AT THE FOOT OF THE HIMALAYAS: PALEONTOLOGY AND ECOSYSTEM DYNAMICS OF THE SIWALIK RECORD

AT THE FOOT OF THE
HIMALAYAS
PALEONTOLOGY AND ECOSYSTEM DYNAMICS OF
THE SIWALIK RECORD

edited by
Catherine Badgley,
Michèle E. Morgan,
and David Pilbeam

JOHNS HOPKINS UNIVERSITY PRESS
BALTIMORE

Johns Hopkins University Press
2715 North Charles Street
Baltimore, Maryland 21218
www.press.jhu.edu

Cover illustration: The western Siwalik sub-Himalayan alluvial plain ecosystem at ~9 Ma (similar in age to Locality Y182, Fig. 24.11). The reconstruction is based on scientific evidence discussed in this book and illustrated by Julius Csotonyi. The view is toward the north, with a large, mountain-sourced river entering from the upper left and a smaller, foothill-sourced tributary entering from the upper right. The scene occurs after the end of the rainy season and shows ponding, sediment deposition, and debris from recent flooding, with smoke from a wildfire visible in a drier area on the upper right. Low hills are visible in the near distance and higher mountains in the far distance.

Library of Congress Cataloging-in-Publication Data

Names: Badgley, C. E. (Catherine E.), editor. | Morgan, Michèle E., 1962–
    editor. | Pilbeam, David R., editor.
Title: At the foot of the Himalayas : paleontology and ecosystem dynamics
    of the Siwalik record / edited by Catherine Badgley, Michèle E. Morgan,
    and David Pilbeam
Description: Baltimore : Johns Hopkins University Press, 2025. | Includes
    bibliographical references and index.
Identifiers: LCCN 2024035202 | ISBN 9781421450278 (hardcover) |
    ISBN 9781421450285 (ebook)
Subjects: LCSH: Paleontology—Siwalik Range. | Fossils—Siwalik Range. |
    Biotic communities—Siwalik Range.
Classification: LCC QE756.S68 A85 2024 | DDC 560.954—dc23/eng/
    20240917
LC record available at https://lccn.loc.gov/2024035202

A catalog record for this book is available from the British Library.

Special discounts are available for bulk purchases of this book. For more
information, please contact Special Sales at specialsales@jh.edu.

To the friends and colleagues in Pakistan and around the world who worked on the Siwalik record and to those who will continue that work

CONTENTS

Preface ix
Acknowledgments xi
Abbreviations xiii
Online Supplement xv

I Preamble to Framing Chapters 1
CATHERINE BADGLEY, MICHÈLE E. MORGAN, AND
DAVID PILBEAM

1 A Long Story 3
DAVID PILBEAM

2 Rocks, Rivers, and Time 12
ANNA K. BEHRENSMEYER AND JOHN C. BARRY

3 Documenting the Siwalik Fossil Record 34
JOHN C. BARRY AND ANNA K. BEHRENSMEYER

4 Siwalik Taphonomy: Fossil Assemblage
Preservation 54
ANNA K. BEHRENSMEYER, CATHERINE BADGLEY,
AND MICHÈLE E. MORGAN

5 Stable Isotopes as a Record of Ecological Change
in the Siwalik Group of Pakistan 86
THURE E. CERLING, ANNA K. BEHRENSMEYER,
JOHN C. BARRY, SHERRY NELSON, MICHÈLE E. MORGAN,
AND JAY QUADE

II Preamble to Faunal Chapters 97
CATHERINE BADGLEY, MICHÈLE E. MORGAN, AND
DAVID PILBEAM

6 Freshwater Molluscan Fossils of the Siwalik Record
of the Potwar Plateau 99
DAMAYANTI GURUNG

7 Fishes in the Siwalik Record 108
CATHERINE BADGLEY AND LOIS ROE

8 Siwalik Reptilia, Exclusive of Aves 117
JASON J. HEAD

9 Siwalik Birds 136
ANTOINE LOUCHART

10 Siwalik Hedgehogs, Shrews, Bats,
and Treeshrews 149
LAWRENCE J. FLYNN

11 Siwalik Lagomorphs and Rodents 163
LAWRENCE J. FLYNN, LOUIS L. JACOBS, YURI KIMURA,
AND EVERETT H. LINDSAY

12 Siwalik Primates 199
JAY KELLEY, MICHÈLE E. MORGAN, ERIC DELSON,
AND DAVID PILBEAM

13 Siwalik Creodonta and Carnivora 213
JOHN C. BARRY

14 Tubulidentata and Pholidota of the Siwalik
Record 240
THOMAS LEHMANN

15 Siwalik Proboscidea 249
PASCAL TASSY AND WILLIAM J. SANDERS

16 Equidae from the Potwar Plateau, Pakistan 260
RAYMOND L. BERNOR, MICHÈLE E. MORGAN,
SHERRY NELSON, AND GINA M. SEMPREBON

17 Siwalik Chalicotheriidae 272
SUSANNE COTE AND MARGERY C. COOMBS

18 Rhinocerotids from the Siwalik Record 281
PIERRE-OLIVIER ANTOINE

19 Siwalik Suidae and Palaeochoeridae 295
    OLIVIER CHAVASSEAU

20 Siwalik Hippopotamoidea 313
    FABRICE LIHOREAU AND JEAN-RENAUD BOISSERIE

21 Siwalik Tragulidae 332
    JOHN C. BARRY

22 Siwalik Giraffoidea 353
    NIKOS SOLOUNIAS AND MELINDA DANOWITZ

23 Siwalik Bovidae 368
    ALAN W. GENTRY, NIKOS SOLOUNIAS, JOHN C. BARRY,
    AND S. MAHMOOD RAZA

III Preamble to Synthetic Chapters 389
    CATHERINE BADGLEY, MICHÈLE E. MORGAN, AND
    DAVID PILBEAM

24 Reconstructing Siwalik Miocene Paleoecology from
    Rocks and Faunas 409
    ANNA K. BEHRENSMEYER AND MICHÈLE E. MORGAN

25 Siwalik Mammalian Community Structure and
    Patterns of Faunal Change 432
    MICHÈLE E. MORGAN, LAWRENCE J. FLYNN,
    AND DAVID PILBEAM

26 Taxonomic and Ecological Dynamics of Siwalik
    Mammalian Faunas 481
    CATHERINE BADGLEY

27 Highlights of the Siwalik Record and Future Research
    Opportunities 527
    CATHERINE BADGLEY, MICHÈLE E. MORGAN,
    AND DAVID PILBEAM, WITH CONTRIBUTIONS
    FROM JOHN C. BARRY, ANNA K. BEHRENSMEYER,
    LAWRENCE J. FLYNN, AND S. MAHMOOD RAZA

Contributors 537
Index 539

PREFACE

The two of us met in Pakistan in November 1973, around 50 years ago. One of us, a young Geological Survey of Pakistan (GSP) geologist, the other, a not quite so young Yale paleontologist, were present at the beginning of a long collaboration between the GSP and a Yale- and later the Harvard-initiated project. Over the past five decades we went from being collaborators to becoming close friends. No matter what our aspirations were five decades ago, neither of us would have predicted all that followed.

The project's first 30 years of fieldwork covered much of fossiliferous Pakistan, but our main focus quickly became the magnificent rocks of the Potwar Plateau with its 18 to 6 Ma sequence of highly productive Miocene exposures. The first field party included just ten of us, but the extended team rapidly grew to include many times that number of paleontologists and geologists—from Pakistan, North America, Europe, and elsewhere.

In addition to the project's scientific contributions, we must acknowledge and salute the very cordial and productive relationships between "hosts" (the people of Pakistan, their government agencies, particularly the Geological Survey of Pakistan and the Pakistan Museum of Natural History) and their "guests" (spanning many different areas of expertise and representing many different nationalities).

After an initial three decades of fieldwork (1973–2000), the 20-plus years since have been invested in descriptions, analyses, and publications, along with many stimulating discussions. This scientific wealth generated multiple research efforts and formed the basis of thesis and PhD research at diverse institutions. All of this research has provided the foundation for the present volume.

The first part of this volume reviews ways in which varying and changing paleolandscapes of the plateau, dominated by rivers, streams, and lakes, were reconstructed from detailed geological studies, and how the fossil record accumulated and is dated. Part II, which comprises the bulk of the book, consists of 18 chapters on faunal groups, the majority of which are mammals. The third part is devoted to analyses and syntheses. What has emerged is a detailed "case study" of a series of faunal communities spanning over 10 million years, the reconstructed environmental context of ongoing habitat and climate change, and discussion of possible causes of the evolutionary and ecological transformations that led to today's biodiversity in southern Asia.

The project's legacies include the extensive networks of collaborations and friendships and a rich and abundant publication record on a fascinating time and place in earth history, ultimately culminating in this book. As important is that the legacy will include the planned Geological Survey of Pakistan Vertebrate Paleontology Research and Collections Center in Islamabad, Pakistan, which will secure the irreplaceable paleontological collections along with all of their essential supporting documentation. There is still much to learn about the forces that shaped the geological past and our modern world, and we hope that the international partnerships that have sustained our efforts over 50 years provide inspiration for a new era of paleontological research and collaboration.

David Pilbeam and S. Mahmood Raza

ACKNOWLEDGMENTS

We have several realms of people to acknowledge both for the creation of knowledge that is the foundation of this book and for the assembly and production of the book. All of them have contributed to this book in essential and meaningful ways.

We begin with colleagues at the Geological Survey of Pakistan. These include our professional collaborators and facilitators, such as the Director General, the Director of the Stratigraphy and Paleontology division, the officers who joined us in the field, and the drivers, cooks, and helpers who made day-to-day life in the field possible, enjoyable, and delicious. Among these many individuals, we highlight Dr. S. M. Ibrahim Shah, who in 1974 became Director of the Paleontology and Stratigraphy section of the Geological Survey of Pakistan and hence a leading collaborator from our second field season in 1975 onward, and Dr. Iqbal U. Cheema, Khalid A. Sheikh, and Dr. Imran A. Khan—each of whom accompanied us in fieldwork for many years and came to the United States for graduate degrees. Special recognition goes to Dr. S. Mahmood Raza, one of the contributors to this book, for his many years, since 1973, of collaboration, guidance, and friendship.

We salute and thank the international group of scholars, students, and staff who conducted and supported fieldwork that provided the foundation for all subsequent analyses of the Siwalik record of the Potwar Plateau. A list of field participants can be found in the Preamble to Part I. We also thank all who have contributed to analyses and laboratory work in universities and institutions around the world—too many to list individually—for their ideas, comraderie, and publications. Both the fieldwork and lab work have stimulated and advanced Siwalik research over the past five decades and are reflected in the chapters of this book.

Among these collaborators are the contributors to this volume, listed separately.

We also note with appreciation colleagues and friends who have died over the course of this research, including Niaz Ahmed, William Bishop, Iqbal Cheema, Will Downs, Andrew Hill, Muhammed Asif Jah, Noye Johnson, Everett Lindsay, Grant Meyer, and Neil Opdyke.

Several funding agencies have supported the work of our collective research effort. These include the Smithsonian Foreign Currency Program, the U.S. National Science Foundation, the American School of Prehistoric Research through the Peabody Museum at Harvard University, the Wenner Gren Foundation, and our home institutions. We thank these organizations and the individuals within them who made our work financially possible.

At the Johns Hopkins University Press, we highlight Tiffany Gasbarrini, our editor, for her support, patience, and advice over the years of preparation of this book, through delays caused by the Covid pandemic as well as the sheer amount of material that we needed to summarize. We have been fortunate to work with an excellent publishing team, including Diem Bloom, Kris Lykke, Ezra Rodriguez, and Jennifer Paulson.

We greatly appreciate the insights of the editing team, including Tatiana Holway and Helen Wheeler at Westchester Publishing Services. We also thank Mary Parrish and Katharine Loughney for their superb skill and artistry with Adobe Illustrator and Photoshop to finalize and format the many figures in this book. Pilar Wyman ably created the index.

We met with several core members of our research group biweekly for several years during the planning, analysis, and writing stages of this book. We thank John Barry, Kay Behrensmeyer, Larry Flynn, and Mahmood Raza for their friendship, ideas, and guidance, beyond their direct contributions to the book.

Finally, we thank our spouses and family members for their support and encouragement over the many years of this project and the writing of this book.

Thanks to you all!

ABBREVIATIONS

INSTITUTIONS AND MUSEUMS

AMNH	American Museum of Natural History, New York
BPI	Bernard Price Institute, Johannesburg
BSM	Bavarian State Museum, Munich
BSP	Bayerische Staatssammlung für Paläontologie und Geologie, Munich
GIU	Geological State Museum, Utrecht
GSI	Geological Survey of India
GSP	Geological Survey of Pakistan
HLMD	Hessisches Landesmuseum, Darmstadt
IMC	Indian Museum, Kolkata (Calcutta)
ISEM	Institut des Sciences de l'Évolution, Université de Montpellier, Montpellier
KNM	Kenya National Museum
MfN	Museum für Naturkunde, Berlin
MNHN	Muséum National d'Histoire Naturelle, Paris
NHM	Natural History Museum, London (formerly British Museum of Natural History)
NMK	National Museums of Kenya, Nairobi
NYIT-COM	Department of Anatomy, New York Institute of Technology College of Osteopathic Medicine
P.I.N.	Palaeontological Institute, Russian Academy of Sciences, Moscow
PMNH	Pakistan Museum of Natural History, Islamabad
PUPC	Punjab University Palaeontological Collection, Lahore, Pakistan
SGT	Service Géologique de Tunisie
SMNS	Staatliches Museum für Naturkunde in Stuttgart
SPGM	Paläontologisches Museum, München
TM	Transvaal Museum, Pretoria
VPL	Vertebrate Palaeontology Laboratory, Geology Department of Panjab University, Chandigarh, India
YPM	Yale Peabody Museum, New Haven, CT

LOCALITY, SPECIMEN, COLLECTION, AND RESEARCH PROJECT ABBREVIATIONS

AJ	Al Jadidah collection (Hofuf Formation, Saudi Arabia) of MNHN Paris
AK	prefix for specimens from Akkasdagi, Turkey in the MNHN paleontological collections, Paris
AM	prefix for specimens at the American Museum of Natural History, New York
AMHH	prefix for specimens at the American Museum of Natural History, New York
BMNH	prefix for specimens at the Natural History Museum, London
BSM	prefix for some specimens in the BSM collections
BSPG	M prefix for some specimens in the BSP collections
CCZ RH	relates to the Khorat deposits and fossils, stored in the Nakhon Ratchasima University collections, Thailand
CH	prefix for some specimens from the Chinji Fm. in the BSP collections
CHC	prefix for some specimens from the Chinji Fm. in the BSP collections
CHD	prefix for some specimens from the Chinji Fm. in the BSP collections
CHJ	prefix for some specimens from the Chinji Fm. in the BSP collections

CHK	prefix for some specimens from the Chinji Fm. in the BSP collections
CHO	prefix for some specimens from the Chinji Fm. in the BSP collections
D	prefix for localities resulting from expeditions of the Dartmouth-Peshawar University project
DP	prefix for fossils resulting from expeditions of the Dartmouth-Peshawar University project
H-GSP	prefix for specimens catalogued by Howard University-Geological Survey of Pakistan (and Utrecht collaborative projects)
Harvard-GSP	Harvard University-Geological Survey of Pakistan research project
Howard-GSP	Howard University-Geological Survey of Pakistan research project
KNM-RU	prefix for collections from Rusinga, Kenya, that are curated at the National Museums of Kenya
NG	prefix for Nagri Formation specimens in the BSP collections
NHML	prefix for some specimens at the Natural History Museum, London
NHML M	prefix for some specimens at the Natural History Museum, London
PAK	prefix for collections made by a joint project of the University of Montpellier, the Muséum National d'Histoire Naturelle in Paris, and the University of Balochistan
PUA	collections of Panjab University Anthropology Department, Chandigarh, India

S-GSP	prefix for collections made by the Yale / Harvard University-GSP Project in the Miocene of Sindh Province
SNSB-BSPG	prefix used for collections in the Bavarian State Collection of Paleontology and Geology, Munich
Y	prefix for fossil localities in the Harvard–GSP research project database (originally Yale-GSP research project)
YGSP	prefix for specimens catalogued by the Harvard–GSP research group (originally Yale-GSP research group)
YPM	prefix for specimens curated at the Yale Peabody Museum, New Haven, CT
YPM-VP	prefix for some specimens curated in Vertebrate Paleontology at the Yale Peabody Museum, New Haven, CT
Z	prefix for localities and fossils from the Zinda Pir area, western Pakistan

ADDITIONAL ABBREVIATIONS

GPTS	geomagnetic polarity time scale
kyr	thousand years
Ma	mega-annum, a specific age in millions of years
MCO	Miocene Climatic Optimum
MN	Mammal Neogene, a series of biostratigraphic zones used to correlate fossil localities with mammals in Europe
myr	million years

ONLINE SUPPLEMENT

*A*t the Foot of the Himalayas: Paleontology and Ecosystem Dynamics of the Siwalik Record is supported by an online supplement. Additional files with data and images relevant to chapters in this book can be found by visiting the specific book page on **press.jhu.edu**.

PREAMBLE TO FRAMING CHAPTERS

CATHERINE BADGLEY, MICHÈLE E. MORGAN, AND
DAVID PILBEAM

These five chapters describe the history and objectives of the Harvard-Geological Survey of Pakistan research project (1973–present) and the ecological theater of the western Siwaliks. Since the beginning of this research project, David Pilbeam has adroitly guided, encouraged, and facilitated field and lab research. The Geological Survey of Pakistan (GSP) has been an invaluable partner throughout. Fieldwork in the Potwar Plateau was conducted over 27 field seasons between 1973 and 2001 and included more than 200 participants. A map (Fig. I.1) details the Potwar Plateau study region and situates it within both the expanse of Siwalik Group sediments south of the Himalayas and the broader geography of South Asia that influences access and egress of biota.

While this project began more than 50 years ago, paleontological research in Siwalik Group rocks spanned centuries and is acknowledged in David Pilbeam's retrospective (Chapter 1). His chapter also provides an overview of the book. In Chapter 2, Anna K. Behrensmeyer and John C. Barry reinterpret the classic Siwalik rock formations of the Potwar Plateau with descriptions of the fluvial sediments and impacts of fluvial change and stability at different scales of time and space on evolution of ecosystems. Project documentation protocols follow in Chapter 3, in which John C. Barry and Anna K. Behrensmeyer review core information for fossils and fossil localities and how this information is carefully integrated with sedimentological, stratigraphic, and taphonomic data. A critical result of this work is the ever-evolving chronostratigraphic framework, which is central to all research. In Chapter 4, Anna K. Behrensmeyer, Catherine Badgley, and Michèle E. Morgan discuss the taphonomic signatures of fossil assemblages and the processes that shaped them. In the final chapter of this group, Chapter 5, Thure E. Cerling, Anna K. Behrensmeyer, John C. Barry, Sherry Nelson, Michèle E. Morgan, and Jay Quade present the discoveries of isotopic paleoecology through the Siwalik sequence based on carbonates in paleosols and fossil mammal teeth and the broader impact of this record on global ecosystem evolution research.

Fieldwork was conducted by dedicated GSP geologists and a multinational team of scientists and students. Throughout, staff of the GSP and residents near the field camps worked toward the common goal of documenting the fossil history of Siwalik biota. Participants include Niaz Ahmed, Zulfiqar Ahmed, Mohammad Amin, Mohammad Anwar, David Archibald, Robert Holt Ardrey, Mohammad Arif, Catherine Badgley, Malik Sikander Bakht, Nazar Mohammad Baloch, Qasim Baluch, Jeff Barndt, John Barry, Wendy Barry, Mohammad Bashir, Rich Beck, Jim Beer, Kay Behrensmeyer, Ray Bernor, Bill Bishop, Melissa Brannon, John Bridge, Bobbie Brown, Michel Brunet, Mike Cassiliano, Thure Cerling, Dan Chaney, George Chaplin, Iqbal U. Cheema, Rich Cifelli, Dave Clark, Melissa Connell, Glenn Conroy, Jane Conroy, Julie Cormack, Dana Coyle, Dixon Cunningham, Allah Dad, John Damuth, Elizabeth Dawson-Saunders, Ilam Din, Todd Disotell, Phil Dormitzer, Kevin Downing, Will Downs, Ali Nasir Fatmi, Regina Figge, Tony Fiorillo, Kathy Flanagan, Francine Flynn, Larry Flynn, Carol French, Hod French, Peter Friend, Carol Frost, Tony Gaston, Jeff Gee, Ted Gehring, Philip Gingerich, Fred Grady, Cliff Gronsetch, Nigel Hacking, Abdul Haq, Munir-ul-Haq, Jessica Harrison, Richard Haskin, Jason Head, Emile Heintz, Jason Hicks, Andrew Hill, Sylvia Hixson, Steve Hochman, Julie Hutt, Masood Idris, Jameel Imtiaz, Mohammed Iqbal, Nina Jablonski, Bonnie Fine Jacobs, Louis Jacobs, Muhammad Asif Jah, Qazi Jahangir, Bakht Jamal, Akhtar Jan, Noye Johnson, Tom Jorstad, John Kappelman, Jay Kelley, Bill Keyser, Ahmed Khan, Imran Ahmed Khan, Mohammad Hashim Khan, Munsif Khan, John Kingston, Ev Lindsay, Bruce MacFadden, Laura MacLatchy, Sandra Madar, Abdul Majid, Lawrence Martin, Wendy Martin, Rich May, Melanie McCollum, Robert McCord, Michael McMurtry, Lee McRae, Gerhard Meyer, Grant Meyer, Pat Monaghan, Marc Monahan, Fateh Mohammad, Noor Mohammad, Kevin Moodie, Michèle Morgan, Mohammad Munir, Sherry Nelson, Neil Opdyke, Stephanie Pangas, Martin Pickford, David Pilbeam, Joyce Pohlman, Bob Poreda, Jay Quade, Kathy Rafferty, A. Rahim Rajpar, Ghulam

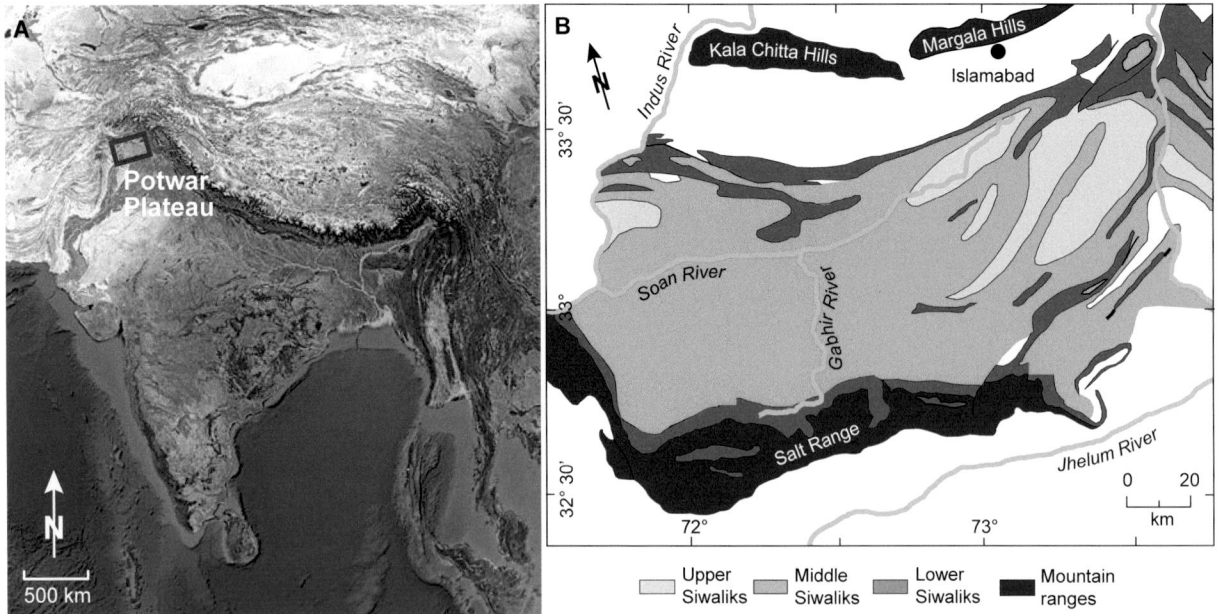

Figure I.1. Map situating Potwar Plateau (A) within the outlined sub-Himalayan foreland basin (from Fig. 2.1) and Potwar Plateau details (B) with primary study areas outlined (Fig. 2.3B).

Raza, Rizwana Raza, S. Mahmood Raza, Robin Renaut, Greg Retallack, Lois Roe, Ghazala Roohi, Mohammad Sarwar, Jeff Saunders, Margaret Schoeninger, Karen Sears, Asrar Shah, Mustafa Shah, S. M. Ibrahim Shah, Khalid Ayub Sheikh, Tanya Sher Khan, Richard Sherwood, Tauqir Ahmed Shuja, Fred Sibley, Mark Solomon, Nikos Solounias, Steve Speyer, David Stern, Steve Sullivan, Daris Swindler, Lisa Tauxe, Lou Taylor, Herbert Thomas, Mike Tiffany, Yuki Tomida, Annie Walton, Steve Ward, Kathy White, Brian Willis, Marion Wingfield, Alisa Winkler, John Wozny, Mike Zaleha, Rukan Zaman, and Mohammad Zareen.

1

A LONG STORY

DAVID PILBEAM

This book documents ecosystem change in a long fossiliferous sequence generated and dominated by the
rising Himalayas. It focuses particularly on diverse and
changing mammal communities during the Miocene in a relatively restricted area of South Asia, an important and interesting Old-World faunal region.

South Asia, the geographical entity today comprising Pakistan, India, Bangladesh, Sri Lanka, and Myanmar, is also a geological entity, the Indian Plate, which for over 100 million years
has been moving steadily north from what is now Antarctica.
Some 50 million years ago the leading edge of the plate began
to contact the Eurasian Plate, and by 35 to 40 million years ago,
it was pinching off and eliminating a seaway, the Tethys Sea.
The process of subducting or sliding the Indian Plate under the
Eurasian Plate raised the southern margins of the latter, forming the mountain barriers surrounding South Asia, the Himalayas running west to east, and lesser but significant north-south
aligned systems marking the western and eastern margins of
the subcontinent.

The uplifting areas were eroded by numerous south-flowing
rivers and streams. The sediment that was generated accumulated particularly in the foredeep or depression to the south of
the Himalayas, forming a thick and layered system of sedimentary rocks along the margins of these emergent mountains.
These sediments are now exposed as foothills along the flanks
of the Himalayas from northern Pakistan through northwestern India and Nepal, with coeval deposits to the west in southern Pakistan and to the east in northeastern India and Myanmar, and are known formally as the Siwalik Group and
informally as "the Siwaliks." Miocene rocks throughout the
northern portion of South Asia contain similar contemporaneous faunas, the "Siwalik faunas"; recording a single faunal
domain, these are precursors of the current Oriental zoogeographical region communities.

One of the most pressing questions in evolutionary biology
concerns the roles of intra- and interspecies competition and
climate and habitat change in the evolution of species. While

this project did not begin with such a focus, that is what it has
become.

Charles Darwin was the first evolutionary biologist, and *On
the Origin of Species by Means of Natural Selection*, published in
1859, was its foundational text. All such foundations necessarily undergo change and modulation. The first involved the
late nineteenth- to early twentieth-century rediscovery of
Mendelian genetics and its linking to Darwinian selectionism.
Beginning a quarter century later, the "modern synthesis"
brought together population genetics, ecology, systematics,
and population biology. Paleontology was soon added to this
mix, providing the necessary temporal dimension (Simpson
1944). The geological and paleontological records had been an
essential part of Darwin's long argument in the *Origin*. "For
my part," he wrote, "following out Lyell's metaphor, I look at
the natural geological record, as a history of the world imper-
fectly kept, and written in a changing dialect" (Darwin 1964,
310). And, he added, "all the chief laws of palaeontology
plainly proclaim, as it seems to me, that species have been produced by ordinary generation: old forms having been sup-
planted by new and improved forms of life, produced by the
laws of variation still acting around us, and preserved by
Natural Selection" (Darwin 1964, 345).

Building to a great extent on Simpson's 1944 contribution,
paleontology became a critical aspect of the modern synthesis.

The fossil record has far surpassed what was available to
Darwin. It is clearly essential to addressing questions about
macroevolution; its role grows in importance as the record of
the past steadily improves. While at the genomic level change
occurs in a quasi-clocklike way, in contrast, the hard-tissue phenotypes making up the fossil record frequently show minimal
change often over hundreds of thousands to millions of years.
Good fossil records document species' lifespans. If dense
enough, records can constrain the time involved in the origination of new species—even on rare occasions capturing the
splitting of an ancestor into descendants. Because at least some

faunal turnover occurs over even short intervals of geological time, combinations of species documented at various spatial and temporal scales are constantly changing. It is sometimes possible to observe whether and how immigrant taxa might have affected resident members of communities. Understanding the dynamics of these paleocommunities requires understanding patterns and inferring interactions (for example, competition, predation, food preferences) underlying ecological interactions within living communities. Until geologically recent time, for many groups, particularly mammals, both species and paleocommunities differed from those in the present, often markedly.

There has long been interest in and argument over the relative roles of biotic and abiotic factors in determining changes at species and community levels. Bennett (2004) notes that Darwin and Lyell had contrasting views, the former advocating the biotic—what came more than a century later to be termed the "Red Queen" hypothesis (Van Valen 1973)—the latter the abiotic—the hypothesis Barnosky (2001) labeled the "Court Jester." Notably, a recent review of over one hundred papers, the great majority summarizing other studies, supports "an important and measurable role of biotic interactions in shaping communities and lineages on long timescales" (Fraser et al. 2021, 61).

This statement echoes what Darwin wrote more than a century and a half ago: "The action of climate seems at first sight to be quite independent of the struggle for existence; but in so far as climate chiefly acts in reducing food, it brings on the most severe struggle between the individuals, whether of the same or of distinct species, which subsist on the same kind of food" (Darwin 1964, 68).

In an important paper commemorating the bicentenary of Darwin's birth and the 150th anniversary of the publication of the *Origin of Species*, Reznick and Ricklefs (2009) noted that because the macroevolutionary consequences of ecological competition and habitat change cannot be addressed over the span of one or even several human lifetimes, long fossiliferous sequences are critical contributors to debates about their relative roles.

Although our research is about a particular place and time interval in the long history of life, it relates fundamentally to these big questions in evolutionary biology, including those first posed by Charles Darwin. Stimulated by Darwin, we reflect on that history and on how those questions guided and informed our interpretations of the forces that shaped the evolution of Siwalik faunas.

The long fossiliferous sequence in the Potwar Plateau of Pakistan described in this book records past biotic events and their dynamics: species originations and extinctions; range extensions and contractions; changing climates, habitats, and paleocommunities. Like other long sequences, it shows a pattern of steady and unsteady change and turnover, with species appearing and disappearing throughout the record and presenting another opportunity to assess the relative contributions of abiotic and biotic factors in shaping species, clades, and communities.

BOOK ORGANIZATION

The book is organized in three sections. The first contains framing chapters that provide information on the geology, paleontology, and habitats of the Potwar Plateau, including how the sequence is dated; the nature of deposition with insights into changing landscapes and habitats; collection, documentation, dating, and analyses of fossils from this sequence; how the fossils accumulated and were affected by taphonomic processes. Studies of these processes, along with more ecological approaches, provide essential insights into faunal paleocommunities and their environments and habitats.

In the following section, 18 chapters cover the fossil record by taxonomic group, all but one describing vertebrates and all but three of those concerning mammals. With varying degrees of attention to new material or to systematics, their primary goal is to summarize the most important paleobiological and paleogeographical aspects of each group. What are significant adaptations as they relate to diet, body size, locomotion, and habitat preference, for example? What might we infer about the provenance of immigrants or of additional biogeographic information provided by emigrants?

The synthetic chapters of the third and final section integrate and analyze changes over time in the ecology and the structure of communities and on occasion individual lineages, as well as suggesting explanations for these macroevolutionary changes. They assess what can be determined about the relative contributions of habitat change, competition, and other biological factors to faunal turnover and discuss the broader significance to paleontology of this unique "case study" spanning 18 million years and more of evolutionary time.

OUR FIELDWORK

While doing some fieldwork on older sequences in southern Pakistan, the main focus since our project began in 1973 has been the Middle and Late Miocene fossiliferous rocks in the Potwar Plateau of northwestern Pakistan, particularly the faunas between 18 and 6 million years ago. The best exposures of Siwalik Group rocks anywhere in the subcontinent are in the plateau, where a combination of uplift, folding, and erosion expose a layered sedimentary sequence deposited by a complex of rivers and streams over the course of almost 20 million years. An added advantage is that the tilt of sediments is relatively low and thus affords a large area for accumulation. The plateau covers around 20,000 km² but the majority of our collections comes from three smaller and spatially separate parts of the plateau. A single deep-sea core from the central Atlantic Ocean captures a good representation of the more widely distributed trans-Atlantic marine biota. Picture the Potwar fossils sampled between 18 and 6 million years from these areas as comprising a kind of composite, calibrated core. The Potwar does not capture everything contemporaneously present throughout the South Asian faunal zone, but we believe that it records a significant fraction. Appearances and disappearances of species reflect the steady movement of species across a constantly changing and evolving landscape. They have their entrances and exits.

Our collections have weaknesses as well as strengths. We find few "complete" specimens where head and body parts are from the same individual or where an entire articulated limb is recovered. We have almost no botanical record. Deposition was effectively continuous, however. The exposures are varyingly fossiliferous throughout and now very well calibrated in time so that fossil localities, circumscribed patches yielding fossil bones and teeth, are well enough dated for most now to be sorted ("binned") into 100,000-year intervals. Such geologically brief intervals, for the Miocene, make it possible to consider the relative roles of climate and habitat change and competitive interactions among species within paleobiologically tractable timeframes. While fossils from the plateau range in age from 18 to 6 million years old, we have focused some of our analyses on the densely sampled sequence spanning around 14 to 7 million years ago, which provides an impressively continuous sequence of species lineages and their association in communities. Plant remains, wood, leaves, fruits, pollen and leaf waxes are known from other Siwalik sequences in India and Nepal, and there is a good Potwar sequence of fossil soils. This means that directly and indirectly we have a reasonable understanding of habitat structure throughout most of the sequence.

South Asia presents a particularly valuable opportunity to examine species evolution and adaptations and their possible links to habitat change. It is, in effect, a laboratory for testing biotic and abiotic causes of faunal change (lineage and community). The region has clearly delimited and variably permeable paleogeographic margins. We are able to review faunal change during times of both global climate change and stability, as well as during episodes of what we infer to be relatively closed versus open boundaries. This has become an important analytical focus of the project because it asks the most integrative of questions and demands the most integrative of approaches. Our analyses of the long Potwar sequence will also make possible comparisons to other long sequences, such as those in Spain, Kenya, and China, and potentially help assess the extent to which change may or may not be synchronous on a global scale.

EARLIER FIELDWORK

A brief summary of earlier work in the Siwaliks provides some background and shows how work like ours is dependent on that of so many others. The *Mahâbhârata*, a long Indian epic poem composed between 400 BCE and 400 CE, tells the history of the descendants of King Bharata and of a great battle waged between his descendants. Van der Geer, Dermitzakis, and de Vos (2008) believe that observation of fossil bones eroding continually throughout the entire west to east range of the Siwalik Series contributed to the construction of a mythical great battle, the fossils being interpreted as the remains of fallen giant humans and other animals. A millennium later, large fossils were recovered and reported during the 1335 CE renovation by Firuz Shah Tughlaq of the Yamuna Canal in northwestern India (pers. comm. Rajesh Kochhar via Michael Witzel).

Before the First World War considerable geological progress had already been made in describing the rocks from which the fossils were recovered in the Indian subcontinent, the most notable contributions being those of Guy Pilgrim (1913). Such studies accelerated between the two world wars. Among others, Robert Anderson (1927), Gerald Cotter (1933), Edwin Colbert (1935), and G. Edward Lewis (1937) continued the work, which was followed after the Second World War by that of William Gill (1951) and more recently A. N. Fatmi (1973) from the Geological Survey of Pakistan (GSP). Pilgrim in particular, and two decades later Lewis, played leading roles in describing and naming the rock units making up the Siwalik Series, the "classic" sequence being based on rock units named from villages in the Potwar Plateau: from oldest to youngest, Kamlial, Chinji, Nagri, Dhok Pathan, and Tatrot Formations.

Siwalik fossils from the subcontinent were well known before the arrival of European traders and eventual colonizers, but among the first "scientific" reports, those of Hugh Falconer beginning in the 1830s stand out (for example, Falconer 1837). Two additional colonial India-based paleontologists were critical in collecting and describing fossils and their geological contexts: Richard Lydekker starting in the 1870s (for example, Lydekker 1877); and Guy Pilgrim whose important paleontological contributions began early in the twentieth century (1910, 1913). These paleontologists, along with their Indian collectors and collaborators, worked in fossiliferous areas of what became India and Pakistan after partition in 1947. Scholars from the United States, including Barnum Brown in the early 1920s (Brown, Gregory, and Hellman 1924), G. Edward Lewis from 1931 to 1933 and later in 1935 with Helmut de Terra (Lewis 1934, 1937; Gregory, Hellman, and Lewis 1938), as well as Richard Dehm from Munich (Dehm, Oettingen-Spielberg, and Vidal 1958; Dehm 1983), collected fossils immediately prior to (1939) and after the Second World War (1952–1953 and 1955), as did Gustaf (Ralph) von Koenigswald from Utrecht in 1966–1967 (von Koenigswald 1983). Elwyn Simons from Yale worked with S. R. K. Chopra in northwestern India in 1967 and 1968 at another classic Siwalik location, Haritalyangar near Chandigarh (Simons and Chopra 1969). There have been continuing field projects led by Indian, Pakistani, and Nepali colleagues from the mid-1900s on: for example, Rajeev Patnaik (2013) for India; Abu Bakr and his students at the University of the Punjab, Lahore (MacFadden and Abu Bakr 1979; Kahn et al. 2020) for Pakistan; and the Nepali-based Gudrun Corvinus with Nepali colleagues (Corvinus 1993; West et al. 1974; Quade et al. 1995; Corvinus and Rimal 2001) for Nepal.

SOME CONTEXT FOR OUR PROJECT

South Asian mammal fossils have long been of interest (Colbert 1935) to paleontologists (including paleoprimatologists and paleoanthropologists). Documenting the early stages in the development of the Indo-Malayan biogeographical realm and sampling subtropical habitats, these fossils are recovered from a temporally long sequence. Although South Asian faunas are relatively endemic, throughout our sequence we document continuing faunal interchanges of both immigrants and emigrants shared with Africa, Southeast Asia, Europe, and China.

The region had been noted among students of primate and human evolution for important fossil ape remains, mainly of jaws and teeth, described until the 1960s as comprising many genera and species (see Simons and Pilbeam 1965). By the 1970s most agreed that the principal South Asian genera were *Sivapithecus*, *Ramapithecus*, and *Indopithecus*. Notable was the recovery of *Ramapithecus brevirostris* at Haritalyangar in India by then-Yale graduate student G. Edward Lewis during the Yale North India Expedition of 1931–1933 and its subsequent description by Lewis in 1934 (also see note about the discovery in Simons 1964). Despite initial and continuing disagreement as to its status, by the early 1960s many paleontologists and paleoanthropologists came to regard *Ramapithecus* as a probable early direct human ancestor ("hominid" in the terminology of the day, now "hominin"), on the order of 14 or 15 million years old (Simons 1961, 1964). Although by the 1960s, comparative genetic studies were estimating that the split of hominins and African apes was less than half that age (Sarich and Wilson 1967), there was continuing robust support for *Ramapithecus*'s hominin status from many primatologists and paleontologists well into the 1970s.

We consider our work in the essentially continuously fossiliferous Potwar sequence as important for two main reasons. It provides well-dated and current descriptions, comparisons, and analyses of mammal groups. It offers insight into the mix of competition and habitat change as determinants of faunal evolution. This, however, was not how the project was initially designed nor why it started in 1973. It began as a search for more *Ramapithecus*, considered to be the "earliest human ancestor" at the time. Some personal reminiscences can set the scene for the genesis of that original intention.

Sometime in the spring of 1962, while I was an undergraduate at Cambridge University, I read Elwyn Simons' brief reanalysis (1961) of Lewis' description of *Ramapithecus brevirostris* (1934), which restated the case for its being an early human ancestor. Barely a year later, I was Elwyn's graduate student at Yale, quickly engaging with him on studies of Miocene fossil apes and putative hominins. I returned to Cambridge in 1965 to teach for three years, but went back to Yale in 1968. Elwyn was at that time in the second of a two-year Siwalik field project at Haritalyangar in India with S. R. K. Chopra, the goal being to find more primate fossils, particularly *Ramapithecus*. But by 1969, he had decided not to continue work in South Asia and turned his energies back to older sediments in Egypt with which he had been engaged since the early 1960s.

Before commencing Siwalik work at Haritalyangar, Elwyn had also explored the possibility of collecting in Siwalik exposures of Pakistan. I was working in Elwyn's laboratory in the Yale Geology and Geophysics Department in the early spring of 1973 when the late Grant Meyer, then his lab and field manager, walked in with a letter from the GSP agreeing to a collaborative project. Elwyn generously encouraged me to follow up the invitation. Even more generously, the National Science Foundation swiftly allowed me to redirect some funds for work in deposits elsewhere of similar age for use in Pakistan. We were also fortunate in having immediate support from the Smithsonian Foreign Currency Program. Together, these enabled us to mount a small expedition to Pakistan with the initial and principal goal of recovering more of "the oldest putative hominin," *Ramapithecus*.

Grant Meyer, along with Glenn Conroy (then a Yale graduate student), embarked on an epic truck journey, driving uninterruptedly overland from Barcelona in Spain to Quetta in Pakistan (this would now be completely impossible!). In 2021 Glenn sent me his recollection of that drive: "The one interesting anecdote I remember about our epic journey was that Grant and I determined to swim in every named body of water between Spain and Pakistan. I particularly remember Lake Van in Turkey and the Caspian Sea in Iran as being interesting plunges!" They arrived in Quetta, then the headquarters of the GSP, in October 1973. We were most fortunate in that first season in having the support of A. N. Fatmi, then director of Paleontology and Stratigraphy of the GSP (to be succeeded in 1974 by the equally supportive S. M. Ibrahim Shah).

We were even more fortunate to be assigned a then-junior GSP officer, S. Mahmood Raza, to work with us as the GSP representative on the joint project. A few years later Mahmood became a graduate student in geology and geophysics at Yale, from which he subsequently received his PhD. He has remained to this day a constant and essential member of the research group. That October, Mahmood joined Grant and Glenn, Philip Gingerich (another Yale graduate student), and Tony Gaston (whom I'd taught at Cambridge more than a decade earlier). The first season involved fieldwork in the Potwar Plateau, first around the village of Dhok Pathan and later at Chinji. While driving back to Rawalpindi on the last day of that first season, we took a detour to examine exposures near the village of Khaur, which turned out to fall in age between the sequences at Dhok Pathan and Chinji. The Khaur deposits subsequently produced fossils of great importance.

During that first season we became aware of two other similar projects. One involved a group of geologists and paleontologists from Arizona, Columbia, and Dartmouth Universities (Everett Lindsay, Neil Opdyke, Noye Johnson, and Gary Johnson), who were collaborating with Rashid Tahirkheli of Peshawar University (Chapter 3, this volume). We quickly developed close, complementary, and eventually symbiotic relationships with this group, and our aims effectively became a single project. A third project, also with the GSP, was initiated by S. Taseer Hussain of Howard University and worked on the west side of the Indus River. (Taseer had collected in the Potwar a decade earlier when an undergraduate at Lahore University, and then later with Ralph von Koenigswald as a Utrecht PhD student.)

Even before the end of that first field season, we realized how fortunate we were to have landed in the Potwar Plateau. The Siwalik deposits throughout northern Pakistan, India, and

Nepal, along with their equivalents in southern Pakistan and in Myanmar are largely "layer cake" because of the nature of their deposition. The Potwar sequence comprises well over 3,000 m of sediment that accumulated over the last 18 million years (see Chapters 2, 3, and 4, this volume). With continual northward Indian Plate movement, the sedimentary pile has been buried, tilted, somewhat deformed, and exposed. Deposited largely from rivers and streams, the rocks range from gravel to coarse sandstone, finer-grained silts, muds, and clays. The sands represent ancient river channels, while the finer sediments are overbank materials laid down on floodplains adjacent to rivers and streams. The sequence is made up of repeated motifs of sandstones fining upward into silts and clays, capped in turn by another sandstone. Fossil soils (paleosols) formed between successive sandstones and, like the latter, can often be traced laterally for tens of kilometers. They formed geologically at the "same" time and hence are very useful isochrons. Many small recent streams have eroded through the ancient sediments perpendicular to their depositional axes, making it possible to explore these rocks through time from deepest and oldest to youngest.

Our second season started in January 1975, and I was fortunate in being able to recruit another Yale graduate student to the team, John Barry. Over almost five decades, John has become very much the "vertebral column" of the project and an essential member of our now Harvard-based group. As well as being more than competent across our entire fossil record, he developed and maintains a very original database for the project, making possible the integration of different categories of data. In that same season I also recruited yet another Yale graduate student, Catherine Badgley, the initiator of this volume and one of its co-editors. It has been realized thanks to her tenacious and persistent energy and enthusiasm. Catherine began documenting the depositional environments of fossil accumulations, developing taphonomic studies aimed at understanding how fossil sites formed and thereby how validly the dead record what was once a living present (Chapter 4, this volume). These studies have continued over the life of the project and have yielded many critically important insights on habitats and community dynamics. Through my colleague and friend the late Bill Bishop, I was fortunate in being able to also recruit Martin Pickford to the team in 1975. (I had invited Bill to join the group as coordinator of geological research, but he very sadly died two years later in 1977.) Martin's energy, fossil-finding prowess, and overall enthusiasm for the project were of inestimable help in those early years, and I regret that our association was relatively brief.

Louis Jacobs, at that time a graduate student member of the Peshawar group, visited our camp in early spring of 1975, and over the following seasons became a member of the research group. Louis, along with his advisor, Everett Lindsay (one of the leaders of the Peshawar group), and later with younger colleagues Larry Flynn and the late Will Downs, began collecting small mammals, which became one of the major foci of the project. Innovative for the Siwaliks, the retrieval of small mammals was based almost entirely on screen-washing, which involved considerable labor and logistical

efforts. Small mammals now comprise almost 12,000 specimens collected from almost 90 screen-washed localities. As discussed in Chapter 4, small-mammal localities are predominantly accumulations of teeth recovered in spatially restricted patches and subsequently buried by fluvial processes—hence representing very brief time intervals. They were mainly concentrated initially by raptorial birds or small carnivorous mammals. In contrast, large-mammal skeletal elements—cranial, dental, postcranial—are scattered and fragmentary. They nevertheless provide adequate sampling of the broader community composition steadily through the sequence and can to some extent provide information about "subhabitat" preferences within always heterogeneous habitats.

After that second season in 1975 I stopped in Paris to meet with a French colleague and friend, the late Yves Coppens, the goal being to recruit specialists in ungulate groups, especially the bovids. I was fortunate to meet Herbert Thomas. This particular French connection led to other collaborators, and not only French: a survey of the authors of the faunal chapters shows just how transnational the project has become. During the third season, we were fortunate again in being able to persuade Kay Behrensmeyer to join. Kay had recently completed her PhD on taphonomy and paleoecology of Plio-Pleistocene sites in East Africa and became a key organizer of a broad range of geological work and fossil and locality documentation methods, a major conduit for recruiting other impressive and talented researchers, and the coauthor of many framing and synthetic chapters in this book.

APPROACHES TO FIELDWORK

With exceptions for particular analytical reasons, we rarely collected all fossil material at a locality. Where possible, we located sites from which earlier published fossils had been recovered, at first particularly hominoids. Over time, with careful attention to stratigraphy, our collecting efforts could focus on more poorly documented time intervals. This became possible because during those first few seasons considerable effort was invested in understanding the classic sequence of rock formations exposed on the Potwar Plateau: Kamlial, Chinji, Nagri, and Dhok Pathan Formations. A three-dimensional geometry of the rocks emerged along with the first steps to dating them (see Chapters 2 and 4, this volume). Earlier I described the sedimentary sequence as "layer cake," but it is more complex than that. Deposition was from a network of major and minor rivers and streams. After episodes of levee buildup, particularly along larger channels, a river could "jump its banks" and shift course, leaving abandoned parts of the river as lakes, ponds, and swamps. Detailed geological work documenting the three-dimensional nature of the sediments allowed reconstruction of contemporaneous landscapes through time, providing a four-dimensional story of change over millions of years.

The ability to reconstruct such a complex history involved studying the paleomagnetism of the sediments, a major research focus of the previously mentioned Dartmouth-Peshawar team. (The principals along with their students are discussed in Chapter 3 of this volume but it is important to recognize some of them here: the late Noye Johnson, Gary Johnson, the late Neil Opdyke, the late Ev Lindsay, John Kappelman, and Lisa Tauxe.) The signatures of normal and reversed magnetic polarity, documented in multiple stratigraphic columns measured through the sediments, can be correlated to a global standard time scale. Over 80 sections were measured, with lengths varying from a few meters to over 3,000 m and almost 60 of them sampled for paleomagnetism. In this way it became possible to calibrate and correlate the rocks and their contained fossils ("calibrate" being a better term than "date" because the ages of magnetic transitions, normal to reversed to normal, have been steadily adjusted as work updates transition ages in the geomagnetic polarity time scale; see Chapter 3, this volume.) By the early- to mid-1980s, through the substantial work of many, we had a good understanding of the ages of the rocks and were able to tie fossil localities into local and magnetically calibrated stratigraphic sequences. The value of these studies to the project cannot be overstated. It became clear that past attempts to disentangle the use of Kamlial, Chinji, Nagri, and Dhok Pathan as names of rock formations from their use as names of faunal chronostratigraphic units were incomplete and at times incorrect. Paleomagnetism was the liberator, as we now had access to a temporal yardstick linked neither to rock formations nor to faunal or temporal units (see Chapter 2, this volume).

Most localities are limited spatially to tens to hundreds of m^2 and temporally to 10^2 to 10^4 years. The analytical quality of the "bins" into which localities could be placed has improved, and their temporal duration has decreased. Our first analyses used 500,000 year intervals. We now have almost 55,000 fossil records from almost 1,300 localities, a significant fraction of them placeable in 100,000 year bins. For such a long sweep of Miocene time, these are unusually and satisfyingly brief. As the fossil record steadily grew, our ability to date localities and their fossils, and hence to analyze faunal change, increased (see Chapter 3, this volume).

Our home base shifted from Yale University to Harvard University in 1981. This was the first field season for another Yale graduate student, John (Jay) Kelley, who has become the project's coordinator for studies of primates, particularly the hominoids.

INTEGRATION OF RESEARCH APPROACHES

By the end of the 1980s, we had become aware that a fossil record of variable quality could at times present confusing patterns of faunal change. Gaps in the record might give the impression that first occurrences clustered close together in time—these could represent either immigrants (range extensions from outside the bioprovince) or endemic originations (speciations). But driving the patterns was often a fossil record of varying quality, and we began using various approaches to infer confidence intervals on observed ranges. When might we comfortably infer the age of a true species appearance relative to its first occurrence or of its disappearance relative to last occurrence? We became interested in the patterns of faunal

turnover. Did episodes of first occurrences (or appearances) coincide with, precede, or follow episodes of last occurrences (or disappearances)? As the analytical intervals or bins became shorter and as the record expanded, we saw more nuanced patterns. There was always turnover, but it might be slow and steady over long intervals and then intermittently sprinkled with briefer intervals showing greater turnover. Our goal became explaining such varying patterns.

As noted earlier, the Potwar sediments themselves contain few botanical remains—trees, leaves, fruits, or pollen—although some plant biomarkers are found. But Siwalik rocks in northwestern India and Nepal do have plant remains indicating that floodplains and the nearby foothills of the mountains were covered with tropical and subtropical forests and woodlands along with varying amounts of nonwoody plants, including cool-season or yearlong growing grasses, all using the C_3 photosynthetic pathway. As reviewed in Chapter 5 of this volume, beginning in the 1980s ground-breaking research by Thure Cerling and his then–graduate student Jay Quade analyzed carbon and oxygen stable isotopes from Potwar paleosols (fossil soils) to reconstruct floodplain habitats (Quade, Cerling, and Bowman 1989). South Asian vegetation was predominantly C_3 until the middle of the Late Miocene. Beginning around 8 million years ago, and depending on location, floodplains rather rapidly became covered by seasonally warm/dry grasses using the C_4 photosynthetic pathway.

In addition to isotopic approaches to reconstructing habitats, other studies contribute to reconstructing the paleobiology of the animals comprising the changing communities. These studies include analyses of carbon and oxygen stable isotopes, as well as tooth wear (micro- and meso-) accumulated just prior to death to monitor physical properties of the most recently chewed items; tooth structure (crown height and morphology) providing further information about the kinds of foods being eaten; and estimates of body size, which are useful indicators of change in community structure. For example, we were able to make plausible inferences about grazing versus browsing, and among browsers about the relative contributions of leaves and fruit. A leader in such studies was Michèle Morgan, who joined the project as a Harvard graduate student in 1987. She is a third editor of this book, and an essential contributor to its production throughout its entire "life."

As the fossil record has grown, its description and analysis have become more mature, and with the ability to package it in ever-briefer "bins," our faunal research has coalesced around a few main themes. First, continuing taphonomic studies showed that the terrestrial vertebrates, most often represented by fragmentary remains, are associated with abandoned channels and that they represent the activities of predators and scavengers around sources of water. They sample the wider animal community and do so repeatedly over long intervals. This approximately "isotaphonomic" sampling of the fauna makes possible the study of individual species lineages and of the clades of which they are members in terms of their dietary, locomotor, and other important paleobiological adaptations. Which new species were immigrants and from where? What impact might immigrants have had on other community members? Which groups waxed or waned? We were able to examine putative or hypothesized competitive interactions, for example, between cricetid and murid rodents, hominoid and cercopithecoid primates, tragulid and bovid artiodactyls, and the impact if any of the Late Miocene insertion of horses into the bioprovince. Our third focus has been on the dynamics of faunal change, culminating most recently on its analysis within a specific interactive model: examining change when the external boundaries to South Asia were relatively closed rather than open and when climate and habitat change could be inferred to be relatively changing or stable (Badgley et al. 2015).

Our final season of work in Pakistan was in 2001. Collecting by colleagues in the Pakistan Museum of Natural History, the GSP, and Punjab University (Khan et al. 2020) has continued when feasible.

We have focused over the past two decades on faunal descriptions and analyses. Although that work will continue after this book is published, what is presented in the faunal chapters of this volume (Chapters 6 through 23) represents the best current statement about the faunal groups. We asked the authors to summarize published systematic studies (some occasionally in press) and to focus wherever possible on paleobiology and adaptations, patterns of change through time, and relevant biogeographical insights. The diversity of the fauna, both in species richness and in the adaptive range of species, is impressive.

These faunal chapters make possible the approaches taken in the final synthetic chapters (Chapters 24 through 27), which integrate faunal data mainly across the 7-million-year analytical interval between 14 and 7 Ma, reviewing several temporally brief "communities" and the intervals of change between them. What has been in equal measure challenging and rewarding is the clearly non-analogue nature of these faunas. These Miocene mammal communities include a very small fraction of species in extant genera; hence the use of a variety of ecomorphological approaches, such as stable isotopes, tooth wear, and dental and postcranial morphologies. Relative to living communities, there is a strikingly high species richness in the smallest mammals in the early samples (analytical focus is mainly on rodents); this proportion declines steadily over our temporal sequence, with notable acceleration after 10 Ma. As impressive is the very high megaherbivore richness across the interval, mainly in rhinocerotids and proboscideans. Within the large-mammal cohort, especially the ungulates, average body size is initially low, but with a significant shift midway through the sequence toward increased body mass. Relative to faunas of similar age in Africa, the species richness of tragulids in our Potwar sequence is higher and of bovids lower; relative to Eurasia, cervids are notably absent from the Potwar Plateau until the Pleistocene.

These patterns of high richness and diversity, relative abundances, and body mass distribution are unmatched in modern communities (Faith et al. 2018; Faith, Rowan, and Du 2019). Also unmatched are the habitats in which the faunas flourished; until very late in our sequence, landscapes were dominated by what must have been highly productive C_3 vegetation.

Back to that original reason for beginning the project in 1973—what happened to the earliest putative human ancestor *Ramapithecus* and to my belief in the importance of *Ramapithe-cus* to human evolution? Three years before the Potwar project began in 1973, the late John T. Robinson visited Yale to examine the original *Ramapithecus* specimens housed there. Robinson was famous for his discoveries, descriptions, and analyses of the australopithecines, undoubted hominins from South Africa. During his Yale visit, Robinson made clear that in his opinion *Ramapithecus* was not a hominin, as he also emphasized in his book (Robinson 1972, 255). He was not alone in expressing skepticism about its hominin status from the viewpoint of morphology or from the genetically-based inference about the timing of the human-ape divergence (Sarich and Wilson 1967).

Although primate fossils are rare in the Siwaliks, after three or four field seasons we had discovered enough new primate material for me also to have doubts, and this can be seen in what I now consider to be my odd published musings about the relationships of *Ramapithecus* and other similar species (Pilbeam et al. 1977). I was trying to accommodate new facts to my a priori assumptions. But as Thomas Henry Huxley noted in his 1870 presidential address to the Liverpool meeting of the British Association, reported in the second year of publication of a new scientific journal, *Nature*: "The great tragedy of science" is "the slaying of a beautiful hypothesis by an ugly fact" (Foster and Lankester 1901, 580).

By the early 1980s, new facts based on more new fossils had arrived, stimulating new interpretations. A reasonably complete face of *Sivapithecus* was discovered in the Potwar on January 23, 1979. It included the palate, lower jaw, and teeth, which showed facial and palatal features resembling the orangutan, *Pongo*, as did on closer analysis the original *Ramapithecus brevirostris* upper jaw fragment. A plausible conclusion was that *Sivapithecus* and *Ramapithecus* were congeneric, perhaps even conspecific (male and female), and that they were best assigned to the lineage leading to *Pongo*. Postcranial remains attributable to *Sivapithecus* continued to accumulate slowly, and most puzzling was that in no way did any of them resemble *Pongo*. Indeed they show surprisingly few resemblances to any living ape. Most view the facial similarities as supporting a link to *Pongo*, while a few prefer to view the matter as unsettled (see Chapter 12, this volume).

We have traveled a long way from "*Ramapithecus* the earliest human ancestor." Along that journey we have come to recognize our collective good fortune: the ability to contribute in some small way to a larger endeavor about the major processes of evolutionary change—biotic, abiotic, or both.

ACKNOWLEDGMENTS

I was extremely fortunate in the support we received from successive program directors at the National Science Foundation, and from Francine Berkowitz of the Smithsonian Institution Foreign Currency Program. I am enormously appreciative of their understanding and flexibility (and funding).

I would especially like to acknowledge Hubert Horatio Humphrey Jr. and Dwight David Eisenhower. Public Law (P. L.) 480, initially proposed by Senator Humphrey and signed into law on July 10, 1954, by President Eisenhower, allowed countries to pay for American food imports in their own currencies, such currencies then being used to pay for a range of projects inside those countries. The Smithsonian Institution was one US entity authorized to distribute such funds.

At Harvard we have benefited greatly from ongoing support from the American School of Prehistoric Research. Our many other collaborators had diverse sources of support.

We are grateful for the essential, ongoing, and continuing support of the Geological Survey of Pakistan and of its successive directors general, particularly our colleagues and friends, S. Mahmood Raza and S. M. Ibrahim Shah, as well as colleagues in the Pakistan Museum of Natural History, in both field and laboratory, including I. U. Cheema.

We deeply appreciate the extraordinary friendliness, generosity, and kindness of the many Pakistani field and laboratory personnel who made our work possible, indeed of the people of Pakistan in general.

At the end of the Preamble to Framing Chapters we present as comprehensive a list as possible of the very many people who have been involved in the field over the past five decades.

We want to thank the 33 contributors to the 27 chapters, whose impressive patience over the past too many years we appreciate.

I must single out a few for special acknowledgment here. Our friend and collaborator, now back in Pakistan, S. Mahmood Raza, who has been essential to our success. My lab group at Harvard, John Barry, Larry Flynn, and Michèle Morgan, along with two other close colleagues in the United States, Catherine Badgley and Kay Behrensmeyer, have been essential to the final production of this book. Thure Cerling, and Jay Kelley have also been great sources of stimulation and argument. Our editor at Johns Hopkins University Press, Tiffany Gasbarrini, has shown great patience with and faith in this project, and she has been critical in bringing it to completion. Last but by no means least, we are grateful to our superb field managers, the late Grant Meyer and the unflappable Hod French.

It has been an enormous privilege and great fun to work with such a talented group; I have learned far more from them than I can ever adequately acknowledge.

One truly final point. It would be mistaken to see the project described here as a grand plan conceived in 1973—it was not. The research, like the group, grew by accretion. A recruit added for a specific reason might well end up following a new direction; unexpected opportunities regularly appeared, usually to be seized. At any particular time there was a plan, but it was always a plan that constantly evolved.

References

Anderson. R. V. V. 1927. Tertiary stratigraphy and orogeny of the Northern Punjab. *Bulletin of the Geological Society of America* 38(4): 665–720.

Badgley, C., M. S. Domingo, J. C. Barry, M. E. Morgan, L. J. Flynn, and D. Pilbeam. 2016. Continental gateways and the dynamics of mammalian faunas. *Comptes Rendu Palevol* 15(7): 763–779.

Barnosky, A. D. 2001. Distinguishing the effects of the Red Queen and Court Jester on Miocene mammal evolution in the northern Rocky Mountains. *Journal of Vertebrate Paleontology* 21(1): 172–185.

Bennett, K. D. 2004. Continuing the debate on the role of Quaternary environmental change for macroevolution. *Philosophical Transactions of the Royal Society of London. Series B: Biological Sciences* 359(1442): 295–303.

Brown, B., W. K. Gregory, and M. Hellman. 1924. On three incomplete anthropoid jaws from the Siwaliks, India. *American Museum Novitates* no. 130.

Colbert, E. H. 1935. Siwalik mammals in the American Museum of Natural History. *Transactions of the American Philosophical Society (New Series)* 26:1–401.

Corvinus, G. 1993. The Siwalik Group of sediments at Surai Khola in western Nepal and its palaeontological record. *Journal of Nepal Geological Society* 9:21–35.

Corvinus, G., and L. N. Rimal. 2001. Biostratigraphy and geology of the Neogene Siwalik Group of the Surai Khola and Rato Khola areas in Nepal. *Palaeogeography, Palaeoclimatology, Palaeoecology* 165(3): 251–279.

Cotter, G. De P. 1933. The geology of the part of the Attock District west of longitude 72° 45′ E. *Memoirs of the Geological Survey of India* 55:63–161.

Darwin, C. 1964. *On the Origin of Species: A Facsimile of the First Edition.* Cambridge, MA: Harvard University Press.

Dehm, R. 1983. Miocene hominoid primate dental remains from the Siwaliks of Pakistan. In *New Interpretations of Ape and Human Ancestry,* edited by Russell L. Ciochon, and Robert S. Corruccini, 527–537. New York: Plenum Press.

Dehm, R., T. Prinzessin zu Oettingen-Spielberg, and H. Vidal.1958. Palaontologische und geologische Untersuchungen im Tertiar von Pakistan. I. Die Munchner Forschungsreise nach Pakistan, 1955–1956. *Abhandlungen Bayerische Akademie der Wissenschaften, Mathematisch-Naturwissenschaft liche Klasse (Neue Folge)* 90:1–13.

Faith, J. T., J. Rowan, and A. Du. 2019. Early hominins evolved within non-analog ecosystems. *Proceedings of the National Academy of Sciences* 116(43): 21478–21483.

Faith, J. T., J. Rowan, A. Du, and P. L. Koch. 2018. Plio-Pleistocene decline of African megaherbivores: No evidence of ancient hominin impacts. *Science* 362(6417): 938–941.

Falconer, H. 1837. Note on the occurrence of fossil bones in the Sewalik Range, eastward of Hardwar. *Journal of the Asiatic Society of Bengal* 6:23.

Fatmi, A. N. 1973. Lithostratigraphic units of the Kohat-Potwar Province, Indus Basin, Pakistan. *Memoirs of the Geological Survey of Pakistan* 10:1–80.

Foster, M., and E. R. Lankester. 1901. *The Scientific Memoirs of Thomas Henry Huxley Vol. III.* London: Macmillan and Co. Limited.

Fraser, D., L. C. Soul, A. B. Tóth, M. A. Balk, J. T. Eronen, S. Pineda-Munoz, A. B. Shupinski, et al. 2021. Investigating biotic interactions in deep time. *Trends in Ecology and Evolution* 36(1): 61–75.

Gill, W. D. 1951. The stratigraphy of the Siwalik Series in the Northern Potwar, Punjab, Pakistan. *Journal of the Geological Society* 107:375–394.

Gregory W. K., M. Hellman, and G. E. Lewis. 1938. Fossil anthropoids of the Yale-Cambridge India expedition of 1935. *Carnegie Institution of Washington* No. 495, 1–26.

Khan, M. A., J. Kelley, L. J. Flynn, M. A. Babar, and N. G. Jablonski. 2020. New fossils of *Mesopithecus* from Hasnot, Pakistan. *Journal of Human Evolution* 145:102818.

Lewis, G. E. 1934. Preliminary notice of new man-like apes from India. *American Journal of Science (Series 5)* 27:161–179.

Lewis, G. E. 1937. A new Siwalik correlation (India). *American Journal of Science (Series 5)* 33:191–204.

Lydekker, R. 1877. Notices of new and other vertebrata from Indian tertiary and secondary rocks. *Records of the Geological Society of India* 10:30–43.

MacFadden, B. J., and A. Bakr. 1979. The horse *Cormohipparion theobaldi* from the Neogene of Pakistan, with comments on Siwalik hipparions. *Palaeontology* 22:439–447.

Patnaik, R. 2013. Indian Neogene Siwalik mammalian biostratigraphy: An overview. In *Fossil mammals of Asia*, edited by X.-M. Wang, Lawrence J. Flynn, and Mikael Fortelius, 423–444. New York: Columbia University Press.

Pilbeam, D., G. E. Meyer, C. Badgley, M. D. Rose, M. H. L. Pickford, A. K. Behrensmeyer, S. M. I. Shah. 1977. Hominoid primates from the Siwaliks of Pakistan and their bearing on hominoid evolution. *Nature* 270(5639): 689–695.

Pilgrim, G. E. 1910. Notices of new mammalian genera and species from the Tertiaries of India. *Records of the Geological Survey of India* 40:63–71.

Pilgrim, G. E. 1913. The correlation of the Siwaliks with mammal horizons of Europe. *Records of the Geological Survey of India* 43:264–326.

Quade, J., J. M. L. Cater, T. P. Ojha, J. Adam, and T. M. Harrison. 1995. Late Miocene environmental change in Nepal and the northern Indian subcontinent: Stable isotopic evidence from paleosols. *GSA Bulletin* 107(12): 1381–1397.

Quade, J., T. E. Cerling, and J. R. Bowman. 1989. Development of Asian monsoon revealed by marked ecological shift during the latest Miocene in northern Pakistan. *Nature* 342(6246): 163–166.

Reznick, D. N., and R. E. Ricklefs. 2009. Darwin's bridge between microevolution and macroevolution. *Nature* 457(7231): 837–842.

Robinson, J. T. 1972. *Early Hominid Posture and Locomotion*. Chicago: University of Chicago Press.

Sarich, V. M., and A. C. Wilson. 1967. Immunological time scale for hominid evolution. *Science* 158(3805): 1200–1203.

Simons, E. L. 1961. The phyletic position of *Ramapithecus*. *Postilla* 57.1–9.

Simons, E. L. 1964. On the mandible of *Ramapithecus*. *Proceedings of the National Academy of Sciences* 51: 528–535.

Simons, E. L., and S. R. K. Chopra. 1969. *Gigantopithecus* (Pongidae, Hominoidea): A new species from North India. *Postilla* 138:1–18.

Simpson, G. G., 1944. *Tempo and Mode in Evolution*. New York: Columbia University Press.

Van der Geer, A., M. Dermitzakis, and J. de Vos. 2008. Fossil folklore from India: The Siwalik Hills and the *Mahâbhârata*. *Folklore* 119(1): 71–92.

Van Valen, L. 1973 A new evolutionary law. *Evolutionary Theory* 1:1–30.

Von Koenigswald, G. H. R. 1983. The significance of hitherto undescribed Miocene hominoids from the Siwaliks of Pakistan in Senckenberg Museum, Frankfurt. In *New Interpretations of Ape and Human Ancestry*, edited by Russell L. Ciochon, and Robert S. Corruccini, 517–526. New York: Plenum Press.

West, R. M., J. R. Lukacs, J. Munthe Jr., and S. T. Hussain. 1974. Vertebrate fauna from Neogene Siwalik Group, Dang Valley, Western Nepal. *Journal of Paleontology* 52(5): 1015–1022.

2 ROCKS, RIVERS, AND TIME

ANNA K. BEHRENSMEYER AND JOHN C. BARRY

INTRODUCTION

In this chapter, we describe and reinterpret the classic Siwalik and Rawalpindi Group formations of the Potwar Plateau in northern Pakistan as the cumulative deposits of rivers, the primary physical agents that shaped the Siwalik record of life over time. Multiple large river systems responded to millions of years of uplift caused by the collision of India with Asia by depositing vast quantities of sediment south of the rising mountains in an arcuate trough called the sub-Himalayan foreland basin. We include a summary of the Zinda Pir region, which lies west of the Indus River and the Potwar Plateau, and briefly mention the Siwalik deposits in Sind, southwestern Pakistan (Fig. 2.1). The fossiliferous marginal marine and fluvial deposits of these regions provide important Late Oligocene and Early Miocene fossils that record faunas and environments older than those of the Potwar Plateau.

Our project has assembled a unique space-time picture of the three-dimensional architecture of these complex fluvial systems from decades of sedimentological field research, documenting both vertical stratigraphy and lateral patterns of deposition in different regions of the Potwar Plateau. Along with other types of geological and geochemical evidence, this information allows us to reconstruct the physiographic structure and substrates of successive Miocene paleo-landscapes, providing environmental context for changing vegetation and faunas at local to regional scales.

None of this would make sense, however, without the detailed chronological framework for the thick rock sequences that our project developed using paleomagnetic stratigraphy. This approach relies on measuring "normal" versus "reversed" polarity in the sedimentary rocks and linking these changes to well-dated changes in Earth's magnetic polarity field over time. The broad synthesis of Potwar geology and geochronology is built on knowledge accumulated and tested over five decades by the dedicated work of many scientists and graduate students. Coordinated field and laboratory research on this scale is a true group effort, and we do our best to fairly represent the major contributions of the extended team.

In this chapter, we first describe the geology of the Potwar Plateau and Zinda Pir regions, then discuss the chronostrati-graphic methods that allow us to place the rock layers and faunas accurately in time. We follow with descriptions of the fluvial sediments and interpretations of the environments they represent at different scales and conclude with an overview of change through time in the depositional systems of the western sub-Himalayan foreland basin. All of this sets the stage for the following chapters, which focus on the extraordinary paleontological record preserved because of the Cenozoic tectonic forces and fluvial depositional processes that shaped this region of southern Asia.

THE SIWALIK GROUP

The Siwalik Group includes the Chinji, Nagri, Dhok Pathan, and Soan Formations, which are extensively exposed in the Potwar Plateau, as well as the more localized Tatrot and Samwal Formations. The Kamlial Formation underlies the Chinji Formation, and with the Murree Formation constitutes the Rawalpindi Group (Cheema, Raza, and Ahmed 1977; Burbank, Beck, and Mulder 1996; Friend et al. 2001; Flynn et al. 2013; Fig. 2.2). These formations record the sedimentary filling of the huge foreland basin formed when Earth's crust buckled downward during the early stages of collision between the Indian subcontinent and Asia to form an arcuate depression that stretches for over 2,500 km south along the plate boundaries (Fig. 2.1a). Dates for the Murree Formation farther north indicate that the foreland basin began to form as early as 55 million years ago (Ma) and extended south over time (Bossart and Ottiger 1989; Burbank, Beck, and Mulder 1996).

The Potwar Plateau is an uplifted block of these sediments, which provides a window into the geological history near the western end of the foreland basin (Fig. 2.1). This window represents only about 2% of the total estimated area of the basin (18,000 km^2/850,000 km^2) but is the best-exposed and well-studied region for the Siwalik sequence, which also extends westward into Afghanistan and eastward across India and Nepal, with similar strata as far as Assam, north of Bangladesh. (See Flynn et al. 2013 for more details.) Our project has focused on the Kamlial through Dhok Pathan Formations (see

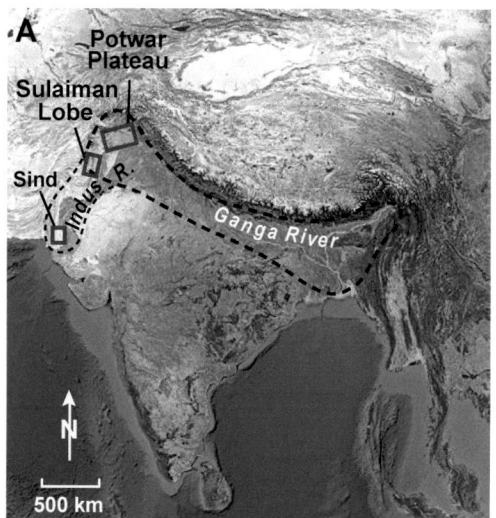

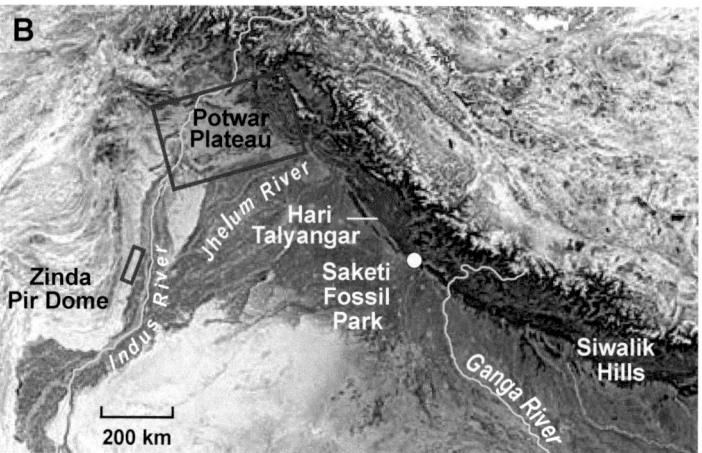

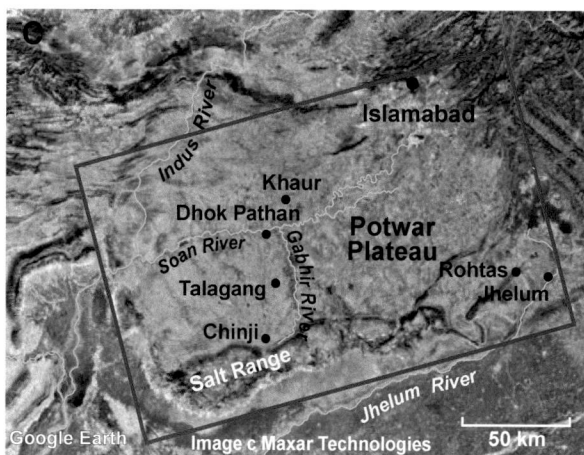

Figure 2.1. A. Map of Indian subcontinent and Tibetan Plateau, with major study areas, showing the extent of the sub-Himalayan foreland basin. Thicker dashed line = Miocene–Pleistocene foreland basin; thinner dashed line = earlier fossiliferous deposits in the lower Indus region (part of Sind). B. View of main region of Siwalik exposures in Pakistan and India. C. Closer view of Potwar Plateau showing the less deformed Soan Synclinorium between more folded strata to the east and west, and landmarks mentioned in the text. See also Figure 2.3. Images: Google Earth, 2020.

Fig. 2.2). These four rock formations together are over 3,000 m thick and represent a relatively continuous record of riverborne sediments between 18 and 3.5 Ma. The Kamlial Formation overlies the fluvial Murree Formation, which is poorly exposed in the Potwar region and has few vertebrate fossils, although it contains the distinctive large foramineran, *Numulites*, reworked by Miocene erosion of shallow marine deposits that formed in the Eocene Tethys Sea.

Our geological and paleontological work on the Potwar Plateau has concentrated on the Khaur, Chinji, Hasnot, and Rohtas areas (Fig. 2.3). The Khaur area is on the northern limb of a large geological structure known as the Soan Synclinorium, and the other areas are distributed along its southern limb, just north of the Salt Range (Fig. 2.3a). Eocambrian salt deposits underlie the Potwar region and allowed a structural detachment ("decollement") of a large "plate" of the overlying sedimentary rocks during the Plio-Pleistocene phase of Himalayan uplift. This thrust plate slid southward over underlying, relatively viscous salt deposits as the Indian subcontinent continued to move under Asia (Fig. 2.3b; Burbank, Beck, and Mulder 1996; Borderie et al. 2018). Many other areas along the major boundary fault suffered intense faulting and compression, but because of the relatively gentle structural deformation that formed the Potwar Plateau, the Miocene–Pliocene stratigraphic sequence

was preserved. Paleomagnetic readings taken at different sites on the plateau show that portions of the block also rotated up to 10 degrees counterclockwise as it slid southward (Opdyke et al. 1982). On the north side of the plateau, strata dip 12–15 degrees southward, while those in the south dip 12–15 degrees northward, with laterally continuous outcrops exposed by streams that follow the strike valleys and cut through them at fairly regular intervals. The landscape is relatively free of vegetation due to intense summer heat, erosion by monsoonal rains, and livestock foraging. This creates ideal topographic conditions for following vertical and lateral stratigraphy over large areas as well as exposing fossils that can be precisely linked to sedimentary context and geochronology.

The different formations of the Rawalpindi Group and the Lower and Middle Siwalik Group are relatively easy to distinguish based on color and proportion of channel-deposited sandstones versus silty to clayey floodplain sediments (Fig. 2.4). This distinction was recognized early in the geological exploration of the Potwar Plateau (Pilgrim 1913, 1926, 1934; Gill 1951; Fatmi 1973; Cheema, Raza, and Ahmed 1977; Hussain et al. 1992), as well as in adjacent areas in India (Siwalik Hills; Prasad 1968; Pillans et al. 2005; Patnaik 2013) and the "trans-Indus" region (Sulaiman Lobe, Zinda Pir Dome, Daud Khel) to the southwest (Hussain et al. 1977; Hussain 1979; Munthe et al. 1979; Burbank,

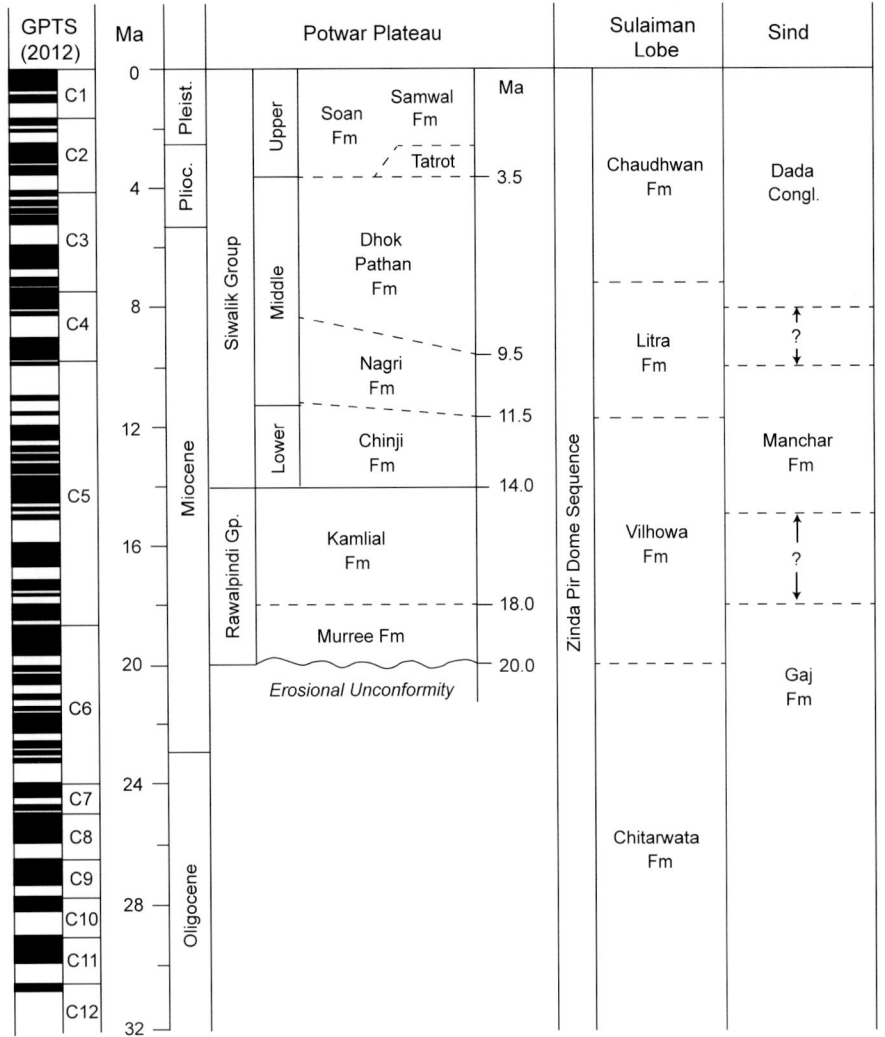

Figure 2.2. Generalized stratigraphic relationships and chronostratigraphy of rock formations in northern Pakistan that preserve the fossil record discussed in this and other chapters. Global Polarity Time Scale (GPTS) from Gradstein et al. (2012). Black = normal, white = reversed, C = Chron, short for Magnetochrons of the GPTS, left column. Potwar Plateau stratigraphy from Cheema, Raza, and Ahmed (1977); chronostratigraphy based on updated information by our team. Sulaiman Lobe stratigraphy and approximate dating (dashed lines) adapted from Lindsay and Flynn (2016) and Sind from Flynn et al. (2013).

Beck, and Mulder 1996; Figs. 2.1, 2.2). The Upper Siwalik Group has fewer fossils and a greater proportion of sand and gravel, representing the coarsening-upward of the overall foreland basin sequence. From older to younger, the Potwar Plateau Formations (Fig. 2.2) can be described as follows:

Rawalpindi Group Formations (Cheema, Raza, and Ahmed 1977)
- Murree—dark red and purple clay and purple, gray and greenish-gray sandstone with subordinate intraformational conglomerate. The basal strata of the formation consist of light greenish-grey calcareous sandstone and conglomerate with abundant derived Eocene large foraminifers (20–18 Ma).
- Kamlial—dark red to purple mudstones with interbedded gray and red, medium- to coarse-grained sandstones and laterally extensive yellow and purple intraformational conglomerates (18–14 Ma).

Siwalik Group Formations
Lower Siwaliks
- Chinji—thick red to orange mudstones and small-scale gray channel sandstones, alternating with laterally extensive gray

sandstone bodies, with multiple fining-upward cycles recorded in the mudstone lithofacies, paleosols, and abandoned channel fills, also with occasional laterally extensive intraformational conglomerates (14–11.5 Ma).

Middle Siwaliks
- Nagri—multistorey gray sandstones up to 30 m thick with interbedded thin intervals of orange, brown, and yellow mudstones, paleosols, and occasional large-scale channel fills (11.5–9.5 Ma).
- Dhok Pathan—yellow to orange mudstones with paleosols, alternating with gray and buff sandstone bodies representing different alluvial source areas (mountain versus foothill), with increased dominance of buff sandstones upward in the sequence (9.5–~3.5 Ma).

Upper Siwaliks
- Soan—buff fluvial sandstones and conglomerates with associated sandy and silty overbank deposits. Only locally exposed in our study area; includes strata assigned to the Tatrot and Samwal Formations, also very locally exposed (Flynn et al. 2013; ~3.5–0.9 Ma).

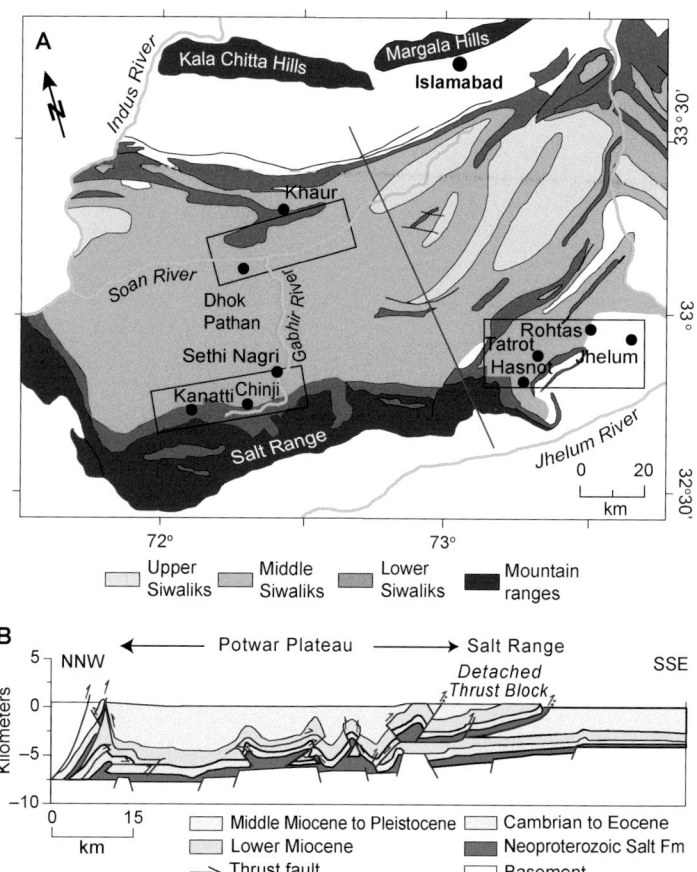

Because the Kamlial Formation is described as part of the Rawalpindi Group (Cheema, Raza, and Ahmed 1977) rather than the Siwalik Group, it is technically incorrect to refer to all the formations targeted for study in the Potwar Plateau sequence as "the Siwaliks." However, in this book we use "the Siwaliks" or "Potwar Siwaliks," as an informal shorthand that includes the Kamlial Formation with Siwalik Group formations. Note also that the descriptions and dates provided for these formations relate to Potwar Plateau exposures and may differ in other areas.

The Potwar Plateau Miocene to Pleistocene rocks are capped by the Potwar Silts. During the later Pleistocene glacial periods, wind-blown and water-transported silt (loess) accumulated in the bowl-like Potwar synclinorium, burying previously eroded topography and creating an extensive plain with fertile soil. The Potwar Silts are now rapidly eroding, but these unconsolidated deposits still cover large areas of the Siwalik and older exposures.

ZINDA PIR DOME AND SULAIMAN FOOTHILLS

The beginning of the story of the Siwalik faunas and paleoenvironments is recorded on the eastern flank of the Sulaiman Range and the Bugti Hills, west of the Indus River and south of the Potwar Plateau (see Fig. 2.1a). The area around Dera Bugti, at the southern end of the Sulaiman Range, is known

for important late Oligocene–early Miocene fossils including the immense perissodactyl *Paraceratherium* (Raza et al. 2002; Métais et al. 2009; Antoine et al. 2013). Early work by our team (Downing et al. 1993) on the Chitarwata and Vihowa Formations exposed farther north in the Zinda Pir Dome recovered assemblages of vertebrate fossils biostratigraphically older than those of the Siwalik Group. Paleomagnetic chronology now places the age of the Chitarwata Formation in the Zinda Pir Dome (Fig. 2.1b) as Oligocene at the formation's base and earliest Miocene at the contact with the overlying Vihowa Formation (Lindsay et al. 2005; Métais et al. 2009). In this age assessment, the Chitarwata Formation overlaps with the Murree Formation, which is exposed in the northwest part of the Potwar Plateau (Abbasi and Friend 1989; Najman et al. 2003; see Fig. 2.2).

Sulaiman Lobe Formations

- Chitarwata Formation—red, gray, and green claystone and subordinate amounts of siltstone and sandstone; highly variable lithology with hard ferruginous sandstones and conglomerates near the top (Hemphill and Kidwai 1973); includes marine invertebrates and trace fossils indicating estuarine and coastal environments (Lindsay and Flynn 2016; Métais et al. 2009; ~30–20 Ma).
- Vihowa Formation—gray and brown sandstone and subordinate amounts of red and brownish-red sandy siltstone. The sandstone is medium to coarse in texture, with massive,

Figure 2.4. Representative images of the four Siwalik formations of the Potwar Plateau that are the source of the vertebrate fossil record and the main focus of the research presented in this volume. A. Kamlial Formation at Locality Y591, with Will Downs for scale. B. Lower Chinji Formation near Chinji Village, with Gabhir Kas (river). C. Nagri Formation looking north into the Nagri Gorge, with Gabhir Kas. D. Dhok Pathan Formation exposed in a high cliff under the "UU" sandstone (right column of Fig. 2.5), Kaulial Kas, east of Khaur Village, small human figure in lower left center for scale.

thick-bedded, and cross-bedded units (Hemphill and Kidwai 1973). This formation is fluvial throughout and is interpreted as the earliest record of an Indus River and its delta (Downing et al. 1993; Downing and Lindsay 2005; 20–~11 Ma).

According to work by Downing and others, these formations represent a coastal to fluvial (river) transition associated with the regression of the Tethys Sea. Fluvial systems drained areas to the northwest and prograded into a basin south of the Potwar region (Métais et al. 2009). The Chitarwata and Lower Vihowa Formations record, in ascending order, estuary, strand plain, tidal flat, tidal channel, and fluvial paleoenvironments (Downing et al. 1993). The Sulaiman/trans-Indus depositional system was associated with tectonic upheaval resulting from the early phases of collision of the Indian subcontinent with Asia, but the area is ~500 km south of the Potwar Plateau and has its own lithostratigraphy and paleoenvironmental history. The Potwar Rawalpindi and Siwalik Groups are fluvial throughout and characterized by relatively continuous deposition on the sub-Himalayan alluvial plain, preserving magnetic reversal patterns that can be confidently matched with the Global Polarity Time Scale (GPTS; Gee and Kent 2007; Gradstein et al.

2012). In contrast, the magnetostratigraphic record of the Zinda Pir sediments contains normal and reversed polarity patterns that are less clearly relatable to magnetostratigraphic time, indicating variable rates of deposition and significant hiatuses (time breaks) in the stratigraphic record (Métais et al. 2009).

MAGNETOSTRATIGRAPHY—THE KEY TO SIWALIK TIME

The Siwalik rocks generally lack volcanic components that would allow direct radiometric dating. Biostratigraphy using the mammalian fossils provided initial chronological controls for earlier research, but rigorous investigation of faunal changes and paleoecology through time requires a dating method that is not dependent on the fossils themselves. The sequential formations (Kamlial, Chinji, Nagri, Dhok Pathan) are clearly represented in different areas of the Potwar Plateau, and earlier researchers assumed that the distinctive rock units also represented discrete blocks of time (Colbert 1935). We now know, however, that these formations interfinger with each other, that they are thus partly contemporaneous, and that the formational boundaries occur at different times in different places (i.e., they are time-transgressive; Flynn et al. 2013). The magnetostratigraphic framework allows us to understand these geological relationships and provides absolute dates for the rich Siwalik fossil record, facilitating reconstructions of evolutionary and ecological changes through time.

Earth's magnetic field has changed polarities many times over the geological past, and a distinctive sequence of "normal" (i.e., aligned with present north) versus "reversed" (a compass would point south) has been established for the GPTS (see Fig. 2.2; Gradstein et al. 2012). The GPTS provides the standard for dating all magnetostratigraphic sequences across Earth's surface, with absolute ages of reversals periodically revised as dating methods improve.

The Siwalik sequence played an important role in the development of magnetostratigraphy in dating terrestrial sedimentary rocks. Neil Opdyke, a leader in demonstrating polarity reversals in deep-sea sediments, helped initiate research in the Siwaliks to test whether it was feasible to obtain polarity records in terrestrial depositional sequences (Opdyke et al. 1979). This work was expanded by Noye Johnson (Johnson et al. 1982b) and his Dartmouth College students (e.g., Bob Raynolds, Jeff Barndt, and Lea McRae), as well as by Gary Johnson, Everett Lindsay, Lisa Tauxe, Jason Hicks, Imran Khan, Khalid Sheikh, John Kappelman, and others who contributed to the field sampling and laboratory analyses required to build a detailed magnetostratigraphy for the Potwar Plateau (Tauxe and Opdyke 1982; Johnson et al. 1985) and the Zinda Pir sedimentary succession (Lindsay et al. 2005; Lindsay and Flynn 2016). Bob Raynolds is also a long-term contributor to documenting Siwalik stratigraphy and dating (Raynolds et al. 2022 and references therein) in Pakistan and established a website compiling published information on strata thicknesses and magnetostratigraphy (Raynolds n.d.).

In sedimentary sequences such as the Siwaliks, magnetostratigraphy depends on rock polarity established during or soon after sediment deposition and subsequently preserved without significant change up to the present. Microscopic particles of hematite and magnetite in sand, silt, and clay impart this signal when they align with Earth's magnetic field, before being immobilized in hardened rocks. Sensitive magnetometers can detect even weak magnetic signals, but because these signals can be reset later by "overprinting" of unstable magnetic components, great care is needed during laboratory analysis to recover the original signals. This involves a chamber shielded from the present-day magnetic field and a stepped-heating process in which each sample is analyzed at increasing temperatures to determine the most stable, presumably original, magnetic signal. The samples used to build a magnetostratigraphic record are small cubes or cylindrical cores cut from rock and marked with compass orientations before being removed from the outcrop. The orientation of the sample in the rock strata is used in combination with the laboratory measurement of polarity to determine the original north versus south orientation of the sediment's magnetic components when it first lithified.

Magnetostratigraphers typically take sequential samples at regular intervals through layers of strata (e.g., every 5 or 10 meters) to construct a vertical record of normal versus reversed polarity through time. This sequential pattern of different stratigraphic thicknesses of normal (black) and reversed (white) "bands" (Fig. 2.5) is then compared with the GPTS (Fig. 2.2) to determine the age of the rock sequence. Variable rates of sediment accumulation can make the correlation challenging, and other age information from fossils or radiometric age dating is often needed to place the rock sequence in the correct interval of geological time. In the Potwar Plateau Siwalik formations, we have two radiometric dates that inform our magnetostratigraphic column. Early in the efforts to develop Siwalik magnetostratigraphy, a thin volcanic ash with zircons was discovered in the Gabhir Kas section (Nagri Fm) and dated at 9.5 ± 0.63 Ma (Johnson et al. 1982a). This age confirmed the presence of C5n, the "long normal" in our sequence, although the original date no longer fits the updated GPTS calibration. A second volcanic ash provides a date of 3.0 ± 0.1 Ma for part of the Rohtas section (Soan Formation; Behrensmeyer et al. 2007).

The drainage patterns in both the Potwar Plateau and Zinda Pir regions, combined with the gentle dips of the tilted strata, continuous vertical exposure, and suitable lithologies, greatly facilitated the development of our chronostratigraphy. Drainages typically form a lattice-like arrangement of small intermittent streams (see Fig. 2.6). The larger streams, which are perpendicular to strike (i.e., the linear trace that a tilted rock layer makes with a horizontal plane) are spaced every 3 to 5 km and have lateral branches parallel to strike that expose continuous bands of outcrop between the larger streams. Many people and field seasons were devoted to sampling vertical sections of strata along the larger streams at different places across the Potwar Plateau, then filling in between these "long sections" to establish lateral correlations and to control for local variation in the magnetostratigraphic record (e.g., cryptic hiatuses, or missing time intervals). These efforts resulted in well-established sequences of normal and reversed polarity that could be assembled into larger chronostratigraphic panel diagrams showing how the fossil-bearing sedimentary sequences correlate in both time and space (Figs. 2.7, 2.8; Chapter 3, this volume). We also traced specific magnetic reversals, or isochrons, laterally by sampling the same short sedimentary intervals over multiple kilometers (e.g., Behrensmeyer and Tauxe 1982; Tauxe and Badgley 1988; Behrensmeyer et al. 2007), which provided added certainty regarding correlation of the vertical magnetostratigraphic records.

Much of the Potwar Plateau Siwalik magnetostratigraphic record is based on sampling fine-grained floodplain sediments, although sandstones can also provide reliable polarity signals. Lisa Tauxe established detrital hematite as the primary source of remnant magnetism in the Siwalik sediments and examined other sources of magnetic overprinting (Tauxe, Kent, and Opdyke 1980; Tauxe et al. 1990). As part of this investigation, she tested whether the signals were "locked in" at an early stage of lithification by collecting paleomagnetic samples from mudstone clasts that had been reworked from the floodplain into

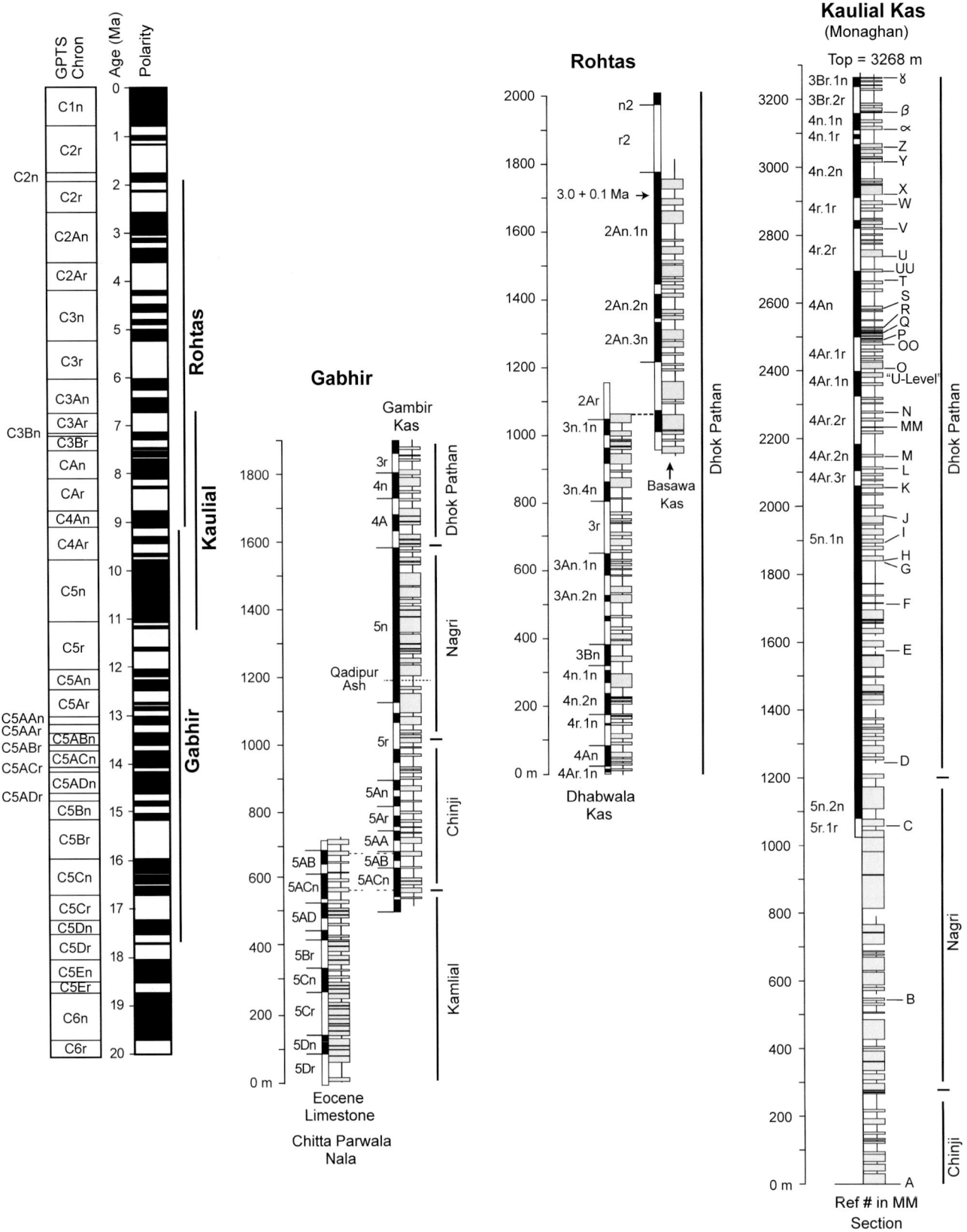

Figure 2.5. Reference stratigraphic sections showing generalized lithology and magnetostratigraphy (GPTS Chrons) for the three major regions of the Potwar Plateau Siwalik sequence (see Fig. 2.3, Table 2.1). Vertical black bars to the right of the GPTS polarity column show where each section correlates to the GPTS. Kaulial Kas lacks magnetostratigraphy for the lower part but extends into the Chinji Formation based on lithofacies descriptions. Colored blocks indicate sandstones; blue indicates blue-gray fluvial system; yellow indicates buff fluvial systems, respectively (Behrensmeyer and Tauxe 1982). The transition between these two systems occurs earlier in the north (Kaulial Kas ~10 Ma) than in the south (Rohtas ~8 Ma) and is not evident in the Gabhir section at ~9 Ma. This pattern is consistent with increasing influence of the foothill-sourced buff system as it prograded into the foreland basin from the north. All sections are drawn to the same vertical scale. Note that "U-level" marked on the Kaulial Kas section is based on an earlier sandstone labeling system.

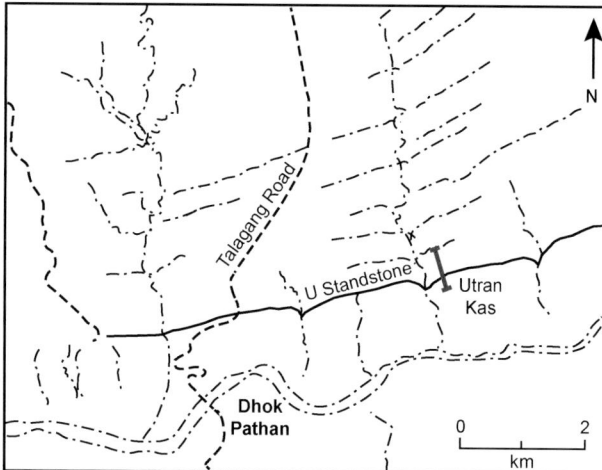

Figure 2.6. Google Earth view of the strata near Dhok Pathan village, the type area of the Dhok Pathan Formation. The stratigraphic dip is to the south on the northern limb of the Soan Synclinorium (Fig. 2.3). The Nagri Formation to the north of the Soan River shows the typical lattice outcrop pattern formed by drainages that have eroded the strata parallel to strike, feeding into larger drainages cutting perpendicular to strike. In this area, the Nagri Formation consists mainly of multistorey sandstones; the overlying Dhok Pathan Formation to the south is composed of interbedded silts and sands. The "U" Sandstone serves as an isochronous marker linking the Nagri and Dhok Pathan Formations (see Fig. 2.7). Bar at Utran Kas indicates position of measured section.

conglomerates in the same strata. She found that the clasts were magnetized before being reworked; that is, they acquired measurable polarity signals, but the north-south orientations were randomized relative to the consistent polarity of the original floodplain deposits from which they were reworked. This confirmed that magnetic signals in these sediments were locked in soon after the strata were deposited. Years of laboratory analysis also demonstrated that Siwalik and underlying fluvial sediments preserve reproducible magnetostratigraphic data, which has added to our confidence that they provide a reliable chronostratigraphic framework for the Potwar Plateau Siwaliks (Barndt 1977; Barndt et al. 1978; Opdyke et al. 1979; Barry, Behrensmeyer, and Monaghan 1980; McMurtry 1980; Behrens-

meyer and Tauxe 1982; Johnson et al.1982b; Stix 1982; Tauxe and Opdyke 1982; Johnson et al. 1985; Tauxe and Badgley 1988; Willis 1993a, 1993b), as well as the Zinda Pir sequence in southwestern Punjab (Lindsay et al. 2005; see Figs. 2.2, 2.6).

The resulting chronostratigraphy for the Potwar Plateau is based on 44 measured sections between 250 m and 3200 m thick, as well as 43 shorter sections. Thirty-one of the 44 long sections and 30 of the 43 short sections have magnetic polarity stratigraphies. These 87 sections (61 total with magnetic polarity stratigraphy) form three regional networks (see Figs. 2.3a, 2.5; Table 3.1, this volume) corresponding to the classic collecting areas of Guy Pilgrim and subsequent authors, with the exception of the Rohtas and Jalalpur areas. Within each region, exposures are vertically and laterally continuous, and individual sections can be correlated lithologically as well as by magnetic polarity. However, between regions the absence of connecting exposures means that longer-distance correlations depend entirely on the magnetostratigraphy. This approach allows us to confidently assign absolute ages (with appropriate error bars) to all of the Potwar Plateau fossil-bearing strata that we have studied; no sections in our record have been correlated on the basis of biostratigraphy.

In the fluvial Siwalik formations of the Potwar region, stratigraphic sections, which are typically about 250 m long in the Potwar region, contain six or seven magnetic transitions that can usually be independently correlated to the GPTS (Johnson and McGee 1983; see Fig. 3.7, this volume). Although sections with fewer magnetic transitions may not be independently matched to the GPTS, they can be placed into the chronostratigraphic framework using laterally continuous lithological horizons by determining their stratigraphic relationships to the longer sections (Behrensmeyer and Tauxe 1982; Johnson et al. 1988; Badgley and Tauxe 1990; McRae 1989, 1990; Kappelman et al. 1991; Sheikh 1984). Our methods for correlating sections to the GPTS are discussed in detail in Barry et al. (2002, 2013) and in Chapter 3 of this volume. These stratigraphic correlations, both between sections and to the GPTS, have not changed, but in this volume, we use the timescale of Gradstein et al (2012) as the reference for magneto-chron boundary ages (or "Chron") boundary ages rather than previous timescales.

In the Zinda Pir sequence, the marginal marine deposits of the Chitarwata Formation (Fig. 2.2) are harder to place in the GPTS due to apparently longer, more irregular cryptic hiatuses. The Zinda Pir chronostratigraphic framework is based on 5 paleomagnetically sampled sections (Friedman et al. 1992; Lindsay et al. 2005; Métais et al. 2009; Lindsay and Flynn 2016) that together are 700 meters thick. In contrast to the wholly fluvial formations of the Potwar Plateau, in the Zinda Pir region the Chitarwata Formation was deposited near the shoreline of a receding seaway and is dominated by coastal sediments representing estuarine, strandplain, and tidal-flat paleoenvironments. The overlying Vihowa Formation is wholly fluvial (Downing et al. 1993, 2005; Lindsay et al. 2005, Métais et al. 2009). As on the Potwar Plateau, the Zinda Pir sections are correlated to each other by tracing lithologic transitions and the magnetic polarity stratigraphy.

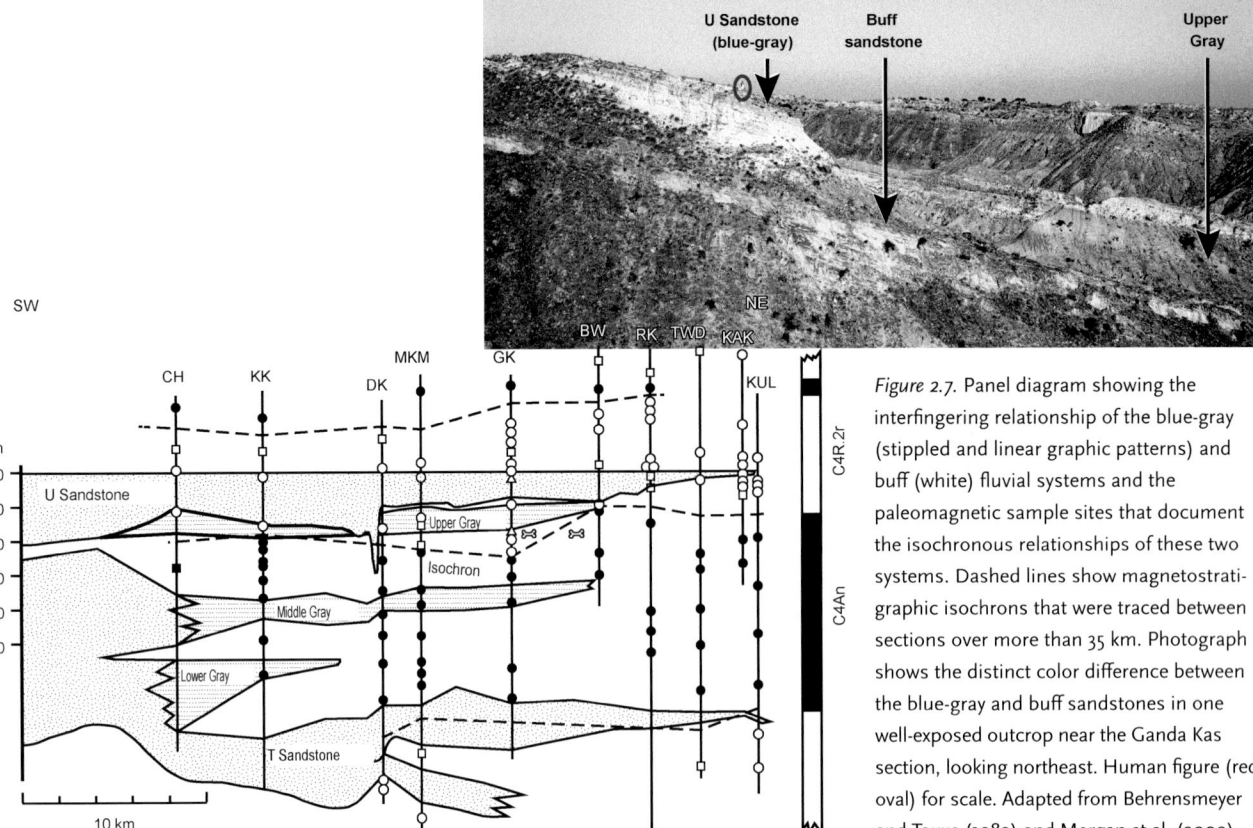

Figure 2.7. Panel diagram showing the interfingering relationship of the blue-gray (stippled and linear graphic patterns) and buff (white) fluvial systems and the paleomagnetic sample sites that document the isochronous relationships of these two systems. Dashed lines show magnetostratigraphic isochrons that were traced between sections over more than 35 km. Photograph shows the distinct color difference between the blue-gray and buff sandstones in one well-exposed outcrop near the Ganda Kas section, looking northeast. Human figure (red oval) for scale. Adapted from Behrensmeyer and Tauxe (1982) and Morgan et al. (2009).

The Manchar Formation in Sind Province (Fig. 2.2) is a Middle and Upper Miocene fluvial sequence overlying the marine Gaj Formation and composed of alternating sandstones and silts, with thin conglomerates. Two sections have been measured (Raza et al. 1984; Khan et al. 1984; Bernor et al. 1988), but paleomagnetic results were reported as preliminary (Khan et al. 1984); thus ages of the contained fossils are only approximately known. Adjusted to the Gradstein et al. (2012) timescale, the preliminary magnetostratigraphic correlation of the Manchar Formation suggests that it is between 15 and 10 Ma, although it may be as old as 18 Ma and extend up to 8 Ma.

In addition to establishing the overall chronostratigraphic framework, combined geochronological and lithological documentation allows us to create detailed panel diagrams for particular time slices within the different Potwar Siwalik formations (see Figs. 2.7, 2.8, Chapter 4, this volume). Using the magnetic isochrons as timelines, we can generate cross-sections of lateral lithofacies relationships to show the geographic patterns of contemporaneous fluvial channels and floodplains at the scale of the original Miocene landscapes occupied by the biota (Behrensmeyer and Tauxe 1982; Behrensmeyer 1987; Behrensmeyer et al. 2007; Morgan et al. 2009). The resulting reconstructions of the three-dimensional architecture of the Potwar Plateau formations allow us to examine how the floras and faunas of the sub-Himalayan ecosystems interacted with fluvial processes and environmental change over different scales of space and time.

SIWALIK FLUVIAL PALEOENVIRONMENTS

The sedimentary rocks of the Siwalik sequence are made up of varying amounts of gravel, sand, silt, and clay washed down from the mountains and foothills to the north, and they also include calcium carbonate, iron oxide, and iron hydroxide deposits that formed in these sediments after deposition on the alluvial plain. The associations of the sediments represent particular environments (lithofacies), and their stratigraphic arrangements record a rich history of fluvial landscapes and climatic conditions spanning millions of years of basin evolution. Potwar Plateau outcrops exposing these strata occur over an area of about 18,000 km² (i.e., ~180 × 100 km), and magnetostratigraphy allows high-resolution correlation of rock layers over kilometers to hundreds of km, as well as to the GPTS. The local to regional correlations provide an essential foundation for examining different areas of the sub-Himalayan alluvial plain, which record infilling of the sub-Himalayan foreland basin during Miocene–Pleistocene time (~18–~2 Ma).

Documentation of the sediments has been done at different scales, from centimeter-scale measured sections spanning meters, to meter-scale resolution over hundreds to thousands of vertical meters (see Figs. 2.5, 2.7, 2.8). High-resolution stratigraphy documents fossil localities, lateral lithofacies patterns, and magnetic isochrons at specific stratigraphic intervals (Behrensmeyer and Tauxe 1982; Badgley and Tauxe 1990; Morgan et al. 2009), while a lower-resolution approach was necessary to build the chronostratigraphic framework for the Potwar sequence as a whole. We integrate these different scales to

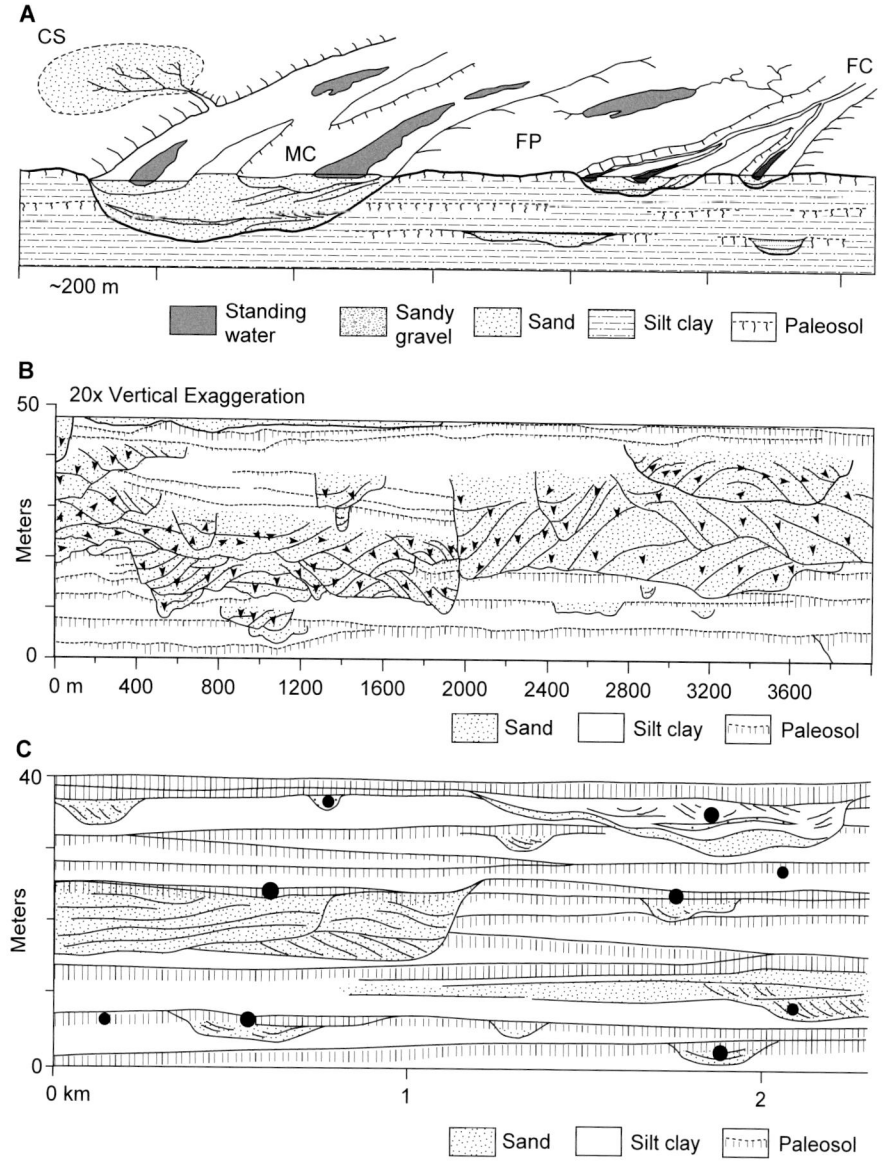

Figure 2.8. A. Schematic diagram representing four major fluvial sub-environments based on sedimentological features and lithofacies relationships. MC = major channel; FC = floodplain channel; CS = crevasse splay; FP = floodplain. Additional subcategories and details are provided in Behrensmeyer et al. (2005) and Chapter 4, this volume. B. Panel diagram showing fluvial architecture of channel and floodplain deposits spanning the Chinji–Nagri Formation boundary near Chinji Village, adapted from Willis (1993a) and Willis and Behrensmeyer (1995). Major multistoried sandstone body (MC in A) is shown with storey boundaries (heavy black lines), dominant bedding dip (lighter lines), and current directions (short arrows, toward the viewer = down arrow, into the plane of the panel = up arrow). Overbank sediments (FP in A) are shown in white, and paleosols (i.e., zones leached of matrix carbonate) appear as dashed lines and light vertical line pattern. C. Schematic cross-section representing the depositional context of fossil localities in the fluvial architecture of the Siwalik formations, with size of black dot proportional to typical numbers of specimens and quality of preservation. This is based on detailed fossil locality studies by Behrensmeyer, Badgley, and Raza and adapted from Behrensmeyer (1987); see also Chapter 4, this volume.

interpret the evolution of Siwalik fluvial systems over time. At the finest space-time scale, for example, we can reconstruct the environment of fossil localities as a paleochannel or a floodplain soil, then expand the scale to place these localities within the interconnected systems of channels of various sizes and flow patterns as they changed over hundreds of thousands of years. The same methods were applied to the three major study areas

(Khaur, Chinji, Rohtas; Fig. 2.3a), expanding the spatial scale to a larger region of the western Miocene foreland basin. For the Potwar Plateau as a whole, future paleoenvironmental reconstructions could compare the same time intervals in the different geographic areas using magnetostratigraphic timelines (Fig. 2.5) to provide a clearer picture of physical differences across the alluvial plain.

Although our main paleontological focus is the time interval between 18 and 6 Ma, uninterrupted deposition continued in the western foreland basin until about 2.5 Ma, the time of major uplift and deformation of the Potwar Plateau and adjacent areas (Burbank, Beck, and Mulder 1996). The eastern parts of the larger foreland basin are still actively subsiding and accumulating sediment from the Ganges, Brahmaputra, and other river systems (Fig. 2.1a).

In describing the fluvial stratigraphy and sedimentology of the Siwalik and Rawalpindi Group formations, we will start with evidence associated with fossil localities, move on to laterally extensive time slices representing the architecture of the fluvial systems, and conclude with an overview of the vertical succession of formations and what this means for the evolution of the western sub-Himalayan river systems over ~20 million years.

Fossil Localities as Samples of the Ancient Landscape and Its Fauna

Interest in how the paleontological record was preserved led to documentation of the "micro-stratigraphy" of short sequences of rocks (i.e., 1–10 m thicknesses described at 10 cm scale) associated with concentrations of fossil vertebrates. Most of the fossil assemblages occur on outcrop surfaces in spatially discrete localities, and these can be linked to source strata where fossils are observed in situ or inferred using adhering rock matrix. When it is not possible to pinpoint a source bed, micro-stratigraphy provides a lower resolution but still useful record of burial context. The work of Badgley (1986; Badgley, Downs, and Flynn 1998), Raza (1983), Raza et al. (1983, 1984), and Behrensmeyer (1987; Behrensmeyer and Barry 2005; Behrensmeyer et al. 2005) resulted in micro-stratigraphic documentation of over 300 fossil localities in the Kamlial, Chinji, Nagri, and Dhok Pathan Formations, with additional information recorded on locality cards (Behrensmeyer and Raza 1984; see Chapters 3 and 4, this volume). This work often required digging to expose fresh outcrop surfaces, which revealed many otherwise hidden sedimentary structures and lithologies useful for reconstructing the paleoenvironments where animal remains were preserved.

The Siwalik formations include gravels to silty clays and carbonate-rich mudstones representing a range of depositional energies and environments (Fig. 2.8). Channels are typically filled with coarse- to medium-grained sand with lenses of intraclast gravel composed of mudclasts and soil carbonate nodules locally derived from contemporaneous sediments. Clast size and the frequency of gravel increase upward through the Dhok Pathan Formation, with increasing proportions of extraformational clasts. The dominance of well-sorted sand and scarcity of allochthonous gravel in the lower formations are evidence that the mountain source areas were some distance away, with generally low-gradient flow across an extensive depositional plain. Overbank deposits consist of fine sand, silt, and clay, as expected in a lower-energy environment, but there are also sand bodies representing crevasse splays that formed when water broke through levees during floods and spread

sediment onto the floodplain. Paleosols formed on all types of sediment but are most clearly developed in the finer-grained floodplain units. They represent hiatuses in deposition when plants and microorganisms colonized the exposed land surfaces and include varying degrees of soil development and time periods (Behrensmeyer 1987; Retallack 1991; see Chapters 4 and 5, this volume, for more about paleosols).

Vertebrate fossils are found in all types of sediment, including depositional units modified by paleosols (Fig. 2.8c), but a notably productive lithofacies is mudclast (or intraclast) conglomerate with reworked, often angular pieces of clay, silt, and carbonate embedded in a sandy matrix (see Chapter 4, this volume). This lithofacies reflects fluvial erosion of previously deposited mudstone from levees, floodplains, or dried-out channel segments, as well as paleosols, which are the main source of carbonate nodules. Some of this erosion likely occurred during flooding and crevasse-splay events. It is no surprise that the reworking process also entrained bones or teeth that had accumulated on floodplains, within soils, or in temporarily inactive channels. All of the Siwalik formations have fossil vertebrates preserved in mudclast conglomerates, and such localities can be very productive. Most of the remains are fragmentary, however, and consist mainly of durable elements such as teeth, jaws, bovid horncores, limb ends, and foot bones.

Lateral Stratigraphy—Time Slices of Siwalik Landscapes

The extensive outcrops of the Siwalik sequence in the Potwar Plateau are ideal for studying lateral patterns in the fluvial strata, especially since these strata are tilted at angles of only 12–15 degrees and stretch for tens of kilometers along the north and south limbs of the Soan Synclinorium (see Figs. 2.1, 2.3, 2.7, 2.8). It is possible to follow sand bodies, overbank deposits, and landscape surfaces marked by paleosols for distances of many kilometers, documenting lateral changes using stratigraphic sections 10–30 m in thickness and spaced hundreds of meters apart. Strata are then correlated to construct "panel" diagrams—slices of the fluvial deposits representing thousands to tens of thousands of years of shifting conditions on the aggrading alluvial plain (see Figs. 2.7, 2.8, Fig. 4.3, this volume).

Stratigraphic panels (e.g., Behrensmeyer 1987; Willis 1993a, 1993b; Willis and Behrensmeyer 1994; Sheikh 1984; Khan 1993, 1995; Khan et al. 1997; Zaleha 1997b) provide windows into the cross-sectional architecture of the Siwalik deposits and show the relationships and scales of channel and floodplain environments (Fig. 2.8). Using this approach, Behrensmeyer and Tauxe (1982) documented the existence of two contemporaneous river systems, termed the "blue-gray" and "buff" systems because of their distinctively colored sandstones (see Fig. 2.7b). These two fluvial packages represent mountain-sourced (blue-gray) versus foothill-sourced (buff) catchment areas of the sub-Himalayan setting, and their patterns of superposition and interfingering reflect the evolution of the foreland basin. In our study area, the Kamlial, Chinji, and Nagri Formations consist of blue-gray system deposits, and the two river systems were contemporaneous only during the lower Dhok Pathan Formation. The upper Dhok Pathan Formation consists solely of buff-system lithologies. There is regional variation, with the blue-gray system persisting later in time in the Gabhir and Rohtas sections (Figs. 2.2b, 2.5) and indicating that mountain-sourced rivers gradually shifted to the southwest as the buff system prograded into the foreland basin from the north.

Interbedded between the channel deposits are finer-grained overbank sediments that formed much of the Miocene alluvial landscape. Detailed lateral description of these sediments through the Chinji, Nagri, and Dhok Pathan Formations (Behrensmeyer and Tauxe 1982; Behrensmeyer 1987; Willis and Behrensmeyer 1994; Morgan et al. 2009; Behrensmeyer, unpublished data) also document the larger-scale context of the fossil localities, most of which are associated with floodplains rather than major channel sandstones. Paleosols in the floodplain sequences represent times when the landscape was occupied by plant and animal communities. Lateral tracing of paleosols in relation to paleomagnetic isochrons shows that stable areas with well-developed paleosols were kilometers to tens of kilometers in scale and contemporaneous with floodplains and channels experiencing active sedimentation (Willis and Behrensmeyer 1994; Behrensmeyer, Willis, and Quade 1995). Thus, we know that the alluvial plain on a large scale was a mosaic of drier and wetter areas, with patches that were stable for hundreds to thousands of years while others were disturbed more frequently by flooding, erosion, and deposition. During early Dhok Pathan Formation time, this was further complicated by the presence of both mountain-sourced (blue-gray) and foothill-sourced (buff) river systems, which had differing flow regimes and possibly asynchronous flood cycles (Behrensmeyer and Tauxe 1982; Morgan et al. 2009). Vegetation on the alluvial plain would have consisted of differing plant communities adapted to varying amounts of disturbance, which in turn affected the distribution of animals and likely kept many of them moving on a seasonal cycle (Morgan et al. 2009). It is possible that some mammals migrated tens to hundreds of km into the sub-Himalayan foothills to the north or to drier areas to the south, returning when floods receded and new plant growth appeared.

We have designated four major types of fluvial environments that make up the Siwalik formations: major channels, floodplain channels, crevasse splays, and floodplains (with paleosols; Fig. 2.8a). These represent different habitats available to the original floras and faunas (See Behrensmeyer and Barry 2005; also Chapter 4, this volume) and vary in relative frequency through time.

Major Channels
The tilted Siwalik outcrops have prominent ridges formed of well-cemented sandstones originally deposited in river channels (Fig. 2.4; Fig. 3.2, this volume). Documentation of internal sedimentary structures, size, and frequency of these deposits through the Potwar Plateau Siwalik formations (Figs. 2.4, 2.8b) allows us to infer that the active major channels varied from ~80–400 m across and 4–30 m deep and were associated with braided channel belts 1–2 km wide (Willis 1993a, 1993b).

Willis and Behrensmeyer (1994, 1995), Khan et al. (1997), Sheikh (1984), and Zaleha (1997a, 1997b) documented laterally continuous sandstones composed of many individual "storeys" representing multiple episodes of channel cutting and filling over time (see Fig. 2.8b). Deposition of these laterally extensive sandstones terminated when the entire channel belt changed course (avulsed) to another floodplain position. Avulsions resulted from the build-up of an alluvial ridge by the active channel, eventually leading to instability that caused water to break out of the elevated banks and switch course to a lower topographic position. Such events were relatively common; according to our magnetochronology, they occurred every 10^3–10^4 years. When a major channel changed course, it left behind an elevated ridge with a linear depression that became a site for temporary ponds, vegetation growth, and animal habitation. The avulsion-driven dynamics of the sub-Himalayan fluvial systems were important for maintaining habitat heterogeneity on the alluvial plain and also for providing settings for fossil preservation (see Chapter 4, this volume).

Floodplain Channels
Throughout the Siwalik formations of the Potwar Plateau there are many smaller-scale channels associated with the finer-grained overbank (floodplain) deposits. These have varying amounts of sand and gravel, the latter composed primarily of mudclasts, pedogenic carbonate, and iron-manganese nodules reworked from contemporaneous paleosols. The coarser-grained sediments are often well-cemented, forming small plateaus and ridges on the outcrops, and paleontologists are drawn to these because they often signal a concentration of fossils. Lateral tracing as well as internal sedimentary structures show that these deposits represent channels, often with well-defined edges cut into contemporaneous floodplain sediments and paleosols. These channels vary from a few meters to tens of meters wide and a few meters deep, and their fill is typically a mix of intraclast conglomerates and sandy to silty sediment, often in a fining upward sequence and sometimes with sand-silt couplets forming lateral accretion deposits. They were tributaries to the major channels that drained the floodplain (Fig. 2.9). Depending on flooding cycles on the alluvial plain versus mountain source areas, some floodplain channels may have reversed flow direction seasonally to become distributaries for floodwaters from major rivers onto the floodplain.

Crevasse Splays
The Kamlial Formation includes numerous sheet-like deposits composed of sand and intraclast conglomerate with soil nodules and (sometimes) fossil bones and teeth. These deposits are usually less than a meter thick and laterally restricted to tens of meters along the outcrop, although some larger-scale examples extend for hundreds of meters. We interpret them as crevasse-splay deposits, which represent flooding events that broke through levees and spread lobe-like sheets of sediment across the floodplain, picking up skeletal remains and components of the soils (e.g., carbonate nodules) and depositing these coarser materials at the base of the crevasse-splay. The basal

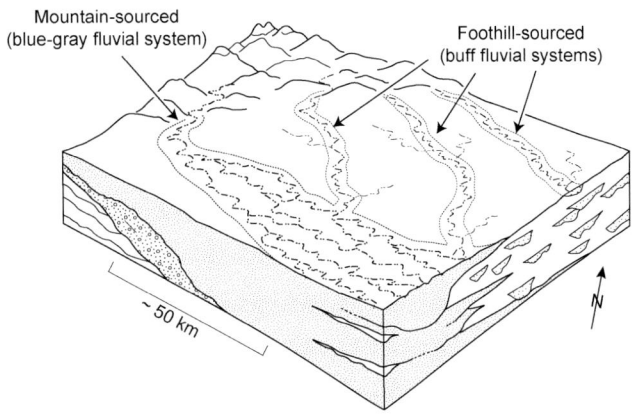

Figure 2.9. Schematic illustration of the relationship of the blue-gray and buff fluvial systems. Petrographic evidence and current-direction measurements show that the blue-gray system was draining higher montane areas, while the buff fluvial systems were sourced in the foothills; adapted from Behrensmeyer and Tauxe (1982). The blue-gray system dominated the southern Potwar Plateau area throughout the Kamlial, Chinji, and Nagri Formations, with increasing influence of buff system deposition in the Dhok Pathan Formation in the eastern and northern areas of the Plateau (see Figs. 2.6, 2.10). The Dhok Pathan buff system channels also include more gravels and cobbles, a consequence of uplift and faulting in the Potwar region that ended deposition in the western foreland basin (Fig. 2.3). Channel belt boundaries are indicated by dotted lines; blue-gray system channels were often braided and anastomosing, while buff system channels were meandering (smaller channels) to braided (larger channels). Scale of block diagram is ~100 km on a side. Reconstruction below the diagram by Julius Csotonyi shows the two fluvial systems at ~ 9 Ma; see Chapter 24 (this volume).

contact is usually horizontal, in contrast to channel deposits that erode irregular troughs into the floodplains when they shift course. Crevasse splays in the Chinji Formation form sand bodies up to 5 m thick and represent longer-term (e.g., multi-flood event) influxes of overbank sand from a major channel. Thinner, sheet-like crevasse-splay deposits are sometimes associated with floodplain channels, indicating that some crevasse-splay events evolved into more established conduits for seasonal water flow.

Floodplains

The thick intervals of red, orange, and buff-colored strata sandwiched between the Siwalik sandstone ridges (see Fig. 2.4) were deposited by flooded rivers that spread silt and clay across the alluvial plain. These accumulated in annual, centimeter-scale increments that built up into thousands of meters as the foreland basin subsided over millions of years (Willis and Behrensmeyer 1994, 1995). The fine-grained sediments generally are not as well cemented as the sandstones and erode more easily, forming slopes and flat areas where fossils can accumulate as they weather from the outcrops. Floodplain strata appear deceptively homogeneous on weathered outcrop surfaces but contain a wealth of detail in fresh exposures—from complexly bedded fine-grained channel fills to laminated floodplain ponds and massive, bioturbated paleosols.

Over thicknesses of a few meters, there are two distinct patterns of floodplain sedimentation—one a fining-upward sequence with sand at the bottom to silt and clay at the top (Fig. 4.2, this volume) and the other a coarsening-upward sequence with finer sediment at the base changing upward to sand. The first of these resulted from the waning influence of flooding as portions of the floodplain aggraded or channels moved farther away. Fining-upward sequences are typically topped by paleosols marking stable landscapes that organisms could inhabit for long periods of time. Coarsening-upward sequences signal the increasing proximity of higher-energy flooding as a channel or a crevasse-splay lobe moved closer. Both sequences contributed to the overall vertical build-up of the floodplain over time. It is likely that much of the well bedded floodplain sediment, which includes calcareous silts (marls), was deposited rapidly in ponds and lakes (Willis and Behrensmeyer 1994). Classic lacustrine deposits with finely laminated, clay-rich sediments or limestones are absent from the Potwar Siwalik sequence. Most of the floodplain water bodies likely were seasonal, with many meters of vertical fluctuation in the water table annually. Average sedimentation rates are lower in the Kamlial and Chinji Formations (Behrensmeyer 1987; Fig. 2.10; Table 2.1), but their deposits have more well-defined floodplain stratification than the Dhok Pathan Formation. This indicates higher water levels relative to subsidence rates in the

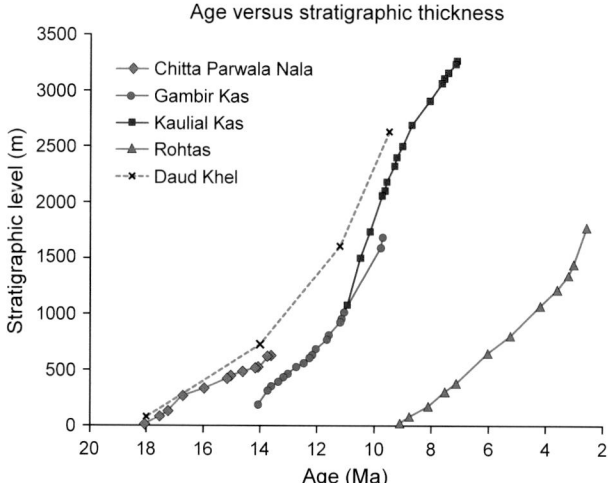

Figure 2.10. Plot of sediment thickness by age for different stratigraphic sections in the Potwar Siwaliks (see Figs. 2.3, 2.5, Table 2.1). Dates are based on boundary ages for magnetostratigraphic chrons, except for Daud Khel (trans-Indus region), where age estimates are from Hussain (1979).

likely incorporated large quantities of organic matter and probably were originally brown rather than red or orange. Preserved free carbon (bulk soil organic matter) is relatively low in the Siwalik rocks, however, which implies that it was oxidized and destroyed by seasonal drying and soil processes and/or after deeper burial and diagenesis. The Dhok Pathan and Nagri Formations are notably less red than the Chinji and Kamlial, which suggests that color intensity in part reflects greater diagenetic iron oxidation in the more deeply buried parts of the Siwalik sequence and contributed to their intense red and purple colors (Krynine 1937). Fortunately, diagenetic alteration of the Siwalik rocks did not seriously affect the original stable isotope composition of fossils and soil carbonates or the biomarker record in the paleosols (Chapters 5 and 24, this volume).

Paleosols

As records of ancient land surfaces, the Siwalik paleosols hold important information about the habitats of the animals and plants that occupied the sub-Himalayan alluvial plain. The floodplain deposits of all Siwalik formations include hundreds of these soils, repeated every few meters through the long vertical sections and traceable laterally in some cases for many kilometers (Fig. 2.8). Although they make up a small proportion of the sediment thickness, they represent much of the time encompassed by the rock record, when sediment accumulation temporarily slowed or stopped and biological processes took over to populate the alluvial landscape with terrestrial life.

The Siwalik paleosols include many characteristic features of soils in varying stages of development and water saturation, from "gleyed" soils representing continuously wet, reducing conditions to highly oxidized, red and orange "vertisols" representing places where the water table was low (at least seasonally). Plants plus soil microbiota stirred and bioturbated the floodplain sediment, sometimes to depths of meters (Retallack 1991; Willis and Behrensmeyer 1995). In general, paleosols in the lower Siwalik formations are better developed and represent more time (up to 10^4 years) than those in the Dhok Pathan Formation (e.g., 10^3 years; Behrensmeyer 1987; Behrensmeyer et al. 2007). Many of the paleosols include a mix of features indicating that they slowly accreted over time (e.g., hundreds to thousands of years), in some cases evolving from one soil type into another (Behrensmeyer et al. 2007). Pedogenic iron-manganese and carbonate nodules are a common component of most (but not all) of the paleosols, with some representing seasonally reducing conditions that mobilized and concentrated iron and manganese and others in which calcium carbonate precipitated during periods of drying and oxidation. Pedogenic carbonate also occurs as root casts and in dense concentrations as incipient caliches (Behrensmeyer et al. 2007). Soil moisture was transpired primarily through plants growing on the soils, and this resulted in a carbon and oxygen stable isotope record preserved in pedogenic carbonate (Quade and Cerling 1995; Cerling et al. 1997) and biomarkers (Freeman and Colarusso 2001).

earlier phases of foreland basin filling, combined with sedimentation events that filled low areas before they formed paleosols, protecting primary bedding from disruption by biological processes. This balance changed as sediment accumulation rates increased relative to subsidence rates later in the Nagri and Dhok Pathan Formations (see Table 2.1, Fig. 2.10), but there is no evidence for unconformities indicating periods of sediment bypassing the Potwar Plateau region. This indicates that accommodation space south of the major boundary thrust kept pace with deposition. A more seasonal climate likely amplified annual fluctuations in the water table as well, contributing to oxidation of floodplain deposits. Slower build-up of overbank sediments meant that floodplain surfaces were more stable and provided longer periods of time for pedogenic processes to destroy primary bedding.

Fossil vertebrates occur in floodplain sediments but are relatively uncommon. The well-bedded sediments that are preserved appear to represent environments subject to rapid flood sedimentation, which land vertebrates likely avoided. The lack of mass bone accumulations (other than occasional fish concentrations) in the Siwalik formations argues against mass mortality as a significant taphonomic process in our vertebrate fossil record. Most of the remains associated with floodplain deposits occur in small channels, as described above, or in paleosols (Fig. 2.8c; Chapter 4, this volume).

The distinctive bright colors of the Siwalik floodplain sediments are caused by oxidized iron, reflecting the transformation of reduced iron (Fe++) to iron hydroxides including goethite (FeOOH, brown to yellow), and hematite (Fe_2O_3, red; Krynine 1937; Abrajevitch et al. 2009). The abundant evidence for biological activity in the floodplain paleosols (see below) suggests that the original Miocene floodplain soils

Table 2.1. Magnetostratigraphic Ages and Stratigraphic Thicknesses for the Major Sections of the Chronostratigraphic Framework for the Potwar and Zinda Pir Siwaliks

Gabhir Kas—Composite Section (Gambir + Chitta Parwala Nala)

Gambir Kas (a)

Age (midpoint)	Strat Level	Chron Boundary	Formation (Fm)	Meters/million yrs	cm/1,000 yrs	Fm Av.—cm/1,000 yrs	
9.721	1,687	4Ar.2n/4Ar.3r	Dhok Pathan				
9.786	1,595	4Ar.3r/5n.1n	Nagri	1,423	142.31		
11.056	1,015	5n.2n/5r.1r	Nagri	456	45.60		
11.146	956	5r.1r/5r.1n	Nagri	666	66.56		
11.188	927	5r.1n/5r.2r	Nagri	671	67.14		
11.592	812	5r.2r/5r.2n	Nagri	286	28.61	70.04	Av. Nagri Fm
11.657	770	5r.2n/5r.3r	Chinji	640	64.00		
12.049	687	5r.3r/5An.1n	Chinji	212	21.20		
12.174	635	5An.1n/5An.1r	Chinji	420	42.00		
12.272	607	5An.1r/5An.2n	Chinji	280	27.96		
12.474	561	5An.2n/5Ar.1r	Chinji	228	22.82		
12.735	526	5Ar.1r/5Ar.1n	Chinji	133	13.26		
13.032	467	5Ar.3r/5AAn	Chinji	201	20.10		
13.183	435	5AAn/5AAr	Chinji	209	20.86		
13.363	394	5AAr/5ABn	Chinji	232	23.17		
13.608	355	5ABn/5ABr	Chinji	158	15.80		
13.739	315	5ABr/5ACn	Chinji	305	30.46	27.42	Av. Chinji Fm
14.07	190	5ACn/5ACr	Kamlial	377	37.73		
			Average	**344**	**34.42**		

Chitta Parwala Nala (b)

Age (midpoint)	Strat Level	Chron Boundary	Formation (Fm)	Meters/million yrs	cm/1,000 yrs	Fm Av.—cm/1,000 yrs	
13.608	627	5ABn/5ABr	Chinji				
13.739	622	5ABr/5ACn	Chinji	40	3.97		
14.070	526	5ACn/5ACr	Chinji	290	29.00	16.49	Av. Chinji Fm
14.163	516	5ACr/5ADn	Kamlial	102	10.22		
14.609	485	5ADn/5ADr	Kamlial	71	7.13		
15.032	448	5Bn.1r/5Bn.2n	Kamlial	87	8.75		
15.160	423	5Bn.2n/5Br	Kamlial	191	19.14		
15.974	335	5Br/5Cn.1n	Kamlial	108	10.79		
16.721	268	5Cn.3n/5Cr	Kamlial	91	9.06		
17.235	131	5Cr/5Dn	Kamlial	266	26.63		
17.533	83	5Dn/5Dr.1r	Kamlial	160	15.97		
18.056	11	5Dr.2r/5En	Kamlial	138	13.77	13.50	Av. Kamlial Fm
			Average	**138**	**13.85**		

Rohtas (c)

Age (Midpoint)	Strat Level	Chron Boundary	Formation (Fm)	Meters/million yrs	cm/1,000 yrs	Fm Av.—cm/1,000 yrs	
2.581	1,777	2r base	Dhok Pathan				
3.032	1,448	2An.1n/2An.2r	Dhok Pathan	729	72.95	39.01	Av. Upper DP Fm
3.207	1,345	2An.2n/2An.3r	Dhok Pathan	589	58.86		
3.596	1,217	2An.3n/2Ar	Dhok Pathan	329	32.90		
4.187	1,073	2Ar/3n.1n, Basawa Kas	Dhok Pathan	244	24.37		
5.235	804	3n.4n/3r	Dhok Pathan	257	25.67		
6.033	650	3r/3An	Dhok Pathan	193	19.30		
7.14	381	3An/3Bn	Dhok Pathan	243	24.30	19.67	Av. Lower DP Fm
7.528	306	3Bn/4n	Dhok Pathan	193	19.33		
8.108	175	4n.2n/4r.1n	Dhok Pathan	226	22.59		
8.771	83	4n1n/4An	Dhok Pathan	139	13.88		
9.105	22	4An/4Ar.1r	Dhok Pathan	183	18.26		
			Average	**269**	**26.90**		

Table 2.1. continued

Kaulial Kas (d)

Age (Midpoint)	Strat Level	Chron Boundary, Notes	Formation (Fm)	Meters/million yrs	cm/1,000 yrs	Fm Av.—cm/1,000 yrs	
	3,268	Top (m)					
7.135	3,268	3Br.1n top(?), (Gamma)	Dhok Pathan				
7.17	3,237	3Br.1n/3Br.2r	Dhok Pathan	886	88.57		
7.432	3,160	3Br.3r/4n.1n (Beta)	Dhok Pathan	294	29.39		
7.562	3,110	4n.1n/4n.1r (Alpha in middle)	Dhok Pathan	385	38.46		
7.65	3,068	4n.1r/4n.2n top (~Z)	Dhok Pathan	477	47.73		
8.072	2,910	4n.2n/4r.1r (~X)	Dhok Pathan	374	37.44		
8.699	2,694	4r.2r/4An (~UU)	Dhok Pathan	344	34.45		
9.025	2,500	4An/4Ar.1r (between P and Q)	Dhok Pathan	595	59.51		
9.23	2,400	4Ar.1r/4Ar.1n ("U level" just below O)	Dhok Pathan	488	48.78	48.04	Av. Upper DP Fm
9.308	2,325	4Ar.1n/4Ar.2r (between N and O)	Dhok Pathan	962	96.15		
9.58	2,182	4Ar.2r/4Ar.2n (between M and MM)	Dhok Pathan	526	52.57		
9.642	2,103	4Ar.2n/4Ar.3r (between K and L)	Dhok Pathan	1274	127.42		
9.74	2,059	4Ar.3r/5n.1n (~K, "Long Normal" top)	Dhok Pathan	449	44.90		
10.159	1,738	Y258	Dhok Pathan	766	76.61		
10.492	1,502	Y-259	Dhok Pathan	709	70.87	58.57	Av. Lower DP Fm
10.94	1,080	5n.2n/5r.1r (just above C)	Nagri	942	94.20		
		Average		**575**	**57.50**		

Daud Khel (e)

Age Estimate	Strat Level	Notes		M/Ma	Cm /1,000 yrs
	3,413	Local Top Dhok Pathan			
9.500	2,634	Dhok Pathan/Nagri			
11.200	1,610	Nagri/Chinji		603	60.28
14.000	729	Chinji/Kamlial		315	31.45
18.000	81	Kamlial/Murree		162	16.21
	0	Eocene Limestone			
		Average	**Average**	**300**	**30.05**

Gabhir Section—Levels of Selected Localities (f)

Age (midpoint)	Strat Level	Fossil Locality	Formation (Fm)	Meters/million yrs	cm/1,000 yrs	Fm Av.—cm/1,000 yrs	
10.063	1,468.5	Y311	Nagri				
10.356	1,335.4	Y729/Y789	Nagri	455	45.52		
10.365	1,331.2	Y728	Nagri	467	46.67		
10.759	1,151.3	Y779	Nagri	456	45.59		
10.810	1,128.1	Y799	Nagri	456	45.59		
10.875	1,098.5	Y646	Nagri	455	45.46		
11.081	997.0	Y798	Nagri	493	49.27		
11.120	971.6	Y804	Nagri	660	65.97		
11.230	915.9	Y791	Nagri	507	50.68	49.34	Av. Nagri Fm
11.410	863.9	Y76/L66	Chinji	288	28.78		
11.429	858.8	Y77	Chinji	276	27.57		
11.450	852.5	Y773	Chinji	293	29.30		
11.457	850.7	Y638	Chinji	277	27.69		

(continued)

Table 2.1. continued

Gabhir Section—Levels of Selected Localities (f)

Age (midpoint)	Strat Level	Fossil Locality	Formation (Fm)	Meters/million yrs	cm/1,000 yrs	Fm Av.—cm/1,000 yrs
12.357	587.6	Y821	Chinji	292	29.22	
12.779	517.8	Y714	Chinji	166	16.55	
12.829	507.8	Y667	Chinji	200	20.00	
12.861	501.3	Y849	Chinji	203	20.31	
12.865	500.6	Y60	Chinji	186	18.57	
12.935	486.5	Y716	Chinji	201	20.07	
12.943	485.0	Y770	Chinji	193	19.33	
12.984	476.7	Y698/Y699	Chinji	200	20.00	
13.013	470.9	Y710	Chinji	200	20.00	
13.036	466.5	Y725	Chinji	191	19.13	
13.050	463.8	Y726/Y774	Chinji	196	19.64	
13.374	393.0	Y848	Chinji	219	21.89	
13.411	386.5	Y724	Chinji	174	17.43	
13.436	382.5	Y842	Chinji	157	15.69	21.73 Av. Chinji Fm

Sources for each section: a. Gambir Kas: Johnson et al. (1985); Willis (1993a); J. C. Barry (pers. comm. 2023). b. Chitta Parwala Nala: McRae (1989); J. C. Barry (pers. comm. 2023). c. Rohtas: Behrensmeyer et al. (2007). d. Kaulial Kas: measured by Marc Monaghan; updated by Behrensmeyer (pers. comm. 2023). e. Daud Khel: Hussain et al. (1979). f. Gabhir Section: J. C. Barry (pers. comm. 2023).

Notes: These data have been reassessed and updated according to the original stratigraphic sections and magnetochron assignments, but in most cases we use the same polarity zonation and correlation to the GPTS as cited references or have made only minor adjustments. Chron boundaries for the Gabhir Composite Section may differ slightly from those shown in Fig. 2.5, which is based on stratigraphic sections by Willis (1993a), McRae (1989), and Johnson et al. (1985). Rates of sediment accumulation are averaged for the formations in the different sections (right column labeled Fm—cm/1,000 yrs) and show a general increase through time (see text and Fig. 2.10). The Gabhir fossil localities provide an alternative dataset for calculating sedimentation rates based on age-assignment methods discussed in Chapter 3, this volume; this subset consists of 31 localities (27 levels) with the most tightly constrained chronostratigraphic ages (see Fig. 3.8, this volume).

Abbreviation Notes: For Chron Boundary abbreviations (numbers and letters), see text and Figs. 2.3, 2.5. For farthest right column, Fm—cm/1,000 yrs is the average depositional rate in cm/1,000 yrs for the thickness of each formation represented in the section.

These provide a critical proxy for understanding changes in Siwalik vegetation and climate through time (Chapters 5, 24, this volume).

Vertical Stratigraphy—Siwalik Rocks through Time

The river deposits that make up the Potwar Plateau Siwalik record add up to over 3,200 meters in the northern Potwar Plateau, where they are thickest in the Kaulial Kas section (see Fig. 2.5). The deposits represent "layer cake" geology of superimposed rock layers spanning nearly 18 million years. Because of the relatively gentle uplift and bending of the sedimentary layers to form the Soan Synclinorium, few faults disrupt the stratigraphic sequence. Variable erosion across the Plateau, however, means that different parts of the stratigraphy are exposed in different regions, and these are assembled into the composite history using magnetostratigraphy as shown in Figures 2.3 and 2.5. This synthesis reflects many years of work by both geologists who described different parts the stratigraphic sequence and geochronologists whose magnetostratigraphic data allowed accurate correlation of geographically separated regions (Figs. 2.3, 2.5).

The succession of formations (Figs. 2.3, 2.5) documents changing dominance of channel versus floodplain deposition during the evolution of the western sub-Himalayan foreland basin after it began to subside in the Early Miocene. The estuarine deposits of the Zinda Pir region record the final marine influence of the Tethys Sea, and from then on, consistent fluvial deposition shows that erosion of the rising Himalayas was approximately in balance with subsidence in the foreland basin. The Potwar Plateau record of the sub-Himalayan alluvial plain is limited in scale (see Fig. 2.1), but its overall stratigraphy (not necessarily the ages of the formation boundaries) is also represented by the Siwalik sequence in India, over 600 km to the southeast (Flynn et al. 2013; Patnaik 2013; Hu et al. 2016). The Kamlial and Chinji Formations are dominated by mudstones representing fluvial floodplains, but these finer-grained sediments alternate with sandstones deposited by river channels similar in scale to large tributaries of the modern Indus and Ganges systems. This pattern shows that the foreland basin between 18 and 12 Ma was underfilled, with subsidence slightly greater than the input of sediment from the eroding mountains. Down-warping of the basin was also steady enough that there are no erosional breaks (hiatuses) of more than ~100,000 years, thus providing a "complete record" of sediment build-up at this level of resolution. Much of the sequence also records time at finer time scales (10^0–10^4 years), with breaks at paleosol surfaces and where channels have eroded into the floodplains.

The Nagri Formation represents a marked change to massive sand deposits in the Potwar Plateau Siwaliks, with stacks of channels interbedded with much thinner and laterally discontinuous floodplain sediments (Willis 1993a; Willis and Behrensmeyer 1995; Zaleha 1997a, 1997c; Figs. 2.6, 2.8, 2.10). This lithological change suggests that a deeper part of the foreland basin shifted into the Potwar region, perhaps from a position farther south, causing regional focusing of major channel axes during Nagri Formation time. Although large (up to 5 km wide), the Nagri channel belts are not considered to represent a major "trunk river" on the scale of the Ganges or Indus, but rather "second order" mountain-sourced rivers that were tributaries to this trunk river (Willis 1993b; Barry et al. 2013). There is no evidence for widespread unconformities that would indicate a period of lower subsidence ("overfilled" conditions leading to erosion) at the scale of the foreland basin. The Nagri channels did erode into their previous deposits, however, as they shifted back and forth within the channel belts, reworking large volumes of sediment along with any organic remains that they contained (Fig. 2.8b).

The balance of sand versus silt and clay shifted again to greater proportions of fine sediment in the Dhok Pathan Formation, which coincides with the gradual disappearance of channels with blue-gray sand and increased dominance of the buff sand system (Figs. 2.5, 2.9). Explanations for this change must take into account the well-documented interfingering of Nagri and Dhok Pathan strata, which is time-transgressive over about 1 million years across the Potwar Plateau (Figs. 2.2, 2.7; Behrensmeyer and Tauxe 1982; Barry et al. 2013). The shift in fluvial systems could have resulted from increased transport of buff system sediment from the chemically weathered foothills farther out into the basin, filling up available accommodation space and displacing the blue-gray channel belt southward. Foreland basin subsidence rates increased over time, as indicated by a change in rates of sediment accumulation from an average of ~23 cm/1,000 years in the Chinji Formation to 40–60 cm/1,000 years in the Dhok Pathan Formation (Fig. 2.10 Table 2.1; Burbank, Beck, and Mulder 1996). The thicker section and high rates of sediment accumulation in the Khaur area (Kaulial Kas) indicate greater rates of subsidence close to the main boundary fault in the north (see Fig. 2.3, 2.5). The top of the Kaulial Kas section averages ~60 cm/1,000 years between 7.4 and 7.1 Ma. The Rohtas section, which is farther south, shows a slower accumulation rate of ~20 cm/1,000 years between 9 and 7 Ma but this rate increases to 73 cm/1,000 years near the top of the section between 3 and 2.6 Ma (see Fig. 2.10). These magnetochron-calibrated sedimentation accumulation rates thus show accelerated down-warping along the main boundary fault prior to tectonic disruption of the Potwar region, which began around 2.5 Ma. Before this time, accommodation space in the basin was more than matched by sediment input. The upper part of the Dhok Pathan Formation is increasingly coarse, as uplift along the boundary fault caused more erosion and transport of sands and gravels onto the alluvial plain. This continued into the Upper Siwalik formations, although subsidence was still sufficient in the lower Soan Formation to preserve thick intervals of fine-grained floodplain sediments between the coarser channel deposits (Khan et al. 1997).

Summary of Potwar Siwalik Fluvial Paleoenvironments

Understanding the lateral and vertical sedimentology and stratigraphy of the Siwalik formations at different scales allows us to reconstruct the three-dimensional architecture of the Potwar Plateau deposits (see Fig. 2.9). This in turn is essential for understanding the role of foreland basin geology in shaping the Siwalik ecosystem and preserving its organic remains as fossils.

Throughout millions of years of continuous deposition during the Miocene, the balance of coarse and fine sediment shifted, rates of sediment accumulation changed, and contexts for fossil preservation varied, but the mountain-sourced river systems in our study area were consistently directed toward the east and southeast, while the foothill-sourced rivers were directed toward the south and southwest as tributaries to the larger rivers (Behrensmeyer and Tauxe 1982; Willis 1993a; Willis and Behrensmeyer 1995; Zaleha 1997b; see Fig. 2.9). This is documented by current directions measured on cross-bedding and other structures preserved in channels and by other research on the fluvial systems (Raynolds 1981; Raynolds and Johnson 1985; Beck and Burbank 1990; Burbank, Beck, and Mulder 1996). The directions of the modern Yamuna and Ganga (Ganges) Rivers of today are similar but differ from the Indus, which today flows toward the south-southwest, passing across the western edge of the Potwar Plateau after exiting the Himalayan mountain front (see Fig. 2.1b). Prior to the late Pliocene uplift of the Salt Range, Potwar Plateau, the eastward-flowing mountain-sourced Siwalik rivers suggest a connection with the upper reaches of the paleo-Ganges system (Burbank, Beck, and Mulder 1996). The Indus River is thought to have existed as early as the Eocene (Bossart and Ottiger 1989; Métais et al. 2009), and Zinda Pir fluvial sediments may have been deposited by a paleo-Indus river. Drill cores from the Indian Ocean document input from a major south-flowing river in the middle Miocene (Routledge et al. 2019; Tauxe and Feakins 2020; Feakins et al. 2020). Thus, there appear to have been two major river systems bringing sediment into the sub-Himalayan alluvial plain during the Miocene. The southeast-directed drainage system was eventually disrupted by uplift of the Salt Range and other parts of the western foreland basin, causing Potwar region rivers to change direction and become southwest flowing tributaries to the modern Indus (Burbank, Beck, and Mulder 1996; Métais et al. 2009).

Although the positions of Miocene river channels and floodplain habitats shifted frequently, driving ecological heterogeneity at spatial scales of tens of kilometers, the overall physiographic conditions across the sub-Himalayan alluvial plain persisted for over 12 million years (see Fig. 2.2, 2.5). This relatively flat physical environment was ~100 km from the Indian Ocean (Burbank 1983) at subtropical paleolatitude ~29 degrees north. It would have looked much the same across hundreds of thousands of square kilometers—a vast ecosystem stretching ~2,000 km from west to east. On this plain, Siwalik floras and faunas could move or otherwise adapt to changes in climate and monsoonal flooding cycles as well as experience species immigration events, evolution, and extinction. Uplift in the western part of the basin combined with broader-scale climate change toward the end of the Miocene likely created an increasing climatic gradient, with drier, more seasonal conditions (Quade and Cerling 1995; Karp, Behrensmeyer, and Freeman 2018) that graded eastward toward moister, more equable environments. This gradient extended into the Malaysian peninsula, where some elements reminiscent of the Miocene Siwalik fauna, such as rhinoceros and tragulids, still live today (Chapters 18, 21, and 27, this volume).

This detailed geological and geochronological knowledge of the Potwar Plateau and Zinda Pir stratigraphic sequences provides the foundation for understanding the ecological setting where the Siwalik fossil record was preserved and where faunal and floral evolution shaped the modern ecology of southern Asia, as laid out in the following chapters.

ACKNOWLEDGMENTS

We thank the editors David Pilbeam, Catherine Badgley, and Michèle Morgan for inviting us to participate in the Siwalik book project. The research represented in this chapter is a result of a long collaboration with the Geological Survey of Pakistan, and we appreciate the efforts of the directors general and members of the survey who have supported the project over the years—especially the interest of Ibrahim Shah, long-standing director of the stratigraphy and palaeontology branch. Financial support included grants from the National Science Foundation, the Smithsonian Foreign Currency Program, the Wenner-Gren Foundation for Anthropological Research, and the American School for Prehistoric Research. We would also like to acknowledge the much-appreciated support of Francine Berkowitz of the Smithsonian Office of International Relations. We are grateful for the generosity and hospitality of the local people of Pakistan throughout our decades of fieldwork—their kind offerings of tea and goodwill often buoyed our spirits during long days on the outcrops.

We dedicate this chapter to our late colleagues Noye Johnson, Will Downs, Asif Jah, Neil Opdyke, and Ev Lindsay. All were instrumental in the work described here and in other chapters.

References

Abbasi, I. A. and P. F. Friend. 1989. Uplift and evolution of the Himalayan orogenic belts, as recorded in the foredeep molasse sediments. *Zeitschrift für Geomorphologie*, Supplement, 76:75–88.

Abrajevitch, A., K. Kodama, A. K. Behrensmeyer, and C. Badgley. 2009. Magnetic signature of moisture availability in ancient subtropical soils: A pilot rock magnetic study of Neogene paleosols in Pakistan. American Geophysical Union, Fall Meeting 2009, abstract id. GP43B-0853.

Antoine, P.-O., G. Métais, M. J. Orliac, J.-Y. Crochet, L. J. Flynn, L. Marivaux, A. R. Rajpar, G. Roohi, and J.-L. Welcomme. 2013. Mammalian Neogene biostratigraphy of the Sulaiman Province, Pakistan. In *Fossil Mammals of Asia: Neogene Biostratigraphy and Chronology*, edited by X. Wang, L. J. Flynn, and M. Fortelius, 400–422. New York: Columbia University Press.

Badgley, C. E. 1986. Taphonomy of mammalian fossil remains from Siwalik rocks of Pakistan. *Paleobiology* 12(2): 119–142.

Badgley, C. E., W. Downs, W., L. J. and Flynn. 1998. Taphonomy of small-mammal fossil assemblages from the Middle Miocene Chinji Formation, Siwalik Group, Pakistan. In *Advances in Vertebrate Paleontology and Geochronology*, edited by Y. Tomida, L. J. Flynn, and L. L. Jacobs, 145–166. Tokyo: National Science Museum Monographs, No. 14.

Badgley, C. E., and L. Tauxe. 1990. Paleomagnetic stratigraphy and time in sediments: Studies in alluvial Siwalik rocks of Pakistan. *Journal of Geology* 98(4): 457–477.

Barndt, J. 1977. The magnetic polarity stratigraphy of the type locality of the Dhok Pathan Faunal Stage, Potwar Plateau, Pakistan. Unpublished MS thesis, Dartmouth College.

Barndt, J., N. M. Johnson, G. D. Johnson, N. D. Opdyke, E. H. Lindsay, D. Pilbeam, and R. A. H. Tahirkheli. 1978. The magnetic polarity stratigraphy and age of the Siwalik Group near Dhok Pathan Village, Potwar Plateau, Pakistan. *Earth and Planetary Science Letters* 41(3): 355–364.

Barry J. C., A. K. Behrensmeyer, C. Badgley, L. J. Flynn, H. Peltonen, I. U. Cheema, D. Pilbeam, et al. 2013. The Neogene Siwaliks of the Potwar Plateau, Pakistan. In *Fossil Mammals of Asia: Neogene Biostratigraphy and Chronology*, edited by X. Wang, L. J. Flynn, and M. Fortelius, 373–399. New York: Columbia University Press.

Barry, J. C., A. K. Behrensmeyer, and M. Monaghan. 1980. A geologic and biostratigraphic framework for Miocene sediments near Khaur Village, northern Pakistan. *Postilla* 183:1–19.

Barry, J. C., M. E. Morgan, L. J. Flynn, D. Pilbeam, A. K. Behrensmeyer, S. M. Raza, I. A. Khan, C. Badgley, J. Hicks, and J. Kelley. 2002. Faunal and environmental change in the late Miocene Siwaliks of northern Pakistan. *Paleobiology Memoirs* No. 3:1–71; Supplement to *Paleobiology* 28 (2).

Beck, R. A., and D. W. Burbank. 1990. Continental-scale diversion of rivers: A control of alluvial stratigraphy. Geological Society of America, Abstracts and Programs 22:238.

Behrensmeyer, A. K. 1987. Miocene fluvial facies and vertebrate taphonomy in northern Pakistan. In *Recent Developments in Fluvial Sedimentology*, edited by F. G. Ethridge, R. M. Flores, and M. D. Harvey, 169–176. Special Publication No. 39: Society of Economic Paleontologists and Mineralogists.

Behrensmeyer, A. K., C. E. Badgley, J. C. Barry, M. E. Morgan, and S. M. Raza. 2005. The paleoenvironmental context of Siwalik Miocene vertebrate localities. In *Interpreting the Past: Essays on Human, Primate, and Mammal Evolution in Honor of David Pilbeam*, edited by D. E. Lieberman, R. J. Smith, and J. Kelley, 47–61. Boston: Brill.

Behrensmeyer, A. K., and J. C. Barry 2005. Biostratigraphic surveys in the Siwaliks of Pakistan: A method for standardized surface sampling of the vertebrate fossil record. *Palaeontologia Electronica* 8(1). https://palaeo-electronica.org/2005_1/behrens15/issue1_05.htm.

Behrensmeyer, A. K., J. Quade, T. E. Cerling, J. Kappelman, I. A. Khan, P. Copeland, L. Roe, et al. 2007. The structure and rate of late Miocene expansion of C_4 plants: Evidence from lateral variation in stable isotopes in paleosols of the Siwalik Group, northern Pakistan. *Geological Society of America Bulletin* 119(11–12): 1486–1505.

Behrensmeyer, A. K., and M. Raza. 1984. A procedure for documenting fossil localities in Siwalik deposits of northern Pakistan. *Memoirs of the Geological Survey of Pakistan* 11:65–69.

Behrensmeyer, A. K., and L. Tauxe. 1982. Isochronous fluvial systems in Miocene deposits of northern Pakistan. *Sedimentology* 29(3):331–352.

Behrensmeyer, A. K., B. J. Willis, and J. Quade. 1995. Floodplains and paleosols of Pakistan Neogene and Wyoming Paleogene deposits: A comparative study. *Palaeogeography, Palaeoclimatology, Palaeoecology* 115(1–4): 37–60.

Bernor, R. L., L. J. Flynn, T. Harrison, S. T. Hussain, and J. Kelley. 1988. *Dionysopithecus* from southern Pakistan and the biochronology and biogeography of early Eurasian catarrhines. *Journal of Human Evolution* 17(3): 339–358.

Borderie, S., F. Graveleau, C. Witt, and B. C. Vendeville. 2018. Impact of an interbedded viscous décollement on the structural and kinematic coupling in fold-and-thrust belts: Insights from analogue modeling. *Tectonophysics* 722:118–137.

Bossart, P., and R. Ottiger. 1989. Rocks of the Murree Formation in Northern Pakistan: Indicators of a descending foreland basin of late Paleocene to Middle Eocene age. *Eclogae Geologicae Helvetiae* 82(1):133–165.

Burbank, D.W. 1983. The chronology of intermontane-basin development in the northwestern Himalaya and the evolution of the Northwest Syntaxis. *Earth and Planetary Science Letters* 64(1): 77–92.

Burbank, D. W., R. A. Beck, and T. Mulder. 1996. The Himalayan foreland basin. In *Tectonic Evolution of Asia*, edited by A. Yin and T. M. Harrison, 149–188. Cambridge: Cambridge University Press.

Cerling, T. E., J. M. Harris, B. J. MacFadden, M. G. Leakey, J. Quade, V. Eisenmann, and J. R. Ehleringer. 1997. Global vegetation change through the Miocene/Pliocene boundary. *Nature* 389:153–158.

Cheema, M. R., S. M. Raza, and H. Ahmed. 1977. Cainozoic. In *Stratigraphy of Pakistan*, edited by S. M. I. Shah. Memoirs of the Geological Survey of Pakistan 12:56–98.

Colbert, E. H. (1935) Siwalik mammals in the American Museum of Natural History. Transactions of the American Philosophical Society (New Series) 26:1–401.

Downing, K. F., E. H. Lindsay, W. R. Downs, and S. E. Speyer. 1993. Lithostratigraphy and vertebrate biostratigraphy of the early Miocene Himalayan foreland Zinda Pir Dome, Pakistan. *Sedimentary Geology* 87(1–2):25–37

Downing, K. F., and E. H. Lindsay, E. H. 2005. Relationship of Chitarwata Formation paleodrainage and paleoenvironments to Himalayan tectonics and Indus River paleogeography. *Palaeontologia Electronica* 8(1): 1–12.

Fatmi, A. N. 1973. Lithostratigraphic units of the Kohat-Potwar Province, Indus Basin, Pakistan. *Memoirs of the Geological Survey of Pakistan* 10:80.

Feakins, S. J., H. M. Liddy, L. Tauxe, V. Galy, X. Feng, J. E. Tierney, Y. Miao, and S. Warny. 2020. Miocene C_4 grassland expansion as recorded by the Indus Fan. *Paleoceanography and Paleoclimatology* 35(6): e2020PA003856.

Flynn, L. J., E. H. Lindsay, D. Pilbeam, S. M. Raza, M. E. Morgan, J. C. Barry, C. E. Badgley, et al. 2013. The Siwaliks and Neogene evolutionary biology in South Asia. In *Fossil Mammals of Asia: Neogene Biostratigraphy and Chronology*, edited by X. Wang, L. J. Flynn, and M. Fortelius, 353–372. New York: Columbia University Press.

Freeman K. H., and L. A. Colarusso. 2001. Molecular and isotopic records of C_4 grassland expansion in the late Miocene. *Geochimica et Cosmochimica Acta* 65(9): 1439–1454.

Friedman, R., J. Gee, L. Tauxe, K. Downing, and E. H. Lindsay. 1992. The magnetostratigraphy of the Chitarwata and lower Vihowa Formations of the Dera Ghazi Khan area, Pakistan. *Sedimentary Geology* 81:253–268.

Friend, P.F., S. M. Raza, G. Geehan, and K. A. Sheikh. 2001. Intermediate-scale architectural features of the fluvial Chinji Formation (Miocene), Siwalik Group, northern Pakistan. *Journal of the Geological Society* 158(1): 163–177.

Gee, J. S., and D. V. Kent. 2007. Source of oceanic magnetic anomalies and the geomagnetic polarity timescale. *Treatise on Geophysics* 5 (12): 455–507. https://www.sciencedirect.com/science/article/abs/pii/B9780444527486000973?via%3Dihub.

Gill, W. D. 1951. The stratigraphy of Siwalik series in the northern Potwar. Punjab, Pakistan. *Quarterly Journal of the Geological Society* (London) 107:375–394.

Google Earth 2020. Version 7.3.4. Southern Asia, Indian sub-continent, Potwar Plateau. Image Date 12/13/2015. Available at https://earth.google.com/A.

Gradstein, F., J. G. Ogg, M. Schmitz, and G. Ogg. 2012. *The Geologic Time Scale.* Elsevier., 1176

Grelaud, S., W. Sassi, D. F. de Lamotte, T. Jaswal, and F. Roure. 2002. Kinematics of eastern salt range and south Potwar Basin (Pakistan): A new scenario. *Marine and Petroleum Geology*19(9): 1127–1139. https://doi.org/10.1016/S0264-8172(02)00121-6.

Hu, X., E. Garzanti, J. Wang, W. Huang, W. An, and A. Webb. 2016. The timing of India-Asia collision onset–Facts, theories, controversies. *Earth-Science Reviews* 160: 264–299.

Hussain, S. T. 1979. *Neogene Stratigraphy and Fossil Vertebrates of the Daud Khel Area, Mianwali District, Pakistan*, Vol. 13. Geological Survey of Pakistan. Islamabad, Pakistan.

Hussain, S. T., J. Munthe, R.M. West, and J. R. Lukacs. 1977. The Daud Khel local fauna: A preliminary report on a Neogene vertebrate assemblage from the Trans-Indus Siwaliks, Pakistan. *Milwaukee Public Museum Contributions in Biology and Geology*, 16:1–17.

Hussain, S. T., G. D. van den Bergh, K. J. Steensma, J. A. de Visser, J. de Vos, M. Arif, J. van Dam, P. Y. Sondaar, and S. B. Malik. 1992. Biostratigraphy of the Plio- Pleistocene continental sediments (Upper Siwaliks) of the Mangla-Samwal Anticline, Asad Kashmir, Pakistan. *Proceedings of the Koninklijke Nederlandse Akademie van Wetenschappen* 95(1): 65–80.

Johnson, G. D., P. Zeitler, C. W. Naeser, N. M. Johnson, D. M. Summers, C. D. Frost, N. D. Opdyke, and R. A. K. Tahirkheli. 1982a. The occurrence and fission-track ages of Late Neogene and Quaternary volcanic sediments, Siwalik Group, northern Pakistan. *Palaeogeography, Palaeoclimatology, Palaeoecology* 37(1): 63–93.

Johnson, N. M., and V. E. McGee. 1983. Magnetic polarlty stratigraphy: Stochastic properties of data, sampling problems, and the evaluation of interpretations. *Journal of Geophysical Research* 88(B2): 1213–1221.

Johnson, N. M., N. D. Opdyke, G. D. Johnson, E. H. Lindsay, and R. A. K. Tahirkheli. 1982b. Magnetic polarity stratigraphy and ages of Siwalik Group rocks of the Potwar Plateau, Pakistan. *Palaeogeography, Palaeoclimatology and Palaeoecology* 37(1): 17–42.

Johnson, N. M., K. A. Sheikh, E. Dawson-Saunders, and L. E. McRae. 1988. The use of magnetic-reversal time lines in stratigraphic analysis: A case study in measuring variability in sedimentation rates. In *New Perspectives in Basin Analysis*, edited by K. L. Kleinspehn and C. Paola, 189–200. New York: Springer.

Johnson, N. M., J. Stix, L. Tauxe, P. F Cerveny, and R. A. K. Tahirkheli. 1985. Paleomagnetic chronology, fluvial processes and tectonic implications of the Siwalik deposits near Chinji Village, Pakistan. *Journal of Geology* 93(1): 27–40.

Kappelman, J, J. Kelley, D. Pilbeam, K. A. Sheikh, S. Ward, M. Anwar, J. C. Barry, et al. 1991. The earliest occurrence of *Sivapithecus* from the Middle Miocene Chinji Formation of Pakistan. *Journal of Human Evolution* 21(1): 61–73.

Karp, A. T., A. K. Behrensmeyer, and K. H. Freeman. 2018. Grassland fire ecology has roots in the Late Miocene. *Proceedings of the National Academy of Sciences* 115(48): 12130–12135. www.pnas.org/cgi/doi/10.1073/pnas.1809758115.

Khan, I. A. 1993. Evolution of Miocene-Pliocene fluvial paleoenvironments in eastern Potwar Plateau, northern Pakistan. Unpublished PhD diss., Binghamton University.

Khan, I. A. 1995. Complexity in stratigraphic division of fluviatile successions. *Pakistan Journal of Mineral Sciences* 7:23–34.

Khan, I. A., J. S. Bridge, J. Kappelman, and R. Wilson. 1997. Evolution of Miocene fluvial environments, eastern Potwar Plateau, northern Pakistan. *Sedimentology* 44(2):221–251. https://doi.org/10.1111/j.1365-3091.1997.tb01522.x.

Khan, M. J., S.T. Hussain, M. Arif, and H. Shaheed. 1984. Preliminary paleomagnetic investigations of the Manchar Formation, Gaj River Section, Kirthar Range, Pakistan. *Geological Bulletin of the University of Peshawar* 17:145–152.

Krynine, P. D. 1937. Petrography and genesis of the Siwalik Series. *American Journal of Science* 5(34): 422–446.

Lindsay, E. H., L. J. Flynn, I. U. Cheema, J. C. Barry, K. F. Downing, A. R. Rajpar, and S. M. Raza. 2005. Will Downs and the Zinda Pir Dome. *Palaeontologia Electronica* 8(1): 1–18. https://palaeo-electronica/2005_1/lindsay19/issue1_05.htm.

Lindsay, E. H., L. J. Flynn, I. U. Cheema, J. C. Barry, K. F. Downing, A. R. Rajpar, and S. M. Raza. 2005. Will Downs and the Zinda Pir Dome. *Palaeontologia Electronica* 8(1): 19A:1-18

Lindsay, E. H. and L. J. Flynn. 2016. Late Oligocene and Early Miocene muroidea of the Zinda Pir Dome. *Historical Biology* 28:1–2, 215–236. https://www.tandfonline.com/doi/full/10.1080/08912963.2015.1027888.

McMurtry, M. G. 1980. Facies changes and time relationships along a sandstone stratum, Middle Siwalik Group, Potwar Plateau, Pakistan. Unpublished BS thesis, Dartmouth College.

McRae, L. E. 1989. Chronostratigraphic variability in fluvial sequences as revealed by paleomagnetic isochrons: Examples in Miocene strata from a central Andean intermontane basin (Salla, Bolivia) and the northwest Himalayan foreland (Lower Siwalik Group Chinji Formation, northern Pakistan). Unpublished PhD diss., Dartmouth College.

McRae, L. E. 1990. Paleomagnetic isochrons, unsteadiness, and nonuniformity of sedimentation in Miocene fluvial strata of the Siwalik Group, northern Pakistan. *Journal of Geology* 98(4):433–456.

Métais, G., P.-O. Antoine, S. R. H. Baqri, J.-Y. Crochet, D. de Franceschi, L. Marivaux, L., and J.-L. Welcomme. 2009. Lithofacies, depositional environments, regional biostratigraphy and age of the Chitarwata Formation in the Bugti Hills, Balochistan, Pakistan. *Journal of Asian Earth Sciences* 34:154–167.

Morgan, M. E., A. K. Behrensmeyer, C. Badgley, J. C. Barry, S. Nelson, and D. Pilbeam. 2009. Lateral trends in carbon isotope ratios reveal a Miocene vegetation gradient in the Siwaliks of Pakistan. *Geology* 37(2): 103–106.

Munthe, J., S. T. Hussain, J. R. Lukacs, R. M. West, and S. M. I. Shah. 1979. Neogene stratigraphy of the Daud Khel Area, Mianwali District, Pakistan. *Milwaukee Public Museum Contributions in Biology and Geology* No. 23: 1–18.

Najman, Y., E. Garzanti, M. Pringle, M. Bickle, J. Stix, and I. Khan. 2003. Early-Middle Miocene paleodrainage and tectonics in the Pakistan Himalayas. *Geological Society of America Bulletin* 115(10):1265–1277.

Opdyke, N. D., N. M. Johnson, G. Johnson, E. Lindsay, and R. A. K. Tahirkheli. 1982. The paleomagnetism of the Middle Siwalik formations of northern Pakistan and the rotation of the Salt Range Decollement. *Palaeogeography, Palaeoclimatology, Palaeoecology* 37(1): 1–15.

Opdyke, N. D., E. H. Lindsay, G. D. Johnson, N. M. Johnson, R. A. K. Tahirkheli, and M. A. Mirza. 1979. Magnetic polarity stratigraphy and vertebrate paleontology of the Upper Siwalik Subgroup of northern Pakistan. *Palaeogeography, Palaeoclimatology, Palaeoecology* 27:1–34.

Patnaik, R. 2013. Indian Neogene Siwalik mammalian biostratigraphy: An overview. In *Fossil Mammals of Asia: Neogene Biostratigraphy and Chronology*, edited by X. Wang, L. J. Flynn, and M. Fortelius, 423–444. New York: Columbia University Press.

Pilgrim, G. E. 1913. The correlation of the Siwaliks with mammal horizons of Europe. *Records of the Geological Survey of India* 43:264–326.

Pilgrim, G. E. 1926. The Tertiary formations of India, and the interrelation of marine and terrestrial deposits. *Proceedings of the Pan-Pacific Science Congress in Australia, 1923*, 893–931.

Pilgrim, G. E. 1934. Correlation of the ossiferous sections in the Upper Cenozoic of India. *American Museum of Natural History Novitates* 704:1–5.

Pillans, B., M. Williams, D. Cameron, R. Patnaik, J. Hogarth, A. Sahni, J. C. Sharma, F. Williams, and R. L. Bernor. 2005. Revised correlation of the Haritalyangar magnetostratigraphy, Indian Siwaliks: Implications for the age of the Miocene hominids *Indopithecus* and *Sivapithecus*, with a note on a new hominid tooth. *Journal of Human Evolution* 48(5): 507–515.

Prasad K. N. 1968 The vertebrate fauna from the Siwalik beds of Haritalyangar, Himachel Pradesh, India. *Memoirs of the Geological Survey of India, Palaeontologia Indica*, n.s., 391–79.

Quade, J., and T. E. Cerling. 1995. Expansion of C_4 grasses in the late Miocene of northern Pakistan: Evidence from stable isotopes in paleosols. *Palaeogeography, Palaeoclimatology, Palaeoecology* 115(1–4): 91–116.

Raza, S. M. 1983. Taphonomy and Paleoecology of Middle Miocene vertebrate assemblages, southern Potwar Plateau, Pakistan. PhD diss., Yale University. 414 pp.

Raza, S. M., J. C. Barry, G. E. Meyer, and L. Martin. 1984. Preliminary report on the geology and vertebrate fauna of the Miocene Manchar Formation, Sind, Pakistan. *Journal of Vertebrate Paleontology* 4(4): 584–599.

Raza, S. M., I. U. Cheema, W. R. Downs, A. R. Rajpar, and S. C. Ward. 2002. Miocene stratigraphy and mammal fauna from the Sulaiman Range, Southwestern Himalayas, Pakistan. *Palaeogeography, Palaeoclimatology, Palaeoecology* 186(3–4): 185–197.

Raza, S. M., J. C. Barry, D. Pilbeam, M. D. Rose, S. I. Shah, and S. Ward. 1983. New hominoid primates from the Middle Miocene Chinji Formation, Potwar Plateau, Pakistan. *Nature* 306(5938): 52–54.

Raynolds, R. G. H. 1981. Did the ancestral Indus flow into the Ganges drainage? *Geological Bulletin of the University of Peshawar* 14:141–150.

Raynolds, R. G. H. n.d. Siwalik stratigraphy. https://www.siwalikstratigraphy.org/.

Raynolds, R. G. H., and G. D. Johnson. 1985. Rate of Neogene depositional processes, north-west Himalayan foredeep margin, Pakistan. In *The Chronology of the Geologic Record*, edited by N. J. Snelling. Geological Society, London, Memoirs 10(1): 297–311.

Raynolds, R. G., G. D. Johnson, C. D. Frost, H. M Keller, M. G. McMurtry, and C. F. Visser. 2022. Magnetic polarity maps and time maps in the eastern Potwar Plateau: Applications of magnetic polarity stratigraphy. *Journal of Himalayan Earth Science* 55(2): 1–20.

Retallack, G. J. 1991. *Miocene Paleosols and Ape Habitats of Pakistan and Kenya*. Oxford: Oxford University Press.

Routledge, C. M., D. K. Kulhanek, L. Tauxe, G. Scardia, A. D. Singh, S. Steinke, E. M. Griffith, and R. Saraswat. 2019. A revised chronostratigraphic framework for International Ocean Discovery Program expedition 355 sites in Laxmi basin, Eastern Arabian Sea. *Geological Magazine* 157(6): 1–18. https://doi.org/10.1017/S0016756819000104.

Sheikh, K. A. 1984. Use of magnetic reversal time lines to reconstruct the Miocene Landscape near Chinji Village, Pakistan. Unpublished MS thesis, Dartmouth College.

Stix, J. 1982. Stratigraphy of the Kamlial Formation near Chinji Village, northern Pakistan. Unpublished BA thesis, Dartmouth College.

Tauxe, L., and C. Badgley. 1988. Stratigraphy and remanence acquisition of a paleomagnetic reversal in alluvial Siwalik rocks of Pakistan. *Sedimentology* 35(4):697–715.

Tauxe, L., C. G. Constable, L. Stokking, and C. Badgley. 1990. Use of anisotropy to determine the origin of characteristic remanence in the Siwalik red beds of northern Pakistan. *Journal of Geophysical Research* 95(B4): 4391–4404.

Tauxe, L., and S. J. Feakins. 2020. A reassessment of the chronostratigraphy of Late Miocene C_3–C_4 transitions. *Paleoceanography and Paleoclimatology* 35(7): e2020PA003857.

Tauxe, L., D. V. Kent, and N. D. Opdyke. 1980. Magnetic components contributing to the NRM of Middle Siwalik red beds. *Earth and Planetary Science Letters* 47(2): 279–284.

Tauxe, L., and N. D. Opdyke. 1982, A time framework based on magnetostratigraphy for the Siwalik sediments of the Khaur area, northern Pakistan. *Palaeogeography, Palaeoclimatology, Palaeoecology* 37(1): 43–61.

Willis, B. J. 1993a. Ancient river systems in the Himalayan foredeep, Chinji Village area, northern Pakistan. *Sedimentary Geology* 88(1–2): 1–76.

Willis, B. J. 1993b. Evolution of Miocene fluvial systems in the Himalayan foredeep through a two kilometer–thick succession in northern Pakistan. *Sedimentary Geology* 88(1–2): 77–121.

Willis, B. J., and A. K. Behrensmeyer. 1994. Architecture of Miocene overbank deposits in northern Pakistan. *Journal of Sedimentary Research* B64:60–67.

Willis, B. J., and A. K. Behrensmeyer. 1995. Fluvial systems in the Siwalik Miocene and Wyoming paleogene. *Palaeogeography, Palaeoclimatology, Palaeoecology* 115(1–4): 13–35.

Zaleha, M. J. 1997a. Fluvial and lacustrine palaeoenvironments of the Miocene Siwalik Group, Khaur area, northern Pakistan. *Sedimentology* 44(2): 349–368.

Zaleha, M. J. 1997b. Intra- and extrabasinal controls on fluvial deposition in the Miocene Indo-Gangetic foreland basin, northern Pakistan. *Sedimentology* 44(2): 369–390.

Zaleha, M. J. 1997c. Siwalik paleosols (Miocene, northern Pakistan): Genesis and controls on their formation. *Journal of Sedimentary Research* 67(5): 821–839.

DOCUMENTING THE SIWALIK FOSSIL RECORD

JOHN C. BARRY AND ANNA K. BEHRENSMEYER

INTRODUCTION

Investigating how environmental forces, especially climate, shape species and ecosystems over geologic time involves documenting temporal patterns in species richness, composition, and ecological structure, then investigating how these patterns are related to environmental changes over the same time periods. The extraordinarily long sequence of ancient river deposits exposed throughout the Indian subcontinent contains abundant and varied assemblages of fossil vertebrates and freshwater invertebrates providing an opportunity to study links between evolution and environmental processes over millions of years in a terrestrial setting. Such studies, however, require detailed knowledge of the stratigraphy, geological age, and depositional context of the fossils. This in turn requires close attention to documenting each of the hundreds of fossil localities and integrating the resulting data on sedimentology, taphonomy, and geochronological age with research on the paleobiology of the preserved species. Our interpretations about evolutionary trends ultimately depend on the quality of the locality-based data that has been carefully accumulated by many people over decades of research.

This chapter describes how we collected and archived locality and specimen data, together with supplementary geological and ecological data. Our collecting and documenting techniques have been adapted to the unique aspects of the Siwalik fossil record and technological advances, specifically to the ability to estimate geological ages arising from the development of paleomagnetic methods. Measuring numerous rock sections with paleomagnetic sampling has produced a well-tested chronostratigraphic framework for the Siwalik exposures and, as a consequence, age estimates for the fossil localities and fossil remains. This temporal framework is the foundation for analyzing in later chapters of this volume the causes of change in faunal assemblages and evolutionary lineages through time. It has also allowed us to examine the complex relationships of the lithostratigraphic facies, both laterally and through time (see Chapters 2 and 4, this volume).

Fossil mammals from the Indian subcontinent first became known to scientists during the 1820s (Pilgrim 1939). Subsequently, during the nineteenth and twentieth centuries, large collections accumulated in museums in India (most notably the Geological Survey of India in Kolkata, Panjab University in Chandigarh, and the Wadia Institute in Dehra Dun) and Pakistan (the Pakistan Museum of Natural History and the Geological Survey of Pakistan, both in Islamabad), as well as in Europe (the Natural History Museum in London, the Bavarian State Museum for Palaeontology and Historical Geology, and the Geological Institute, State University of Utrecht) and in North America (the American Museum of Natural History and the Yale Peabody Museum). More recently, beginning in 1973, collections have been made by collaborative projects involving Harvard University, Howard University, the Geological Survey of Pakistan (GSP), and the Pakistan Museum of Natural History, as well as between the University of Peshawar and a consortium including Dartmouth College, the Lamont-Doherty Geological Observatory, the University of Arizona, and the Smithsonian Institution.

As might be expected for collections assembled over nearly 200 years, the fossils in the older collections are often without useful stratigraphic data, although they contain many of the type specimens for Siwalik taxa, especially in the Natural History Museum in London and the Indian Geological Survey. Because of better stratigraphic control on fossils collected since 1973, for our work we have primarily relied on the collections held by the GSP and the Pakistan Museum of Natural History. In addition, some fossils collected by Barnum Brown and G. Edward Lewis in 1922 and 1932 and now in the American Museum of Natural History and the Yale Peabody Museum are well enough documented to include in our fossil and locality databases. Because previous collections from the nineteenth and early twentieth centuries cannot be included, most type specimens and other significant fossils described by Falconer, Lydekker, and Pilgrim are not incorporated into our biostratigraphic analyses.

Note that throughout this chapter, we use "Siwalik(s)" as a term of convenience to refer to the Miocene to Pleistocene rock formations and fossils that technically belong both to the Siwalik and Rawalpindi Groups. See Chapter 2, this volume, for further explanation.

POTWAR PLATEAU STRATIGRAPHY, DEPOSITIONAL ENVIRONMENTS, AND FOSSIL OCCURRENCES

Exposures of Siwalik sediments are known from the southern margin of the Himalayas in both Pakistan and India, but are notably well exposed on the Potwar Plateau in northern Pakistan (Fig. 3.1a). The Potwar sediments, which formed in a variety of fluvial environments, are of Early Miocene through Pleistocene age. Significant additional exposures of earliest Miocene and latest Oligocene fluvial and fluvio-deltaic deposits are also known in southwestern Punjab, Baluchistan, and Sind (Fig. 3.1b, c; Chapter 2, this volume). A critical attribute of the Siwalik deposits is that they form thick (over 3,200 m), unbroken sequences of long duration (up to 10 million years in some instances) with only short, episodic hiatuses. A second

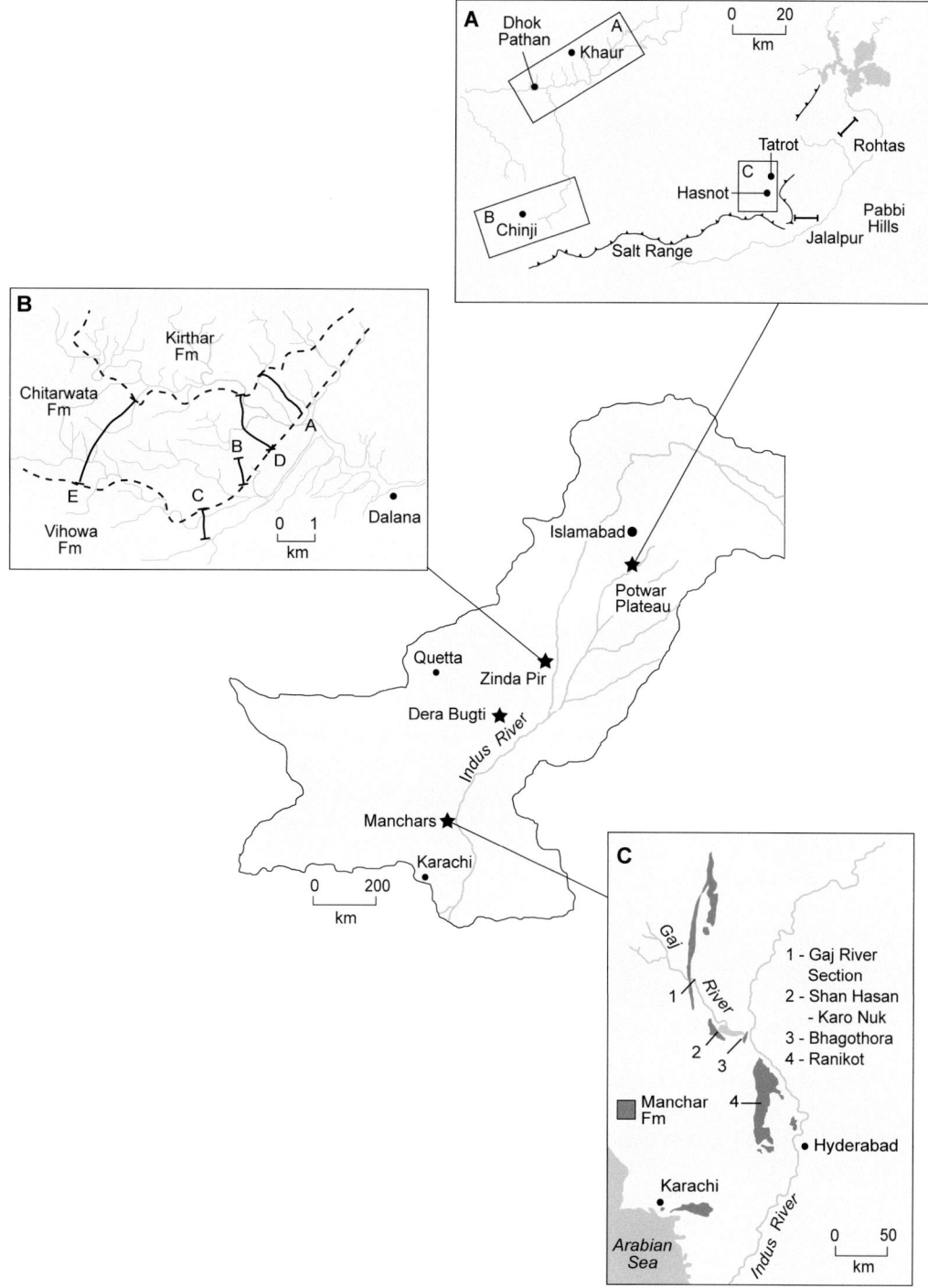

Figure 3.1. Map of Pakistan showing source areas for the Miocene formations and fossils discussed in this and other chapters: A. the Potwar Plateau in northern Pakistan, B. the Zinda Pir Dome in southwestern Punjab, and C. the Manchars in Sind.

Figure 3.2. Typical view of Chinji Formation exposures near the Chinji Rest House on the Potwar Plateau, looking northeast toward the villages of Bhilomar and Qadipur. The gray resistant beds forming the tilted layers on the left are examples of the major channels discussed in Chapters 2 and 4, this volume. Photograph taken in 1975 by J. Barry.

significant attribute is that the outcrops occur as extensive lateral exposures that may extend for kilometers or even tens of kilometers (Fig. 3.2; Chapter 2, this volume), which has allowed us to trace distinctive lithological horizons that are isochronous within time intervals of less than 10,000 years. Since the strata have clear superpositional relationships that easily establish the temporal succession of both rocks and fossils, they form a unique terrestrial sequence that allows detailed and temporally resolved study of paleoenvironments and faunas through time. Fossil productivity over the whole sequence is uneven, however, with some time intervals having very few sites or fossils while others have many (Fig. 3.3a, b, c). This varying abundance of fossils, which is a result both of outcrop availability and fluvial processes affecting fossil preservation, can bias the observed stratigraphic ranges of species and may distort or obscure patterns of faunal change.

Although relatively few Siwalik fossils are found in situ, they do frequently occur in low density surface concentrations of eroded bones, teeth, and occasional shells surrounded by barren or nearly barren exposures. A typical concentration may contain between 5 and 200 fossils, although there are concentrations that contain a thousand or more specimens. The fossils are normally found as isolated teeth, jaws, and postcranials, most of which are not only disarticulated but also fragmentary due both to pre-burial taphonomic processes and weathering on the outcrops. Although there are partial skulls in the collections, very few cranial and skeletal elements belonging to a single individual are known. The fragmentary, disarticulated state of the fossils makes it difficult to link dental, cranial, and postcranial morphologies and thus to fully characterize species. Although abundant calcified roots and very rare leaf impressions are known, we have not been able to recover pollen or other diagnostic plant remains from the sediments. Analysis of biomarkers preserved within the paleosols documents the existence of some surviving molecular evidence for the ancient vegetation (Freeman and Colarusso 2001; Karp, Behrensmeyer, and Freeman 2018), suggesting an approach that could be further developed in the future. Ichnofossils are also found and include abundant burrow fills and root traces in paleosols; invertebrate and vertebrate trackways are so far unknown. Calcareous stromatolites occur in some abandoned channel deposits.

Depositional environments are discussed in more detail in Chapters 2 and 4 of this volume. Briefly, the sedimentary sequences are a mix of alternating sandstone, siltstone, and mudstone lithologies, with rare marls (muddy limestones). The

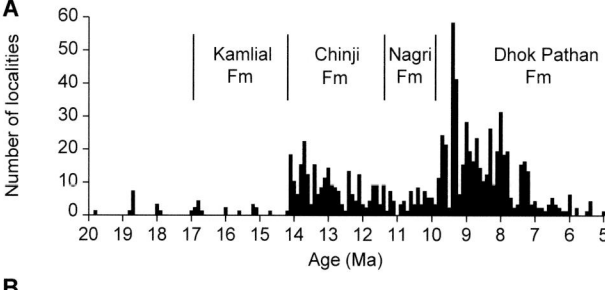

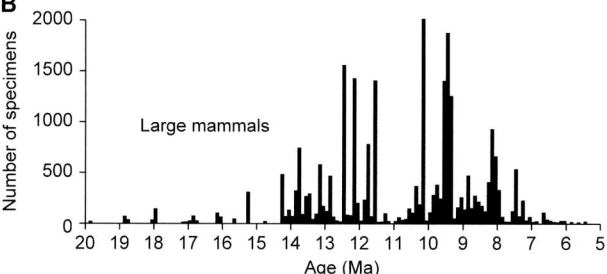

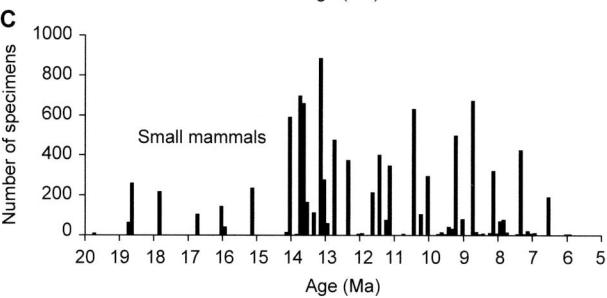

Figure 3.3. Plots of fossil frequencies per 100,000 year-long intervals between 20 and 5 Ma.: A. number of localities, B. number of fossils of large mammals, and C. number of fossils of small mammals. Graphed data include only Class I, II, and III localities as defined in the text and Figure 3.7. The boundaries of the Potwar Plateau Siwalik formations in Figure 3.3a are schematic and approximate. Both Figures 3.3b and 3.3c have very small counts that are not visible. The maximum value on Figure 3.3b is 4,478 in the 10.1 Ma interval (Locality Y311).

most common depositional environments include large or small, active or abandoned channels, levees, crevasse splays, and paleosols (Chapters 2 and 4, this volume). These occur in all Siwalik formations but differ in their frequency of occurrence. Such differences in frequency contribute to the overall uneven-ness in fossil preservation over time (Fig. 3.3a). The Kamlial and Nagri Formations are formed predominately from sandstones deposited in large channels, while the Chinji and Dhok Pathan Formations have more floodplain deposits including floodplain channels and paleosols. Fossils of both vertebrates and inverte-brates occur in nearly all lithologies, but vertebrates are most common in the fills of abandoned large channels and medium to small size floodplain channels (Chapter 4, this volume). Because the Nagri and Kamlial Formations are predominantly composed of large-scale channel deposits with few of the more productive abandoned channel or floodplain channel deposits, they have poorer fossil productivity. Also noteworthy is the re-duction in number of localities after 7.2 Ma (Fig. 3.3a), probably largely due to a decrease in the area of rock exposures.

Fossil Localities and Survey Blocks

Here we use two terms to refer to the occurrence and sam-pling protocol for fossils: locality and survey block. As we use the term, a locality is a concentration of fossils with very limited dimensions of time and space (Fig. 3.4a, b, c, d, e). In the Si-waliks, such fossil concentrations typically occur on small areas of outcrop a few tens or hundreds of square meters in extent, with one sedimentary unit as the source of the fossils. Between such patches there usually are only infrequent, widely dispersed fossils on the outcrops, so that localities are spatially distinct, natural entities. The duration of accumulation of the fossil material in a given locality cannot be precisely known; estimates involve both time averaging of the original bone as-semblage and time averaging of eroded fossils on the modern outcrop surfaces. Thus, localities may mix materials from dif-ferent, closely spaced stratigraphic levels. The amount of time represented by a fossil assemblage depends on the sedimento-logical and taphonomic processes involved in forming the lo-cality. In most cases this time interval probably is brief, ranging from tens to thousands of years (Behrensmeyer 1982; Badgley 1986; Behrensmeyer 1987; Badgley et al. 1995; Behrens-meyer, Willis, and Quade 1995; Chapter 4, this volume). The amount of time represented by surface accumulations of fos-sils eroded from different source beds could be much longer, probably up to tens to hundreds of thousands of years.

Because it is more extensive in area, spans a greater strati-graphic thickness, and is likely to include fossils from multiple sedimentary layers, a survey block, or collecting level, encom-passes more space and time than a single locality (Fig. 3.5a, b, c, d; Behrensmeyer and Barry 2005). Survey blocks can span substantial stratigraphic thicknesses (up to 110 m) with the tem-poral duration of individual blocks estimated as typically be-tween 30,000 and 350,000 years. The survey blocks may extend laterally 1–2 km and generally are demarcated by over- and underlying sandstone bodies. Many of these survey blocks were systematically searched for all fossils, including unidentifiable scraps of bone, as an additional measure of fossil productivity. During such surveys we kept a record of the number of hours spent searching and all finds were counted and tabulated. Fos-sils were only collected if they were identifiable to taxon and relatively informative or of taphonomic interest. In some cases a concentration of fossils or a notable specimen on a survey block was designated as a locality in order to preserve details of the occurrence (e.g., Locality Y665 on Fig. 3.5d).

Initially designed as a means of standardizing surface col-lections (Behrensmeyer and Barry 2005), survey blocks also proved to be a useful alternative to designating individual localities for isolated fossils of biostratigraphic importance, such as single equid molars. Data collected in these surveys produce less biased estimates of the relative abundance of fossils on outcrop surfaces (fossil productivity) and the rela-tive proportions of taxa and skeletal parts than other proto-cols (Behrensmeyer and Barry 2005). Fossils collected from both survey blocks and localities are used in the analyses of this volume.

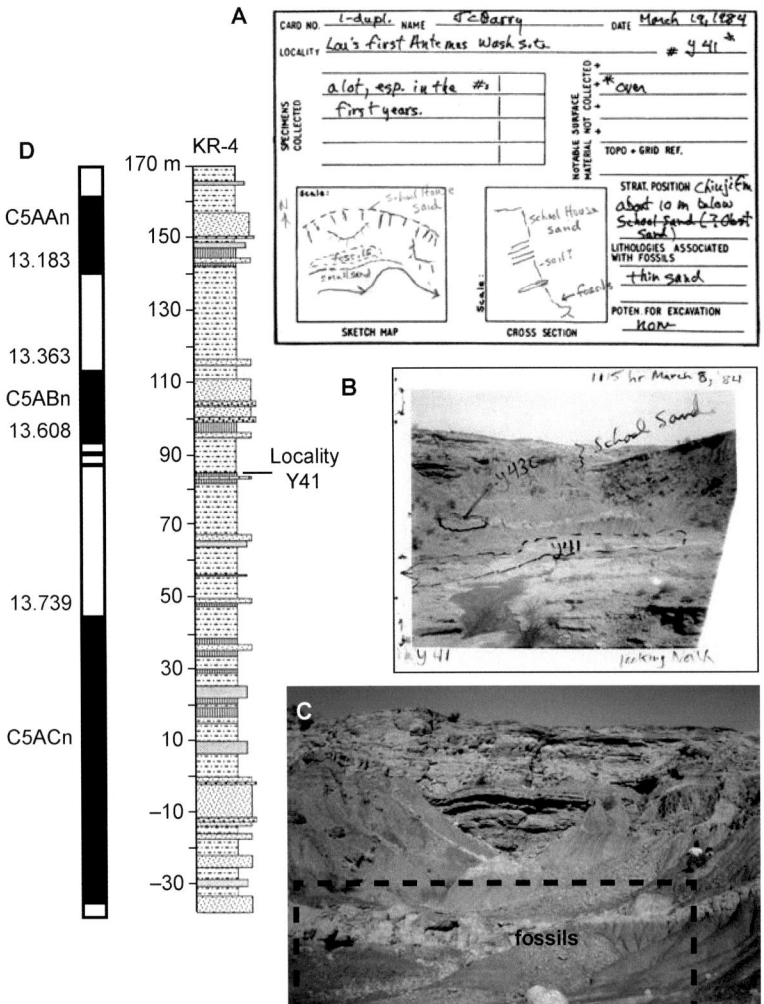

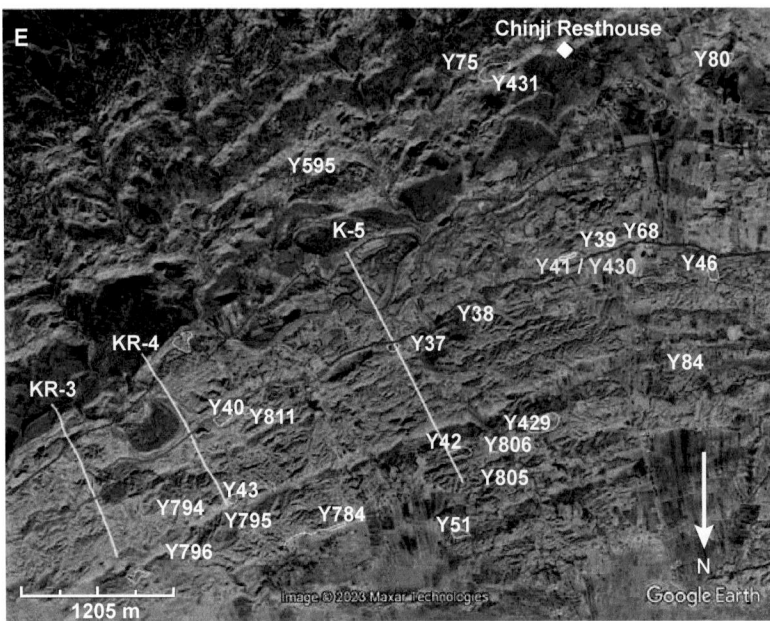

Figure 3.4. Documentation of a fossil locality as exemplified by Locality Y41 in the lower Chinji Formation.: A. locality card, B. annotated Polaroid, C. ground photo of Locality Y41 taken in 1975 by W. C. Barry, D. lithological section modified from McRae (1989) with local magnetic zonation and its correlation to GPTS chrons C5AAn through C5ACn, and E. Google Earth image (January 12, 2003) of exposures with Localities Y41 and Y430. Red, purple and blue lines trace outcrop of major sands. Yellow lines trace three stratigraphic sections (K5, KR4, and KR3) with paleomagnetic sampling (Table 3.1). Localities Y41 and Y430 in yellow; other localities in white.

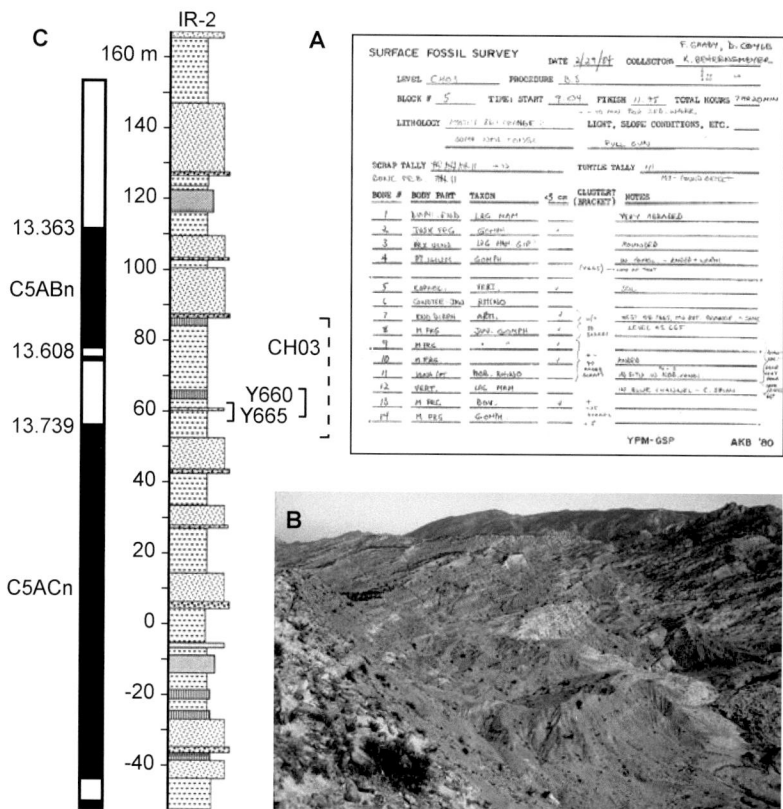

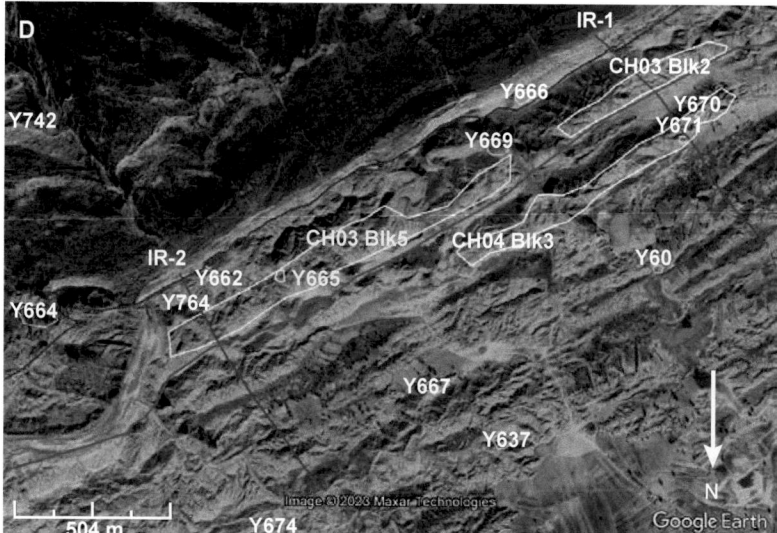

Figure 3.5. Documentation of a survey block as exemplified by CH03 Blk5 in the lower Chinji Formation: A. survey tally card, B. ground photo of typical lower Chinji Formation exposures, C. lithological section modified from McRae (1989), with local magnetic zonation and correlation to the GMTS, and D. Google Earth image (January 12, 2003) of exposures with Survey Block CH03 Blk5. Red and purple lines trace outcrop of major sands. Blue lines trace two stratigraphic sections (IR-1 and IR-2) with paleomagnetic sampling (Table 3.1). CH03 Blk5 and Locality Y665 in yellow, others in white. The difference between the estimated minimum and maximum ages for CH03 is 207,000 years. Locality Y665 is a productive concentration screened for small mammals.

Recent collecting in Pakistan combined with records for fossils in the American Museum of Natural History and the Yale Peabody Museum has produced over 54,700 records in our project database from 1,297 different localities. As discussed in the following section, a large number of the localities, and thus fossils, have ages estimated to within 100,000 years or less. However, for reasons explained above, the specific localities and ages of most type specimens and many referred specimens of named species are not known.

Fossil Collection and Documentation

Our Siwalik fossil collections were made using a variety of techniques. Natural erosion of the sediments has exposed fossils that are scattered unevenly across outcrop surfaces. Larger specimens, and sometimes very small fossils as well, were collected from these exposures. Although excavation can result in more complete specimens, it is very time-intensive relative to the rate of fossil recovery. Therefore, few localities have been excavated to any extent, but excavated localities have provided

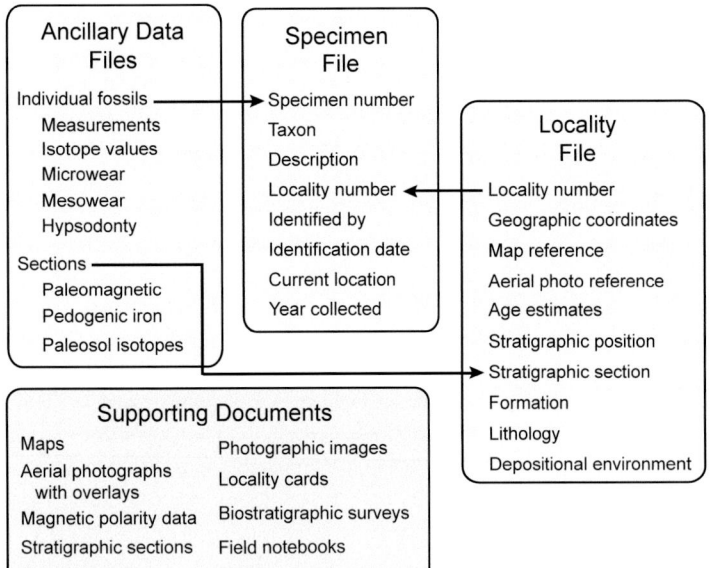

Ancillary Data Files

Individual fossils
  Measurements
  Isotope values
  Microwear
  Mesowear
  Hypsodonty

Sections
  Paleomagnetic
  Pedogenic iron
  Paleosol isotopes

Specimen File

Specimen number
Taxon
Description
Locality number
Identified by
Identification date
Current location
Year collected

Locality File

Locality number
Geographic coordinates
Map reference
Aerial photo reference
Age estimates
Stratigraphic position
Stratigraphic section
Formation
Lithology
Depositional environment

Supporting Documents

Maps
Aerial photographs
  with overlays
Magnetic polarity data
Stratigraphic sections
Fence diagrams

Photographic images
Locality cards
Biostratigraphic surveys
Field notebooks
Publications

Figure 3.6. Simplified diagram showing the content and relationships of our primary and ancillary digital files, as well as paper documents recording information from our years of geological and paleontological research. Note that only a selection of fields in the specimen and locality files is listed here, while the ancillary files are separate files, each containing data for either individual specimens or stratigraphic sections. Arrows show fields in common that are used to link files. In individual ancillary files these would be either specimen numbers or stratigraphic section names. For example, a file with measurements on dentitions would also have the specimen number, thus allowing the measurements to be linked to the taxonomic identification and age of the fossil. The files under "Sections" are all connected to the measured lithostratigraphic sections included in the supporting documents.

some important fossils and also allowed documentation of taphonomic processes associated with fossil burial and preservation (Fig. 4.7, Table 4.4, this volume). Screening of bulk sediment samples to recover rodents and other small species has been very successful, resulting in almost 11,800 specimens of small mammals (rodents, lagomorphs, insectivores, lorisids, tupaiids, and bats) from 89 screen-washed sites as well as many other small vertebrates such as snakes, lizards, fish, and, very rarely, amphibians. Most of the screen-washed localities have larger fossils as well, but since screen washing was not introduced until the mid-1970s, the older collections in European, American, and Indian museums rarely include rodents or other small vertebrates. As mentioned previously, standardized surveys documented the number of fossils and the relative abundance of taxa at different stratigraphic levels. These surveys entailed diligent searching for and recording of all occurrences of fossil bone along belts of outcrop within defined stratigraphic intervals (survey blocks).

A large proportion of fossils in collections made since 1973 are elements of the postcranial skeleton. This contrasts with the older collections and reflects a shift in focus from collecting taxonomically identifiable cranial material to understanding more about the paleobiology of the individual species as well as preservation processes. Because of our concern for gathering accurate and reliable geographic and stratigraphic information, we avoided buying fossils from villagers. We have, however, benefited from their local knowledge and interest in our work, and we employed them to help with excavating and transporting bulk rock samples and large specimens with camels and donkeys. This proved mutually beneficial and we are very grateful for their generosity and hospitality throughout the decades of fieldwork.

The geographic positions of newly discovered localities and survey blocks have been documented with topographic maps,

aerial photograph overlays, and more recently (for a small subset) imagery available through Google Earth, while their stratigraphic positions have been recorded on measured stratigraphic sections (Fig. 3.4d). We also assembled a file of locality cards (Behrensmeyer and Raza 1984), each with a brief lithological description of the locality, sketch maps and cross sections, and annotated Polaroid ground photographs (Fig. 3.4a, b). Although most of our collecting occurred prior to the availability of high-resolution Global Positioning System (GPS) technology, geographic coordinates for many localities have been determined from maps, hand-held GPS devices, or Google Earth imagery (Figs. 3.4e, 3.5d).

Fossil specimens are cataloged with a unique number, taxonomic identification, coded description of the specimen, and reference to a locality or survey block (Barry 1984). The specimen and locality catalogs are separate but linked digital files (Fig. 3.6). There are also important ancillary files, both paper and digital. Paper documents include maps, stratigraphic sections, locality cards, and biostratigraphic survey cards, as well as field notebooks, aerial photographs, and 35 mm slides. Many of these have been converted to digital formats and archived. Others already in digital format include a project bibliography, lists of taxonomic names and their authors, and a limited number of scanned images of fossils and fossil localities. The entire specimen database now holds over 54,700 records for fossils from 1,297 localities and survey blocks, 89% of which are well documented localities discovered since 1973. Our project database also includes approximately 1,080 records for selected fossils in the American Museum of Natural History and the Yale Peabody Museum collections. These are fossils from localities that can in many cases be confidently placed in the stratigraphic framework, although their estimated ages may be poorly determined. Data presented in subsequent chapters are derived from these databases.

CHRONOSTRATIGRAPHY: ESTIMATING THE AGE OF THE FOSSILS

Ordering the Siwalik fossils in time is critical to all of our interpretations of evolution and paleoenvironmental change and is not always easy or straightforward. Various protocols for age determination must be followed in order to provide the highest possible temporal resolution without introducing additional sources of error. We are fortunate that the field of magnetostratigraphy developed during the early years of our Siwalik research, and we benefited from collaborating with pioneers in applying their techniques to fluvial sediments (Chapter 2, this volume). Many decades of dedicated field and laboratory work gave our project a secure chronostratigraphic framework, but determining the age of localities and fossils still depends on a number of assumptions and decisions that affect what we can and cannot do with the faunal and geological data.

The Chronstratigraphic Framework

The thickness and continuity of Siwalik deposition have allowed us to construct a detailed paleomagnetic reversal stratigraphy, providing a reliable chronostratigraphic framework for much of the Potwar Plateau (Barndt 1977; Barndt et al. 1978; Opdyke et al. 1979; Barry, Behrensmeyer, and Monaghan 1980; McMurtry 1980; Behrensmeyer and Tauxe 1982; G. D. Johnson et al. 1982; N. M. Johnson et al. 1982; Stix 1982; Tauxe and Opdyke 1982; Sheikh 1984; Johnson et al. 1985; Tauxe and Badgley 1988; Badgley and Tauxe 1990; McRae 1989, 1990; Kappelman et al. 1991; Willis 1993a, 1993b; Khan et al. 1997) as well as for the Zinda Pir Dome in southwestern Punjab (Lindsay et al. 2005) and Manchars in Sind (Khan et al. 1984). The drainage patterns in the three Potwar regions form a lattice-like arrangement of small intermittent streams that have cut ravines in the uplifted sedimentary rocks (Fig. 2.6, this volume). The larger streams, which are perpendicular to strike and spaced 3–5 km apart, have lateral tributary branches parallel to strike that expose continuous bands of outcrop between the larger streams. The gentle structural dip of the strata, combined with continuous vertical exposure, relatively little vegetation cover, and suitable lithologies, facilitated the development of the Potwar Siwaliks chronostratigraphy. (For details see Figs. 2.2 and 2.5, this volume.)

The relative thickness of the normal and reversed zones in a stratigraphic section provides the means for correlation to the Geomagnetic Polarity Time Scale (GPTS). Because the accumulation of sediments was constant-enough over long intervals of time, the relative thickness of a magnetostratigraphic zone reflects the duration of time over which it accumulated. However, over shorter time periods there may be inconsistencies in the patterns of normal and reversed polarities caused by variation in fluvial depositional rates and sampling density, which can lower confidence in section correlations (Johnson and McGee 1983; Johnson et al. 1988). Table 3.1 lists the thickness and published references for the sections used to construct the chronostratigraphic framework. In most cases we use the same polarity zonation and correlation to the GPTS as the references or have made only small changes.

The Potwar Plateau chronostratigraphic framework is constructed from 44 measured sections between 250 m and 3,270 m thick, as well as another 43 shorter sections (see Table 3.1). Thirty-one of the sections 250 meters or more and 30 of the shorter sections have documented magnetic polarity stratigraphy. With the exception of Rohtas, Jalalpur, and the Pabbi Hills, these sections form three regional networks corresponding to three of the areas (Chinji, Dhok Pathan, and Hasnot) that Pilgrim and other authors used to construct "faunal zones" (Pilgrim 1913, 1934; Cotter 1933; Colbert 1935; Lewis 1937; Barry et al. 2013; Flynn et al. 2013). In each regional network (A, B, and C on Fig. 3.1a; see also Fig. 2.3, this volume) exposures are both vertically and laterally continuous (e.g., Fig. 3.2), and individual sections can be correlated lithologically as well as by their magnetostratigraphy. However, in the absence of continuous exposures between regions, correlations among the three regional networks depend entirely on the magnetostratigraphy. None of our sections on the Potwar Plateau has been correlated using biostratigraphy.

The Zinda Pir chronostratigraphic framework is based on five paleomagnetically sampled sections in the Chitarwata and Vihowa Formations that together encompass approximately 700 m (Fig. 3.1b; Friedman et al. 1992; Lindsay and Downs 2000; Lindsay et al. 2005; Lindsay and Flynn 2016). Unlike the wholly fluvial formations of the Potwar Plateau, the Chitarwata Formation of Zinda Pir was deposited on the shore of a receding seaway and is dominated by sediments of estuary, strandplain, and tidal flat paleoenvironments. The overlying Vihowa Formation is wholly fluvial (Downing et al. 1993; Downing and Lindsay 2005; Lindsay et al. 2005). The Zinda Pir sections can be correlated to each other by tracing lithologic units laterally and by magnetic polarity stratigraphy.

The Manchar Formation (Fig. 3.1c) is a Middle and Upper Miocene fluvial formation in Sind. The formation, which at its base is gradational with the underlying marine Gaj Formation, is composed of alternating sandstones and silts, with thin conglomerates. The top is also gradational with the overlying Dada Conglomerate. Two sections have been measured (Raza et al. 1984; Khan et al. 1984; Bernor et al. 1988) and some preliminary paleomagnetic results were obtained by Khan et al. (1984) for one (see Table 3.1). While the age of the sediments and fossils is only approximately established, when adjusted to the Gradstein et al. (2012) timescale, the preliminary correlation of Khan et al. (1984) indicates the lower 1,700 m of Manchar Formation is probably between 15 and 10 Ma (Fig. 2.2, this volume).

In the fluvial Siwalik formations, stratigraphic sections containing six or more magnetic transitions can usually be independently correlated to the GPTS (Johnson and McGee 1983). Such sections are typically at least 250 m thick in the Potwar region. While sections with fewer magnetic transitions may not be directly correlated with the GPTS, they can still be placed in the chronostratigraphic framework by establishing their stratigraphic relationships to adjacent longer sections by tracing lithological horizons between them (e.g., Tauxe and Opdyke 1982; Sheikh 1984). The marginal marine deposits of the Chitarwata Formation are more problematic, with apparently

Table 3.1. Names, Thicknesses, and References for the Stratigraphic Sections Used to Construct the Chronostratigraphic Framework

Region	Section	Thickness (m)	Paleomagnetic sampling	References	Comments
Rohtas	Dhabwala Kas (Rohtas)	1,050	yes	Opdyke et al. 1979; Behrensmeyer et al. 2007	
	Sanghoi/Basawa	1,850	yes	Opdyke et al. 1979; Behrensmeyer et al. 2007	
Jalalpur	Jarmaghal Kas	2,100	yes	Opdyke et al. 1979; Johnson, G. D. 1982	Chambal Ridge in Opdyke et al. (1979)
Hasnot Area	Andar Kas	1,300	yes	Johnson, G. D. 1982	
	Kotal Kund	1,800	yes	Johnson, G. D. 1982	
	Lower Dhala Nala	862	yes	Johnson, G. D. 1982, Barry et al. 2002	Lower 100 m without paleomag
	Dhok Saira	250	yes	Barry et al. 2002	
	Padhri	183	no	Barry et al. 2002	
	Y910	68	no	Barry et al. 2002	
	Hasnot SW	68	no	Barry et al. 2002	
	East of Bhandar	100	no	Barry et al. 2002	
	Pillar I	64	yes	McMurtry 1980	
	Pillar II	71	yes	McMurtry 1980	
	Pillar III	66	yes	McMurtry 1980	
	Pillar IV	55	yes	McMurtry 1980	
	Pillar V	80	yes	McMurtry 1980	
	Pillar VI	90	yes	McMurtry 1980	
	Tatrot-Bhandar	52	yes	Johnson, G. D. 1982, McMurtry 1980	Tatrot in McMurtry (1980)
	Tatrot	63	yes	Opdyke et al. 1979	
Khaur Area	Kundval Kas	500	yes	Barndt 1977	
	Dhok Pathan Rest House	497	yes	Barry, Behrensmeyer, and Monaghan 1980; Barry et al. 2002	
	Bhianwala Kas	1,692	no	Barry, Behrensmeyer, and Monaghan 1980	Shown as lower part of Dhok Pathan section
	Utran Kas	1,300	yes	Barndt 1977; Barndt et al. 1978; Zaleha 1997b	
	Utran (UN)	70	no	Behrensmeyer and Tauxe 1982	Remeasured segment
	Kot Maliaran	600	yes	Barndt 1977; Barry, Behrensmeyer, and Monaghan 1980; Barndt et al. 1978	
	Kot Maliaran (KM)	60	no	Behrensmeyer and Tauxe 1982	Remeasured segment
	Chouktriwali Kas	1,708	no	Barry, Behrensmeyer, and Monaghan 1980	
	Chouktriwali Kas (CH)	100	yes	Behrensmeyer and Tauxe 1982	Remeasured and resampled upper segment
	Dhok Mila	1,000	yes	Barndt 1977; Barndt et al. 1978	
	Khaur Kas (KK)	270	yes	Barry, Behrensmeyer, and Monaghan 1980; Behrensmeyer and Tauxe 1982	Paleomag sampling in lower 75 m
	Hasal Kas (HL)	2,425	yes	Barndt 1977; Barndt et al. 1978; Behrensmeyer and Tauxe 1982; Tauxe and Opdyke 1982 Zaleha 1997b	Bora Kas in Tauxe and Opdyke (1982)
	Hasal Kas	2,850	no	Barry, Behrensmeyer, and Monaghan 1980	Section of Barndt (1977) remeasured by Raza and Meyer
	Y227	75	no	Behrensmeyer and Tauxe 1982	
	Dinga Kas (DK)	120	yes	Barry, Behrensmeyer, and Monaghan 1980; Behrensmeyer and Tauxe 1982; Tauxe and Opdyke 1982	
	Malhuwala Kas	1,200	yes	Barry et al. 2002; Barndt et al. 1978	
	Malhuwala Kas (MKM)	2,230	yes	Barry, Behrensmeyer, and Monaghan 1980; Behrensmeyer and Tauxe 1982; Tauxe and Opdyke 1982	Upper 838 m of paleomag sampling of Tauxe and Opdyke (1982) transfered to measured section of Barry, Behrensmeyer, and Monaghan (1980)

Table 3.1. continued

Region	Section	Thickness (m)	Paleomagnetic sampling	References	Comments
	Jabbi	203	yes	Barry, Behrensmeyer, and Monaghan 1980; Tauxc and Opdykc 1982	
	Gandakas Road (GKR)	70	no	Behrensmeyer and Tauxe 1982	
	Gandakas (GK)	160	yes	Barry, Behrensmeyer, and Monaghan 1980; Barndt et al. 1978; Behrensmeyer and Tauxe 1982; Tauxe and Opdyke 1982	Section 3 of Tauxe and Badgley (1988) is close to GK
	Bhagwa Kas South (BWS)	75	yes	Behrensmeyer and Tauxe 1982; Tauxe and Badgley 1988	Section 5 in Tauxe and Badgley (1988)
	Bhagwa Kas (BW)	110	yes	Barry, Behrensmeyer, and Monaghan 1980; Behrensmeyer and Tauxe 1982; Tauxe and Opdyke 1982; Tauxe and Badgley 1988	Section 6 in Tauxe and Badgley (1988)
	Y269	75	no	Behrensmeyer and Tauxe 1982	
	Silt Bridge (SB)	30	no	Behrensmeyer and Tauxe 1982	
	Ratha Kas South (RKS)	35	no	Behrensmeyer and Tauxe 1982	
	Ratha Kas (RK)	1,966	yes	Barry, Behrensmeyer, and Monaghan 1980; Behrensmeyer and Tauxe 1982; Tauxe and Opdyke 1982	
	Thagenwali Dhok (TWD)	170	yes	Behrensmeyer and Tauxe 1982; Tauxe and Opdyke 1982	
	Y260 (KAK)	75	yes	Behrensmeyer and Tauxe 1982	
	Kaulial Kas	3,262	yes	Barry, Behrensmeyer, and Monaghan 1980; Tauxe and Opdyke 1982	Combined KUL and LK of Tauxe and Opdyke (1982)
	Kaulial Kas (KUL)	70	yes	Behrensmeyer and Tauxe 1982	Additional sampling added to Tauxe and Opdyke (1982)
Chinji Area	Gambhir	2,100	yes	Johnson, G. D. 1982; Johnson et al. 1985; Willis 1993a, 1993b	Paleomag sites of Johnson et al. (1982b, 1985) placed onto section of Willis (1993a, 1993b); Various spellings, including Gabhir
	Chita Parwala	745	yes	Johnson et al. 1985; Stix 1982	
	Trimman Nala	580	no	Stix 1982	
	I	276	no	Raza 1983	
	II	830	no	Raza 1983	
	Huch Nala	712	yes	Raza 1983; Kappelman 1986	Section III of Willis (1993a, 1993b); H in Stix (1982)
	IV	268	no	Raza 1983	
	V	470	no	Raza 1983	
	Rata Dala Nala	654	yes	I. Khan and Sheikh, unpublished data	
	K1	195	yes	Sheikh 1984; McRae 1989, 1990	KR1 in McRae (1989, 1990)
	K2	225	yes	Sheikh 1984; McRae 1989, 1990	KR2 in McRae (1989, 1990)
	K3	248	yes	Sheikh 1984; McRae 1989, 1990	KR3 in McRae (1989, 1990)
	K4	207	yes	Sheikh 1984; McRae 1989, 1990	KR4 in McRae (1989, 1990)
	K5	117	yes	Sheikh 1984	
	K6	119	yes	Sheikh 1984	
	K7	129	yes	Sheikh 1984	
	McRae's Chita Parwala	230	yes	McRae 1989, 1990	
	IR-1	225	yes	I. Khan, unpublished data; McRae 1989, 1990	
	IR-2	220	yes	I. Khan, unpublished data; McRae 1989, 1990	
	IR-3	207	yes	I. Khan, unpublished data; McRae 1989, 1990	
	R	311	no	Kappelman 1986	
	Q	343	no	Kappelman 1986	
	Parrawali	366	yes	Kappelman 1986; Kappelman et al. 1991	Section P in Kappelman (1986)
	M	257	no	Kappelman 1986	
	L	267	no	Kappelman 1986	

(*continued*)

Table 3.1. continued

Region	Section	Thickness (m)	Paleomagnetic sampling	References	Comments
	Kanatti	478	yes	Kappelman 1986; Kappelman et al. 1991	Section K in Kappelman (1986)
	Samriala Nala	402	yes	Kappelman et al. 1991	
	Bhaun	1,750	yes	Opdyke et al. 1979; Johnson, G. D. 1982	Also called Chakwal-Bhaun
Pabbi Hills	2	620	yes	Opdyke et al. 1979; Keller 1975; Keller et al. 1977; Johnson et al. 1979	
	15	650	yes	Opdyke et al. 1979; Keller 1975; Keller et al. 1977; Johnson et al. 1979	
	16	100	yes	Opdyke et al. 1979; Keller 1975; Keller et al. 1977; Johnson et al. 1979	
Mangla Samwal Anticline	Jhel Kas	840	yes	Opdyke et al. 1979; Johnson et al. 1979; Johnson, G. D. 1982 Hussain et al. 1992	
	Jhelawala Kas	850	no	Hussain et al. 1992	
	Samwal	830	no	Hussain et al. 1992	
	Dhok Dara	150	no	Hussain et al. 1992	
	Rakh	145	no	Hussain et al. 1992	
Campbellpore Basin	Haro River	61	yes	Opdyke et al. 1979; Johnson 1982; Saunders and Dawson 1998	
Mahesian Anticline	Chakwala Kas	1,200	yes	Khan 1993; Khan et al. 1997	
Sind	Gaj River	2,220	yes	Raza et al. 1984; Khan et al. 1984; Bernor et al. 1988	
	Bhagothoro	430	no	Raza et al. 1984; Bernor et al. 1988	Sehwan in Bernor et al. (1988)
Zinda Pir Dome	A	494	yes	Friedman et al. 1992; 35	DGA in Friedman et al. (1992)
	B	212	yes	Friedman et al. 1992; 35	DGB in Friedman et al. (1992)
	C	320	yes	Friedman et al. 1992; Lindsay et al. 2005	DGC in Friedman et al. (1992)
	D	412	yes	Lindsay et al. 2005	
	E	458	yes	Lindsay et al. 2005	

longer and more irregular undetected pauses in sedimentation resulting in missing intervals of time. Our methods for correlating sections to the GPTS were discussed in detail in Barry et al. (2002, 2013). The stratigraphic correlations among the various sections and to the GPTS have not changed from previous project publications, but we now use the time scale of Gradstein et al. (2012) to determine magnetochron boundary ages rather than previous calibrations of the GPTS.

Ages of Localities and Assignment to Specific Time Intervals

On the Potwar Plateau the fossil localities occur in exposures at most 2 or 3 kilometers (usually less) from local stratigraphic sections and can be correlated to them by tracing lithological horizons or marker beds. These laterally continuous horizons might be paleosols or sandstone bodies such as crevasse splays, which are more or less isochronous over distances of 2–3 km (Behrensmeyer and Tauxe 1982; see Fig. 2.8c, this volume). However, the lithological correlation has a degree of uncer-

tainty specific for each locality. To preserve this uncertainty, we record the stratigraphic position for a locality as both the highest and lowest reasonable positions in the reference section (Fig. 3.7). We have been conservative in these judgments, and as a result localities have a range of possible ages, some longer than others. Even when confident that a locality is neither above nor below these limits, however, we do not make any judgments about where it is within the bracketed time interval. Instead, we assume that it could be at any level in the designated interval with equal likelihood.

The relative stratigraphic positions of localities correlated to the same section are established by superposition alone. But to order localities on different sections or a network of sections, each locality must be assigned a relative age that can be determined from its stratigraphic position relative to known marker horizons or timelines shared by all the sections under consideration (Fig. 3.7). In our case we estimate numeric ages using the accepted magnetochronology, and these ages have varying degrees of accuracy. The algorithm used to determine them is based on interpolation between two points of known age. In our case all points of known age are magnetic polarity transitions (reversed to normal polarity or normal to reversed polarity), but in principle radiometric dates or events of known age in the stratigraphic sequence could also be used. The formula for finding an age is

$$Age = UAge + ((LAge - UAge)/T)SD,$$

where $UAge$ and $LAge$ are the ages of the top (Upper) and bottom (Lower) of the geomagnetic zone, T is the stratigraphic thickness of the magnetozone in the local section, and SD is the stratigraphic distance between the locality and the top of the magnetozone. In this formula, the quantity $(LAge - UAge)/T$ is the average rate of sediment accumulation in the section during the geomagnetic interval (e.g., Table 2.1, this volume). Some localities lie in truncated topmost or bottommost magnetozones, and interpolation cannot be used to estimate their ages. In such cases, ages are extrapolated, using the rate of sediment accumulation of the sub- or superjacent magnetic interval and the stratigraphic distance between the locality and the underlying or overlying magnetic transition.

Data used for the calculations include: (1) the stratigraphic position of the locality in a local lithological section, (2) the stratigraphic positions of the magnetic transitions in the same section, and (3) the ages of the magnetic transitions, as derived from correlation of the section to the GPTS (Gradstein et al. 2012).

We use two further refinements. The first is to derive separate age estimates for the possible upper and lower limits of the stratigraphic position for each locality that can be traced into a local or reference section (Fig. 3.7). This allows our estimate to incorporate the uncertainty in the correlation of the locality, which usually depends on tracing lithological markers to a stratigraphic section. When the stratigraphic position is precisely determined, these age estimates will be the same. Second, although magnetic polarity transitions are conventionally placed midway between sites of differing polarities, we instead use the highest and lowest possible positions of the magnetic transitions in the stratigraphic section. Doing so incorporates uncertainties in the stratigraphic positions of the magnetic transitions in a section by generating two age estimates, one for the highest position and one for the lowest. When the magnetic transitions are tightly constrained stratigraphically, again these age estimates converge. Taken together, these two refinements produce estimates for a pair of maximum and minimum ages for each locality—a conservative approach that provides the largest reasonable range of possible ages. In the analyses of subsequent chapters in this book, we frequently average the oldest and youngest ages for each locality to produce a "midpoint age" and calculate the difference between the oldest and youngest ages to produce an "age span."

Interpolation and extrapolation both assume a constant rate of sediment accumulation within an interval of constant polarity. Fluvial sequences are by nature episodic, with many short periods of rapid sedimentation alternating with periods of nondeposition or erosion that result in hiatuses. Consequently,

that assumption is certainly wrong, but as long as the hiatuses are short relative to the duration of the geomagnetic polarity interval and are evenly distributed within it, the approach will produce reliable results (Johnson and McGee 1983; Johnson et al. 1988).

Of the 1,297 localities with cataloged fossils in our database, 79% have an age estimate, although the precision or temporal resolution of the estimates, expressed as the difference between the maximum and minimum estimates, varies considerably. The greatest difference between the oldest possible age and the youngest is over 1.3 million years (Locality Y853). However, the ages of most localities are much better resolved, with 54% of the dated localities having a temporal resolution equal to 100,000 years or less, while an additional 26% have a resolution between 100,000 and 200,000 years. An additional 99 localities have associated estimated ages, but for various other reasons fossils from these sites are not cataloged in the project database. (They are largely localities found by G. E. Lewis and Barnum Brown.) Placing a fossil within a 100,000-year time interval is a relatively high level of temporal resolution. For example, if a fossil is 10 million years old, 100,000 years is only 1% of its age.

The youngest locality in our database is D54, the Haro River quarry of Saunders and Dawson (1998), a site originally excavated by de Terra and Teilhard de Chardin in 1935. The midpoint age of the host rock is 1.425 Ma, although the fossil deposits are younger (Saunders and Dawson 1998). (Note that in our project database, midpoint ages are expressed to the nearest thousand years, but this is only to preserve the superposition of localities in the measured sections, not because the age estimates have such a high degree of precision. For example, the oldest and youngest estimated ages of the host rock of D54 are 1.778 Ma and 1.072 Ma, a difference of more than 700,000 years.) The oldest dated locality is Z157 in the Chitarwata Formation, which is about 24 to 25 Ma, but there are other undated localities that are older (Lindsay et al. 2005), including several in Chron 10 that range in age from 28 to 29 Ma (i.e., in the middle of the Oligocene; Lindsay and Flynn 2016). Although our faunal analyses concentrate on localities older then 5 Ma (Fig. 3.3a), there are noteworthy younger localities. Most are in the Tatrot Formation and younger than 4.0 Ma. Five at Rohtas range between 4.5 and 4.9 Ma (Behrensmeyer et al. 2007).

In a previous analysis of faunal change (Barry et al. 2002), we proposed the following protocol to assign localities to 100,000 year intervals (e.g., 7.5 Ma) and to identify a subset of localities that cannot be assigned to a single interval. (Interval boundaries are set at +/- .05 so that, for example, the 7.5 Ma interval has upper and lower boundaries at 7.45 and 7.55 Ma.) With this protocol we refer to Class I those localities having both maximum and minimum age estimates in the same 100,000 year interval and can therefore be unambiguously assigned to that interval (Fig. 3.8). They make up approximately 19% of the 1297 localities with fossils. Class II localities have their minimum and maximum ages in two adjacent intervals, although the difference between their maximum and minimum

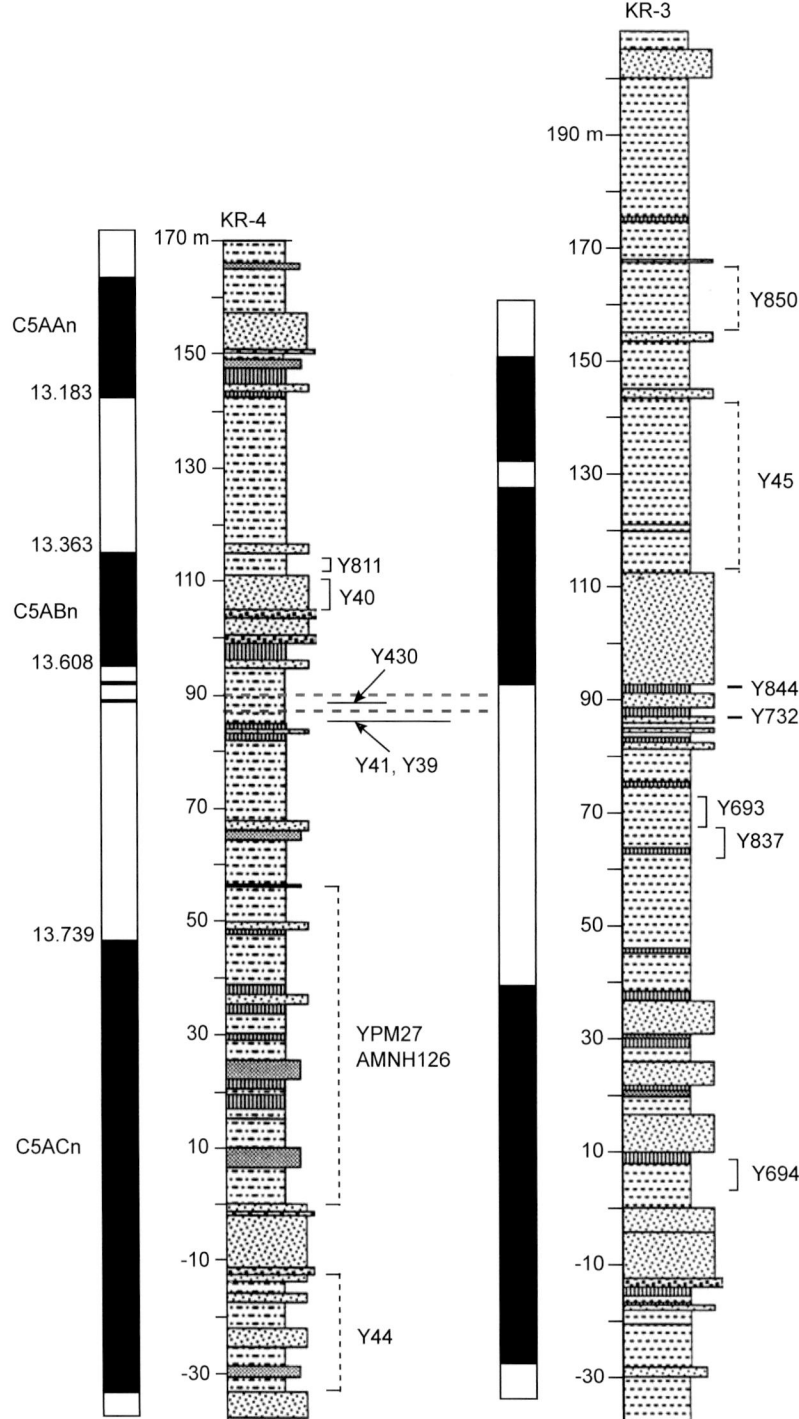

Figure 3.7. Two adjacent short stratigraphic sections in the lower Chinji Formation. The two sections are 0.8 km apart, as shown on Figure 3.4d; both sections modified from McRae (1989). The red and blue dashed connecting lines show levels of two paleosols traced between the sections. (These are not the sandstones traced on Figs. 3.4e and 3.5d.) Correlation of the local magnetostratigraphy of section KR-4 to the GPTS shown on left. Correlated localities are shown by letters and numbers, with dashes and brackets to indicate the upper and lower bounds of their stratigraphic positions. YPM designates a locality for fossils found by G. E. Lewis in 1932 and now in the Yale Peabody Museum. AMNH designates fossils found by Barnum Brown in 1922, now in the American Museum of Natural History. All the others are Harvard-Geological Survey of Pakistan localities.

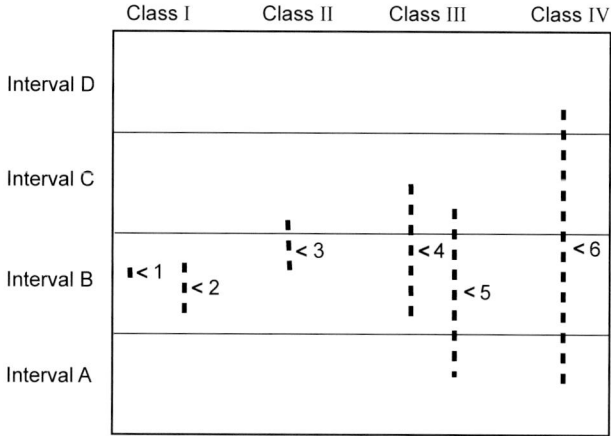

Figure 3.8. Examples illustrating different classes of age estimates for six hypothetical localities (1–6), showing how we determine levels of certainty in dating our fossil localities. Intervals A–D are each of 100,000 years duration and the midpoint ages of the six localities (numbers 1 through 6), as indicated by small arrows, are all in Interval B. Locality 1: maximum estimated age the same as the minimum estimated age; Locality 2: difference between maximum and minimum estimated ages less than 100,000 years with both in the same 100,000 year interval; Locality 3: difference between maximum and minimum estimated ages also less than 100,000 years, but falling in different intervals; Locality 4: difference between maximum and minimum estimated ages greater than 100,000 years and less than 200,000 years, but only spanning two intervals; Locality 5: difference between maximum and minimum estimated ages greater than 100,000 years and less than 200,000 years, but spanning three intervals; and Locality 6: difference between maximum and minimum estimated ages greater than 200,000 years and spanning three or more intervals. Localities 1, 2, 3, 4, and 5 would be assigned to Interval B. Locality 6 would not be assigned to any interval, although its midpoint lies in Interval B.

ages is less than 100,000 years' duration. These localities cannot be unambiguously assigned to a single interval. This is also true for a third group, Class III localities, which have differences between their minimum and maximum ages greater than 100,000 and less than 200,000 years, thus straddling two or three intervals. Nevertheless, all the 815 Class II and III localities (64%) must have at least 50% of their span in a single 100,000-year interval (Fig. 3.8). If the percentage of overlap is equivalent to the probability that the true age lies within an interval, then localities of both Classes II and III can be assigned to the interval with the greatest probability (or overlap). That interval will be the one in which the midpoint age falls. If a locality does not overlap with any 100,000 year interval by 50%, then an age is not assigned. This will be the case for all localities with age spans greater than 200,000 years, which may therefore be excluded from the syntheses of later chapters. The unassigned localities may have important fossils that enter into discussions of specific taxa, but because of poor temporal resolution, they are not as useful for our investigations of faunal change dynamics.

DATA QUALITY IN THE SIWALIK FOSSIL RECORD

Reliability of Age Estimates for Fossil Sites

Many of the syntheses of subsequent chapters depend on our age estimates for the fossils. We have taken as comprehensive and consistent an approach to the dating as possible, but it is still important to consider various other factors that may affect the reliability of the dates. As previously noted, the algorithm used to estimate ages of the localities makes several assumptions. The most basic is that the rate of sediment accumulation remains constant within each magnetic interval. The assumption is surely violated (Johnson et al. 1988), since sedimentation in fluvial systems is stochastically episodic with many short hiatuses that collectively may represent more time than the accumulated sediment (Sadler 1981). Even so, if no single hiatus is large relative to a magnetozone's duration and if the stratigraphic distribution of the many small hiatuses within it is random, then interpolation is still justifiable (Johnson and McGee 1983; Badgley, Tauxe, and Bookstein 1986). Assuming a constant rate of sediment accumulation is consistent with correlations between our individual sections, which ordinarily, but not always, show the magnetozones with the same relative stratigraphic thickness in laterally adjacent sections (e.g., Behrensmeyer and Tauxe 1982 and Figs. 3.7 and 2.7).

We also assume a uniform, rectangular distribution for the probability that a locality lies somewhere between its upper and lower stratigraphic limits, with zero probability above and below the marked range. This is a conservative assumption, because it produces a wider spread between the maximum and minimum age estimates than other probability distributions, such as a normal distribution around the center of the interval.

A third assumption concerns the age estimates of the magnetic transitions of the GPTS. These ages are derived from interpolations between tie points of known age and assume constancy in oceanic ridge–spreading rates. The age interpolations have been reassessed many times giving rise to a series of published geomagnetic polarity time scales, each reflecting different choices of tie points, spreading rates, and inflection points. Each variant timescale gives different estimates for the ages of Siwalik localities, with maximum differences of over 1 million years. In previous publications we have used the successive timescales of Mankinen and Dalrymple (1979), Berggren et al. (1985), Cande and Kent (1992, 1995), and Gradstein, Ogg, and Smith (2004). In this and subsequent chapters we use the timescale of Gradstein et al. (2012). When discussing older publications using geomagnetic derived ages, it is important to note the timescale used and make necessary adjustments.

Other potential sources of error in our age estimates are not included in the above uncertainties. Mistakes in field correlations between the localities and sections are most important in this regard, but there may also be errors in the measurement of stratigraphic thicknesses. When placing localities stratigraphically, we have tried to preserve the uncertainties in

Table 3.2. Siwalik Project Research Areas, Topics, and References

Research Area	Research Topic	References
Paleosols and sediments	Carbon and oxygen isotopic signals	Quade, Cerling, and Bowman 1989; Quade et al. 1992 Morgan et al. 1994; Quade and Cerling 1995; Behrensmeyer et al. 2007; Cerling et al. Chapter 5
	Biomarkers	Freeman and Colarusso 2001; Karp, Behrensmeyer, and Freeman 2018
	Paleosol morphology	Retallack 1991; Behrensmeyer et al. 1995; Abrajevitch et al. 2009
	Lithological facies	McMurtry 1980; Behrensmeyer and Tauxe 1982; Behrensmeyer 1987; Downing et al. 1993; Khan 1993; Willis 1993a; Willis and Behrensmeyer 1994; Behrensmeyer, Willis, and Quade1995; Behrensmeyer et al. 2005; Khan et al. 1997; Zaleha 1997a, 1997b; Downing and Lindsay 2005
Paleobiology of species	Functional morphology	Rose 1984, 1986, 1989; Flynn 1985, 1986; Kappelman 1988; Scott, Kappelman, and Kelley 1999; DeSilva et al. 2010; Kimura, Jacobs, and Flynn 2013; Lihoreau et al. 2007
	Body-size estimates	Kappleman 1986; Morgan 1994; Barry et al. 2002; Scott 2004; Lihoreau et al. 2007; Bibi 2007a
	Diet: microwear	Nelson 2005a; Badgley et al. 2008; Bibi 2007b
	Diet: mesowear	Belmaker et al. 2007; Morgan et al. 2015
	Diet: isotopes	Morgan 1994; Morgan, Kingston, and Marino 1994; Morgan et al. 2009; Nelson 2005a, 2007 Bibi 2007b Badgley et al. 2008 Kimura et al. 2013
	Diet: hypsodonty	Flynn and Jacobs 1982; Flynn 1986; Bibi 2007a, 2007b
Taphonomy	Fossil sites	Behrensmeyer 1982, 1987; Behrensmeyer, Willis, and Quade 1995; Behrensmeyer et al. 2005
	Faunal assemblages	Badgley 1986; Badgley and Behrensmeyer 1995; Badgley et al. 1995; Badgley, Downs, and Flynn 1998
Seasonality	Paleosol formation	Retallack 1991; Behrensmeyer, Willis, and Quade 1995; Behrensmeyer et al. 2007
	Invertebrates	Gurung Chapter 6, this volume Nora Loughlin, pers. comm.
	Vertebrates	Nelson 2007

our correlations, but the potential remains for undetected errors. Finally, our correlations between the local magnetic sections and the geomagnetic time scale may be incorrect. Correlations are fundamentally interpretations of data, and we could have misidentified magnetic transitions for which there are plausible alternative correlations. Additional fieldwork will someday resolve these problems.

Quality of Fossil Occurrence Data

Throughout the Siwalik Miocene nearly all stratigraphic intervals have some fossils, and in many intervals the record is very good. Nevertheless, there are considerable differences in the number of fossils among intervals. This is illustrated in Fig. 3.3a, b, and c, which show, respectively, the number of localities and the number of cataloged large and small mammal specimens in each 100,000 year interval between 20 and 5 Ma.

Variation in fossil recovery through the sequence results from multiple factors including (1) the area and quality of exposure, together with the amount of vegetation and ease of access to the outcrops, (2) the types and relative frequency of the depositional environments, and (3) differences in the time spent prospecting at different stratigraphic horizons. Note, though, that the relationship between the number of surveying hours and the number of fossils encountered during biostratigraphic surveys (Behrensmeyer and Barry 2005) demonstrates that the greater abundance of fossils in some intervals is not due solely to collecting effort or outcrop area. Rather, greater fossil abundance is also due to the intrinsic productivity of the lithologies. The variety of depositional environments and their relative frequencies have the greatest influence in determining the patterns of locality and fossil occurrences over time, as documented by a number of taphonomic studies (Raza 1983; Badgley 1986; Behrensmeyer 1987; Behrensmeyer et al. 2005). These factors are examined in Chapter 4 in this volume.

The Siwalik faunal record is best between about 14.1 and 7.2 Ma, with all but one interval having some fossils (Fig. 3.3a). The details differ depending on whether number of localities or number of fossils is considered, but the record is notably poorer before 14.1 Ma and importantly after 7.2 Ma, while there is also a less productive period between 11.4 and 10.5 Ma. The bias in the abundance of the fossils could in itself account for differences in the patterns of first and last occurrences and thus affect our assessment of the number of species in each stratigraphic interval (Badgley and Gingerich 1988; Badgley 1990; Barry et al. 2002).

Other biases affecting the reliability of the fossil occurrence data result from diverse taphonomic processes and the methods used to make the collections. The fragmentary nature of almost all Siwalik fossils also creates a bias that makes easily identified taxa, those that can be identified from more skeletal elements, or those with more durable parts appear to be more common. All these biases can limit paleoecological inferences, such as assessing relative abundance of taxa or recognizing habitat specificity. We assume in the following analyses (Chapters 25 and 26, this volume) that these biases are the same for

all intervals under study and do not affect comparisons between intervals (see also Chapter 4, this volume).

LINKING LOCALITY AND FAUNAL DATA TO ISOTOPIC AND OTHER PALEOECOLOGICAL DATA

In subsequent chapters of this volume, a primary focus of our inquiries is documenting patterns of both change in the fauna and change in the depositional systems and local habitats. To this end, we and others have conducted a variety of parallel and mutually dependent studies on Siwalik fossils and sediments (Table 3.2), studies that aimed to investigate aspects of paleoecology as related to larger scale change in climate, tectonics, and environment. Such research includes (1) investigations of the paleosols and sediments, especially analyses of carbon and oxygen isotopic signals in the carbonate components, biomarkers, paleosol morphology, and the vertical and lateral relationships of the lithological facies; (2) investigations of the paleobiology of individual species, including functional morphology, body size estimates, and diet; (3) investigations of the formation of the fossil sites and faunal assemblages; and (4) limited work on seasonality.

Key to integrating these data is our ability to confidently link the localities (or lithological unit in the case of sedimentary studies) and the individual fossils. Both linkages place the data in our lithologic and chronostratigraphic framework, allowing localities and fossils to be related not just to each other, but also to broad-scale tectonic, environmental, or climatic events (e.g., Chapter 26, this volume) or to faunal change throughout the Old World. Other auxiliary digital files that can be linked include dental and postcranial measurements, carbon and oxygen isotope values for both paleosols and fossils (including invertebrates), pedogenic iron content and biomarkers, micro- and mesowear data as well as measures of crown height among ruminants, and excavation data tied to both locality and fossils (Fig. 3.6).

SUMMARY

The thick sequence of ancient Siwalik deposits contains a dense fossil record, offering an opportunity to improve our understanding not just of the evolution of individual species, but also aspects of faunal change and paleoecology related to climate, tectonics, and environment. Such research demands knowledge of the geological age and depositional contexts of the fossils, which in turn entails not only collecting but also integrating data from diverse fields such as geochemistry, sedimentology, taphonomy, and geochronology with research on the systematics and paleobiology of extinct species.

This chapter has described some of the collecting techniques adapted to unique aspects of the Siwaliks, as well as discussing dating the fossil localities and archiving the locality and specimen data along with supplementary ecological and geological data. Paleomagnetic sampling in numerous rock sections has produced a chronostratigraphic framework that dates the fossil localities and fossil remains. This temporal framework forms the basis for analyzing change in lineages and faunal assemblages and even the lateral and vertical complexity of the lithological facies. Furthermore, the temporal framework allows us to connect those changes to climatic and tectonic events. In all these endeavors, special concerns have been uneven sampling through time, differential preservation of the fossils, errors in age estimates, and collecting methods.

There are 54,700 records for individual fossils in the project database, the majority of which are postcranial elements of mammals. These fossils come from 1,297 localities, of which 79% have an age estimate, although the precision of these estimates varies considerably. There are also considerable differences in the number of fossils from different parts of the Siwalik Miocene sequence. The fossil record is best between about 14.1 and 7.2 Ma, while poor both before 14.1 and after 7.2 Ma. There is also a less productive period between 11.4 and 10.5 Ma. Significantly there are also vertebrates from the Oligocene and Pliocene, although the Oligocene record is meager.

The differences in fossil recovery throughout the sequence result from at least three factors: the extent of the exposures, combined with vegetation cover; the ease of access to outcrops and effort spent prospecting them; and the relative frequency of different depositional environments. Data from our biostratigraphic surveys indicate that the greater or lesser abundance of fossils is not due solely to collecting effort or outcrop area. Rather, greater fossil abundance primarily reflects the intrinsic productivity of the lithologies, which is a result of the mix of depositional environments and their associated taphonomic processes. All these biases can limit paleoecological inferences and should be kept in mind in the following chapters.

ACKNOWLEDGMENTS

We thank the editors David Pilbeam, Catherine Badgley, and Michèle Morgan for inviting us to contribute this chapter, which is based on decades of careful documentation of the Siwalik paleontological record. The research on the Siwaliks is a result of a long collaboration with the Geological Survey of Pakistan and we appreciate the efforts of the Directors General and members of the Survey who have over the years supported the project—and especially the interest of Ibrahim Shah, long Director of the GSP Stratigraphy and Palaeontology Branch. Financial support has included grants from the National Science Foundation, the Smithsonian Foreign Currency Program, the Wenner-Gren Foundation for Anthropological Research, and the American School for Prehistoric Research. We would also like to acknowledge the abiding support of Francine Berkowitz of the Smithsonian Office of International Relations. Finally, we are grateful for the generosity and hospitality of the local people of Pakistan throughout our decades of fieldwork.

This chapter is dedicated to our late colleagues Noye Johnson, Will Downs, Neil Opdyke, and Ev Lindsay. They were instrumental in the work describe here and in other chapters.

References

Abrajevitch, A., K. Kodama, A. K. Behrensmeyer, and C. Badgley. 2009. Magnetic signature of moisture availability in ancient subtropical soils: A pilot rock magnetic study of Neogene Paleosols in Pakistan. American Geophysical Union, Fall Meeting 2009, abstract id. GP43B-0853

Badgley, C. 1986. Taphonomy of mammalian fossil remains from Siwalik rocks of Pakistan. *Paleobiology* 12(2):119–142.

Badgley, C. 1990. A statistical assessment of last appearances in the Eocene record of mammals. In *Dawn of the Age of Mammals in the Northern Part of the Rocky Mountain Interior, North America*, edited by T. M. Bown and K. D. Rose. Geological Society of America Special Paper 243:153–167.

Badgley, C., J. C. Barry, M. E. Morgan, S. V. Nelson, A. K. Behrensmeyer, T. E. Cerling, and D. Pilbeam. 2008. Ecological changes in Miocene mammalian record show impact of prolonged climatic forcing. *Proceedings of the National Academy of Sciences* 105(34): 12145–12149.

Badgley, C., W. S. Bartels, M. E. Morgan, A. K. Behrensmeyer, and S. M. Raza. 1995. Taphonomy of vertebrate assemblages from the Paleogene of northwestern Wyoming and the Neogene of northern Pakistan. In *Long Records of Continental Ecosystems*, edited by C. Badgley and A. K. Behrensmeyer. *Palaeogeography, Palaeoclimatology, Palaeoecology* 115(1–4):157–180.

Badgley, C., and A. K. Behrensmeyer. 1995. Preservational, paleoecological and evolutionary patterns in the Paleogene of Wyoming-Montana and the Neogene of Pakistan. *Palaeogeogrctphy, Palaeoclimatology, Palaeoccology* 115(1–4):319–340.

Badgley, C., W. Downs, and L. J. Flynn. 1998. Taphonomy of small-mammal fossil assemblages from the Middle Miocene Chinji Formation, Siwalik Group, Pakistan. In *Advances in Vertebrate Paleontology and Geochronology*, edited by Y. Tomida, L. J. Flynn, and L. L. Jacobs. *National Science Museum Monographs* No. 14:145–166.

Badgley, C., L. Tauxe, and F.L. Bookstein. 1986. Estimating the error of age interpolation in sedimentary rocks. *Nature* 319:139–141.

Badgley, C. E., and P. D. Gingerich. 1988. Sampling and faunal turnover in Early Eocene mammals. *Palaeogeography, Palaeoclimatology, Palaeoecology* 63(1–3):141–157.

Badgley, C. E., and L. Tauxe. 1990. Paleomagnetic stratigraphy and time in sediments: Studies in alluvial Siwalik rocks of Pakistan. *Journal of Geology* 98(4):457–477.

Barndt, J. 1977. The magnetic polarity stratigraphy of the type locality of the Dhok Pathan faunal stage, Potwar Plateau, Pakistan. Unpublished MA thesis, Dartmouth College.

Barndt, J., N. M. Johnson, G. D. Johnson, N. D. Opdyke, E. H. Lindsay, D. Pilbeam, and R.A.K. Tahirkheli. 1978. The magnetic polarity stratigraphy and age of the Siwalik Group near Dhok Pathan village, Potwar Plateau, Pakistan. *Earth and Planetary Science Letters* 41(3):355–364.

Barry, J. C. 1984. A summary of data file structure and data standards for the Yale-Geological Survey of Pakistan Project-Siwalik fossil collections. In *Contribution to the Geology of Siwaliks of Pakistan*, edited by S. M. I Shah and D. Pilbeam. *Memoirs of the Geological Survey of Pakistan* No. XI:71–74.

Barry, J. C., A. K. Behrensmeyer, and M. Monaghan. 1980. A geologic and biostratigraphic framework for Miocene sediments near Khaur Village, northern Pakistan. *Postilla* 183:1–19.

Barry, J. C., C. Badgley, A. K. Behrensmeyer, L. J. Flynn, H. Peltonen, I. U. Cheema, D. Pilbeam, S. Mahmood Raza, A. R. Rajpar, and M. E. Morgan. 2013. The Neogene Siwaliks of the Potwar Plateau and other regions of Pakistan. In *Fossil Mammals of Asia: Neogene Biostratigraphy and Chronology*, edited by X.-M. Wang, L. J. Flynn, and M. Fortelius, 373–399. New York: Columbia University Press.

Barry, J. C., M. E. Morgan, L. J. Flynn, D. Pilbeam, A. K. Behrensmeyer, S. M. Raza, I. A. Khan, C. Badgley, J. Hicks, and J. Kelley. 2002. Faunal and environmental change in the late Miocene Siwaliks of northern Pakistan. *Paleobiology Memoirs* No. 3:1–71 (supplement to *Paleobiology* 28[2]).

Behrensmeyer, A. K. 1982. Time resolution in fluvial vertebrate assemblages. *Paleobiology* 8(3):211–227.

Behrensmeyer, A. K. 1987. Miocene fluvial facies and vertebrate taphonomy in northern Pakistan. In *Recent Developments in Fluvial Sedimentology*, edited by F. G. Ethridge, R. M. Flores, M. D. Harvey, SEPM Special Publication No. 39:169–176.

Behrensmeyer, A. K., C. E. Badgley, J. C. Barry, M. E. Morgan, and S. M. Raza. 2005. The paleoenvironmental context of Siwalik Miocene vertebrate localities. In *Interpreting the Past: Essays on Human, Primate, and Mammal Evolution in Honor of David Pilbeam*, edited by D. E. Lieberman, R. J. Smith, and J. Kelley, 47–61. Boston: Brill.

Behrensmeyer, A. K. and J. C. Barry. 2005. Biostratigraphic surveys in the Siwaliks of Pakistan: A method for standardized surface sampling of the vertebrate fossil record. Special issue in honor of W. R. Downs, *Palaeontologica Electronica* 8(1): 15A. https://palaeo-electronica.org/2005_1/behrens15/issue1_05.htm.

Behrensmeyer A. K., J. Quade, T. E. Cerling, I. A. Kahn, P. Copeland, L. Roe, J. Hicks, P. Stubblefield, B. J. Willis, and C. Lalorre. 2007. The structure and rate of late Miocene expansion of C_4 plants: Evidence from lateral variation in stable isotopes in paleosols of the Siwalik Group, northern Pakistan. *Geological Society of America Bulletin* 119:1486–1505.

Behrensmeyer, A. K., and L. Tauxe. 1982. Isochronous fluvial systems in Miocene deposits of northern Pakistan. *Sedimentology* 29(3):331–352.

Behrensmeyer, A. K., and M. Raza. 1984. A procedure for documenting fossil localities in Siwalik deposits of northern Pakistan. In *Contribution to the Geology of Siwaliks of Pakistan*, edited by S. M. I. Shah and D. Pilbeam. *Memoirs of the Geological Survey of Pakistan* No. XI:65–69.

Behrensmeyer, A. K., B. J. Willis, and J. Quade. 1995. Floodplains and paleosols of Pakistan Neogene and Wyoming Paleogene deposits: Implications for the taphonomy and paleoecology of faunas. In *Long Records of Continental Ecosystems*, edited by C. Badgley and A. K. Behrensmeyer. *Palaeogeography, Palaeoclimatology, Palaeoecology* 115(1–4):37–60.

Belmaker, M., S. Nelson, M. Morgan, J. Barry, and C. Badgley. 2007. Mesowear analysis of ungulates in the Middle to Late Miocene of the Siwaliks, Pakistan: Dietary and paleoenvironmental implications. *Society of Vertebrate Paleontology Meeting Program and Abstracts* 2007:46A.

Berggren, W. A., D. V. Kent, J. J. Flynn, and J. A. Van Couvering. 1985. Cenozoic geochronology. *Geological Society of America Bulletin* 96:1407–1418.

Bernor, R. L., L. J. Flynn, T. Harrison, S. T. Hussain, and J. Kelley. 1988. *Dionysopithecus* from southern Pakistan and the biochronology and biogeography of early Eurasian catarrhines. *Journal of Human Evolution* 17(3):339–358.

Bibi, F. 2007a. Origin, paleoecology, and paleobiogeography of early bovini. *Palaeogeography, Palaeoclimatology, Palaeoecology* 248(1–2):60–72.

Bibi, F. 2007b. Dietary niche partitioning among fossil bovids in late Miocene C_3 habitats: Consilience of functional morphology and stable isotope analysis. *Palaeogeography, Palaeoclimatology, Palaeoecology* 253(1–2):529–538.

Cande, S. C., and D. V. Kent. 1992. A new geomagnetic polarity timescale for the Late Cretaceous and Cenozoic. *Journal of Geophysical Research* 97(B10):13917–13951.

Cande, S. C., and D. V. Kent. 1995. Revised calibration of the geomagnetic polarity timescale for the Late Cretaceous and Cenozoic. *Journal of Geophysical Research* 100(B4):6093–6095.

Colbert, E. H. 1935. Siwalik mammals in the American Museum of Natural History. *Transactions of the American Philosophical Society* n.s. 26:1–401.

Cotter, G. de P. 1933. The geology of the part of the Attock District west of longitude 72° 45′ E. *Memoirs of the Geological Survey of India* 55:63–161.

DeSilva, J. M., M. E. Morgan, J. C. Barry, and D. Pilbeam. 2010. A hominoid distal tibia from the Miocene of Pakistan. *Journal of Human Evolution* 58(2):147–154.

Downing, K. F., and E. H. Lindsay. 2005. Relationship of Chitarwata Formation paleodrainage and paleoenvironments to Himalayan tectonics and Indus River paleogeography. *Palaeontologia Electronica* 8(1): 20A. https://palaeo-electronica.org/2005_1/downing20/issue1_05.htm.

Downing, K. F., E. H. Lindsay, W. R. Downs, and S. E. Speyer. 1993. Lithostratigraphy and vertebrate biostratigraphy of the early Miocene Himalayan Foreland Zinda Pir Dome, Pakistan. *Sedimentary Geology* 87(1–2):25–37.

Flynn, L. J. 1985. Evolutionary patterns and rates in Siwalik Rhizomyidae. *Acta Zoologica Fennica* 170:141–144.

Flynn, L. J. 1986. Species longevity, stasis and stairsteps in rhizomyid rodents. In *Vertebrates, Phylogeny, and Philosophy: Contributions to Geology Special Paper 3*, edited by K. M. Flanagan and J. A. Lillegraven. University of Wyoming Special Paper 3:273–285.

Flynn, L. J., and L. L. Jacobs.1982. Effects of changing environments on Siwalik rodent faunas of northern Pakistan. *Palaeogeography, Palaeoclimatology, Palaeoecology* 38(1–2):129–138.

Flynn, L. J., E. H. Lindsay, D. Pilbeam, S. Mahmood Raza, M. E Morgan, J. C. Barry, C. Badgley, et al. 2013. The Siwaliks and Neogene evolutionary biology in South Asia. In *Fossil Mammals of Asia: Neogene Biostratigraphy and Chronology*, edited by X.-m. Wang, L. J. Flynn, M. Fortelius, 353–372. New York: Columbia University Press.

Freeman, K. H., and L. A. Colarusso. 2001. Molecular and isotopic records of C_4 grassland expansion in the late Miocene. *Geochimica et Cosmochimica Acta* 65(9):1439–1454.

Friedman, R., J. Gee, L. Tauxe, K. Downing, and E. Lindsay. 1992. The magnetostratigraphy of the Chitarwata and lower Vihowa Formations of the Dera Ghazi Khan area, Pakistan. *Sedimentary Geology* 81(3–4):253–268.

Gradstein, F. M., J. G. Ogg, M. D. Schmitz, and G. M. Ogg, 2012. *The Geological Time Scale 2012*. Amsterdam: Elsevier.

Gradstein, F. M., J. G. Ogg, and A. G. Smith. 2004. *A Geologic Time Scale 2004*. Cambridge: Cambridge University Press.

Hussain, S. T., G. D. van den Bergh, K. J. Steensma, J. A. de Visser, J. de Vos, M. Arif, J. van Dam, P. Y. Sondaar, and S. B. Malik. 1992. Biostratigraphy of the Plio-Pleistocene continental sediments (Upper Siwaliks) of the Mangla-Samwal Anticline, Asad Kashmir, Pakistan. *Proceedings of the Koninklijke Nederlandse Akademie van Wetenschappen* 95(1):65–80.

Johnson, G. D., P. Zeitler, C. W. Naeser, N. M. Johnson, D. M. Summers, C. D. Frost, N. D. Opdyke, and R. A. K. Tahirkheli. 1982. The occurrence and fission-track ages of Late Neogene and Quaternary volcanic sediments, Siwalik Group, northern Pakistan. *Palaeogeography, Palaeoclimatology, Palaeoecology* 37(1):63–93.

Johnson, G. D., N. M. Johnson, N. D. Opdyke, and R. A. K. Tahirkheli. 1979. Magnetic reversal stratigraphy and sedimentary tectonic history of the Upper Siwalik Group, Eastern Salt Range and Southwestern Kashmir. In *Geodynamics of Pakistan*, edited by A. Farah and K. A. de Jong, 149–165. Quetta: Geological Survey of Pakistan.

Johnson, N. M., and V. E. McGee. 1983. Magnetic polarity stratigraphy: Stochastic properties of data, sampling problems, and the evaluation of interpretations. *Journal of Geophysical Research* 88(B2):1213–1221.

Johnson, N. M., N. D. Opdyke, G. D. Johnson, E. H. Lindsay, and R. A. K. Tahirkheli. 1982. Magnetic polarity stratigraphy and ages of Siwalik Group rocks of the Potwar Plateau, Pakistan. *Palaeogeography, Palaeoclimatology, Palaeocology* 37(1):17–42.

Johnson, N. M., K. A. Sheikh, E. Dawson-Saunders, and L. E. McRae. 1988. The use of magnetic-reversal time lines in stratigraphic analysis: a case study in measuring variability in sedimentation rates. In *New Perspectives in Basin Analysis*, edited by K L. Kleinspehn and C. Paola, 189–200. New York: Springer.

Johnson, N. M., J. Stix, L. Tauxe, P. F. Cerveny, and R. A. K.Tahirkheli. 1985. Paleomagnetic chronology, fluvial processes and tectonic implications of the Siwalik deposits near Chinji Village, Pakistan. *Journal of Geology* 93(1):27–40.

Karp, A. T., A. K. Behrensmeyer, and K. H. Freeman. 2018. Grassland fire ecology has roots in the Late Miocene. *Proceedings of the National Academy of Sciences* 115:12130–12135.

Kappelman, J. W. 1986. The paleoecology and chronology of the Middle Miocene hominoids from the Chinji Formation of Pakistan. Unpublished PhD diss., Harvard University.

Kappelman, J. W. 1988. Morphology and locomotor adaptations of the bovid femur in relation to habitat. *Journal of Morphology* 198(1):119–130.

Kappelman, J. W., J. Kelley, D. Pilbeam, K. A. Sheikh, S. Ward, M. Anwar, J. C. Barry, et al. 1991. The earliest occurrence of *Sivapithecus* from the Middle Miocene Chinji Formation of Pakistan. *Journal of Human Evolution* 21(1):61–73.

Keller, H. M. 1975. The magnetic polarity stratigraphy of an Upper Siwalik sequence in the Pabbi Hills of Pakistan. Unpublished MS thesis, Dartmouth College.

Keller, H. M., R. A. K. Tahirkheli, M. A. Mirza, G. D. Johnson, N. M. Johnson, and N. D. Opdyke. 1977. Magnetic polarity stratigraphy of the Upper Siwalik deposits, Pabbi Hills, Pakistan. *Earth and Planetary Science Letters* 36(1):187–201.

Khan, I. A. 1993. Evolution of Miocene-Pliocene fluvial paleoenvironments in eastern Potwar Plateau, northern Pakistan. Unpublished PhD diss., Binghamton University.

Khan, I. A., J. S. Bridge, J. Kappelman, and R. Wilson. 1997. Evolution of Miocene fluvial environments, eastern Potwar Plateau, northern Pakistan. *Sedimentology* 44(2):221–251. https://doi.org/10.1111/j.1365-3091.1997.tb01522.x.

Khan, M. J., S. T. Hussain, M. Arif, H. Shaheed. 1984. Preliminary paleomagnetic investigations of the Manchar Formation, Gaj River section, Kirthar Range, Pakistan. *Geological Bulletin University of Peshawar* 17:145–152.

Kimura, Y., L. L Jacobs, T. E. Cerling, K. T. Uno, K. M. Ferguson, L. J. Flynn, and R. Patnaik. 2013. Fossil mice and rats show isotopic evidence of niche partitioning and change in dental ecomorphology related to dietary shift in Late Miocene of Pakistan. *PLOS ONE* 8(8): e69308.

Kimura, Y., L. L. Jacobs, and L. J. Flynn. 2013. Lineage-specific responses of tooth shape in murine rodents (Murinae, Rodentia) to Late Miocene Dietary Change in the Siwaliks of Pakistan. *PLOS ONE* 8(10): e76070.

Lewis, G. E. 1937. A new Siwalik correlation (India). *American Journal of Science*, Series 5, 33(195):191–204.

Lihoreau, F., J. Barry, C. Blondel, Y. Chaimanee, J.-J.Jaeger, and M. Brunet. 2007. Anatomical revision of the genus *Merycopotamus* (Artiodactyla; Anthracotheriidae): Its significance for Late Miocene mammal dispersal in Asia. *Palaeontology* 50(2):503–524.

Lindsay, E. H., and W. R. Downs. 2000. Age assessment of the Chitarwata Formation. *Himalayan Geology* 21:99–107.

Lindsay. E. H., and L. J. Flynn. 2016. Late Oligocene and Early Miocene Muroidea of the Zinda Pir Dome. *Historical Biology* 28(1–2): 215–236. https://doi.org/10.1080/08912963.2015.1027888.

Lindsay, E. H., L. J. Flynn, I. U. Cheema, J. C. Barry, K. F. Downing, A. R. Rajpar, and S. M. Raza. 2005. Will Downs and the Zinda Pir Dome. *Palaeontologia Electronica* 8(1). https://palaeo-electronica.org/2005_1/lindsay19/issue1_05.htm.

McMurtry, M. G. 1980. Facies changes and time relationships along a sandstone stratum, Middle Siwalik Group, Potwar Plateau, Pakistan. Unpublished BA thesis, Dartmouth College.

McRae, L. E. 1989. Chronostratigraphic variability in fluvial sequences as revealed by paleomagnetic isochrons: Examples in Miocene strata from a central Andean intermontane basin (Salla, Bolivia) and the northwest Himalayan foreland (Lower Siwalik Group Chinji Formation, northern Pakistan). Unpublished PhD diss., Dartmouth College.

McRae, L. E. 1990. Paleomagnetic isochrons, unsteadiness, and nonuniformity of sedimentation in Miocene fluvial strata of the Siwalik Group, northern Pakistan. *Journal of Geology* 8(4):433–456.

Mankinen, E. A., and G. B. Dalrymple. 1979. Revised geomagnetic polarity time scale for the interval 0–5 m.y. B.P. *Journal of Geophysical Research* 84(B2):615–626.

Morgan, M. E. 1994. Paleoecology of Siwalik Miocene hominoid communities: Stable carbon isotope, dental microwear, and body size analyses. Unpublished PhD diss., Harvard University.

Morgan, M. E., A. K. Behrensmeyer, C. Badgley, J. C. Barry, S. Nelson, and D. Pilbeam. 2009. Lateral trends in carbon isotope ratios reveal a Miocene vegetation gradient in the Siwaliks of Pakistan. *Geology* 37:103–106.

Morgan, M. E., J. D. Kingston, and B. D. Marino. 1994. Carbon isotopic evidence for the emergence of C_4 plants in the Neogene from Pakistan and Kenya. *Nature* 367:162–165.

Morgan, M. E., L. Spencer, E. Scott, M. S. Domingo, C. Badgley, J. C. Barry, L. J. Flynn, and D. Pilbeam. 2015. Detecting differences in forage in C_3-dominated ecosystems: Mesowear and stable isotope analyses of Miocene bovids from northern Pakistan. *Society of Vertebrate Paleontology Meeting Program and Abstracts* 2015:184.

Nelson, S. V. 2005a. Habitat requirements and the extinction of the Miocene ape, *Sivapithecus*. In *Interpreting the Past: Essays on Human, Primate, and Mammal Evolution in Honor of David Pilbeam*, edited by D. E. Lieberman, R. J. Smith, and J. Kelley, 145–166. Boston: Brill.

Nelson, S.V. 2007. Isotopic reconstructions of habitat change surrounding the extinction of *Sivapithecus*, a Miocene hominoid, in the Siwalik Group of Pakistan. *Palaeogeography, Palaeoclimatology, Palaeoecology* 243(1–2):204–222.

Opdyke, N. D., E. H. Lindsay, G. D. Johnson, N. M. Johnson, R. A. K. Tahirkheli, and M. A. Mirza. 1979. Magnetic polarity stratigraphy and vertebrate paleontology of the Upper Siwalik Subgroup of northern Pakistan. *Palaeogeography, Palaeoclimatology, Palaeoecology* 27:1–34.

Pilgrim, G. E. 1913. The correlation of the Siwaliks with mammal horizons of Europe. *Records of the Geological Survey of India* 43:264–326.

Pilgrim, G. E. 1934. Correlation of the ossiferous sections in the Upper Cenozoic of India. *American Museum of Natural History Novitates* 704:1–5.

Pilgrim, G. E. 1939. The fossil bovidae of India. *Memoirs of the Geological Survey of India, Palaeontologia Indica* n.s. 26:1–356.

Quade, J., and T. E. Cerling. 1995. Expansion of C_4 grasses in the late Miocene of northern Pakistan: Evidence from stable isotopes in paleosols. *Palaeogeography, Palaeoclimatology, Palaeoecology* 115(1–4): 91–116.

Quade, J., T. E. Cerling, J. C. Barry, M. E. Morgan, D. R. Pilbeam, A. R. Chivas, J. A. Lee-Thorp, and N. J. van der Merwe. 1992. A 16-Ma record of paleodiet using carbon and oxygen isotopes in fossil teeth from Pakistan. *Chemical Geology* 9(3):183–192.

Quade J., T. E. Cerling, and J. R. Bowman. 1989. Development of Asian monsoon revealed by marked ecological shift during the latest Miocene in northern Pakistan. *Nature* 342:163–166.

Raza, S. M. 1983. Taphonomy and paleoecology of Middle Miocene vertebrate assemblages, southern Potwar Plateau, Pakistan. Unpublished PhD diss., Yale University.

Raza, S. M., J. C. Barry, G. E. Meyer, and L. Martin. 1984. Preliminary report on the geology and vertebrate fauna of the Miocene Manchar Formation, Sind, Pakistan. *Journal of Vertebrate Paleontology* 4(4): 584–599.

Retallack, G. J. 1991. *Miocene Paleosols and Ape Habitats of Pakistan and Kenya*. Oxford: Oxford University Press.

Rose, M. D. 1984. Hominoid postcranial specimens from the Middle Miocene Chinji Formation, Pakistan. *Journal of Human Evolution* 13(6): 503–516.

Rose, M. D. 1986. Further hominoid postcranial specimens from the Late Miocene Nagri Formation of Pakistan. *Journal of Human Evolution* 15(5):333–367.

Rose, M. D. 1989. New postcranial specimens of catarrhines from the Middle Miocene Chinji Formation, Pakistan: Descriptions and a discussion of proximal humeral functional morphology in anthropoids. *Journal of Human Evolution* 18(2):131–162.

Sadler, P. M. 1981. Sediment accumulation rates and the completeness of stratigraphic sections. *Journal of Geology* 89(5):569–584.

Saunders, J. J., and B. K. Dawson. 1998. Bone damage patterns produced by extinct hyena, *Pachycrocuta brevirostris* (Mammalia:carnivora), the Haro River Quarry, northwestern Pakistan. In *Advances in Vertebrate Paleontology and Geochronology*, edited by Y. Tomida, L. J. Flynn, and L. L. Jacobs. *National Science Museum Monographs* No. 14:215–242.

Scott, R. S. 2004. The comparative paleoecology of Late Miocene Eurasian hominoids. Unpublished PhD diss., University of Texas at Austin.

Scott, R. S., J. W. Kappelman, and J. Kelley. 1999. The paleoenvironment of *Sivapithecus parvada*. *Journal of Human Evolution* 36(3):245–274.

Sheikh, K. A. 1984. Use of magnetic reversal time lines to reconstruct the Miocene landscape near Chinji Village, Pakistan. Unpublished MS thesis, Dartmouth .

Stix, J. 1982. Stratigraphy of the Kamlial Formation near Chinji Village, northern Pakistan Unpublished BA thesis, Dartmouth College.

Tauxe, L., and C. Badgley. 1988. Stratigraphy and remanence acquisition of a paleomagnetic reversal in alluvial Siwalik rocks of Pakistan. *Sedimentology* 35(4):697–715.

Tauxe, L., and N. D. Opdyke. 1982. A time framework based on magnetostratigraphy for the Siwalik sediments of the Khaur area, northern Pakistan. *Palaeogeography, Palaeoclimatology, Palaeoecology* 37(1):43–61.

Willis, B. J. 1993a. Ancient river systems in the Himalayan foredeep, Chinji Village area, northern Pakistan. *Sedimentary Geology* 88(1–2):1–76.

Willis, B. J. 1993b. Evolution of Miocene fluvial systems in the Himalayan foredeep through a two kilometer–thick succession in northern Pakistan. *Sedimentary Geology* 88(1–2):77–121.

Willis, B. J., and A. K. Behrensmeyer. 1994. Architecture of Miocene overbank deposits in northern Pakistan. *Journal of Sedimentary Research* 64(1B):6460–6467.

Zaleha, M. J. 1997a. Fluvial and lacustrine palaeoenvironments of the Miocene Siwalik Group, Khaur area, northern Pakistan. *Sedimentology* 44(2):349–368.

Zaleha, M. J. 1997b. Intra- and extrabasinal controls on fluvial deposition in the Miocene Indo-Gangetic foreland basin, northern Pakistan. *Sedimentology* 44(2):369–390.

SIWALIK TAPHONOMY
Fossil Assemblage Preservation

ANNA K. BEHRENSMEYER, CATHERINE BADGLEY,
AND MICHÈLE E. MORGAN

INTRODUCTION

This chapter focuses on the taphonomy of Siwalik vertebrate fossil assemblages—how they were preserved and what the processes of preservation reveal about life and death in the sub-Himalayan fluvial ecosystems. The specimens we have available to study are informative but limited samples of the original biota, and they may be biased for or against particular taxa or environments. Taphonomic processes that controlled the preservation of these samples affect what we can and cannot know about the original plant and animal communities that inhabited the vast sub-Himalayan plains. These communities changed over time along with shifts in climate and fluvial depositional systems. Understanding the patterns of fossil occurrence in relation to the evolving Siwalik environments provides a unique opportunity to investigate how the story of fossil preservation both limits and enhances our reconstructions of South Asian faunas and their paleoecology.

The Harvard-Geological Survey of Pakistan (Harvard-GSP) project documented 1,018 fossil localities on the Potwar Plateau. These localities, combined with biostratigraphic survey blocks, and previously recorded localities by Barnum Brown,

G. E. Lewis, and the Dartmouth-Peshawar team, result in a total of 1,315 geographically designated localities with at least formation assignments. These localities represent many different environmental settings for burial and preservation for vertebrates (teeth and bones) and some invertebrates as well (shells and gastropod opercula). The quality of fossil vertebrate preservation, as evaluated using a subset of 834 Potwar Siwalik localities with <200 kyr (i.e., <200,000 years) temporal resolution (Fig. 4.1; see also Chapter 3), fluctuated through time because of changing depositional and taphonomic conditions and is also influenced by the outcrop area available for fossil collecting.

Plant fossils (leaf impressions, pollen, and phytoliths) are rare and poorly preserved in the Potwar record, reflecting loss during decomposition on the Miocene landscapes or through later diagenesis. In contrast, Siwalik sequences of India and Nepal, 400–1200 km to the east, preserve microfloral and some macrofloral remains (e.g., Hoorn, Ohja, and Quade 2000; Srivastava et al. 2018), a record of vegetation that likely characterized the Miocene fluvial ecosystems of Pakistan as well. In the absence of actual fossil plants or pollen, we use paleoenviron-

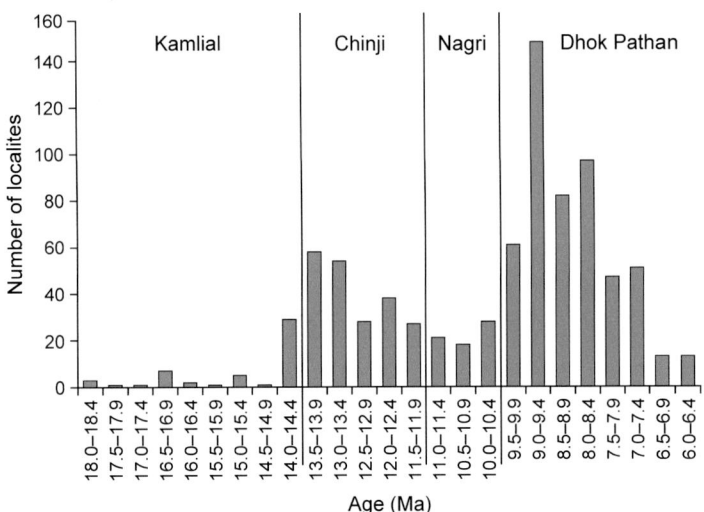

Figure 4.1. Fluctuating frequency of fossil localities in four different Siwalik formations between 18.0 and 6.0 Ma, including 834 localities with <200 kyr temporal resolution, binned here in 500 kyr intervals.

mental proxies, such as stable isotopes of carbon and molecular biomarkers, as well as the inferred ecological requirements of the preserved animals, as evidence for the vegetation and physical habitats (aquatic and nonaquatic) that once covered the alluvial landscape of our study area (Chapters 5 and 24, this volume). We also refer to fluvial ecosystems in India and Bangladesh as modern analogues for the Miocene habitats of the Potwar Siwalik record.

Most Siwalik vertebrate fossils are fragmentary. Associated or articulated specimens are rare both in situ and on the surface, where recent erosion also contributes to their breakdown. The condition of Siwalik fossils before they emerge onto modern outcrop surfaces is a consequence of many taphonomic processes that affected the remains before, during, and after burial. Their fragmentary state affects the identifiability of fossils and limits what we can say about the anatomical details of the living animals. Taphonomic processes, however, also helped to concentrate remains from different species in discrete localities representing specific depositional settings. A great strength of the Siwalik sequence is that it preserves evidence of animal communities from different parts of the landscape in samples resulting from similar taphonomic processes, and these "iso-taphonomic" samples (Behrensmeyer and Hook 1992) can be followed through millions of years.

The physical settings of Siwalik fossil localities within the larger context of the sub-Himalayan alluvial plain provide important evidence for the ecology as well as the taphonomic processes that led to fossil preservation. For example, many Siwalik fossil concentrations were buried in abandoned channels, which often contained ponded water and likely were surrounded by abundant vegetation. Modern ecosystems suggest that such places would have attracted animals to drink and feed—herbivores as well as the carnivores that preyed on them (Haynes 1985a; Crosmary et al., 2012). Over time, the remains of animals that died in this setting accumulated and were buried by trampling and influxes of sediment that gradually filled the abandoned-channel depression. Other types of fossiliferous deposit, such as channel bars, crevasse splays, and floodplain land surfaces, provide samples of the Miocene communities with somewhat different taphonomic histories and paleoecological interpretations.

The geological background for the fluvial environments in the Siwalik record is discussed in Chapter 2 of this volume. Here we present sedimentological and taphonomic evidence as a basis for interpreting how the Siwalik fossil assemblages represent the original animal habitats and communities. This knowledge helps us understand preservational biases and enables realistic reconstruction of many aspects of Miocene paleoecology, which underpin the history of profound ecological change during the Late Miocene (Chapters 24, 25, and 26, this volume).

VISUALIZING AN ANCIENT ALLUVIAL LANDSCAPE

The geological evidence presented in Chapter 2 of this volume combined with knowledge of the Siwalik fossil record and mod-

ern fluvial systems enable us to visualize the landscape of the Miocene sub-Himalayan ecosystem and its influence on the life habits and accumulations of animal remains. Imagine that you are flying a small airplane or piloting a drone and looking down over the vegetated surface of a vast alluvial plain, laced with waterways flowing from a line of hills and high mountains visible in the distance (Fig. 24.4, this volume). Braided river channels up to 1 km across and spaced around 10 km apart drain the foothills and mountains, and smaller streams feed into these rivers from the adjacent floodplains. These river systems are tributaries to a larger river flowing from northwest to southeast—a precursor of the modern Ganges River. Flying closer to the ground, you can see patches of woodland and strips of forest along the streams, areas of thickets and grassland, and gleaming wetlands between the streams. It is the end of the rainy season, and the vegetation is lush and green. Aprons of fresh sand and mud from recent flooding cover some areas, partially burying plant debris near the channels. Groups of large animals mill about on higher ground—proboscideans, rhinoceroses, and giraffes, which are returning to the floodplain from the northern foothills as the flood waters recede. Herds of smaller antelopes and suids feed in mixed shrubland and grassy areas, and the shadowy form of a bear-like, carnivorous amphicyonid lurks near a dense thicket. Scattered flocks of water birds are visible in the wetlands, with groups of tragulids and anthracotheres along the margins. Crocodiles bask on the exposed sandbars of the river channels, and one of them has grabbed a huge, thrashing catfish.

As far as your eye can see to the northwest and southeast, this landscape stretches along the foothills of the Himalayas. At the largest scale, the vast alluvial plain was created by the slow collision of the Indian continental plate with Eurasia. An arcuate subsiding trough called a foreland basin (see Fig. 2.1, this volume) formed where the smaller Indian tectonic plate was and still is being forced under the much bigger Asian plate. The basin filled with sediment carried by rivers as they eroded the rising mountains and foothills over millions of years, forming a verdant plain extending from Afghanistan to Myanmar. The balance of sediment input and basin subsidence controlled not only the channel patterns and the topography of this plain, but also the local environments where organic remains were concentrated and preserved. These places became the fossil localities that provide the long record of Miocene ecosystems and faunal change in southern Asia.

DIFFERENT PATHWAYS TO PRESERVATION

Most fossil occurrences in the Potwar Plateau are concentrated in restricted areas on the outcrops rather than being distributed widely across the eroding landscape. These fossil localities have been the focus of attention through decades of collecting and geological field research and are the primary source of information about the Siwalik faunas (Chapter 3, this volume). Magnetostratigraphic geochronology provides absolute ages for fossil localities within the Siwalik stratigraphic succession, and sedimentology enables us to assign them to specific

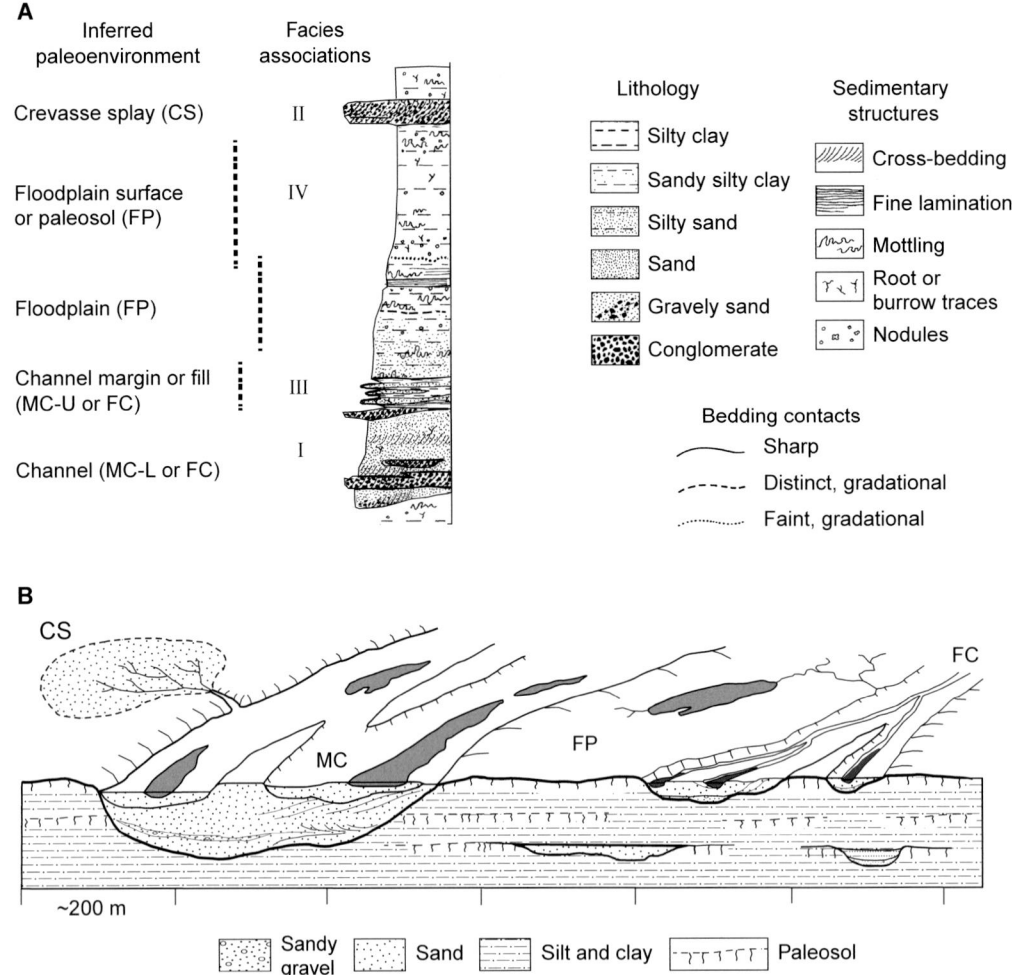

Figure 4.2. A. Lithofacies and inferred paleoenvironments associated with the Siwalik vertebrate fossil record of the Potwar Plateau. Facies associations represent the typical lithologies and sedimentary structures found in these lithofacies (Badgley 1986b; indicated in Table 4.1 by initials CEB-FA); and inferred paleoenvironment shows how these relate to fluvial depositional settings (see Table 4.1). B. Schematic diagram representing four major fossil-bearing fluvial sub-environments used in this chapter, also indicated under A. MC = major channel, MC–U or MC–L indicate upper or lower; FC = floodplain channel; CS = crevasse splay; FP = floodplain. Additional subcategories and details are provided in Behrensmeyer et al. (2005) and Table 4.1. See also Fig. 2.10 and aerial view reconstruction in Fig. 24.4, this volume.

fluvial environments (Chapter 2, this volume). While other paleontological expeditions have traditionally recorded the map location, stratigraphic position, and geological age of fossil localities, we expanded our documentation to include systematic study of depositional environments and taphonomic attributes of the fossil assemblages. This practice has resulted in ~60% (784/1,315) of our Potwar Plateau localities being documented on standardized locality cards designed for our field research; these cards also provide essential geographic and topographic information that allows paleontologists to relocate and recollect specific localities over multiple years (Behrensmeyer and Raza 1987; Chapter 3, this volume).

Detailed stratigraphic sections of fossil localities document lithologies associated with the fossils and show vertical lithofacies relationships at decimeter to meter scales (Fig. 4.2a). This information enables us to assign many of the fossil localities to specific environmental settings. Systematic collecting and

recording of all taxa and skeletal elements (including fragments) found on the outcrop surface at a subset of localities minimize distortions from collecting biases and facilitate reconstruction of the taphonomic history of fossil vertebrates preserved at each locality. From a total of 1,315 localities, 316 have been assigned to fluvial sub-environments (Behrensmeyer et al., 2005), and 63 from the Dhok Pathan and Chinji Formations also have detailed information on taphonomy, based mainly on standardized surface collecting (Badgley 1982; Raza 1983; Badgley 1986a, 1986b; Morgan 1994). Surface fossil collections are subject to additional processes that affect taphonomic variables, however. A limited set of excavations at Nagri Formation Locality Y311 offers the most reliable information about pre-burial taphonomic processes and faunal composition (Behrensmeyer, Gordon, and Yanagi 1989; Burke et al. 2014; Loughlin, Behrensmeyer, and Badgley 2018). All of these well-documented subsets of our fossil localities contribute to our taphonomic and paleoecological interpretations of the Siwalik faunas.

Pathways to preservation of Siwalik fossils were determined in part by fluvial processes that characterized the sub-Himalayan alluvial plain. In any depositional system controlled by rivers with annual flooding cycles, different areas receive varying amounts of sediment when water inundates the floodplain. Flooding rivers deposit sand, silt, and clay near their channels, in levees, or in crevasse splays (sheet floods that transport sediment out of channels and across the floodplain). Gradually the channel becomes elevated above the surrounding terrain to form an alluvial ridge (Willis and Behrensmeyer 1994; Behrensmeyer, Willis, and Quade 1995). This ridge eventually reaches a point of instability such that floodwaters break through the levees and establish a new channel across the floodplain, a process called "avulsion." The many major and minor channels of the Siwalik alluvial system were subject to frequent avulsion, as is evident in the complex stratigraphic architecture of interfingering channel and floodplain sediments (Chapter 2, this volume). Continuous subsidence of the sub-Himalayan foreland basin contributed to this "avulsion-prone" depositional system, and abandoned channels were common. Since rapid, permanent burial is a key factor in fossil preservation, this actively aggrading system contributed to preserving Siwalik localities with their diverse vertebrate remains.

The depositional environments of the Siwalik paleo-river systems can be classified into four major categories: major channels, floodplain channels, crevasse splays (sheet deposits), and floodplains (Behrensmeyer et al. 2005; Chapter 2, this volume). Paleoenvironmental interpretations are based on lithology, sedimentary structures, and lateral and vertical facies patterns showing preserved channel banks and fining-upward stratification. The alluvial plain included emergent areas with low rates of sediment accumulation where preexisting deposits were modified by soil development to become paleosols (Fig. 4.2B; Table 4.1; see also Chapter 2, this volume). Major channels were hundreds of meters across and concentrated within channel belts 1–2 km across, similar to tributaries of the modern Indus and Ganges Rivers. Between the channel belts were floodplains with many smaller channels, tens to hundreds of meters across. These drained water into the major channels from the floodplains as well as reversing direction seasonally to act as distributaries when the major channels experienced overbank flooding (see Figs. 2.9 and 24.4, this volume). While all these fluvial environments have produced fossils, the majority of our vertebrate localities occur in abandoned-channel fills of varying scales and lithologies (i.e., gravels, sands, mudstones) (Figs. 4.2, 4.3, Tables 4.1, 4.2; see Fig. 2.5, this volume).

The four major fossil-bearing fluvial environments (see Fig. 4.2) are major channels, floodplain channels, crevasse splays, and floodplains.

Major Channels
Major channel (MC) deposits consisting of sand and gravel—that is, representing active, sustained channel flow—have produced relatively few fossil assemblages. We attribute this to high energy and frequent bedload reworking, which tended to abrade and disperse skeletal remains. Although we have recovered a few important specimens including proboscidean skulls from this context, most fossils in these major channel deposits consist of fragmentary, abraded skeletal parts and rounded bone pebbles representing multiple cycles of transport and reworking. Well-preserved bones and teeth indicate that some were entrained by channel processes and rapidly buried with minimal reworking, such as could occur immediately before an avulsion event. The build-up of multistorey sandstones (see Fig. 2.8b, this volume), which form when channels erode into and rework their previous deposits, would have subjected most vertebrate remains buried in these sands to multiple phases of erosion and redeposition. Although the sand-dominated, active channel deposits are not the source of most Siwalik fossils, finer-grained channel fills within major channels are highly productive settings for fossil quantity and quality (see Fig. 4.2). When a major Siwalik channel changed position during an avulsion event, it left behind a curvilinear depression on the alluvial landscape. This area would have hosted permanent ponds or small lakes in the early stages after abandonment, and plants and animals could rapidly populate the newly stable substrates, which became linear habitats dense with trees and ground vegetation (Behrensmeyer et al. 2007). Renewed flow in the abandoned channel during flood seasons would mobilize sediment and bury organic remains. As an abandoned channel filled slowly over time, it would have provided a reliable source of forage and water for the vertebrate community, which resulted in localized samples of individuals that died in this environment. Oxidation and biological recycling removed any direct record of the plants that inhabited these settings other than root traces and rhizoliths, but channel-fill sedimentation and favorable post-burial chemical conditions preserved bones and teeth as well as occasional invertebrate remains. Time-averaging of vertebrate remains in these major channel fills would have been on the order of thousands to at most ~10,000 years (Behrensmeyer 1982; 1987), and the accumulated remains were autochthonous—that is, they represent places where the animals lived and died, even if transported over short distances.

Floodplain Channels
Floodplain channels (FC) are an important source of fossil vertebrates in the Siwalik formations of the Potwar region (see Figs. 4.2, 4.3a). These deposits often form "shoe-string" sand bodies, some of which are bounded by levee deposits indicating periods of stability within the aggrading alluvial system. The fossils preserved in these settings were derived from two major sources—bones and teeth from local erosion of floodplain surfaces and soils, including bank collapse, and surface remains that accumulated in or near the channels during times when these provided a localized source of water and food (e.g., during dry seasons). Fossils from floodplain channel localities therefore represent samples of the original faunal community from near the channel and the nearby floodplain. These assemblages would have been time-averaged over the lifetime of the channel, estimated at hundreds to thousands of years based on

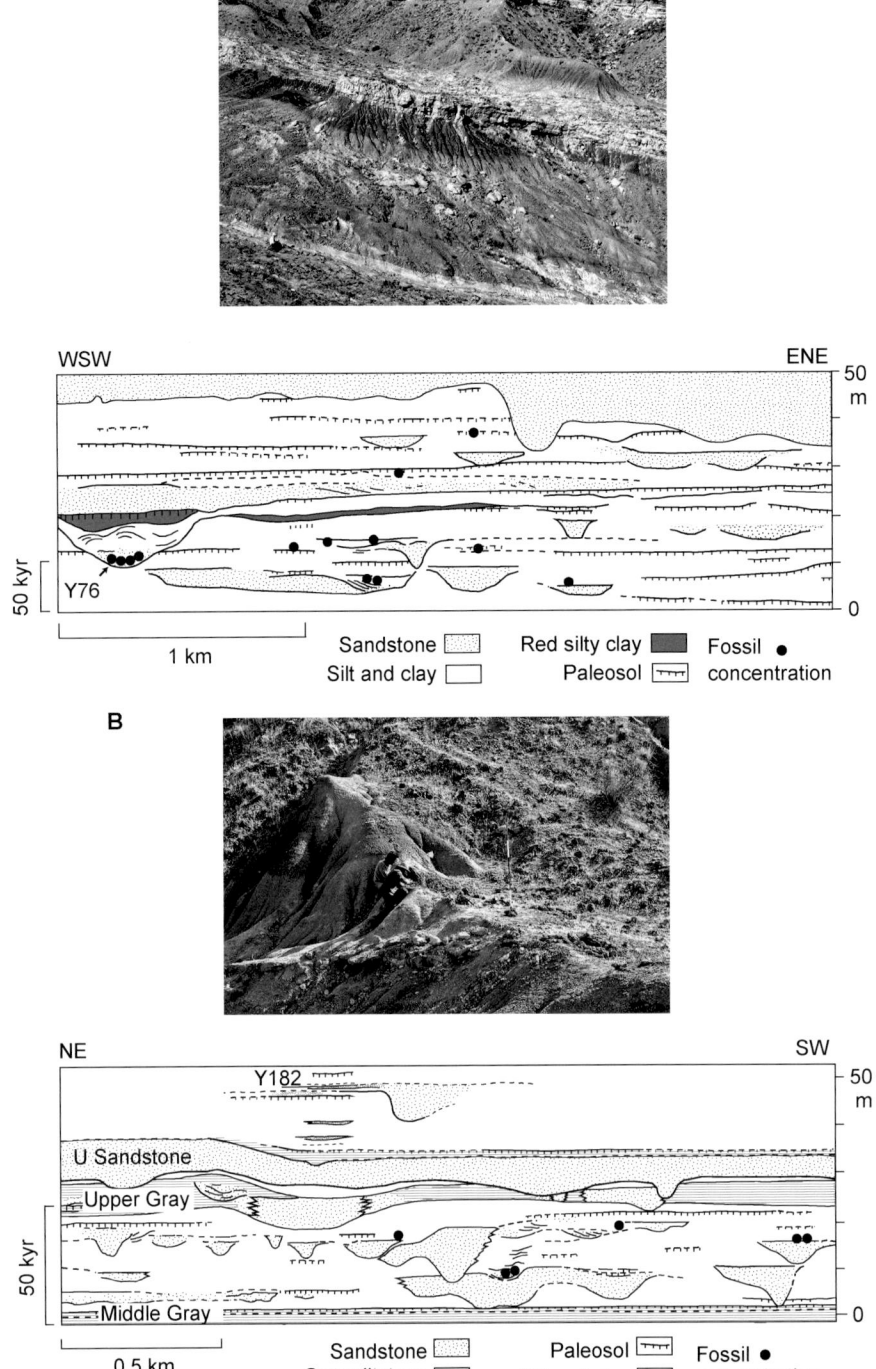

Figure 4.3. Examples of fossil locality contexts within the architecture of Siwalik fluvial deposits. A. Top: Overview of Locality Y76, a highly fossiliferous, fining-upward floodplain channel fill of interbedded sand and silt, capped by a red silty clay paleosol followed by a gray, sandy crevasse-splay deposit. Human figure for scale in lower left. Photo by A. K. Behrensmeyer. Bottom: Panel diagram (50 m thickness) of part of the upper Chinji Fm showing the position of fossil-producing localities in relation to the fluvial channel and floodplain deposits, with Y76 indicated in the lower left. Number of dots is roughly proportional to fossil productivity. Vertical exaggeration = 20×. Black vertical bar on left indicates 50,000 years, represented in sediment accumulation and hiatuses on stable land surfaces (paleosols). B. Top: Part of Locality Y182, showing fining-upward lithologies representing the crevasse-splay and levee deposits associated with a nearby buff system channel. Andrew Hill and Catherine Badgley are excavating in situ fossil vertebrates preserved in these deposits. Photo by A K. Behrensmeyer. Bottom: Panel diagram of part of the Dhok Pathan Formation in the northern part of the Potwar Plateau (Behrensmeyer 1987). Vertical exaggeration = 44×; areas lateral to Y182 above the U Sandstone are not documented at this scale. The U sandstone and the Middle and Upper Gray units are laterally extensive channel and overbank deposits, respectively, of the blue-gray fluvial system, while the fossiliferous channel and floodplain deposits represent the buff fluvial system (Behrensmeyer and Tauxe 1982). The black reference bar for 50,000-year shows more rapid sediment accumulation and fewer hiatuses (paleosols) in the Dhok Pathan Formation compared with the Chinji Formation in Fig. 4.3a.

Table 4.1. Classification Scheme for Depositional Environments and Sub-Environments That Preserve Siwalik Fossils of the Potwar Plateau

Context	Abbrev.	Sub-Context	Fluvial Deposits	Fossiliferous Lithofacies	Paleoenvironment	Processes Affecting Vertebrate Remains
Major channel	MC		Sand bodies usually 5–20 m thick and laterally extensive over hundreds of meters	Sand, gravel, mixed to finer-grained in upper parts (CEB-FA I)	Main channel, either active or abandoned	Fluvial transport and abrasion of skeletal remains from the alluvial plain at large
	MC-L	Lower 2/3	High energy deposits of active channel bars	Channel lag—gravel, intraclast conglomerate (CEB-FA I, II)	Channel base and barforms, reactivation surfaces, storey boundaries	Fluvial transport, reworking representing large areas of alluvial plain
	MC-U	Upper 1/3	Mixed lithologies of sand, intraclast conglomerate, silt and clay, generally fining-upward, indicating waning channel energy	Heterolithic inclined strata (alternating sand, silt, clay), mudstone, intraclast conglomerate, paleosols	Abandoned channel, may have ponds or wetlands during early fill phase	Local, attritional mortality, minor transport and reworking
Floodplain channel	FC		Laterally discontinuous, generally tabular, or shoe-string sand bodies defining channels within floodplain sequences	Sand, silt, and intraclast lenses, cross-cutting fluvial strata within erosional channel margins	Floodplain tributaries and distributaries	Fluvial transport and reworking from local area of adjacent floodplain and channel
	FC-C	Complex Fill	Sediment body 1–5 m thick and usually <~100 m in cross-section with multiple, cross-cutting sub-units	Channel lag and channel margin deposits, interfingering coarse and fine-grained lenses showing multiple episodes of channel activity (CEB-FA II, III)	Active channel, seasonally variable flow, channel margin, levee, pond, deposits fine upward as channel becomes abandoned	Minor transport and reworking, mostly local and attritional
	FC-S	Simple fill	U-shaped in cross-section, representing fill of single channels	Coarse to fine sediment lenses, fining upward, some with only fine-grained fill (CEB-FA III, IV)	Abandoned channel representing one cycle of erosion and filling	Attritional and local
Crevasse splay	CS		Thin, laterally extensive coarse-grained deposits (sheet deposits) caused by flood-generated overbank deposition	Mixed sand and intraclast gravel, channel margin (CEB-FA II)	Levees, sediment lobes formed by sheet flow on the floodplain	Mix of transported, reworked, and floodplain surface
Floodplain	FP		Laterally extensive, bedded to massive, fine-grained deposits with paleosols.	Silt to clay, with vertical sequences from bedded to massive, pedogenically modified, laterally continuous silty clays, often with pedogenic carbonate (CEB-FA IV)	Soil surfaces—incremental build-up of sediment due to seasonal flood events	Local, attritional deaths; rare predator den and burrow concentrations

Source: Adapted from Behrensmeyer et al. (2005).

Note: The fossiliferous lithofacies or Facies Associations of Badgley (1986b) may occur in several of these depositional settings. These are indicated by CEB-FA in the table and refer to categories described in Fig. 4.2.

the overall sediment accumulation rate of the Siwalik formations (Behrensmeyer 1987). Vertebrate remains were also primarily autochthonous, although some were transported and reworked locally by the floodplain channels.

Crevasse Splays
Crevasse splays (CS) form when water breaks through channel banks and levees during floods and spreads a broad sheet of

sediment onto the floodplain (see Figs. 4.2b, 4.3b). The energy of such floods can erode sediments from the floodplain surface, forming thin, laterally extensive mud- and pedogenic-clast conglomerates. These tabular deposits should therefore preserve vertebrates that died on the relatively limited areas of land surface affected by this fluvial process. Crevasse-splay fossil assemblages also should represent shorter periods of time-averaging than floodplain channels (e.g., tens to hundreds of

Table 4.2. Depositional Environments of 316 Potwar Plateau Fossil Localities in Four Siwalik Formations

	Formations					
Environments	Kamlial	Chinji	Nagri	Dhok Pathan	Total	Proportion
Major channel (MC)	6	37	11	40	94	0.30
Floodplain channel (FC)	9	57	12	57	135	0.43
Crevasse splay (CS)	4	3	1	13	21	0.07
Floodplain (FP)	4	23	2	37	66	0.21
Total localities	23	120	26	147	316	

	Proportions			
Environments	Kamlial	Chinji	Nagri	Dhok Pathan
Major channel (MC)	0.261	0.308	0.423	0.272
Floodplain channel (FC)	0.391	0.475	0.462	0.388
Crevasse splay (CS)	0.174	0.025	0.038	0.088
Floodplain (FP)	0.174	0.192	0.077	0.252

Note: Data graphed in Fig. 4.4.

years) and localized accumulations of skeletal remains. Transport was likely minimal, as floodplain vegetation would have impeded entrainment of bones and teeth.

Floodplains

Fossil vertebrates in fine-grained floodplain (FP) deposits were mostly preserved in small channels or within the upper parts of soils (see Figs. 4.2, 4.3a) rather than in well-bedded floodplain silts and clays. The latter represent low-lying areas subject to rapid flood sedimentation in the Siwalik alluvial system, which land vertebrates likely avoided. Our paleoecological reconstructions would benefit if a larger sample of vertebrates were preserved in soils representing the land surfaces where these animals spent most of their lives, but fossils are relatively rare in this context in the Siwalik record. Those that have been discovered in paleosols consist of isolated specimens or clusters of associated elements, often encrusted with soil carbonate, concentrations of microvertebrate remains (Badgley, Downs, and Flynn 1998), and rare occurrences of partial skeletons in burrow fills. The pedogenic processes that formed the soils were apparently efficient at recycling organic material, including bones, leaving little vertebrate or plant remains to be fossilized. Notably, however, these paleosols contain biomolecules that preserve information about floodplain vegetation (Freeman and Colarusso 2001) as well as natural fires and conditions generating repeated wildfires over time (Karp, Behrensmeyer, and Freeman 2018).

Siwalik formations generally lack dense bone accumulations (other than occasional fish concentrations), which argues against mass mortality as a significant taphonomic process contributing to our vertebrate fossil record. The evidence points instead to attritional accumulations representing animals that died from normal causes and were modified by a variety of taphonomic processes prior to final burial and mineralization.

Within the four major categories of depositional environment listed above are subcategories determined by the type of sediment, the spatial scale and complexity of the deposit, and (sometimes) the distribution and abundance of fossils (Figs. 4.2, 4.3; Tables 4.1, 4.2, 4.3; Behrensmeyer et al. 2005). Classification of a given fossil locality within this system depends on how much information we have about context—that is, the extent and exposure of the sedimentary outcrops and time spent documenting the locality's geology. The scale of paleoenvironmental interpretation also depends on whether documentation is based on microstratigraphic sections or on the larger-scale context within the fluvial architecture. For example, fossils described as occurring in "channel-margin" deposits (Badgley 1986b) could be associated with abandoned channel fills or levee environments, but it may not be possible to distinguish between these settings from the evidence at hand.

A compilation of paleoenvironmental settings of the 316 well-documented fossil localities showed that channels are the dominant context for fossil preservation throughout the Siwalik sequence (Behrensmeyer et al. 2005; Fig. 4.4; Table 4.2). Gradual filling of abandoned channels of various sizes (major and floodplain) created particularly favorable settings for preserving vertebrate remains, and these are the most common productive environments in all four Siwalik formations. This pattern is consistent with enhanced preservation in situations that both attracted animals and experienced sporadic lower-energy flow. Fluvial dynamics changed over time in the Potwar Siwalik sequence (Behrensmeyer 1987), with more abandoned channels in the Chinji and Dhok Pathan Formations contrasting with laterally continuous channel sand bodies in the Kamlial and Nagri Formations (Behrensmeyer and Tauxe 1982; see Fig. 2.8, this volume). This contrast directly affected the quality of the fossil record, with greater numbers and better-preserved vertebrate remains in the formations with more abandoned-channel fills, from either major or floodplain channels. The Siwalik fossil record therefore bears a strong depositional signature of preservation that reflects the ecological and taphonomic processes typical of abandoned-channel settings. Although the distributions of fossil-bearing environments in the different formations (Fig. 4.4) are statistically similar (chi-square = 6.97, p <0.0729; Kruskal-Wallis H = 5.16, df = 3, p = 0.1604), floodplain and crevasse-splay localities are fewer and more variable through time than major and floodplain channels; both of these environments offer faunal samples that differ taphonomically and ecologically from those of the major and floodplain channels (see Taphonomic Indicators section below).

The physical processes characterizing each fluvial sub-environment would be expected to leave distinctive taphonomic signatures on the fossil assemblages, such as variable evidence for reworking and transport. At the same time, biological processes such as predator bone modification or trampling by ungulates should affect animal remains across all

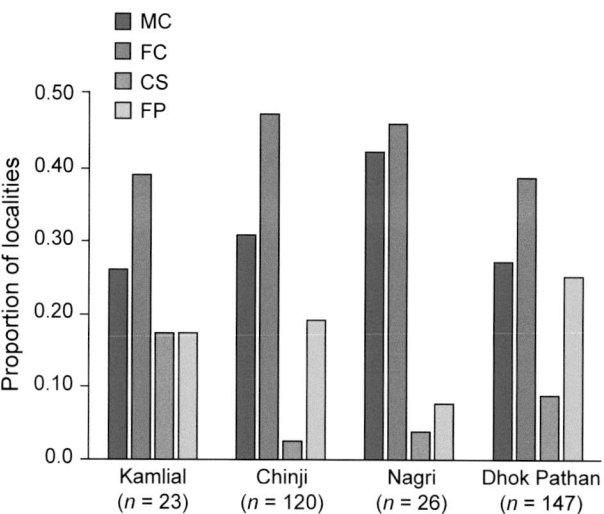

Figure 4.4. Distribution of depositional contexts across Lower and Middle Siwalik Formations for 316 localities that can be assigned with confidence to one of these four categories. See Figure 4.2b for diagram showing the different environments in the sub-Himalayan fluvial system. A chi-square test on the distribution of environments in the finer-grained Chinji and Dhok Pathan formations shows no significant difference (chi-square = 6.97; p = 0.073), although floodplain localities are slightly more common in the latter. A nonparametric Kruskal-Wallis test shows no significant difference in the distribution of fossil-preserving environments among the four formations (H = 5.16, df = 3, p = 0.160). Although the proportions of different fossil-preserving environments do not differ statistically through time, observed changes are consistent with differences between the major channel sand-dominated formation (Nagri) and those with high proportions of floodplain channels and floodplain environments (Chinji, Dhok Pathan).

environments. Based on this reasoning, we might expect settings characterized by lower-energy sedimentary processes to show more biological bone modification compared with higher-energy settings. Careful study of the fossil assemblages themselves tests this idea with taphonomic evidence for different physical, chemical, and biological processes that affected vertebrate remains, as discussed in the following section.

TAPHONOMIC INDICATORS

Five empirical attributes of vertebrate fossil assemblages, elaborated below, have proven useful for inferring the taphonomic history of Siwalik localities and are the basis for characterizing the vertebrate assemblages of the Potwar Plateau (see Fig. 4.5; Table 4.3). Comparable research has not yet been conducted in Siwalik sequences of India and Nepal, so whether similar taphonomic patterns occur in these regions is currently unknown.

Bone Modification
Biological and physical processes can leave distinctive marks on the surfaces of bones before and after they have lost soft tissue. Carnivores and scavengers may leave tooth punctures or deep scratches, as well as chewed ends, and may break or crush bones, leaving diagnostic fracture patterns (Behrensmeyer, Gordon, and Yanagi 1989; Shipman 1993; Marean et al. 1992; Lyman and Lyman 1994; Fosse et al. 2012; Faith, Marean, and Behrensmeyer 2007; Fernandez-Jalvo and Andrews 2016). Termites can chew through bone, leaving distinctive shallow pits, holes, and deeper burrow structures. Snails also sometimes scour the outer surface of bones. Fungal hyphae associated with plant roots may produce a lacey pattern of shallow grooves. Trampling may scratch, break, or crush bones on the surface or after shallow burial in loose sediment. Digestive acids from crocodilians or hyenas dissolve the surfaces of teeth, sometimes completely removing the enamel, and etch bone fragments or whole bones. All these features have been identified in Siwalik fossils (see Fig. 4.5) and are most reliably recorded on excavated specimens that have not been further modified by erosion and breakage on modern outcrops.

Weathering is a form of bone modification that occurs when defleshed bones are exposed to varying temperature, humidity, and sunlight on land surfaces. The breakdown of the bone structure follows a characteristic sequence over periods of years in a modern semi-arid savannah ecosystem (Behrensmeyer 1978; Behrensmeyer and Miller 2012), although rates are highly variable depending on body size, microclimate, and vegetation cover. Weathering stages of Siwalik fossils, along with other features, were documented for samples of bones excavated from a large, abandoned channel (MC-U in Table 4.1) in the Nagri Formation (Y311, dated to 10.1 Ma, with >4,000 catalogued specimens). A large proportion of the combined excavated sample (67%; Table 4.4b) shows little evidence of post-mortem, pre-burial surface weathering. About 40% of the excavated fossils have damage patterns indicating carnivore activity, such as spiral fractures, chewed break edges, and

Table 4.3. Taphonomic Attributes of Potwar Plateau Fossil Assemblages

A. Artiodactyl Skeletal Elements by Formation and Depositional Environment

	Specimen Counts (Including Scapula and Pelvis)								Proportion					
Formation	Tooth	Cranial	Forelimb	Hindlimb	Metapodial	Pod-Phal	Total	Formation	Tooth	Cranial	Forelimb	Hindlimb	Metapodial	Pod-Phal
Kamlial	117	37	33	33	63	238	521	Kamlial	0.22	0.07	0.06	0.06	0.12	0.46
Chinji	1,345	882	734	670	673	2,494	6,798	Chinji	0.20	0.13	0.11	0.10	0.10	0.37
Nagri	874	533	459	485	412	1500	4,263	Nagri	0.21	0.13	0.11	0.11	0.10	0.35
Dhok Pathan	969	829	350	363	304	823	3,638	Dhok Pathan	0.27	0.23	0.10	0.10	0.08	0.23
Total Specimens	3,305	2,281	1,576	1,551	1,452	5,055	15,220							

	Specimen Counts (Including Scapula and Pelvis)								Proportion					
Environment	Tooth	Cranial	Forelimb	Hindlimb	Metapodial	Pod-Phal	Total	Environment	Tooth	Cranial	Forelimb	Hindlimb	Metapodial	Pod-Phal
MC	1,607	1,047	868	900	829	2,997	8,248	MC	0.195	0.127	0.105	0.109	0.101	0.363
FC	1,101	751	502	485	469	1,566	4,874	FC	0.226	0.154	0.103	0.100	0.096	0.321
CS	283	237	112	104	92	288	1,116	CS	0.254	0.212	0.100	0.093	0.082	0.258
FP	314	246	94	62	62	204	982	FP	0.320	0.251	0.096	0.063	0.063	0.208
Total Specimens	3,305	2,281	1,576	1,551	1,452	5,055	15,220							

Notes: Counts are based on catalogued specimen numbers (records) and include scapulae with forelimb elements and pelves with hindlimb elements. "Cranial" includes maxillae, mandibles, horn cores, and crania; Pod-Phal = Podials and Phalanges. See Figs 4.6, 4.7.

B. Catalogued Artiodactyl Records of Juveniles and Adults, by Formation and Depositional Environment

Catalogued Artiodactyls (All Environments)

Formation	Juv	Adult	Total	J/Total
Dhok Pathan	262	3,495	3,757	0.070
Nagri	339	4,035	4,374	0.078
Chinji	493	6,427	6,920	0.071
Kamlial	44	522	566	0.078
Total	1,138	14,479	15,617	0.073

Catalogued Artiodactyls (All Formations)

Environment	Juv	Adult	Total	J/Total
MC	659	7,788	8,447	0.078
FC	297	4,678	4,975	0.060
CS	66	1,076	1,142	0.058
FP	106	908	1,014	0.105
Total specimens	1,128	14,450	15,578	0.073
Mandible and maxilla only	192	636	828	0.232

Dhok Pathan: 21/42 Systematically Collected Localities

Lithofacies	Juv	Total	J/Total
MC	12	822	0.015
FC	18	366	0.049
CS	4	271	0.015
FP	18	286	0.063
	52	1,745	0.030

Y311 Excavated Samples—Dental Age Only (Mandibles, Maxillae, Teeth)

	Y311-DS4	Y311-DS4ME	Y311-W1	Y311-W2	Total
Adult	2	5		10	17
Juv, young ad	1		1	2	4
Total	3	5	1	12	21

J/Total = 0.190

Notes: Excavation names: Y311-DS4, Y311-DS4ME, Y311-West-1, Y311-West-2. Counts are based on unfused versus fused postcranial skeletal elements, juvenile (J) versus adult dentitions in mandibles and maxillae, and adul: versus deciduous teeth. Dhok Pathan comparison is based on 21 localities with controlled surface fossil collections (Badgley 1986b, Table 3). Table with excavated Y311 mammals (including many artiodactyls) is based on dental age for mandibles, maxillae, and teeth only. For Y311 excavated samples, using all elements where age could be determined, the ratio of J/Total is 14/100 = 0.140.

Table 4.3. continued

C. Body-Size Estimates for All Catalogued Artiodactyls

Formation	All Artiodactyl Data						Proportions				Fig. 4.8A: Excluding 10–1,000 kg data				
	1–10 kg	10–100 kg	10–1,000 kg	100–1,000 kg	1,000+ kg	Total	1–10 kg	10–100 kg	100–1,000 kg	1,000+ kg	1–10 kg	10–100 kg	100–1,000 kg	1,000+ kg	Total
Kamlial	144	269	40	68	0	521	0.299	0.559	0.141	0.000	144	269	68	0	481
Chinji	717	3,605	278	2,197	1	6,798	0.110	0.553	0.337	0.000	717	3,605	2,197	1	6,520
Nagri	425	2,624	90	892	232	4,263	0.102	0.629	0.214	0.056	425	2,624	892	232	4,173
Dhok Pathan	347	1,941	116	952	282	3,638	0.099	0.551	0.270	0.080	347	1,941	952	282	3,522
Total records	1,633	8,439	524	4,109	515	15,220	0.111	0.574	0.280	0.035	1,633	8,439	4,109	515	14,696

Environment	All Artiodactyl Data						Proportions				Fig. 4.8B: Excluding 10–1,000 kg data				
	1–10 kg	10–100 kg	10–1,000 kg	100–1,000 kg	1,000+ kg	Total	1–10 kg	10–100 kg	100–1,000 kg	1,000+ kg	1–10 kg	10–100 kg	100–1,000 kg	1,000+ kg	Total
MC	854	4,881	232	2,018	263	8,248	0.107	0.609	0.252	0.033	854	4,881	2,018	263	8,016
FC	568	2,461	206	1,528	111	4,874	0.122	0.527	0.327	0.024	568	2,461	1,528	111	4,668
CS	151	585	40	252	88	1,116	0.140	0.544	0.234	0.082	151	585	252	88	1,076
FP	60	512	46	311	53	982	0.064	0.547	0.332	0.057	60	512	311	53	936
Total records	1,633	8,439	524	4,109	515	15,220	0.111	0.574	0.280	0.035	1,633	8,439	4,109	515	14,696

Notes: Counts based on records of identified taxa in four Siwalik formations and depositional environments (see Fig. 4.8). The broad category 10–1,000 kg is excluded from the main tables and figure, but the number of records in this category is given to the far right; these are animals with uncertain weight estimates.

tooth marks, punctures, deep grooves, and scoring marks (see Table 4.4c,d, Fig. 4.5b). A significant fraction (40%) of the excavated sample shows abrasion of break edges and limb ends (Table 4.4f), and 13% shows parallel scratch marks and striations indicative of trampling (Table 4.4e) (Behrensmeyer, Gordon, and Yanagi 1986; Behrensmeyer,

Table 4.3. continued

D. Comparison of Taxonomic Occurrences in Chinji Formation Localities Y639 and Y699

Major Group	MNI Y639	MNI Y699	Proportions Y639	Y699
Fish	2	0	0.12	0.00
Turtle	1	0	0.06	0.00
Croc	7	0	0.41	0.00
Anthracothere	1	0	0.06	0.00
Tragulid	4	0	0.24	0.00
Suid	0	2	0.00	0.22
Bovid	2	2	0.12	0.22
Giraffoid	0	1	0.00	0.11
Aardvark	0	1	0.00	0.11
Rodent	0	2	0.00	0.22
Hyaenodont	0	1	0.00	0.11
Total MNI	17	9		

Notes: Comparison is based on estimates of the minimum number of individuals (MNI) represented by skeletal remains recorded in the Siwalik Catalogue for Chinji Formation localities Y639 (MC) and Y699 (FP). Both aardvark and suid were represented by multiple elements associated with single individuals. See Fig. 4.9.

Gordon, and Yanagi 1989). Dissolution from acid etching, which we attribute to carnivore digestion, is apparent on about 18% of the specimens (Table 4.4g).

These data suggest that each of the Y311 excavated fossil assemblages likely represents decades to centuries of attritional accumulation, with cumulative taphonomic modification resulting from carnivores, scavengers, trampling, and fluvial abrasion before final burial. Data from several different excavations indicate that these processes operated somewhat differently across similar fluvial settings at Y311, with more variation in biotic modification (carnivore damage, trampling, dissolution) than physical bone modification (abrasion, weathering). Y311 as a whole was preserved in an unusually large and complex abandoned channel, and the combined sample of surface and excavated fossils likely represents 10^3–10^4 years of time-averaging.

Surface-collected fossils from taphonomically documented localities in the Dhok Pathan Formation offer additional evidence for pre-burial modification and indicate that polish and abrasion are more common on specimens from channel and crevasse-splay environments in this formation, compared with punctures and chewed edges that are more frequent on specimens from channel-margin and floodplain environments (Badgley 1986a, Table 2). This difference is consistent with expectations based on the greater impact of physical processes in the higher-energy depositional environments versus biological processes in lower-energy environments.

E. Major Mammalian Groups Represented in the Four Depositional Environments

	Carnivore	Rodent	Artiodactyl	Equid	Total	Carnivore (N=321)	Rodent (N=5,113)	Artiodactyl (N=7,502)	Equid
	Fig. 4.10a: Pre-10.8 Ma					*Proportions: Pre-0.8 Ma*			
MC	148	1,067	3,552	0	4,767	0.46	0.21	0.47	0.00
FC	154	2,948	3,539	0	6,641	0.48	0.58	0.47	0.00
CS	10	931	218	0	1159	0.03	0.18	0.03	0.00
FP	9	167	193	0	369	0.03	0.03	0.03	0.00
Total Records	321	5,113	7,502	0	12,936	1.00	1.00	1.00	0.00

	Carnivore	Rodent	Artiodactyl	Equid			Carnivore	Rodent	Artiodactyl	Equid
	Fig. 4.10b. Post-10.8 Ma						*Proportions: Post-10.8 Ma*			
MC	309	786	4,895	803	6,793	MC	0.51	0.29	0.61	0.41
FC	114	1,080	1,436	597	3,227	FC	0.19	0.40	0.18	0.31
CS	57	503	924	281	1,765	CS	0.09	0.19	0.11	0.14
FP	129	324	821	262	1,536	FP	0.21	0.12	0.10	0.13
Total Records	609	2,693	8,076	1,943	13,321		1.00	1.00	1.00	1.00

Notes: Based on number of catalogued records of taxa in two time intervals, before and after 10.8 Ma, which is the first appearance datum (FAD) of equids in our Siwalik record. Data plotted in Fig. 4.10.

Table 4.3. continued

F. Mammalian Specimen Numbers and Taxonomic Richness by Locality and Environment (212 Localities with >9 specimens)

Major Channel (MC)—65 Localities			Floodplain Channel (FC)—93 Localities			Crevasse Splay (CS)—18 Localities			Floodplain (FP)—36 Localities		
Locality	# Specimens	# Unique Taxa	Locality	# Specimens	# Unique Taxa	Locality	# Specimens	# Unique Taxa	Locality	# Specimens	# Unique Taxa
D12	12	1	Y33	22	5	Y182	401	29	Y17	165	11
D13	93	12	Y41	105	12	Y211	230	23	Y20	79	9
Y34	51	7	Y58	24	7	Y221	118	14	Y196	51	10
Y37	27	5	Y59	147	19	Y239	19	6	Y218	12	3
Y38	172	14	Y60	79	11	Y317	359	22	Y236	44	7
Y42	14	4	Y61	41	7	Y406	42	10	Y240	30	6
Y43	34	5	Y66	35	6	Y418	10	2	Y258	140	13
Y51	38	6	Y67	23	9	Y452	172	9	Y262	128	12
Y53	71	9	Y69	39	5	Y491	148	13	Y270	96	11
Y83	65	12	Y71	26	6	Y592	99	15	Y322	14	3
Y107	14	3	Y140	49	7	Y644	15	4	Y327	63	10
Y155	14	3	Y159	37	5	Y665	16	5	Y328	16	8
Y160	144	8	Y161	16	6	Y678	76	10	Y363	25	1
Y171	19	2	Y165	11	1	Y721	15	3	Y366	19	2
Y176	135	8	Y173	10	2	Y833	19	3	Y367	10	2
Y214	19	2	Y174	40	8	Y908	45	10	Y387	17	1
Y225	51	10	Y191	42	9	Y913	14	4	Y389	21	1
Y243	90	11	Y193	28	8	Y921	15	5	Y399	45	7
Y254	30	5	Y207	32	9				Y403	33	4
Y260	193	20	Y213	25	4				Y429	20	4
Y310	525	26	Y224	65	12				Y439	13	3
Y330	138	9	Y226	21	9				Y448	22	5
Y362	22	5	Y227	342	24				Y479	52	2
Y413	20	5	Y259	105	13				Y544	16	3
Y430	30	5	Y261	54	10				Y545	204	17
Y450	354	20	Y269	164	20				Y630	12	3
Y457	102	14	Y285	11	5				Y663	44	10
Y478	392	29	Y286	14	5				Y680	12	3
Y494	83	17	Y309	241	18				Y692	14	3
Y496	755	28	Y312	128	12				Y699	34	5
Y499	555	21	Y314	135	18				Y783	16	5
Y500	483	20	Y315	21	9				Y795	15	4
Y503	117	14	Y336	35	4				Y809	16	3
Y515	325	19	Y337	82	13				Y837	16	3
Y539	134	9	Y350	90	14				Y941	37	6
Y589	15	2	Y359	29	6				Y986	13	4
Y599	84	10	Y388	37	9						
Y639	17	4	Y454	40	9						
Y642	114	15	Y495	174	14						
Y647	200	25	Y498	76	10						
Y682	106	14	Y502	179	21						
Y690	85	13	Y504	374	19						
Y691	26	7	Y541	49	10						
Y694	80	12	Y578	86	11						
Y711	53	9	Y591	51	7						
Y714	80	8	Y631	18	7						
Y724	12	5	Y634	110	17						
Y728	19	6	Y640	232	17						
Y731	12	6	Y641	94	9						
Y735	43	10	Y650	107	10						
Y786	15	2	Y651	27	5						
Y802	56	9	Y660	58	5						
Y811	17	5	Y661	48	4						
Y822	74	15	Y666	23	3						

Table 4.3. continued

Major Channel (MC)—65 Localities			Floodplain Channel (FC)—93 Localities			Crevasse Splay (CS)—18 Localities			Floodplain (FP)—36 Localities		
Locality	# Specimens	# Unique Taxa	Locality	# Specimens	# Unique Taxa	Locality	# Specimens	# Unique Taxa	Locality	# Specimens	# Unique Taxa
Y825	71	9	Y668	42	4						
Y826	21	5	Y669	20	4						
Y831	24	4	Y675	12	8						
Y834	35	5	Y695	76	13						
Y875	15	3	Y698	50	11						
Y883	24	3	Y710	28	6						
Y889	13	3	Y716	15	5						
Y906	35	6	Y718	76	3						
Y932	24	5	Y719	19	3						
Y935	52	6	Y725	37	6						
Y943	18	6	Y726	39	6						
			Y738	34	8						
			Y739	13	4						
			Y744	13	7						
			Y747	143	16						
			Y750	282	24						
			Y767	33	11						
Not plotted in Fig. 4.11			Y771	13	2						
Y311	4229	45	Y772	77	14						
			Y775	29	10						
			Y776	35	9						
			Y779	40	6						
			Y782	15	4						
			Y784	46	8						
			Y791	11	2						
			Y797	64	12						
			Y801	16	5						
			Y810	10	3						
			Y824	25	5						
			Y861	14	4						
			Y870	12	3						
			Y898	19	5						
			Y909	16	4						
			Y910	39	6						
			Y925	10	1						
			Y927	79	4						
			Y948	11	5						
			Y950	24	9						
			Y953	31	4						
			Not plotted in Fig. 4.11:								
			Y76	1369	35						
	Total Specimens	Total Unique Taxa		Total Specimens	Total Unique Taxa		Total Specimens	Total Unique Taxa		Total Specimens	Total Unique Taxa
	6,761	620		5,719	789		1,813	187		1,564	204
Average # Specimens/ Locality	~104		**Average # Specimens/ Locality**	~62		**Average # Taxa/ Locality**	~101		**Average # Specimens/ Locality**	~43	
Average # Taxa/ Locality	9.5		**Average # Taxa/ Locality**	8.5		**Average # Taxa/ Locality**	10.4		**Average # Taxa/ Locality**	5.7	

Notes: These data were used to generate the rarefaction curves in Fig. 4.11 using specimen numbers versus mammal taxonomic richness per locality; all body sizes, for 212 localities with >9 specimens. MC, FC, and CS localities with 300–400 specimens begin to level out at a maximum of 20–25 taxa. In contrast, FP localities approach a maximum of ~15 taxa at 200 specimens. Two very productive localities, Y76 (1369 specimens and 35 taxa) and Y311 (4229 specimens, 45 taxa), are not plotted on Fig. 4.11.

Table 4.3. continued

G. Distribution of Ecological Categories of Taxa across Four Environmental Settings

| Major Group | Taxon | Counts of Records by Taxon | | | | | Ecological Category |
		MC	FC	CS	FP	Total	
Bird	Bird	78	49	10	11	148	Volant
Fish	Fish	532	603	74	51	1,260	Aquatic
Invertebrate	Invertebrate	241	239	83	34	597	Aquatic
Mammal	Aardvark/ pangolin	11	54	3	78	146	Terrestrial
Mammal	Anthracothere	374	206	40	23	643	Near-aquatic
Mammal	Bovid	4,187	2,169	637	575	7,568	Terrestrial
Mammal	Carnivore	457	268	67	138	930	Terrestrial
Mammal	Chalicothere	35	40	5	4	84	Terrestrial
Mammal	Equid	803	597	281	262	1,943	
Mammal	Giraffoid	802	838	115	91	1,846	Terrestrial
Mammal	Primates	130	42	51	26	249	Arboreal
Mammal	Proboscidean	248	256	47	53	604	Terrestrial
Mammal	Rhinocerotid	800	697	140	103	1,740	Terrestrial
Mammal	Rodent	1,853	4,028	1,434	491	7,806	Scansorial
Mammal	Other small mammal	142	269	85	30	526	Scansorial
Mammal	Suid	683	632	169	202	1,686	Terrestrial
Mammal	Tragulid	2,138	883	127	74	3,222	Near-aquatic
Reptile	Crocodilia	472	353	77	44	946	Semi-aquatic
Reptile	Lizard	47	51	10	18	126	Arboreal
Reptile	Snake	1,134	621	68	83	1,906	Semi-aquatic
Reptile	Turtle	251	256	58	91	656	Semi-aquatic
	Total	15,418	13,151	3,581	2,482	34,632	

| | Proportions of Ecological Categories across Four Depositional Environments | | | | | | Proportions of Ecological Categories within Each Depositional Environment | | | |
	MC	FC	CS	FP	Total		MC	FC	CS	FP
Aquatic	0.416	0.453	0.085	0.046	1,857	Aquatic	0.050	0.064	0.044	0.034
Semi-aquatic	0.529	0.351	0.058	0.062	3,508	Semi-aquatic	0.120	0.094	0.057	0.088
Near-aquatic	0.650	0.282	0.043	0.025	3,865	Near-aquatic	0.163	0.083	0.047	0.039
Scansorial	0.239	0.516	0.182	0.063	8,332	Scansorial	0.129	0.327	0.424	0.210
Terrestrial	0.485	0.335	0.088	0.091	16,547	Terrestrial	0.521	0.422	0.409	0.607
Arboreal	0.472	0.248	0.163	0.117	375	Arboreal	0.011	0.007	0.017	0.018
Volant	0.527	0.331	0.068	0.074	148	Volant	0.005	0.004	0.003	0.004
						Total	15,418	13,151	3,581	2,482

Notes: Data plotted in Fig. 4.12, based on number of catalogued records for each taxon in 316 documented localities. Key: Aquatic = invertebrates (mollusks) and fish; semi-aquatic = snakes, crocodiles, turtles; near-aquatic = anthracotheres, tragulids; scansorial = rodents and other small mammals; arboreal = primates and lizards; volant = birds; all remaining taxa = terrestrial (see Fig. 4.12).

Data on bone modification from the Dhok Pathan localities and excavated sites in the Nagri (Y311) and Chinji Formations show differing proportions of weathering, rounding, and carnivore damage but support our overall interpretation that these types of pre-burial taphonomic modification were typical of Siwalik fossil assemblages. The combined evidence indicates that most of the remains preserved in the Siwalik deposits were subject to predation and scavenging by large vertebrate carnivores (mammalian and reptilian) but were less affected by physical and chemical taphonomic processes before being permanently stabilized by burial and mineralization.

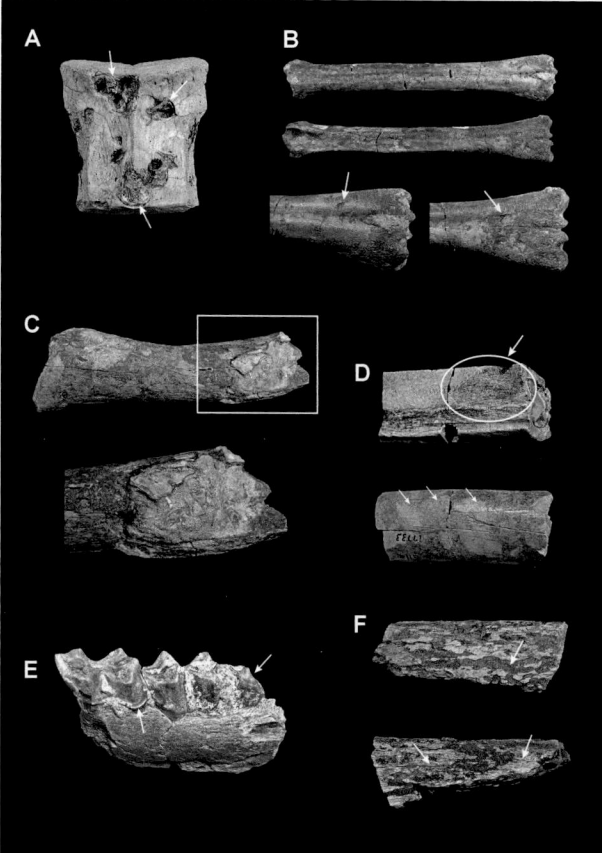

Figure 4.5. Examples of bone-modification features in Siwalik fossils—see Table 4.4 for abbreviations such as Y311-W2, relating to excavated specimens. A. Multiple tooth marks (arrows) on the lumbar vertebra of a small ungulate (YGSP 42030 [Geological Survey of Pakistan Catalogue Number], Locality Y516), possibly from crocodilian predation; B. Juvenile bovid metapodial with enlarged view of both sides of distal end, showing scoring marks (left, white arrow) and tooth punctures (right, white arrow), also indicating possible crocodilian predation (Locality Y311, YGSP 17731, W2-516); C. Mammal limb shaft (humerus) showing abrasion, surface weathering and erosion, plus crushing of proximal end (enlarged) attributed to trampling (Locality Y311, W2-502); D. Bovid radius shaft section showing carnivore tooth mark and flake on posterior shaft (pale oval with arrow) and fine, parallel scratches on anterior side (white arrows), interpreted as the consequence of trampling; note otherwise smooth, unweathered bone surface (YGSP 17733, Locality Y311, W2-511a); E. Tragulid mandible fragment (excavated), showing dissolution of enamel around tooth crowns (white arrows) attributed to digestion by crocodilian predator (Locality Y311, YGSP 20182); F. Pre-burial surface weathering indicated by cracking parallel to original bone structure (lower image, examples indicated by arrows) (Locality Y690, YGSP 26833), also deeper grooving attributed to termites (upper image, example indicated by single arrow); sediment infill in these features shows that damage occurred prior to final burial and mineralization.

Skeletal Elements

The average intact mammal has over 200 skeletal elements. Reptiles, birds, fishes, and amphibians have more or fewer, depending on taxonomic group. Some mammal groups, such as bovids and equids, have reduced numbers of distal limb elements, (metapodials, podials, and phalanges); otherwise, numbers are similar for crania, long bones of the forelimb, hindlimb, and pre-caudal vertebrae. Deviations from the expected frequency or proportion of elements reflect processes of mortality, scavenging, weathering, surface scattering, or fluvial transport and winnowing (Behrensmeyer 1982; Badgley 1986b; Shipman 1993; Lyman and Lyman 1994). Actualistic and experimental studies of mammal skeletons, as well as flume experiments with bones and teeth, have identified suites of elements that characterize different taphonomic pathways (Voorhies 1969; Hanson 1980; Behrensmeyer 1991; Aslan and Behrensmeyer 1996).

Catalogued skeletal elements provide one source of comparative information for this taphonomic feature. We focus on artiodactyls since most of their identifiable skeletal elements were collected, including all astragali, and this approach was applied consistently across all Siwalik formations. Comparison of skeletal-element representation for a sample of 15,220 artiodactyl specimens from 316 localities (Fig. 4.6; Table 4.3a) shows that proportions of major body parts were similar across four formations (Fig. 4.6a), with slightly lower representation of more fragile elements in Kamlial Formation localities and a change toward increased cranial elements in the Dhok Pathan Formation. These variations could reflect a combination of increase through time in artiodactyl body size and changes in predator impact on the pre-burial bone assemblage. In the different fluvial environments, artiodactyl distal limb elements are most frequently preserved in channel fills, while cranial remains are more often associated with crevasse-splay deposits and floodplains (including teeth in the latter). Podials and phalanges would have survived trampling in channel-fill settings but could have been subject to increased scavenging in floodplain settings. Cranial remains may be better preserved in floodplain and crevasse splays because of lower levels of destructive trampling than were typical of channel-fill environments.

We can compare these patterns with fossil localities in which all skeletal elements were systematically documented during surface collection. Fossil assemblages from the four major Dhok Pathan facies associations (Fig. 4.2a) are more similar to each other in skeletal-element composition than to an average whole Miocene ungulate (Badgley 1986a). This pattern is consistent across all four Dhok Pathan environments, supporting the hypothesis that the primary taphonomic transformation of skeletal proportions resulted from early post-mortem processes such as predation, scavenging, weathering, and trampling rather than from subsequent fluvial transport and winnowing.

Excavated fossil assemblages in Locality Y311 provide a further test for skeletal-element bias (see Table 4.4a). All assemblages include abundant axial elements (vertebrae and ribs) as well as appendicular elements (limb parts including podials and

Table 4.4. Taphonomic Attributes of Specimens from Y311 Excavations

A. Skeletal Elements Recorded in Four Excavations at Y311

	Y311 Excavated Skeletal Elements (Fig. 4.7a)										Comparison with Catalogued Artiodactyls (Fig. 4.7b)			
	Numbers of Specimens in Each Excavation					Proportions					Environment, All Formations			
	East 1	West 1+2	DS-4+ DS-4ME	GM Pod	Total	East-1 n=73	West-1+2 n=209	DS-4+DS-4ME n=84	GM Pods n=106	Total	Y311 Excav	MC	Y311 Excav n=297	MC n=4910 (w/o Y311)
Tooth	11	29	7	0	47	0.15	0.14	0.08	0.00	0.100	47	956	0.158	0.195
Cranial	10	31	9	21	71	0.14	0.15	0.11	0.20	0.150	71	654	0.239	0.133
Forelimb	5	18	9	2	34	0.07	0.09	0.11	0.02	0.072	34	514	0.114	0.105
Hindlimb	4	13	13	14	44	0.05	0.06	0.15	0.13	0.093	44	512	0.148	0.104
Metapodial	3	7	7	10	27	0.04	0.03	0.08	0.09	0.057	27	502	0.091	0.102
Pod-Phal	13	25	12	24	74	0.18	0.12	0.14	0.23	0.157	74	1772	0.249	0.361
Axial	27	86	27	35	175	0.37	0.41	0.32	0.33	0.371				
	73	209	84	106	472						297	4910 excluding Y311		

Notes: See Fig. 4.7. Data for four excavations (East-1, West 1 + 2, DS-4+DS-4ME, GM Pods) and comparison of the combined excavation sample with proportions of catalogued records of artiodactyl skeletal elements from all other major channel localities (excluding Y311). Axial remains were consistently recorded in the excavations but not in the surface assemblages, so are not included in the comparison with artiodactyls. Differences in proportions of skeletal elements in excavated versus surface collections indicate greater fragility of cranial remains on outcrop surfaces, contrasting with greater durability of podials and phalanges.

Table 4.4. continued

B. Bone-Weathering at Four Y311 Excavations

	Y311-DS4	Y311-DS4ME	Y311-W1	Y311-W2	Total
WS 0	27	59	19	137	242
WS 1	10	16	6	79	111
WS 2	0	4	0	6	10
Total	37	79	25	222	363
WS 0 proportion	0.730	0.747	0.760	0.617	0.667 ~67% Unweathered

Notes: Shows number of specimens that can be confidently assigned to weathering stages (WS) 0, 1, and 2 (Behrensmeyer 1978), based on the presence or absence of cracking and roughness on bone surfaces. Weathering Stage 0 indicates minimal time of exposure to surface conditions prior to burial and is the dominant group (67%) in this excavated sample. None of these excavations had bones showing greater degrees of surface weathering (WS 3–5).

C. Overall Carnivore Damage

	Y311-DS4	Y311-DS4ME	Y311-W1	Y311-W2	Total
No	28	51	17	116	212
Yes	7	21	7	104	139
Total	35	72	24	220	351
Yes proportion	0.200	0.292	0.292	0.473	0.396 ~40% with Carnivore damage

Notes: Evidence for carnivore damage includes spiral breaks, a hallmark of breakage while fresh; frayed or chewed bone ends; and tooth marks such as punctures, scores, and v-shaped grooves. All of these are included in the tallies on bones with well-preserved surfaces from four excavation sites at Y311.

D. Tooth Marks

	Y311-DS4	Y311-DS4ME	Y311-W1	Y311-W2	Total
No	32	68	25	182	307
Yes	5	9	0	39	53
Total	37	77	25	221	360
Yes proportion	0.135	0.117	0.000	0.176	0.147 ~15% Tooth-marked

Notes: Carnivore tooth marks (see Fig. 4.5a, b, d) provide reliable evidence for predator or scavenger bone modification. The tallies in this table are a subset of marks included in Table 4.3c, from the same Y311 excavations.

E. Trample Damage

	Y311-DS4	Y311-DS4ME	Y311-W1	Y311-W2	Total
No	35	66	19	185	305
Yes	0	6	5	35	46
Total	35	72	25	220	351
Yes proportion	0.000	0.083	0.200	0.159	0.131 ~13% Trampled

Notes: Trample damage is indicated by parallel striations on bone surfaces and by crushing and breakage that depress and flake bone edges and surfaces. See Fig. 4.5c, d.

Table 4.4. continued

F. Rounding

	Y311-DS4	Y311-DS4ME	Y311-W1	Y311-W2	Total
Angular (A)	9	14	4	36	63
Sub-angular (SA)	12	30	11	108	161
Sub-rounded (SR)	15	20	7	71	113
Rounded (R)	1	16	3	10	30
Total	37	80	25	225	367
R + SR	16	36	10	81	143
Proportion R+SR	0.432	0.450	0.400	0.360	0.390
					~40% Rounded

Notes: Physical abrasion (rounding) is most clearly expressed by rounding of breaks and bone edges that were originally sharp and angular. The four categories, from angular to rounded, are similar to those used for sedimentary grain description.

G. Dissolution

Code	Y311-DS4	Y311-DS4ME	Y311-W1	Y311-W2	Total
0	34	39	24	171	268
1	1	13	0	16	31
2	1	19	0	31	53
3	1	6	0	3	13
Total	37	77	24	221	365
2 + 3	2	25	0	34	66
Proportion 2 + 3	0.054	0.325	0.000	0.154	0.181
					~18% Dissolved

Notes: Code: 0–3 = no evidence (0) to highly dissolved (3). Dissolution from water, soil or digestive acids is evident in pitted, roughened bone surfaces but may also be recorded in the removal of enamel from teeth and smoothing of bone and tooth edges where these were exposed to carnivore digestive processes. See Fig. 4.5e.

phalanges; Fig. 4.7a). Comparison between the excavated ungulates and catalogued artiodactyls (Fig. 4.7b) indicates a significant bias (chi-square = 32.73; $p < .0001$) in the surface-collected sample against cranial remains (other than teeth) and toward podials and phalanges. This likely reflects increased survival rates for more compact and durable distal limb elements on outcrop surfaces, since collecting biases would have favored cranial remains (maxillae and mandibles).

Age Distribution

The age structure of individuals in a death assemblage provides information about agents and processes of mortality. For example, a sample from catastrophic mortality due to floods, droughts, or epidemics can preserve unusual age distribution samples of the population, including an abundance of healthy adult individuals (Berger 1983; Haynes 1985b). In contrast, attritional mortality from predation, disease, or old age preferentially affects juvenile and old individuals (Lyman and Lyman 1994), and the resulting death assemblage thus should have fewer prime-age adults. Mammalian carnivores today often selectively prey on juveniles of their target species (Kruuk 1972; Miller 2012), although prey selection varies with season and hunting strategies (e.g., group versus individual, pursuit versus ambush; Van Valkenburgh 1999). For example, bone accu-

mulations collected by modern spotted hyena and leopard have a higher proportion of juveniles than in the living populations (Hill 1989; de Ruiter and Berger 2000).

Siwalik fossil assemblages include relatively few juveniles. For 15,220 catalogued artiodactyls, juveniles constitute an average of 7% of catalogued specimens across all four formations (see Table 4.3b). Floodplain and major channel environments have 11% and 8% juveniles, respectively, while floodplain channel and crevasse splay both have 6%. Taphonomically documented surface assemblages in the Dhok Pathan Formation have only 3% juveniles, with fewer in channel and crevasse-splay environments than in channel margins and floodplains (Badgley 1986a, Table 3). Fluvial reworking likely reduced representation of juveniles relative to the original populations, but there may also be an overall bias against the survival of juvenile remains in surface fossil assemblages. Similarities across all 21 of the Dhok Pathan surface-collected assemblages (Badgley 1986b, Fig. 5) indicate that predators had the strongest overall influence on skeletal-part destruction, although other taphonomic processes (e.g., trampling and fluvial reworking) likely contributed to pre-burial loss of juveniles. At Nagri Locality Y311, of 100 excavated mammalian skeletal remains that can be classified by age, 14% (14/100) are juvenile. A subsample of 21 cranial and dental elements includes four juvenile (immature)

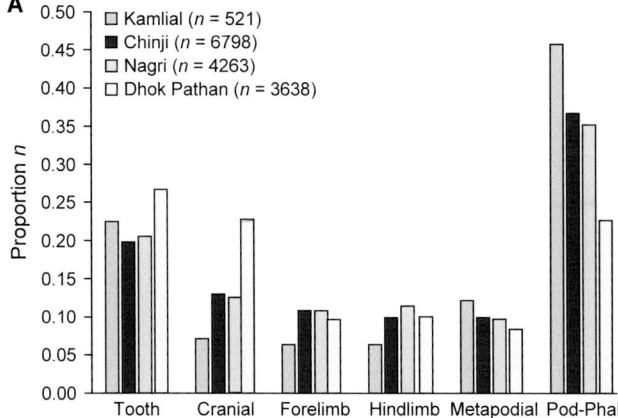

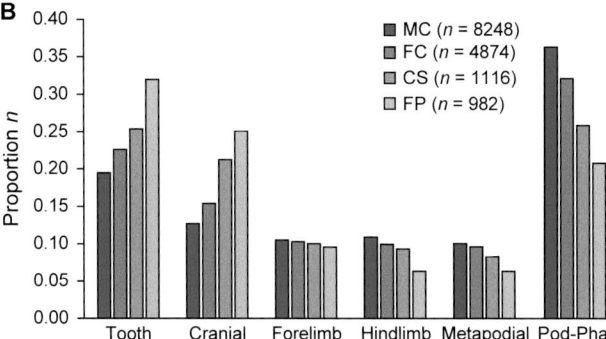

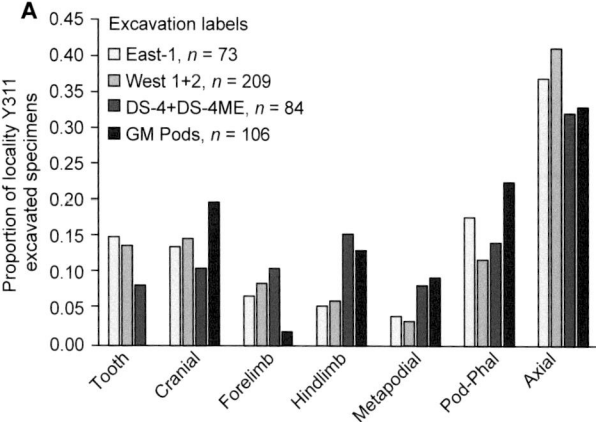

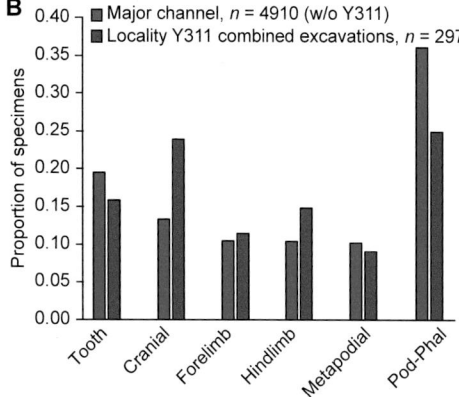

Figure 4.6. Skeletal-part representation for 15,220 catalogued artiodactyl specimens, including scapulae and pelves in forelimb and hindlimb counts (see data in Table 4.3). Note that all body parts identifiable as bovid were collected, thus representing a standardized protocol that minimizes biases across formations and depositional environments. A. Similar distributions of skeletal-part preservation across four Siwalik formations. From the Kamlial to the Dhok Pathan Formation, the main trends are the decrease in the proportion of Pod-Phal (Podials and Phalanges) and the increase in cranial remains. Differences in Dhok Pathan Formation preservation may be attributable to increased body size or changes in predation. B. Skeletal-part preservation across all four depositional environments, which is generally consistent but with better preservation of crania and teeth in crevasse splays and floodplains likely reflecting lower energy and/or more rapid burial in these settings. Trampling in major channels and floodplain channels may be a factor in the survival of podials and phalanges versus cranial remains. Based on a one-way ANOVA, the skeletal-part distributions differ significantly across the four environments ($F = 8.62$; $p = 0.001$), and the Tukey HSD Test shows that the major channel mean is significantly different from those of both crevasse splays and floodplains ($p < 0.01$). (Key: MC = major channel [upper], FC = floodplain channel, CS = crevasse splay, FP = floodplain.)

Figure 4.7. A. Skeletal-part distribution based on four excavations at Locality Y311, showing overall similarity in different parts of an abandoned channel fill (MC-U) in the Nagri Formation. Excavation names: East-1, West 1 + 2, DS-4+DS-4ME, GM Pods. Note high relative abundance of axial elements (vertebrae, ribs). B. Comparison of combined Y311 excavated samples (all mammals) with skeletal-part data from artiodactyls preserved in the major channel depositional environment (mostly abandoned-channel fills). Y311 data are not included in the major channel skeletal-part totals (all formations), and axial remains are excluded because these were not consistently collected or recorded in the catalogued surface assemblages. A chi-square test comparing cranial and pod-phal (podials + phalanges) in the excavated versus surface assemblage shows a significant difference (*Phi* = -0.11, Pearson chi-square = 32.73, $p = < 0.0001$). Data in Table 4.4.

individuals (20%). This figure, compared with 6–10% juvenile artiodactyls across all Siwalik formations and environments and 3% in systematically collected Dhok Pathan samples, suggests that juveniles were more common in the original Siwalik death assemblages than recorded in our surface-collected fossil samples.

Taphonomic explanations for the relatively low representation of juveniles in the Siwalik vertebrate assemblages indicate that attritional rather than catastrophic mortality was the primary mode of bone accumulation. Lower proportions of adult versus advanced-age individuals would provide an additional test, but this assessment is not feasible due to limited sample sizes and the rarity of complete dentitions.

Body Size Range Represented by Fossil Specimens
Two aspects of size pertain to preservation history: the inferred adult body size (weight) of individuals represented in a fossil assemblage and the size range of their individual skeletal elements. The body mass of preserved individuals reflects many

different aspects of ecological and preservation history, including size variation in species and individuals, habitats in the original ecosystem, processes of mortality (e.g., predator preferences for prey of a particular size range), and transportability of skeletal elements by fluvial or biological agents. Skeletal part size and density affect how remains interact with taphonomic processes including predation, scavenging, and fluvial transport, with smaller and more fragile elements less likely to be preserved than large, dense, or otherwise robust ones.

Siwalik fossil vertebrate assemblages show a general pattern of preservation favoring larger-bodied species, although concentrations of small bones and teeth also occur and record somewhat different taphonomic and sedimentary circumstances (Badgley, Downs, and Flynn 1998). Examination of the body-size distribution for artiodactyls shows dominance of the 10–100 kg body size through time, with a post-Kamlial decline in smaller body sizes and a later increase of megaherbivores >1000 kg (Fig. 4.8a). The size distribution is quite similar in all four major environments (Fig. 4.8b; One-way ANOVA: $F = 2.02$; p = 0.1897), however, which indicates that this body-size structure was characteristic of the >1 kg mammalian community across the sub-Himalayan alluvial landscape (see Fig. 4.8b).

The size distribution of all mammals (not just artiodactyls) varies among a subset of systematically sampled localities from different lithofacies associations (Fig. 4.2) in the Dhok Pathan Formation. For all mammals greater than 1 kg in estimated body weight recovered in surface collections, smaller species (1–15 kg) are quite rare (2–4% of individuals) in channel-lag and crevasse-splay deposits but an order of magnitude more common (12–16% of individuals) in channel-margin and floodplain deposits (Badgley 1986a, Table 9; see also Fig. 4.2a, Table 4.1). Assemblages affected by fluvial transport in active channel environments thus appear biased against smaller species in the Dhok Pathan Formation, although there is no similar pattern for catalogued artiodactyls in the 1–10 kg size range (Fig. 4.8a). The size range of 10–100 kg (i.e., small gazelle to large suid) is best represented in terms of number of individuals in Dhok Pathan localities from all depositional environments except crevasse splays (Badgley 1986a). This size range also encompasses much of the taxonomic diversity of ungulates throughout the Siwalik record of the Potwar Plateau (Fig. 4.8a; Chapters 2 and 5, this volume; Part III, Table III.1).

Taxonomic Composition

Certain aspects of taxonomic composition provide evidence for taphonomic history and depositional environment. In Siwalik localities, the frequency of aquatic vertebrate remains indicates the extent to which flowing or standing water dominated the depositional environment. Fossils of fish, turtles, and crocodilians are common in channel-lag assemblages (see Table 4.1), which formed in active river channels, and in some larger abandoned channels with evidence for ponded water. In such assemblages, aquatic-vertebrate remains may outnumber those of terrestrial mammals. In contrast, aquatic vertebrates and water-dependent mammals are rare to absent in floodplain

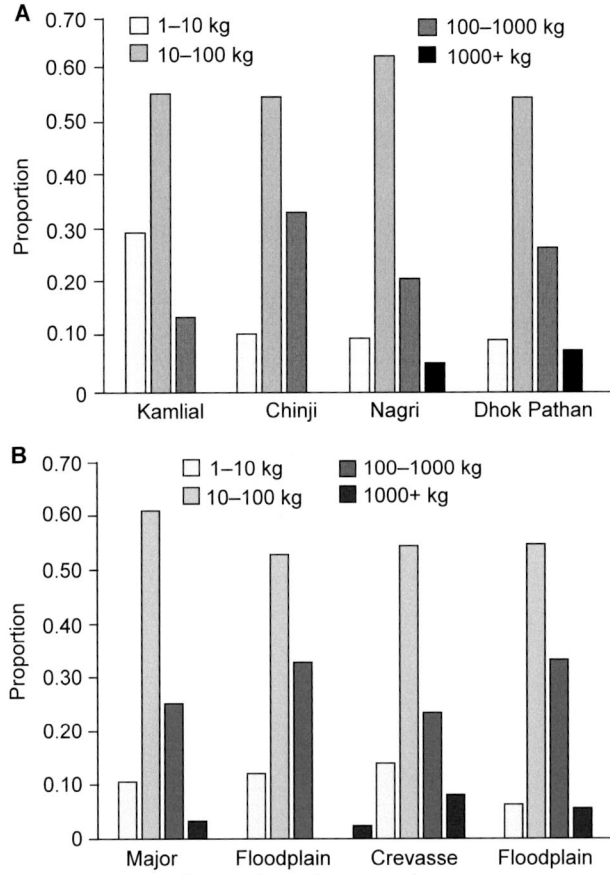

Figure 4.8. A. Comparison of artiodactyl body sizes across Siwalik formations and depositional environments, using the same dataset as in other figures, but excluding 534 records for taxa with uncertain body size (Table 4.3c). Body-size distributions for the formations (excluding >1,000 kg) are not significantly different based on a one-way ANOVA ($F = 2.02$; $p = 0.190$). Other groups (e.g., Perissodactyla, Proboscidea) include species with large body sizes across these formations and environments but are not represented in this analysis. B. Body-size distributions (artiodactyls) across four environmental settings (all formations), showing no significant differences (one-way ANOVA, $F = 2.27$, $p = 0.133$). Data in Table 4.3c.

assemblages, which are characterized by cursorial ungulates (e.g., equids) and rodents, as well as rare burrowing animals, such as aardvarks. Comparison of two Chinji Formation localities, Y639 and Y699, of similar-age (~13 Ma) provides an example of environmental controls on faunal composition, with nearly complete separation of taxa in a major channel–fill deposit versus a floodplain deposit (Fig. 4.9).

For the major mammalian groups, relative abundance based on specimen counts varies across depositional environments (Fig. 4.10), with significant differences in the proportions of rodents, carnivores, and artiodactyls (see statistical tests in caption for Fig. 4.10) before and after the appearance of equids (Fig. 4.10b), which perhaps also reflects the shift toward the

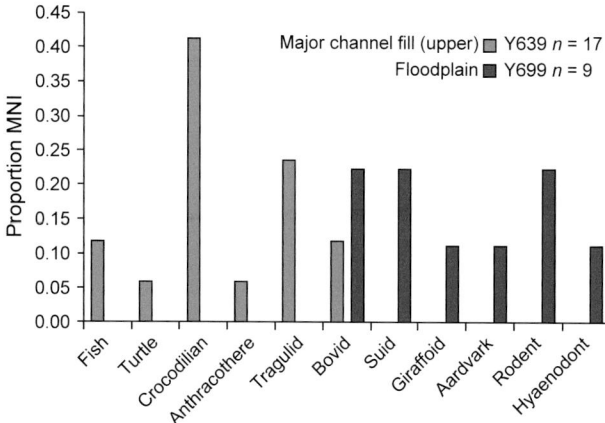

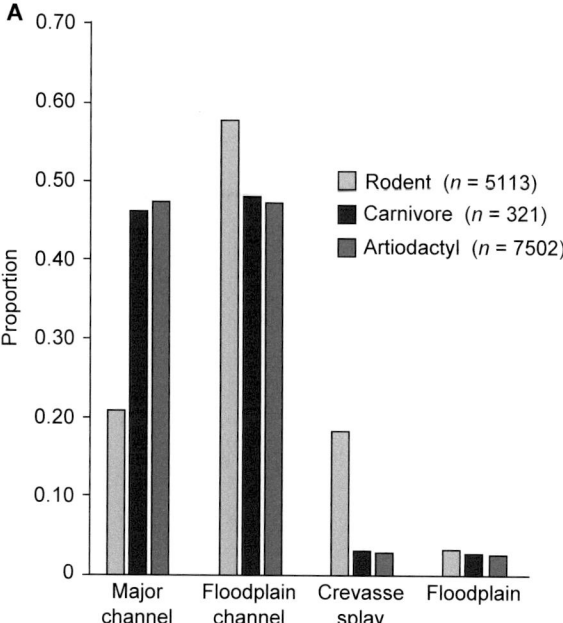

Figure 4.9. Comparison of relative abundance using estimated minimum number of individuals (MNI) of taxa in two Chinji Formation localities, both ~13 Ma, which represent two distinct paleoenvironmental contexts based on sedimentological evidence. Locality Y639 is in a large-scale channel fill near the top of a major channel sand, while Locality Y699 is a fossiliferous patch associated with floodplain paleosols. Taxa are ordered from more aquatic to more terrestrial, left to right. Only bovids occur in both sedimentary environments. Adapted from Behrensmeyer et al. (2005, Fig.). Data in Table 4.3d.

buff-system increase in floodplain habitats (Chapter 2, this volume). Overall, these patterns are consistent with ecological variation in the original habitats occupied by these mammals as well as differences in taphonomic processes characterizing the fluvial settings. For example, localities with rodent fossils occur more frequently in floodplain channel and crevasse splay contexts relative to larger taxa, while the opposite is true for the infillings of abandoned major channels (e.g., Fig. 4.9). This contrast suggests more frequent association of micromammal predators with riparian floodplain habitats than with abandoned major channels, although it also may reflect higher preservation probability for small bones and teeth in floodplain habitats. Carnivore and artiodactyl relative abundances are similar in all environments prior to 10.8 Ma, but carnivores are more common in floodplain settings after the appearance of equids, possibly as a consequence of more den-associated remains in open habitats. When equids become part of the mammal community, they occur in all depositional contexts but are notably more abundant than artiodactyls in floodplain channels (Fig. 4.10). Their abundance in post-10.8 Ma fossil assemblages, relative to artiodactyls and carnivores (except in major channels), indicates success in populating the alluvial landscape and probably also the high preservation potential of their robust skeletal elements (Behrensmeyer and Barry 2005). This shift is also consistent with a marked taxonomic turnover shortly after 10.8 Ma (see Fig. 25.6, this volume). It is possible that equid interaction with vegetation structure contributed to the period of faunal change by promoting more open habitats (i.e., via "grazing lawns") during the C_3 to C_4 transition (Potts et al., 2020; see Chapters 25 and 26, this volume).

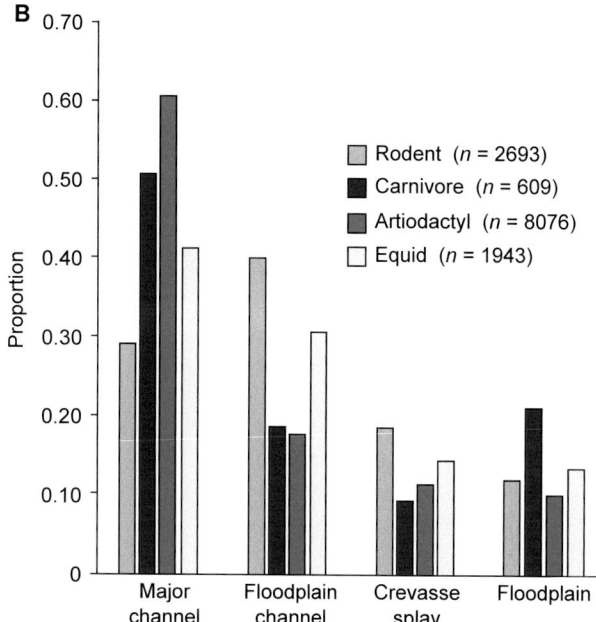

Figure 4.10. Distribution of major mammal groups across four major depositional environments before (A) and after (B) the arrival of equids in the Siwalik fauna at about 10.8 Ma. Numbers are based on catalog entries for 316 localities with well-documented environments, as for other figures (see Table 4.3). Note the striking difference in proportion of carnivores in floodplain deposits before and after the appearance of equids in the fauna and higher numbers of artiodactyls and carnivores in floodplain channels before equids arrived. These differences are significant at the $p < 0.0001$ level using chi-square tests for comparisons before and after 10.8 Ma, excluding equids (rodents: $X^2 = 372.4$, $p < 0.0001$; carnivores: $X^2 = 122.58$; $p < 0.0001$; artiodactyls: $X^2 = 1,909.33$, $p < 0.0001$), indicating a shift in taphonomic and ecological processes across all depositional settings. Data in Table 4.3e.

In the Dhok Pathan Formation, finer scale analysis of faunas associated with different depositional contexts indicates variation resulting from (1) original habitat associations in the Miocene ecosystems and (2) the durability and transportability of skeletal elements of different size and shape (Badgley 1986a, 1986b). At the family level, most taxonomic groups occur in all major depositional environments, but the relative abundance of taxa based on specimen counts varies considerably, and these differences are even greater using abundance of individuals (Badgley 1986a, Table 4). Bovids and equids are the most numerous taxa across all Dhok Pathan depositional environments, especially in assemblages influenced by fluvial transport. Hominoids, suids, and tragulids are more abundant in channel-margin (including abandoned channel assemblages). This second pattern may indicate original habitat preference for places with water, tree-canopy cover, and shade—that is, it reflects animals that died in the same habitats where they lived.

Analysis of body-size representation of all mammals >1 kg in individual locality samples from the four major depositional contexts (Fig. 4.8b) shows that floodplain channels, major channels, and crevasse splays are similar in sampling large mammals and megaherbivores from the Siwalik paleocommunity. The rarefaction curves (summarized as power functions; Fig. 4.11) quantify the increase in taxa per locality recovered with increasing specimen numbers; these increases level off at a maximum of 25–30 taxa (excluding small mammals and carnivorans) per locality for floodplain channels, major channels, and crevasse splays. The floodplain curve has a notably lower trajectory, with maximum taxonomic richness below 15 taxa. This pattern is consistent with reduced preservation rates per locality on the floodplain (likely due to lower probability of permanent burial or more rapid recycling of organic remains) and shorter periods of time-averaging in floodplain assemblages. Two especially productive fossil localities, Y76 and Y311 (Table 4.3f), preserve 35 and 45 distinct mammalian taxa (respectively), indicating that higher local community richness was occasionally sampled by unusual taphonomic and depositional circumstances.

INTERPRETING SIWALIK TAPHONOMIC HISTORIES

Taphonomic analyses presented above indicate that Siwalik fossil assemblages of the Potwar Plateau were subjected to similar early post-mortem processes (damage and fragmentation due to predation, scavenging, trampling) across all four environmental settings (Figs. 4.5, 4.6, 4.7). Subsequent taphonomic modification by fluvial transport, sorting, and reworking of the skeletal remains varied, however, due to different depositional processes among these settings. Many of the fragmentary, well-preserved fossils represent attritional mortality in or near the places where the animals lived (autochthonous or para-autochthonous), but substantial parts of our record were also subjected to fluvial processes that combined remains from different habitats or time intervals. Inferences about taphonomic history described in this section pertain mainly to remains

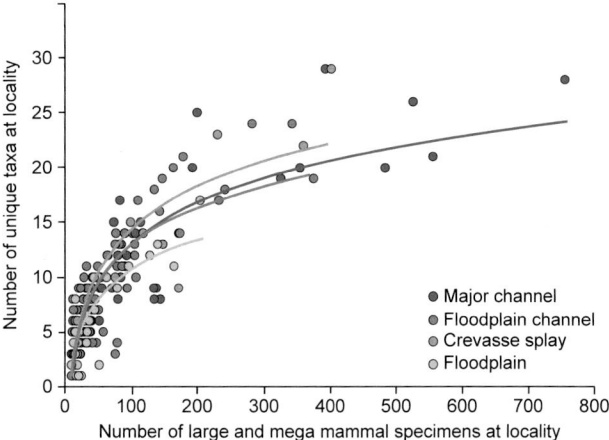

Figure 4.11. Rarefaction curves based on the number of unique taxa plotted against sample size (see Table 4.3f) show that the major channels (mostly upper) and floodplain channels are similar in terms of how they represent large to mega-mammal diversity (1–5,000 kg adult body weight). Power functions show that floodplain localities have fewer unique taxa for a given sample size, while crevasse splay localities have a slightly higher number of taxa for a given sample size. Our two most productive localities, Y76 (N = 1,369 specimens and 35 taxa) and Y311 (N = 4,229 specimens and 45 taxa) are not included in the plot but show that higher taxonomic richness can be recorded in more complex, spatially enlarged, or time-averaged taphonomic settings.

of the larger vertebrates (>1 kg adult body size); microvertebrate fossil assemblages and a small subset of localities represent unusual circumstances such as predator concentrations and mass mortality and are discussed separately in "Unusual Fossil Concentrations" at the end of this section.

Attritional Mortality

Animals lived in great numbers across the Siwalik alluvial plain, and their normal cycles of life and death resulted in a constant "rain" of soft tissues and skeletons onto the land surfaces. Predation, scavenging, and other ecological processes recycled most of the remains, but fluvial deposition sometimes led to burial and preservation of bones and teeth, with variable effects of fluvial transport or reworking. Sedimentological as well as taphonomic evidence shows that most of our fossils are associated with specific environments of the alluvial plain, particularly abandoned-channel settings (Figs. 4.2, 4.4, 4.6; Tables 4.2, 4.3), rather than the more extensive floodplains. Fossiliferous lithofacies in abandoned channels typically include thin beds of sandstone, gravel lenses, and mudstones with root traces, bioturbation structures, and indications of fluctuating water tables, evidence for subaerial land surfaces, saturated substrates (swamps or ponds), and episodic, flood-related deposits of sand to fine conglomerate (see also Chapter 2, this volume).

Attritional mortality as the dominant pattern of accumulation is supported by properties of a set of taphonomically documented Dhok Pathan fossil assemblages characterized by spatial clustering of associated skeletal elements, a preponderance of unweathered bone surfaces, and relatively high frequencies of juvenile remains and carnivores (Badgley 1986a). This evidence indicates autochthonous (local) processes of accumulation that were primarily biological, suggesting that diverse species were attracted to specific areas over multiple years. A longer-term component is evident from a low proportion of bones in advanced weathering stages (Badgley 1986a; Table 4.3). Both aquatic (crocodilian) and terrestrial (amphicyond, felid, hyaenid) predators could have contributed to attritional mortality of their prey. Time-averaging was likely low to moderate (10^0–10^3 years), based on the concentration of fossils and the number of productive horizons at a given locality. The number of mammalian taxa represented by fossils at individual channel-related localities (e.g., average taxa/locality for major channels was 9.5 and for floodplain channels, 8.5; Table 4.3f) is based on the remains that survived destructive taphonomic processes to accumulate over time in these depositional settings.

In contrast to the abundance of fossils in channel-related deposits, only ~20% of our well-documented localities (total $N = 316$) occur in floodplain sediments that formed the vast ecosystems where most of the fauna and flora actually lived (Table 4.2, Fig. 4.4), and these localities generally have fewer specimens and individuals than other environments (average number of taxa/locality on floodplains was 5.7; Table 4.3f). Fine-grained, laterally extensive floodplain sediments, including those that became paleosols, do contain fossil assemblages with isolated skeletal elements and rare partial skeletons, but these usually represent few species and individuals (e.g., Fig. 4.9). Most of these likely represent isolated deaths scattered across the land surface, which happened to become buried and preserved by flooding events before the remains were destroyed by natural recycling processes. A few localities with associated, well-preserved skeletal parts belonging to multiple individuals are interpreted as unusual, biologically mediated accumulations (see "Unusual Fossil Concentrations" at the end of this section).

Fluvial Transport

Sediments representing energetic water flow, such as coarse lenses within channel deposits (channel-lag), contain notable concentrations of vertebrate remains (Table 4.1, Fig. 4.2). Most specimens are incomplete skeletal elements with fresh to highly abraded surfaces, including bone pebbles, and are approximately hydraulically equivalent to the grain-size distribution of the sedimentary matrix, which typically varies from medium-grained sandstone to pebble-grade conglomerate (Badgley 1986b). The isolated, usually fragmentary fossils in these deposits represent vertebrate taxa across the entire size range of large mammals (1–5,000 kg), with species at the lower end of this range (1–15 kg) relatively uncommon (Fig. 4.8).

Channel-lag assemblages occur in both major channel and floodplain channel depositional contexts (Table 4.1, Figs. 4.2, 4.3). Because they represent higher-energy fluvial processes of erosion, reworking, and transport, these assemblages are likely to include multiple sources of skeletal input, such as carcasses entrained by active river currents and bones harvested from bank collapse as channels reworked their floodplain deposits (Tauxe, Kent, and Opdyke 1980; Badgley 1986b; Aslan and Behrensmeyer 1996). The inorganic clasts in the fossiliferous deposits include angular to sub-rounded pieces of mudstone and calcareous nodules from eroded paleosols, providing additional evidence for reworking of previously buried skeletal remains. These assemblages likely represent time averaging on the order of 10^3–10^4 years (Behrensmeyer 1982, 1987) and could include material reworked from other depositional contexts such as floodplains. Channel-lag fossil assemblages may preserve the best overall "census" of animals inhabiting the western Siwalik alluvial plain, but they also sample long periods of ecological time, which could increase species richness relative to that of the faunal community in any narrow interval of time.

Crevasse-splay assemblages occur in thin sheets of sandstone and conglomerate deposited over floodplain surfaces (Badgley 1986b), often directly above a paleosol. These depositional events were likely of short duration (days to weeks) but could potentially have incorporated vertebrate remains from different sources, including material entrained in the channel, buried skeletal parts exhumed when the channel broke through its banks, and bones and teeth on the floodplain surface that were mobilized during the splay event. Crevasse splay assemblages have an average of 10.5 taxa/locality (Table 4.3f), nearly twice the average for floodplain localities, which suggests that remains preserved within the floodplain deposits (including in paleosols) were significantly affected by post-burial destruction.

In the Dhok Pathan Formation, a small fraction (<10%) of skeletal elements in these crevasse-splay assemblages exceeds hydraulic equivalence with the inorganic conglomerate clasts (Badgley 1986b); these specimens may represent rapidly buried, "first-cycle" skeletal remains that were not subject to extensive sorting and winnowing during the splay event. Although fragmentary, they tend to be well-preserved, with minimal evidence of sedimentary abrasion. The resulting assemblages combine skeletal material with variable degrees of transport across limited areas of the floodplain and represent moderate amounts of time-averaging (e.g., 10^2–10^3 years).

Micro-Vertebrate Assemblages

Small-vertebrate faunas are a critical component of Siwalik species diversity and were recovered separately from fossil assemblages of larger animals (>1 kg). Assembling a record for the Siwalik small mammals and other "micro-vertebrate" remains required special collection methods, notably a long-term, large-scale program of wet-screening bulk sediment from sites where very small bones were identified on the outcrop surface (Chapters 3, 10, and 11, this volume). Detailed taphonomic study of Siwalik small-mammal assemblages focused on 9 localities out of a total of 153. These represent a narrow stratigraphic interval in the Lower Chinji Formation (Badgley, Downs, and Flynn 1998), and further study will be necessary to understand how they compare with small-mammal assemblages from the other Siwalik formations of the Potwar Plateau.

The productive layers of the micro-vertebrate localities from the Lower Chinji Formation have diverse lithologies ranging from clayey silt to conglomerate. For six of the nine sites, the depositional environments were swales or ponds associated with abandoned channels. Three other sites were mudstones with strong pedogenic overprint: two were floodplain deposits, and the third was part of an abandoned-channel fill. These latter lithologies had the fewest fossils. Within the abandoned channels, several productive horizons of thin siltstone to gravelly sandstone units 5–20 cm thick were laterally restricted, representing single depositional events—such as deposits from a storm or annual flood. A few fossil-rich assemblages occurred in shallow ponds with visible swale topography and stratification, also within abandoned channels. Small-mammal teeth were hydraulically equivalent to grains in the enclosing matrix, and most teeth were not abraded. Some showed signs of acid etching, which could indicate partial digestion from predator consumption or post-burial damage from soil acids.

The frequency of small-mammal teeth and bones, in combination with the restricted lateral extent of the sedimentary units to a few tens of square meters, implies that a concentration of small-mammal remains existed prior to assimilation into the sediment matrix (Badgley, Downs, and Flynn 1998). Such concentrations typically occur under a raptor perch (e.g., owl-pellet accumulations) or in latrines of carnivorous mammals (Andrews 1990; Reed 2007; Fernandez-Jalvo and Andrews 2016). In addition to teeth and postcrania of small mammals, assemblages in sedimentary contexts interpreted as ponds contain fish teeth and bones of fishes and frogs. Postcrania of small mammals are greatly underrepresented compared to cheek teeth, and may have been destroyed during predation, dispersed by flowing water, or degraded during screen-washing.

In terms of taxonomic composition, small-mammal assemblages are dominated by rodents. Bats and insectivores are uncommon, which likely reflect the preferences of the predatory agents of accumulation, since both groups are typically abundant in modern subtropical terrestrial ecosystems. There is no indication of the predator identities, although small carnivores, snakes, and birds of prey are all plausible candidates. Bird fossils are rare in our Siwalik assemblages, and of raptors that could have been responsible for small-vertebrate accumulations, we have only an eagle recorded at Nagri Locality Y311.

Unusual Fossil Concentrations

A small number of Siwalik fossil localities can be attributed to sedimentological and taphonomic processes that differ from the major modes of accumulation described above. One indicator of such differences is the number of skeletal elements that can be assigned to single individuals, representing partial carcasses that were less altered by early post-mortem taphonomic destruction and dispersion. Only 43 of our localities (~3%) have eight or more skeletal elements that can be attributed to single (or in some cases multiple) mammal individuals, out of the total of 1,315 localities documented at least to for-

mation. In a subset of 30 that can be assigned to depositional environment, such localities are predominantly floodplain (40%) and major channel (33%), with 17% in floodplain channels and 10% in crevasse splays (Table 4.5). The floodplain depositional environment is even more dominant (8/13, or 62%) for localities with 15 or more associated skeletal elements. This provides evidence for particular circumstances leading to lower rates of taphonomic destruction on the Siwalik floodplains and suggests the impact of increased floodplain prevalence and deposition in the Dhok Pathan Formation on increased carnivore and rodent preservation after 10.8 Ma (Figs. 4.4, 4.10).

Predator Dens and Burrows

A subset of the localities with associated skeletal remains is interpreted as dens or burrows based on contextual and taphonomic evidence (Table 4.5). Spatially concentrated remains of multiple individuals of medium-sized mammals, including partial skeletons, indicate accumulation by one or more predatory or scavenging animals that brought carcasses to a burrow, den, or nest (Hill 1989; Lansing et al. 2009) or fossorial animals that died in their burrows (e.g., Papa, Santis, and Togo 2017; Weaver et al. 2021). Articulated or closely associated skeletal elements, chewed edges, puncture marks, uniform weathering stage, a relatively high frequency of juvenile and carnivore specimens, and lack of hydraulic equivalence between bones and the sediment matrix all point to biological agents of accumulation.

Such localities are rare in the Siwalik record of the Potwar Plateau—only six likely dens or burrows have been documented in 25 years of fieldwork by the Harvard-GSP group. These six are noteworthy, however, because associated skeletal material, sometimes with dental remains as well, has high taxonomic value. One locality (Y270), excavated from a small patch of floodplain sediment, contains at least two partial carnivore skeletons (of different species) along with fragmentary remains of fish, turtle, lizard, snake, and bird, in addition to associated elements of several mammalian herbivores. This mix of different taxa, and associated remains of perhaps as many as ten individuals in a small volume of sediment, suggests bone col-

Table 4.5. Environmental Distribution of Localities with Multiple Specimens Per Individual

Environment	At Least 8 Associated Elements	At Least 15 Associated Elements	8≥	15≥
Major Channel (MC)	10	1	0.333	0.077
Floodplain Channel (FC)	5	2	0.167	0.154
Crevasse Splay (CS)	3	2	0.100	0.154
Floodplain (FP)	12	8	0.400	0.615
TOTAL	30	13		

Note: Localities with multiple specimens per individual, showing numbers and proportions of localities associated with each of the four major depositional environments. Most represent the skeletal parts of single individuals, but a few localities have associated remains from multiple individuals. One Crevasse Splay (CS) locality (Y481) included here was not used in other taphonomic analyses of environments.

lecting by one or more carnivore species and possibly occupation by multiple species over time. Five other localities have associated remains, four with aardvark and one with pangolin fossils. Both groups today are known to occupy burrows, and their associated remains indicate that the animals likely died underground and were protected from additional taphonomic processing by burrow collapse or rapid infilling. Short intervals of time (e.g., 10^0–10^1 years) likely are represented in all these assemblages. Of the four localities (out of six) with documented sedimentary context, three are in floodplain silts and one occurs in a small channel fill (Fig. 4.2, Table 4.1).

Sediment "Pods"

Enigmatic bone concentrations that may represent a different kind of predator accumulation occur in "pods" at Nagri Locality Y311. There are multiple occurrences of these sharply defined, sand- and bone-filled sedimentary structures (roughly cylindrical, or trough-shaped) that protrude downward as much as 1 m into a thick bed of pedogenically modified clayey silt bed, all at approximately the same stratigraphic level. The pods contain a wide variety of herbivore taxa, including a partial deinothere skull and fragmentary small to medium-sized artiodactyl bones showing digestive dissolution and other evidence of carnivore bone modification. The pattern suggests carnivore dens dug into fine-grained sediments within the larger context of the abandoned channel that formed Y311 (Loughlin, Behrensmeyer, and Badgley 2018). It is possible that original "piping structures" caused by subterranean removal of floodplain sediment by ground water initiated these structures, which later were occupied or enlarged for stashing carcasses by carnivores (including crocodilians) or served as traps for fluvially transported skeletal remains.

Mass Accumulations of Fish

Several localities in our Siwalik record have dense concentrations of disarticulated fish bones within thin, laterally restricted lenses of silt (see Chapter 6, this volume). We interpret these as mass deaths of aquatic animals caused by the drying up of small bodies of water that became disconnected from larger aquatic habitats (e.g., major channels). One of these, Locality Y719 (Chinji Formation, ~14 Ma), occurs in the fill of a complex floodplain channel and has at least five fish taxa, with many complete and some articulated skeletal elements, as well as fragmentary remains of the land snail *Pila*, crocodile, snake, turtle, and mammals (mostly teeth). Given the widely different states of preservation of the fish compared to the other faunal elements, the fragmentary remains likely represent longer-term accumulation in the floodplain channel, with the mass mortality event primarily affecting fish. Locality Y388 (Dhok Pathan Formation, 8.8 Ma) preserves abundant fish remains, including delicate elements, dispersed in poorly sorted silt lenses above a thick sandstone unit. This locality also had numerous small mammal remains and represents intermittent fill of an abandoned channel.

The scarcity of mass bone accumulations and complete skeletons in the Siwalik fossil record contrasts with other fluvial paleoecosystems in which these accumulations are more common, such as the Judith River Formation in the Cretaceous of North America (Rogers, Eberth, and Fiorillo 2010), the Cretaceous Maevarano Formation of Madagascar (Rogers 2005), the Triassic Ischigualasto Formation of Argentina (Columbi, Rogers, and Alcober 2012), and the Miocene Kipsaramon bonebed (Muruyur Beds) in Kenya (Behrensmeyer et al. 2002). It is possible that more of these types of taphonomic concentrations occurred originally on the Siwalik alluvial plain and were reworked by fluvial processes. Their low numbers may also indicate that environmental stresses were buffered by the large, mountain-sourced fluvial systems and rarely became extreme enough to cause mass mortality events in either the aquatic or terrestrial animal communities. In modern floodplain ecosystems, resident animals are highly sensitive to changing water levels, moving in and out of the floodplain seasonally (Sheppe and Osborne 1971; Andrews, Grove, and Horne 1975). Seasonal migration may also have characterized mobile aquatic and terrestrial members of the Siwalik vertebrate paleocommunity, thus reducing the likelihood of mass mortality.

CHANGES IN FOSSIL PRESERVATION THROUGH THE SIWALIK RECORD

The overall frequency of fossil localities and their representation in different fluvial environments change through the four formations of the Lower and Middle Siwalik record of the Potwar Plateau (Figs. 4.1, 4.4). The formations in which thick, multistorey sandstones (representing major channels) are the dominant lithology (Kamlial, Nagri) have fewer fossil localities than the two formations in which mudstones are dominant (Chinji, Dhok Pathan). Long-term increase in the rate of sediment accumulation (Chapter 2, this volume) occurs in the Siwalik sequence and may provide large-scale control on this pattern of fossil productivity, via increased survival of floodplain and associated channel deposits and decreased fluvial reworking by major channels through time. At the subenvironment scale, higher sediment accumulation rates should increase the chances for bone burial and preservation.

The relative abundance of fossil localities among four major fluvial contexts is statistically similar (at the $p < 0.0001$ level) in all four formations (Fig. 4.4). Floodplain channels and major channels are the primary contexts for fossil preservation in all formations. There are nearly equal numbers of crevasse splay and floodplain localities in the Kamlial Formation, and floodplain localities are most numerous in the Dhok Pathan Formation. Together, these four fluvial sub-environments made up the paleo-landscape of the Siwalik alluvial system (see Fig. 4.2, Table 4.1; Fig. 24.4, this volume). Although the drainage areas for the major channels changed from mountain to lowland source areas (Behrensmeyer and Tauxe 1982; see also Chapter 2, this volume), it is remarkable that the distribution of fossil localities across sub-environments of the two different fluvial systems remained similar over millions of years.

Floodplain channels occur within sequences of finer-grained floodplain sediments (Fig. 4.2) and are distinct from major perennial drainage channels. The floodplain channels were more numerous and smaller in size (depth and width) than major channels and were bordered by channel-margin deposits (including levees) that graded laterally into extensive mudstone-dominated floodplains, with both well-drained and poorly drained soils developed between the channels. Floodplain channels served as tributaries or distributaries of the major channels and represent variable flow regimes, some of which were seasonal and probably ephemeral. Both kinds of channels include localities with channel-lag and abandoned-channel deposits, with the most productive fossil assemblages associated with fine-grained fills of large floodplain channels or the tops of major channels (Figs. 4.2, 4.3, 4.4, Tables 4.1, 4.2). This is a common pattern of preservation in fluvial depositional systems throughout the vertebrate fossil record (Behrensmeyer 1988).

In a monsoonal ecosystem, both active and abandoned channels (especially large ones), as well as the floodplains, would have been inundated during the wet season. As a result of rising floodwaters, it is likely that populations of larger terrestrial species moved to higher ground, possibly for months at a time. During falling water levels, a flush of new vegetation would have attracted terrestrial animals back into lower areas, where water remained available during the dry season. Thus, cycles of life and death modulated by seasonal climate and variable river flow helped to sustain the Miocene ecosystems and generate the Siwalik fossil record.

Variation in the frequency of fossil localities over 12 million years of Siwalik formations (Fig. 4.1) has implications for our record of mammal diversity over time. Changes through the sequence in the number of fossil localities influence the number of documented species (a sampling effect) and their associated first and last occurrences in the Siwalik record (Chapter 3, this volume). Within the spatial and temporal window provided by the Potwar Plateau, the changing proportions of major channel versus floodplain deposits (Figs. 4.4, 4.10) and the habitats they represent affect what we know about faunal composition, turnover, and evolution. For example, in the Kamlial and Nagri Formations, which are sand-dominated and have fewer productive localities than the mudstone-dominated Chinji and Dhok Pathan Formations, we expect that *actual* first occurrences of species could be significantly older than *observed* first occurrences. Likewise, actual last occurrences could be younger than observed last occurrences (Chapters 2, 3, and 26, this volume).

The prevalence of fossil localities in abandoned major channels and floodplain channels (Table 4.2, Fig. 4.4) throughout the Potwar Plateau record, along with the similar spectrum of environments sampled throughout that record, suggest that, in spite of the variation in number of localities among the different formations (Fig. 4.1), taphonomic biases resulting from depositional environment and mode of accumulation are generally uniform through all four formations, for both large and small mammals (Fig. 4.10). This is fortunate from the standpoint of paleoecological reconstructions because the record provides us with similar (i.e., "isotaphonomic") samples through time. Nevertheless, comparisons of taxa recovered

from different contexts (e.g., major channel and floodplain) show marked differences reflecting the preferred life habitats of the preserved species (see Fig. 4.9). For example, Locality Y639 in a major-channel fill has mostly aquatic vertebrate remains along with anthracothere, tragulid, and bovid fossils; anthracotheres and tragulids were likely forest browsers with an affinity for riparian habitats. In contrast, Locality Y699 in floodplain mudstones preserved artiodactyl, aardvark, rodent, and creodont fossils, but no aquatic vertebrate remains; the artiodactyl families indicate a range of closed- and open-canopy habitats.

Trends in taxonomic diversity versus specimen number show that floodplain localities preserve lower mammalian diversity than the other major contexts (Fig. 4.11). This contrast demonstrates why it is important to control for depositional setting in assessing biodiversity trends across space and through time. For example, using only floodplain channel or major channel localities to trace artiodactyl diversity provides taphonomically similar samples through time, which reduces environmental biases in the taxa that are represented. But including floodplain assemblages could introduce biases, such as better representation of carnivorous mammals than in other depositional environments (Fig. 4.10). Since floodplain localities represent only ~20% of our locality sample (and contribute less than 10% of catalogued specimens), it is likely that carnivore diversity, though impressive (Chapter 13, this volume), is under-sampled for the Siwalik faunal record as a whole.

PALEOCOMMUNITY RECONSTRUCTION

The different taphonomic pathways that characterize the Siwalik fossil record have important consequences for paleoecological interpretations of the faunas as well as the evolution of the component taxa. To characterize the Miocene faunal communities and compare them through time, we need to estimate original richness (number of species), species abundances, and functional diversity (e.g., dietary and size categories, substrate use), which are standard measures used to characterize present-day communities (Chapters 24, 25, and 26, this volume). Taphonomic processes and biases affect all these ecological variables, and knowledge about the preservation history of the Siwalik fossil assemblages helps us understand and account for these biases.

The number and abundance of species are critical features of community structure. These variables, as recorded at a fossil locality, are linked to sample size and to the preservability of the component taxa, depending in part on the original durability of their skeletal remains. Reconstructing the relative abundance of different species in a fossil assemblage is a prerequisite for inferring abundances (or rank abundance) of once-living populations and relies on accurate estimates of individuals belonging to each taxon. To do this, several methods have been applied to Siwalik fossil assemblages.

Taphonomic studies in modern ecosystems show that animal body size is generally correlated with the probability of preservation—durable remains of larger animals are more likely to survive early post-mortem taphonomic processes, increasing their potential for burial and fossilization (e.g., Behrensmeyer, Western, and Boaz 1979; Behrensmeyer, Kidwell, and Gastaldo 2000). In attritional land-surface assemblages, bones of mammals between 1–15 kg body weight are likely to be consumed by predators or destroyed by scavenging, and they also weather more rapidly than those of mammals >15 kg body weight (Behrensmeyer 1978; Behrensmeyer, Western, and Boaz 1979). Assuming that these fates also applied to the Siwalik fossil record, for the Dhok Pathan fauna we have estimated original abundance of species in a community by applying the inverse of the probability of preservation to their relative frequencies in the fossil assemblage (Badgley 1986a). If there are few individuals of a smaller-sized species (< 15 kg), which we assume had a lower probability of preservation, then we can estimate a higher abundance for this species in the original death assemblage. In contrast, a larger-sized species that is abundant in the fossil assemblage would have had a higher probability of preservation, and we estimate its original abundance as low to moderate in the original death assemblage. Dhok Pathan ungulates of 100–250 kg body weight (e.g., large antelopes, equids) had a relatively high probability of preservation, and we can infer that the actual relative abundances of these species in the original ecosystem likely were lower than their proportional abundances in the fossil assemblages. Estimates based on higher rates of destruction for mammals <15 kg characterize the Chinji and Nagri Formations but do not fit the pattern of greater representation of small artiodactyls in the Kamlial Formation (Fig. 4.8a). There was higher relative species richness and likely also larger populations of animals in this size group during the earlier part of the Siwalik sequence, and it is also possible that bone consumption by predators was less destructive than later, when hyenas became members of the paleocommunity.

The general pattern of lower representation of smaller body sizes across formations and environments (Fig. 4.8) does not include other types of bone concentrations, such as owl-pellet assemblages, accumulations in predator dens, or mass-death events. Notably, Siwalik micro-vertebrate localities represent taphonomic processes that preserved large numbers of delicate remains of animals <1 kg in body weight. Different recovery methods used for large and small vertebrates in the Siwalik record provide complementary datasets that help to overcome taphonomic biases against smaller species.

Research in modern mammal communities indicates that for body-size ranges subjected to similar pre-burial taphonomic processes, live-dead relative abundance fidelity can be high (Western and Behrensmeyer 2009). We can use taphonomic criteria to specify which subsets of our fossil assemblage data were affected by similar processes as a basis for judging whether relative abundances within these subsets reflect those of the original source communities. Clearly, macro- and micro-vertebrate assemblages are too different in this respect to provide credible estimates of rodent versus bovid abundances, but within each of these two categories, specific

body-size ranges could represent similar pre-burial taphonomic biases. Given the consistent representation of the 10–100 kg artiodactyls in all four depositional environments (Fig. 4.8b), it is likely that numbers of component taxa within this weight group reflect relative abundance in the original communities with moderate to high levels of live-dead fidelity. For example, the ratio of bovid to tragulid catalogue records for 10–100 kg taxa remains at ~3:1 through the Kamlial, Chinji, and Nagri formations but jumps to 16:1 in the Dhok Pathan Formation. This change likely reflects a community shift accompanying the increase of grassy floodplains and a decrease of near-aquatic habitats preferred by tragulids after ~8 Ma (Chapter 25, this volume).

Sample size (the number of localities per time interval) is uneven through the Siwalik fossil record (Fig. 4.1), as is typical of fossil records in general. Paleontologists have developed various ways to control for sample-size biases and their effects on species appearances and disappearances (Chapter 3, this volume). Sampling also controls what we can and cannot infer regarding species diversity and abundance. The larger the sample from a locality, the more likely it is that originally uncommon species will be present in the assemblage. For a combined sample of large to mega-mammals (excluding carnivorans), the rarefaction curves in Fig. 4.11 show positive correlations between numbers of specimens and identified taxa per locality. The curves level off at approximately 20–25 taxa per locality (~15 for floodplain localities), indicating that individual localities preserve only a fraction of the total Siwalik mammalian taxonomic diversity. Most localities record 15 or fewer taxa, however, which may be a reasonable subsample of contemporaneous species in the local community. In the combined sample for all formations, the total number of all large-mammal taxa including carnivorans is over 190, with a maximum of 70 and 71 taxa (excluding carnivorans) per 200 kyr interval in the Chinji and Dhok Pathan Formations, respectively (see Chapters 25 and 26, this volume). For small mammals (<1 kg), a separate analysis (not figured) showed considerably more variation in the number of taxa captured by increasing the number of specimens. This variation relates to changes in diversity over time (see Chapters 25 and 26, this volume), since rodent diversity drops sharply after 10.1 Ma, and there are many localities with fewer taxa that document this change. It is also likely that taphonomic processes responsible for the micromammal assemblages, e.g., different biological agents (predators) that formed the original concentrations, contributed to differences in the numbers of taxa represented for a given sample size.

Different spatial and temporal scales of taphonomic fidelity also affect the extent to which the Siwalik fossil assemblages preserve information about faunas in different subenvironments of the original alluvial ecosystem. Assemblages subjected to fluvial transport and reworking—whether in active channels or crevasse splays—could show lower fidelity to local habitats because skeletal material was mixed from different sources. If the habitats changed over time (e.g., drying ponds in an abandoned-channel swale), then taxa with different ecologies could be mixed due to time-averaging. On the other hand, transported and time-averaged assemblages record taxa from different areas of the drainage basin and should have higher fidelity to species richness at the landscape scale. Autochthonous assemblages—whether from abandoned channels or floodplains—record associations at the scale of home ranges of individual animals—for example, on the order of 10–100 km² for large and mega-sized mammals (10–5000 kg—including large giraffoids and proboscideans; Morgan et al. 2009). In combination with paleoecological information about individual species (e.g., from stable carbon isotopes or morphometrics, Chapter 24, this volume), taphonomic analysis of the faunal assemblages presented in this chapter indicates that variation in taxonomic composition of most Siwalik fossil assemblages generally reflects fidelity to different habitats and resource gradients in the original ecosystem (e.g., Morgan et al. 2009; Behrensmeyer et al. 2005; Figs. 4.9, 4.10; Figs. 24.8, 24.9, and 24.10, this volume). Plots of the generalized ecology of fossil animals, inferred from functional anatomy and based on catalogued specimen numbers in taxa from 316 documented localities, show their relative abundance (Fig. 4.12a) and distribution across the four depositional environments (Fig. 4.12b; see also Table 4.3g). These patterns support ecologi-

cal live-dead fidelity in different Siwalik fluvial habitats, with interesting variations, such as scansorial mammals being less common in wetter major channel environments than in floodplain channels. These patterns also underscore the importance of major and floodplain channels for preserving animals with terrestrial adaptations.

CONCLUSION

Documenting and interpreting the depositional context and taphonomy of the Siwalik fossil localities establish the essential foundation for defining strengths and limitations in our faunal and environmental data. The fossil record did not provide us with random samples of the original Miocene ecosystems through time and across space. Rather, varying physical and biological processes generated different types of fossil assemblages that affect what we can say about the Miocene faunas in important ways—primarily though environmental and body size-related controls on taxonomic representation at locality to landscape and formation scales.

Excavated localities as well as systematic surface collections provide essential information on taphonomic features not evident in collections from more traditional paleontological survey

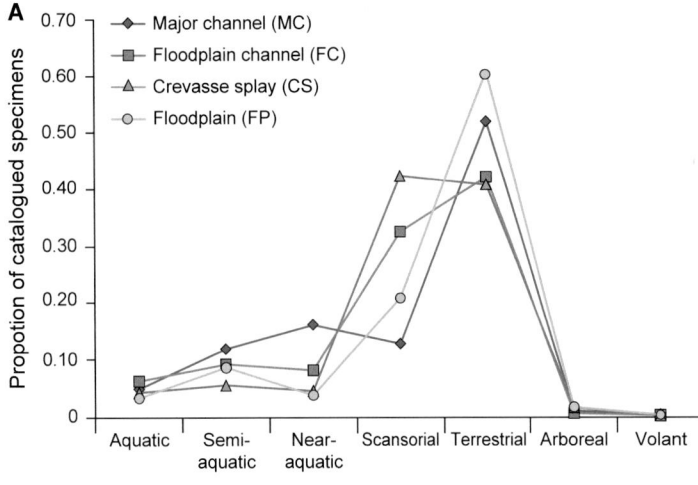

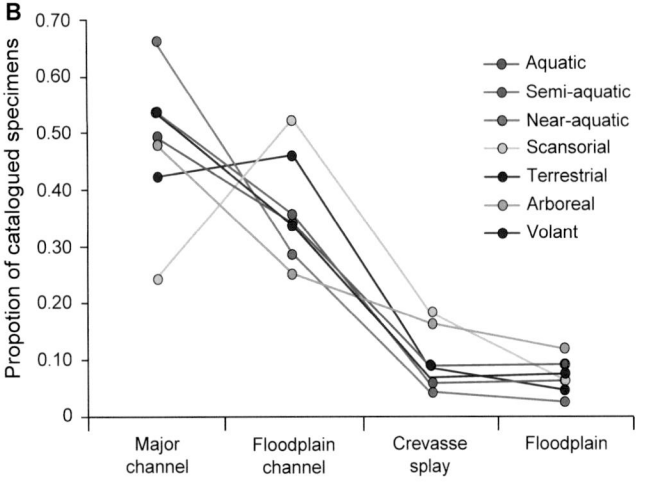

Figure 4.12. Distribution of fossil animals across depositional environments, based on catalogued specimen numbers and a taxon's generalized ecology, inferred from its functional anatomy (see Table 4.3g). A. Plot of the proportion of catalogued specimens in four depositional environments, for each ecological category. B. Plot of the proportion of catalogued specimens in each ecological category, across the four depositional environments, from wetter on the left to drier on the right. Aquatic = invertebrates (mollusks) and fish; semi-aquatic = snakes, crocodiles, turtles; near-aquatic = anthracotheres, tragulids; scansorial = rodents and other small mammals; arboreal = primates and lizards; volant = birds; all remaining taxa = terrestrial (see Table 4.3g). The distribution pattern of arboreal taxa suggests that there were trees on the floodplain, and low numbers of aquatic taxa indicate lack of perennial water bodies there as well.

methods. Parallel efforts to recover both macro- and micro-vertebrate records, which represent different accumulation processes as well as collecting strategies, offer complementary taphonomic and faunal data for reconstructing both components of the Siwalik vertebrate communities. Our small sample of unusual fossil concentrations (associated skeletal remains, burrow, pod, and mass-death assemblages) reveals interesting differences from preservation in the four dominant modes of accumulation in major and floodplain channels, crevasse splays, and floodplains. All types of preservation contribute to our story, and consistency through time in these depositional settings represents a fortuitous balance of skeletal input and sediment accumulation that resulted in the rich Siwalik fossil record.

Taphonomic study reveals changes in the quantity and quality of fossil preservation over time; these were controlled primarily by large-scale physical processes—the shifting proportions of channel versus floodplain environments accompanied by increased rates of sediment accumulation over time. It allows us to distinguish the roles of biological versus physical processes in shaping our fossil record and to reconstruct processes affecting mortality and bone destruction versus survival that operated across different environments through millions of years.

Taphonomic analyses described here have four major implications for community reconstruction of the Siwalik record of the Potwar Plateau. First, consistent representation of similar environments of preservation through the four formations of our fossil record provides generally similar (isotaphonomic) samples of vertebrate faunas for paleocommunity analysis and artistic reconstructions (Chapters 24, 25, and 26, this volume). Second, our finest level of ecological resolution for most localities is on the order of areas of <10 km^2 and hundreds to thousands of years. Fossil assemblages that formed over centuries to millenia of time-averaged attritional remains, combined with those that experienced fluvial reworking and transport of skeletal remains, provide species associations representing different habitat types and paleocommunity censuses over significant areas of the Siwalik alluvial plain. Third, across the body-size spectrum of mammals in the original Miocene ecosystems, species of intermediate size (50–250 kg, including equids after 10.8 Ma) had the highest probability of preservation in all depositional environments. Taxa in this size group also likely were dominant elements of the mammalian fauna, and their taxonomic diversity and relative abundance thus have greater fidelity than for smaller or larger species. Fourth, the large number of individual localities with fossils representing different environments and degrees of time-averaging combine to provide samples of Siwalik large and mega-mammals that likely approached their original Miocene diversity in the western part of the sub-Himalayan foreland basin. These insights all contribute to the paleoenvironmental and paleocommunity analyses presented in Chapters 24, 25, and 26 of this volume.

Comprehensive documentation and analysis of Siwalik taphonomy and depositional context provide evidence for how fluvial systems preserve records of terrestrial ecosystems through time. Tectonic forces driving foreland basin subsidence and the rise of the Himalaya Mountains, along with climate change and biological processes across a range of scales, combined to generate a rich vertebrate fossil record. Future comparisons with other fluvial sequences could help to distinguish the varying roles of tectonics, climate, and biology in preserving these valuable chronicles of terrestrial life throughout the Phanerozoic.

ACKNOWLEDGMENTS

We thank the Pakistan Geological Survey and its staff, as well as many helpful assistants from the local communities of the Potwar Plateau, for field support that contributed to documenting the paleoenvironments and taphonomy of the Siwalik fossil localities. Bill Keyser provided field support in stratigraphic section measuring as well as carefully engineered Jacob's staffs. We gratefully acknowledge the pioneering work of S. Mahmood Raza on Chinji Formation taphonomy. Excavations at Nagri Formation Locality Y311 involved many of our Pakistani and other colleagues, and we especially thank Jay Kelley, Fred Grady, and Will Downs for their contributions to excavation-based taphonomic data summarized in this chapter, and Jana Burke and Nora Loughlin for their preliminary analyses of these data.

References

Andrews, P. 1990. *Owls, Caves and Fossils: Predation, Preservation and Accumulation of Small Mammal Bones in Caves, with an Analysis of the Pleistocene Cave Faunas from Westbury-Sub-Mendip, Somerset, UK*. Chicago: University of Chicago Press.

Andrews, P., C. P. Grove, and J. F. M. Horne. 1975. Ecology of the lower Tana River floodplain (Kenya). *Journal of the East African Natural History Society and National Museum* 151:1–31.

Aslan, A., and A. K. Behrensmeyer. 1996. Taphonomy and time resolution of bone assemblages in a contemporary fluvial system: the East Fork River, Wyoming. *Palaios* 11(5): 411–421.

Badgley, C. 1986a. Counting individuals in mammalian fossil assemblages from fluvial environments. *Palaios* 1(3): 328–338.

Badgley, C. 1986b. Taphonomy of mammalian fossil remains from Siwalik rocks of Pakistan. *Paleobiology* 12(2): 119–142.

Badgley, C., W. Downs, and L. J. Flynn. 1998. Taphonomy of small-mammal fossil assemblages from the Middle Miocene Chinji Formation, Siwalik Group, Pakistan. *National Science Museum Monographs* 14: 145–166.

Badgley, C. E. 1982. *Community reconstruction of a Siwalik mammalian assemblage*. PhD diss., Yale University.

Behrensmeyer, A. K. 1978. Taphonomic and ecologic information from bone weathering. *Paleobiology* 4(2): 150–162.

Behrensmeyer, A. K. 1982. Time resolution in fluvial vertebrate assemblages. *Paleobiology* 8(3): 211–227.

Behrensmeyer, A. K. 1987. Miocene fluvial facies and vertebrate taphonomy in northern Pakistan. In *Recent Developments in Fluvial Sedimentology*, edited by F. G. Ethridge, R. M. Flores, and M.D. Harvey, 169–176. Tulsa, OK: Society for Economic Paleontology and Mineralogy, Special Publication 39.

Behrensmeyer, A. K. 1988. Vertebrate preservation in fluvial channels. *Palaeogeography, Palaeoclimatology, Palaeoecology* 63(1–3): 183–199.

Behrensmeyer, A. K. 1991. Terrestrial vertebrate accumulations. *Taphonomy: Releasing the Data Locked in the Fossil Record*, edited by P. Allison, P. and D. E. G. Briggs, 291–335. New York: Plenum.

Behrensmeyer, A. K., C. E. Badgley, J. C. Barry, M. Morgan, and S. M. Raza. 2005. The paleoenvironmental context of Siwalik Miocene vertebrate localities. In *Interpreting the Past: Essays on Human, Primate, and Mammal Evolution in Honor of David Pilbeam*, edited by Daniel E. Lieberman, Richard J. Smith, and Jay Kelley, 48–62. Boston: Brill.

Behrensmeyer, A. K., A. L. Deino, A. Hill, J. Kingston, and J. J. Saunders. 2002. Geology and chronology of the Middle Miocene Kipsaramon fossil site complex, Muruyur Beds, Tugen Hills, Kenya. *Journal of Human Evolution* 42(1–2): 11–38.

Behrensmeyer, A. K., K. D. Gordon, and G. T. Yanagi. 1986. Trampling as a cause of bone surface damage and pseudo-cutmarks. *Nature* 319(6056): 768–771.

Behrensmeyer, A. K., K. D. Gordon, and G. T. Yanagi. 1989. Non-human bone modification in Miocene fossils from Pakistan. In *Bone Modification*, Proceedings of First International Conference on Bone Modification, edited by R. Bonnichsen and M. H. Sorg, 99–120. Orono, ME: Center for the Study of the First Americans.

Behrensmeyer, A. K., and R. W. Hook. 1992. Paleoenvironmental contexts and taphonomic modes in the terrestrial fossil record. In *Terrestrial Ecosystems Through Time*, edited by A. K. Behrensmeyer, J. D. Damuth, W. A. DiMichele, R. Potts, H.-D. Sues, and S. L. Wing, 15–136. Chicago: University of Chicago Press.

Behrensmeyer, A. K., S. M. Kidwell, and R. A. Gastaldo. 2000. Taphonomy and paleobiology. *Paleobiology* 26(S4): 103–147.

Behrensmeyer, A. K., and J. H. Miller. 2012. Building links between ecology and paleontology using taphonomic studies of recent vertebrate communities. In *Paleontology in Ecology and Conservation*, edited by J. Louys, 69–91. Berlin: Springer.

Behrensmeyer, A. K., J. Quade, T. E. Cerling, J. Kappelman, I. A. Khan, P. Copeland, L. Roe, et al. 2007. The structure and rate of late Miocene expansion of C$_4$ plants: Evidence from lateral variation in stable isotopes in paleosols of the Siwalik Group, northern Pakistan. *Geological Society of America Bulletin* 119(11–12): 1486–1505.

Behrensmeyer, A. K., and S. M. Raza. 1987. A procedure for documenting fossil localities in Siwalik deposits of northern Pakistan. *Memoirs of the Geological Survey of Pakistan* 9:65–69.

Behrensmeyer, A. K., and L. Tauxe. 1982. Isochronous fluvial systems in Miocene deposits of northern Pakistan. *Sedimentology* 29(3): 331–352.

Behrensmeyer, A. K., D. Western, and D. E. D Boaz. 1979. New perspectives in vertebrate paleoecology from a recent bone assemblage. *Paleobiology* 5(1): 12–21.

Behrensmeyer, A. K., B. J. Willis, and J. Quade. 1995. Floodplains and paleosols in the Siwalik Neogene and Wyoming Paleogene: Implications for the taphonomy and paleoecology of faunas. *Palaeogeography, Palaeoclimatology, Palaeoecology* 115(1–4): 37–60.

Berger, J. 1983. Ecology and catastrophic mortality in wild horses: Implications for interpreting fossil assemblages. *Science* 220(4604): 1403–1404.

Burke, J. E., A. K. Behrensmeyer, C. Badgley, J. C. Barry, and S. K Lyons. 2014. Assessing the impact of time-averaging on a Miocene vertebrate fauna from northern Pakistan. Abstract Book, Paleontological Society Special Publication, Vol. 13:41, 10th North American Paleontological Convention, Gainesville, FL.

Colombi, C. E., R. R. Rogers, and O. A. Alcober. 2012. Vertebrate taphonomy of the Ischigualasto Formation. *Journal of Vertebrate Paleontology* 32(suppl. 1): 31–50.

Crosmary, W. G., M. Valeix, H. Fritz, H. Madzikanda, and S. D. Côté. 2012. African ungulates and their drinking problems: Hunting and predation risks constrain access to water. *Animal Behaviour* 83(1): 145–153.

De Ruiter, D. J., and L. R. Berger. 2000. Leopards as taphonomic agents in dolomitic caves—Implications for bone accumulations in the hominid-bearing deposits of South Africa. *Journal of Archaeological Science* 27(8): 665–684.

Faith, J. T., C. W. Marean, and A. K. Behrensmeyer. 2007. Carnivore competition, bone destruction, and bone density. *Journal of Archaeological Science* 34(12): 2025–2034.

Fernandez-Jalvo, Y., and P. Andrews. 2016. *Atlas of Taphonomic Identifications: 1001+Images of Fossil and Recent Mammal Bone Modification*. London: Springer

Fosse, P., N. Selva, A. Wajrak, J. B. Fourvel, S. Madelaine, M. Esteban-Nadal, J. Yravedra, J. P. Brugal, A. Prucca, and G. Haynes. 2012. Bone modification by modern wolf (*Canis lupus*): A taphonomic study from their natural feeding places. *Journal of Taphonomy* 10(3–4): 197–217.

Freeman, K. H., and L. A. Colarusso. 2001. Molecular and isotopic records of C4 grassland expansion in the late Miocene. *Geochimica et Cosmochimica Acta* 65(9): 1439–1454.

Hanson, C. B. 1980. Fluvial taphonomic processes: Models and experiments. In *Fossils in the Making*, edited by A. K. Behrensmeyer and A. Hill, 156–181. Chicago: University of Chicago Press.

Haynes, G. 1985a. On watering holes, mineral licks, death, and predation. In *Environments and Extinctions: Man in Late Glacial North America*, edited by J. Mead and D. Meltzer, 53–71. Orono, ME: Center for the Study of Early Man.

Haynes, G. 1985b. Age profiles in elephant and mammoth bone assemblages. *Quaternary Research* 24(3): 333–345.

Hill, A. 1989. Bone modification by modem spotted hyenas. In *Bone Modification*, Proceedings of First International Conference on Bone Modification, edited by R. Bonnichsen and M. H. Sorg, 169–178. Orono, ME: Center for the Study of the First Americans.

Hoorn, C., T. Ohja, and J. Quade. 2000. Palynological evidence for vegetation development and climatic change in the sub-Himalayan zone (Neogene, central Nepal). *Palaeogeography, Palaeoclimatology, Palaeoecology* 163(3–4): 133–161.

Karp, A. T., A. K. Behrensmeyer, and K. H. Freeman, K.H., 2018. Grassland fire ecology has roots in the Late Miocene. *Proceedings of the National Academy of Sciences* 115(48): 12130–12135.

Kruuk, H. 1972. Surplus killing by carnivores. *Journal of Zoology* 166(2): 233–244.

Lansing, S. W., S. M. Cooper, E. E. Boydston, and K. E. Holekamp. 2009. Taphonomic and zooarchaeological implications of spotted hyena (*Crocuta crocuta*) bone accumulations in Kenya: A modern behavioral ecological approach. *Paleobiology* 35(2): 289–309.

Loughlin, N., A. K. Behrensmeyer, and C. Badgley. 2018. Deciphering carnivore impact via taphonomic analysis of surface and excavated bones in a Miocene fossil assemblage from the Siwalik sequence of Pakistan. Society of Vertebrate Paleontology Meeting, October, Albuquerque, NM.

Lyman, R. L., and C. Lyman. 1994. *Vertebrate Taphonomy*. Cambridge: Cambridge University Press.

Marean, C. W., L. M. Spencer, R. J. Blumenschine, and S. D. Capaldo. 1992. Captive hyaena bone choice and destruction, the Schlepp effect and Olduvai archaeofaunas. *Journal of Archaeological Science* 19(1): 101–121.

Miller, J. H., 2012. Spatial fidelity of skeletal remains: Elk wintering and calving grounds revealed by bones on the Yellowstone landscape. *Ecology* 93(11): 2474–2482.

Morgan, M. E. 1994. Paleoecology of Siwalik Miocene hominoid communities: Stable carbon isotope, dental microwear, and body size analyses. PhD diss., Harvard University.

Morgan, M. E., A. K. Behrensmeyer, C. Badgley, J. C. Barry, S. Nelson, and D. Pilbeam. 2009. Lateral trends in carbon isotope ratios reveal a Miocene vegetation gradient in the Siwaliks of Pakistan. *Geology* 37(2): 103–106.

Papa, L. M. D., L. D. Santis, and J. Togo, 2017. The fossorial faunal record at the Beltrán Onofre Banegas-Lami Hernandez archaeological site (Santiago del Estero province, Argentina): A taphonomic approach. In *Zooarchaeology in the Neotropics*, edited by M. Mondini, A. Munoz, and P. Fernandez, 137–156. Edinburgh: Springer Cham.

Potts, R., R. Dommain, J. Moerman, A. K. Behrensmeyer, A. Deino, R. B. Owen, E. Beverly, et al. 2020. Increased ecological resource variability during a critical transition in hominin evolution. *Science Advances* 6(43): 1–14. www.science.org/doi/10.1126/sciadv.abc8975.

Raza, S. M. 1983. Taphonomy and paleoecology of Middle Miocene vertebrate assemblages, southern Potwar Plateau, Pakistan. PhD diss., Yale University.

Reed, D. N. 2007. Serengeti micromammals and their implications for Olduvai paleoenvironments. In *Hominin Environments in the East African Pliocene: An Assessment of the Faunal Evidence*, edited by R. Bobe, Z. Alemseged, A. K. Behrensmeyer, 217–255. Dordecht: Springer.

Rogers, R. R. 2005. Fine-grained debris flows and extraordinary vertebrate burials in the Late Cretaceous of Madagascar. *Geology* 33(4): 297–300.

Rogers, R. R., D. A. Eberth, and A. R. Fiorillo, eds. 2010. *Bonebeds: Genesis, Analysis, and Paleobiological Significance*. Chicago: University of Chicago Press.

Sheppe, W., and T. Osborne. 1971. Patterns of use of a flood plain by Zambian mammals. *Ecological Monographs* 41(3): 179–205.

Shipman, P. 1993. *Life History of a Fossil: An Introduction to Taphonomy and Paleoecology*. Cambridge, MA: Harvard University Press.

Srivastava, G., K. N. Paudayal, T. Utescher, and R. C. Mehrotra. 2018. Miocene vegetation shift and climate change: Evidence from the Siwalik of Nepal. *Global and Planetary Change* 161:108–120.

Tauxe, L., V. D. Kent, and N. D. Opdyke. 1980. Magnetic components contributing to the NRM of Middle Siwalik red beds. *Earth and Planetary Science Letters* 47(2): 279–284.

Van Valkenburgh, B. 1999. Major patterns in the history of carnivorous mammals. *Annual Review of Earth and Planetary Sciences* 27(1): 463–493.

Voorhies, M. R. 1969. Taphonomy and population dynamics of an Early Pliocene vertebrate fauna, Knox County, Nebraska. *University of Wyoming Contributions to Geology*, Special Paper 1:1–69.

Weaver, L. N., D. J. Varricchio, E. J. Sargis, M. Chen, W. J. Freimuth, and G. P. Wilson Mantilla. 2021. Early mammalian social behaviour revealed by multituberculates from a dinosaur nesting site. *Nature Ecology & Evolution* 5(1): 32–37.

Western, D., and A. K. Behrensmeyer. 2009. Bone assemblages track animal community structure over 40 years in an African savanna ecosystem. *Science* 324(5930): 1061–1064.

Willis, B. J., and A. K. Behrensmeyer. 1994. Architecture of Miocene overbank deposits in northern Pakistan. *Journal of Sedimentary Research* B64(1): 60–67.

STABLE ISOTOPES AS A RECORD OF ECOLOGICAL CHANGE IN THE SIWALIK GROUP OF PAKISTAN

THURE E. CERLING, ANNA K. BEHRENSMEYER, JOHN C. BARRY, SHERRY NELSON, MICHÈLE E. MORGAN, AND JAY QUADE

INTRODUCTION

The expansion of summer grasslands (C_4 grasses) in the late Cenozoic certainly ranks among the most important ecological developments in the evolution of vascular plants since they first appeared in the Silurian Period ~420 million years ago (Ma). Until sometime in the Cenozoic, all vascular plants used the Rubisco enzyme to fix carbon dioxide during what is termed C_3 photosynthesis. During the mid-Cenozoic, a new form of photosynthesis—C_4 photosynthesis—first evolved from C_3 plants in the form of a series of biochemical changes that allowed plants to concentrate CO_2 around the carboxylating enzyme Rubisco (Sage 2004). Today, C_3 plants include nearly all trees and shrubs, as well as grasses favored by cool growing seasons, such as at high latitudes and high elevations, or in shaded settings. C_4 plants are dominated by grasses specially adapted to warm summers and are found today mainly in the mid-latitudes to tropics. The latest evidence points to two key steps in the evolution of C_4 from C_3 plants. These include the first appearance of C_4 plants in the mid-Cenozoic, as suggested by molecular clock studies and fossil plants (Sage 2004), followed by the globally transgressive radiation of C_4 biomass, starting at the equator 10 to 8 Ma (Uno et al. 2011; Polissar et al. 2019) and expanding across the mid-latitudes 8 to 4 Ma (Cerling et al. 1997).

The isotopic records from the Siwalik Group sediments occupy center stage in our understanding of the dramatic late Miocene (10–4 Ma) radiation of C_4 plants. Two groundbreaking studies, one by Cerling (1984) and the second by Lee-Thorp and van der Merwe (1987), established the feasibility of using the carbon isotopic composition of carbonate in ancient soils and fossil teeth to reconstruct the proportion of C_3 versus C_4 plants on paleolandscapes and in the diets of fossil mammals, respectively. At that time, sediments and fossils of the Siwalik Group presented an ideal testing ground for these new methods. The Siwalik Group is exposed widely across the northern subcontinent, including Pakistan, and encompasses one of the most continuous, highest resolution, and best-dated terrestrial sedimentary records on Earth. The Siwalik Group had been under study for over a decade prior to 1989, and the

age of fossils and soil carbonates were by then well dated, setting the stage for subsequent isotopic studies.

In this chapter, we summarize and review the evidence from the Siwalik Group as a key example for early testing of stable isotopes in paleosols and fossil tooth enamel. We also describe the basis for the stable isotope analyses of pedogenic (soil and paleosol) carbonate and the rationale for diet reconstruction of mammals using stable isotopes.

The natural variation in carbon and oxygen isotopes is small and we use standard notation methods to discuss these differences. We use the "permil" notation, whereby

$$\delta^{13}C = (R_{sample}/R_{standard} - 1) \star 1{,}000,$$

and

$$\delta^{18}O = (R_{sample}/R_{standard} - 1) \star 1{,}000,$$

where R_{sample} and $R_{standard}$ are the $^{13}C/^{12}C$ (or $^{18}O/^{16}O$) ratios in the sample and standard, respectively. In this chapter, V-PDB is the standard used for both isotopes (Coplen 1988).

Isotope enrichment $\varepsilon^{\star}$ describes the isotope difference between two related phases:

$$\varepsilon^{\star} = ((R_a/R_b) - 1) \star 1{,}000 = ((1{,}000 + \delta_a)/(1{,}000 + \delta_b) - 1) \star 1{,}000.$$

CO_2 AND $\delta^{13}C$ OF THE ATMOSPHERE: MIOCENE TO PRESENT

The concentration history of carbon dioxide in the Earth's atmosphere during the Cenozoic still has high uncertainty, with alkenone barometry and boron-isotope barometry giving conflicting results. Pagani, Arthur, and Freeman (1999) and Pagani, Freeman, and Arthur (1999) favor low CO_2 concentrations (<350 parts per million by volume [ppmV]) throughout most of the Miocene, although recent revisions suggest that these early estimates using alkenone barometry were too low and that CO_2 concentrations in the early Miocene may have been considerably higher (Zhang et al. 2013). Reevaluations of both the alkenone and boron-isotope proxies suggest an important decline in atmospheric CO_2 between about 6 and 8 Ma (Herbert et al. 2016), possibly passing through the 500 ppmV threshold during this interval. Holbourn et al. (2018) propose a CO_2 shift around 7 Ma. Polissar et al. (2019) examined leaf waxes as an indicator of paleoaridity in northwest and northeast Africa and concluded that CO_2 decline was the likely driver of C_4 expansion globally beginning around 10 Ma rather than aridity as had been proposed by other research groups (e.g., Pagani et al. 2005). The proposal by Cerling, Wang, and Quade (1993) and Cerling et al. (1997) that the global C_4 expansion, first recognized in the Siwaliks by Quade, Cerling, and Bowman (1989a), was related to a decline in atmospheric CO_2 concentration through a critical threshold of about 500 ppmV remains unsettled.

It is now well recognized that the $\delta^{13}C$ value of the modern atmosphere has changed since the beginning of the Industrial Revolution. Thus, the isotope values of modern samples must be adjusted to the preindustrial value of the atmosphere (e.g., Cerling and Harris 1999; Cerling et al. 2015) in order to compare fossil ecosystems to a constant reference condition (the modern $\delta^{13}C$ value for atmospheric CO_2 has changed by 2‰ in the past 200 years). It is less well appreciated that the preindustrial $\delta^{13}C$ ratio of the atmosphere also changed significantly in the Miocene compared to the Pliocene and the Pleistocene. The $\delta^{13}C$ value for the preindustrial atmosphere was about −6.3‰, according to the ice-core records of Francey et al. (1999); compiled studies of marine foraminifera (e.g., Zachos et al. 2001) show that the Miocene ocean-atmosphere system was quite different from the modern one. Cerling (1997) attributed this change, in part, to the greater abundance of global C_4 biomass in the Plio-Pleistocene compared to the earlier C_3-dominated world. The overall effect is that the $\delta^{13}C$ of the atmosphere was essentially the same from the preindustrial atmosphere to 6.5 Ma; it was 1‰ enriched in ^{13}C from 6.5 to 12.5 Ma; it was 1.5‰ enriched in ^{13}C from 12.5 to 17.0 Ma; and it was 1‰ enriched in ^{13}C from 17.0 to 20.0 Ma (Passey et al. 2002; Cerling et al. 2010; Tipple, Meyers, and Pagani 2010). These enrichments have important implications for the interpretation of the contribution of C_4 biomass to diets of any mammals or to ecosystem biomass prior to 6.5 Ma when using the $\delta^{13}C$ of tooth enamel or of pedogenic carbonate, respectively, and therefore we consider them in our analysis below.

The change in the $\delta^{13}C$ value of the atmosphere was a "baseline shift" because photosynthesis uses atmospheric CO_2 as the basis for structural plant carbon. This shift affects the interpretations of isotope ratios of plant carbon (e.g., bulk carbon, biomarkers); soil carbonate whose CO_2 is ultimately derived from plant carbon; tooth enamel carbon, which is derived from animal metabolism based on plant consumption; and, likewise, any other material whose $\delta^{13}C$ values are related to the coeval atmosphere. In the discussion below, we use the preindustrial value of −6.3‰ as the reference value and recognize that plant and animal tissue values have progressively decreased since the beginning of the Industrial Revolution in ca. 1750. The shift has been accelerating in the past two decades

and the $\delta^{13}C$ of the atmosphere is now almost 2‰ lower than the preindustrial value of −6.3‰. Modern plants or animal tissues collected in the past 40 or more years need to be corrected to the preindustrial atmospheric value to account for this effect (see Cerling and Harris 1999; Cerling et al. 2015).

C_3 AND C_4 PHOTOSYNTHESIS

Today there are two principal photosynthetic pathways used by plants: the C_3 and C_4 photosynthetic pathways; a third pathway, CAM metabolism, makes up a minor proportion of net primary productivity in most ecosystems and will not be discussed further, unless needed. Both the C_3 and C_4 pathways use the Rubisco enzyme to convert CO_2 to photosynthate products; the difference is that the C_3 pathway operates in "open conditions" at the leaf level (i.e., isotope discrimination is allowed), whereas the C_4 pathway operates under "closed" conditions at the leaf level (isotope discrimination not allowed; Ehleringer and Monson 1993; Tipple and Pagani 2007). The net result is that the $\delta^{13}C$ value of C_3 plants ranges between about −30 to −25‰, with higher values being found in more xeric conditions and the more negative values in mesic conditions (Cerling et al. 1997). Closed-canopy conditions can result in values even more negative than −30‰ (van der Merwe and Medina 1989; Cerling, Hart, and Hart 2004; Blumenthal et al. 2016). C_4 plants have $\delta^{13}C$ values ranging from about −15‰ to −10‰, with more negative values being associated with more xeric conditions and the more positive values being associated with more mesic conditions (Cerling, Harris, and Passey 2003). Thus, the ^{13}C-isotope effect with respect to water stress is opposite for C_3 and C_4 plants, respectively, and different mixing lines between C_3 and C_4 plants are expected for xeric versus mesic conditions (see Cerling et al. 2015).

C_4 photosynthesis is a response to the low CO_2 concentrations in the atmosphere, which prevailed in the Neogene (Ehleringer and Monsoon 1993; Ehleringer, Cerling, and Helliker 1997); C_4 plants have $\delta^{13}C$ values between about −11 and −15‰ (Cerling et al. 1997). Photorespiration occurs at low atmospheric CO_2 conditions in C_3 plants, and this process takes place with increasing rates relative to photosynthesis at elevated temperatures. The temperature at which photorespiration exceeds photosynthesis is known as the *cross-over temperature*, and the competition between C_3 and C_4 photosynthesis is related to both leaf temperature and CO_2 concentrations. C_4 photosynthesis occurs largely independent of atmospheric CO_2 concentrations because C_4 plants have higher intercellular concentrations of CO_2 than the ambient atmosphere due to a concentration mechanism. At concentrations above approximately 500 to 600 ppm V CO_2, C_3 photosynthesis should be favored under virtually all conditions amenable to photosynthesis on Earth.

For both C_3 and C_4 plants, different plant parts may vary in $\delta^{13}C$ values (e.g., Codron et al. 2005), and likewise (as discussed above) there are environmental factors that influence the final $\delta^{13}C$ value of plants. Thus, for the fossil record, reconstructing the precise $\delta^{13}C$ of plants contributing to an ecosystem or to dietary intake is challenging; however, the clear ~12 to

15‰ difference between C_3 and C_4 plants makes the C_3 and C_4 distinction a very important tracer of past ecosystems so that the small 1 to 2‰ differences related to other environmental factors is of minor significance.

THE SIWALIK GROUP

The Siwalik Group refers to a belt of Neogene-age, mainly fluvial sedimentary rocks that extends from Pakistan to eastern India. The Siwalik Group found on the Potwar Plateau in Pakistan spans about 17.5 myr (18–0.5 Ma) and is up to 5 km in thickness (Johnson et al. 1982). It is composed of fluvial sediments deposited by large river systems ancestral to the modern Ganges and Brahmaputra, and in the case of Pakistan, the Indus River or several of its large paleo-tributaries (Behrensmeyer and Tauxe 1982). Regional shortening in response to Indo-Asian collision starting in the mid?-Eocene led to the development of the foreland basin that contains the Siwalik Group. The Siwalik Group itself represents the most recent infilling of that evolving basin, which is derived from a wide range of bedrock types eroded from the ancestral Himalaya to the north. The stratigraphy of the Siwalik Group is reviewed in Chapter 2 of this volume. For the purposes of this chapter, the Siwalik Group encompasses five formations from base to top: Kamlial Formation (18–14 ka), Chinji Formation (14–11.5 Ma), Nagri Formation (11.5–9.5 Ma), Dhok Pathan Formation (9.5–3.5 Ma), and the Soan Formation (or "Upper Siwaliks"; 3.5 to 0.5 Ma).

SIWALIK PALEOSOLS

The Siwalik Group is composed of two main textural facies: sand-filled, large-scale channel deposits and clay and silt-dominated floodplain deposits. During pauses in deposition, both facies were subjected to alterations that formed soils. Soils develop through physical and chemical alteration of surface sediments and rock. Soils require extended surface stability to develop, and do so during pauses in deposition in the geologic record. After burial and isolation from further surface alteration, soils are referred to as *paleosols*. Jenny (1994) suggested that modern soil properties were related to climate, biota, relief, parent material, and time (with the implication that others factors, too, might be important). Bedrock in the Himalaya is dominated by silicate rocks, but marble, limestone, and dolostone is also widespread. The alluvial derivatives of these rocks therefore contain significant detrital carbonate in the 3–20% range. Upon weathering of both the silicate but especially the detrital carbonate fraction, significant Ca^{2+}, Mg^{2+}, and HCO_3^- is dissolved in soil water. In drier climates such as characterize the modern Gangetic and Indus floodplains, evapotranspiration exceeds infiltration, and as a result low-Mg calcite accumulates in the lower half of paleosols, now and throughout the Siwalik paleosol record (Fig. 5.1).

Also associated with soil formation is the neo-formation of clay minerals and translocation of elements downwards in the soil profile, as well as changes in oxidation state that may assist in translocating metals such as iron and manganese. This en-

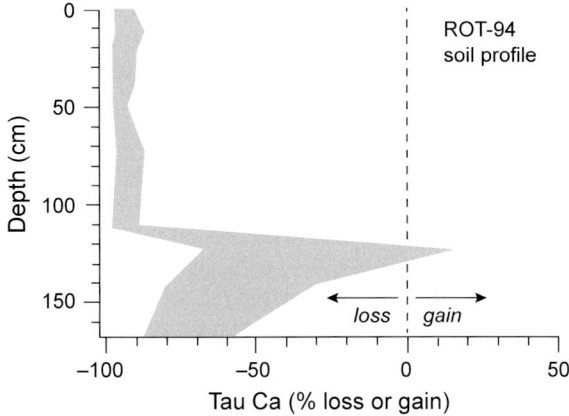

Figure 5.1. Estimated net loss or gain of calcium ("Tau Ca") from paleosol ROT-194 using the method of Brimhall and Dietrich (1987) and Chadwick et al. (1990). Hatched area shows range of leaching assuming siltstone parent material (± 1 sigma for Ca and Ti). CaO comprises 10.5 ± 4.2% of parent material siltstone ($n = 27$). Figure from Quade et al. (1995).

semble of processes leads to a diagnostic set of paleosol features recognizable both in hand specimen and thin section (Fig. 5.2). These include (1) massive, unbedded appearance, (2) distinct zone in the upper 50–100 cm of paleosols that is completely leached of detrital carbonate (see Fig. 5.1) and depleted in other cations such as Mg and Na, (3) generally reddish to orangish colors often with some local gray mottling, (4) fine-grained matrix due to the vertical translocation of clays, (5) frequent presence of pellet-size secondary Fe-Mn nodules, and (6) transitional lower boundaries and abrupt upper contacts with overlying sediment. Nodular (2–5 cm) carbonate is almost always present in paleosols developed on fine-grained floodplain sediment (Fig. 5.2) and sparse to absent where developed in sandy channel deposits. The early, pedogenic origin of carbonate nodules is verified by their abundance in contemporaneous channel deposits at all levels in the Siwalik Group, where they were reworked during bank collapse that exposed slightly older paleosols (Fig. 5.3a).

Carbon and Oxygen Isotope Values of Soil Carbonate

Systematic studies presented in Cerling (1984), Cerling et al. (1989) and Quade, Cerling, and Bowman (1989b) show that the $\delta^{13}C$ and $\delta^{18}O$ values of pedogenic carbonate can record aspects of vegetation and climate, respectively. These reconstructions require that paleosol carbonate be sampled, and not other forms of secondary carbonate, based on the list of criteria given previously. Once securely identified as paleosol carbonate, there are several other key sampling considerations. One is that it should be verified that the matrix of paleosols is leached of inherited detrital carbonate in the zone of pedogenic carbonate accumulation; otherwise, detrital carbonate would otherwise contaminate the analysis. Another key issue is that paleosols should be texturally homogeneous, indicating that pedogen-

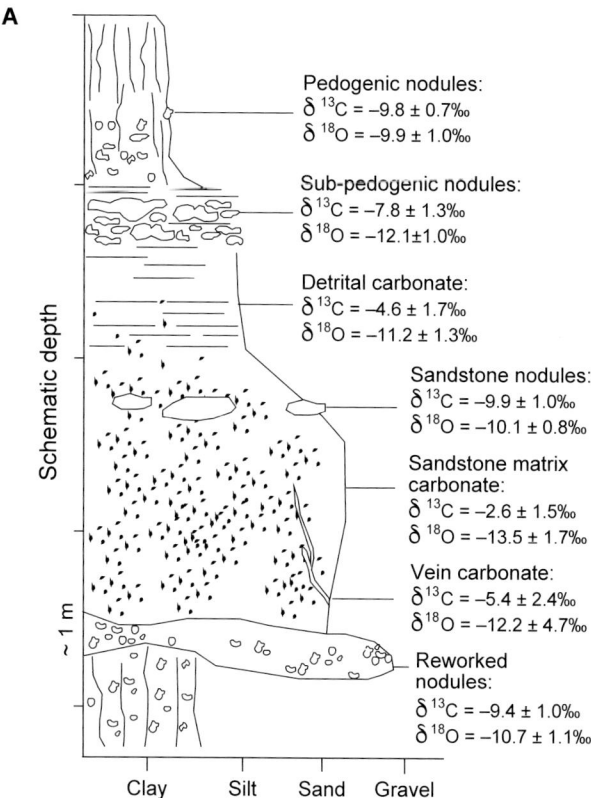

Figure 5.2. A. Modern soil showing surface-leached A- and B- soil horizons to a depth of about 30 cm with abundant large CaCO₃ nodules below 30 cm depth. Geological hammer is ~30 cm in length. B. Paleosol in Dhok Pathan Formation: above top of "paleo-pick" is overlying bedded carbonate-rich (~10%) fluvial sediments; bioturbated leached sediments are below hammer top and extend to near the bottom of the hammer handle; CaCO₃ nodule zone begins near base of hammer (~1 m) and matrix is leached to about 1.5 meters in this soil. Paleo-pick is 65 cm in length. C. Closeup of paleosol with overlying brown unleached siltstone; gray A-horizon is leached of carbonate; carbonate nodules begin in yellow B-horizon. Geological hammer is ~30 cm in length. D. Dipping paleosol showing abrupt carbonate nodular horizon beginning at top of "paleo-pick." Paleo-pick is 65 cm in length.

esis therefore followed a single major (>1 m) depositional event. Many paleosols are texturally heterogenous and appear to be more cumulic in nature, and in these cases it is very hard to determine the original depth of secondary carbonate formation. Care must be taken that soil carbonate formed well below the soil-air interface, to avoid contamination by atmospheric CO_2 in the upper few tens (<30 cm) of centimeters in soils.

Other forms of non-pedogenic carbonate are commonly observable in the Siwalik Group (see Fig. 5.3a), These include diagenetic cements in sandstones (Quade and Roe 1999; Sanyal, Bhattacharya, and Prasad 2005) that can also record valuable

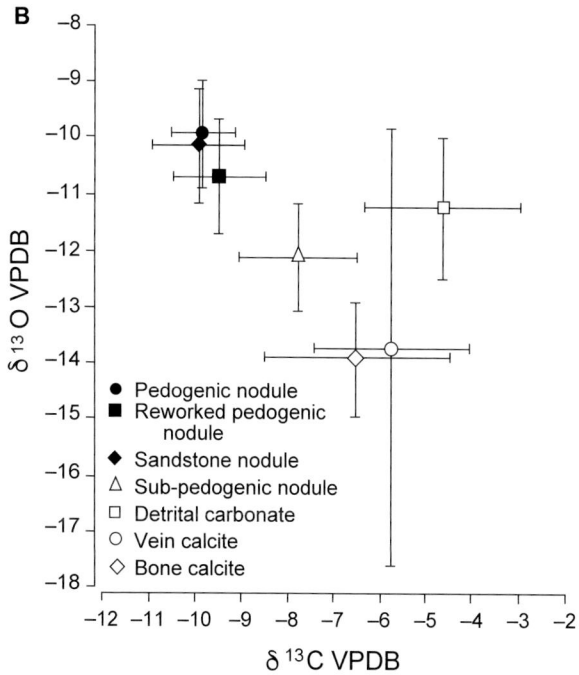

Figure 5.3. Different carbonate types found in Siwalik sediments. A. Schematic drawing of the different carbonate types found in the Chinji and Nagri Formations, along with their average $\delta^{13}C$ and $\delta^{18}O$ values. B. $\delta^{13}C$ and $\delta^{18}O$ of different carbonate types in the Siwaliks sediments from the Chinji Formation. Paleosol carbonates, reworked carbonate nodules, and sandstone nodules have lower $\delta^{13}C$ values and higher $\delta^{18}O$ values than other carbonate types.

paleoenvironmental information; massive cementation of floodplain sediment in the sub-paleosol zone; and late-stage veins filled with coarse calcite and filling voids such as in bone (Quade et al. 1992).

$\delta^{13}C$ and $\delta^{18}O$ values of carbonates in the Siwalik Group span a wide range depending on type of material (Table 5.1; see Fig. 5.3b). Volumetrically, probably the most abundant carbonate consists of reworked detrital grains of calcite and dolomite shed from the Himalaya and found in floodplain siltstone and channel sands, where carbonate content averages $10.2 \pm 4.7\%$ by weight. Samples of this detrital carbonate from the Chinji and Nagri Formations have values of -4.6 ± 1.7 and $-11.2 \pm 1.3\text{‰}$ ($n = 10$), respectively. Pedogenic carbonates from the Chinji Formation have $\delta^{13}C$ and $\delta^{18}O$ values of -9.8 ± 0.7 and $-9.9 \pm 1.0\text{‰}$ ($n = 36$), respectively. These samples were obtained entirely from the lower half of paleosol profiles in which the matrix is generally leached of carbonate, indicating that no detrital carbonate was present prior to pedogenic carbonate formation. Deeper in profiles below the zone of active weathering and bioturbation, nodular carbonate is often observed to cement unaltered and well-bedded floodplain sediment (Fig. 5.3a). These "sub-pedogenic" nodules have $\delta^{13}C$ and $\delta^{18}O$ values of -7.8 ± 1.3 and $-12.1 \pm 1.0\text{‰}$ ($n = 7$), respectively (Fig. 5.3b). These values reflect mixing primary pedogenic cement with detrital carbonate. The sub-pedogenic carbonates tend to be platey in form, mimicking the floodplain sediment that they cement. This contrasts with the more rounded appearance of pedogenic nodules. Carbonate nodules are also common in basal channel conglomerates of the Chinji and Nagri Formations. They are associated with rounded mudstone clasts and likely experienced brief transport following bank collapse of floodplain deposits. They have $\delta^{13}C$ and $\delta^{18}O$ values of -9.4 ± 1.0 and $-10.7 \pm 1.1\text{‰}$ ($n = 9$), respectively, which is virtually identical to pedogenic carbonate (Fig. 5.3b). This similarity verifies that carbonate of this composition is in fact pedogenic and formed very early, prior to deep burial of sediments.

Late diagenetic carbonate in the Chinji Formation occurs principally in two forms: coarse sparry calcite veins that cross-cut depositional bedding features and large calcite crystals infilling bone cavities. The late diagenetic calcite veins have $\delta^{13}C$ and $\delta^{18}O$ values of -5.4 ± 2.4 and $-12.2 \pm 4.7\text{‰}$ ($n = 10$), respectively, and the bone-cavity filling calcite spar -7.6 ± 2.0 and $-13.9 \pm 3.0\text{‰}$ ($n = 7$), respectively (see Fig. 5.3, Table 5.1); late-stage vein carbonates may be similar in their time of formation to the bone-cavity carbonates. These data for the different calcite forms from the Chinji Formation show that the $\delta^{13}C$ and $\delta^{18}O$ values of the pedogenic carbonates are distinct from detrital carbonate as well as from later diagenetic carbonates.

A final carbonate phase of note is secondary carbonate nodules found in sandstones. These nodules are typically quite large (≤50 cm) and ovoid in shape, cementing sandstone along bedding planes. Like pedogenic carbonate, they are often observed as reworked clasts in contemporaneous paleochannels, demonstrating their very early diagenetic origin. Quade and Roe (1999) studied the isotope systematics of these "sandstone nodules" and report averages of $\delta^{13}C$ and $\delta^{18}O$ values of -9.9 ± 1.0 and $-10.2 \pm 0.8\text{‰}$ ($n = 10$) from the Chinji Formation (see Fig. 5.3, Table 5.1). These averages are virtually identical to that of pedogenic nodules and thus show that both phases can be used for paleoecological reconstruction described below.

Long-Term Carbon Isotope Record from Paleosols

We have compiled >400 stable isotope analyses from all five formations in the Siwalik Group (Fig. 5.4; data from Quade, Cerling, and Bowman 1989a; Quade and Cerling 1995; Behrensmeyer et al. 2007; and unpublished data). The data density is much higher for the period 5–8 Ma, with results from at

Table 5.1. $\delta^{13}C$ and $\delta^{18}O$ Values of Carbonates in the Siwalik Group

	n (samples)	$\delta^{13}C$‰ (mean)	$\delta^{13}C$‰ (standard deviation)	$\delta^{18}O$‰ (mean)	$\delta^{18}O$‰ (SD)
Soil carbonate nodules	39	−9.8	1.1	−8.9	1.1
Reworked nodules	10	−9.9	1.0	−10.1	0.8
Weak diagenetic cement	4	−2.6	1.5	−13.5	1.7
Parent siltstone	4	−5.4	0.9	−11.5	0.9
Bone calcite	8	−6.2	1.8	−14.8	2.2
Vein carbonate	5	−5.2	1.9	−15.5	1.6

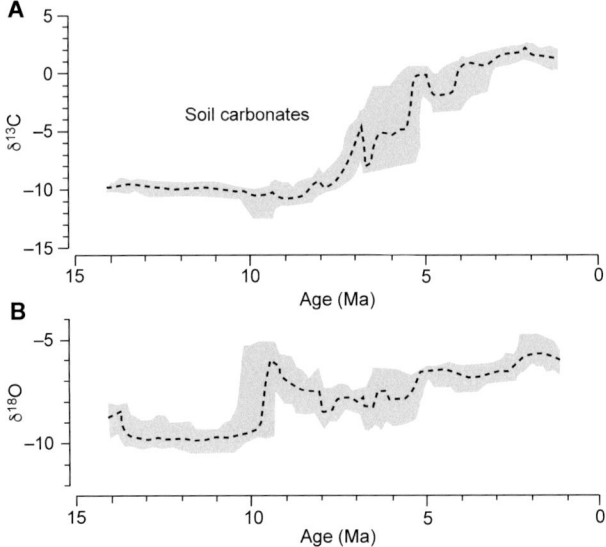

Figure 5.4. $\delta^{13}C$ and $\delta^{18}O$ of pedogenic carbonates from the Siwalik group in northern Pakistan; boundaries of shaded area are the 20th and 80th percentiles of >400 carbonates using a 25-point running sample mean; dashed line is the median of 25-point running sample size. A. $\delta^{13}C$ for paleosols from Siwaliks, northern Pakistan; B. $\delta^{18}O$ for paleosols from Siwaliks, northern Pakistan.

least five separate sections. By contrast, the very youngest (uppermost Siwaliks) and oldest samples (Kamlial Formation) are represented by samples from only a single section.

The most significant changes in $\delta^{13}C$ values occur between 8.5 and 6.5 Ma, when the average changes from -9 to $-10‰$ for sediments ≥ 8.5 Ma to values of 0 to $+2‰$ for sediments ≤ 6.5 Ma. The decrease in $\delta^{13}C$ values marks the change from C_3-dominated ecosystems prior to 8.5 Ma, to a mixed C_3-C_4 ecosystem between 8.5 and 6.5 Ma, and to a C_4-dominated ecosystem after 6.5 Ma. $\delta^{13}C$ analysis of bulk organic matter (Quade and Cerling 1995) and n-alkanes in Siwalik paleosols (Freeman and Collaruso 2001) confirm this increase in C_4 biomass in the late Miocene.

Behrensmeyer et al. (2007) document variation in $\delta^{13}C$ in lateral transects extending away from major channels at specific time intervals across this ecologic transition.

Subsequent studies have shown that this shift in C_4 biomass during this same time period (8.5 to 6.5 Ma) is visible in the Siwalik Group paleosols of Nepal (Quade et al. 1995) and India (Sanyal et al. 2004, 2010; Roy, Ghosh, and Sanyal 2020), as well as in bulk organic matter in the Bengal Fan (France-Lanord and Derry 2001). A study by Hoorn, Ojha, and Quade (2000) of pollen from the Siwalik Group of Nepal revealed that grass pollen is uncommon in the pre–8 Ma record, but abundant thereafter. This suggests that in lowland Nepal C_4 grassland replaced C_3 forests, not C_3 grassland, ~6 to 8 Ma. Uplands surrounding floodplains remained forested throughout this period, as they are today, although forest composition shifted from evergreen toward drier, semi-deciduous trees after ~7 Ma (Awashi and Prasad 1989).

Further evidence from paleosols and fossil teeth in the Americas (Latorre, Quade, and McIntosh 1997; Fox and Koch 2003) and globally (Cerling et al. 1997) show that the C_4 expansion was time-transgressive, first appearing in the tropics as early as ~10 Ma (Uno et al. 2011; Polissar et al. 2019) and making first clear appearances as late as the early Pliocene at higher latitudes.

This dramatic expansion in C_4 biomass between 10 and 4 Ma should not be confused with the actual first development of C_4 plants. Unfortunately, few fossil plants known to exhibit C_4 characteristics (e.g., Kranz anatomy and $\delta^{13}C$ values) exist, and so the record of C_4 plants is a "shadow record" that has been best documented using stable isotopes rather than macrofossils. Stromberg (2005) recognized chloridoid (arid-adapted grasses) phytoliths in 19 Ma North American sediments, and, based on phylogenetic analyses, Christin et al. (2008, 2014) argued for the origins of C_4 photosynthesis in the Oligocene. Thus, the origins of C_4 photosynthesis in grasses predate the expansion of C_4 grasses by more than 10 myr. The impetus behind the expansion between 8.5 and 6.5 Ma, or perhaps earlier, remains unresolved: the competing hypotheses of a CO_2 threshold (Cerling et al. 1997; Ehleringer, Cerling, and Helliker 1997) versus global aridity (Pagani, Freeman, and Arthur 1999) are still not resolved, although Polissar et al. (2019) make a strong case for CO_2 falling below a critical threshold around 10 Ma. Fire in the late Miocene has also been proposed as an important driver of the expansion of C_4 grasslands (Bond,

Midgley, and Woodward 2003; Keeley and Rundell 2005; Scheiter et al. 2012). Karp, Behrensmeyer, and Freeman (2018) show an increase in biomarkers associated with fire in Siwalik sediments, starting at 10 Ma and increasing during the same period as the soil carbonate and molecular-isotope transition (Freeman and Collaruso 2001).

Long-Term Oxygen Isotope Record from Paleosols

$\delta^{18}O$ values of paleosol carbonate in the Siwalik Group of Pakistan display an increase of 3–4‰ at ~9 Ma, smaller in magnitude than the ~10‰ shift in $\delta^{13}C$ values one to two million years later (see Fig. 5.4). This increase in $\delta^{18}O$ values is also seen in some but not all paleosol records outside Pakistan, notably in Nepal (Quade et al. 1995), East Africa (Levin et al. 2004), and South America (Latorre, Quade, and McIntosh 1997). As with the carbon isotope record, no single explanation is accepted for this large isotopic shift. One explanation that can be rejected is the role of soil temperature, since the shift is much too large to be explained by soil temperature change alone, which would have to decrease by 12–16°C to account for the 3–4‰ increase in $\delta^{18}O$ values. There is no evidence for a global decrease in temperature of this magnitude at ~9 Ma.

Another explanation offered in Quade, Cerling, and Bowman (1989a) for the late Miocene increase in $\delta^{18}O$ (and $\delta^{13}C$) values in the Siwalik Group is intensification of the Asian monsoon. This explanation can now be rejected based on three arguments: (1) the increase in $\delta^{18}O$ and $\delta^{13}C$ values is now known to be global and not confined to the Indian subcontinent, thus requiring a global explanation and not just a local climate phenomenon such as the Asian monsoon, (2) other non-isotopic and paleofloral evidence from Pakistan and Nepal points to a weakening, not strengthening of the Asian monsoon (Quade et al. 1995) post ~8 Ma, and (3) evidence of changes in seasonality patterns from $\delta^{18}O$ values of fossil mollusks also point to less rain and a stronger dry season in regional climate post ~8 Ma (Dettman et al. 2001).

These three lines of evidence combine to suggest that the increase in $\delta^{18}O$ values was the result of increased soil evaporation rates, either somehow connected to the change from forest to grass cover (and the resultant increases in solar radiation reaching the soil) or to aridification of late Miocene climate (e.g. Pagani et al. 1997). This aridification would need to have been global in extent to explain the global nature of the C_4 grass expansion across the low to mid-latitudes. While some records outside the Indian subcontinent do clearly record a late Miocene increase in $\delta^{18}O$ values (East Africa: Levin et al. 2004; South America: Latorre, Quade, and McIntosh 1997; Kleinert and Strecker 2001; Greece: Quade, Solounias, and Cerling 1994), others do not (e.g., North America: Fox and Koch 2003; East Africa: Polissar et al. 2019).

STABLE ISOTOPE RECORD IN FOSSIL MAMMAL ENAMEL

From a historical perspective, we began work on isotopes in Siwalik fossil teeth as a test of the fidelity of stable isotopes in tooth enamel. Lee-Thorp and van der Merwe (1987; Lee-Thorp, Sealy, and van der Merwe 1989; Lee-Thorp, van der Merwe, and Brain 1989) had done pioneering work to establish that tooth enamel was a robust indicator of diet in the geological record. However, confusion between the fidelity of "bone" and "enamel" apatite for stable isotope analysis remained controversial due to earlier work on fossil bone (Sullivan and Krueger 1981; Schoeninger and DeNiro 1982; Nelson et al. 1986); fossil enamel was equated with the demonstrably unreliable fossil bone. Since isotopic analysis of pedogenic carbonate showed that the Siwalik sequence had undergone a nearly complete conversion from C_3- to C_4-dominated floodplain ecosystems beginning around 8 Ma, this provided a basis for comparisons with fossil bone and enamel to resolve the fidelity question for fossil enamel. The earliest work on fossil enamel from the Siwaliks (Quade et al. 1992) was done with this test in mind, and the results of that study concluded that fossil tooth enamel retained its original isotope value through the diagenetic process but that the isotope values of fossil bone was likely altered by diagenesis. Therefore, fossil enamel, but not bone, could be used for reconstructing ancient diets of mammals over millions of years. Concurrently, Koch, Zachos, and Gingerich (1992) demonstrated a shift at the Paleocene-Eocene Thermal Maximum in both the marine carbon record and the terrestrial paleosol carbonate, as well as in mammalian tooth enamel records. This indicated fidelity back to more than 50 Ma for tooth enamel.

The results of these preliminary studies of stable isotopes in tooth enamel opened new possibilities for paleoecological research and hypothesis testing. The paleosol carbonate record provided a testing ground for studies of how animal lineages respond to changes in dietary resources: the earlier C_3-dominated ecosystems were replaced over a few million years by C_4-dominated grasslands, and therefore a major diet shift might be expected in herbivorous mammals as the diet resources changed. Using isotopes alone, it is not possible to distinguish between C_3 grasses or other C_3 herbaceous plants and C_3 woody vegetation, and questions about such dietary distinction awaits other methods. It is likely that isotope studies of biomarkers will be informative on C_3 herbaceous versus C_3 woody plant abundance for the period before C_4 plants became abundant.

Following this early work with fossils in South Africa, Pakistan, and the western United States (Lee-Thorp, van der Merwe, and Brain 1989; Quade et al. 1992; Koch, Zachos, and Gingerich 1992), stable isotope studies of tooth enamel eventually became widely accepted as an ecological and paleoecological tool (e.g., MacFadden 2000; Morgan, Kingston, and Marino 1994; Morgan et al. 2009; Cerling, Harris, and Passey 2003; Badgley et al. 2005; MacFadden 2005; Nelson 2005; Koch 2007; Nelson 2007; Uno et al. 2011; Cerling et al. 2015; Suraprasit et al. 2018; Sun et al. 2019).

Tooth enamel is now an accepted source of information on the diet of extinct organisms and also on their sources of water. For carbon, there is an isotope enrichment between diet and tooth enamel, which can vary somewhat from species to species. Basically, the energy source (carbohydrates for herbivores) is converted to CO_2 by respiration:

$$CH_2O + O_2 = CO_2 + H_2O,$$

and a large fractionation (isotope enrichment) of ^{13}C in bicarbonate occurs compared to CO_2 (Friedman and O'Neil 1977). Passey et al. (2005) showed that the isotope enrichment between breath and enamel was about 11.4‰ across taxa with widely ranging digestive physiology (suids, rodents, bovids, lagomorphs). Passey et al. (2005) also suggested that mammals that produce more methane have higher ^{13}C enrichment in CO_2 going from diet to breath; methane production results from the very large isotope enrichment (ca. 51‰ at 37°C) between CH_4 and CO_2 in ruminants (Klevenhusen et al. 2010). Thus, isotope enrichment values $\varepsilon^\star_{\text{diet-enamel}}$ range from about +9 to +14‰ (Lee-Thorp, Sealy, and van der Merwe 1989; Cerling and Harris 1999; Passey et al. 2005), with rodents having the smallest enrichment and ruminants the highest enrichment.

Tejada et al. (2018) evaluated all published diet-enamel isotope enrichment studies and concluded that body size and digestive physiologies were the principal factors in determining the isotopic enrichment between diet and bioapatite for herbivorous mammals. For ungulate mammals ranging from 100 to 4,000 kg in that study, the $\varepsilon^\star$ (isotope enrichment) was 13.8 ± 0.6‰ ($n = 9$ taxa), which is similar to the value of 14.1‰ suggested by Cerling and Harris (1999) for ruminant and other large ungulate mammals. Thus, we use the 14.1‰ value for diet to bioapatite enrichment for the large Artiodactyla, Perissodactyla, and Proboscidea lineages for the Siwalik fauna. Smaller values for $\varepsilon^\star$ have been established for rodents and lagomorphs (DeNiro and Epstein 1978; Tieszen and Fagre 1993; Passey et al. 2005). Below, we discuss primarily the Artiodactyla, Perissodactyla, and Proboscidea (APP) taxa with body mass > 100 kg.

The isotope differences between diet and tooth enamel for herbivores is likely related to CH_4 / CO_2 production rates and perhaps other factors; additional work is needed to determine values that apply to the fossil records. There may be additional limitations for mammals with no living analogs (e.g., *Anthracotheres, Notoungulates*). However, the total range in isotope enrichment factors between diet and tooth enamel is on the order of 4‰ (i.e., from 10‰ to 14‰) or so for herbivores, hence the large-scale trends over time will stand even when precise enrichment values are not known.

Figure 5.5 shows the long-term trend of stable isotopes in mammals in this study ($n > 500$ fossil specimens, APP taxa, >100 kg). We use only the single isotope enrichment value of 14.1‰ determined for large ruminants (Cerling and Harris 1990) because of the restricted dietary physiologies and body mass restriction of >100 kg. Expected isotope ranges for C_4 grazers, C_3–C_4 mixed feeders, C_3 foragers, and C_3 hyperforagers are shown in Figure 5.5, following definitions in Cerling et al. (2015), with provision made for expected values during periods of Earth history when the $\delta^{13}C$ of the atmosphere changed with the resultant cascade to the food webs. Because the end-member C_3- and C_4-values are not precisely known,

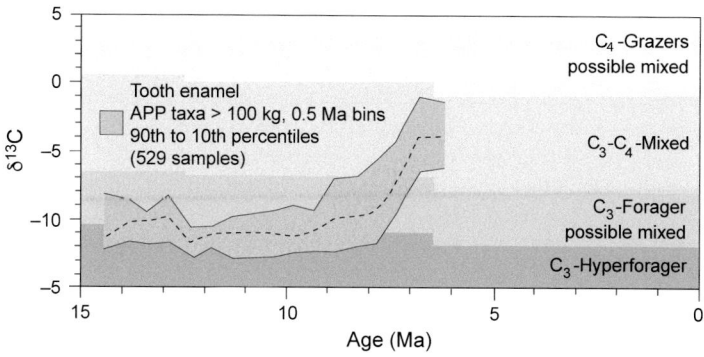

Figure 5.5. $\delta^{13}C$ of large mammals through Miocene Siwaliks binned in 0.5 myr increments; dashed heavy line is average values and gray shaded area includes 10th through 90th percentiles of data. Fields for C_4-grazers, C_3–C_4 mixed feeders, C_3-browsers, and C_3-hyperbrowsers shown with orange, yellow, light green, and dark green shaded areas, respectively; the shaded areas account for change in the $\delta^{13}C$ of the atmosphere at 6.5 Ma (1‰ higher than the preindustrial $\delta^{13}C$ value) and at 12.5 Ma (1.5‰ higher than the preindustrial value).

and also because isotope enrichments are not precisely known, in our discussion C_4 grazers are considered to be those herbivores whose $\delta^{13}C$ values >-1‰ and are likely to have diets comprised of ~75 to 100% C_4 plants. Mixed feeders (sub-equal amounts of C_3 and C_4 plants comprising the diet) have $\delta^{13}C$ values between -8‰ and -1‰. Predominantly C_3 foragers, with diets between ~75 to 100% C_3 plants have $\delta^{13}C$ values between -8 and -12‰, and C_3 foragers (i.e., unlikely to measurable C_4 contribution) have $\delta^{13}C$ values <-12‰. These values are considered useful between the preindustrial atmosphere back to about 6.5 Ma, when there was a shift in the $\delta^{13}C$ of the atmosphere as discussed above; for samples older than 6.5 Ma these values must be shifted accordingly, as also noted above.

Our compilation of all results shows that diets are solidly in the C_3-dominated diet range for samples older than 9 Ma with $\delta^{13}C$ values averaging between -10 and -12‰ (see Fig. 5.5). The first significant isotopic change begins between 9.0 and 8.5 Ma with continuing increase in $\delta^{13}C$ values through 6.0 Ma; Quade et al. (1992) documented continued increase in $\delta^{13}C$ values into the Pleistocene. Thus, in a period of a few million years, the diets of herbivores in the Siwalik record change from a virtually pure C_3-dominated diet to a C_4-dominated one. Given that the floodplain paleosols indicate an almost complete turnover from C_3 to C_4 plants, this change is not unexpected. However, it provides a good opportunity to study the response of different mammalian groups to this change in ecosystem structure. Some of these changes are discussed in other chapters of this volume.

References

Awashi, N., and M. Prasad. 1989. Siwalik plant fossils from the Surai Khola area, western Nepal. *Palaeobotanist* 38:298–318.

Badgley, C., S. Nelson, J. Barry, A. K. Behrensmeyer, and T. E. Cerling. 2005. Testing models of faunal turnover with Neogene mammals from Pakistan. In *Interpreting the Past: Essays on Human, Primate, and Mammal Evolution*, edited by D. E. Lieberrman, R. J. Smith, and J. Kelly, 29–46. Boston: Brill.

Behrensmeyer, A.K., J. Quade, T. E. Cerling, I. A. Kahn, P. Copeland, L. Roe, J. Hicks, P. Stubblefield, B.J. Willis, and L. Catorre. 2007. The structure and rate of late Miocene expansion of C_4 plants: Evidence from lateral variation in stable isotopes in paleosols of the Siwalik Group, northern Pakistan. *Geological Society of America Bulletin* 119(11–12): 1486–1505.

Behrensmeyer, A. K., and L. Tauxe. 1982. Isochronous fluvial systems in Miocene deposits of northern Pakistan. *Sedimentology* 29(3): 331–352.

Blumenthal, S. A., J. M. Rothman, K. L. Chritz, and T. E. Cerling. 2016. Stable isotopic variation in tropical forest plants for applications in primatology. *American Journal of Primatology* 78(10): 1041–1054. https://doi.org/10.1002/ajp.22488.

Bond, W. J., G. F. Midgley, and F. I. Woodward, F.I., 2003. The importance of low atmospheric CO_2 and fire in promoting the spread of grasslands and savannas. *Global Change Biology* 9(7): 973–982.

Brimhall, G. H., and W. E. Dietrich. 1987. Constitutive mass balance relations between chemical composition, volume, density, porosity, and strain in metasomatic hydrochemical systems: Results on weathering and pedogenesis. *Geochimica et Cosmochimica Acta* 51(3): 567–587.

Cerling, T. E. 1984. The stable isotopic composition of modern soil carbonate and its relationship to climate. *Earth and Planetary Science Letters* 71(2): 229–240.

Cerling, T. E. 1997. Late Cenozoic vegetation change, atmospheric CO_2, and tectonics. In *Tectonic Uplift and Climate Change*, edited by W. F. Ruddiman, 313–327. New York: Plenum.

Cerling, T. E., S. A. Andanje, S. A. Blumenthal, F. H. Brown, K. L. Chritz, J. M. Harris J. A. Hart, et al. 2015. Dietary changes of large herbivores in the Turkana Basin, Kenya from 4 to 1 million years ago. *Proceedings of the National Academy of Sciences* 112(37): 11467–11472.

Cerling, T. E., and J. M. Harris. 1999. Carbon isotope fractionation between diet and bioapatite in ungulate mammals and implications for ecological and paleoecological studies. *Oecologia* 120(Aug): 347–363.

Cerling, T. E., J. M. Harris, M. G. Leakey, B. H. Passey, and N. E. Levin. 2010. Stable carbon and oxygen isotopes in East African mammals: Modern and fossil. In *Cenozoic Mammals of Africa*, edited by L. Werdelin and W. Sanders, 949–960. Berkeley: University of California Press.

Cerling, T. E., J. M. Harris, B. J. MacFadden, M. G. Leakey, J. Quade, V. Eisenmann, and J. R. Ehleringer. 1997 Global change through the Miocene/Pliocene boundary. *Nature* 389(6647): 153–158.

Cerling, T. E., J. M. Harris, and B. H. Passey. 2003. Dietary preferences of East African bovidae based on stable isotope analysis. *Journal of Mammalogy* 84(2): 456–471.

Cerling, T. E., J. A. Hart, and T. B. Hart. 2004. Stable isotope ecology in the Ituri Forest. *Oecologia* 138(Jan): 5–12.

Cerling, T. E., J. Quade, Y. Wang, and J. R. Bowman. 1989. Carbon isotopes in soils and paleosols as ecologic and paleoecologic indicators. *Nature* 341(6238): 138–139.

Cerling, T. E., Y. Wang, and J. Quade. 1993. Expansion of C_4 ecosystems as an indicator of global ecological change in the late Miocene. *Nature* 361(6410): 344–345.

Chadwick, O.A., G. H. Brimhall, and D. M. Hendricks. 1990. From a black to a gray box—A mass balance interpretation of pedogenesis. *Geomorphology* 3(3–4): 369–390.

Christin, P.A., G. Besnard, E. Samaritani, M. R. Duvall, T. R. Hodkinson, V. Savolainen, and N. Salamin. 2008. Oligocene CO_2 decline promoted C_4 photosynthesis in grasses. *Current Biology* 18(1): 37–43.

Christin, P.A., E. Spriggs, C. P. Osborne, C. A. Strömberg, N. Salamin, and E. J. Edwards. 2014. Molecular dating, evolutionary rates, and the age of the grasses. *Systematic Biology* 63(2): 153–165.

Codron, J., D. Codron, J. A. Lee-Thorp, M. Sponheimer, W. J. Bond, D. de Ruiter, and R. Grant. 2005. Taxonomic, anatomical, and spatio-temporal variations in the stable carbon and nitrogen isotopic composition of plants from an African savanna. *Journal of Archaeological Science* 32(12): 1757–1772.

Coplen, T. B. 1988. Normalization of oxygen and hydrogen isotope data. *Chemical Geology: Isotope Geoscience Section* 72(4): 293–297.

DeNiro, M. J., and S. Epstein. 1978. Influence of diet on the distribution of carbon isotopes in animals. *Geochimica et Cosmochimica Acta* 42(5): 495–506.

Dettman, D. L., M. J. Kohn, J. Quade, F. J. Ryerson, T. P. Ojha, and S. Hamidullah. 2001. Seasonal stable isotope evidence for a strong Asian Monsoon throughout the past 10.7 Ma. *Geology* 29(1): 31–34.

Ehleringer, J. R., T. E. Cerling, and B. Helliker. 1997 C_4 photosynthesis, atmospheric CO_2, and climate. *Oecologia* 112(Oct): 285–299.

Ehleringer, J. R., and R. K. Monson. 1993. Evolutionary and ecological aspects of photosynthetic pathway variation. *Annual Reviews of Ecology and Systematics* 24(1): 411–439

Fox, D. L., and P. L. Koch. 2004. Carbon and oxygen isotopic variability in Neogene paleosol carbonates: Constraints on the evolution of C_4-grasslands of the Great Plains, USA. *Palaeogeography, Palaeoclimatology, Palaeocology* 207(3–4): 305–329.

France-Lanord, C., and L. Derry. 1994. $\delta^{13}C$ of organic carbon in the Bengal Fan: source, evolution, and transport of C3 and C4 plant carbon to marine sediments. *Geochimica et Cosmochimica Acta* 58(21): 4809–4814.

Francey, R. J., C. E. Allison, D. M. Etheridge, C. M. Trudinger, I. G. Enting, M. Leuenberger, R. L. Langenfelds, E. Michel, E., and L. P. Steele. 1999. A 1000-year high precision record of $\delta^{13}C$ in atmospheric CO_2. *Tellus B*, 51(2): 170–193.

Freeman, K. H., and L. A. Colarusso. 2001. Molecular and isotopic records of C_4 grassland expansion in the late Miocene. *Geochimica et Cosmochimica Acta* 65(9): 1439–1454.

Friedman, I., and J. R. O'Neil. 1977. Compilation of stable isotope fractionation factors of geochemical interest. US Geological Survey Professional Paper 440-KK. Washington, DC: US Government Printing Office.

Herbert, T. D., K. T. Lawrence, A. Tzanova, L. C. Peterson, R. Caballero-Gill, and C. S. Kelly. 2016. Late Miocene global cooling and the rise of modern ecosystems. *Nature Geoscience* 9(11): 843–847.

Holbourn, A.E., W. Kuhnt, S. C. Clemens, K. G. Kochhann, J. Jöhnck, J. Lübbers, and N. Andersen. 2018. Late Miocene climate cooling and intensification of southeast Asian winter monsoon. *Nature Communications* 9(1): 1–13.

Hoorn, C., T. P. Ojha, and J. Quade. 2000. Palynologic evidence for paleo-vegetation changes in the sub-Himalayan zone of Nepal during the Late Neogene. *Palaeoecology. Palaeogeography. Palaeoclimatology* 163(3–4): 133–161.

Jenny, H. 1994. *Factors of Soil Formation: A System of Quantitative Pedology.* New York: Dover.

Johnson, N. M., N. D. Opdyke, G. D. Johnson, E. H. Lindsay, and R. A. K. Tahirkheli. 1982. Magnetic polarity stratigraphy and ages of Siwalik Group rocks of the Potwar Plateau, Pakistan. *Palaeogeography, Palaeoclimatology, Palaeocology* 37(1): 17–42.

Karp, A. T., A. K. Behrensmeyer, and K. H. Freeman. 2018. Grassland fire ecology has roots in the late Miocene. *Proceedings of the National Academy of Sciences* 115(48): 12130–12135.

Keeley, J. E., and P. W. Rundel. 2005. Fire and the Miocene expansion of C_4 grasslands. *Ecology Letters* 8(7): 683–690.

Kleinert, K., and M. R. Strecker. 2001. Climate change in response to orographic barrier uplift: Paleosol and stable isotope evidence from the late Neogene Santa Maria basin, northwestern Argentina. *Geological Society of America Bulletin* 113(6): 728–742.

Klevenhusen, F., S. M. Bernasconi, M. Kreuzer, and C. R. Soliva. 2010. Experimental validation of the Intergovernmental Panel on Climate Change default values for ruminant-derived methane and its carbon-isotope signature. *Animal Production Science* 50(3): 159–167.

Koch, P. L. 2007. Isotopic study of the biology of modern and fossil vertebrates. *Stable Isotopes in Ecology and Environmental Science* 2:99–154.

Koch, P. L., J. C. Zachos, and P. D. Gingerich. 1992. Correlation between isotope records in marine and continental carbon reservoirs near the Palaeocene/Eocene boundary. *Nature* 358(6384): 319–322.

Latorre, C., J. Quade, and W. C. McIntosh. 1997. The expansion of C_4 grasses and global change in the late Miocene: Stable isotope evidence from the Americas. *Earth and Planetary Science Letters* 146(1–2): 83–96.

Lee-Thorp, J., and N. J. van der Merwe. 1987. Carbon isotope analysis of fossil bone apatite. *South African Journal of Science* 83(11): 712–715.

Lee-Thorp, J. A., J. C. Sealy, and N. J. van der Merwe. 1989. Stable carbon isotope ratio differences between bone collagen and bone apatite, and their relationship to diet. *Journal of Archaeological Science* 16(6): 585–599.

Lee-Thorp, J. A., N. J. Van der Merwe, and C. K. Brain. 1989. Isotopic evidence for dietary differences between two extinct baboon species from Swartkrans. *Journal of Human Evolution* 18(3): 183–189.

Levin, N., J. Quade, S. Simpson, S. Semaw, and M. Rogers. 2004. Isotopic evidence for Plio-Pleistocene environmental change at Gona, Ethiopia. *Earth and Planetary Science Letters* 219(1): 93–100.

MacFadden, B. J. 2000. Cenozoic mammalian herbivores from the Americas: Reconstructing ancient diets and terrestrial communities. *Annual Review of Ecology and Systematics* 31(1): 33–59.

MacFadden, B.J. 2005. Diet and habitat of toxodont megaherbivores (Mammalia, Notoungulata) from the late Quaternary of South and Central America. *Quaternary Research* 64(2): 113–124.

Morgan, M. E., A. K. Behrensmeyer, C. Badgley, J. C. Barry, S. Nelson, and D. Pilbeam. 2009. Lateral trends in carbon isotope ratios reveal a Miocene vegetation gradient in the Siwaliks of Pakistan. *Geology* 37(2): 103–106.

Morgan, M. E., J. D. Kingston, and B. D. Marino. 1994. Carbon isotopic evidence for the emergence of C_4 plants in the Neogene from Pakistan and Kenya. *Nature* 367(6459): 162–165.

Nelson, B. K., M. J. DeNiro, M. J. Schoeninger, D. J. De Paolo, and P. E. Hare. 1986. Effects of diagenesis on strontium, carbon, nitrogen and oxygen concentration and isotopic composition of bone. *Geochimica et Cosmochimica Acta* 50(9): 1941–1949.

Nelson, S. V. 2005. Paleoseasonality inferred from equid teeth and intra-tooth isotopic variability. *Palaeogeography, Palaeoclimatology, Palaeoecology* 222(1–2): 122–144.

Nelson, S. V. 2007. Isotopic reconstructions of habitat change surrounding the extinction of *Sivapithecus*, a Miocene hominoid, in the Siwalik Group of Pakistan. *Palaeogeography, Palaeoclimatology, Palaeoecology* 243(1–2): 204–222.

Pagani, M., M. A. Arthur, and K. H. Freeman. 1999. Miocene evolution of atmospheric carbon dioxide. *Paleoceanography* 14(3): 273–292.

Pagani, M., K. H. Freeman, and M. A. Arthur. 1999. Late Miocene atmospheric CO_2 concentrations and the expansion of C4 grasses. *Science* 285(5429): 876–879.

Pagani, M., J. C. Zachos, K. H. Freeman, B. Tipple, and S. Bohaty. 2005. Marked decline in atmospheric carbon dioxide concentrations during the Paleogene. *Science* 309(5734): 600–603.

Passey, B. H., M. E. Perkins, M. R. Voorhies, T. E. Cerling, J. M. Harris, and S. T. Tucker. 2002. Timing of C_4 biomass expansion and environmental change in the Great Plains: An isotopic record from fossil horses. *Journal of Geology* 110(2): 123–140.

Passey, B. H., T. F. Robinson, L. K. Ayliffe, T. E. Cerling, M. Sponheimer, M. D. Dearing, B. L. Roeder, and J. R. Ehleringer. 2005. Carbon isotopic fractionation between diet, breath, and bioapatite in different mammals. *Journal of Archaeological Science* 32(10): 1459–1470.

Polissar, P. J., C. Rose, D. T. Uno, S. R. Phelps, and P. B. de Menocal, 2019. Sychronous rise of the African C_4 ecosystems 10 million years ago in the absence of aridification. *Nature Geoscience* 12(8): 657–660.

Quade, J., J. M. Cater, T. P. Ojha, J. Adam, and T. Mark Harrison. 1995. Late Miocene environmental change in Nepal and the northern Indian subcontinent: Stable isotopic evidence from paleosols. *Geological Society of America Bulletin* 107(12): 1381–1397.

Quade, J., and T. E. Cerling. 1995 Expansion of C_4 grasses in the late Miocene of northern Pakistan: Evidence from stable isotopes in paleosols. *Palaeogeography, Palaeoclimatology, Palaeoecology* 115(1–4): 91–116.

Quade, J., T. E. Cerling, J. C. Barry, M. E. Morgan, D. R. Pilbeam, A. R. Chivas, J. A. Lee-Thorp, and N. J. van der Merwe. 1992. A 16-Ma record of paleodiet using carbon and oxygen isotopes in fossil teeth from Pakistan. *Chemical Geology* 94(3): 183–192.

Quade, J., T. E. Cerling, and J, R. Bowman. 1989a. Development of Asian monsoon revealed by marked ecological shift during the latest Miocene in northern Pakistan. *Nature* 342(6246): 163–16.

Quade, J., T. E. Cerling, and J. R. Bowman. 1989b. Systematic variations in the carbon and oxygen isotopic composition of pedogenic carbonate along elevation transects in the southern Great Basin, United States. *Geological Society of America Bulletin* 101(4): 464–475.

Quade, J., and L. Roe. 1999. The stable isotopic composition of early diagenetic cements from sandstone in paleoecological reconstruction. *Journal of Sedimentary Research* 69(3): 667–674.

Quade, J., N. Solounias, and T. E. Cerling. 1994. Stable isotopic evidence from paleosol carbonates and fossil teeth in Greece for forest or woodlands over the past 11 Ma. *Palaeogeography, Palaeoclimatology, Palaeoecology* 108(1–2): 41–53.

Roy, B., S. Ghosh, and P. Sanyal. 2020. Morpho-tectonic control on the distribution of C_3–C_4 plants in the central Himalayan Siwaliks during Late Plio-Pleistocene. *Earth and Planetary Science Letters* 535:116–119.

Sage, R. 2004. The evolution of C_4 photosynthesis. *New Phytologist* 161(2): 341–370.

Sanyal, P., S. K. Bhattacharya, R. Kumar, S. K. Ghosh, and S. J. Sangode. 2004. Mio-Pliocene monsoonal record from Himalyan foreland basin (Indian Siwaliks) and its relation to vegetation change. *Palaeoecology. Palaeogeography. Palaeoclimatology* 205(1–2): 23–41.

Sanyal, P., S. K. Bhattacharya, and M. Prasad. 2005. Chemical diagenesis of Siwalik sandstone: Isotopic and mineralogical proxies from the Surai Khola section. *Sedimentary Geology* 180(1–2): 57–74.

Sanyal, P., A. Sarkar, S. K. Bhattacharya, R. Kumar, S. K. Ghosh, and S. Agrawal. 2010. Intensification of monsoon, microclimate, and asynchronous C_4 appearance: Isotopic evidence from the Indian Siwalik

sediments. *Palaeoecology. Palaeogeography. Palaeoclimatology* 296(1–2): 165–173.

Scheiter, S., S. I. Higgins, C. P. Osborne, C. Bradshaw, D. Lunt, B. S. Ripley, L. L. Taylor, D. J. and Beerling. 2012. Fire and fire-adapted vegetation promoted C_4 expansion in the late Miocene. *New Phytologist* 195(3): 653–666.

Schoeninger, M. J., and M. J. DeNiro. 1982. Carbon isotope ratios of apatite from fossil bone cannot be used to reconstruct diets of animals. *Nature* 297(5867): 577–578.

Strömberg, C. A. E. 2005. Decoupled taxonomic radiation and ecological expansion of open-habitat grasses in the Cenozoic of North America. *Proceedings of the National Academy of Sciences* 102(34): 11980–11984.

Sullivan, C. H., and H. W. Krueger. 1981. Carbon isotope analysis of separate chemical phases in modern and fossil bone. *Nature* 292(5821): 333–335.

Suraprasit, K., H. Bocherens, Y. Chaimanee, S. Panha, and J. J. Jaeger. 2018. Late Middle Pleistocene ecology and climate in Northeastern Thailand inferred from the stable isotope analysis of Khok Sung herbivore tooth enamel and the land mammal cenogram. *Quaternary Science Reviews* 193(Aug): 24–42.

Sun, F., Y. Wang, Y. Wang, C. Z. Jin, T. Deng, and B. Wolff. 2019. Paleoecology of Pleistocene mammals and paleoclimatic change in South China: Evidence from stable carbon and oxygen isotopes. *Palaeogeography, Palaeoclimatology, Palaeoecology*, 524(June): 1–12.

Tejada-Lara, J. V., B. J. MacFadden, L. Bermudez, G. Rojas, R. Salas-Gismondi, and J. J. Flynn. 2018. Body mass predicts isotope enrichment in herbivorous mammals. *Proceedings of the Royal Society B: Biological Sciences* 285(1881): 20181020.

Tieszen, L. L., and T. Fagre. 1993. Effect of diet quality and composition on the isotopic composition of respiratory CO_2, bone collagen, bioapatite, and soft tissues. In *Prehistoric Human Bone*, edited by Joseph B. Lambert and Gisela Grupe, 121–155. Berlin: Springer.

Tipple, B. J., S. R. Meyers, and M. Pagani. 2010. Carbon isotope ratio of Cenozoic CO_2: A comparative evaluation of available geochemical proxies. *Paleoceanography* 25(3). https://doi.org/10.1029/2009PA001851.

Tipple, B. J. and M. Pagani. 2007. The early origins of terrestrial C_4 photosynthesis. *Annual Review of Earth and Planetary Sciences* 35(1): 435–461.

Uno, K. T., T. E. Cerling, J. M. Harris, Y. Kunimatsu, M. G. Leakey, M. Nakatsukasa, and H. Nakaya. 2011. Late Miocene to Pliocene carbon isotope record of differential diet change among East Africa herbivores. *Proceedings of the National Academy of Sciences* 108(16): 6509–6514.

van der Merwe, N. J., and E. Medina. 1989. Photosynthesis and $^{13}C/^{12}C$ ratios in Amazonian rain forests. *Geochimica et Cosmochimica Acta* 53(5): 1091–1094

Zachos, J., M. Pagani, L. Sloan, E. Thomas, and K. Billups. 2001. Trends, rhythms, and aberrations in global climate 65 Ma to present. *Science* 292(5517): 686–693.

Zhang, Y.G., M. Pagani, Z. Liu, S. M. Bohaty, and R. DeConto. 2013. A 40-million-year history of atmospheric CO_2. *Philosophical Transactions of the Royal Society A: Mathematical, Physical and Engineering Sciences* 371(2001): 20130096.

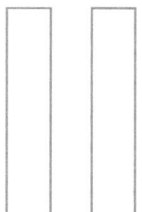

PREAMBLE TO FAUNAL CHAPTERS

CATHERINE BADGLEY, MICHÈLE E. MORGAN, AND
DAVID PILBEAM

The eighteen chapters in this section present specimen-based research by an international team of scientists on the rich terrestrial and aquatic faunas of the western Siwalik sequence in the Potwar Plateau. These chapters reflect current understanding of the Siwalik fauna and do not formally name new taxa. For many groups the taxonomy is well resolved, while for others new species identified during research for this book will result in future systematic papers.

All chapters report ages of fossil localities using the magnetostratigraphic framework described in Chapters 2 and 3 and aligned to the timescale of Gradstein et al. (2012). However, some older Siwalik fossil collections include important diagnostic fossils that can only be placed within formations. The chronostratigraphic relationships of Potwar Plateau rock formations are shown in Figure II.1 for reference and may facilitate reader comparisons to fauna from other geographic regions. Chapters follow a similar organizational structure—introduction to the group, summary of Siwalik taxa, ecomorphological traits, evolutionary patterns and biogeography, specialist insights—and include a standardized stratigraphic range chart depicting observed first and last occurrence dates of taxa between 18 and 6 Ma in the Potwar Plateau region of northern Pakistan.

Nearly every mammalian group documented between 18 and 6 Ma in the Potwar Plateau is addressed here at the ordinal or family level, the exceptions being the Sanitheriidae, suoid-like artiodactyls present in the Middle Miocene Siwaliks (van der Made and Hussain 1992), with a last occurrence in the Potwar at 14.1 Ma, and Pecora *incertae sedis* from the basal Kamlial Formation, dating to 17.9 Ma. Siwalik fishes, non-avian reptiles, and birds each have their own chapter. Amphibians, poorly represented in the fossil record, are not included. The chapter on invertebrates addresses two groups of Mollusca—bivalves and gastropods. Arthropods are not addressed; the sparse record includes attributions to Brachyura (crabs), and, more specifically Panopeidae (mud crabs) in the late Miocene (Rodney Feldmann, pers. comm. 2006), as well as ostracods.

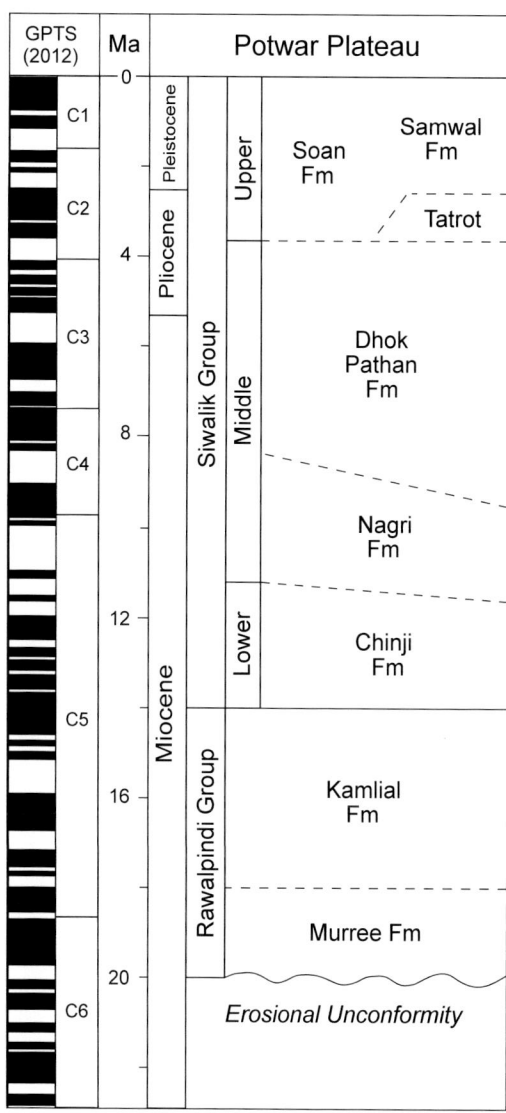

Figure II.1. Potwar Plateau rock-formation stratigraphy calibrated to the Gradstein et al. 2012 geological time scale (modified from Fig. 2.2; see Chapter 2 for details).

This section does not include a chapter on Siwalik flora. Few in situ fragments of fossil wood and leaf impressions have been documented in the Potwar; neither pollen nor phytoliths were detected in Potwar sediments in multiple attempts. In the absence of well-preserved flora, we rely on indirect evidence; geochemical analyses of paleosols, mammal teeth, and marine sediments; along with pollen and plant fossil records in the central and eastern Siwaliks to infer vegetation type and change through time (see Framing and Synthetic chapters).

References

Gradstein, F. M., J.G . Ogg, M. D. Schmitz, and G. M. Ogg, 2012. *The Geological Time Scale 2012*. Amsterdam: Elsevier.

Van der Made, J., and T. Hussain. 1992. Sanitheres from the Miocene Manchar Formation of Sind, Pakistan and remarks on sanithere taxonomy and stratigraphy. In *Proceedings of the Koninklijke Nederlandse Akademie van Wetenschappen* (1990). 95(1): 81–95.

6

FRESHWATER MOLLUSCAN FOSSILS OF THE SIWALIK RECORD OF THE POTWAR PLATEAU

DAMAYANTI GURUNG

INTRODUCTION

The Potwar Plateau in northern Pakistan is part of the Himalayan foreland basin that filled with the alluvial sediments of Neogene to Quaternary age. This sedimentary sequence is called the Siwalik Group. The sediments are deposits of ancient rivers with vertebrate, invertebrate, and trace fossils, along with some pollen and macroflora, and extend along the southern margin of the Himalayas from Pakistan east to Myanmar. Siwalik invertebrate fossils are dominated by freshwater molluscs.

Miocene to Pliocene Siwalik sediments contain molluscan fossils in various states of preservation. Collection began in the nineteenth and early twentieth centuries in India, Pakistan, and Myanmar (Hislope 1860; Forbes 1868; Vredenburg and Prashad 1921; Annandale 1921a, 1924; Prashad 1925a, 1927; Vokes 1935, 1936), with more recent reports beginning around 1970 (Modell 1969; Bhatia and Mathur 1971, 1973; Bhatia 1974; Srivastava and Arora 1983; Varun 2013; Bhandari, Köhler, and Tiwari 2014; Sharma, Singh, and Vipina 2015). Freshwater molluscan fossils from the Siwalik Group in Nepal were reported by West et al. (1975) and have been comprehensively reviewed by Takayasu (1992), Takayasu, Gurung, and Matsuoka (1995), Gurung, Takayasu, and Matsuoka (1997), and Gurung (1998).

The extant freshwater molluscs of the Indian subcontinent and surrounding areas are relatively well documented (Annandale 1920, 1921b; Prashad 1930; Brandt 1974; Subba Rao 1989; Nesemann et al. 2005; Ramakrishna and Dey 2007; Afshan et al. 2013; Kakar et al. 2017). Diverse molluscan faunas inhabit large and small rivers, lakes and ponds, floodplain lakes, and permanent and seasonal wetlands and marshes (Annandale and Prashad 1921a, 1921b; Annandale and Srinivasa Rao 1923; Prashad 1925b).

The freshwater molluscs consist of bivalves (clams, mussels) and gastropods (snails). Freshwater bivalves belong in six orders: Arcoida, Mytiloida, Unionida, Veneroida, Myoida and Anomalodesmata (Graf and Cummings 2007; Carter et al. 2011). A majority of freshwater bivalve species belong to the order Unionida, with six families occupying distinct geographical regions. The extant freshwater mussel family Unionidae is the most species-rich, with more than 600 species distributed in the freshwater ecosystems of Asia, Europe, North America, and Africa (Graf and Cummings 2007; Bogan and Roe 2008; Lopes-Lima et al. 2017). In the Indian subcontinent, freshwater mussels are mainly unionids or members of the order Veneroida (families Corbiculidae, Sphaeriidae and Etheriidae; see Brandt 1974; Subba Rao 1989; Graf 2013). Gastropoda are worldwide in distribution, with more than 4,000 species reported (Strong et al. 2008). Siwalik gastropod taxa represent the three subclasses: Caenogastropoda, Opisthobranchia, and Pulmonata (Subba Rao 1989; Boucher and Rocroi 2005; Strong et al. 2008). Fourteen freshwater caenogastropod families are represented on the Indian subcontinent: Ampullariidae, Viviparidae, Paludomidae, Pachychilidae, Pleuroceridae, Thiaridae, Amnicolidae, Bithyniidae, Hydrobiidae, Pomatiopsidae, Valvatidae, and the Pulmonate families Lymnaeidae, Planorbidae and Physidae (Subba Rao 1989; Boucher and Rocroi 2005; Ramakrishna and Dey 2007; Strong et al. 2008).

Freshwater molluscan fossils have been reported in the Siwalik Group from northern Pakistan to Nepal. The gastropod fossils reported from the Siwaliks belong to two orders: Prosobranchia (represented by the families Ampullariidae, Viviparidae, Valvatidae, Pleuroceridae, Thiaridae, Melanopsidae, Hydrobiidae, Bithyniidae) and Pulmonata (represented by the families Lymnaeidae, Ancylidae, Planorbidae, Physidae). Most of the Siwalik freshwater bivalve fossils belong to the order Unionida (family Unionidae) with a few in the order Veneroida (families Corbiculidae, Pisidiidae).

Freshwater molluscan fossils are important indicators of sedimentary environments because they are generally preserved close to their original habitat. Furthermore, freshwater molluscan fossils document past geographical distributions as they have restricted dispersal ability due to their constrained environments and limited mobility (Green and Figuerola 2005; Daraio, Weber, and Newton 2010; Kappes and Haase 2012; Van Leeuwen et al. 2013). Freshwater gastropods and bivalves generally disperse through connected waterways or when carried by aquatic or terrestrial animals (Rees 1965; Graf 1997; Dillon

2000; Bauer 2001; Wächtler, Mansur, and Richter 2001; Green and Figuerola 2005; Kappes and Haase 2012; Pfeiffer and Graf 2015). Studies of the fossil record of these molluscs are useful for detecting paleoenvironmental changes and ways in which freshwater taxa have colonized evolving river systems. Hence, the study of the fossil molluscan fauna of the Potwar Plateau is informative biogeographically, given the plateau's location at the boundary of three major zoogeographical regions: the northern Palearctic realm, the southeastern Indo-Malayan realm, and the western Ethiopian realm.

FRESHWATER MOLLUSCS OF THE SIWALIK GROUP, POTWAR PLATEAU

Freshwater molluscan fossils from the Siwalik Group in the Potwar Plateau were first described by H. E. Vokes (1935, 1936). He carried out detailed identifications and descriptions of the fossils collected by the first and second Yale North India Expeditions. He described eight new species of Unionidae, belonging to three genera, *Lamellidens*, *Parreysia*, and *Indonaia*, from the Siwalik record of the Potwar Plateau (Table 6.1).

During the Munich research program in Pakistan from 1955–1956, numerous bivalve molluscan fossils were collected from the Siwalik Group in the Potwar Plateau, as well as from the Murree Formation and Nagpur, Central Deccan Traps, India. Freshwater molluscan fossils from the Siwalik Group were analyzed by Modell (1969), who described 28 new species be-

longing to 13 genera (see Table 6.1). (Some of these were later synonymized with other genera, as noted in parentheses in Table 6.1.) The most common freshwater bivalves in the Siwalik sediments were species in the genera *Lamellidens*, *Parreysia*, and *Indonaia*, which have also been reported from Siwalik Group rocks of India and Nepal. The entire molluscan fauna of the Siwalik Group of the Potwar is more diverse, with additional species belonging to genera *Rectidens*, *Physunio*, *Pseudodon*, and *Trapezoideus* (Table 6.2), found to occur earlier than in India and Nepal.

No published reports to date have focused on gastropod fossils from the Siwalik Group of the Potwar Plateau of Pakistan. Specimens currently on loan to Harvard University from the Geological Survey of Pakistan (GSP) contain many gastropod shells not yet studied in detail. They are identified as belonging to the families Ampullariidae, Viviparidae, Thiaridae, and Planorbidae (Table 6.2). The fossil gastropods from the Siwalik sediments of India and Nepal include the above-mentioned four families and, in addition, gastropods from the families Melanopsidae, Hydrobiidae, Bithyniidae, Lymnaeidae, and Physidae (Prashad 1925a; Bhatia 1974; Srivastava and Arora 1983; Gurung, Takayasu, and Matsuoka 1997; Gurung 1998; Bhandari, Köhler, and Tiwari 2014). This difference in taxonomic composition may be due to collection bias and preservation differences, because many gastropod shells are thin and small, as well as differences in alluvial environments.

Table. 6.1. Freshwater Molluscan Fossils from the Siwalik Group of the Potwar Plateau

Sub-group	Formations	Age	Author	Molluscan Taxa
Upper Siwalik	Tatrot	Upper Pliocene	Vokes (1935, 1936)	*Lamellidens lewisi, Lamellidens* sp. B, *L. jammuensis* (Parshad 1927), *L. turgidus, Parreysia tatrotensis, P. deterrai, P. near corrugata, Indonaia mittali* (Parshad 1927), *Indonaia* sp. B
Middle Siwalik	Dhok Pathan	Upper Miocene to Lower Pliocene (~3.5–9.5 Ma)	Vokes (1935, 1936) Modell (1969)	*Lamellidens wadiai, P. tatrotensis, Indonaia hasnotensis, I. mittali* (Prashad 1927) *Lamellidens marginalis vokesi, L. marginalis feueri, Parreysia dhokpathana, P. nearchi, P. gnanami, P. discus, , P. Nannonaia* (=*Indonaia*) *parlewalensis, Nannonaia* (=*Indonaia*) *hasnotensis* (Vokes 1935), *Nannonaia* (=*Indonaia*) *mittalli* (Prashad1927), *Leguminaia* (=*Lamellidens*) *soanensis, Rectidens haquei, R. datwalensis, Physunio* (=*Lamellidens*) *pakistanus, Cosmopseudodon* (=*Lamellidens*) *indicus, Unio schlagintweitianus*
	Nagri	Upper Miocene (9.5–11.5 Ma)	Modell (1969)	*Bellamya* sp., *Pila* sp., *Brotia* sp. B, *Indoplanorbis* sp. *Lamellidens marginalis vokesi, Parreysia praecursor, P. dayi, P. gnanami, Nannonaia* (=*Indonaia*) *pachysomoides, Leguminaia* (=*Lamellidens*) *nagriensis, Hemisolasma* (=*Indonaia*) *annandalei, Rectidens haquei* *Bellamya* sp., *Pila* sp.
Lower Siwalik	Chinji	Upper Miocene (11.5–14.0 Ma)	Vokes (1935, 1936) Modell (1969)	*Lamellidens subparallelus, Lamellidens* sp. A, *Lamellidens* sp. C, *Parryesia tatrotensis, Indonaia prashadi* *Lamellidens marginalis vokesi, L. lamellatus feueri, Parreysia praecursor, P. raniensis, Dehminaia dehmi, Nannonia* (= *Indonaia*) *blandfordi, N.* (=*Indonaia*) *pachysomoides, N.* (=*Indonaia*) *pinfoldi, Pseudodon oettingenae, Monodontina* (=*Lamellidens*) *mogul, Hemisolasma* (=*Indonaia*) *ahmedi, Amblemella gundae, Limnoscapha* (=*Pseudodon*) *vidali, , Trapezoideus* (=*Lamellidens*) *zoebeleini* *Pila* sp., *Brotia* sp. A,

Notes: Bivalves are described by Vokes (1935, 1936) on the First and Second Yale North India Expedition (1931–1933) and Modell (1969) on the Munich research trip to the Potwar Plateau (1955–1956). Gastropod fossils were studied by this author; the fossils are currently on loan to Harvard University from the Geological Survey of Pakistan. Some genera are synonymized with others and are noted in parentheses.

Table 6.2. Taxonomy of Freshwater Molluscs of the Siwalik Group of the Potwar Plateau

Class: Gastropoda
Subclass: Prosobranchia Milne-Edwards, 1848
Superorder: Caenogastropoda Cox, 1960
Order: Architaenioglossa
　　Superfamily Ampullarioidea
　　　Family Ampullariidae
　　　　Genus: *Pila* Röding, 1798
　　Superfamily Viviparoidea
　　　Family Viviparidae
　　　　Genus: *Bellamya* Jousseaume, 1886
Order: Sorbeoconcha
　　Superfamily Cerithioidea
　　　Family Pachychilidae
　　　　Genus: *Brotia* Adams, 1866
　　Superfamily Planorboidea
　　　Family Planorbidae
　　　　Genus: *Indoplanorbis* Annandale and Prashad, 1921
Class: Bivalvia Linnaeus, 1758
Order: Unionoida
　　Superfamily Unionoidea Rafinesque, 1820
　　　Family Unionidae Fleming, 1828
　　　　Subfamily Parreysiinae Henderson, 1935
　　　　　Genus: *Parreysia* Conrad, 1853
　　　　　Genus: *Lamellidens* Simpson, 1900,
　　　　　Genus: *Radiatula* (=Indonaia) Simpson, 1900
　　　　Subfamily Goniderinae Ortmann, 1916
　　　　　Genus: *Pseudodon* Gould, 1844
　　　　　Genus: *Rectidens* Simpson, 1900

STRATIGRAPHIC RANGES OF MOLLUSCAN FOSSILS FROM THE POTWAR PLATEAU

The molluscan fossil record of the Siwalik Group, including the Potwar Plateau, is not as well studied as the vertebrate fossil record. Stratigraphic ranges of bivalve fossil taxa of the Potwar Plateau are presented as much as possible at the species level, but many fossil shells are in fragments, while gastropod fossil taxa are presented at the generic level. Many molluscan fossils were difficult to identify due to poor preservation. Despite this, stratigraphic distributions of molluscan taxa from the Potwar Plateau (Figs. 6.1a, b) are based on published records (Vokes 1935, 1936; Modell 1969) and this author's study of collections currently on loan to Harvard University from the GSP. The Siwalik sediments of the Potwar Plateau are divided, based on lithofacies, into Kamlial, Chinji, Nagri and Dhok Pathan Formations (Badgley and Behrensmeyer 1995; Zaleha 1997; Barry et al. 2002; Behrensmeyer and Barry 2005). Chronostratigraphy is based mainly on paleomagnetism, supplemented by isotopic dating of rare volcanic ash beds (Tauxe and Opdyke 1982; Johnson et al. 1982; see Chapter 2, this volume).

The fossil gastropod fauna belongs to families Ampullariidae, Viviparidae, Thiaridae and Planorbidae and is represented by four genera: *Pila, Bellamya, Brotia* and *Indoplanorbis* (Figs. 6.1, 6.2). Specimens of *Pila* sp., family Ampullariidae, were documented by occurrences of their calcareous opercula near the base of the Kamlial Formation; this is the earliest known record

of this genus. The fossilized opercula become more abundant between 11.5 and 9.1 Ma in the upper part of the Chinji Formation through the Nagri Formation. The youngest occurrence of *Pila* in the Potwar Plateau is around 8.0 Ma (Locality Y547) in the Dhok Pathan Formation. The only other report of opercula of *Pachylabra* (= *Pila*) *prisca* from the Lower Siwaliks is from Poonch, Kashmir (Prashad 1925a). Fossils of the genus *Bellamya* in the family Viviparidae first occur in the Potwar Siwaliks at around 10.1 Ma (Locality Y311) in the Nagri Formation. Their occurrences increase from 7.0 (Locality Y908) to 6.4 Ma (Locality Y913) in the Dhok Pathan Formation. Fossil gastropods belonging to the genus *Brotia*, family Thiaridae, first occur around 11.4 Ma (Locality Y76) in the Chinji Formation and again around 7.4 Ma (Locality Y944) in the Dhok Pathan Formation. Because the shell morphology of the fossils from these two localities is different, they are identified as *Brotia* sp. A and *Brotia* sp. B, respectively (Figs. 6.1a,b). First occurrence of the genus *Indoplanorbis*, family Planorbidae, is around 7.4 Ma (Locality Y926) with a last occurrence around 6.4 Ma (Locality Y913).

According to reports by Vokes (1935, 1936) and Modell (1969), fossil bivalves in the Siwalik Group of the Potwar Plateau include seven genera of Unionidae: *Lamellidens, Parreysia, Indonaia, Pseudodon, Rectidens, Physunio,* and *Trapezoideus* (Fig. 6.3). Close examination of shells possibly representing *Physunio* and *Trapezoideus* was inconclusive, and these two genera are therefore not included in Fig. 6.3. The large *Lamellidens* sp. appears in the fossil record around 12.4 Ma (Locality Y496). The genus *Parreysia* sp. appears around 13.0 Ma (Locality Y499). These genera, *Lamellidens* sp. and *Parreysia* sp., are recorded at around the same time in the Chinji Formation and become more common after 11 Ma in the Nagri and Dhok Pathan Formations. The genus *Indonaia* appears around 14.1 Ma (Locality Y758) in the Chinji Formation and becomes common between 12 and 11 Ma, then rare between 11 and 9 Ma; some occurrences in the Dhok Pathan Formation are noted after 9 Ma. Large-shelled *Pseudodon* first appears at around 13.9 Ma (Locality Y694) in the Chinji Formation and is rare after 11 Ma. The genus *Rectidens* was reported by Modell (1969) in the Middle and Upper Siwaliks. The stratigraphic distribution of all fossil molluscan taxa from the Potwar Plateau is presented in Fig. 6.1.

There are currently more documented molluscan fossils from the Siwalik record of the Potwar Plateau than from Siwalik deposits in India and Nepal. Those from the Potwar Plateau (spanning 16 to 6 Ma) were large-shelled bivalve species of the Unionidae with Southeast Asian affinities (see fig. 6.1a). Opercula of the genus *Pila* from around 16.8 Ma (Locality Y802) are among the earliest known; another record of opercula from the Lower Siwalik series (Chinji-equivalent) from Poonch, Kashmir (Prashad 1925a) is of indeterminate absolute age. In India and Nepal, most freshwater molluscan fossils are from the Middle and the Upper Siwaliks (Prashad 1927; Srivastava and Arora 1983; Takayasu, Gurung, and Matsuoka 1995; Gurung, Takayasu, and Matsuoka 1997; Gurung 1998; Bhandari, Köhler, and Tiwari 2014);

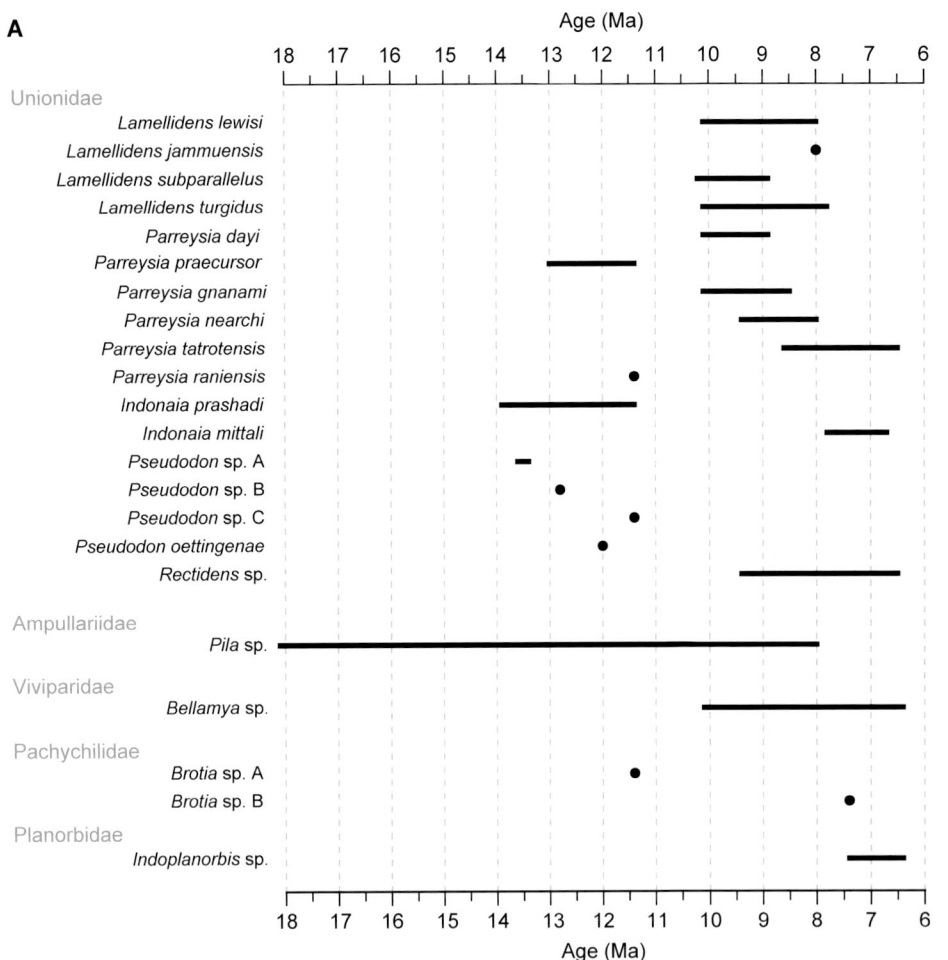

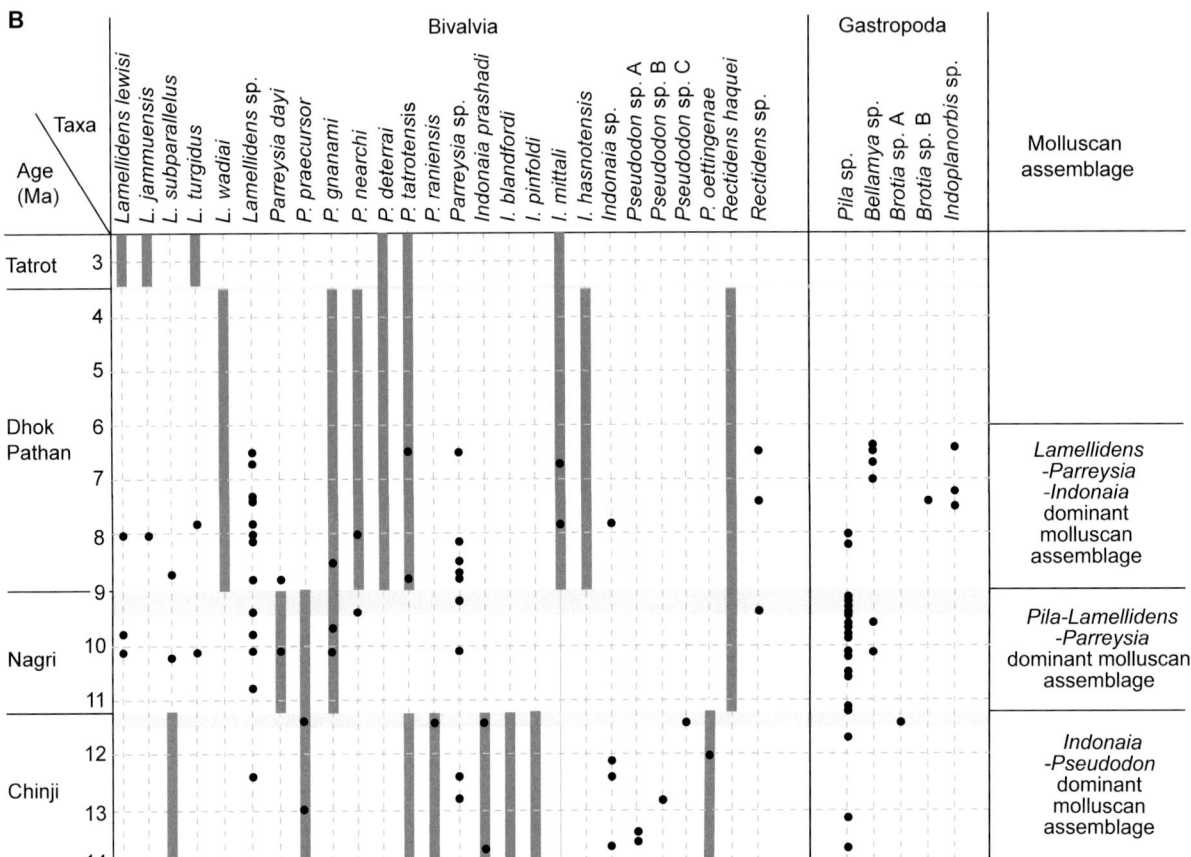

Figure 6.1. Stratigraphic range of freshwater molluscan fossil lineages in the Siwalik Group of the Potwar Plateau, Pakistan, as documented in the Harvard-GSP database (A) and in greater detail (B), with stratigraphic division into formations and their age range based on Chapter 2 (this volume). Bands indicate time-transgressive nature of formation boundaries. Black dots indicate molluscan occurrences with age of the localities. Vertical gray lines indicate molluscan faunal range as reported by Vokes (1935, 1936) and Modell (1969).

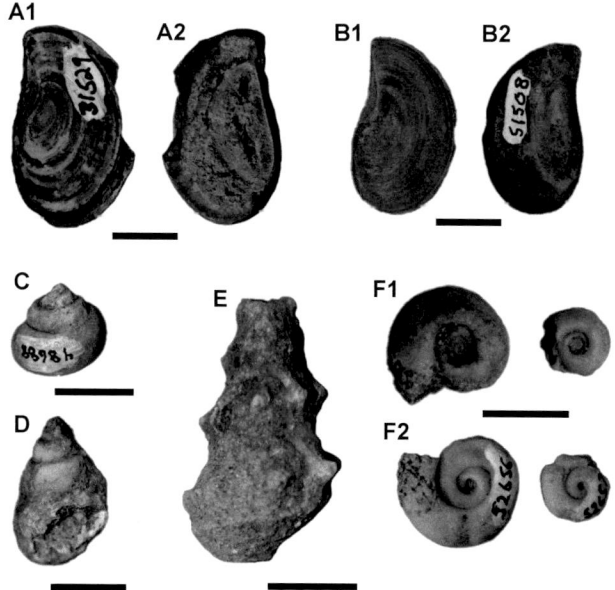

Figure 6.2. Freshwater gastropod fossils from the Siwalik Group, Potwar Plateau, from the Harvard-GSP collection: A1 operculum of *Pila* sp. (outer surface) (YGSP 31529 from Locality Y537); A2. operculum of *Pila* sp. (inner surface); B1. operculum of *Pila* sp. (outer surface) (YGSP 51508 from Locality Y991); B2. operculum of *Pila* sp. (inner surface); C. *Bellamya* sp. (YGSP 48688 from Locality Y311); D. *Bellamya* sp. (YGSP 52272 from Locality Y919); E. *Brotia* sp. (YGSP 46590 from Locality Y311); F1. *Indoplanorbis* sp. (apical view) (YGSP 52656 and YGSP 52657 from Locality Y970); F2. *Indoplanorbis* sp. (umbilical view). Scale bar represents 1 cm.

few molluscan fossils are reported from the Lower Siwaliks (Varun 2013).

ENVIRONMENTAL IMPLICATIONS OF SIWALIK MOLLUSCS

The fossil molluscan genera from the Potwar Plateau belong to the extant molluscan faunas of the warm humid regions of the Indian subcontinent and Southeast Asia (Brandt 1974; Subba Rao 1989; Graf and Cummings 2007).

Species in the genus *Pila*, family Ampullariidae, are common and widely distributed in tropical Asia and Africa today (Subba Rao 1989; Hayes et al. 2015). The genus *Pila* possesses a calcified operculum, forming a seal during aestivation, when the snail becomes dormant in response to drought or cold temperatures (Seuffert and Martin 2010; Giraud-Billoud et al. 2011; Hayes et al. 2015; Glasheen et al. 2017). In general, species of *Pila* inhabit shallow, slow-moving rivers, lakes, ponds, canals, rice paddies, and swamps with aquatic vegetation (Prashad 1925b). Many ampullariids are generalist feeders, capable of feeding on a variety of items such as biofilms, periphyton, and floating, submerged, and emergent aquatic vegetation. In the Potwar Plateau, fossil opercula of genus *Pila* occur in all four Miocene Siwalik formations (Fig. 6.1a, b) and are common between 11.5 and 9.1 Ma (Fig. 6.1b). Abundant occur-

Figure 6.3. Freshwater bivalve fossils from the Siwalik Group, Potwar Plateau: A. *Lamellidens lewisi* Vokes, 1935, right valve (Holotype, YPM 15030); B *Parreysia tatrotensis* Vokes, 1935, right valve (Holotype, YPM 15036); C. *Lamellidens subparallelus* Vokes, 1936, left valve (YPM 15046); D. *Indonaia hasnotensis* Vokes, 1936, left valve (Holotype YPM 15033); E. *Pseudodon* sp. B, right valve (YGSP 26740 from Locality Y714); F. *Parreysia gnanami* Modell, 1969, left valve (YGSP 12505 from Locality Y330); G. *Pseudodon oettingenae* Modell, 1969, left valve (Type, BSP 1561I-156, 1178); H. *Lamellidens vokesi* Modell, 1969, right valve (BSP 1956II-169, 1110) I. *Rectidens haquei* Modell, 1969, right valve (BSP 1956II-193, 1320). Scale bar represents 1 cm.

rences of *Pila* sp. in the Nagri Formation likely indicate the prevalence of freshwater environments with seasonal changes in flow and probably periodic drought.

The viviparid genus *Bellamya* occurs today in Africa, the Indian subcontinent, and Southeast Asia (Subba Rao 1989; Ramakrishna and Dey 2007). Species inhabit permanent shallow, slow-flowing water bodies, such as rivers, lakes, ponds, canals, rice paddies. and swamps, and are mostly detritivores and filter feeders, which ingest bacteria with detritus (Dillon 2000). Occurrences of fossil *Bellamya* species (Fig. 6.2c, 6.2d) suggest the presence of semi-permanent, shallow aquatic habitats and less seasonal change; they are common between 6.9 and 6.4 Ma in the Dhok Pathan Formation (Fig. 6.1b).

Species belonging to the genus *Brotia* live in freshwater rivers from northeastern India (Ganges-Brahmaputra) to Sumatra

(Kohler and Glaubrecht 2006; Kohler and Dames 2009). The gastropods of this genus are mostly algivores (scrapers), which inhabit fast-flowing water in mountain streams as well as slow-flowing rivers; these snails occur on rocky substrates as well as on soft surfaces of mud, sand, and plant root systems (Kohler and Glaubrecht 2001). In the Siwalik record of the Potwar Plateau, fossils of *Brotia* sp. are found at just two localities, in the upper Chinji Formation and in the Dhok Pathan Formation (Fig. 6.2e). However, these occurrences suggest a far wider geographic distribution than known at present.

The family Planorbidae, genus *Indoplanorbis*, includes common snails across the Indian subcontinent and Southeast Asia (Brandt 1974; Subba Rao 1989). They live close to the surface, avoid turbulent waters, and inhabit ponds and semi-permanent wetlands with aquatic vegetation where they can survive the dry season buried in mud (van Damme 1984; Raut, Rahman, and Samanta 1992; Islam et al. 2012). Planorbid fossils were found at localities between 7.4 and 6.4 Ma in the Dhok Pathan Formation (Fig. 6.2f). Their occurrences suggest the presence of shallow water bodies that experienced seasonal drought.

The unionid bivalves of the Potwar Plateau include the genera *Lamellidens*, *Parreysia*, *Indonaia*, *Pseudodon*, and *Rectidens* (Fig. 6.3). Extant species of these genera inhabit more permanent aquatic environments (Brandt 1974; Subba Rao 1989; Strong et al. 2008; Graf 2013). Adult unionids live partially buried in sediments of large rivers and floodplain lakes, streams, and lakes. They are infaunal filter feeders, which ingest inorganic and organic particles (diatoms, algae cells, bacteria) (Watters 1996; Dillon 2000). Unionids reproduce sexually and include a parasitic life-cycle stage. Larvae (glochidia) attach to the external surface of a suitable host, such as fishes or amphibians (Dillon 2000; Barnhart, Haag, and Roston 2008; Graf 2013; Reis, Collares-Pereira, and Araujo 2014). Consequently, dispersal in freshwater environments is determined by the movements of the host animal. In the Siwalik Group of the Potwar Plateau, bivalve fossils were present from 14.1 to 6.5 Ma. Their common presence indicates the prevalence of large, stable freshwater environments.

MOLLUSCAN FAUNAL CHANGES IN THE SIWALIK GROUP, POTWAR PLATEAU

Although freshwater molluscan fossils are found throughout much of the Siwalik Group of the Potwar Plateau, more work on species-level identifications is needed, along with additional specimen collection from well-documented localities. The stratigraphic distribution and occurrences of freshwater molluscan taxa begin to reveal assemblage patterns. In general, three molluscan assemblage zones can be distinguished, from older to younger: *Indonaia-Pseudodon* dominant molluscan assemblage; *Pila-Lamellidens-Parreysia* dominant molluscan assemblage; and *Lamellidens-Parreysia-Indonaia* dominant molluscan assemblage (Fig. 6.1b).

The earliest molluscan assemblage, *Indonaia-Pseudodon* dominant molluscan assemblage, occurs almost entirely in the Chinji Formation between 14.0 and 11.2 Ma. Large-shelled

Pseudodon was represented by four species: *P. oettingenae*, *Pseudodon* sp. A, *Pseudodon* sp. B, *Pseudodon* sp. C, and *Indonaia* by *Indonaia parshadi*, *I. blandfordi*, *I. pinfoldi*. A few taxa belonging to *Lamellidens* and *Parreysia* were also associated with the assemblage. Bivalve fossils were more abundant around 12 to 11 Ma. Gastropods, *Pila* sp. and *Brotia* sp., occur rarely. Common occurrences of *Pseudodon* sp. and *Indonaia* sp. indicate the presence of permanent bodies of water in rivers and floodplain lakes.

The *Pila-Lamellidens-Parreysia* dominant molluscan assemblage occurs between 11.2 and 9.0 Ma in the Nagri and Dhok Pathan Formations. In the Potwar Plateau, the Nagri-Dhok Pathan Formation boundary is time-transgressive between 9.5 and 9.0 Ma (Flynn et al. 2013; see Chapter 2, this volume). The assemblage comprises the gastropod *Pila* sp. and bivalves *Lamellidens lewisi*, *L. jammuensis*, *L. subparallelus*, *L. turgidus*, *Parreysia dayi*, *P. praecursor*, *P. gnanami*, and *P. nearchi*, along with a few *Rectidens haquei*. In the upper part of the Chinji Formation these genera were rare. The gastropod *Pila* sp. increased in abundance in fossil localities between 11 and 9 Ma. The abundance of *Pila* indicates extensive occurrence of shallow water environments with aquatic vegetation, which were susceptible to periodic drought and experienced seasonal changes in flow. At the same time, the large bivalves, such as species of *Lamellidens* and *Parreysia*, indicate the presence of larger permanent freshwater environments.

The youngest molluscan assemblage, *Lamellidens-Parreysia-Indonaia* dominant molluscan assemblage, is found between 9.0 and 6.0 Ma in the Dhok Pathan Formation. This assemblage was dominated by bivalves *Parreysia gnanami*, *P. nearchi*, *P. deterrai*, and *P. tatrotensis*, along with *Lamellidens lewisi*, *L. turgidus*, and *L. wadiai*, as well as *Indonaia mittali* and *I. hasnotensis* and *Rectidens haquei*. Gastropod taxa *Bellamya* sp., *Brotia* sp., and *Indoplanorbis* sp. were rare. This assemblage indicates the presence of permanent freshwater habitats, such as large rivers or shallow floodplain lakes.

The freshwater fossil molluscan assemblages of the Potwar Plateau record are more diverse than at present, which indicates that there were more favorable and extensive freshwater habitats in the past. The occurrences of taxa *Brotia* sp., *Pseudodon* sp., and *Rectidens* sp. (Fig. 6.1a, b) in the Potwar region indicate a wide geographical distribution of this Southeast Asian fauna (Brandt 1974; Subba Rao 1989; Kohler and Glaubrecht 2002) and thus suggest that the paleo-drainage of the Potwar Plateau may have been connected to a southeastern river system, such as the Ganges-Brahmaputra Rivers. Studies on paleocurrent directions in the Siwalik of Potwar Plateau have suggested a southeastward flow toward the Ganges drainage before 5 Ma (Cerveny et al. 1989; Burbank, Beck, and Mulder 1996; Clift and Blusztajn 2005). The species of *Indonaia* are more common today in northeastern India, in Assam, West Bengal, and Myanmar (Brandt 1974; Subba Rao 1989).

CONCLUSION

This chapter has reviewed the stratigraphic distribution and occurrence of the freshwater molluscan fossil fauna of the Siwalik

Group of the Potwar Plateau, using published reports and recent fossil collections. The molluscan fossil distribution documents three assemblages, between 14 and 6 Ma; from older to younger, they are *Indonaia-Pseudodon* dominant molluscan assemblage, from 14.0 to 11.2 Ma; *Pila-Lamellidens-Parreysia* dominant molluscan assemblage, from 11.2 to 9.0 Ma; and *Lamellidens-Parreysia-Indonaia* dominant molluscan assemblage, from 9.0 to 6.0 Ma. More work on species-level identifications, with additional specimens from well-documented localities, will provide a more comprehensive understanding of lineage occurrences.

The molluscan assemblages indicate successive environmental changes. The mostly Middle Miocene *Indonaia-Pseudodon* assemblage dominated the Chinji Formation. Common occurrence of these unionids between 14.0 and 11.2 Ma indicates the presence of large permanent rivers and lakes with little fluctuation in freshwater habitats. Between 11.2 and 9.0 Ma, the molluscan assemblages consisted mainly of *Pila*, *Lamellidens*, and *Parryesia*; the abundant presence of the gastropod *Pila* sp. suggests that freshwater environments were more susceptible to seasonal fluctuations with droughts. The youngest molluscan assemblage, *Lamellidens-Parreysia-Indonaia* present between 9.0 and 6.0 Ma in the Dhok Pathan Formation, indicates the return of more stable large fluvial environments.

The Siwalik record of the Potwar Plateau includes freshwater molluscan taxa present in the Indian subcontinent and Southeast Asia today. Most species, including *Lamellidens* sp., *Parreysia* sp., *Indonaia* sp., *Pila* sp., *Bellamya* sp., and *Indoplanorbis* sp., are common at present in the Ganges-Brahmaputra river system and in rivers in Myanmar (Brandt 1974; Subba Rao 1989). In addition, the occurrence in the Potwar Plateau of *Pseudodon* sp., *Rectidens* sp., and *Brotia* sp., of Southeast Asian faunal affinity, indicates a far wider geographical distribution of these lineages during the Miocene. Their distribution suggests connection of the river systems of the Potwar Plateau to an eastward-flowing river system, such as the Ganges-Brahmaputra, where these genera are at present abundant (Brandt 1974; Subba Rao 1989; Kohler and Glaubrecht 2001). Further study of the Siwalik molluscan fossils will greatly contribute to understanding past changes in the environment and distribution of this fauna.

ACKNOWLEDGMENTS

I would like to express my appreciation to all for providing me with the opportunity to write this chapter. I greatly appreciate the editorial advice and reviews and also thank Harvard University, Yale University, and also Prof. Alexander Nützel, Ludwig-Maximilians-Universität München, for access to their molluscan fossil collections.

References

Afshan, K., M. A. Beg, I. Ahmad, M. M. Ahmad, and M. Qayyum. 2013. Freshwater snail fauna of Pothwar Region, Pakistan. *Pakistan Journal of Zoology* 45(1): 227–233.

Annandale, N. 1920. XV. Materials for a generic revision of the freshwater gastropod molluscs of the Indian Empire. Part II. The Indian genera of Viviparidae. *Records of the Indian Museum* 19:107–115.

Annandale, N. 1921a. Indian fossil Viviparae. *Records of the Geological Survey of India* 51:362–367, plate 11.

Annandale, N. 1921b. The genus *Temnotaia* (Viviparidae). *Records of the Indian Museum* 22:293–295.

Annandale, N. 1924. Fossil molluscs from the oil-measures of the Dawna Hills, Tenasserim. *Records of the Geological Survey of India* 55:97–101, plates 1–3.

Annandale, N., and B. Prashad. 1921a. II. Materials for a generic revision of the freshwater gastropod molluscs of the Indian Empire. Part IV. The Indian Ampullariidae. *Records of the Indian Museum* 22:7–12.

Annandale, N., and B. Prashad. 1921b. The aquatic and amphibious Mollusca of Manipur. The Prosobranchia. *Records of the Indian Museum* 22:529–631, plates 4–8.

Annandale, N., and H. Srinivasa Rao. 1923. The molluscs of the Salt Range, Punjab. *Records of the Indian Museum* 25:387–397, plate 9.

Badgley, C., and A. K. Behrensmeyer. 1995. Two long geological records of continental ecosystems. *Palaeogeography, Palaeoclimatology, Palaeoecology* 115(1–4): 1–11.

Barnhart, M. C., W. R. Haag, and W. N. Roston. 2008. Adaptations to host infection and larval parasitism in Unionoida. *Journal of North American Benthological Society* 27(2): 370–394.

Barry, J. C., M. E. Morgan, L. J. Flynn, D. Pilbeam, A. K. Behrensmeyer, S. M. Raza, I. A. Khan, C. Badgley, J. Hicks, and J. Kelley. 2002. Faunal and environmental change in the late Miocene Siwaliks of northern Pakistan. *Paleobiology* 28(S2): 1–71.

Bauer, G. 2001. Factors affecting Naiad occurrence and abundance. In *Ecological Studies. 145—Ecology and Evolution of the Freshwater Mussels Unionoida*, edited by G. Bauer and K. Wächtler, 155–162. Berlin: Springer.

Behrensmeyer, A. K., and Barry, J. C. 2005. Biostratigraphic surveys in the Siwaliks of Pakistan: A method for standardized surface sampling of the vertebrate fossil record. *Palaeontologia Electronica* 8 (1): 15A–24A.

Bhandari, A., F. Köhler, F., and B. N. Tiwari. 2014. Neogene freshwater gastropods from the Siwalik Group in Uttarakhand, northern India: First fossil record of the protoconch of a freshwater cerithioidean snail (Mollusca, Caenogastropoda). *Himalayan Geology* 35(2): 182–189.

Bhatia, S. B. 1974. Some Pleistocene molluscs from Kashmir, India. *Himalayan Geology* 4:371–395.

Bhatia, S. B., and A. M. Mathur. 1971. Late Pleistocene gastropods from Nalagarh Tehsil, Himachal Pradesh. *Journal of the Geological Society of India* 12(3): 280–285.

Bhatia, S. B., and A. M. Mathur. 1973. Some Upper Siwalik and Late Pleistocene molluscs from Punjab. *Himalayan Geology* 3:24–58.

Bogan, A. E., and K. J. Roe. 2008. Freshwater bivalve (Unioniformes) diversity, systematics, and evolution: Status and future directions. *Journal of the North American Benthological Society* 27(2): 349–369.

Boucher, P., and J. P. Rocroi, 2005. Classification and nomenclature of gastropod families. *Malacologia* 47(1–2): 1–397.

Brandt, R. A. M. 1974. The non-marine aquatic Mollusca of Thailand. *Achiv für Molluskenkunde* 105:1–423, plates 1–30.

Burbank, D. W., R. A. Beck, T. and Mulder. 1996. The Himalayan foreland basin. In *The Tectonic Evolution of Asia*, edited by A. Yin and T. M. Harrison, 149–188. Cambridge: Cambridge University Press.

Carter, J. G., C. R. Altaba, L. C. Anderson, R. Araujo, A. S. Biakov, A. E. Bogan, D. C. Campbell, et al. 2011. A synoptical classification of the Bivalvia (Mollusca). *Paleontological Contributions* 4:1–47.

Cerveny, P. F., N. M. Johnson, R. A. K. Tahirkheli, and N. R. Bonis. 1989. Tectonic and geographic implications of Siwalik Group heavy minerals, Potwar Plateau, Pakistan. Special paper, *Geological Society of America* 232:129–137.

Clift, P. D., and J. Blusztajn. 2005. Reorganization of the western Himalayan river system after five million years ago. *Nature* 438:1001–1003.

Daraio, J. A., L. J. Weber, and T. J. Newton. 2010. Hydrodynamic modeling of juvenile mussel dispersal in a large river: The potential effects of bed shear stress and other parameters. *Journal of the North American Benthological Society* 29(3): 838–851.

Dillon, R. T. 2000. *The Ecology of Freshwater Mollusks*. Cambridge: Cambridge University Press.

Flynn, L. J., E. H. Lindsay, D. Pilbeam, S. M. Raza, M. E. Morgan, J. C. Barry, C. E. Badgley, et al. 2013. The Siwaliks and Neogene evolutionary biology in South Asia. In *Fossil Mammals of Asia: Neogene Biostratigraphy and Chronology*, edited by X. Wang, L. J. Flynn, and M. Fortelius, 353–372. New York: Columbia University Press.

Giraud-Billoud, M., M. Abud, J. Cueto, J., I. Vega, and A. Castro-Vazquez. 2011. Uric acid deposits and estivation in the invasive apple snail, *Pomacea canaliculata*. *Comparative Biochemistry and Physiology* 158:506–512.

Glasheen, P. M., C. Calvo, M. Meerhoff, K. A. Hayes, and R. L. Burks. 2017. Survival, recovery, and reproduction of apple snails (Pomacea spp.) following exposure to drought conditions. *Freshwater Science* 36(2): 316–324.

Graf, D. I. 1997. Sympatric speciation of freshwater mussels (Bivalvia: Unionoida): A model. *American Malacological Bulletin* 14:35–40.

Graf, D. L. 2013. Patterns of freshwater bivalve global diversity and the state of phylogenetic studies on the Unionoida, Sphaeriidae, and Cyrenidae. *American Malacological Bulletin* 31(1): 135–153.

Graf, D. L., and K. S. Cummings. 2007. Review of the systematics and global diversity of freshwater mussel species (Bivalvia: Unionoida). *Journal of Molluscan Studies* 73:291–314.

Green, A. J., and J. Figuerola. 2005. Recent advances in the study of long distance dispersal of aquatic invertebrates via birds. *Diversity and Distributions* 11:149–156.

Gurung, D. 1998. Freshwater molluscs from the Late Neogene Siwalik Group, Surai Khola, western Nepal. *Journal of Nepal Geological Society* 17: 728.

Gurung, D., K. Takayasu, and K. Matsuoka. 1997. Middle Miocene-Pliocene freshwater gastropods of the Churia Group, west-central Nepal. *Paleontological Research* 1(3): 166–179.

Hayes, K. A., R. L. Burks, A. Castro-Vazquez, P. C. Darby, H. Heras, P. R. Martin, J. Qiu, J., et al. 2015. Insights from an integrated view of the biology of apple snails (Caenogastropoda: Ampullariidae). *Malacologia* 58(1–2): 245–302.

Hislope, S. 1860. On the Tertiary deposits, associated with trap-rock, in the East-Indies. With description of the fossil shells. *Quaternary Journal of the Geological Society of London* 16:154–182, plates 5–10.

Islam, Z., M. Z. Alam, S. Akter, B. C. Roy, and M. M. H. Mondal. 2012. Distribution patterns of vector snails and Trematode cercaria in their vectors in some selected areas of Mymensingh. *Journal of Environmental Science and Natural Resources* 5(2): 37–46.

Johnson, N. M., N. D. Opdyke, G. D. Johnson, E. H. Lindsay, and R. A. K. Tahirkheli. 1982. Magnetic polarity stratigraphy and ages of Siwalik Group rocks of the Potwar Plateau, Pakistan. *Palaeogeography, Palaeoclimatology, Palaeoecology* 37:17–42.

Kakar, S., K. Kamran, S. A. Essote, and A. Iqbal. 2017. New locality report on freshwater clam, *Corbicula striatella* (Deshayes, 1854) (Bivalvia: Mollusca) with reference to freshwater bivalves from Baluchistan Province, Pakistan. *Journal of Entomology and Zoology Studies* 5(3): 231–235.

Kappes, H., and P. Haase. 2012. Slow, but steady: Dispersal of freshwater molluscs. *Aquatic Sciences* 74:1–14.

Kohler, F., and C. Dames. 2009. Phylogeny and systematics of the Pachychilidae of mainland South-East Asia—Novel insights from morphology and mitochondrial DNA (Mollusca, Caenogastropoda, Cerithioidea). *Zoological Journal of the Linnean Society* 157:679–699.

Kohler, F., and M. Glaubrecht. 2001. Toward a systematic revision of the Southeast Asian freshwater gastropod *Brotia* H. Adams, 1866 (Cerithioidea: Pachychilidae): An account of species from around the South China Sea. *Journal of Molluscan Studies* 67:281–318.

Kohler, F., and M. Glaubrecht. 2002. A new speices of *Brotia* H. Adams, 1866 (Caenogastropoda: Cerithioidea: Pachychilidae). *Journal of Molluscan Studies* 68:353–357.

Kohler, F, and M. Glaubrecht. 2006. A systematic revision of the Southeast Asian freshwater gastropod *Brotia* (Caenogastropoda: Pachychilidae). *Malacologia* 48(1–2): 159–251.

Lopes-Lima, M., E. Froufe, V. T. Do, M. Ghamizi, K. E. Mock, U. Kebapci, O. Klishko, et al. 2017. Phylogeny of the most species-rich freshwater bivalve family (Bivalvia: Unionida: Unionidae): Defining modern subfamilies and tribes. *Molecular Phylogenetics and Evolution* 106:174–191.

Modell, H. 1969. Paläontologische und geologische Untersuchungen im Tertiär von Pakistan. 4. Die tertiären Najden des Punjab und Vorderindiens. *Abhandlungen der Bayerischen Akademie Der Wissenschaften, Mathematisch-naturwissenschaftliche Klasses, neue Folge.* 135:1–49, 4 plates.

Nesemann, H., S. Sharma, G. Sharma, and R. K. Sinha. 2005. Illustrated checklist of large freshwater bivalves of the Ganga River System (Mollusca: Bivalvia: Solecurtidae, Unionidae, Amblemidae). *Nachrichenblatt der Ersten Vorarlberger Malakologischen Gesellschaft* 13:1–51.

Pfeiffer, J. M. III, and D. L. Graf. 2015. Evolution of bilaterally asymmetrical larvae in freshwater mussels (Bivalvia: Unionoida: Unionidae). *Zoological Journal of the Linnean Society* 175:307–318.

Prashad, B. 1925a. On a fossil Ampullariid from Poonch, Kashmir. *Records of Geological Survey of India* 56(3): 210–212.

Prashad, B. 1925b. Revision of the Indian Ampullariidae. *Memoirs of the Indian Museum* 8:69–89, plates 13–15.

Prashad, B. 1927. On some fossil Indian Unionidae. *Records of the Geological Survey of India* 60:308–312, plate 25.

Prashad, B. 1930. On some undescribed freshwater molluscs from various parts of India and Burma. *Records of the Geological Survey of India* 63:428–433, plate 19.

Ramakrishna, and A. Dey. 2007. *Handbook on Indian Freshwater Molluscs*. Calcutta: Zoological Survey of India.

Raut, S.K., M. S. Rahman, and S. K. Samanta. 1992. Influence of temperature on survival, growth and fecundity of the freshwater snail *Indoplanorbis exustus* (Deshayes). *Memorias do Instituto Oswaldo Cruz* 87(1): 15–19.

Rees, W. J. 1965. The aerial dispersal of Mollusca. *Proceedings of the Malacological Society of London* 36:269–282.

Reis, J., M. J. Collares-Pereira, and R. Araujo. 2014. Host specificity and metamorphosis of the glochidium of the freshwater mussel *Unio tumidiformis* (Bivalvia: Unionidae). *Folia Parasitologica* 61(1): 81–89

Seuffert, M. E., and P. R. Martin. 2010. Dependence on aerial respiration and its influence on microdistribution in the invasive freshwater snail *Pomacea canaliculata* (Caenogastropoda, Ampullariidae). *Biology Invasions* 12:1695–1708.

Sharma, K. M., S. S. Singh, and P. V. Vipina. 2015. Microfossil assemblage from Dhok Pathna Formation (Middle Siwaliks) exposed near Polian Prohita, Una District, Himachal Pradesh, India. *Earth Science India* 8(1): 15–31.

Srivastava, J. P., and R. K. Arora. 1983. On the occurrence of fossil molluscs in the Tatrot Rocks (Pliocene) near Saketi, Sirmur District, Himachal Pradesh. *Journal of the Palaeontological Society of India* 28:112–113, plate 1.

Strong, E. E., O. Gargominy, W. F. Ponder, and P. Bouchet. 2008. Global diversity of gastropods (Gastropoda; Mollusca) in freshwater. *Hydrobiologia* 595:149 166.

Subba Rao, N. V. 1989. *Handbook Freshwater Molluscs of India.* Calcutta: Zoological Survey of India.

Takayasu, K. 1992. Palaeoenvironmental aspects of the freshwater molluscs from the Siwalik Group in the Arung Khola Area, West Central Nepal. *Bulletin of the Department of Geology,* Tribhuvan University, Kathmandu, Nepal, *Proceedings of the Symposium on Geodynamics of the Nepal Himalaya* 2(1): 107–115.

Takayasu, K., D. Gurung, and K. Matsuoka. 1995. Some new species of freshwater bivalves from the Mio-Pliocene Churia Group, west-central Nepal. *Transactions and Proceedings of the Palaeontological Society of Japan* 179:157–168.

Tauxe, L., and N. D. Opdyke. 1982. A time framework based on magnetostratigraphy for the Siwalik sediments of the Khaur area, northern Pakistan. *Palaeogeography, Palaeoclimatology, Palaeoecology* 37:43–61.

Van Damme, D. 1984. The freshwater Mollusca of Northern Africa: Distribution, biogeography and palaeoecology. In *Developments in Hydrobiology,* edited by H. J. Dumont, 25:1–164. The Hague: Junk.

Van Leeuwen, C. H. A., N. Huig, G. van der Velde, T. A. van Alen, C. A. M. Wagemaker, C. D. H. Sherman, M. Klaassen, and J. Figuerola. 2013. How did this snail get here? Several dispersal vectors inferred for an aquatic invasive species. *Freshwater Biology* 58:88–99.

Varun, P. 2013. Fossil Molluscs from the Middle Miocene Lower Siwalik deposits of Jammu, India. *International Research Journal of Earth Sciences* 1(1): 16–23.

Vokes, H. E. 1935. Unionidae of the Siwalik Series. *Memoirs of the Connecticut Academy of Arts and Sciences* 9:37–48, plate 3.

Vokes, H. E. 1936. Siwalik Unionidae from the collections of the second Yale North India Expedition. *Quarterly Journal of the Geological, Mining and Metallurgical Society of India* 8:133–141, plate 10.

Vredenburg, E., and B. Prashad. 1921. Unionidae from the Miocene of Burma. *Records of the Geological Survey of India* 51:371–374, plate 12.

Wächtler, K., M. C. Dreher-Mansur, and T. Richter. 2001. Larval types and early postlarval biology in Najads (Unionoida). In *Ecology and Evolution of the Freshwater Mussels Unionoida,* edited by G. Bauer and K. Wachtler, 93–125. Berlin: Springer.

Watters, G. T. 1996. Small dams as barriers to freshwater mussels (Bivalvia, Unionoida) and their hosts. *Biological Conservation* 75:79–85.

West, R. M., J. Munthe, J. R. Lukacs, and T. B. Shrestha. 1975. Fossil Mollusca from the Siwaliks of eastern Nepal. *Current Science* 44(14): 497–498.

Zaleha, M.J. 1997. Intra- and extrabasinal controls on fluvial deposition in the Miocene Indo-Gangetic foreland basin, northern Pakistan. *Sedimentology* 44:369–390.

7 FISHES IN THE SIWALIK RECORD

CATHERINE BADGLEY AND LOIS ROE

INTRODUCTION

The Siwalik fossil record documents vast riverine ecosystems that occupied the foreland basin of the Himalaya Mountains—a tectonically maintained lowland along its southern margin. The present-day Ganges River system flows along the foreland basin where the Eurasian plate overrides the Indian plate, and Siwalik sediments document the extensive, dynamic river system that was its precursor. While the Punjab region today is drained by the south-flowing Indus River, in the Miocene it was drained by the southeast-flowing ancestral Ganges until ~5 million years ago (Ma), when rivers of the Punjab were captured by the Indus (Clift and Blusztajn 2005).

Siwalik Group sediments record a variety of depositional environments, including active river channels, abandoned channels and oxbow ponds, shallow floodplain lakes, and seasonally dry floodplains in river belts that were tens of kilometers wide (Chapter 2, this volume; Willis 1993a, 1993b). Extensive floodplains and channels provided year-round habitats for a diverse aquatic fauna, including molluscs (Chapter 6), fishes (this chapter), frogs and turtles (Chapter 8), crocodilians (Chapter 8), water snakes (Chapter 8), water birds (Chapter 9), and a few semi-aquatic mammals, such as otters and some anthracotheres (Chapters 13, 20). Aquatic vertebrates are a major component of the Siwalik vertebrate record in terms of fossil specimens, taxonomic diversity, and occurrences in Siwalik sequences across the Indian subcontinent.

Although fossil fishes were recognized early in the study of Siwalik paleontology (e.g., Lydekker 1886), they remain critically understudied. Fishes were important members of fluvial ecosystems—as predators, prey, and nutrient cyclers—and serve as indicators of the size and persistence of paleo-habitats. This chapter focuses on fishes of the Siwalik record of the Potwar Plateau, with reference to fishes from other fossil sequences and modern ichthyofaunas. It is based largely on the unpublished master's thesis of Lois Roe (1990) with additional contributions from G. R. Smith (pers. comm., 2017–2021).

Many of the specimens described here were collected during the 1988 field season of the Harvard-Geological Survey of Pakistan (Harvard-GSP) research project. Most were retrieved by surface collecting, with some in situ preparation. Small delicate fossils, such as the teeth of dasyatid rays and cyprinids, were obtained by screen-washing at numerous localities throughout the sequence.

Comparative material includes prepared dry skeletons of recent fishes from the American Museum of Natural History, the Central African Museum in Tervuren (Belgium), the California Academy of Sciences (CAS), the Natural History Museum of Los Angeles County, and the University of Michigan Museum of Zoology (UMMZ). Fossil fish material from the Siwalik collections of the Natural History Museum, London (NHM) were also examined for comparison.

THE FOSSIL RECORD OF SIWALIK FISHES

Fish fossils have been recovered from Siwalik sequences throughout the Himalayan foreland basin. In the Potwar Plateau, more than 1,100 specimens have been catalogued from 205 localities spanning the interval from ~18.0 Ma to ~2.6 Ma (Flynn et al. 2016).

Many Siwalik fossil localities in the Potwar Plateau preserve a small amount of fish remains, but a few have notably large quantities. Locality Y719 (14.1 Ma) at the base of the type Chinji Formation preserves hundreds of fragmentary fish fossils, with 115 specimens catalogued. This locality is a rare example of a catastrophic assemblage in which fish fossils account for 83% of catalogued specimens, whereas at most localities they make up fewer than 5% of specimens. For example, Locality Y311 (10.1 Ma) has yielded >4,800 catalogued vertebrate fossils, of which 59 (~1%) are fish fossils.

Badgley et al. (1995) noted a consistent relationship between dominant lithology and mammalian fossil productivity in the Siwaliks of Pakistan: the mudstone-dominated Chinji and Dhok Pathan Formations have more localities and higher taxonomic richness than the sandstone-dominated Kamlial and Nagri Formations (also Chapter 4, this volume). Documented fish localities, however, occur primarily in Lower Siwalik sediments. This distribution reflects the fact that collecting efforts focused on these formations when a fish paleontologist (Lois Roe) was in the field party and does not accurately reflect the distribution of fish remains in Middle and Upper Siwalik sediments.

Fossil localities that commonly preserve fish remains occur in depositional environments that represent persistent aquatic habitats, such as river channels, abandoned channels with standing water, and perennial wetlands (Chapter 4, this volume). Willis and Behrensmeyer (1994) noted that the majority of Siwalik fossil remains occur in fills of floodplain channels, often after their abandonment, and are less common from localities representing high-energy depositional environments. Most fish fossils were retrieved in surface collecting, with additional small fish teeth and other elements obtained from screen-washing.

Nanda, Sehgal, and Chauhan (2018) provided a detailed synopsis of Siwalik and Siwalik-equivalent deposits in Pakistan, India, Nepal, and Myanmar, noting their lithologic units and their biostratigraphic correlation to the Siwalik formations in the Potwar Plateau (their Table 12). In India, the majority of Siwalik fish fossils are from Haritalyangar (Himachal Pradesh), where Sahni and Khare (1977) and Sahni and Mitra (1980) documented 20,000 specimens; in Jammu and Kashmir (Gaur and Kotlia 1987; Parmar and Prasad 2012); and in Ladakh (Sahni et al. 1984). Gaur and Kotlia (1987) documented (but did not illustrate) >4,600 fish fossils from the Karewa Intermontane Basin of Kashmir. Of these, they were able to refer 3,500 to 5 genera (*Cyprinus, Diptychus, Oreinus, Schizopygopsis,* and *Schizothorax*), all of which occur in the Ganges today but not in the Indus. This is unusual in that all referred taxa are cyprinids.

In Nepalese Siwalik sequences, fish fossils comprise ~33% of collected vertebrate specimens (Munthe et al. 1983; West, Hutchison, and Munthe 1991; Paudayal 2015), as compared to ~2% of catalogued specimens from the Potwar Plateau. The fossiliferous sequences are dated to the Middle to early Late Miocene (West, Hutchison, and Munthe 1983; Paudayal 2015). Fish fossils identified at the family level include Cyprinidae, Clariidae, and Channidae (Paudayal 2015). To date, fossil fishes have not been described from the Siwaliks of Myanmar (Nanda, Sehgal, and Chauhan 2018).

In aggregate, Siwalik fish collections from India and Nepal include six families: Cyprinidae, Bagridae, Clariidae, Siluridae, Sisoridae, and Channidae. With the exception of the African clariid genus, *Heterobranchus,* all of the Indian Siwalik fish taxa are present in the modern Ganges-Brahmaputra River system. The Pakistan Siwalik record also includes the Dasyatidae and Chacidae.

TAXONOMIC DIVERSITY

The taxonomic diversity of Siwalik fishes from the Potwar Plateau is summarized in Table 7.1 with biostratigraphic ranges shown in Figure 7.1. Four orders are represented.

Myliobatiformes

The Dasyatidae are a euryhaline group of rays, with several obligate freshwater taxa in the modern Ganges-Brahmaputra rivers and a separate group of freshwater dasyatid taxa in Thai-

Table 7.1. Fish Taxa, Major Skeletal Elements, and Representative Specimen Numbers Recovered from Siwalik Localities of the Potwar Plateau

Taxon	Major Skeletal Elements	Abundance
Class Chondrichthyes		
Order Myliobatiformes	Isolated teeth	Rare
Family Dasyatidae		
Class Osteichthyes		
Order Cypriniformes	Teeth, dentaries, fragmentary	Uncommon
Family Cyprinidae	skull elements, vertebrae	
Labeo	YGSP 35037, YGSP 35038	
Cyprinid indet.		
Order Siluriformes	Spines, skull elements,	Uncommon
Family Bagridae	dentaries, pectoral girdles,	
	vertebrae	
Sperata (Mystus)	YGSP 35482, YGSP 35050,	
Rita	YGSP 35180, YGSP 35412	
Family Chacidae	Spines, skull elements, pectoral	Uncommon
Chaca sp. nov. 1	girdles, vertebrae	
Chaca sp. nov. 2	YGSP 46617, YGSP 27867	
Family Clariidae	Spines, skull elements, dentaries,	Abundant
Clarias	pectoral girdles, vertebrae	
Clariid indet.	YGSP 35133	
Family Siluridae	Dentaries, skull elements	Rare
Wallago	YGSP 35127	
Family Sisoridae	Skull elements, pectoral girdles	Rare
cf. *Bagarius*	YGSP 35132	
Order Anabantiformes	Dentaries, skull elements,	Common
Family Channidae	vertebrae	
Channa spp.	YGSP 35191	

Notes: Specimens included are in the Harvard-Geological Survey of Pakistan Fossil Collection. Tentative identification of genera from Roe (1990). Abundance refers to frequency of fossil occurrences among the source fossil localities.

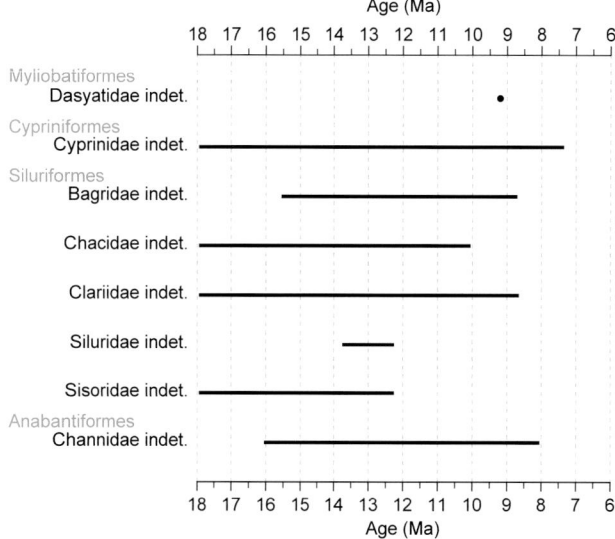

Figure 7.1. Biostratigraphic range chart for major groups of freshwater fishes at the family level.

land, Laos, and Vietnam (Roberts 1982). Dasyatids are notably absent from the Indus River today.

Isolated teeth of dasyatid rays (Fig. 7.2a) occur in a few localities in the Potwar Plateau (Roe 1990) and are also reported

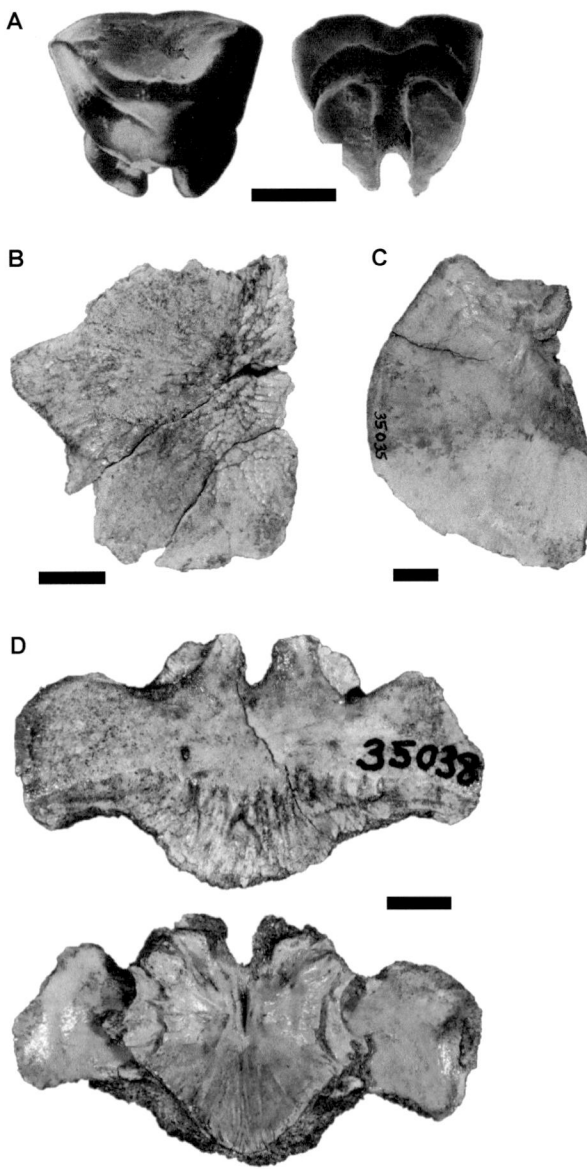

Figure 7.2. Representative specimens of Dasyatidae and Cyprinidae. A. dasyatid teeth. B. cyprinid right frontal, ventral view (YGSP 35037). C. cyprinid left opercule (YGSP 35035). D. cyprinid mesethmoid in two views (YGSP 35038). Scale bar is 1 mm in A, 1 cm in B–D.

from the Oligocene Chitarwata Formation of Baluchistan (Adnet et al. 2007). Two types of teeth—cuspate and pavement—are found, but as rays have a heterodont dentition (Underwood et al. 2015), both types of teeth may belong to the same species. Dasyatids date back at least to the Paleogene of Rajasthan, India (Rajkumari and Guntupalli 2020) and Madagascar (Samonds et al. 2015), and may be descendants of Gondwanan taxa, as Jayaram (1977) suggested for several modern taxa.

Cypriniformes

The order Cypriniformes (minnows, carp, suckers, and loaches) is the most diverse clade of predominantly freshwater fishes, with over 4,000 valid species that include minnows, suckers,

loaches, and hillstream loaches (Mayden et al. 2009; Conway et al. 2010; Fricke et al. 2021). They are native everywhere except South America, Australia, and Antarctica and are most diverse in Asia (>3,300 species). Siwalik representatives are all referred to the family Cyprinidae.

Siwalik cyprinid fossils from the Potwar Plateau include 28 catalogued specimens from 8 localities ranging in age from 17.9 Ma (Locality Y747) at the base of the type Kamlial Formation to 7.4 Ma (Locality Y457) in the type Dhok Pathan Formation. Cyprinid fossils include pharyngeal teeth from screen-washed samples and skull and pectoral-girdle elements from surface collections. Roe (1990) attributed a frontal (YGSP 35037, Fig. 7.2b) and several associated elements from Locality Y777 (8.7 Ma) to *Labeo* cf. *calbasu* based on the square outline in dorsal view, the shape of the postorbital process, the supraorbital margin, and the articular surface with the supraoccipital. Several other well-preserved, large cyprinid skull and pectoral-girdle specimens from Locality Y777 include two left opercles (YGSP 35035, Fig. 7.2c, and YGSP 35036) with different shapes and likely representing different species, three infraorbitals (YGSP 35040, YGSP 35041, and YGSP 35042), a mesethmoid (YGSP 35038, Fig. 7.2d) that shares features with modern representatives of *Labeo* and *Catla*, and a cleithrum (YGSP 35477). Both *Labeo* and *Catla* occur in the modern Indus and Ganges and in rivers of Southeast Asia (e.g., Bangladesh, Cambodia; Rainboth 1991). Several pharyngeal teeth are tentatively referred to *Barbus*.

The oldest cyprinid fossils are from the early Eocene of Kazakhstan (Systchevskaya 1986). In South Asia, the oldest cyprinids are from the Eocene of Pakistan (Gayet 1987; Murray and Thewissen 2008) and Ladakh, India (Verma, Maurya, and Bajpai 2020). Early Miocene cyprinids occur in Saudi Arabia (Otero 2001). Cyprinids also occur in Miocene-Pliocene Siwalik sequences of Jammu and Kashmir (Gaur and Kotlia 1987) and Nepal (West, Hutchison, and Munthe 1991; Paudayal 2015).

Siluriformes

The Siluriformes (catfishes) are the second most diverse clade of fishes, with over 2,800 recognized living species in more than 30 families (Lundberg and Friel 2003). These fishes have a worldwide distribution on the continents, and most species occupy freshwater habitats. Siwalik catfish fossils have been referred to five families: Bagridae, Chacidae, Clariidae, Siluridae, and Sisoridae (Table 7.1).

Bagridae

This family is known as the naked catfishes because they lack scales. Bagrid fossils in the Potwar Plateau include 23 specimens from 13 localities. Roe (1990) assigned bagrid fossils to two extant genera, *Rita* and *Sperata* (formerly *Mystus*). Notable specimens include a partial ethmoid (YGSP 35482), two partial supraoccipitals (YGSP 35050 and YGSP 35180) and several partial or complete dorsal or pectoral spines (Table 7.1, Fig. 7.3a,b). The partial ethmoid (YGSP 35482, Locality Y798) has a distinctive shape. The partial supraoccipital (YGSP 35050) is assigned on the basis of granular surface texture, longitudinal grooves on the sides of the fontanelle, and transverse grooves posterior to the longitudinal grooves. A large (16 cm) and nearly complete pectoral spine (YGSP 35412, Locality Y784) came from a fish ~1 meter long. Its striated texture and stout cross-sectional proportions, as well as the presence of a ridge with denticles on the anterior surface are similar to the pectoral spines of *Sperata*. The large size implies an expansive aquatic habitat. Other bagrid specimens include humeral processes of the cleithrum, nuchal plates, and additional spine fragments.

The oldest known bagrid fossils are from the Middle Eocene Kuldana Formation of Pakistan (Gayet 1987) and the Late Eocene Birket Qarun Formation in the Fayum Depression of Egypt (El-Sayed et al. 2020). Bagrid fossils from the Potwar Plateau range in age from 15.5 Ma (Locality Y689) to 8.75 Ma (Locality Y388). Sahni and Khare (1977) reported bagrid pectoral spines from Middle Siwalik deposits near Haritalyangar, India, and Parmar and Prasad (2012) documented bagrid spines from the Lower Siwalik Mansar Formation of Jammu/Kashmir.

Chacidae

Known as squarehead or angler catfishes, chacids are ambush predators that spend most of their time buried in the substrate waiting for prey. They use their barbels to lure prey, then engulf them into their wide mouths (Roberts 1982). They have flat bodies, with short, strong depressed pectoral fins and short dorsal fins with short spines. Just four species are placed in one genus (*Chaca*). They occur in the Ganges-Brahmaputra and Irrawaddy Rivers of India, Nepal, Bangladesh, and Burma, as well as in Thailand, Malaysia, and Indonesia. They are notably absent from the modern Indus river system.

Chacid fossils from the Potwar Plateau include 43 skull elements, pectoral spines, girdle elements, and vertebrae from 17 localities (see Table 7.1), ranging in age from 17.9 Ma (Locality Y747) to 10.1 Ma (Locality Y311). Two species are recognized from two different skulls with diagnostic characters of the mesethmoids (Fig. 7.3c, YGSP 46617) and two different types of well-preserved partial pectoral and dorsal spines with different textures (Fig. 7.3d). A distinctive characteristic of chacids is the large, straight cornua of the mesethmoid, a synapomorphy of the family (Brown and Ferraris 1988), which makes them easily recognizable in the fossil record. The dorsal side of skull elements and dorsal and pectoral spines have a distinctive honeycomb texture that is one of the most recognizable features of the group and may be related to their keratinized sensory protuberances (Mistri et al. 2018). Based on the skull and pectoral spine dimensions, these Siwalik chacids were 20 and 75 cm in body length.

Clariidae

Known as air-breathing catfishes, clariids are mostly freshwater fish with about 115 extant species in 16 genera (Nelson, Grande, and Wilson 2016). They have a widespread but disjunct distribution, occurring throughout Africa and in South and Southeast Asia, the Philippines, and Indonesia. Living species occur in the Ganges (Sarkar et al. 2012) but not in the Indus (Rafique 2000).

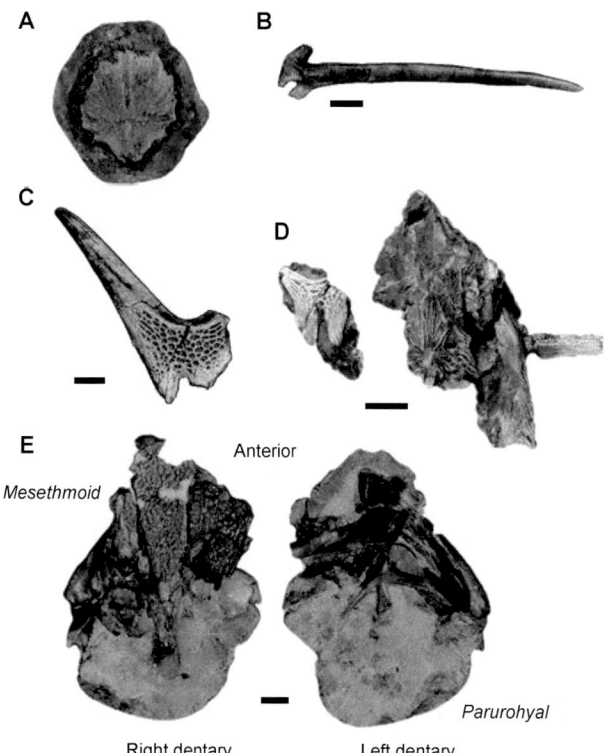

Figure 7.3. Representative specimens of Bagridae, Chacidae, and Clariidae. A. cf. *Bagrus*, posterior of skull (YGSP 35475). B. bagrid pectoral spine (YGSP 35067). C. chacid mesethmoid (YGSP 46617). D. chacid skull, dorsal view (YGSP 35484). E. durophagous clariid, right and left dentaries (YGSP 17543). Scale bar is 1 cm.

Clariids have long dorsal fins with no leading spine, long anal fins, and small (or absent) pectoral fins without spines. They breathe air by means of a labyrinthine organ attached to the gill arches. Some bury themselves in mud and are capable of moving short distances over land; these are known as walking catfish. Their fossil record in Asia begins in the Middle Eocene Kuldana Formation (Gayet 1987) and includes Siwalik sequences in India (Sahni and Khare 1977) and Nepal (West, Hutchison, and Munthe 1991), as well as Jammu/Kashmir (Parmar and Prasad 2012).

Clariid fossils are abundant in Lower Siwalik deposits of the Potwar Plateau, with over 158 catalogued specimens from 31 localities (Fig. 7.3). Roe (1990) recognized three species. One proposed new taxon was based on a dentary and a prevomer covered with a tooth patch for many blunt and villiform teeth. A similar specimen was identified in the Nepal Siwaliks by J. Howard Hutchinson (pers. comm. 1988–1989) but not published. The urohyal attributed to this taxon resembles that of *Clarias batrachus* and *C. bythipogon*. A second taxon was based on partial skulls with prevomerine tooth patches, a separate supraoccipital, dentaries, and a cleithrum. The third new taxon was recognized on the basis of a partial skull roof (YGSP 35133, Locality Y691, 13.1 Ma). The posterior margin formed by the supraoccipital and epiotic bones is concave in outline, resembling the posterior margin in *Clarias ngamensis*, an extant African species (Roe 1990, p. 38). An incomplete supraorbital was also referred to this new taxon. Clariid fossils are known from localities spanning about nine million years of the Potwar Siwalik record; the oldest catalogued specimens occur at Locality Y747 (17.9 Ma) and the youngest at Locality Y388 (8.7 Ma).

Siluridae

This group, known as sheatfishes, includes extant species that inhabit fresh and brackish waters in rivers and lakes in Europe and Asia. In species with dorsal fins, these are small and lack spines. Pelvic fins are small or absent, and the anal fin extends along much of the ventral body length. *Wallago*, to which Siwalik specimens are referred, is large, reaching 1 meter in length and weighing around 45 kg. Of the 107 extant species of Siluridae, 6 occur in the Indus River (Rafique 2000) and 7 in the Ganges (Sarkar et al. 2012). *Wallago attu* has a disjunct distribution (Roberts 2014). It inhabits all major rivers on the Indian subcontinent and has a second population in Southeast Asia. The gap between the two populations appears to be the Salween and Tenasserim river drainages in Burma (Roberts 2014). This species is an aggressive predator that lives in deep pools of rivers and lakes.

Siwalik silurids are documented by only three specimens from three localities. The oldest locality (Y640) is 13.6 Ma, the youngest (Y775) 12.3 Ma. Roe (1990) tentatively referred a partial maxilla and two partial dentaries to the extant genus *Wallago*. A partial dentary (YGSP 35127, Locality Y640) has a ventral deflection of the alveolar surface, large lingual tooth sockets, and placement of mandibular pores similar to the anatomy in *Wallago*. *Wallago* is also reported from a Miocene locality in northern Thailand (Roberts 2014).

Sisoridae

Over 200 species of sisorid catfishes occur today in Asian rivers in montane and lowland settings from Turkey to Indochina (Nelson, Grande, and Wilson 2016) with substantial diversity in South Asia and China. About a dozen species in six genera inhabit the Indus River system (Rafique 2000), and representatives of the same genera also occur in the Ganges River (Sarkar et al. 2012). The oldest report of sisorid fossils is from Lydekker (1886) from the Siwalik Hills (India) with no age information.

Sisorid fossils from the Potwar Plateau are documented from four specimens from two localities, Y747, 17.9 Ma, and Y647, 12.3 Ma. Roe (1990) tentatively attributed one fossil to the extant genus *Bagarius*. The specimen is a partial quadrate (YGSP 35132) from Locality Y647. Resemblance to *Bagarius* includes a deep dorsal notch for insertion of the hyomandibular, the pattern of pores on the lateral surface, and overall surface texture.

Molecular analyses of Chinese sisorids have been the basis for the hypothesis that much of the diversification and external morphology of this group occurred in major fast-flowing drainages that became established during uplift of the Qinghai-Tibetan Plateau (Guo, He, and Zhang 2005; Zhou et al. 2016). The Siwalik fossil record in Pakistan and India points to their

origin being more likely to the west, with subsequent diversification in the Ganges-Brahmaputra rivers and later diversification in the large rivers draining the eastern Tibetan Plateau, with river captures providing dispersal routes followed by allopatric speciation. These alternative biogeographic hypotheses underscore the importance of the Siwalik fossil fish record.

Anabantiformes

Channids are obligate air-breathing freshwater fishes (Graham 1997) with long, eel-like bodies and long, continuous dorsal and anal fins. Their body shape, large scales, large gape, protruding lower jaw, villiform teeth, and dorso-laterally placed eyes give them a snakelike appearance—hence their nickname "snakeheads." Channids are prized as food fish and aquarium fish. They are also a keystone species that can grow up to a meter in length and, where introduced, often devastate prey populations in aquatic ecosystems. They can live out of water as long as they stay moist (Courtenay and Williams 2004), sometimes encase themselves in mud, and can move up to 400 meters over land. These behaviors enable them to survive in ephemeral environments. Two primary genera and 44 valid species are recognized (Fricke, Eschmeyer, and van der Laan 2021), 4 in the African genus *Parachanna* and 40 in the Asian genus *Channa*.

Fossil channids are known from several sequences throughout Asia and Afro-Arabia. Siwalik sequences with channid fossils include Jammu and Kashmir (Parmar and Prasad 2012), Haritalyangar, India (Sahni and Khare 1997), and Nepal (West, Hutchison, and Munthe 1991; Paudayal 2015).

There are earlier records of channid material from Eocene sediments in Pakistan (Gayet 1987; Roe 1991; Murray and Thewissen 2008), Jammu and Kashmir (Khare 1976), and Ladakh (Sahni et al. 1984). Fossil channids have also been described from Eocene, Oligocene, and Neogene sediments in Egypt (Murray 2004, 2006, 2016; Murray, Argyriou, and Cook 2014) and from the Miocene-Pleistocene of Ethiopia (Stewart and Murray 2017).

Channid fossils from the Potwar Plateau include a partial skull, dentaries, premaxillae, supraoccipitals, and anguloarticulars, which are readily identified by their double facets, and vertebrae. These fossils are fairly common (187 identified specimens from 31 localities). The oldest channid fossils are from the Kamlial Formation at Locality Y592, dated at 16.0 Ma, the youngest from the Dhok Pathan Formation at Locality Y545, dated at 8.1 Ma. Roe (1990) tentatively referred dentary specimens to four extant species of *Channa* (Fig. 7.4) based on dentary shape, sensory pores, and tooth sockets, but definitive identifications require more complete associated material. Referred fossils are summarized in Table 7.1. Two of the referred species, *Channa marulius* and *Channa striata*, are among the 19 species of *Channa* documented from the modern Ganges-Brahmaputra rivers. Five of these 19 species occur in the Indus River as well.

It has been suggested that Channiformes originated in Asia and that *Parachanna* represents a dispersal of taxa to Africa (e.g., Böhme 2004). An alternative hypothesis is supported by a newly documented family, Aneigmachannidae, and genus,

Figure 7.4. Representative specimens of Sisoridae and Channidae. A. sisorid symphyseal end of left dentary (YGSP 35337). B. channid dentary (YGSP 35219). C. channid skull, dorsal (right) and ventral (left) views (YGSP 40957). Scale bar is 1 cm.

Aneigmachanna, from subterranean aquifers of Kerala, south India. Britz et al (2020) suggest that these are living fossils that originated in Gondwana.

DIVERSITY AND BIOGEOGRAPHY

The major groups of freshwater fishes documented for the Potwar Plateau are similar to those preserved in other Siwalik sequences. For example, Parmar and Prasad (2012) describe the same families of cypriniforms, siluriforms, and channids from Siwalik rocks of Jammu. Three of the families documented in the Potwar Plateau (Clariidae, Bagridae, Channidae) are also reported from a rich Siwalik sequence at Haritalyangar, India (Sahni and Khare 1977). The Siwalik record of Nepal also includes fossils of Cyprinidae, Clariidae, and Channidae (West, Hutchison, and Munthe 1991). In contrast, the ichthyofaunas of the present-day Ganges and Indus Rivers exhibit much greater taxonomic diversity. A survey of freshwater fish diversity of the Ganges River drainage reported 133 native (non-exotic) species in 32 families and 11 orders (Sarkar et al. 2012). Species of Cyprinidae, Bagridae, and Channidae dominated the survey. In Pakistan, 165 native species in 24 families are documented in the Indus River drainage (Rafique 2000), with the most speciose families including Cyprinidae (70 species) and Sisoridae (13 species). Thus, the

Siwalik record preserves representatives of families with high numbers of extant species in the Ganges and Indus river systems but lacks many of the native families present in one or both drainages today.

The most likely factors that contribute to the discrepancy between the taxonomic diversity of fossil and present-day ichthyofaunas are taphonomic bias, collection bias, and lack of study. For many Cenozoic alluvial sequences that preserve a robust record of vertebrates or plants, the representation of fish diversity is anomalously low (Smith, Stearley, and Badgley 1988). The fossil fishes preserved in these settings are dominated by a few species of relatively large body size and robust skeletal elements. In contrast, fish faunas in modern river systems around the world have high taxonomic diversity. A combination of taphonomic factors contributes to loss of this diversity after death. Most cyprinid fishes are consumed by predators or scavengers, often leaving only small pharyngeal arches and teeth to become fossilized. Larger benthic or semi-benthic fishes with high bone density, such as gar and *Amia* in North American Cenozoic sequences and catfishes and channids in South Asian sequences, contribute disproportionately to the fossil record. Even for taxa with robust skeletal elements, the preserved material is often fragmentary and disarticulated, presumably from scavenging. Collection bias may play a role, since most alluvial fossiliferous records preserve notable crocodilians, turtles, or mammals and attract specialists who focus their collecting efforts on those groups. Fish fossils may be ignored or under-reported if their diagnostic skeletal features are not recognized. Screen-washing may abrade the remains of small vertebrates, including small fishes, but the main goal of screen-washed samples is typically small mammals. The residue of screen-washed samples from the Siwalik record of the Potwar Plateau contains hundreds of tiny bones and teeth of fishes and amphibians, most of which await study.

We report residence times at the family level (Fig. 7.1), since so much material remains to be identified and taxonomic resolution is uncertain for most genera and species. Table 7.2 illustrates family-level composition and the number of catalogued specimens in each family for the richest fish localities in the four formations that span much of the well-studied Potwar Siwalik sequence. The number of reported specimens is a subset of the collected material, much of which awaits further study. The pattern in Fig. 7.1 and Table 7.2 illustrates persistence of most families over most of the sequence.

Siwalik fish fossils from the Potwar Plateau reinforce the biogeographic pattern of the major groups from older Cenozoic sequences and across South Asia. The cyprinids, catfish families, and channids belong to old South Asian lineages that persist today in large and small rivers of the Indian subcontinent and other parts of Asia. Among the bony fishes, several families occur in the oldest localities at the base of the Kamlial Formation, often with precursors in older continental deposits of Pakistan, and persist into the late Miocene. Only one family, the Siluridae, has a short duration (1.3 million years, Myr), as surmised from currently identified material. All other families are present for at least 7 Myr. Cyprinids have the longest

Table 7.2. Taxonomic Composition of the Richest Fish Locality from Each of Four Siwalik Formations of the Potwar Plateau

Locality Number	Formation	Age in Myr	Families Represented (no. catalogued specimens)
Y689	Kamlial	15.50	Bagridae (2), Chacidae (3), Clariidae (19), Siluriformes (5), Channidae (8), Osteichthyes indet. (2)
Y719	Chinji	14.06	Chacidae (3), Clariidae (63), Siluriformes (23), Channidae (24), Osteichthyes indet. (3)
Y311	Nagri	10.06	Bagridae (3), Chacidae (1), Clariidae (1), Siluriformes indet. (4), Channidae (3), Osteichthyes indet. (57)
Y388	Dhok Pathan	8.75	Bagridae (1), Clariidae (1), Siluriformes (6), Channidae (15), Osteichthyes indet. (82)

Notes: Age estimates based on calibration to timescale of Gradstein et al. (2012). At the family level, the taxonomic composition is stable over time, although proportional representation varies through the sequence.

duration of 10.5 Myr. More detailed taxonomic resolution is needed before ideas about endemic evolution, dispersal, and extinction can be evaluated.

SUMMARY

Siwalik fish fossils from the Potwar Plateau provide critical information about environments and freshwater fishes of South Asia. They record the first occurrence in Neogene freshwater sediments in South Asia of the freshwater ray family Dasyatidae and the first fossil record of the catfish family Chacidae. The presence of the Dasyatidae and Chacidae in the Pakistan Siwaliks and the modern Ganges-Brahmaputra rivers but not in the modern Indus is consistent with the isolation of the ancestral Indus at that time. It also suggests an origin of the family Chacidae in the ancestral Ganges. The other dominant families present—Cyprinidae, Bagridae, Clariidae, Siluridae, Sisoridae, and Channidae—are all regarded as Asian in origin and are known from Eocene sequences in Pakistan, Afro-Arabia, and Eurasia.

The presence of large individuals (up to 1 meter in length) indicates that large bodies of water were persistent elements of the landscape. The abundant presence of two obligate air-breathing taxa, *Clarias* and *Channa* (Graham 1997), suggests that ephemeral environments such as abandoned channels may have been favorable to the preservation of these taxa because of their tendency to encase themselves in mud. Similarly, the benthic sedentary nature of *Chaca*, which spends most of its time buried into substrate, may also have contributed.

Identifications at the family level capture critical aspects of the Siwalik fish record, which reflects the ichthyofauna of the modern Ganges but not the modern Indus. Further study of uncatalogued Siwalik fish fossils is needed to confirm taxonomic identifications and better assess diversity. Comparisons with Neogene and earlier faunas from South Asia and Afro-Arabia could answer critical questions about the origin, diversification, and biogeography of several major fish clades.

ACKNOWLEDGMENTS

We thank the many collectors of these fish fossils, especially Will Downs for the screen-washed samples; John Barry for assistance with cataloguing; and Gerald Smith for assistance with identifications and comments on the text.

References

Adnet, S., P.-O. Antoine, S. R. Hassan Baqri, J.-Y. Crochet, L. Marivaux, J.-L. Welcomme, and G. Métais. 2007. New tropical carcharhinids (Chondrichthyes, Carcharhiniformes) from the late Eocene–early Oligocene of Balochistan, Pakistan: Paleoenvironmental and paleogeographic implications. *Journal of Asian Earth Sciences* 30(2): 303–323.

Badgley, C., W. S. Bartels, M. E. Morgan, A. K. Behrensmeyer, and S. M. Raza. 1995. Taphonomy of vertebrate assemblages from the Paleogene of northwestern Wyoming and the Neogene of northern Pakistan. *Palaeogeography, Palaeoclimatology, Palaeoecology* 115(1–4): 157–180.

Böhme, M. 2004. Migration history of air-breathing fishes reveals Neogene atmospheric circulation patterns. *Geology* 32(5): 393–396.

Britz, R., N. Dahanukar, V. K. Anoop, S. Philip, B. Clark, R. Raghavan, and L. Rüber. 2020. Aenigmachannidae, a new family of snakehead fishes (Teleostei: Channoidei) from subterranean waters of South India. *Scientific Reports* 10, https://doi.org/10.1038/s41598-020-73129-6.

Brown, B. A., and C. J. Ferraris. 1988. Comparative osteology of the Asian family Chacidae: With the description of a new species from Burma. *American Museum Novitates* No. 2907.

Clift, P. D., and J. Blusztajn. 2005. Reorganization of the western Himalayan river system after five million years ago. *Nature* 438:1001–1003.

Conway, K. W., M. V. Hirt, L. Yang, R. L. Mayden, and A. M. Simons. 2010. Cypriniformes: Systematics and paleontology. In *Origin and Phylogenetic Interrelationships of Teleosts,* edited by J. S. Nelson, H.-P. Schultze, and M. V. H. Wilson, 295–316. Munich: Verlag Dr. Friedrich Pfeil.

Courtenay, Jr., W. R., and J. D. Williams. 2004. Snakeheads (Pisces, Channidae)—A biological synopsis and risk assessment. U.S. Geological Survey Circular 125, 143 pp.

El-Sayed, S., A. M. Murray, M. A. Kora, G. A. Abuelkheir, M. S. Antar, E. R. Seiffert, and H. M. Sallam. 2020. Oldest record of African Bagridae and evidence from catfishes for a marine influence in the Late Eocene Birket Qarun Locality 2 (BQ-2), Fayum Depression, Egypt. *Journal of Vertebrate Paleontology* 40:3, https://doi.org/10.1080/02724634.2020.1780248.

Flynn, L. J., D. Pilbeam, J. C. Barry, M. E. Morgan, and S. M. Raza. 2016. Siwalik synopsis: A long stratigraphic sequence for the Later Cenozoic of South Asia. *Comptes Rendus Palevol* 15(7): 877–887.

Fricke, R., W. N. Eschmeyer, and R. van der Laan, eds. 2021. Eschmeyer's Catalog of Fishes: Genera, Species, References. Database. San Francisco: California Academy of Sciences. https://researcharchive.calacademy.org/research/ichthyology/catalog/fishcatmain.asp.

Gaur, R., and B. S. Kotlia. 1987. Paleontology, age, palaeoenvironment and palaeoecology of the Karewa Intermontane Basin of Kashmir (J&K, India). *Rivista Italiana di Paleontologia e Stratigrafia* 93(2): 237–250.

Gayet, M. 1987. Lower vertebrates from the Early-Middle Eocene Kuldana Formation of Kohat (Pakistan): Holostei and Teleostei. *Contributions of the University of Michigan Museum of Paleontology* 27(7): 151–193.

Graham, J. B. 1997. *Air-Breathing Fishes: Evolution, Diversity, and Adaptation.* Cambridge, MA: Academic Press.

Guo, X., S. He, Y. Zhang. 2005. Phylogeny and biogeography of Chinese sisorid catfishes re-examined using mitochondrial cytochrome b and 16S rRNA gene sequences. *Molecular Phylogenetics and Evolution* 35(2): 344–362.

Jayaram, K. C. 1977. Zoogeography of Indian freshwater fishes. *Proceedings of the Indian Academy of Sciences* 86B: 265–274.

Khare, S. K. 1976. Eocene fishes and turtles from Subathu Formation, Beragua Coal Mine, Jammu and Kashmir. *Journal of the Palaeontological Society of India* 18: 36–43.

Lundberg, J. G., and J. P. Friel. 2003. Siluriformes. Catfishes. The Tree of Life Web Project. www.tolweb.org/Siluriformes/15065/2003.01.20.

Lydekker, R. 1886. Indian Tertiary and post-Tertiary vertebrata. *Memoirs of the Geological Survey of India (Palaeontologica Indica)* 10(3): xxiv, 1–264.

Mayden, R. L., W. J. Chen, H. L. Bart, M. H. Doosey, A. M. Simons, K. L. Tang, R. M. Wood, et al. 2009. Reconstructing the phylogenetic relationships of the earth's most diverse clade of freshwater fishes—order Cypriniformes (Actinopterygii: Ostariophysi): A case study using multiple nuclear loci and the mitochondrial genome. *Molecular Phylogenetics and Evolution* 51(3): 500–514.

Mirza, M. R. 1975. Freshwater fishes of Pakistan. *Bijdragen tot der Dierkunde* 45(2): 143–180.

Mistri, A., U. Kumari, S. Mittal, and A. Mittal. 2018. Keratinization and mucogenesis in the epidermis of an angler catfish *Chaca chaca* (Siluriformes, Chacidae): A histochemical and fluorescence microscope investigation. *Zoology* 131:10–19.

Munthe, J., B. Dongol, J. H. Hutchison, W. F. Kean, K. Munthe, and R. M. West. 1983. New fossil discoveries from the Miocene of Nepal include a hominoid. *Nature* 303:331–333.

Murray, A. M. 2004. Late Eocene and early Oligocene teleost and associated ichthyofauna of the Jebel Qatrani Formation, Fayum, Egypt. *Palaeontology* 47(3): 711–724.

Murray, A. M. 2006. A new channid (Teleostei: Channiformes) from the Eocene and Oligocene of Egypt. *Journal of Paleontology* 80(6): 1172–1178.

Murray, A. M. 2012. Relationships and biogeography of the fossil and living African snakehead fishes (Percomorpha, Channidae, *Parachanna*). *Journal of Vertebrate Paleontology* 32(4): 820–835.

Murray, A. M., T. Argyriou, and T. D. Cook. 2014. Palaeobiogeographic relationships and palaeoenvironmental implications of an earliest Oligocene Tethyan ichthyofauna from Egypt. *Canadian Journal of Earth Sciences* 51(10): 909–918.

Murray, A. M. and J. G. M. Thewissen. 2008. Eocene actinoptyerigian fishes from Pakistan, with the description of a new genus and species of channid (Channiformes). *Journal of Vertebrate Paleontology* 28(1): 41–52.

Nanda, A. C., R. K. Sehgal, and P. R. Chauhan. 2018. Siwalik-age faunas from the Himalayan foreland basin of South Asia. *Journal of Asian Earth Sciences*, 162:54–68.

Nelson, J. S., T. C. Grande, and M. V. H. Wilson. 2016. *Fishes of the World*, 5th ed. Hoboken, NJ: John Wiley and Sons.

Otero, O. 2001. The oldest-known cyprinid fish of the Afro-Arabic Plate, and its paleobiogeographical implications. *Journal of Vertebrate Paleontology* 21(2): 386–388.

Parmar, V., and G. V. R. Prasad. 2012. Fossil fish fauna from the Lower Siwalik beds of Jammu. *Journal of the Palaeontological Society of India* 57(1): 43–52.

Paudayal, K. N. 2015. The Cenozoic vertebrate fossils from the Nepal Himalaya: A review. *Journal of Natural History Museum* 27:120–131.

Rafique, M. 2000. Fish diversity and distribution in Indus River and its drainage system. *Pakistan Journal of Zoology* 32(4): 321–332.

Rajkumari, P., and V. R. Prasad Guntupalli. 2020. New chondrichthyan fauna from the Paleogene deposits of Barmer district, Rajasthan, western India; Age, paleoenvironment and intercontinental affinities. *Geobios* 58:55–172.

Roberts, T. R. 1982. A revision of the South and Southeast Asian Angler-Catfishes (Chacidae). *Copeia* 1982(4): 895–901.

Roberts, T. R. 2014. *Wallago* Bleeker, 1851, and *Wallagonia* Myers, 1938 (Ostariophysi, Siluridae), distinct genera of tropical Asian catfishes, with description of †*Wallago maemohensis* from the Miocene of Thailand. *Bulletin of the Peabody Museum of Natural History* 55(1): 35–47.

Roe, L. J. 1990. Neogene freshwater fishes from the Siwalik Group of Pakistan: A preliminary report. Unpublished MA thesis, University of Michigan.

Roe, L. J. 1991. Phylogenetic and ecological significance of Channidae (Osteichthyes, Teleostei) from the Early Eocene Kuldana Formation of Kohat, Pakistan. *Contributions from the University of Michigan Museum of Paleontology* 28(5): 93–100.

Sahni, A., and S. K. Khare. 1977. A Middle Siwalik fish fauna from Ladhyani (Haritalyangar), Himachal Pradesh. *Biological Memoirs, Vertebrate Paleontology Series–1*, vol. 2 (1 & 2): 187–221.

Sahni, A., and H. C. Mitra. 1980. Neogene palaeobiogeography of the Indian subcontinent with special reference to fossil vertebrates. *Palaeogeography, Palaeoclimatology, Palaeoecology* 31:39–62.

Sahni, A., S. V. Srikantia, T. M. Ganesan, and C. Wangdus. 1984. Tertiary fishes and mollusks from the Kuksho Formation of the Indus Group, near Nyoma, Ladakh. *Journal of the Geological Society of India* 25(11): 744–747.

Samonds K. E., T. H. Andrianavalona, L. A. Wallett, I. S. Zalmout, and D. J. Ward. 2019. A middle–late Eocene neoselachian assemblage from nearshore marine deposits, Mahajanga Basin, northwestern Madagascar. *PLOS ONE* 14(2): e0211789.

Sarkar, U. K., A. K. Pathak, R. K. Sinha, K. Sivakumar, A. K. Pandian, A. Pandey, V. K. Dubey, and W. S. Lakra. 2012. Freshwater fish biodiversity in the River Ganga (India): Changing pattern, threats and conservation perspectives. *Reviews in Fish Biology and Fisheries* 22(1): 251–272.

Smith, G. R., R. F. Stearley, and C. E. Badgley. 1988. Taphonomic bias in fish diversity from Cenozoic floodplain environments. *Palaeogeography, Palaeoclimatology, Palaeoecology* 63(1–3): 263–273.

Stewart, K. M., and A. M. Murray. 2017. Biogeographic implications of fossil fishes from the Awash River, Ethiopia. *Journal of Vertebrate Paleontology* 37:1. https://doi:10.1080/02724634.2017.1269115.

Systchevskaya, E. K. 1986. Paleogene freshwater fish fauna of the USSR and Mongolia. The Joint Soviet-Mongolian paleontological expedition. *Transactions* 29:5–157.

Verma, S., A. Maurya, and S. Bajpai. 2020. Cyprinid fishes from Eocene molasse deposits (Liyan Formation), India-Asia collision zone,

Eastern Ladakh, NW Himalaya, India. Implications for non-marine aquatic connections between India and Mainland Asia. *Historical Biology* 33(12): 3215–3223. https://doi.org/10.1080/08912963.2020.1855636.

West, R. M., J. H. Hutchison, and J. Munthe. 1991. Miocene vertebrates from the Siwalik Group, western Nepal. *Journal of Vertebrate Paleontology* 11(1): 108–129.

Willis, B. J. 1993a. Ancient river systems in the Himalayan foredeep, Chinji Village area, northern Pakistan. *Sedimentary Geology* 88: 1–76.

Willis, B. J. 1993b. Evolution of Miocene fluvial systems in the Himalayan foredeep through a two-kilometer thick succession in northern Pakistan. *Sedimentary Geology* 88:77–121.

Willis, B. J., and A. K. Behrensmeyer. 1994. Architecture of Miocene overbank deposits in northern Pakistan. *Journal of Sedimentary Research* B64(1): 60–67.

Zhou, C., X. Wang, X. Gan, Y. Zhang, D. M. Irwin, R. L. Mayden, and S. He. 2016. Diversification of Sisorid catfishes (Teleostei: Siluriformes) in relation to the orogeny of the Himalayan Plateau. *Scientific Bulletin* 61(13): 991–1002.

SIWALIK REPTILIA, EXCLUSIVE OF AVES

JASON J. HEAD

INTRODUCTION

The Cenozoic reptile fossil record is important for understanding the evolution of modern ecosystems because it provides historical data on climate and environment from the perspective of poikilothermic metabolism (e.g., Hutchison 1982; Böhme 2003; Head et al. 2009; Head et al. 2013) and because it documents histories of diversification and radiation of the most speciose tetrapod clades. In the context of the Siwalik Group, the reptile fossil record is unique in sampling both aquatic and terrestrial faunas, especially from diverse fluvio-lacustrine communities, and as a basis for inferring biogeographic histories, especially the assembly of South Asian vertebrate ecosystems.

The Siwalik fossil reptile record here includes Squamata, Testudines, and Archosauria limited to Crocodilia. Historic discovery and documentation of this record are concurrent with the earliest discoveries of fossil mammals. Early descriptions are based on the collecting expeditions of William Theobald and colleagues from the Geological Survey of India to the eastern and northern Punjab, including the Potwar Plateau, during the mid-nineteenth century, and of Proby T. Cautley and Hugh Falconer in the Upper Siwaliks of northwestern India near Chandigarh between 1834 and 1837 (Jukar, Sun, and Bernor 2018). Early documentation is limited to brief descriptions of emydine and testudinid turtles (Theobald 1877, 1879); the crocodilians *Crocodilus bombifrons*, *Crocodilus crassidens*, *Crocodilus leptodus*, and *Leptorhynchus gangeticus*; the turtles *Colossochelys atlas* and *Emys tecta*; and the varanid lizard *Varanus sivalensis* (Falconer 1868). Lydekker (1885, 1886a, 1886b, 1889a, 1889b) revised the taxonomy and identifications of Siwalik reptiles based on the collections of Theobald, Falconer, and Cautley, as well as collections from the Siwaliks of Piram Island in the Gulf of Khambhat, Sind. These studies greatly increased the taxonomic richness of the record, including the first descriptions of Siwalik snakes and trionychine turtles. Subsequent studies in the Indian Upper Siwaliks (e.g., Tewari and Badam 1969; Badam 1981; Rage, Gupta, and Prasad 2001; Verma, Mishra, and Gupta 2002; Patnaik et al. 2014; Nanda, Schleich, and Kotlia 2016) and the Nepalese Middle and Upper Siwaliks

(West, Hutchison, and Munthe 1991; Corvinus and Schleich 1994) have continued to expand reptile diversity within the higher-order clades identified by Lydekker.

Aside from the early collections of Theobald, Siwalik reptile fossils from the Potwar Plateau were collected primarily during the course of five field research programs: the Geological Survey of India expeditions led by G. E. Pilgrim during the first two decades of the twentieth century, the American Museum of Natural History Siwalik Hills Indian Expedition in 1922, the Yale University North India expedition in the 1930s, the Bayerische Staatssammlung für Paläontologie und Historische Geologie (BSP Munich) expeditions in 1955–1956 (Dehm, Oettingen-Spielberg, and Vidal 1958), and the collaborative field research program begun in the 1970s between the Geological Survey of Pakistan and multiple US institutions, including Yale University, Harvard University, Dartmouth College, the University of Arizona, and the Smithsonian Institution (e.g., Pilbeam et al. 1979; Barry, Behrensmeyer, and Monaghan 1980; Barry, Lindsay, and Jacobs 1982; Flynn et al. 1990; Badgley and Behrensmeyer 1995; Barry et al. 1995, 2002, 2013). Although substantial reptile collections have been recovered from these projects, there was limited research conducted on them prior to the 1990s (e.g., Mook 1932, 1933; Lull 1944; Hoffstetter 1964; Norell and Storrs 1989).

Field surveys specifically targeting Siwalik Group reptiles were conducted as part of continuing research on the Potwar Plateau from 1996–2000 in three collecting regions: Pahdri-Hasnot, Khaur, and Chinji-Bhilomar (e.g., Flynn et al. 2013). These surveys sampled primarily the Middle Siwalik Dhok Pathan and Nagri Formations but additionally sampled the Lower Siwalik Chinji and Kamlial Formations. Where possible, assemblage composition at the locality level was documented, and diagnostic specimens were collected. These collections were combined with reptile specimens derived from screen-washing of sediments for micromammal fossils (e.g., Flynn et al. 1998) to provide a preliminary biostratigraphic history of reptiles from the Dhok Pathan Formation (Head 1997) and a synthesis of the snake faunas of the Potwar Plateau

Siwalik Group (Head 2005). Results of these surveys and collections allow for a comprehensive assessment of the taxonomic diversity and biostratigraphic and biogeographic histories of Siwalik reptile faunas of the Potwar Plateau.

Lower-order taxonomic ambiguity in osteological anatomy among reptiles combined with taphonomic modes of preservation in Siwalik vertebrates limits our ability to delineate taxa at generic or specific levels within much of the reptile record. Although reptiles exhibit higher species richness than mammals globally, especially within tropical and subtropical environments, most species-level apomorphies are soft-tissue, integumentary characters. By comparison, skeletal morphology is conservative, subject to growth and remodeling throughout ontogeny, and lacks durable histological structure comparable to prismatic enamel of mammalian dentitions. As a result, inferences of reptile species identity and richness based on fossils almost certainly underestimate original richness, especially relative to coeval mammalian records.

Despite these limitations, the Siwalik record of the Potwar Plateau preserves diverse terrestrial and aquatic reptile assemblages throughout the sequence and is the only fossil record of the origins and initial assembly of diverse modern South Asian reptile faunas. The majority of the Siwalik reptile record remains undescribed, and the following synthesis provides the first biostratigraphic framework for most clades. In establishing first occurrences (FOs) and last occurrences (LOs), reported ages for localities are median estimates from Barry et al. (2002) updated to the time scale of Gradstein et al. (2012) (see Chapter 3, this volume). For many taxa, the reported LO is 6.5 million years ago (Ma), which is the median age estimate for Locality D13. This locality represents the youngest taxonomically diverse sample recovered between 1996 and 2000. LOs based on D13 should therefore be considered simply the top of the well-sampled section and not be misconstrued as the boundary of stratigraphic duration.

SQUAMATA OPPEL, 181

Squamates have been documented from the Siwalik Group of northern India (Himachal Pradesh, Punjab, Jammu, and Kashmir; Falconer 1968; Lydekker 1886a; Rage, Gupta, and Prasad et al. 2001), Nepal (West, Hutchison, and Munthe 1991; Patnaik and Schleich 1998), and Pakistan (Hoffstetter 1964; Hussian et al. 1979; Head 2005). Previously described lizard taxa consist of the genus *Varanus*, including indeterminate vertebral elements from the Middle to Upper Siwaliks of Jammu (Rage, Gupta, and Prasad 2001), and *V. sivalensis* from the Upper Siwaliks of northern India (Lydekker 1886a; Falconer 1968), as well as Agamidae indeterminate from the middle Miocene Chinji Formation near Del Khaul in Pakistan (Hussain et al. 1979), and records of *Uromastyx* sp., *Calotes* sp., and Scincidae indet. from the Pliocene Siwaliks near Chandigarh (Patnaik and Schleich 1998).

The squamate record from the Potwar Plateau is derived from both surface collection of isolated and occasionally associated postcranial elements of larger taxa and screen-washing,

which has produced teeth, tooth-bearing elements, and vertebrae of small-bodied taxa (Head 2005). For screen-washed specimens, the majority of taxa are represented by single specimens or small sample sizes. This limitation is especially true for lizards, which are known only from fragmentary tooth-bearing elements. As a result, the record provides a minimum FO for many taxa but is not informative for detailed histories of immigration, extirpation, or extinction.

Among lizards, the majority of extant South Asian high-order diversity is present in the Siwalik Group. Acrodontans (dragons, forest lizards, mastigures) have been recovered from near the base to near the top of the Potwar Plateau section. They are represented by uromastycines and the two major clades of Agamidae, all of which include extant species that occur locally in Pakistan (Minton 1966). Among agamids, dracoines are represented by a single fragmentary record of cf. *Bronchocela*, identified by the large accessory tooth cusps and strong overlap between adjacent posterior tooth positions (Fig. 8.1a; Smith et al. 2011), from 15.1 Ma (YGSP 48110, Locality Y678), and by specimens similar to *Calotes* and *Draco*, based on the presence of more widely spaced teeth and deep and narrow interdental grooves. Specimens with these morphologies first appear at around 14.3 Ma (Locality Y709) and are recovered up to 7.2 Ma (Locality Y581). Agamines, identified by acrodont teeth lacking accessory cusps and by low, wide posterior teeth (Smith et al. 2011), first enter the Potwar Siwalik record at 13.2 Ma (Locality Y718) and are last recovered at 11.4 Ma (Locality Y76). An anterior right dentary that compares favorably with extant species of *Laudakia* (Baig et al. 2012; Fig. 8.1b), on the basis of simple successional teeth and two anterior pleurodont teeth with a greatly enlarged second tooth position, occurs at 11.4 Ma (YGSP 24404, Locality Y76).

Two diminutive, incomplete maxillae compare favorably to Uromastycinae on the basis of tall posterior teeth, strongly decreasing tooth size anteriorly, and tooth-wear patterns most similar to those described for fossil uromastycines (Holmes et al. 2010). The two specimens are otherwise distinct, with one (YGSP 39507) being more gracile than either the extant uromastycines *Uromastyx* and *Leiolepis* or the other Siwalik specimen (Fig. 8.1c). The specimens occur high in the Potwar Plateau sequence, at 8.7 Ma (Locality Y388) and 7.2 Ma (Locality Y581). Gekkotans, represented by fragmentary dentary elements with high numbers of small pleurodont teeth and completely enclosed Meckelian grooves (e.g., Müller and Mödden 2001; Fig. 8.1d), first occur at 11.4 Ma (Locality Y76) and are last recovered at 7.8 Ma (Locality Y898). Scincids are represented by fragmentary pleurodont maxillae and dentaries. Both Scincinae and Lygosominae (*sensu* Pyron, Burbrink, and Wiens 2013) are represented by specimens possessing distinct morphologies of the Meckelian groove (Čerňanský and Syromyatnikova 2021). Lygosominae are represented by a partial left dentary from 8.7 Ma (YGSP 27072, Locality Y388), which includes a dorsally angled posterior ramus with four preserved teeth and enclosure of the Meckelian groove along the preserved length of the element (Fig. 8.1e). Scincinae are represented by three mandibles possessing a well-developed

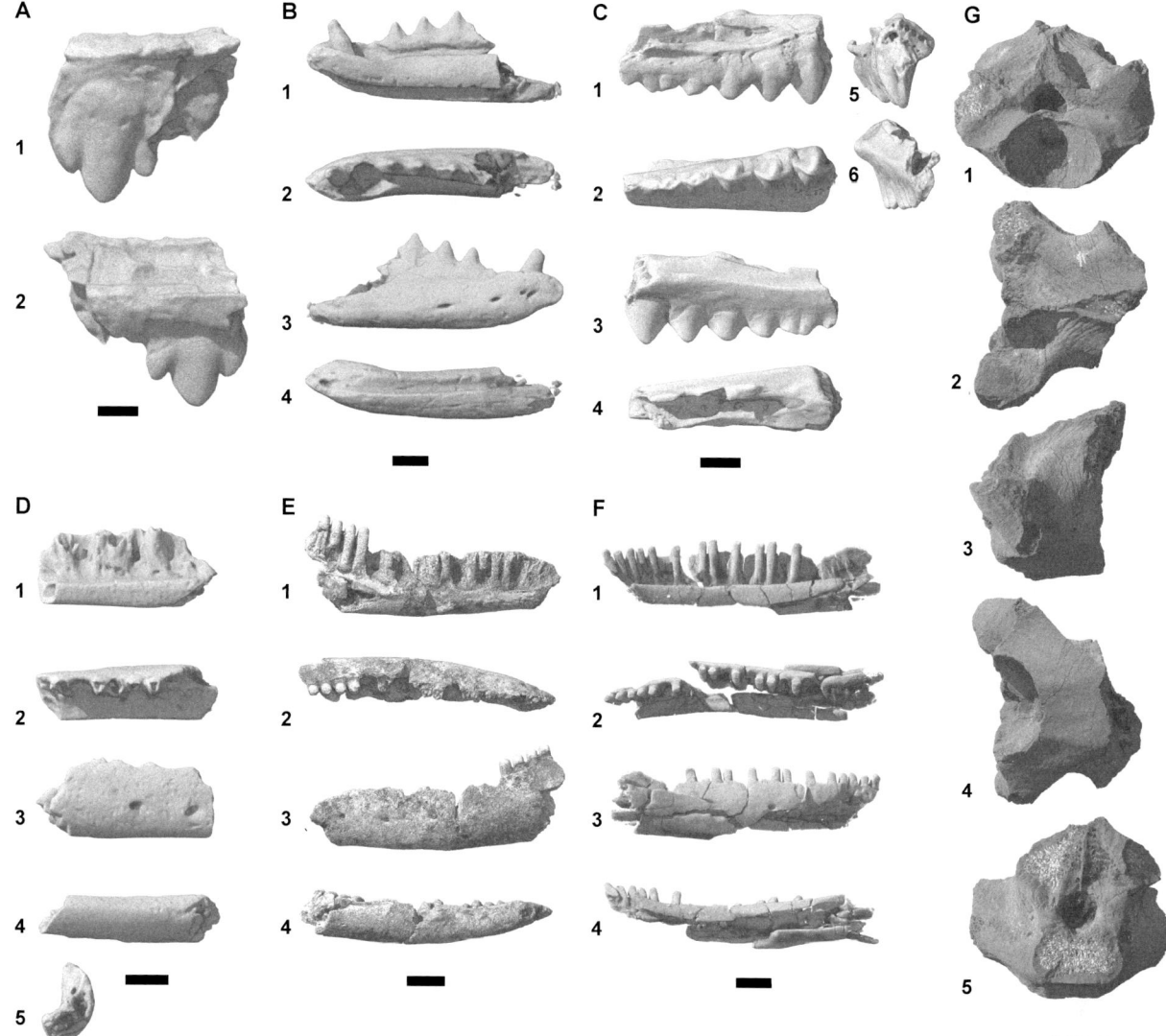

Figure 8.1. Siwalik Group lizards, high resolution computed-tomographic images. A. Cf. *Bronchocela* sp. left posterior maxilla fragment (YGSP 48110) in (1) medial and (2) lateral views. Scale equals. 0.3 mm. B. cf. *Laudakia* sp., anterior right maxilla (YGSP 24404) in (1) medial, (2) dorsal, (3) lateral, and (4) ventral views; scale equals 0.5 mm. C. cf. Uromastycinae sp. posterior right maxilla (YGSP 39507) in (1) medial, (2) ventral, (3) lateral, (4) ventral, (5) posterior, and (6) anterior views; scale equals 1 mm. D. Gekkota sp. right dentary fragment (YGSP 27116) in (1) medial, (2) dorsal, (3) lateral, (4) ventral, and (5) posterior views; scale equals 0.5 mm. E. Lygosominae sp. left dentary (YGSP 27072, reconstructed) in (1) medial, (2) dorsal, (3) lateral, and (4) ventral views; scale equals 1 cm. F. Scincinae sp. right dentary (YGSP 39415, reconstructed) in (1) medial, (2) dorsal, (3) lateral, and (4) ventral views; scale equals 1 cm. G. *Varanus* sp. posterior precloacal vertebra (YGSP 46663) in (1) anterior, (2) dorsal, (3) left lateral, and (4) ventral, 5 posterior views; scale equals 1 cm.

Meckelian groove, including a partial right dentary from 13.1 Ma (YGSP 39415, Locality Y726; Fig. 8.1f) and two dentary fragments from 13.2 Ma (Locality Y718). Additional fragmentary specimens that cannot be identified beyond Scincidae occur between 13.2 (Locality Y718) and 7.8 Ma (Locality Y906).

The fossil record of *Varanus* in Pakistan precedes deposition of Siwalik Group sediments, with fragmentary cranial remains referred to cf. *Varanus* (Z805) recovered from the earliest Miocene Chitarwata Formation of the Zinda Pir Dome (Z127, 23 Ma; Lindsay et al. 2005). In the Potwar Siwalik record, *Varanus* is represented by isolated and occasionally articulated precloa-

cal vertebrae recovered from surface collecting (e.g., Villa and Delfino 2022), with an FO at approximately 15.1 Ma (Locality Y678). The most complete specimen is an undescribed partial postcranial skeleton from the Dhok Pathan Formation (Locality Y989, undated). Vertebral morphology of Siwalik *Varanus* is consistent with that of extant Asian species (Conrad, Balcarcel, and Mehling 2012), and vertebral sizes are consistent with those in *V. salvator* and large specimens of *V. niloticus* (Pianka and King 2004; Conrad, Balcarcel, and Mehling 2012). A single, incomplete precloacal vertebra (YGSP 46663) from 6.8 Ma (Locality Y856; see Fig. 8.1e) represents an individual that is

larger than most extant *Varanus* species, consistent with vertebral sizes in large specimens of *V. salvator* (Hocknull et al. 2009),
and approximately 75% the size of precloacal vertebrae referred
to *V. sivalensis*, a large-bodied species with skeletal dimensions
similar in size to extant *V. komodoensis*. Though larger than any
other *Varanus* recovered from the Potwar record, the specimen
cannot be unambiguously assigned to a unique species in the
absence of apomorphic vertebral morphology.

SERPENTES LINNEAEUS, 1758

Snakes are the most thoroughly studied Siwalik reptiles.
Lydekker (1880, 1882, 1886a) first documented snake taxa, identifying a small collection of precloacal vertebrae from the
Siwaliks of Sind and the Punjab as *Python* cf. *molurus*. This
collection actually includes two taxa, *Python* sp. and *Acrochordus dehmi*, with specimens of the latter erroneously referred to
the former. Hoffstetter (1964) provided the first description
of snakes of the Potwar Plateau, based on collections from the
BSP Munich expeditions. From this record, Hoffstetter recognized *Python* and erected the taxon *Acrochordus dehmi*. West,
Hutchison, and Munthe (1991) described *A. dehmi*, including a
distal quadrate, which represents the only documented snake
cranial elements then recovered from any Siwalik section, from
the Middle Siwaliks of Nepal. Rage, Gupta, and Prasad (2001)
described the fossil snake record from Jammu and Kashmir,
which consists of *A. dehmi* from the upper Miocene Ramnagar
Member of the Mansar Formation and indeterminate colubroid fossils from the upper Pliocene Labli Member of the
Uttarbaini Formation. Head (2005) documented a diverse snake
record from the Potwar Plateau, based on approximately 1,600
isolated or associated vertebrae derived from screen-washing
and surface collecting at approximately 140 localities throughout the Siwalik sequence. Overview of the Siwalik snake record is based primarily on that work; taxonomic reassessments
were made subsequently.

Relative to extant South Asian higher-order clades, only sco-
lecophidians are absent from the revised snake record of the
Potwar Plateau, and only *Acrochordus* and, potentially, part of
the Siwalik elapoid record represent extralimital distributions
for northern South Asia.

Booid snakes include *Python*, represented by precloacal vertebrae recovered from surface collections ranging from 16.8
Ma (Locality Y802) to 7.0 Ma (Locality Y908). A suite of vertebral characters unambiguously assigns specimens to the genus
(Hoffstetter 1964; Head 2005), and specimens can be differentiated from coeval species of fossil *Python*. The Siwalik record
is distinct from *Python europaeus* from the Early to Middle Miocene of Europe in lacking apomorphically short neural spines
(Szyndlar and Rage 2003) and differs from *P. marus* from the
Middle Miocene of Morocco on the basis of synapophyseal
morphology (Rage 1976; Head and Müller 2020). Vertebral
sizes are consistent with those of extant South Asian large-
bodied pythonids, *Python molurus* and *Malayopthon reticulatus*.
Species-level comparisons and assignment based on vertebral
morphology are hampered by the aforementioned sources of

osteological variation; however, a single incomplete right maxilla of *Python* (YGSP 50165, Locality Y908, 7.0 Ma, Fig. 8.2a)
possesses a suborbital excavation and palatine process that are
more similar to *P. molurus* than to *Malayopython*.

The other records of booid snakes include a single precloacal vertebra previously referred (Head 2005) to Boidae? indet.
(Locality Y41, 13.6 Ma) and a caudal vertebra assigned to cf.
Erycidae indet. (Locality Y641, 13.7 Ma). Morphology of both
specimens is consistent with referral to Erycidae. The precloacal
element is anatomically distinct from coeval *Python* specimens
and is similar to erycids in possessing a larger, less triangular
neural canal; slightly concave interzygapophysial ridges; a
more elongate, relatively large neural spine; and more ovoid
zyagpophyseal articular facets. As in erycids, the caudal vertebra possesses the bases of wide lateral processes and a vertically oriented pterapophysis (e.g., Szyndlar 1994). Among
extant erycids, *Eryx* is locally present in Pakistan, but the Siwalik record is too incomplete to be informative on the duration
of the genus in South Asia.

In both total number of localities and specimens per locality, *Acrochordus dehmi* (Fig. 8.2b) is the most abundant squamate
within the Siwalik sequence. The oldest occurrence of the genus predates the Siwalik record, with an unnamed species distinct from *A. dehmi* on the basis of neural spine morphology
and the absence of parazyosphenal foramina recovered from
the Chitarwata Formation (Locality Z127, 23 Ma; Lindsay et al.
2005). On the Potwar Plateau, approximately 1,300 specimens
of *Acrochordus dehmi* have been recovered from 140 localities,
ranging from 17.9 Ma (Locality Y747) to 6.5 Ma (Locality D13;
Head 2005).

Colubroid snakes include the first Siwalik records of viperids based on reassessment of previously described specimens.
Three large precloacal vertebrae previously referred to Elapidae?
indet., known from 10.1 Ma (Locality Y311) to 7.0 Ma (Locality
Y908), are reassigned to Viperidae on the basis of enlarged
cotyle and condyle; tall, well-developed, and deep dorsoventrally oriented hypapophyses; prominent parapophyseal an-
teroventral projections; depressed neural arches; and elongate,
narrow vertebral centra with straight lateral margins (e.g.,
Szyndlar 1991; Szyndlar and Rage 1999; Head, Sánchez-Villagra,
and O. Aguilera 2006; Bailon et al. 2010). While the large size of
the specimens is consistent with that of extant South Asian *Daboia russelii*, the neural spine of the Siwalik specimens is much
shorter than the tall, wide spine of *Daboia* and the neural arch
is more elevated, reaching a level well dorsal to the dorsal edge
of the zygosphene. Similarly, the specimen is distinct from the
Neogene "Oriental vipers" complex of *Vipera* (Szyndlar and
Rage 1999; Bailon et al. 2010) in possessing a high angle of the
posterodorsal edge of the neural arch at the base of the neural
spine (YGSP 51107, Locality Y311, Fig. 8.2c).

A second, distinct viperid record is a single precloacal vertebra from the Dhok Pathan Formation (Locality Y908, 7.0 Ma).
This specimen (YGSP 49900) was identified as the holotype for
the taxon *Gansophis potwarensis* erected by Head (2005). Previously assigned to Colubroidea without greater precision, *Gansophis* is here referred to Viperidae on the basis of an extremely

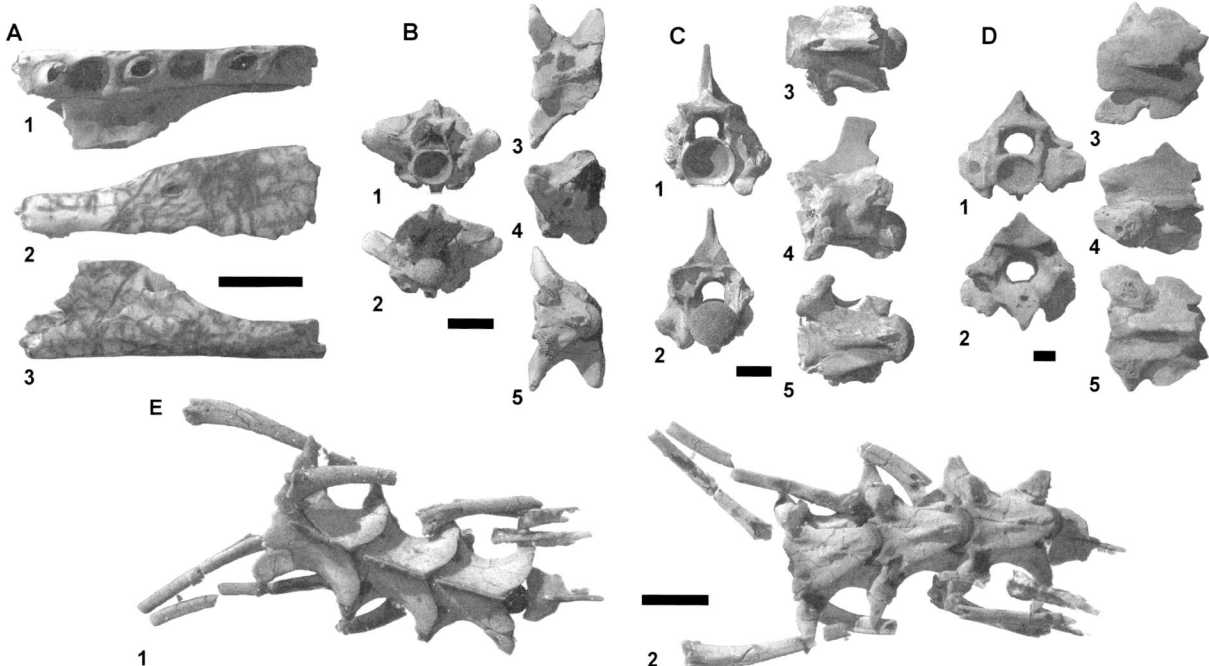

Figure 8.2. Siwalik Group snakes, high resolution computed-tomographic images. A. *Python* sp. right maxilla (YGSP 50165) in (1) ventral, (2) lateral, and (3) dorsal views; scale equals 5 mm. B. *Acrochordus dehmi* precloacal vertebra (YGSP 47568) in (1) anterior, (2) posterior, (3) dorsal, (4) left lateral, and (5) ventral views; scale equals 1cm. C. Viperidae gen. et sp. indet. precloacal vertebra (YGSP 51107) in (1) anterior, (2) posterior, (3) dorsal, (4) left lateral, and (5) ventral views; scale equals 5 mm. D. *Bungarus* sp. posterior precloacal vertebra (YGSP 52568) in (1) anterior, (2) posterior, (3) dorsal, (4) left lateral, and (5) ventral views; scale equals 1 mm. E. Elapoid morph "A," articulated precloacal vertebrae (YGSP 26666) in (1) dorsal and (2) ventral views; scale equals 5 mm.

large, posteroventrally oriented hypapophyseal base and depressed neural arch. *Gansophis* is distinct from the larger viperid specimens, which indicates the existence of multiple, coeval viperids in Upper Siwalik ecosystems.

Elapoidea is represented by *Bungarus* and two vertebral morphs previously allocated to Colubroidea (Head 2005). The record of *Bungarus* consists of precloacal vertebrae recovered primarily from screen-washing and extends from 10.2 Ma (Locality Y450) to 7.8 Ma (Locality Y898; Fig. 8.2d). Vertebral morphology is highly diagnostic within the genus, and the presence of well-developed pre- and postzygapophyseal wings is the basis for assigning the Siwalik specimens to a clade of *Bungarus* species more closely related to each other than to either *B. flaviceps* or *B. bungaroides* (Hoffstetter and Gasc 1969; McDowell 1970; Slowinski 1994). The Potwar Plateau record is approximately one million years older than the only other fossil record of the genus ("*Xenodermus* Reinhardt, 1836 or *Bungarus* Daudin, 1803," Lapparent De Broin et al. 2020, 422) from the Vallesian-Turolian boundary of Pul-E Charkhi, Afghanistan.

The two precloacal vertebral morphs are assigned to Elapoidea on the basis of well-developed, posteriorly oriented hypapophyses and wide central bodies with concave lateral margins relative to Colubridae *sensu* Pyron, Burbrink, and Wiens (2013; Szyndlar and Zerova 1990). Beyond these diagnostic characters, the morphs are taxonomically unresolved but distinct from each other. Morph "A" is represented by two specimens from

the Chinji Formation, including a partial skeleton (YGSP 26666, Locality Y726, 13.1 Ma; Fig. 8.2e) and an isolated precloacal vertebra (Locality Y698, 13.0 Ma). Morph "B" is represented by 25 precloacal vertebrae with a temporal range from 9.0 Ma (Locality Y367) to 7.0 Ma (Locality Y908). The two morphs are differentiated on the basis of relative centrum length. Morph "A" is more elongate and gracile, whereas morph "B" is comparatively stout beyond the range of individual or intracolumnar morphological variation within a single taxon (e.g., Hoffstetter and Gasc 1969; Gasc 1974, 1976).

Although precise assignment of the two morphs to exclusive clades within Elapoidea is hampered by poor documentation of axial osteology for most taxa, which limits comparative analyses and potential referral, the Siwalik taxa can be excluded from Asian elapids including *Naja* on the basis of relatively smaller cotylar-condylar articulations and less robust centra (e.g., Szyndlar 1991), from *Calliophis* based on the absence of short neural spines, and from *Bungarus* based on the absence of zygapophyseal accessory processes. With the exception of the aforementioned genera, the majority of extant elapoid taxa are either endemic to Africa or have distributions spanning Africa, the Middle East, and southern Europe. Thus, the Siwalik record may represent a past expansion of the clade into South Asia.

Colubroidea is represented by three taxa in the Siwalik record of the Potwar Plateau. The record of Natricinae consists of three precloacal vertebrae, spanning a temporal range of

8.0 Ma (Locality Y547) to 7.4 Ma (Locality Y457). The fossil record of Colubrinae is relatively rich and includes two named taxa and multiple indeterminate records ranging from about 14.3 Ma (Locality Y709) to 6.5 Ma (Locality D13; Head 2005). Of the named taxa, *Sivaophis downsi* is represented by 23 isolated precloacal vertebrae recovered from surface and screen-washed collections by both the BSP Munich and Harvard-GSP expeditions (Head 2005). This species spans a temporal range from 13.7 Ma (Locality Y59) to 6.5 Ma (Locality D13). The second taxon, *Chotapohis padhriensis*, is represented by two isolated precloacal vertebrae from 9.0 Ma (Locality Y367) to 7.4 Ma (Locality Y457).

TESTUDINES BATSCH, 1788

Turtles have been documented from all sampled Siwalik sections, including those in the Potwar Plateau (Theobald 1879; Pilbeam et al. 1979; Head 1997), Nepal (West, Hutchison, and Munthe 1991), and northern India (Falconer 1868; Lydekker 1885, 1886b; Tewari and Badam 1969; Prasad 1974; Vasishat, Gaur, and Chopra 1978; Nanda, Schleich, and Kotlia 2016). The record from the Potwar Plateau is dense throughout the Siwalik sequence, reflecting persistent populations, active habitation in depositional environments, and high taphonomic durability of turtle elements. Nonetheless, the record consists primarily of partial to fragmentary shells with some isolated postcranial remains and only two well-preserved skulls. Among higher-order clades, trionychoids (pig-nosed, softshell, and flapshell turtles) are effectively ubiquitous in most fluvial environments of the Chinji, Nagri, and Dhok Pathan Formations and are the most taxonomically diagnostic on the basis of fragmentary shell morphologies. Testudinoids (tortoises and Old World pond turtles) are less common and more taxonomically ambiguous from fragments.

The trionychoid record consists of rare carettochelyid specimens and abundant trionychids. The carettochelyid record is restricted to three specimens, two of which have resolved stratigraphic occurrences, with one from the Kamlial Formation at 16.8 Ma (Locality Y802) and a second from the Dhok Pathan Formation at 6.3 Ma (Locality Y930). All are peripheral elements that lack scute sulci but include well developed digitate anterior and posterior articular facets that form a continuous bony margin of the carapace, which articulates with the peripheral series of the carapace and the hyo-hypoplastral elements of the plastron (Fig. 8.3a). The combination of these characters is diagnostic for Carettocheylinae within the clade (e.g., Meylan 1987, 1988; Joyce 2014). Peripheral elements possess high-angled pleural articular surfaces indicating a tall shell, with a roughly radial arrangement of shallow, slightly anastomosing ridges, similar to extant *Carettochelys insculpta* and some elements of Paleogene *Burmemys* (Hutchison, Holroyd, and Ciochon 2004), but distinct from other fossil carettochelyid records (Joyce, Klein, and Mörs 2004).

The sole extant carettochelyid, *Carettochelys insculpta*, occurs in freshwater aquatic environments in Papua New Guinea and northern Australia. Extralimital fossil records indicate a more cosmopolitan distribution during the Neogene, with occurrences in Europe, Africa, and the Arabian Peninsula (Joyce 2014). Carettochelyids have a wide range in the Paleogene of Asia, and the overall Cenozoic record of the clade suggests that it should occur widely in the Asian Neogene. This is not the case, however, and the carettochelyid record from the Potwar Plateau is the only record for both the greater Siwalik Group and the entire Neogene of Asia (Havlik et al. 2014).

Trionychids have the densest record of any turtle clade within the Siwalik Group and include the two primary extant South Asian clades, Trionychinae and the cyclanorbine genus *Lissemys*. *Lissemys* is one of the most common reptiles recovered throughout the Potwar Plateau sequence and is represented by all shell elements, including partial carapaces and all elements of the plastron. Shell elements possess a distinct pustulate ornamentation on external surfaces. Peripheral elements, which compose an array along the posterolateral edge of the carapace in extant *L. punctata* and *L. scutata*, are the most common isolated elements of *Lissemys*. Plastral elements with greatly expanded dermal callosities have been recovered from well-sampled localities ranging throughout the Potwar Plateau Siwalik sequence. An isolated prenuchal element (YGSP 39518) with pustulations restricted to the dorsal surface, as in extant *Lissemys* (Delfino et al. 2010), was recovered from the Dhok Pathan Formation (Locality Y935, 7.4 Ma).

Two distinct nuchal-element morphologies referable to *Lissemys* are present. One morph, recovered from the Dhok Pathan Formation (FO is 9.6 Ma, Y262; LO is 6.5 Ma, D13; YGSP 39516, Fig. 8.3b), is a robust, domed element with ornamentation extending onto the visceral surface of a thickened anterior rim of the carapace and with well-developed pore channels for Rathke's glands (e.g., Deraniyagala 1939) at the sutural contact for the first pleurals. The second morph is a comparatively flat, elongate element with pustulate ornamentation restricted to the dorsal surface, recovered from the Chinji (Locality Y496, 12.4 Ma) and Dhok Pathan (Locality D13, 6.5 Ma) Formations (YGSP 46949, Fig. 8.3c). Both morphologies possess fan-like costiform processes (Meylan 1987; Meylan, Weig, and Wood 1990; Head, Raza, and Gingerich 1999), but are distinct in the lateral extent of the process. In the first nuchal morph, the lateral edges of the processes are contained within the lateral edges of the element. In the second morph, the dorsolateral edge extends laterally past the sutural contact with the first pleural. Maximum size estimates for shells of both nuchal morphologies indicate carapace lengths of up to 50 cm, which greatly exceed maximum lengths of approximately 25–30 cm in extant *L. punctata* and *L. scutata* (Gramentz 2011).

Lydekker (1885) described *Lissemys punctata* and the fossil species *L. palaeindica*, *L. sivalensis*, and *L. lineata* based on fragmentary specimens collected by Theobald and colleagues from the Siwaliks of the eastern Punjab and from a nearly complete shell referred to *L. punctata* and attributed to the Siwalik Hills with no additional provenance information. Two distinct *Lissemys* morphologies were included in Siwalik faunal lists for the Potwar Plateau by Pilbeam et al. (1979), with no additional discussion. Most characters used by Lydekker (1885) to erect fossil

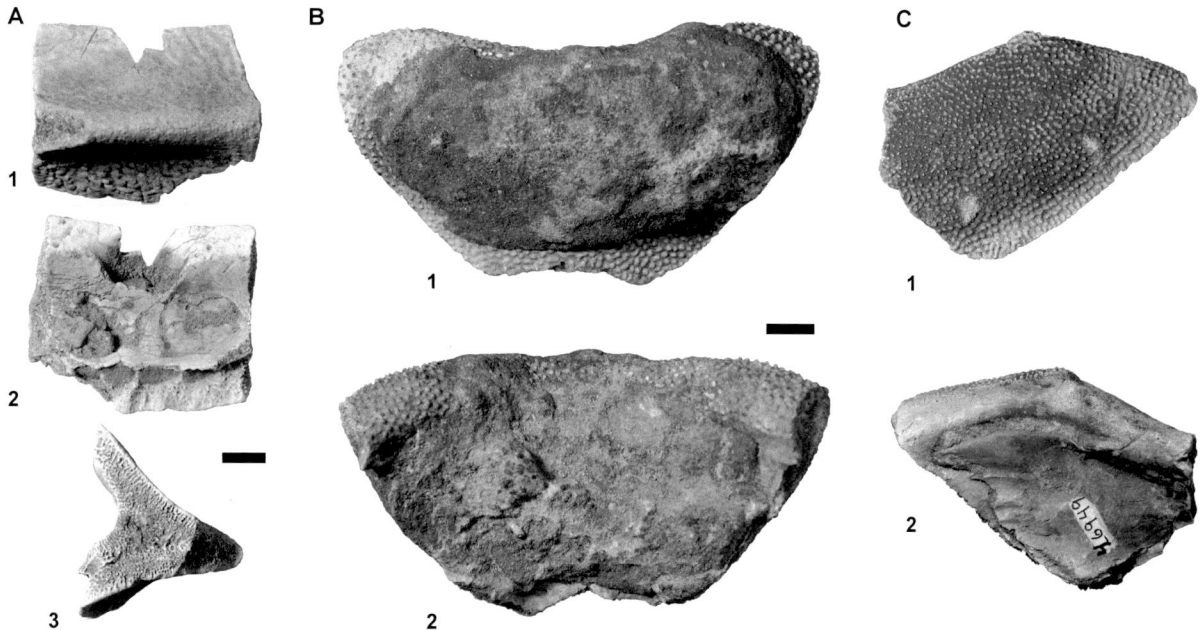

Figure 8.3. Siwalik Group trionychoid turtles. A. Carettocheylinae indet., right seventh peripheral element (YGSP 39515) in (1) lateral, (2) medial, and (3) posterior views. B. *Lissemys* sp. morph "A," complete nuchal (YGSP 39516) in (1) dorsal and (2) visceral views. C. *Lissemys* sp. morph "B," nuchal right ramus (YGSP 46949) in (1) dorsal and (2) visceral views. Scales equal 1 cm.

species are subtle (pustulate pattern, degree of excavation on anterior edges of peripheral elements) and problematic with respect to variation within extant *Lissemys* species; however, nuchal morphologies confirm two distinct taxa, albeit with different diagnostic characters from those of Lydekker (1885).

Other trionychid records include fragmentary shell remains with well-developed, strongly anastomosing ridges, often forming distinct punctate pockets, on the external surface of the shell, and reduced or absent dermal callosities on anterior plastral elements (e.g., Meylan 1987). Within this record, trionychine taxa are diagnosable on the basis of shell shape and discrete apomorphies (e.g., Meylan 1987).

Trionychines include *Chitra* and the *Nilssonia* clade. Among Asian trionychines, *Chitra* is recognizable based on very large, comparatively flat carapacial elements, elongate costiform processes, partial subdivision of the eighth pleural elements by the terminal neural element, and more distinctly punctate ornamentation than in other taxa. Large trionychid fragments that compare favorably to *Chitra* occur throughout the Chinji and Kamlial Formations, with a partial carapace from 13.1 Ma (Locality Y726) providing a cf. *Chitra* FO on the basis of shell ornamentation. Specimens possessing specific apomorphies of *Chitra* were recovered in the Chinji Formation and throughout the Dhok Pathan Formation; these include a partial nuchal with a deep medial excavation for the first pleural and an anterior position of the costiform process occurring at 12.4 Ma (Locality Y695) and nuchals with elongate costiform processes from a locality dated between 8.0 and 8.5 Ma and a second locality at 7.4 Ma (Locality Y34 and Locality Y943, respectively). A posterior carapace with partially separated eighth pleurals occurs

at 7.8 Ma (YGSP 50031, Locality Y917; Fig. 8.4a), and the LO for *Chitra* is 6.5 Ma (Locality D13) based on two large pleural elements representing a bony carapace with a diameter in excess of 70 cm.

The fossil record of *Chitra* is restricted to the Neogene and Pleistocene of South Asia. Extant *C. indica* was described from the "Siwalik Hills" by Lydekker (1885), based on two partial carapaces. These specimens were presented by P. T. Cautley and are therefore likely from the late Pliocene Upper Siwaliks near Chandigarh. Srivastava and Patnaik (2002) similarly described *C. indica* on the basis of several fragmentary pleural elements from the Upper Siwaliks of Himachal Pradesh. West, Hutchison, and Munthe (1991) described *C.* cf. *indica* from the middle Miocene Siwaliks of Nepal on the basis of a partial hypoplastron, and Claude et al. (2011) reported *Chitra* sp. from poorly dated middle Miocene (?) to Pleistocene localities in northeastern Thailand.

The fossil record of *Chitra* from the Potwar Plateau record provides the most precise FO. If the cf. *Chitra* record from the Chinji Formation does represent the genus, then the Pakistan record is equivalent in age to the oldest estimates for the more poorly constrained Nepalese and Thai records (West, Hutchison, and Munthe 1991; Claude et al. 2011). At a minimum, the Siwalik record of the Potwar Plateau unambiguously constrains the appearance of the genus to no younger than 12.4 Ma.

Nilssonia, the most speciose clade of extant trionychids, is represented by diagnostic cranial and shell remains from the Chinji and Dhok Pathan Formations. A fragmentary shell occurring at 11.4 Ma (Locality Y76) possesses a straight, approximately horizontal hyo-hypoplastral suture, an interdigitate posteromedian hypoplastral margin, and a large singular hypoplastral medial process, as in extant *N. gangeticus*, and is the FO for the genus based on these characters (YGSP 39520, Fig. 8.4b). A mostly complete, articulated carapace and plastron of a juvenile *Nilssonia*, possessing a straight hyo-hypoplastral suture and a large preneural (Meylan 1987), occurs at 7.8 Ma (Locality Y898).

Only two well-preserved turtle skulls have been recovered from the Potwar Plateau Siwalik record. Both are from the Chinji Formation and are assignable to *Nilssonia* based on the presence of a medially constricted basisphenoid (Meylan 1987). One specimen dated at 13.7 Ma (Locality Y882) is complete and is nearly indistinguishable from *N. gangeticus* in overall shape, size, and articulations of skull-roof and palatal elements (YGSP 47215, Fig. 8.4c). The second specimen is dated at 12.0 Ma (Locality Y502) and is nearly complete, missing only the posterior braincase, supraoccipital process, and lateral margins of the otic region (YGSP 41000, Fig. 8.4d). This specimen preserves the anatomy of the trigeminal foramen with exclusion of the quadrate from the foramen rim by the epipterygoid and pterygoid as in extant *Nilssonia* (Meylan 1987). The quadratojugal appears to participate in the otic trochlea in both specimens, which suggests closer affinities to *N. gangeticus* than to other extant *Nilssonia* species (Meylan 1987).

The Potwar Plateau record extends the temporal range of *Nilssonia* and places the divergence of the *N. gangeticus* lineage from the rest of the clade at no younger than 11.4 Ma. All other records for the genus are either Pliocene, consisting of *Nilssonia* sp. from the Upper Siwaliks of India (*Nilssonia* cf. *gangeticus*; Srivastava and Patnaik 2002), the Plio-Pleistocene of Nepal (Corvinus and Schleich 1994), or of indeterminate provenance within the Siwalik Group (*N. hurum sivalensis* from the "Siwalik Hills"; Lydekker 1885, 1889a).

Testudinoid turtles from the Potwar Plateau Siwalik Group include common but fragmentary records of geoemydids (pond turtles) and rare testudinid (tortoise) records. The majority of the geoemydid record is too incomplete to assign to generic or specific levels, although several partial shells can be assigned to exclusive clades. A partial carapace and plastron from 11.6 Ma (Locality Y504, YGSP 23368, Fig. 8.5a) are referrable to *Geoclemys* based on the presence of wide, tall carapacial ridges with a saw-toothed lateral profile and without a tectiform carapacial cross-section (Hirayama 1985; Joyce and Bell 2004), as in other fossils of the genus (Tewari and Badam 1969; Das 1991). A single third neural, with a tall posterior ridge edge positioned anteriorly and the base of a second ridge posteriorly as in *"Clemmys (Geoclemys) palaeindica"* (Lydekker 1885; synonymized with *Geoclemys hamiltonii* by Garbin, Bandyopadhyay, and Joyce 2020), was recovered from 10.1 Ma (Locality Y311).

Melanochelys is represented by plastra and a partial carapace referable to the genus. Plastra possessing scute sulci patterns identical to those of extant *M. trijuga*, including elongate gular scutes, strongly anterior concave pectoral scute margins, and a nearly imbricate marginal-pectoral sulcus pattern (e.g., Bourret 1941; Fig. 8.5b), were recovered from the Dhok Pathan Formation (FO at 8.7 Ma, Y388; 7.4 Ma Y457). A partial carapace from 7.7 Ma (Locality Y900) includes a narrow second neural scute with strongly concave posterolateral margins and a low median ridge, as in extant *M. trijuga* (Fig. 8.5c). Taxonomically ambiguous remains that are distinct from either *Geoclemys* or *Melanochelys* indicate a diversity of small-bodied geoemydids within the Siwalik sequence. A posterior plastron with a deep, narrow anal notch and an elongate, anteromedially angled sulcus for the anal scute was recovered from 7.2 Ma (Locality Y581), and articulated epi- and entoplastra with squared anterior and

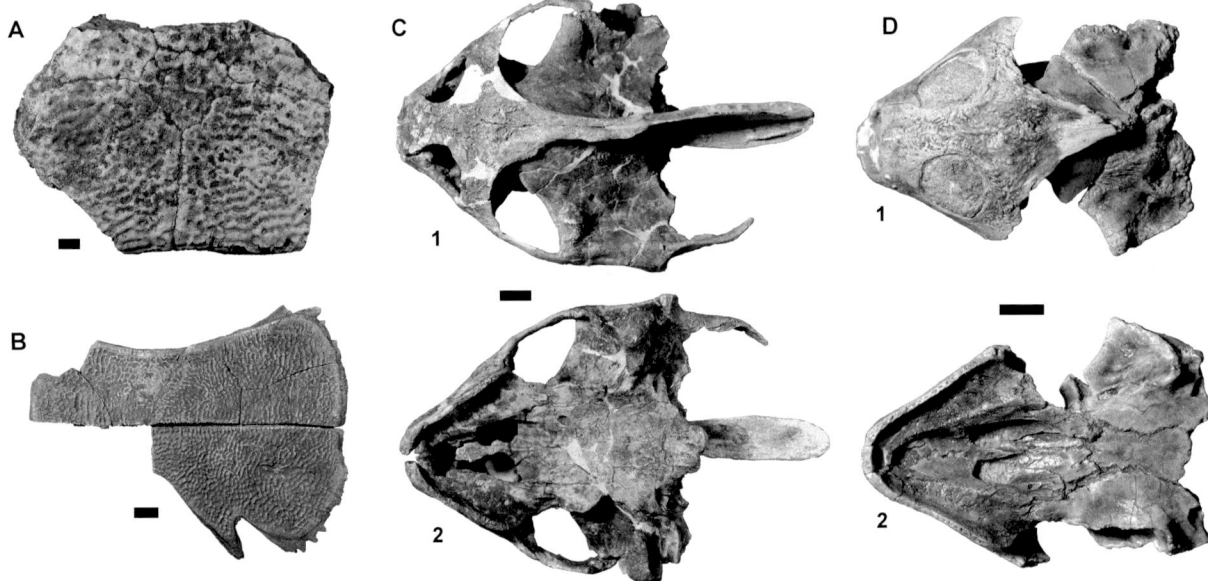

Figure 8.4. Siwalik Group trionychine turtles. A. *Chitra* sp., posterior carapace (YGSP 50031) in dorsal view. B., *Nilssonia* cf. *gangeticus* lineage, left hyo-hypoplastron (YGSP 39520) in ventral view. C. *Nilssonia* cf. *gangeticus* lineage, skull (YGSP 47215) in (1) dorsal and (2) ventral views. D. *Nilssonia* sp., skull (YGSP 41000) in (1) dorsal and (2) ventral views. Scales equal 1 cm.

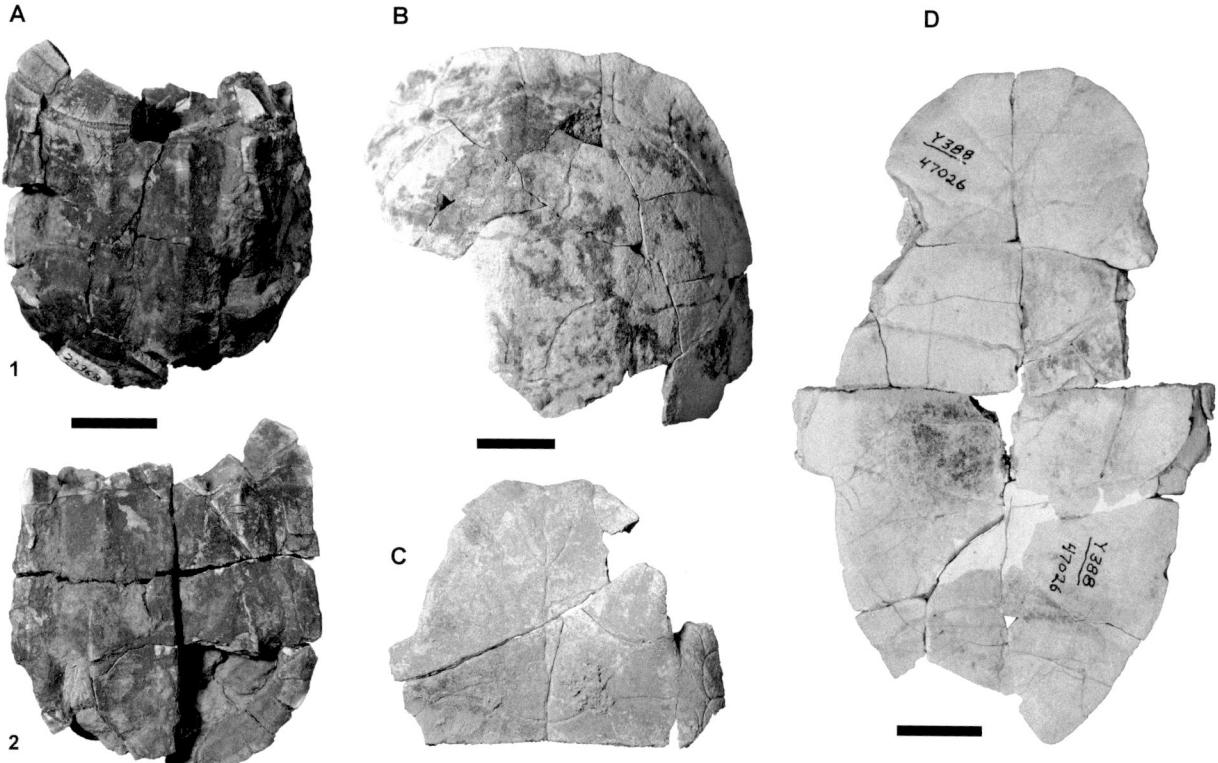

Figure 8.5. Siwalik Group geoemydid turtles. A. *Geoclemys* sp. partial shell (YGSP 23368) in (1) dorsal and (2) ventral views. B. *Melanochelys* sp. anterior carapace (YGSP 49739) in dorsal view. C. *Melanochelys* sp. anterior plastron (YGSP 51130) in ventral view. D. *Melanochelys* sp. plastron (YGSP 47026) in ventral view. Scales equal 2 cm.

lateral margins favorably comparable to *Cyclemys* were recovered from 7.8 Ma (Locality Y917). Fragmentary shell elements of large-bodied geoemydids, similar in size to large individuals of extant *Batagur*, *Callagur*, and *Hardella* and fossils assigned to them by Lydekker (1885), are present throughout the Potwar Plateau sequence from at least 14.1 Ma (Locality Y478). Large neural elements from a single individual recovered from 7.3 Ma (Locality Y937) are elongate and narrow, similar to the first three neurals of fossil *Batagur* species erected by Lydekker (1885), but unlike those of most other large geoemydids.

Diverse geoemydids have been described from the Siwaliks of northwestern India. Lydekker (1885) described multiple species of *Clemmys* (*Melanochelys*), *Batagur*, and *Pangshura* from the Cautley and Falconer collections, as well as a specimen collected by Theobald from "Asnot, Punjab" (Lydekker 1885, 188), which likely corresponds to the Padhri-Hasnot collecting area. Multiple records of these genera, as well as the previous FO for *Geoclemys*, have been subsequently described from the Upper Siwaliks (e.g., Nanda, Schleich, and Kotlia 2016). West, Hutchison, and Munthe (1991) documented multiple indeterminate geoemydid shell morphologies as well as tentative records of *Kachuga*, *Batagur*, and *Cuoura/Chinemys* from the Middle Siwaliks of Nepal. From the Potwar Plateau, Theobald (1877) described "*Bellia* (*Siebenrockiella*) *sivalensis*" from the Upper Siwaliks (*Elephas planifrons* Interval Zone, approximately 2.5–1.8 Ma; Barry, Lindsay, and Jacobs 1982) near Jhand, and Joyce and Lyson (2010) described *Pangshura tatrotia* from the

Pliocene Tatrot Formation near Padhri, both on the basis of nearly complete shells.

Taxonomies of Siwalik geoemydids collected and described by Cautley and Falconer were revised by Garbin, Bandyopadhyay, and Joyce (2020) based on collections of the Natural History Museum, London, and the Indian Museum, Kolkata. Provenance data for these specimens is limited, but most were considered by Garbin, Bandyopadhyay, and Joyce (2020) to span the Miocene-Pliocene based on assumptions that Cautley and Falconer collected specimens on the Potwar Plateau. The majority of collections were made in the younger Siwaliks of northern India, however (Jukar, Sun, and Bernor 2018), and the preservation of most specimens is more consistent with geoemydid fossils from the Pliocene-Pleistocene Pinjor and pre-Pinjor beds between the Sutlej and Yamuna rivers (Nanda, Schleich, and Kotlia 2016) than with specimens from the Pakistan Siwaliks. The geoemydid records definitively from the Potwar Plateau extend the histories for most previously recognized taxa, including middle Late Miocene FOs for *Geoclemys* and *Melanochelys*, and occurrences of the large-bodied geoemydid clade that predate the Nepalese and Indian records.

Testudinids are sparsely represented in the Siwalik record. Small-bodied specimens (carapace length approximately 50 cm or less) are represented by shell fragments throughout the Potwar Plateau. Large-bodied taxa (carapace length approaching or exceeding 1 m) increase in occurrence from approximately 9.3 Ma (Locality Y439) to younger intervals.

Among small-bodied taxa, the *Indotestudo elongata* lineage is represented at 8.7 Ma (Locality Y388) by a posterior plastron with a wide, concave anal notch and short anal scute that extends anteriorly just past the apex of the notch (YGSP 47029, Fig. 8.6a). Cf. *Indotestudo*, consisting of a badly fragmented shell, including an anterior plastron with a dorsally overhanging epiplastral lip, was recovered at 12.8 Ma (Locality Y825). A poorly preserved, badly abraded shell of a specimen approximately 50 cm long was recovered at 7.4 Ma (Locality Y926).

Large-bodied testudinids with estimated carapace lengths exceeding 1.5 meters are represented by fragmentary pleural elements, partial plastral elements, and peripheral elements relatively high in the Potwar Plateau sequence. Because of their size and weight, large testudinids have not been systematically collected, and fragmentary remains deposited in the Geological Survey of Pakistan remain the primary evidence for constraining the biostratigraphic record. The FO for the largest testudinid specimens is 9.3 Ma (Locality Y439), and the youngest collected specimen is a fragmentary peripheral from 6.3 Ma (Locality Y930). Large-bodied testudinids were disproportionally recovered from the Khaur collecting area, including localities where specimens were left in situ (Locality Y1011, approximately 9.7 Ma, Fig. 8.6b).

Testudinids have been documented from the Upper Siwaliks along the Himalayan foredeep. *Colossochelys atlas*, a giant taxon represented by shell and appendicular elements, was recovered from the Upper Siwaliks of Northern India (Falconer 1868; Lydekker 1885; Badam 1981). Theobald (1879) identified a large testudinid peripheral consistent in size with specimens recovered from the Khaur area on the Potwar Plateau and dated between 11 and 8.5 Ma, based on co-occurrence with *Sivapithecus sivalensis* (Morgan et al. 2015). West, Hutchison, and Munthe (1991) described *Geochelone* (?*Hersperotestudo*) on the basis of osteoderms and fragmentary shell pieces from the Middle Siwaliks of Nepal.

CROCODYLIA GMELIN, 1788

Crocodilians are well-documented in the middle and upper Siwalik Group of Pakistan, India, and Nepal. Three clades have been recognized: *Gavialis*, *Crocodylus*, and Tomistominae, including *Rhamphosuchus* and indeterminate remains (e.g., Falconer 1868; Lull 1944; West, Hutchison, and Munthe 1991; Patnaik and Schleich 1993; Head 2001; Nanda, Schleich, and Kotlia 2016). Indeterminate crocodilian remains, including generalized conical teeth, vertebrae, and fragmentary osteoderms, are ubiquitous throughout the section. *Gavialis* is the most commonly recovered crocodilian from the Potwar Plateau and occurs throughout the Siwalik sequence. The oldest sampled remains are isolated elongate, recurved teeth and partial rostra, which occur from the Kamlial Formation (Locality Y807, 16.9 Ma). Cranial and dental remains have been recovered throughout the section, and extant *G. gangeticus* has been locally extirpated from the northern Punjab only during the twentieth century.

Multiple fossil *Gavialis* species have been described from the Siwalik Group, including *Gavialis leptodus* Falconer (1868), *G. hysudricus* Lydekker (1886a), *G. curvirostris* Lydekker (1886a), *G. pachyrhynchus* Lydekker (1886a), *G. leptodus* Lydekker (1886a), *G. browni* Mook (1932), and *G. lewisi* Lull (1944). Several of these taxa are distinct from *Gavialis* or Gavialidae. The type materials of *Gavialis curvirostris* are referrable to Tomistominae; *G. pachyrhynchus*, a massive rostral tip, is indistinguishable from the possible tomistomine *Rhamphosuchus* (Head 2001; Martin 2019), and *G. leptodus* is either a tomistomine (Martin 2019) or a gavialid distinct from *Gavialis* on the basis of rostral morphology. The remaining taxa are based on specimens referable to *Gavialis*.

Of the previously named species from the Potwar Plateau, *G. browni* was recovered from the Dhok Pathan Formation (Mook 1932) and *G. lewisi* from the Nagri Formation (Lull 1944). The validity of these taxa remains uncertain. Norell and Storrs (1989) recognized *G. lewisi* as marginally distinct from *G. gangeticus*, and Brochu (1997) combined all Siwalik Group *Gavialis* into a single informal taxon as the sister taxon to *G. gangeticus*. Martin (2019) revised the Siwalik gavialoids based on the collections housed in the Natural History Museum, London and referred *G. hysudricus* to *G. gangeticus*, while leaving the validity of *G. lewisi* and *G. browni* unresolved. Head (1997) reported *Gavialis* cf. *G. curvirostris* on the basis of a distally recurved partial right dentary (YGSP 51126) from the Dhok Pathan Formation at approximately 7.4 Ma (Locality Y926); the specimen is unambiguously *Gavialis*, whereas *G. curvirostris* is not, and the curvature used for reference is most likely the result of taphonomic distortion.

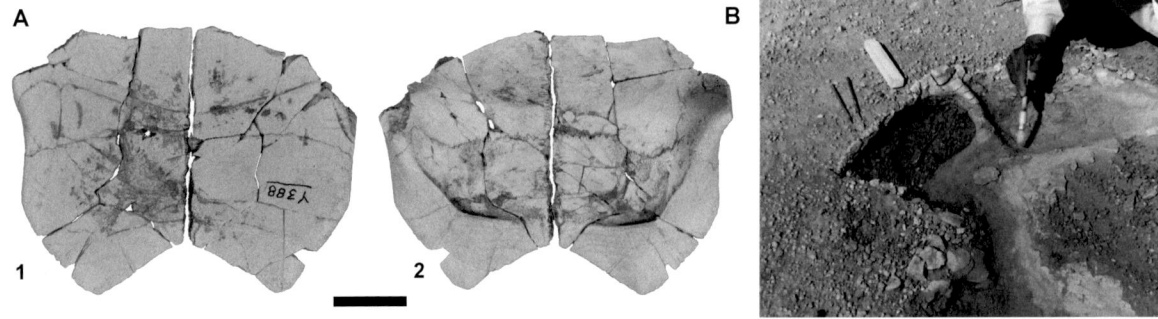

Figure 8.6. Siwalik Group testudinid turtles. A. Cf. *Indotestudo* sp. posterior plastron (YGSP 47029) in (1) ventral and (2) visceral views. Scale equals 2 cm. B. Indeterminate testudinid plastron exposed in situ by Geological Survey of Pakistan officer Munir el Haq, Locality Y1011, near Khaur, 2000.

Referral of the *Gavialis* record to species is limited both by the fragmentary nature of specimens and to ambiguities in specific diagnostic apomorphies of *G. lewisi* and *G. browni* relative to each other and to *G. gangeticus*. Both *G. lewisi* and *G. browni* are recognized as distinct on the basis of a deeper skull than in *G. gangeticus*, with *G. browni* possessing a strongly angled rostrum, though this may represent taphonomic distortion (Lull 1944; Norell and Storrs 1989). *G. lewisi* differs from *G. gangeticus* by several additional characters, including smaller basioccipital tubera and supratemporal fenestrae (Norell and Storrs 1989). None of these characters has been considered diagnostically robust by their respective authors, and none can be unambiguously recognized in the *Gavialis* record of the Potwar Plateau. As a result, only generic-level assignments are applied here.

The Siwalik Group spans the majority of the *Gavialis* fossil record. However, the oldest records of the genus based on cranial elements are from the early Miocene of South Asia. Piras and Kotsakis (2005) described an indeterminate *Gavialis* skull from the early Miocene of Gaj, and undescribed cranial remains have been recovered from the middle to upper Miocene Kharinadi Formation at Pasuda in Gujarat, India (Bhandari et al. 2018), from the same horizon that produced records of *Acrochordus* (Head, Mohabey, and Wilson 2007). The youngest fossil records of *Gavialis* include Pleistocene specimens from modern Thailand and Indonesia (Dubois 1908; Delfino and De Vos 2010; Martin et al. 2012).

Recognition of tomistomines within the Siwalik Group crocodilian record is comparatively recent (West, Hutchison, and Munthe 1991, Head 2001). Specimens recovered from late Paleogene and Neogene strata of Pakistan have been misidentified as fossil gavialoids (*Gavialis curvirostris*, *G. curvirostris* var. *gangeticus*, and *G. breviceps*; Pilgrim 1908, 1912; Lydekker 1886a; Lull 1994; Martin 2019) since the earliest field explorations. Within the Potwar Plateau, two tomistomine specimens have been recovered from the type Chinji Formation. The first is a partial rostrum including premaxillae and maxillae (YGSP 23417) from 13.3 Ma (Locality Y670). The second is a complete skull and mandible (YGSP 39521) recovered from 13.0 Ma (Locality Y43; Fig. 8.7b), tentatively referred to *Rhamphosuchus crassidens* (Head 2001), which was previously known from partial cranial elements from the Upper Siwaliks of Chandigarh (Falconer 1868; Nanda, Schleich, and Kotlia 2016). Both Chinji Formation specimens represent the same taxon, based on the shape of the external nares, the spacing and relative sizes of premaxillary tooth positions, and the degree of maxillary longirostry. Prior referral of YGSP 39521 to *R. crassidens* was similarly based on the morphology of rostral elements, as well as extremely large body size (Falconer 1868; Lydekker 1886a). Palatal exposure of the vomer medial to the palatine (Brochu 1997) on YGSP 39521 was used to assign the specimen unambiguously to Tomistominae (Head 2001).

Subsequent examination of the type and referred materials of *Rhamphosuchus crassidens* demonstrates important differences in relative longirostry and indicates that the specimens from the Chinji Formation represent a new taxon. In particular, the rostrum is proportionately more elongate in

R. crassidens, including higher tooth counts and a more elongate symphseal portion of the mandible (Lydekker 1888). Additionally, the only specimen referred to *R. crassidens* (British Museum of Natural History 39804) that preserves a visible palate does not appear to possess palatal exposure of the vomer, which suggests that either the taxon is not a tomistomine or that the hypodigm of *Rhamphosuchus* is a chimera. The Chinji Formation taxon also possesses a dorsally inflated narial margin and a slightly more dorsoventrally shallow maxillary rostrum than *R. crassidens*. These features are similar to the enigmatic tomistomine *"Gavialis" breviceps* from the early Miocene Kumbi site in the Chitarwata Formation at Dera Bugti (Pilgrim 1908, 1912; Antoine and Welcomme 2000; Antoine et al. 2013) and, potentially, to tomistomine remains from the Early Miocene of Kutch (Patnaik et al. 2014). The Chinji Formation taxon is distinct from *"Gavialis" breviceps*, however, in possessing a flatter interorbital skull roof, fewer tooth positions, and an even shorter mandibular symphysis.

Crocodylus is represented by cranial elements from the Chinji, Nagri, and Dhok Pathan Formations. A fragmentary right anterior maxilla indistinguishable from *C. palaeindicus* (e.g., Lydekker 1886a; Fig. 8.7a) represents the FO for the genus in the Siwalik record at 12.4 Ma (Locality Y496, YGSP 46952). The LO for the genus is a partial left dentary from 6.4 Ma (Locality Y913, YGSP 49862). As with the LO for *Gavialis*, this occurrence simply represents the youngest sampled locality and not the end of its duration, as *Crocodylus* occurs throughout wetland environments in modern Pakistan (e.g., Minton 1966).

The fossil record of *Crocodylus* within the Siwalik Group is generally referred to *C. palaeindicus* (Chabrol et al. 2024). Exceptions include several nominally erected species (e.g., Mook 1933; Suneja, Singh, and Chopra 1977; Garg 1988) and extant *Crocodylus* cf. *palustris* and *C. porous* (*biporcatus*), which have been reported from the Pliocene Siwaliks of northwestern India at Moginand and Naipli, Himachal Pradesh, respectively (Badam 1973; Patnaik and Schleich 1993). Although the Siwalik record of *Crocodylus* on the Potwar Plateau cannot be assigned to species on the basis of discrete apomorphies, recovered cranial fragments are effectively identical to specimens of upper Siwalik *C. palaeindicus*, which suggests that the lineage extended into at least the Middle Miocene. The Potwar Plateau record may additionally represent the FO for the genus. The interrelationships of *Crocodylus palaeindicus* are controversial. It is either part of an unresolved polytomy with *C. niloticus* (and African fossil taxa; Brochu 2017), New World *Crocodylus*, and Indopacific *Crocodylus*, (Brochu 2000, 2003 2017; Brochu and Storrs 2012), sister taxon to the *Crocodylus* crown (Conrad et al. 2013), or sister taxon to *C. palustris* (Chabrol et al. 2024). Only the latter hypothesis unequivocally unites the Siwalik record with extant Asian species.

DISCUSSION

Biostratigraphy and Faunal Change

Low sample sizes of diagnostic elements at generic and specific levels for many taxa limit the ability to interpret the

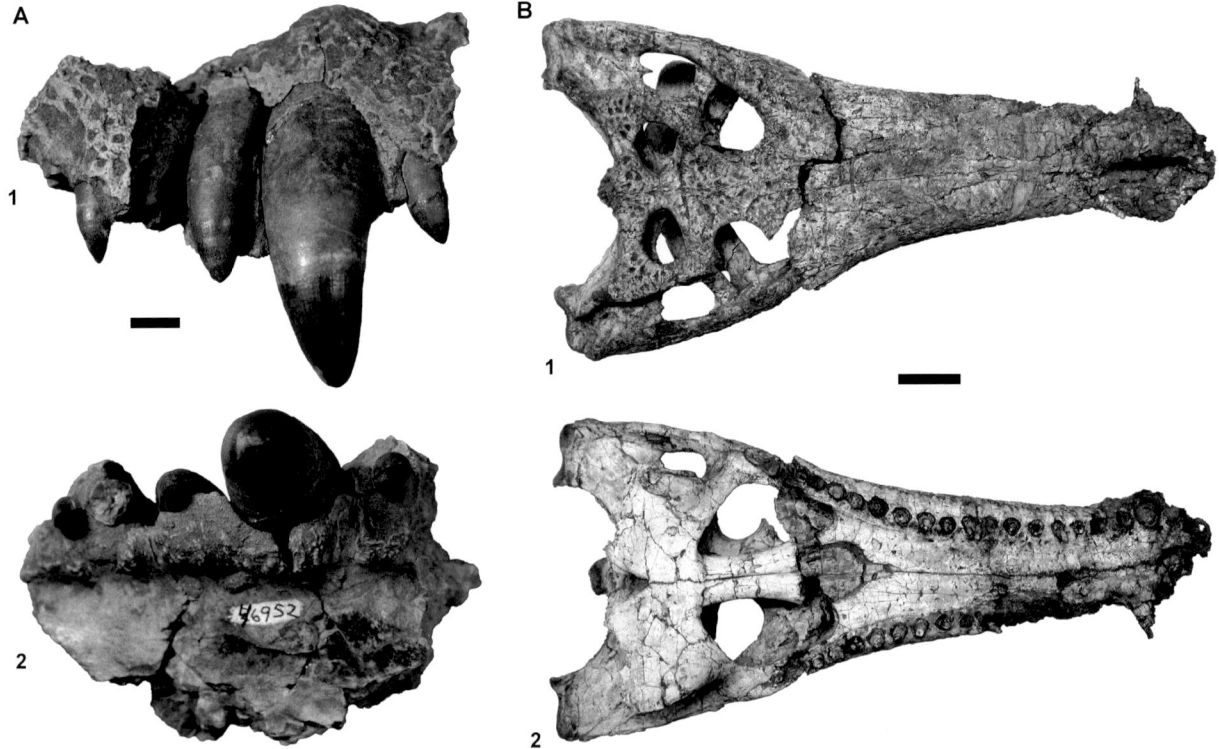

Figure 8.7. Siwalik Group crocodylians. A. *Crocodylus* cf. *palaeindicus* left anterior maxilla (YGSP 46952) in (1) lateral and (2) ventral views. Scale equals 1 cm. B. Tomistominae gen. et sp. nov. (YGSP 39521) complete skull in (1) dorsal and (2) ventral views. Scale equals 10 cm.

record of first and last occurrence of Siwalik reptiles as literal historical events. For example, the isolated records of Boidae? indet. and Erycidae indet. and the last occurrences of most agamine lizard records within the Chinji Formation, as well as LOs for gekkotans and scincids at approximately 7.8 Ma (Localities Y898, Y906), are almost all preservation or sampling artefacts since extant species of those clades are abundant throughout South Asia, including the modern Potwar Plateau. In contrast, the stratigraphic ranges of extant turtle genera such as *Geoclemys* (11.6–10.1 Ma, Localities Y504–Y311) and *Melanochelys* (8.7–7.7 Ma, Localities Y388–Y900) represent instances of relatively good preservation within less diagnostic, fragmentary records that span the investigated section (Fig. 8.8).

Despite these limitations, the Siwalik record of the Potwar Plateau is important for extending the temporal ranges for younger Siwalik reptile taxa, as well as extant generic and higher-order lineages, back to at least the Middle Miocene. This record provides the oldest precise FOs for *Crocodylus*, all described geoemydid genera, and *Nilssonia* and *Chitra*. The FOs for large testudinids and uromastycines may additionally represent local immigration due to regional-scale environmental change. Additionally, the record demonstrates the persistence of Carettochelyinae beyond Southeast Asia and Australia-New Guinea until at least the late Miocene.

The increasing number of first occurrences upsection among colubroid snakes likely approximates immigration at local geographic scales within northern South Asia, as these patterns postdate the larger global diversification of Colubroi-

dea during the Late Paleogene and Early Neogene (e.g., Head, Mahlow, and Müller 2016). The taxonomic and morphotypic richness of colubroids increases through the Potwar Plateau section, beginning at approximately 10.2 Ma (Locality Y450). Colubroid FOs increase upsection, with no evidence of preservational or sampling biases driving changes in richness or record quality (Head 2005).

Thus, some of the reptile biostratigraphic record contains historical events approximately synchronous with environmental changes recorded in Siwalik pedogenic and biogenic proxy data. Lithofacies, isotopic records, paleosol composition, or pollen records taken in situ from terrestrial sections combined with isotopic and hydrocarbon environmental proxies derived from coeval Siwalik sediments shed into the Bengal Fan (Karp et al. 2021; Polissar et al. 2021) demonstrate consistent, though apparently asynchronous, environmental transitions within the Siwalik Group (e.g., Barry et al. 2002; Srivastava et al. 2018; Karp et al. 2021; Polissar et al. 2021). Hydrologic changes included a shift from more equable climates to increased seasonality beginning at approximately 10.7 Ma in western Nepal (Dettman et al. 2001) and 10.0-9.5 Ma in the Potwar Plateau (Polissar et al. 2021), with evidence of shifts in fire regimes from low-fire to variable fire stages beginning at 10.2 Ma (Karp et al. 2021). Floral changes include increase in wet deciduous woodlands from evergreen forests by 12 Ma in Nepal based on plant fossils, whereas the earliest evidence of floral transitions in the Pakistan Siwaliks is the first isotopic evidence for C_4 diets in equids at approximately 9.5 Ma (Polissar et al. 2021).

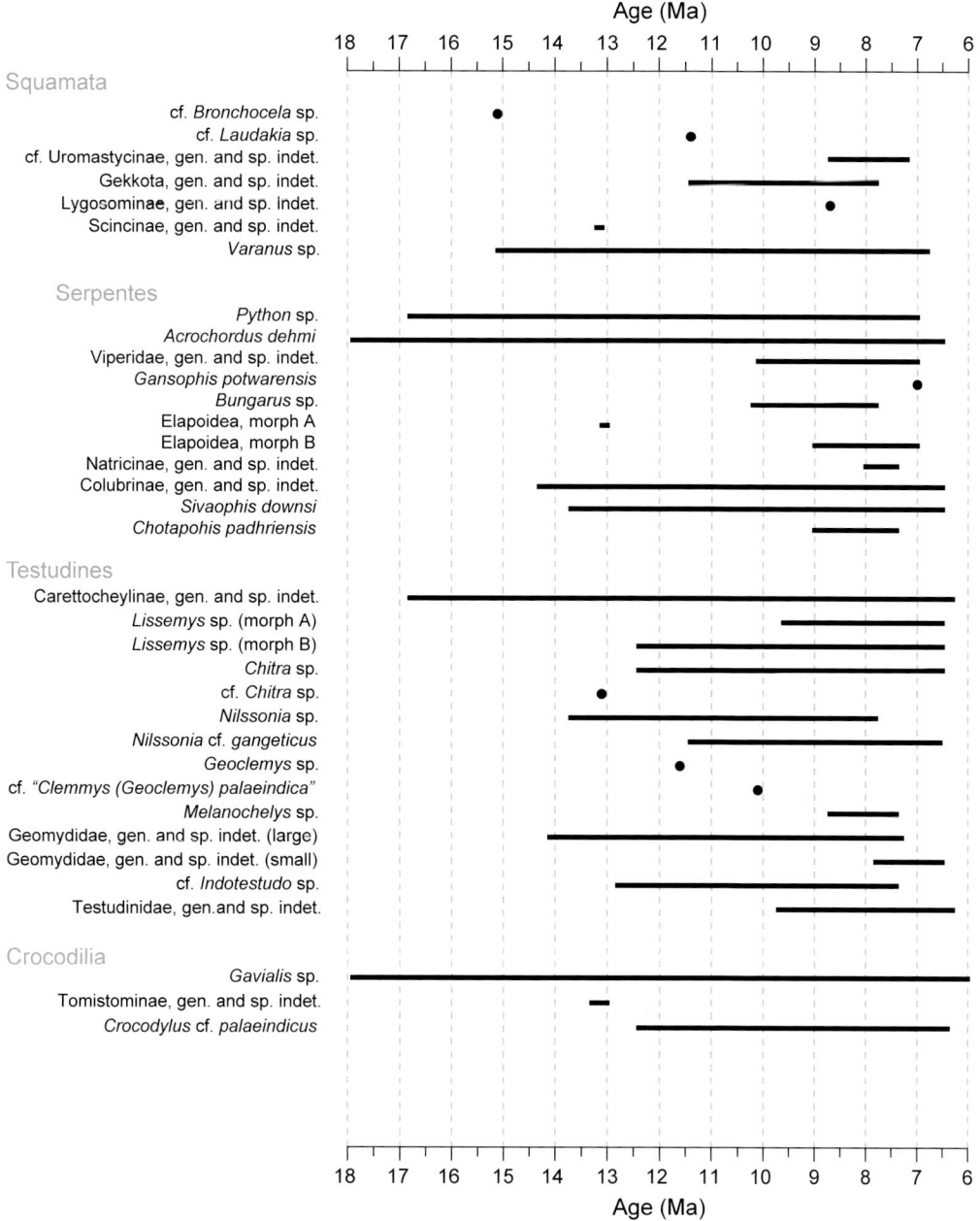

At local scales, changes in fluvial architecture from the Nagri Formation to the Dhok Pathan Formation starting at approximately 9.5 Ma resulted in a wider range of habitats surrounding smaller river systems with greater seasonal drainage than during the time of Nagri lithofacies deposition (Willis 1993a, 1993b; Behrensmeyer, Willis, and Quade 1995; Willis and Behrensmeyer 1995; Barry et al. 2002; Morgan et al. 2009; Barry et al. 2013).

Increasing taxonomic diversity over time among snakes, the increasing occurrence of large testudinids, the FO for Uromastyincae, and the young record of large *Varanus* are consistent with ecosystem shifts to more heterogenous, open environments. The transition from more closed-canopy wetland forests associated with trunk river systems to more patchy vegetation habitats associated with smaller, more seasonal river systems would have supported a wider range of ecomorphs and physiologies by increasing basking opportunities (e.g., Huey et al. 2009) for squamates and terrestrial turtles and would have provided more opportunities for grazing in terrestrial testudinids.

Alternatively, increase in the taxonomic richness of colubroid snakes could represent shifts in agents of accumulation through time. A lack of correlation between specimen abundance of micromammals and sedimentary particle sizes in the localities that produced screen-washed microvertebrates implicated non-sedimentary (biotic) accumulation processes, such as predator accumulation (Badgley, Downs, and Flynn 1998). Conversely, examination of nonmammalian vertebrate concentrations from representative screen-washed localities through the Chinji, Nagri, and Dhok Pathan Formations reveals the continued presence, albeit in low abundances, of aquatic taxa on

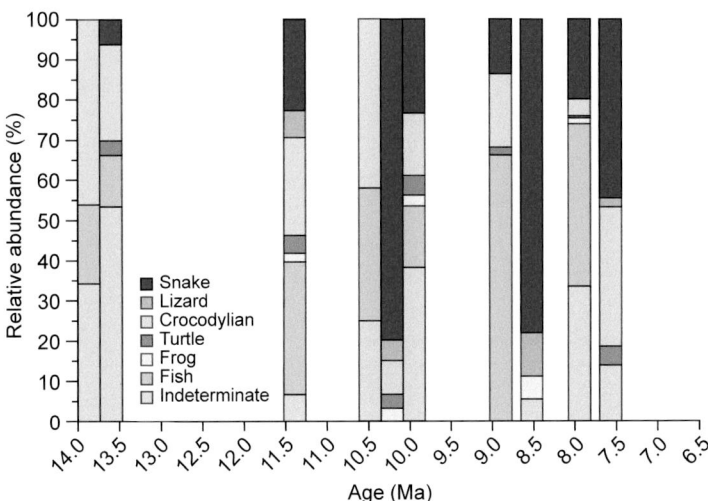

Figure 8.9. Relative abundance of nonmammalian specimens from select Siwalik vertebrate microsite concentrations. Specimens were obtained by bulk screen-washing of sediments by William R. Downs III.

the basis of shed crocodilian teeth, juvenile turtle shell elements, and fish vertebrae (Fig. 8.9, Table 8.1). These records indicate that at least some of the Siwalik microvertebrate locality composition may be derived from hydrologic deposition of autchthonous remains in addition to predator-transported elements. Colubroid snakes often possess semi-aquatic habits, and concomitant increases of larger colubroids, including *Bungarus* and large viperid elements in surface-collected samples beginning at approximately 10 Ma, suggest increasing local habitation as opposed to shifts in taphonomic mode.

Environmentally mediated faunal changes within Siwalik mammalian ecosystems, including the presence of equids in large herbivore communities from 10.8 until 6.5 Ma and the loss of arboreal hominoids by 8.5 Ma (Barry et al. 2002; Nelson 2007), appear to have been controlled by increasing aridity driven by large-scale Himalayan tectonics and regional climate change. Conversely, the increase in snake diversity predates the local sedimentological and isotopic evidence for the major late Miocene climatic transition by approximately 1 million years and is approximately coeval with fluvially mediated habitat shifts at the Nagri-Dhok Pathan transition. Late Miocene increases in occurrence of testudinids and the FO for uromastycines are ecologically consistent with more open habitats, but the relative influences of local fluvial reorganization or regional

climate change in facilitating these faunal changes cannot be differentiated with the limited record quality for both taxa.

Even with reductions in river size and increasing seasonality through time, the reptile record indicates the persistence of large, permanent aquatic ecosystems throughout the sampled sequence (Head 1997, 1998). The surface-collected reptile record is dominated by large-bodied, fully aquatic, obligate piscivores, including *Acrochordus* and *Gavialis*, and aquatic carnivorous trionychids are abundant. The only potential change in the aquatic fauna with environmental change is the absence in the Nagri or Dhok Pathan Formations of the large-bodied tomistomine, which would have inhabited larger fluvial systems. Body-size estimates for specimens recovered from the Chinji Formation exceed 8 m (Head 2001), and the absence of this taxon upsection may represent a combination of limited sampling (Nagri Formation) and a transition to smaller river systems (Dhok Pathan Formation; Zaleha 1997) that would be more favorable to smaller-bodied *Crocodylus*.

Biogeography and Assembly of South Asian Reptile Faunas

The Siwalik reptile record of the Potwar Plateau includes taxa shared at generic and more inclusive phylogenetic levels with other Neogene, Old World assemblages. Together with the

Table 8.1. Number of Nonmammalian Specimens from Microvertebrate Sites by Taxon

Locality	Estimated Age (Ma)	Snake	Lizard	Crocodylian	Turtle	Frog	Fish	Indeterminate	Total
Y906	7.8	19	1	15	2	0	0	6	43
Y547	8.0	29	0	6	1	2	60	50	148
Y388	8.7	14	2	0	0	1	0	1	18
Y367	9.0	13	0	18	2	0	66	0	99
Y311	10.1	9	0	6	2	1	6	15	39
Y450	10.2	47	3	5	2	2	0	0	59
Y259	10.5	0	0	5	0	0	4	3	12
Y76	11.4	10	3	11	2	1	15	3	45
Y491	13.8	3	0	13	2	0	7	29	54
Y589	14.0	0	0	53	0	0	23	40	116

upper Siwalik records of northern India, the record of the Potwar Plateau minimally constrains the assembly of components of modern reptile faunas within the larger Indo-Malayan biogeographic province and demonstrates a greater past distribution of taxa that are now segregated into either sub-Himalayan South Asia or peninsular or insular Southeast Asia.

Multiple Siwalik taxa were part of greater Miocene transcontinental distributions, including *Varanus*, *Python*, colubroid snakes, tomistomine crocodylians, and carettochelyid turtles, with records spanning southern Europe, Africa, and Asia (e.g., Piras et al. 2007; Havlik et al. 2014; Georgalis et al. 2016; Head and Müller 2020; Villa et al. 2018). Other taxa are either the oldest or only records for extant lineages with wide geographic distributions. *Crocodylus* cf. *palaeindicus* from the Chinji Formation represents the generic FO, predating the next oldest Asian records from Piram Island by approximately 2 to 3 million years (Chabrol et al. 2024—see discussion in Nanda, Sehgal, and Chauhan (2017) for ages of Piram Island Siwaliks), and African and southern European records by approximately five million years (Brochu 2000; Brochu and Storrs 2012). This pattern of oldest occurrences suggests an Asian, rather than African (Delfino and Rook 2008), origin of the genus. Siwalik *Bungarus* belongs to a derived species complex within the genus and minimally constrains the divergence and western dispersal of extant lineages from eastern Asia to no younger than approximately 10 Ma. The Neogene records of *Chitra*, *Indotestudo*, and potentially *Cyclemys* are restricted to South Asia, despite cosmopolitan distributions of these taxa through much of modern Indo-Malaysia (Rhodin et al. 2017). Despite its high abundance throughout the Siwalik record and widespread distribution of extant species throughout Southeast Asia to Northern Australia (McDowell 1979; Shine 1986), the definitive fossil record of *Acrochordus* is restricted to India, Pakistan, and Nepal from at least the Early Miocene to the Late Pliocene (Hoffstetter 1964; West, Hutchison, and Munthe 1991; Rage, Gupta, and Prasad 2001; Head 1997, 1998, 2005; Head, Mohabey, and Wilson 2007; Kapur et al. 2021; Singh et al. 2022).

The majority of aquatic turtle records suggest Miocene establishment of South Asian provincial turtle faunas. *Geoclemys*, *Melanochelys*, *Lissemys*, and *Nilssonia* are restricted to the Siwalik Group without any evidence of extralimital distributions from their extant South Asian ranges (Rhodin et al. 2017). There is no evidence of extant East or Southeast Asian endemics in the Potwar Plateau turtle record.

Although the upper Siwalik reptile record of northern India demonstrates Pliocene continuation of the reptile communities from the Miocene Potwar Plateau sequence, several differences indicate faunal change within the temporal gap between the youngest well-sampled Potwar records at 6.5 Ma and the oldest records from the "Pre-Pinjor Beds" (Saketi or Tatrot Formations; Verma, Mishra, and Gupta 2002; Nanda, Schleich, and Kotlia 2016) of the Chandigarh area, dated to 3.6–2.6 Ma (Sao et al. 2016; Jukar, Sun, and Bernor 2018). Carettocheylines, though rare in the Miocene Siwaliks, are absent from the Upper Siwaliks, despite multiple studies of the turtle record. The Pliocene giants *Varanus sivalensis*, *Colossochelys atlas*, and potentially *Rhamphosuchus crassidens* from Saketi are

closely related to taxa from the Potwar Plateau, but all first occur in the "Pre-Pinjor Beds" and are distinct from the Miocene records in their extremely large body sizes. This size increase in both fully terrestrial and fully aquatic reptile species does not correspond to any obvious local environmental change between the late Miocene record of the Potwar Plateau and the Upper Siwaliks but does correspond to similar patterns of size increase among tropical reptile records during the late Neogene (e.g., Brochu and Storrs 2012).

Except for the majority of freshwater turtles, the segregation of modern reptile faunas from western South Asia and insular and peninsular Southeast Asia within the larger Indo-Malayan province appears to be a Quaternary history. *Acrochordus*, Tomistominae, Carettochelyinae, and possibly *Bronchocela*, which are all present in the Siwalik record, are absent from modern sub-Himalayan South Asia. Marine *Acrochordus granulatus* is present in coastal habitats of southern India, but freshwater aquatic taxa more closely related to *A. dehmi* (Sanders et al. 2010) are Southeast Asian through Australasian. *Tomistoma* is restricted to peninsular Malaysia and Borneo. Post-Pleistocene *Carettochelys* is now extirpated from Indo-Malaysia, with distributions limited to Australasia, and *Bronchocela* is restricted to Southeast Asia (Diong and Lim 1998).

Thus, despite the preservational and sampling limitations relative to the mammalian record, the Siwalik reptile record documents the initial herpetofaunal assembly of the Indo-Malayan biogeographic realm by at least the middle Miocene. The record consists mainly of taxa that are now primarily or wholly South Asian, with a smaller number of taxa that are cosmopolitan or have become predominantly restricted to equatorial Southeast Asia or Australasia since at most the late Pliocene. This geographic pattern is generally consistent with the mammalian record, in which modern equatorial Southeast Asian taxa, including pongines, tragulids, and lorisids, were present north to the foothills of the Himalaya during the Miocene. But the Southeast Asian timing of extinction or extirpation differs for reptilian and mammalian faunas from the Siwalik Group and reflects distinct biogeographic histories.

ACKNOWLEDGMENTS

This work is based primarily on study of Siwalik reptile specimens collected through 40 years of field research on the Potwar Plateau, to which I am privileged to have contributed in small part from 1996 to 2000. I am grateful for the efforts of numerous field participants in surface collection, as well as the screen-washing collections generated by Everett Lindsay and his students Louis Jacobs and Larry Flynn, and especially by William R. Downs III. For both education and tolerance over two decades through this research, I am grateful to Catherine Badgley, John Barry, Kay Behrensmeyer, Will Downs, Larry Flynn, Doc Lindsay, Michèle Morgan, and David Pilbeam. Officers of the Geological Survey of Pakistan were crucial for facilitating field research, especially Ibrahim Shah, Munir ul Haq, Mohammad Anwar, and Khalid Sheikh. Many of the aforementioned, as well as Jay Kelley, Nina Jablonski, and Sherry Nelson provided field education and company on the

Potwar. I am grateful to Louis Jacobs, Dale Winkler, and Christopher Bell for their mentorship during the early stages of this research. Funding for both the field and laboratory components of this study came from the Saurus Institute, the Institute for the Study of Earth and Man, and the Department of Geological Sciences (Now Huffington Department of Earth Sciences), Southern Methodist University, as well as more recently by National Science Foundation EAR-1338028, and Natural Environment Research Council Award (NE/S000739/1). I thank Keturah Smithson, Cambridge Biotomography Centre, University of Cambridge, for C-T scanning of specimens. Portions of this chapter were written and edited at the Live and Let Live Public House, Cambridge, UK.

References

Antoine, P.O., G. Métais, M. J. Orliac, J. Y. Crochet, L. J. Flynn, L. Marivaux, A. R. Rajpar, G. Roohi, and J.-L. Welcomme. 2013. Mammalian Neogene biostratigraphy of the Sulaiman Province, Pakistan. In *Fossil Mammals of Asia: Neogene Biostratigraphy and Chronology*, edited by X. Wang, L. J. Flynn, and M. Fortelius, 400–422. New York: Columbia University Press.

Antoine, P. O., and J.-L. Welcomme. 2000. A new rhinoceros from the Lower Miocene of the Bugti Hills, Baluchistan, Pakistan: The earliest elasmotheriine. *Palaeontology* 43(5): 795–816.

Badam, G. L. 1981. *Colossochelys atlas*, a giant tortoise from the upper Siwaliks of North India. *Bulletin of the Deccan College Research Institute* 40(1981): 149–153.

Badgley, C., and A. K. Behrensmeyer. 1995. Two long geological records of continental ecosystems. *Palaeogeography, Palaeoclimatology, Palaeoecology* 115(1–4): 1–12.

Badgley, C., W. R. Downs, and L. J. Flynn. 1998. Taphonomy of small-mammal fossil assemblages from the Middle Miocene Chinji Formation, Siwalik Group, Pakistan. In *Advances in Vertebrate Paleontology and Geochronology*, edited by Y. Tomida, L. J. Flynn, and L. L. Jacobs, 145–166. Tokyo: National Science Museum Monograph 14.

Baig, K. J., P. Wagner, N. B. Ananjeva, and W. Böhme. 2012. A morphology-based taxonomic revision of *Laudakia* Gray, 1845 (Squamata: Agamidae). *Vertebrate Zoology* 62(2): 213–260.

Bailon, S., P. Bover, J. Quintana, and J. A. Alcover. 2010. First fossil record of *Vipera* Laurenti 1768 "Oriental vipers complex" (Serpentes: Viperidae) from the Early Pliocene of the western Mediterranean islands. *Comptes Rendus Palevol* 9(4): 147–154.

Barry, J. C., A. K. Behrensmeyer, C. E. Badgley, L. J. Flynn, H. Peltonen, I. U. Cheema, D. Pilbeam, et al. 2013. The Neogene Siwaliks of the Potwar Plateau and other regions of Pakistan. In *Fossil Mammals of Asia: Neogene Biostratigraphy and Chronology*, edited by X. Wang, L. J. Flynn, and M. Fortelius, 373–399. New York: Columbia University Press.

Barry, J. C., A. K. Behrensmeyer, and M. Monaghan. 1980. A geologic and biostratigraphic framework for Miocene sediments near Khaur Village, Northern Pakistan. *Postilla* 183:1–19.

Barry, J.C., E. H. Lindsay, and L. L. Jacobs. 1982. A biostratigraphic zonation of the middle and upper Siwaliks of the Potwar Plateau of northern Pakistan. *Palaeogeography, Palaeoclimatology, Palaeoecology* 37(1): 95–130.

Barry, J. C., M. E. Morgan, L. J. Flynn, D. Pilbeam, A. K. Behrensmeyer, S. M. Raza, I. A. Khan, et al. 2002. Faunal and environmental change in the late Miocene Siwaliks of northern Pakistan, *Paleobiology* 28(Suppl. 2): 1–71.

Barry, J. C., M. E. Morgan, L. J. Flynn, D. Pilbeam, L. L. Jacobs, E. H. Lindsay, S. M. Raza, and N. Solounias. 1995. Patterns of faunal turnover and diversity in the Neogene Siwaliks of northern Pakistan. *Palaeogeography, Palaeoclimatology, Palaeoecology* 115(1–4): 209–226.

Batsch, A. J. G. K. 1788. *Versuch einer Anleitung, zur Kenntniß und Geschichte der Thiere und Mineralien*. Akademische Buchhandlung, Jena.

Behrensmeyer, A. K., B. J. Willis, and J. Quade. 1995. Floodplains and paleosols of Pakistan Neogene and Wyoming Paleogene deposits: Implications for the taphonomy and paleoecology of faunas. *Palaeogeography, Palaeoclimatology, Palaeoecology* 115(1–4): 37–60.

Bhandari, A., R. F. Kay, B. A. Williams, B. N. Tiwari, S. Bajpai, and T. Hieronymus. 2018. First record of the Miocene hominoid *Sivapithecus* from Kutch, Gujarat state, western India. *PLOS ONE* 13(11): e0206314.

Böhme, M. 2003. The Miocene climatic optimum: Evidence from ectothermic vertebrates of Central Europe. *Palaeogeography, Palaeoclimatology, Palaeoecology* 195(3–4): 389–401.

Bourret, R. 1941. *Les tortues de l'Indochine avec une note sur la pêche et l'elevage des tortues de mer par F. Le Poulain*. Édité part l'Institut Océanographique de l'Indochine.

Brochu, C. A. 1997. Morphology, fossils, divergence timing, and the phylogenetic relationships of *Gavialis*. *Systematic Biology* 46(3): 479–522.

Brochu, C. A. 2000. Phylogenetic relationships and divergence timing of *Crocodylus* based on morphology and the fossil record. *Copeia* 2000(3): 657–673.

Brochu, C. A. 2003. Phylogenetic approaches toward crocodilian history. *Annual Review of Earth and Planetary Sciences* 31(1): 357–397.

Brochu, C. A. 2017. Pliocene crocodiles from Kanapoi, Turkana Basin, Kenya. *Journal of Human Evolution* 117(April): 33–43. https://doi:10.1016/j.jhevol.2017.10.003.

Brochu, C. A., and G. W. Storrs. 2012. A giant crocodile from the Plio-Pleistocene of Kenya, the phylogenetic relationships of Neogene African crocodylines, and the antiquity of *Crocodylus* in Africa. *Journal of Vertebrate Paleontology* 32(3): 587–602.

Čerňanský, A., and E. V. Syromyatnikova. 2021. The first pre-Quaternary fossil record of the clade Mabuyidae with a comment on the enclosure of the Meckelian canal in skinks. *Papers in Palaeontology* 7(1): 195–215.

Chabrol, N., A. M. Jukar, R. Patnaik, and P. D. Mannion. 2024. Osteology of *Crocodylus palaeindicus* from the late Miocene-Pleistocene of South Asia and the phylogenetic relationships of crocodyloids. *Journal of Systematic Palaeontology* 22(1): 2313133.

Claude, J., W. Naksri, N. Boonchai, E. Buffetaut, J. Duangkrayom, C. Laojumpon, P. Jintasakul, et al. 2011. Neogene reptiles of northeastern Thailand and their paleogeographical significance. *Annales de Paléontologie* 97(3–4): 113–131.

Conrad, J. L., A. M. Balcarcel, and C. M. Mehling. 2012. Earliest example of a giant monitor lizard (*Varanus*, Varanidae, Squamata). *PLOS ONE* 7(8): p.e41767.

Conrad, J. L., K. Jenkins, T. Lehmann, F. K. Manthi, D. J. Peppe, S. Nightingale, A. Cossette, et al. 2013. New species of '*Crocodylus* pigotti (Crocodylidae) from Rusinga Island, Kenya, and generic reallocation of the species. *Journal of Vertebrate Paleontology* 33(3): 629–646.

Corvinus, G., and H. H. Schleich. 1994. An upper Siwalik reptile fauna from Nepal. *Courier Forschungsinstitut Senckenberg* 173:239–259.

Das, I. 1991. The taxonomic status of the Pleistocene turtle *Geoclemys sivalensis*. *Journal of Herpetology* 25(1): 104–107.

Dehm R., T. Oettingen-Spielberg, and H. Vidal. 1958. Paläontologische und geologische Untersuchungen im Tertiär Pakistans. I. Die Münchner Forschungsreise nach Pakistan. *Bayerische Akaddemie der Wissenschaften Mathematisch-naturwissenschaftliche Klasse, Abhandlungen, Neue Folge Heft, München* 90:1–13.

Delfino, M., and J. de Vos. 2010. A revision of the Dubois crocodylians, *Gavialis bengawanicus* and *Crocodylus ossifragus*, from the Pleistocene *Homo erectus* beds of Java. *Journal of Vertebrate Paleontology* 30(2): 427–441.

Delfino, M., and L. Rook. 2008. African crocodylians in the Late Neogene of Europe: A revision of *Crocodylus bambolii* Ristori, 1890. *Journal of Paleontology* 82(2): 336–343.

Delfino, M., T. M. Scheyer, U. Fritz, and M. R. Sánchez-Villagra. 2010. An integrative approach to examining a homology question: Shell structures in soft-shell turtles. *Biological Journal of the Linnean Society* 99(2): 462–276.

Deraniyagala, P. E. P. 1939. *The Tetrapod Reptiles of Ceylon: Volume 1 Testudinates and Crocodilians.* Colombo Museum Natural History Series. London: Dulau.

Dettman, D. L., M. J. Kohn, J. Quade, F. J. Ryerson, T. P. Ojha, and S. Hamidullah. 2001. Seasonal stable isotope evidence for a strong Asian monsoon throughout the past 10.7 Ma. *Geology* 29(1): 31–34.

Diong, C. H. and Lim, S. S. 1998. Taxonomic review and morphometric description of *Bronchocela cristatella* (Kuhl, 1820) (Squamata: Agamidae) with notes on other species in the genus. *Raffles Bulletin of Zoology* 46(2): 345–359.

Dubois E. 1908. Das Geologische alter der Kendengoder Trinilfauna. *Tijdschrift van het Koninklijk Nederlandsch Aardrijkskundig Genootschap* 2(25): 1235–1270.

Falconer, H. P. 1868. Descriptions by Dr. Falconer of the fossil remains of crocodiles from Ava, Perim Island, and the Nerbudda, in the catalogue of the Museum of the Asiatic Society of Bengal. In *Paleontological Memoirs and Notes of the Late Hugh Falconer, A. M., M. D., with a Biographical Sketch of the Author*, Vol. 1, *Fauna Antiqua Sivalensis*, edited by C. Murchison, 357–358. London: Robert Hardwickle Publishers.

Flynn, L. J., W. Downs, M. E. Morgan, J. C. Barry, and D. Pilbeam. 1998. High Miocene species richness in the Siwaliks of Pakistan. In *Advances in Vertebrate Paleontology and Geochronology*, edited by Y. Tomida, L. J. Flynn, and L. L. Jacobs, 167–180. Tokyo: National Science Museum, Monograph 14.

Flynn, L. J., E. H. Lindsay, D. Pilbeam, S. Mahmood Raza, M. E. Morgan, J. C. Barry, C. Badgley, et al. 2013. The Siwaliks and Neogene evolutionary biology in South Asia. In *Fossil Mammals of Asia: Neogene Biostratigraphy and Chronology*, edited by X. Wang, L. J. Flynn, and M. Fortelius, 353–372. New York: Columbia University Press.

Flynn, L. J., D. Pilbeam, L. L. Jacobs, J. C. Barry, A. K. Behrensmeyer, and J. W. Kappelman. 1990. The Siwaliks of Pakistan: Time and faunas in a Miocene terrestrial setting. *Journal of Geology* 98(4): 589–604.

Garbin, R. C., S. Bandyopadhyay, and W. G. Joyce. 2020. A taxonomic revision of geoemydid turtles from Siwalik-age of India and Pakistan. *European Journal of Taxonomy* 652:1–67.

Garg, R. L. 1988. New fossil reptile from the Siwalik Fossil Park, Saketi, Sirmur District, Himachal Pradesh. *Geological Survey of India Special Publication* 11:207–211.

Gasc, J.-P. 1974. L'interprétation fonctionnelle de l'appareil musculosquelettique de l'axe vertébral chez les serpents (Reptilia). *Mémoires du Muséum National d'Historie Naturelle, Série A* 83:1–182.

Gasc, J.-P. 1976. "Snake vertebrae—A mechanism or merely a taxonomist's toy?" In *Morphology and Biology of Reptiles*, edited by A'dA Bellairs, and C. B. Cox, 177–190. Linnean Society Symposium Series 3. London: Academic Press.

Georgalis, G. L., A. Villa, E. Vlachos, and M. Delfino. 2016. Fossil amphibians and reptiles from Plakias, Crete: A glimpse into the earliest late Miocene herpetofaunas of southeastern Europe. *Geobios* 49(6): 433–444.

Gmelin, J. F. 1788. *Systema Naturae, per regna tria Natura: secundum Classes, Ordines, Genera, Species, cum Characteribus, Differentiis, Synonymis, Locis*, tome 1, pars I: 1–500. Lyon.

Gramentz, D. 2011. *Lissemys Punctata: The Indian Flap-\Shelled Turtle.* Frankfurt: Edition Chimaira.

Havlik, P. E., W. G. Joyce, and M. Böhme. 2014. *Allaeochelys libyca*, a new carettochelyine turtle from the Middle Miocene (Langhian) of Libya. *Bulletin of the Peabody Museum of Natural History* 55(2): 201–214.

Head, J. J. 1997. Reptile paleontology of the Dhok Pathan Formation, Siwalik Group: Preliminary results. In *Siwaliks of South Asia. Proceedings of the Third GEOSAS Workshop*, edited by M. I . Ghaznavi, S. M. Raza, and M. T. Hasan, 49–56. Islamabad: Geological Survey of Pakistan.

Head, J. J. 1998. Relative abundance as an indicator of ecology and physiology of the giant filesnake *Acrochordus dehmi*, from the Miocene of Pakistan. *Journal of Vertebrate Paleontology* 18(suppl. to 3): 49A.

Head, J. J. 2001. Systematics and body size of the gigantic, enigmatic crocodyloid *Rhamphosuchus crassidens*, and the faunal history of Siwalik Group (Miocene) crocodylians. *Journal of Vertebrate Paleontology* 21(suppl. to 3): 59A.

Head, J. J. 2005. Snakes of the Siwalik Group (Miocene of Pakistan): Systematics and relationship to environmental change. *Palaeontologia Electronica* 8(1),16A: 1–32.

Head, J. J., J. I. Bloch, A. K. Hastings, J. R. Bourque, E. A. Cadena, F. A. Herrera, P. D. Polly, and C. A. Jaramillo. 2009. Giant boid snake from the Paleocene neotropics reveals hotter past equatorial temperatures. *Nature* 457(7230): 715–717.

Head, J. J., G. F. Gunnell, P. A. Holroyd, J. H. Hutchison, and R. L. Ciochon. 2013. Giant lizards occupied herbivorous mammalian ecospace during the Paleogene greenhouse in Southeast Asia. *Proceedings of the Royal Society of London, Series B* 280(1763): 20130665.

Head, J. J., K. Mahlow, and J. Müller. 2016. Fossil calibration dates for molecular phylogenetic analysis of snakes 2: Caenophidia, Colubroidea, Elapoidea, Colubridae. *Palaeontologia Electronica* 19(2), 2FC: 1–21.

Head, J. J., D. Mohabey, and J. A. Wilson. 2007. *Acrochordus* Hornstedt (Serpentes: Caenophidia) from the Miocene of Kachchh, western India: Neogene radiation of a derived snake. *Journal of Vertebrate Paleontology* 27(3): 720–723.

Head, J. J., and J. Müller. 2020. Squamate reptiles from Kanapoi: Faunal evidence for hominin paleoenvironments. *Journal of Human Evolution* 140(March): 102451.

Head, J. J., S. M. Raza, and P. D. Gingerich. 1999. *Drazindaretes tethyensis*, a new large trionychid (Reptilia: Testudines) from the marine Eocene Drazinda Formation of the Suliman Range, Punjab (Pakistan). *Contributions from the Museum of Paleontology, The University of Michigan* 30(7): 199–214.

Head, J. J., M. R. Sánchez-Villagra, and O. Aguilera. 2006. Fossil snakes from the Neogene of Venezuela (Falcón State). *Journal of Systematic Palaeontology* 4(3): 233–240.

Hirayama, R. 1985. Cladistic analysis of batagurine turtles. *Studia Palaeocheloniologica* 1:140–157.

Hocknull, S. A., P. J. Piper, G. D. van den Bergh, R. A. Due, M. J. Morwood, and I. Kurniawan. 2009. Dragon's paradise lost: Palaeobiogeography, evolution and extinction of the largest-ever terrestrial lizards (Varanidae). *PLOS ONE* 4: p.e7241.

Hoffstetter, R. 1964. Les serpents du Néogène du Pakistan (couches des Siwaliks). *Bulletin de la Société Géologique de France, Série 7*(6): 467–474.

Hoffstetter, R., and J.-P. Gasc. 1969. Vertebrae and ribs of modern reptiles. In *Biology of the Reptilia*, edited by C. Gans, 1: 201–310. London: Academic Press.

Holmes, R. B., A. M. Murray, P. Chatrath, Y. S. Attia, and E. L. Simons. 2010. Agamid lizard (Agamidae: Uromastycinae) from the lower Oligocene of Egypt. *Historical Biology* 22(1–3): 215–223.

Huey, R. B., C. A. Deutsch, J. J. Tewksbury, L. J. Vitt, P. E. Hertz, H. J. Álvarez Pérez, and T. Garland Jr. 2009. Why tropical forest lizards are vulnerable to climate warming. *Proceedings of the Royal Society of London, Series B* 276(1664): 1939–1948.

Hussain, S. T., R. M. West, J. Munthe, and J. R. Lukacs. 1979. The Daud Khel local fauna: A preliminary report on a Neogene vertebrate assemblage from the Trans-Indus Siwaliks, Pakistan. *Milwaukee Public Museum Contributions in Biology and Geology* 16(1): 1–17.

Hutchison, J. H. 1982. Turtle, crocodilian, and champsosaur diversity changes in the Cenozoic of the north-central region of western United States. *Palaeogeography, Palaeoclimatology, Palaeoecology* 37(2–4): 149–164.

Hutchison, J. H., P. A. Holroyd, and R. Ciochon. 2004. A preliminary report on Southeast Asia's oldest Cenozoic turtle fauna from the late Middle Eocene Ponduang Formation, Myanmar. *Asiatic Herpetological Research* 10:38–52.

Joyce, W. G. 2014. A review of the fossil record of turtles of the clade *Pan-Carettochelys*. *Bulletin of the Peabody Museum of Natural History* 55(1): 3–33.

Joyce, W. G., and C. J. Bell. 2004. A review of the comparative morphology of extant testudinoid turtles (Reptilia: Testudines). *Asiatic Herpetological Research* 10:53–109.

Joyce, W. G., N. Klein, and T. Mörs. 2004. Carettochelyine turtle from the Neogene of Europe. *Copeia* 2004(2): 406–411.

Joyce, W. G., and T. R. Lyson. 2010. *Pangshura tatrotia*, a new species of pond turtle (Testudinoidea) from the Pliocene Siwaliks of Pakistan. *Journal of Systematic Palaeontology* 8(3): 449–458.

Jukar, A., B. Sun, and R. L. Bernor. 2018. The first occurrence of *Plesiohipparion huangheense* (Qiu, Huang & Guo, 1987) (Equidae, Hipparionini) from the late Pliocene of India. *Bollettino della Società Paleontologica Italiana* 57:125–132.

Kapur, V. V., M. Pickford, G. Chauhan, and M. G. Thakkar. 2021. A Middle Miocene (~14 Ma) vertebrate assemblage from Palasava, Rapar Taluka, Kutch (Kachchh) District, Gujarat State, western India. *Historical Biology* 33(5): 595–615.

Karp, A. T., K. T. Uno, P. J. Polissar, and K. H. Freeman. 2021. Late Miocene C4 grassland fire feedbacks on the Indian Subcontinent. *Paleoceanography and Paleoclimatology* 36(4): e2020PA004106.

Lapparent de Broin, F., S. Bailon, M. L. Augé, and J.-C. Rage. 2020. Amphibians and reptiles from the Neogene of Afghanistan. *Geodiversitas* 42(22): 409–426.

Lindsay, E. H., L. J. Flynn, I. U. Cheema, J. C. Barry, K. F. Downing, A. R. Rajpar, and S. M. Raza. 2005. Will Downs and the Zinda Pir dome. *Palaeontologia Electronica* 8(1),19A: 1–19.

Linnaeus, C. von. 1758. *Systema naturae per regna tria naturae, secundum classes, ordines, genera, species, cum characteribus, differentiis, synonymis, locis*, 10th ed. Stockholm: Laurentii Salvii.

Lull, R. S. 1944. Fossil gavials from north India. *American Journal of Science* 242(8): 417–430.

Lydekker, R. 1880. A sketch of the history of the fossil vertebrata of India. *Journal of the Asiatic Society of Bengal* 49(1): 4–36.

Lydekker, R. 1882. Note on some Siwalik and Narbada fossils. *Records of the Geological Survey of India* 15:106–107.

Lydekker, R. 1885. Indian Tertiary and Post-Tertiary Vertebrata. Volume iii. Siwalik and Narbada Chelonia. *Memoirs of the Geological Survey of India, Paleontographica Indica* 10(6): 155–208, plates 18–28.

Lydekker, R. 1886a. Indian Tertiary and Post-Tertiary Vertebrata. Volume iii. Siwalik Crocodilia, Lacertilia, & Ophidia. *Memoirs of the Geological Survey of India, Paleontographica Indica* 10(7): 209–240, plates 28–35.

Lydekker, R. 1886b. On a new emydine chelonian from the Pliocene of India. *Quarterly Journal of the Geological Society of London* 42:540–541, plate 15.

Lydekker, R. 1888. *Catalogue of the Fossil Reptilia and Amphibia in the British Museum of Natural History. Part I, Containing the Order Ornithosauria,* Crocodilia, Dinosauria, Squamata, Rhynchocephalia, and Proterosauria. London: Trustees of the British Museum.

Lydekker, R. 1889a. Notes on Siwalik and Narbada Chelonia. *Records of the Geological Survey of India* 22(1): 56–58.

Lydekker, R. 1889b. On the land tortoises of the Siwaliks. *Records of the Geological Survey of India* 22:209–212.

Martin, J. E. 2019. The taxonomic content of the genus *Gavialis* from the Siwalik Hills of India and Pakistan. *Papers in Palaeontology* 5(3): 483–497.

Martin, J. E., E. Buffetaut, W. Naksri, K. Lauprasert, and J. Claude. 2012. *Gavialis* from the Pleistocene of Thailand and its relevance for drainage connections from India to Java. *PLOS ONE* 7(9): p.e44541.

McDowell, S. B. 1970. On the status and relationships of the Solomon Island elapid snakes. *Journal of Zoology* 161(2): 145–190.

McDowell, S. B. 1979. A catalogue of the snakes of New Guinea and the Solomons, with special reference to those in the Bernice P. Bishop Museum. Part III: Boinae and Acrochordoidea (Reptilia, Serpentes). *Journal of Herpetology* 13(1): 1–92.

Minton, S. A., Jr. 1966. A contribution to the herpetology of West Pakistan. *Bulletin of the American Museum of Natural History* 134(2): 27–34.

Meylan, P. A. 1987. The phylogenetic relationships of soft-shelled turtles (Family Trionychidae). *Bulletin of the American Museum of Natural History* 186(1): 1–101.

Meylan, P. A. 1988. *Peltochelys* Dollo and the relationships among the genera of the Carettocheylidae (Testudines: Reptilia). *Herpetologica* 44(4): 440–450.

Meylan, P. A., B. S. Weig, and R. C. Wood. 1990. Fossil soft-shelled turtles (Family Trionychidae) of the Lake Turkana Basin, Africa. *Copeia* 1990(2): 508–528.

Mook, C. C. 1932. A new species of fossil Gavial from the Siwalik Beds. *American Museum Novitates* 514:1–5.

Mook, C. C. 1933. A skull with jaws of *Crocodilus sivalensis* Lydekker. *American Museum Novitates* 670:1–10.

Morgan, M. E., A. K. Behrensmeyer, C. Badgley, J. C. Barry, S. Nelson, and D. Pilbeam. 2009. Lateral trends in carbon isotope ratios reveal a Miocene vegetation gradient in the Siwaliks of Pakistan. *Geology* 37(2): 103–106.

Morgan, M. E., K. L. Lewton, J. Kelley, E. Otárola-Castillo, J. C. Barry, L. J. Flynn, and D. Pilbeam. 2015. A partial hominoid innominate from the Miocene of Pakistan: Description and preliminary analyses. *Proceedings of the National Academy of Sciences* 112(1): 82–87.

Müller, J., and C. Mödden. 2001. A fossil leaf-toed gecko from the Oppenheim/Nierstein Quarry (Lower Miocene, Germany). *Journal of Herpetology* 35(3): 529–532.

Nanda, A. C., H. H. Schleich, and B. S. Kotlia. 2016. New fossil reptile records from the Siwalik of North India. *Open Journal of Geology* 6(8): 673–691.

Nanda, A. C., R. K. Sehgal, and P. R. Chauhan. 2018. Siwalik-age faunas from the Himalayan foreland basin of South Asia. *Journal of Asian Earth Sciences* 162(August): 54–68.

Nelson, S. V. 2007. Isotopic reconstructions of habitat change surrounding the extinction of *Sivapithecus*, a Miocene hominoid, in the Siwalik Group of Pakistan. *Palaeogeography, Palaeoclimatology, Palaeoecology* 243(1–2): 204–222.

Norell, M. A., and G. W. Storrs. 1989. Catalogue and review of the type fossil crocodilians in the Yale Peabody Museum. *Postilla* 203:1–28.

Oppel, M. 1811. *Die ordnungen, familien und gattungen der reptilien als prodrom einer naturgeschichte derselben*. München: Lindauer.

Patnaik, R., K. Milankumar Sharm, L. Mohan, B. A. William, R. F. Kay, and P. R. Chatrath. 2014. Additional vertebrate remains from the Early

Miocene of Kutch, Gujarat. *Special Publication of the Paleontological Society of India* 5:335–351.

Patnaik, R., and H. H. Schleich. 1993. Fossil crocodile remains from the Upper Siwaliks of India. *Mitteilungen der Bayersiche Staatssammlung für Paläontologie und Historie Geologie* 33:91–117.

Patnaik, R., and H. H. Schleich. 1998. Fossil micro-reptiles from Pliocene Siwalik sediments of India. *Veroffentlichungen aus dem Fuhlrott Museum* 4:295–300.

Pianka, E. R. and King, D. R., eds. 2004. *Varanoid Lizards of the World.* Indianapolis: Indiana University Press.

Pilbeam, D. R., A. K. Behrensmeyer, J. C. Barry, and S. M. Ibrahim Shah, eds. 1979. Miocene sediments and faunas of Pakistan. *Postilla* 179:1–45.

Pilgrim, G. E. 1908. The tertiary and post-tertiary freshwater deposits of Baluchistan and Sind with notices of new vertebrates. *Records of the Geological Survey of India* 37:139–166.

Pilgrim, G. E. 1912. The vertebrate fauna of the Gaj Series in the Bugti Hills and the Punjab. *Memoirs of the Geological Survey of India (Palaeontologia Indica) New Series* 4(2): 1–83.

Piras, P., M. Delfino, L. del Favero, and T. Kotsakis. 2007. Phylogenetic position of the crocodylian *Megadontosuchus arduini* and tomistomine palaeobiogeography. *Acta Palaeontologica Polonica* 52(2): 315–328.

Piras, P., and T. Kotsakis. 2005. A new gavialid from Early Miocene of south-eastern Pakistan (preliminary report). *Rendiconti Società Paleontologia Italiana* 2:201–207.

Polissar, P. J., K. T. Uno, S. R. Phelps, A. T. Karp, K. H. Freeman, and J. L. Pensky. 2021. Hydrologic changes drove the late Miocene expansion of C4 grasslands on the Northern Indian subcontinent. *Paleoceanography and Paleoclimatology* 36(4): e2020PA004108.

Prasad, K. N. 1974. The vertebrate fauna from Piram Island, Gujarat, India. *Memoirs of the Geological Survey of India, New Series* 41:1–23.

Pyron, R. A., F. T. Burbrink, and J. J. Wiens. 2013. A phylogeny and revised classification of Squamata, including 4161 species of lizards and snakes. *BMC Evolutionary Biology* 13:93.

Rage, J.-C. 1976. Les Squamates du Miocène de Bèni Mellal, Maroc. *Géologie Méditerranéenne* 3(2): 57–70.

Rage, J.-C., S. G. Gupta, and G. V. R. Prasad. 2001. Amphibians and squamates from the Neogene Siwalik beds of Jammu and Kashmir, India. *Paläontologische Zeitschrift* 75(2): 197–205.

Rhodin, A. G. J., J. B. Iverson, R. Bour, U. Fritz, A. Georges, H. B. Shaffer, and P. P. Van Dijk. 2017. Turtles of the world: Annotated checklist and atlas of taxonomy, synonymy, distribution, and conservation status. *Chelonian Research Monographs* 7:1–292.

Sanders, K. L., A. Hamidy, J. J. Head, and D. J. Gower. 2010. Phylogeny and divergence times of filesnakes (*Acrochordus*): Inferences from morphology, fossils and three molecular loci. *Molecular Phylogenetics and Evolution* 56(3): 857–867.

Sao, C. C., S. Abdessadok, A. D. Malassé, M. Singh, B. Karir, V. Bhardwaj, S. Pal, et al. 2016. Magnetic polarity of Masol 1 Locality deposits, Siwalik Frontal Range, northwestern India. *Comptes Rendus Palevol* 15(3–4): 407–416.

Shine, R. 1986. Ecology of low-energy specialists: Food habits and reproductive biology of the Arafura filesnake, *Acrochordus arafurae. Copeia* 1986(2): 424–437.

Singh, N. P., R. Patnaik, A. Čerňanský, K. M. Sharma, N. A. Singh, D. Choudhary, and R. K. Sehgal. 2022. A new window to the fossil herpetofauna of India: Amphibians and snakes from the Miocene localities of Kutch (Gujarat). *Palaeobiodiversity and Palaeoenvironments* 102:419–435.

Slowinski, J. B. 1994. A phylogenetic analysis of *Bungarus* (Elapidae) based on morphological characters. *Journal of Herpetology* 28(4): 440–446.

Smith, K. T., S. F. K. Schaal, S. Wei, and L. Chun-Tian. 2011. Acrodont iguanians (Squamata) from the Middle Eocene of the Huadian Basin of Jilin Province, China, with a critique of the taxon *"Tinosaurus." Vertebrata PalAsiatica* 49(1): 69–84.

Srivastava, R., and R. Patnaik. 2002. Large soft-shelled turtles from the Upper Pliocene rocks of Saketi (District, Sirmaur), Himachal Pradesh, India. *Journal of the Palaeontological Society of India* 47:65–76.

Srivastava, G., K. N. Paudayal, T. Utescher, and R. C. Mehrotra. 2018. Miocene vegetation shift and climate change: Evidence from the Siwalik of Nepal. *Global and Planetary Change* 161:108–120.

Suneja, I. J., G. Singh, and S. R. K. Chopra. 1977. A new species of *Crocodylus* Laurenti from Jammu Hills, India. *Journal of the Indian Academy of Geoscience* 20:1–6.

Szyndlar, Z. 1991. A review of Neogene and Quaternary snakes of Central and Eastern Europe. Part II: Natricinae, Elapidae, Viperidae. *Estudios Geológicos* 47(3–4): 237–266.

Szyndlar, Z. 1994. Oligocene snakes of southern Germany. *Journal of Vertebrate Paleontology* 14(1): 24–37.

Szyndlar, Z., and J.-C. Rage. 1999. Oldest fossil vipers (Serpentes: Viperidae) from the Old World. *Darmstädter Beiträge zur Naturgeschichte* 8:9–20.

Szyndlar, Z., and J.-C. Rage. 2003. *Non-erycine Booidea from the Oligocene and Miocene of Europe.* Krakow: Institute of Systematics and Evolution of Animals, Polish Academy of Sciences.

Szyndlar, Z., and G. A. Zerova. 1990. Neogene cobras of the genus *Naja* (Serpentes: Elapidae) of East Europe. *Annalen des Naturhistorischen Museums in Wien. Serie A für Mineralogie und Petrographie, Geologie und Paläontologie, Anthropologie und Prähistorie* 91:53–61.

Tewari, B. S., and G. L. Badam. 1969. A new species of fossil turtle from the Upper Siwaliks of Pinjore, India. *Palaeontology* 12(4): 555–558.

Theobald, W. 1877. Description of a new Emydine from the Upper Tertiaries of the Northern Punjab. *Records of the Geological Society of India* 10:43–45.

Theobald, W. 1879. On a marginal bone of an undescribed tortoise from the Upper Siwaliks near Nila in the Potwár, Punjab. *Records of the Geological Society of India* 12:186–187.

Vasishat, R. N., R. Gaur, and S. R. K. Chopra. 1978. Geology, fauna and palaeoenvironments of Lower Sivalik deposits around Ramnagar, India. *Nature* 275(5682): 736–737.

Verma, B. C., V. P. Mishra, and S. S. Gupta. 2002. Pictorial catalogue of Siwalik vertebrate fossils from Northwest Himalaya. Islamabad: *Geological Survey of India, Catalogue Series No. 5.*

Villa, A., J. Abella, D. M. Alba, S. Almécija, A. Bolet, G. D. Koufos, F. Knoll, et al. 2018. Revision of *Varanus marathonensis* (Squamata, Varanidae) based on historical and new material: Morphology, systematics, and paleobiogeography of the European monitor lizards. *PLOS ONE* 13(12): p.e0207719.

Villa, A., and M. Delfino. 2022. First fossil of *Varanus* Merrem, 1820 (Squamata: Varanidae) from the Miocene Siwaliks of Pakistan. *Geodiversitas* 44(7): 229–235.

West, R. M., J. H. Hutchison, and J. Munthe. 1991. Miocene vertebrates from the Siwalik Group, western Nepal. *Journal of Vertebrate Paleontology* 11(1): 108–129.

Willis, B. 1993a. Ancient river systems in the Himalayan foredeep, Chinji Village area, northern Pakistan. *Sedimentary Geology* 88(1–2):1–76.

Willis, B. 1993b. Evolution of Miocene fluvial systems in the Himalayan foredeep through a two-kilometer-thick succession in northern Pakistan. *Sedimentary Geology* 88(1–2): 77–121.

Willis, B. J., and A. K. Behrensmeyer. 1995. Fluvial systems in the Siwalik Miocene and Wyoming Paleogene. *Palaeogeography, Palaeoclimatology, Palaeoecology* 115(1–4): 13–35.

Zaleha, M. J. 1997. Fluvial and lacustrine palaeoenvironments of the Miocene Siwalik Group, Khaur area, northern Pakistan. *Sedimentology* 44(2): 349–368.

SIWALIK BIRDS

ANTOINE LOUCHART

INTRODUCTION

Fossil birds from the Cenozoic of South and Southeast Asia are rarer than those from the rest of Eurasia and other continents. The representation of fossil avifaunas is also low in this region compared with, for example, the mammalian record. This paucity of fossils makes it difficult to decipher aspects of bird evolution in southern Asia. The Neogene Siwalik record constitutes the major window for the exploration of this evolutionary history in southern Asia. Hitherto, few avian fossil remains have been described. Since the second half of the nineteenth century, and prior to 1982, several authors described a handful of poorly dated fossils (see Stidham 2014), which led to the recognition of several different taxa from the Late Miocene or the Pliocene (references in Harrison and Walker 1982). These authors studied 39 avian fossil remains from more recent excavations recording precise stratigraphy and spanning about 10 million years in the Middle to Late Miocene and yielding 11 taxa in 10 families (Harrison and Walker 1982).

This chapter synthesizes the results of the investigations of 101 new fossil bird remains found in the same strata of the Potwar Plateau during fieldwork done in the last four decades, together with reinterpreted earlier studies. Bird eggshells are also included (see also Stern, Johnson, and Chamberlain 1994) for a total of 34 eggshell fragments, essentially of ratites. The present work, based on a total of 141 well-dated fossil bones, plus eggshells, allows recognition of at least 29 species in 18 families, based on 80 bones identifiable at least to the family level (plus a dozen ratite eggshell fragments types A and S). The taxa are primarily non-passerines (except for one species with two fossils). These identifications make it possible to improve our understanding of the evolution of certain avian taxa and to reveal new records and patterns for otherwise poorly represented fossil groups at a global level. The Late Miocene Siwalik birds open a wider window on what was hitherto largely terra incognita.

Miocene birds belong to the crown group generally referred to as Neornithes. Presumably in the late Cretaceous, the Neornithes evolved into (1) the Palaeognathae (including ratites and tinamous) and (2) the Neognathae (i.e., all other modern birds). The Neognathae are split into (a) the Galloanserae (Anseriformes, Galliformes and allies) and (b) the Neoaves. The Neoaves differentiated during the Early Cenozoic into several major clades that were only recently partially resolved using phylogenomics (Jarvis et al. 2014; Prum et al. 2015). Most avian orders are now more confidently grouped with one another. The Siwalik avian record of the Potwar Plateau ranges from 16.8 to at least 0.5 Ma, with an additional fossil from the Late Oligocene. By these times, avian families were already well differentiated, with a complex history of evolution and distribution across the world (Blondel and Mourer-Chauviré 1998). The Potwar Plateau fossil birds yield new biogeographic, biostratigraphic, and evolutionary history of the bird fauna of this region and beyond. For some of the taxa, the first and last dates of occurrence can be interpreted as constraining either speciation or range extension or as extinction or range contraction. Paleoecological and paleoenvironmental indices also emerge from these results, as well as information on the evolution of communities and landscapes in the Siwaliks and the history of certain avian groups.

SUMMARY OF TAXA AND TEMPORAL RANGES

What follows are the species accounts and analyses from the present original study and some reinterpretations of Harrison and Walker (1982), as well as of other records of poorly dated fossils described in the nineteenth and early twentieth centuries. All identifications are based on comparisons using extant skeletal specimens and some fossil specimens, as well as those in the literature. Data on extant taxa (systematics, taxonomy, distribution, and ecology) are derived mainly from *Handbook of the Birds of the World Alive* (del Hoyo et al. 2017).

Fossil eggshells from several localities represent the Ratites (Palaeognathae), belonging either to aepyornithoid or struthioid types. Aepyornithoid-type eggshells range from 9.3 to 3.3 Ma (Localities D24, D33, D56, D70, Y10, Y11, Y221, Y317, Y318,

Y325). Struthioid-type eggshells range from 9.4 to 2.1 Ma (Localities D67, KL1, Y221, Y331; see also Stern, Johnson, and Chamberlain 1994). Among the struthioid eggshells, one exhibits pores grouped in disks with diameters and densities as well as eggshell thickness corresponding to extant ostrich *Struthio camelus* and differing from other *Struthio* species and oospecies (such as *S. asiaticus* and *S. molybdophanes*; see Pickford, Senut, and Dauphin 1995; Mourer-Chauviré et al. 1996a, 1996b; Harrison and Msuya 2005; Bibi et al. 2006; Mourer-Chauviré and Geraads 2008).

Also present are neognath birds, among which are Phasianidae (Galliformes), represented by species of *Pavo* (peafowl) and at least one other phasianid species, a pheasant in the genus cf. *Lophura*. A first species of *Pavo*, sp. A, is present at 12.9 Ma (Locality Y60; Fig. 9.1a). A second species of cf. *Pavo*, smaller (sp. B) is present at 13.6 Ma (Y59). A possible interpretation is that they represent large males and small females of a single species, but the size dimorphism would then greatly exceed what we observe in the two extant *Pavo* species. The species cf. *Lophura* sp. occurs from 11.6 to 9.4 Ma (Localities Y64, Y227, Y310, Y311). Among the fossils of cf. *Lophura* are those considered to represent a new species ("*L. wayrei*") by Harrison and Walker (1982), but considered here insufficiently diagnostic.

The family Anatidae (Anseriformes) is represented by three taxa. A first species (A) belongs to *Anser* or *Branta* (geese, brant geese) and is present at 9.4 Ma (Y269). Another anatid is tentatively assigned to the tribe Tadornini (shelducks) at 7.0 Ma (Y581). An indeterminate duck is present at 13.0 Ma (Y691). A putative merganser "*Mergus* (?) sp.*" was tentatively identified by Lydekker (1884) based on a single vertebra from a horizon situated between 10.0 and 3.3 Ma (Stidham et al. 2014). Such a tentative identification as a piscivorous, diving duck is in need of reassessment based on a comprehensive comparison with all genera of Anatinae. Until then, it is better considered as Anatinae indet.

Flamingos (Phoenicopteriformes, Phoenicopteridae) are represented by *Phoenicopterus* cf. *roseus* from 10.1 to 9.4 Ma (Y269, Y311), and *Phoeniconaias* cf. *minor* at 12.0 Ma (Y500).

An indeterminate rallid (Gruiformes, Rallidae), intermediate in size between a Eurasian coot (*Fulica*) and an extant *Porphyrio* species, is present at 9.4 Ma (Y310). The distal tibiotarsus concerned was considered to represent a new species ("*Porphyrio parvus*") by Harrison and Walker (1982), but this fragment is insufficiently diagnostic.

The extinct family of didactyl cursorial gruiforms Ergilornithidae (Gruiformes) is represented by several fossils, principally distal tibiotarsi, of *Amphipelargus cracrafti* (Harrison and Walker 1982) that range from 16.0 Ma (tentative, Y592) and 13.4 to 10.1 Ma (Y40, Y311, Y499; Fig. 9.1b, c, d, e). Initially, Harrison and Walker (1982) ascribed a distal tibiotarsus (and tentatively two partial femurs) to their new species named *Urmiornis cracrafti*, which they compared with *Urmiornis maraghanus* from the Late Miocene of Iran (de Mecquenem 1908, 1925), but not with other contemporaneous ergilornithids. Noteworthy is *Amphipelargus majori* Lydekker, 1891, from the Pliocene of Samos (Greece), which was transferred from the Ciconiidae

to the Ergilornithidae by Harrison (1981). *A. cracrafti* differs slightly from *Urmiornis* (de Mecquenem 1908, 1925) and corresponds to *Amphipelargus* according to the characters emphasized in Harrison (1981). Furthermore, *A. cracrafti* differs from *A. majori*, principally in being smaller than the latter species; it is among the smallest late Neogene ergilornithids (see Kurochkin 1985; Karhu 1997). Recently, all the other Middle to Late Miocene and Pliocene ergilornithids have been placed in *Urmiornis* (e.g., Karhu 1997; Zelenkov 2016; Zelenkov, Boev, and Lazaridis 2016). This practice should be treated with caution; whatever the interpretation, the Siwalik species differs and is best placed in *Amphipelargus* alongside *A. majori*.

A distal tibiotarsus belongs to a crane, *Grus* sp. (Gruiformes, Gruidae), at 16.8 Ma (Y802; Fig. 9.1e).

The family Otididae (Otidiformes), the bustards, is represented by three fossils. One belongs to a male of a large species in the genus *Otis* or *Ardeotis*, from the 7.0 Ma Locality Y581. Another is compatible in size with a female of the same species, from the 13.6 Ma Locality Y640. A third fossil from Locality Y52 at 13.2 Ma belongs to a much smaller species, compatible with extant *Chlamydotis* bustards, for instance (Fig. 9.1f).

Storks and marabou storks of the family Ciconiidae (Ciconiiformes) are represented by several species. A marabou stork cf. *Leptoptilos dubius* is present at 12.0 Ma (Y515), while a similar large species (either *L. dubius* or the extinct *L. falconeri*) is present at 10.5 Ma (Y259). *Leptoptilos falconeri* (Milne-Edwards 1867–1871; Davies 1880; Lydekker 1884) was described from poorly dated Siwalik horizons (Late Miocene or Pliocene; see Stidham et al. 2014) and is generally larger than *L. dubius*, with relatively short wings; it is also known from the Pliocene of northern and eastern Africa and tentatively Ukraine (Louchart et al. 2005). A smaller species, cf. *Leptoptilos javanicus*, is present at 13.1 Ma (Y513). One or more species of saddle-billed stork, *Ephippiorhynchus* sp., is represented from 8.2 to 7.0 Ma (Y24, Y384, Y910). Two of the preceding fossils were described as the basis for a new extinct species, *E. pakistanensis*, by Harrison and Walker (1982). However, these fossils are not sufficiently diagnostic and are only assignable to *Ephippiorhynchus* sp. (Louchart et al. 2005). Other fossils assignable to the large stork tribe Leptoptilini, gen. sp. indet. (some tentatively) come from Localities Y317, Y500, Y650, Y776, Y918, Y926, ranging from 13.1 to 7.3 Ma. Among these fossils is an earlier tentative assignment by Harrison and Walker (1982) to an extinct species (*Leptoptilos siwalicensis* Harrison, 1974, erroneously spelled *siwalikensis* in Harrison and Walker [1982]). But this fossil, as well as the original (Late Miocene/Pliocene) fossils on which *L. siwalicensis* is based, are not sufficiently diagnostic and are assignable only to Leptoptilini, gen. sp. indet. (Louchart et al. 2005). One further fossil was assigned to a new extinct genus and species, *Cryptociconia indica*, by Harrison (1974), but is now assignable to *Leptoptilos dubius/falconeri* (Louchart et al. 2005); it derives from unknown strata, Late Miocene or Pliocene in age. Smaller stork(s) are represented by fossils assignable to the genus *Ciconia* or *Mycteria* from the 12.3 Ma Locality Y496. In addition, one fossil of a Ciconiidae indet., the size of those from Locality Y496, comes from the 9.3 Ma Locality Y269.

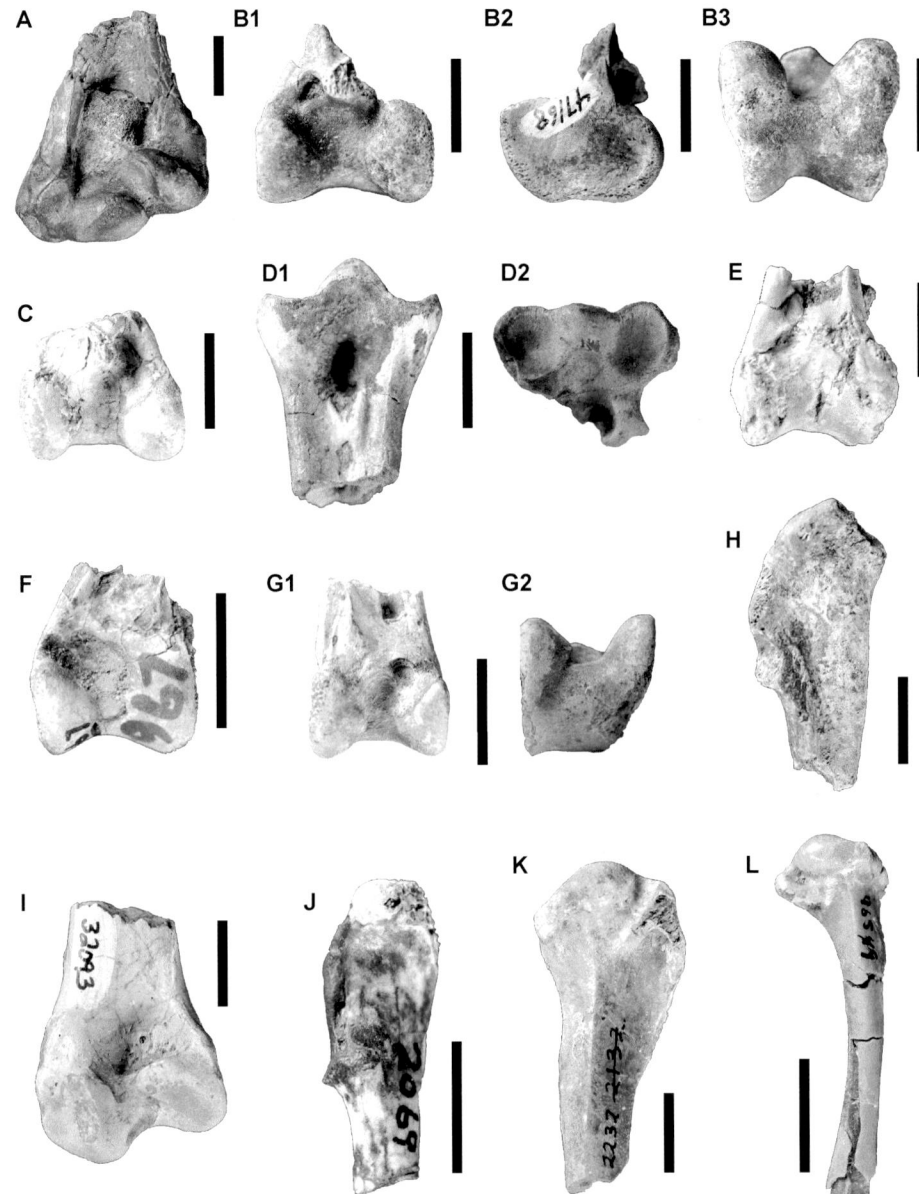

Figure 9.1. Selected avian fossils from the Potwar Plateau. A. *Pavo* sp. A, distal left humerus YGSP 32563, cranial view; B. *Amphipelargus cracrafti*, distal left tibiotarsus YGSP 47168, cranial (B1), medial (B2) and distal (B3) views; C. *A. cracrafti*, distal right tibiotarsus YGSP 25500, cranial view; D. cf. *A. cracrafti*, proximal right tarsometatarsus YGSP 17533, dorsal (D1) and proximal (D2) views; E. *Grus* sp., distal left tibiotarsus YGSP 32723, cranial view; F. Otididae sp. B size of *Chlamydotis macqueeni*, distal left tibiotarsus YGSP 967, cranial view; G. cf. *Platalea* size of *P. leucorodia*, distal right tibiotarsus YGSP 50464, cranial (G1) and distal (G2) views; H. *Ardea* cf. *A. goliath*, proximal left coracoid YGSP 46683, medial view; I. *Pelecanus* cf. *P. cautleyi*, distal left femur YGSP 32093, caudal view; J. *Phalacrocorax* cf. *P. fuscicollis*, proximal left coracoid Z 2069, medial view; K. *Anhinga* cf. *A. melanogaster*, proximal left humerus YGSP 2232, caudal view; L. *Ceryle rudis*, right humerus lacking distal end YGSP 46549, caudal view. Scale bars 10 mm.

In the family Threskiornithidae (Pelecaniformes), a probable spoonbill, genus *Platalea*, is present at 10.1 Ma (Y311). It is based on a distal tibiotarsus, the size of extant *P. leucorodia* (Fig. 9.1g).

A heron (Pelecaniformes, Ardeidae) is present in the 5.9 Ma Locality Y861. In the genus *Ardea*, it is tentatively assignable to the extant species *A. goliath*. This large heron is represented by a proximal coracoid (Fig. 9.1h).

A shoebill (Pelecaniformes, Balaenicipitidae), different from the unique extant African species *Balaeniceps rex*, was recognised by Harrison and Walker (1982) based on a distal tarsometatarsus from the 10.2 Ma Locality Y450. It was named *Paludiavis richae*. Concomitantly, a distal tibiotarsus from the Late Miocene of Tunisia was tentatively assigned to this genus and species (Harrison and Walker 1982). Both are likely to belong to Balaenicipitidae, even though they are fragmentary.

A pelican (Pelecaniformes, Pelecanidae) is represented by five fossils, three of which are assignable to the extinct *P. caut-leyi* or *P. sivalensis* (Fig. 9.1i) and originally described (Davies 1880; Lydekker 1884, 1891) from poorly dated Late Miocene or Pliocene horizons in the Siwaliks (see Stidham et al. 2014 for age). Harrison and Walker (1982) referred to these three fossils as *P.* cf. *sivalensis*. They come from Localities Y159, Y162, and Y269, all three dating to around 9.4 Ma. Based on size and proportions, another fossil is more precisely assignable to *P. caut-leyi*, to which a fifth one can be tentatively assigned; these come from the 12.0 Ma Locality Y500 and the 11.1 Ma Locality Y798, respectively. Furthermore, it is possible that *P. cautleyi* and *P. siv-alensis* are synonymous, considering likely sexual dimorphism (Olson 1985) and recognizing that only a handful of fragmen-tary specimens is known for these species. For the time being, either one or two species can be recognized, the size of these pelicans falling within the size range of smaller extant species (e.g., Afrotropical *P. rufescens* or Australian *P. conspicillatus*).

Several fossils represent one or more cormorants (Suli-formes, Phalacrocoracidae). The oldest locality that yielded a bird fossil, the 23–25 Ma Locality Z129, produced a proximal coracoid of a relatively small cormorant, *Phalacrocorax* sp., the size of extant South Asian *P. fuscicollis* (Fig. 9.1j). Part of the omal extremity is abraded, so that the fossil is hardly sufficient for a confident generic assignment (*Phalacrocorax* versus a more basal, extinct genus), but as preserved nothing contradicts this assignment. In younger strata, fossils assigned to *Phalacrocorax* cf. *fuscicollis* range from 11.6 to 7.8 Ma (Localities Y83 and Y906). An extinct new genus and species of cormorant, *Valen-ticarbo praetermissus*, was erected by Harrison (1979) based on a proximal tarsometatarsus (cast) from an unknown horizon (Late Miocene/Pliocene) of the Siwaliks. However, this fossil (initially identified as *Phalacrocorax* sp. by Lydekker [1884]) of-fers no convincing character that would differentiate it from *Phalacrocorax* and was insufficiently compared with extant spe-cies. A deeper groove, concavity, and muscle attachment are likely due to older individual age at death than the average ex-tant museum specimens (see Louchart, Tourment, and Carrier 2011). Therefore, this poorly dated specimen probably belongs to *Phalacrocorax* and is roughly the size of *P. fuscicollis*.

Six fossils belong to a darter, *Anhinga* cf. *melanogaster* (Suli-formes, Anhingidae), ranging from 14.0 to 7.4 Ma (Localities Y86, Y227, Y457, Y498, Y733, Y735; Fig. 9.1k). They include the fossils that were assigned by Harrison and Walker (1982) to the extinct large species *A. pannonica*, but that fall within the met-ric variability of *A. melanogaster* and are smaller or less robust than *A. pannonica* (see Louchart et al. 2008); Harrison and Walker (1982) had not compared the bones with *A. melanogas-ter*, but they did compare them with the smaller *A. anhinga*. Incidentally, fossils assigned to *Anhinga* sp., interpreted as dif-ferent from *A. melanogaster*, from the latest Pliocene of the In-dian Siwaliks (2.6 Ma; Stidham et al. 2017) offer no convincing evidence of being specifically distinct from *A. melanogaster* (and see above under *Phalacrocorax*).

Two large birds of prey in the family Accipitridae (Accipi-triformes) are represented, by one fossil each. A vulture, *Gyps* cf. *fulvus*, is present from the 11.4 Ma Locality Y76 (Harrison and Walker 1982). An eagle, cf. *Aquila* sp., the size of the larg-est species (Royal *A. chrysaetos* or Imperial *A. heliaca*), is pre-sent from the 10.1 Ma Locality Y311.

Among the few smaller fossils, a proximal humerus from the 10.1 Ma Locality Y311 belongs to a kingfisher (Coraci-iformes, Alcedinidae; Fig. 9.1l), similar to the extant Afrotropi-cal and Oriental species *Ceryle rudis* (a male).

Finally, two distal tibiotarsi, unfortunately poorly diag-nostic, belong to a single species among the Passeriformes (songbirds), probably to a corvid (Corvidae) the size of a crow (such as *Corvus corone*), although with a notably narrow dis-tal epiphysis, it is not possible to identify them further. They derive from the 12.3 Ma Locality Y502 and the 12.0 Ma Lo-cality Y499.

PALEOECOLOGICAL AND PALEOENVIRONMENTAL INFERENCES

Considering estimated body size, larger birds are much better represented than smaller ones (Table 9.1). A tiny distal tibio-tarsus of Aves indet., possibly a small passerine, belongs to a bird around 15 or 20 g. Next in size is the kingfisher, ~85 g. The size category from 500 to 700 g comprises the medium-sized passerines—probable corvid, the rallid, and the cormorant. From 1,000 to 1,200 g are the duck, the shelduck, and the *Lophura* pheasant. From 1,400 to 1,900 g are the darter, the smaller bustard, the smaller flamingo, and the probable spoon-bill. From 2,500 to 3,000 g are the goose/brant goose, the smaller stork, the ergilornithid *A. cracrafti*, the smaller peafowl, and the larger flamingo. From 4,400 to 6,500 g are the heron, the eagle, most of the leptoptiline storks, the pelicans, the larger peafowl, the crane, and the shoebill. From 7,000 to 9,000 g are the largest leptoptiline storks (*L. dubius/falconeri*), the vulture, and the larger bustard. Finally, around 120,000 g are *Struthio* sp. and *Struthio camelus*. There is no particular change in body size across localities and their ages.

Carnivorous and piscivorous habits dominate a diverse range of diets. Herbivorous birds are represented by ostriches. More generally, herbivorous (or mainly herbivorous) birds in-clude the shelduck, the goose/brant goose, and the bustards, as well as the smaller flamingo (*Phoeniconaias* cf. *minor*), which primarily filters phytoplankton. Omnivorous birds include the phasianids (*Lophura*, *Pavo*), the rallid, and the crane, as well as the larger flamingo (*Phoenicopterus* cf. *roseus*), which filters small invertebrates, and possibly the cf. Corvidae. Carnivorous taxa include the eagle (fully carnivorous) and the storks and heron (carnivorous, invertivorous and aquatic carnivorous). Aquatic carnivores also include the threskiornithid, a probable spoon-bill. Scavenger or scavenger/carnivorous birds include the vulture and the *Leptoptilos* storks. Finally, piscivorous taxa in-clude the pelican, shoebill, cormorant, darter, and kingfisher. Unfortunately, the diet of *Amphipelargus* (and ergilornithids in general) is not known. From 16–12 Ma to 12–7 Ma, there is a

Table 9.1. Bird Taxa from the Potwar Plateau Fossil Record

Group Name	Genus	Species	Locality FO	FO (Ma)	LO (Ma)	Locality LO	Diet	Body size (g)	Probable Habitat	Common Name
Struthionidae	*Struthio*	cf. *camelus*	KL01	9.2	—	—	Herbivorous	120,000	Open country	Ostrich
Struthionidae	*Struthio*	sp.	Y221	9.4	2.1	D67	Herbivorous	120,000	Open country	Ostrich
Pavoninae	*Pavo*	sp. A (larger)	Y60	12.9	12.9	Y60	Omnivorous	4,400–6,500	Forest/stream	Peafowl
Pavoninae	cf. *Pavo*	sp. B	Y59	13.6	13.6	Y59	Omnivorous	2,500–3,000	Forest/stream	Peafowl
Phasianinae	cf. *Lophura*	sp.	Y64	11.6	9.4	Y310	Omnivorous	1,000–1,200	Forest	Pheasant
Anatidae	A (*Anser/Branta*)	sp. indet.	Y269	9.4	9.44	Y269	Herbivorous	2,500–3,000	Wetlands	Goose
Anatidae	B (cf. *Tadornini*)	sp. indet.	Y581	7.0	7.0	Y581	Herbivorous	1,000–1,200	Wetlands	Shelduck
Anatidae	C	sp. indet.	Y691	13.0	13.0	Y691	Herbivorous	1,000–1,200	Water bodies	Duck
Phoenicopteridae	*Phoenicopterus*	cf. *roseus*	Y311	10.1	9.4	Y269	Omnivorous	2,500–3,000	Water bodies; saline or alkaline	Flamingo
Phoenicopteridae	*Phoeniconaias*	cf. *minor*	Y500	12.0	12.0	Y500	Herbivorous	1,400–1,900	Water bodies; saline or alkaline	Flamingo
Gruiformes, Rallidae	indet.		Y310	9.4	9.4	Y310	Omnivorous	500–700	Wetlands	Rallid
Gruiformes, Ergilornithidae	cf. *Amphipelargus*	(*cracrafti*)	Y592	16.0	16.0	Y592	?	2,500–3,000	Open country	—
Gruiformes, Ergilornithidae	*Amphipelargus*	*cracrafti*	Y40	13.4	10.1	Y311	?	2,500–3,000	Open country	—
Gruiformes, Gruidae	*Grus*	sp.	Y802	16.8	16.8	Y802	Omnivorous	4,400–6,500	Open country/humid	Crane
Otidiformes, Otididae	Size *Otis/Ardeotis*	sp. indet.	Y640	13.6	13.6	Y640	Herbivorous	7,000–9,000	Open to semi-open	Bustard
Otidiformes, Otididae	A (*Otis/Ardeotis*)	sp. indet.	Y581	7.0	7.0	Y581	Herbivorous	7000–9000	Open to semi-open	Bustard
Otidiformes, Otididae	B (size *Chlamydotis*)	sp. indet.	Y52	13.2	13.2	Y52	Herbivorous	1,400–1,900	Open to semi-open	Bustard
Ciconiidae	cf. *Leptoptilos*	A (cf. *dubius*)	Y515	12.0	12.0	Y515	Scavenger/carnivorous	7,000–9,000	Open country/humid	Stork
Ciconiidae	*Leptoptilos*	*dubius/falconeri*	Y259	10.5	10.5	Y259	Scavenger/carnivorous	7,000–9,000	Open country/humid	Stork
Ciconiidae	cf. *Leptoptilos*	B (cf. *javanicus*)	Y513	13.1	13.1	Y513	Scavenger/carnivorous	4,400–6,500	Open country/humid	Stork
Ciconiidae	*Ephippiorhynchus*	sp.	Y24	8.2	7.0	Y910	Carnivorous/invertivorous/aquatic carnivorous	2,500–3,000	Open country/humid	Stork
Ciconiidae	*Ciconia/Mycteria*		Y496	12.3	12.3	Y496	Carnivorous/invertivorous/aquatic carnivorous	2,500–3,000	Open country/humid	Stork
Pelecaniformes, Threskiornithidae	cf. *Platalea*		Y311	10.1	10.1	Y311	Aquatic carnivorous	1,400–1,900	Wetlands	Spoonbill
Pelecaniformes, Ardeidae	*Ardea*	cf. *goliath*	Y861	6.0	6.0	Y861	Carnivorous/invertivorous/aquatic carnivorous	4,400–6,500	Wetlands	Heron

(continued)

Table 9.1. continued

Group Name	Genus	Species	Locality FO	FO (Ma)	LO (Ma)	Locality LO	Diet	Body size (g)	Probable Habitat	Common Name
Pelecaniformes, Balaenicipitidae	*Paludiavis*	*richae*	Y450	10.2	10.2	Y450	Piscivorous	4,400–6,500	Wetlands	Shoebill
Pelecanidae	*Pelecanus*	cf. *cautleyi*	Y798	11.1	11.1	Y798	Piscivorous	3,900–7,000	Water bodies	Pelican
Pelecanidae	*Pelecanus*	*cautleyi*	Y500	12.0	12.0	Y500	Piscivorous	3,900–7,000	Water bodies	Pelican
Pelecanidae	*Pelecanus*	*sivalensis/ cautleyi*	Y162	9.4	9.4	Y159	Piscivorous	3,900–7,000	Water bodies	Pelican
Suliformes, Phalacrocoracidae	*Phalacrocorax*	cf. *fuscicollis*	Y83	11.6	7.8	Y906	Piscivorous	500–700	Water bodies	Cormorant
Suliformes, Phalacrocoracidae	*Phalacrocorax*	sp. (size *fuscicollis*)	Z129	25–23	25–23	Z129	Piscivorous	500–700	Water bodies	Cormorant
Suliformes, Anhingidae	*Anhinga*	cf. *melanogaster*	Y733	14.0	7.4	Y457	Piscivorous	1,400–1,900	Water bodies	Darter
Accipitriformes, Accipitridae	*Gyps*	cf. *fulvus*	Y76	11.4	11.4	Y76	Scavenger/ carnivorous	7,000–9,000	—	Vulture
Accipitriformes, Accipitridae	cf. *Aquila*		Y311	10.1	10.1	Y311	Carnivorous	4,400–6,500	—	Eagle
Coraciiformes, Alcedinidae	*Ceryle*	*rudis*	Y311	10.1	10.1	Y311	Piscivorous	85	Water bodies	Kingfisher
Passeriformes, cf. Corvidae			Y502	12.3	12.0	Y499	?Omnivorous	500–700	—	Cf. crow

Notes: Information on various levels of avian identification derives from knowledge of extant representatives, principally from del Hoyo et al. (2017).

fourfold decrease in the proportion of piscivorous taxa compared with taxa with other diets, linked to a decrease in aquatic taxa between these two time intervals. The reasons for such a decrease might be merely taphonomic; hence, it is not possible to make environmental inferences. There is otherwise no particular change over time in the distribution of taxa with different diets.

Some taxa (family, genus, species level) are habitat indicators, as reasonably interpreted from the range of habitats occupied by the same taxa today. Locomotion is closely tied to habitat. Long-legged birds are the most common and represent either open or semi-open–habitat landbirds or wading birds. Among open-country taxa, two are remarkable in terms of locomotion: the ostrich (*Struthio*) and the ergilornithid (*Amphipelargus*). These two flightless taxa (likely for the second) are the only known didactyl (two pedal digits) families of birds in the world, with a cursorial adaptation (de Mecquenem 1908, 1925; Olson 1985). The presence of forest is indicated by the phasianid birds, *Pavo* spp. and *Lophura* sp. (Table 9.2); at Localities Y59 and Y60, *Pavo* indicates more precisely proximity of a stream. The presence of open to semi-open dry habitats is indicated by the Struthionidae, Ergilornithidae and Otididae. The Gruidae and Ciconiidae signify mesic environments—among the Ciconiidae, the smaller storks indicate more humid settings than Leptoptilini. Water bodies (stream, lake, lagoon) are indicated by aquatic birds, such as ducks, Pelecanidae, Phalacrocoracidae and Anhingidae (see Table 9.2). The kingfisher (Y311) also signifies at least small patches of open water within its

habitat. Wetland settings (marshes, swamps, or humid grasslands) are indicated by the shoebill (rather closed habitats such as swamps), the probable spoonbill (shallow wetlands), the rallid, the heron, the shelduck and the probable anserine anatid. A more specific wetland setting, namely saline or alkaline lakes or lagoons, is indicated by the flamingos (Phoenicopteridae) from Localities Y269, Y311, and Y500. Finally, the two large raptors (cf. *Gyps* and cf. *Aquila*, respectively, from Localities Y76 and Y311), while not indicative of precise habitats, nevertheless require open areas for hunting (cf. *Gyps* however points to rather dry open habitat; see Table 9.2).

The different avian habitats are represented at localities that span ~16 Ma to 7 Ma, with the exception of forested and saline habitats. Birds of forest or saline habitats occur only at localities from 12.8 to 9.3 Ma. Generally paleoenvironments appear heterogeneous, with different habitat preferences represented by different birds of a single locality (even with few taxa). The best example of multiple contemporaneous habitats is provided by Locality Y311, which is the richest in bird (and all vertebrate) remains. Birds from Y311 are indicative of open, semi-open, forested, and humid (including saline or alkaline) habitats, but do not include aquatic taxa (duck, darter, cormorant, pelican) or any gruid or ciconiid bird incidentally, in contrast to a number of other localities. Apart from Y311, the few localities that yield more than one species often represent diverse habitats. For example, Y581 has a shelduck (aquatic) and a bustard (open dry). Y227 preserves a darter (aquatic) and a pheasant (forest, dry). Y269 and Y500 have water birds and

Table 9.2. Taxa Groupings According to Localities of Occurrence and Habitats

Taxa	Localities	Type of Habitat
Pavo spp., *Lophura* sp.	Y59, Y60, Y64, Y227, Y310, Y311	Forested
Struthionidae, Ergilornithidae, Otididae, cf. *Gyps*	Y40, Y52, Y76, Y221, Y311, Y331, Y499, Y581, Y592, Y640, KL1, D67	Open to semi-open dry
Gruidae, Ciconiidae	Y24, Y259, Y269, Y317, Y384, Y496, Y500, Y513, Y515, Y650, Y776, Y802, Y910, Y918, Y926	Open to semi-open mesic
Anatini, Pelecanidae, Phalacrocoracidae, Anhingidae	Y83, Y86, Y159, Y162, Y227, Y269, Y457, Y498, Y500, Y691, Y733, Y735, Y798, Y906, and the late Oligocene Z129	Stream, lake, or lagoon
Balaenicipitidae, cf. *Platalea*, Rallidae, Ardeidae, cf. Tadornini, *Anser/Branta*	Y310, Y311, Y450, Y861, Y581, Y269	Marsh, swamp, or humid grassland
Phoenicopteridae	Y269, Y311, and Y500	Saline or alkaline lake/lagoon

taxa of riparian habitats or diverse humid settings. Y310 includes a rallid (wetland) and a pheasant. The other localities have only a single species of birds (often a single avian fossil).

PATTERNS IN REGIONAL DIVERSITY, BIOGEOGRAPHY, AND TEMPORAL RANGES

Considering the taphonomic biases and the rarity of avian fossils throughout the Potwar Plateau record (even though this record is important and unusual for South Asia), changes in diversity through time are difficult to interpret. There is no discernible trend that attests to genuine patterns in the evolution of ornithocoenoses. However, taxon by taxon, several findings are important, especially in terms of the dates of first and last appearance (see Table 9.1, Figure 9.2) and their plausible interpretations (Table 9.3). Ostriches (*Struthio*) appear after the disappearance of the ergilornithid (*Amphipelargus*), a possible regional replacement of one didactyl cursorial flightless species by an unrelated but convergent species. *Amphipelargus cracrafti* derives from a stock of ergilornithid species that inhabited Central to South Asia from the Late Eocene to the Oligo-Miocene, which in turn derives from non-didactyl, Eurasian ancestors in the Paleogene (Zelenkov, Boev, and Lazaridis 2016). Phylogenetic relationships among the different Miocene species are obscure, but *A. cracrafti* from the Potwar Plateau represents the southernmost species of ergilornithid known and is remarkable for being widely separated by the Himalayas from Central Asian steppe species. Closer biogeographic continuity would be with species from the Early Miocene of Kazakhstan and Late Miocene species from Iran. The whole family reached its peak in diversity in the Late Miocene, having been present from Greece to western Mongolia (and Pakistan until 10.1 Ma) and then becoming extinct with the last Early Pliocene species in western Mongolia (Zelenkov, Boev, and Lazaridis 2016). The appearance and disappearance of *A. cracrafti* therefore likely reflect range extension (end of Burdigalian) and range contraction (beginning of Tortonian) of ergilornithids. The oldest dates of either 16.0 or 13.4 Ma for *A. cracrafti* correspond, respectively, to tentative and firm identification of the Siwalik ergilornithid, interpretable as an extension into Pakistan of the ergilornithid range, concomitant with differentiation of *A. cracrafti*, found only in the Siwaliks.

The last appearance of *A. cracrafti* reflects the contraction of the range of the whole family Ergilornithidae. It is notable that ergilornithids and ostriches were not contemporaneous in the Siwaliks, in contrast with Maragheh in Iran, where they coexisted from 9.5 to 7.0 Ma.

The appearance of ostrich in the Siwaliks around 9.4–9.2 Ma, although indirectly (because it is based on eggshell), constitutes the earliest record of *Struthio* (*S. camelus* at 9.2 Ma) in Asia east of Turkey. Indeed, the *Struthio* lineage evolved in Africa, and the earliest record outside of Africa dates to ~12 Ma in Turkey (Mourer-Chauviré et al. 1996a, 1996b). Since 11 Ma, *Struthio* expanded its range to southeastern Europe, with various species described generally up to 50% larger than *S. camelus* (Ukraine, Moldova and Greece; Boev and Spassov 2009). *Struthio* eggshell from the Potwar Plateau attests to colonization of South Asia from southwestern Asia around 9.4 Ma. Iran hosted a large *Struthio* sp. (Maragheh, ~9.5-7.0 Ma; de Mecquenem 1925; Bernor 1986), which is sometimes assigned to the extinct species *S. asiaticus*. This latter species was present in the Late Miocene or Pliocene of the Siwaliks, where it was the first ostrich named based on bones (Milne-Edwards 1867–1871; Davies 1880; Lydekker 1884). The age of these first-named remains is not known precisely; although they were previously considered to be Pliocene, they must now be considered Late Miocene or Pliocene, as is the case for other early-named Siwalik taxa (Stidham et al. 2014). *Struthio asiaticus* must be placed within the larger context of numerous extinct species, often contemporaneous and based on limited material, from different areas of Central, South, and West Asia and southeastern Europe, as well as several sites in Africa. Some of these species are placed in synonymy with *S. asiaticus* by various authors, in which case this species occurred over wide regions of Eurasia and even tentatively North Africa (Pliocene of Morocco; Mourer-Chauviré and Geraads 2008) and persisted until the Late Pleistocene and even into the Holocene of China and Mongolia (Boev and Spassov 2009). The disappearance of *S. asiaticus* from the Siwaliks is not dated precisely, but might be at the end of the Pliocene (Stidham et al. 2014). These large extinct ostriches constitute at least one lineage contemporaneous with the extant *S. camelus* lineage.

More remarkably, the Siwaliks eggshell specimen that clearly corresponds to the extant *S. camelus* ootype is potentially the

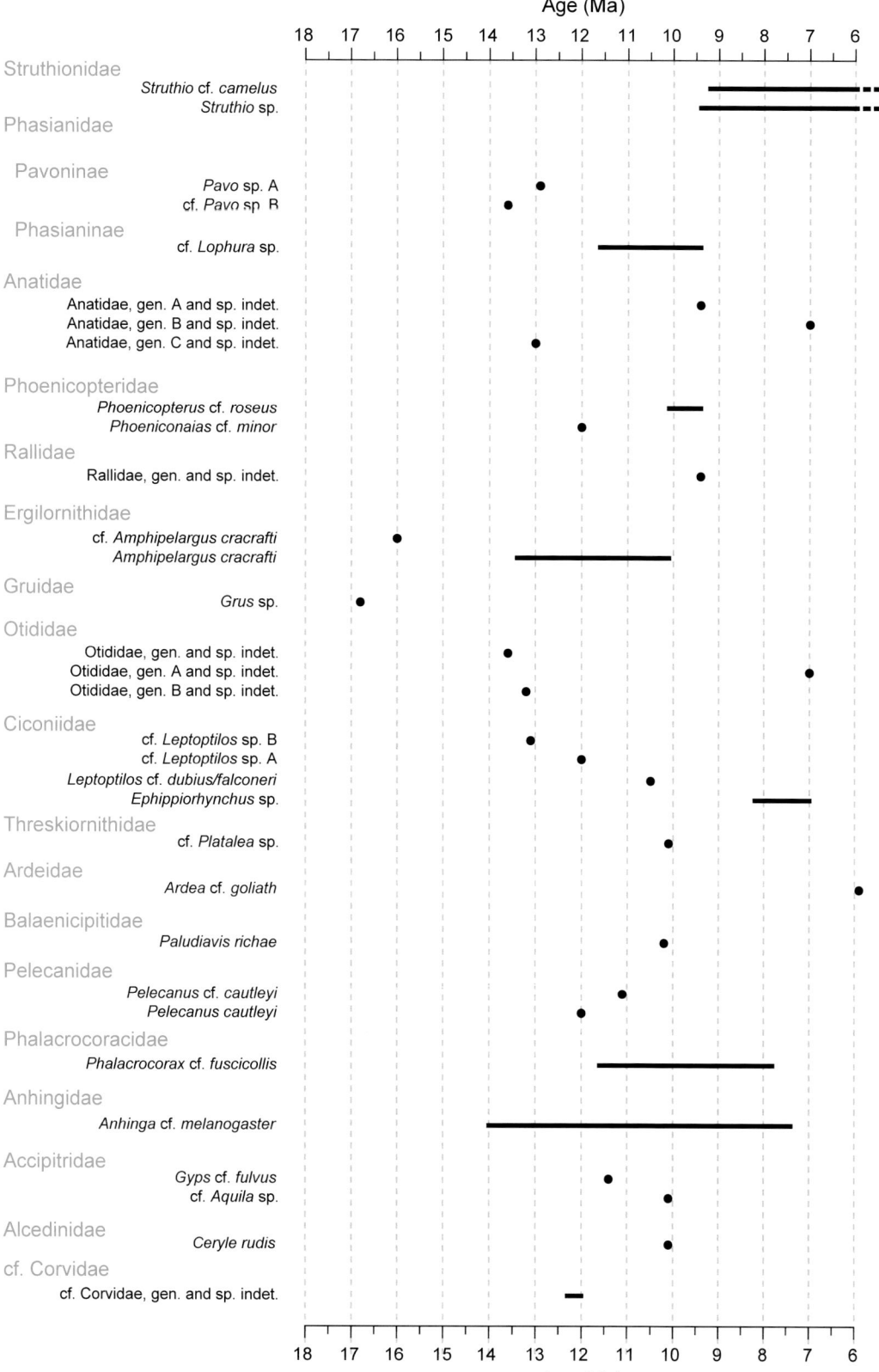

Age (Ma)

Figure 9.2. Biostratigraphic range chart generated from Table 9.1 data.

earliest record of the species at 9.2 Ma. *S. camelus* is otherwise known from the Middle Pliocene (3.6–3.8 Ma) in East Africa (Harrison and Msuya 2005; Mourer-Chauviré and Geraads 2008). *S. camelus* might therefore have evolved in Asia, in the Middle Tortonian, and later recolonized Africa in the Middle

Pliocene, where the earliest species of the *Struthio* lineage had evolved. To evaluate such a scenario, more remains from the Middle and Late Miocene of Central and South Asia are needed, including skeletal material. *S. camelus* eggshells attest to the presence of this extant species in India and Pakistan until 0.5 Ma

Table 9.3. Interpretations Regarding First and Last Occurrence of the Main Potwar Plateau Avian Taxa

Taxon[1]	First Appearance			Last Appearance		Still Extant	
	Origination (=speciation, anagenesis/cladogenesis[2])	Unknown origin	Immigration (range extension)	Extinction	Regional extirpation (range contraction)	Without geographic continuity	With geographic continuity[3]
Struthio cf. *camelus*	?[4](9.2 Ma)	—	—	—	Regionally extinct (Pakistan, India) ~25 Ka; until 0.5 Ma in the Siwaliks	X[4]	—
Struthio (sp. B, larger)[5]	—	—	X (late Miocene/Pliocene)	—	X (late Miocene-Pliocene to Plio-Pleistocene)	—	—
Pavo sp. A[5]	X (12.9 Ma)	—	—	X (12.9 Ma)	—	—	—
cf. *Pavo* sp. B	?(13.6 Ma)	—	—	—	—	—	X (today *P. cristatus* very near: northwestern India)
cf. *Lophura* sp.	?(11.6 Ma)	—	—	—	—	—	X
Phoenicopterus cf. *roseus*	?(10.1 Ma)	?(10.1 Ma)	?(10.1 Ma)	—	—	—	X
Phoeniconaias cf. *minor*	?(12 Ma)	?(12 Ma)	?(12 Ma)	—	—	—	X (today not far: south Pakistan, parts of India)
Amphipelargus cracrafti[5]	Both ?(16 Ma/13.4 Ma)	Both ?(16 Ma/13.4 Ma)	Both ?(16 Ma/13.4 Ma)	X (10.1 Ma)	—	—	—
Ergilornithidae[5]	—	—	X (16 Ma/-13.4 Ma)	—	X (10.1)	—	—
Grus sp.	?(16.8 Ma)	—	—	—	—	—	X
Ardea cf. *goliath*	?(5.9 Ma)	?(5.9 Ma)	?(5.9 Ma)	—	X (5.9 Ma) (today closest occurrences: Iran and eastern India)	X	—
Paludiavis richae[5]	X (10.2 Ma)	X (10.2 Ma)	X (10.2 Ma)	X (10.2 Ma or somewhere between 11.2 and 7.1 Ma)	X (10.2 Ma or somewhere between 11.2 and 7.1 Ma)	—	—
Balaenicipitidae	—	—	X (10.2 Ma)	—	X (10.2 Ma)	X	—
Pelecanus cautleyi[5]	?(12 Ma)	—	—	X (9.4 Ma)	—	—	—
Phalacrocorax cf. *fuscicollis*	?(11.6 Ma)	?(11.6 Ma)	?(11.6 Ma)	—	—	—	X
Phalacrocorax	?(25–23 Ma)	?(25–23 Ma)	?(25–23 Ma)	—	—	—	X
Anhinga cf. *melanogaster*	?(14 Ma)	?(14 Ma)	?(14 Ma)	—	—	—	X
Gyps cf. *fulvus*	?(11.4 Ma)	?(11.4 Ma)	?(11.4 Ma)	—	—	—	X
Ceryle rudis	—	—	?(10.1 Ma)	—	—	—	X

[1]Only the species (and higher taxa when pertinent) for which interpretations can be proposed are listed here. For other taxa the identification is too imprecise. Data based on Table 9.1 (first- and last-occurrence ages) and text.

[2]For cases of origination (speciation), the ancestor is considered to have been present in the region previously (even if not precisely identified) based on the widespread nature of the record of the whole group; it was not possible, however, to distinguish between cladogenesis and anagenesis, the data for surrounding regions and periods and related taxa being too rare.

[3]Even if there is a gap between the last-appearance date in the Potwar Plateau record and today, this does not affect interpretation of continuity between these fossil occurrences and the presence today of the same taxa in the same region. The gap is explained by both the limits of the Siwalik record for birds (most recent fossils being 7 or 5 Ma, except for a few Pliocene taxa and for ratite eggshell) and the scarcity of avian remains in general due to taphonomic bias.

[4]More hypothetical interpretations are marked with " ?," especially in cases where fossils are rare for the whole group (e.g., family) concerned, in contrast with interpretations marked with "X," which point to firmer interpretations.

[5]Extinct taxon. See text for further explanation.

in the Siwaliks (present study; Stern, Johnson, and Chamberlain 1994), then at least from before 60,000 years ago and 25,000 years ago in India (Blinkhorn, Achyuthan and Petraglia 2015; Jain et al. 2017). *S. camelus* gradually disappeared from Asia and southeastern Europe (apparently due to human impact), remaining only in Africa. A remnant population, present until recently in the Arabian Peninsula, Syria, and Iraq (*S. c. syriacus*), probably became extinct in the late twentieth century (del Hoyo et al. 2017). Today ostriches live only in Africa (with *S. camelus* in most parts, and *S. molybdophanes* in the Horn of Africa).

The enigmatic aepyornithoid eggshells are difficult to interpret, but might represent a separate lineage of ratite, possibly a small *Struthio* lineage (see Mourer-Chauviré et al. 1996a, 1996b). These eggshells are widespread in the Neogene of Africa, Eurasia, and one locality in Spain and were hitherto unknown after the Pliocene (Louchart, Bibi, and Stewart 2022). However, aepyornithoid eggshells occur in the Potwar Plateau from 11.0 to 1.3 Ma (present study; Stern, Johnson, and Chamberlain 1994) and provide the youngest ages known for this ootype. Interpretations are further complicated in various Neogene sites by the coexistence within a single eggshell of aepyornithoid-type and struthioid-type morphologies, depending on their position in the egg (more equatorial or more polar; Mourer-Chauviré and Geraads 2008).

Among phasianids, remains of the peafowl, cf. *Pavo* sp. B (13.6 Ma) and *Pavo* sp. A (12.9 Ma), are the earliest known fossils of the genus, which likely evolved within the region, where two extant species live. Younger fossils are known from southern Europe and East Africa in the Pliocene (Louchart 2003; Pickford, Senut, and Mourer-Chauviré 2004), probably attesting to an extension before a contraction toward the present.

Two species of pelicans, known exclusively from the Siwaliks, are extinct: *Pelecanus cautleyi* and *P. sivalensis*. *P. cautleyi* is present between at least 12.0 and 11.1 and probably 9.4 Ma. The last appearance date for *P. cautleyi* (9.4 Ma) is actually that of three fossils assigned to *P. cautleyi/sivalensis*, while the fossil from 12 Ma is *P. cautleyi* and an intermediate-aged fossil from 11 Ma is *P.* cf. *cautleyi*. Contemporaneous or younger *P. cautleyi* is present in undated Siwalik localities, from which it was initially described (Davies 1880; Lydekker 1891) and initially claimed to be Pliocene, but is now considered Late Miocene or Pliocene (Stidham et al. 2014). In the absence of precise ages for the original fossils of *P. cautleyi*, 9.4 Ma represents the last plausible date for that species. The other extinct Siwalik pelican, *P. sivalensis*, is known from poorly dated (Late Miocene/Pliocene) horizons and tentatively from a Late Pliocene Indian locality (2.6 Ma; Stidham et al. 2014).

Paludiavis richae (10.2 Ma; Balaenicipitidae) is close to cf. *Paludiavis richae* from the Late Miocene of Tunisia (~11.2–7.1 Ma) and attests to an extension of the family's geographic range from Africa to southern Asia (before 10.2 Ma), followed by a contraction (after 10.2 Ma), which led to its current Afrotropical distribution (see Table 9.3). The first-appearance age of *Paludiavis richae* (and concomitantly, the Balaenicipitidae) likely corresponds to expansion from Africa, since the oldest record of the family is from the Oligocene of Egypt (Olson and Rasmussen 1987). *P. richae* is known tentatively from between 11.2 and 7.1 Ma in Tunisia (Rich 1972, 1974; Harrison and Walker 1982), and therefore it is possible that the date of 10.2 Ma in the Potwar Plateau corresponds to immigration from Africa. Alternatively, a different ancestor evolved into *Paludiavis* in Asia, and the Tunisian record, if more recent, might correspond to a re-immigration of this genus into Africa. Similarly, the last occurrence for *Paludiavis* could correspond to global or regional extinction. In both cases, it is only a regional extinction of the Balaenicipitidae, since this family is still extant in sub-Saharan Africa.

Flamingos, tentatively represented by the extant species *Phoenicopterus* cf. *roseus* at 10.1–9.4 Ma and *Phoeniconaias*

cf. *minor* at 12.0 Ma, constitute potentially the earliest remains for these extant lineages and would imply a significantly greater age for crown flamingos than indicated by molecular studies (Pliocene; Torres et al. 2015). More basal taxa, generally considered stem flamingos, are known from the Oligocene and Early Miocene of France and the Early and Middle Miocene of Kenya, while an Early Miocene flamingo from Thailand is questioned as being either crown group—*Phoeniconaias*—or stem group (Cheneval et al. 1991; Mayr 2017). If the Thailand species is indeed a *Phoeniconaias* (it was named *Phoeniconaias siamensis*), then, at ~21 Ma, it precedes the record of the genus from the Potwar Plateau.

Among cranes, the *Grus* sp. reported here (16.8 Ma) represents the earliest occurrence of this extant genus, the next youngest being ~12–11 Ma from Romania, and then ~8 Ma from Afghanistan, which is roughly contemporaneous with material from Hungary and Greece (see Mourer-Chauviré et al. 1985; Zelenkov 2015).

The bustard family Otididae probably has its earliest representatives to date in the Potwar Plateau (at 13.6 Ma), with a second smaller species at 13.2 Ma, which indicates an earlier diversification. There are earlier possibilities elsewhere, but all are either putative or undescribed (Eocene/Oligocene of Quercy, France: Mourer-Chauviré 2006 in Mayr 2009; Oligocene of Kazakhstan: Kurochkin 1976 in Mayr 2009; 16-15 Ma from Kenya: Harrison 1980; Pickford 1986; Mayr 2014). Contemporaneous with the early record from the Potwar Plateau are probable Otididae from Mongolia (undescribed, 13.5–11.0 Ma; Zelenkov 2016).

The genus *Phalacrocorax* identified at 25–23 Ma (Late Oligocene) from Locality Z129 of the Zinda Pir region is the earliest occurrence of crown Phalacrocoracidae. All earlier fossils assigned to the family are actually stem representatives (several genera), undescribed, or doubtful (Mayr 2009). Hitherto crown phalacrocoracids were known only from the Middle Miocene, from various localities in the world (Mayr 2017). The early Siwalik record therefore represents a major extension of the lineage.

Other Potwar Plateau fossils are assignable to extant species and include the earliest known occurrences of these species. This is the case for *Ardea* cf. *goliath* (5.9 Ma), *Phalacrocorax* cf. *fuscicollis* (11.6 Ma), *Anhinga* cf. *melanogaster* (14 Ma; *A. melanogaster* in the strict sense, i.e., Oriental darter, today restricted to South and Southeast Asia), *Gyps* cf. *fulvus* (11.4 Ma), and *Ceryle rudis* (10.1 Ma). Concerning the darter, only two earlier records of the genus *Anhinga* exist in the Old World, one from ~16 Ma (*A. pannonica*, Germany; Dalsätt, Mörs, and Ericson 2006), and one from ~21 Ma (*Anhinga* sp., size of *A. melanogaster*, Thailand; Cheneval et al. 1991). The Early Miocene darter from Thailand was initially proposed as *A. pannonica* (Cheneval et al. 1991) but is actually smaller and compatible with *A. melanogaster* (personal observation). This Thailand darter could well be ancestral to the Siwalik and extant lineage, whereas the German record attests to a second lineage that arose before 16 Ma and became extinct during the Pliocene (*A. pannonica*). For the kingfisher at 10.1 Ma (*Ceryle rudis*; monospecific genus *Ceryle*; today in Africa and South Asia), this relatively early age for an extant small species is compatible with its basal phylogenetic position among extant kingfishers (Moyle 2006; Andersen et al. 2018). It is also one of the earliest known Alcedinidae, with another Alcedinidae gen. sp. indet. from 14–15 Ma in Kenya (Mayr 2014).

The plausible processes that would account for the first- and last-appearance ages of various taxa (see Table 9.3) can be placed into different categories. Concerning first-appearance dates, speciation pertains to between 3 and 15 taxa, whereas immigration applies to between 5 and 13 taxa (between 2 and 10 of these are possibly in both categories—i.e., speciation concomitant with immigration). The three firmer speciation events range in age from 16.0 to 10.2 Ma, whereas the five firmer immigrations extend from 16.0 or 13.4 Ma to 10.2 Ma. Concerning the last-appearance dates, four are extinctions (from 12.9 Ma to 9.4 Ma) and six are regional extirpations (range contractions; from 10.2 Ma to 0.5 Ma). In addition, a number of Siwalik avian taxa are extant, including eight species and four genera (sp. indet.)—the earliest being the Late Oligocene *Phalacrocorax*. Among these extant taxa, ten still occur today in the same area (six at the species level, four at the genus level). This continuity (taxonomic, and in part geographic) is the most prominent aspect of the paleoavifaunas, and as such attests to the absence of notable evolutionary changes for these taxa.

Turnovers (disappearances and appearances) in the avifauna are infrequent, but several of the well-constrained and significant events lie between 10.2 and 9.4 Ma. These include the disappearance of *Amphipelargus cracrafti* (and Ergilornithidae as a whole), the appearance of ostriches (*Struthio*), the single occurrence of Balaenicipitidae, the disappearance (extinction) of *Pelecanus cautleyi*, and the appearance of *Ceryle rudis*. These events do not seem to be correlated with climatic or environmental change, and they might have resulted principally from immigration and emigration events, the origins of which are unknown, and perhaps related mainly to events outside the paleotropical biogeographic region. Comparisons with contemporaneous sequences in other regions are not feasible because of the sparse fossil avifaunas; occurrences are rare or poorly described, except for parts of Europe, the New World, and Australia—locations that are generally too distant to provide valuable insight. The Siwalik record (essentially Miocene) is therefore remarkable for birds but remains an exception compared with the rest of Asia and Africa.

Geographic links between the Siwaliks and other regions are primarily with Africa. At least the following taxa arrived from Africa via southwest Asia: *Struthio*, aepyornithoid ootaxon, Balaenicipitidae, and *Ceryle rudis* (the latter being the only lineage still present in South Asia). The Ergilornithidae originated in Central Asia and immigrated probably via a route west of the Himalaya Mountains. Finally, one or two possible originations are from Southeast Asia (*Phoeniconaias, Anhinga*).

PROMINENT ASPECTS OF THE AVIAN RECORD IN THE POTWAR PLATEAU

Among the most interesting aspects of the sequence in the Siwalik record are the changes observed between 10.2 Ma and 9.4 Ma, which involve several species and families, including an apparent replacement of ergilornithids by struthionids. These changes cannot be positively linked to regional environmental change. Another striking aspect of the record is the presence of a large majority of extant families and extant genera, as well as many extant species. This is quite common for birds in general since the Middle Miocene (e.g., Louchart et al. 2008; Louchart, Tourment, and Carrier 2011). There is no clear example of phyletic evolution and no obvious replacement within a family. Overall, the avifaunal succession shows a mixture of continuity along with faunal change involving mainly incursions and excursions from and to surrounding regions, as far as Africa.

CONCLUSION

Siwalik avian remains yield important information concerning a number of orders and families (of rather large birds owing to taphonomic biases): ostrich(es) (Ratitae, Struthionidae), landfowl (Galliformes, Phasianidae), waterfowl (Anseriformes, Anatidae), storks and allies (Ciconiidae), ibis (Threskiornithidae), darters and cormorants (Anhingidae, Phalacrocoracidae), pelican(s) (Pelecanidae), cf. shoebill (cf. Balaenicipitidae), flamingo (Phoenicopteridae), heron (Ardeidae), bustards (Otididae), rallid (Rallidae), vulture (Accipitridae), kingfisher (Alcedinidae), and passerines (Passeriformes). In addition, the avian fossils include representatives of the extinct family Ergilornithidae (Gruiformes), which were unique flightless cursorial birds with two-toed feet, a characteristic otherwise known only in ostriches. For most of these groups, the Neogene Siwalik record provides the first or earliest occurrence for the region or the subcontinent and for some in the entire world. Among the more interesting and crucial groups are the ostriches (eggshells and bones; two species), as well as the aepyornithoid eggshell ootype. More fossils from the Miocene and beyond, when found in South Asia, will help resolve the mysteries surrounding these and other avian groups that once lived at the foot of the Himalayas.

ACKNOWLEDGMENTS

I would like to thank David Pilbeam, Michèle Morgan, Larry Flynn, John Barry, and Catherine Badgley for having invited me years ago to study the Siwalik birds then housed at Harvard University, and for having supported me in this task. I also thank the curators of comparative osteological collections at MCZ Harvard (USA), UCBL Lyon and MNHN Paris (France), NHM Tring (UK), IRSNB Brussels and Africa Museum Tervuren (Belgium).

References

Andersen, M. J., J. M. McCullough, W. M. Mauck III, B. T. Smith, and R. G. Moyle. 2018. A phylogeny of kingfishers reveals an Indomalayan origin and elevated rates of diversification on oceanic islands. *Journal of Biogeography* 45:269–281.

Bernor, R. L. 1986. Mammalian biostratigraphy, geochronology, and zoogeographic relationships of the Late Miocene Maragheh fauna, Iran. *Journal of Vertebrate Paleontology* 6(1): 76–95.

Bibi, F., A. B. Shabel, B. P. Kraatz, and T. A. Stidham. 2006. New fossil Ratite (Aves: Palaeognathae) eggshell discoveries from the Late Miocene Baynunah Formation of the United Arab Emirates, Arabian Peninsula. *Palaeontologia Electronica* 9 (1): 2A. https://palaeo-electronica.org/paleo/2006_1/eggshell/issue1_06.htm.

Blinkhorn, J., H. Achyuthan, and M. D. Petraglia. 2015. Ostrich expansion into India during the Late Pleistocene: Implications for continental dispersal corridors. *Palaeogeography, Palaeoclimatology, Palaeoecology* 417:80–90.

Blondel, J., and C. Mourer-Chauviré. 1998. Evolution and history of the western palearctic avifauna. *TREE* 13(12): 488–492.

Boev, Z., and N. Spassov. 2009. First record of ostriches (Aves, Struthioniformes, Struthionidae) from the late Miocene of Bulgaria with taxonomic and zoogeographic discussion. *Geodiversitas* 31(3): 493–507.

Cheneval, J., L. Ginsburg, C. Mourer-Chauviré, and B. Ratanasthien. 1991. The Miocene avifauna of the Li Mae Long Locality, Thailand: Systematics and paleoecology. *Journal of Southeast Asian Earth Sciences* 6(2): 117–126.

Dalsätt, J., T. Mörs, and P. G. P. Ericson. 2006. Fossil birds from the Miocene and Pliocene of Hambach (NW Germany). *Palaeontographica Abteilung A* 277(1–6): 113–121.

Davies, W. 1880. On some fossil bird-remains from the Siwalik Hills in the British Museum. *Geological Magazine* 7(1): 18–27.

Del Hoyo, J., A. Elliott, J. Sargatal, D. A. Christie, and E. de Juana, eds. 2019. *Handbook of the Birds of the World Alive*. Barcelona: Lynx Edicions. https://www.hbw.com/.

De Mecquenem, R. 1908. Contribution à l'étude du gisement des vertébrés de Maragha et de ses environs. In *Annales d'Histoire Naturelle. Ministère de l'instruction publique et des beaux-arts (Délégation en Perse)*, edited by de Morgan. Ernest Leroux (éditeur), Paris, Tome 1(1): 27–79.

De Mecquenem, R. 1925. Contribution a l'étude des fossiles de Maragha. *Annales de Paléontologie* 14:135–160.

Harrison, C. J. O. 1974. A re-examination of the extinct marabou stork *Leptoptilos falconeri*; with descriptions of some new species. *Bulletin of the British Ornithologists' Club* 34:42–49.

Harrison, C. J. O. 1979. The Pliocene Siwalik cormorant. *Tertiary Research* 2:57–58.

Harrison, C. J. O. 1980. Fossil birds from Afrotropical Africa in the collection of the British Museum (Natural History). *Ostrich* 51(2): 92–98.

Harrison, C. J. O. 1981. A re-assignment of *Amphipelagus* [sic] *majori* from Ciconiidae (Ciconiiformes) to Ergilornithidae (Gruiformes). *Tertiary Research* 3(3): 111–112.

Harrison, C. J. O., and C. A. Walker. 1982. Fossil birds from the Upper Miocene of Pakistan. *Tertiary Research Special Papers* 4(2): 53–69.

Harrison, T., and C. P. Msuya. 2005. Fossil struthionid eggshells from Laetoli, Tanzania: Taxonomic and biostratigraphic significance. *Journal of African Earth Sciences* 41(4): 303–315.

Jain, S., N. Rai, G. Kumar, P. Aggarwal Pruthi, K. Thangaraj, S. Bajpai, and V. Pruthi. 2017. Ancient DNA reveals late Pleistocene existence of ostriches in Indian sub-continent. *PLOS ONE* 12(3): e0164823.

Jarvis, E. D., S. Mirarab, A. J. Aberer, B. Li, P. Houde, C. Li, S. Y. W. Ho, et al. 2014. Whole-genome analyses resolve early branches in the tree of life of modern birds. *Science* 346(6215): 1320–1331.

Karhu, A. A. 1997. A new species of *Urmiornis* (Gruiformes: Ergilornithidae) from the Early Miocene of Western Kazakhstan. *Paleontological Journal* 31(1): 102–107.

Kurochkin, E. N. 1985. Birds of Central Asia in the Pliocene. *Proceedings of the Joint Soviet-Mongolian Paleontological Expedition, Transactions* 26:1–120 (in Russian).

Louchart, A. 2003. A true peafowl in Africa. *South African Journal of Science*. 99(7): 368–371.

Louchart, A., F. Bibi, and J. Stewart. 2022. Chapter 9: Birds from the Baynunah Formation. In *Sands of Time: Ancient Life in the Late Miocene of Abu Dhabi, United Arab Emirates*, edited by F. Bibi, B. Kraatz, M. J. Beech, and A. Hill, 125–139. New York: Springer.

Louchart, A., Y. Haile-Selassie, P. Vignaud, A. Likius, and M. Brunet. 2008. Fossil birds from the Late Miocene of Chad and Ethiopia and zoogeographical implications. *Oryctos* 7:147–167.

Louchart, A., N. Tourment, and J. Carrier. 2011. The earliest known pelican reveals 30 million years of evolutionary stasis in beak morphology. *Journal of Ornithology* 152(1): 15–20.

Louchart, A., P. Vignaud, A. Likius, M. Brunet, and T. D. White. 2005. A large extinct marabou stork in African Pliocene hominid sites, and a review of the fossil species of *Leptoptilos*. *Acta Palaeontologica Polonica* 50(3): 549–563.

Lydekker, R. 1884. Siwalik birds. *Memoirs of the Geological Survey Of India* 10:135–147.

Lydekker, R. 1891. *Catalogue of the Fossil Birds in the British Museum (Natural History)*. London: British Museum.

Mayr, G. 2009. *Paleogene Fossil Birds*. Heidelberg: Springer.

Mayr, G. 2014. On the Middle Miocene avifauna of Maboko Island, Kenya. *Geobios* 47(3): 133–146.

Mayr, G. 2017. *Avian Evolution*. Chichester: John Wiley & Sons.

Milne-Edwards, A. 1867–1871. *Recherches anatomiques et paléontologiques pour servir à l'histoire des oiseaux fossiles de la France*, 4 vols. Paris: Victor Masson et Fils.

Mourer-Chauviré, C., J. -C. Balouet, Y. Jehenne, and E. Heintz. 1985. Une nouvelle espèce de grue, *Grus afghana* (Aves, Gruiformes), du Miocène supérieur de Molayan, Afghanistan. *Bulletin du Muséum national d'histoire natuirelle. Section C. Sciences de la terre* 7(3): 179–187.

Mourer-Chauviré, C., and D. Geraads. 2005. The Struthionidae and Pelagornithidae (Aves: Struthioniformes, Odontopterygiformes) from the late Pliocene of Ahl al Oughlam, Morocco. *Oryctos* 7:169–194.

Mourer-Chauviré, C., B. Senut, M. Pickford, and P. Mein. 1996a. Le plus ancien représentant du genre *Struthio* (Aves, Struthionidae), *Struthio coppensi* n. sp., du Miocène inférieur de Namibie. *Comptes rendus de l'Académie des Sciences de Paris. série 2a. Sciences de la terre et des planètes* 322(4): 325–332.

Mourer-Chauviré, C. B. Senut, M. Pickford, P. Mein, and Y. Dauphin. 1996b. Ostrich legs, eggs and phylogenies. *South African Journal of Science* 92:492–495.

Moyle, R. G. 2006. A molecular phylogeny of kingfishers (Alcedinidae) with insights into early biogeographic history. *Auk* 123(2): 487–499.

Olson, S. L. 1985. The fossil record of birds. In *Avian biology*, edited by Donald S. Farner, James R. King, and Kenneth C. Parkes, 8:79–238. New York: Academic Press.

Olson, S. L., D. Tab Rasmussen, and Elwyn Simons. 1987. Fossil birds from Oligocene Jebel Qatrani Formation, Fayum Province, Egypt. *Smithsonian Contributions to Paleobiology* 62:1–20.

Pickford, M. 1986. Cainozoic palaeontological sites of western Kenya. *Münchner Geowissenschaftliche Abhandlungen* 8:1–151.

Pickford, M., B. Senut, and Y. Dauphin. 1995. Biostratigraphy of the Tsondab Sandstone (Namibia) based on gigantic avian eggshells. *Geobios* 28(1): 85–98.

Pickford, M., B. Senut, and C. Mourer-Chauviré. 2004. Early Pliocene Tragulidae and peafowls in the Rift Valley, Kenya: Evidence for rainforest in East Africa. *Comptes Rendus Palevol* 3(3): 179–189.

Prum, R. O., J. S. Berv, A. Dornburg, D. J. Field, J. P. Townsend, E. M. Lemmon, and A. R. Lemmon. 2015. A comprehensive phylogeny of birds (Aves) using targeted next-generation DNA sequencing. *Nature* 526(7574): 569–573.

Rich, P. V. 1972. A fossil avifauna from the upper Miocene Beglia Formation of Tunisia. *Notes du Service géologique de Tunisie* 35:29–66.

Rich, P. V. 1974. Significance of the Tertiary avifaunas from Africa (with emphasis on a mid to late Miocene avifauna from southern Tunisia). *Annals of the Geological Survey of Egypt* 4:167–209.

Stern, L. A., G. D. Johnson, and C. Page Chamberlain. 1994. Carbone isotope signature of environmental change found in fossil ratite eggshells from a South Asian Neogene sequence. *Geology* 22(5): 419–422.

Stidham, T. A., K. Krishan, B. Singh, A. Ghosh, and R. Patnaik. 2014. A pelican tarsometatarsus (Aves: Pelecanidae) from the latest Pliocene Siwaliks of India. https://doi.org/10.1371/journal.pone.0111210.

Stidham, T. A., R. Patnaik, K. Krishan, B. Singh, A. Ghosh, A. Singla, and S. S. Kotla. 2017. The first darter (Aves: Anhingidae) fossils from India (late Pliocene). *PLOS ONE* 12(5): e0177129. https://doi.org/10.1371/journal.pone.0177129.

Torres, C. R., V. L. de Pietri, A. Louchart, and M. van Tuinen. 2015. New cranial material of the earliest filter feeding flamingo *Harrisonavis croizeti* (Aves, Phoenicopteridae) informs the evolution of the highly specialized filter feeding apparatus. *Organisms. Diversity and Evolution* 15:609–618.

Zelenkov, N. V. 2015. The fossil record and evolutionary history of cranes. *Cranes of Eurasia (Biology, Distribution, Captive Breeding)* 5:83–90.

Zelenkov, N. V. 2016. Evolution of bird communities in the Neogene of Central Asia, with a review of the Neogene fossil record of Asian birds. *Paleontological Journal* 50(12): 1421–1433.

Zelenkov, N. V., Z. Boev, and G. Lazaridis. 2016. A large ergilornithine (Aves, Gruiformes) from the late Miocene of the Balkan Peninsula. *Paläontologische Zeitschrift* 90:145–151.

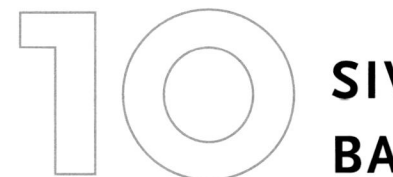

10

SIWALIK HEDGEHOGS, SHREWS, BATS, AND TREESHREWS

LAWRENCE J. FLYNN

INTRODUCTION

A century of exploration in the extensive Siwalik deposits had yielded few small mammal fossils when Colbert (1935) produced his magnum opus on the mammal collections from the Indian subcontinent housed in the American Museum of Natural History. To explain the rarity of small mammal remains, Colbert reasoned that most small vertebrates could have been destroyed during the deposition of the tremendously thick strata. We now recognize that, rather than fragility, rarity reflects bias against recovery due to the small size of the fossils. Early reports, such as Lydekker (1884), did note larger species of rodents, such as bamboo rats (Rhizomyinae) and a porcupine (Hystricidae, *Hystrix sivalensis*). Hinton (1933), Colbert (1933), and Wood (1937) added several more rhizomyines, a cane rat (Thryonomyidae), gundis (Ctenodactylidae), and the archaic hystricid *Sivacanthion*. Thus, up until the 1930s, nearly all known small mammal fossils were rodents of relatively large body size. The paleontology of Siwalik small mammals could not advance without a new approach to fossil collection.

That advance came in the 1970s with the application of new techniques for wet-screening large volumes of fossiliferous rock for small specimens not easily seen on the surface of the ground. Two teams independently adopted such screening programs. Jacobs (1977, 1978, 1980) began systematic screening of multiple horizons in the classical Chinji and Dhok Pathan formations of the Potwar Plateau and the Upper Siwaliks of the adjacent Pabbi Hills. His work produced shrews, hedgehogs, a rabbit, and diverse rodents, including mice, cricetids (hamster relatives), squirrels, and dormice. Whereas surface finds had produced mostly large species, screening made clear that small mice were truly dominant in the paleofauna. At about the same time, associates of S. Taseer Hussain (Howard University), especially Jens Munthe and Robert M. West, began productive screening programs elsewhere in Pakistan. At Daud Khel, near the Indus River, Munthe (1980) and Munthe and West (1980) recovered a diverse microfauna of four insectivores, a treeshrew, and six rodent groups (squirrels, gundis, dormice, bamboo rats, hamsters, and mice). In Kohat, west of the Indus River, colleagues from the University of Utrecht re-covered rich faunas from the Early Miocene Murree Formation (De Bruijn, Hussain, and Leinders 1981) and the Middle Miocene Chinji Formation (Wessels et al. 1982). These studies revealed consistently diverse small mammal faunas that showed both change and stability over various intervals of time.

By late in the twentieth century, the Potwar Siwaliks had proven to be highly productive of rodent material, with superposed assemblages representing a relatively complete record from multiply sampled horizons for the interval between 18 and 6 million years ago (Ma; see for example Flynn et al. 1995, 1998). In Sind Province of southern Pakistan, Wessels (1996) documented a series of horizons with Miocene muroids. Coupled with her monograph (Wessels 2009), she showed that the small mammal assemblages of the Miocene were largely the same from southern to northern Pakistan. Lindsay et al. (2005) discovered similar assemblages in the Early Miocene deposits of western Pakistan. With careful attention to stratigraphic control and a magnetostratigraphic framework for age interpolation, the western and northern successions of sites combine to yield a generally well-dated composite record that spans the Miocene Epoch.

Today the Siwalik small mammal record is therefore far different from that perceived a century ago. Small mammals are diverse and enrich our view of the paleocommunities from which they derive. On a backdrop of faunal stability they show changes—and surprises—throughout the Miocene. As part of the Indo-Malayan paleogeographic region and in partial isolation by physical barriers, the taxonomic composition of Siwalik assemblages is distinctive. Many elements are shared eastward with Southeast Asia and South China. Certain non-rodent groups, such as moles and various shrew clades, were blocked from entry into the theater of evolution of the Indian subcontinent. Exotics, such as marsupials and dermopterans that occurred much earlier to the southwest in the Bugti Hills, Pakistan (Marivaux et al. 2006; Crochet et al. 2007), are not recorded in the Miocene of the Potwar area. On the other hand, treeshrews flourished in the Potwar Siwaliks, showing greater diversity than expected based on modern South Asian mainland tropical forest faunas.

Siwalik micromammal samples screened by modern methods are dominated by Rodentia. The rodent sister group Lagomorpha adds a small fraction of the specimens. These two orders are treated together as Cohort Glires in Chapter 11 of this volume. Small primates are included in Chapter 12. Other elements found in Siwalik small mammal assemblages are hedgehogs, shrews, bats, and treeshrews. These important faunal elements inform our analyses about the paleobiology of Siwalik mammalian communities. Of these, hedgehogs were once surprisingly abundant, and treeshrews were, for a time, diverse. The distinctive and changing composition of Siwalik small mammal faunas exclusive of rodents and lagomorphs is the subject of this chapter.

AGES OF FOSSILS

Terrestrial sediments of earliest Miocene age occur in western and southern Pakistan, but in the Potwar Plateau of northern Pakistan the Siwalik sedimentary record begins later, about 18 Ma. In western Pakistan, a few small mammal fossil sites between 23 and 18 Ma (Fig. 10.1) occur in the Chitarwata and Vihowa formations of the Zinda Pir Dome. Magnetic reversal stratigraphy provides age estimates (Lindsay et al. 2005), although significant hiatuses of geological time in the Early Miocene without fossil data make age assignments uncertain. For example, Z113 occurs in sediments that lack magnetic resolution, and Z135 could be several hundred thousand years older than interpolated here (21 Ma) if the normal magnetic interval with which it is associated is older than here proposed. It is correlated with chron C6An.2n but could match C6AAr.1n. Flynn et al. (2013) revised age estimates for Zinda Pir sites and these are readjusted to the current time scale (see Fig. 10.1).

There are a few much older rodent sites in the Chitarwata Formation near Bugti and Dalana, Pakistan. Early Oligocene Localities Y417, Z108, and Z144, sampled in collaboration with paleontologists of the Geological Survey of Pakistan (GSP) and the Pakistan Museum of Natural History, are approximately as old as the rich Early Oligocene Paali Nala locality south of Bugti, discovered by the French Paleontological Mission and the University of Baluchistan (Marivaux, Vianey-Liaud, and Welcomme 1999).

Every formation of the fossiliferous Potwar sequence produces microfauna, and in contrast to western Pakistan, Potwar biostratigraphy is based on dense fossil sampling. The Kamlial Formation and the Chinji Formation constitute the Early and Middle Miocene age Lower Siwaliks. The earliest Potwar sediments, mainly the less intensively surveyed Kamlial Formation, yield relatively few microfauna sites. The superjacent Chinji Formation is richer, both in site quality and in density of microsites throughout the sequence (see discussion in Barry et al. 2013). In the succeeding Late Miocene Middle Siwaliks, the Nagri Formation is sporadically productive in contrast to the richer Dhok Pathan Formation lying above it. The microfauna of the Plio-Pleistocene Upper Siwaliks, not a focus of this

study, has been worked in Pakistan by our group (Jacobs 1978; Cheema, Raza, and Flynn 1997; Cheema, Flynn, and Rajpar 2003; Khan et al. 2020) and is well studied in India (Kotlia 1991; Patnaik 1997; Kotlia et al. 1999; Patnaik 2001; Patnaik and Nanda 2010).

SIWALIK SMALL MAMMAL SITES

Figure 10.1 includes all Neogene localities that have yielded small mammals for the YGSP group. These localities differ in number and quality of specimens. Some represent surface finds of single specimens, as is the case for D70 at 3.3 Ma, where a bamboo rat was found by Neil Opdyke while conducting a magnetostratigraphic survey. Most sites represent finds resulting from careful prospecting. Well over 100 sites were sampled in bulk for microfauna, but those found to be unproductive were not aggressively screened. From 61 sites, at least 300 kg of fossiliferous sediment were screen-washed. About 42 of the more productive localities were selected for processing larger samples, over 500 kg of matrix. Uneven sampling resulted from differing site productivity and screening effort. One influence on productivity was ease of wet-screening. Some sites were characterized by a combination of clays and silts that disintegrated readily in water. Other lithologies needed encouragement to break down. We used diluted acetic acid to remove calcareous cement from some of the matrix, while other samples required presoaking in kerosene for about 15 minutes prior to wet-screening to facilitate dispersal and flocculation of clays when the samples were immersed in water.

The screened residue from sites was rewashed at least once. We used tandem screen boxes: a coarse-mesh inside box and a fine-screen outer box to catch small fossils in a mesh of 0.5 mm. The coarse material of the inner box was normally sorted in the field. Much of the finer concentrate was processed by our skillful colleague, the late William R. Downs, who perfected a method of flotation in the laboratory: the concentrate was treated with a heavy liquid (sodium polytungstate) to separate denser fossils from the matrix; then the residue was washed and sorted by microscope. Will became very proficient at reliable extraction of quite small fossils.

In its entirety, the collection of small mammals from sites shown in Figure 10.1 numbers over 11,500 specimens. These are primarily isolated cheek teeth, although some jaw fragments, anterior dentition, or postcrania are present. Inevitably we have had to consider whether isolated teeth in a single sample might represent the same individual. In a few cases this is evident (based on similar distinctive preservation or stage of wear), but generally we consider each specimen to represent a unique individual.

The consistent taxonomic composition of species and genera throughout our long series of Middle Miocene localities shows stability through time and represents an evolving biota comprising successive metacommunities (Leibold et al. 2004; Eronen 2007; Flynn et al. 2023). Despite species substitution and some additions or deletions, the stability indicates that faunal content of individual sites reflects snapshots or censuses

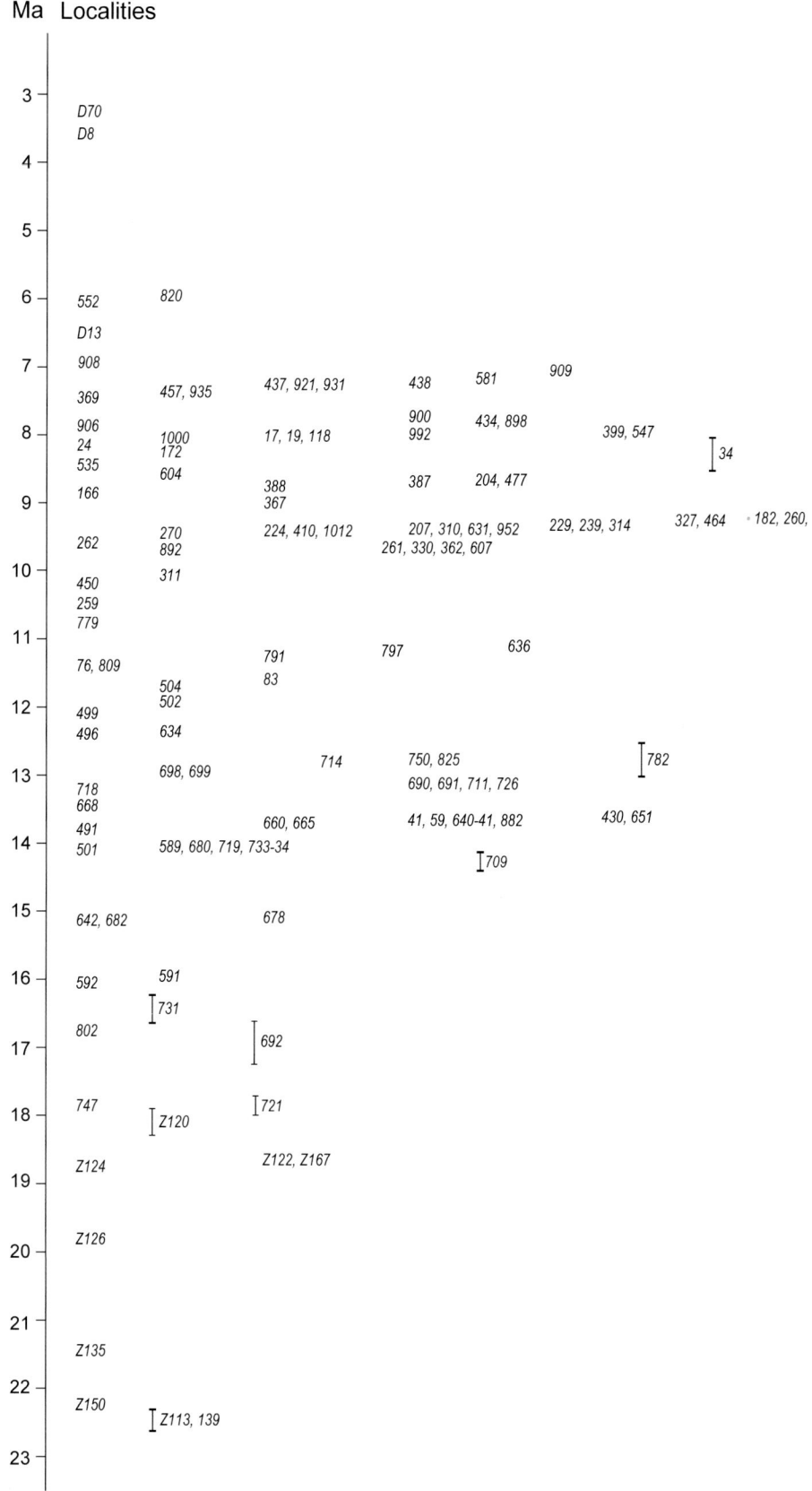

Ma Localities

Figure 10.1. Temporal distribution of sampled Potwar small mammal localities. Localities are ordered by observed stratigraphy; a few with imprecise correlation have vertical bars indicating maximum likely range of age. Prefixes D and Z refer to localities of the Dartmouth-Peshawar and the Zinda Pir projects, respectively; numbers without prefix designate Y localities. To the left of the timeline, age in millions of years (Ma) is estimated by magnetic reversal stratigraphy. Note the nonuniform distribution of localities through time. Productivity also varies greatly, with some sites as surface finds producing one or a few fossils and others yielding hundreds of molars collected by screening.

through time of the resident assemblage and that microsites repeatedly sample similar ecology. Well-represented Middle Miocene age localities and intervals yield ~30 rodent species and appear to capture a significant proportion of the past microfauna. That spatially adjacent sites have comparable assemblages suggests that they formed under similar conditions with comparable ecology (Badgley, Downs, and Flynn 1998). An analogous study in North America by Rogers et al. (2017) observed high faunal similarity for coeval microsites in the Late Cretaceous Judith River Formation and inferred minimal differential preservation bias affecting faunal composition of assemblages within that system.

The fossil representation of successive Potwar localities can be compared because sources of taphonomic bias among them are limited. Isolated rodent teeth between localities are comparable hydraulically. Fossils from successive samples do not vary greatly in shape, which otherwise could affect recovery (Blob and Fiorillo 1996).

Siwalik small mammal localities formed typically in riparian settings, in overbank floodplain deposits (Badgley, Downs, and Flynn 1998). The fossils are predominantly isolated teeth and are not heavily corroded. Fossil assemblages were likely concentrated by avian predators in small areas, for example under roosts (Wood 1949, Denys et al. 2018). Owls and raptors that feed during the day are indicated by the light degree of enamel corrosion, which we observe in preserved assemblages (Andrews 1990a; Rey-Rodríguez et al. 2019). Multiple avian predators and mammalian carnivores contribute to fossil microfaunas, but snakes and crocodilians are not known to pass significant identifiable remains (Andrews 1990a). That postcrania do not dominate assemblages suggests that remains were

not buried immediately after deposition. Subaerial trampling and disintegration of most skeletal elements would result in loss of bone and disproportionate survival of teeth. The seasonal or annual accumulations of dental remains were buried by silt from seasonal flooding.

Taking owls as likely concentrators of our larger microfaunal assemblages, we can estimate the scale of area of intense hunting. Andrews (1990a) indicates that a large owl ranges 3–4 km (from a presumed home roost). Using this as a radius, we surmise that discrete fossil microsites can contain fauna from an area of up to 50 km². Faunal diversity comparisons should be done at that scale (and similar 30° latitude; Rosenzweig 1995). While specialist predators today accumulate remains of favored prey at proportions that may not reflect relative abundances in the paleocommunity, generalists concentrate most prey species in proportion to their relative abundances in the present communities (Andrews 1990b; Terry 2010).

Small mammal fossil assemblages enrich the vision of mammal life in the Miocene of the Indian subcontinent. They shed light on aspects of mammalian interactions in a dynamic community setting as they do for modern communities. They reveal unexpected patterns through time in faunal composition and diversity. Much of small mammal diversity other than that of the ubiquitous rodents is surveyed in this chapter.

The Siwalik microfauna documentation in this chapter and the following one includes body mass estimates for all species-level taxa of the Potwar Plateau. The estimates shown in Table 10.1 are based mainly on dental comparisons, assuming that tooth dimensions in fossils correlate with body mass as do molars of living species of known weight. Regression equations are also applied in Chapter 11 for species-rich groups.

Table 10.1. Lipotyphlan (Erinaceid, Soricid) and Scandentian Species in the Miocene Record of the Potwar Plateau, Pakistan

Species	Author	Specimen	Age Range	Mass
Erinaceidae				
Galerix wesselsae	Zijlstra and Flynn 2015	YGSP 52202, right M1 (type)	18.8–14.0 Ma	80–90 g
Galerix rutlandae	Munthe and West 1980	Howard-GSP 2371, right M1 (type)	14.3–11.6 Ma	60–70 g
Schizogalerix sp.	Zijlstra and Flynn 2015	YGSP 39496, left M2	11.4–10.2 Ma	>100 g
Erinaceinae, gen. A and sp. indet.		YGSP 52204	(Locality Z122) 18.7 Ma	60 g
Erinaceinae, gen. B and sp. indet.		YGSP 34899	(Locality Y691) 13.1 Ma	60 g
Erinaceinae, gen. C and sp. indet.		YGSP 39462	(Locality Y931) 7.3 Ma	50 g
Soricidae				
Indeterminate		Z 774 incisor	22.5–18.7 Ma	8 g
cf. *Crocidura* sp.	Flynn et al. 2020	YGSP 40515, Y182 mandible	14.0–6.5 Ma	10 g
Suncus honeyi	Flynn et al. 2020	YGSP 40539, right M2 (type)	10.5–6.5 Ma	80 g
Beremendiini, gen. and sp. indet.	Flynn et al. 2020	YGSP 40538 M1	8.7 Ma	50 g
Tupaiidae				
cf. *Dendrogale* sp. 1		YGSP 48097 M2	18.7–10.5 Ma	60 g
cf. *Dendrogale* sp. 2		YGSP 36684 m3	18.7–13.1 Ma	40 g
cf. *Dendrogale* sp. 3		YGSP 33139 m2	14.0–7.3 Ma	80 g
cf. *Tupaia* sp.		YGSP 34948 M2	12.8–11.2 Ma	80 g
Tupaiidae, n. gen. and sp.		YGSP 8089 skull fragment	11.6–8.7 Ma	300 g
Tupaiidae, gen. and sp. indet. (tiny)		YGSP 49011 frag	8.7 Ma	~30 g

SMALL MAMMAL ORDERS EXCLUSIVE OF GLIRES

Fieldwork in Pakistan since 1975 and subsequent analysis of its Neogene stratigraphy and faunas have built a rich fossil record of Siwalik small mammals. Rather than the earlier view of a small mammal record dominated by large bamboo rats and porcupines, we now perceive murids and cricetids as the most diverse and abundant members of the rodent fauna. We also find a wealth of other groups complementing the microfauna, notably bats and other mammals of small body size. Bats, order Chiroptera, are expected to be diverse and locally abundant in the Miocene Siwalik paleocommunity, as they are in modern faunas. Working against their recovery are taphonomic processes, because their volant locomotion and restricted roosts make fossil preservation less likely in typical Siwalik small mammal localities than for many rodents. Also expected in Siwalik assemblages are lipotyphlan insectivorans. Both groups are reviewed below. Another distinct kind of mammal represented in the Siwaliks is order Scandentia, the treeshrews. The largely arboreal treeshrews occur at many localities, which indicates that the same biological agents that concentrated the rodents also gathered treeshrews as well as insectivorans. Lagomorpha have a restricted Siwalik record and are treated in Chapter 11 of this volume with rodents as Cohort Glires. Almost any mammalian component of the paleocommunity (small carnivores, artiodactyls, primates) has the potential to emerge in the screened concentrate from fossiliferous sites. Here, the fossil record that we have built for Lipotyphla, Chiroptera, and Scandentia is reviewed for clues about faunal diversity, turnover, and paleohabitat.

Lipotyphla Haeckel, 1866

The core group of living insectivoran mammals is Lipotyphla Haeckel, 1866. The rank and content (e.g., McKenna and Bell 1997) of the Lipotyphla vary among authorities, but usage here is unambiguous, because only two extant families occur in the Siwalik fossil record, hedgehogs (Erinaceidae) and shrews (Soricidae). A remarkable characteristic of Siwalik mammal faunas is that the third widespread lipotyphlan group, the moles, family Talpidae, apparently never reached South Asia. Parmar, Norboo, and Magotra (2023) reported an erinaceid and possible mole from Ramnagar, India. They compared the latter extensively with bats, but preferred talpid identification. Chiropteran assignment of the "mole" appears more likely, given its pre- and postcingulum. Restricted mole distribution reflects the semi-isolated nature of the Indian subcontinent, which limited entry of certain terrestrial groups. Although present in the Late Miocene of South China, moles did not disperse to the Indian subcontinent, perhaps being blocked from westward expansion by major north-south waterways such as the Irrawaddy River.

Family Erinaceidae Fischer, 1814

The dominant hedgehogs of the Siwaliks, and in southeastern Asia today, are the gymnures, subfamily Galericinae named for

the fossil genus *Galerix*. Frost, Wozencraft, and Hoffman (1991) felt that Hylomyinae is the appropriate subfamily name, but McKenna and Bell (1997) did not agree. Gymnures lack the spiny dorsal hairs familiar in living European and North Asian *Erinaceus*. They eat insects, other invertebrates, and some plant matter. Gymnures are terrestrial and occur in humid forests (Nowak 1999). Siwalik hedgehogs are small in size, comparable to *Hylomys* (the smallest living genus). Using *Hylomys suillus* as a model, the common mid-Miocene *Galerix rutlandae* of similar dental size would have had a body mass of 60–70 g.

Galerix rutlandae Munthe and West, 1980 (Chinji Formation, holotype Howard-GSP 2371) was recognized in the Siwaliks over 40 years ago. In the Chinji Formation, common *G. rutlandae* accounts for over 5% of the small mammal specimens at several sites. Hedgehogs have a long history in Europe, where early records of *Galerix* show antiquity comparable to that in Pakistan (MN3 age, 19–20 Ma; Ziegler 1999), and there are older fossils from Turkey. Given its abrupt appearance in the Indian subcontinent, *Galerix* is considered an immigrant, dispersing presumably from the west.

Hedgehogs are unknown in the earliest Miocene of Pakistan. Their first local record in western Pakistan is at Z124, top of chron C6n, about 18.8 Ma. The Siwalik hedgehog story is largely that of the long-lived and geographically widespread galericines. The analysis of Zijlstra and Flynn (2015) revealed two successive species, *Galerix wesselsae* (holotype from Kamlial Locality Y642, 15.2 Ma) and the later *G. rutlandae* (Figs. 10.2, 10.3). *Galerix wesselsae* was somewhat larger than *G. rutlandae*, with estimated body mass of 80–90 g, but the older *G. wesselsae* was more derived dentally and apparently not ancestral to

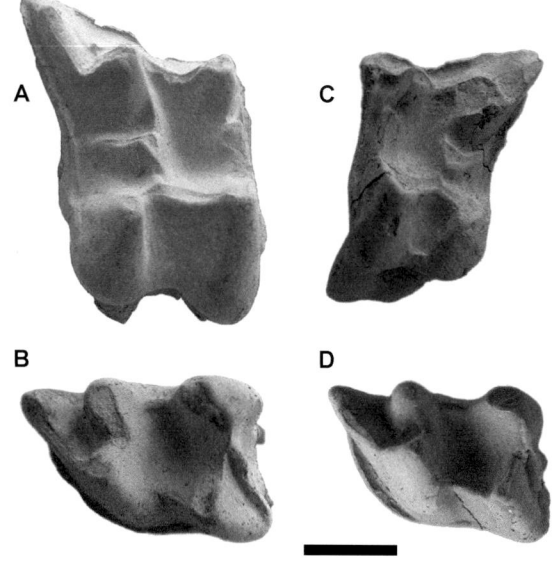

Figure 10.2. Upper and lower first molars of two hedgehogs (Erinaceidae). *Galerix wesselsae* A. right M1 YGSP 52202, holotype, from Locality Y642; B. left m1 YGSP 24542 from Locality Y733; *G. rutlandae*, C. left M1 YGSP 40010 from Locality Y718, D. left m1 YGSP 34925 from Locality Y718. Bar scale = 1 mm.

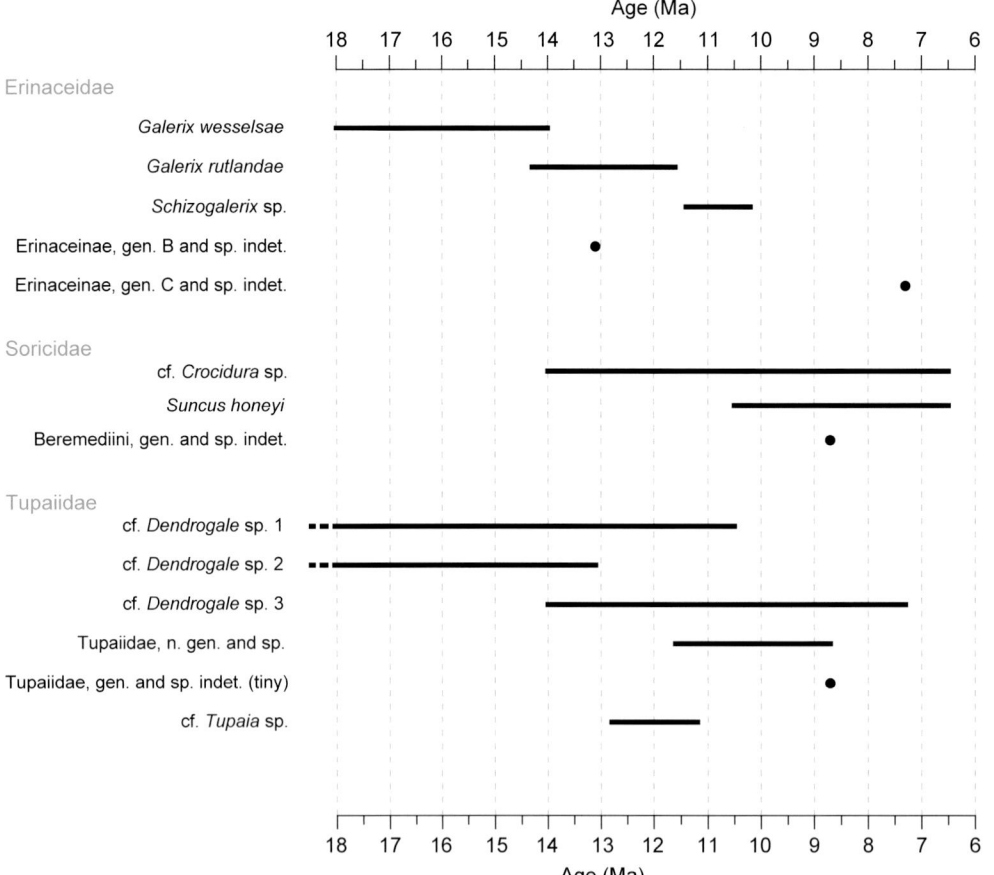

Figure 10.3. Temporal ranges of shrews, hedgehogs, and treeshrews on the Potwar Plateau, 18–6 Ma.

G. rutlandae, which implies separate immigrations. The pattern suggests replacement of *G. wesselsae* by *G. rutlandae,* but the two overlapped for over 100 k.y., perhaps even 14.3 to 14.0 Ma. *Galerix* was characteristic of Middle Miocene Siwalik faunas, with its last record at 11.6 Ma.

Galerix was succeeded by an indeterminate species of *Schizogalerix,* which survived in low numbers for over one m.y. Linear dimensions of *Schizogalerix* teeth are ~10% larger than those of *Galerix rutlandae,* suggesting a body mass estimate near 100 g. *Schizogalerix* has a long record in the Mediterranean region (Engesser 1980), which suggests that the Siwalik appearance is another probable immigration event from the west. This species was less common than Chinji hedgehogs and is last recorded at 10.2 Ma. A renewed overview by Crespo et al. (2021) suggests that Siwalik species previously considered *Galerix* in fact represent forms of *Schizogalerix.* Elsewhere in southern Asia, *Hylomys*-like hedgehogs occurred in the Middle Miocene of Thailand (Mein and Ginsburg 1997) and at the Late Miocene Lufeng locality in Yunnan, China (Storch and Qiu 1991). Galericines apparently disappeared from the Potwar before 10 Ma.

Three indeterminate teeth represent the distinctive spiny hedgehog subfamily Erinaceinae at scattered Potwar Miocene localities (see Table 10.1) dating to 18.7, 13.1, and 7.3 Ma. Each tooth probably represents a different species. The latest record of Potwar hedgehogs is the youngest of these, at Late Miocene

Locality Y931. In addition, Munthe and West (1980) note a large erinaceine (molar Howard-GSP 2380, estimated body mass about 200 g) in the late Middle Miocene Daud Khel fauna, which they named *Amphechinus kreuzae.* Furthermore, Wazir et al. (2022) describe a late Oligocene erinaceine in the high-elevation Ladakh fauna of northern India. These finds indicate under-appreciated hedgehog diversity, hinting at temporary range expansions and contractions of erinaceines rather than undetected continuous occupation.

Family Soricidae Fischer, 1814

Before galericines appeared in the Indian subcontinent record, shrews were evident in the earliest Miocene (Locality Z113, ~22.5 Ma). Unlike hedgehogs, shrews were uncommon throughout the Miocene Siwaliks. Their fossil record is limited, and they are unknown from the Oligocene of the Indian subcontinent. Generally uncommon in Potwar fossil assemblages, shrews became more frequent in the Late Miocene. At Localities Y182 and Y388 (9.2 and 8.7 Ma), soricid abundances reach 1% of the small mammal fauna. Most fossils represent white-tooth shrews, subfamily Crocidurinae. Due to poor shrew representation, alpha taxonomy is difficult.

The oldest shrew records are a few teeth from Localities Z113 and Z150, ~22 Ma, including a specimen of the diagnostic soricid hooked upper incisor (Fig. 10.4a). A small shrew

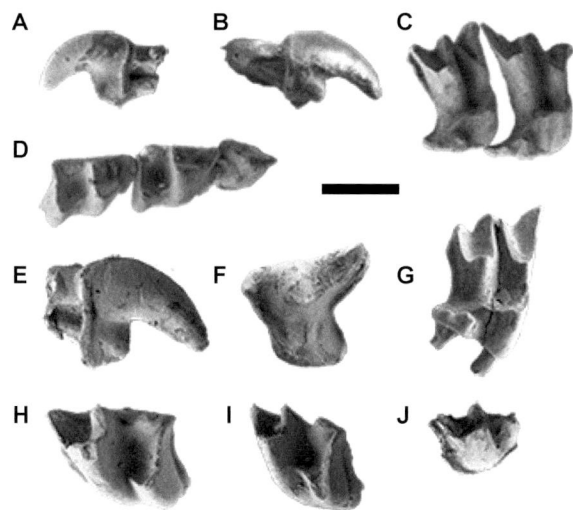

Figure 10.4. Potwar Soricidae. A. Subfamily indet., left upper incisor Z774 from Locality Z113. Crocidurine shrew cf. *Crocidura* sp. represented by B. right upper incisor YGSP 40527 from Locality Y182, C. right maxilla with M1-2 YGSP 40572 from Locality Y547, D. right dentary YGSP 40515 with p4m1-2 from Locality Y182, occlusal view. Large crocidurine *Suncus honeyi* represented by E. right upper incisor YGSP 54201 from Locality Y311, F. left P4 YGSP 40535 from Locality Y388, G. holotype right M2 YGSP 40539 from Locality Y388, H. left m1 YGSP 40566 from Locality D13, I. right m2 (reversed image) YGSP 40528 from Locality Y182, J. left m3 YGSP 54277 from Locality Y921. Bar scale = 1 mm.

(estimated about 8 g) occurs in the Early Miocene Zinda Pir area deposits of the upper Chitarwata Formation and the Vihowa Formation. Several younger shrew specimens recovered from Z124 and Z122 (18.8–18.7 Ma) probably represent two species (Flynn et al. 2020), one a stem crocidurine and another shrew that differs in that molar enamel shows red iron (goethite) pigmentation, clearly indicating a different subfamily, perhaps Crocidosoricinae Reumer, 1987.

For the Potwar Plateau, fossil assemblages of the Kamlial Formation (roughly 18–14 Ma) lack shrews entirely. The original shrew stock may have been extirpated, or its absence may reflect small sample sizes and few localities. In younger Potwar faunas, Crocidurinae, the subfamily of white-tooth shrews, constitutes a minor but persistent small mammal component beginning at 14 Ma (Locality Y733). Unpigmented teeth of that age are consistent morphologically with extant Crocidurinae and constitute the oldest definitive records of the subfamily in the world. From the entire superposed Chinji Formation (30 wash sites), 10 shrew teeth were recovered. Some Chinji sites are very rich in fossils so shrew rarity may be real and significant ecologically. Five Nagri Formation sites from 11.2 to 10.1 Ma produced 13 specimens. The fossils resemble living *Crocidura* of about 10 g mass. This record is succeeded by similar samples of small white-toothed shrew comprising 34 specimens (Fig. 10.4b, c, d) from sites of the Dhok Pathan Formation (9.4–6.5 Ma). The 14–6.5 Ma record is considered to represent a

single lineage because members cannot be differentiated morphologically. Given the limitations of the record, the parsimonious interpretation is that the Middle and Late Miocene localities record one small, uncommon crocidurine shrew lineage (cf. *Crocidura* spp.) spanning 14 to 6.5 Ma (Fig. 10.3).

Crocidura may persist in Late Neogene deposits of the Indian subcontinent. Raghavan and Pathmanathan (2007) announce a small Pliocene shrew from northern India (under the name *Chandisorex*) with an m1 length of ~1 mm. In lateral view the fossil shows a cingulum, which may indicate that it represents a bat.

A larger crocidurine makes a striking first appearance as an immigrant in the Late Miocene. *Suncus honeyi* Flynn et al., 2020 (80 g; see Table 10.1), first recorded in the Nagri Formation, 10.5–10.1 Ma (5 specimens), is smaller than the living house shrew *Suncus murinus*, which has a body mass near 100 g (Nowak 1999). *S. honeyi* is represented throughout the Dhok Pathan Formation by 40 specimens of variable size (Fig. 10.4e, f, g, h, i, j). Some Late Miocene *Suncus* teeth, particularly larger specimens high in the sequence, rival *S. murinus* in size. At 10.5 Ma, the Potwar Late Miocene *S. honeyi* calibrates the split of Asian *Suncus* from *Crocidura* at an age greater than the 6.8 Ma genetic-based estimate of Dubey et al. (2008).

Cheema, Raza, and Flynn (1997) and Nanda (2002) previously noted *Suncus* in the Pliocene of the northern part of the Indian subcontinent. Patnaik (2001) found *Suncus* in the small mammal fauna at about 2.5 Ma in northern India. At the Pabbi Hills Pleistocene Locality D24, the screening efforts of Jacobs (1978) recovered a smaller *Suncus* species.

Sahni and Khare (1976) published a large shrew from strata in the 9.0–8.5 Ma interval of the Haritalyangar area as *Siwalikosorex prasadi*, and Sahni and Kotlia (1985) named *Indosuncus bhatiai* for limited material from the Early Pleistocene of Kashmir. Both are similar in size to *Suncus murinus*, but were not differentiated from that species. These records attest to a large crocidurine (*Suncus*) in the Late Neogene of India, with later individuals rivaling in size the extant *S. murinus*.

The small shrew lineage cf. *Crocuidura* sp. remained a minor resident component of small mammal assemblages into the Late Miocene when it was joined by larger immigrant *Suncus*, a genus now widespread throughout the Indo-Malayan biogeographic province. Given its abrupt appearance in the Siwaliks, *Suncus honeyi* is considered an immigrant to the Indian subcontinent. The Potwar specimens are relevant to molecular studies because they date the appearance of the subfamily at 14 Ma and the *Suncus-Crocidura* split by 10.5 Ma. The geographic source, first of crocidurine shrews and later of *Suncus*, appears to be Palaearctic (Dubey et al. 2008) implying two phases of eastward range expansion via southwestern Asia.

Because soricids are generally uncommon in the fossil record, the true diversity of the group is likely underestimated. A distinctive non-crocidurine upper molar from Dhok Pathan Locality Y388 (YGSP 40538, with flared hypoconal flange) may represent a soricine of the Neogene age tribe Beremendiini. Strong red pigmentation of the metacone and paracone distinguish it from any crocidurine. YGSP 40538 resembles Chinese

Lunanosorex Jin and Kawamura, 1996, but other than the Si-walik record, the tribe is known only from Plio-Pleistocene localities. This medium-size shrew (body mass 50 g) in the Late Miocene Siwalik fauna is smaller than a Pleistocene beremen-diine from Kashmir reported by Kotlia (1991).

Order Chiroptera Blumenbach, 1779

Bat teeth have been recovered from more than 30 Potwar locali-ties, but always in low numbers, with very few sites yielding more than two teeth. Based on modern diversity, however, as well as their presence in faunas all over the globe, bats are ex-pected to have been a usual component of Miocene faunas of the subtropical Indian subcontinent. They would be expected to occur in every paleocommunity with appreciable diversity, but with considerable bias against their preservation and retrieval due to their volant lifestyle and special roosting habit. In the modern mammal fauna of Pakistan, Roberts (1977) lists 36 spe-cies of bats for the entire country. We would not expect a number like that in any setting sampled by the Potwar fossil record, but multiple lineages should be represented in the col-lection. The dental diversity of recovered Potwar fossils does indicate several genera; none are fruit bats (Megachiroptera); all belong to the predominantly insectivorous Microchiroptera.

Bats occur in the Siwalik record with the earliest small shrews. At least two species appear at the debut of the Mio-cene in the Zinda Pir area of western Pakistan at Locality Z113 (~22.5 Ma). Chiropterans occur in the younger Vihowa For-mation at Localities Z124 and Z122 (18.8, 18.7 Ma) and Kam-lial Formation sites of the Potwar Plateau. Chiropteran repre-sentation improves in the superjacent deposits of the Potwar Plateau. Most of the Siwalik evidence of bats (50 of 74 speci-mens) comes from the Chinji Formation. Modest diversity is indicated by remains of often broken upper and lower molars, and canines, which are relatively robust. The several species of Vespertilionidae (Fig. 10.5) were small with body mass under 10 g (see study by Moyers Arévalo et al. 2020). A few larger bats, probably Rhinolophidae, would have been over 10 g in mass.

Nagri Formation localities, stratigraphically above Chinji levels, yield six teeth (at least three species), but only four speci-mens come from Dhok Pathan sites, which are otherwise rich in rodent fossils. That bat diversity on the Potwar Plateau likely did not decline dramatically during the Miocene suggests in-stead a change in the Late Miocene mode of concentration of microvertebrates prior to fossilization. Taphonomic changes might include typical location of bone accumulations relative to streams or roosting sites (bluffs or banks of trees), the geo-morphology of riverways, or species of raptor. Vegetation cover could be important. Multiple factors could have changed the Late Miocene pattern of preservation.

Order Scandentia Wagner, 1855

The order Scandentia, treeshrews, contains the family Tu-paiidae Gray, 1825, and the monotypic family Ptilocercidae Lyon, 1913. These are small arboreal mammals that today are restricted to south and southeastern Asia and islands of the Indo-Malayan region. On mainland Asia today, *Tupaia*

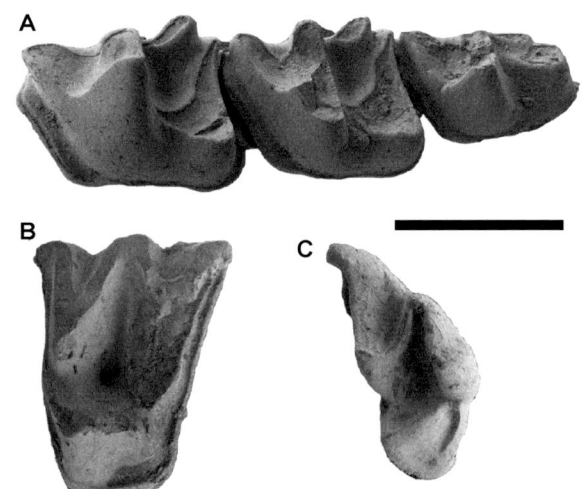

Figure 10.5. Chiroptera: A. right dentary with m1-3 YGSP 53957 from Locality Y809, B. left M1 YGSP 40600 from Locality Y76, C. left M3 YGSP 40586, from Locality Y76, B. and C. possibly same species. A. is reversed so that all appear with anterior to the left. Bar scale = 1 mm.

minor and *Ptilocercus lowii* exist in proximity on peninsular Malaysia, and *Tupaia belangeri* co-occurs with *Dendrogale mu-rina* in Southeast Asia; *Anathana ellioti* has a patchy distribu-tion in India.

Treeshrews are squirrel-like in locomotion, but differ in dentition, as evidenced by the lack of a large rodent dia-stema and retention of a near-full dentition that lacks only one upper incisor and the first premolar relative to the primi-tive placental formula (Butler 1980). Living species range from 50 to over 300 g and show no size dimorphism (Em-mons 2000). They are scansorial with scampering arboreal ability. Being small and unconstrained by climbing adapta-tions, they are capable of efficient terrestrial locomotion. Larger species spend more time on the ground than smaller species (Emmons 2000).

Emmons (2000) revealed the world of treeshrews (she pre-fers a single word form rather than "tree shrews") as observed from a Bornean perspective. The large island of Borneo in-cludes some relatively undisturbed forests of varied elevation that are presumably a close model for some Siwalik paleohabi-tats. Emmons (2000) studied an area of <1 km², home to *Ptilo-cercus lowii* and five species of *Tupaia*. As a model, the forests of Borneo suggest that treeshrew species diversity today in mainland southern Asia is reduced from that which its forests might be expected to support. The study area forest is varied, and its tupaiid species assort vertically, with small species rang-ing higher into the canopy. Emmons (2000) documented the diets of species, inclusive of variation over time due to masting events and annual patterns of fruit ripening. Many kinds of fruit are exploited, and invertebrates, which are hunted, sup-plement the diet. Emmons (2000) notes mean annual rainfall of 2.82 m for the period of 1986 to 1990. These observations on natural history provide clues to the ecology of extinct Siwa-lik treeshrews.

South Asian Tupaiid Fossils

As a group, order Scandentia is a characteristic if infrequent component of Siwalik assemblages. Siwalik treeshrews comprise several species of Tupaiidae; there are no Siwalik Ptilocercidae. Treeshrews occur throughout the Potwar composite stratigraphic section up to Locality Y931 at 7.3 Ma. Because they are uncommon, generalities about patterns of occurrence are made cautiously. Nonetheless, the heyday for Siwalik tupaiids appears to have been the Middle and early Late Miocene, when three species co-occur at some sites. Treeshrews occur in the Sindh Province Manchars and are found throughout the mid-Cenozoic western Pakistan Zinda Pir sequence, from the Oligocene Chitarwata Formation Locality Z108 to the earliest Miocene Locality Z113 at the base of the Vihowa Formation and the higher Z122, where, at about 18.7 Ma, there are at least two species distinguished by size. Early tupaiid fossils demonstrate the antiquity of the group in South Asia, and diverse tooth fragments attest to more than one contemporaneous species for the Oligocene pre-Siwalik fauna.

Tupaiidae of the Potwar Plateau are represented mainly by isolated teeth, some only fragments. A major difficulty for species level systematics is association of upper and lower canines, premolars, and molars collected from multiple localities. About 80 fossils are scattered among dozens of localities. Considering the multiple pitfalls in postulating phyletic affinity based on single teeth (see Selig et al. 2020), taxonomic decisions seem premature. In the following discussion, size differences are interpreted to indicate species diversity.

Kamlial Formation records are few; only four specimens of small body size are known, in keeping with the overall low fossil representation from those deposits. Throughout the Chinji Formation a medium size species (~40 specimens) co-occurs with both a smaller and a larger treeshrew based on scattered teeth. Most specimens indicate a species with dentition about the size of *Tupaia minor* or *Dendrogale melanura*, mass estimated near 60 g (comparison with data in Emmons 2000). The poorly represented small species (about 40 g) and an uncommon larger form of 80–90 g show that three possibly congeneric treeshrews resembling *Dendrogale* occurred in Chinji forests. In addition, a very large genus and another treeshrew resembling *Tupaia* join them high in the Chinji Formation, thereby increasing Late Miocene diversity (see Table 10.1).

The little cf. *Dendrogale* sp. is at the lower end of the size range of observed extant treeshrews. Sehgal et al. (2022) publish and name as *Sivatupaia ramnagarensis* a beautiful yet smaller lower tooth from late middle Miocene Ramnagar, India. It is very difficult to compare with Potwar teeth, being but a single lower molar (the authors suggest it represents m2). Cited features do not shed light on comparisons but suggest unappreciated diversity of insectivorans in the Miocene of the Indian subcontinent.

Thirteen Nagri Formation specimens include a jaw fragment from Y607 (9.7 Ma) and the anterior portion of a skull studied by Jacobs (1980) from Y450 (10.2 Ma). The latter is exceptionally large, more robust than *Tupaia tana*, and matching in size older teeth dating back to 11.6 Ma. Thus, we add to the Siwalik roster of three cf. *Dendrogale* treeshrews, a species resembling *Tupaia* and an early Late Miocene form that is larger than living Bornean tupaiids, near the size of Philippine *Urogale* (mass about 300 g according to Nowak [1999]). Cf. *Tupaia* sp. continues into the Nagri Formation (Locality Y791, 11.2 Ma, Fig. 10.6a), where upper molar YGSP 34948 presents a robust morphology of squared tooth with strong crests rarely seen among Siwalik specimens (discussed in the next section).

Only seven treeshrew teeth have been recorded from Late Miocene sites younger than 10 Ma. This includes the mid-size cf. *Dendrogale* sp. and a very small species late in the record (8.7 Ma). Again, poor representation limits systematic conclusions. After 8.7 Ma, a single specimen occurs at Locality Y931 (7.3 Ma). Given the wealth of Dhok Pathan small mammal fossils from some localities, treeshrew rarity may indicate true decline in abundance.

The 80 specimens under study (molars, premolars, anterior dentition, tooth fragments) include a Chinji Formation tupaiid molar reported by Cheema, Raza, and Rajpar (1996) and a Jalalpur tooth fragment from the Nagri Formation (Cheema et al. 2000). A partial tupaiid tooth was recovered by Munthe and West (1980) from the late Middle Miocene Daud Khel locality, and a Sehwan molar represents treeshrews in southern Pakistan (Locality HGSP 8227, Middle Miocene; Wessels 2009).

In northern India, exceptionally fine treeshrew specimens (Panjab University Anthropology [PUA] collection I-3, I-5, I-6) from Haritalyangar, *Palaeotupaia sivalicus* Chopra and Vasishat, 1979, date to over 9 Ma (paleomagnetic correlation of Pillans et al. 2005). The trivial name may have formal priority for some *Dendrogale*-like Potwar fossils of similar age, but the taxon is poorly differentiated at the generic level from other tupaiids

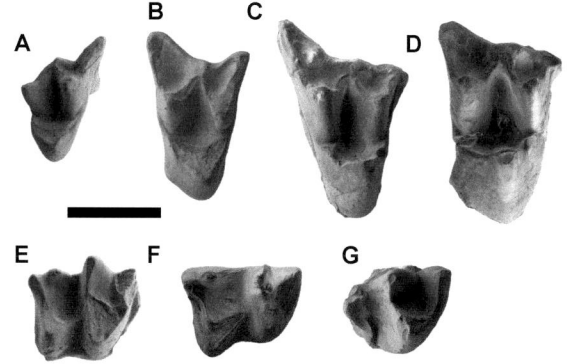

Figure 10.6. Scandentia isolated teeth, cf. *Dendrogale* sp. (A., B., C., E., F.) and cf. *Tupaia* sp. (D. G.). Cf. *Dendrogale* sp.: A. left M3, YGSP 49007 from Locality Y714, B. right M2, YGSP 48097 from Locality Y750, C. right M1, YGSP 33154 from Locality Y76, E. left m2 YGSP 33139 from Locality Y718, F. left m1 YGSP 8090 from Locality Y182 described by Jacobs (1980). Cf. *Tupaia* sp., D. left molar YGSP 34948 from Locality Y791, G. right p4 YGSP 34949 from Locality Y791. Specimens A., D., E., F. reversed so that all teeth appear as anterior to right. All at same 1 mm scale.

(Luckett and Jacobs 1980). To the east in the Middle Miocene of Thailand, Mein and Ginsburg (1997) recovered an upper molar with a small hypocone, which they named *Tupaia miocenica*. Potwar upper molars differ in lacking a distinct hypocone cusp.

Affinities and Biogeographic Significance of Potwar Tupaiids

The Siwalik fossil treeshrews display an array of features that are a mosaic of the dental structures seen in the living genera *Dendrogale*, *Tupaia*, and *Anathana*, as well as hypothetical stem genera. None of the fossil specimens presents characters seen in *Ptilocercus*, which is classified in its own family of Paleogene origin. The molecular study of Roberts et al. (2011) uses the ancient separation of *Ptilocercus* from Tupaiidae to infer Early Oligocene divergence of *Dendrogale* from other tupaiids. Divergence of *Anathana* and then radiation within the *Tupaia* cluster of species were mainly Middle and Late Miocene phenomena. Recalling that three or more contemporaneous fossil species occur in Middle and early Late Miocene Potwar localities, as predecessors of modern species, some Siwalik treeshrews might belong to living genera. They could represent early *Dendrogale*, *Anathana*, or *Tupaia*, or, alternatively, extinct genera.

Most of the smaller Siwalik fossil species present conservative features as in *Dendrogale*: gracile upper molars without distinct hypocones (Fig. 10.6a, b, c), rather low crests, and trigonids with relatively low but wide-angled metalophid and protolophid (Fig. 10.6e, f). Three sizes of *Dendrogale*-like tupaiids are apparent; body mass would have been about 40, 60, and 80 g by comparison with living species.

While these specimens resemble *Dendrogale* and may represent that genus, a few lower molars resemble *Tupaia* in their high trigonids with large cusps making a strongly acute angle. Unlike most Siwalik treeshrews, upper molars of *Tupaia* are squarer, and M1 shows a small hypocone; M2 has a less prominent hypocone, more of a bead of enamel in that position. Only a few Siwalik upper molars show a minute enamel bulge in the position of the hypocone. The morphological outlier molar from Locality Y791 mentioned above (YGSP 34948, Fig. 10.6d) dates to 11.2 Ma and is most like *Tupaia* in its tiny "hypocone" and squarer dimensions. Comparing molar size with species of living *Tupaia*, the estimated mass of this treeshrew was about 80 g. A few Potwar fossils resemble *Anathana* in P3 and P4, having protocones well separated from the buccal sectorial blade (meaning lingually extended, unlike *Tupaia*). Closer study may clarify phylogenetic relationships.

Central to systematics is the status of the species *Sivatupaia ramnagarensis* (Sehgal et al. 2022) and *Palaeotupaia sivalicus* (Chopra and Vasishat 1979), which their authors propose as distinct (despite absence of differential diagnoses). The Siwalik South Asian Tupaiidae were species-rich and differentiated during the Miocene into extant groups presently considered as distinct genera. How the multiple fossil species might be linked with *Dendrogale*, *Tupaia*, and *Anathana* remains uncertain.

South China provides fascinating complementary information on tupaiid evolution. While a purported Eocene tupaiid record from China (Tong 1988) is doubtful (Ni and Qiu 2012),

highly informative Oligocene fossils from Yunnan indicate a species there close to living *Ptilocercus* (Li and Ni 2016). For the Late Miocene of Yunnan, 16 specimens from Lufeng (~7 Ma) were described and named *Prodendrogale yunnanica* Qiu, 1986. This record was complemented by fossils from the somewhat older Yuanmou Basin, Yunnan, described as *Prodendrogale engesseri* Ni and Qiu, 2012. Three other Yuanmou specimens were distinguished as *Tupaia storchi* Ni and Qiu, 2012. As with the Siwalik fossils, most Late Miocene Yunnan specimens resemble *Dendrogale*, although a few tooth characters suggest *Tupaia*. *Prodendrogale* from Yunnan is gracile with thin crests (before wear), lower premolars have less prominent talonids with smaller cusps than Siwalik fossils, and P4 has a protocone less lingually placed. The Siwalik tupaiids differ in these features and do not seem to be assignable to Yunnan *Prodendrogale*.

Treeshrews are not only a characteristic element of the Miocene fossil record of the Indian subcontinent, but also a hallmark of the Oligocene to extant faunas of the Indo-Malayan biogeographic region. The records from the Middle Miocene Li Mae Long Basin, Thailand (Mein and Ginsburg 1997), and Late Miocene of Yunnan, China (Qiu 1986; Ni and Qiu 2012; Li and Ni 2016), demonstrate the occurrence of the group in the fossil record across southern Asia. Scandentia were endemic to this region at least since the Oligocene, possibly long before. It remains for us to tie the scanty fossil record to the treeshrew radiation that produced the modern species.

OVERVIEW OF SMALL MAMMALS OTHER THAN GLIRES

Shrews, hedgehogs, bats ,and treeshrews are components of Siwalik assemblages. Treeshrews and probably bats were present in South Asia during the Oligocene. Each group lends distinctive character to the South Asian Miocene biota, and hedgehogs became abundant in the Middle Miocene small mammal community. On balance, these small mammals are consistent with the moist, forested conditions inferred on other criteria for Middle Miocene Siwalik habitats. The Late Miocene disappearance of gymnure hedgehogs, coupled with decline of treeshrews but persistence of shrews, is consistent with increased climatic seasonality and vegetation change hypothesized for the Potwar Plateau after 10 Ma.

The family Erinaceidae (hedgehogs) has Late Oligocene first records in Europe (Hugueney and Maridet 2022) and in northern India (Wazir et al. 2022). In Pakistan, the gymnure hedgehog *Galerix wesselsae* is known from the Early Miocene and postdates somewhat the first records of *Galerix* in Turkey. By 18.7 Ma both *Galerix* and a spiny erinaceine hedgehog occur in Pakistan. The erinaceines have a spotty record in the Siwaliks, in contrast to the well-represented *Galerix*. The Early Miocene species *Galerix wesselsae* was replaced by the common *G. rutlandae*, which persisted until about 11.6 Ma. *Galerix* was succeeded by a larger gymnure, *Schizogalerix*, which spanned about one million years of the early Late Miocene. Erinaceids show repeated immigration of separate species into the Potwar region. *Galerix rutlandae* was not derived from *G. wesselsae*, but was more closely related to European species, and the affinity of *Schizogalerix* is with Anatolian predecessors. The source of hedgehog immigrants was likely from the west, possibly from Anatolia. Later the species *Galerix rutlandae* spread eastward and was recovered with the hedgehog *Lantanotherium* in the Middle Miocene of Mae Moh, Thailand (Cailleux et al. 2020).

Shrews were present from the earliest Miocene onward, apparently before erinaceids arrived. An early Miocene fossil from the Zinda Pir area indicates a small shrew with goethite (red iron) enamel pigment, which distinguishes it from the white-tooth lineage, cf. *Crocidura* sp. Shrews were never common, which is a cause of uncertainty for both species-level systematics and for calculating temporal ranges. The small Potwar shrews were morphologically conservative crocidurines, comparable to living *Crocidura*, and they show little size variation over their Miocene record of 14 to 6.5 Ma. The distinctive large crocidurine *Suncus* appeared in northern Indo-Pakistan by 10.5 Ma and was more common throughout the Late Miocene. An immigrant from the west, *Suncus* became widespread throughout the Indo-Malayan biogeographic province. Larger Late Miocene individuals of the *Suncus* lineage converged on the size of the extant South Asian house shrew, *S. murinus*. The Potwar shrew record includes the oldest evidence of the subfamily Crocidurinae (14 Ma) and provides at 10.5 Ma an early calibration date for the origin of the crown genus *Suncus*.

Chiroptera were probably diverse in Siwalik and pre-Siwalik faunas and, although underrepresented in the fossil record, include multiple species. Despite selective bias against preservation, some inferences are possible. First, no fruit bats are found—only diverse small microchiropterans. Second, in contrast to the rare Middle Miocene shrews, bats are present at many Chinji Formation localities. That they occur much less commonly in the Late Miocene Dhok Pathan Formation may suggest a change in sampled habitat or mode of concentration or preservation after 10 Ma.

Tupaiidae were characteristic of South Asian faunas during the Miocene, reflecting faunal connections throughout the Indo-Malayan biogeographic province. Siwalik Tupaiidae were diverse, and the several co-occurring species in the early Late Miocene may have been stem members of enduring lineages. Their diversity was greater than that in mainland Asia today, but is in line with that observed in the relatively undisturbed forests of Borneo (Emmons 2000). Potwar treeshrews appear to represent distinct lineages that include morphologies seen in the extant genera *Dendrogale*, *Anathana*, and *Tupaia*, which is consistent with the molecular-based inference that these lineages were differentiating at that time. Siwalik treeshrews were not common but remained elements of Potwar Plateau ecosystems through about 10 Ma. Thereafter, their occurrences declined, which may indicate less favorable Potwar habitats for tupaiids in the Late Miocene.

As with the tupaiids, some other small mammal elements are distributed widely in the Indo-Malayan biogeographic province and indicate a degree of faunal exchange throughout that biogeographic province. Among gymnures, a small hedgehog similar to *Hylomys* is recorded in the Siwaliks, Thailand, and Yunnan. A species of *Suncus* immigrating from the west by 10.5 Ma spread throughout South Asia and was succeeded by living *Suncus murinus*, the widespread house shrew of today. These South Asian and Southeast Asian distributions reflect the antiquity and persistence of several mammal lineages of the Indo-Malayan biogeographic province, which is duplicated among several rodent groups as seen in the following chapter.

ACKNOWLEDGMENTS

The Siwalik small mammal record has been built through the efforts of many. It originated thanks to field collectors' conscientious attention to details. The backbone of the biostratigraphy is the careful stratigraphic control of each fossil find, primarily under John Barry's direction, that allows precise temporal placement of most fossils. Many individuals contributed to this effort, while others retrieved important fossils as surface finds. In our group, small mammal fossil collections began to grow in earnest under the hands of Louis Jacobs, who early on with Everett Lindsay saw the potential for retrieval of fossils, mainly rodents, from the Siwaliks. These efforts began 50 years ago, but were launched into high production primarily by the late and lamented Will Downs, who organized the survey of many fossil prospects and then systematized fossil retrieval by heavy-liquid separation. Will and our longtime colleague, Dr. Yukimitsu Tomida, worked initially on many of the materials discussed here. Colleagues of the GSP and of the Pakistan Museum of Natural History, I.U. Cheema and A.R. Rajpar, improved the micromammal collection. Numerous other individuals participated in screening fossiliferous sediment, including the "Buckettes" Catherine Badgley, Michèle Morgan, and Alisa Winkler. In the laboratory, John Barry and Jelle Zijlstra worked to identify many of the incomprehensible micromammal fossil fragments. Wilma Wessels kindly allowed access to unstudied Manchars fossils form southern Pakistan. I thank Eric Sargis and Chris Gilbert for their advice. The Museum of Comparative Zoology at Harvard University was a resource for identification. Finally, the analysis presented here benefited from the efforts of the editors and suggestions of Louis Jacobs.

References

Andrews, P. 1990a. *Owls, Caves, and Fossils*. Chicago: University of Chicago Press.

Andrews, P. 1990b. Small mammal taphonomy. In *European Neogene Mammal Chronology*, edited by E. H. Lindsay, V. Fahlbusch, and P. Mein, 487–494. New York: Plenum Press.

Badgley, C., W. R. Downs, and L. J. Flynn. 1998. Taphonomy of small-mammal fossil assemblages from the Middle Miocene Chinji Formation, Siwalik Group, Pakistan. In *Advances in Vertebrate Paleontology and Geochronology*, edited by Y. Tomida, L. J. Flynn, and L. L. Jacobs, 145–166. Tokyo: National Science Museum Monograph 14.

Barry, J. C., A.K. Behrensmeyer, C. E. Badgley, L. J. Flynn, H. Peltonen, I. U. Cheema, D. Pilbeam, et al. 2013. The Neogene Siwaliks of the Potwar Plateau and other regions of Pakistan. In *Fossil Mammals of Asia: Neogene Biostratigraphy and Chronology*, edited by X. Wang, L. J.

Flynn, and M. Fortelius, 373–399. New York: Columbia University Press.

Blob, R. W., and A. R. Fiorillo. 1996. The significance of vertebrate microfossil size and shape distributions for faunal abundance reconstructions: A Late Cretaceous example. *Paleobiology* 22(3): 422–435.

Blumenbach, J. R. 1779. Handbuch der Naturgeschichte. Göttingen: Johann Christian Dietrich.

Butler, P. M. 1980. The tupaiid dentition. In *Comparative Biology and Evolutionary Relationships of Tree Shrews*, edited by W. Patrick Luckett, 171–204. New York: Plenum.

Cailleux, F., Y. Chaimanee, J.-J. Jaeger, and O. Chavasseau. 2020. New Erinaceidae (Eulipotyphla, Mammalia) from the middle Miocene of Mae Moh, northern Thailand. *Journal of Vertebrate Paleontology*, e1783277. https://www.tandfonline.com/doi/full/10.1080/02724 634.2020.1783277.

Cheema, I. U., L. J. Flynn, and A. R. Rajpar. 2003. Late Pliocene murid rodents from Lehri, Jhelum District, northern Pakistan. In *Advances in Vertebrate Paleontology "Hen to Panta": A Tribute to Constantin Radulescu and Petre Mihai Samson*, edited by A. Petculescu and E. Ştiucă, 85–92. Bucharest: Romanian Academy.

Cheema, I. U., S. M. Raza, and L. J. Flynn. 1997. Note on Pliocene small mammals from the Mirpur District, Azad Kashmir, Pakistan. *Géobios* 30(1): 115–119.

Cheema, I. U., S. M. Raza, L. J. Flynn, A. R. Rajpar, and Y. Tomida. 2000. Miocene small mammals from Jalalpur, Pakistan, and their biochronologic implications. *Bulletin of the National Science Museum* C26(1, 2): 57–77.

Cheema, I. U., S. M. Raza, and A. R. Rajpar. 1996. Late Astaracian small mammal fauna of the Miocene Siwaliks of Kallar Kahar-Dhok Tahlian, District Chakwal. *Pakistan Journal of Zoology* 28(3): 231–243.

Chopra, S. R. K., and R. N. Vasishat. 1979. Śivalik fossil tree shrew from Haritalyangar, India. *Nature* 281:214–215.

Colbert, E. H. 1933. Two new rodents from the lower Siwalik beds of India. *American Museum Novitates* 633:1–6.

Colbert, E. H. 1935. Siwalik mammals in the American Museum of Natural History. *Transactions American Philosophical Society* 27:1–401.

Crespo, V. D., P. Montoya, R. Marquina-Blasco, and F. J. Ruiz-Sánchez. 2021. *Parasorex ibericus* from Venta del Moro (late Miocene, Iberian Peninsula): A new light on the tribe Galericini. *Palaeontographica, Abteilung A* 320:65–86.

Crochet, J.-Y., P.-O. Antoine, M. Benammi, N. Iqbal, L. O. Marivaux, G. Métais, and J.-L. Welcomme. 2007. A herpetotheriid marsupial from the Oligocene of Bugti Hills, Balochistan, Pakistan. *Acta Palaeontologica Polonica* 52(3): 633–637.

De Bruijn, H., S. Taseer Hussain, and J. J. M. Leinders. 1981. Fossil rodents from the Murree Formation near Banda Daud Shah, Kohat, Pakistan. *Proceedings of the Koninklijke Nederlandse Akademie van Wetenschappen* B84(1): 71–99.

Denys, C., P. Andrews, S. Bailon, A. Rihane, J., Huchet, Y. Fernandz-Jalvo, and V. Laroulandie. 2018. Taphonomy of small predators multi-taxa accumulations: paleoecological implications. *Historical Biology* 30(6): 868–881. https://doi.org/10.1080/08912963.2017.1347647.

Dubey, S., N. Salamin, M. Ruedi, P. Barrière, M. Colyn, and P. Vogel. 2008. Biogeographic origin and radiation of the Old World crocidurine shrews (Mammalia: Soricidae) inferred from mitochondrial and nuclear genes. *Molecular Phylogenetics and Evolution* 48:953–963.

Emmons, L. H. 2000. *Tupai: A Field Study of Bornean Treeshrews*. Berkeley: University of California Press.

Engesser, B. 1980. Insectivora und Chiroptera (Mammalia) aus dem Neogen der Türkei. *Schweizerische Paläontologische Abhandlungen* 102:45–149.

Eronen, J. T. 2007. Locality coverage, metacommunities and chronofauna: Concepts that connect paleobiology to modern population biology. *Vertebrata PalAsiatica* 45(2): 137–144.

Fischer, G. 1814. Zoognosia. Volumen III. Quadrupeda reliqua, Ceti, Monotrymata. Moscow: Nicolai Sergeidis Vsevolozsky.

Flynn, L. J., J. C. Barry, M. E. Morgan, D. Pilbeam, L. L. Jacobs, and E. H. Lindsay. 1995. Neogene Siwalik mammalian lineages: Species longevities, rates of change, and modes of speciation. *Palaeogeography, Palaeoclimatology, Palaeoecology* 115:249–264.

Flynn, L. J., W. Downs, M. E. Morgan, J. C. Barry, and D. Pilbeam. 1998. High Miocene species richness in the Siwaliks of Pakistan. In *Advances in Vertebrate Paleontology and Geochronology*, edited by Y. Tomida, L.J. Flynn, and L. L. Jacobs, 167–180. Tokyo: National Science Museum Monograph 14.

Flynn, L. J., L. L. Jacobs, Y. Kimura, L. H. Taylor, and Y. Tomida. 2020. Siwalik fossil Soricidae: A calibration point for the molecular phylogeny of *Suncus*. *Paludicola* 12(4): 274–258.

Flynn, L. J., E. H. Lindsay, D. Pilbeam, S. Mahmood Raza, M. E. Morgan, J. C. Barry, C. E. Badgley, et al. 2013. The Siwaliks and Neogene evolutionary biology in South Asia. In *Fossil Mammals of Asia: Neogene Biostratigraphy and Chronology*, edited by X. Wang, L. J. Flynn, and M. Fortelius, 353–372. New York: Columbia University Press.

Flynn, L. J., M. E. Morgan, J. C. Barry, S. M. Raza, I. U. Cheema, and D. Pilbeam. 2023. Siwalik rodent assemblages for NOW. In *Evolution of Cenozoic Land Mammal Faunas and Ecosystems. 25 years of the NOW database of fossil mammals*, edited by I. Casanovas-Vilar, C. Janis, L. Van den Hoek Ostende, and J. Saarinen , 43–58. Cham: Springer.

Frost, D., W. C. Wozencraft, and R. S. Hoffman. 1991. Phylogenetic relationships of hedgehogs and gymnures (Mammalia: Insectivora: Erinaceidae). *Smithsonian Contributions to Zoology* 518:1–69.

Haeckel, E. 1866. Generelle Morphologie der Organismen. Allgemeine Grundzüge der Organischen Formen-Wissenschaft, Mechanisch Begründet durch die von Charles Darwin Refirmte Descendenz-Theorie. Band 1: Allgemeine Anatomie der Organismen. Berlin: Georg Reimer.

Hinton, M. A. C. 1933. Diagnoses of new genera and species of rodents from the Indian Tertiary deposits. *Annals and Magazine of Natural History* 10:620–622.

Hugueney, M., and O. Maridet. 2022. The Erinaceidae from the late Oligocene of Coderet-Bransat and Peublanc (Allier, France; MP 30): New data on *Amphechinus pomeli* (Schlosser, 1925–1926) and *Galerix minor* (Filhol, 1880). *Revue de Paléobiologie* 41(2): 267–279.

Jacobs, L. L. 1977. A new genus of murid rodent from the Miocene of Pakistan and comments on the origin of the Muridae. *PaleoBios* 25:1–12.

Jacobs, L. L. 1978. Fossil rodents (Rhizomyidae and Muridae) from Neogene Siwalik deposits, Pakistan. *Museum of Northern Arizona Press Bulletin Series* 52:1–103.

Jacobs, L. L. 1980. Siwalik fossil tree shrews. In *Comparative Biology and Evolutionary Relationships of Tree Shrews*, edited by W. Patrick Luckett, 205–216. New York: Plenum.

Jin, C.-Z., and Y. Kawamura. 1996. A new genus of shrew from the Pliocene of Yinan, Shandong Province, northern China. *Transactions and Proceedings of the Paleontological Society of Japan* new series 182(1): 478–483.

Khan, M. A., F. Y. Dar, L. J. Flynn, S. G. Abbas, M. A. Babar, M. Akhtar, H. Zahra, and H. Amanat. 2020. The Upper Siwaliks east of Jhelum, Pakistan: A new large Quaternary mammal assemblage from the Pabbi Hills and Reconnaissance near Bhimbar. *Quaternary International* 568:20–35.

Kotlia, B. S. 1991. Pliocene Soricidae (Insectivora, Mammalia) from Kashmir Intermontane Basin, northwestern Himalaya. *Journal of the Geological Society of India* 38(3): 253–275.

Kotlia, B. S., K. Nakayama, B. Phartiyal, S. Tanaka, M. S. Bhalla, T. Tokuoka, and R. N. Pande 1999. Lithology and magnetostratigraphy of the upper Siwaliks. *Memoirs of the Geological Society of India* 44:209–220.

Leibold, M. A, M. Holyoak, N. Mouquet, P. Amarasekare, J. M. Chase, M. F. Hoopes, R. D. Holt, et al. The metacommunity concept: A framework for multi-scale community ecology. *Ecology Letters* 7:601–613.

Li, Q. and X. Ni. 2016. An early Oligocene fossil demonstrates treeshrews are slowly evolving "living fossils." *Scientific Reports* 6 (January): 18627 https://doi.org/10.1038/srep18627.

Lindsay, E. H., L. J. Flynn, I. U. Cheema, J. C. Barry, K. F. Downing, A. R. Rajpar, and S. Mahmood Raza. 2005. Will Downs and the Zinda Pir Dome. *Palaeontologia Electronica* 8(1), 19A:1–19, 1MB. https://palaeo-electronica.org/2005_1/lindsay19/issue1_05.htm.

Luckett, W. P., and L. L. Jacobs. 1980. Proposed fossil tree shrew genus *Palaeotupaia*. *Nature* 28:104.

Lydekker, R. L. 1884. Rodents and new ruminants from the Siwaliks and synopsis of Mammalia. *Paleontographica Indica* 10(3): 134–185.

Marivaux, L., L. Bocat, Y. Chaimanee, J.-J. Jaeger, B. Marandat, P. Srisuk, P. Tafforeau, C. Yamee, and J.-L. Welcomme. 2006. Cynocephalid dermopterans from the Palaeogene of South Asia (Thailand, Myanmar, and Pakistan): Systematic, evolutionary and palaeobiogeographic implications. *Zoologica Scripta* 35(4): 395–420.

Marivaux, L., M. Vianey-Liaud, and J.-L. Welcomme. 1999. Première découverte de Cricetidae (Rodentia, Mammalia) oligocènes dans le synclinal sud de Gandoï (Bugti Hills, Balochistan, Pakistan). *Comptes Rendus de l'Académie des Sciences de la Terre et des Planètes* 329:839–844.

McKenna, M. C., and S. K. Bell. 1997. *Classification of Mammals above the Species Level*. New York: Columbia University Press.

Mein, P., and L. Ginsburg. 1997. Les mammifères du gisement miocène inférieur de Li Mae Long, Thaïlande: Systématique, biostratigraphie et paléoenvironnement. *Geodiversitas* 19(4): 783–844.

Moyers Arévalo, R. L., L. I. Amador, F. C. Almeida, and N. P. Giannini. 2020. Evolution of body mass in bats: Insights from a large supermatrix phylogeny. *Journal of Mammalian Evolution* 27(1): 123–138.

Munthe, J. 1980. Rodents of the Miocene Daud Khel local fauna, Mianwali District, Pakistan. Part II. Sciuridae, Gliridae, Ctenodactylidae and Rhizomyidae. *Milwaukee Public Museum, Contributions in Biology and Geology* 34:1–36.

Munthe, J., and R. M. West. 1980. Insectivora of the Miocene Daud Khel local fauna, Mianwali District, Pakistan. *Milwaukee Public Museum, Contributions in Biology and Geology* 38:1–17.

Nanda, A. C. 2002. Upper Siwalik mammalian faunas of India and associated events. *Journal of Asian Earth Sciences* 21:47–58.

Ni, X.-J., and Z.-D. Qiu. 2012. Tupaiinae tree shrews (Scandentia, Mammalia) from Yuanmou *Lufengpithecus* locality of Yunnan, China. *Swiss Journal of Palaeontology* 131:51–60.

Nowak, R. M. 1999. *Walker's Mammals of the World*, 6th ed. Baltimore: Johns Hopkins University Press.

Parmar, V., R. Norboo, and R. Magotra. 2023. First record of Erinaceidae and Talpidae from the Miocene Siwalik deposits of India. *Historical Biology* 35(2): 276–282. https://doi.org/10.1080/08912963.2022.2034806.

Patnaik, R. 1997. New murids and gerbillids (Rodentia, Mammalia) from Pliocene Siwalik sediments of India. *Palaeovertebrata* 26(1–4): 129–165.

Patnaik, R. 2001. Late Pliocene micromammals from Tatrot Formation (Upper Siwaliks) exposed near Village Saketi, Himachal Pradesh, India. *Palaeontographica* Abteilung A 261:55–81.

Patnaik, R., and A. C. Nanda. 2010. Early Pleistocene mammalian faunas of India and evidence of connections with other parts of the world. In *Out of Africa I: The First Hominin Colonization of Eurasia*, edited by

J. Fleagle, J. Shea, F. E. Grine, A. L. Baden, and R. Leakey, 129–143. Dordrecht: Springer.

Pillans, B., M. Williams, D. Cameron, R. Patnaik, J. Hogarth, A. Sahni, J. C. Sharma, F. Williams, and R. L. Bernor. 2005. Revised correlation of the Haritalyangar magnetostratigraphy, Indian Siwaliks: Implications for the age of the Miocene hominids *Indopithecus* and *Sivapithecus*, with a note on a new hominid tooth. *Journal of Human Evolution* 48:507–515.

Qiu, Z.-D. 1986. Fossil tupaiid from the hominoid locality of Lufeng, Yunnan. *Vertebrata PalAsiatica* 24(4): 308–319 (in Chinese with English summary).

Raghavan, P., and G. Pathmanathan. 2007. Discovery of an Upper Pliocene insectivore *Chandisorex minuta* from the basal Pinjor Formation of the sub-Himalayas, India. *Indian Journal of Physical Anthropology and Human Genetics* 26:7–16.

Reumer, J. W. F. 1987. Redefinition of the Soricidae and the Heterosoricidae (Insectivora, Mammalia), with the description of the Crocidosoricinae, a new subfamily of Soricidcae. *Revue de Paléobiologie* 6(2): 189–192.

Rey-Rodríguez, I., E. Stoetzel, J. M. López-García, and C. Denys. 2019. Implications of modern barn owls pellets analysis for archaeological studies in the Middle East. *Journal of Archaeological Science* 111:105029.

Roberts, T. E., H. C. Lanier, E. J. Sargis, and L. E. Olson. 2011. Molecular phylogeny of treeshrews (Mammalia: Scandentia) and the timescale of diversification in Southeast Asia. *Molecular Phylogenetics and Evolution* 60:358–372.

Roberts, T. J. 1977. *The Mammals of Pakistan*. London: Ernest Benn Limited.

Rogers, R. R., M. T. Carrano, K. A. Curry Rogers, M. Perez, and A. K. Regan. 2017. Isotaphonomy in concept and practice: An exploration of vertebrate microfossil bonebeds in the Upper Cretaceous (Campanian) Judith River Formation, north-central Montana. *Paleobiology* 43(2): 248–273.

Rosenzweig, M. L. 1995. *Species Diversity in Space and Time*. Cambridge: Cambridge University Press.

Sahni, A., and S. K. Khare. 1976. Siwalik Insectivora. *Journal of the Geological Society of India* 17:114–116.

Sahni, A., and B. S. Kotlia. 1985. Karewa microvertebrates: Biostratigraphical and palaeoecological implications. In *Climate and Geology of Kashmir, the Last 4 Million Years: Proceedings of the International Workshop on the Late Cenozoic Palaeoclimatic Changes in Kashmir and Central Asia*, edited by D. P. Agrawal, S. Kusumgar, and R. V. Krishnamurthy, 29–43. New Delhi: Today & Tomorrow's Printers and Publishers.

Sehgal, R. K., A. P. Singh, C. C. Gilbert, B. A. Patel, C. J. Campisano, K. R. Selig, R. Patnaik, and N. P. Singh. 2022 A new genus of treeshrew and other micromammals from the middle Miocene hominoid locality of Ramnagar, Udhampur District, Jammu and Kashmir, India. *Journal of Paleontology* 96(6): 1318–1335. https://doi.org/10.1017/jpa.2022.41.

Selig, K. R., E. J. Sargis, S. G. B. Chester, and M. T. Silcox. 2020. Using three-dimensional geometric morphometric and dental topographic analyses to infer the systematics and paleoecology of fossil treeshrews (Mammalia, Scandentia). *Journal of Paleontology* 94(6): 1202–1212. https://doi.org/10.1017/jpa.2020.36.

Storch, G., and Z.-D. Qiu. 1991. Insectivores (Mammalia: Erinaceidae, Soricidae, Talpidae) from the Lufeng hominoid locality, late Miocene of China. *Géobios* 24(5): 601–621.

Terry, R. C. 2010. On raptors and rodents: Testing the ecological fidelity and spatiotemporal resolution of cave death assemblages. *Paleobiology* 36(1): 137–160.

Tong, Y. 1988. Fossil tree shrews from the Eocene Hetaoyuan Formation of Xichuan, Henan. *Vertebrata PalAsiatica* 26(3): 214–220.

Wagner, J. A. 1855. Die Säugethiere in Abbildungen Nach Der Natur. Weiger, Leipzig.

Wazir, W. A., F. Cailleux, R. K. Seghal, R. Patnaik, N. Kumar, and L. W. van den Hoek Ostende. 2022. First record of insectivore from the Late Oligocene, Kargil Formation (Ladakh Molasse Group), Ladakh Himalayas. *Journal of Asian Earth Sciences* 8(December): 100105. https://doi.org/10.1016/j.jaesx.2022.100105.

Wessels, W. 1996. Myocricetodontinae from the Miocene of Pakistan. *Proceedings of the Koninklijke Nederlandse Akademie van Wetenschappen* 99(3–4): 253–312.

Wessels, W. 2009. Miocene rodent evolution and migration: Muroidea from Pakistan, Turkey, and northern Africa. *Geologica Ultraiectina* 307:1–290.

Wessels, W., H. Bruijn, S. T. Hussain, and J. J. M. Leinders. 1982. Fossil rodents from the Chinji Formation, Banda Daud Shah, Kohat, Pakistan. *Proceedings of the Koninklijke Nederlandse Akademie van Wetenschappen* B85(3): 337–364.

Wood, A. E. 1937. Fossil rodents from the Siwalik beds of India. *American Journal of Science* 36:64–76.

Wood, H. E. 1949. Oligocene faunas, facies, and formations. *The Geologisal Society of America* Memoir 39:83–90.

Ziegler, R. 1999. Order Insectivora. In *The Miocene Land Mammals of Europe*, edited by G.E. Rössner and K. Heissig, 53–74. Munich: Verlag.

Zijlstra, J., and Flynn, L. J. 2015. Hedgehogs (Erinaceidae, Lipotyphla) from the Miocene of Pakistan, with the description of a new species of *Galerix*. *Palaeobiodiversity and Palaeoenvironments* 95(3): 477–495.

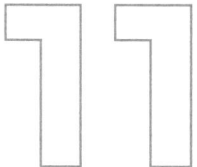

SIWALIK LAGOMORPHS AND RODENTS

LAWRENCE J. FLYNN, LOUIS L. JACOBS, YURI KIMURA, AND
EVERETT H. LINDSAY

INTRODUCTION

The diminutive rodents were the dominant members of Neogene mammalian paleofaunas of the Indian subcontinent—
dominant in both abundance and species richness. As in regional
faunas today, Miocene rodent species were diverse, showing
striking numbers for some clades. True abundances in paleofaunas are difficult to determine, because they are filtered by
taphonomic factors and by predator preference. Many Potwar
fossil samples are, however, large enough (those of at least 200
specimens; Flynn et al. 1998) to capture at least one individual
of most species, so paleodiversity can be estimated. Larger
samples also allow us to judge relative abundances. Well-dated
Siwalik paleofaunas reflect changes through time in the diversity and major characteristics, including relative abundances,
of Miocene micromammal assemblages of South Asia. As we
shall see, screened samples from rich localities can shed light
on the changing microfauna of the past.

The stratigraphically continuous Potwar Siwaliks document
12 million years (~18–6 Ma) of small mammal history. Older
and younger strata (see Fig. 10.1, this volume) from adjacent
areas can be related to this sequence by magnetostratigraphy,
yielding a paleobiological history that spans the Neogene. The
Potwar Miocene interval records nine rodent families, several
with high diversity at the genus and species levels. If Oligocene–
Early Miocene data from noncontiguous deposits of the Sindh,
Bugti, and Zinda Pir areas and the Murree Formation at Banda
Daud Shah are added, then the total rodent diversity over an
additional 10 million years increases by one or two families.

The fossil record of the Potwar Plateau yields key details
of the histories of many rodent groups. Its power of resolution flows from the many superposed, productive fossil horizons, which reflect intense effort over many field seasons to
extract and document the paleobiological evidence needed
to interpret and understand major small mammal events in
the Neogene Siwaliks.

Among small mammals, the record of rodents is particularly dense, but as a record comprised primarily of isolated
teeth, its interpretation is limited by the morphological characters found in the dentition. Biostratigraphic richness, system-

atic studies, quantitative morphology of rodent taxa through
time, molecular contributions to the phylogeny of extant taxa,
and stable isotope analyses of fossil tooth enamel (Kimura et al.
2013)—together, these document the great significance of the
Siwalik record for chronology, biogeography, paleoecology, and
evolution. It is important to recognize that stratigraphic ranges
are built on a strong foundation of successive samples of highly
comparable material: virtually all Siwalik rodent fossils are
teeth. Biostratigraphy and comparative studies of successive
samples is developed directly on the fossils in hand. Ranges may
be more difficult to establish for macromammals, whose records may consist of diverse osteological and dental fossils that
may at times limit direct comparison.

The Potwar Plateau, currently at about 33° latitude north,
was located south of 30° paleolatitude during the Miocene
(Tauxe and Opdyke 1982; Scotese 1997), and its paleofaunas
including rodent assemblages were subtropical in character,
particularly through 11 Ma. Late Miocene shifts to increased
seasonality of precipitation and more open habitats affected
small mammal faunas, as documented below.

The biogeographical role played by small mammals in the
Siwaliks is revealed through global comparisons. Families such
as dipodids (jerboas) and castorids (beavers) found elsewhere
throughout Eurasia are not known to have ever entered the Indian subcontinent, but that theater of evolution preserves
chapters in the history of other groups such as sciurids (squirrels) and ctenodactylids (gundis). The premier example of a biological success story among Siwalik small mammals is the
origin and progressive domination of the rodent fauna by murine mice, which evolved and diversified in the Indian subcontinent over millions of years prior to their widespread dispersal beyond South Asia in the early Late Miocene. South Asia,
as seen through the Siwalik lens of the Potwar Plateau, was a
center of evolution for burrowing rhizomyine rodents, whose
history as discussed below plausibly responded to regional
tectonic events and culminated in full fossorial exploitation of
underground food resources. Rodent families were either
barred from the Indian subcontinent or thrived there.

The physical geography of the Indian subcontinent contributes to its isolation and controls the provincial distribution of its mammals. The Indo-Pakistan area is limited on the north by high mountain ranges towering over the vast plateau of Tibet. Major forests and north-south rivers are barriers to the east. Mountainous terrain isolates the arid area to the west. Together, these physical barriers constitute a filter on land mammal distribution, and the small mammals discussed in Chapters 10 and 11 of this volume illustrate the effects on the biogeography of these Siwalik groups.

The Indian subcontinent today is part of the Indo-Malay biogeographic province, and its fossils indicate that the roots of this biogeographic region extend over 20 million years into the past. The paleofaunas of the Potwar Plateau represent a subregion, the northwestern portion of the Indo-Malay biogeographic province. The biogeographic pattern overprinting the endemism of the Siwalik subtropical rodent fauna reflects both Indo-Malay elements to the east across southern Asia and interchange with the west. Ancient Siwalik clades of hamsters and gerbils share Asiatic and African roots beyond the Potwar subregion. Various Siwalik tree squirrels are predominantly Southeast Asian, but other rodents, such as thryonomyid "cane rats," are strictly African today. Distinctive immigrants to the Indian subcontinent, such as the dormouse and porcupine, appear without precedent in the Miocene, dispersing presumably from southwest Asia. The evolutionary histories of South Asian rodents, especially the prominent mouse and bamboo rat groups are well-represented in the Siwaliks.

Muroid rodents have a long Miocene history on the Indian subcontinent, with a pattern of high diversity. Multiple congeneric hamster species co-occur within the habitat, and stem gerbils became increasingly diverse. Early and Middle Miocene myocricetodontin gerbils evolved dental structures heralding the radiation of murines, the "true mice." Significantly, both gerbils and hamsters declined precipitously after 12.5 Ma as the true mice rose in prominence.

The success of true mice relative to hamsters and gerbils suggests an example of biotic interaction in deep time (see Fraser et al. 2021): replacement due to competition. Late Miocene mice subsequently dominated the small mammal fauna and established the roots of modern Indian rodent communities. As mice became abundant, two temporally very long, dominant murine lineages differentiated. The Potwar record captures graded details of murine divergence because it is well-represented and well-dated, and it offers minimum age estimates for several crown group (tribe level) divergences. Murine dental structures and stable isotopes of molar enamel indicate partitioning of food sources (Kimura et al. 2013; Kimura, Jacobs, and Flynn 2013) as lineages diverged morphologically.

Following a different evolutionary pattern, the burrowing rhizomyine rodents became species rich and at times quite numerous, with species widely distributed to the east and west. Their history can be tied to geobiological events of mountain-building and uplift of the Tibetan highlands to the north, which drove waxing and waning of monsoonal weather

(López-Antoñanzas et al. 2015). Crown bamboo rats emerged after 10 Ma, exploiting underground food sources by becoming fully fossorial. Rhizomyines declined during Pliocene time, in parallel with increasing seasonal aridity.

Rodents do not stand alone in the record of Siwalik microfaunas. Small-bodied insectivorans, bats, and treeshrews are discussed in Chapter 10 of this volume. Herein, we focus on Rodentia and companion Lagomorpha as sister orders of the larger Cohort Glires. Glires are one of the best-supported, best-defined superordinal clades among Mammalia (Meng, Hu, and Li 2003; Asher et al. 2005; Kriegs et al. 2007). Sister taxon to the rodents, lagomorphs include living pikas (Ochotonidae) and rabbits plus hares (Leporidae). The limited South Asian Neogene fossils of both living lagomorph families are reviewed here, followed by accounts of the diverse Siwalik rodent families.

ORDER LAGOMORPHA BRANDT, 1855

Order Lagomorpha comprises pikas (family Ochotonidae) and rabbits and hares (family Leporidae). Beyond the Indian subcontinent, the middle Cenozoic of China and Central Asia contains a rich Paleogene lagomorph record. Although most of that record is absent from the Indian subcontinent, Rose et al. (2008) report lagomorphs in the Eocene Vastan coal fields of India. While younger lagomorphs failed to penetrate the geographic barriers around Indo-Pakistan until Late Miocene time, there is rare evidence of an Early Miocene ochotonid from Sindh and from the Potwar Plateau. The partial isolation of the Indian subcontinent from the rest of Asia contributed to the long absence of Leporidae from Siwalik assemblages.

Family Ochotonidae Thomas, 1897
A relatively primitive, but indeterminate ochotonid tooth was recovered from Potwar Plateau Locality Y747 and dated to 17.9 Ma. Anterior and posterior moieties of the upper molariform (YGSP 48048) are nearly symmetrical (Flynn et al. 1997) and are therefore derived with respect to basal ochotonids such as *Sinolagomys*. This tooth is perhaps best designated as a small species of *Alloptox*. That surprising occurrence is bolstered by an ochotonid incisor from Sindh of about the same age (Wessels 2009). Based on tooth size comparisons, this fossil pika was near the size of larger species of living *Ochotona*, probably about 250 g in body mass.

Family Leporidae Fischer, 1814
Crown Leporinae appeared in the Potwar area in the Late Miocene at several localities in the upper part of the Dhok Pathan Formation. The first occurrence at 7.4 Ma (Y457) was succeeded closely in time by others; once rabbits appeared, they became a frequently encountered, if minor, part of the small mammal fauna. No small mammal localities are known immediately preceding 7.4 Ma, but absence of the family at well-represented sites dating to 7.8 Ma and older constrains immigration of family Leporidae between 7.4 and 7.8 Ma. Dispersion into the Indian subcontinent followed the invasion of

modern leporids (crown group Leporinae) into northern Asia from North America by 8 Ma (Flynn et al. 2014b).

Winkler, Flynn, and Tomida (2011) based the Siwalik species *Alilepus elongatus* on a right p3 (YGSP 50951, Fig. 11.1) from Y921, "Bunny Hill," with age estimated as 7.34 Ma. This species is distinctive in its enlarged trigonid on p3 and crenellated enamel on cheek teeth, unlike most other *Alilepus*. Sen (2020) observed that these features may distinguish this rabbit as a new genus. The Potwar *Alilepus* survived until at least 6.5 Ma (Locality D13). About the size of extant *Oryctolagus*, its mass was about 2 kg. (Mass estimates for this and other taxa with close living relatives are based on dental comparisons, and reference to Nowak [1999].)

Leporid species of similar age in China differ from those found in the Potwar. In the Late Miocene of Yunnan Province, South China, the very different *Nesolagus longisinuosus* is known from many specimens from Lufeng, ~7 Ma (Qiu and Han 1986), and from the younger Zhaotong Basin in Yunnan (Flynn et al. 2019). Modern *Nesolagus*, the extant striped rabbit, and apparently fossil representatives in China prefer wet habitat. The Late Miocene Shanxi Province *Alilepus annectens* with simple, unwrinkled enamel, may have preferred drier habitat. The small *A. lii* Jin, 2004, from Anhui Province, shows enamel crenellation, as does *A. elongatus*.

With no possible ancestors recorded in South Asia, *Alilepus elongatus* appears to be an immigrant to the Siwaliks and therefore seems to represent a case of dispersal of small mammals into the Indian subcontinent. *Alilepus lii* and other *Alilepus* from North China, are 3,500 km distant from the Potwar. An undetermined leporine in the Late Miocene of Afghanistan (Heintz

and Brunet 1982), older than 7 Ma, is geographically closer but separated from the Siwaliks by mountainous terrain. Does the affiliation of *A. elongatus* lie with a species from distant North China or closer Afghanistan? The Potwar appearance of *Alilepus* could represent a rare case for the Siwaliks of species dispersion from Afghanistan across mountain passes into the subcontinent, rather than from the east or west at lower latitude. Winkler, Flynn, and Tomida (2011) proposed that by ~6 Ma, a species of *Alilepus* with crenellated enamel dispersed westward from the Indian subcontinent to East Africa.

Leporidae remain part of the fossil record through the Pliocene and Pleistocene Upper Siwaliks. Indeterminate leporid teeth occur at Tatrot Locality D8 (~3.5 Ma) in the Potwar Plateau. For fossils with complexly folded enamel from Kanthro, Late Pliocene of northern India, Patnaik (2001) defined the rabbit *Pliosiwalagus whitei* (type specimen VPL/RP-SM-89). Patnaik (2002) transferred the fossil species *Caprolagus sivalensis* Major, 1899 (BM 16529, unknown locality) and an Early Pliocene p3 from Moginand, India, to *Pliosiwalagus*; he postulated sister relationship of *Pliosiwalagus* with derived extant *Caprolagus*. The early Pleistocene Locality D24 in the Pabbi Hills, Pakistan, yielded fragments of similarly complex teeth (Winkler, Flynn, and Tomida 2011).

The modern mammalian faunas of the Indian subcontinent include Holarctic *Lepus* and geographically restricted *Caprolagus* with a complex enamel pattern on p3. *Lepus* is a widespread immigrant, but the origin of *Caprolagus* may involve Siwalik Neogene rabbits.

ORDER RODENTIA BOWDICH, 1821

Globally, rodents have been diverse for 50 million years. Clearly related to Lagomorpha in Cohort Glires, the orders share derived features, including enlarged, paired upper and lower incisors, which are ever-growing with enamel restricted mesially, and are followed by a large diastema (other incisors and canine and anterior premolars absent). Modern rodent diversity is extraordinary, accounting for nearly half of living mammal species. Siwalik faunal assemblages capture an appropriate scale of diversity for the region. Well-represented single fossil assemblages from the Middle Miocene Chinji Formation record ~30 rodent species (Flynn et al. 1998). While even rich fossil assemblages may not capture every species in a paleocommunity, the species richness found at many Chinji Formation localities is great—and greater than that observed today in Pakistan. Excluding species restricted to extreme habitats of high elevation or the desert southwest, Roberts (1977) records 28 species in total for the country.

Tóth, Lyons, and Behrensmeyer (2014) find greater diversity in historical data for modern communities in large equatorial game parks of East Africa, but richness of any faunas, living or past, should be evaluated on an equivalent basis of paleolatitude and area sampled (Rosenzweig 1995), which in this case would be 30° latitude north and ~50 km². The area estimate is derived on the hypothesis that single sites are concentrated by avian predators such as owls (Denys et al. 2018;

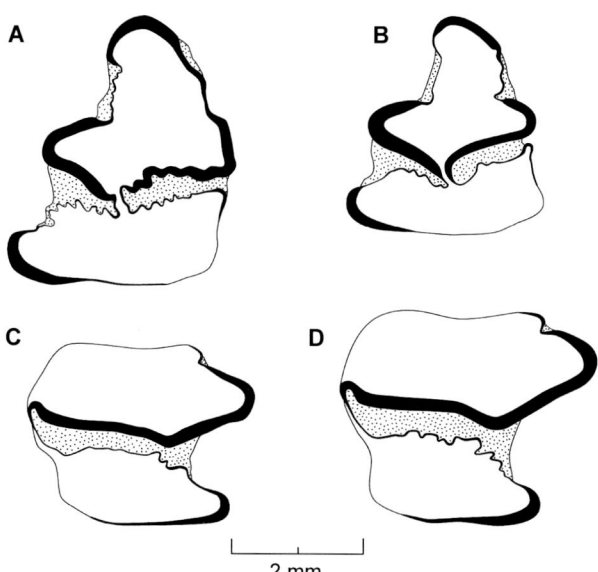

Figure 11.1. The leporid *Alilepus elongatus*. A. Right p3 (reversed so it appears as a left) YGSP 50951 from Locality Y921, B. left p3 (YGSP 34504 from Locality D13), C. right m1 YGSP 34508 from Locality D13, D. right m1 (YGSP 50952 from Locality Y921); after Winkler et al. (2011). Scale = 2 mm.

	Holotype	Age Range	Mass (g)
Lagomorpha			
Family Ochotonidae			
Alloptox sp.		(~18 Ma)	250
Family Leporidae			
Alilepus elongatus Winkler et al., 2011	YGSP 50951, right p3	(7.4–6.5 Ma)	2,000
Rodentia			
Family Diatomyidae (Indo–Malay)			
Marymus dalanae Flynn, 2009	Z 2353, left dp4	(~22.5–22 Ma)	60
cf. *Diatomys shantungensis*		(18.7–18.6 Ma)	>100
Diatomys chitaparwalensis Flynn, 2006	YGSP 53256, left M2	(17.9 Ma)	80
Willmus maximus Flynn and Morgan, 2005	YGSP 33100, right dp4	(11.2 Ma)	120
Family Ctenodactylidae (gundis)			
Subfamily Ctenodactylinae			
Sayimys flynni (Baskin, 1996)	Z 295, left m3	(~22.5 Ma)	40
Sayimys minor De Bruijn et al., 1981	Howard-GSP 116/313, right m1	(21–16.0 Ma)	60
Sayimys intermedius (Sen and Thomas, 1979)	AJ 545, left mandible, p4-m2	(18.7–15.1 Ma)	70
Sayimys baskini (López-Antoñanzas and Sen, 2003)	YGSP 48125, right P4	(17.9–16.0 Ma)	60
Sayimys cf. *S. giganteus*		(16.8–16.5 Ma)	120
Sayimys sivalensis (Hinton, 1933)	GSI D284, left mandible, m2–3	(15.2–12.7 Ma)	80
Sayimys chinjiensis Baskin, 1996	YGSP 45186, partial left mandible, p4m1	(12.4–10.1 Ma)	90
Sayimys sp. B		(12.4–11.2 Ma)	80
Sayimys perplexus Wood, 1937	YPM 13800 left mandible, p4m1–3	(7.4 Ma)	<100
Family Thryonomyidae cane rats (African)			
Kochalia geespei De Bruijn and Hussain, 1985	Manchar left p4	(16.0–13.4 Ma)	<100
Kochalia n. sp.		(17.9–16.8 Ma)	80
Paraulacodus indicus Hinton, 1933	GSI D283, right maxilla, P4M1–2	(13.1–12.7 Ma)	500
Family Hystricidae (porcupines)			
Atherurus complicatus (Colbert, 1933)	AMNH 19626 L and R dentary	Middle Miocene	3,000
Hystrix primigenia Wagner, 1848		(8.0–3.3 Ma)	20–25 kg
Hystrix indica Kerr, 1792		(Pleistocene)	15–20 kg
Family Gliridae (dormice)			
Subfamily Myomiminae			
Myomimus sumbalenwalicus Munthe, 1980	Howard-GSP 1986, right M1	(13.8–12.3 Ma)	30
Myomimus n. sp. A		(11.6–10.1 Ma)	35
Myomimus n. sp. B		(9.4–8.0 Ma)	40
Family Sciuridae (squirrels)			
Subfamily Pteromyinae (gliding)			
Aliveria sp.		(16.8–16 Ma)	160
Pteromyinae, gen. and sp. indet.		(17.9–15.2 Ma)	
Subfamily Ratufinae			
Ratufini n. gen. and sp.		(21 Ma)	<100
cf. *Ratufa* n. sp.		(16.8; 14.3–13.6)	350
Ratufa sylva Flynn, 2003	YGSP 17260, left m1	(10.5–9.4 Ma)	350
Subfamily Callosciurinae			
Tamiops sp., cf. *T. asiaticus*		(18.6 Ma)	160
Tamiops large sp.		(14.3–8.7 Ma)	75
Tamiops small sp. (isolated early occurrence)		(16.8; 13.7–10.1)	55
Tamiops giant sp.		(8.0 Ma)	>200
Callosciurus sp. (+isolated late occurrence)		(14.3–13.2, 9.2)	185
Dremomys sp.		(14.3–7.4 Ma)	135
Subfamily Sciurinae			
Oriensciurus n. sp.		(18.6–17.9 Ma)	370
Subfamily Marmotinae (ground squirrels)			
Large gen. and sp. indet.		(~22 Ma)	450
Tribe Xerini			
Heteroxerus sp.		(14.3–13.2 Ma)	300
Atlantoxerus sp.		(14.3–13.2 Ma)	190

Table 11.1. continued

	Holotype	Age Range	Mass (g)
Tribe Sciurotamiini			
Sciurotamias Kamlial sp. (+ isolated late record)		(16.0–15.2, 12.8)	170
Sciurotamias sp. A (small)		(14.3–11.4 Ma)	120
Sciurotamias sp. B (large)		(11.2–7.2 Ma)	170
Tribe Tamiini			
Tamias "pre-urialis"		(14.3–13.6 Ma)	90
Tamias urialis (Munthe, 1980)	Howard-GSP 2013, left dentary m1–2	(13.2–11.2 Ma)	65–70
Tamias sp., cf. *T. gilaharee* Patnaik et al., 2022		(10.5–8.0 Ma)	170
Family Anomaluridae (scaly tailed gliders)			
cf. *Downsimys* sp.		(Oligocene to 18.6 Ma)	10
Anomaluridae, n. gen. and sp. indet.		(16.8)	<100
Family Spalacidae (burrowing rats)			
Subfamily Rhizomyinae			
Tribe incertae sedis			
Prokanisamys kowalskii (Lindsay, 1996)	Z 679, right M1	(~22.5–21 Ma)	39
Prokanisamys arifi De Bruijn et al., 1981	Howard-GSP 116/22, right m1	(19.7–17.9 Ma)	27
Prokanisamys major Wessels and De Bruijn, 2001	Howard-GSP 81.14/4522, left M1	(18.6–15.1 Ma)	37
Prokanisamys sp. indet. (large)		(18.6 Ma)	75
Prokanisamys sp. indet. (small)		(18.7–18.2 Ma)	15
Prokanisamys n. sp. (small)		(16.8–13.1 Ma)	20
Prokanisamys n. sp.		(14.3–12.3 Ma)	45
Tribe Tachyoryctini			
Kanisamys indicus Wood, 1937	YPM 13810, right dentary	(16.8–11.4 Ma)	70
Kanisamys potwarensis Flynn, 1982	YGSP 8379, left m3	(14.3–13.4 Ma)	100
Kanisamys nagrii Prasad, 1970	GSI 18086, left dentary	(11.6–9.7 Ma)	125
Kanisamys sivalensis Wood, 1937	YPM 13801, left dentary	(9.4–7.3 Ma)	140
Protachyoryctes tatroti Hinton, 1933	GSI D272, right dentary	(7.8–7.1 Ma)	330
Eicooryctes kaulialensis Flynn, 1982	YGSP 15320, palate	(7.8–7.3 Ma)	180
Rhizomyides punjabiensis (Colbert, 1933)	AMNH 19762, right dentary	(11.2–10.1 Ma)	120
Rhizomyides sp. indet.		(~8.2)	180
Rhizomyides sivalensis (Lydekker, 1884)	GSI D97, left dentary	(7.3–6.5 Ma)	700
Tribe Rhizomyini (Asian bamboo rats)			
Miorhizomys nagrii (Hinton, 1933)	GSI D273, left dentary	(9.7–9.2 Ma)	410
Miorhizomys tetracharax (Flynn, 1982)	YGSP 4810, partial skull	(9.4–7.9 Ma)	1,500
Miorhizomys cf. *M. pilgrimi* (in Flynn 1982)	GSI D278, right dentary	(9.5–7.4 Ma)	3,500
Miorhizomys choristos (Flynn, 1982)	YGSP 4053, partial skeleton	(8.4–6.8 Ma)	2,000
Miorhizomys micrus (Flynn, 1982)	YGSP 5494, left dentary	(9.3 Ma)	200
Miorhizomys blacki (Flynn, 1982)	YGSP 15560, right dentary	(9.3 Ma)	400
Anepsirhizomys opdykei Flynn, 1982	DP 678, dentary, left m1–2	(3.3 Ma)	4,000
Family Cricetidae			
Subfamily incertae sedis			
Primus microps De Bruijn et al., 1981	Howard-GSP 116/135, left m1	(~22.5–18.1 Ma)	5
Primus cheemai Lindsay and Flynn, 2015	PMNH 2001, right m1	(~22.5–18.1 Ma)	11
Spanocricetodon lii De Bruijn et al., 1981	Howard-GSP 116/254, left M1	(~20 Ma)	8
Subfamily Cricetinae (hamsters)			
Democricetodon khani (De Bruijn et al., 1981)	Howard-GSP 116/221, left m2	(21–18.1 Ma); cf. 22 Ma	10
Democricetodon sp. A		(17.9–13.4 Ma)	12
Democricetodon sp. B, C		(16.0–10.5 Ma)	16
Democricetodon sp. E		(14.3–10.1 Ma)	32
Democricetodon kohatensis Wessels et al., 1982	Howard-GSP 50301, left m1	(17.9–10.1 Ma)	24
Democricetodon fejfari Lindsay, 2017	YGSP 19321, left dentary, m1–3	(13.8–10.1 Ma) (8.7 Ma)	41
Subfamily Megacricetodontinae			
Megacricetodon aguilari Lindsay, 1988	YGSP 19562, left M1	(17.9–12.8 Ma)	15
Megacricetodon mythikos Lindsay, 1988	YGSP 19556, right M1	(17.9–12.7 Ma)	26
Megacricetodon sivalensis Lindsay, 1988	YGSP 19231, left M1	(16.8–12.7 Ma)	12
Megacricetodon daamsi Lindsay, 1988	YGSP 19525, left M1	(15.2–12.8 Ma)	12
Punjabemys mikros Lindsay, 1988	YGSP 19524, left m1	(16.0–12.7 Ma)	16
Punjabemys leptos Lindsay, 1988	YGSP 19517, right m1	(16.0–12.3 Ma)	23
Punjabemys downsi Lindsay, 1988	YGSP 19119, right m1	(16.0–13.2 Ma)	33

Table 11.1. continued

	Holotype	Age Range	Mass (g)
Family Muridae			
Subfamily Gerbillinae (gerbils)			
Myocricetodon tomidai Lindsay and Flynn, 2015	PMNH 394, left M1	(~21–18.1 Ma)	11
Myocricetodon sivalensis Lindsay, 1988	YGSP 19576, left M1	(17.9–12.8 Ma)	13
Myocricetodon large sp.		(16.8–12.8 Ma)	34
Dakkamys barryi Lindsay, 1988	YGSP 24717, right maxilla M1–3	(15.2–12.7 Ma)	39
Dakkamys n. sp.		(14.3–11.4 Ma)	
Dakkamys asiaticus Lindsay, 1988	YGSP 19541, left M1	(13.2–10.5 Ma)	48
Paradakkamys chinjiensis Lindsay, 1988	YGSP 24656, right M1	(12.4–10.5 Ma)	15
Mellalomys lavocati (Lindsay, 1988)	YGSP 19229, right M1	(16.8–12.4 Ma)	24
Mellalomys perplexus (Lindsay, 1988)	YGSP 24651, left M1	(17.9–10.5 Ma)	23
Abudhabia pakistanensis Flynn and Jacobs, 1999	YGSP 27649, right M1	(8.7 Ma)	60
Subfamily Murinae (true mice)			
Stem Murinae			
Potwarmus mahmoodi Lindsay and Flynn, 2015	PMNH 515, left m1	(18.8–18.1 Ma)	11
Potwarmus primitivus (Wessels et al., 1982)	Howard-GSP 5521, left M1	(17.9–12.3 Ma)	11
Potwarmus minimus Lindsay, 1988	YGSP 19555, left M1	(17.9–14.0 Ma)	7
Antemus chinjiensis Jacobs, 1977	YGSP 7649, right M1	(13.8–12.7 Ma)	18
Crown Murinae			
"Post-*Antemus*" n. sp.		(12.4–12.3 Ma)	23
"Pre-*Progonomys* n. sp. X"		(11.6–11.4 Ma)	21
Murinae, gen. D and n. sp. (Y809 sp. nov. in Flynn et al. 2020)	YGSP 54088, right M1	(11.6–10.2 Ma)	29
Murinae, gen. C and n. sp. (large sp. in Flynn et al. 2020)	YGSP 36822, right M1	(11.2 Ma)	50
Progonomys morganae Kimura et al., 2017	YGSP 33180, left M1	(11.6–10.1 Ma)	17
Progonomys hussaini Cheema et al., 2000	PMNH 5062, left M1	(10.5–10.1 Ma)	24
Karnimata fejfari Kimura et al., 2017	YGSP 34546, left M1	(10.5–10.1 Ma)	26
"*Karnimata–Progonomys*" n. sp.		(11.2 Ma)	19
Progonomys debruijni Jacobs, 1978	YGSP 7739, left M1	(9.7–8.2 Ma)	15
Progonomys n. sp.		(8.0–7.2 Ma)	22
Parapodemus badgleyi Kimura et al., 2017	YGSP 33936, left M1	(10.5–10.1 Ma)	22
Parapodemus hariensis Vasishat, 1985	PUA 70-13, left dentary, m1–3	(9.2–8.0 Ma)	43
Karnimata darwini Jacobs, 1978	YGSP 7720, right M1	(9.7–8.0 Ma)	39
Karnimata huxleyi Jacobs, 1978	DP 253, left M1	(7.8–6.5 Ma)	29
Karnimata n. sp. (large) (see Jacobs 1978)		(9.2–8.0 Ma)	70
Parapelomys cf. *P. robertsi*		(7.8–7.3 Ma)	75
Parapelomys robertsi Jacobs, 1978	DP 250, right M1	(7.2–6.5 Ma)	80
Mus absconditus Flynn and Kimura, 2023	YGSP 52920, left M1	(8.0–7.2 Ma)	11
Mus auctor Jacobs, 1978	DP 210, left M1	(6.5 Ma)	15
Murinae, gen. A and n. sp. (see Jacobs 1978)	DP 310, right M1	(6.5 Ma)	
Golunda kelleri Jacobs, 1978	DP 197, right dentary	(~1.8 Ma)	110
Hadromys loujacobsi Musser, 1987	DP 204, right M1	(~1.8 Ma)	>100

Notes: Authorship and holotype given for named taxa, as well as age ranges for all. Mass estimates are derived by reference to living taxa through side-by-side comparisons with museum specimens of known mass and by regressions (Martin 1996) of cheek tooth dimensions on known body weights. Squirrel body-mass estimates utilize p4 lengths. Rhizomyine, cricetine, and murine masses are based on m1 length, although Martin (1996) does not claim that his regression applies to non-cricetines. Intraspecific variation in body mass is unknown; we use average lengths for species known by multiple individuals. *Abbreviations:* AJ = Al Jadidah collection (Hofuf Formation, Saudi Arabia) of the Muséum national d'Histoire naturelle, Paris; AMNH = American Museum of Natural History; D, DP = prefixes for localities (D) and fossils (DP) resulting from expeditions of the Dartmouth-Peshawar University project; GSI = Geological Survey of India; Howard-GSP = prefix for localities resulting from expeditions of the Howard-Geological Survey of Pakistan-Utrecht project; PMNH = Pakistan Museum of Natural History; PUA—collections of Panjab University Anthropology Department, Chandigarh, India; YPM = Yale Peabody Museum.

Denys, Reed, and Dauphin 2023); large owls range up to 3 or 4 km (Andrews 1990). Sampling at this scale reveals species-rich rodent communities of the Middle Miocene Siwaliks.

Each small mammal fossil locality represents an assemblage from a limited point in space and time that can be compared through geologic time to assemblages in consecutively older and younger localities. The Tóth, Lyons, and Behrensmeyer (2014) data for mammalian species richness in Kenyan game parks show increases through shifts in species distribution among neighboring parks on a scale of 10^2 years. However, evo-lutionary and ecological changes for a sequence of events calibrated through geologic time elucidate a larger pattern than do relatively local distributions. Change through time involves essential nonrepetitive components such as origination and extinction as well as broader scale biogeographic events such as faunal dispersal and immigration.

Systematic phylogenetic studies of Siwalik rodents based on dental morphology are generally limited because most nominal species are represented by isolated teeth that are associated as taxonomic groups by similarity and size into composite dentitions and then interpreted to define genera and species. Despite these limitations, because most relevant living families and genera of mammals can be distinguished by their dentitions, this practice has yielded a reproducible scheme. Siwalik rodents can be linked with modern relatives according to molecular-based phylogenetic hypotheses. Family-level systematic relationships for most Neogene rodents are relatively stable given that we can use genera and species of living crown groups to infer them. Problems remain for some extinct stem genera (e.g., the muroid *Primus*) that are not clearly allied to a living group, but these are few for Siwalik taxa.

Starting from the beginning of the twenty-first century, molecular phylogenetics has clarified the interrelationships of extant families (Huchon et al. 2000; Blanga-Kanfi et al. 2009; Honeycutt 2009; Gatesy et al. 2017). Living rodents cluster as three groups, one named formally by Huchon, Catzeflis, and Douzery (2000) as suborder Ctenohystrica. A mouse-like clade of rodent families gained recognition as suborder Supramyomorpha by D'Elía, Fabre, and Lessa (2019). Flynn et al. (2019) reviewed the status of high-level rodent taxa and named the clade of squirrel-like rodents suborder Eusciurida. The distribution of incisor enamel microstructure features (Von Koengswald 2004) is consistent with these subordinal clusters.

Molecular data are not likely to address the entirety of rodent evolution because the highly diverse fossil record extends globally throughout the Cenozoic and there are many extinct taxa currently beyond the reach of molecular analysis. For the Indian subcontinent, some extinct Paleogene groups may not fall into living higher level clades recognized currently based on dental and skeletal features. For example, certain stem rodents from Pakistan (Hussain et al. 1978; De Bruijn, Hussain, and Leinders 1982; Hartenberger 1982) that existed with primitive ctenohystricans in Eocene assemblages may have deep roots with no surviving relatives. In contrast all Neogene Siwalik fossil rodents can be assigned to crown suborders, and all three living rodent suborders are represented by Siwalik fossils.

Although the focus of this chapter is the Miocene record of the Potwar Plateau, the Glires story in South Asia must be placed in its biogeographic and temporal context, particularly with respect to the Oligocene age assemblages of Baluchistan and the Zinda Pir Dome in western Pakistan. The Early Oligocene rodent faunas from the lower part of the Chitarwata Formation of western Pakistan are mainly ctenohystrican assemblages, with other minor rodent components. Bugti Localities Paali Nala and Y417 in Baluchistan and Zinda Pir Localities Z108 and Z144 (Early Oligocene, ~28 Ma), preserve rodent assemblages containing primitive "ctenodactyloid" hystricognaths, the diatomyid *Fallomus*, an anomaluroid, and rare muroids and squirrels (Marivaux, Vianey-Liaud, and Welcomme 1999; Marivaux 2000). Flynn, Jacobs, and Cheema (1986) considered an array of rodents from the Bugti (Y417) assemblage as "baluchimyines" derived from Eocene chapattimyids, but subsequent work (e.g., Marivaux, Vianey-Liaud, and Jaeger 2004) disproved the monophyly of the Bugti assemblage. The Oligocene faunas differed taxonomically and, by inference, ecologically from Siwalik microfaunas of the earliest Miocene and younger. This led Flynn

(2000) to propose a revolutionary turnover among Rodentia from a "Bugti" microfauna to an Early Miocene "Siwalik" microfauna. The turnover was indeed a "revolution" in the sense that it was substantial, but we now know it was spread over 6 million years. The faunal change for the Zinda Pir Dome is displayed in its chronological context (Fig. 11.2). The rich Paali Nala fauna of Baluchistan adds uncommon faunal elements, including the large hystricognath *Bugtimys* Marivaux et al., 2002, now recognized at Y417 and Zinda Pir Locality Z108.

We lack a Late Oligocene record of small mammals in Pakistan. The record resumes higher in the Chitarwata Formation with earliest Miocene Zinda Pir sites (~22.5 Ma, see Fig. 11.2). From then on, the Miocene Siwalik rodent assemblages, distinct from Oligocene predecessors, are dominated by diverse Muroidea and Sciuridae; only a few ctenohystricans persist.

SUBORDER CTENOHYSTRICA HUCHON ET AL., 2000

Family Diatomyidae Mein and Ginsburg, 1997

Diatomyidae are medium-sized rodents (Table 11.1) with four cheek teeth and hystricomorphous jaw musculature (Dawson et al. 2006). An enigmatic group, the species *Laonastes aenigmamus* survives into the modern world and is now perceived as a basal taxon in crown Ctenohystrica. Diatomyids have bilophodont dentition although early genera retain distinct cusps until late wear.

Marymus dalanae Flynn, 2007, from Z113 (~22.5 Ma) is a derived holdover from Oligocene ctenohystrican-dominated assemblages. Its predecessor (see Fig. 11.2) is the Oligocene *Fallomus*. *Marymus* had cuspidate premolars, but molars dominated by a pair of transverse lophs. In occlusion, teeth abraded food by a fully proal power stroke (mandible drawn anteriorly), without trace of a lateral component in tooth wear (Flynn 2007).

In contrast to the abundant *Marymus*, younger Siwalik diatomyids were uncommon and found at only a few sites. Single teeth at Localities Z124 and Z122, 18.8 and 18.7 Ma, record large, high-crowned, bilophodont cf. *Diatomys shantungensis* (Fig. 11.3a), indistinguishable from the type species from the Early Miocene of Shandong Province, China. At a younger Potwar Locality Y747 (17.9 Ma), derived *Diatomys chitaparwalensis* is known from only a few teeth (Flynn 2006). *Diatomys* succeeded *Marymus* (see Fig. 11.2), but an ancestor-descendant relationship is not demonstrated.

The rich Middle Miocene faunas of the Siwaliks do not record diatomyids. The family is absent from younger strata, except for a single occurrence at Locality Y797, 11.2 Ma, where *Willmus maximus* was established for two large teeth (Flynn and Morgan 2005; Fig. 11.3b). Thereafter, there is no record of the family anywhere in the world, until extant *Laonastes* was discovered in Laos (Jenkins et al. 2005). The long absences of the family from the Siwalik record may reflect true absences from the subregion, with reintroduction of *Willmus* occurring by range extension.

This pattern of occurrences and absences provides biogeographical as well as paleoecological information. First, Diato-

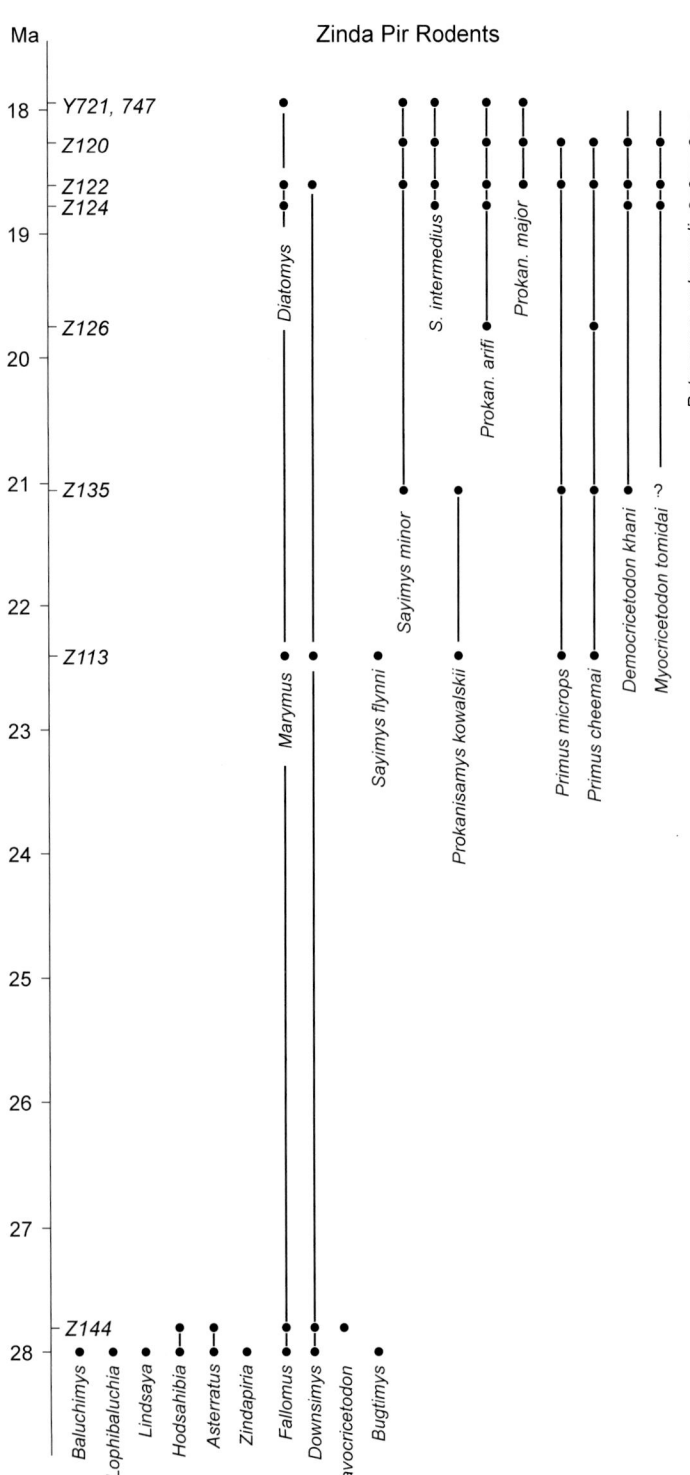

Zinda Pir Rodents

Figure 11.2. Ranges of key Zinda Pir rodents from Locality Z108 at ~28 Ma to Locality Z120 at 18.2 Ma, including early Potwar Plateau occurrences at 17.9 Ma. Zinda Pir Localities of ~28 Ma produce an array of Oligocene ctenohystrican genera. A gundi and muroid species of modern aspect appear by the Early Miocene. Different species of *Democricetodon*, *Myocricetodon* and *Potwarmus* occur at 17.9 Ma and younger Potwar sites.

myidae are South Asian in origin, although they occur as far west as Romania during the Paleogene (Marković et al. 2017); Neogene diatomyids are characteristic of the Indo-Malay biogeographic province. Second, early diatomyids are found in humid habitats, as judged from their diatomaceous type locality in China with reconstructed wetland habitat where diatomyids are richly preserved, for *Fallomus* in the Bugti Oligocene

(Flynn, Jacobs, and Cheema 1986; Marivaux and Welcomme 2003) and for *Diatomys liensis* in the Middle Miocene Li Mae Long Basin, Thailand (Mein and Ginsburg 1997). The youngest fossil diatomyid is the uncommon Siwalik *Willmus* at 11.2 Ma. The decline of Siwalik diatomyids possibly reflects loss of dependably moist habitat. The habitat of *Laonastes* today (humid forest developed on limestone hills in Laos and Viet Nam;

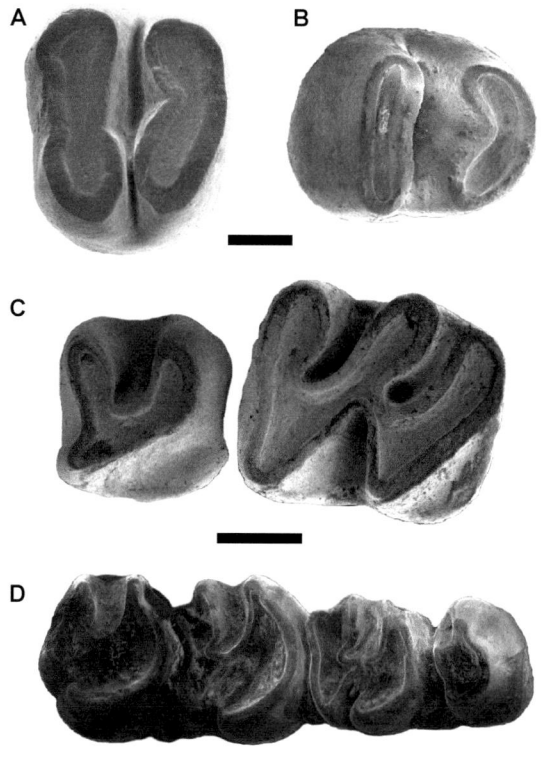

A

B

C

D

Figure 11.3. Ctenohystrican cheek teeth: Diatomyidae (A, B) and
Ctenodactylidae (C, D). A. *Diatomys* sp. right M2 Z595 from locality Z122;
B. *Willmus maximus* right dp4 YGSP 33100 from Locality Y797 at same
scale. C. *Sayimys chinjiensis* YGSP 45186, left p4m1 from Locality Y634,
anterior to left; D. *S. chinjiensis* YGSP 48411, right maxilla with P4M1-3
from Locality Y311, anterior to right. All scales = 1 mm.

Jenkins et al. 2005; Nguyen et al. 2014) contrasts with the low-
lands associated with its predecessors.

Family Ctenodactylidae Gervais, 1853

Ctenodactylidae are related to Diatomyidae but they are earliest Miocene immigrants to the Indian subcontinent, not derived from earlier Paleogene faunas of the region. They have
four lophodont cheek teeth with large deciduous premolars being replaced ontogenetically by smaller adult teeth. The few
living ctenodactylid species, known in the vernacular as "gundis," are restricted now to dry areas of Africa, a fraction of their
former broad Old World distribution. Paleogene ctenodactylids
of Asia, generally small and diverse, indicate a group that used
a broad range of habitats, including humid conditions, probably through most of Miocene time.

The oldest modern gundi, subfamily Ctenodactylinae, is
early Miocene *"Prosayimys" flynni* Baskin, 1996, from western
Pakistan Locality Z113 (22.5 Ma). It was proposed as distinguishable from later gundis by an extra crest (metalophulid II)
on lower teeth and deep folds on upper molars. Hartman et al.
(2019) found that the crest is not unique but occurs in later
ctenodactylines of the genus *Sayimys* and therefore submerged

Prosayimys into the long-lived Siwalik genus *Sayimys* Wood,
1937. (Oliver et al. [2023] accept *Prosayimys* in their analysis.) As
a basal ctenodactyline, *Sayimys* is an immigrant from elsewhere
in Asia, which joined the distinctive Miocene "Siwalik" fauna.

Sayimys is found at many localities in Pakistan throughout
the Early and Middle Miocene. De Bruijn et al. 1981 named the
small *Sayimys minor*, though larger than *S. flynni*, for Early Miocene specimens from Howard-GSP 116 at Banda Daud Shah in
the Murree Formation. Representatives of the genus the size of
S. minor, some with metalophulid II, occur at Locality Z135,
about 21 Ma, and superposed Localities Z124, Z122, Z120 (18.8
to 18.2 Ma), and in lower Manchars Localities, such as Howard-
GSP 81-07 (Wessels 2009). López-Antoñanzas and Sen (2003)
questioned the validity of *Sayimys minor* because the original
sample was inadequate to document variation. They proposed
S. minor as a junior synonym of *S. intermedius* (a mandible holotype from Saudi Arabia; Sen and Thomas 1979). We believe
S. minor is now based on sufficient samples to retain the species.

Somewhat younger than the Zinda Pir sequence, Potwar
Plateau Kamlial Formation Localities Y721 and Y747 (17.9 Ma,
postdating the Zinda Pir sequence) yield gundis similar in size
to *S. minor* and some larger specimens. Like Baskin (1996), we
consider these to represent the species *Sayimys minor* and *S. intermedius*. López-Antoñanzas and Sen (2003) also noted small specimens with variant premolar morphology at Y747 for which they
proposed *Sayimys baskini* (holotype YGSP 48125). We agree with
the parsimonious view of Hartman et al. (2019) that *S. minor*
appears to be valid for small, variable late Early Miocene *Sayimys*. We also agree with Baskin (1996) that there are two size
classes in these gundi samples, the larger being *S. intermedius*,
which was later replaced by *S. sivalensis* around 15 Ma.

An alternative interpretation for ctenodactylines (Hartman
et al. 2019) restricted species *S. intermedius* (from Al Jadidah,
Saudi Arabia) and the later *S. sivalensis* (from near Chinji Village)
to their holotypes. This proposal left a large array of Siwalik
Sayimys from numerous localities without a species name. Hartman et al. (2019) used Early Miocene Manchars specimens to
create *Sayimys hintoni* (holotype from Howard-GSP 81.14a).
This seems unnecessary because both Baskin (1996) and De
Bruijn, Boon, and Hussain (1989) documented diverse *Sayimys*
samples from throughout the Lower Siwaliks and the Manchars
series, thus placing *S. intermedius* and *S. sivalensis* on firm bases.

The Middle Miocene series of gundi localities of the Potwar
Plateau (15.2 to 10.1 Ma) is dominated by a single non-
splitting lineage (Fig. 11.4). For Baskin (1996) *Sayimys* continued
as successive species *S. sivalensis* and *S. chinjiensis* (Fig. 11.3c, d),
with increasing body size and crown height through time. De
Bruijn, Boon, and Hussain (1989) also documented this size
change in the Manchar series of southern Pakistan, showing
replacement of *Sayimys intermedius* by larger *S. sivalensis*.

The revision of López-Antoñanzas and Sen (2003) and
López-Antoñanzas and Knoll (2011) endorsed a simple phyletic
interpretation of the *Sayimys* lineage, submerging named samples into a long-ranging *Sayimys sivalensis*. The Potwar Plateau
sequence confirms an uninterrupted *Sayimys* lineage showing
increasing size and hypsodonty, with the body mass of larger

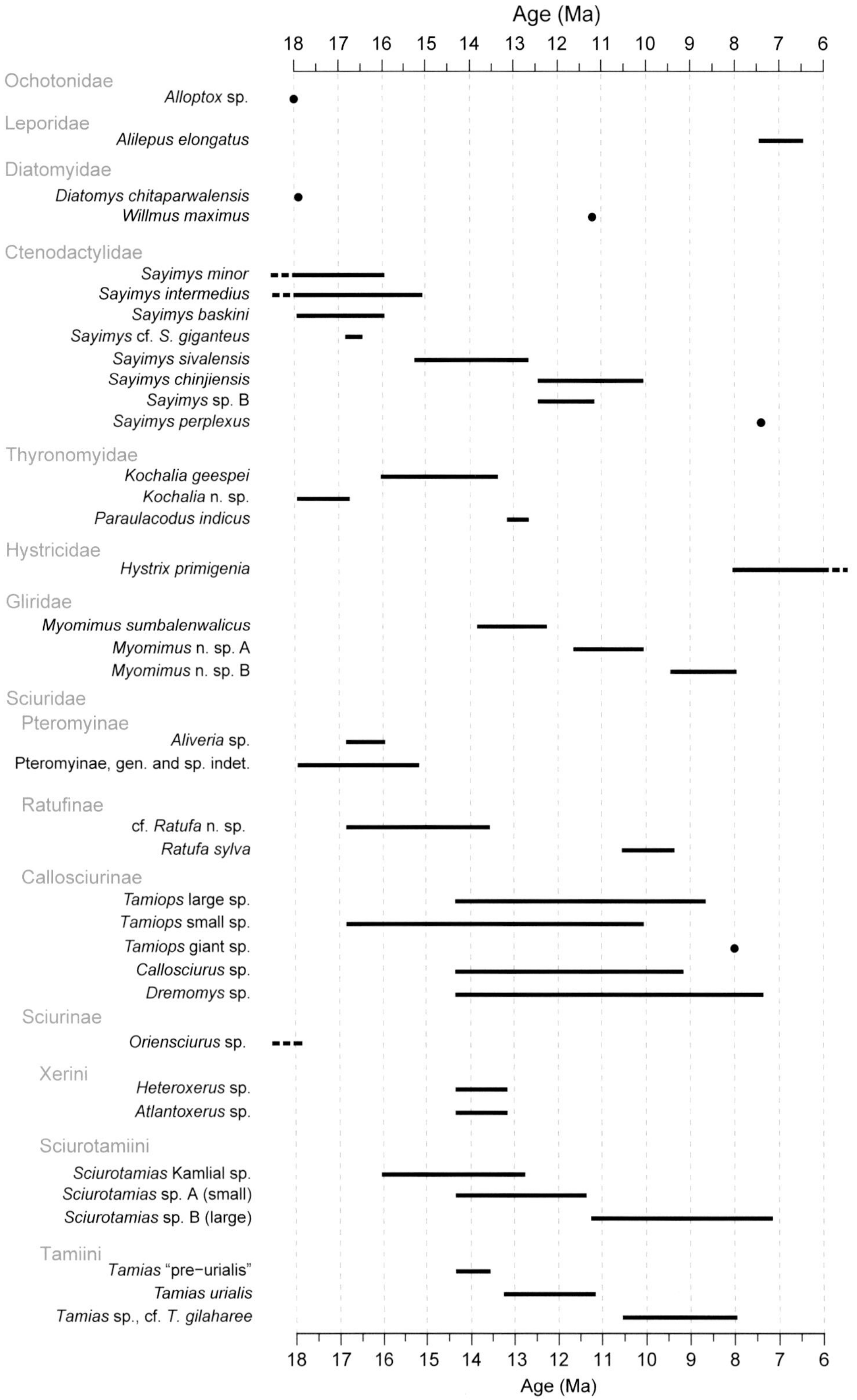

Figure 11.4. 18–6 Ma temporal range chart of Potwar rodent species.

S. sivalensis reaching 100 g (near the size of living *Petromus typicus* or *Ctenodactylus gundi*; Honigs and Greven 2003; *Encyclopedia of Life* 2017). The succession of *Sayimys* species can be used for Potwar biostratigraphic correlation.

All researchers interpret the evolution of *Sayimys* in the Tertiary of South Asia as a long anagenetic lineage, but a

few localities attest to some diversity. Several teeth of a large species occur briefly in the late Early Miocene of both the Potwar (16.8–16.5 Ma) and the Manchars series. We follow Hartman et al. (2019) to apply the name "cf. *Sayimys giganteus*" for this large, uncommon gundi. A high-crowned *Sayimys* sp. B, clearly more hypsodont than *Sayimys chinjiensis*, is

Age (Ma)

Figure 11.4. continued

represented by a few teeth between 12.4 and 11.2 Ma (Baskin 1996).

On the Potwar Plateau *Sayimys* has no local record after 10.1 Ma until later in the Miocene, at which time one tooth from Y457 (YGSP 45126, 7.4 Ma) indicates presence of the genus in Pakistan. In India the *Sayimys* lineage continued with Late Miocene genotypic species *Sayimys perplexus* Wood, 1937. *S. perplexus* from Haritalyangar (~9 Ma, Patnaik 2013) has been considered a large end-member of the *S. intermedius–S. sivalensis* lineage. Munthe (1980) regarded *Sayimys perplexus* as a synonym of *S. sivalensis*. For fossils from higher in the Haritalyangar section, Vasishat (1985) proposed *Sayimys badauni* as a larger species. A recent review (Patnaik 2020) shows that the several referred specimens are close in size to *S. perplexus* (and *S. sivalensis*). We consider *Sayimys perplexus* as the appropriate name for the post-10 Ma Late Miocene fossil gundis in the northern Indian subcontinent. Why *Sayimys* persisted later in India than in Pakistan remains an important question.

The biogeographic history of *Sayimys* includes at least one subtropical latitude dispersal event. Early Middle Miocene *Sayimys* spread westward from the Indian subcontinent, as evidenced by Miocene *S. intermedius* and *S. assarrarensis* of the Arabian Peninsula and *S. giganteus* and *S. obliquidens* of Anatolia and North China. *S. intermedius* (Sen and Thomas 1979) is one of the few Siwalik rodent species known outside the

Indian subcontinent. While named for a jaw from the Miocene of Al Jadidah, Saudi Arabia, Siwalik records range from approximately 19 to 16 Ma. The cladistic analysis of López-Antoñanzas and Knoll (2011) places older *Sayimys flynni* at the base of the Ctenodactylinae, so we propose a westward direction of dispersal from the Indian subcontinent. Later from the Arabian-Anatolian region, *Sayimys* may have spread secondarily to the north and eastward across the Old World and into North China. Under this scenario, *Sayimys* bypassed Tibet rather than expanding directly northward from the Indian subcontinent (Flynn and Wessels 2013, Flynn et al. 2023b).

Family Thryonomyidae Pocock, 1922
This ctenohystrican family with derived hystricognathous jaw morphology includes two living species of *Thryonomys*, the African grasscutters, which weigh up to several kilograms. The Tertiary African record shows the family to be a long-time resident of that continent since the Paleogene, with occasional dispersals eastward. The two genera of Thryonomyidae found in the Siwalik fossil record fall separately among African cane rats in a cladistic analysis (Flynn and Winkler 1994). This implies two immigration events dispersing eastward from Africa. The less derived *Kochalia* appeared in the Siwaliks at ~18 Ma, while later *Paraulacodus* arrived independently ~13.1 Ma in the late Middle Miocene.

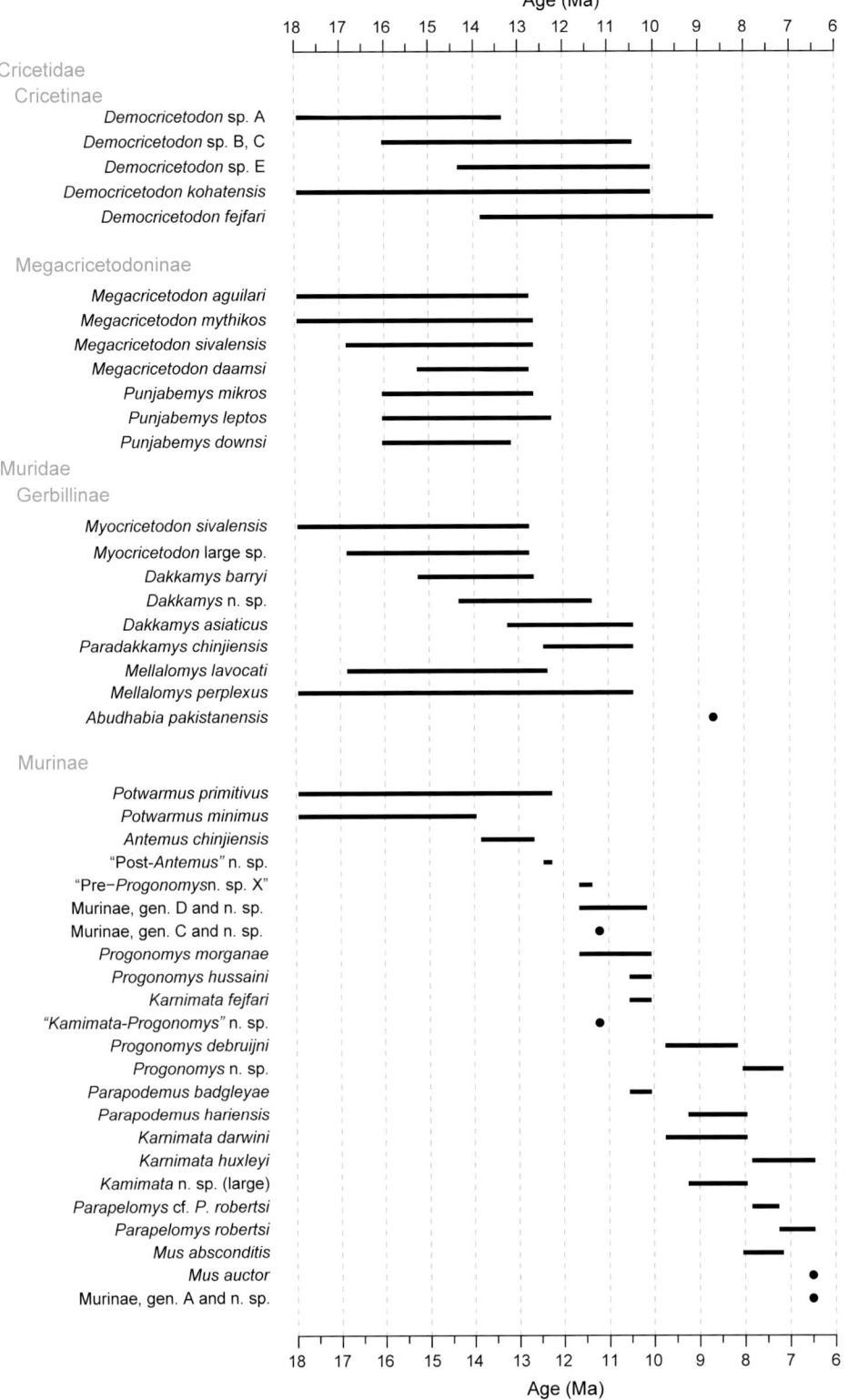

Figure 11.4. continued

Kochalia geespei De Bruijn and Hussain, 1985, is based on a Middle Miocene sample from the Manchar Formation. The species ranges from 16.0 to 13.4 Ma in the Potwar area. *K. geespei* is preceded by similar *Kochalia* sp. in both the Manchar Formation (Wessels 2009) and the Kamlial Formation (17.9 to 16.8 Ma) of the Potwar Plateau.

Paraulacodus indicus Hinton, 1933, is derived with respect to *Kochalia* and closer to modern *Thryonomys* (Flynn and Winkler 1994). Its younger congener from the Late Miocene (8.0–8.5 Ma) of Ch'orora, Ethiopia, is *P. johanesi* Jaeger, Michaux, and Sabatier 1980; both *Paraulacodus* species have upper incisors with double grooves. Multiple grooves characterize modern

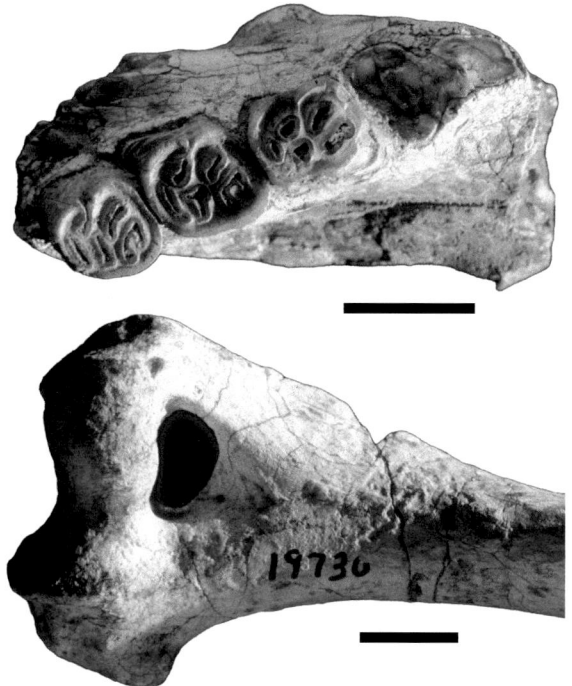

Figure 11.5. Hystrix primigenia. Above, left dentary with m1–m3, YGSP 19722 from Locality Y434; below, right distal humerus YGSP 19730 from Locality Y547. Scale = 1 cm.

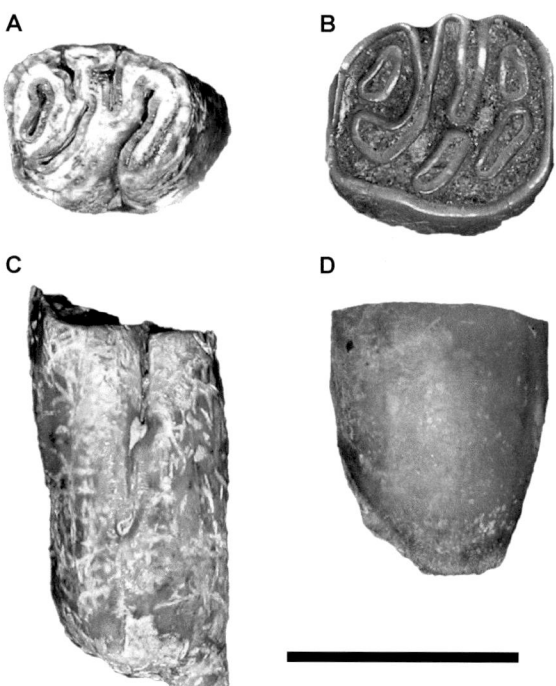

Figure 11.6. Occlusal and lateral views of high crowned *Hystrix indicus* (A, C, left m3 YGSP 34831 from Haro River, Locality Y54) compared with larger *H. primigenia* (B, D, right M2 YGSP 51000 from Locality Y935). Scale = 1 cm.

Thryonomys. Africa, being the center of diversity for thryonomyids since the Paleogene, was the source for the immigrant *Paraulacodus.* After Middle Miocene dispersal from Africa, *P. indicus* (holotype from an unknown locality) had a short temporal duration on the Potwar Plateau (13.1–12.7 Ma). Living thryonomyids are herbivores that prefer moist habitat. Whereas *Kochalia* was of modest size, near 100 g, *Paraulacodus* had a larger body mass of about 500 g.

Family Hystricidae Linnaeus, 1758

Hystricidae, the familiar Old-World porcupines, are both hystricomorphous (jaw musculature) and hystricognathous (dentary structure). Porcupines have four lophodont, high-crowned cheek teeth and tend to large body size. The only Potwar rodent species recorded and published that our group has not recovered, and a large one at that, was originally named *Sivacanthion complicatus* by Colbert (1933), who recognized this primitive porcupine among fossils attributed to the Chinji area. Only one other specimen has been recovered from Pakistan, a larger jaw fragment found by a Yale party in 1935 in the Chinji Formation near Kanatti. Comparative study (Landry 1957; Van Weers 2005) shows that *Sivacanthion* is a junior synonym of *Atherurus,* the brush-tailed porcupine, and further that the species *A. complicatus* represents a predecessor of Indian Pleistocene *Atherurus karnuliensis.* The species was either uncommon or restricted temporally to a brief residence in the Potwar area. However, *A. complicatus* is also known from the Middle and Late Miocene of India (Patnaik 2013). Larger than living *Atherurus africanus,* its body mass would have been about 3 kg. *Atherurus* is

a forest animal, mainly terrestrial but capable of climbing, and nocturnal (Nowak 1999).

Lydekker (1884) recognized a derived Siwalik porcupine assignable to the living genus *Hystrix* Linnaeus, 1758. *Hystrix* (Fig. 11.5) appeared on the Potwar Plateau at about 8 Ma, without predecessor. Lydekker (1884) named this Siwalik porcupine (Geological Survey of India [GSI] D96) *Hystrix sivalensis,* but Van Weers and Rook (2003) show it to be indistinguishable from *Hystrix primigenia* Wagner, 1848, a member of the Pikermian chronofauna (Bernor et al. 1996; Steininger and Papp 1979) to the west of Pakistan. The same species persists into the Pliocene at Tatrot (Khan and Flynn 2017). This large Siwalik fossil species had a body mass in the range of 20–25 kg. A younger Late Miocene porcupine from Yunnan (*H. lufengensis* Wang and Qi, 2005) resembles the Potwar species.

Matthew (1929) and Black (1972) noted a higher-crowned *Hystrix* (American Museum of Natural History [AMNH] 19909, "2.5 miles north of Hasnot") among fossils from Pakistan and India. A similar hypsodont molar (YGSP 34831) was recovered from D54, the Haro River Quarry (De Terra and Teilhard de Chardin 1936) reopened by Saunders (Early Pleistocene, ~1.4 Ma; Opdyke et al. 1979; Saunders and Dawson 1998). The molar is much like AMNH 19909, and these resemble molars of extant *Hystrix indica,* a species less massive (15–20 kg) than *H. primigenia* (Fig. 11.6). This change in *Hystrix* species is part of Plio-Pleistocene faunal turnover in the Indian subcontinent. Modern *Hystrix* has broad ecological tolerances.

Accepting the attribution of the Potwar *Atherurus complicatus* type locality as Middle Miocene in age (Chinji Formation), *A. complicatus* represents an early record for the family. A reasonable hypothesis is that Hystricidae at the family level were of South Asian origin. Yet Sen (2001) notes that the oldest known *Hystrix* fossils are European (MN 10, >9 Ma). While the family possibly has evolutionary roots in South Asia, the genus *Hystrix* is an immigrant to Indo-Pakistan at ~8 Ma. It is distinct from Late Miocene porcupines of North China such as *H. gansuensis* but, based on resemblance to *Hystrix primigenia* and *H. aryanensis* from Afghanistan and Iran (Sen 2001), the Late Miocene Siwalik porcupine likely was a Pikermian chronofauna immigrant from the direction of Iran or Afghanistan. This porcupine lineage may have dispersed from the Potwar to South China as *H. lufengensis*.

SUBORDER EUSCIURIDA FLYNN ET AL., 2019

Eusciurida include squirrels with sciurognathous jaw structure and uniserial incisor enamel microstructure, and they are defined genetically by unique nuclear and mitochondrial sequences (Montgelard et al. 2002). The core of this clade of rodent families is the diverse Sciuridae (squirrels). Other members include the dormice (Gliridae) and the Aplodontiidae. Aplodontiids, a Holarctic group of higher latitude represented by the extant "mountain beaver" *Aplodontia*, never spread to South Asia. The diminutive dormice (Gliridae) appeared abruptly in the Middle Miocene but were neither diverse nor abundant in the Siwaliks.

Family Gliridae Muirhead, 1819

A single dormouse species has been described from the Siwaliks, a member of the extant genus *Myomimus* Ognev, 1924. Living *Myomimus* is terrestrial (not arboreal; Storch 1978), omnivorous, hibernates underground, with a body mass for fossil dormouse of 30–40 g (Burildağ and Kurtonur 2001). The earliest occurrence of *Myomimus* at 13.8 Ma in the Chinji Formation is preceded by rich assemblages lacking dormice, which thus constrains their entry into the Potwar area to 14.0–13.8 Ma. Dormice thereafter comprise a minor component of Middle and Late Miocene Siwalik faunas. The small size of dormice teeth introduces bias against retrieval in screen-washing, but the morphology of upper and lower cheek teeth is distinctive and easily recognized.

The Chinji Formation dormouse is *Myomimus sumbalenwalicus*, first recognized by Munthe (1980) at Daud Khel, west of the Potwar Plateau. *M. sumbalenwalicus* extends through Middle Miocene levels, but in younger beds a change in morphology signals replacement by *Myomimus* sp. A. This slightly larger dormouse (11.6–10.1 Ma) has more nearly square upper molars and more elongate lower molars. Younger sites reveal yet another morphology, *Myomimus* sp. B (9.4–8.0 Ma), with larger teeth and stronger ridges (Fig. 11.7). This youngest Potwar dormouse last occurs at Locality Y547 (8.0 Ma).

The dormouse succession can be interpreted as a single evolving lineage from 13.8 to 8.0 Ma. Biogeographically it signals a lineage of dormouse that penetrated the geographic bar-

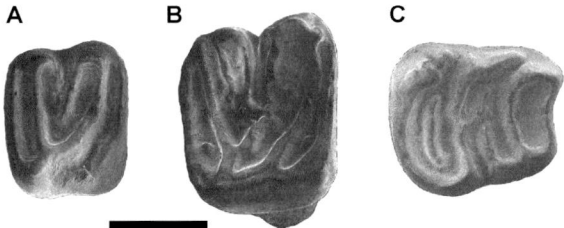

Figure 11.7. Siwalik dormouse molars with 1 mm scale. *Myomimus sumbalenwalicus* (A,YGSP 21631, M1 from Locality Y491) and *Myomimus* sp. B (B, C: YGSP 21602 M1 and 21610 m1, both from Locality Y388).

riers of the subcontinent from the west once only, after the Middle Miocene Climatic Optimum and the subsequent cooling event ~13.9 Ma (Zachos et al. 2001). Thereafter, the lineage apparently evolved in isolation from dormouse populations of northern Asia and the Middle East.

Family Sciuridae Fischer, 1814

For a group of rodents that is not abundant in the Siwalik fossil record, squirrels in aggregate show high species richness. Present in low numbers early in the Neogene terrestrial fossil record of Pakistan, squirrels became more common and diverse through the Middle Miocene, with a peak of ten coexisting species. Ground and tree squirrel groups dominated in the Siwalik record of that time.

The oldest squirrel in Pakistan is the unusual *Oligopetes* from the pre-Siwalik Early Oligocene locality Paali Nala, Baluchistan (Welcomme et al. 2001). Rugose cheek tooth enamel suggests it was a flying squirrel (Pteromyinae). A younger putative flying squirrel *Aliveria* sp. with a body mass of ~160 g (Table 11.1, estimated by comparison with livings species of similar molar size), occurs at Early Miocene Potwar Localities Y802 and Y591 (16.8 and 16.0 Ma). A few indeterminate teeth from Kamlial Formation Localities Y721, Y592, Y642, and Y682 (17.9-15.2 Ma) may represent a different flying squirrel, but we see no evidence for younger Pteromyinae.

Neogene squirrel diversity in Pakistan lies among nongliding groups. In the Early Miocene, a large indeterminate ground squirrel occurs at Zinda Pir Locality Z150, ~22 Ma, and at Banda Daud Shah (De Bruijn, Hussain, and Leinders 1981). Tribe Ratufini is recorded at Z135 (21 Ma, Fig. 11.8a). This relatively small tree squirrel, undefined genus, supports early differentiation of the tribe as postulated by Mercer and Roth (2003). Evidence of a larger ratufin (YGSP 48043, 350 g), closer to living *Ratufa* in morphology, comes from younger Locality Y802 (16.8 Ma). This species is smaller than extant *Ratufa*, as is the younger Middle Miocene *Ratufa maelongensis* from Li Basin in Thailand (Mein and Ginsburg 1997).

At Locality Z122 (18.7 Ma) the medium-size tree squirrels *Tamiops* and *Oriensciurus* suggest significant forest cover. *Tamiops* cf. *T. asiaticus* would have had a mass of about 160 g (see Table 11.1), while *Oriensciurus* sp. was larger (~370 g). Several younger specimens from high in the Kamlial Formation (Y642 and Y682, 15.2 Ma) indicate *Sciurotamias* sp., a 170-g ground

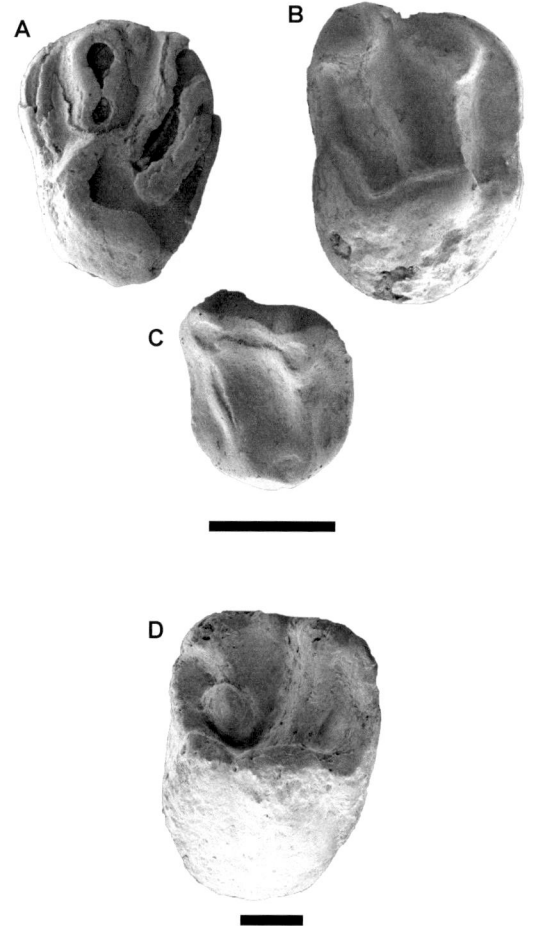

Figure 11.8. Siwalik squirrel upper cheek teeth (A–C share the same 1 mm scale). A. Ratufini n. gen. and sp. (YGSP Z757 P4 from Locality Z135), B. *Atlantoxerus* sp. (YGSP 17196 from Locality Y491), C. *Tamiops* large sp. (YGSP 36565 from Locality Y797), D. *Heteroxerus* sp. (YGSP 36421 from Locality Y718), much larger with its own 1 mm scale.

squirrel related to the living Chinese rock squirrel. This species recurs later at 12.8 Ma but is not found at productive sites of intervening age in the Chinji Formation. These late Early Miocene squirrels resemble species from the diatomites of Shanwang, eastern China (Qiu and Yan 2005).

Squirrel diversity in Chinji Formation microsites increases through about 13 Ma. Small mammal assemblages show consistently high species richness among squirrels, coincident with the global cooling trend following the Middle Miocene Climatic Optimum (Zachos et al. 2001). At 14.3 Ma *Ratufa* sp. appears with a small *Sciurotamias* (~120 g) that ranges to 11.4 Ma, as well as two species of *Tamiops* that co-occur at several sites. Middle Miocene assemblages record other genera new to the Siwaliks. The xerine ground squirrels *Atlantoxerus* (Fig. 11.8b) and the more massive *Heteroxerus* (Fig. 11.8d, both 14.3–13.2 Ma) are represented by nine and five specimens, respectively. Table 11.1 body mass estimates are in line with those of Freudenthal and Martín-Suárez (2013) for other species of those genera. The callosciurines *Callosciurus* (185 g) and *Dremomys*

(135 g) are contemporary tree squirrels, the latter continuing into the Late Miocene. The chipmunk *Tamias* appears by 14 Ma and ranges through the Nagri and Dhok Pathan Formations as a succession of species. The early *Tamias* species (body weight about 90 g) persists for less than 1 million years. It is succeeded by a smaller species that Munthe (1980) named for specimens from Daud Khel (originally *Eutamias urialis*). *Tamias urialis* may have evolved endemically in the Potwar area. This ubiquitous Chinji chipmunk survived to at least 11.2 Ma. By 10.5 Ma, a larger species (170 g) of chipmunk with inflated cusps with stronger connections appears and ranges to 8.0 Ma. Patnaik et al. (2002) name *T. gilaharee* for a sample of similar morphology from Kutch, India, about 10 Ma. Distinct differences from *T. urialis* imply immigration of this chipmunk.

The squirrel assemblage begins to decline in richness in the early Late Miocene. The small *Sciurotamias* terminates at 11.4 Ma, but the larger *Sciurotamias* persists to 7.2 Ma. *Dremomys* sp. and both species of *Tamiops* continue (Fig. 11.8c). After 10.5 Ma (Y259) the squirrel guild shows change. In addition to the large immigrant chipmunk *Tamias*, the distinct tree squirrel *Ratufa sylva* appears. Weighing ~350 g, as estimated from dentition, *R. sylva* is not as large as extant species of the genus.

Younger assemblages contain the common large *Tamias*. At Y182 (9.2 Ma), *Ratufa sylva* is still present, and *Callosciurus* reappears (same size as the Chinji species). The small-sized *Tamiops* is absent after 10.1 Ma, but the larger *Tamiops* survives to 8.8 Ma. Both *Sciurotamias* and *Dremomys* continue into the upper levels of the Dhok Pathan Formation. A single upper milk tooth from Y547 indicates a different large species of *Tamiops* (>200 g) at 8 Ma. After 9.2 Ma, Dhok Pathan sites show decreased squirrel diversity and abundance, which may imply less forest cover.

An intriguing aspect of Siwalik faunas is the general lack of flying squirrels (Pteromyinae), except for the few Early Miocene records. Today the flying squirrels of Pakistan occur only in Himalayan moist temperate forests (Roberts 1977). Pteromyines are unknown in the Middle Miocene Li Mae Long Basin of Thailand but occur in Late Miocene sites of Yunnan (Qiu and Ni 2005).

Middle Miocene Sciuridae of the Indian subcontinent were diverse, and the family constitutes up to 10% of small mammal teeth at some sites. Most species are represented by few teeth at any single locality. Up to ten species coexisting during the Middle Miocene indicate forested habitats, although some squirrels were likely terrestrial. Based on tooth size, a few Early Miocene individuals were large (300–400 g), compared with most later squirrel species, which are small to medium in size. Many Siwalik squirrel genera are also recorded in Miocene assemblages of southern China (Qiu and Ni 2005), which indicates evolution of those lineages within the forests of the Indo-Malay biogeographic province. One prominent exception is the chipmunk *Tamias*, which has no records in South China. *Tamias* possibly spread southward into the Potwar area from Central Asia in two events; the lineage that appeared at 14 Ma differs considerably from the larger immigrant at 10.5 Ma.

SUBORDER SUPRAMYOMORPHA D'ELÍA ET AL., 2019

Mitochondrial and nuclear sequences support monophyly of Supramyomorpha (Nedbal, Honeycutt, and Schlitter 1995; Montgelard et al. 2002). The core of the suborder is the classical Muroidea, many of which show myomorphy, a secondarily derived jaw musculature feature in which the masseter muscle extends onto the snout, partially through the expanded infraorbital foramen. Sister clades Muroidea and Dipodoidea are associated in Supramyomorpha with North American geomyoids, Holarctic castorids, and African pedetids and anomalurids (Blanga-Kanfi et al. 2009; Honeycutt 2009). The Castoridae (beavers), mainly northern but extending into Indo-Malaya as far as Thailand during the Miocene (Suraprasit et al. 2011), never entered the Indian subcontinent. Anomaluridae ranged throughout subtropical Asia and Africa in the earlier Cenozoic (Dawson et al. 2003), including for a time into the Indian subcontinent.

Family Anomaluridae Gervais, 1849

Living anomalurids, known as the "scaly-tailed flying squirrels," do not fly and are not squirrels. They are small to large arboreal rodents restricted to Africa today. Two of the three extant genera have a gliding membrane (see Fabre et al. 2018). Anomalurids are hystricomorphous and sciurognathous and have uniserial incisor enamel (Von Koenigswald 2004). They have four cheek teeth with a crown structure of transverse lophs.

The oldest anomalurid from the Indian subcontinent is *Downsimys margolisi* Flynn et al., 1986, which when described from the Oligocene of Bugti, Baluchistan, was unassigned at the family level. Subsequently, Marivaux (2000) studied a large sample of *Downsimys* from the Early Oligocene of Paali Nala and reconstruction of the dentition supported inclusion in Anomaluridae. *Downsimys* was considered limited to the Oligocene, its extinction part of the small mammal revolution that ushered in the predecessors of Siwalik microfaunas (Flynn 2000). De Bruijn, Hussain, and Leinders (1981) had illustrated two minute, round, taxonomically undetermined rodent teeth from the Early Miocene Murree assemblage, with a diameter of just over a millimeter, the size of *Downsimys*. Our field team working in the Zinda Pir area also found small Early Miocene teeth that are difficult to assign: molar Z640 from Locality Z113 (~22.5 Ma) and molars Z587 and YGSP 45123 from Z122 (18.7 Ma). We now consider these three specimens and the two from the Murree assemblage reported by De Bruijn et al. (1981) as cf. *Downsimys* sp., a small species of perhaps 10 g body mass. These specimens indicate that the family Anomaluridae survived the end-Oligocene rodent revolution in the Indian subcontinent (see Fig. 11.2).

In addition, Flynn (2003) described a single tooth from Locality Y802 (16.8 Ma). This 2 mm lophodont specimen (YGSP 36144) was proposed as a platacanthomyid (a muroid family), but without matching any known genus. Mörs (2006) suspected that the specimen does not represent that family. An alternate proposal, perhaps as likely as the platacanthomyid hypothesis, is that this tooth represents another anomalurid of ~100 g body

mass, as different from *Downsimys* as is today's *Anomalurus* from living *Idiurus*.

Whether or not these fossils represent late-surviving anomalurids in Asia, the family was a late Paleogene element of the Indian subcontinent microfauna. During the Paleogene, the family was broadly distributed across low-latitude South Asia and extended into Africa (Dawson et al. 2003). Anomalurids are indicators of forest habitats, a consistent signal for the mid-Cenozoic of the Indian subcontinent. They became quite uncommon in Asia after the Oligocene, with shrinking range, but we conclude that they persisted locally in the Early Miocene of the Indian subcontinent.

Superfamily Muroidea, Illiger, 1811

Throughout the Miocene, the diverse clade Muroidea dominated Siwalik microfaunas in numbers of both individuals and genera. Crown Muroidea have only three cheek teeth in the upper and lower jaws. The superfamily was differentiated by the Late Eocene. All but the most basal species are united morphologically by loss of all premolars. Primitively bunodont, crown morphology is based on four cusps joined by crests, plus a fifth anterocone (anteroconid) on first molars. Derived muroids have myomorphous jaw musculature, but early members retain primitive hystricomorphy. Some early stem genera do not sort readily into extant families. Today's rich Muroidea embrace half a dozen crown families; three of these, Spalacidae, Cricetidae, and Muridae, dominated the Siwaliks.

Family Spalacidae Gray, 1821

Bamboo rats (Spalacidae: Rhizomyinae) are a hallmark of Siwalik fossil small mammal faunas. Living bamboo rats are medium- to large-size burrowers (see Table 11.1), with adaptations for subterranean life: fusiform body with robust humerus and claws and skull modified to dissipate stress associated with chisel-incisor digging. Molars are lophodont, with thick enamel and crests elevated to cusp level. Third molars are large. Chewing produces a pronounced wear gradient along the tooth row (heavier wear anteriorly).

Siwalik fossil rhizomyines include both primitive bamboo rats tribe Rhizomyini and their sister tribe Tachyoryctini. Tachyoryctins such as *Kanisamys* showed fewer burrowing features, but were at times common. The Siwalik record of Rhizomyinae is rich in species diversity and number of individuals throughout the Miocene.

We follow current usage supported by molecular studies (e.g., Steppan and Schenk 2017) that unites in family Spalacidae bamboo rats with blind mole rats (*Spalax*) and north Asian zokors (*Myospalax*) plus the forerunners of each. Their basal divergence dates to the Paleogene. Previously we considered monophyly of the three to be unlikely (Flynn et al. 1985), and De Bruijn et al. (2022) continue to suspect that similarities are homoplasies rather than homologies. Flynn (1982, 1985, 2009) placed Siwalik bamboo rats at the family level, but here they are a subfamily under Spalacidae, with later genera as members of living tribes Rhizomyini (bamboo rats) and Tachyoryctini (East African mole rats). Spalacines, myospalacines, and rhizomyines have distinctive distributions: spalacines in the Middle East, Asia Minor, Caucasus, and North Africa; myospalacines in China, Russia, Kazakhstan, and Mongolia; rhizomyines in southern Asia and Africa. Most members of these burrowing groups share similar lower incisor enamel microstructure, specifically Type 10 of Kalthoff (2000).

The oldest rhizomyine rat known is *Prokanisamys kowalskii* (Lindsay, 1996) from the upper member of the Chitarwata Formation, Zinda Pir Dome (Locality Z113, ~22.5 Ma). *P. kowalskii* is well represented at Z113, Z139 and Z150, and the lineage continues at Z135 (21 Ma). All known early rhizomyines are from the Indian subcontinent, and we believe that the clade originated in the Indo-Malay biogeographic province. *Prokanisamys*, plesiomorphic with low-crowned molars, is not attributed to tribe, but it shows the subfamily features of lophodonty, thick enamel, and crests elevated to cusp levels. Its lower incisor is robust and rounded and bears a pair of longitudinal ridges.

The congeneric successor of *P. kowalskii* is *Prokanisamys arifi*, type species of the genus. Based on a sample from the Murree Formation at Banda Daud Shah (De Bruijn, Hussain, and Leinders 1981), *P. arifi* appears at Locality Z126 (19.7 Ma). As a smaller species (~27 g), it is less likely derived from *P. kowalskii*, so it may have originated outside the local area but within the biogeographic province.

Beginning at about 18.8 Ma, multiple species of *Prokanisamys* are recorded in the Zinda Pir Dome. In addition to *P. arifi*, the larger *P. major* coexists with a much larger and a very small species, both unnamed (Lindsay and Flynn 2015). *Prokanisamys major* (its type locality, Manchar Formation Howard-GSP 8114; Wessels and De Bruijn 2001) continues through the Early Miocene. The diminutive *P. arifi* lineage shows slight increase in size and persists (informally as *Prokanisamys* "chinjiarifi"; see Table 11.1) in low numbers until about 13 Ma.

Prokanisamys is joined by *Kanisamys indicus* Wood, 1937 at the end of the Early Miocene (Potwar Locality Y802, 16.8 Ma). The long-lived type species *K. indicus* is higher crowned with elongate m3 and was probably similar to *Prokanisamys* in habitat and diet. Both were terrestrial, capable of burrowing but not fully fossorial (Flynn 2009). A larger, unnamed *Prokanisamys* species at Y718 occurs through much of the Chinji Formation; it was erroneously called *Prokanisamys benjavuni* by Flynn et al. (1995), before the large topotypic sample of *P. benjavuni* from Li Basin, Thailand, became available (Mein and Ginsburg 1997). The Thai species *P. benjavuni* is distinctive and Middle Miocene in age. The pattern of coexistence of multiple species within the genera *Prokanisamys* and *Kanisamys* continues as a characteristic aspect of Siwalik assemblages for the next five million years.

Kanisamys indicus is the characteristic mid-Miocene Siwalik rhizomyine and shows the derived increased crown height of tribe Tachyoryctini. The higher-crowned *Kanisamys nagrii* appears toward the top of the Chinji Formation at Y504 (11.6 Ma). The lower incisor of *K. nagrii* is round in cross-section with a single longitudinal ridge. The species co-occurs with *K. indicus* for 200,000 years and continues after the 11.4 Ma local termi-

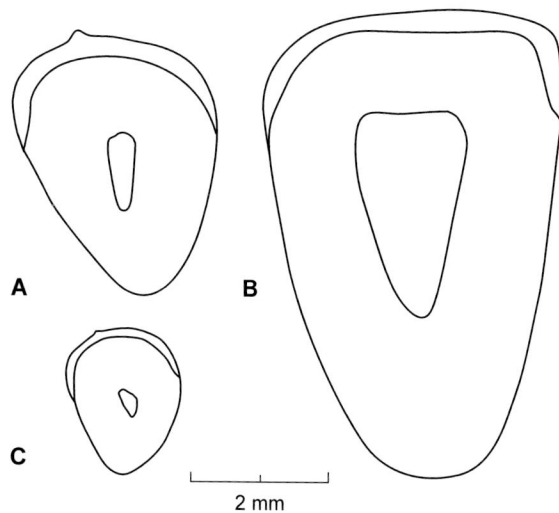

Figure 11.9. Rhizomyinae lower incisor cross sections in anterior view. Right incisors of the tachyoryctins *Rhizomyides sivalensis* (A, Punjab University Palaeontological Collection [PUPC] 09/30) and *Kanisamys sivalensis* (C, YGSP 8358 from Locality Y387) compared to bamboo rat *Miorhizomys choristos* (B. PUPC 00/37), all drawn to the same scale. Note the longitudinal ridge somewhat medial to the midline in rounded cross sections A and C; the much larger *Miorhizomys* (B) has flattened enamel without a ridge.

nation of *K. indicus*. The replacement by higher-crowned *K. nagrii* may signal a change in some aspect of diet. Throughout the Chinji and Nagri strata, rhizomyines are abundant, sometimes amounting to over 10% of total small mammal specimens at localities.

The Middle Miocene Potwar record also documents the robust *Kanisamys potwarensis*, resident from 14.3 to 13.4 Ma. *K. potwarensis* brings to four the number of small rhizomyine contemporaries in the lower part of the Chinji Formation. Unlike higher crowned species of the genus, *K. potwarensis* may have originated elsewhere in the Indo-Malay region.

In the Late Miocene high-crowned *Kanisamys* is joined by a larger species, *Rhizomyides punjabiensis* (11.2 to 10.1 Ma). *R. punjabiensis*, the smallest species of its long-lived genus (see Table 11.1), appears without a clear South Asian predecessor, although *Kanisamys potwarensis* may prove related if more shared characters are recognized. Species of this lineage, apparently native to the Indo-Malay region, occupied the Potwar subregion in successive immigration events.

The Potwar record lacks *Kanisamys* fossils between 9.7 and 9.4 Ma. By 9.4 Ma the hypsodont *K. sivalensis* is present, and *K. nagrii* is no longer found. Molars are high-crowned but the teeth are rooted. Two partial skeletons from Locality Y387 (8.7 Ma) were preserved within a burrow. This confirms that *K. sivalensis* could burrow, but associated postcrania show no special digging adaptations, and unmodified skull anatomy indicates that *K. sivalensis* was not committed to fossorial life (Flynn 1982, 2009). Its round incisor with a single enamel ridge is robust but otherwise unmodified for digging (Fig. 11.9). *K. sivalensis* was a successful terrestrial herbivore and quite common at some

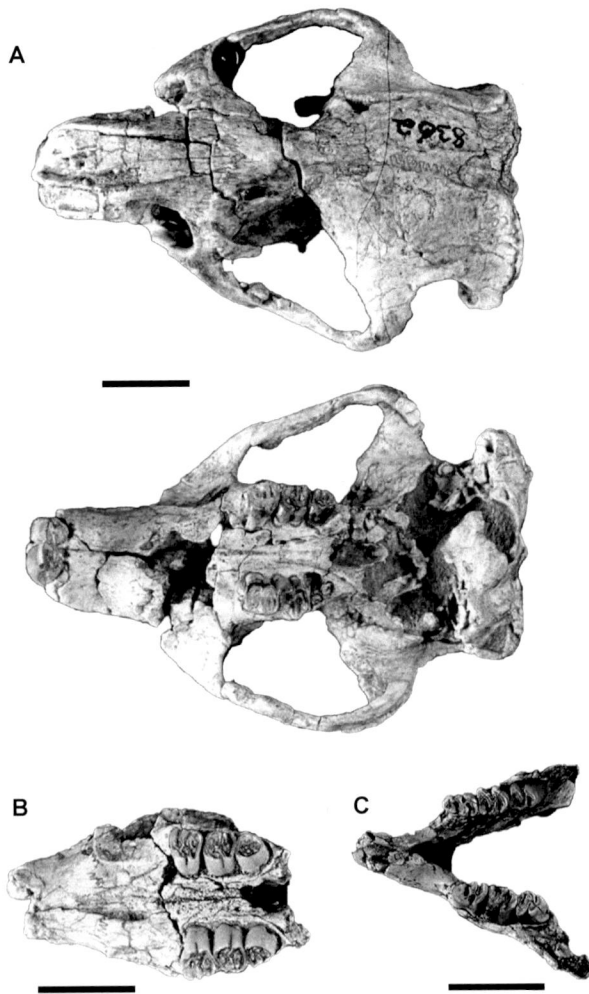

Figure 11.10. Siwalik Rhizomyinae. A. Bamboo rat, rhizomyin *Miorhizomys nagrii* (YGSP 8362 skull in dorsal and ventral views from Locality Y270); B. tachyoryctins *Kanisamys sivalensis* (YGSP 8358 from Locality Y387, snout and palate in ventral view), and C. *Protachyoryctes tatroti* (mandible YGSP 15336 from Locality Y434). Photos by George Marcek; scales = 1 cm.

localities, comprising up to an impressive one-third of small mammal specimens at some sites in the lower part of the Dhok Pathan Formation.

Kanisamys sivalensis continues to 7.3 Ma, overlapping the 8–7 Ma ranges of two derived tachyorctin genera, the fully hypsodont *Eicooryctes kaulialensis* and the larger *Protachyoryctes tatroti* (Fig. 11.10). Based on limited remains, neither appears to have been committed to fossorial life, both probably spending time on the surface. The crown height of *Eicooryctes* suggests exploitation of a particularly abrasive diet. The uneven quality of the fossil record limits assessment of the precise timing of termination of these genera, but the occurrence of *Protachyoryctes* near Bhandar (Khan and Flynn 2017) shows that *P. tatroti* continued to at least 6.8 Ma. These tachyoryctins are considered endemic, because they are known only from the Potwar area and their closest relative is *Kanisamys sivalensis*.

Rhizomyides punjabiensis (11.2-10.1 Ma) was followed two million years later by larger *Rhizomyides* sp. (single m3, YGSP 8222, from Y034) dating to ~8 Ma, a pattern representing brief reappearance of this lineage on the Potwar Plateau. Following another million-year absence, *Rhizomyides* returned by 7.2 Ma in the form of *Rhizomyides sivalensis*, type species of the genus. *R. sivalensis* was large, with the tachyoryctin high-crowned molars and robust incisors (round in cross-section with a longitudinal ridge; see Fig. 11.9). In contrast to the shorter ranges of earlier species of the genus, *R. sivalensis* was probably present continuously in northern Indo-Pakistan, as indicated by discoveries from the late Neogene of India (e.g., Black 1972; Vasishat 1985; Patnaik 2001). Later species of *Rhizomyides* appeared in Afghanistan in the Pliocene (Brandy 1979; Flynn et al. 1983), which indicates a surprising dispersal westward through mountain passes of ~1000 m elevation and implies broad habitat tolerance.

The foregoing summarizes the Miocene Siwalik history of basal Rhizomyinae and Tribe Tachyoryctini. None of the Asian tachyorctins has adaptations showing commitment to subterranean life; although capable of burrowing, they probably remained largely surface feeders. The group shows multiple dispersals to Africa (López-Antoñanzas, Flynn, and Knoll 2013). After Early Miocene westward emigration of *Prokanisamys* sp. to North Africa (Wessels 2009), later Miocene records in East Africa (*Pronakalimys* and *Nakalimys*) represent another dispersal event (Winkler et al. 2010). Later Neogene East African root rats phylogenetically closer to fully fossorial living *Tachyoryctes* (López-Antoñanzas, Flynn, and Knoll 2013) represent an additional late Miocene westward dispersal from South Asia of the *Kanisamys-Protachyoryctes-Eicooryctes* group. The earliest recognized species, Ethiopian Awash *Tachyoryctes makooka* at 5.7 Ma (López-Antoñanzas and Wesselman 2013), retains features of *Protachyoryctes* (Patnaik 2020).

In contrast to Tachyoryctini, Siwalik members of the extant tribe Rhizomyini evolved fully fossorial adaptations during the Late Miocene. True bamboo rats displaying an array of derived burrowing features appear in the Siwalik record after 10 Ma. No suitable endemic ancestor for the tribe has been recovered in Pakistan prior to its first appearance, but based on the distribution of living and fossil crown bamboo rats, the group likely originated within the Indo-Malay biogeographic province. The earliest known rhizomyin is *Miorhizomys nagrii* (see Fig. 11.10) from low in the Dhok Pathan Formation (9.7 Ma). Rhizomyini are recognized by derived features including a dorsally restricted infraorbital foramen, deep dentary with shortened anterior extension of the masseteric crest, an enlarged lower incisor with flattened enamel lacking a longitudinal ridge (see Fig. 11.8), small posterolingual reentrant on m3, and forelimb and skull showing fossorial adaptations (Flynn 2009). Skulls from Y270 (9.45 Ma) and somewhat older dentaries unequivocally demonstrate that *Miorhizomys* was fossorial. Potwar habitats by then record the prevalence of fire (Karp, Behrensmeyer, and Freeman 2018; Karp et al. 2021), which would promote the protective strategy of underground life for *Miorhizomys* and the investment of resources in subterranean structures such as rhizomes for plants. Rhizomyini began to exploit a subterranean food source that was either new or not accessed before 10 Ma.

Bamboo rat diversification following the first occurrence of Rhizomyini is impressive. Within a few hundred thousand years of the appearance of *Miorhizomys*, several species of the genus co-occur on the Potwar Plateau, some of >1 kg mass (see Table 11.1). By 9.3 Ma, four species are known, two documented by single specimens. Multiple fossils of *M. tetracharax* span 9.4 to 7.9 Ma, and fossils of the largest species, cf. *M. pilgrimi*, persist from 9.5 to 7.4 Ma. The later large-bodied *M. choristos* ranges from 8.4 Ma to 6.8 Ma (Khan and Flynn 2017). Several of these Potwar species occur also in the Late Miocene of northern India, but *Miorhizomys harii* (Prasad 1970) is a distinctive endemic of the Haritalyangar area. The appearance of multiple *Miorhizomys* congeners led Flynn (1985) to observe that fossoriality associated with burrowing lifestyles of small populations in isolation would be consistent with splitting speciation.

Flynn and Qi (1982) recognized Potwar bamboo species at the Late Miocene Lufeng site, Yunnan Province, South China. At the time Flynn (1982) considered the appropriate generic name for these bamboo rats to be *Brachyrhizomys* Teilhard de Chardin, 1942, type species from the Pliocene of Yushe Basin in Shanxi Province, China. Subsequently, based on a fine Yushe skull that had become available, Flynn (2009) realized that Yushe *Brachyrhizomys shansius* presents features of *Rhizomys* (notably, as a synapomorphy, the infraorbital foramen restricted dorsally as a round aperture) and is a subgenus of *Rhizomys*. The fossorial Miocene bamboo rats of Yunnan and the Indian subcontinent constitute genus *Miorhizomys*.

Bamboo rats show continued Pliocene diversity in South Asia, although the Upper Siwaliks are incompletely surveyed in Pakistan. The 4-kg *Anepsirhizomys opdykei* occurs at a 3.3 Ma horizon near Jalalpur. In India, several rhizomyine species thrived, including particularly large *Rhizomyides* and *Anepsirhizomys pinjoricus* (see Flynn et al. 1990).

Miorhizomys represents a radiation of burrowing rodents that ranged throughout South Asia and characterized the Indo-Malay biogeographic province. Subsequently *Rhizomys* (*Brachyrhizomys*) extended northward into Shanxi Province in the late Neogene (6–4 Ma; Flynn 2009). Modern *Rhizomys* species exploit subterranean food resources in moist thickets from 1,000 to 4,000 m elevation. Using *Rhizomys* as a model, the radiation of bamboo rats after 10 Ma (oldest fossil from Locality Y892, 9.7 Ma) points to subterranean foraging as a new behavior for rhizomyines. This is consistent with Late Miocene habitat change, including vegetation with underground nutrient-storage organs evolved in response to increasingly seasonal precipitation and increased incidence of fire. Other than their Late Neogene northward spread into Shanxi Province, tribe Rhizomyini remained Indo-Malay.

Families Cricetidae Fischer, 1814, and Muridae Illiger, 1811

These muroid sister families form a monophyletic crown group sharing the derived cheek tooth dentition consisting only of molars (M1–3, m1–3); all premolars are absent. Aside from the divergent platacanthomyids and spalacids, the grouping in families of muroid genera and tribes has been unclear when based on morphology alone. There was little consensus among specialists on the content of families and higher taxa until the thorough review of Musser and Carleton (2005) and consistent resolution from molecular data (e.g., Steppan, Adkins, and Anderson 2004). As groupings of living genera became clear and supported by repeated analyses, the content of crown families Cricetidae and Muridae became better understood. Early cricetids are hamster-like rodents of small body-size; murids, also generally small, include gerbils and the true mice and rats. Most muroid genera in the Neogene Siwalik fossil record other than rhizomyines represent Cricetidae or Muridae, but some of the earliest Miocene species may be stem taxa that do not fall clearly into either family. An example is the small muroid genus *Primus*.

Living Cricetidae embrace cricetine hamsters, the derived arvicolines (voles, lemmings, muskrats), and a large array of New World mice. Muridae include Murinae and Otomyinae as traditionally understood, as well as the Deomyinae and Gerbillinae. Molecular data clarify a radiation of African muroids, not known from the Siwaliks, and distinguish them at the family level; these are the Nesomyidae of Madagascar and continental Africa. One African nesomyid clade is Dendromurinae, represented by the highly derived *Dendromus* and *Malacothrix*. Both genera have extremely elongate first molars: m1 with greatly alternating cusps, and M1 with a small enterostyle. Although some Siwalik mice have been published as dendromurines, the resemblance is superficial, and we agree with Wessels (1998) that those fossil species are, at least in part, myocricetodontin gerbils.

The few known muroids of Oligocene age from low in the Chitarwata Formation of western Pakistan (Marivaux, Vianey-Liaud, and Welcomme 1999) are primitive and excluded from extant families. Lindsay and Flynn (2015) illustrate *Atavocricetodon* from Locality Z144. By the early Miocene, stem mice appear alongside the rhizomyine *Prokanisamys* and the gundi *Sayimys*. As part of the Oligocene-Miocene Siwalik rodent turnover (Flynn 2000), these appearances signal the beginnings of the "Siwalik" rodent fauna. These Early Miocene muroids from the upper Chitarwata and Vihowa formations of the Zinda Pir Dome share derived features that set them apart from their Oligocene predecessors as crown muroids (Flynn et al. 1985).

Primus microps De Bruijn et al., 1981 (Early Miocene Murree Formation of Banda Daud Shah), although a stem taxon, exemplifies the emergence of muroids of modern aspect in the Siwaliks. In the Zinda Pir area *Primus*, *Prokanisamys*, and *Sayimys* are first recorded at the earliest Miocene Locality Z113 (~22.5 Ma; see Fig. 11.2). Subsequent Zinda Pir rodent assemblages add the cricetid *Democricetodon* and the early murids *Potwarmus mahmoodi* and *Myocricetodon tomidai*, all present at Z124 (18.8 Ma; Lindsay and Flynn 2015). *Myocricetodon* and Murree Formation *Spanocricetodon* show the distribution of early gerbils (myocricetodontins) from Pakistan to South China (Li 1977; De Bruijn, Hussain, and Leinders 1981; Lindsay and Flynn 2015). All are small mice (see Table 11.1).

The Zinda Pir appearance of the early stem-hamster (Criceti-dae) *Democricetodon* is biogeographically important. *Democricetodon* appears at about the same time across the continent, from central Asia (Xinjiang Province, China, 21–22 Ma; Maridet et al. 2011) to the Early Miocene of Turkey (Theocaropoulos 2000). In Pakistan we recovered *Democricetodon khani* (De Bruijn, Hussain, and Leinders 1981) from Early Miocene Locality Z135 (21 Ma) with possible older damaged specimens at Z150 (~22 Ma). These early stem-hamster records push back the minimum age of the root of the cricetine tree, making it earlier than the 20 Ma estimate used recently to calibrate a molecular phylogeny (Neumann et al. 2006). *D. khani* continues into the Vihowa Formation. *Democricetodon* is an immigrant to South Asia, there being no indication of a related predecessor in the region, but the precise, presumed Asiatic origin of the genus (and of cricetine hamsters) is unknown.

Democricetodon remains characteristic of the Siwalik rodent fauna for 12 million years after its appearance and recurs briefly at 8.7 Ma. This common stem-hamster is represented by multiple species. Wessels et al. (1982) documented this genus in the Middle Miocene at Banda Daud Shah. At that time the limited sample (25 molars) for which her new species *Democricetodon kohatensis* was named showed considerable size variation. A much larger collection (hundreds of teeth) from numerous Chinji Formation localities of the Potwar area showed that the *D. kohatensis* hypodigm included two species, which Lindsay (1996) distinguished independently as *Democricetodon* D and E. *Democricetodon* D corresponds to *D. kohatensis* (Lindsay 1996, Fig. 3).

Multiple *Democricetodon* species occur in the lower part of the Potwar Siwaliks, beginning at Locality Y747 (17.9 Ma). A small species is cited in several publications (e.g., Lindsay 1996) as *Democricetodon* A, and it ranges up to 13.4 Ma. Its larger companion (see Table 11.1) *D. kohatensis* Wessels et al., 1982, spans almost 8 million years, from Y747 to Y311 at 10.1 Ma. A third Lower to Middle Siwalik hamster *Democricetodon* B-C coexists at many localities from Y592 (16.0 Ma) to Y259 (10.5 Ma).

The species *Democricetodon* E, larger than *D. kohatensis*, joins the cohort of hamsters at the base of the Chinji Formation (~14.3 Ma) and runs to 10.1 Ma in the Nagri Formation. The even larger cricetid *Democricetodon fejfari* Lindsay, 2017 (see Table 11.1), brings to five the number of similar congeneric species. *D. fejfari* includes variants previously enumerated as F, G, and H (see Lindsay 1996) and spans 13.8 to 10.1 Ma (at Y491 and Y311). By 10.1 Ma, only three *Democricetodon* species remain: *D. fejfari* and the last records of *D. kohatensis* and *Democricetodon* E. These ranges supersede those of Fig. 1 in Lindsay (2017).

In contrast to its previous abundance and species diversity, *Democricetodon* is notably absent from rich localities of 9.4 to 9.0 Ma. We interpret this as local extirpation. The single species *Democricetodon fejfari* recurs in abundance (~9% of rodent teeth) at 8.7 Ma (Localities Y387 and Y388; Fig. 11.11a). We consider this pattern to reflect a long residence time with a true temporary absence of this species from the Potwar Plateau, followed by reoccupation of the local habitat.

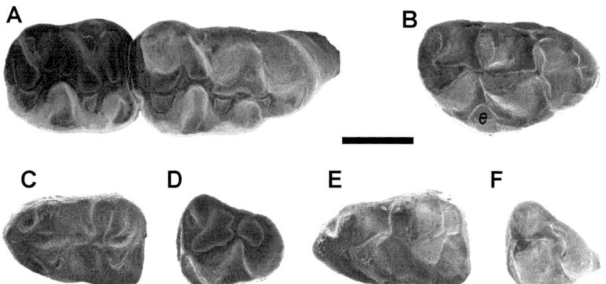

Figure 11.11. Siwalik muroids compared. Cricetid *Democricetodon fejfari* (A) right maxilla fragment with M1-2, YGSP 19285 from Locality Y388, next to (B) right M1 of the gerbil *Dakkamys barryi*, YGSP 43811 from Locality Y718. Anterior to the right for both. *Democricetodon fejfari* (C, D) left m1 and m3 (YGSP 19298, 19329 from Locality Y388). *Dakkamys barryi* (E, F) left m1 and m3 (YGSP 24720 from Locality Y726, 44951 from Locality Y718). Anterior to left for C–F. All teeth at same 1 mm scale. *Dakkamys* M1 is more lophodont and bears a lingual enterostyle (e); *Dakkamys* m1 is narrow anteriorly and its cusps alternate; its m3 is reduced in size.

The Potwar sediments preserve no record of cricetid rodents after 8.7 Ma. Vole-like distant relatives of *Democricetodon*, two species of the arvicoline *Kilarcola*, are known from the beginning of the Pleistocene at >2,000 m elevation in Kashmir (~2.4 Ma; Kotlia and von Koenigswald 1992).

The distinct megacricetodontines co-occur with *Democricetodon* in the Siwaliks. *Megacricetodon* is a classical European Early and Middle Miocene muroid of modern grade that retains the "cricetid plan" cusp pattern shared with the ancestors of gerbillines and murines (Flynn, Jacobs, and Lindsay 1985). Lindsay (1988) named four species of Siwalik *Megacricetodon* for large collections from Kamlial and Chinji beds of the Potwar Plateau and differentiated them by crown height and size. We do not recognize *Megacricetodon* in the Vihowa Formation; it appears at the base of the Kamlial Formation as *M. aguilari* and *M. mythikos*, which coexist from 17.9 to 12.8 Ma. Other late Early Miocene *Megacricetodon* species are *M. sivalensis* and *M. daamsi*, also terminating at 12.8 Ma (see Table 11.1). The Siwalik species resemble gerbils in some features, which led Wessels (1996) to differentiate them from European *Megacricetodon* as new taxon *Sindemys*, a proposal not followed here.

Punjabemys Lindsay, 1988, is based on species related to *Megacricetodon* that are derived in having inflated cusps. Most *Punjabemys* species are larger than *Megacricetodon*, except for *M. mythikos* (table 11.1). The earliest Potwar *Megacricetodon* predates *Punjabemys*, which is not securely known before 16.0 Ma. *Punjabemys leptos* persists until 12.3 Ma as the last record of the genus.

SUBFAMILY GERBILLINAE GRAY, 1825

Gerbillinae, sister to Murinae, share dental features with the true mice: reduced longitudinal connections along the midlines of molars, doubling of anterocones, and posterior reduction of the tooth rows (small third molars). We refer primitive gerbils to Myocricetodontini Lavocat, 1961, at the tribe level, within Gerbillinae. The earliest Siwalik gerbil is Early Miocene *Myocricetodon tomidai* from the Vihowa Formation of the Zinda Pir region (see Fig. 11.2).

Potwar Myocricetodontini include *Myocricetodon* and *Mellalomys*, as well as genera formerly listed by us as Dendromurinae. As noted above, Dendromurinae are a small group, well-defined genetically and morphologically and currently restricted to sub-Saharan Africa. Myocricetodontini, in contrast, were widespread from North Africa to China. *Myocricetodon sivalensis* Lindsay, 1988, is much less common in micromammal samples of fossil teeth than its contemporaries *Democricetodon*, *Megacricetodon*, or *Punjabemys*. *Myocricetodon sivalensis* occurs from 17.9 to 14 Ma and less commonly in younger beds up to 12.8 Ma. A larger unnamed Potwar *Myocricetodon* spans almost the same time range. To the south in Kutch, India, Patnaik et al. (2022) recognize a late-surviving member of the genus.

Two species of *Mellalomys* occur in the Siwaliks. They were published originally (Lindsay 1988) as *Dakkamyoides*, which was subsequently shown (Wessels 1996) to be a junior synonym of *Mellalomys* Jaeger, 1977. *Mellalomys perplexus*, at the base of the Potwar sequence (17.9 Ma), continues through Siwalik formations to 10.5 Ma. *M. lavocati* is less abundant, and ranges from 16.8 to 12.4 Ma.

Dakkamys Jaeger, 1977, is another widespread myocricetodontin and, like *Mellalomys*, is known from North Africa to the Indian subcontinent. *Dakkamys* is distinctive in its isolated enterostyle on the cingulum of upper molars and its reduced longitudinal crest that joins alternating cusps. There are two named Potwar species of *Dakkamys*: the abundant *Dakkamys barryi* (Fig. 11.11c) ranging from high in the Kamlial Formation (15.2 Ma) to 12.7 Ma in the Chinji Formation, and a larger but less common *Dakkamys asiaticus*, which occurs from 13.2 Ma to Nagri Formation Locality Y259 at 10.5 Ma. Extending the Siwalik record, Vasishat and Chopra (1981) named a still larger species "*?Dakkamys nagrii*" from near Haritalyangar, India (~9 Ma; Patnaik 2013), indicating later survival in northern India of a larger species of *Dakkamys* with body size ~80 g.

Paradakkamys Lindsay, 1988, is a derived myocricetodontin found between 12.4 and 10.5 Ma. It has elongated first molars with a prominent enterostyle on M1-2 and an ectostylid on m1. The small *Paradakkamys chinjiensis* was not abundant, but the fact that its short range overlaps that of *Dakkamys asiaticus* makes it useful for biochronologic correlation.

What habitat was preferred by myocricetodontin gerbils? Whereas modern gerbils occupy dry, sparsely vegetated habitats, there is no indication that such conditions characterized the Early to Middle Miocene Siwaliks. On the contrary, the once-abundant Myocricetodontini declined to local extirpation by 10.5 Ma, when other evidence indicates that Potwar habitats were becoming at least seasonally drier. We suspect that modern gerbils survived by tolerating dry habitats, but that Miocene myocricetodontin diversity and abundance depended on moist, vegetated conditions.

The Late Miocene crown gerbil *Abudhabia pakistanensis* Flynn and Jacobs, 1999, a member of the extant tribe Taterillini,

occurs at Locality Y387 (8.7 Ma). Related to living *Tatera indica*, which is adapted to dry habitat from India to the Middle East, the Siwalik *Abudhabia* suggests open-canopy vegetation if not dry terrain. Both the decline of fossil myocricetodontins and rise of taterillin gerbils may have been related to increasing seasonality of precipitation. *Abudhabia* is an immigrant without predecessor, and the direction of its introduction into the Potwar Plateau is undetermined, but the genus occurs earlier in the Late Miocene of Shaanxi Province and Inner Mongolia, China (Qiu and Li 2016). *Abudhabia* was named for fossils to the west of the Indian subcontinent in Abu Dhabi (De Bruijn and Whybrow 1994). Later occurrences of the genus are in the Pliocene of Himachal Pradesh and Afghanistan (Patnaik 2001; Flynn et al. 2003).

SUBFAMILY MURINAE ILLIGER, 1811: STEM GENERA

Biogeographically, it appears that the radiation of Muroidea that resulted in the differentiation of subfamily Murinae centers in southern Asia. Fossiliferous deposits of Pakistan record parts of the Early Miocene phase of murid evolution. Crown Murinae emerged by the beginning of the Middle Miocene (Aghová et al. 2018). Primitive Murinae of this age are found only in South Asia. Fortuitously the later Middle Miocene Potwar Siwalik strata capture a record of key events in early murine differentiation.

Molecular data place Gerbillinae as the sister taxon to Murinae within the family Muridae. The Early to Middle Miocene genus *Potwarmus* Lindsay, 1988, displays special murine cusp character states, including the lingual enterostyle on M1 and M2. *Potwarmus* resembles its immediate Potwar successor, the larger *Antemus* Jacobs, 1977, and both are considered stem murines. Key innovations for *Antemus* include addition on the upper molars of the lingual anterostyle and linkage of cusps in transverse chevrons; lower molars develop a double anteroconid on m1 and loss of a mesolophid. Together there are eight major cusps on M1 and six on m1, autapomorphies for crown Murinae (Fig. 11.12).

Potwarmus, widespread and common in the Indo-Malay biogeographic region, may be considered an Early Miocene immigrant to the Potwar subregion. *P. mahmoodi* appears by 18.7 Ma in western Pakistan. On the Potwar Plateau two species, *Potwarmus primitivus* (Wessels et al. 1982) and *P. minimus* Lindsay, 1988, range through most of the Kamlial and Chinji Formations. Both are identified at the base of the Potwar series at Locality Y747 (17.9 Ma). *P. minimus* is abundant up to 14 Ma (Y733), whereas *P. primitivus*, also common, persists in low numbers to 12.3 Ma. *Potwarmus thailandicus* was abundant east of the Potwar in the Middle Miocene Li Basin, Thailand (Mein and Ginsburg 1997). By the late Early Miocene *Antemus mancharensis* Wessels, 2009, of southern Pakistan displays morphology intermediate between *Potwarmus* and *Antemus chinjiensis*.

By 13.8 Ma, the stem murine *Potwarmus* was replaced by the larger *Antemus*. Within 200,000 years after 14 Ma, this abundant mouse transformed as the murine *Antemus*. Specimens occurring during the transition have been regarded by us (Jacobs and Downs 1994; Jacobs and Flynn 2005) as "near

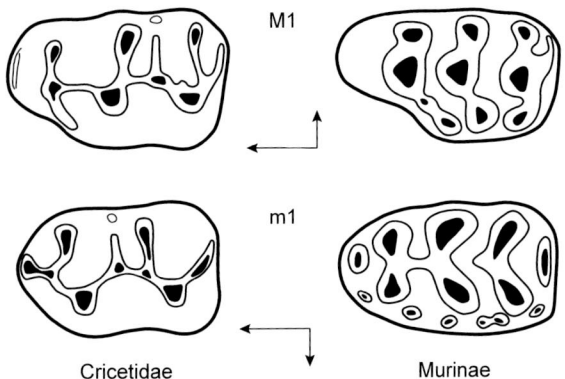

Figure 11.12. Generalized morphology of muroid first molars, M1 above and m1 below; the cricetid plan on the left, derived murine condition on the right. Orientation arrows show anterior to the left, and labial (buccal, short arrow) at right angle. Subfamily Murinae upper molars are derived in the addition of lingual cusps linked transversely as chevron crests. The simple anteroconid of the cricetid m1 illustrated here is twinned in murines, and an extra medial cuspid is often present; this murine m1 adds small buccal cusps and retains a partial midline longitudinal connection of paired cusps.

Antemus." We no longer think it useful to consider the transition as arbitrarily stepped, as implied by that term. Rather, we see *Potwarmus primitivus* as the predecessor of a lineage, individuals of which incrementally acquired variable murine autapomorphies. As derived features in lineage members became stable and co-varied, the population became recognizable and diagnosable as *Antemus chinjiensis*. In the recovered murine samples, a few individuals with *Potwarmus* morphology persisted until 12.3 Ma as variants among common *Antemus* (leading to late database records of "*Potwarmus*"). The murine autapomorphies are fully developed chevrons on upper molars and correspondingly derived lower molars (see Fig. 11.12). *Antemus chinjiensis* displays the modern murine phenotype of 8-cusped M1 with transverse chevrons and similarly modified M2; and m1 with twin anteroconids, oval paired and opposed cusps, and buccal cuspules on a cingulum.

Murinae: Abundant from the Outset

The Potwar record samples evolving Murinae lineages serially through time. We have the luxury of tracing phyletic change in small increments, but as alluded to above, this leads to challenges in applying binomial nomenclature to samples of populations with variable intermediate character states. When fossil samples are spaced widely in time, species level taxonomy can be based on defining features; when spaced closely in time, variation blurs distinction between successive samples. To address such a problem Crusafont-Pairó and Reguant (1970) and Harrison (1978) adopted hybrid compound species names for intermediates. We adopt their approach with minor modification.

Stratigraphically dense sampling also poses a special problem for distinguishing and calibrating splitting events because genetic divergence, which is used in "molecular" phylogenetic analyses, begins before morphological divergence accumulates

and becomes detectable in daughter species. The common origin of two Potwar species lineages recognized as the genera *Progonomys* and *Karnimata* documents one such splitting event. Incremental accumulation of characteristic morphologies unfolded over hundreds of thousands of years. Early in their differentiation, the lineages cannot be distinguished reliably, and only individuals showing morphological extremes can be separated; other individuals overlap in key features. How do we distinguish by nomenclature such early phases of divergence? Below, we apply distinct species names for those murine taxa that can be morphologically supported. If morphological differences are not definitive, we use a compound name to signal that the lineages are either not yet fully split genetically, or if cryptic, have not yet diverged morphologically.

We resume the Potwar murine story with *Antemus chinjiensis*. The oldest record of *A. chinjiensis* is a large collection of mouse teeth from Locality Y491 (13.8 Ma). As with *Potwarmus* in older deposits, *Antemus chinjiensis* was abundant, comprising more than one-third of the small mammal fauna at Y491. Large samples of *A. chinjiensis* show considerable morphological variation (Jacobs, Flynn, and Downs 1989). The M1 of *A. chinjiensis* was 17% longer on average than that of *P. primitivus*, corresponding to 50% greater body mass (see Table 11.1). The successful early stem murine *Antemus* appeared and flourished after 14 Ma during global climatic cooling following the Middle Miocene Climatic Optimum (Zachos et al. 2001). *Antemus chinjiensis* remained common but declined in relative abundance by 13.2 Ma and is last recorded at Y750 (12.7 Ma).

The next younger murine ("post-*Antemus*") sampled at Localities Y496 and Y634 (12.4–12.3 Ma), is derived in having relatively stronger upper molar chevrons, particularly a connection, although low and weak, from the enterostyle to the protocone. A few specimens lack the connection, so remain *Antemus*-like in that trait. Strong chevrons, an apomorphic complex linked to herbivory (Coillot et al. 2013), are shared with all later murines including *Progonomys*. This variant became extraordinarily abundant, accounting for 40% of all rodent specimens at 12.4 Ma. Under the rubric "Near *Progonomys*" Jacobs and Flynn (2005) portrayed these fossils as representing a transitional species, but although advanced beyond *Antemus*, this taxon is equally like all later murines, without unique relationship to *Progonomys*. The Y496 and Y634 murine samples are more derived than *Antemus* and are better described as "post-*Antemus*." We interpret this Y496-Y634 murine to represent one species, despite the few atavistic individuals with *Antemus*-like morphology.

Modern Murinae: The Rise of Living Tribes

Although abundant, basal mice were not diverse compared with later Neogene and living murines; one or two species co-existed in the Middle Miocene Siwaliks. As documented in the Siwalik record, the extraordinary diversification of Murinae is rooted in the Late Miocene. Stem taxa *Potwarmus* and *Antemus* had new apomorphic lingual cusps in upper molars and buccal cingula in lowers—but as stem taxa lacked some of the features of later murine groups; they were effectively aberrant genera among cricetids and gerbils of the time. Morphometric and wear facet analyses of molars (Coillot et al. 2013) show that *Potwarmus* and *Antemus* lacked inclined cusps and strong transverse crests (chevrons) with facets for shearing vegetation and retained a lateral component to jaw movement in mastication. These stem murines likely included insects in their diet. High crests joining inclined cusps accompanied a shift to the fully back to front (proal) chewing of extant murine tribes, and since then (~12 Ma), the modern Murinae diversified explosively. Portions of that diversification are recorded in the Potwar sequence.

Local Record of Murine Tribe Origins on the Potwar

Our interpretation of mouse evolution is based on available fossils, predominantly isolated molar teeth. The first upper molar is the most complex dental element of the murine cheek-tooth series. (Other molars present some useful features but most distinctions are based on M1.)

Very few rodent fossils were recovered in the interval between the dominance of "post-*Antemus*" at 12.3 Ma and the next younger screened localities (Y504, Y83) at 11.6 Ma. By that time murine samples show evidence of more than one species. A few teeth represent a small murine with narrow molars that are referred to *Progonomys morganae* Kimura et al., 2017, named originally for younger samples (10.5–10.1 Ma). Several larger specimens represent a mouse with broad molars ("Genus D" in Flynn, Kimura, and Jacobs 2020).

Most specimens in the age range of 11.6 to 11.4 Ma represent a murine with derived cusp arrangement but considerable variability in occlusal features, including the shape of upper molar chevrons. Currently, we interpret as intraspecific the observed variation in features of this murine. Molars of this age (Y83 and Y504 at 11.6 Ma; Y76 and Y809 at 11.4 Ma) precede the distinct Middle Siwalik *Karnimata* and *Progonomys* and cannot be attributed consistently as species of either of those two genera. Some sampled M1s present morphology like that of *Progonomys*, others of *Karnimata*, but variation is not bimodal, and other tooth positions show inconsistent variation. Connections between cusps are not as marked as in the genotypic species *Progonomys cathalai* and *Karnimata darwini*. This early mouse presents a variable, primitive morphological grade less derived than in either *Progonomys* or *Karnimata*, and we apply informally the descriptor "pre-*Progonomys* sp. X" (simply pre-*Progonomys* in Kimura, Flynn, and Jacobs 2021). Although some specimens in the 11.6–11.4 Ma range show apomorphies that herald later *Karnimata* or *Progonomys*, these are interpreted as variants within a single species, phenotypes characteristic of later lineages after they fully split (Flynn, Kimura, and Jacobs 2020; Kimura, Flynn, and Jacobs 2021). Many molars lack distinctive features of either genus, so cannot be identified with certainty. The fossils apparently represent one interbreeding population with variants that only later became characteristic of the successor genera.

Stratigraphically superposed samples at 11.2 Ma show more consistently modern murine morphology and clearer *Karnimata* and *Progonomys* M1 morphotypes (but still with continuous

variation). Some M1 in these samples have a more symmetrical anterior chevron; others have more inclined cusps with an asymmetrical chevron (the anterostyle displaced posteriorly and laterally compressed). These phenotypes are, respectively, the earliest evidence of the long *Karnimata* Jacobs, 1978, and *Progonomys* Schaub, 1938, lineages (Fig. 11.13). Some molars resemble younger named species so could be called "cf. *K. fejfari*" or "cf. *P. hussaini*." Kimura, Jacobs, and Flynn (2013), Kimura, Flynn, and Jacobs (2015), and Kimura et al. (2015) cited these variants as "?*Karnimata*" and "*Progonomys* sp." But at 11.2 Ma most of the recovered teeth, particularly second and third molars, still show no apomorphic features of either *Progonomys* or *Karnimata*.

These fossils probably sample a species complex at a time after genomic splitting began but before morphologies of the dentition definitive of the crown tribes Murini and Arvicanthini to which *Progonomys* and *Karnimata* are respectively assigned were fully established. This suggests at least two scenarios: the murine sample represents two cryptic species that have split genetically but have not yet diverged greatly in morphology, or the sample represents a single population with high variation, some individuals with phenotypes that are later characteristic

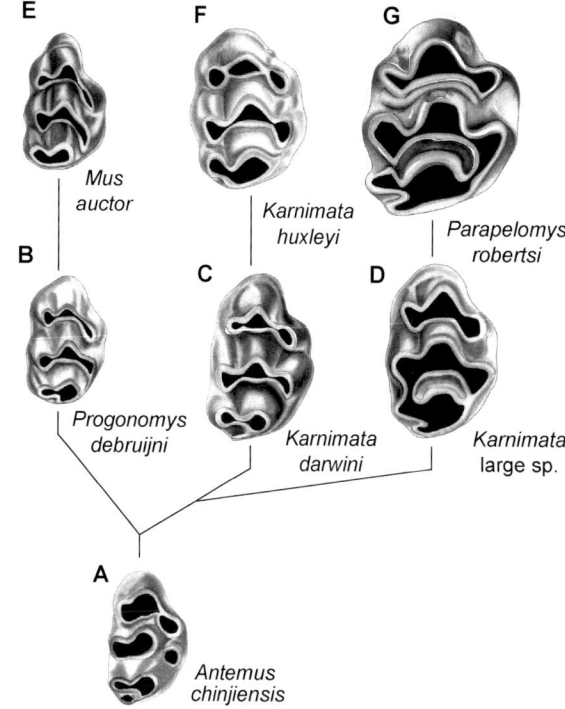

Figure 11.13. A scenario of fossil murine evolution with taxa from the Potwar Plateau Siwaliks represented by upper first molars. The *Antemus* lineage (A at ~13 Ma) is continued by *Progonomys* and *Karnimata*, stem members of tribes Murini and Arvicanthini. The murin *Progonomys* lineage (B, *P. debruijni* at 9.4 Ma) continues to *Mus auctor* at 6.5 Ma (E). The arvicanthin *Karnimata* diversifies in the Late Miocene; at 9.4 Ma *K. darwini* (C) coexists with a large *Karnimata* sp. (D); by 6.5 Ma, *K. darwini* is succeeded by *K. huxleyi* (F) and the large *Karnimata* by *Parapelomys robertsi* (G).

of daughter *Progonomys* and *Karnimata* lineages. The lineages may have diverged in features other than those later captured by molar morphology. The ultimate differentiation of the murine tribes was possibly tied to partitioning of different resources by stem lineages, which Kimura, Jacobs, and L.J. Flynn (2013) argue was reflected in tooth morphology after 10 Ma.

In either case the 11.2 Ma fossils sample abundant mice in the process of splitting and diverging to become two long-enduring lineages representing extant tribes. Their splitting process was not a brief event in generational or ecological time, but unfolded over hundreds of thousands of years. Introgression of the two nascent lineages was likely. Because most individuals of the 11.2-Ma sample cannot be distinguished as two species, we label them as a single, abundant, incompletely differentiated taxon at the morphological grade "cf. *K. fejfari*-cf. *P. hussaini*" ("indeterminate *Progonomys-Karnimata* grade" in Flynn, Kimura, and Jacobs [2020] and Kimura, Flynn, and Jacobs [2021]). We do not know the exact point in time of genetic separation, but the acquisition of distinctive morphologies in the two lineages required at least 400 ky between the 11.6–11.2 Ma sites.

The large fossil collection at 11.2 Ma includes a minority of teeth representing other murines, *Progonomys morganae* and the robust "Genus D," plus a single large M1 designated "Genus C." This pattern of uncommon species co-occurring with one or two abundant murine lineages is characteristic of Potwar murine assemblages throughout the Late Miocene.

Younger Miocene species of *Progonomys* and *Karnimata* diverged in distinct features with respect to samples of 11.2-Ma age. Specimens from sites in the Nagri Formation dating from 10.5 to 10.1 Ma (Y259, Y450, Y311) with morphological differences characterizing the two genera are formally named *Progonomys hussaini* and *Karnimata fejfari* (Cheema et al. 2000; Kimura, Flynn, and Jacobs 2017). Kimura et al. (2015) showed that *Progonomys* and *Karnimata* are respectively early members of the crown groups Murini and Arvicanthini; hence *P. hussaini* and *K. fejfari* demonstrate separation of these tribes prior to 10.5 Ma.

The Potwar Late Miocene 10.5–10.1-Ma localities Y259, Y450, and Y311 with abundant *Progonomys hussaini* and *Karnimata fejfari* also contain the uncommon, small *P. morganae* and robust Genus D, as well as the stephanodont *Parapodemus badgleyae* (Kimura, Flynn, and Jacobs 2017). Stephanodonty is a linkage of buccal cusps on upper molars, presumably enhancing longitudinal shearing crests, a character of tribe Apodemini. At 10.5 Ma, *Parapodemus badgleyae* provides a minimum age for the split of the Apodemini-Murini clades.

"U-Level" Mice
The murine pattern of high relative abundance of two species dominating two or three other mice continues through 9.7 Ma and younger deposits. The productive 9.4 to 9.2–Ma "U-level" of the Dhok Pathan Formation produces common *Progonomys debruijni* (small with strongly inclined cusps) and the even more abundant *Karnimata darwini* (large with broad molars; Jacobs 1978). Phenetic differences between *P. debruijni* and *K. darwini* are greater than those between their predecessors one million

years earlier (Kimura, Flynn, and Jacobs 2017). Kimura, Jacobs, and Flynn (2013) concluded that early *Progonomys* and *Karnimata* evolved to exploit different food sources. By 9.4 Ma, size difference was distinct, in concert with an interpretation of competition for food (Bowers and Brown 1982).

Two uncommon mice from "U-level" microsites are *Karnimata* large sp. (unnamed, the first sample of a long lineage) and *Parapodemus hariensis*, first named for coeval material from Haritalyangar, northern India (Vasishat 1985) and now recognized in the Potwar Plateau (Kimura, Flynn, and Jacobs 2017). Vasishat (1985) also named "*Progonomys chopraii*" for a partial skull (Panjab University Anthropology Department [PUA] 71-48) found east of Haritalyangar but restudy shows synonymy of *P. chopraii* with *Progonomys debruijni* (Patnaik 2020).

The "U-level" assemblage persists to younger horizons of the Dhok Pathan Formation. The same murines are encountered over the succeeding million years in Localities Y367 at 9.0 Ma, Y387 and Y388 at 8.7 Ma, and Y24 at 8.2 Ma. *Karnimata darwini* and *Progonomys debruijni* are accompanied by the unnamed large *Karnimata* species; *Parapodemus hariensis* occurs sporadically. Probably falling in this time range is the Haritalyangar "*Siwalkomus nagrii*" of Vasishat (1985). Its maxilla, PUA 18–75, confirms identity with *Karnimata darwini*, and a lower jaw included in "*Siwalkomus nagrii*" represents *Parapodemus hariensis* (Patnaik 2020). These finds indicate distribution of the murine community across the northern part of the subcontinent.

Late Miocene Change
At Locality Y547 (8.0 Ma) Potwar *Karnimata darwini* persists with the unnamed large species of the genus, but the sample of *Progonomys* shows a slight size increase (Kimura et al. 2013). This new *Progonomys* species, not yet named, persists until at least 7.2 Ma (Locality Y581). Y547 also records the first appearance of a much smaller mouse, a representative of the extant genus *Mus*. As in living species of *Mus*, the posterior cingulum is lacking, and its M1 is strongly asymmetrical, with sloping anterobuccal margin extended forward and a correspondingly retracted, compressed lingual anterostyle. This species, *Mus absconditus* Flynn and Kimura, 2023, is the oldest representative of the genus from anywhere. Other named fossils of the genus are larger on average: Siwalik (6.5 Ma) *Mus auctor* Jacobs, 1978; Afghan Pliocene (~5 Ma) *Mus elegans* Sen, 1983.

The murine assemblage continues to change in the Late Miocene. At sites of 7.8 to 7.4 Ma, a tiny *Mus absconditus* appears alongside *Progonomys*, and the *Karnimata* lineages are represented by derived species. *Karnimata huxleyi* replaces *K. darwini*, and in place of the large, unnamed *Karnimata* is the larger murine designated "cf. *Parapelomys robertsi*." Specimens of the latter are not as big as the type series of *P. robertsi* from Locality D13 (6.5 Ma). Localities Y921 and Y931 (7.3 Ma) produce large samples of *Karnimata huxleyi*, *Mus absconditus*, cf. *Parapelomys robertsi*, and *Progonomys* n. sp. One hundred thousand years later (Y581, 7.2 Ma), the assemblage persists with larger *Parapelomys robertsi*.

Our hypothesis of murine relationships is developed for these Siwalik taxa (Fig. 11.14) as a tree with dates at nodes for

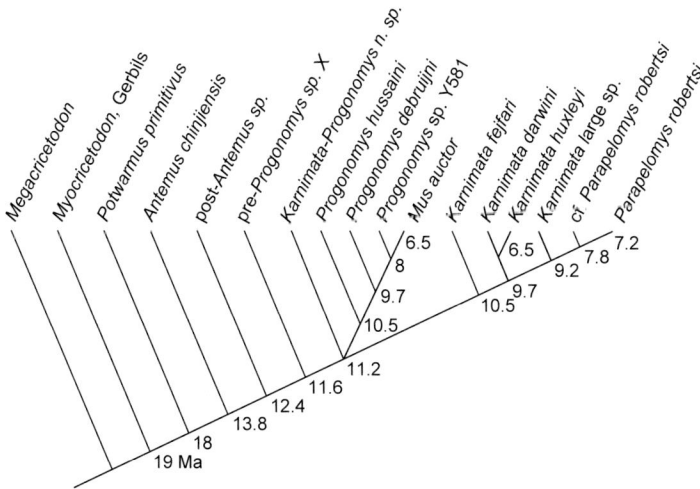

Figure 11.14. Tree of relationships proposed for Siwalik murids with *Megacricetodon* as outgroup. *Myocricetodon* represents primitive gerbillines; *Potwarmus* is a stem Murinae. Potwar Plateau fossil murines diversify mainly as species in two clades, the Tribe Murini (*Progonomys* and relatives) and Tribe Arvicanthini (*Karnimata* and relatives). Node ages are estimates for localities of first occurrences.

first occurrences in the fossil record. There is good internal concordance of age and placement on the tree and of agreement with recent molecular-based age estimates of clade origins (Lecompte et al. 2008; Aghová et al. 2018; Flynn, Kimura, and Jacobs 2020).

The Late Miocene (6.5 Ma) Locality D13 is the type locality (Jacobs 1978) for *Karnimata huxleyi*, *Parapelomys robertsi*, and *Mus auctor*. The site is dominated by murines, with *K. huxleyi* and *M. auctor* each about twice as abundant as *P. robertsi*. A few teeth of a fourth unidentified murine occur with them (Genus A; Jacobs 1978). By this time both the derived *Progonomys* and small *Mus* species are no longer present. *Mus auctor* has a less distinctive M1 chevron, more like that of *Progonomys* and not as asymmetrical and projected. M1 length of *Mus auctor* is 7% less than that of preceding *Progonomys* but 12% greater than that of the older *Mus* species (Flynn and Kimura 2023).

In our interpretation, the Dhok Pathan Formation localities of 8–7 Ma record an advanced *Progonomys* and small *Mus* of modern grade. Clearly the *Mus* species is an immigrant. By 6.5 Ma (D13) both *Mus absconditus* and *Progonomys* are gone; common *Mus auctor*, not derived from earlier Siwalik *Mus*, may also be an immigrant.

At 6.5 Ma, Locality D13 is the youngest Potwar site that produces a large sample of micromammals. It occurs in the Dhok Saira paleomagnetic section of Barry et al. (2002), at the top of a local succession of superposed fossiliferous localities. Younger murids occur elsewhere in Pakistan, notably at Mirpur and Lehri (Cheema, Raza, and Flynn 1997; Cheema, Flynn, and Rajpar 2003). The Samwal Formation of the Mirpur area (~3 Ma) preserves *Golunda kelleri*, *Hadromys* sp., and *Mus* sp. The Soan Formation Lehri site estimated at 2 Ma occurs below the Olduvai subchron in the local paleomagnetic section. It yields specimens of *Golunda kelleri*, *Hadromys loujacobsi*, a small *Mus* sp. and a larger murine, cf. *Cremnomys* sp.

Pleistocene deposits of the Pabbi Hills adjacent to the Potwar Plateau produce two distinctive murines (Jacobs 1978). Just above the Olduvai Subchron (paleomagnetic section of Keller et al. 1977), about 1.75 Ma, Locality D24 yields the hypodigms of *Golunda kelleri* and *Hadromys loujacobsi*. Murines remain a

major part of the small mammal fauna of the Indian subcontinent, thereafter including extant species of *Golunda* and *Hadromys* (Musser 1987).

PALEOECOLOGY, BIOGEOGRAPHIC DISTRIBUTIONS, AND EVOLUTIONARY TRENDS

Paleoecology and Communities

The composition of Siwalik microfaunas indicates warm, humid conditions in the Early and Middle Miocene of Pakistan. The diverse squirrels, bamboo rats, early gerbils, and mice, as well as the more restricted records of cane rats, scaly-tailed flying squirrels, and diatomyids reflect subtropical conditions through the Middle Miocene. The high species richness (near 30 species; Flynn et al. 1998) recorded at single Chinji Formation localities (~13 Ma) also reflects ecological diversity in the Middle Miocene Siwaliks. The early Late Miocene Siwaliks of the Potwar Plateau remained seasonally moist (Flynn 2003), but by then species turnover and declining richness among small mammals (Barry et al. 1991) signal significant change.

The decreased species richness after 10.5 Ma probably reflects increasing seasonality of precipitation and diminishing food sources for small mammals during dry seasons. Several Late Miocene appearances suggest open habitat: genera with greater crown height, a taterillin gerbil, and the *Karnimata* murine lineage with molar crests adapted for tough vegetation. Additionally, δ13C molar enamel isotopic values show that *Karnimata* ingested some proportion of C4 grasses, likely exploiting vegetation not consumed by other mice (Kimura et al. 2013).

By 9.7 Ma early bamboo rats (*Miorhizomys*), burrowers of modern grade, appear in the record and diversify over the next few hundred thousand years, signaling ecological change probably including increased seasonality of moisture. An inferred dry season and increased fire incidence (Karp, Behrensmeyer, and Freeman 2018; Karp et al. 2021; see also Chapter 24, this volume) may have selected some plant species to

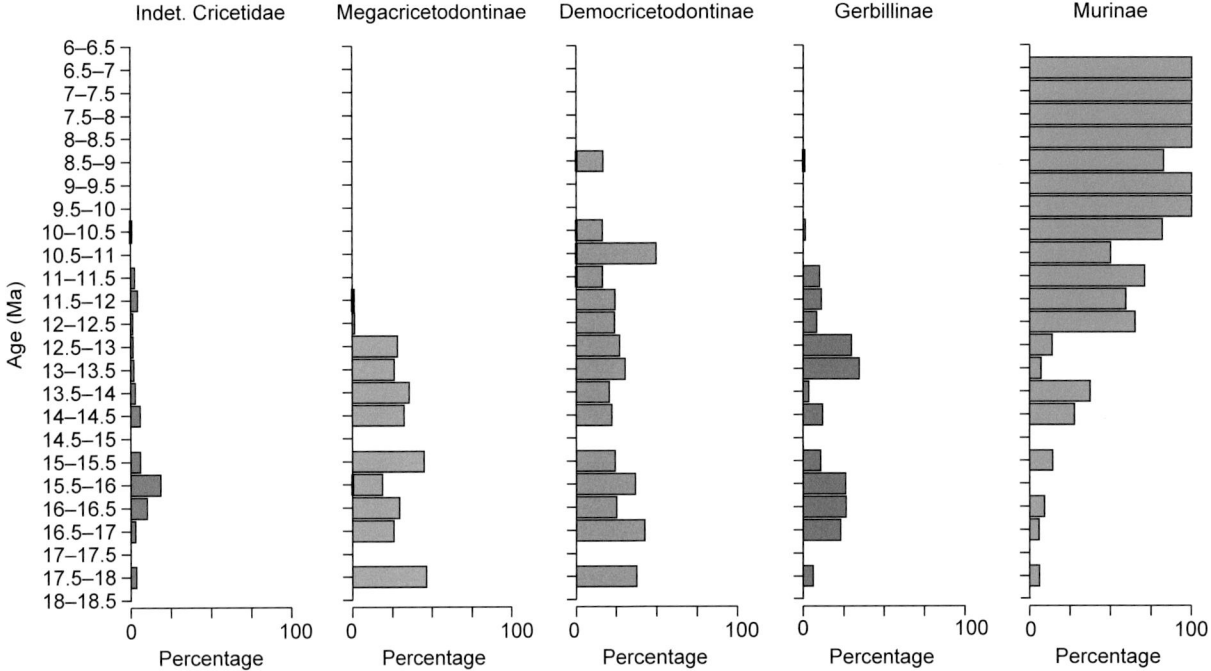

Figure 11.15. Bar graph showing replacement among derived muroids. Murine rodents replace cricetids plus gerbils in numbers of individuals during the Late Miocene. Here abundances are estimated by numbers of teeth in samples, presented as percentages of total muroids from localities aggregated into 0.5 m.y. time bins. The graph is based on a database upload of July 2017, counting all molars as individuals. Age bins: any site with a date matching the younger age (on the left) is included in the next younger interval. By ~10 Ma, replacement of cricetines is nearly complete, but note the brief reappearance of *Democricetodon* at 8.7 Ma. The late Early Miocene *Potwarmus*, preceding *Antemus*, is included in the murine count.

evolve subterranean organs to store nutrition. The success of *Miorhizomys*, committed to fossorial life, suggests exploitation of such a previously unavailable food source. Subfamily Rhizomyinae later declined in abundance on the Potwar Plateau and in species richness by the Pleistocene, presumably due to loss of dependable resources.

The paramount change among Miocene small mammals is replacement of cricetids and gerbils by murines. Late in the Middle Miocene, Murinae came to greatly exceed Cricetidae plus Gerbillinae in numbers of individuals (Fig. 11.15). Cricetid and gerbil species richness also declined especially after 10.5 Ma. The success over genera with the old dental "cricetid plan" by mice with the novel murine cusp arrangement and especially full proal mastication, was nearly complete after 10 Ma (Lindsay 2017).

Subtropical habitats supported similar microfauna over a large portion of the Indian subcontinent, judging from the wide distribution of Potwar taxa from the trans-Indus in the west, to northern India, and southward to the Manchars of Sindh and the Kutch District in Gujarat (Wessels et al. 1982; Patnaik 2013; Bhandari et al. 2021; Patnaik et al. 2022). Similarity of assemblages through time (most taxa the same) indicates stable associations of species distributed over a large area, and these document a succession of metacommunities (Leibold et al. 2004; Flynn et al. 2023a).

Flynn et al. (2023a) distinguish successive Miocene metacommunities in the Potwar fossil record based on characteris-

tic taxa and the change in muroid abundances noted above. A mid-Miocene *Antemus chinjiensis* metacommunity was followed by a crown Murini metacommunity of greatly increased murine abundance at the expense of cricetids and gerbils in both numbers of individuals and of species. This community transformation took place within 300,000 years. The early crown murine metacommunity was succeeded after 10 Ma by the Late Miocene *Karnimata darwini* metacommunity with very abundant murines (mainly *K. darwini*) and numerous rhizomyines but near total absence of cricetids or gerbils. Here we add earlier and later small mammal associations and propose five successive metacommunities. Periods of relatively stable ecology promoted widespread Potwar metacommunities during the Miocene (Table 11.2), which were distinctly South Asian in character and contrasted in structure with communities elsewhere in Eurasia.

Biogeography

The prevailing paleobiogeographic signature for Siwalik small mammals is predominantly endemic but with episodic introductions of lineages from the wider Indo-Malay region, as well as the Middle East and Africa. Because few Siwalik small mammal species are known from beyond Indo-Pakistan, strict application of fossil species occurrences could indicate nearly complete endemism. The successive assemblages of the Potwar Plateau suggests that the proportion of the fauna that is truly endemic varies through time.

Table 11.2. Potwar Small-Mammal Metacommunities

Metacommunity	Age Range	Characteristics
Parapelomys spp.	7.8–6.5 Ma	Abundant Murinae *Parapelomys*, *Mus*, *Karnimata huxleyi*; no cricetids; immigrant *Hystrix* and Leporidae
Karnimata darwini	9.7–8.0 Ma	Abundant *Karnimata darwini* and *Progonomys debruijni*; diverse *Miorhizomys*, abundant *Kanisamys*; *Sayimys* absent; cricetids and gerbils rare
Crown Murine Radiation	12.4–10.1 Ma	Abundant crown Murinae; cricetids and gerbils reduced proportionately
Antemus chinjiensis	13.8–12.7 Ma	*Antemus chinjiensis*; Cricetidae and Gerbillinae common, abundant hedgehog *Galerix rutlandae*
Potwarmus primitivus	16.0–14.0 Ma	Abundant *Potwarmus*; diverse Cricetidae; *Sayimys* spp. exceeding 5% of individuals; abundant hedgehogs (*Galerix* spp.)

To more closely evaluate the contribution of immigrants, Flynn et al. (2014a) made a series of explicit assumptions, which we follow here. First, the Potwar Plateau sequence provides a reasonable sampling of communities distributed throughout the northern part of the Indo-Pakistan biogeographic subprovince of the wider Indo-Gangetic floodplains and the regional Indo-Malay biogeographic province. Second, if for a particular Potwar species an earlier potential Potwar ancestor can be identified, either through formal cladistic analysis or by prior presence of a congeneric species, then the inferred lineage is scored as "endemic." By viewing fossil assemblages from well-sampled intervals as adequate representatives of communities, species within assemblages can be sorted into three categories: "residents," or those already earlier present; new "endemics," as defined here and in Flynn et al. (2014a); and new "immigrants," or Potwar species for which there are no plausible ancestors within the sub-region. Some examples follow.

The species *Alilepus elongatus* appears at 7.4 Ma in the Potwar Plateau, without predecessor, the first leporine in the long Neogene record. It is an immigrant. The closest relatives (geographically and morphologically) are from Afghanistan and China. The species *Alilepus elongatus* is considered a Late Miocene Siwalik immigrant, while the genus *Alilepus* is an earlier immigrant to Asia from North America (Flynn et al. 2014b).

Diatomyid rodents were present in the subcontinent since the Early Oligocene, the oldest known member being Bugti *Fallomus*. The next younger diatomyid, Early Miocene *Marymus*, is known only from Pakistan (Flynn 2007); later *Diatomys* occurred from Pakistan to Japan. The family, primarily South Asian, ranged westward to Romania (Marković et al. 2017). *Marymus* may be endemic, because a possible evolutionary predecessor occurs in the region, but the origin of *Diatomys* is not yet constrained due to its wide geographic range. As a family, Diatomyidae is a characteristic Indo-Malay element that spread beyond that province.

The Early Miocene ctenodactyline *Sayimys flynni* (Z113, itself an immigrant) at the base of the *Sayimys* radiation (López-Antoñanzas and Knoll 2011) is plausibly ancestral to later *Sayimys*, the genus considered to have evolved endemically. One species in the lineage, *S. intermedius* of Saudi Arabia, indicates westward dispersal from Pakistan in the late Early Miocene. Subsequent east-northeastward migration of *Sayimys* produced *S. obliquidens* in China (Flynn and Wessels 2013).

Cane rats (Thryonomyidae) have a long evolutionary history in Africa and dispersed eastward to South Asia twice. Immigration of Early Miocene *Kochalia* was followed in the mid-Miocene by derived genus *Paraulacodus* (not closely related to *Kochalia*; Flynn and Winkler 1994). Both dispersed from East Africa. Late Miocene reverse spread is possible, given the different species of *Paraulacodus* in Ethiopia at 8–8.5 Ma (Suwa et al. 2015).

The porcupine genus *Hystrix* has its oldest records in Europe and became widespread across Eurasia in the Late Miocene. *Hystrix primigenius* (no local ancestor) is a Siwalik immigrant at 8.0 Ma from the Pikermian chronofauna to the west (Van Weers and Rook 2003).

The tiny dormouse *Myomimus* suggests a similar direction of immigration. The genus is absent from China until the Late Miocene (Qiu et al. 2013), but present by 13.8 Ma in Pakistan, which indicates that its source is likely from the west at subtropical latitude. Successive dormouse populations of the Potwar Plateau appear to represent in situ evolution. Therefore, the Late Miocene lineage is interpreted as endemic.

The squirrel family was diverse in Potwar assemblages. Most genera are shared with Late Miocene faunas of southern China and illustrate Siwalik faunal affinity with the Indo-Malay biogeographic province. These squirrels are immigrants to the Potwar, but the Asian fossil record is not yet documented well enough to suggest the source area for most individual genera. The Middle Miocene chipmunk, however, is not found in the fossil record of Southeast Asia. Ancient *Tamias* was widespread throughout Holarctica, with Early Miocene presence in Inner Mongolia (Qiu and Li 2016). The ground squirrels *Atlantoxerus* and *Heteroxerus*, widespread across southern Europe and North Africa and therefore not endemic, suggest mid-Miocene faunal influence from the west.

After the earliest Miocene appearance of *Prokanisamys*, the evolution of rhizomyines on the Potwar Plateau was mainly endemic. Westward emigration events can be documented: *Prokanisamys* sp. found in North Africa in the late Early Miocene (Wessels 2009) and *Pronakalimys* in the Middle Miocene of East Africa (Tong and Jaeger 1993). The derived tachyoryctins *Protachyoryctes* and *Eicooryctes* are endemic to South Asia. Younger *Tachyoryctes* in East Africa represents late Neogene emigration from South Asia of a hypsodont burrower (Black 1972; López-Antoñanzas and Wesselman 2013). The Mio-Pliocene appearance of *Rhizomyides* in Afghanistan is a rare example of a small mammal dispersing westward across mountain passes from the Potwar Plateau. The fossorial bamboo rat *Miorhizomys* appears in the Potwar after 10 Ma, an immigrant from elsewhere in the Indo-Malay province. Later bamboo rats (*Rhizomys*) spread into

Shanxi Province at about 6 Ma, an unusual example of a small mammal emigrating to North China from the Indo-Malay biogeographic province.

For much of their histories, *Rhizomyides* and the Diatomyidae were rare or absent from the Potwar Siwaliks, appearing only briefly and sporadically. They may represent rodents of the Indo-Malay region that were occasional residents of the Potwar when ecological conditions were favorable. These peripheral rodent lineages expanded into the Potwar subregion opportunistically.

The diminutive muroid *Primus* of the Early Miocene Murree Formation and the Zinda Pir Dome, derived with respect to Oligocene muroids, appears without antecedent and is therefore considered an immigrant, preceding the cricetid immigrant *Democricetodon khani*. Early *Democricetodon* species of about the same age (21–22 Ma) are widespread from Xinjiang, China, to Turkey and the eastern Mediterranean. The points of origin of neither *Primus* nor *Democricetodon* can be specified at present. *D. khani* is the first member of a long, possibly endemic, cluster of hamster species in the Potwar Plateau. By about 19 Ma two murids are present, the early gerbil *Myocricetodon* and the mouse *Potwarmus*; neither of these is readily derived from local predecessors, so both are inferred to be immigrants. *Myocricetodon* and its allies diversified in the Middle and Late Miocene of North Africa and Early Miocene of Turkey (Wessels 2009) and east China (*Spanocricetodon* Li, 1977). *Potwarmus* (known in the Middle Miocene of Thailand) and *Myocricetodon* could have originated elsewhere in Indo-Malaya or in the eastern Mediterranean region. The later endemic Potwar species *Myocricetodon sivalensis* persisted until the late Middle Miocene. The early gerbils *Mellalomys* and *Dakkamys* appear in the Potwar as immigrants, both having Middle Miocene records in North Africa (Jaeger 1977). Early *Dakkamys* in southern France predates the earliest Potwar record of the genus (Lazzari, Michaux, and Aguilar 2007).

Megacricetodon represents a distinctive species group from the Miocene of Europe and northern Asia. The genus appears in the Potwar sequence at 17.9 Ma, roughly the same time as in Europe (Oliver and Peláez-Campomanes 2013) and persists until about 12.3 Ma. As for other cricetid genera, the center of origin is unclear. After the immigration of *M. aguilari* and *M. mythikos*, Potwar species of *Megacricetodon* and the closely related *Punjabemys* are interpreted as endemic.

Despite ranking as a genus distinct from *Potwarmus*, the stem murine *Antemus* probably originated endemically from it. Successive Middle Miocene and early Late Miocene populations of derived murines are known nowhere else in the world, and we propose that most evolved within the Indo-Malay realm, leaving a record in the Potwar biogeographic subregion (Flynn, Kimura, and Jacobs 2020). Siwalik murines after *Antemus* until 11.2 Ma comprise an endemic succession of murines under the names "post-*Antemus*," "pre-*Progonomys* sp. X," and "cf. *Progonomys hussaini–Karnimata fejfari*." Currently these are interpreted as a single lineage that split after 11.2 Ma and became diagnosable as the long-ranging genera *Progonomys* and *Karnimata* (see Fig. 11.13). Other Potwar murines

(Genus D, *Progonomys morganae*, *Parapodemus badgleyi*) are immigrants from elsewhere in the Indo-Malay province. The Late Miocene lineages *P. hussaini–P. debruijni*, *K. fejfari–K. darwini*, and *Parapodemus badgleyae–P. hariensis* are endemic.

One of the most striking emigration events for the Siwalik record is the early Late Miocene dispersal of *Progonomys* from the Indian subcontinent to the circum-Mediterranean region (Wessels 2009; Casanovas-Vilar et al. 2016). The modern murin *Progonomys* spread westward after the genus differentiated, occupying murine-free territory. Both *Progonomys* and the genus *Parapodemus* have records in the Middle East and the Mediterranean region by 10 Ma. Siwalik *Progonomys hussaini* (10.5 to 10.1 Ma) is similar morphologically to early *Progonomys cathalai* (type species of the genus) of the circum-Mediterranean region (Wessels 2009). Also similar is the earliest *Progonomys* in North China (*P. sinensis* Qiu et al., 2004), well-dated paleomagnetically to 10.0 Ma (Kaakinen 2005). Flynn and Wessels (2013) proposed a scenario of early westward dispersal of *Progonomys* and then spread of *Progonomys* northeastward to China. López-Antoñanzas et al. (2019) placed the dispersal of *Progonomys* to the Levant near 11 Ma. Westward dispersal of *Parapodemus* is recorded at Buzhor 1, Moldova (~10 Ma, Sinitsa and Delinchi 2016).

Late Miocene Potwar Murinae sample long-lived lineages. Successive *Karnimata* and *Progonomys* species are endemic and the former lead to *Parapelomys robertsi*. The immigrant *Mus absconditus* at 8.0 Ma is an early member of the living species-rich genus, not closely related to the *Progonomys-Mus auctor* lineage (Kimura et al. 2013). Possibly *Mus absconditus* evolved elsewhere in Southeast Asia, and its lineage immigrated to the Potwar Plateau. *Mus auctor* may also be an immigrant, arriving after *Mus absconditus*. It remains to be determined exactly the pattern of decline of *Progonomys* and rise of early *Mus auctor*. The evolution of modern Murini (*Mus* and close relatives) likely centered in southeastern Asia and the peripheral Potwar sampled the edges of that radiation.

Sorting Potwar Plateau murines into endemics or immigrants is possible because of sampling density, especially between 14 and 6.5 Ma. Youngest Miocene and Pliocene rodent samples are few, so the appearances of later genera (e.g., *Golunda*) are cautiously interpreted as immigrations from elsewhere in the Indo-Malay region.

Immigrant sources for the Siwalik biogeographic subregion, as represented in the Potwar Plateau, appear to have been primarily from parts of the Indo-Malay province to the east and from either southwestern Asia or Africa to the west. Few taxa are likely to have spread into the area from north of the Himalayas. Dispersal routes for emigrants from the Potwar appear to have been largely westward along a subtropical belt (Flynn and Wessels 2013). Some rodents spread eastward to South China, but few (notably *Miorhizomys*) spread secondarily northward from South China into the Palaearctic biogeographic province. *Sayimys* and *Myocricetodon* occurred in the Siwaliks and later in North China; the hypothetical route of entry to North China was first westward from Indo-Pakistan and then northeastward across Asia (Flynn and Wessels 2013).

Evolutionary Trends

We have documented evolutionary change in rodent groups over millions of years across multiple species within Potwar Siwalik lineages. *Sayimys intermedius–S. chinjiensis*, for example, increased in crown height and size over its long 18–10-Ma age range (Baskin 1996). The dormouse *Myomimus* exhibited slight lengthening of molars. The chipmunk *Tamias urialis* showed slight size reduction from its predecessor.

A comprehensive cladistic analysis of Rhizomyinae (López-Antoñanzas, Flynn, and Knoll 2013) makes it possible to assess relationships of genera and rates of diversification. These rodents diversified mainly endemically, with occasional dispersals east and west. Rhizomyine muroids increased in size and showed cumulative changes in molar crown morphology, accompanied by punctuated increments in hypsodonty (Flynn 1982). The bamboo rat tribe (Rhizomyini) exhibited increased body mass, incisor size, and depth of the dentary. Evolutionary rates in this group appear related to changes in the intensity of the Asian monsoon system complex (López-Antoñanzas et al. 2015), perhaps linked to tectonic uplift to the north. A pulse in rhizomyine evolution culminated in the radiation of crown Rhizomyini after 10 Ma. The overall pattern of bamboo rat evolution accompanied exploitation of a new food source available after 10 Ma. By 9.7 Ma, these bamboo rats evolved fossorial characteristics to exploit underground plant nutrient storage organs. Seasonality of precipitation, perhaps forced by regional tectonic events, led to increased incidence of fire and changes in plant nutritional resources. The fully fossorial bamboo rats diversified despite a decline in optimal forest habitat by exploiting underground food resources.

Set on the Indian Peninsula with mountain and river barriers on three sides and bounded by the Indian Ocean, the Potwar Siwaliks preserved a sequence of resident mouse populations sampled by fossil teeth (Flynn, Kimura, and Jacobs 2020). Murines left a fascinating record of evolution showing trends that span multiple species. *Antemus chinjiensis* originated from gerbil-like *Potwarmus primitivus*, as documented by novel adaptations of the murine chevron with extra lingual cusps in stable position on upper molars (see Fig. 11.10) and coordinated changes in lower molars. The stem murine *Antemus* lineage documents strengthened cusp connections toward a *Progonomys*-like dentition.

Early in murine history in the 14–12 Ma interval, dental evidence indicates a single abundant lineage, *Antemus chinjiensis*, followed by species informally designated post-*Antemus* and pre-*Progonomys*. Early murines underwent great dental and masticatory transformation: stronger chevrons with crosscutting crests and fully proal mastication gave them great competitive advantage over contemporary muroids (Coillot et al. 2013; Flynn et al. 2023a).

Later murine assemblages from single sites show within-species variation in an abundant mouse species accompanied by two or three uncommon murines. By 11.2 Ma, rich murine samples include variant morphologies that represent the roots of the genera *Progonomys* and *Karnimata*. By 10.5 Ma these were distinct as *P. hussaini* and *K. fejfari*. Kimura, Flynn and Jacobs (2015, 2017). Kimura et al. (2015) documented gradual divergence of these lineages, stem members of extant murine crown groups, the *Mus* and the *Arvicanthis* tribes, respectively. A *Parapodemus* lineage of 10.5–8.0 Ma represents the somewhat younger stem of Tribe Apodemini.

It is precisely because we benefit from a long, densely sampled record with later easily recognized members that we can perceive the subtle origins of early *Karnimata-Progonomys* grade lineages. Large samples and a morphometric approach demonstrate the diverging trajectories of the genera after 11.2 Ma (Kimura, Jacobs, and Flynn 2013; Kimura, Flynn, and Jacobs 2021). Earlier samples (11.4 and 11.6 Ma) include some individuals having phenotypes (presumably genetically based) that became characteristic of the distinct daughter taxa. Genetically determined morphological variation preceded species separation, the point at which the paleontological record can resolve distinct lineages (Steiper and Young 2008; Kimura, Flynn, and Jacobs 2015). This high-quality Potwar fossil record can be used to gauge the time between the phenotypic variation in a single parent lineage and ultimate species divergence. Our data indicate that this particular splitting event unfolded over hundreds of thousands of years.

The pattern of succession of Potwar mouse assemblages indicates the origin of several lineages within the biogeographic subprovince and provides minimum dates for the separation of important crown lineages. The stem murine *Antemus chinjiensis* phyletically replaces *Potwarmus primitivus* about 14 Ma and is in turn replaced by a currently unnamed murine before 12 Ma. Following the cladogenic split of *Karnimata-Progonomys* (~11.2 Ma, conservative date for Arvicanthini and Murini tribes), dual parallel sequences evolved (see Fig. 11.13): *Progonomys* cf. *P. hussaini*→*P. hussaini*→*P. debruijni*→*Progonomys* n. sp. at Y581→*Mus auctor*; *Karnimata* cf. *K. fejfari*→*K. fejfari*→*K. darwini*→*K. huxleyi*. The tribe Apodemini differentiated by 10.5 Ma, and the *Parapodemus badgleyae*→*P. hariensis* succession occurred by 9.4 Ma. Another Late Miocene replacement involved origin of *Parapelomys* about 7.4 Ma: *Karnimata* unnamed large sp.→cf. *Parapelomys robertsi*→*P. robertsi*.

Siwalik murine data are in good accord with recent estimates of lineage splitting times that assume clocklike genomic evolution. The trees of Rowe et al. (2008), Fabre et al. (2012), and Steppan and Schenk (2017) propose age estimates for nodes that are consistent with the Siwalik fossil record. New fossil data (Flynn, Kimura, and Jacobs 2020) show that the phylogenetic analysis and time tree of Lecompte et al. (2008), while concordant with the Potwar record, likely underestimates slightly the antiquity of murine tree nodes. These various analyses use the age of fossil *Antemus* as a calibration point, so general concordance is not surprising. Aghová et al. (2018) add multiple calibration points from Siwalik murines in their study and find not only better agreement of the fossil record with estimated origins of extant tribes, but also general proportionality to degree of molecular divergence.

CONCLUSION

What was the "Siwalik" small mammal fauna, and where did it originate? The pre-Siwalik, Oligocene age "Bugti" fauna was distinct from subsequent assemblages at high taxonomic levels. Oligocene age rodents were dominantly archaic ctenohystricans, most lineages of which did not continue into the Miocene. The Siwalik small mammal fauna differed fundamentally from Oligocene assemblages in its non-rodent component as well as the rodents. In addition to the "Bugti-holdover" treeshrews, sivaladapid primates, and bats, the Early Miocene Siwalik micromammal record added immigrant shrews and hedgehogs (documented by 22 Ma) and lorises. Bugti diatomyids and uncommon anomalurids persisted into the Early Miocene, but otherwise the Siwalik rodent assemblage was almost entirely new. Siwalik rodent immigrants included several squirrels, the gundi *Sayimys*, the early rhizomyine *Prokanisamys*, and the higher muroids *Primus*, followed by the gerbil *Myocricetodon* and the mouse *Potwarmus*. Five million years after the Early Oligocene Bugti small mammal fauna (~28 Ma, Welcomme et al. 2001; Lindsay et al. 2005) the Siwalik fauna was in place and differed at high taxonomic levels. The paleobiogeographic origins of many early Siwalik groups, although likely subtropical, are still not clear.

Later Siwalik small mammal assemblages changed through the Miocene by both endemic evolution and immigration. Early and Middle Miocene cane rats came from Africa, Middle Miocene dormice likely from the west, the Late Miocene porcupine *Hystrix* from the west, and the Late Miocene rabbit *Alilepus* via China or Afghanistan. Introductions of taxa were dominantly at subtropical latitude, given that sister taxa of most immigrants occurred in regions of low latitude to the west or to the east.

Siwalik rodents reflect moist, vegetated conditions during the Early and Middle Miocene. By the Late Miocene, more open woodland habitat with greater seasonality of precipitation resulted in declining small mammal diversity. Change in dominant types of vegetation likely led to partitioning of resources among some rodents, likely reflected in different ratios of stable isotopes in tooth enamel among mice. For fossorial bamboo rats, a new strategy of exploitation of subterranean food sources was adopted after 10 Ma. A replacement among higher groups of mice involved the Late Miocene rise of Murinae at the expense of other muroid groups (see Fig. 11.15). Opportunistic Late Miocene diversification of Murinae in the Indian subcontinent is seen in local species-level differentiation as in Kutch, India (Patnaik et al. 2022).

The Siwalik rocks of the Potwar Plateau have provided a profoundly important and informative record of a wide variety of mammals including rodents. The record is unparalleled in its sequence of superposed fossil localities with ages calibrated by magnetic polarity stratigraphy. Rich assemblages enable estimates of relative abundances at the species level; we can begin to perceive community structure in these ancient microfaunas. The series of evolving species, each with special attributes, contributes to understanding the ecological and biogeographical evolution of southern Asia. For several groups, the Siwaliks contain an early record of evolution leading to living representatives, thereby facilitating the melding of paleontological and biological data sets. Perhaps this is demonstrated no better than with murine rodents, the most species-rich group of mammals in the modern world, now documented from its origin to its living descendants. But it is the combined knowledge of all members of evolving Siwalik communities, viewed together—their ways of life, their interactions, and their relation to their environments—that is most enlightening.

ACKNOWLEDGMENTS

Our work stands on the shoulders of friends no longer with us. We take this opportunity to thank three scientists for their contributions to and discovery of the riches of Siwalik fossil small mammals. Coauthor and cricetid rodent specialist Ev Lindsay was able to see the project to fruition. Our dear colleague Dr. Iqbal Umer Cheema recently passed and would have also been a coauthor. His initiative and enthusiasm in the project's early days were essential and we think fondly of him. He was also one of the founding pillars of the PMNH as it developed and grew into a public resource. We also thank and think of Will Downs (1950–2002) who built much of the collection upon which our analysis depends. All three inspire us as we interpret the fabulous small mammal story of Pakistan. Many other scientists have contributed in varied ways to our long study of Siwalik Glires, especially J. A. Baskin, J. A. Harrison, A. R. Rajpar, J. J. Saunders, A. J. Winkler and the "Buckettes." Fossils were retrieved and prepared under a series of grants to David Pilbeam from the National Science Foundation (NSF) and Public Law 480 programs, from NSF and National Geographic Society grants to Everett Lindsay, and under NSF grant DEB-7709539 to Louis Jacobs and BNS-9196211 to Jay Kelley. Fossils were analyzed most recently through NSF grant EAR 0958178 and its supplement to David Pilbeam, Catherine Badgley and Thure Cerling and with support from the American School of Prehistoric Research. Japan Society for the Promotion of Science (JSPS) KAKENHI grants JP15H06884 and 18K13650 funded the work of Y. Kimura. Michèle Morgan assembled species abundance data and taxon ranges from the Harvard University database maintained by John Barry; the contributions of both made our analysis possible. This chapter is greatly improved by the careful reading of editors David Pilbeam, Catherine Badgely, and Michèle Morgan. Imaging was performed in part at the Harvard University Center for Nanoscale Systems, member of the National Nanotechnology Coordinated Infrastructure Network, which is supported by the NSF under award ECCS-2025158.

References

Aghová, T., Y. Kimura, J. Bryja, G. Dobigny, and G. J. Kergoat. 2018. Fossils know it best: Using a new set of fossil calibrations to improve the temporal phylogenetic framework of murid rodents (Rodentia: Muridae). *Molecular Phylogenetics and Evolution* 128:98–111.

Andrews, P. 1990. *Owls, Caves, and Fossils*. Chicago: University of Chicago Press.

Asher, R. J., J. Meng, J. R. Wible, M. C. McKenna, G. W. Rougier, D. Dashzeveg, and M. J. Novacek. 2005. Stem Lagomorpha and the antiquity of Glires. *Science* 307(5712): 1091–1094.

Barry, J. C., M. E. Morgan, L. J. Flynn, D. Pilbeam, A. K. Behrensmeyer, S. M. Raza, I. A. Khan, et al. 2002. Faunal and environmental change in the Late Miocene Siwaliks of Northern Pakistan. *Paleobiology Memoir* 3:1–71.

Barry, J. C., M. E. Morgan, A .J. Winkler, L. J. Flynn, E. H. Lindsay, L. L. Jacobs, and D. Pilbeam. 1991. Faunal interchange and Miocene terrestrial vertebrates of southern Asia. *Paleobiology* 17(3): 231–245.

Baskin, J. A. 1996. Systematic revision of Ctenodactylidae (Mammalia, Rodentia) from the Miocene of Pakistan. *Palaeovertebrata* 25(1): 1–49.

Bernor, R. L., N. Solounias, C. C. Swisher III, and J. A. Van Couvering. 1996. The correlation of three classical "Pikermian" mammal faunas—Maragheh, Samos, and Pikermi—with the European MN Unit system. In *The Evolution of Western Eurasian Neogene Mammal Faunas*, edited by R. L. Bernor, V. Fahlbusch, and H.-V. Mittmann, 137–154. New York: Columbia University Press.

Bhandari, A., S. Bajpai, L. J. Flynn, B. N. Tiwari, and N. Mandal. 2021. First Miocene rodents from Kutch, western India. *Historical Biology* 33(12): 3471–3479.

Black, C. C. 1972. Review of fossil rodents from the Neogene Siwalik beds of India and Pakistan. *Palaeontology* 15(2): 238–266.

Blanga-Kanfi, S., H. Miranda, O. Penn, T. Pupko, R. W. DeBry, and D. Huchon. 2009. Rodent phylogeny revised: analysis of six nuclear genes from all major rodent clades. *BMC Evolutionary Biology* 9:71.

Bowdich T. E. 1821. *An Analysis of the Natural Classifications of Mammalia, for the Use of Students and Travelers*. Paris: J. Smith

Bowers, M. A., and J. H. Brown. 1982. Body size and coexistence in desert rodents: Chance or community structure? *Ecology* 63(2): 391–400.

Brandt, J. F. 1855. Beitrage der Nähern Kenntniss der Säugethiere Russland's. Kaiserlichen Akademie der Wissenschaften, Mathématiques, Physiques et Naturelles, 7:1–365. Mémoires de L'Académie Impériale des Sciences de Saint Pètersbourg. Eggers et Companie.

Brandy, L. D. 1979. Rongeurs nouveaux du Neogene d'Afghanistan. *Comptes Rendus Hebdomadaires des Séances de l'Académie des Sciences. Série D* 289(2): 81–83.

Burildağ, E., and C. Kurtonur. 2001. Hibernation and postnatal development of the mouse-tailed dormouse, *Myomimus roachi*, reared outdoor's in a cage. *Trekya University Journal of Scientific Research* B2(2): 179–186.

Casanovas-Vilar, I., A. Madern, D. M. Alba, L. Cabrera, I. García-Paredes, L. W. van den Hoek Ostende, D. DeMiguel, et al. 2016. The Miocene mammal record of the Vallès-Penedès Basin (Catalonia). *Comptes Rendus de l'Académie des Sciences Palevol* 15(7): 791–812.

Cheema, I. U., L. J. Flynn, and A. R. Rajpar. 2003. Late Pliocene murid rodents from Lehri, Jhelum District, northern Pakistan. In *Advances in Vertebrate Paleontology "Hen to Panta": A Tribute to Constantin Radulescu and Petre Mihai Samson*, edited by A. Petculescu and E. Ştiucă, 85–92. Bucharest: Romanian Academy.

Cheema, I. U., S. M. Raza, and L. J. Flynn. 1997. Note on Pliocene small mammals from the Mirpur District, Azad Kashmir, Pakistan. *Géobios* 30(1): 115–119.

Cheema, I. U., S. M. Raza, L. J. Flynn, A. R. Rajpar, and Y. Tomida. 2000. Miocene small mammals from Jalalpur, Pakistan, and their biochronologic implications. *Bulletin of the National Science Museum* C26(1, 2): 57–77.

Coillot, T., Y. Chaimanee, C. Charles, H. Gomez-Rodrigues, J. Michaux, P. Tafforeau, M. Vianey-Liaud, L. Viriot, and V. Lazzari. (2013). Correlated changes in occlusal pattern and diet in stem Murinae during the onset of the radiation of Old World rats and mice. *Evolution* 67(11): 3323–3338.

Colbert, E. H. 1933. Two new rodents from the lower Siwalik beds of India. *American Museum Novitates* 633:1–6.

Crusafont-Pairó, M., and S. Reguant. 1970. The nomenclature of intermediate forms. *Systematic Zoology* 1970(3): 254–257.

Dawson, M. R., T. Tsubamoto, M. Takai, N. Egi, S. T. Tun, and C. Sein. 2003. Rodents of the family Anomaluridae (Mammalia) from Southeast Asia (middle Eocene, Pondaung Formation, Myanmar). *Annals of Carnegie Museum* 72(3): 203–213.

De Bruijn, H., E. Boon, and S. Taseer Hussain. 1989. Evolutionary trends in *Sayimys* (Ctenodactylidae, Rodentia) from the lower Manchar Formation (Sind, Pakistan). *Proceedings of the Koninklijke Nederlandse Akademie van Waterscape* B92(3): 191–214.

De Bruijn, H., and S. Taseer Hussain. 1985. Thryonomyidae from the Lower Manchar Formation of Sind, Pakistan. *Proceedings of the Koninklijke Nederlandse Akademie van Wetenschappen* B88(2): 155–166.

De Bruijn, H., S. Taseer Hussain, and J. J. M. Leinders. 1981. Fossil rodents from the Murree Formation near Banda Daud Shah, Kohat, Pakistan. *Proceedings of the Koninklijke Nederlandse Akademie van Wetenschappen* B84(1): 71–99.

De Bruijn, H., S. Taseer Hussain, and J. J. M. Leinders. 1982. On some Early Eocene rodent remains from Barbara Banda, Kohat, Pakistan, and the early history of the order Rodentia. *Proceedings of the Koninklijke Nederlandse Akademie van Wetenschappen* B85(3): 249–258.

De Bruijn, H., Z. Markovic, W. Wessels, and A. A. van de Weerd. 2022. On the antiquity and status of the Spalacidae, new data from the late Eocene of South-East Serbia. *Palaeobiodiversity and Palaeoenvironments* 103:433–445.

De Bruijn, H., and P. J. Whybrow. 1994. A Late Miocene rodent fauna from the Baynunah Formation, Emirate of Abu Dhabi, United Arab Emirates. *Proceedings of the Koninklijke Akademie van Wetenschappen* 97(4): 407–422.

D'Elía, G., P.-H. Fabre, and E. P. Lessa. 2019. Rodent systematics in an age of discovery: Recent advances and prospects. *Journal of Mammalogy* 100(3): 852–871.

Denys, C., D. N. Reed, and Y. Dauphin. 2023. Deciphering alterations of rodent bones through in vitro digestion: An avenue to understanding pre-diagenetic agents? *Minerals* 13:124. https://doi.org/10.3390/min13010124.

Denys, C., E. Stoetzel, P. Andrews, S. Bailon, A. Rihane, J. B. Huchet, Y. Fernandez-Jalvo, and V. Laroulandie. 2018. Taphonomy of small predators multi-taxa accumulations: Paleoecological implications. *Historical Biology*, 30:868–881.

De Terra, H., and P. Teilhard de Chardin. 1936. Observations on the Upper Siwalik Formation and later Pleistocene deposits in India. *Proceedings of the American Philosophical Society* 76(6): 791–822.

Encyclopedia of Life. https://eol.org/pages/8702.

Fabre, P. H., Hautier, L., Dimitrov, D., and Douzery, E. J. 2012. A glimpse on the pattern of rodent diversification: A phylogenetic approach. *BMC Evolutionary Biology* 12:88.

Fabre, P.-H., M.-K. Tilak, C. Denys, P. Gaubert, V. Nicolas, E. J. P. Douzery, and L. Marivaux. 2018. Flightless scaly-tailed squirrel never learned how to fly: A reappraisal of Anomaluridae phylogeny. *Zoologica Scripta* 47(4): 404–417.

Fischer, G. 1814. Zoognosia. Volumen III. Quadrupeda reliqua, Ceti, Monotremata. Moscow: Nicolai Sergidis Vsevolozsky.

Flynn, L. J. 1982. Systematic revision of Siwalik Rhizomyidae (Rodentia). *Géobios* 15(3): 327–389.

Flynn, L. J. 1985. Evolutionary patterns and rates in Siwalik Rhizomyidae. *Acta Zoologica Fennica* 170:141–144.

Flynn, L. J. 2000. The great small mammal revolution. *Himalayan Geology* 21(1–2): 39–42.

Flynn, L. J. 2003. Small mammal indicators of forest paleoenvironment in the Siwalik deposits of the Potwar Plateau, Pakistan. *Deinsea* 10:183–195.

Flynn, L. J. 2006. Evolution of the Diatomyidae, an endemic Asian family of rodents. *Vertebrata PalAsiatica* 44: 182–192.

Flynn, L. J. 2007. Origin and Evolution of the Diatomyidae, with clues to paleoecology from the fossil record. *Bulletin of Carnegie Museum* 39:173–181.

Flynn, L. J. 2009. The antiquity of *Rhizomys* and independent acquisition of fossorial traits in subterranean muroids. In *Systematic Mammalogy: Contributions in Honor of Guy G. Musser*, edited by R. S. Voss and M. D. Carleton. *Bulletin of the American Museum of Natural History* 331:128–156.

Flynn, L. J., J. C. Barry, W. Downs, J. A. Harrison, E. H. Lindsay, M. E. Morgan, and D. Pilbeam. 1997. Only ochotonid from the Neogene of the Indian subcontinent. *Journal of Vertebrate Paleontology* 17(3): 627–628.

Flynn, L. J., J. C. Barry, M. E. Morgan, D. Pilbeam, L. L. Jacobs, and E. H. Lindsay. 1995. Neogene Siwalik mammalian lineages: Species longevities, rates of change, and modes of speciation. *Palaeogeography, Palaeoclimatology, Palaeoecology* 115(1–4): 249–264.

Flynn, L. J., W. R. Downs, M. E. Morgan, J. C. Barry, and D. Pilbeam. 1998. High Miocene species richness in the Siwaliks of Pakistan. In *Advances in Vertebrate Paleontology and Geochronology*, edited by Y. Tomida, L. J. Flynn, and L.L. Jacobs, 167–180. Tokyo: National Science Museum Monograph 14.

Flynn, L. J., E. Heintz, S. Sen, and M. Brunet. 1983. A new Pliocene tachyoryctine (Rhizomyidae, Rodentia) from Lataband, Sarobi Basin, Afghanistan. *Proceedings of the Koninklijke Nederlandse Akademie van Wetenschappen* B86(1): 61–68.

Flynn, L. J., and L. L. Jacobs. 1999. Late Miocene small mammal faunal dynamics: The crossroads of the Arabian Peninsula. In *Vertebrate Fossils of the Arabian Peninsula*, edited by P. Whybrow and A. Hill, 410–419. New Haven, CT: Yale University Press.

Flynn, L. J., L. L. Jacobs, and I. U. Cheema. 1986. Baluchimyinae, a new ctenodactyloid rodent subfamily from the Miocene of Baluchistan. *American Museum Novitates* 2841:1–58.

Flynn, L. J., L. L. Jacobs, Y. Kimura, and E. H. Lindsay. 2019. Rodent suborders. *Fossil Imprint* 75(3–4): 292–298.

Flynn, L J., L .L. Jacobs, and E. H. Lindsay. 1985. Problems in muroid phylogeny: Relationships to other rodents and origin of major groups. In *Evolutionary Relationships among Rodents: A Multidisciplinary Analysis*, edited by W. P. Luckett and J. L. Hartenberger, 589–616. New York: Plenum Press.

Flynn, L. J., Jin C.-Z., J. Kelley, N.G. Jablonski, X.-P. Ji, D. F. Su, Deng T., and Li Q. 2019. Late Miocene fossil calibration from Yunnan Province for the striped rabbit *Nesolagus. Vertebrata PalAsiatica* 57(4): 214–224.

Flynn, L. J., and Y. Kimura. 2023. One more Siwalik surprise: The oldest record of *Mus* (Mammalia, Rodentia) from the late Miocene of northern Pakistan. In *Windows into Sauropsid and Synapsid Evolution. Essays in honor of Prof. Louis L. Jacobs*, edited by Y.-N. Lee, 264–272. Seoul: Dinosaur Science Center Press.

Flynn, L. J., Y. Kimura, and L. L. Jacobs. 2020. The murine cradle. In *Biological Consequences of Plate Tectonics: New Perspectives on Post-Gondwana Break-Up. A Tribute to Ashok Sahni*, edited by G. V. R. Prasad and R. Patnaik, 347–362. Cham: Springer.

Flynn, L. J. and M. E. Morgan. 2005. An unusual diatomyid rodent from an infrequently sampled late Miocene interval in the Siwaliks of Pakistan. *Palaeontologica Electronica*, 8(1): 17A. https://palaeo-electronica.org/2005_1/flynn17/issue1_05.htm.

Flynn, L. J., M.E. Morgan, J. C. Barry, S. M. Raza, I. U. Cheema, and D. Pilbeam. 2023a. Siwalik rodent assemblages for NOW. In *Evolution of Cenozoic Land Mammal Faunas and Ecosystems. 25 Years of the NOW Database of Fossil Mammals*, edited by I. Casanovas-Vilar, C. Janis, L. Van den Hoek Ostende, and J. Saarinen, 43–58. Cham: Springer.

Flynn, L. J., M. E. Morgan, D. Pilbeam, and J. C. Barry. 2014a. "Endemism" relative to space, time, and taxonomic level. *Annales Zoologici Fennici* 51(1–2): 245–258.

Flynn, L. J., A. Sahni, J.-J. Jaeger, B. Singh, and S. B. Bahtia. 1990. Additional fossil rodents from the Siwalik beds of India. *Proceedings of the Koninklijke Nederlandse Akademie van Wetenschappen* B93(1): 7–20.

Flynn, L. J., and G. Qi. 1982. Age of the Lufeng, China, hominoid locality. *Nature* 298(5876): 746–747.

Flynn, L. J., and W. Wessels. 2013. Paleobiogeography and South Asian small mammals: Neogene latitudinal faunal variation. In *Fossil Mammals of Asia: Neogene Biostratigraphy and Chronology*, edited by X. Wang, L. J. Flynn, and M. Fortelius, 445–460. New York: Columbia University Press.

Flynn, L J., and A. J. Winkler. 1994. Dispersalist implications of *Paraulacodus indicus*, a South Asian rodent of African affinity. *Historical Biology* 9:223–235.

Flynn, L J., A. J. Winkler, M. Erbaeva, N. Alexeeva, U. Anders, C. Angelone, S. Čermák, et al. 2014b. The leporid datum: A Late Miocene biotic marker. *Mammal Review* 44:164–176.

Flynn, L. J., A. Winkler, L. L. Jacobs, and W. R. Downs. 2003. Tedford's gerbils from Afghanistan. In *Vertebrate Fossils and Their Context: Contributions in Honor of Richard H. Tedford*, edited by L. J. Flynn, *Bulletin of the American Museum of Natural History* 279:603–624.

Flynn, L. J., W. Wu, L. Li, and Z. Qiu. 2023b. New material of *Sayimys* (Rodentia, Ctenodactylidae) from China. In *Windows into Sauropsid and Synapsid Evolution Essays in honor of Prof. Louis L. Jacobs*, edited by Yuong-Nam Lee, 273–289. Seoul: Dinosaur Science Center Press.

Fraser, D., L. C. Soul, A. B. Tóth, M. A. Balk, J. T. Eronen, S. Pineda-Munoz, A. B. Shupinski, et al. 2021. Investigating biotic interactions in deep time. *Trends in Ecology and Evolution* 36:61–75.

Freudenthal, M., and E. Martín-Suárez. 2013. Estimating body mass of fossil rodents. *Scripta Geologica* 145:1–130, online appendix.

Gatesy, J., R. W. Meredith, J. E. Janecka, M. P. Simmons, W. J. Murphy, and M. S. Springer. 2017. Resolution of a concatenation/coalescence kerfuffle: Partitioned coalescence support and a robust family-level tree for Mammalia. *Cladistics* 33(3): 295–332.

Gervais, F. L. P. 1841. Mammifères. In *Zoologie*, edited by J. F. T. Eydoux and L. F. A. Souleyet, 1–68. Voyage autour du Monde exécuté pendant les années 1836 et 1837 sur la corvette La Bonite 4:1–334, pls. 1–12. Paris: Arthus Bertrand,

Gervais, M.P. 1853. Description ostéologique de *l'Anomalurus* et remarques sur la classification naturelle des rongeurs. *Annales des Sciences Naturelles Zoologie* 20(3): 238–246.

Gray, J. E. 1821. On the natural arrangement of vertebrose animals. *London Medical Repository*, 15(1): 296–310.

Gray, J. E. 1825. Outline of an attempt at the disposition of the Mammalia into tribes and families with a list of the genera apparently appertaining to each tribe. *Annals of Philosophy*, n.s. 2(10): 337–344.

Harrison, J. A. 1978. Mammals of the Wolf Ranch local fauna, Pliocene of the San Pedro Valley, Arizona. *University of Kansas Museum of Natural History Occasional Paper* 73:1–18.

Hartenberger, J.-L. 1982. A review of the Eocene rodents of Pakistan. *Contributions from the Museum of Paleontology, University of Michigan* 26 (2): 19–35.

Hartman, J., A. A. van de Weerd, H. De Bruijn, and W. Wessels. 2019. An exceptional large sample of the Early Miocene ctenodactyline rodent *Sayimys giganteus*, specific variation and taxonomic implications. *Fossil Imprint* 75(3–4): 359–382.

Heintz, E., and M. Brunet. 1982. Stand der Kenntnisse über die fossilen Wirbeltierfaunen von Afghanistan. *Wissenschaftliche Zeitschrift der Humboldt-Universität zu Berlin. Mathematisch-Naturwissenschaftliche Reihe* 31:135–141.

Hinton, M.A.C. 1933. Diagnoses of new genera and species of rodents from Indian Tertiary deposits. *Annals and Magazine of Natural History* 10(12): 620–622.

Honeycutt, R. L. 2009. Rodents (Rodentia). In *The Time Tree of Life*, edited by S. B. Hedges and S. Kumar, , 490–494. Oxford: Oxford University Press.

Honigs, S., and H. Greven. 2003. Biology of the gundi, *Ctenodactylus gundi* (Rodentia: Ctenodactylidae), and its occurrence in Tunisia. *Kaupia* 12:43–55.

Huchon, D., F. M. Catzeflis, and J. P. Douzery. 2000. Variance of molecular datings, evolution of rodents, and the phylogenetic affinities between Ctenodactylidae and Hystricognathi. *Proceedings Royal Society of London* B 267(1441): 393–402.

Hussain, S. Taseer, H. De Bruijn, and J. J. M. Leinders. 1978. Middle Eocene rodents from the Kala Chitta Range (Punjab, Pakistan). *Proceedings of the Koninklijke Nederlandse Akademie van Wetenschappen* B81(1): 74–112.

Illiger, C. D. 1811. Prodromus systematis mammalium et avium additis terminis zoographicis uttriusque classis. Berlin: Salfeld,

Jacobs, L. L. 1977. A new genus of murid rodent from the Miocene of Pakistan and comments on the origin of the Muridae. *PaleoBios* 25:1–12.

Jacobs, L. L. 1978. Fossil rodents (Rhizomyidae & Muridae) from Neogene Siwalik deposits, Pakistan. *Museum of Northern Arizona Press, Bulletin Series* 52:1–103.

Jacobs, L. L., and W.R. Downs. 1994. The evolution of murine rodents in Asia. In *Rodent and Lagomorph Families of Asian Origins and Diversification*, edited by Y. Tomida, C. K. Li, and T. Setoguchi, 149–156. Tokyo: National Science Museum Monograph 8.

Jacobs, L. L., and L. J. Flynn. 2005. Of mice . . . again: The Siwalik rodent record, murine distribution, and molecular clocks. In *Interpreting the Past: Essays on Human, Primate, and Mammal Evolution in Honor of David Pilbeam*, edited by D. E. Lieberman, R. J. Smith, and J. Kelley, 63–80. Boston: Brill.

Jacobs, L. L., L. J. Flynn, and W. R. Downs. 1989. Neogene rodents of southern Asia. In *Papers on Fossil Rodents in Honor of Albert Elmer Wood* edited by C. C. Black and M. R. Dawson, 157–177. Los Angeles: Natural History Museum.

Jaeger, J. J. 1977. Rongeurs (Mammalia, Rodentia) do Miocène de Beni-Mellal. *Palaeovertebrata* 7(4): 91–125, 2 plates

Jaeger, J. J., J. Michaux, and M. Sabatier. 1980. Premières données sur les rongeurs de la formation de Ch'orora (Ethiopie) d'age Miocène supérieur. I. Thryonomyidés. *Palaeovertebrata, Mémoire Jubilaire en Hommage à René Lavocat*, 365–374.

Jenkins P. D., C. W. Kilpatrick, M. F. Robinson, and R. J. Timmins. 2005. Morphological and molecular investigations of a new family, genus and species of rodent (Mammalia: Rodentia: Hystricognatha) from Lao PDR. *Systematics and Biodiversity* 2(4): 410–454.

Jin, C.-Z. 2004. Fossil leporids (Mammalia, Lagomorpha) from Huainan, Anhui, China. *Vertebrata PalAsiatica* 42(3): 230–245.

Kaakinen, A. 2005. A long terrestrial sequence in Lantian—a window into the late Neogene palaeoenvironments of northern China. *Publications of the Department of Geology* D4: 1–49, University of Helsinki, Finland.

Kalthoff, D. C. 2000. Die Schmelzmikrostruktur in den Incisiven der hamsterartigen Nagetiere und anderer Myomorpha (Rodentia Mammalia). *Palaeontographica A* 259:1–193.

Karp, A. T., A. K. Behrensmeyer, and K. H. Freeman. 2018. Grassland fire ecology has roots in the late Miocene. *Proceedings of the National Academy of Sciences* 115(48): 12130–12135.

Karp, A. T., K. T. Uno, P. Polissar, and K. H. Freeman. 2021. Late Miocene C_4 grassland fire feedbacks on the Indian Subcontinent. *Paleoceanography and Paleoclimatology* 36(4): e2020PA004106. https://doi.org/10.1029/2020PA004106.

Keller, H. M., R. A. K. Tahirkheli, M. A. Mirza, G. D. Johnson, N. M. Johnson, and N. D. Opdyke. 1977. Magnetic polarity stratigraphy of the Upper Siwalik deposits, Pabbi Hills, Pakistan. *Earth and Planetary Science Letters* 36:187–201.

Khan, M. A., and L. J. Flynn. 2017. New small mammals from the Hasnot-Tatrot area of the Potwar Plateau, northern Pakistan. *Paläontologische Zeitschrift* 91(4): 589–599.

Kimura, Y., L. J. Flynn, and L. L. Jacobs. 2015. A paleontological case study for species delimitation in diverging fossil lineages. *Historical Biology* 28(1–2): 189–198.

Kimura, Y., L. J. Flynn, and L. L. Jacobs. 2017. Early Late Miocene murine rodents from the upper part of the Nagri Formation, Siwalik Group, Pakistan, with a new fossil calibration point for the Tribe Apodemurini (*Apodemus / Tokudaia*). *Fossil Imprint* 73(1–2): 197–212.

Kimura, Y., L. J. Flynn, and L. L. Jacobs. 2021. Tempo and mode: Evidence on a protracted split from a dense fossil record. *Frontiers in Ecology and Evolution* 9:64281, 15 pp.

Kimura, Y., L. L. Jacobs, T. E. Cerling, K. T. Uno, K. M. Ferguson, L. J. Flynn, and R. Patnaik. 2013. Fossil mice and rats show isotopic evidence of niche partitioning and change in dental ecomorphology related to dietary shift in Late Miocene of Pakistan. *PLOS ONE* 8(8). https://doi.org/10.1371/journal.pone.0069308.

Kimura, Y., L. L. Jacobs, and L. J. Flynn. 2013. Lineage-specific responses of tooth shape in murine rodents (Murinae, Rodentia) to Late Miocene dietary change in the Siwaliks of Pakistan. *PLOS ONE* 8(10). https://doi.org/10.1371/journal.pone.0076070.

Kimura, Y., M. T. R. Hawkins, M. M. McDonough, L. L. Jacobs, and L. J. Flynn. 2015. Corrected placement of *Mus-Rattus* fossil calibration forces precision in the molecular tree of rodents. *Scientific Reports* 5:14444, 9 pp.

Kotlia, B. S. and W. von Koenigswald. 1992. Plio-Pleistocene arvicolids (Rodentia, Mammalia) from Kashmir intermontane basin, northwestern India. *Palaeontographica A* 223:103–135.

Kriegs, J. O., G. Churakov, J. Jurka, J. Brosius, and J. Schmitz. 2007. Evolutionary history of 7SL RNA-derived SINEs in Supraprimates. *Trends in Genetics* 23(4): 158–61.

Landry, S. 1957. The interrelationships of the New and Old World hystricomorph rodents. *University of California Publications in Zoology* 56:1–118.

Lavocat, R. 1961. Étude systématique de la faune de mammifères et conclusions générals., *Éditions du Service Géologique du Maroc* 155:29–145.

Lazzari, V., J. Michaux, and J. P. Aguilar. 2007. First occurrence in Europe of Myocricetodontinae (Rodentia, Gerbillidae) during the lower Middle Miocene in the karstic locality of Blanquetère 1 (southern France): Implications. *Journal of Vertebrate Paleontology* 27:1062–1065.

Lecompte, E., K. Aplin, K. Denys, F. Catzeflis, M. Chades, and P. Chevret. 2008. Phylogeny and biogeography of African Murinae based on mitochondorial and nuclear gene sequences, with a new tribal classification of the subfamily. *BMC Evolutionary Biology* 8(199): 21 pp.

Leibold, M. A., M. Holyoak, N. Mouquet, P. Amarasekare, J. M. Chase, M. F. Hoopes, R. D. Holt, et al. 2004. The metacommunity concept: A framework for multi-scale community ecology. *Ecology Letters* 7:601–613.

Li, C.-K. 1977. A new cricetodont rodent of Fangshan, Nanking. *Vertebrata PalAsiatica* 15(1): 67–75.

Lindsay, E. H. 1988. Cricetid rodents from Siwalik deposits near Chinji Village. Part 1, Megacricetodontinae, Myocricetodontinae, and Dendromurinae. *Palaeovertebrata* 18(2): 95–154.

Lindsay, E. H. 1996. A new eumyarionine cricetid from Pakistan. *Acta Zoologica Cracoviensia* 39(1): 279–288.

Lindsay, E. H. 2017. *Democricetodon fejfari* sp. nov. and replacement of Cricetidae by Muridae in Siwalik deposits of Pakistan. *Fossil Imprint* 73(3–4): 445–453.

Lindsay, E. H., and L. J. Flynn. 2015. Late Oligocene and early Miocene Muroidea of the Zinda Pir Dome. *Historical Biology* 28(1–2): 215–236.

Lindsay, E. H., L. J. Flynn, I. U. Cheema, J. C. Barry, K. F. Downing, A. R. Rajpar, and S. M. Raza. 2005. Will Downs and the Zinda Pir Dome. *Palaeontologia Electronica* 8(1): 19A, 19 pp. https://palaeo -electronica.org/2005_1/lindsay19/issue1_05.htm.

Linnaeus, C. 1758. *Systema naturae per regna tria naturae, secudum classes, ordines, genera, species, cum characteribus, differentiis, synonymis, locis. Vol. I Regnum Animale.* Editio decima, reformata. Stockholm: Laurentii Salvii.

López-Antoñanzas, R., L. J. Flynn, and F. Knoll. 2013. A comprehensive phylogeny of extinct and extant Rhizomyinae (Rodentia): Evidence for multiple intercontinental dispersals. *Cladistics* 29(3): 247–273.

López-Antoñanzas, R., and F. Knoll. 2011. A comprehensive phylogeny of the gundis (Ctenodactylinae, Ctenodactylidae, Rodentia). *Journal of Systematic Palaeontology* 9(3): 379–398.

López-Antoñanzas, R., F. Knoll, S. Wan, and L. J. Flynn. 2015. Causal evidence between monsoon and evolution of rhizomyine rodents. *Scientific Reports* 5:9008, 8 pp.

López-Antoñanzas, R., S. Renaud, P. Peláez-Campomanes, D. Azar, G. Kachala, and F. Knoll. 2019. First Levantine fossil murines shed new light on the earliest intercontinental dispersal of mice. *Scientific Reports* 9:11874. https://doi.org/10.1038/s41598-019-47894-y.

López-Antoñanzas, R., and S. Sen. 2003. Systematic revision of Mio-Pliocene Ctenodactylidae (Mammalia, Rodentia) from the Indian subcontinent. *Eclogae Geologicae Helvetiae* 96:521–529.

López-Antoñanzas, R., and H. B. Wesselman. 2013. *Tachyoryctes makooka* (Tachyoryctini, Spalacidae, Rodentia) and its bearing on the phylogeny of the Tachyoryctini. *Palaeontology* 56(1): 157–166.

Lydekker, R. L. 1884. Rodents and new ruminants from the Siwaliks and synopsis of Mammalia. *Paleontographica Indica* 10(3): 105–134.

Maridet, O., W. Wu, J. Ye, S. Bi, X. Ni, and J. Meng. 2011. Earliest occurrence of *Democricetodon* in China, in the early Miocene of the Junggar Basin (Xinjiang), and comparison with the genus *Spanocricetodon*. *Vertebrata PalAsiatica* 49(4): 393–405.

Marivaux, L. 2000. Les rongeurs de l'Oligocène des Collines Bugti (Balouchistan, Pakistan): ouvelles données sur la phylogénie des Rongeurs paléogènes, implications biochronologiques et paléobiogéographiques. PhD diss., Université Montpellier II.

Marivaux, L., M. Vianey-Liaud, and J.-J. Jaeger. 2004. High-level phylogeny of Early Tertiary rodents: Dental evidence. *Zoological Journal of the Linnaean Society* 142:105–134.

Marivaux, L., M. Vianey-Liaud, and J.-L. Welcomme. 1999. Première découverte de Cricetidae (Rodentia, Mammalia) oligocènes dans le synclinal sud de Gandoï (Bugti Hills, Balouchistan, Pakistan). *Comptes Rendus de l'Académie des Sciences. Série IIA* 329(11): 839–844.

Marivaux, L., M. Vianey-Liaud, J.-L. Welcomme, and J.-J. Jaeger. 2002. The role of Asia in the origin and diversification of hystricognathous rodents. *Zoologica Scripta* 31(1): 225–239.

Marivaux, L., and J.-L. Welcomme. 2003. New diatomyid and baluchimyine rodents from the Oligocene of Pakistan (Bugti Hills, Balochistan): Systematic and paleobiogeographic implications. *Journal of Vertebrate Paleontology* 23(2): 420–434.

Marković, Z., W. Wessels, A. Van de Weerd, and H. De Bruijn. 2017. On a new diatomyid (Rodentia, Mammalia) from the Paleogene of S. E. Serbia, the first record of the family in Europe. *Palaeobiodiversity and Palaeoenvironments* 98:459–469.

Martin, R. A. 1996. Tracking mammal body size distributions in the fossil record: A preliminary test of the "rule of limiting similarity." *Acta Zoologica Cracoviensia* 39(1): 321–328.

Matthew, W. D. 1929. Critical observations upon Siwalik mammals. *Bulletin of the American Museum of Natural History* 61:437–560.

Mercer, J. M., and V. L. Roth. 2003. The effects of Cenozoic global change of squirrel phylogeny. *Science* 299:1568–1572.

Mein, P., and L. Ginsburg. 1997. Les mammifères du gisement miocène inférieur de Li Mae Long, Thaïlande: Systématique, biostratigraphie et paléoenvironnement. *Geodiversitas* 19(4): 783–844.

Meng, J., Y. Hu, and C.-K. Li. 2003. The osteology of *Rhombomylus* (Mammalia, Glires): Implications for phylogeny and evolution of Glires. *Bulletin of the American Museum of Natural History* 275:1–247.

Montgelard, C., S. Bentz, C. Tirard, O. Verneau, and F. M. Catzeflis. 2002. Molecular systematics of Sciurognathi (Rodentia): The mitochondrial cytochrome *b* and 12Sr RNA genes support the Anomaluroidea (Pedetidae and Anomaluridae). *Molecular Phylogenetics and Evolution* 22(2): 220–233.

Mörs, T. 2006. The platacanthomyine rodent *Neocometes* Schaub and Zapfe, 1953 from the Miocene of Hambach (NW Germany). *Beiträge zur Paläontologie* 30:329–337.

Muirhead, L. 1819. Mazology. In *The Edinburgh Encyclopaedia*, edited by Brewster, (D):393–486 [pls. 353–358]. 4th edition, Edinburgh: William Blackwood.

Munthe, J. 1980. Rodents of the Miocene Daud Khel local fauna, Mianwali District, Pakistan. Part 1. Sciuridae, Gliridae, Ctenodactylidae, and Rhizomyidae. *Milwaukee Public Museum Contributions in Biology and Geology* 34:1–36.

Musser, G. G. 1987. The occurrence of *Hadromys* (Rodentia: Muridae) in early Pleistocene Siwalik strata in northern Pakistan and its bearing on biogeographic affinities between Indian and northeastern African murine faunas. *American Museum Novitates* 2883:1–36.

Musser, G. G. and M. D. Carleton. 2005. Superfamily Muroidea. In *Mammal Species of the World: A Taxonomic and Geographic Reference*, edited by D. E. Wilson and D. M. Reeder, 894–906. Baltimore: Johns Hopkins University Press.

Nedbal, M. A., R. L. Honeycutt, and D. A. Schlitter. 1996. Higher-level systematics of rodents (Mammalia, Rodentia): Evidence from the mitochondrial 12S rRNA gene. *Journal of Mammalian Evolution* 3(3): 201–237.

Neumann, K., J. Michaux, V. Lebedev, N. Yigit, E. Colak, N. Ivanova, A. Poltoraus, et al. 2006. Molecular phylogeny of the Cricetinae subfamily based on the mitochondrial cytochrome b and 12S rRNA genes and the nuclear vWF gene. *Molecular Phylogeny and Evolution* 39(1): 135–148.

Nguyen, D., N. Nguyen, T. Dinh, D. Le, and D. Dinh. 2014. Distribution and habitat of the Laotian Rock Rat *Laonastes aenigmamus* Jenkins,

Kilpatrick, Robinson and Timmins, 2005 (Rodentia: Diatomyidae) in Viet Nam. *Biodiversity Data Journal* 2:e4188. https://doi.org/10.3897/BDJ.2.e4188.

Nowak, R. M. 1999. *Walker's Mammals of the World*, 6th ed. Baltimore: Johns Hopkins University Press.

Ognev, S. I. 1924. Priroda Okhota Ukraine [Nature and Hunting in Ukraine], Kharkov, 1-2:115.

Oliver, A., P. M. Carro-Rodríguez, P. López-Guerrero, G. Daxner-Höck. 2023. A new framework of the evolution of the ctenodactylids (Mammalia: Rodentia) in Asia: New species and phylogenetic status of distylomyins. *Zoological Journal of the Linnean Society*.

Oliver, A., and P. Peláez-Campomanes. 2013. *Megacricetodon vandermeuleni*, sp. nov. (Rodentia, Mammalia), from the Spanish Miocene: A new evolutionary framework for *Megacricetodon. Journal of Vertebrate Paleontology* 33(4): 943–955.

Opdyke, N. D., E. H. Lindsay, G. D. Johnson, N. M. Johnson, R. A. K. Tahirkheli, and M. A. Mirza 1979. Magnetic polarity stratigraphy and vertebrate paleontology of the Upper Siwalik Subgroup of northern Pakistan. *Palaeogeography, Palaeoclimatology, Palaeoecology* 27(1):1–34.

Patnaik, R. 2001. Late Pliocene micromammals from Tatrot Formation (Upper Siwaliks) exposed near Village Saketi, Himachal Pradesh, India. *Palaeontographica* Abt. A 261:55–81.

Patnaik, R. 2002. Pliocene Leporidae (Lagomorpha, Mammalia) from the Upper Siwaliks of India: Implications for phylogenetic relationships. *Journal of Vertebrate Paleontology* 22(2): 443–452.

Patnaik, R. 2013. Indian Neogene Siwalik mammalian biostratigraphy: An overview. In *Fossil Mammals of Asia: Neogene Biostratigraphy and Chronology*, edited by X. Wang, L. J. Flynn, and M. Fortelius, 423–444. New York: Columbia University Press.

Patnaik, R. 2020. New data on the Siwalik murines, rhizomyines and ctenodactylines (Rodentia) from the Indian Subcontinent. In *Biological Consequences of Plate Tectonics: New Perspectives on Post-Gondwana Break-Up—A Tribute to Ashok Sahni*, edited by G. V. R. Prasad and R. Patnaik, 363–391. Cham: Springer.

Patnaik, R., N. P. Singh, K. M. Sharma, N.A . Singh, D. Choudhary, Y. P. Singh, R. Kumar, W. A. Wazir, and A. Sahni. 2022. New rodents shed light on the age and ecology of late Miocene ape locality of Tapar (Gujarat, India). *Journal of Systematic Palaeontology* 20(1): 2084701. https://doi:10.1080/14772019.2022.

Pocock, R. I. 1922. On the external characters of some hystricomorph rodents. *Proceedings of the Zoological Society of London*, 365–427.

Prasad, K. N. 1970. The vertebrate fauna from the Siwalik beds of Haritalyangar, Himachal Pradesh, India. *Memoirs of the Geological Survey of India, Palaeontologia Indica*, Delhi new series 39:1–56.

Qiu, Z.-X., T. Deng, Z.-D. Qiu, C.-K. Li, Z.-Q. Zhang, B.-Y. Wang, and X. Wang. 2013. Neogene Land Mammal Ages of China. In *Fossil Mammals of Asia: Neogene Biostratigraphy and Chronology*, edited by X. Wang, L. J. Flynn, and M. Fortelius, 29–90. New York: Columbia University Press.

Qiu, Z.-D., and D. Han. 1986. Fossil Lagomorpha from the hominoid locality of Lufeng, Yunnan. *Acta Anthropologica Sinica* 5(1): 41–53, (in Chinese with English summary).

Qiu, Z.-D., and Q. Li. 2016. Neogene rodents from central Nei Mongol, China. *Palaeontologia Sinica* 198, n.s. C30:1–684.

Qiu, Z.-D., and X.-J. Ni. 2005. Small mammals. In *Lufengpithecus hudiensis Site*, edited by Guo-Qin Qi and Wei Dong, 113–130. Beijing: Science Press (English summary).

Qiu, Z.-D., and C.-L. Yan. 2005. New sciurids from the Miocene Shanwang Formation, Linqu, Shandong. *Vertebrata PalAsiatica* 43(3): 194–207.

Qiu, Z.-D., S.-H. Zheng, and Z.-Q. Zhang. 2004. Murids from the late Miocene Bahe Formation, Lantian, Shaanxi. *Vertebrata PalAsiatica* 42(1): 67–76.

Roberts, T. J. 1977. *The Mammals of Pakistan*. London: Ernest Benn Limited.

Rose, K. D., V. Burke DeLeon, P. Missiaen, R. S Rana, A. Sahni, L. Singh, and T. Smith, 2008. Early Eocene lagomorph (Mammalia) from western India and the early diversification of Lagomorpha. *Proceedings B of the Royal Society* 275:1203–1208.

Rosenzweig, M. L. 1995. *Species Diversity in Space and Time*. Cambridge: Cambridge University Press.

Rowe, K. C., M. L. Reno, D. M. Richmond, R. M. Adkins, and S. J. Steppan. 2008. Pliocene colonization and adaptive radiations in Australia and New Guinea (Sahul): Multilocus systematics of the old endemic rodents (Muroidea: Murinae). *Molecular Phylogenetics and Evolution* 47:84–101.

Saunders, J. J., and B. K. Dawson. 1998. Bone damage patterns produced by extinct hyena, *Pachycrocuta brevirostris* (Mammalia: Carnivora) at the Haro River Quarry, northwestern Pakistan. In *Advances in Vertebrate Paleontology and Geochronology*, edited by Y. Tomida, L. J. Flynn, and L. L. Jacobs, 215–242. Tokyo: National Science Museum Monograph 14.

Schaub, S. 1938. Tertiäre und Quartäre Murinae. *Abhandlungen des Schweizerischen paläontologischen Gesellschaft* 61:1–38.

Scotese, C. R. 1997. Paleogeographic atlas. Paleomap Project. University of Texas, Arlington.

Sen, S. 1983. Rongeurs et Lagomorphes du gisement pliocène de Pul-e Charkhi, bassin de Kabul, Afghanistan. *Bulletin du Museum national d'Histoire naturelle*, 5e Séries, 5C(1): 33–74.

Sen, S. 2001. Rodents and insectivores from the upper Miocene of Molayan, Afghanistan. *Palaeontology* 44:913–932.

Sen, S. 2020. Lagomorphs (Mammalia) from the early Pliocene of Dorkovo, Bulgaria. *Fossil Imprint* 76:99–117.

Sen, S., and H. Thomas. 1979. Découverte de Rongeurs dans le Miocène moyen de la Formation Hofuf (Province du Hasa, Arabie Saoudite). *Compte Rendu Sommaire Des Séances de la société Géologique de France* 1:34–37.

Sinitsa, M. V., and A. Delinschi. 2016. The earliest member of *Neocricetodon* (Rodentia: Cricetidae): A redescription of *N. moldavicus* from Eastern Europe, and its bearing on the evolution of the genus. *Journal of Paleontology* 90(4): 771–784.

Steininger, F. F., and A. Papp. 1979. Current biostratigraphic and radiometric correlations of Late Miocene Central Paratethys stages (Sarmatian s.str., Pannonian s.str. and Pontian) and Mediterranean stages (Tortonian and Messinian) and the Messinian event in the Paratethys. *Newsletters on Stratigraphy* 8(2): 100–110.

Steiper, M., and N. Young. 2008. Timing primate evolution: Lessons from the discordance between molecular and paleontological estimates. *Evolutionary Anthropology* 17:179–188.

Steppan, S. J., R. M. Adkins, and J. Anderson. 2004. Phylogeny and divergence-date estimates of rapid radiations in muroid rodents based on multiple nuclear genes. *Systematic Biology* 53(4): 533–553.

Steppan, S. J., and J. J. Schenk. 2017. Muroid rodent phylogenetics: 900-species tree reveals increasing diversification rates. *PLOS ONE* 12(8): e0183070.

Storch, G. 1978. Familie Gliridae Thomas, 1897-Schläfer. In *Handbuch der Säugetiere Europas*, edited by J. Niethammer and F. Krapp, 201–280. Wiesbaden: Akademische Verlagsgesellschaft.

Suraprasit, K., Y. Chaimanee, T. Martin, and J.-J. Jaeger. 2011. First castorid (Mammalia, Rodentia) from the Middle Miocene of Southeast Asia. *Naturwissenschaften* 98:315–328.

Suwa, G., Y. Beyene, H. Nakaya, R. L. Bernor, J.-R. Boisserie, F. Bibi, S. H. Ambrose, S. Katoh, and B. Asfaw. 2015. Newly discovered cercopithecid, equid and other mammalian fossils from the Chorora Formation, Ethiopia. *Anthropological Science* 123:19–39.

Tauxe, L., and N. D. Opdyke. 1982. A time framework based on magnetostratigraphy for the Siwalik sediments of the Khaur area, northern Pakistan. *Palaeogeography, Palaeoclimatology, Palaeoecology* 37:43–61.

Teilhard de Chardin, P. 1942. New rodents of the Pliocene and Lower Pleistocene of North China. Institut de Géo-Biologie de Pekin, 9:101 pp.

Theocaropoulos, C. D. 2000. Late Oligocene–Middle Miocene *Democricetodon, Spanocricetodon,* and *Karydomys* n. gen. from the eastern Mediterranean area. *Gaia* 8:1–103.

Thomas, O. 1896 [1897]. On the genera of rodents: An attempt to bring up to date the current arrangement of the order. *Proceedings of the Zoological Society of London*, 1012–1028.

Tong, H., and J.-J. Jaeger. 1993. Muroid rodents from the Middle Miocene Fort Ternan locality (Kenya) and their contribution to the phylogeny of muroids. *Palaeontographica A* 229(1–3): 51–73.

Tóth, A. R., S. K. Lyons, and A. K. Behrensmeyer. 2014. A century of change in Kenya's mammal communities: Increased richness and decreased uniqueness in six protected areas. *PLOS ONE* 9(4). https://doi.org/10.1371/journal.pone.0093092.

Van Weers, D. J. 2005. A taxonomic revision of the Pleistocene *Hystrix* (Hystricidae, Rodentia) from Eurasia with notes on the evolution of the family. *Contributions to Zoology* 74(3–4): 301–312.

Van Weers, D. J., and L. Rook. 2003. Turolian and Ruscinian porcupines (genus *Hystrix,* Rodentia) from Europe, Asia, and North Africa. *Paläontologische Zeitschrift* 77(1): 95–113.

Vasishat, R. N. 1985. *Antecedents of Early Man in Northwestern India: Paleontological and Paleoecological Evidences.* New Delhi: Inter-India Publications.

Vasishat, R. N., and S. R. K. Chopra. 1981. First record of fossil cricetid (Rodentia) from the Indian Siwaliks. *Bulletin of the Indian Geological Association* 14:59–63.

Von Koenigswald, W. 2004. The three basic types of schmelzmuster in fossil and extant rodent molars and their distribution among rodent clades. *Palaeontographica A* 270(4–6): 95–132.

Wagner, A. 1848. Urweltliche saugetierreste aus Griechenland. *Abhandlungen der Bayerische Akademie der Wissenschaften* 5(2): 333–378.

Wang, B.-Y., and G.-Q. Qi. 2005. A porcupine (Rodentia, Mammalia) from *Lufengpithecus* site, Lufeng, Yunnan. *Vertebrata PalAsiatica* 43(1): 11–23.

Welcomme, J. L., M. Benammi, J.-Y. Crochet, L. Marivaux, G. Métais, P.-O. Antoine, and I. Baloch. 2001. Himalayan forelands: Palaeontological evidence for Oligocene detrital deposits in the Bugti Hills, (Balochistan, Pakistan). *Geological Magazine* 138:397–405.

Wessels, W. 1996. Myocricetodontinae from the Miocene of Pakistan. *Proceedings of the Koninklijke Nederlandse Akademie van Wetenschappen* B99(3–4): 253–312.

Wessels, W. 1998. Gerbillidae from the Miocene and Pliocene of Europe. *Mitteilungen der Bayerischen Staatsamlung für Paläontologie und historische Geologie* 38:187–207.

Wessels, W. 2009. Miocene rodent evolution and migration: Muroidea from Pakistan, Turkey, and northern Africa. *Geologica Ultraiectina* 307:1–290.

Wessels, W., and H. De Bruijn. 2001. Rhizomyidae from the lower Manchar Formation (Miocene, Pakistan). *Annals of Carnegie Museum* 70(2): 143–168.

Wessels, W., H. De Bruijn, S. Taseer Hussain, and J. J. M. Leinders. 1982. Fossil rodents from the Chinji Formation, Banda Daud Shah, Kohat, Pakistan. *Proceedings of the Koninklijke Nederlandse Akademie van Wetenschappen* B83(3): 337–364.

Winkler, A. J., Christiane Denys, and D. Margaret Avery. 2010. Rodentia. In *Cenozoic Mammals of Africa* edited by L. Werdelin and W.J. Sanders, 263–304. Berkeley: University of California Press.

Winkler, A. J., L. J. Flynn, and Y. Tomida. 2011. Fossil lagomorphs from the Potwar Plateau, northern Pakistan. *Palaeontologia Electronica* 14(3), 38A. https://palaeo-electronica.org/2011_3/17_winkler/index.html.

Wood, A. E. 1937. Fossil rodents from the Siwalik beds of India. *American Journal of Science* 36:64–76.

Zachos, J., M. Pagani, L. Sloan, E. Thomas, and K. Billups. 2001. Trends, rhythms, and aberrations in global climate 65 Ma to present. *Science* 292:686–693.

12

SIWALIK PRIMATES

JAY KELLEY, MICHÈLE E. MORGAN, ERIC DELSON,
AND DAVID PILBEAM

INTRODUCTION

Primates are present throughout the Siwalik sequence of the Potwar Plateau. At higher taxonomic levels they are highly diverse, with at least six superfamilies represented, but with superfamily and family representation changing through time. The early part of the record (Early to early Middle Miocene) contains only lower primates of small body size, sivaladapids and lorisids, both known from multiple species, and a small, unspecified primitive catarrhine known only from a few teeth. Sivaladapids were almost surely present in the region at the onset of Siwalik sedimentation, as evidenced by their presence near the base of the Kamlial Formation, and in the Oligocene/Early Miocene faunas from Bugti and the Zinda Pir Dome in west-central Pakistan (Marivaux et al. 2002; Lindsay et al. 2005). The earliest lorisid remains postdate those of sivaladapids by more than 1.5 myr and it seems most likely that they immigrated into the region from Africa subsequent to the beginning of Siwalik sedimentation (Phillips and Walker 2000, 2002). Sivaladapids disappear from the Potwar Siwalik record during the Middle Miocene, but some lorisids persist well into the Late Miocene.

The Middle Miocene saw the appearance of the first large-bodied primate, the hominid *Sivapithecus*. *Sivapithecus* was clearly an immigrant to the region, but whether from somewhere in Eurasia, Afro-Arabia, or perhaps even elsewhere in the subcontinent is unknown. Its craniofacial anatomy is distinctive among all known Miocene primates and bears many features otherwise found only in the living orangutan, *Pongo*. In its entirety, its postcranium is also distinctive among Miocene large primates, with a unique combination of ape-like and more primitive features. The postcranium is even less like that of *Pongo* than it is of other Miocene apes. These issues are discussed further below, but they are raised here to underscore the difficulty in determining both the geographic and phyletic origins of *Sivapithecus*. For most of its history in the Potwar Plateau, *Sivapithecus* is one of the more uncommon taxa, but it is nearly always found when total mammalian specimen counts are sufficiently high. A large, very rare, and somewhat enigmatic hominid, *Indopithecus*, is represented by a single tooth from the Potwar Siwaliks, but it is of uncertain provenance. It is known from a more complete specimen from the Indian Siwaliks and is most likely derived from *Sivapithecus*. Another enigmatic primate from the late Middle Miocene is represented by a small proximal humerus that most resembles those of primitive catarrhines or platyrrhines.

The early colobine monkey *Mesopithecus* first appears in the Siwalik record in the Late Miocene, almost one million years after the local extinction of *Sivapithecus* and in the context of progressive faunal change. *Mesopithecus* is widespread throughout southern and central Eurasia in the Late Miocene and spread rapidly from what is either its likely center of origin, or point of entry into Eurasia from Africa, in southwestern Europe. Its first appearance in the Siwaliks is one of the earlier records of its occurrence (Alba et al. 2015). Temporal ranges of all primate taxa recovered from the Potwar Plateau are shown in Figure 12.1.

Primates are also known from several locations in the Siwaliks of India, and a few specimens have been recovered from Early-Middle to perhaps Late Miocene sediments in the Manchar Formation in southern Pakistan. Not surprisingly, these represent many of the same genera as in the Potwar Plateau sequence, but in some cases different species. From Ramnagar and Haritalyangar in India, there are also rare representatives of groups unknown in the Potwar: respectively, Hylobatidae and Pliopithecoidea. Finally, from Early-Middle Miocene sediments in the Manchar Formation is a representative of the Tarsiidae (Zijlstra, Flynn, and Wessels 2013), unknown from the Siwaliks of either Pakistan or India.

STREPSIRRHINI É. GEOFFROY SAINT-HILAIRE, 1812

Fossil Lorisidae and Sivaladapidae have been recovered almost exclusively from screen-washing bulk sediments and are represented primarily by isolated teeth. The exception is a partial skeleton of *Nycticeboides simpsoni* (Jacobs 1981). Despite small samples, the Siwalik small-primate fossils hint at considerable diversity through the sequence. At one locality (Y682), dated

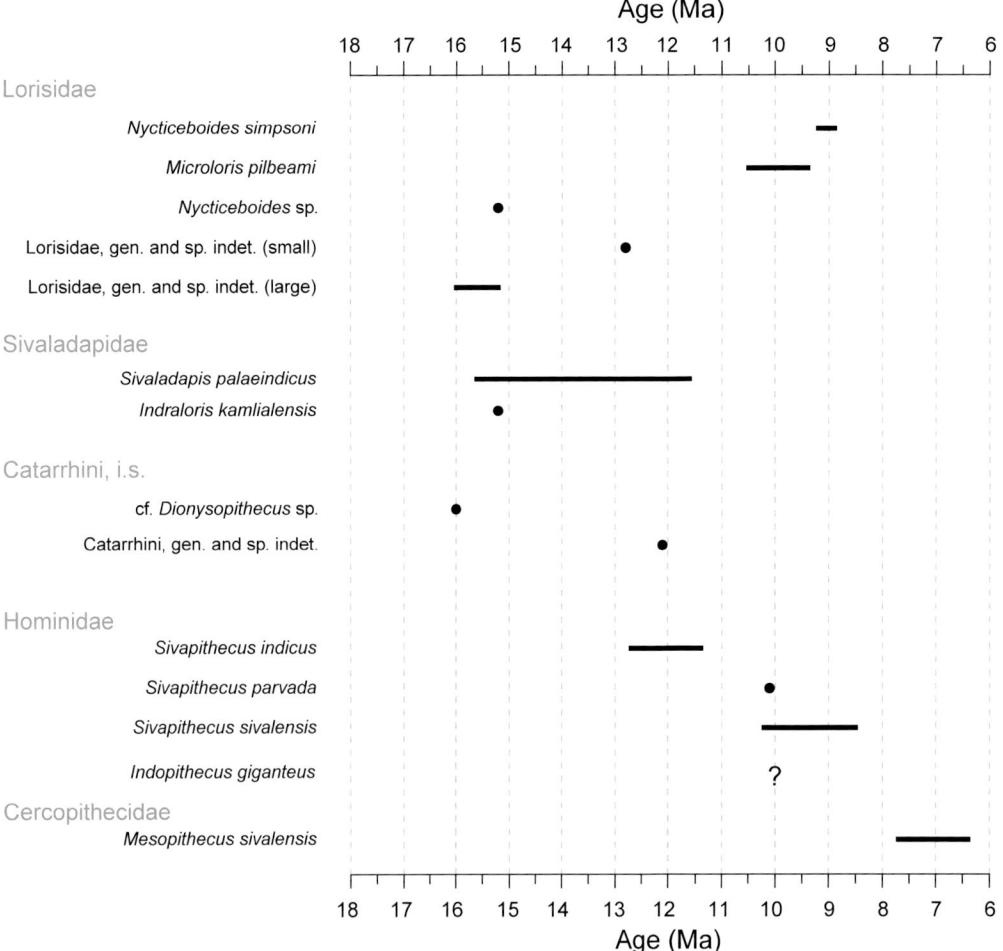

Age (Ma)

Figure 12.1. Stratigraphic ranges of primate species in the Potwar Plateau, Pakistan.

to 15.2 Ma (million years ago), four species, two lorisids and two sivaladapids, have been identified.

Lorisoidea Gray, 1821

Nycticeboides simpsoni. The type specimen of *Nycticeboides simpsoni*, YGSP 8091 (Figure 12.2), is a fragmentary cranium, mandible, and partial skeleton (Jacobs 1981; MacPhee and Jacobs 1986). *N. simpsoni* was probably similar in size to extant *Nycticebus pygmaeus* at around 400 g (MacPhee and Jacobs 1986; Ravosa 1998). Jacobs (1981) reconstructed *N. simpsoni* as a slow-moving arboreal quadruped primarily on the basis of features of the humerus. MacPhee and Jacobs (1986) inferred from tooth morphology and body size that *N. simpsoni* likely consumed insects, exudates, and fruit.

Microloris pilbeami. The type specimen of *Microloris pilbeami* is YGSP 26006, a left M2 (Flynn and Morgan 2005). *M. pilbeami* is much smaller than living *Nycticebus*. Its teeth are similar in size to those of *Galagoides demidovii*, which weighs less than 100 g (Nowak and Paradiso 1983). The molar morphology of *M. pilbeami* suggests an insectivorous diet, although a concave wear facet on the upper P2 might indicate lateral stripping of vegetation (Flynn and Morgan 2005). Sussman, Rasmussen, and

Raven (2012) note that among extant species, even small lorises are omnivorous.

At least four additional lorisoid species are represented in the Siwaliks. These have not been formally named due to the fragmentary material but are distinct from both *Nycticeboides simpsoni* and *Microloris pilbeami* in morphology or size (Flynn and Morgan 2005).

Nycticeboides sp. YGSP 48029, a left m1 from Locality Y682 (15.2 Ma), is similar to *Nycticeboides simpsoni* but much smaller. This specimen considerably extends the temporal range of *Nycticeboides* in the Siwaliks.

cf. *Nycticeboides* sp. YGSP 26009, a left p4 from Locality Y182 (9.2 Ma), is somewhat similar to *Nycticeboides simpsoni*, also represented at this locality, but it clearly represents a different species (MacPhee and Jacobs 1986; Flynn and Morgan 2005).

Lorisidae, unnamed small genus and species. YGSP 53266, a left p2 from Locality Y714 (12.8 Ma), is smaller than *Microloris pilbeami* and has a blunt cusp (Flynn and Morgan 2005). The specimen extends the temporal range of small lorisids into the Chinji Formation.

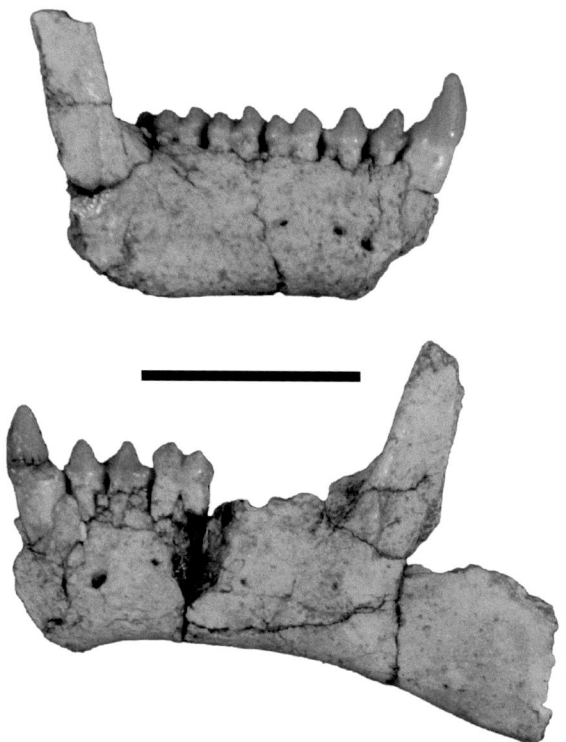

Figure 12.2. Mandible of the type specimen of *Nycticeboides simpsoni*, YGSP 8091, from Locality Y363 dated to about 9 Ma. Scale bar = 1 cm.

Lorisidae, unnamed large genus and species. Two lower molars from the Kamlial Formation, YGSP 19069, a left m2 from Locality Y592 (16.0 Ma), and YGSP 33156, a right m3 from Locality Y682 (15.2 Ma), represent a large lorisid (Flynn and Morgan 2005), larger dentally than living *Nycticebus coucang* and therefore probably weighing more than 2 kg (Nowak and Paradiso 1983). The size and bulbous shape of the molar cusps suggest a largely frugivorous diet.

Lorisoid Evolution and Biogeography

Comparative genetic data suggest a late Middle Eocene origin for Lorisoidea (Yoder and Yang 2004). Several lorisoid genera have been described from the Early and Middle Miocene of East Africa (Phillips and Walker 2002) and, based on this record, Seiffert (2007) and Seiffert, Simons, and Attia (2003) have also suggested a late Middle Eocene Afro-Arabian origin for the group. A late Early Miocene specimen from Thailand attributed to "*?Nycticebus*" (Mein and Ginsburg 1997) is the earliest record of Lorisoidea in southern Asia. Lorisoids were present for at least 7 million years in the Potwar Plateau (minimum range of 16.0 to 8.9 Ma). Dental and postcranial morphology of the Potwar fossil specimens, as well as the size range, suggest similarities with both galagids and lorisids (Flynn and Morgan 2005).

Sivaladapidae Thomas and Verma, 1979

Sivaladapis palaeindicus. The type specimen of *Sivaladapis palaeindicus*, GSI D224, is a partial mandible from an unknown locality near Chinji Village (Pilgrim 1932; Gingerich and Sahni 1979). Body size likely exceeded 3 kg, similar to large species of *Lemur* and *Varecia* (Flynn and Morgan 2005).

Indraloris kamlialensis. The type specimen of *Indraloris kamlialensis*, YGSP 44443, is a lower right molar, probably m1, from Locality Y682 (15.2 Ma). A handful of other teeth from this and nearby localities at the same stratigraphic level are about 20% smaller than specimens of *Indraloris himalayensis* from the Late Miocene at Haritalyangar, India (Gingerich and Sahni 1979; Flynn and Morgan 2005).

Indraloris large species. Two specimens of a large *Indraloris* have been identified from the Kamlial Formation (Flynn and Morgan 2005). These are larger than *I. kamlialensis* and similar in size to *I. himalayensis* (Gingerich and Sahni 1979).

Sivaladapid Evolution and Biogeography

Sivaladapids are also present in the Indian Siwaliks at Ramnagar and Haritalyangar. Ramnagar has not been paleomagnetically dated but is faunally equivalent to the Chinji Formation in the Potwar Plateau (~14–11 Ma), perhaps correlating to the lower half of that range (Gilbert et al. 2019), but most likely in the middle of the range. Primate-bearing sediments at Haritalyangar are dated to ~9.2–8.6 Ma (Pillans et al. 2005). Ramnagar contains *S. palaeindicus* and the newly named *Ramadapis sahnii* (Gilbert et al. 2017). *Sivaladapis nagrii* from Haritalyangar differs from *S. palaeindicus* in having a more derived molar morphology, including higher cusps and more molariform lower premolars, and is interpreted to have been a mixed frugivore-folivore (Patnaik et al. 2014). It is also smaller than *S. palaeindicus* and is plausibly derived from it (Gingerich and Sahni 1979, 1984; Flynn and Morgan 2005). *Indraloris himalayensis* is represented by two specimens from Haritalyangar (Gingerich and Sahni 1979). Thus, of the six sivaladapid species known from Siwalik sediments, *S. palaeindicus* is the only one common to both the Potwar Plateau and the Indian Siwaliks, despite Ramnagar and Haritalyangar being, respectively, only ~150 and 400 km to the east of the Potwar. The presence of *Sivaladapis* and *Indraloris* at Haritalyangar reveals that both genera were present for several million years in the Indian subcontinent and persisted later in the Indian Siwaliks than in the Potwar Plateau.

Sivaladapids are known from elsewhere in Asia during the Miocene—*Siamodapis* from Thailand, dated to 13.3–13.1 Ma (Chaimanee et al. 2008), and *Sinoadapis* from Lufeng in southern China, dated to 6.9–6.2 Ma (Pan and Wu 1985; Wu and Pan 1986; Yue and Zhang 2006), the latter being the latest record of the family. They are also known from Eocene and Oligocene localities in China, Myanmar, and Pakistan (Beard et al. 2007; Marivaux et al. 2002, 2008; Qi and Beard 1998), which gives the group a temporal range of more than 30 million years. The known geographic range of sivaladapids indicates that they were likely endemic to southern and eastern Asia.

CATARRHINI É. GEOFFROY SAINT-HILAIRE, 1812

Siwalik fossil catarrhines include primitive catarrhines of uncertain status, a pliopithecoid, a hylobatid, hominids, and cercopithecids, all unevenly distributed through space and time. By far the most abundant remains are those of the hominid *Sivapithecus*, while primitive catarrhines, the hylobatid, and the pliopithecoid are known from only one or a few specimens each.

Sivapithecus Pilgrim, 1910

The taxonomy of *Sivapithecus* has proven quite vexing from its earliest description (see Kelley 2002), and there is still no consensus about the number of species or the allocation of specimens to species (Kelley 2005). This is a consequence of *Sivapithecus* being relatively uncommon and the sample comprising, apart from a partial skull, YGSP 15000 (Figure 12.3), mostly fragmentary remains. There is general agreement about the validity of two of the three first-named Siwalik *Sivapithecus* species, *S. sivalensis* Lydekker, 1879 and *S. indicus* Pilgrim, 1910, but not about their hypodigms. The types of both of these species

are from the Potwar Plateau, while that of the third, *S. ("Ramapithecus") punjabicus*, now regarded as a junior synonym of *S. sivalensis*, is from Haritalyangar in India. The traditional conception is of a large (*S. indicus*) and small (*S. sivalensis*) species pair (e.g., Simons and Pilbeam 1965; Greenfield 1979; Kay 1982), but there is no consistency among those using this taxonomy in the assignment of specimens in the middle of the overall size range. Moreover, acknowledgment of the implications of this taxonomic scheme for dental, and by extension, body-size sexual dimorphism, is rare; these would have been quite low by extant ape standards, given the overall size range in each of these species. An alternative taxonomy followed here sees time-successive species (Kelley 1986, 1988, 2005)—the earlier *S. indicus* from the Chinji Formation and the later *S. sivalensis* from the Nagri and Dhok Pathan Formations (see Fig. 12.1). In this taxonomy, both species contain large (male) and small (female) specimens, and dental sexual dimorphism in both is substantial, more or less equivalent to that found in extant *Gorilla* or *Pongo* (Kelley 2005). The two species are nearly identical in known dento-gnathic morphology, but there are differences in molar proportions, and the teeth of *S. indicus* are, on average, slightly smaller

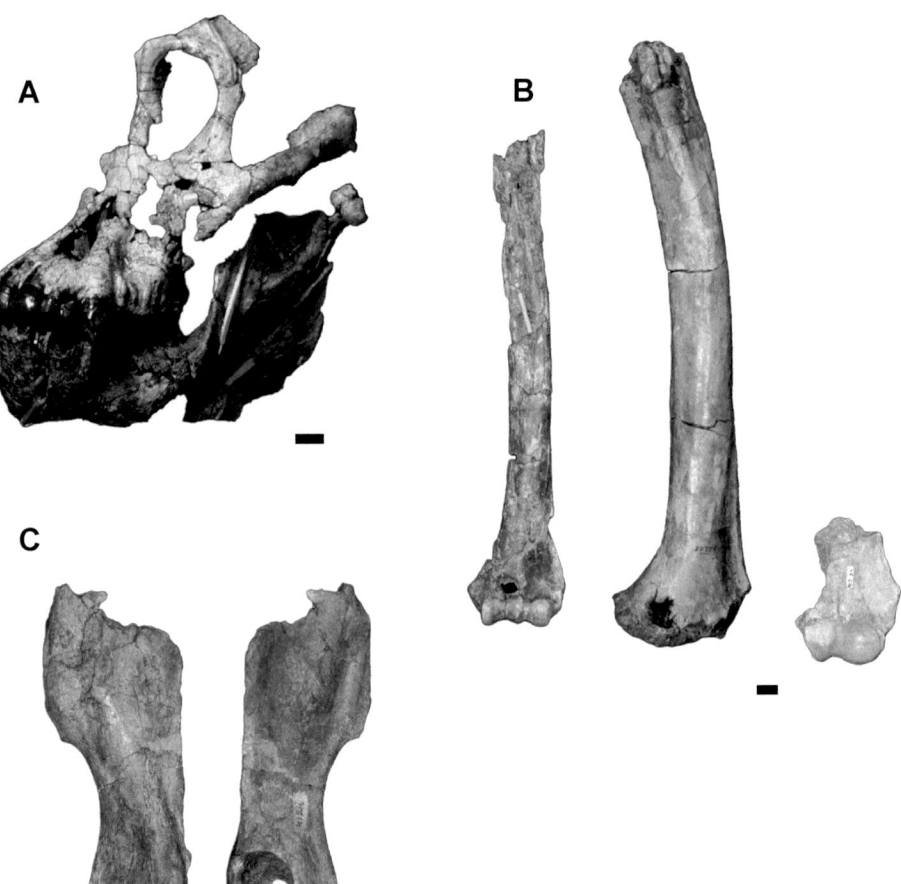

Figure 12.3. Sivapithecus: A. partial skull of *S. sivalensis*, YGSP 15000 from Locality Y410 dating to 9.4 Ma; B. humeri, from left to right: YGSP 30730, left humerus of *S. indicus* from Locality Y76 dated to 11.4 Ma, YGSP 30754 (left humerus) and YGSP 12271 (right humerus, reversed) of *S. parvada* from Locality Y311 dated to 10.1 Ma; C. partial left innominate of *S. indicus*, YGSP 41216 from Locality Y647 dated to 12.4 Ma. Scale bars = 1 cm.

than those of *S. sivalensis* (Kelley 2005). There are also differences between the one partial palate of *S. indicus* and those of *S. sivalensis* (Begun and Güleç 1998), but whether these reflect inter-individual variation or species-specific morphology is not clear. There are few shared elements among the postcranial samples of the two species, and those that are shared (humerus, calcaneus) show no meaningful differences.

There is also general agreement about a third, very large species of *Sivapithecus*, *S. parvada* Kelley, 1988, known from a single Potwar locality, Y311, and not known from outside the Potwar Plateau. In addition to the inferred body mass of *S. parvada* being substantially greater than that of both *S. indicus* and *S. sivalensis* (Table 12.1 and below), *S. parvada* also differs from both in having relatively large premolars in relation to molars and different upper central incisor morphology (Kelley 1988; Kelley et al. 1995).

While the temporal range of *S. parvada* is confined to the depositional period of the Y311 locality, perhaps several tens of thousands of years at ~10.1 Ma, the narrow stratigraphic interval that has produced most of the remains spans perhaps no more than a few hundred years (Behrensmeyer 1987). Both *S. indicus* and *S. sivalensis* are known from numerous localities, but their relative rarity, combined with the low quality of the Potwar fossil record during long stretches of time (Barry et al. 1995, 2002), make it difficult to precisely delimit their temporal ranges. The first appearance of *S. indicus* (12.7 Ma) and the last appearance of *S. sivalensis* (8.5 Ma) are preceded and followed, respectively, by several 100,000-year intervals[1] of high record quality (large numbers of mammal specimens and numerous species represented) in which they are not found, making these dates highly reliable. In contrast, the quality of the record over the interval in which *S. indicus* last appeared and *S. sivalensis* first appeared (or during which anagenesis occurred—see below) is relatively poor and specimens are few. Thus, the precise temporal ranges of *S. indicus* and *S. sivalensis* remain unknown.

Finally, there is a long history of discussion regarding the possible presence of a fourth, very small species of *Sivapithecus* (see Kelley 2002, 2005; Bhandari et al. 2018). The few suggested hypodigms for this species all contain only a handful of specimens, some from the Potwar Plateau and some from Haritalyangar in India, and all include specimens that extend over much of the temporal and geographic range of *Sivapithecus* as a whole. While there are certainly some quite small dental and mandibular specimens in the greater Siwalik sample (including India), there are no obvious breaks in the continuity of dental metric distributions by which this species, if it exists, might be easily delimited. If a very small species is present, naming it will depend on which specimens are included in the hypodigm (Kelley 2005; Bhandari et al. 2018).

Body Size and Body-Size Dimorphism

Postcranial remains are known for all three *Sivapithecus* species, and for all three, most or all postcranial elements broadly fall

into two size groups corresponding to males and females (see Table 12.1; see also DeSilva et al. 2009). Based on comparisons to extant primates, male and female body-size estimates for the three *Sivapithecus* species are: *S. parvada*, ~60–70 kg and ~30–35 kg; *S. sivalensis*, ~50–55 kg and ~25–30 kg; *S. indicus*, ~30–45 kg and ~20–25 kg. Body-size dimorphism in all three species is therefore estimated to have been about 2:1 or similar to that in *Pongo pygmaeus*.

Locomotion and Substrate Use

As shown in Table 12.1, there is not a great deal of overlap in postcranial elements among the three *Sivapithecus* species. However, where it occurs (humerus, calcaneus), the morphology is remarkably consistent among species, in spite of the considerable size disparity between *S. parvada* and the other two species. These and all other elements present a consistent picture regarding positional behavior, locomotion, and substrate use in *Sivapithecus*: a generalized, above-branch, arboreal quadruped that was capable of engaging to some degree in a variety of other arboreal locomotor modes such as climbing and certain forms of suspension (Rose 1984, 1986, 1989, 1994, 1997; Pilbeam et al. 1980, 1990; Madar et al. 2002; DeSilva et al. 2009; Morgan et al. 2015). Even occasional terrestriality may have been practiced (Spoor, Sondar, and Hussain 1991; Madar et al. 2002), although we find evidence that *Sivapithecus* was a habitual knuckle-walker (Begun and Kivell 2011) unconvincing. That *Sivapithecus* engaged primarily in monkey-like quadrupedal locomotion rather than the suspensory behaviors common in modern apes is particularly borne out by the morphology of the humeral diaphysis and pelvis (Figure 12.3).

In addition, it has been proposed that *Sivapithecus* relied on a mechanism of balance that may have been peculiar to a number of other mostly quadrupedal arboreal Miocene apes, but not utilized by any living anthropoid. It is presumed that *Sivapithecus*, like other apes, did not possess a tail, so that balance was maintained by maneuvering the torso and the center of gravity from secure hand and foot holds held in a variety of orientations, rather than by reliance on a tail as a counterweight or counter force (Kelley 1997; Morgan et al. 2015; see also Larsen and Stern 2006). Such a mechanism is compatible with postcranial features in *Sivapithecus* that are associated with enhanced limb mobility, especially distally (Rose 1986; Pilbeam et al. 1980, 1990; Madar et al. 2002), and with powerfully grasping hands and feet, enhanced by hallux and pollex elements that are both relatively long and robust (Pilbeam et al. 1980; Madar et al. 2002).

Diet

Tooth morphology, microwear, and carbon and oxygen isotopes from enamel support a predominantly frugivorous diet derived largely from upper-canopy forage (Nelson 2003, 2007). Intraspecific variation in *Sivapithecus sivalensis* microwear suggests that the fruits consumed varied in hardness, including both fruits softer in consistency, like those ordinarily consumed by chimpanzees, and some harder fruits, as often consumed by orangutans (Teaford and Walker 1984; Nelson 2003).

---

1. 100,000-year intervals are the standard for analysis of the Potwar Plateau faunal record.

Table 12.1. Body Mass Estimation in *Sivapithecus*

	Element	Comparison to Extant Primate Species
S. indicus		
YGSP 1656	Distal tibia	Similar to male proboscis monkey: ~20 kg
YGSP 17118	Ectocuneiform	Similar to small male chimpanzee: ~45–50 kg
YGSP 17119	Capitate	Slightly smaller than male & female bonobos: ~30–40 kg
YGSP 28230	Calcaneus	Similar to male proboscis monkey: ~20 kg
YGSP 30730	Humerus	Similar to male baboon: ~25 kg
YGSP 41216	Innominate	Similar to male proboscis monkey & baboon: ~20–25 kg
YGSP 46459	Navicular	Similar to male proboscis monkey: ~20 kg
S. parvada		
NG 933	Metacarpal I	Head slightly larger than male chimpanzee, shaft as stout as male gorilla: ~65–70 kg
NG 940	Hamate	Slightly larger and stouter than male chimpanzee: ~60 kg
YGSP 6454	Int. cuneiform	Similar to male baboon: ~25 kg
YGSP 6663	Distal humerus	Similar to male baboon; slightly smaller than female chimpanzee: ~25–30 kg
YGSP 6664	Pollical prox. phalanx	Slightly larger than male gorilla; relationship of phalanges to body mass uncertain
YGSP 12271	Distal humerus	Similar to male orangutan: ~70 kg
YGSP 17152	Calcaneus	Similar to male chimpanzee: ~60 kg
YGSP 17606	Calcaneus	Similar to small female chimpanzee: ~30–35 kg
YGSP 19905	Cuboid	Slightly larger than male baboon: ~30 kg
YGSP 30754	Humerus	Slightly smaller than male orangutan: ~65 kg
YGSP 47700	Hallucal dist. phalanx	Similar to male gorilla; relationship of phalanges to body mass uncertain
YGSP 51344	Pollical prox. phalanx	Similar to male gorilla; relationship of phalanges to body mass uncertain
S. sivalensis		
YGSP 4664	Calcaneus	Larger than siamang; smaller than male baboon: ~15–20 kg
YGSP 6178	Femoral head	Intermediate between siamang and female chimpanzee: ~20–25 kg
YGSP 10785	Talus	Larger than siamang; much smaller than male baboon: ~15–18 kg
YGSP 11867	Femoral head	Slightly larger than male baboon: ~25 kg
YGSP 13420	Femoral shaft	Similar to male baboon: ~25 kg
YGSP 13506	Distal humerus	Similar to small female chimpanzee but less robust: ~30 kg
YGSP 13929	Femoral head	Intermediate between siamang and female chimpanzee: ~20–25 kg
YGSP 14046	Hallux	Metatarsal head slightly smaller than male chimpanzee; phalanges larger than female chimpanzee: ~50 kg
YGSP 15782	Femoral head	Slightly larger than male baboon: ~25–30 kg
YGSP 47025	Distal femur	Slightly larger than male baboon: ~25–30 kg

Habitat Preference

The Potwar Siwalik sequence preserves very few plant remains and lacks pollen. Reconstruction of habitats associated with *Sivapithecus* between ~13 and ~8 Ma relies therefore on carbon and oxygen isotopic analyses of paleosols and mammalian teeth, on mammalian ecomorphology, and on inferences from sedimentology. Habitats over the interval when *Sivapithecus* was part of the mammal community are reconstructed as mosaic, with probable gallery forests, woodlands of varying density, and some grassland, while climates would have been subtropical and seasonal although not strongly so (Behrensmeyer, Willis, and Quade 1995; Scott, Kappelman, and Kelley 1999; Barry et al. 2002; Nelson 2005, 2007; Behrensmeyer 2007; Badgley et al. 2008; Morgan et al. 2009). Given the inferred positional repertoire of *Sivapithecus*, together with indications that a substantial amount of feeding took place in the upper canopy (Nelson 2005, 2007), it is likely that *Sivapithecus* was highly dependent on forests and denser woodlands.

Life History

Life history in *Sivapithecus parvada* has been examined using the chronology of dental development derived from the remains of a juvenile, YGSP 11536 and YGSP 46460, that died shortly after the first molars erupted. From the incremental growth features preserved on the upper incisors (Figure 12.4), it was possible to estimate a likely age at first molar emergence of 40–41 months, which falls within the ranges of extant great apes (Kelley 1997). Since age at first-molar emergence is a useful proxy for the overall schedule of development in primates (Smith 1989; Kelley and Schwartz 2010), the estimated age at death for this individual reveals that life history in *Sivapithecus* would have been broadly the same as the prolonged life histories of extant apes.

Indopithecus von Koenigswald, 1950
Indopithecus giganteus (Pilgrim 1915) is known from only two specimens, of which the type, an m2, is from the Potwar Plateau but is of uncertain provenance. The other specimen, a relatively

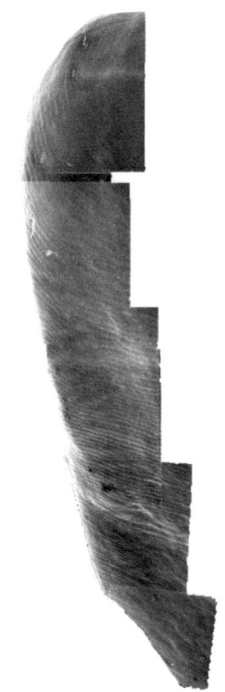

Figure 12.4. Incremental growth features on upper incisor enamel of juvenile *Sivapithecus parvada* (YGSP 46460, from Locality Y311 dated to 10.1 Ma).

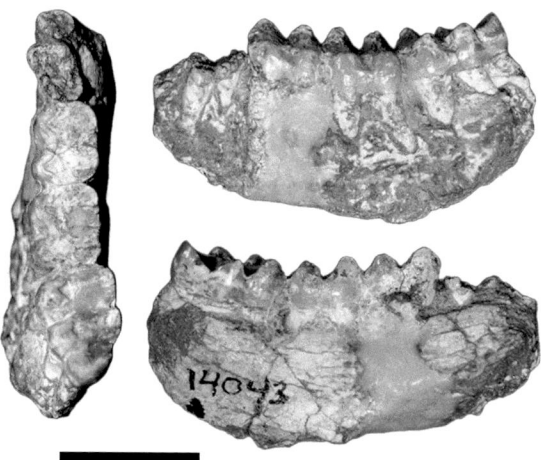

Figure 12.5. Partial mandible of *Mesopithecus* (YGSP 14043 from Locality Y370 dated to 7.2 Ma) in occlusal, lingual and labial views. Scale bar = 1 cm.

complete mandible, is from Haritalyangar in India. The mandible was described as representing a species of *Gigantopithecus*, *G. bilaspurensis* (Simons and Chopra 1969). Szalay and Delson (1979) combined the two specimens as *G. giganteus*, but they were returned to the resurrected *Indopithecus* by Cameron (2001). At ~8.6 Ma (Pillans et al. 2005), the Haritalyangar mandible is the youngest hominid specimen recovered from that sequence and is among the youngest hominids from the Siwaliks.

Cf. *Dionysopithecus* Li, 1978/Primitive Catarrhine

A small upper molar from ~16.0 Ma in the Kamlial Formation was regarded by Barry, Jacobs, and Kelley (1986) as taxonomically indeterminate, being equally similar to the early Middle Miocene *Dionysopithecus* from China and Early Miocene *Micropithecus* from East Africa. *Dionysopithecus* is possibly a primitive pliopithecoid (Harrison and Gu 1999), while *Micropithecus* is generally regarded as a primitive catarrhine and is placed in the superfamily Dendropithecoidea by Harrison (2010).

A small proximal humerus from the Chinji Formation bears similarities to a variety of primitive catarrhines or even platyrrhines such as *Cebus* (Rose 1989). Whether it represents the same taxon as the Kamlial molar cannot be determined, but it is additional evidence for the rare presence of small, non-hominoid, non-cercopithecoid catarrhines in the early Potwar record.

Mesopithecus Wagner, 1839

Over decades of collecting, a small number of cercopithecid primates have been recovered from the Late Miocene of the Potwar Plateau. Only four of these have reliable provenance, three recovered by the Harvard-GSP project from localities in the Kaulial Kas area and one collected by G. E. Lewis in the 1930s from Locality L082 near Hasnot, and range in age from 7.7 to 7.0 Ma. The best preserved of these is YGSP 14043, a partial mandible (Figure 12.5). Several other fragmentary specimens were also recovered by earlier researchers, mainly from deposits in the Hasnot area, estimated to range in age from 7.2 to 6.4 Ma (J. Barry, pers. comm., revised from Barry 1987). Two species were named from this earlier material, the first being *Macaca sivalensis* Lydekker, 1878, but this and the other remains were eventually all recognized to be colobine (Simons 1970). Delson (1975, 1994) and Szalay and Delson (1979) combined all the specimens into a single species, which Harrison and Delson (2007) later tentatively assigned to the cosmopolitan Late Miocene genus *Mesopithecus*, as *M. sivalensis*. However, the available material is too fragmentary and insufficiently studied to say anything about the paleobiology of this species.

More recently, additional cercopithecid dento-gnathic specimens were recovered from the Hasnot area, dated to ~7.9–7.0 Ma and assigned to cf. *Mesopithecus* sp., pending comparison to the earlier material (Khan et al. 2020).

Catarrhine Primates from the Indian Siwaliks, Southern Pakistan, and Nepal

Catarrhine primates are also known from the more easterly regions of the Siwaliks in India. Among hominids, these include, in addition to *Indopithecus*, both *S. indicus* from the Ramnagar area and *S. sivalensis* from the vicinity of Haritalyangar and farther east (Kelley 2002 and references therein; Cameron, Patnaik, and Sahni 1999; Patnaik 2013; Gilbert et al. 2019). The sample from the Haritalyangar area contains several specimens that are larger than any *S. sivalensis* specimens from the Potwar Plateau, although more similar in size to the latter than to *S. parvada* (Kelley 1988), as well as some that are among the smallest, which makes the taxonomy of the sample difficult to interpret (Kelley 2005; Scott, Schrein and Kelley 2009). If the largest specimens are in fact *S. sivalensis*, they appear to represent a larger regional variant than in the Potwar Plateau. Both Ramnagar and Haritalyangar also preserve catarrhines unknown from the Potwar Siwalik sequence. A single m3 from Ramnagar, representing a new genus and species, *Kapi ramnagarensis*, has been assigned to Hylobatidae (Gilbert et al. 2020). If this assignment is accurate (see Ji et al. 2022; Gilbert et al. 2023), it would nearly double the temporal range of fossil hylobatids to between ~12 and ~13 Ma and represent the first record of the family outside southeastern Asia (Pan 2006; Harrison 2016). The pliopithecoid from Haritalyangar, *Krishnapithecus krishnaii* Chopra and Kaul, 1979 is highly derived and, dentally, is one of the largest pliopithecoids (Sankhyan, Kelley, and Harrison 2017).

Six cercopithecid dento-gnathic specimens, one partial maxilla and five partial mandibles, have also been recovered from (younger) Indian Siwalik sediments. The maxilla was the first fossil primate to be described from the subcontinent and the first fossil to have been recognized as primate (Baker and Durand 1836); it was named *Semnopithecus subhimalayanus* in a catalog by von Meyer (1848) but is now generally assigned to *Procynocephalus*. The mandibular fragments, three of which were described by Falconer and Cautley (1837), have a checkered taxonomic history (Lydekker 1884, 1886; Verma 1969; Delson 1973, 1975; Szalay and Delson 1979). There is still disagreement about species attribution and, for two specimens, subfamily attribution (Delson 1973, 1975; Jablonski 2002).

A single hominid molar has been recovered from sediments in Nepal that appear to be faunally equivalent to the Chinji Formation of the Potwar Plateau (Munthe et al. 1983).

Primate teeth have also been recovered from the (mostly) early Middle Miocene Manchar Formation at Sehwan in southern Pakistan (Raza et al. 1984; Bernor et al. 1988). The smaller teeth were provisionally assigned to *Dionysopithecus*, but this has been questioned by Harrison and Gu (1999). In size and morphology, they are compatible with the unassigned molar from the Kamlial Formation in the Potwar Plateau described by Barry, Jacobs, and Kelley (1986) and offer further evidence for the widespread occurrence of small, primitive catarrhine primates in South Asia at this time. Two of the Manchar Formation teeth, however, are larger. One (GSP S214) is a large hominid molar from near the top of the local section, which also produced teeth of *Hipparion*, so it is Late Miocene in age and contemporaneous with later hominid-bearing sediments in the Potwar Plateau. The other (GSP S100) is a worn molar fragment that cannot be definitively identified as hominid or even hominoid. If it is, its position from much lower in the section would establish the presence of large hominids in the subcontinent considerably earlier than the first appearance of *Sivapithecus* in the Potwar Plateau at 12.8 Ma, and perhaps among the earliest in Eurasia (see Raza et al. 1984).

Finally, a poorly preserved maxillary fragment from Kutch, Gujarat state in western India, assigned to *Sivapithecus* sp., is dated faunally to younger than 10.8 Ma (Bhandari et al. 2018). This is the only definitive occurrence of *Sivapithecus* outside the Siwaliks and extends its range southwestward by roughly 1,000 km, although much less than this if either of the two large catarrhine teeth from Sehwan in southern Pakistan is *Sivapithecus*.

Siwalik Catarrhine Evolution and Biogeography

There are no obvious precursors to *Sivapithecus* in the fossil record, although the record of hominoids in Eurasia is quite sparse prior to its first appearance at 12.8 Ma. The only earlier sample with more than a handful of specimens is from Paşalar, at 14–15 Ma, while the earliest hominoids from the Vallès Penedès Basin in Spain are ~12 Ma (Casanovas-Vilar et al. 2011). The two species in the large Paşalar sample, *Kenyapithecus kizili* and *Griphopithecus alpani*, are known almost entirely from teeth (Alpagut, Andrews, and Martin 1990; Martin and Andrews 1993; Gençturk, Alpagut, and Andrews 2008), which, along with the few cranial and mandibular remains, are largely uninformative with respect to *Sivapithecus* ancestry. One specimen reveals a rather large palatal incisive foramen (Kelley, Andrews, and Alpagut 2008), unlike the more restricted foramen in *Sivapithecus*, which is part of a derived subnasal morphological complex that is distinctive for *Sivapithecus* and the later *Khoratpithecus* from Thailand and Myanmar (Chaimanee et al. 2019) among Miocene hominoids. None among the very large number of catarrhine species from the Early and Middle Miocene of East Africa shows any particular affinity with *Sivapithecus*.

If the origins of *Sivapithecus* are uncertain, its phylogenetic position within Hominoidea seems clearer, but even that comes with caveats. As noted, *Sivapithecus* uniquely shares with extant *Pongo* a suite of craniofacial features that are widely interpreted as confirming membership in the Ponginae (see Figure 12.3). These include vertically elongate orbits; a narrow interorbital septum lacking ethmoid sinuses; absence of a frontal sinus; a continuously concave upper and midfacial profile combined with pronounced nasoalveolar prognathism; a subnasal morphology characterized by a premaxilla that extends well into the nasal cavity; a smooth transition from the premaxilla to the palate; a reduced incisive fossa, narrow incisive canal, and a slit-like incisive foramen (Andrews and Cronin 1982; Pilbeam 1982; Ward and Kimbel 1983, Ward and Pilbeam 1983; Ward and Brown 1986, McCollum et al. 1993, Brown, Kappelman, and Ward 2005). In contrast, the teeth and mandible do not bear any particular resemblance to those of *Pongo*, with the possible exception of upper central incisor morphology and premolar/molar proportions in *S. parvada* (Kelley 1988; Kelley et al. 1995). Likewise, while many individual features among the *Sivapithecus* postcranial elements are broadly like those of extant great apes (and many more that are not), none is exclusively like those of *Pongo* (Madar et al. 2002 and references therein; Morgan et al. 2015), nor, as noted, is overall postcranial morphology at all like that of extant great apes. While we tend to favor the view that the shared craniofacial features of *Sivapithecus* and *Pongo* represent homologies while much of the shared postcranial morphology of extant great apes results from homoplasy, we acknowledge that there are no clear biological criteria for deciding such questions (Morgan et al. 2015).

Regarding other Miocene apes, there is little morphological similarity between *Sivapithecus* and most other fossil taxa, especially cranially, even among taxa considered by some to be in the *Pongo* clade. *Ankarapithecus* from Turkey bears some similarity in a few cranial features, but overall has a fundamentally different morphology and mostly lacks the numerous *Pongo*-like features evident in *Sivapithecus* (Begun and Guleç 1998; Alpagut et al. 1996; Kappelman et al. 2003; Brown, Kappelman, and Ward 2005). Species of *Lufengpithecus* from southern China are even more dissimilar and show no compelling evidence to support membership in the *Pongo* clade (Kelley and Gao 2012; Ji et al. 2013). Other Eurasian fossil apes with more than a few postcranial remains (*Hispanopithecus* and *Pierolapithecus* from Spain; *Rudapithecus* from Hungary) show a fundamentally different body plan, one more compatible with orthogrady and the frequent use of suspensory behaviors (Moyà Solà and Köhler 1996; Moyà Solà et al. 2004; Kivell and Begun 2009; Begun 2010; Alba 2012; Alba et al. 2012; Begun, Nargolwalla, and Kordos 2012; Ward et al. 2019).

An exception to this lack of similarity is *Khoratpithecus*, which is known from the late Middle and Late Miocene of Thailand and Myanmar. While the first published remains of *Khoratpithecus* were limited to two mandibles and a small number of teeth that are either unlike those of *Sivapithecus* or show no particular affinity (Chaimanee et al. 2003, 2004, 2006; Jaeger et al. 2011), a more recently recovered partial maxilla (Chaimanee et al. 2019) broadly displays the same diagnostic subnasal morphology otherwise evident only in *Pongo* and *Sivapithecus* (Ward and Kimbel 1983).

Beyond origins and phylogeny, some of the more intriguing questions regarding *Sivapithecus* concern the occurrence and abundance of the different species in space and through time, particularly *S. parvada*. Given that species of *Sivapithecus* were probably continuously present in the region for more than 4 myr, the restriction of *S. parvada* to the Y311 locality is noteworthy. Y311 is unusual both for its abundance of fossils and its depositional environment within the Potwar sequence, as the facies associated with fossil concentrations in the Chinji and Dhok Pathan Formations are uncommon in the Nagri Formation (Behrensmeyer 1987: Y311 designated as Level 3; Willis 1993a, 1993b). The massive channel system represented by the Nagri Formation would almost surely have offered a different mix of habitats, or at least in different proportions, than those most frequently associated with other formations. Whether or not the presence of *S. parvada* was dependent on such factors is unknown, and its restriction to this single locality is a conundrum.

The Y311 locality is also unusual for the abundance of *Sivapithecus* specimens relative to other mammalian fossils. For the 25 hundred-thousand-year intervals from the 12.8 to 8.5 Ma *Sivapithecus* time range that have produced at least 150 total mammalian specimens (24 of these intervals have produced at least 200), the average proportion of *Sivapithecus* specimens among all specimens is 0.22 (range: 0–0.66); the value for Y311, which accounts for nearly all the mammal remains in the 10.1–10.0 Ma interval, is 1.18. There are three other hundred-thousand-year intervals with a high relative abundance of *Sivapithecus* remains

(*S. sivalensis*), all clustered around the U-level sandstone at ~9.4-9.2 Ma. The values for these are, respectively, 1.20, 4.89 and 3.49. Since these comparisons are restricted to intervals in which record quality is good, the exceptional abundance of *Sivapithecus* within these few time intervals, particularly at 9.3–9.2 Ma, would appear to be real. While the Y311 locality is somewhat unusual as noted above, there is nothing obvious about either the depositional environments or the faunas from localities clustered around the U-level that might explain the exceptional abundance of *Sivapithecus* during this period.

Regarding relationships among species of *Sivapithecus*, it was suggested earlier that *S. indicus* and *S. sivalensis* might be anagenetic species, with *S. parvada* showing no closer relationship to one or the other. However, this is based on little more than the two former species being essentially the same size while the latter is substantially larger. There is nothing in the morphology of the three species to favor any particular hypothesis of relationships among them. Resolving these relationships is made even more intractable by the near absence of any *Sivapithecus* remains between 11.4 and 9.4 Ma, much of this period being of low record quality, other than those of *S. parvada*.

There is little that can be inferred about the evolution and biogeography of other Siwalik catarrhines. Few morphological features distinguish the dentition of *Indopithecus* from that of *Sivapithecus*, although dental proportions differ, with *Indopithecus* having relatively very small canines and incisors (Simons and Chopra 1969). Further, despite its great size, the *Indopithecus* mandible is closely matched morphologically by some mandibles of *S. sivalensis*. It is most reasonable, therefore, to regard *Indopithecus* as a derived taxon within the *Sivapithecus* radiation. The affinities of the small, primitive catarrhine teeth from the Kamlial and Manchar Formations could as easily be with some Early Miocene catarrhine from East Africa as with *Dionysopithecus* from eastern Asia (Barry et al. 1986). Seemingly, a single species of *Mesopithecus*, *M. pentelicus*, ranges from Spain to southwestern China during the Late Miocene (Alba et al. 2015; Jablonski et al. 2020). Further comparative study of the Siwalik *Mesopithecus* will be necessary to determine if it does in fact represent a different species.

PRIMATE EXTINCTIONS IN THE POTWAR PLATEAU

The local extinction of sivaladapids in the Potwar Plateau sequence at 11.6 Ma occurred during a period of low faunal extinction overall, but this was followed by a period of significantly higher rates of extinction that peaked at around 11 Ma (Badgley et al. 2016). It may be that sivaladapids were particularly sensitive to whatever conditions led to this subsequent wave of extinctions.

The conditions that led to the local extinction of other primate groups are clearer. Marked change in habitat from predominantly forests and woodlands to more open woodlands and grasslands, inferred primarily from stable isotope analysis of paleosol carbonates and mammalian enamel, began around 8 Ma on the Potwar Plateau (Behrensmeyer et al. 2007). This change is correlated with significant faunal change, in which most closed-habitat frugivorous and browsing mammals disappear from the record, or with shifts in dietary habits among herbivorous mammals from consuming browse and other C_3 vegetation to increasing reliance on C_4 grasses (Barry et al. 2002; Behrensmeyer et al. 2007; Badgley et al. 2008). A similar pattern and timing of habitat change, from predominantly closed to more open habitats, is documented in the paleobotanical and palynological records of India and Nepal (Hoorn, Ohja, and Quade 2000).

Notably, isotopic evidence, especially from mammalian tooth enamel, also indicates that these habitat changes began much earlier, by 10 Ma or even earlier (Quade, Cerling and Bowman 1989; Morgan, Kingston and Marino 1994; Behrensmeyer et al. 2007), but the isotopic shifts were initially subtle, with as-yet uncertain effects on faunal composition. *Sivapithecus* and other ripe-fruit dependent or extensive closed-canopy-dependent species appear to have been the earliest affected as these changes accelerated (Nelson 2003, 2005, 2007; Badgley et al. 2008), with lorisoids last recorded at 8.9 Ma and *Sivapithecus* at 8.5 Ma. The first appearance of *Mesopithecus* is at 7.7 Ma, roughly at the midpoint of the accelerated period of isotope transition, and postdates the last record of *Sivapithecus* by nearly 1 myr. Thus, the "replacement" of apes by monkeys in the Siwaliks is driven entirely by the differing ecological requirements of the two groups and owes nothing to competition between them.

PRIMATE FAUNAS OF THE POTWAR PLATEAU AND THE INDIAN SIWALIKS

One of the more interesting aspects of the Siwalik primate record is the differences between the primate faunas of the Potwar Plateau and those of the Indian Siwaliks throughout the entirety of this record, despite the relatively small distances between the various Indian sites and the Potwar. These include mostly different sivaladapid species and the persistence of the group in India for more than 2 myr longer than in the Potwar; the absence of *S. parvada* in the Indian record (although possibly being due to an unrepresented time interval); apparent differences in the body size of *Sivapithecus* in the two regions ~9 Ma, perhaps reflecting the appearance of a different, larger species in the Indian Siwaliks at this time; and the presence of both a pliopithecoid and a probable hylobatid in the Indian Siwaliks but not in the Potwar. The absence of lorisoids from the Indian Siwalik record is likely an artifact of infrequent sediment screen-washing by groups who have collected there. The reasons for the other differences have not been explored and remain a potential informative line of inquiry into the history of Siwalik primates and into overall regional differences in the biogeography of the Siwalik sequence of South Asia.

References

Alba, D. M. 2012. Fossil apes from the Vallès-Penedès Basin. *Evolutionary Anthropology* 21(6): 254–269.

Alba, D. M., S. Almécija, I. Casanovas-Vilar, J. M. Méndez, and S. Moyà-Solà. 2012. A partial skeleton of the fossil great ape *Hispanopithecus laietanus* from Can Feu and the mosaic evolution of crown-hominoid positional behaviors. *PLoS ONE* 7(6): e39617.

Alba, D. M., P. Montoya, M. Pina, L. Rook, J. Abella, J. Morales, and E. Delson. 2015. First record of *Mesopithecus* (Cercopithecidae, Colobinae) from the Miocene of the Iberian Peninsula. *Journal of Human Evolution* 88:1–14

Alpagut, B., P. Andrews, M. Fortelius, J. Kappelman, I. Temizsoy, H. Çelebi, and W. Lindsay. 1996. A new specimen of *Ankarapithecus meteai* from the Sinap Formation of central Anatolia. *Nature* 382(6589): 349–351.

Alpagut, B., P. Andrews, and L. Martin. 1990. New hominoid specimens from the middle Miocene site at Paşalar, Turkey. *Journal of Human Evolution* 19(4–5): 397–422.

Andrews, P., and J. Cronin. 1982. The relationships of *Sivapithecus* and *Ramapithecus* and the evolution of the orang-utan. *Nature* 297(5867): 541–546.

Badgley, C., J. C. Barry, M. E. Morgan, S. V. Nelson, A. K. Behrensmeyer, and T. E. Cerling. 2008. Ecological changes in Miocene mammalian record show impact of prolonged climatic forcing. *Proceedings of the National Academy of Sciences* 105(34): 12145–12149.

Badgley, C., M. Soledad Domingo, J. C. Barry, M. E. Morgan, L. J. Flynn, and D. Pilbeam. 2016. Continental gateways and the dynamics of mammalian faunas. *Comptes Rendus Palevol* 15(7): 763–779.

Baker, W. E., and N. H. Durand. 1836. Sub-Himalayan fossil remains of the Dadoopoor collection. *Journal of Asiatic Society of Bengal* 5:739–741.

Barry, J. C. 1987. The history and chronology of Siwalik cercopithecids. *Human Evolution* 2:47–58.

Barry, J. C., L. L. Jacobs, and J. Kelley. 1986. An early middle Miocene catarrhine from Pakistan with comments on the dispersal of catarrhines into Eurasia. *Journal of Human Evolution* 15(6): 501–508.

Barry, J. C., M. E. Morgan, L. J. Flynn, D. Pilbeam, A. K. Behrensmeyer, S. K. Raza, I. A. Khan, et al. 2002. Faunal and environmental change in the Late Miocene Siwaliks of northern Pakistan. *Paleobiology* 28(S2): 1–71.

Barry, J. C., M. E. Morgan, L. J. Flynn, D. Pilbeam, L. L. Jacobs, E. H. Lindsay, S. M. Raza, and N. Solounias. 1995. Patterns of faunal turnover and diversity in the Neogene Siwaliks of northern Pakistan. *Palaeogeography, Palaeoclimatology, Palaeoecology* 115(1–4): 209–226.

Beard, K. C., L. Marivaux, S. T. Tun, A. N. Soe, Y. Chaimanee, W. Htoon, B. Marandat, A. H. Aung, and J.-J. Jaeger. 2007. New sivaladapid primates from the Eocene Pondaung Formation of Myanmar and the anthropoid status of Amphipithecidae. *Bulletin of Carnegie Museum of Natural History* 39:67–76.

Begun, D. R. 2010. Miocene hominids and the origins of the African apes and humans. *Annual Review of Anthropology* 39:67–84.

Begun D. R., and E. Güleç. 1998. Restoration of the type and palate of *Ankarapithecus meteai*: Taxonomic and phylogenetic implications. *American Journal of Physical Anthropology* 105(3): 279–314.

Begun, D. R., and T. L. Kivell. 2011. Knuckle-walking in *Sivapithecus*? The combined effects of homology and homoplasy with possible implications for pongine dispersals. *Journal of Human Evolution* 60(2): 158–170.

Begun, D. R., M. C. Nargolwalla, and L. Kordos, L. 2012. European Miocene hominoids and the origin of the African ape and human clade. *Evolutionary Anthropology* 21(1): 10–23

Behrensmeyer, A .K. 1987. Miocene fluvial facies and vertebrate taphonomy in northern Pakistan. In *Recent Developments in Fluvial Sedimentology*, edited by F. A. Ethridge, R. M. Flores, and M. D. Harvey, 169–176. Special Publication 39, Society of Economic Paleontologists and Mineralogists, Tulsa, Oklahoma.

Behrensmeyer, A. K., J. Quade, T. E. Cerling, J. Kappelman, I. A. Khan, P. Copeland, L. Roe, et al. 2007. The structure and rate of Late Miocene expansion of C$_4$ plants: evidence from lateral variation in stable isotopes in paleosols of the Siwalik Group, northern Pakistan. *Geological Society of America Bulletin* 119(11–12): 1486–1505.

Behrensmeyer, A. K., B. J. Willis, and J. Quade. 1995. Floodplains and paleosols in the Siwalik Neogene and Wyoming Paleogene deposits: Implications for taphonomy and paleoecology of faunas. *Palaeogeography, Palaeoclimatology, Palaeoecology* 115(1–4): 37–60.

Bernor, R. L., L. J. Flynn, T. Harrison, S. T. Hussain, and J. Kelley. 1988. *Dionysopithecus* from southern Pakistan and the biochronology and biogeography of early Eurasian catarrhines. *Journal of Human Evolution* 17(3): 339–358.

Bhandari, A., R. F. Kay, B. A. Williams, B. N. Tiwari, S. Bajpai, and T. Hieronymus. 2018. First record of the Miocene hominoid *Sivapithecus* from Kutch, Gujarat state, western India. *PLoS ONE* 13, e0206314.

Brown, B., J. Kappelman, and S. Ward. 2005. Lots of faces from different places: What craniofacial morphology does(n't) tell us about hominoid phylogenetics. In *Interpreting the Past: Essays on Human, Primate, and Mammal Evolution in Honor of David Pilbeam*, edited by D. E. Lieberman, R. H. Smith, and J. Kelley, 167–188. Boston: Brill.

Cameron, D. W. 2001. The taxonomic status of the Siwalik Late Miocene hominoid *Indopithecus* (=*Gigantopithecus*). *Himalayan Geology* 22:29–34.

Cameron, D., R. Patnaik, and A. Sahni. 1999. *Sivapithecus* dental specimens from Dhara locality, Kalgarh District, Uttar Pradesh, Siwaliks, northern India. *Journal of Human Evolution* 37(6): 861–68.

Casanovas-Vilar, I., D. M. Alba, M. Garcés, J. M. Robles, and S. Moyà Solà. 2011. Updated chronology for the Miocene hominoid radiation in western Eurasia. *Proceedings of the National Academy of Sciences* 108(14): 5554–5559.

Chaimanee, Y., D. Jolly, M. Benammi, P. Tafforeau, D. Duzer, I. Moussa, and J.-J. Jaeger. 2003. A Middle Miocene hominoid from Thailand and orangutan origins. *Nature* 422(6927): 61–65.

Chaimanee, Y., V. Lazzari, K. Chaivanich, and J.-J. Jaeger. 2019. First maxilla of a late Miocene hominid from Thailand and the evolution of pongine derived characters. *Journal of Human Evolution* 134: 102636.

Chaimanee, Y., V. Suteethorn, P. Jintasakul, C. Vidthayanon, B. Marandat, and J.-J. Jaeger. 2004. A new orang-utan relative from the late Miocene of Thailand. *Nature* 427(6973): 439–441.

Chaimanee, Y., C. Yamee, P. Tian, O. Chavasseau, and J.-J. Jaeger, J.-J. 2008. First Middle Miocene sivaladapid primate from Thailand. *Journal of Human Evolution* 54(3): 434–443.

Chaimanee, Y., C. Yamee, P. Tian, K. Khaowiset, B. Marandat, P. Tafforeau, C. Nemoz, and J.-J. Jaeger, J.-J. 2006. *Khoratpithecus piriyai*, a Late Miocene hominoid of Thailand. *American Journal of Physical Anthropology* 131(3): 311–323.

Chopra, S. R. K., and S. Kaul. 1975. New fossil *Dryopithecus* material from the Nagri beds at Haritalyangar (H. P.), India. In *Contemporary Primatology*, edited by S. Kondo, M. Kawai, and A. Ehara, 2–11. Basel: Karger.

Delson, E. 1973. Fossil colobine monkeys of the circum-Mediterranean region and the evolutionary history of the Cercopithecidae (Primates, Mammalia). PhD diss., Columbia University.

Delson, E. 1975. Evolutionary history of the Cercopithecidae. *Contributions to Primatolology* 5:167–217.

Delson, E. 1994. Evolutionary history of the colobine monkeys in palaeoenvironmental perspective. In *Colobine Monkeys: Their Ecology, Behaviour and Evolution*, edited by G. Davies and J. F. Oates, 11–43. Cambridge: Cambridge University Press.

DeSilva, J. M., M. E. Morgan, J. C. Barry, and D. Pilbeam. 2010. A hominoid distal tibia from the Miocene of Pakistan. *Journal of Human Evolution* 58(2): 147–154.

Falconer, H., and P. T. Cautley. 1838. On additional fossil species of the order Quadrumana from the Siwalik Hills. *Journal of Asiatic Society of Bengal* 6:354–361.

Flynn, L. J., and M. E. Morgan. 2005. New lower primates from the Miocene Siwaliks of Pakistan. In *Interpreting the Past: Essays on Human, Primate, and Mammal Evolution in Honor of David Pilbeam*, edited by D. L. Lieberman, R. J. Smith, and J. Kelley, 81–101. Boston: Brill.

Gençturk, I., B. Alpagut, and P. Andrews. 2008. Interproximal wear facets and tooth associations in the Paşalar hominoid sample. *Journal of Human Evolution* 54(4): 480–493.

Geoffroy Saint-Hilaire. É. 1812. Tableau des quadrumanes. I. Ordre Quadrumanes. *Annales du Museum d'Histoire Naturelle Paris* 19:85–122.

Gilbert, C. C., A. Ortiz, K. D. Pugh, C. J. Campisano, B. A. Patel, N. P. Singh, J. G. Fleagle, and R. Patnaik. 2020. New Middle Miocene ape (Primates: Hylobatidae) from Ramnagar, India, fills major gaps in the hominoid fossil record. *Proceedings of the Royal Society B* 287(1934): 20201655.

Gilbert, C. C., A. Ortiz, K. D. Pugh, C. J. Campisano, B. A. Patel, N. P. Singh, J. G. Fleagle, and R. Patnaik. 2023. Reanalysis of *Kapi ramnagarensis* supports its classification as a stem hylobatid. *Journal of Human Evolution* 180, S 75: 62 (abstr.).

Gilbert, C. C., B. A. Patel, N. P. Singh, C. J. Campisano, J. G. Fleagle, K. L. Ruse and R. Patnaik. 2017. New sivaladapid primate from Lower Siwalik deposits surrounding Ramnagar (Jammu and Kashmir State), India. *Journal of Human Evolution* 102:21–41.

Gilbert, C. C., R. K. Sehgal, K. D. Pugh, C. J. Campisano, E. May, B. A. Patel, N. P. Singh, and R. Patnaik. 2019. New *Sivapithecus* specimen from Ramnagar (Jammu and Kashmir), India, and a taxonomic revision of Ramnagar hominoids. *Journal of Human Evolution* 135:102665.

Gingerich, P. D., and A. Sahni. 1979. *Indraloris* and *Sivaladapis*: Miocene adapid primates from the Siwaliks of India and Pakistan. *Nature* 279(5712): 415–416.

Gingerich, P.D., and A. Sahni. 1984. Dentition of *Sivaladapis nagrii* (Adapidae) from the Late Miocene of India. *International Journal of Primatology* 5:63–79.

Greenfield, L. O. 1979. On the adaptive pattern of *Ramapithecus*. *American Journal of Physical Anthropology* 50(4): 527–548.

Harrison, T. 2010. Dendropithecoidea, Proconsuloidea and Hominoidea (Catarrhini, Primates). In *Cenozoic Mammals of Africa*, edited by L. Werdelin, and W. J. Sanders, 429–469. Berkeley: University of California Press.

Harrison, T. 2016. The fossil record and evolutionary history of hylobatids. In *Evolution of Gibbons and Siamang: Phylogeny, Morphology, and Cognition*, edited by U. H. Reichard, H. Hirai, and C. Barelli, 91–110. New York: Springer.

Harrison, T., and E. Delson. 2007. *Mesopithecus sivalensis* from the Late Miocene of the Siwaliks. *American Journal of Physical Anthropology* 132, Supp. S44: 126 (abstr.).

Harrison, T., and Y. Gu. 1999. Taxonomy and phylogenetic relationships of Early Miocene catarrhines from Sihong, China. *Journal of Human Evolution* 37(2): 225–277.

Hoorn, C., T. Ohja, and J. Quade. 2000. Palynological evidence for vegetation development and climatic change in the sub-Himalayan zone (Neogene, central Nepal). *Palaeogeography, Palaeoclimatology, Palaeoecology* 163:133–161.

Jablonski, N. G. 2002. Fossil old world monkeys: The late Neogene radiation. In *The Primate Fossil Record*, edited by W. C. Hartwig, 255–300. Cambridge: Cambridge University Press.

Jablonski, N. G, X. Ji, J. Kelley, L. J. Flynn, C. Deng, and D. F. Su. 2020. *Mesopithecus pentelicus* from Zhaotong, China, the easternmost representative of a widespread Miocene cercopithecoid species. *Journal of Human Evolution* 146:102851.

Jacobs, L .L. 1981. Miocene lorid primates from the Siwaliks of Pakistan. *Nature* 289(5798): 585–587.

Jaeger, J-J., A. N. Soe, O. Chavasseau, P. Coster, E.-G. Emonet, F. Guy, R. Lebrun, et al.. 2011. First hominoid from the Late Miocene of the Irrawaddy Formation (Myanmar). *PLoS ONE* 6(4): e17065.

Ji, X., T. Harrison, Y. Zhang, Y. Wu, C. Zhang, J. Hu, D. Wu, Y. Hou, S. Li, G. Wang, and Z. Wang. 2022. The earliest hylobatid from the Late Miocene of China. *Journal of Human Evolution* 171:102251.

Ji, X., N. G. Jablonski, D. F. Su, C. Deng, L. J. Flynn, Y. You, and J. Kelley. 2013. Juvenile hominoid cranium from the terminal Miocene of Yunnan, China. *Chinese Science Bulletin* 58:3771–3779.

Kappelman, J., B. Richmond, E. R. Seiffert, M. Maga, and T. Ryan. 2003. Fossil primates from the Sinap Formation. In *Geology and Paleontology of the Sinap Formation*, edited by M. Fortelius, J. Kappelman, R. Bernor, and S. Sen, 90–124. New York: Columbia University Press.

Kay, R. F. 1982. *Sivapithecus simonsi*, a new species of Miocene hominoid with comments on the phylogenetic status of the Ramapithecinae. *International Journal of Primatology* 3:113–73.

Kelley, J. 1986. Paleobiology of Miocene hominoids. Ph.D. diss., Yale University.

Kelley, J. 1988. A new large species of *Sivapithecus* from the Siwaliks of Pakistan. *Journal of Human Evolution* 17:305–324.

Kelley, J. 1997. Paleobiological and phylogenetic significance of life history in Miocene hominoids. In *Function, Phylogeny and Fossils: Miocene Hominoid Evolution and Adaptations*, edited by D. R. Begun, C. V. Ward, and M. D. Rose, 173–208. New York: Plenum.

Kelley, J. 2002. The hominoid radiation in Asia. In *The Primate Fossil Record*, edited by W. Hartwig, 369–384. Cambridge: Cambridge University Press.

Kelley, J. 2005. Twenty-five years contemplating *Sivapithecus* taxonomy. In *Interpreting the Past: Essays on Human, Primate, and Mammal Evolution in Honor of David Pilbeam*, edited by D. E. Lieberman, R. H. Smith, and J. Kelley, 123–143. Brill Academic Publishers, Boston.

Kelley, J., P. Andrews, and B. Alpagut. 2008. A new hominoid species from the Middle Miocene site of Paşalar, Turkey. *Journal of Human Evolution* 54(4): 455–479.

Kelley, J., M. Anwar, M. A. McCollum, and S. C. Ward. 1995. The anterior dentition of *Sivapithecus parvada*, with comments on the phylogenetic significance of incisor heteromorphy in Hominoidea. *Journal of Human Evolution* 28(6): 503–517.

Kelley, J., and F. Gao. 2012. Juvenile hominoid cranium from the Late Miocene of southern China and hominoid diversity in Asia. *Proceedings of the National Academy of Sciences* 109(18): 6882–6885.

Kelley, J., and G. T. Schwartz. 2010. Dental development and life history in living African and Asian apes. *Proceedings of the National Academy of Sciences* 107(18): 1035–1040.

Khan, M. A., J. Kelley, L. J. Flynn, M. A. Babar, and N. G. Jablonski. 2020. New fossils of *Mesopithecus* from Hasnot, Pakistan. *Journal of Human Evolution* 145:102818.

Kivell, T. L., and D. R. Begun. 2009. New primate carpal bones from Rudabánya (Late Miocene, Hungary): Taxonomic and functional implications. *Journal of Human Evolution* 57(6): 697–709.

Larson, S. G., and J. T. Stern. 2006. Maintenance of above-branch balance during primate arboreal quadrupedalism: Coordinated use of forearm rotators and tail motion. *American Journal of Physical Anthropology* 129(1): 71–81.

Li, C. 1978. A Miocene gibbon-like primate from Shihhung, Kangsu Province. *Vertebrata PalAsiatica* 16:187–192.

Lindsay, E. L., L. J. Flynn, U. Cheema, J. C. Barry, K. Downing, A. R. Rajpar, and S. M. Raza. 2005. Will Downs and the Zinda Pir Dome. *Palaeontologia Electronica* 8 (1): 19A. https://palaeo-electronica.org/2005_1/lindsay19/issue1_05.htm.

Lydekker, R. 1879. Further notices of Siwalik mammals. *Records of the Geological Survey of India* 12:33–52.

Lydekker, R. 1884. Rodents and new ruminants from the Siwaliks, and synopsis of Mammalia. *Memoirs of the Geological Survey of India: Palaeontologia Indica Ser.* 10, 3:105–134.

Lydekker R. 1886. Siwalik Mammalia. *Memoirs of the Geological Survey of India* 4:1–18.

MacPhee, R. D. E., and L. L. Jacobs. 1986. *Nycticeboides simpsoni* and the morphology, adaptations, and relationships of Miocene Siwalik Lorisidae. In *Contributions to Geology*, Special Paper 3, edited by K. M. Flanagan and J. A. Lillegraven, 131–161. Laramie: University of Wyoming.

Madar, S. I., M. D. Rose, J. Kelley, L. MacLatchy, and D. Pilbeam. 2002. New *Sivapithecus* postcranial specimens from the Siwaliks of Pakistan. *Journal of Human Evolution* 42(6): 705–752.

Marivaux, L., K. C. Beard, Y. Chaimanee, J.-J. Jaeger, B. Marandat, A. N. Soe, S. T. Tun, and A. A. Kyaw. 2008. Proximal femoral anatomy of a sivaladapid primate from the late Middle Eocene Pondaung Formation (Central Myanmar). *American Journal of Physical Anthropology* 137(3): 262–273.

Marivaux, L., J.-J. Welcomme, S. Ducrocq, and J.-J. Jaeger. 2002. Oligocene sivaladapid primate from the Bugti Hills (Balochistan, Pakistan) bridge the gap between Eocene and Miocene adapiform communities in southern Asia. *Journal of Human Evolution* 42(4): 379–388.

Martin, L., and P. Andrews. 1993. Species recognition in Middle Miocene hominoids. In *Species, Species Concepts, and Primate Evolution*, edited by W. H. Kimbel and L. B. Martin, 393–427. New York: Plenum.

McCollum, M. A., F. E. Grine, S. C. Ward, and W. H. Kimbel. 1993. Subnasal morphological variation in extant hominoids and fossil hominids. *Journal of Human Evolution* 24(2): 87–111.

Mein, P., and L. Ginsburg. 1997. Les mammifères du gisement miocène inférieur de Li Mai Long, Thaïlande: Systématique, biostratigraphie et paléoenvironnement. *Geodiversitas* 19:783–844.

Morgan, M. E., A. K. Behrensmeyer, C. Badgley, J. C. Barry, S. Nelson, and D. Pilbeam. 2009. Lateral trends in carbon isotope ratios reveal a Miocene vegetation gradient in the Siwaliks of Pakistan. *Geology* 37(2): 103–106.

Morgan, M. E., J. D. Kingston, and B. D. Marino. 1994. Carbon isotopic evidence for the emergence of C_4 plants in the Neogene from Pakistan and Kenya. *Nature* 367(6459): 162–165.

Morgan, M. E., K. L. Lewton, J. Kelley, E. Otárola-Castillo, J. C. Barry, L. J. Flynn, and D. Pilbeam. 2015. A partial hominoid innominate from the Miocene of Pakistan: Description and preliminary analysis. *Proceedings of the National Academy of Sciences* 112(1): 82–87.

Moyà-Solà, S., and M. Köhler. 1996. A *Dryopithecus* skeleton and the origins of great-ape locomotion. *Nature* 379(6561): 156–159.

Moyà-Solà, S., M. Köhler, D. M. Alba, I. Casanovas-Vilar, and J. Galindo. 2004. *Pierolapithecus catalaunicus*, a new Middle Miocene great ape from Spain. *Science* 306(5700): 1339–1344.

Munthe, J., B. Dongo, J. H. Hutchison, W. F. Kean, K. Munthe, and R. M. West. 1983. New fossil discoveries from the Miocene of Nepal include a hominoid. *Nature* 303(5915): 331–333.

Nelson, S. 2003. *The Extinction of Sivapithecus: Faunal and Environmental Changes in the Siwaliks of Pakistan.* American School of Prehistoric Research Monographs, vol. 1. Boston: Brill.

Nelson, S.V. 2005. Paleoseasonality inferred from equid teeth and intra-tooth isotopic variability. *Palaeogeography, Palaeoclimatology, Palaeoecology* 222(1–2): 122–144.

Nelson, S.V. 2007. Isotopic reconstructions of habitat change surrounding the extinction of *Sivapithecus*, a Miocene hominoid, in the Siwalik Group of Pakistan. *Palaeogeography, Palaeoclimatology, Palaeoecology* 243(1–2): 204–222.

Nowak, R. M., and J. L. Paradiso. 1983. *Walker's Mammals of the World*, 4th ed. Baltimore: Johns Hopkins University Press.

Pan, Y. 2006. Primates Linnaeus: 1758. In *Lufengpithecus Hudienensis Site*, edited by G. Qi and W. Dong, 131–148, 320–322. Beijing: Science Press.

Pan, Y., and R. Wu. 1986. A new species of *Sinoadapis* form the Lufeng hominoid locality. *Acta Anthropologica Sinica* 5:31–40.

Patnaik, R. 2013. Indian Neogene Siwalik mammalian biostratigraphy: An overview. In *Fossil Mammals of Asia: Neogene Biostratigraphy and Chronology*, edited by X. Wang, L. J. Flynn, and M. Fortelius, 423–444. New York: Columbia University Press.

Patnaik, R., T. E. Cerling, K. T. Uno, and J. G. Fleagle. 2014. Diet and habitat of Siwalik primates *Indopithecus*, *Sivaladapis* and *Theropithecus*. *Annales Zoologici Fennici* 51(1–2): 123–142.

Phillips, E. M., and A. Walker. 2000. A new species of fossil lorisid from the Miocene of East Africa. *Primates* 41:367–372.

Phillips, E. M., and A. Walker. 2002. Fossil lorisoids. In *The Primate Fossil Record*, edited by H. W. Hartwig, 83–95. Cambridge: Cambridge University Press.

Pilbeam, D. 1982. New hominoid skull material from the Miocene of Pakistan. *Nature* 295(5846): 232–234.

Pilbeam, D., M. D. Rose, C. Badgley, and B. Lipschutz. 1980. Miocene hominoids from Pakistan. *Postilla* 181:1–94.

Pilbeam, D., M. D. Rose, J. C. Barry, and S. M. I. Shah. 1990. New *Sivapithecus* humeri from Pakistan and the relationship of *Sivapithecus* and *Pongo*. *Nature* 348(6298): 237–239.

Pilgrim, G. E. 1910. Notices of new mammalian genera and species from the Tertiaries of India. *Records of the Geological Survey of India* 40:63–71.

Pilgrim, G. E. 1915. New Siwalik primates and their bearing on the question of the evolution of man and the Anthropoidea. *Records of the Geological Survey of India* 45:1–74.

Pilgrim, G. E. 1932. The fossil Carnivora of India. *Memoirs of the Geological Survey of India* 18:1–232.

Pillans, B., M. Williams, D. Cameron, R. Patnaik, J. Hogarth, A. Sahni, J. C. Sharma, F. Williams, and R. L. Bernor. 2005. Revised correlation of the Haritalyangar magnetostratigraphy, Indian Siwaliks: Implications for the age of the Miocene hominids *Indopithecus* and *Sivapithecus*, with a note on a new hominid tooth. *Journal of Human Evolution* 48(5): 507–515.

Qi, T., and K. C. Beard. 1998. Late Eocene sivaladapid primate from Guangxi Zhuang Autonomous Region, People's Republic of China. *Journal of Human Evolution* 35(3): 211–220.

Quade, J., T. E. Cerling, and J. R. Bowman. 1989. Development of Asian monsoon revealed by marked ecological shift during the latest Miocene in northern Pakistan. *Nature* 342(6246): 163–165.

Ravosa, M. 1998. Cranial allometry and geographic variation in slow lorises (*Nycticebus*). *American Journal of Primatology* 45(3): 225–243.

Raza, S. M., J. C. Barry, G. E. Meyer, and L. Martin. 1984. Preliminary report on the geology and vertebrate fauna of the Miocene Manchar Formation, Sind, Pakistan. *Journal of Vertebrate Paleontology* 4(4): 584–599.

Rose, M. D. 1984. Hominoid postcranial specimens from the Middle Miocene Chinji Formation, Pakistan. *Journal of Human Evolution* 13(6): 503–516.

Rose, M. D. 1986. Further hominoid postcranial specimens from the Late Miocene Nagri Formation of Pakistan. *Journal of Human Evolution* 15(5): 333–367.

Rose, M. D. 1989. New postcranial specimens of catarrhines from the Middle Miocene Chinji Formation, Pakistan: Descriptions and a discussion of proximal humeral functional morphology in anthropoids. *Journal of Human Evolution* 18(2): 131–162.

Rose, M. D. 1994. Quadrupedalism in some Miocene catarrhines. *Journal of Human Evolution* 26(5–6): 387–411.

Rose, M. D. 1997. Functional and phylogenetic features of the forelimb in Miocene hominoids. In *Function, Phylogeny and Fossils: Miocene Hominoid Evolution and Adaptations*, edited by D. R. Begun, C. V. Ward, and M. D. Rose, 79–100. New York: Plenum.

Sankhyan, A., J. Kelley, and T. Harrison. 2017. A highly derived pliopithecoid from the late Miocene of Haritalyangar, India. *Journal of Human Evolution* 105:1–12.

Scott, J. E., C. M. Schrein, and J. Kelley. 2009. Beyond *Gorilla* and *Pongo*: Alternative models for evaluating variation and sexual dimorphism in fossil hominoid samples. *American Journal of Physical Anthropology* 140(2): 253–264.

Scott, R. S., J. Kappelman, and J. Kelley. 1999. The paleoenvironment of *Sivapithecus parvada*. *Journal of Human Evolution* 36(3): 245–274.

Seiffert, E. R. 2007. Early evolution and biogeography of lorisiform strepsirrhines. *American Journal of Primatology* 69(1): 27–35.

Seiffert, E. R., E. L. Simons, and Y. Attia. 2003. Fossil evidence for an ancient divergence of lorises and galagos. *Nature* 422(6930): 421–424.

Simons, E. L. 1970. The deployment and history of Old World monkeys (Cercopithecidae, Primates). In *Old World Monkeys*, edited by J. R. Napier and P. H. Napier, 97–137. London: Academic Press.

Simons, E. L., and S. R. K. Chopra. 1969. *Gigantopithecus* (Pongidae, Hominoidea) a new species from North India. *Postilla* 138:1–18.

Simons, E., and D. Pilbeam. 1965. Preliminary revision of the Dryopithecinae (Pongidae, Anthropoidea). *Folia Primatologica* 3(2–3): 81–152.

Spoor, C. F., P. Y. Sondar, and S. T. Hussain. 1991. A new hominoid hamate and first metacarpal from the late Miocene Nagri Formation of Pakistan. *Journal of Human Evolution* 21(6): 413–424.

Sussman, R. W., D. T. Rasmussen, and P. H. Raven. 2012. Rethinking primate origins again. *American Journal of Primatology* 75(2): 95–106.

Szalay, F. S., and E. Delson. 1979. *Evolutionary History of the Primates*. New York: Academic Press.

Teaford, M. F., and A. Walker. 1984. Quantitative differences in dental microwear between primate species with different diets and a comment on the presumed diet of *Sivapithecus*. *American Journal of Physical Anthropology* 64(2): 191–200.

Thomas, H., and S. N. Verma. 1979. Découverte d'un Primate Adapiforme (Sivaladapinae sub. fam. nov.) dans le Miocène moyen des Siwaliks de la région de Ramnagar (Jammu et Cachemire, Inde). *Comptes Rendus de l'Académie des Sciences. Série D* 289(12): 833–836.

Verma, B. C. 1969. *Procynocephalus pinjorii*, sp. nov. A new fossil primate from Pinjor beds (lower Pleistocene), east of Chandigarh. *Journal of the Paleontological Society of India* 13:53–57.

Von Koenigswald, G. H. R. 1950. Bemerkungen zu *"Dryopithecus giganteus"* Pilgrim. *Eclogae Geologicae Helvetiae* 42:515–519.

Von Meyer, H. 1848. In *Index Palaeontologicus, Abteilung I, 2nd half, N-Z*, edited by H. G. Bronn, 1133. Stuttgart: Schweizerbart.

Wagner, A. 1839. Fossile Überreste von einem Affenschadel und anderen Säugtieren aus Griechenland. *Gelehrte Anzeigen* 8:306–311.

Ward, C. V., A. S. Hammond, J. M. Plavcan, and D. R. Begun. 2019. A late Miocene hominid partial pelvis from Hungary. *Journal of Human Evolution* 136:102645.

Ward, S., and W. Kimbel. 1983. Subnasal alveolar morphology and the systematic position of *Sivapithecus*. *American Journal of Physical Anthropology* 61(2): 157–171.

Ward, S. C., and B. Brown. 1986. Facial anatomy of Miocene hominoids. In *Comparative Primate Biology*, Vol. 1: *Systematics, Evolution, and Anatomy*, edited by D. Swindler and J. Erwin, 413–452. New York: Liss.

Ward, S. C., and D. R. Pilbeam. 1983. Maxillo-facial morphology of Miocene hominoids from Africa and Indo-Pakistan. In *New Interpretations of Ape and Human Ancestry*, edited by R. L. Ciochon and R. S. Corruccini, 211–238. New York: Plenum.

Willis, B. J. 1993a. Ancient river systems in the Himalayan Foredeep, Chinji Village area, northern Pakistan. *Sedimentary Geology* 88(1–2): 1–76.

Willis, B. J. 1993b. Evolution of Miocene fluvial systems in the Himalayan foredeep through a two-kilometer-thick succession in northern Pakistan. *Sedimentary Geology* 88(1–2): 77–121.

Wu, R., and Y. Pan. 1985. A new adapid primate from the Lufeng Miocene, Yunnan. *Acta Anthropologica Sinica* 4(1): 1–6.

Yoder, A. D., and Z. Yang. 2004. Divergence dates for Malagasy lemurs estimated from multiple gene loci: Geological and evolutionary context. *Molecular Ecology* 13(4): 757–773.

Yue, L., and Y. Zhang. 2006. Paleomagnetic dating of *Lufengpithecus hudienensis* localities. In *Lufengpithecus Hudienensis Site*, edited by G. Qi and W. Dong, 245–255. Beijing: Science Press.

Zijlstra, J. S., L. J. Flynn, and W. Wessels. 2013. A westernmost tarsier: A new genus and species from the Miocene of Pakistan. *Journal of Human Evolution* 65(5): 544–550.

13

SIWALIK CREODONTA AND CARNIVORA

JOHN C. BARRY

INTRODUCTION

Although the status of many species is ambiguous and they are always rare, creodonts and carnivores are diverse in the Neogene Siwalik faunas of the Indian subcontinent. Moreover, Carnivora were part of the first fossil collections made from the Siwaliks, with the discovery of Pliocene and Miocene species that included canids, ursids, mustelids, viverrids, hyaenids, and felids (Falconer and Cautley 1836a, 1836b; Lydekker 1876, 1877, 1878, 1884; Bose 1880). These earliest collections, as well as many additional holotypes and referred specimens discovered later, are mostly held by the Natural History Museum in London (NHM) and the Geological Survey of India (GSI) in Kolkata (Calcutta) and were subject to a monographic treatment by Pilgrim (1932).

Ensuing years added carnivore fossils to collections at the Panjab University in Chandigarh, the University of the Punjab in Lahore, the Wadia Institute of Himalayan Geology in Dera Dun, the American Museum of Natural History (AMNH) in New York, the Yale Peabody Museum (YPM) in New Haven, the Bavarian State Collection of Palaeontology and Geology (BSP), and the Geological Institute at the State University of Utrecht (GIU). And most recently, additional collections with many specimens of interest have been made by collaborative projects involving Harvard University, Howard University, the Geological Survey of Pakistan (GSP), and the Pakistan Museum of Natural History (PMNH), as well as between the University of Peshwar and a consortium including Dartmouth College, the Lamont-Doherty Geological Observatory, and the University of Arizona. It is these most recent collections that are of principal interest in this chapter.

Fossil vertebrates have been found in the fluvial Siwalik sediments along the southern margin of the Himalayas in Pakistan, India, and Nepal (Barry et al. 2013; Flynn et al. 2013; Patnaik 2013; Chapter 2, this volume). While Pliocene deposits are present everywhere, the Miocene Siwaliks are particularly well exposed on the Potwar Plateau of northern Pakistan. Significant exposures of the fluvial and fluvio-deltaic sediments of the Vihowa and Chitarwata Formations are also known in southwestern Punjab and Baluchistan in Pakistan, as well as in the Manchar Formation of Sind. The Vihowa, Chitarwata, and Manchar Formations are important because they contain fossils of Oligocene and earliest Miocene age (Raza et al. 1984, 2002; Lindsay et al. 2005; Antoine et al. 2013) while the oldest Potwar localities are only Early Miocene.

The provenance of the fossils in the older collections, and unfortunately of many of the type specimens collected in the late nineteenth or early twentieth centuries, is almost always unknown, and consequently their ages are unknown. However, material collected since 1973 (and some in the American Museum of Natural History and Yale Peabody Museum collected in 1926 and 1932) can generally be placed in a reliable chronostratigraphic framework (Chapters 2 and 3, this volume), which makes these fossils of special interest. Using estimates based on the 2012 Geomagnetic Polarity Time Scale (GPTS) (Gradstein et al. 2012), the oldest Siwalik carnivores and creodonts are from the Chitarwata Formation and range in age between about 22 and 28 Ma. Most fossil specimens, however, are between 18 and 4 Ma, with the best record being restricted to between ~14 and 6 Ma (Fig. 13.1). As discussed later, this variation in sampling biases the perceived stratigraphic ranges of the species, especially so for taxa rare to begin with.

Many of Lydekker's and Pilgrim's holotypes came from near the village of Hasnot (or "Asnot"), and these pose a special problem. First, while the most extensive and most fossiliferous outcrops nearest Hasnot are in the Dhok Pathan Formation, outcrops of the older Nagri and Chinji Formations are not far distant and in rare cases potentially were the source of important fossils. Furthermore, while Dhok Pathan Formation outcrops near Hasnot are between 9.2 and 6.3 Ma, nearly 80% of the localities are younger than 8.0 Ma, and about 50% are younger than 7.4 Ma. Thus, most fossiliferous outcrops of the Dhok Pathan Formation near Hasnot are younger than the deposits along the Soan River near Dhok Pathan, Nila, and Jabi. Most Hasnot localities are between about 8 and 6.5 Ma,

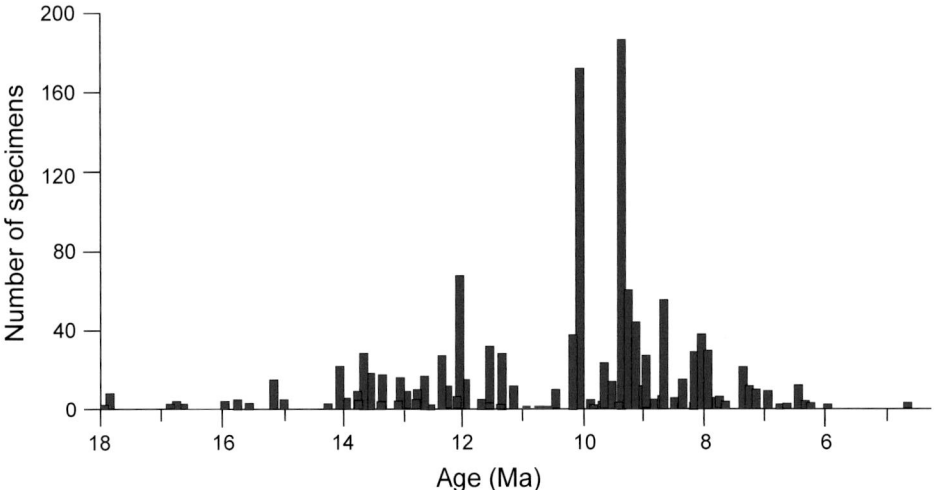

Figure 13.1. Number of carnivore and creodont specimens versus time. The counts are grouped into 100,000-year intervals. The time scale on this and subsequent figures is based on Gradstein et al. (2012).

while localities along the Soan River are between 9.3 and 7.8 Ma (Barry et al. 2002, with ages adjusted for the 2012 GPTS).

As of August 2022, the Harvard Siwalik project database has records for 85 creodonts and 1,246 carnivores older than 4.5 Ma. In both groups there are about equal numbers of dental (isolated teeth or jaws) and disarticulated postcranial elements, nearly all of which are fragmentary. Apart from one extraordinary 9.4 Ma site (Locality Y270, found by Andrew Hill in 1977), there are very few teeth and skeletal elements belonging to a single individual. Such fragmentary, disarticulated preservation makes it difficult to link dental and postcranial morphologies in order to identify or characterize species. These difficulties are compounded by the large number of taxa, many of the same size and all rare. Inadequate type specimens are also a persistent problem, and use of local names obscures the zoogeographic and phylogenetic relationships of the species in the absence of detailed studies.

Table 13.1 lists the species discussed in this chapter and Figure 13.2 shows their observed chronostratigraphic ranges. Fifty-three species are represented in the GSP collections, plus an additional two from the YPM collections. The table also includes three species recognized by Pilgrim (1932), but are not present in the more recent collections. There are, in addition, many unidentified skeletal elements that may belong to otherwise unrecognized species, which would increase the species counts. A few named species that are problematic are discussed at the end of the "Species Accounts" below. (*Note*: A (?) before the genus and species name, as in (?)*Phoberogale bugtiensis*, means that the fossil material may not be assigned to the correct genus or species.)

SPECIES ACCOUNTS

Creodonta

The specimen records in the database for creodonts represent approximately 70 different individuals from 53 localities. *Dissopsalis carnifex* is by far the most common taxon, with 49 records from 35 localities.

Dissopsalis Pilgrim, 1910
Dissopsalis carnifex Pilgrim, 1910
Includes *D. ruber* Pilgrim, 1910

This is a relatively common species, with collections in the American Museum, Yale University, and especially the Geological Survey of Pakistan AMNH, YPM, and especially GSP containing additional dental and postcranial specimens (Colbert 1933a; Barry 1988). The locality and age of the lectotype (GSI D 143) are not known, but the lectotype was said by Pilgrim to come from the lower portion of the Chinji Formation near Chinji Village, which would place it between about 14 and 12 Ma. The oldest and youngest securely dated and identified specimens are 16.0 and 11.4 Ma. However, there is a fragmentary maxilla from Zinda Pir which would be at least 17 Ma and an ulna fragment that is about 22 Ma (Lindsay et al. 2005). Barry (1988) very tentatively identified a 10.1 Ma molariform tooth (age adjusted to the 2012 GPTS) as the youngest dental material, while a damaged cuboid of a slightly larger creodont is 10.0 Ma. Sehgal (2015) also reported a late-occurring *Dissopsalis carnifex* at Nurpur in India, although Patnaik (2013) gave an older age for the sequence.

Hyainailouros Biedermann, 1863
Hyainailouros sulzeri Biedermann, 1863

A truly gigantic species, represented in the recent collections by two partial maxillas with associated lower molar and premolar fragments, several isolated teeth, and a number of postcranial elements, most notably including a set of vertebrae associated with a distal humerus and proximal radius complex. Ginsburg (1980) described a juvenile skeleton to which the Siwalik material closely compares.

Hyainailouros bugtiensis (Pilgrim, 1908) and *Sivapterodon lahirii* (Pilgrim, 1932) are junior synonyms of *H. sulzeri*. Pilgrim (1932) noted features of the m3 that he thought separated *H. bugtiensis* from *H. sulzeri*. They included m3 larger and posteriorly broader with a more vertical anterior face of the paraconid, a more horizontal and blade-like postparaconid

Table 13.1. Species of Siwalik Creodonts and Carnivora

Taxon	Author and Date	Holotype / Collections[1]	Age (Ma)[2]	Est. Mass (kg)[3]	Pre 1935[4]
Creodonta					
Dissopsalis carnifex	Pilgrim, 1910	GSI D 143	>17–10.0?	30–50	p
Hyainailouros sulzeri	Biedermann, 1863	GSI, YGSP, etc	~ 19–11.4	430	p
(?)*Metapterodon* unnamed sp.		YGSP, AMNH	13.7–10.1	ca. 10–20	n
Hyaenodontidae, genus and species indet.		GSI D 217, GSI D 218	Middle Miocene	ca. 40	p
Carnivora					
Family Incerta Sedis					
Unnamed genus and species		YGSP	10.1	5?	n
Arctoidea					
Arctoidea, genus and species indet.		YGSP	11.6	50?	n
Amphicyonidae					
Amphicyonidae?, genus and species indet.		S–GSP, H–GSP	?24–22?	50	n
Amphicyon palaeindicus	Lydekker, 1876	GSI D 26	17.9–11.6	150–200	p
Amphicyon lydekkeri	Pilgrim, 1910	GSI D 133	12.7–7.9	200–300	p
Amphicyon shahbazi	(Pilgrim, 1908)	GSI D 110 (lectotype)	ca. 22–18.5	120	p
Amphicyon giganteus	Schinz, 1825	PAK / YGSP	ca. 19–17.9	300–400	n
cf. *Agnotherium antiquum*	Kaup, 1832	AMNH, YGSP	?15.8–10.1	170–275	n
Ursidae					
(?)*Phoberogale bugtiensis*	(Forster Cooper, 1923)	NHML M 12338	Early Miocene?	ca. 60	p
Indarctos punjabiensis	Lydekker, 1884	GSI D 6	8.7–7.8?	150–250?	p
Agriotherium palaeindicum	(Lydekker, 1878)	GSI D 16	Latest Miocene?	200–300?	p
Ailuridae					
(?)*Alopecocyon sp.*		PMNH	between 27.9 and 28.3	ca. 1–2	n
Mustelidae					
Eomellivora necrophila	Pilgrim, 1932	GSI D 243	10.1–9.4	30–50	p
Cernictis lydekkeri	(Colbert, 1933)	AMNH:FM 19407	?14.0–12.1	3.0–6.0	p
Plesiogulo crassa	Teilhard, 1945	YPM	6.7	50	p
cf. *Anatolictis* sp. A		YGSP	9.4–6.5	0.3–1.5	n
cf. *Anatolictis* sp. B		YGSP	14.1	0.3	n
cf. *Anatolictis* sp. C		YPM	12.3	1.0–1.5	n
cf. *Hoplictis anatolicus*	(Schmidt–Kittler, 1976)	BSPG M Nr. 646	9.7–9.2?	12–25	n
Circamustela sp.		YGSP	?9.3–7.4	2.5–3.0	n
Enhydriodon falconeri	Pilgrim, 1931	NHML M 4847	6.4	10–20	p
Sivaonyx bathygnathus	(Lydekker, 1884)	GSI D 33	9.4–6.3	10–20	p
Vishnuonyx chinjiensis	Pilgrim, 1932	GSI D 223	13.7–11.6	2.5–10	p
Vishnuonyx large species		YGSP	?11.6–10.1	10–20	n
cf. *Torolutra* sp.		YGSP	7.4–7.2	5–10	n
Herpestidae					
Urva small species		YGSP	13.6–7.0	0.3–0.9	n
Urva medium species		YGSP	15.1–7.0	1.3	n
Urva large species		YGSP	13.1–7.2	2–5	n
Viverridae					
Large, indet. genus and species		YGSP	18.0–12.1	10–30	n
Vishnuictis salmontanus	Pilgrim, 1932	GSI D 160	10.1–7.9	2–4	p
Vishnuictis chinjiensis	Pilgrim, 1932	GSI D 214	?14.3–10.5?	3–5	p
Vishnuictis hasnoti	(Pilgrim, 1913)	GSI D 135	10.5–6.5	9–20	p
Mioparadoxurus meini	Morales and Pickford, 2011	GSI K 29–469	10.1–7.3	15–20	n
Paradoxurine genus and species indet.		YGSP	?12.1–9.2?	5–10?	n
Percrocutidae					
Percrocuta carnifex	(Pilgrim, 1913)	GSI D 172	14.0–9.6	40–60	p
Percrocuta grandis	(Kurtén, 1957)	GSI D 162	9.4–6.3	70–100?	p
Hyaenidae					
Proctitherium sp		YGSP	9.7	10–15	n
"*Thalassictis*" *proava*	(Pilgrim, 1910)	GSI D 126	?13.6–10.2	ca. 15	p
cf."*Thalassictis*" *montadai*	(Villalta and Crusafont, 1943)	YPM and YGSP	?13.0–11.6	30–60?	n
Lepthyaena sivalensis	Lydekker, 1877	GSI D 38a	12.0–8.2	15–25?	p
Lepthyaena (?) pilgrimi	(Werdelin and Solounias, 1991)	GSI D 53	9.2–7.3	25–35?	p
Lycyaena macrostoma	(Lydekker, 1884)	GSI D 44	8.7–6.4	60?	p
Adcrocuta eximia	(Roth and Wagner, 1854)	GSI collections	?8.5–8.1?	60–100?	p

(continued)

Table 13.1. continued

Taxon	Author and Date	Holotype / Collections[1]	Age (Ma)[2]	Est. Mass (kg)[3]	Pre 1935[4]
Barbourofelidae					
(?)*Sansanosmilus serratus*	Pilgrim, 1932	GSI D 165	15.2–11.6	30–40	p
(?)*Barbourofelis piveteaui*	(Ozansoy, 1965)	YGSP	?11.4–8.6	50–70	n
Felidae					
Paramachaerodus sivalensis	(Lydekker, 1877)	GSI 95	?10.1–7.0?	50–60?	p
Sivasmilus copei	Kretzoi, 1929	GSI D 151	?14.1–11.4?	30–50	p
cf. *Miomachairodus* sp.		YGSP	10.2–8.0	70–175	n
(?)*Amphimachairodus giganteus*		YGSP	?9.4–4.7?	200–450	n
cf. *Lokotunjailurus* sp.		YGSP	8.1	50–60	n
Sivaelurus chinjiensis	Pilgrim, 1910	GSI D 150	13.8–12.4?	30–50	p
Small Felidae, genus and species indet.			13.9–7.8	5–10	
Medium Felidae, genus and species indet.			22.0–6.5	10–30	
Large Felidae, genus and species indet..			13.8–7.4	30–50	

[1] Institutional Abbreviations:
H-GSP: Geological Survey of Pakistan, Islamabad, collections made by Howard University-GSP Project
S-GSP: Geological Survey of Pakistan, Islamabad, collections made by Yale/Harvard University-GSP Project in the Miocene of Sind
YGSP: Geological Survey of Pakistan, Islamabad, collections made by Yale/Harvard University-GSP Project mostly on The Potwar Plateau
AMNH: American Musuem of Natural History, New York
GSI: Geological Survey of India, Kolkata
NHML: Natural History Museum London
PMNH: Pakistan Museum of Natural History, Islamabad
PAK: Collections made by a joint project of the University of Montpellier 2, the Muséum National d'Histoire Naturelle in Paris, and the University of Balochistan
[2] Observed first and last occurrences (Ma). "?" indicates uncertainties in either the estimated ages or uncertainties in identications.
[3] Body-mass estimates based on comparisons to skulls of related, extant species with either known body mass or averages taken from the literature.
[4] Known in Siwalik collections before 1935 (p) or discovered after 1935 (n).

cristid, and a stronger and more trenchant protoconid. However, all of these differences are the result of differences in size, stage of wear, or simply individual variation. Savage (1973) also previously suggested that *S. lahirii* is a junior synonym of *H. sulzeri*. Pilgrim (1932) diagnosed *S. lahirii* primarily on the large but short and wide molars, small m3 talonid, and lingual displacement of the m2 paraconid and talonid. These characters are largely artifacts as the holotype and only specimen is badly damaged and crudely reconstructed.

The material described by Pilgrim (1908, 1912) came from the Chitarwata Formation at Dera Bugti, while Ginsburg and Welcomme (2002) reported remains of a gigantic creodont from Levels 6 through 6 sup in the Vihowa Formation (Antoine et al. 2013). These remains should be between approximately 22 and 18.5 Ma. Material from the Potwar Plateau spans 17.9 to 11.6 Ma. Pilgrim's *Sivapterodon lahirii* holotype is from near the local base of the Miocene sediments on the Potwar Plateau, where they overlie Eocene marine sediments. The specimen should be on the order of 18 Ma, if not older.

Metapterodon Stromer, 1926
(?)*Metapterodon* unnamed species

Four specimens of this taxon are known. The youngest is about 10.1 Ma (Barry 1980; age adjusted to the 2012 timescale), while the oldest is 13.7 Ma. AMNH specimens, which might be slightly older, include two associated lower molars (AMNH 19344) from a locality that Brown thought was near the base of the Chinji Formation (Colbert 1935), while the exact locality of a maxilla (AMNH 96593) is not known. The museum

label states that the fossil came from near Sosinwallah, a small village about 1.5 km south of Chinji Village. Sosinwallah sits on the lower half of the Chinji Formation.

Metapterodon has, in the words of Lewis and Morlo (2010, 551), "proven to be quite problematic," and addition of the Middle and Late Miocene specimens does not reduce the confusion. In briefly describing the Siwalik material, Barry (1980) discussed the possible generic referral of the specimens, noting that only the maxilla could be directly compared to Stromer's *M. kaiseri*, while the reference of mandibular material had to be based on size and a presumed functional correlation between upper and lower dentitions. Subsequently Morales, Pickford, and Soria (1998) described circa 20 million-year-old material from Namibia that included both upper and lower teeth. Their species, *M. stromeri*, was characterized as being larger than *Metapterodon kaiseri*, with a stronger, more isolated parastyle and more anteriorly directed protocone. Although similar in size, the Siwalik specimens differ in the degree of fusion of the paracone and metacone, the strength of the metastyle, and more labial position of the lower molar talonids. Holroyd (1999) also described Paleogene material from Egypt, recognizing additional species. One, *M. schlosseri*, is about the same size as the Siwalik species, but differs in the ratio of m2/m3 length, the size of the m2 talonid, and the lack of diastemata between the premolars.

The two AMNH lower molars are probably from the lower half of the Chinji Formation and might be as old as 14 Ma. The GSP specimen from the overlying Nagri Formation is the youngest well-dated creodont and gives the species an ap-

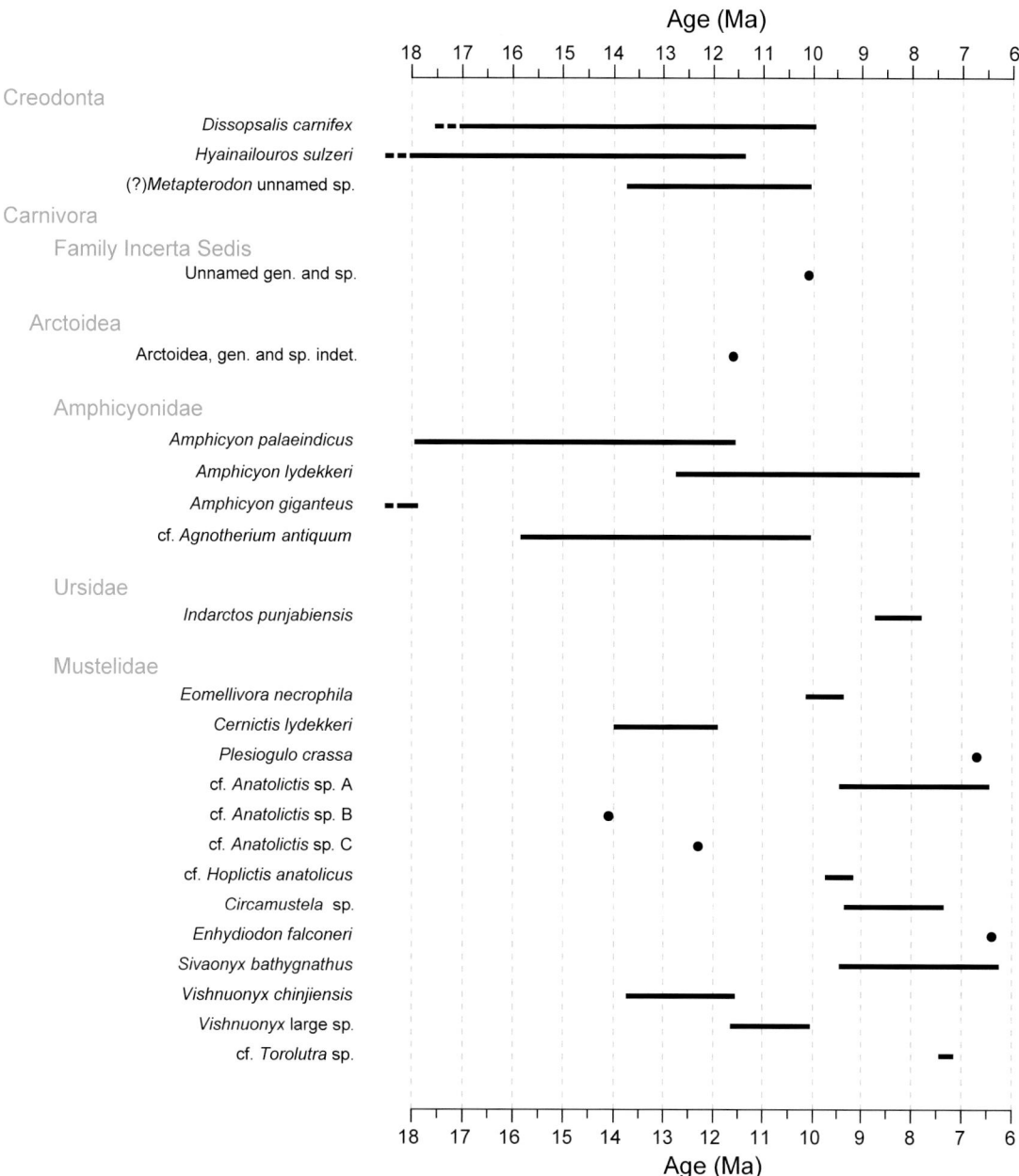

Figure 13.2. Observed chronostratigraphic ranges of species, based on first and last occurrences. Data from Table 13.1.

proximate temporal range of at least 13.7 to 10.1 Ma, which is much younger than the known ages of the African species (Morales, Pickford, and Soria 1998; Holroyd 1999).

Hyaenodontidae, genus and species indet.
This is a taxon that is not represented in the new collections, but is noteworthy. In 1932 Pilgrim named the genus *Vinayakia* with two species, *V. noctura* and *V. sarcophaga*. The genotype species, *V. noctura*, was based on a mandible (GSI D 221) and a referred maxilla (GSI D 218) with P3 and P4, while *V. sarcophaga* was based on a single specimen, a maxilla (GSI D 217) with P4 and M1. The holotype mandible (GSI D 221) is an indeter-

minate hyaenid, but the two maxillae may belong to a medium-size creodont different from the others discussed here. Both maxillae were described by Pilgrim (1932), who put a different interpretation on the teeth. It is important to realize, however, that the enamel has been stripped from the teeth by a post-mortem process. This was apparently not recognized by Pilgrim, which is understandable as the specimens are dark and have a satiny, enamel-like appearance. The loss of enamel accounts for the apparent dilation and height of the roots, the slope of the premolar crowns, the small diastemata between the teeth, the absence of anterior and posterior accessory cusps, and the unusual "wear" on the tips of the cusps.

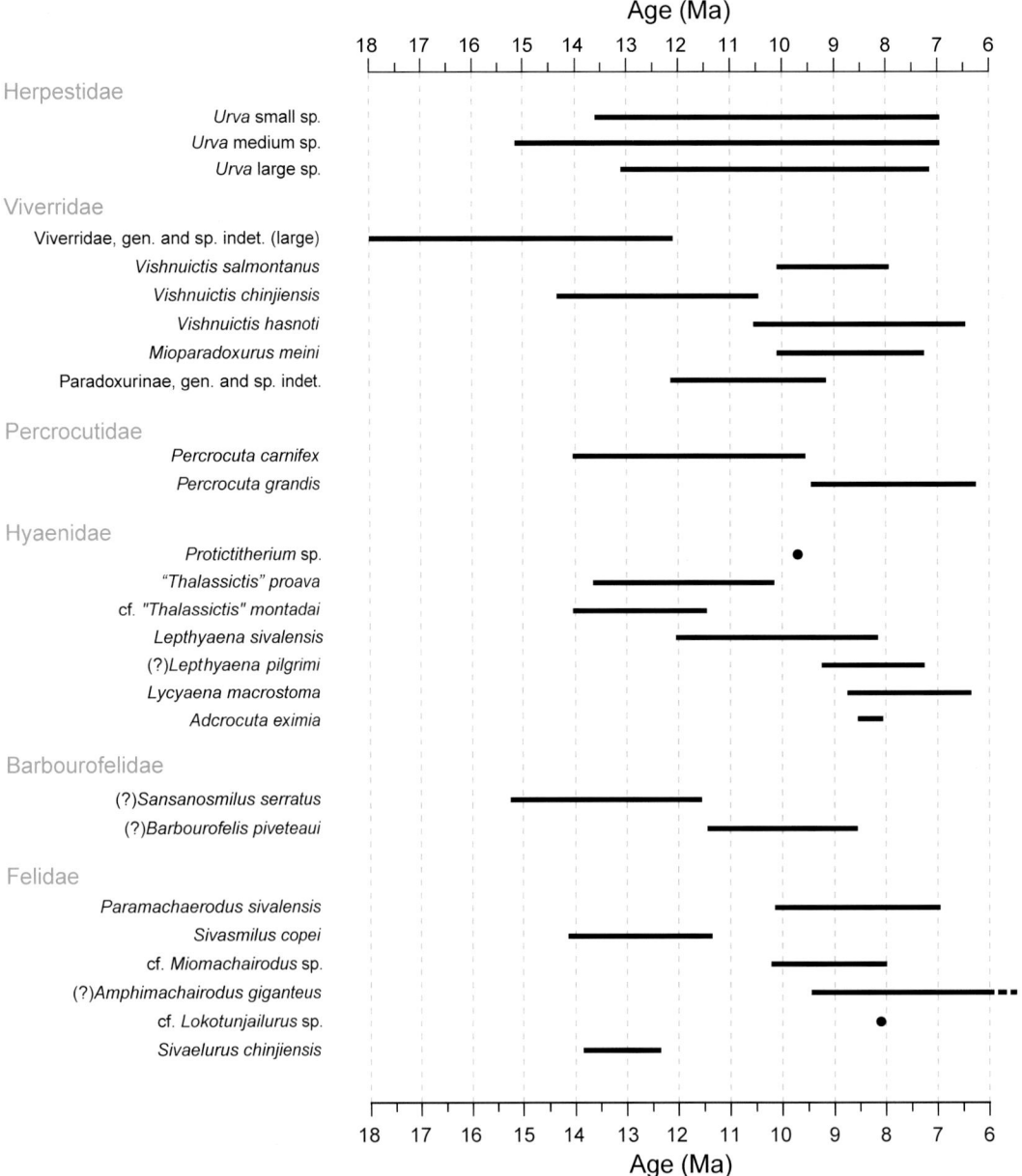

Figure 13.2. (Continued)

Pilgrim (1932) considered his new genus to be a primitive felid, but characteristic features that suggest the maxillae, or at least the GSI D 217 specimen, might be a creodont are the three-rooted P4 and blade-like construction of the P4 talon. In addition, the P4 paracone of GSI D 217 is taller than the M1 amphicone (paracone + metacone), while the M1 protocone is large, extends in front of the amphicone, and does not broaden at the point of contact with the amphicone. The P3 of GSI D 218 also has a distinct internal lobe and in both specimens the infraorbital canal opens above the P4. The exact localities of the GSI specimens are not known. Pilgrim (1932) noted that D 218 was collected near Bahitta in Jhelum District from what he considered to be the Dhok Pathan Formation, while D 217 is from the Chinji Formation south of Kotal Kund. Neither

specimen is referrable to any other Siwalik creodont. Size rules out *Hyainailouros sulzeri* and (?)*Metapterodon* unnamed species, while morphology excludes *Dissopsalis carnifex.*

Carnivora

The 1,246 records in the Harvard Siwalik project database represent approximately 1,100 different individuals from 335 localities.

CARNIVORA INCERTAE SEDIS
Unnamed Genus and species
A single 10.1 million-year-old mandible belongs to an unnamed genus and species (Barry 1989, age adjusted to the 2012 timescale). This taxon differs from other carnivores in having an

unusual canine morphology, while the relative size and morphological details of its premolars and the construction of the carnassial differ from those of other carnivores. Although autapomorphies and symplesiomorphies are present, no certain synapomorphies shared with other groups can be identified. Consequently, this new taxon's relationships within the Carnivora are ambiguous and it cannot be referred to any currently recognized family.

ARCTOIDEA

Arctoidea, Genus and species indet.

A fragment of a calcaneum belongs to a large (ca. 50 kg), unidentified arctoid (noted in Table 13.1 as Arctoidea, genus and species indet.). With a broad and flat posterior calcaneal facet, the specimen is about the size of *Simocyon*, but is morphologically different. While too small to be any of the following Middle Miocene amphicyonid species, a similar size arctoid has also been reported from Middle Miocene deposits in northern Thailand (Ducrocq et al. 1995; Peigné et al. 2006).

Amphicyonidae

There are two distinct groups of amphicyonids in the Middle and Late Miocene Siwaliks: species allied to *Amphicyon* and one species of the thaumastocyoninae *Agnotherium*. There are, in addition, small isolated molars from the Manchar, Chitarwata, and Gaj Formations that probably belong to small amphicyonids with at least two species being present. Among these teeth is an m1 trigonid from the Gaj Formation in the Gaj River section listed as "cf. *Cynelos* sp." in Raza et al. (1984). These specimens are best left as indeterminate. The oldest amphicyonid in the GSP collections is approximately 23 Ma. Generic diagnoses and descriptions of various amphicyonids are given by Kuss (1965), Viranta (1996), Hunt (1998), and Werdelin and Peigné (2010).

Typically amphicyonids are strongly sexually dimorphic in size with relatively uniform morphology between species. Dimorphism and variability within species make it difficult to define the limits of the time-successive species, and consequently assigning individual specimens is arbitrary to some extent.

Amphicyonidae?, Genus and species indet.

Fragmentary remains of small (ca 50 kg), presumed amphicyonids are known from the base of the Manchar Formation, uppermost part of the Gaj Formation, and upper member of the Chitarwata Formation. The fossils, which may belong to one species, are all poorly dated, but probably are from between about 24 and 22 Ma.

Amphicyon Lartet, 1836

Five Siwalik species have been attributed to *Amphicyon*, but several authors have expressed doubts about not only the number of species but also the generic attribution (among others, Matthew 1929, Pilgrim 1932, and Peigné et al. 2006). So far as known, Siwalik species allied to *Amphicyon* are characterized by (1) widely spaced, blunt lower premolars with p1 probably absent, one- or two-rooted p2, and p4 with weak posterior accessory cusp; (2) m1 paraconid much lower than protoconid,

metaconid present but small and slightly displaced posteriorly, and the talonid wide with tall centrally located hypoconid and entoconid reduced to narrow shelf or rim; (3) rectangular m2 large relative to m1, with distinct and large protoconid, metaconid, and hypoconid; (4) m3 one- or two-rooted with indistinct cusps; and (5) P4 with small parastyle and shelf-like protocone.

Previously named Siwalik species attributed to *Amphicyon* include *A. palaeindicus* Lydekker, 1876; *A. shahbazi* (Pilgrim, 1908); *A. pithecophilus* Pilgrim, 1932; *A. cooperi* Pilgrim, 1932; and *A. sindiensis* Pilgrim, 1932. To these, as discussed below, I would add *Arctamphicyon lydekkeri* (Pilgrim, 1910). These six species were based on the following specimens: an M2 (*A. palaeindicus*, GSI D 26), a mandible with p3–m2 and roots of p1–p2 and m3 (*A. shahbazi*, GSI D 110), an M2 (*A. pithecophilus*, GSI D 129), an m1 (*A. cooperi*, NHML M 12341), a mandible with m2 and alveolus of m3 (*A. sindiensis*, GSI D 25), and what is probably an M2 (*Arctamphicyon lydekkeri*, GSI D 133). This combination of different, and mostly unassociated, upper or lower molars makes it difficult to compare the species. Consequently, the relationships and validity of the multiple named Siwalik species are to a great extent unresolved, but there are probably only four: *A. palaeindicus*, *A. lydekkeri*, and *A. shahbazi*, plus *A. giganteus* Schinz, 1825—a species otherwise known from the Lower Miocene of Europe.

Arctamphicyon was based by Pilgrim (1932) on his species *Amphicyon lydekkeri*, as typified by a tooth he considered to be an M1 (GSI D 133) and a referred M2 (GSI D 134). (GSI D 133 was identified as "M₁" in the original description, but Pilgrim was not following the subscript/superscript conventions of recent practice as can be seen from his preceding description of *Dissopsalis carnifex*. The context makes clear that he was discussing an upper molar.) The generic distinction was the quadrangular shape of the supposed M1, which differs from the more triangular shape typical of *Amphicyon*, combined with the smaller, quadrangular M2 GSI D 134. (Pilgrim believed the two molars were probably from the same individual.) Both Matthew (1929) and Peigné et al. (2006), however, noted that GSI D 133 has a typical M2 morphology and if it is an M2, then Pilgrim's generic distinction fails. The assertion that GSI D 133 is an M2 is supported by a recently recovered 10.1 Ma specimen which is morphologically similar to GSI D 133. This specimen has an elongated interdental wear facet on its anterior margin, as would be expected of an M2, instead of the restricted anterolabial facet of an M1. The assertion is further supported by discovery of a large M1 of *Amphicyon lydekkeri* with a typical triangular shape.

Amphicyon palaeindicus Lydekker, 1876

Includes most specimens referred to *Amphicyon pithecophilus* Pilgrim, 1932. Lydekker's identification of GSI D 26 as an M2 has generally been accepted, but differences between it and other known M2's leave unsettled questions about how many species are represented. Most of the specimens previously attributed to *A. palaeindicus* are from the Chinji Formation and thus between about 14 and 11 Ma, while GSI D 133, the holotype of *A. lydekkeri*, is said to be from near Padhri and is probably between 8 and 7 Ma (Chapter 3, this volume). In the new collections there are now many isolated amphicyonid teeth and skeletal elements between 18 and 7 Ma. However, apart from a difference in size, the only morphological difference suggesting there is more than one species is the triangular shape of the M2 of GSI D 26 by comparison to the more rectangular shape of GSI D 133 and other intact M2's. The various skeletal elements belong to individuals of a range of sizes (Table 13.1), but generally are about the size of *A. major*.

Amphicyon lydekkeri (Pilgrim, 1910)

Includes GSI D 23, referred to *Amphicyon pithecophilus* by Pilgrim (1932). There is a difference in size between the smaller *A. palaeindicus* and larger *A. lydekkeri*, which is best seen in the length of the second lower molar (Fig. 13.3a, b). It is also apparent in the relative sizes of skeletal elements as exemplified by the width of the proximal radius (Fig. 13.4). *Amphicyon lydekkeri* is a very large species (Table 13.1), being in some instances nearly the size of *Amphicyon giganteus*.

Amphicyon shahbazi (Pilgrim, 1908)

Includes *Amphicyon cooperi* Pilgrim, 1932 and *Amphicyon sindiensis* Pilgrim, 1932. The mandible MNHL M 12339 is also tentatively included (Fig. 13.3a, b), although it is smaller than

the other specimens (Forster Cooper 1923). The ages of the holotype and other remains studied by Pilgrim are not known, but Ginsburg and Welcomme (2002) list specimens from the upper Chitarwata and lower Vihowa Formations (Levels 4 through 6 sup), which should be between approximately 22 and 18.5 Ma.

Dimensions of m1 and m2 as well as of the proximal radius show that *Amphicyon shahbazi* was slightly smaller than *Amphicyon palaeindicus* (Figs. 13.3a,b; 13.4), although there is overlap in the samples. Ginsburg and Welcomme (2002) suggested that *A. shahbazi* might be the ancestor to *A. major*, a common Middle Miocene species in Europe.

Amphicyon giganteus Schinz, 1825

Ginsburg and Welcomme (2002) report postcranial remains of *Amphicyon giganteus* from Levels 5 through 6 sup in the Vihowa Formation at Dera Bugti (see also Antoine et al. 2013). These remains should be between approximately 19 and 18.5 Ma. There is also an M3 of this species from the base of the Potwar Plateau sequence, extending its range to 17.9 Ma. Viranta, Hussain, and Bernor (2004) discuss a humerus from the base of the Manchar Formation which might be even younger. On Figure 13.3a and 13.3b, the data for this species are all from European specimens (Ginsburg and Antunes 1968).

Agnotherium Kaup, 1832
cf. *Agnotherium antiquum* Kaup, 1832

Includes YPM VP 19530, identified as aff. *Magericyon* sp. by Jiangzuo, Yu, and Flynn (2019), and possibly *Vishnucyon chinjiensis* Pilgrim, 1932. An intact 10.5 Ma mandible shows that this species is unusual in that all the anterior premolars have been lost, leaving a long diastema between the canine and p4, while the p4 and carnassial display a "thaumastocyonine" wear pattern dominated by shear. The p4 has a large posterior accessory cusp, but only a slight swelling at the anterolingual base of the main cusp. There is no trace of a metaconid on the narrow m1, which has a short talonid (0.23 and 0.27 of m1 length in two examples) with a tall, central hypoconid and raised

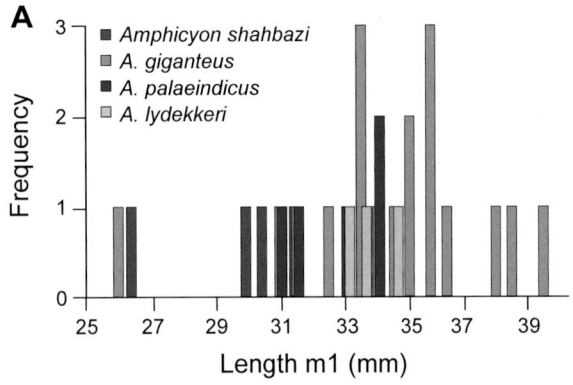

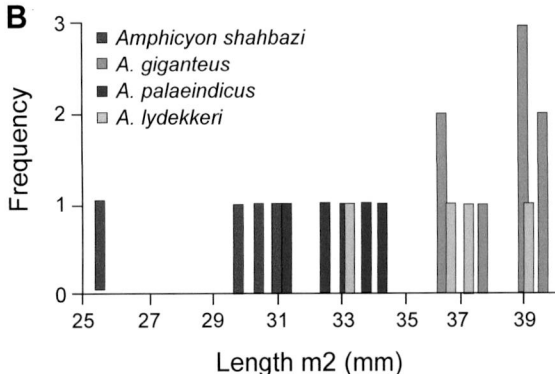

Figure 13.3. Molar dimensions for four species of *Amphicyon*. A. m1 length, B. m2 length. The values for *A. giganteus* are from Ginsburg and Antunes (1968), that for other species from Pilgrim (1932), Forster Cooper (1923), Colbert (1935), and my own observations.

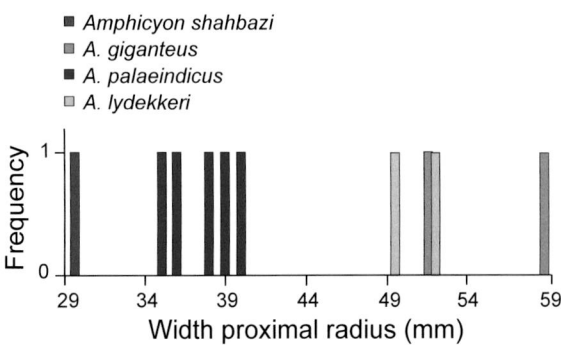

Figure 13.4. Width of proximal radius for four species of *Amphicyon*. Values are from Ginsburg and Antunes (1968) for *A. giganteus* and from Ginsburg and Welcomme (2002) for *A. shahbazi*. Values for *A. palaeindicus* and *A. lydekkeri* are my own observations.

lingual rim. The m1 paraconid has a vertical anterior edge, and there is a faint vertical crest on the posterolingual face of the protoconid. The 10.5 Ma mandible has alveoli for a short m2 (0.43 of m1 length) and a minute, single-rooted m3. An isolated m2 has a trenchant protoconid, the paraconid reduced to a lingually directed cristid, and a short talonid supporting a low, trenchant hypoconid. The metaconid is absent on this tooth and there is a posterolingually directed vertical crest on the protoconid. The YPM m2 (YPM VP 19530) has a similar structure, which was identified as a reduced metaconid by Jiangzuo, Yu, and Flynn (2019). Two moderately compressed canines with sharp, unserrated posterior edges and rounded anterior faces are the oldest dental remains (ca.14.1 Ma), but a small amphicyonid calcaneum questionably referred to this species is considerably older (ca.15.8 Ma).

Apart from its smaller size, "*Vishnucyon chinjiensis*" was characterized by a tall paracone on M1 and M2, features that contribute to the "thaumastocyonine" wear pattern of the lower molars. (See Hunt [1998] for a discussion of similar wear in *Pseudocyon* Lartet, 1851.)

Agnotherium is a difficult taxon, because it is so poorly known (Morlo et al. 2020). The absence of all premolars anterior to p4 and the absence of the anterior accessory cusp on p4, as well as the absence of metaconids on the molars, the very short talonid, and a minute m3 favor attribution of the Siwalik material to *Agnotherium* among other thaumastocyonines.

Canidae

Canids are not known from the Miocene Siwaliks, although early collections have specimens that are presumably from the Pinjor Formation or younger sites (Pilgrim 1932). The oldest canid in the GSP collections is about 2.6 Ma.

Ursidae

Ursids were among the first fossil species to be documented in the Siwaliks (Falconer and Cautley 1836b). Three Miocene species are recognized.

Phoberogale Ginsburg and Morales, 1995
(?)*Phoberogale bugtiensis* (Forster Cooper, 1923)

Forster Cooper (1923) based *Cephalogale bugtiensis* on a weathered maxillary fragment with P3 through M2 (NHML M 12338) that was obtained from the Bugti Hills in Baluchistan. Pilgrim (1932) subsequently suggested it was a species of *Metarctos*, which he placed in the Amphicyonidae. More recently, Wang et al. (2009) placed the specimen in *Phoberogale* as a larger, earlier species relative to those known from the European Early Miocene. De Bonis (2013) has suggested that it is best referred to as Cephalogalini, gen. indet., but here I have included it in *Phoberogale*.

In the absence of more specific information, Forster Cooper's type specimen has been considered to be Early Miocene in age. However, because there are also Oligocene deposits in the Bugti Hills (Antoine et al. 2013), an older age is possible. Unfortunately, no additional material has been found, and this species is not in the new GSP collections.

Indarctos Pilgrim, 1913
Indarctos punjabiensis Lydekker, 1884

Pilgrim (1932) identified three ursid species based on specimens from the "vicinity" of Hasnot. These were *Agriotherium palaeindicum* (Lydekker, 1878) (a referred mandible GSI D 8), *Indarctos punjabiensis* (the holotype maxilla GSI D 6), and *Indarctos salmontanus* Pilgrim, 1913 (the holotype maxilla GSI D 158 and a referred mandible GSI D 159). Matthew (1929) and Petter and Thomas (1986) regard *I. salmontanus* as a junior synonym of *I. punjabiensis*, a conclusion that I accept here. There have also been reservations expressed about the referral of the GSI D 8 mandible to *Agriotherium palaeindicum*, since it and other GSI specimens show a mix of characters reflecting both sexual dimorphism and a transitional position between species of the two genera (see especially Hendy 1980, 82–84, but also Pilgrim 1932, 44–46).

A partial skull in the YPM collections (YPM VP 014246) was identified as *Indarctos atticus* (Weithofer, 1888) by Jiangzuo, Yu, and Flynn (2019). The specimen was said to have come from G. E. Lewis's locality 76 (YPM Locality VP 04872 in Jukar and Brinkman [2021]), but it lacks a proper Lewis field number, which would have been written in ink on the specimen by G. E. Lewis, and it was not with the Siwalik material at YPM in the late 1970s. I think it was probably misplaced from another part of the Yale collections and is certainly not a specimen from the Siwaliks.

The GSP material, as well as that in the YPM and AMNH collections, ranges in age from 8.1 to 7.8 Ma. As noted in the introduction, the holotype from near Hasnot could be older, but is most likely between 8 and 6.5 Ma, as there are extensive outcrops of this age nearer to Hasnot. *Indarctos* is exceptionally widespread, with species throughout Eurasia, North Africa, and North America (Petter and Thomas 1986; Barysknikov 2002). I follow Barysknikov (2002) and not Petter and Thomas (1986) in accepting *Indarctos punjabiensis* as having priority over Weithofer's *I. atticus*.

Agriotherium Wagner, 1837
Agriotherium palaeindicum (Lydekker, 1878)

This taxon is not represented in the GSP collections but is certainly a valid Late Miocene species. The holotype maxilla of *A. palaeindicum* (GSI D 16) can be distinguished from *Indarctos punjabiensis*, but the specimen is of unknown age. It was assumed by Pilgrim (1932) to be from the Dhok Pathan Formation. Other specimens (GSI D 9, D 10, and D 11) came from "Jabi" and may also have been from the Dhok Pathan Formation. The species of *Agriotherium* are typically very large and sexually dimorphic.

Ailuridae

Alopecocyon Camp and Vanderhoof, 1940
(?)*Alopecocyon* sp.

A small m2 from the middle of the Chitarwata Formation (Table 13.1) is similar to species of *Alopecocyon* or *Amphictis* Pomel, 1853. The specimen has very low, blunt cusps; a short trigonid with the metaconid taller than the protoconid and connected to it by a transverse ridge; a very long, wide talonid with a labially positioned hypoconid with subdued cristid oblique and posthypocristid; and the lingual structures suppressed. I have identified this specimen as *Alopecocyon* because the trigonid is only about one-third the length of the talonid, although some species of *Amphictis* also have long talonids (de Beaumont 1976; Morlo 1996; Morlo and Peigné 2010). The specimen differs from the m2 of *Magerictis* Ginsburg, Morales, Soria, and Herraez, 1997 in the absence of the longitudinal valley separating the cusps of the trigonid as well as those of the talonid.

What are taken to be primitive ailurids, the red panda group, have a wide distribution in the Miocene of Eurasia and North America (Baskin 1998; Morlo and Peigné 2010; Salesa et al. 2011). *Alopecocyon* appears to be restricted to the Miocene, but species of *Amphictis* occur in the Oligocene. The Chitarwata Formation locality has been correlated to Geomagnetic Polarity Chron C10n (Lindsay et al. 2005; Lindsay and Flynn 2016), which would place it in the middle of the Oligocene with an age of between 27.9 and 28.3 Ma on the GPTS of Gradstein et al. (2012). This is consistent with the biochronological correlations of Welcomme et al. (2001) and Antoine et al. (2013).

Mustelidae

Pilgrim (1932) listed 11 Miocene taxa he considered to be mustelids, 6 of which are represented in the GSP, YPM, and AMNH collections. Of the remainder, one (GSI D 135) is a viverrid, another (GSI D 22) is indistinguishable from *Eomellivora necrophila*, while two (GSI D 227 and D 128) might be amphicyonids. The fifth, GSI D 207, is indeterminate.

Worldwide a number of small, hypercarnivorous fossil mustelids have been recognized. There are a number of small forms in the Siwaliks as well, although the available material is often too fragmentary for satisfactory identification. I have not recognized any badger-like forms nor any skunks (Mephitidae), although both are known as fossils and as extant species in Southeast and southern Asia.

Eomellivora Zdansky, 1924
Eomellivora necrophila Pilgrim, 1932

The holotype, a m1 (GSI D 243), is said to have come from the upper part of the Chinji Formation near Bhilomar (Pilgrim 1932) and could be between 12 and 11 Ma. Three isolated premolars (an upper GSI D 256a and two lower D 255b and D 256b) and a broken m1 (GSI D 255a) from the same site were referred to the species by Pilgrim (1932). He also included an edentulous mandible (GSI D 254), also from the Chinji Formation "south of Chinji." Pilgrim noted that the associated lower premolars (both thought to be left p4s) must have been from different individuals than the holotype m1.

This is a large mustelid (Table 13.1), which Pilgrim (1932) distinguished from *Eomellivora wimani* Zdansky, 1924, by its slightly smaller size (m1 lengths: 20.4 mm versus 24.5 mm). However, Pilgrim's diagnosis and comparison to *Eomellivora wimani* were based primarily on the referred edentulous mandible GSI D 254, which he thought to be older and had doubts

about belonging to the same species as the *E. necrophila* holotype (Pilgrim 1932). I am retaining Pilgrim's name pending further study, but have removed the query. The GSP material is younger than the probable age of the holotype (Table 13.1). This species is also very similar to *E. piveteaui* Ozansoy, 1965.

E. necrophila has a complicated nomenclatural history. It is frequently referred to *Hoplictis* Ginsburg, 1961 (Schmidt-Kittler 1976; Petter 1987; de Bonis et al. 2009), while Kretzoi (1942) made it the type of *Sivamellivora*, an attribution which has been followed by other authors (Wolsan and Semenov 1996; Valenciano et al. 2017, 2019; Alba et al. 2022). These attributions were usually without explanation or elaboration, although Kretzoi justified creation of a new genus as ". . . especially due to the shape and basic plan of the premolars . . ." (Kretzoi 1942, 322). The premolars Kretzoi was referring to (D 255b and D 256a), however, are respectively a P2 and a P3 of a small hyaenid such as *Lepthyaena sivalensis*, while D 256b might be a worn anterior lower premolar of a small suoid, such as *Schizochoerus* or *Microbunodon*.

New cranial and mandibular material justifies attribution to *Eomellivora*. The new specimens show that the species is characterized by a deep symphysis; m1 with a very narrow and trenchant talonid and complete absence of metaconid; slender, crowded lower premolars with salient anterior and larger posterior accessory cusps and a posterolingual expansion of the crown; single-rooted p1 present; p4 with backward inclination of the principal cusp; P3 with 3 roots, a single small posterior accessory cusp, and posterolingual shelf bounded by a low cingulum; P4 with a massive protocone slightly in advance of the paracone and a short carnassial blade; and M1 with broad, flat stylar shelf, very reduced metacone, and protocone with preprotocrista connected to the paracone.

E. necrophila includes *Eomellivora* (?)*tenebrarum* Pilgrim, 1932 (GSI D 22) from "Niki." In Pilgrim's bovid monograph (Pilgrim 1939) a list of localities has a "Niki," which is presumably where the fossil came from, about 33 km northwest of Chinji village, but the locality's age is not known. There are postcranial remains in the GSP collections of a large mustelid older than 12 Ma, which might belong to *Eomellivora*. There are, however, no dental remains that would confirm their identity.

Cernictis Hall, 1935
Cernictis lydekkeri (Colbert, 1933)

I follow a suggestion of Camille Grohé (pers. comm. February 4, 2011) that Colbert's species, first described as *Mustela lydekkeri*, is best placed in *Cernictis* rather than *Martes*. If Colbert's interpretation of Brown's stratigraphic position is correct (Colbert 1935—the original maps in the AMNH archives do not have the locality marked), then the holotype should be between about 14.2 and 13.8 Ma. Known only from mandibles, *C. lydekkeri* has crowded anterior premolars with p1 possibly absent, p2 and p3 slightly obliquely rotated, and p4 with a strong posterior accessory cusp, wide talonid, and posterior cingulum. The m1 has a relatively short paraconid, a large metaconid, and a relatively long, basined talonid with a prominent hypoconid and a low, but distinct, lingual rim. The m1 is short relative to the length of the p4 and the corpus is very deep beneath m1.

Pilgrim (1932) discussed a mandible (GSI D 239) from the Chinji Formation, considering it to be a new species of otter. Unable to specify a genus, he referred to the mandible as "Genus indet. *furtivus*." The teeth of GSI D 239 are battered, with only the p4 and m1 retaining any structure. The specimen, however, is similar to *C. lydekkeri*, including the absolute and relative size of the teeth and depth of the mandible. In addition, the p4 of GSI D 239 has a prominent posterior accessory cusp and wide heel, while the m1 has a large and probably tall metaconid. The talonid is about 75% the length of the trigonid. Pilgrim (1932) thought the specimen had a p1, but it is not readily apparent that it did. I think *C. lydekkeri* is probably the same species, but it is a problem that will need more study to be resolved.

Colbert (1933b) noted a similarity of his new species to *Pekania palaeosinensis* (Zdansky, 1924) as shown by the ratio of the talonid length to total m1 length. The Siwalik form, however, has a larger, taller metaconid. There are mustelid postcranial remains in the GSP collections with a general marten-like morphology that probably belong to *Cernictis lydekkeri*.

Plesiogulo Zdansky, 1924
Plesiogulo crassa Teilhard, 1945

This species is known from a mandible in the YPM collections (Lewis 1933; Kurtén 1970; Jiangzuo, Yu, and Flynn 2019). Its premolars are robust, without anterior or posterior accessory cusps, and short and wide in comparison to the molar. The m1 is low, but elongated relative to the premolars, with a large, distinct metaconid and a long, wide talonid. The hypoconid is low.

The holotype of *Promellivora punjabiensis* (Lydekker, 1884; GSI D 20) is too poor to confidently determine whether it is a distinct species or what its relations to others might be. Particularly noteworthy are the crowded short and very wide premolars. Although smaller, it may be same species as the YPM mandible, and as it is said to come from Hasnot (Lydekker 1884; Pilgrim 1932), it is probably approximately the same age as the YPM specimen. This is another problem that will need more study to be resolved.

Anatolictis Schmidt-Kittler, 1976
cf. *Anatolictis* sp. A

Among the small, hypercarnivorous mustelids, some Siwalik fossils are most similar to *Anatolictis laevicaninus* Schmidt-Kittler, 1976, although only about two-thirds the size of the Turkish species. The premolars are closed, with p2 and p3 slightly rotated. A small, simple, single-rooted p1 is present, while the principal cusps of p2 and p3 are situated anteriorly. The p4 has a more centrally situated principal cusp, a low posterior accessory cusp, and weak anterior and posterior cingula. The m1 metaconid is low but distinct, while the short talonid (about one-third the length of the tooth) has a centrally located bulbous hypoconid and a very small entoconid. The talonid has a low lingual rim and labial cingulum. The m2 has the proto-

cone and metacone of similar size and a low shelf-like talonid lacking the hypoconid. M1 has a distinct paracone with a smaller metacone and a crescent-shaped, anteriorly positioned protocone. The molar lacks the anterior expansion of the M1 of *Mustela* and has an enlarged stylar area. There may be two species present, one smaller and the other slightly larger, although the difference in size is possibly due to sexual dimorphism. One 9.4 Ma locality has associated skeletal elements together with mandibles.

This species is about the same size as and very similar to *Aragonictis* Valenciano, Morales, Azanza, and Demiguel, 2022, but differs in the larger p4 posterior accessory cusp and more centrally located and bulbous m1 hypoconid. It differs from *Circamustela* in the closed premolar series, shorter narrower talonid, and central hypoconid. It also differs from a small, extant ferret badger, such as *Melogale moschata*, in the more crowded premolars, small p1, lower m1 metaconid, and the structure and size of the m1 talonid.

cf. *Anatolictis* sp. B

A minute mandible fragment with m1 (length = 0.51 mm) possibly belongs to a different species of *Anatolictis*, although the identification is unsatisfactory. It is, however, different from the larger cf. *Anatolictis* sp. A in several features. As with other species of *Anatolictis*, the specimen has an open trigonid with an anterior paraconid forming a long carnassial, but the metaconid is taller and, although worn and damaged, the talonid appears to be shorter. The specimen is also much older (Table 13.1).

cf. *Anatolictis* sp. C

A mandible of a small hypercarnivorous mustelid in the YPM collections was tentatively identified by Jiangzuo, Yu, and Flynn (2019) as aff. *Iberictis* sp., but might better be placed in *Anatolictis*. The specimen has only p4 and m1 preserved. The p4 has a very subdued posterior accessory cusp (contra Jiangzuo, Yu, and Flynn 2019), while the m1 metaconid is very slightly behind the protoconid and the short talonid has a centrally located and trenchant hypoconid. The lingual rim and labial cingulum of cf. *Anatolictis* sp. A are absent. The specimen is slightly smaller than the holotype of *Anatolictis laevicaninus* and has a stouter p4 with a wider heel and weaker posterior accessory cusp.

The identification by Jiangzuo, Yu, and Flynn (2019) was based on the structure of the p4, notably the apparent absence of a posterior accessory cusp, the size of the m1 metaconid, and the absence of a talonid basin and entoconid. The specimen, however, lacks the distinct lingual bulge and surrounding cingulum of an *Iberictis* p4 (Valenciano et al. 2018). It differs from species of *Circamustela* in the smaller and posteriorly placed m1 metaconid and shorter talonid.

Hoplictis Ginsburg, 1961
cf. *Hoplictis anatolicus* (Schmidt-Kittler, 1976)

A weathered, slender P4 with a small, anteriorly placed protocone and low shelf or platform at the base of the paracone is most similar to *H. anatolicus,* although slightly larger. The width (as measured through the base of the paracone) to length ratio is very close to that reported by Schmidt-Kittler (1976) for *Hoplictis anatolicus.* An m1 fragment with a very flat talonid basin and low hypoconid is similar to *Pekania pennanti.*

Circamustela Petter, 1967
Circamustela sp.

This species is best known from a mandible bearing teeth that have lost some enamel, especially on the labial faces. Although not retaining their original lengths, the premolars are separated by short spaces, and there is an alveolus for a single-rooted p1. The premolars are tall and slender, p4 being nearly the same height as the m1 protoconid. The cusps of the unicuspid p2 and p3 are situated anteriorly, while that of the p4 is more central and bears a salient posterior accessory cusp. The m1 has an open trigonid, but with short postpara- and preprotocristids and a low bulbous metaconid set slightly behind the protoconid. The talonid is about one-third of the m1 length and relatively wide, with a shallow basin bounded by a low entocristid that terminates lingual to the base of the metaconid. The low hypoconid is at the labial edge and there may have been a small entoconid.

Compared to other hypercarnivorous mustelids, this species differs most notably in the tall premolars, short postparacristid and preprotocristid of the trigonid, and the wider and shallower talonid with a low, labial hypoconid.

Enhydriodon Falconer, 1868
Enhydriodon falconeri Pilgrim, 1931

This species is known from just three specimens: the holotype P4 (NHML M 4847), a large m1 (GSI D 161), and a P4 in the GSP collections. The m1 is said to have come from near Hasnot (Pilgrim 1932), while the holotype is noted as from the "Siwalik Hills" (Pilgrim 1931, 1932). The GSP specimen is from a locality about 3 km northeast of Hasnot and is 6.4 Ma. The upper carnassials are very distinctive, but other than for its size (Fig. 13.5a), the lower first molar is similar to those of *Sivaonyx.* (See Morales, Pickford, and Soria 2005, 54–55, for a discussion of the two taxa.)

Sivaonyx Pilgrim, 1931
Sivaonyx bathygnathus (Lydekker, 1884)

This species is known from several specimens in the older GSI and NHM collections (Pilgrim 1932; Pickford 2007) as well as a substantial number in the new GSP collections. Among the GSP specimens, two are particularly significant, a 9.4 Ma skull and a 9.2 Ma mandible, so that all the teeth except the incisors and m2 are now known. *S. bathygnathus* is characterized by single-rooted P1 and p1 and crowded premolars, with P1 positioned lingually to the canine and p1 and p2 rotated. The premolars have a salient cingulum surrounding the principal cusp and lack anterior and (except for p4) posterior accessory cusps. The P4 has a much enlarged anterior protocone and smaller hypocone surrounded by a cingulum, a bulbous short metastyle, and large, but low, parastyle. The unworn m1 metaconid

is slightly taller than the unworn protoconid, and both are taller than the paraconid. The talonid and trigonid have a labial cingulum that extends to the lingual side of the paraconid. The talonid is as long as and wider than the trigonid and may have a beaded lingual rim with a small cusp at the base of the protoconid separate from the hypoconid.

Pickford (2007) based a new species, *S. gandakasensis,* on a mandible (YGSP 4225), diagnosing it primarily on smaller size and supposed more bunodont upper and lower carnassials. The holotype mandible plus three other GSP specimens he referred to the new species are 9.2 Ma and presumably older than Pilgrim's specimens from near Hasnot—which we think are most likely between about 8 and 6.5 Ma (Barry et al. 2002). While the older fossils are generally smaller than the younger ones (Fig. 13.5a, b), *S. gandakasensis* is problematic, as there are small *Sivaonyx* specimens from younger horizons and no clear separation between them and the *S. gandakasensis* materials (Fig. 13.5a, b). In addition, some of the distinguishing characters cited by Pickford are variable. For instance, the lingual cingulum of a 9.3 Ma P4 is heavy with an inflated hypocone filling more of the central basin. In addition, the P4 may or may not have a very small cusp between the protocone and hypocone, with the protocone and hypocone consequently being farther apart. Finally, other than for the AMNH specimen discussed by Colbert (1935), the m2 of all other specimens has only one root.

The low, wide skull with a narrow and long postorbital constriction is much like that of an extant river otter and probably indicates aquatic adaptation. Fossils of this species are common, as would be expected in fluvial deposits. Numerous species of *Sivaonyx* have been described from Europe and Africa, as well as Asia (Grohé et al. 2013; Ghaffar and Akhtar 2016).

Vishnuonyx Pilgrim, 1932

Compared to *Sivaonyx,* the upper carnassial of *Vishnuonyx* has a relatively longer, more sectorial metastyle, which gives the tooth a more triangular shape. Although variable, the parastyle, which may be separated from the protocone by a rounded embrasure, is less prominent and the inner face of the hypocone less bulbous so that the hypocone is more sectorial in appearance. The P3 of *Vishnuonyx* is only slightly expanded posterolingually in contrast to that of *Sivaonyx.* Also, unlike *Sivaonyx,* the unworn m1 protoconid of *Vishnuonyx* is the tallest cusp and the metaconid the shortest. The trigonid overall has a more sectorial aspect with the paraconid having a more acute anterior edge and a more distinct lingual paracristid. The labial cingulum is not as salient and in some cases is absent.

Vishnuonyx chinjiensis Pilgrim, 1932

In the lower jaw the canine, m1, and m2 are now known plus the roots of the premolars, while in the upper jaw P2 through P4 are known. *Vishnuonyx chinjiensis* had crowded premolars, with a single-rooted p1, p2 and p3 rotated, and a cingulum surrounding the premolars. The m1 talonid and trigonid have a labial cingulum that extends around to the lingual side of the paraconid. The talonid may have a beaded lingual rim or

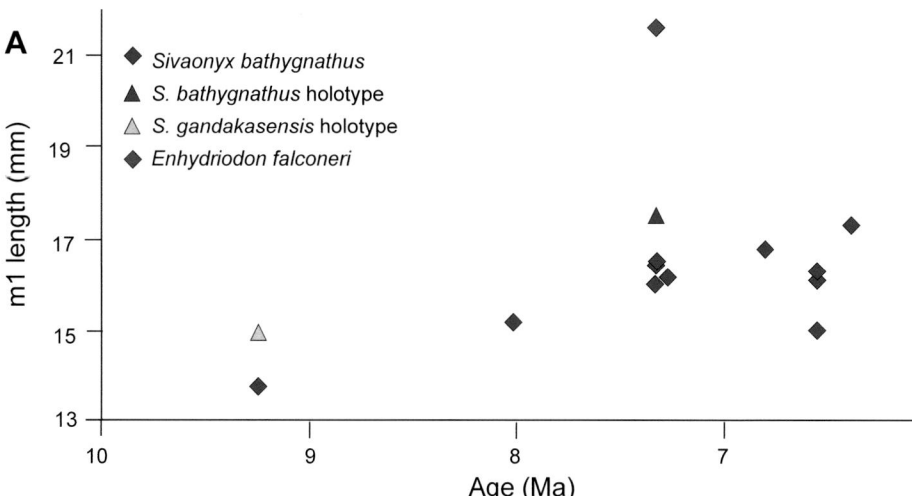

A. length of m1 with axes m1 length (mm) from 13 to 21 versus Age (Ma) from 10 to 6.

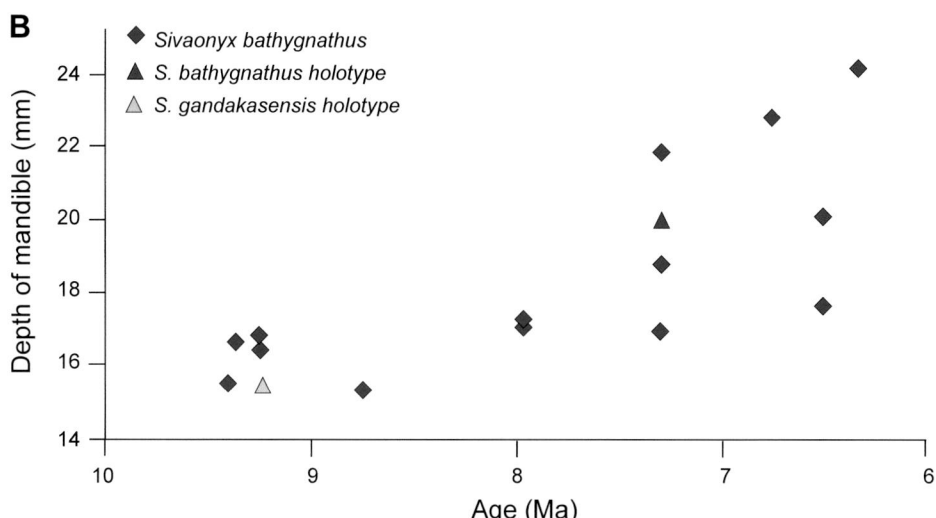

B. depth of mandible with axes Depth of mandible (mm) from 14 to 24 versus Age (Ma) from 10 to 6.

Figure 13.5. Dental dimensions of *Sivaonyx* and *Enhydriodon* versus time. A. length of m1, B. depth of mandible below m1. Specimens from near Hasnot are arbitrarily assigned an age of 7.3 Ma, while the Haritalyangar specimen is assigned an age of 8.0 Ma. Values from Lydekker (1884), Pilgrim (1932), Colbert (1935), Pickford (2007), and my own observations.

single crest, while the talonid cusp at the base of the protoconid is very small.

There are mustelid skeletal elements in the GSP collections that may belong to this species but do not have lutrine morphologies that would confirm their identity. The one exception is a distal humerus with an enlarged lateral supracondyloid ridge and lateral epicondyle which are similar to an otter's.

Vishnuonyx was reported from the Ngorora Formation of northern Kenya by Hill et al. (1985). Morales and Pickford (2005) referred the specimen to *Vishnuonyx chinjiensis*, but the shorter metastyle, more inflated hypocone, and division of protocone into two small cusps suggest a different species.

Vishnuonyx large species

A much larger species of *Vishnuonyx* is known from a mandible, upper P4, and skull fragments in the GSP collections. It is the size of *Sivaonyx*, but differs in the conformation of the P4, chiefly in the long metastyle, triangular shape, and smaller sectorial hypocone. The m1 has an acute anterior edge. A

single-rooted p1 was present, and p2 and p3 were rotated. Apart from size, this species differs from *Vishnuonyx chinjiensis* in that the P4 protocone is not a distinct cusp, but rather a crest connected to the paracone with a small cusp between the paracone and protocone. A distal humerus with a greatly enlarged lateral supracondylar crest suggests aquatic adaptation. Although of unknown age, the mandible NHMG 4 reported by Pickford (2007) from Chhorwali Katas, Pakistan, is probably this species.

Torolutra Petter, Pickford, and Howell, 1991
cf. *Torolutra* sp.

An indeterminate Late Miocene fish-otter known from an m1 and a p2 most closely resembles species of *Lutra* and *Torolutra*. The m1 has tall, trenchant trigonid cusps, with the protoconid the tallest and the paraconid the lowest. The metaconid is set slightly behind the protoconid and with the protoconid forms a vertical posterior wall, while the paraconid has an acute anterior edge and a distinct lingual paracristid. The talonid is narrower than the trigonid and less than one-third the length of m1. The talonid is similar to that of *Lutra* with the hypoconid forming a slender cristid that ends in a tiny cusp at the base of the protocone. A barely discernible rim encircles the posterior and lingual parts of the talonid, which has a flat, shallow basin. A faint cingulum runs anteriorly from the base of the protoconid around to the lingual face of the paraconid. Although the labial side of the talonid is damaged, this cingulum may have extended posteriorly and, as in *Lutra,* formed an apron-like structure below the hypoconid. The m1 is widest through the protoconid. The two-rooted p2 is narrow relative to its length, with a posterolingually expanded heel and strong lingual cingulum.

The most distinctive features of this species are the tall and trenchant trigonid and the short talonid of the m1. The short talonid, which is narrower than the trigonid, is suggestive of *Torolutra ougandensis* (Petter, Pickford, and Howell 1991) from the African Pliocene, but the Siwalik form differs in the crest-like hypoconid and relative heights of the trigonid cusps—both to each other and to the talonid. The structure of the talonid is similar to species of *Lutra*, both extant and extinct, but they have relatively longer talonids and lower trigonids with the metaconid being much lower than the protoconid. Species of *Paralutra* Roman and Viret, 1934 and *Tereulictis* Salesa, Antón, Siliceo, Pesquero, Morales, and Alcalá, 2013 have longer talonids as well as posterior cristids on the metaconid and protoconid (Salesa et al. 2013).

AELUROIDEA
Herpestidae

While mongooses are a diverse group in Africa, with as many as thirteen genera, there is only one extant genus in Asia. I follow Patou et al. (2009) in using *Urva* in place of *Herpestes*, assuming that the fossils are most likely related to the extant Asian species.

Urva Hodgson, 1837

Barry (1983) recognized Late Miocene species of *Urva* (referred to as *Herpestes* Illiger, 1811), the common mongoose of southern Asia. The Siwalik material is rare and fragmentary, but appeared to represent three different size classes, which correspond to differences among the extant Asian species. Since 1983, a few additional specimens have been discovered, which extend the range of Siwalik herpestids into the Middle Miocene. Although the size classes are distinct, the material is too poor to allow any judgments about whether any size class is a single, stable species. It does, however, demonstrate that, although rare, at least by 15 Ma, herpestids are part of the Siwalik fauna.

Urva small species

The fossils of this class are the size of the smallest of the extant Indian mongoose species, *U. auropunctata*, but the fossils are too scarce to determine whether it belongs to *U. auropunctata* or a different species.

Urva medium species

At about 1.3 kg, these fossils are from an animal the size of the brown Indian mongoose, *U. fuscus*, and morphologically similar to it, but may be a different species as the available material has a longer, narrower P3.

Urva large species

A second, much older mandible (13.1 Ma) is very similar to the 8.0 Ma mandible described by Barry (1983). Both have a slender p4 with a more trenchant anterior accessory cusp, which differs from that of the smaller species of *Urva*.

Viverridae

The extant viverrids of Asia are divided into three subfamilies, the paradoxurines with five genera, the hemigalines with four, and the viverrines with just two, including *Viverra* with four species and the monotypic *Viverricula*. These taxa are diagnosed primarily on the basis of soft tissue and basicranial characters (Veron 2010), but to a large degree they can also be recognized on the basis of dentitions.

Three genera of Siwalik viverrids have been described: the viverrines *Viverra* and *Vishnuictis* and a paradoxurine *Mioparadoxurus*. Pilgrim (1932) used *Viverra* and *Vishnuictis* for two large Pliocene species (*Viverra bakerii* Bose, 1880, and *Vishnuictis durandi* Lydekker, 1884) and established two smaller Middle and Late Miocene species (*Viverra ?chinjiensis* and *Vishnuictis salmontanus*). Prasad (1968) subsequently named two Late Miocene species (*Viverra nagri* and *Vishnuictis hariensis*). The distinction between *Vishnuictis* and *Viverra*, however, is equivocal. The principal characters of *Vishnuictis* that Pilgrim (1932) cited were an elongate and narrow skull that gradually narrows behind the post-orbital processes; a long, slender muzzle; and the presence of a sharp ridge along the nasal maxillary suture that extends onto the frontal, creating a shallow depression along the midline. Possible distinguishing dental characters of *Vishnuictis* include M1 and M2 both triangular, with the protocone on the lingual margin and without any trace of a lingual cingulum; m1 metaconid much lower than the protoconid and paraconid; and m1 talonid much shorter than the trigonid. The variety of postcranial remains in the collections that cannot be assigned to known fossil species suggests there are other unrecognized taxa present.

Large indet. Genus and species

An edentulous mandible from the Kamlial Formation (18.0 Ma) belongs to a large viverrid of about 20 to 30 kg. It is the oldest well-dated viverrid in the GSP collections and documents the presence of large viverrids in the Early Miocene. A Middle Miocene specimen belongs to a smaller individual of about 10–20 kg, while a navicular of a yet smaller and different species (ca. 2–5 kg) occurs in the Vihowa Formation near 19.7 Ma.

Vishnuictis Pilgrim, 1932

The Siwalik Miocene members of this genus include two of Pilgrim's species, plus an additional much larger species referred to as "Lutrine genus indet, *hasnoti*" by Pilgrim (1932). Morales and Pickford (2008) have also reported a species from the Middle Miocene of Kenya. I have removed the query from Pilgrim's attributions.

Vishnuictis salmontanus Pilgrim, 1932

Includes *Vishnuictis hariensis* Prasad, 1968, and *Viverra nagri* Prasad, 1968. Recently recovered GSP material includes skulls and mandibles plus associated postcranials of at least four individuals, including a juvenile, from a single 9.4 Ma locality. Although crushed and incomplete, the skull material seems to show the differentiating characters of *Vishnuictis*, and I therefore provisionally retain the genus. *Vishnuictis hariensis* and *Viverra nagri* are known from mandibles, while the type of *Vishnuictis salmontanus* (GSI D 160) is a skull with attached mandible and worn dentition, which makes direct comparison difficult. The small sample of new material of this species, however, shows that there is substantial variation in the details of the construction of the teeth, variation that encompasses differences seen in the GSI Haritalyangar specimens. Some of the new GSP material is about 2 million years older and approximately 20% smaller than the holotype of *V. salmontanus*. These specimens were referred to as "*Viverra*, sm. sp." in Barry (1995).

Vishnuictis chinjiensis (Pilgrim, 1932)

The species was based on a mandible and fragmentary M1 and named as a species of *Viverra* by Pilgrim (1932). However, it has characters of *Vishnuictis*, principally a short m1 talonid. The holotype (GSI D 214) has a large m2 relative to the length of m1 and is larger and substantially older than the GSP material of *Vishnuictis salmontanus* (Table 13.1).

Vishnuictis hasnoti (Pilgrim, 1913)

One of the most common viverrids in the collections, this species was initially referred to *Potamotherium*, which was thought at the time to be an Early and Middle Miocene otter. It was referred to as "*Viverra*, lge. sp." in Barry (1995). The new material includes a crushed, incomplete skull and associated mandible as well as postcranials of the same individual. While the crushing of the skull has obscured characters of the skull that differentiate *Vishnuictis* from *Viverra*, dental characters indicate it is a species of *Vishnuictis*. These include m1 with a short talonid and low metaconid as well as a triangular M2 with a lingual protocone. This species has large M2 and m2 relative to M1 and m1.

Mioparadoxurus Morales and Pickford, 2011
Mioparadoxurus meini Morales and Pickford, 2011

Morales and Pickford (2011) based this taxon on a mandible in the GSI collections from Haritalyangar. Although morphologically close to species of *Paradoxurus*, this taxon is very much larger than the extant species, with a less molarized p4 and simpler m1. Exposures at Haritalyangar are thought to be between approximately 9.4 and 8.1 Ma (Pillans et al. 2005; Patnaik 2013; ages adjusted to the Gradstein et al. [2012] timescale). Early

members of this species from the Potwar Plateau (ca. 10.5 Ma) are smaller than the holotype mandible and may represent a different species.

Paradoxurine genus and species indet.

A maxilla with intact P4 and broken M1 is much smaller than the Potwar material of *Mioparadoxurus meini* and is similar to *Paradoxurus*, although the parastyle is relatively much larger and the carnassial blades more trenchant. The P4 has a strong lingual and anterior cingulum and a weaker labial one. A mandible of a juvenile with dp4 and an erupting m1 may go with this taxon. The maxilla is poorly dated, but is between 10.1 and 9.3 Ma, while the mandible is 12.1 Ma. These ages are older than the estimated age of the divergence of *Paradoxurus* from *Paguma* (Patou et al. 2010).

Percrocutidae

Although dentally and in some postcranial features hyena-like, the species of *Percrocuta* belong to a separate family (Werdelin and Solounias 1991).

Percrocuta Kretzoi, 1938

Species of *Percrocuta* can be recognized by a very reduced P4 protocone; absence of a metaconid and presence of a minute talonid on m1; p4 and m1 nearly the same length; and anterior premolars that are wide relative to their length with p2 much shorter than p3 and p3 shorter than p4. P1 and p1 are present in some but not all specimens. The two Siwalik species differ primarily in size and possibly on the position of the P4 protocone relative to the parastyle (Kurtén 1957). However, separating the two is difficult and somewhat arbitrary, especially for isolated teeth. Presumably this is principally due to strong sexual dimorphism and the existence of material from intermediate stratigraphic levels that was not available to Pilgrim (1932) and Kurtén (1957).

Kurtén (1957) also accepted a third species, which he grouped with *Percrocuta*. It is now placed with the true hyenas as *Adcrocuta eximia*.

Percrocuta carnifex (Pilgrim, 1913)

In addition to the holotype (GSI D 172) and the GSP specimens, this species includes GSI D 168, D 169, D 170, D 171, D 173, AMNH 19346, and AMNH 19405. Several of the GSI fossils may belong to a single individual. All are from either the upper or middle of the Chinji Formation (contra Colbert 1935, 113), which would make them approximately 12.5 to 11.3 Ma. Most of the new GSP material comes from a site that is 10.1 Ma, while the youngest is 9.6 Ma.

Percrocuta grandis (Kurtén, 1957)

Apart from the holotype (GSI D 162) and many GSP specimens, this species includes GSI D 137, AMNH 19888, and probably the holotype of *Crocuta talyangari* Prasad, 1968 (GSI 18072). The sole example of an intact *P. carnifex* P4 (GSI D 168) has the protocone adjacent to the parastyle, while several P4s of *P. grandis* have the protocone more posterior. These specimens are also 25% to 50% larger than carnassials of *P. carnifex*. The P3s of *P. grandis* lack the internal lobe present in the GSI D 173 *P. carnifex* specimen.

Hyaenidae

The generic nomenclature of hyenas is complicated, and I generally follow Werdelin and Solounias (1991), although the multiple generic attributions obscure the potential relationships among the Siwalik species and those of other regions. In any case, the remains are often too fragmentary to confidently determine to which genus a species belongs. Measurements of the lower teeth (Fig. 13.6a, b, c, d) separate the Siwalik material into clusters, although the posterior teeth of the two species of *Lepthyaena* are not very distinct.

Protictitherium Kretzoi, 1938
Protictitherium sp.

Known from a pair of mandibles comparable in morphology and size to material from Turkey designated *P.* aff. *gaillardi* by Schmidt-Kittler (1976), *Protictitherium* sp. is very much smaller than other hyaenids and is not included in Figure 13.6.

Thalassictis Gervais, 1850 ex Nordmann
"Thalassictis" proava (Pilgrim, 1910)

This species was assigned to *Thalassictis* by Werdelin (1988) on the basis of the short, low m1 paraconid and tall protoconid. Both p1 and m2 were present. It is another example of a species of uncertain affinities and over the course of 80 years has been placed in 6 genera. (See the synonymy given by Werdelin and Solounias [1991], 30.) The quotation marks follow Werdelin and Solounias (1991) and indicate their doubts about the monophyletic status of the genus.

cf. *"Thalassictis" montadai* (Villalta and Crusafont, 1943)

A mandible in the YPM collections is very similar to material from Turkey (Schmidt-Kittler 1976; Viranta and Werdelin 2003). The specimen is said to be from the Chinji Formation and would be between 14 and 11.5 Ma. The m1 has a low paraconid and tall protoconid combined with a metaconid and large talonid with separate ento- and hypoconids. The status of both m2 and p1 is unknown. A few specimens of a hyaenid larger than either of the following species of *Lepthyaena* are present at 10 Ma. Whether they belong to *"T." montadai* or a different taxon is not evident.

Lepthyaena Lydekker, 1884
Lepthyaena sivalensis Lydekker, 1877

The most common of the small hyenas, this species includes material included by Pilgrim as well as numerous GSP specimens and both GSI D 233, the holotype of *Lycyaena ?chinjiensis* (Pilgrim, 1932), and GSI 18088, the holotype of *Ictitherium nagrii* Prasad, 1968. This species is smaller than (?)*Lepthyaena pilgrimi* (Fig. 13.6a, b, c, d) with a more anterior P4 protocone and short P4 metastyle. Both the p1 and m2 are usually, but not always, absent.

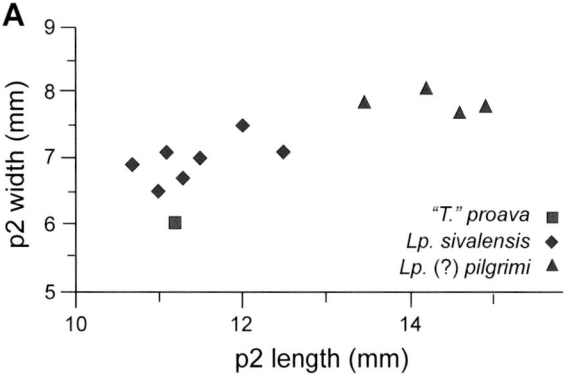

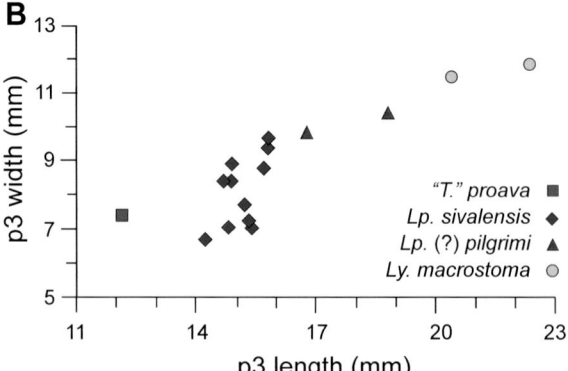

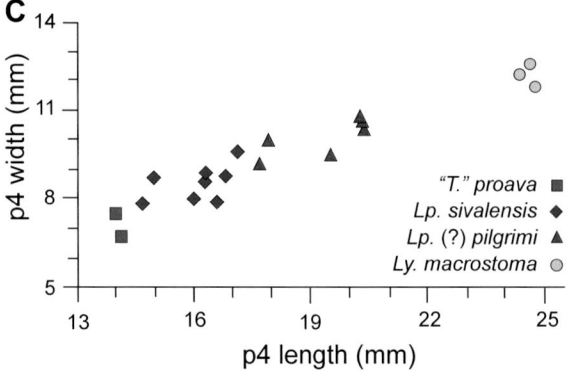

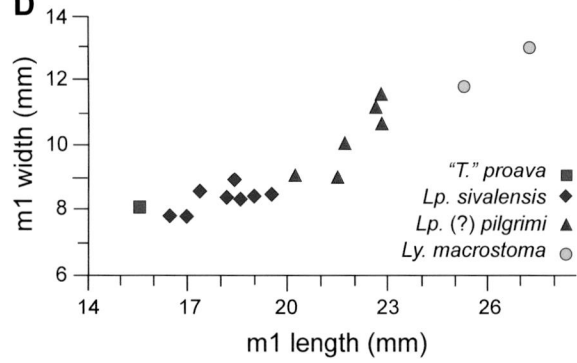

Figure 13.6. Length versus width of lower premolars and molars of *"Thalassictis" proava, Lepthyaena sivalensis,* (?)*Lepthyaena pilgrimi,* and *Lycyaena macrostoma.* A. p2, B. p3, C. p4, and D. m1. Values from Pilgrim (1932) and my own observations.

Lepthyaena (?)*pilgrimi* (Werdelin and Solounias, 1991)

Werdelin and Solounias (1991) have argued that *Palhyaena indica* Pilgrim, 1910 (referred to as *Ictitherium indicum* in Pilgrim [1932]), is a *nomen dubium*, as there is no known holotype. When first establishing the species, Pilgrim (1910) stated that it was based on a maxilla, but he subsequently designated a mandible (GSI D 53) as the holotype (Pilgrim 1932, 120), claiming that he had erred in referring to a maxilla in the original 1910 note. While there could be a misplaced type maxilla in the GSI collections, I am more inclined to think Pilgrim simply made a mistake and had a mandible in mind when creating his species and that there is no lost maxilla. Nevertheless, I accept the solution of Werdelin and Solounias (1991) in creating a new species based on the GSI D 53 mandible. GSI D 138 is the anterior part of the holotype specimen, GSI D 53, but my measurements differ from those of Pilgrim (Pilgrim 1932, 111).

Werdelin and Solounias (1991) placed this species in *"Hyaenictitherium"* because it has a more posterior P4 protocone than *Lepthyaena sivalensis* (Pilgrim 1932, 121), but it is hardly morphologically distinct from *Lepthyaena sivalensis*. Where known, all specimens have p1 and a small m2.

Lycyaena Hensel, 1862
Lycyaena macrostoma (Lydekker, 1884)

The holotype specimen (GSI D 44) is from "Jabi, Punjab" (Lydekker 1884, 298). This designation is ambiguous, as Jabi or a variant is a common village name in the Punjab. It is possible that a village (now "Jabbi") on the south side of the Soan River surrounded by fossiliferous outcrops of the Dhok Pathan Formation was intended. Nearby sediments are around 7.5 Ma, but on the north side of the river there are progressively older and highly fossiliferous outcrops (Barry, Behrensmeyer, and Monaghan 1980), and the holotype could have come from them. In addition to GSP and GSI material, this species includes NHML M 15703 from Hasnot.

Apart from being larger than species of *Lepthyaena*, the P4 has a tall parastyle and paracone and a slender metastyle, while the m1 talonid is short without a distinct hypoconulid.

Adcrocuta Kretzoi, 1938
Adcrocuta eximia (Roth and Wagner, 1854)

A p3 and edentulous mandible of a large hyaenid belong to this species, but otherwise it is not represented in the GSP collections. I include in this species GSI D 206, D 208, D 204/205, and D 163 from the older Siwalik collections. Distinguishing characters of *A. eximia* are the ratio of m1/p4 (greater than 110%) and a long talonid relative to the total length of m1. With the exception of GSI D 204/205, the GSI material is from near Hasnot and is therefore probably between 8 and 6.5 million years old. The GSP fossils are older (Table 13.1).

Barbourofelidae

Barbourofelids are rare in the Siwaliks, with fewer than 30 identified specimens, including the holotype of *Sansanosmilus* (?) *serratus* (GSI D 165) and GSI D 153. I follow Morlo, Peigné, and Nagel (2004) in recognizing barbourofelids as being distinct from nimravids, although a different view has been presented by Barrett, Hopkins, and Price (2021).

Sansanosmilus Kretzoi, 1929
(?)*Sansanosmilus serratus* Pilgrim, 1932

Pilgrim (1932) named two Siwalik species he tentatively referred to *Sansanosmilus*: *Sansanosmilus* (?) *serratus* and *Sansanosmilus* (?) *rhomboidalis*. While the status of *S.* (?) *rhomboidalis* is problematic, the holotype of *S.* (?) *serratus* is the mandible of a small barbourofelid as may also be the P4 (GSI D 153) referred to *S.* (?) *rhomboidalis*. The ratio of m1 to p4 length (1.26) and vestigial metaconid, as well as the smaller size, indicate that the species belongs to *Sansanosmilus* rather than to either *Barbourofelis* or *Albanosmilus*. The very reduced P4 protocone and the more vestigial m1 metaconid preclude assignment to *Prosansanosmilus* Heizmann, Ginsburg, and Bulot, 1980. The GSP collections include fragmentary dental remains, including three upper carnassials and a mandible with p4.

Barbourofelis Schultz, Schultz, and Martin 1970
(?)*Barbourofelis piveteaui* (Ozansoy, 1965)

The GSP collections include fragmentary dental remains of this species, most notably a mandible with the crowns of i2, p4, and m1, as well as an isolated p3 and two upper carnassials. Skeletal remains have also been tentatively identified, mostly of the forelimb. I have not identified any additional specimens of the species in the GSI collection.

I follow Geraads and Gülec (1997) and Robles et al. (2013) in tentatively placing this species in *Barbourofelis*, although there is a case made by Morlo (2006) for *Albanosmilus* Kretzoi, 1929. The absence of an anterior accessory cusp on P3, the medial position of M1 relative to P4, and the ratio of P4 to p4 lengths suggest reference of the Turkish material to *Barbourofelis*, while an enlarged ectoparastyle on P4 and subequal i3 and lower canine favor *Albanosmilus*.

As noted by Robles et al. (2013), if *Barbourofelis* originated in North America from a dispersal of *Albanosmilus*, which is the conventional view and reflected in the assignment of the oldest North American species to *Albanosmilus*, then (?)*B. piveteaui* is presumably a later dispersal from North America into Eurasia. Alternatively, as suggested by Morlo (2006), the Eurasian species of *Barbourofelis* (including *B. vallesiensis* as well as *B. piveteaui*) may have evolved from *Albanosmilus jourdani* in parallel to the North American *Barbourofelis* radiation and should be excluded from *Barbourofelis*.

Felidae

Felids make up about 17% of the Miocene carnivore specimens in the GSP Siwalik Project database. However, while for most carnivores the GSP collections have better material than what was available to Pilgrim (1932), this is not true for the felids. This is because unassociated skeletal elements comprise the greater part of the recent felid material and, with few exceptions, only edentulous mandibles and fragmentary isolated teeth of felids have been found. The named Miocene Siwalik felids are, of course, based exclusively on dentitions (Pilgrim 1932) and consequently identification of the skeletal elements is difficult and frequently unsatisfactory.

Machairodontinae

With a welter of names and opinions about relationships, the Siwalik sabertooths are confusing - which is not unusual for machairodontines (Matthew 1929; Salesa et al. 2010; Werdelin et al. 2010).

Paramachaerodus Pilgrim, 1913
Paramachaerodus sivalensis (Lydekker, 1877)

Paramachaerodus sivalensis is a medium-size species (Table 13.1). A 7.4 Ma maxilla with I1 through P4 is the most significant new specimen. Characters of the dentition include enlarged I3; long and very slender canine with crenulated anterior and posterior keels; P3 without an anterior accessory cusp and with only a slight posterolingual expansion; P4 with an ectostyle that is hardly more than a low cingulum and a reduced protocone between parastyle and paracone; M1 possibly single-rooted.

Apart from the holotype (GSI D 95), the species possibly includes D 140 and D 141 (both mandibles) as well as D 261 and D 262 (upper and lower canines). The holotype is from an unknown locality that is probably Late Miocene, while the others are from the Dhok Pathan Formation in the area near Hasnot (Pilgrim 1932; Colbert 1935) and are most likely between 8 and 6.5 Ma.

Lydekker (1877) based *Pseudaelurus sivalensis* on the GSI D 95 mandible. The specimen, which preserves only a broken m1, has had a convoluted nomenclatural history. (See Table 13.2 for a list of referrals for this one specimen.) Pilgrim (1913) established *Paramachaerodus* for European machairodontines, to which genus he also referred two Siwalik mandibles (GSI D 140 and GSI D 141). These he later (Pilgrim 1915) identified as *Paramachaerodus* cf. *schlosseri,* while the GSI D 95 mandible was retained as a second species, *Paramachaerodus sivalensis.* Subsequently Kretzoi (1929) created a new genus for the D 95 mandible (*Propontosmilus sivalensis*) and two new species for the D 140 and D 141 mandibles (*Paramachaerodus pilgrimi* and *Pontosmilus indicus*). These species (but not the generic assignment of D 141) were reluctantly accepted by Pilgrim (1931, 1932).

Salesa et al. (2010) have argued that the GSI D 141 mandible lacks the characters of a machairodontine and have also questioned the placement of GSI D 140 in *Paramachaerodus.* The D 141 specimen, however, is weathered and very damaged, as noted by Pilgrim (1932). This damage has obscured the

Table 13.2. Generic References for GSI D 95

Pseudaelurus sivalensis	Lydekker 1877
Aelurogale sivalensis	Lydekker 1884
Aelurictis sivalensis	Pilgrim 1910
Sivaelurus sivalensis	Pilgrim 1913
Paramachaerodus sivalensis	Pilgrim 1915
Machaerodus sivalensis	Matthew 1929
Propontosimilus sivalensis	Kretzoi 1929 and Pilgrim 1932
Paramachaerodus sivalensis	This chapter

original machairodontine-like conformation of the symphysis, which was more vertical with a lateral mental crest and deeper ramus than appears in the illustrations.

Sivasmilus Kretzoi, 1929
Sivasmilus copei Kretzoi, 1929

The holotype (GSI D 151) is a poorly preserved mandible from the Chinji Formation with the canine root, intact crown of p3, and a broken p4. The specimen was at first tentatively referred to *Sivaelurus chinjiensis* by Pilgrim (1915) and subsequently recognized as a separate genus and species by Kretzoi (1929). As far as known, the species is relatively small (Table 13.1), with a steeply ascending symphysis and moderately developed mental crest; a long diastema between the canine and p3 with a very small, single-rooted p2 midway between them; a short, slender, and low-crowned p3 with a weak posterior accessory cusp and cingulum but lacking the anterior accessory cusp; and a much larger p4 with a large posterior accessory cusp and broad cingulum surrounding it. A mandible in the GSP collection has a broken canine with a distinct posterior crest, two-rooted p3, and a large, blade-like anterior accessory cusp on p4. The specimen has no trace of a p2.

Miomachairodus Schmidt-Kittler, 1976
cf. *Miomachairodus* sp.

This is both a larger species than the preceding two (Table 13.1) and the most common machairodontine in the GSP collections. The species differs from *Paramachaerodus sivalensis,* but it is not obvious to which genus it should be assigned. Other than for brief comments, Schmidt-Kittler (1976) did not explicitly compare *Miomachairodus* to *Paramachaerodus,* but features of the GSP material suggest attribution to *Miomachairodus* although the Siwalik fossils are smaller ($\sim$ 80%). These features include I2 nearly the size of I3; upper canines long and very compressed with crenulated ("geperlte" or "beaded") anterior and posterior keels; P3 with lingually positioned anterior accessory cusp and modest posterolingual expansion; M1 two-rooted; and m1 with very small talonid. However, these features are seen in unassociated teeth only. Only a few postcranials have diagnostic attributes and most are attributed to this species on the basis of size. These postcranials are the same size as or slightly smaller than those of a lion (*Panthera leo*).

Amphimachairodus Kretzoi, 1929
(?)*Amphimachairodus giganteus* (Wagner, 1857)

A gigantic species (200 to 450 kg), characterized by compressed and very long upper canines with finely crenulated anterior and posterior edges; lower canine reduced in height; P3, p3, and p4 reduced in size relative to the P4 and m1, but with well-developed cusps; upper carnassial with small ectoparastyle and moderately reduced protocone; m1 with very small talonid. Probably includes AMNH 19935, identified as *Megantereon* sp. by Colbert (1935).

The holotype lower carnassial of Ghaffar and Akhtar's (2004) *Sivapanthera padhriensis* is probably the same individual as material in the GSP collections from Locality Y457, a locality located 0.5 km south of the village of Padhri. The GSP material includes a lower left carnassial in the same stage of wear as (PUPC No. 2001/12, a lower p3, massive upper canines, and some skeletal fragments. The locality is 7.4 Ma. *A. giganteus* is widely distributed in Europe.

Lokotunjailurus Werdelin, 2003
cf. *Lokotunjailurus* sp.

A single edentulous mandible closely resembles *Lokotunjailurus emageritus* Werdelin, 2003. Although smaller than the African species, the Siwalik specimen has a long post-canine diastema combined with a very short, single-rooted p3 and much longer p4 and m1. The mandibular body is slender and shallow, with a strong mental crest and steeply ascending symphysis, but lacks a mental process. Possibly also belonging to this taxon is a maxilla with P4 and the alveolus of M1 (GSI D 232) from Haritalyangar on which Pilgrim (1932) based the species *Megantereon* (?) *praecox*. The GSI specimen has an elongated paracone-metastyle blade, very reduced protocone, and a distinct ectoparastyle. GSI D 232 is smaller than a referred P4 of *Lokotunjailurus emageritus* (Pilgrim 1932, 172; Werdelin 2003, Table 7.19).

Felinae
Sivaelurus Pilgrim, 1913
Sivaelurus chinjiensis Pilgrim, 1910

Discussions of Miocene felines have largely neglected *Sivaelurus*, undoubtedly because the known material is scarce, fragmentary, and inaccessible to researchers without visiting India. The holotype maxilla (GSI D 150) has roots of the canine and P2 (or dP2) and crowns of P3-M1. It is similar to *Pseudaelurus* Gervais, 1850, to which Pilgrim first attributed it. *S. chinjiensis* has been distinguished by a tall, short face; upper canine with oval cross section and apparently lacking a posterior keel; the absence of a diastema in the tooth row; an "elongated" single-rooted P2; P3 small relative to P4 with indistinct anterior accessory cusp and weak posterolingual expansion; P4 with small but distinct ectoparastyle ("anteroexternal cingular cusp"), protocone opposite to and close to the parastyle, and the paracone much longer and taller than the metacone; and M1 with separate paracone and metacone and a prominent protocone.

The species is rare and known for certain from only the upper dentition, making comparisons to other small Miocene felids unsatisfactory. It differs from species of *Metailurus* Zdansky, 1924, *Styriofelis* Kretzoi, 1929, *Stenailurus* Crusafont-Pairó, 1972, *Fortunictis* Pons-Moyà, 1987, and *Pristifelis* Salesa et al., 2021, primarily in combining a less compressed canine having a rounded posterior with a one-rooted P2 and a P3 with an indistinct anterior accessory cusp and slight posterolingual expansion. An edentulous mandible in the GSP collections with the roots of the canine through p4 may belong to this species. It has a minute p2 (or dp2) alveolus. *S. chinjiensis* is about the same size as *Sivasmilus copei*, which makes separating their postcranials difficult.

Indeterminate Felids
There are many postcranials and isolated teeth of felids that cannot be identified to species. However, they can be sorted into three size classes (Table 13.1). Two complete upper canines

of the two smaller classes ("less than 10 kg" and "10 to 30 kg") have salient posterior keels and flattened lingual faces, indicating they might be metailurines. Fossils of the larger class (ca. 30 to 50 kg) that are older than 11 Ma probably are either *Sivaelurus chinjiensis* or *Sivasmilus copei*, but a few younger fossils indicate the presence in the Siwaliks of an otherwise unknown felid the size of a cheetah or small leopard. Lydekker's "Felis non det. allied to *Felis lynx*" (GSI D 90) (Lydekker 1884; Pilgrim 1932) is in the medium size class.

Taxa of Uncertain Status
The earliest collections from the Siwaliks, principally those described by Lydekker, included fossils of uncertain affinity. These specimens were thought by Pilgrim to be either aberrant mustelids or primitive felids. The former includes *Sivalictis natans* Pilgrim, 1932 (GSI D 227), and two fossils referred to as Genus indet. sp. indet (GSI D 128 and GSI D 207), while the supposed felids include *Mellivorodon palaeindicus* Lydekker, 1884 (GSI D 21), *Aeluropsis annectans* Lydekker, 1884 (GSI D 41), and *Vishnufelis laticeps* Pilgrim, 1932 (GSI D 266).

Our more recent collections do not contain material referable to any of these problematic taxa. The size and form of GSI D 227 and D 128 suggest that they are amphicyonids, with D 227 being a right M3 and perhaps a species of *Amphicyon*. GSI D 207 is poorly preserved and indeterminate. Lydekker (1884) described *Mellivorodon palaeindicus* as a mustelid, but both Matthew (1929) and Pilgrim (1932) thought it to be a felid. The only specimen is badly damaged, having lost enamel from the teeth (not noted by Matthew or Pilgrim). This accounts for similarities that Pilgrim saw to "*Vinayakia*" (which he also considered to be a felid), such as the large dilated premolar roots, the absence of accessory cusps, and narrow tooth crowns. The specimen, from near Hasnot, is most likely a felid, but the type is inadequate for further determination. The Late Miocene mandible of *Aeluropsis annectans* was described as a felid (Lydekker 1884), but the presence of m2, wide posterior m1 trigonid, and deep mandible all suggest a small hyaenid. The specimen, however, does not fit with other Siwalik species. Finally, *Vishnufelis laticeps* is based on skull fragments lacking the post-canine dentition. Other than for the low and apparently broad face, comparisons to the contemporaneous *Sivaelurus* cannot be made and the validity of the species is in question.

DISCUSSION

Apart from the indeterminate felids, 55 species (or likely species) from 12 families are listed in Table 13.1. Of these nearly half (26) were not known in the Siwaliks prior to 1935, the year E. H. Colbert published a major review of Siwalik mammals. In addition, the recent GSP collections differ from the older collections by having many more small forms and especially those under 10 kg (Fig. 13.7).

Extant carnivores can have very extensive geographic ranges, as for example *Felis concolor* (Currier 1983) and *Mustela erminea* (King 1983). Predictably then, many genera and even species of Siwalik carnivores are found in surrounding regions of Asia, Europe, and in several cases Africa. This is evident from Table 13.3

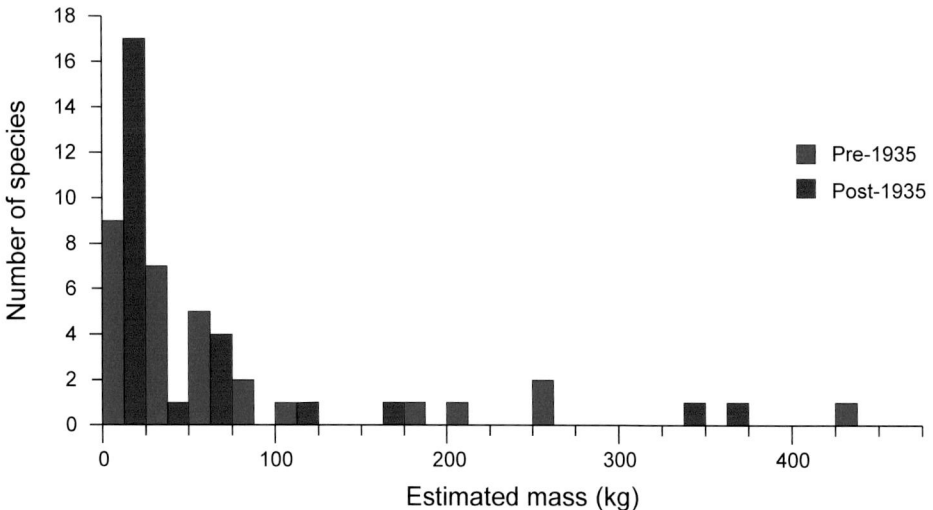

Figure 13.7. Number of species and their estimated mass, separated into those known before 1935 and those recognized since 1935.

Table 13.3. Siwalik Genera Present in Other Regions

Genus	South East Asia and Yunnan	Northern and Western Asia	Eastern Europe and Anatolia	Central Europe	Western Europe	North Africa and Arabia	Sub-Saharan Africa
Dissopsalis	—	—	—	—	—	—	x
Hyainailouros	—	—	—	—	x	x	x
Metapterodon	—	—	—	—	—	x	x
Amphicyon	x	x	x	x	x	x	x
Agnotherium	?	—	—	x	x	?	—
Phoberogale	—	—	—	—	x	—	—
Agriotherium	x	x	x	x	x	x	x
Indarctos	x	x	x	x	x	x	—
Alopecocyon	—	x	x	x	x	—	—
Eomellivora	x	x	x	x	x	—	x
Cernictis	—	—	?	—	—	—	—
Plesiogulo	—	x	x	x	x	—	x
Anatolictis	—	—	x	—	—	—	—
Hoplictis	—	x	x	x	x	—	—
Circamustela	—	—	—	?	x	—	—
Enhydriodon	—	—	x	—	x	—	x
Sivaonyx	x	x	x	x	x	x	x
Vishnuonyx	x	—	—	x	—	—	x
Torolutra	—	—	—	—	—	—	x
Urva/Herpestes	—	—	—	—	x	—	x
Vishnuictis	x	—	—	—	—	—	?
Mioparadoxurus	—	—	—	—	—	—	—
Percrocuta	—	x	x	—	x	x	x
Protictitherium	—	x	x	x	x	x	?
Thalassictis	—	x	x	x	x	—	—
Lepthyaena	—	—	—	—	—	—	—
Lycyaena	—	x	x	—	x	x	—
Adcrocuta	x	x	x	x	x	x	—
Sansanosmilus	—	x	x	x	x	—	—
Barbourofelis	—	—	x	?	—	—	—
Paramachaerodus	x	x	x	x	x	—	—
Sivasmilus	—	—	—	—	—	—	—
Miomachairodus	—	—	x	—	—	—	—
Amphimachairodus	?	x	x	x	x	x	x
Lokotunjailurus	—	—	—	—	—	—	x
Sivaelurus	—	—	—	—	—	—	—

Notes: Information based on a combination of the primary literature and a May 2022 extract from the New and Old Worlds database (NOW Community, undated). *Mioparadoxurus* is probably endemic to the Siwaliks.

where 32 of the 36 Siwalik genera are present in at least one neighboring region. This geographic pattern is especially evident for species over 100 kg, such as *Hyainailouros sulzeri*, the amphicyonids, the ursids, and the sabertooths, which on average have relatives in five neighboring regions. Unfortunately, in the absence of comprehensive phylogenetic analyses, we do not yet know how those taxa relate to the Siwalik species, making it difficult to identify immigrants among the newly appearing taxa. Of the four taxa that are absent from surrounding regions, probably only *Mioparadoxurus* is truly endemic, while the systematics of the other three need more detailed study to determine whether they are in fact restricted to the Siwaliks.

The suite of Siwalik carnivores ranges in size from more than 400 kg to less than 1 kg. The very large species were presumably wholly terrestrial, while many of the small mustelids, viverrids, and felids might have been largely arboreal, at least judging by their extant relatives. The very small species, those less than 2 or 3 kg, would need cover or otherwise have adopted cryptic lifestyles. Aquatic otters were present, but badger-like fossorial species have not yet been documented, although pangolins, aardvarks, and burrowing rodents are well known in the Siwaliks. And the past always has surprises (e.g., Hunt, Xue, and Kaufman 1983).

The diets of the Siwalik species span a broad spectrum. There were large and small hypercarnivores (*Hyainailouros sulzeri*, (?)*Metapterodon* unnamed sp., cf. *Agnotherium antiquum*, (?)*Sansanosmilus serratus*, (?)*Barbourofelis piveteaui*, felids, mustelids, and possibly hyaenids), as well as species capable of crushing bone (principally *Percrocuta*, the larger hyaenids, and *Dissopsalis*, but perhaps also amphicyonids). A few species probably ate significant quantities of fruit (*Mioparadoxurus meini*, (?)*Alopecocyon* sp.) or insects (species of *Urva*), while one of the otters (cf. *Torolutra* sp.) may have consumed fish. Judging from their teeth, the diet of *Vishnuonyx* spp., *Sivaonyx bathygnathus*, and *Enhydriodon falconeri* must have also included crabs or shellfish (Pickford 2007). Omnivores with varied diets, some of which might have also been active predators, include viverrids, amphicyonids, and the ursids.

For species as rare as carnivores, the observed chronostratigraphic ranges (Fig. 13.2) established by the oldest and youngest known fossils (occurrences) are typically unreliable. That is, when a species first appeared or became locally extinct may be anything from slightly to very different from the oldest and youngest occurrences (Barry et al. 2002; Chapter 26, this volume), and in particular the ranges of the smaller species are the most likely to be underestimated. The observed ranges or durations vary from single occurrences (treated as being 0.1 million years) to as long as 8.1 million years (*Urva* medium species). Excluding the less reliable occurrence data and the "indeterminate felids" as well as other taxa likely to contain multiple species, the average duration for all species in these two orders is 2.2 million years (Table 13.4). As might be expected because of taphonomic factors, species larger than 100 kg have longer average durations, approximately 3.7 million years. Apart from species with uncertain Early Miocene first occurrences and taxa that are very likely to encompass

more than one species (e.g. "Large, indet. viverrid," the two smaller size classes of *Urva*, and "Paradoxurine genus and species indet."), there is no clear relationship evident between the age and observed duration of a species (Fig. 13.8).

Patterns of change and faunal dynamics are discussed in greater detail in Chapters 25 and 26 of this volume. Fossils of carnivorous species occur throughout the Siwalik Miocene, but their distribution is uneven and is notably concentrated in the period between 14 and 6 Ma (Fig. 13.1). This unevenness biases not only the perceived chronostratigraphic durations of the species, but also the tabulations of species richness through time as well as the distribution of first and last occurrences. For instance, the total estimated number of carnivore species per 100,000-year interval varies from 11 at 14.0 Ma to a high of 20 at 11.6 Ma and then after 8.1 Ma rapidly decreases to 1 at 6.0 Ma (Fig. 13.9). Total estimated richness for an interval, however, is the sum of the number of species observed (those actually found in the interval) plus the number of range-through species (those found in earlier and later intervals, but not in the focal interval) (Maas et al. 1995). The number of species observed in each interval is much more volatile than the total estimated richness and can fall to low numbers or even zero in some instances. This can happen even while total richness remains high and stable—as between 11.3 and 10.2 Ma (Fig. 13.9).

Turnover of species results from the appearance and disappearance of individual species, the patterns of which can also be artifacts of preservation if only the raw occurrence data are used. (Chapter 26 of this volume addresses this critical issue.) As might be expected, first occurrences of carnivores and creodonts become more frequent after 14 Ma as the overall number of specimens rises, reaching maximums coinciding with the maximum number of specimens at 10.1 and 9.5 Ma (compare Fig. 13.10a and Fig. 13.1). Similarly last appearances become more frequent after 10.1 Ma as the number of specimens declines (compare Fig. 13.10b and Fig. 13.1). The last occurrences at 11.6 Ma might seem anomalous, but they reflect the slight decrease in the modest number of specimens after 12.1 Ma. What does emerge from the record of appearances and disappearances is that in the Late Miocene, a diverse set of large sabertooths and eventually ursids appeared, while amphicyonids and creodonts became increasing rare and eventually disappeared. At the same time, the small mustelids and viverrids maintained a consistent presence, although the species composition changed. The same pattern of replacement is seen with hyaenids, although the appearance between 9.4 and 8.5 Ma of the larger *A. eximia* and *L. macrostoma* together with *P. grandis* is significant.

SUMMARY

Carnivorous species of the Siwaliks include members of two orders: creodonts with 4 species and carnivores with 51, nearly doubling the number of species known prior to 1935. At least 12 families are present, with a wide range of body sizes and feeding adaptations. No one family dominates the fossil collections, although mustelids are the most species-rich and hyaenid

Table 13.4. First and Last Occurrences of Species, Durations, and Estimated Mass

Taxon	First Occurrence (Ma)	Last Occurrence (Ma)	Duration (myr)	Estimated Mass (kg)
Creodonta				
Dissopsalis carnifex	—	10.0	—	30–50
Hyainailouros sulzeri	—	11.4	—	430
(?)*Metapterodon* unnamed sp.	13.7	10.1	3.6	ca. 10–20
Carnivora				
Unnamed genus and species	10.1	10.1	0.1	5?
Arctoidea, genus and species indet.	11.6	11.6	0.1	50?
Amphicyon palaeindicus	17.9	11.6	6.3	150–200
Amphicyon lydekkeri	12.7	7.9	4.8	200–300
Amphicyon giganteus	—	17.9	—	300–400
cf. *Agnotherium antiquum*	—	10.1	—	170–275
Indarctos punjabiensis	8.7	—	—	150–250?
Eomellivora necrophila	10.1	9.4	0.7	30–50
Cernictis lydekkeri	—	12.1	—	3.0–6.0
Plesiogulo crassa	6.7	6.7	0.1	50
cf. *Anatolictis* sp. A	9.4	6.5	2.9	0.3–1.5
cf. *Anatolictis* sp. B	14.1	14.1	0.1	0.3
cf. *Anatolictis* sp. C	12.3	12.3	0.1	1.0–1.5
cf. *Hoplictis anatolicus*	9.7	—	—	12–25
Circamustela sp.	—	7.4	—	2.5–3.0
Enhydriodon falconeri	6.4	6.4	0.1	10–20
Sivaonyx bathygnathus	9.4	6.3	3.1	10–20
Vishnuonyx chinjiensis	13.7	11.6	2.1	2.5–10
Vishnuonyx large species	—	10.1	—	10–20
cf. *Torolutra* sp.	7.4	7.2	0.2	5–10
Urva small species	13.6	7.0	6.6	0.3–0.9
Urva medium species	15.1	7.0	8.1	1.3
Urva large species	13.1	7.2	5.9	2–5
Large, indet. viverrid	—	12.1	—	10–30
Vishnuictis salmontanus	10.1	7.9	2.2	2–4
Vishnuictis hasnoti	10.5	6.5	4.0	9–20
Mioparadoxurus meini	10.1	7.3	2.8	15–20
Percrocuta carnifex	14.0	9.6	4.4	40–60
Percrocuta grandis	9.4	6.3	3.1	70–100?
Protictitherium sp.	9.7	9.7	0.1	10–15
"*Thalassictis*" *proava*	—	10.2	—	ca. 15
cf. "*Thalassictis*" *montadai*	—	11.6	—	30–60?
Lepthyaena sivalensis	12.0	8.2	3.8	15–25?
Lepthyaena (?)*pilgrimi*	9.2	7.3	1.9	25–35?
Lycyaena macrostoma	8.7	6.4	2.3	60?
Adcrocuta eximia	8.5	8.1	0.4	60–100?
(?)*Sansanosmilus serratus*	15.2	11.6	3.6	30–40
(?)*Barbourofelis piveteaui*	—	8.6	—	50–70
Paramachaerodus sivalensis	10.1	7.0	3.1	50–60?
cf. *Miomachairodus* sp.	10.2	8.0	2.2	70–175
(?)*Amphimachairodus giganteus*	9.4	4.7	4.7	200–450
cf. *Lokotunjailurus* sp.	8.1	8.1	0.1	50–60
Sivaelurus chinjiensis	13.8	12.4	1.4	30–50

(13%) and felid (17%) fossils are the most common finds. By comparison to the numbers of fossil species reported here, there are at least 14 species of extant carnivores in Kanha National Park in central India (https:www.Kanha-National-Park.com/wildlife-in-Kanha.html, April 2024). Kanha has a mix of habitats that are probably similar to those of the Middle and Late Miocene Siwaliks.

There are some very large carnivores present and some very small, but most (34 species) weigh between 5 and 100 kg. The very large, those greater than 150 kg, include the creodont *Hyainailouros sulzeri* and an array of amphicyonids, sabertooth felids, and in the Late Miocene ursids, while the smaller species, those less than 5 kg, include mustelids, mongooses, and a few viverrids.

Most Siwalik carnivores and creodonts were undoubtedly carnivorous and predatory to a degree, although only some were exclusively so. Others included a wide range of additional food items in their diets, such as invertebrates and fruit, with

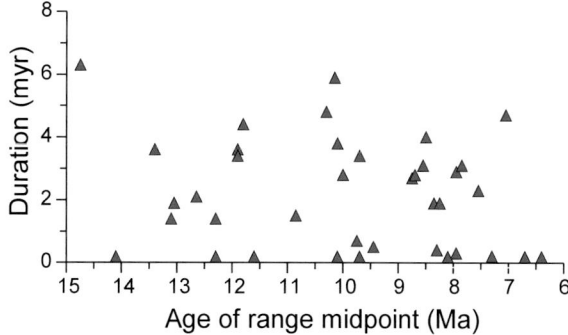

Figure 13.8. Observed duration versus age of 42 species. Ages of first and last occurrences from Table 13.4. Taxa with an uncertain age of first occurrence in the Early Miocene and taxa that are very likely to encompass more than one species are excluded.

in some few cases these foods likely being predominant. Of the carnivorous species, some might have been primarily scavengers (or if like lions, "pirates" might be a more accurate term). One of the otters likely subsisted in part on fish, while the others relied more on aquatic invertebrates. The large more predatory species were likely to have been terrestrial with a variety of hunting techniques, while the smaller ones could have been active in the canopy as well as on the ground. The otters were all aquatic, but no species is known to have been fossorial to the same degree as extant badgers.

Without detailed comparisons to non-Siwalik species and comprehensive phylogenetic analyses, it is difficult to determine the zoogeographic relations of the Siwalik carnivores. However, at least at the generic level, localities in Asia, Europe, and Africa have related taxa, and apparently nearly all the Siwalik genera are also widely found elsewhere. Given our lack of knowledge of the relative ages of the various taxa both in and outside the Siwalik area, we cannot be certain about the direction or timing of any exchanges between regions. However, it is likely that few of the Siwalik species were strictly endemic.

Observed chronostratigraphic ranges can be determined from the oldest and youngest known fossils of a species, but the occurrence data are typically poor indicators of when a species actually appeared and when it went locally extinct. This then has an effect on the estimates of species durations and species richness over time. The overall pattern of turnover, as determined by the appearance and disappearance of species, is also obscured. In the Siwaliks, fossils of carnivores and creodonts are concentrated in the period between 14 and 6 Ma, at which time the estimated number of species per 100,000-year interval rises from 11 to 20 and then, after holding steady until 8.1 Ma, rapidly collapses. Although we know little about the interval before 14 Ma, it is apparent that with the exception of the ursids, the major groups of both carnivores and creodonts, and importantly many of the genera, were already present by 14 Ma. Furthermore, although the timing and magnitude of turnover events between 14 and 6 Ma are unreliable, it is evident that at the end of the Middle Miocene and continuing into the Late Miocene, large sabertooths and eventually ursids appeared, while amphicyonids and creodonts disappeared. This undoubtedly marks a major shift in the interactions between the carnivores and their prey.

ACKNOWLEDGMENTS

I thank the editors, Catherine Badgley, Michèle Morgan, and David Pilbeam, for inviting me to participate. The research on the Siwalik sediments and fossils is a result of a long collaboration with the Geological Survey of Pakistan and I appreciate the efforts of the Directors General and members of the survey who have over the years supported the project. I also thank curators and collection managers at the Geological Survey of India Museum, the Natural History Museum in London, the Bavarian State Collection of Palaeontology and Geology, the Geological Institute at the State University of Utrecht, as well as of the American Museum, the Yale Peabody Museum, and

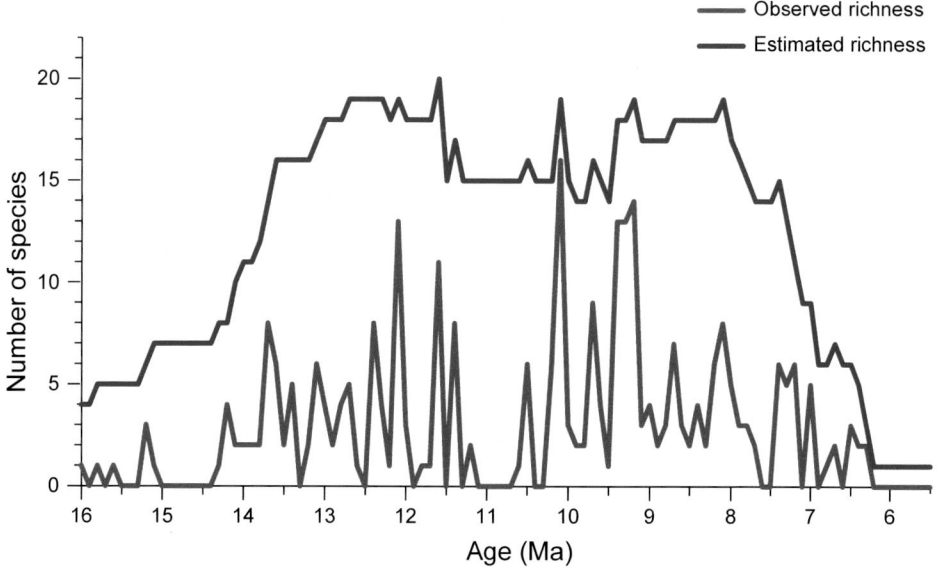

—— Observed richness
—— Estimated richness

Figure 13.9. Observed and estimated numbers of species per 100,000-year interval. The estimated richness is the sum of the number of species observed (those actually found in the focal interval) and the number of range-through species (those found in earlier and later intervals, but not in the focal interval).

A

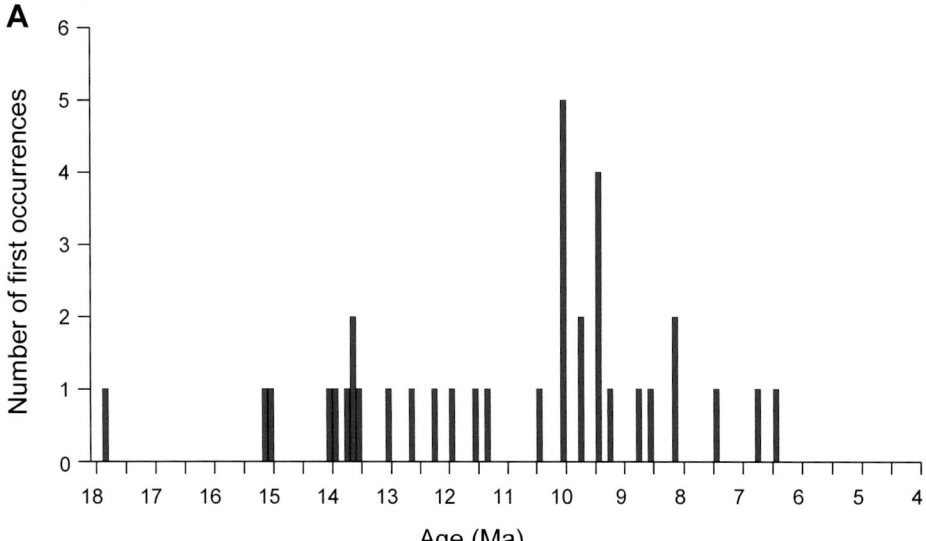

B

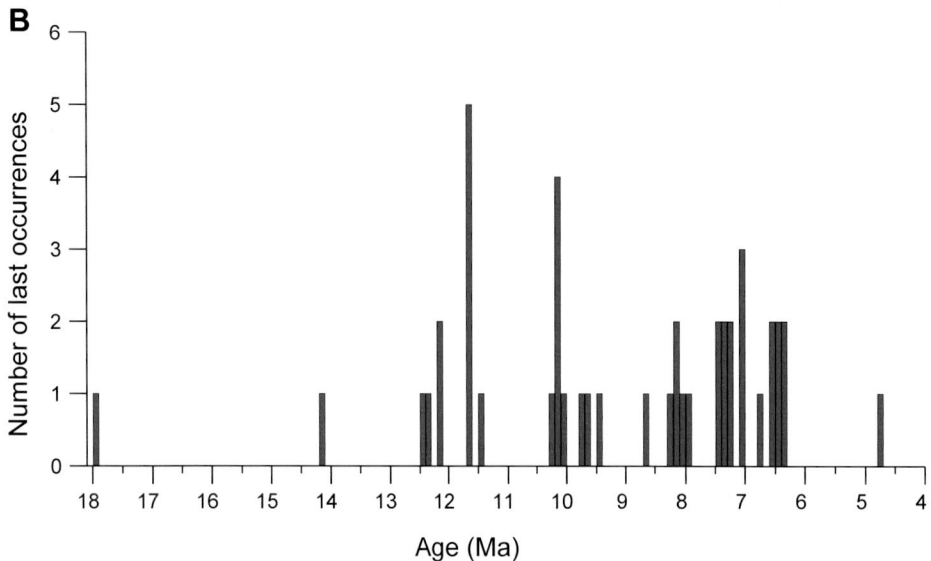

Figure 13.10. Observed first and last occurrence events. A. first occurrences, B. last occurrences.

the Harvard Museum of Comparative Zoology for access to collections and their assistance. Financial support has come from many sources, including grants from the National Science Foundation, the Smithsonian Foreign Currency Program, and the American School for Prehistoric Research. I also acknowledge the past support of Ms Francine Berkowitz of the Smithsonian Office of International Relations and the infectious enthusiasm of many colleagues, most importantly David Pilbeam and S. Mahmood Raza.

References

Alba, D. M., J. M. Robles, A. Valenciano, J. Abella, and I. Casanovas-Vilar. 2022. A new species of *Eomellivora* from the latest Aragonian of Abocador de Can Mata (NE Iberian Peninsula). *Historical Biology* 34:694–703. https://doi.org/10.1080/08912963.2021.1943380.

Antoine, P.-O., G. Métais, M. J. Orliac, J.-Y. Crochet, L. J. Flynn, L. Marivaux, A. R. Rajpar, G. Roohi, and J.-L. Welcomme. 2013. Mammalian Neogene biostratigraphy of the Sulaiman Province, Pakistan. In *Fossil Mammals of Asia: Neogene Biostratigraphy and Chronology*, edited by X. Wang, L. J. Flynn, and M. Fortelius, 400–422. New York: Columbia University Press.

Barrett, P. Z., S. S. B. Hopkins, and S. A. Price. 2021. How many sabertooths? Reevaluating the number of carnivoran sabertooth lineages with total evidence Bayesian techniques and a novel origin of the Miocene Nimravidae. *Journal of Vertebrate Paleontology* 41(1). https://doi.org/10.1080/02724634.2021.1923523.

Barry, J. C. 1980. Occurrence of a hyaenodontine creodont (Mammalia) in the late Miocene of Pakistan. *Journal of Paleontology* 54:1128–1131.

Barry, J. C. 1983. *Herpestes* (Viverridae, Carnivora) from the Miocene of Pakistan. *Journal of Paleontology* 57:150–156

Barry, J. C. 1988. *Dissopsalis*, a Middle and Late Miocene Proviverrine creodont (Mammalia) from Pakistan and Kenya. *Journal of Vertebrate Paleontology* 8:25–45.

Barry, J. C. 1989. A New Late Miocene Carnivore from Pakistan. Talk presented at the annual meeting of the Society of Vertebrate Paleontology, Austin, TX, November 2–4.

Barry, J. C. 1995. Faunal turnover and diversity in the Terrestrial Neogene of Pakistan. In *Paleoclimate and Evolution with Emphasis on Human Origins*, edited by E. S. Vrba, G. H. Denton, T. C. Partridge, and L. H. Burckle, 115–134. New Haven, CT: Yale University Press.

Barry J. C., A. K. Behrensmeyer, C. Badgley, L. J. Flynn, H. Peltonen, I. U. Cheema, D. Pilbeam, E. H. Lindsay, et al. 2013. The Neogene Siwaliks of the Potwar Plateau, Pakistan. In *Fossil Mammals of Asia: Neogene Biostratigraphy and Chronology*, edited by X. Wang, L. J. Flynn, and M. Fortelius, 373–399. New York: Columbia University Press.

Barry, J. C., A. K. Behrensmeyer, and M. Monaghan, M. 1980. A geologic and biostratigraphic framework for Miocene Sediments near Khaur Village, northern Pakistan. *Postilla* 183:1–19.

Barry, J. C., M. E. Morgan, L. J. Flynn, D. Pilbeam, A. K. Behrensmeyer, S. M. Raza, I. A. Khan, C. Badgley, J. Hicks, and J. Kelley. 2002. Faunal and environmental change in the late Miocene Siwaliks of northern Pakistan. *Paleobiology Memoirs* 3:1–71, supplement to *Paleobiology* 28(2).

Baryshnikov, G. F. 2002. Late Miocene *Indarctos punjabiensis atticus* (Carnivora, Ursidae) in Ukraine with survey of *Indarctos* records from the former USSR. *Russian Journal of Theriology* 2:83–89.

Baskin, J. A. 1998. Procyonidae, In *Evolution of Tertiary Mammals of North America*, Vol. 1: *Terrestrial Carnivores, Ungulates, and Ungulate-Like Mammals*, edited by C. M. Janis, K. M. Scott, and L. L. Jacobs, 144–151. Cambridge: Cambridge University Press.

Bose, P. N. 1880. Undescribed fossil Carnivora from the Siválik Hills in the collections of the British Museum. *Quarterly Journal of the Geological Society of London* 36:119–136,

Colbert, E. H. 1933a. The skull of *Dissopsalis carnifex* Pilgrim, a Miocene creodont from India. *American Museum Novitates* 603:1–8.

Colbert, E. H. 1933b. A new mustelid from the Lower Siwalik Beds of northern India. *American Museum Novitates* 605:1–3.

Colbert, E. H. 1935. Siwalik mammals in the American Museum of Natural History. *Transactions of the American Philosophical Society* n. s. 26:1–401.

Currier, M. J. P. 1983. *Felis concolor*. *Mammalian Species* 200:1–7. https://doi.org/10.2307/3503951.

De Beaumont, G. 1976. Remarques préliminaires sur le genre Amphictis Pomel (Carnivore). *Bulletin de la Société Vandoise des Sciences Naturelles* 73:171–180.

De Bonis, L. 2013. Ursidae (Mammalia, Carnivora) from the Late Oligocene of the "Phosphorites du Quercy" (France) and a reappraisal of the genus *Cephalogale* Geoffroy, 1862. *Geodiversitas* 35:787–814. https://www.bioone.org/doi/full/10.5252/g2013n4a4.

De Bonis, L., S. Peigné, F. Guy, A. Likius, H. T. Makaye, P. Vignaud, and M. Brunet. 2009. A new mellivorine (Carnivora, Mustelidae) from the Late Miocene of Toros Menalla, Chad. *Neues Jahrbuch für Geologie und Paläontologie* 252:33–54.

Ducrocq, S., Y. Chaimanee, V. Suteethorn, and J.-J. Jaeger. 1995. Mammalian faunas and the ages of the continental tertiary fossiliferous localities from Thailand. *Journal of Southeast Asian Earth Sciences* 12:65–78.

Falconer, H., and P. T. Cautley. 1836a. Note on the *Felis cristata*, a new fossil tiger from the Siválik Hills. *Asiatic Researches* 19:135–142.

Falconer, H., and P. T. Cautley. 1836b. Note on the *Ursus sivalensis*, a new fossil species, from the Siválik Hills. *Asiatic Researches* 19:193–200.

Flynn, L. J., E. H. Lindsay, D. Pilbeam, S. M. Raza, M. E. Morgan, J. C. Barry, C. E. Badgley, et al. 2013. The Siwaliks and Neogene evolutionary biology in South Asia. In *Fossil Mammals of Asia: Neogene Biostratigraphy and Chronology*, edited by X. Wang, L. J. Flynn, and M. Fortelius, 353–372. New York: Columbia University Press.

Forster Cooper, C. 1923. Carnivora from the Dera Bugti deposits of Baluchistan. *Annals and Magazine of Natural History* ser. 9, 12:259–263.

Geraads, D., and E. Gülec. 1997. Relationships of *Barbourofelis piveteaui* (Ozansoy, 1965, a Late Miocene nimravid (Carnivora, Mammalia) from central Turkey. *Journal of Vertebrate Paleontology* 17:370–375.

Ghaffar, A., and M. Akhtar. 2004. A new addition to the Siwalik Carnivora from the Tertiary rocks of Pakistan. *Pakistan Journal of Biological Sciences* 7:328–330.

Ghaffer, A., and M. Akhtar 2016. New material of *Sivaonyx bathygnathus* (Lutrinae, Mustelidae) from the early Pliocene of Siwaliks, Pakistan. *Revista Brasileira de Paleontologia* 19:347–356.

Ginsburg, L. 1980. *Hyainailouros sulzeri*, mammifère créodonte du Miocène européen. *Annales de Paléontologie* 66:19–73.

Ginsburg, L., and M. T. Antunes. 1968. *Amphicyon giganteus* Carnassier Géant du Miocène. *Annales de Paléontologie* 54:3–32.

Ginsburg, L., and J.-L. Welcomme. 2002. Nouveaux restes de Creodontes et de Carnivores des Bugti (Pakistan). *Symbioses* 7:65–68.

Gradstein, F. M., J. G. Ogg, M. D. Schmitz, and G. M. Ogg. 2012. *The Geological Time Scale 2012*. Amsterdam: Elsevier.

Grohé, C., L. de Bonis, Y. Chaimanee, C. Blondel, and J.-J. Jaeger. 2013. The oldest Asian *Sivaonyx* (Lutrinae, Mustelidae): A contribution to the evolutionary history of bunodont otters. *Palaeontologia Electronica* (16): 3: 29A. palaeo-electronica.org/content/2013/551-oldest-asian-bunodont-otters.

Hendy, Q. B. 1980. *Agriotherium* (Mammalia, Ursidae) from Langebaanweg, South Africa, and relationships of the genus. *Annals of the South African Museum* 81:1–109.

Hill, A., R. Drake, L. Tauxe, M. Monaghan, J. Barry, A. Behrensmeyer, G. Curtis, B. Jacobs, et al.. 1985. Neogene palaeontology and geochronology of the Baringo Basin, Kenya. *Journal of Human Evolution* 14:749–773.

Holroyd, P. A. 1999. New Pterodontinae (Creodonta: Hyaenodontidae) from the Late Eocene-Early Oligocene Jebel Qatrani Formation, Fayum Province, Egypt. *PaleoBios* 19:1–18.

Hunt, R. M., Jr. 1998. Amphicyonidae. In *Tertiary Mammals of North America*, edited by C. Janis, K. Scott, and L. Jacobs, 196–227. London: Cambridge University Press.

Hunt, R. M., X.-X. Xue, and J. Kaufman. 1983. Miocene burrows of extinct bear dogs: Indication of early denning behavior of large mammalian carnivores. *Science* n.s. 4608:364–366.

Jiangzuo, Q., C. Yu, and J. J. Flynn. 2019. *Indarctos* and other Caniformia fossils of G. E. Lewis' YPM collection from the Siwaliks. *Historical Biology* 33:543–557. https://doi.org/10.1080/08912963.2019.1648449

Jukar, A. M., and D. L. Brinkman. 2021. An introduction to the G. Edward Lewis 1932 fossil vertebrate collection from British India and a discussion of its historical and scientific significance. *Bulletin of the Peabody Museum of Natural History* 62: 81–96.

King, C. M. 1983 Mustela erminea. *Mammalian Species* 195:1–8. https://doi.org/10.2307/3503967.

Khana National Park. Undated. "Wildlife." https:www.Kanha-National-Park.com/wildlife-in-Kanha.html.

Kretzoi, M. 1929. Materialen zur phylogenetischen Klassification der Aeluroideen. *Proceedings of the 10th Congrès International de Zoologie*, 1293–1355.

Kretzoi, M. 1942. *Eomellivora* von Polgárdi und Csákvár. *Földtani Közlöny* 72: 318–323.

Kurtén, B. 1957. *Percrocuta* Kretzoi (Mammalia, Carnivora), a group of Neogene hyaenas. *Acta Zoologica Cracoviensia* 2:375–404.

Kurtén, B. 1970. The Neogene wolverine *Plesiogulo* and the origin of *Gulo* (Carnivora, Mammalia). *Acta Zoologica Fennica* 131:1–22.

Kuss, S. E. 1965. Revision der europäischen Amphicyoninae (Canidae, Carnivora, Mamm.) ausschließlich der voroberstampischen Formen. *Sitzungsberichte der Heidelberger Akademie der Wissenschaften, Mathematisch-Naturwissenschaftliche Klasse. Jahrgang 1965, 1. Abhandlung*:1–168.

Lewis, G. E. 1933. Notice of the discovery of *Plesiogulo brachygnathus* in the Siwalik measures of India. *American Journal of Science* 151:80.

Lewis, M. E. and M. Morlo. 2010. Creodonta. In *Cenozoic Mammals of Africa*, edited by L. Werdelin and W. J. Sanders, 543–560. Berkeley: University of California Press.

Lindsay, E. H., and L. J. Flynn 2016. Late Oligocene and Early Miocene Muroidea of the Zinda Pir Dome, *Historical Biology* 28:1–2, 215–236. https://doi.org/ 10.1080/08912963.2015.1027888.

Lindsay E. H., L. J. Flynn, I. U. Cheema, J. C. Barry, K. Downing, A. R. Rajpar, and S. M. Raza. 2005. Will Downs and the Zinda Pir Dome. *Palaeontologia Electronica* 8(1): 19A. https://palaeo-electronica.org /2005_1/lindsay19/issue1_05.htm.

Lydekker, R. 1876. Indian Tertiary and Post-Tertiary Vertebrata. Descriptions of the molar teeth and other remains of Mammalia. *Memoirs of the Geological Survey of India, Palaeontologia Indica* Series 10, I, part 2:19–87.

Lydekker, R. 1877. Notices of new and other Vertebrata from Indian Tertiary and Secondary rocks. *Records of the Geological Survey of India* 10:30–43.

Lydekker, R. 1878. Notices of Siwalik Mammals. *Records of the Geological Survey of India* 11:64–104.

Lydekker, R. 1884. Indian Tertiary & Post-Tertiary Vertebrata. Siwalik and Narbada Carnivora. *Memoirs of the Geological Survey of India, Palaeontologia Indica* Series 10, II, part 6:178–363.

Maas, M. C., M. R. L. Anthony, P. D. Gingerich, G. F. Gunnell, and D. W. Krause. 1995. Mammalian generic diversity and turnover in the Late Paleocene and Early Eocene of the Bighorn and Crazy Mountains Basins, Wyoming and Montana (USA). *Palaeogeography, Palaeoclimatology, Palaeoecology* 115:181–207.

Matthew, W. D. 1929. Critical observations upon Siwalik mammals. *Bulletin of the American Museum of Natural History* 56:437–560.

Morales, J., and M. Pickford. 2005. Carnivores from the Middle Miocene Ngorora Formation (13–12 Ma), Kenya. *Estudios Geológicos* 61:271–284.

Morales, J., and M. Pickford. 2008. Creodonts and carnivores from the Middle Miocene Muruyur Formation at Kipsaraman and Cheparawa, Baringo District, Kenya. *Comptes Rendus Palevol* 7:487–497.

Morales, J., and M. Pickford. 2011. A new paradoxurine carnivore from the Late Miocene Siwaliks of India and a review of the bunodont viverrids of Africa. *Geobios* 44:271–277.

Morales, J. M., M. Pickford, and D. Soria. 1998. A new credodont *Metapterodon stromeri* nov. sp. (Hyaenodontidae, Mammalia) from the Early Miocene of Langental (Sperrgebiet, Namibia). *Comptes rendus de l'Académie des Sciences, Paris, Sciences de la terre et des planètes* 327:633–638.

Morales, J. M., M. Pickford, and D. Soria. 2005. Carnivores from The Late Miocene tand Basal Pliocene of the Tugen Hills, Kenya. *Revista de la Sociedad Geológica de España* 18:39–61.

Morlo, M. 1996. Carnivoren aus dem Unter-Miozän des Mainzer Beckens. *Senckenbergiana Lethaea*, 76:193–249.

Morlo, M. 2006. New remains of Barbourofelidae (Mammalia, Carnivora) for the Miocene of southern Germany: Implications for the history of barbourofelid migrations. *Beiträge zur Paläontologie* 30:339–346.

Morlo, M., K. Bastl, J. Habersetzer, T. Engel, B. Lischewsky, H. Lutz, H., A. von Berg, R. Rabenstein, and D. Nagel. 2020. The apex of amphicyonid hypercarnivory: Solving the riddle of *Agnotherium antiquum* Kaup, 1833 (Mammalia, Carnivora). *Journal of Vertebrate Paleontology* 39(5). https://doi.org/ 10.1080/02724634.2019.1705848.

Morlo, M., and S. Peigné. 2010. Molecular and morphological evidence for Ailuridae and a review of its genera. In *Carnivoran Evolution: New Views on Phylogeny, Form, and Function*, edited by A. Goswami and A. Friscia, 92–140. Cambridge: Cambridge University Press.

Morlo, M., S. Peigné, and D. Nagel. 2004. A new species of *Prosansanosmilus*: Implications for the systematic relationships of the family Barbourofelidae new rank (Carnivora, Mammalia). *Zoological Journal of the Linnean Society* 140:43–61.

Patnaik, R. 2013. Indian Neogene Siwalik mammalian biostratigraphy: An overview. In *Fossil Mammals of Asia: Neogene Biostratigraphy and Chronology of Asia*, edited by X. Wang, L. J. Flynn, and M. Fortelius, 423–444. New York: Columbia University Press.

Patou, M.-L., P. A. Mclenachan, C. G. Morley, A. Couloux, A. P. Jennings, and G. Veron, 2009. Molecular phylogeny of the Herpestidae (Mammalia, Carnivora) with a special emphasis on the Asian *Herpestes*. *Molecular Phylogenetics and Evolution* 53:69–80

Patou, M.-L., A. Wilting, P. Gaubert, J. A. Esselstyn, C. Cruaud, A. P. Jennings, J. Fickel, and G. Veron. 2010. Evolutionary history of the *Paradoxurus* palm civets—A new model for Asian biogeography. *Journal of Biogeography* 37:2077–2097.

Peigné, S., Y. Chaimaneeb, C. Yameeb, P. Tianb, and J.-J. Jaeger. 2006. A new amphicyonid (Mammalia, Carnivora, Amphicyonidae) from the late Middle Miocene of northern Thailand and a review of the amphicyonine record in Asia. *Journal of Asian Earth Sciences* 26:519–532.

Petter, G. 1987. Small carnivores (Viverridae, Mustelidae, Canidae) from Laetoli. *Laetoli, a Pliocene Site in Northern Tanzania*, edited by M. D. Leakey and J. M. Harris, 194–234. Oxford: Clarendon Press.

Petter, G., M. Pickford, and F. C. Howell. 1991. La loutre piscivore du Pliocène de Nyaburogo et de Nkondo (Ouganda, Afrique orientale): *Torolutra ougandensis* n.g., n.sp. (Mammalia, Carnivora). *Comptes Rendus de l'Académie des Sciences de Paris* 312, serie 2: 949–955.

Petter, G., and H. Thomas. 1986. Les Agriotheriinae (Mammalia, Carnivora) Néogènes de l'ancien monde présence du genre *Indarctos* dans la faune de Menacer (ex-Marceau), Algérie. *Geobios* 19:573–586.

Pickford, M. 2007. Revision of the Mio-Pliocene bunodont otter-like mammals of the Indian subcontinent. *Estudios Geológicos* 63:83–127.

Pilgrim, G. E. 1908. The Tertiary and Post-Tertiary freshwater deposits of Baluchistan and Sind with notices of new vertebrates. *Records of the Geological Survey of India*, 37:139–166.

Pilgrim, G. E. 1910. Notices of new mammalian genera and species from the Tertiaries of India. *Records of the Geological Survey of India* 40:63–71.

Pilgrim G. E. 1912. The vertebrate fauna of the Gaj Series in the Bugti Hills and the Punjab. *Memoirs of the Geological Survey of India, Palaeontologia Indica* n. s. 4:1–83.

Pilgrim, G. E. 1913. The Correlation of the Siwaliks with mammal horizons of Europe. *Records of the Geological Survey of India* 43:264–326.

Pilgrim, G. E. 1915. Note on the new feline genera *Sivaelurus* and *Paramachaerodus* and on the possible survival of the subphyllum in modern times. *Records of the Geological Survey of India* 45:138–157.

Pilgrim, G. E. 1931. *Catalogue of the Pontian Carnivora of Europe in the Department of Geology*. London: British Museum (Natural History) Geology.

Pilgrim, G. E. 1932. The fossil Carnivora of India. *Memoirs of the Geological Survey of India, Palaeontologia Indica* n. s. 18:1–232.

Pilgrim, G. E. 1939. The fossil Bovidae of India. *Memoirs of the Geological Survey of India, Palaeontologia Indica* n. s. 26:1–356.

Pillans, B., M. Williams, D. Cameron, R. Patnaik, J. Hogarth, A. Sahni, J. C. Sharma, F. Williams, and R. Bernor. 2005. Revised correlation of the

Haritalyangar magnetostratigraphy, Indian Siwaliks: Implications for the age of the Miocene hominids *Indopithecus* and *Sivapithecus*, with a note on a new hominid tooth. *Journal of Human Evolution* 48:507–515.

Prasad K. N. 1968. The vertebrate fauna from the Siwalik beds of Haritalyangar, Himachel Pradesh, India. *Memoirs of the Geological Survey of India, Palaeontologia Indica* n. s. 39:1–79.

Raza, S. M., J. C. Barry, G. E. Meyer, and L. Martin. 1984. Preliminary report on the geology and vertebrate fauna of the Miocene Manchar Formation, Sind, Pakistan. *Journal of Vertebrate Paleontology* 4:584–599.

Raza S. M., I. U. Cheema, W. R. Downs, A. R. Rajpar, and S. C. Ward. 2002. Miocene stratigraphy and mammal fauna from the Sulaiman Range, Southwestern Himalayas, Pakistan. *Palaeogeography, Palaeoclimatology, Palaeoecology* 186:185–197.

Robles, J. M., D. M. Alba, J. Fortuny, S. De Esteban-Trivigno, C. Rotgers, J. Balaguer, R. Carmona, J. Galindo, S.Almecija, J. V. Berto, and S. Moya-Sola. 2013. New craniodental remains of the barbourofelid *Albanosmilus jourdani* (Filhol, 1883) from the Miocene of the Valles-Penedes Basin (NE Iberian Peninsula) and the phylogeny of the Barbourofelini. *Journal of Systematic Palaeontology* 11(8). https://doi.org/10.1080/14772019.2012.724090.

Salesa, M. J., M. Antón, G. Siliceo, M. D. Pesquero, J. Morales, and L. Alcalá. 2013. A non-aquatic otter (Mammalia, Carnivora, Mustelidae) from the Late Miocene (Vallesian, MN 10) of La Roma 2 (Alfambra, Teruel, Spain): Systematics and functional anatomy. *Zoological Journal of the Linnean Society* 169: 448–482.

Salesa, M. J., M. Anton, A. Turner, L. Alcala, P. Montoya, and J. Morales. 2010. Systematic revision of the Late Miocene sabre-toothed felid *Paramachaerodus* in Spain. *Palaeontology* 53:1369–1391.

Salesa, M. J., S. Peigné, M. Antón, and J. Morales. 2011. Evolution of the family Ailuridae: Origins and Old-World fossil record. In *Red Panda: Biology and Conservation of the First Panda*, edited by A. R. Glatston, 27–41. London: Elsevier.

Savage, R. J. G. 1973. *Megistotherium*, gigantic hyaenodont from Miocene of Gebel Zelten, Libya. *Bulletin of the British Museum (Natural History), Geology* 22:484–511.

Schmidt-Kittler, N. 1976. Raubtiere aus dem Jungtertiär Kleinasiens. *Palaeontographica. Abteilung A, Paläozoologie, Stratigraphie* 155:1–131.

Sehgal, R. K. 2015. Mammalian faunas from the Siwalik sediments exposed around Nurpur, District Kangra (H.P.): Age and palaeobiogeographic implications. *Himalayan Geology* 36:9–22.

Valenciano, A., J. Abella, D. M. Alba, J. M. Robles, M. A. Álvarez-Sierra, and J. Morales. 2018. New Early Miocene material of *Iberictis*, the oldest member of the wolverine lineage (Carnivora, Mustelidae, Guloninae). *Journal of Mammalian Evolution* 27:73–93.

Valenciano, A., J. Abella, U. B. Göhlich, M. Á. Álvarez-Sierra, and J. Morales. 2017. Re-evaluation of the very large *Eomellivora fricki* (Pia, 1939) (Carnivora, Mustelidae, Mellivorinae) from the Late Miocene of Austria. *Palaeontologia Electronica* 20(1)17A:1–22. palaeo-electronica.org/content/2017/1830-the-large-eomellivora-fricki.

Valenciano, A., Q. Jiangzuo, S. Wang, C. Li, X. Zhang, and J. Ye. 2019. First Record of *Hoplictis* (Carnivora, Mustelidae) in East Asia from the Miocene of the Ulungur River Area, Xinjiang, Northwest China. *Acta Geologica Sinica* (English ed.) 93:251–264

Valenciano, A., J. Morales, B. Azanza, and D. Demiguel. 2022. *Aragonictis araid*, gen. et sp. nov., a small-sized hypercarnivore (Carnivora, Mustelidae) from the late Middle Miocene of the Iberian Peninsula (Spain). *Journal of Vertebrate Paleontology* 41(5). https://doi.org/10.1080/02724634.2021.2005615.

Veron, G. 2010. Phylogeny of the Viverridae and "viverrid-like" feliforms. In *Carnivoran Evolution: New Views on Phylogeny, Form, and Function*, edited by A. Goswami and A. Friscia, A., 64–91. Cambridge: Cambridge University Press.

Viranta, S. 1996. European Miocene Amphicyonidae—Taxonomy, systematics and ecology. *Acta Zoologica Fennica* 204:1–61.

Viranta, S., S. T. Hussain, and R. L. Bernor 2004. The anatomical characteristics of a giant Miocene amphicyonid (Carnivora) humerus from Pakistan. *Pakistan Journal of Zoology* 36:1–6.

Viranta, S. and L. Werdelin. 2003. Carnivora. In *Geology and Paleontology of the Miocene Sinap Formation, Turkey*, edited by M. Fortelius, J. Kappelman, S. Sen, and R. L. Bernor, 178–193. New York: Columbia University Press.

Wang, X., R. M. Hunt, R. H. Tedford, and E. B. Lander. 2009. First record of immigrant *Phoberogale* (Mammalia, Ursidae, Carnivora) from Southern California. *Geodiversitas* 31:753–772. https://doi.org/10.5252/g2009n4a753.

Welcomme, J.-L., M. Benammi, J.-Y. Crochet, L. Marivaux, G. Métais, P.-O. Antoine, and I. Baloch. 2001. Himalayan forelands: Palaeontological evidence for Oligocene detrital deposits in the Bugti Hills (Balochistan, Pakistan). *Geological Magazine* 138:397–405.

Werdelin, L. 1988. Studies of fossil hyaenas: The genera *Thalassictis* Gervais ex Nordmann, *Palhyaena* Gervais, *Hyaenictitherium* Kretzoi, *Lycyaena* Hensel and *Palinhyaena* Qiu, Huang and Guo. *Zoological Journal of the Linnean Society* 92:211–265.

Werdelin. L. 2003. Mio-Pliocene Carnivora from Lothagam, Kenya. In *Lothagam: The Dawn of Humanity in Eastern Africa*, edited by M. G. Leakey and J. D. Harris, 261–238. New York: Columbia University Press.

Werdelin, L., and S. Peigné. 2010. Carnivora. In *Cenozoic Mammals of Africa*, edited by L. Werdelin and W. J. Sanders, 603–657. Berkeley: University of California Press.

Werdelin, L., and N. Solounias. 1991. The Hyuaenidae: Taxonomy, systematics and evolution. *Fossils and Strata* 30:1–104.

Werdelin, L., N. Yamaguchi, W. E. Johnson, and S. J. O'Brien. 2010. Phylogeny and evolution of cats (Felidae). In *Biology and Conservation of Wild Felids*, edited by D. Macdonald and A. Loveridge, 59–82. New York: Oxford University Press.

Wolsan, M., and Y. A. Semenov. 1996. A revision of the Late Miocene mustelid carnivoran *Eomellivora*. *Acta Zoologica Cracoviensia* 39:593–604.

TUBULIDENTATA AND PHOLIDOTA OF THE SIWALIK RECORD

THOMAS LEHMANN

PART 1: TUBULIDENTATA

INTRODUCTION

The order Tubulidentata, which belongs to the clade Afrotheria, consists of a single family, the Orycteropodidae, and is currently represented by a single extant species. *Orycteropus afer*, the aardvark, is a medium- to large-sized mammal (40–80 kg) that inhabits a wide range of habitats, including grasslands, savannas, woodlands, and forests. Nocturnal animals, aardvarks sleep in deep burrows during the day and forage for ants and termites during the night (Taylor 2013). The fossil record shows a greater number of species; there is a peak in the Late Miocene, but with relatively limited morphological variation among taxa. In total, 5 genera and at least 15 species have been described so far (for a complete review, see Lehmann 2006, 2007, 2009; Pickford 2019).

The geographic distribution of the Orycteropodidae once spanned Africa and Eurasia but is presently restricted to sub-Saharan Africa. Conspicuously absent from the Paleogene fossil record, the order Tubulidentata makes its first appearance in the Early Miocene of Africa, while the oldest known fossil aardvarks from Eurasia date from the Middle Miocene of Turkey (Çandir, Inönü, and Paşalar; see Fortelius 1990; Tekkaya 1993; van der Made 2003) and Pakistan (e.g., Colbert 1933; Lewis 1938; Pickford 1978).

AARDVARKS OF THE SIWALIK POTWAR PLATEAU

Several species of aardvarks (Tubulidentata, Orycteropodidae) are documented from northern Pakistan during the Middle and Late Miocene. These include two species of *Amphiorycteropus* and a species of *Orycteropus*.

Amphiorycteropus n. sp.

Several specimens of aardvark (Tubulidentata, Orycteropodidae), including, among others, partial skeletons (YGSP 27864, YGSP 31091, YGSP 41936, YGSP S 234), a fragment of mandible (AMNH 101259), and a left first lower molar (YGSP 31391), recovered from the Chinji Formation, represent a new species of the genus *Amphiorycteropus* Lehmann, 2009. This new species roamed the Potwar Plateau (Pakistan) during the Middle Miocene (14.1–12.0 Ma) (Chapter 3, this volume).

Discussion: This species, known only from the Siwaliks, represents the first appearance of Tubulidentata in Pakistan. Along with penecontemporaneous taxa in Turkey, *A.* n. sp. probably arose from the first dispersal of the order out of Africa. This small species shows a few similarities to some of the earliest fossil aardvarks known in Africa (i.e., genus *Myorycteropus*), such as the medio-dorsal to latero-ventral orientation of the head of the femur, the rather vertical cotylar fossa on the astragalus, and its crescent-shaped, medio-ventrally to latero-dorsally oriented condyle for the navicular (Fig. 14.1). Nevertheless, this new species is closer to *Amphiorycteropus* as it displays two synapomorphies of the genus: molars that are trapezoidal in shape and a ventral position of the articular facet for the sesamoid bone on the femur. Additional features, such as the triangular and not proximally bounded olecranon fossa on the humerus, as well as the oblique orientation of the articulation axis of the ulnar trochlear notch, are also shared by the younger, middle-sized type species of the genus, *A. gaudryi*. Hence, *A.* n. sp. is one of the oldest known members of the genus *Amphiorycteropus* (Lehmann, forthcoming-b). Although few elements can be directly compared, this small species is also morphologically similar to *A. browni* (see below), which replaces it in Pakistan by the Late Miocene.

Ecomorphology: Based on fragments of its humerus (Fig. 14.1c), this species was about 50% to 65% the size of extant *O. afer* and equivalent in size to the African *M. africanus*. Its teeth lack enamel and consist of vertical tubes of dentine bound by cementum, as is typical for Tubulidentata, and its humerus and autopodes share all the features present in other *Amphiorycteropus* species. It is thus likely that this animal was also, to a significant extent, myrmecophagous and fossorial.

A

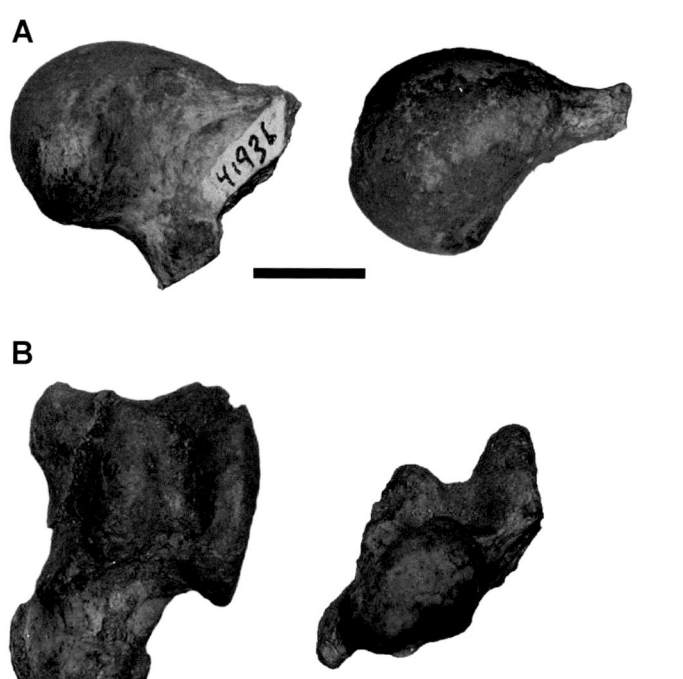

B

C

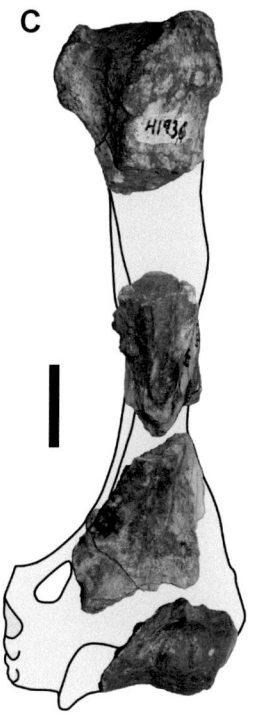

Figure 14.1. Amphiorycteropus n. sp. from the Potwar Plateau. A. YGSP 41936: left femoral head in dorsal and proximal view; B. YGSP 31091: left astragalus in dorsal and distal view; C. YGSP 41936: left humerus in ventral view (reconstruction). Scale bars = 1 cm.

The limited number of specimens makes an evaluation of the degree of dimorphism in *A.* n. sp. challenging. But sexual dimorphism between male and female modern aardvarks is low. Likewise, it is difficult to suggest a habitat preference for fossil Tubulidentata because the living *O. afer* dwells in a broad variety of environments, ranging from savannah to rainforest, and can even be present in altitudes up to 3,200 m (Taylor 2013).

Amphiorycteropus browni

Another set of aardvark specimens, including, among others, two maxillae (AMNH 29840, AMNH 29999), a left first upper molar (AMNH 29997), and elements of the forelimb (YGSP 16405), found in the Chinji, Nagri and Dhok Pathan Formations, belongs to the species *Amphiorycteropus browni* (Colbert, 1933). This species is documented from the Middle to the Late Miocene (12.7–9.4 Ma; Chapter 3, this volume).

Discussion: First described by Colbert (1933) as belonging to the genus *Orycteropus*, this taxon is known from a limited number of specimens, all from Pakistan. In his phylogenetic revision, Lehmann (2009) transferred this species to the genus *Amphiorycteropus* based on dental characters such as the trapezoid shape and shallow lingual groove of the upper molars. But within the genus, *A. browni* stands out with unique features of its cranium, including the rostral position of both the postpalatine foramen (at the level of M2) and the root of the zygomatic arch (at the level of the mesial lobe of M2), which might indicate a snout somewhat shorter than in the other *Amphiorycteropus* species. Both character states are inferred to be plesiomorphic. *A. browni*, with *A.* n. sp. is the basal-most taxon of

the genus *Amphiorycteropus* in the phylogenetic analysis of Lehmann and Vatter (forthcoming). Larger than the slightly older new Siwalik aardvark (see above), *A. browni* overlaps stratigraphically and geographically with the species that it replaces in Pakistan. But in view of the scarcity of comparable material between the two species, it is premature to assume an ancestor-descendant relationship between them. Besides one younger and ambiguous fossil from the Dhok Pathan Formation (see below), *A. browni* disappears without descendants in Pakistan. After a brief temporal hiatus in the fossil record, fossil *Amphiorycteropus* resume their presence in Eurasia from the Late Miocene (Maragheh, Iran) to the Early Pliocene (Perpignan, France) (Lehmann 2009). In parallel, *Amphiorycteropus* species occur from the middle to latest Miocene of Africa.

Ecomorphology: Cranial and postcranial fragments suggest that *Amphiorycteropus browni* was an animal about 60% to 70% the size of the extant aardvark. Except for the possibly shorter snout, the cranium of *A. browni* resembles that of other species of the genus, and its overall morphology is compatible with a myrmecophagous diet. The presence of a median keel on the distal epiphysis of the metapodials, as well as the broad muscle insertion area on the distal epicondyle of the humerus, suggest that the animal was able to dig. As with *A.* n. sp., it is difficult to evaluate the degree of dimorphism and the habitat preference for this species.

Amphiorycteropus sp. indet.

An isolated skull and mandible (YPM 13901) belonging to an unknown species of the genus *Amphiorycteropus*, were found in the Dhok Pathan Formation and are Late Miocene in age (8.0 Ma) (Chapter 3, this volume).

Discussion: Lewis (1938) reported this specimen from a locality close to Mathrala (Attock District) and identified it as *"Orycteropus pilgrimi."* What is left of the cranium and mandible has been supplemented with plaster, which renders the observation of certain characters difficult. A recent reanalysis of this specimen shows that it belongs to the genus *Amphiorycteropus*, and is close to *A. browni*, yet differs from this taxon in having a shallower lingual groove on the upper molars and, more importantly, in sharing a more caudal position of its zygomatic arch with more derived *Amphiorycteropus* species (Lehmann and Vatter, forthcoming). Anatomically intermediate between *A. browni* and younger Eurasian *Amphiorycteropus*, this specimen represents the last occurrence of the genus in Pakistan and also the last known Tubulidentata east of the Middle East.

Ecomorphology: According to the cranial remains, this specimen was about 60% to 70% the size of the extant aardvark, likely in the range of 30 to 50 kg. It was probably myrmecophagous, but nothing can be said about its locomotor adaptations or favored habitat.

Orycteropus sp.

Finally, postcranial specimens (YGSP 5154, YGSP 11604, YGSP 14949, YGSP 53549) and a fragment of mandible (YGSP 46511), which were found in the Nagri and Lower Dhok Pathan Formations, belong to an unknown species of the genus *Orycteropus*. They are dated from the early late Miocene (10.2–8.2 Ma).

Discussion: Although known from only five fragments, this taxon presents striking differences from the older *Amphiorycteropus* species discussed earlier. For instance, its lower molars are rectangular, not trapezoidal, in shape, as in *Orycteropus*

(Fig. 14.2); they are also significantly larger than in any *Amphiorycteropus* species and are comparable only with teeth of the largest *Orycteropus* species (Table 14.1). Moreover, the dimensions of the postcranial elements are either intermediate or closer to those of *Orycteropus* than to those of *Amphiorycteropus* (Lehmann, forthcoming-b). This species is more similar to the African Plio-Pleistocene *Orycteropus* than to the *Amphiorycteropus* from Pakistan, Europe, and Africa. Specimens are, however, too fragmentary to warrant separate species status. As suggested by Lehmann (2009), *Amphiorycteropus* and *Orycteropus* display too many derived characters to be ancestor-descendant. Therefore, the appearance of *O*. sp. in Pakistan more likely results from an immigration event. In that regard, a slightly older representative of the genus *Orycteropus* from Turkey has been recently reexamined (Lehmann and Vatter, forthcoming). The disappearance of *O*. sp. from Pakistan coincides roughly with the extinction of the genus throughout Eurasia and its first occurrence in Africa.

Ecomorphology: This taxon is larger than other Siwalik aardvarks and was about 85% the size of the extant *O. afer*. It is not possible to describe the ecology of this animal on the basis of the available material, but the preserved mandible is similar to that of the extant aardvark, suggesting a similar diet.

IMPORTANT EVOLUTIONARY EVENTS

The order Tubulidentata experienced an important species radiation during the middle and especially the upper Miocene in Eurasia (Lehmann 2006, 2009). It is during this period that aardvarks reached Pakistan, as evidenced by the Siwalik remains found in several localities between 14.1 and 8.0 Ma (Fig. 14.3). These taxa represent the eastern-most record of the order. All known Pakistani Tubulidentata are endemic to this region, and at least three species can be distinguished, belonging to two

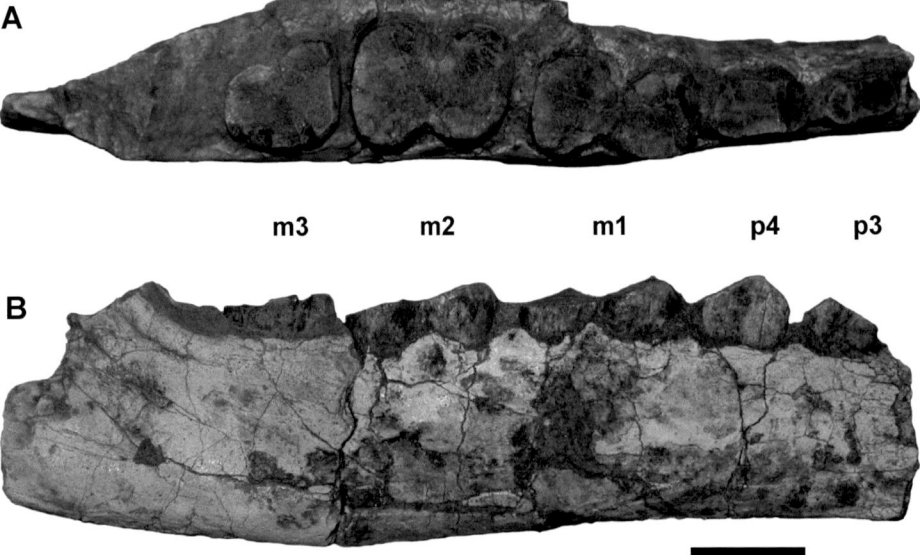

A

m3   m2   m1   p4   p3

B

Figure 14.2. Orycteropus sp. from the Potwar Plateau. YGSP 46511: left hemi-mandible in A. occlusal view; B. buccal view. Scale bar = 1 cm.

Table 14.1. Teeth Measurements (in mm) in Selected Extant and Fossil Tubulidentata Species

Lower Teeth	m1			m2			m3		
	L	B	R	L	B	R	L	B	R
Orycteropus afer	11.6±1.0 (93)	7.9±1.1 (90)	67.4%±7.4 (90)	12.6±1.0 (93)	9.2±1.0 (90)	72.9%±5.6 (90)	10.1±1.1 (90)	7.9±0.7 (87)	78.2%±6.6 (87)
O. crassidens (Holotype)	13.1	8.8	67.2%	14.3	10.6	74.1%	12.0	8.9	74.2%
O. djourabensis (Holotype)	14.7	9.0	61.2%	14.7	10.0	68.0%	14.1	8.9	63.1%
O. pottieri (Holotype)	11.1	6.5	58.6%	11.3	7.0	61.9%	7.6	5.4	71.1%
O. sp. (YGSP 46511)	14.4	8.5	59.0%	13.8	10.2	73.9%	10.0	8.5	85.0%
Amphiorycteropus browni (GSI K13/322)*	10.0	7.0	70.0%	9.9	5.9	59.6%			
A. gaudryi	11.1±0.8 (21)	7.1±0.5 (20)	64.2%±3.5 (20)	11.5±0.6 (19)	7.8±0.5 (18)	67.8%±3.9 (18)	9.3±1.0 (20)	6.5±0.6 (19)	70.9%±5.2 (19)
aff. *A. seni* (Holotype)*	10.6	6.4	60.4%	11.3	6.7	59.3%	8.1	5.6	69.1%
A. n. sp.									
(AMNH 101259)	9.6	5.0	52.1%	~8.7	5.3				
(YGSP 31091)	9.8	5.4	55.1%						
(YGSP 31391)	9.0	5.1	56.7%						
(YGSP 41936)				9.4	6.2	66.0%	6.2	5.2	83.9%
Myorycteropus africanus (Holotype)	7.6	4.0	52.6%	8.0	4.6	57.5%	6.2	4.5	72.6%

Upper Teeth	M1			M2			M3		
	L	B	R	L	B	R	L	B	R
O. afer	11.3±1.0 (93)	7.6±1.1 (90)	67.1%±8.0 (90)	12.0±1.0 (92)	8.7±0.9 (89)	72.2%±5.8 (89)	9.3±1.0 (89)	7.4±0.8 (86)	79.4%±6.8 (86)
A. browni									
(AMNH 29997)	10.2	7.1	69.6%						
(AMNH 29999)	10.4	6.9	66.3%	10.6	8.0	75.5%	7.0	6.6	94.3%
A. gaudryi	11.1±0.7 (15)	7.1±0.4 (15)	64%±3.3 (15)	11.8±1.0 (16)	7.9±0.5 (16)	66.7%±4.0 (16)	7.8±0.8 (16)	6.5±0.5 (16)	83%±5.0 (15)
A. n. sp.									
(YGSP 27864)				8.4	6.2	73.8%	7.0	6.1	87.1%
(YGSP 41936)									
M. africanus (Holotype)	7.1	4.7	66.2%	6.2	5.1	82.3%	4.9	4.8	98.0%

Notes: L = mesio-distal length; B = maximum bucco-lingual breadth; R = robustness index (B/L×100); * = after Pickford (1978) and Tekkaya (1993). Descriptive statistics: mean ± standard deviation; number of observations indicated in brackets. *O. afer* specimens from the AMNH, NHM, BPI, MfN, SMNS, and the TM collections. *A. gaudryi* specimens from the AMNH and the HLMD collections. AMNH = American Museum of Natural History, New York; NHM = Natural History Museum, London; BPI = Bernard Price Institut, Johannesburg; HLMD = Hessisches Landesmuseum, Darmstadt; MfN = Museum für Naturkunde in Berlin; SMNS = Staatliches Museum für Naturkunde in Stuttgart; TM = Transvaal Museum, Pretoria.

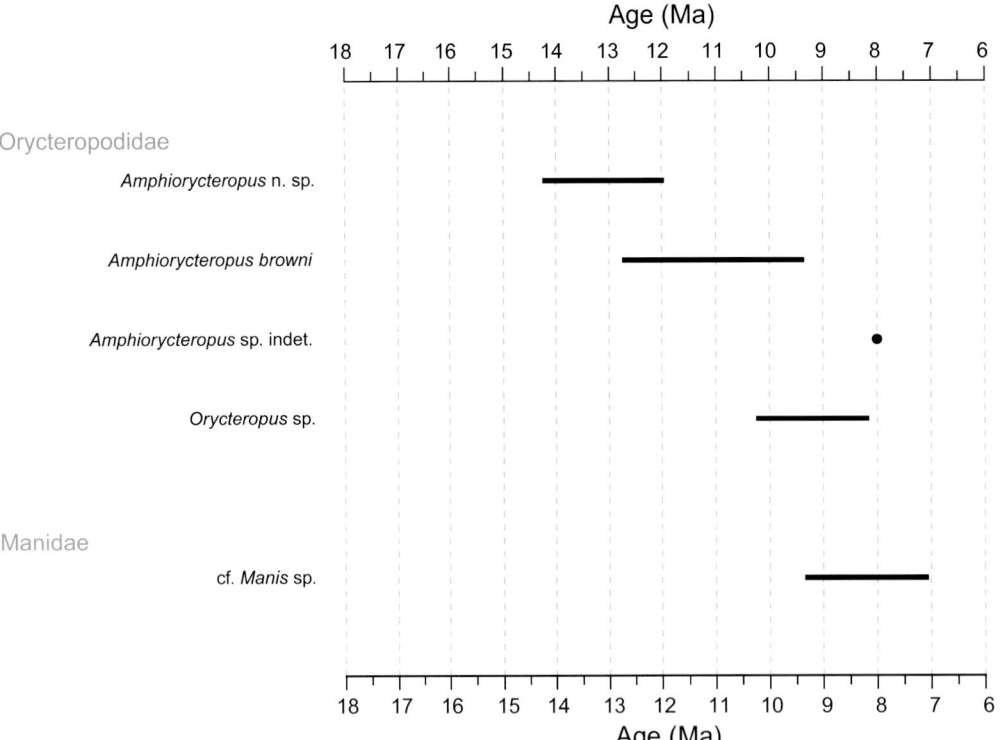

Figure 14.3. Biostratigraphic range chart of aardvarks and pangolins.

distinct lineages with overlapping biostratigraphic ranges. These lineages represent relatively high species richness for the order.

The oldest species of Tubulidentata found in the Siwaliks, *Amphiorycteropus* n. sp., appears suddenly in the fossil record. It shows ancestral character states reminiscent of some older African lineages, and is geographically isolated. Its relationships with slightly older and larger Eurasian fossil aardvarks (i.e., aff. *A. seni* in Turkey; Mammal Neogene units (MN) 5–6) are difficult to establish from the scarcity of material (see Lehmann 2009; Lehmann and Vatter, forthcoming). Nonetheless, both taxa represent the first appearance datum (FAD) of the genus *Amphiorycteropus* worldwide.

Contemporaneous with *A. browni* and *A.* sp. indet. (YPM 13901), the *Orycteropus* specimens from Pakistan count among the oldest representatives of their lineage and could be the first appearance of this genus. Thus Pakistan apparently played a role in the emergence of modern aardvarks (Lehmann, forthcoming-b). Like the new *Amphiorycteropus* species, this taxon appears suddenly in the Siwalik fossil record and is geographically isolated from the coeval and smaller *O. pottieri* (Greece, Turkey), the only other member of the genus in Eurasia (see Lehmann and Vatter, forthcoming).

It is striking that shortly after 8 Ma, all Tubulidentata went extinct in Pakistan, although as many as two (perhaps three with YPM 13901) taxa coexisted at that time. This extinction event is in line with the late Miocene phase of faunal turnover experienced in the Siwaliks and coincides with the reduction in C_3-dominated landscapes (Barry et al. 2002, Chapter 24, this volume).

PATTERNS OF FAUNAL CHANGE

The oldest fossil Tubulidentata are around 20 Ma and are recorded in East Africa (e.g., Napak, Songhor, Rusinga Island). The oldest Eurasian fossil aardvarks are known between 14 and 15 Ma from Turkey (Paşalar), followed shortly after by the ~14 Ma Siwalik aardvarks discussed here, suggesting that African emigrants reached Eurasia between 20 and 15 Ma. Two major events of marine regression could allow passage for mammals between Africa and Eurasia (see Rögl 1999; Sen 2013). The first event, or "*Gomphotherium* bridge," happened around 20 to 16 Ma (MN3–4) and consists of the collision of the Afro-Arabic tectonic plate with Eurasia, closing the Tethys Ocean and creating a land bridge between both continents. It enabled the elephant genus *Gomphotherium*, among other taxa, to enter Anatolia. After a period of marine transgression, dispersal was again possible around 13.8 to 12.5 Ma (MN6). Orycteropodids probably entered Eurasia during either or both of these two regression events. The Siwalik aardvarks of the Potwar Plateau are likely to have evolved from this pool of pioneers.

The genus *Amphiorycteropus* was present in Pakistan without interruption for over 6 million years between the Middle and the Late Miocene, with *A.* n. sp., then *A. browni*, and *A.* sp. indet (YPM 13901). In Europe, this genus is continuously represented from the Middle Miocene to the Pliocene, except for a short gap at the Vallesian-Turolian boundary (between 10.1 and 7.5 Ma). Interestingly, *Amphiorycteropus* reappears in the European fossil record (e.g., *A. gaudryi*) at the same time that it disappears in Pakistan. However, it is speculative to infer close relationships between Pakistani and other Eurasian forms. In

Africa, *Amphiorycteropus* specimens are known from the Middle Miocene of Namibia to the late Miocene of Chad and Algeria (Lehmann 2009).

The sympatric emergence of the *Orycteropus* lineage within the range of the endemic Pakistani *Amphiorycteropus* during the early Late Miocene contrasts with the more drastic replacement of aff. *A. seni* by *O. pottieri* in Turkey. In Africa, the oldest remains of *Orycteropus* come from younger deposits in East Africa (Lothagam, 7–6 Ma) and differ little from the new Siwalik *O. sp.* (Lehmann, forthcoming-b.). Of interest is that the fauna associated with the aardvarks at Lothagam is in large part of Eurasian origin (Leakey et al. 1996). The dispersal of some land mammals from Eurasia to Africa is recorded around MN9 (11.5 to 9.7 Ma), marked in particular by the arrival of *Hipparion* in Africa (Bernor, Tobien, and Woodburne 1990), and is contemporaneous with global paleoclimatic changes, such as major rearrangement of the oceanic circulation and high-latitude cooling trends (Garcés et al. 1997). Thus, Eurasian *Orycteropus* might have been one of the cohort of mammals (e.g., Elephantoidea) that entered Africa during that time (see Kalb, Froehlich, and Bell 1996).

Several authors (e.g., de Bonis et al. 1992; Merceron et al. 2004, 2007) pointed out differences between late Miocene faunal communities from Greece, Iran, and Afghanistan (called the GIA province) on the one hand and that from the Potwar Plateau on the other hand, probably reflecting paleoenvironmental differences, with a more arid and open landscape in the GIA province compared to the increasingly seasonal precipitation and extensive floodplains in the Siwaliks (e.g., Barry et al. 2002; Merceron et al. 2004, 2007; Flynn et al. 2016). Although tubulidentates were present in both provinces during the late Miocene, they had no species in common. Moreover, the Pakistani *Amphiorycteropus* taxa were smaller and the *Orycteropus* were larger than their GIA province counterparts.

The ecological niche of a myrmecophagous burrowing mammal was probably filled in Europe and Asia by Pholidotamorpha until their extinction following the arrival of Tubulidentata (Gaudin, Emry, and Wible 2009), perhaps because of competition with aardvarks. But the Neogene fossil record for this order is very sparse (see below) and there is no record of a coeval pholidotan anywhere in Asia. In the present, only one species of pangolin (*Manis crassicaudata*) occupies this niche in Pakistan.

CONCLUSION

Fossil Tubulidentata are known from localities throughout Africa, but their geographic range historically also includes Europe and Asia. In that respect, the three fossil aardvark taxa from the Siwaliks represent the eastern-most record for the order. Moreover, these specimens are among the oldest Eurasian forms and are important to Tubulidentata evolution.

Among the Siwalik aardvarks, the oldest, *A. n. sp.* represents a first appearance of the genus *Amphiorycteropus* on the Eurasian continent. It is probably the descendant of African pioneers that reached the continent in the Early or Middle Mio-

cene. This species shows affinity with a second form of that genus recognized in the Siwaliks—*A. browni*. Both taxa are morphologically more plesiomorphic than fossil *Amphiorycteropus* found in other regions such as Chad or Greece. The third record in the Siwaliks is the first appearance of the genus *Orycteropus.* The relationships between this form and the older Eurasian aardvarks are not determined yet. However, it is conceivable that the origin of the extant genus is to be found in Eurasia and not in Africa, as previously thought.

PART 2: PHOLIDOTA

INTRODUCTION

The order Pholidota belongs to the clade Laurasiatheria, where it is the sister taxon of the order Carnivora (clade Ferae). Still, Palaeanodonta, an extinct group of fossorial mammals with a reduced dentition, is increasingly supported as the closest relative of Pholidota, with which it is regrouped in the Pholidotamorpha clade (e.g., Rose et al. 2005; Gaudin, Emry, and Wible 2009).

The pangolin fossil record is meager and shows low taxonomic diversity (Gaudin, Emry, and Wible 2009). Unlike Tubulidentata, however, a number of reasonably well-known Paleogene pholidotans exist, such as the oldest known genera *Eomanis* and *Euromanis* (Middle Eocene of Messel, Germany; see Storch and Martin 1994; Gaudin, Emry, and Wible 2009), the only North American representative *Patriomanis* (latest Eocene; Gaudin, Emry, and Morris 2016), and the central European *Necromanis* (Oligocene–Miocene; see von Koenigswald and Martin 1990; Crochet, Hautier, and Lehmann 2015). In Africa, a pair of isolated ungual phalanges from the Early Oligocene of Egypt was identified as Pholidota (Gebo and Rasmussen 1985), and in Asia, *Cryptomanis* was recovered from the late Eocene of Mongolia (Gaudin, Emry, and Pogue 2006). Later fossil pangolins are not known until the Pliocene and Pleistocene in Africa, Europe, and Asia (see Gaudin, Emry, and Pogue 2006; Botha and Gaudin 2007; Terhune et al. 2021). This gap in the fossil record hinders the determination of clear phylogenetic relationships between fossil and extant taxa and in turn makes it difficult to localize the biogeographic origin of modern pangolins (Gaudin, Emry, and Wible 2009). Fortunately, fossil Pholidotans have been found in the Late Miocene Siwalik record of the Potwar Plateau (Pickford 1976), including new material under study that could help address these questions (Lehmann, forthcoming-a).

The eight living species of pangolin form a small group of mammals in a single family, Manidae, which is restricted to the Old World tropics and distributed across Africa, China, the Indian subcontinent, and Southeast Asia. Most pangolins are nocturnal terrestrial mammals with a conical-shaped skull, edentulous jaws, and a long, sticky tongue adapted to a specialized myrmecophagous diet. They are armed with powerful long claws on their forelimb for digging active burrows of ants or termites. Their body is famously covered with a unique external armor of overlapping keratinous scales (Heath 1992, 1995; Lim and Ng 2008). Modern manids are classified in three genera: *Smutsia* (African ground pangolins), *Phataginus* (African arboreal pangolins), and *Manis* (Asian pangolins; Gaudin, Emry, and Wible 2009).

Among the four known Asian species, only the largest, the Indian pangolin (*Manis crassicaudata*), is currently present in Pakistan (Baillie et al. 2014). It is a medium-sized mammal (11 to 20 kg body weight; 60 to 85 cm head–body length), with a long (40 to 85 cm), thick, muscular prehensile tail (Heath 1995; Irshad, Mahmood, and Nadeem 2015). Males are generally larger than females (Roberts 1997). The Indian pangolin lives in various types of tropical forests as well as open land and grasslands, sometimes up to 2,300 m elevation, as long as there is an abundance of ants and termites (Baillie et al. 2014; Mahmood, Irshad, and Hussain 2014). Although it spends its day in a burrow and is considered a "ground pangolin," it is an agile climber that will follow ants into trees (Prater 1980).

SIWALIK POTWAR PLATEAU PANGOLINS

Cf. *Manis* sp.

Specimens of pangolin (Pholidota, Manidae), including a partial skeleton (YGSP 51612) as well as several hand and feet bones (YGSP 2965, YGSP 17077, YGSP 50058, YGSP 52524), were found in the Dhok Pathan Formation. They belong to an unknown species, somewhat comparable to members of the extant genus *Manis*, which lived on the Potwar Plateau (Pakistan) during the Late Miocene (9.3–7.1 Ma; see Chapter 3, this volume).

Discussion: These specimens represent the first appearance of Pholidota in the Indian subcontinent. They are the only Miocene fossil pangolins from Asia known so far and fill the temporal gap between the Late Eocene *Cryptomanis* and the Plio-Pleistocene forms. Most specimens are too fragmentary to allow precise taxonomic identification, but specimen YGSP 51612 includes several diagnostic elements. For instance, the distal phalanges bear a triangular-shaped subungueal process with grooves on each side and are fissured, which is diagnostic of the Manoidea (Gaudin, Emry, and Wible 2009).

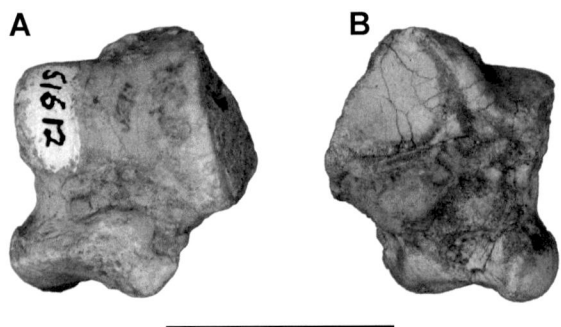

Figure 14.4. Pangolin specimen (cf. *Manis* sp.) from the Potwar Plateau. YGSP 51612: left astragalus in A. dorsal view; B. ventral view. Scale bar = 2 cm.

Moreover, the distal keel of the metapodials extends from the ventral to the dorsal side, which is characteristic of the Manidae (Gaudin, Emry, and Wible 2009). Remarkably, the left astragalus (YGSP 51612; Fig. 14.4) is similar in size to that of *Cryptomanis*; however, unlike the Eocene taxon, the head of this astragalus is concave and its ectal facet is round as in modern pangolins. Furthermore, a groove (for the tendon of M. tibialis posterior) on the distal epiphysis of the associated tibia and the medial extension of the trochlea on its distal humerus suggest affinities to the Asian genus *Manis*. But a more detailed analysis is necessary to confirm this hypothesis and to determine the phylogenetic position of the Siwalik material (Lehmann, forthcoming-a).

Ecomorphology: A preliminary study of YGSP 51612, the most complete Siwalik specimen found so far, indicates that this pangolin was a medium-sized mammal, smaller than *Manis crassicaudata* and closer in size to *M. javanica* but with stouter limbs. The dimensions of several elements of the Siwalik pangolin are quite close to those of the Mongolian *Cryptomanis*. The absence of dental remains suggests that, like other fossil and modern Pholidota, the Siwalik pangolin was edentulous and thus probably myrmecophagous. No cranial material is known for *Cryptomanis* to date (Gaudin, Emry, and Pogue 2006). The strong distal keel on the metapodials, the enlarged distal phalanges, and the broad distal epiphysis of the humerus resemble the locomotor adaptations in modern pangolins. But a more thorough examination of the remains is necessary before making firm inferences about the habitat preference or dimorphism of the Siwalik pangolin.

EVOLUTION AND BIOGEOGRAPHY

The order Pholidota currently consists of an African clade and an Asian clade. But Paleogene and early Neogene fossil records are only from Europe and North America, with a single contemporaneous lineage found in Asia, north of the current distribution of the extant species. This cohort of extinct taxa is followed by a long gap in the record, which makes it difficult to establish the origin of modern pangolins. Gaudin, Emry, and Wible (2009) cautiously hypothesized that a European form, maybe close to *Necromanis*, the only Neogene record for Europe, may be at the origin of the modern pangolins and that it dispersed first into sub-Saharan Africa (oldest fossil of a modern form from the Pliocene) and from there into southern Asia (oldest fossil of a modern form from the Pleistocene). But the recent discovery of a Pleistocene species of *Smutsia* (the African ground pangolin) in Romania (Terhune et al. 2021) has led to the reassessment of this biogeographic scenario. According to Terhune et al. (2021), either pangolins were cryptically present in Europe during the Mio-Pliocene and may have given rise to the African modern forms, or they genuinely disappeared from Europe during the Miocene while African *Smutsia* representatives dispersed back into Europe later.

The Siwalik pholidotes bridge the gap in their fossil record (see Fig. 14.3) and are the oldest representatives found in the geographic distribution area of extant pangolins. If all of the specimens represent a single species, then the Pakistani fossil pangolin is morphologically distinct from *Necromanis* and shares several apomorphies with the Manidae. The state of several characters diagnostic of each extant genus (*Manis*, *Phataginus*, or *Smutsia*)—on the astragalus, the tibia, the humerus, and the axis—still need to be studied in the new Siwalik material from the Potwar Plateau (Lehmann, forthcoming-a). But similarities with the Asian genus *Manis* have already been found. Accordingly, African and Asian pangolins would have diverged as early as at least the Middle Miocene, and the Asian taxa were resident in their current range at least since that time, hence longer than the African forms (see Gaudin, Emry, and Wible 2009, 275 for discussion).

Although pangolins are quite rare in the Siwalik record, it is interesting to note that aardvarks and pangolins have been found together at one site (Locality Y260, 9.3 Ma), which is the oldest record for Pholidota in Pakistan. The biochronological ranges of both orders overlap for about a million years (see Fig. 14.3), which suggests a gradual replacement of species in the ecological niche of myrmecophagous burrowing mammals in fluvial ecosystems during the Late Miocene, until the local extinction of the Tubulidentata. In the present, aardvarks and pangolins coexist only in Africa.

CONCLUSION

Fossil Pholidota are known on several continents since the Paleogene, yet there is a gap in their fossil record for most of the Miocene. Siwalik fossil pangolins partially fill this gap with a first occurrence at 9.3 Ma. Nonetheless, the material is scarce and known only from the Dhok Pathan Formation (Late Miocene). At least one partial skeleton (YGSP 51612) is known, which may bear important information relative to the biogeographic origin of the modern genera.

ACKNOWLEDGMENTS

I would like to thank David Pilbeam, Michèle Morgan, and Catherine Badgley for inviting me to contribute this chapter. My involvement began almost 20 years ago with a visit to study the Harvard-GSP collections at Harvard University, where I was warmly welcomed by John Barry, Lawrence Flynn, and Susanne Cote. They are gratefully thanked.

References

Baillie, J., D. Challender, P. Kaspal, A. Khatiwada, R. Mohapatra and H. Nash. 2014. *Manis crassicaudata*. *The IUCN Red List of Threatened Species 2014*, e.T12761A45221874. https://doi.org/10.2305/IUCN.UK.2014-2.RLTS.T12761A45221874.en.

Barry J. C., M. E. Morgan, L. J. Flynn, D. Pilbeam, A. K. Behrensmeyer, S. M. Raza, I. A. Khan, C. Badgley, J. Hicks, and J. Kelley. 2002. Faunal and environmental change in the Late Miocene Siwaliks of Northern Pakistan. *Paleobiology Memoirs* 28(sp3): 1–72.

Bernor, R. L., H. Tobien, and M. O. Woodburne. 1990. Patterns of Old World hipparionine evolutionnary diversification and biogeographic

extension. In *European Neogene Mammal Chronology*, edited by E. Lindsay, V. Fahlbusch, and P. Mein, 263–319. Plenum Press, New York.

Botha J., and T. J. Gaudin. 2007. A new pangolin (Mammalia: Pholidota) from the Pliocene of Langebaanweg, South Africa. *Journal of Vertebrate Paleontology* 27(2): 484–491.

Colbert, E. H. 1933. The presence of tubulidentates in the Middles Siwalik beds of northern India. *American Museum Novitates* 604:1–10.

Crochet J.-Y., L. Hautier, and T. Lehmann. 2015. A pangolin (Manidae, Pholidota, Mammalia) from the French Quercy phosphorites (Pech du Fraysse, Saint-Projet, Tarn-et-Garonne, late Oligocene, MP 28). *Palaeovertebrata* 39:c4.

De Bonis, L., M. Brunet, E. Heintz, and S. Sen. 1992. La province gréco-irano-afghane et la répartition des faunes mammaliennes au Miocène supérieur. *Paleontologia i Evolucio* 24–25:103–112.

Flynn, L. J., D. Pilbeam, J. C. Barry, M. E. Morgan, and S. M. Raza. 2016. Siwalik synopsis: A long stratigraphic sequence for the Later Cenozoic of South Asia. *Comptes Rendus Palevol* 15(7): 877–887.

Fortelius M. 1990. Less common ungulate species from Paşalar, middle Miocene of Anatolia (Turkey). *Journal of Human Evolution* 19(4–5): 479–487.

Garcés, M., L. Cabrera, J. Agustí, and J. M. Parés. 1997. Old World first appearance datum of "*Hipparion*" horses: Late Miocene large-mammal dispersal and global events. *Geology*. 25(1): 19–22.

Gaudin T. J., R. J. Emry, and J. Morris. 2016. Description of the skeletal anatomy of the North American pangolin *Patriomanis americana* (Mammalia, Pholidota) from the latest Eocene of Wyoming (USA). *Smithsonian Contributions to Paleobiology* 98:1–102.

Gaudin, T.J., R. J. Emry, and B. Pogue. 2006. A new genus and species of pangolin (Mammalia, Pholidota) from the late Eocene of Inner Mongolia, China. *Journal of Vertebrate Paleontology* 26(1): 146–159

Gaudin, T. J., R. J. Emry, and J. R. Wible. 2009. The phylogeny of living and extinct pangolins (Mammalia, Pholidota) and associated taxa: A morphology based analysis. *Journal of Mammalian Evolution* 16:235–305.

Gebo, D. L., and D. T. Rasmussen. 1985. The earliest fossil pangolin (Pholidota: Manidae) from Africa. *Journal of Mammalogy* 66(3): 538–540.

Heath, M. E. 1992. *Manis pentadactyla*. *Mammalian Species* 414:1–6.

Heath, M. E. 1995. *Manis crassicaudata*. *Mammalian Species* 513:1–4.

Irshad, N., T. Mahmood, and M. S. Nadeem. 2016. Morpho-anatomical characteristics of Indian pangolin (*Manis crassicaudata*) from Potohar Plateau, Pakistan. *Mammalia* 80(1): 103–110.

Kalb, J. D., D. J. Froehlich, and G. L. Bell. 1996. Palaeobiogeography of Late Neogene African and Eurasian Elephantoidea. In *Evolution and Palaeoecology of Elephants and Their Relatives*, edited by J. Shoshani, and P. Tassy, 117–123. Oxford: Oxford University Press.

Leakey, M. G., C. Feibel, R. L. Bernor, J. M. Harris, T. E. Cerling, K. M. Stewart, G. W. Storrs, A. Walker, L. Werdelin, and A. Winkler. 1996. Lothagam: A record of faunal change in the Late Miocene of East Africa. *Journal of Vertebrate Paleontology* 16(3): 556–570.

Lehmann, T. 2006. The biodiversity of the Tubulidentata over geological time. *Afrotherian Conservation* 4: 6–11.

Lehmann, T. 2007. Amended Taxonomy of the order Tubulidentata (Mammalia, Eutheria). *Annals of the Transvaal Museum* 44:178–196.

Lehmann, T. 2009. Phylogeny and systematics of the Orycteropodidae (Mammalia, Tubulidentata). *Zoological Journal of the Linnaean Society* 155(3): 649–702.

Lehmann T. Forthcoming-a. New fossil pangolins (Mammalia, Pholidota) from the early late Miocene of Pakistan.

Lehmann T. Forthcoming-b. New specimens of fossil Aardvarks (Mammalia, Tubulidentata) from Pakistan: The Eurasian origin of the extant aardvark?

Lehmann, T., and M. Vatter. Forthcoming. The two lost fossil aardvark skulls from Eurasia—Systematic and palaeobiogeographical implications.

Lewis, G. E. 1938. A skull of *Orycteropus pilgrimi*. *American Journal of Science* 36(5): 401–405.

Lim, N. T., and P. K. Ng. 2008. Home range, activity cycle and natal den usage of a female Sunda pangolin Manis javanica (Mammalia: Pholidota) in Singapore. *Endangered Species Research* 4(1–2): 233–240.

Mahmood, T., N. Irshad, and R. Hussain. 2014. Habitat preference and population estimates of Indian pangolin (Manis crassicaudata) in District Chakwal of Potohar Plateau, Pakistan. *Russian Journal of Ecology* 45(1): 70.

Merceron G., C. Blondel, M. Brunet, S. Sen, N. Solounias, L. Viriot, and E. Heintz. 2004. The Late Miocene paleoenvironment of Afghanistan as inferred from dental microwear in artiodactyls. *Palaeogeography, Palaeoclimatology, Palaeoecology* 207(1–2): 143–163.

Merceron G., E. Schulz, L. Kordos, and T. M. Kaiser. 2007. Paleoenvironment of *Dryopithecus brancoi* at Rudabánya, Hungary: Evidence from dental meso- and micro-wear analyses of large vegetarian mammals. *Journal of Human Evolution* 53(4): 331–349.

Pickford, M. 1976. An Upper Miocene Pholidote from Pakistan. *Pakistan Journal of Zoology* 8(1): 21–24.

Pickford, M. 1978. New evidence concerning the fossil Aardvarks (Mammalia, Tubulidentata) of Pakistan. *Tertiary Research* 2(1): 39–44.

Pickford, M. 2019. Orycteropodidae (Tubulidentata, Mammalia) from the Early Miocene of Napak, Uganda. *Münchner Geowissenschaftliche Abhandlungen Reihe A: Geologie und Paläontologie* 47:1–101.

Prater, S. H. 1980. *The Book of Indian Animals*, 3rd ed. Bombay: Bombay Natural History Society.

Roberts, T. J. 1997. *Mammals of Pakistan*. Karachi, Pakistan: Oxford University Press.

Rögl, F. 1999. Circum-Mediterranean Miocene Paleogeography. In *Land Mammals of Europe*, edited by G. Rössner, and K. Heissig, 39–48. Munich: Dr. Friedrich Pfeil Verlag.

Rose, K. D., R. J. Emry, T. J. Gaudin, and G. Storch. 2005. Xenarthra and Pholidota. In *The Rise of Placental Mammals: Origins and Relationships of the Major Extant Clades*, edited by K. D. Rose and J. D. Archibald, 106–126. Baltimore: Johns Hopkins University Press.

Sen S. 2013. Dispersal of African mammals in Eurasia during the Cenozoic: Ways and whys. *Geobios* 46(1–2): 159–172.

Storch, G., and T. Martin. 1994. *Eomanis krebsi*, ein neues Schuppentier aus dem Mittel-Eozän der Grube Messel bei Darmstadt (Mammalia: Pholidota). *Berliner geowissenschaftliche Abhandlungen Reihe E (Paläobiologie)* 13:83–97.

Taylor, A. 2013. *Orycteropus afer* aardvark. In *Mammals of Africa*, Vol. 1: *Introductory Chapters and Afrotheria*, edited by J. Kingdon, D. Happold, T. Butynski, M. Hoffmann, and M. Happold, 210–215. London, Bloomsbury.

Tekkaya, I. 1993. Türkiye fosil Orycteropodidae'leri. T. C. Kültür Bakanligi Anitlar ve Müzeler Genel Müdürlügü. *Arkeometri Sonuçlari Toplantisi* 1992:275–289.

Terhune C. E., T. J. Gaudin, S. Curran, and A. Petculescu. 2021. The youngest pangolin (Mammalia, Pholidota) from Europe. *Journal of Vertebrate Paleontology*. https://doi.org/10.1080/02724634.2021.1990075.

Van der Made, J. 2003. The aardvark from the Miocene hominoid locality Çandir, Turkey. *Courier Forschungsinstitut Senckenberg, Frankfurt* 240:133–147.

Von Koenigswald, W., and T. Martin. 1990. Ein Skelett von *Necromanis franconica*, einem Schuppentier (Pholidota, Mammalia) aus dem Aquitan von Saulcet im Allier-Becken (Frankreich). *Eclogae Geologicae Helvetiae* 83(3): 845–864

15

SIWALIK PROBOSCIDEA

PASCAL TASSY AND WILLIAM J. SANDERS

INTRODUCTION

The mammalian order Proboscidea extends in time from its origins in the Late Paleocene to the present day (Sanders et al. 2010). Proboscidean diversity culminated during the Early to Middle Miocene, although proboscidean assemblages from any particular time interval and region document relatively few taxa. Thus far, about 160 proboscidean species are known. Today, only two threatened species of elephant remain in Africa, the savanna elephant, *Loxodonta africana*, and the forest elephant, *L. cyclotis*, and one endangered species survives in Asia and South Asia, the Asian elephant, *Elephas maximus*. Proboscideans have been among the largest land animals for much of their existence, including during the time represented by the Lower and Middle Siwalik record (Miocene) of the Potwar Plateau. Miocene deinotheres, mammutids, stegodonts, and gomphotheres routinely ranged to heights of more than 3 m at the shoulder, with massive weights exceeding 3,000 kg, comparable to or exceeding extant elephants in size (Christiansen 2004; Larramendi 2016). Thus, despite their modest diversity in any fauna, proboscideans likely had major impacts on their ecosystems as major primary consumers, dominant shapers of the landscape, and substantial seed and fertilizer dispersers—thus, as "super keystone species," similar to the ecological role filled by extant elephants (Shoshani 1993).

The oldest proboscideans, from the Late Paleocene–Early Eocene of North Africa, are represented by the species *Eritherium azzouzorum* and *Phosphatherium escuilliei* (Gheerbrant et al. 2005; Gheerbrant, Bouya, and Amaghzaz 2012). The initial proboscidean succession in Africa subsequently gave rise to several lophodont and eventually bunolophodont taxa (Gheerbrant and Tassy 2009; Sanders et al. 2010; Sanders 2023). Among them, the Elephantiformes, represented by the genera *Palaeomastodon* and *Phiomia*, differentiated during the latest Eocene-Oligocene (Andrews 1906; Matsumoto 1922, 1924; Savage 1971; Wight 1980), while the possible forerunner of the Deinotheriidae, *Chilgatherium harrisi*, is known from the earliest Late Oligocene (Sanders, Kappelman, and Rasmussen 2004), and indisputable deinotheres were established in Africa by the end of the Oligocene (Rasmussen and Gutíerrez 2009; Abbate

et al. 2012). The earliest known elephantiform proboscideans are from the Middle Eocene of Togo, which suggests an early division of lophodont and bunolophodont proboscidean lineages (Hautier et al. 2021). An early dispersal of probable elephantiforms from Africa toward Asia occurred during the Late Oligocene to Early Miocene (Tassy 1989; Antoine et al. 2003). Later dispersals occurred during the Early and Middle Miocene due to Afro-Arabian-Eurasian land connections made possible by oscillations of the eastern closure of the Tethys Sea. These dispersals included deinotheres and other elephantiforms that belong to the Elephantimorpha—that is, Mammutida ("true mastodons") and Elephantida (the more diverse group of elephants, stegodonts, and various subfamilies of gomphotheres such as amebelodonts and choerolophodonts).

The Siwalik fossil record from the Potwar Plateau exemplifies the Miocene radiation of Elephantimorpha with several lineages of mammutids, choerolophodonts, amebelodonts, gomphotheriines, anancine gomphotheres, tetralophodontines, and stegodonts. Along with deinotheres, the Lower and Middle Siwalik record contains at least 12 lineages of proboscideans (see Tassy 1983a, 1983b, 1983c), which represent high Old World Neogene diversity for proboscideans. Multiple localities in the Potwar Plateau exhibit evidence for the sympatry of five or more proboscidean taxa—an extraordinary co-occurrence of megaherbivores that raises questions about the primary productivity and vegetation composition needed to sustain their viable populations, unlike that of any current ecosystem.

Miocene Siwalik proboscideans were described as early as 1836 (Cautley 1836), initially based on isolated molars in a typological approach. Falconer's (1857) gradistic concepts of trilophodonty and tetralophodonty introduced massive taxonomic confusion to the numerous species described from the Siwalik Hills. Osborn's (1936) first synthesis of proboscideans only added to the confusion. An initial taxonomic revision based on specimens recovered during fieldwork in the Potwar Plateau between 1973 and 1979 was published in 1983 (Tassy 1983a, 1983b, 1983c). Detailed synonymies of the currently recognized species can be found in these papers. The present chapter is an updated summary of the proboscidean taxa from the Lower and Middle Siwaliks (Kamlial to Dhok Pathan formations) of the Potwar Plateau, with remarks about body size, ecology, and biogeography.

TAXONOMY OF SIWALIK PROBOSCIDEA

Deinotheriidae Bonaparte, 1845
Prodeinotherium pentapotamiae (Lydekker, 1876)

Deinotheres occur through much of the Lower and Middle Siwalik sequence of the Potwar Plateau, except in latest Miocene sediments. They are absent from localities dated 7.4–5.4 Ma, which yield other proboscideans. Deinotheres are morphologically conservative, bilophodont (except for trilophed dP4/dp4-M1/m1) species with tapiroid, or chisel-like, molars. Their molars are interpreted as having been adapted for vertical, predominantly phase-I chewing, or the shear-compacting phase of chewing without a substantial accompanying phase-II lateral grinding phase (von Koenigswald 2016). Deinotheres are viewed as having occupied forested environments (Harris 1978). The termination of their stratigraphic distribution in the Potwar Plateau (and in the Indian subcontinent more generally) is probably due to the shift from woodland to open woodland and C_4 grassland beginning ~8.0–7.0 Ma and the absence of woodland in the latest Miocene, contrary to the pattern in western Europe during the same interval (but similar to conditions in Africa; Cerling, Harris, and Leakey 1999; Badgley et al. 2008; Chapter 24 this volume).

The small deinothere *Prodeinotherium pentapotamiae* is recognized from the early Miocene (YGSP 17458 from Locality Y590, 18.0 Ma, type Kamlial Formation) to the early late Miocene (YPM 19322 from Locality L34, 11.7 Ma, type Chinji Formation). As is characteristic for all members of the family, this species lacks upper tusks, has downturned lower tusks, and has a retracted nasal opening (suggesting the presence of at least a short stout trunk). Its cheek teeth are distinguishable from those of larger *Deinotherium* species by their smaller size and more extensive development of postcingulae(ids), particularly in m3 (Fig. 15.1a). All adult teeth were in use simultaneously—a pattern that differs from the horizontal tooth replacement typical of elephantimorph proboscideans (Sanders 2017). Reconstruction of body size for European representatives of *Prodeinotherium* indicate shoulder heights of 2.50–2.75 m and body masses of ~2,700 to greater than 4,000 kg (Larramendi 2016). Dental carbon-isotopic values for Siwalik deinotheres routinely lie at the low end of values for all proboscideans and suggest that they were dedicated browsers (Chapter 25, this volume). Microwear analysis of deinothere molars from the Siwaliks correspondingly shows that they have few pits and scratches, which is indicative of more leaf-browsing than fruit-browsing, similar to microwear of the extant tapir *Tapirus bairdii* (Semprebon et al. 2015).

Deinotherium indicum Falconer, 1845

The large deinothere *Deinotherium indicum* (Fig. 15.1b) occurs from the Middle Miocene (YPM 19374 from Locality L20, 13.1–13.5 Ma) to the Late Miocene (YGSP 16968 from Locality Y545, 8.1 Ma). Specimens are present in the Chinji, Nagri, and Dhok Pathan Formations. Members of the genus reached immense body sizes, exceeding 4 meters in height at the shoulder and estimated body masses well over 9,000 kg (Larramendi 2016).

The interrelationships of deinothere lineages are unresolved (Huttunen 2002). Relatively smaller-bodied species in Africa, Europe, and Asia are allocated to *Prodeinotherium*, while larger-bodied species in those regions are placed in *Deinotherium* (e.g., Harris 1978), which is likely to be polyphyletic. In the Potwar Plateau, the brief temporal overlap of *Prodeinotherium pentapotamiae* and *Deinotherium indicum* in the Late Miocene matches the situation in western Europe 2 million years earlier with *Prodeinotherium bavaricum* and *Deinotherium levius/giganteum* (Duranthon et al. 2007; Aiglstorfer et al. 2014). Small-bodied *Prodeinotherium sinensis* with a primitive p3 (the only deinothere from China) is dated to the Late Miocene (Qiu et al. 2007).

Mammutidae Hay, 1922
Zygolophodon metachinjiensis (Osborn, 1929)

Mammutids—or "true" mastodons—are rare in the Siwalik record of the Potwar Plateau. *Zygolophodon metachinjiensis* is confined to Middle Miocene strata of the type Chinji Formation between 13.8 Ma (AMNH 19414) and 12.4 Ma (YGSP 23963 from Locality Y647). It is represented mainly by isolated gnathodental specimens (Fig. 15.1c), including the holotype right dentary (AMNH 19414) with m2–3. The molars are of the robust morphotype with relatively massive zygodont crests. The two molar morphotypes, robust and gracile, are displayed by species allocated to *Zygolophodon* and *Mammut*, respectively. They represent the limits of what is judged to be significant continuous variation. The molars of *Z. metachinjiensis* are comparable to the robust morphotype of the contemporaneous European species, *Zygolophodon turicensis*. The same morphology characterizes Miocene zygodont molars from China, which are allocated to several species, and also to the African Early Miocene species *Z. aegyptensis* (Sanders and Miller 2002).

Because dental remains are the primary material available for comparative study of *Z. metachinjiensis* and other Miocene species and are more limiting for systematic study than crania, nothing can be said about speciation events in the history of the group. What can be determined is that Early, Middle, and Late Miocene Eurasian species differ from the early Miocene African species *Eozygodon morotoensis* (>20 Ma in age), which has more primitive molars (small, trilophodont M3s) (Pickford and Tassy 1980; Tassy 2019) and from the diminutive, even more primitive Late Oligocene to earliest Miocene African mammutid *Losodokodon losodokius* (from 27–24 to about 23 Ma), which has molars with weakly developed crescentoids and zygodont crests (Rasmussen and Gutíerrez 2009). Molar size of *Z. metachinjiensis* is generally larger than that of *Losodokodon*, *Eozygodon* and *Z. aegyptensis*, but its molars are indistinguishable in size and proportions from molars of Eurasian species of *Zygolophodon* (Sanders et al. 2010; Sanders 2023). A rough estimate of body size of *Z. metachinjiensis*, based on tooth-size

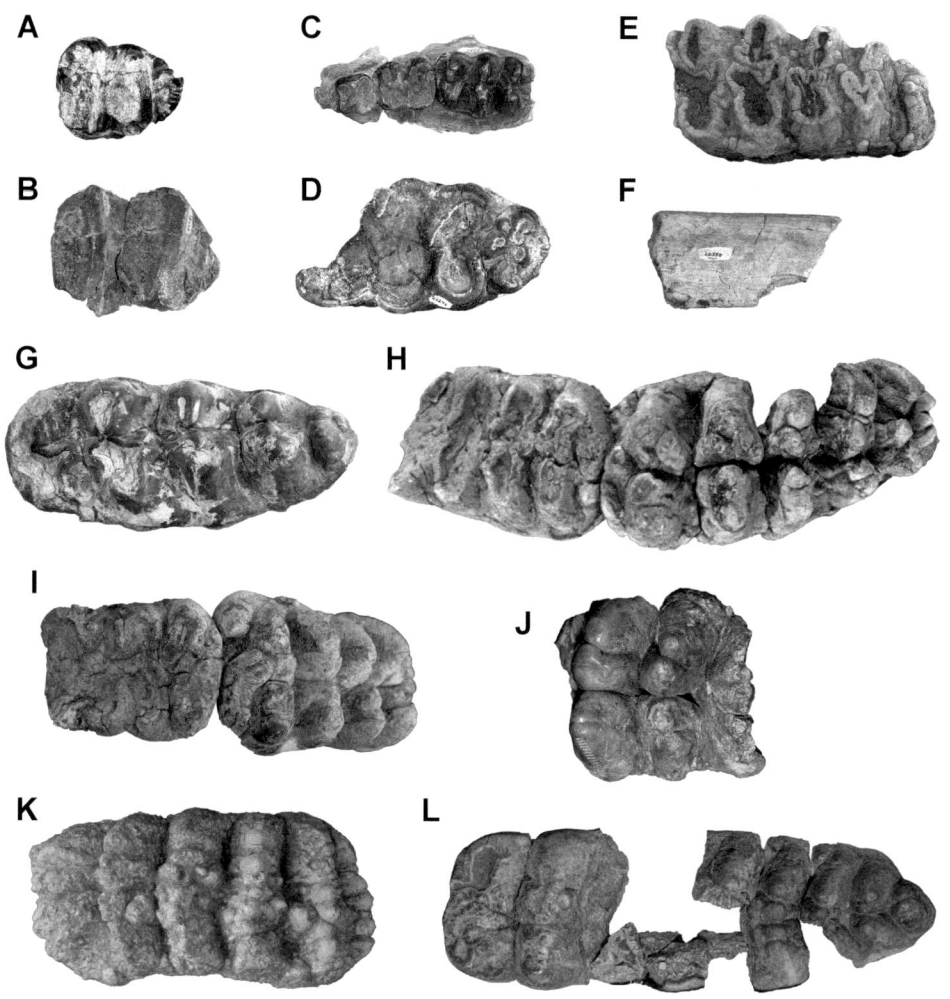

Figure 15.1. Teeth of diverse proboscidean taxa from the Siwalik record of Pakistan. Anterior is to the left, and all images are to the same scale. A. YGSP 42024, m3, *Prodeinotherium pentapotamiae*, Kamlial Formation; B. YGSP 42281, m3, *Deinotherium indicum*, Chinji Formation; C. YGSP 23963, dP2-4, *Zygolophodon metachinjiensis*; D. YGSP 42246, incomplete m3, *Choerolophodon corrugatus*, Chinji Formation; E. S 293, M3, *Protanancus chinjiensis*, Manchar Group; F. YGSP 20550, lower tusk fragment, *Protanancus chinjiensis*, Chinji Formation; G. YGSP 47216, m3, *Gomphotherium browni*, Chinji Formation; H. YGSP 15032, M2-3, *Paratetralophodon hasnotensis*; I. YGSP 41582, M1-2, *Stegolophodon* sp., cf. *S. progressus*; J. YGSP 50197, M3 fragment, *Stegolophodon* cf. *stegodontoides*, Dhok Pathan Formation; K. YGSP 50037, M3, *Stegolophodon maluvalensis*, Dhok Pathan Formation; L. YGSP 52892, m3, *Stegolophodon maluvalensis*, Dhok Pathan Formation.

comparison with *Mammut americanum*, suggests shoulder height between 2.5–3.0 m. Masticatory facets on zygolophodont molars in moderate wear are usually subvertical in orientation, indicating a strong phase-I vertical shearing component to chewing, which, by analogy with deinotheres, is consistent with leaf-browsing (von Koenigwsald 2016). The results of microwear study on Z. *metachinjiensis* molars support this interpretation (Semprebon et al. 2015).

Gomphotheriidae Hay, 1922

The term "gomphothere" has varied meanings for different researchers. Here, it refers to all taxa in the family Gomphotheriidae, which is comprised of choerolophodonts, amebelodonts, and gomphotheriines.

Choerolophodontinae Gaziry, 1976
Choerolophodon corrugatus (Pilgrim, 1913)

The most common species of trilophodont-grade gomphotheres in the Middle and Late Miocene of the Potwar Plateau is *Choerolophodon corrugatus*. The species occurs from 14.1 Ma (YGSP 47138 from Locality Y878) to 7.4 Ma (YPM 19716 from Locality L82) and is most common in Late Miocene localities

of the Dhok Pathan Formation. Choerolophodont molars (Fig. 15.1d) are variably characterized by wrinkled enamel (ptychoidy), the conical shape of cusps of loph(id)s that tend to be pillar-like, reorganization of mesoconelets of half-loph(id)s and pretrite conules to form a V-shape structure (chevroning), multiplication of secondary cusps, and projection of the mesoconelet of the pretrite half-loph(id) anterior to the posttrite half-loph(id). The morphological variation of choerolophodont molars, including those of *C. corrugatus*, is substantial, with simple and complex morphotypes. More complex molars exhibit stronger chevroning with more pronounced projection of pretrite half-loph(id) and greater multiplication and distribution of secondary cusps throughout the crown. Continuity of the lineage is warranted based on the pattern of morphological variation of these molars, which is maintained for more than 7 million years in the Potwar Plateau.

Choerolophodon corrugatus is a large choerolophodont, distinguished by numerous characters, including a cranium with a high facial region, orbits retracted behind the last functioning molar (facial area very elongated), high basicranium correlated with a tall mandibular ramus, and strong downward angulation of the mandibular symphysis (Tassy 1983b). Molars have abundant coronal cementum and exhibit considerable morphological variation. For example, molars with smooth enamel (a primitive feature) can be found in specimens from the Late Miocene Dhok Pathan Formation, while molars with wrinkled enamel (derived condition) are present in specimens from Middle to Late Miocene localities of the Chinji Formation. Molars of *C. corrugatus* are similar to those of contemporaneous eastern Mediterranean and Iranian *C. pentelici*, only slightly larger. The latter species also has a horizontal mandibular symphysis (possibly a paedomorphic feature). In addition, the ramus is more angled on the mandibular corpus in *C. corrugatus* than in *C. pentelici* (Abbas et al. 2018). In the Potwar Plateau, choerolophodont molars with a simple morphotype are present in the Kamlial Formation (e.g., YGSP 1770, Locality Y80, perhaps older than 15 Ma; see Abbas et al. 2018), which possibly extends the *C. corrugatus* lineage back in time or represents a similar, earlier species.

In Pakistan, the earliest choerolophodonts are found in the Early Miocene Bugti Beds of Baluchistan and belong to a small stem species of the entire group, "*Afrochoerodon kisumuensis/ palaeindicus*." In Africa, this species is Early to Middle Miocene in age and is known from sites as old as 19–18 Ma, such as Wadi Moghara, Egypt (Sanders and Miller 2002). This geographical distribution means that choerolophodonts are already present on the northern and southern shores of the Tethys Sea after the first eastern temporary closure, with plausible vicariant speciation during the Middle to Late Miocene. In this scenario, in the northern Tethys area, *C. pentelici* and *C. corrugatus* may have diverged in the West and East, respectively, from an early Miocene precursor (such as *C. chioticus* from Thymiana, Chios Island, Greece, or *C. guangheensis* from the Lingxian Basin, China [Wang and Deng 2011]), while *C. ngorora* evolved in Africa.

An alternative scenario involves a Middle Miocene dispersal of an African ancestor. According to Sanders et al. (2010) and Konidaris et al. (2016), *C. ngorora* can be separated into groups with (1) a simple molar morphology of Middle Miocene age and (2) a complex molar morphotype from the Late Miocene. The comprehensive phylogenetic analysis of all taxa included in the genus *Choerolophodon* by Konidaris et al. (2016; Fig. 10e) is not entirely conclusive. *Choerolophodon ngorora* (simple molar morphology) is the sister to an unresolved clade with a polytomy of *C. ngorora* (complex molar morphology), *C. anatolicus*, *C. pentelici*, and *C. corrugatus*, (*C. anatolicus* has the simple morphotype of *C. pentelici*.) From a chronological and biogeographical viewpoint, this result supports the hypothesis of a Middle Miocene African differentiation and subsequent northern dispersal (including to the Siwaliks) by *C. ngorora* (complex molar morphology).

Choerolophodon corrugatus was, with *C. pentelici*, the largest choerolophodont, being as tall as 2.8–3.0 m at the shoulder. Its molars were adapted for browsing or mixed feeding, with main and central conelets of half-loph(id)s of equal height and chevroning providing additional occlusal surface for phase-II lateral grinding of the compressed bolus. Enamel carbon isotope values for the species are most consistent with a diet dominated by browsing (Chapter 25, this volume).

Amebelodontinae Barbour, 1927
Protanancus chinjiensis (Pilgrim, 1913)

Amebelodonts are commonly known as "shovel-tusked" proboscideans because of the flattened, transversely broadened shape of their lower tusks (Fig. 15.1e; e.g., Osborn and Granger 1931), although the oldest species in the group, such as *Progomphotherium maraisi* from the Early Miocene of Africa, have rounded or modestly flattened lower tusks (Pickford 2003). *Protanancus chinjiensis* is the most common shovel-tusker in the Middle and Late Miocene of the Potwar Plateau. It ranges in age from 14.0 Ma (B 19398 from Locality B52) to 6.5 Ma (YGSP 50758 from Locality D13) in the Chinji and Dhok Pathan Formations, respectively. The mandible has a lengthened symphysis, and its molars have posttrite in addition to pretrite accessory conules, a pseudo-anancoid pattern (main conelets of half-loph[id] pairs offset transversely from one another), and wrinkled enamel. The pseudo-anancoid patten of the molars involves contact between pretrite and posttrite trefoils of subsequent loph(id)s, accentuated by strong pretrite accessory conules angled obliquely across the crown (Fig. 15.1f). Lower tusks are flat with a concentric dentinal structure, and the molar crowns belong to the bunolophodont trilophodont grade (which is plesiomorphic among amebelodonts).

Molars of *P. chinjiensis* are slightly more derived than those of *P. macinnesi* in Africa, whose temporal range extends earlier (Tassy 1986), and are clearly more derived than those of *P. tobieni* from China (Wang et al. 2015) and eastern European cf. *Protanancus* (Markov and Vergiev 2010). In fact, *P. tobieni* does not display the pseudo-anancoid pattern characteristic of the genus. *Protanancus chinjiensis* is possibly derived from a Middle Miocene dispersal of *P. macinnesi* or a species close to it. New finds from Buluk, Kenya, at ~17 Ma constitute the oldest African occurrence of *Protanancus*, which is characterized by five lophids in m3, pseudo-anancoidy, and salient development of accessory conules (Sanders 2023). Alternatively, stem amebelodontines (that is, *Archaeobelodon* with simple molars and a medium-sized mandibular symphysis) dispersed from Africa to Eurasia in the Early to Middle Miocene and could have provided the ancestry for South Asian amebelodonts (Tassy 1989). Moreover, the early Miocene Chinese brevirostrine species, *P. brevirostris*, with primitive non-pseudo-anancoid molars (Wang et al. 2015) could also be a stem species for both *P. macinnesi* and *P. chinjiensis*. This scenario is, however, unlikely as the African record of amebelodonts is older than 21 million years and evidence from Buluk suggests that at least African *Protanancus* was derived regionally from *Archaeobelodon* in the Early Miocene. The role of geographic isolation in the branching pattern of early amebelodontines is not fully resolved, and the relationships of the species placed in *Protanancus* remain to be better elucidated.

Based on the size of the best-preserved material (a complete mandible) and comparison with skeletons of *A. filholi* and *Platybelodon grangeri*, *P. chinjiensis* was at least 2.2–2.4 m high at the shoulder, although its molars are much larger than those of *Archaeobelodon* molars. The body mass of *A. filholi* has been estimated to be 2,350–3,500 kg (Christiansen 2004), and *P. chinjiensis* was almost certainly heavier than that. Dental carbon isotope values for *P. chinjiensis* are higher than those of *C. corrugatus* and indicate that the species was a mixed feeder (Chapter 25, this volume). Microwear analysis on a limited sample of individuals indicates that *P. chinjiensis* was a leaf-browser (Semprebon et al. 2015).

cf. '*Konobelodon*' or '*Torynobelodon*' sp.

A species close to *Konobelodon atticus* (Wagner, 1857) *sensu* Konidaris et al. (2014) is present in the type Dhok Pathan Formation of the Potwar Plateau at Locality Y199, dated to 8.7 Ma. It is represented by partial flat lower tusks with dentinal tubules (YGSP 4772). Both the cross-section and internal structure match the tusk morphology of *Konobelodon atticus* from Maragheh, Iran, and Pikermi, Greece (Konidaris et al. 2014), but not the tusks of *Platybelodon*. In the latter genus, lower tusks are more enlarged and flatter, the dorsal surface is hollowed, the peripheral layer of concentric dentine is thinner, and the internal tubules are more numerous.

The binomen *K. atticus* is problematic (see discussion in Tassy 2016), which poses a minor challenge. Whatever the correct nomen is for this taxon, it is a species of tetralophodont grade, among the most divergent and youngest amebelodontines known in the Old World. Isolated teeth of tetralophodont grade found throughout the late Miocene of the Siwalik Hills, India could belong either to this taxon or to a tetralophodont gomphotheriine. More material is needed to resolve the identification of the Potwar species. The relationship of the Siwalik cf. '*Konobelodon*' or '*Torynobelodon*' sp. with the North American genera *Konobelodon* and *Torynobelodon* (questionable synonyms) is also unclear.

Gomphotheriinae Hay, 1922
Gomphotherium browni (Osborn 1926)

Gomphotheriinae includes the genus *Gomphotherium*, which is something of a waste-basket taxon as it has been used to include all trilophodont gomphotheres with anterior and posterior accessory conules incorporated in enamel wear figures ("trefoils") of pretrite half-loph(id)s, usually without accompanying posttrite accessories, chevroning, pseudo-anancoidy, or flattened lower tusks. Within Elephantida, the subfamily represents taxa with relatively conservative cheek-tooth, incisor, and cranial morphology. The subfamily may extend at least to the early part of the late Oligocene, based on dental evidence from Chilga, Ethiopia (Kappelman et al. 2003; Sanders, Kappelman, and Rasmussen 2004), and depending on the phylogenetic interpretation of the Eritrean proboscidean, *Eritreum melakeghebrekristosi* (see Shoshani et al. 2006; Sanders et al. 2010). The best documented species of the subfamily is the European Middle Miocene *G. angustidens* (Tassy 2013), a nomen applied too broadly both geographically and temporally (e.g., Sanders and Miller 2002: "*G. angustidens libycum*" [Sanders,

2023]). Nonetheless, a gomphothere species in the Siwalik sequence clearly exhibits the conservative occlusal morphology usually associated with *Gomphotherium* and lacks the apomorphies of choerolophodonts and amebelodonts.

Gomphotherium browni is a Siwalik gomphotheriine of trilophodont grade restricted to the Middle to Late Miocene between 13.9 Ma (YGSP 42716 from Locality Y694) and 11.4 Ma (YPM 19405 from Locality L21)). Its bunolophodont molars are simple (Fig. 15.1g) with weak central accessory conules on the upper molars (plesiomorphic) associated with an enlarged and complex postcingulum (apomorphic). The enamel is especially thick, and the brachydonty of the molars is remarkable. The most salient derived feature is the short mandibular symphysis (brevirostry). The cross-section of the lower tusks is oval to circular. The downward curvature of upper tusks and their lateral enamel bands are primitive features. This combination of primitive and derived characters of *G. browni* makes this species distinctive among trilophodont gomphotheriines. Contemporaneous European gomphotheriines, such as *G. angustidens*, display more derived states for all features except their long mandibular symphysis. Molars exhibiting the derived morphotype of *G. angustidens sensu stricto* (Tassy 2014) have not been recovered in the Potwar Plateau.

Gomphotherium browni may be an endemic species from the Indian subcontinent. Its origin from stem gomphotheres such as *G. cooperi* could have occurred in South Asia, but this is hypothetical. *Gomphotherium cooperi* is known from the early Miocene Bugti Beds of Baluchistan. It belongs to the so-called *Gomphotherium annectens* group, consisting of species sharing the most primitive conditions of the cranio-mandibular and dental characters of gomphotheres. This "*annectens* group" is Early Miocene and has a broad geographic distribution, from Portugal to Japan on the northern side of the Tethys Sea and from northern to eastern Africa on the southern side. A direct connection between *G. cooperi* and *G. browni* would imply the existence of a ghost lineage from the Early to Middle Miocene, when *G. subtapiroideum* was present from Europe to China and *G. connexum* was present in China (Wang, Duangkrayom, and Yang 2015). The upper tusks and simple molar pattern of *G. subtapiroideum* are similar to those of *G. browni*, but the peculiar traits of the latter species (extreme brachydonty, thick enamel, and symphyseal brevirostry) are absent from *G. subtapiroideum*. The relatively brachydont *G. connexum* from the Early to Middle Miocene of China is a better candidate for ancestry of *G. browni*, although its mandible is not brevirostrine. As a consequence, an exclusive relationship between *G. cooperi* and *G. browni* cannot be firmly supported, and a close sister-group relationship between *G. connexum* and *G. browni* remains a tentative hypothesis.

Gomphotherium browni is a relatively medium-sized gomphotheriid, such as *G. productum*. Body weight estimates for other congenerics range from ~4,000 kg for *G. angustidens* to between 2,000 and 6,000 kg for *G. productum* (depending on which limb element is used for regression analysis; Christiansen 2004), and shoulder height is reconstructed as 2.5 m for *G. productum* (Larramendi 2015). Although *G. browni* molars are morphologically consistent with dedicated browsing rather than grazing, microwear analysis indicates that they were mixed feeders, probably including some C₃ grasses (Semprebon et al. 2015), as enamel carbon isotope values for the species do not show evidence for eating C₄ grasses (Chapter 25, this volume). In this regard, the species differs from *P. chinjiensis*, for which there is isotopic evidence for mixed feeding.

Subfamily incertae sedis
Paratetralophodon hasnotensis (Osborn, 1929)

Paratetralophodon hasnotensis is a gomphothere of tetralophodont grade, (Fig. 15.1h), known from the Dhok Pathan Formation of the Potwar Plateau between 8.7 Ma (YGSP 15032 from Locality Y407) and 7.4 Ma (YGSP 52892 from Locality Y943). It is characterized by an elevated cranium (high basicranium, facial region, and fronto-parietal vault) and relatively high-crowned tetralophodont molars with coronal cement and posttrite accessory conules variably developed. These traits are derived compared with those of the Late Miocene European species *Tetralophodon longirostris* and, in the case of the molars, are closer to *Tetralophodon* sp. from the Middle to Late Miocene of Africa (Nakaya et al. 1984, 1987; Tsujikawa 2005). *Paratetralophodon hasnotensis* also differs from the contemporaneous European taxon described from the Late Miocene (Turolian) of Spain, "*Tetralophodon* cf. *longirostris* 'grandincisivoid form'" by Mazo and Montoya (2003). The molars of this taxon exhibit plate-like loph(id)s with pre- and posttrite trefoils and an anancoid contact, a morphology clearly different from that of *P. hasnotensis*. The lower tusks are peg-type, with a piriform to round cross-section with concentric dentine.

Although the cranial anatomy of *P. hasnotensis* is well documented (Tassy 1983b), the mandible as well as the lower tusks are poorly known (see also Abbas et al. 2021). As a consequence, the relationship of *P. hasnotensis* to other gomphotheres is unresolved.

Tetralophodont gomphotheres are stem anancines and stem elephantids. *Anancus* is a brevirostrine genus lacking lower tusks. No "anancoid" trait is seen in the molars of *P. hasnotensis* (or in *Tetralophodon* sp. from Africa). Early elephants reach the pentalophodont grade. Tetralophodont molars can also be seen in the basal elephants *Stegotetrabelodon syrticus* (northern Africa) and *S. emiratus* (Arabian Peninsula), ~8–5 Ma. But the molar construction of *Stegotetrabelodon* is transformed toward a plate-like elephantine morphology (absent in *P. hasnotensis*; hence the inclusion of *Stegotetrabelodon* spp. but not *Paratetralophodon hasnotensis* in the Elephantidae), and its massive downturned mandibular symphysis possesses very long lower tusks (unknown in *P. hasnotensis*).

From the size and proportions of known crania, *P. hasnotensis* was approximately 3.5 m tall at the shoulders. Dental carbon isotope values for the species are similar to those for *C. corrugatus* and indicate that it was a browser (Chapter 25, this volume). Microwear data from molars of an indeterminate Siwalik tetralophodont species that could be *P. hasnotensis*

accord with the stable carbon-isotope findings (Semprebon et al. 2015).

Anancinae
Anancus perimensis (Falconer and Cautley 1846)

This species is a stem or primitive-group anancine (Tassy 1986; Metz-Muller 2000), based on its primitive molar morphology. *Anancus perimensis* has tetralophodont intermediate molars with five to six loph(id)s in its third molars, very weak anancoidy (alternating offset of pre- and posttrite half-loph(id)s), no enamel folding, and simple molar crown morphology (Metz-Muller 2000; Konidaris and Roussiakis 2018). Anancoidy is primarily established by contact between pre- and posttrite accessory conules and secondarily by offset of the main conelets of corresponding pre- and posttrite half-loph(id)s. Anancine gomphotheres generally are brevirostrine and lack lower incisors. A mandibular specimen attributed to *A. perimensis* from the Punjab of India lacks lower incisors and has a short symphysis with a narrow symphyseal gutter, whereas a second mandibular specimen from the Dhok Pathan Formation is broken anteriorly, revealing two small lower tusks, which is plesiomorphic for the subfamily (Metz-Muller 2000).

The species is represented in collections since the 2000s from the Potwar Plateau by only a single specimen (YGSP 52745 from Locality Y980 dated to 8.9 Ma) but is reported to occur in South Asia in the time interval 8.6–4.6 Ma (Kalb, Froehlich, and Bell 1996; Konidaris and Roussiakis 2018). This would make this species geologically older than the earliest African anancine gomphotheres from Lothagam and the Mpesida Beds, Kenya, and Toros-Menalla, Chad (see Sanders et al. 2010). Tassy (1986) hypothesized that African anancines were derived from *A. perimensis* dispersing into Africa from South Asia. This hypothesis remains viable, given the age of the species and the more primitive morphology of *A. perimensis* than in the oldest anancines of Africa.

Nothing has been established about the body size of *A. perimensis*. A single enamel carbon isotope value for the species indicates a browsing diet (Chapter 25, this volume).

Stegodontidae Osborn, 1918
Stegolophodon sp. (cf. *S. progressus* Osborn, 1929)

The family Stegodontidae comprises an archaic genus, *Stegolophodon* ("*S.*") and its more derived successor, *Stegodon* ("*St.*"). The oldest records of the family include *S. nasaiensis* from the Early Miocene of Thailand (Tassy et al. 1992). Although a Late Miocene dispersal to Africa is known, the family is predominantly Asian in geography and faunal associations (Coppens et al. 1978; Saegusa, Thasod, and Ratanasthien 2005). Thus, it is likely that the stegodontid taxa in the Siwaliks derive from Southeast Asia, probably recording a series of immigration events.

Stegodontids have been considered a sister taxon to tetralophodont gomphotheres and elephants (e.g., Tassy and Darlu 1986; Tassy 1990, 1996; Kalb and Mebrate 1993; Shoshani 1996; Shoshani, Golenberg, and Yang 1998). Alternatively, given the early divergence of stegodontids from the morphological pattern in gomphotheres, it is possible that stegodontids are the sister taxon to all gomphotheriids + elephants (Sanders et al. 2010). There is strong similarity of molar morphology (formation of plates) and cranial form between elephants and advanced stegodontids. Monophyly of the family has been supported by Saegusa (1987, 1996), Tassy (1990, 1996), Shoshani (1996), and Shoshani, Golenberg, and Yang (1998).

The genus *Stegolophodon* is present in the Middle Miocene Chinji Formation with a species allocated to "*Stegolophodon* sp./ *Stegolophodon progressus*" (Fig. 15.1i) recognized from 13.7 Ma (YGSP 20493 from Locality Y641) to perhaps 7.3 Ma (YPM 19844 from Locality L72). In the Chinji Formation (Localities Y76, B59), molars with nonlinear loph(id)s are less derived than those of *S. stegodontoides*. Previously undescribed molars from the Chinji and Nagri formations that date to 11.4–10.1 Ma are similar to *S. nasaiensis* molars in having a small number of lamellae, few mesoconelets per lamella, massively thick enamel (M2s with enamel thickness of 7.5–7.7 mm), very low lamellar frequency, posterior accessory conules associated with lamellae 1–2 in M2, and low crown height (Johnston and Sanders 2017). A juvenile cranium (AM 19446, the holotype of *S. progressus*, at the AMNH) has tusks with an enamel band, which is also a plesiomorphic feature. This Middle Miocene stegodontid may be ancestral to *S. stegodontoides*, but the relationship is not well supported. The interrelationships of the different species of the genus are unresolved, even when grouped according to the revision of stegodontids by Saegusa (1996) and Saegusa, Thasod, and Ratanasthien (2005).

On the basis of dimensions of the juvenile skull with dP3-M1 (AM 19446), adults of this species could have reached 3.0 m at the shoulder. Stegodontids were heavy-bodied proboscideans, probably heavier than extant elephants of the same height. Larramendi (2016) reconstructs *Stegodon zdanski* as being 3.87 m tall at the shoulder and ~12,000 kg in body mass, and the large, thick-enameled molars of *S. progressus* suggest that it was similarly large. Dental carbon isotope values of a few individuals of the species suggest that it was a dedicated browser (Chapter 25, this volume).

Stegolophodon stegodontoides (Pilgrim 1913)

Stegolophodon stegodontoides occurs in the Potwar Plateau between 9.4 Ma (YGSP 6868 from Locality Y227) and 7.3 Ma (YGSP 50197 from Locality Y933). It includes "*Stegolophodon* cf. *stegodontoides*" described by Tassy (1983c). If the fusion of the mesoconelets and anterior central conules (Fig. 15.1j) is a synapomorphy of Stegodontidae, then the fusion of the mesoconelets and posterior central conules is variable and not fully achieved in most teeth allocated to "*S.* cf. *stegodontoides*" compared to *S. stegodontoides* and likely represents intraspecific variation. As indicated above, stegolophodonts form the stem group of *Stegodon*, and the oldest stegolophodonts are known from the Early Miocene of Japan and Thailand (Tassy et al. 1992; Saegusa 2008). *Stegolophodon stegodontoides* is among the species that are close to the genus *Stegodon*. This species belongs to "*Stegolophodon* group 4" of Saegusa, Thasod, and

Ratanasthien (2005), a group that includes the lectotype of *S. cautleyi* from Perim Island, India. Based on new material from Thailand, Saegusa, Thasod, and Ratanasthien (2005) recognize significantly more intraspecific molar variability than previously thought. As a consequence, the appropriate name for the species from the Potwar Plateau could well be *S. cautleyi*.

Whatever its name, this species is not the sister species of *Stegodon*. Saegusa, Thasod, and Ratanasthien (2005) recognize "*Stegolophodon* group 6" as the stem of the genus *Stegodon*, a group that includes *S. maluvalensis* (see below; Sarwar 1977) from the Dhok Pathan Formation (initially considered an elephantid) together with *S. licenti* and *S. primitium* from China. *Stegolophodon stegodontoides* has large, very brachydont molars with thick enamel, low lamellar frequency, accessory conules as posterior as lophid 4, and a lower third molar formula of x6x (six lophids with pre- and postcingulids) (Johnston and Sanders 2017).

Stegolophodon stegodontoides was large, probably 3.5 m at the shoulder. Enamel carbon isotope values for a large sample from the Potwar Plateau indicate that it was a browsing to mixed-feeding proboscidean (Chapter 25, this volume).

Stegolophodon maluvalensis (Sarwar 1977)

This species occurs in the Potwar Plateau at Locality Y926 dated to 7.4 Ma (YGSP 50236) and is contemporaneous with the youngest recognized specimens of *S. stegodontoides*. As a member of "*Stegolophodon* group 6," *S. maluvalensis* is derived in the direction of the genus *Stegodon* (Saegusa, Thasod, and Ratanasthien 2005). Consequently, a lineage that includes *S. stegodontoides*-*S. maluvalensis*-*Stegodon* is plausible. Other candidates for the origin of *Stegodon* included in "*Stegolophodon* group 6" are the species *S. primitium* and *S. licenti* from southern and northern China, respectively. Thus, the lineage is likely a radiation, notwithstanding the fact that an unequivocal early *Stegodon* is present in East Africa between 7 and 6 Ma (Tassy 1994; Sanders 1999; Mackaye 2001), when only *Stegolophodon* occurs in the Potwar Plateau. The first appearance of *Stegodon* in the Potwar Plateau is documented between 5 and 4 Ma (Barry and Flynn 1989).

Molar morphology of this taxon is derived (Fig. 15.1k, l), with a proliferation of conelets in lamellae, accessory conules throughout the crown, and pentalophodonty of intermediate molars. Lower third molars may have as many as eight lamellae (lamellar formula of x8x) and exhibit an odd distribution of accessory conules on the lateral margin of transverse valleys (Johnston and Sanders 2017). The morphology observed in this species anticipates the crown structure of molars of *St. zhaotongensis* from the late Miocene of China.

DIVERSITY AND PALEOBIOLOGY

Proboscideans are well represented in the Lower and Middle Siwalik record of the Potwar Plateau. They include deinotheres and elephantimorphs comprised of mammutids, stegodonts, and a host of gomphotheres represented by choerolophodonts, gomphotheriines, amebelodonts, and anancines, as well as tetralophodonts. The diversity of taxa from any particular time interval in the South Asian Miocene is modest compared with some other mammalian groups but is extraordinary measured against the situation for extant proboscideans, in which three living elephant species live in isolation from one another. In the Siwaliks, conversely, it is not unusual to find three or more proboscidean taxa at the same locality and six or more contemporaneous species in Siwalik faunas (Fig. 15.2).

Body size of these species is similar to or exceeds that of extant elephants, which means that there were multiple sympatric proboscideans as well as other mammalian megaherbivores throughout the Siwalik sequence, representing an ecosystem of remarkable carrying capacity unlike any that we observe today. The co-occurrence of these proboscideans is even more impressive, considering that dental isotopic and microwear studies indicate a strong overlap of diets among contemporaneous proboscideans in Siwalik habitats, mostly involving leaf-browsing, with a few taxa (amebelodonts, gomphotheriines, and Dhok Pathan stegolophodonts) exploiting a mixed-feeding diet of browse and either C_3 or C_4 grasses. A shift to a strong component of grazing within mixed feeding, and to pure C_4 grazing, among proboscideans did not occur until 8 Ma or later (Quade et al. 1992; Cerling, Harris, and Leakey 1999; Semprebon et al. 2015). Ameen et al. (2022) correlate the incidence of enamel hypoplasy in Siwalik proboscidean lineages with environmental stress due to a more arid environment and finally with extinction (e.g., *Protanancus chinjiensis* ~7 Ma; *Gomphotherium browni* ~9 Ma, and *Choerolophodon corrugatus* ~3 Ma, the two latter dates being more recent than those recorded in this chapter).

The coexistence of multiple proboscidean taxa that characterized localities in the Potwar Plateau (and elsewhere in South Asia) was also common in other Old World Miocene ecosystems. For example, new evidence from the late Early Miocene site of Buluk, Kenya, shows sympatry of six proboscidean genera representing deinotheres, mammutids, and gomphotheres.

The biogeographic origin of most Siwalik proboscideans is problematic. Although most of the taxa represented had their earliest occurrence in Africa, except for the Asian stegodontids, it is not clear if they arrived in South Asia circuitously via either Europe or Asia. Examination of other and earlier collections from the Indian Siwaliks indicates that diversity estimates based on proboscidean assemblages from the Potwar Plateau are almost certainly low. (They represent but the tip of the diversity iceberg.) However, these older collections generally lack adequate stratigraphic information, which makes incorporating them into these discussions of the Potwar Plateau taxa difficult. Yet names, emended diagnoses, and hypodigms are related to these early collections, specimens and their taxonomic burden. They require updated systematic study.

ACKNOWLEDGMENTS

We are grateful to David Pilbeam, John Barry, Michèle Morgan, and Catherine Badgley for the invitation to participate in this project. WJS thanks Catherine Badgley for generous

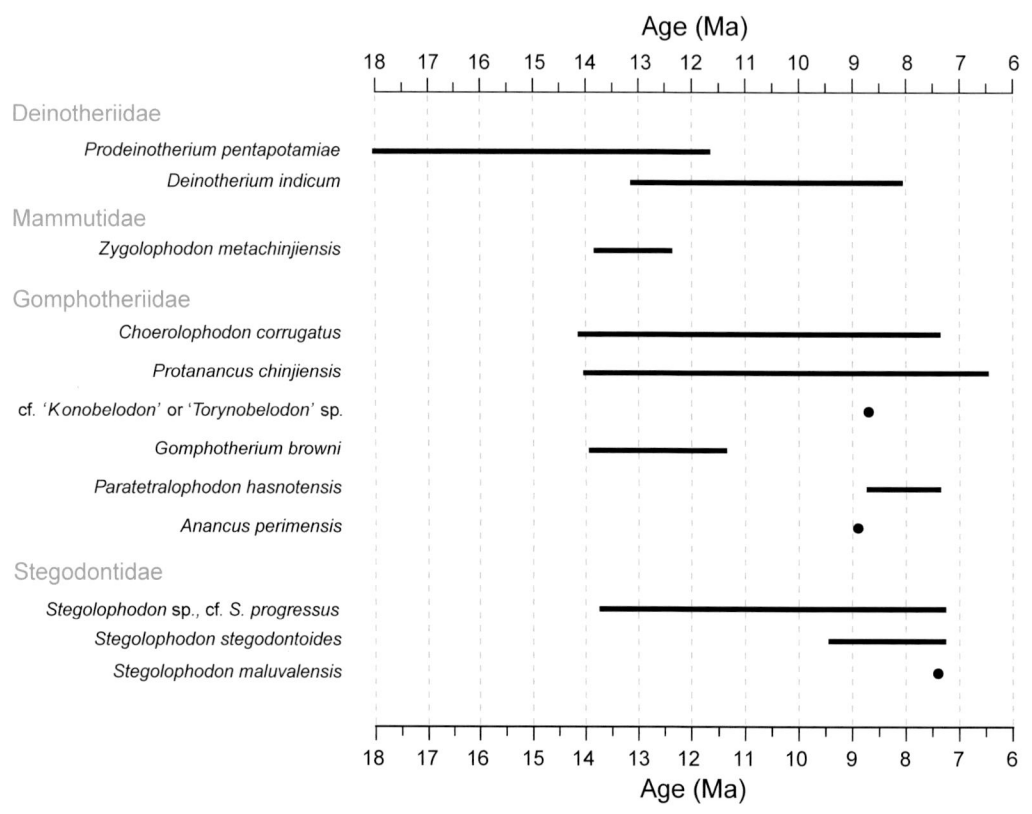

Figure 15.2. Biostratigraphic range of Siwalik proboscideans.

support to study comparative collections housed at the Natural History Museum, London. PT thanks Yves Laurent for his goodwill. Pip Brewer, Adrian Lister (Natural History Museum, London), Emma Mbua, Meave Leakey, Fredrick Manthi, Mary Muungu, and Rose Nyaboke (National Museums of Kenya, Nairobi) provided access to proboscidean collections in their care. In addition, we thank Carol Abraczinskas for expert preparation of Figure 15.1, Gina Semprebon for providing results of microwear analysis of molar specimens, Natalie Laudicina for database assistance, and Michèle Morgan for results of dental carbon isotope analyses.

References

Abbas, S. G., M. A. Babar, M. A. Khan, B. U. Nisa, M. K. Nawaz, and M. Akhtar. 2021. New proboscidean material from the Siwalik Group of Pakistan with remarks on some species. *PaleoBios* 38(1) ucmp_paleobios_54113. https://doi.org/10.5070/P9381054113.

Abbas, S. G., M. A. Khan, M. A. Babar, M. Hanif, and M. Akhtar. 2018. New materials of *Choerolophodon* (Proboscidea) from Dhok Pathan Formation of Siwaliks, Pakistan. *Vertebrata Palasiatica* 56(4): 295–305.

Abbate, E., P. Bruni, A. Coppa, D. Aria, M. P. Ferretti, Y. Libsekal, L. Rook, and M. Sagri. 2012. A new Oligo-Miocene mammal-bearing site from a sedimentary intercalation in the Trap basalts of central Eritrea. *Rivista Italiana di Paleontologia e Stratigrafia* 118:545–550.

Aiglstorfer, M., U. B. Göhlich, M. Böhme, and M. Gross. 2014. A partial skeleton of *Deinotherium* (Proboscidea, Mammalia) from the late Middle Miocene Gratkorn locality (Austria). *Paleobiology and Palaeoenvironment* 94:49–70.

Ameen, M., A. Majid Khan, R. Manzoor Ahmad, M. Ulmar Ijaz, and M. Imran. 2022. Tooth marker of ecological abnormality: The interpretation of stress in extinct mega herbivores (proboscideans) of the Siwaliks of Pakistan. *Ecology and Evolution* 12: e9432. https://doi.org/10.1002/ece3.9432.

Andrews, C. W. 1906. *A Descriptive Catalogue of the Tertiary Vertebrata of the Fayum, Egypt.* London: British Museum of Natural History.

Antoine, P.-O., J.-L. Welcomme, L. Marivaux, I. Baloch, M. Bennami, and P. Tassy. 2003. First record of Paleogene Elephantoidea (Mammalia, Proboscidea) from the Bugti Hills of Pakistan. *Journal of Vertebrate Paleontology* 23(4): 978–981.

Badgley, C., J. C. Barry, M. E. Morgan, S. V. Nelson, A. K. Behrensmeyer, T. E. Cerling, and D. Pilbeam. 2008. Ecological changes in Miocene mammalian record show impact of prolonged climatic forcing. *Proceedings of the National Academy of Sciences, USA* 105 (34): 12145–12149.

Barry, J. C., and L. J. Flynn. 1989. Key stratigraphic events in the Siwalik sequence. In *European Neogene Mammal Chronology*, edited by E. H. Lindsay, V. Fahlbusch, and P. Mein, 557–571. New York: Plenum Press.

Cautley, P. T. 1836. Note on the teeth of the Mastodon à dents étroite of the Siwalik Hills. *Journal of the Asiatic Society of Bengal* 5:294–296.

Cerling, T. E., J. M. Harris, and M. G. Leakey. 1999. Browsing and grazing in elephants: The isotope record of modern and fossil proboscideans. *Oecologia* 120:364–374.

Christiansen, P. 2004. Body size in proboscideans, with notes on elephant metabolism. *Zoological Journal of the Linnean Society* 140:523–549.

Coppens, Y., V. J. Maglio, C. T. Madden, and M. Beden. 1978. Proboscidea. In *Evolution of African Mammals*, edited by V. J. Maglio and H. B. S. Cooke, 336–367. Cambridge, MA: Harvard University Press.

Duranthon, F., P. O. Antoine, D. Laffont, and M. Bilotte. 2007. Contemporanéité de *Prodeinotherium* et *Deinotherium* (Mammalia, Proboscidea)

à Castelnau-Magnoac (Hautes Pyrénées, France). *Revue de Paléobiologie* 26(2): 403–411.

Falconer, H. 1857. On the species of Mastodon and Elephant occurring in the fossil state in Great Britain. *Quarterly Journal of the Geological Society of London* 13:307–360.

Gheerbrant, E., B. Bouya, and M. Amaghzaz. 2012. Dental and cranial anatomy of *Eritherium azzouzorum* from the Paleocene of Morocco, earliest known proboscidean mammal. *Palaeontographica* A 297(5–6): 151–183.

Gheerbrant, E., J. Sudre, P. Tassy, M. Amaghzaz, B. Bouya, and M. Iarochène. 2005. Nouvelles données sur *Phosphatherium escuilliei* (Mammalia, Proboscidea) de l'Eocène inférieur du Maroc, apports à la phylogénie des Proboscidea et des ongulés lophodontes. *Geodiversitas* 27(2): 239–333.

Gheerbrant, E., and P. Tassy. 2009. L'origine et l'évolution des éléphants. *Comptes Rendus Palevol* 8:281–294.

Harris J. M. 1978. Deinotherioidea and Barytherioidea. In *Evolution of African Mammals*, edited by V. J. Maglio and H. B. S. Cooke, 315–332. Cambridge, MA: Harvard University Press.

Hautier, L., R. Tabuce, M. J. Mourlam, K. E. Kassegne, Y. Z. Amoudji, M. Orliac, F. Quillevere, A.-L. Charruault, A. L. C. Johnson, and G. Guinot. 2021. New Middle Eocene proboscidean from Togo illuminates the early evolution of the elephant-like dental pattern. *Proceedings of the Royal Society B* 288:20211439.

Huttunen, K. 2002. Systematics and anatomy of the European Deinotheriidae (Proboscidea, Mammalia). *Annalen des Naturhistorischen Museum in Wien* 103A:237–250.

Johnston, S., and W. J. Sanders. 2017. Age, affinity, and succession of stegodontid proboscideans from the Middle Miocene–Late Pliocene formations of the Siwalik Sequence in South Asia. *Journal of Vertebrate Paleontology, Program and Abstracts* 2017:138.

Kalb, J. E., D. J. Froehlich, and G. L. Bell. 1996. Phylogeny of African and Eurasian Elephantoidea of the late Neogene. In *The Proboscidea: Evolution and Palaeoecology of Elephants and Their Relatives*, edited by J. Shoshani and P. Tassy, 101–116. Oxford: Oxford University Press.

Kalb, J. E., and A. Mebrate. 1993. Fossil elephantoids from the hominid-bearing Awash Group, Middle Awash Valley, Afar Depression, Ethiopia. *Transactions of the American Philosophical Society* 83(1): 1–114.

Kappelman, J., D. T. Rasmussen, W. J. Sanders, M. Feseha, T. Bown, P. Copeland, J. Crabaugh, et al. 2003. Oligocene mammals from Ethiopia and faunal exchange between Afro-Arabia and Eurasia. *Nature* 426:549–552.

Konidaris, G. E., G. D. Koufos, D. S. Kostopoulos, and G. Merceron. 2016. Taxonomy, biostratigraphy and palaeoecology of *Choerolophodon* (Proboscidea, Mammalia) in the Miocene of SE Europe-SW Asia: Implications for phylogeny and biogeography. *Journal of Systematic Palaeontology* 14(1): 1–27.

Konidaris, G.E., and S. Roussiakis. 2018. The first record of *Anancus* (Mammalia, Proboscidea) in the late Miocene of Greece and reappraisal of the primitive anancines from Europe. *Journal of Vertebrate Paleontology* 38:6, e 1534118. https://doi.org/10.1080/02724634.2018.1534118.

Konidaris, G. E., S. Roussiakis, G. Theodorou, and G. Koufos. 2014. The Eurasian occurrence of the shovel-tusker *Konobelodon* (Mammalia, Proboscidea) as illuminated by its presence in the Late Miocene of Pikermi (Greece) *Journal of Vertebrate Paleontology* 34(6): 1437–1453. https://doi.org/10.1080/02724634.2014.873622.

Larramendi, A. 2016. Shoulder height, body mass, and shape of proboscideans. *Acta Palaeontologica Polonica* 61(3): 573–574.

Makaye, H. T. 2001. Les Proboscidiens du Mio-Pliocène du Tchad: Biodiversité, biochronologie, paléoecologie et paléobiogeographie. Unpublished PhD diss., Ingénierie Chimique Biologique et Géologique, Université de Poitiers.

Markov, G. N., and S. Vergiev S. 2010. First report of cf. *Protanancus* (Mammalia, Proboscidea, Amebelodontidae) from Europe. *Geodiversitas* 32(3): 493–500.

Matsumoto, H. 1922. Revision of *Palaeomastodon* and *Moeritherium*. *Palaeomastodon intermedius*, and *Phiomia osborni*, new species. *American Museum Novitates* 51:1–6.

Matsumoto, H. 1924. A revision of *Palaeomastodon* dividing it into two genera, and with descriptions of two new species. *Bulletin of the American Museum of Natural History* 50:1–58.

Mazo, A. V. and P. Montoya. 2003. Proboscidea (Mammalia) from the Upper Miocene of Crevillente (Alicante, Spain). *Scripta Geologica* 126:79–109.

Metz-Muller, F. 2000. La population *d'Anancus arvernensis* (Proboscidea, Mammalia) du Pliocène de Dorkovo (Bulgarie); Étude des modalités évolutives *d'Anancus arvernensis* et phylogénie du genre *Anancus*. Unpublished PhD diss., Museum National d'Histoire Naturelle, Paris.

Nakaya H., M. Pickford, Y. Nakano, and H. Ishida. 1984. The Late Miocene large mammal fauna from the Namurungule Formation, Samburu Hills, northern Kenya. *African Study Monographs* suppl. 2:87–131.

Nakaya, H., M. Pickford, K. Yasui, and Y. Nakano. 1987. Additional large mammalian fauna from the Namurungule Formation, Samburu Hills, northern Kenya. *African Study Monographs* suppl. 5:79–129.

Osborn, H. F. 1936. *Proboscidea Volume 1: Moeritherioidea, Deinotherioidea, Mastodontoidea*. New York: American Museum Press.

Osborn, H. F., and W. Granger. 1931. The shovel-tuskers, Amebelodontinae, of Central Asia. *American Museum Novitates* 470:1–12.

Pickford, M. 2003. New Proboscidea from the Miocene strata in the lower Orange River Valley, Namibia. *Memoir Geological Survey of Namibia* 19:207–256.

Pickford, M. and P. Tassy. 1980. A new species of *Zygolophodon* (Mammalia, Proboscidea) from the Miocene hominoid localities of Meswa Bridge and Moroto (East Africa). *Neues Jahrbuch für Geologie und Paläontologie, Monatshefte* 4:235–251.

Qiu, Z.-X., B.-Y. Wang, H. Li, T. Deng, and Y. Sun. 2007. First discovery of deinothere in China. *Vertebrata PalAsiatica* 45(4): 261–277.

Quade, J., T. E. Cerling, J. C. Barry, M. E. Morgan, D. R. Pilbeam, A. R. Chivas, J. A. Lee-Thorp, and N. J, van der Merwe. 1992. A 16-Ma record of paleodiet using carbon and oxygen isotopes in fossil teeth from Pakistan. *Chemical Geology (Isotope Geoscience Section)* 94(3): 183–192.

Rasmussen, D. T. and M. Gutierrez. 2009. A mammalian fauna from the late Oligocene of Northwestern Kenya. *Palaeontographica* A 288(1–3): 1–52.

Saegusa, H. 1987. Cranial morphology and phylogeny of the stegodonts. *The Compass* 64(4): 221–243.

Saegusa ,H. 1996. Stegodontidae: Evolutionary relationships. In *The Proboscidea: Evolution and Palaeoecology of Elephants and Their Relatives*, edited by J. Shoshani and P. Tassy,178–190. Oxford: Oxford University Press.

Saegusa, H. 2008. Dwarf *Stegolophodon* from the Miocene of Japan: Passengers on sinking boats. *Quaternary International* 182(1): 49–62.

Saegusa, H., Y. Thasod, and B. Ratanasthien. 2005. Notes on Asian stegodontids. *Quaternary International* 126–128:31–48.

Sanders, W. J. 1999. Oldest record of *Stegodon* (Mammalia, Proboscidea). *Journal of Vertebrate Paleontology* 19(4): 793–797.

Sanders, W. J. 2017. Horizontal tooth displacement and premolar occurrence in elephants and other elephantiform proboscideans. *Historical Biology* 30(1–2): 137–156. https://doi.org/10.1080/08912963/2017.1297436.

Sanders, W. J. 2023. *Evolution and Fossil Record of African Proboscidea*. Boca Raton. FL: CRC Press.

Sanders, W. J., E. Gheerbrant, J. M. Harris, H. Saegusa, and C. Delmer. 2010. Proboscidea. In *Cenozoic Mammals of Africa*, edited by L. Werdelin and W. J. Sanders, 161–251. Berkeley: University of California Press.

Sanders, W.J., J. Kappelman, and D. T. Rasmussen. 2004. New large-bodied mammals from the late Oligocene site of Chilga, Ethiopia, *Acta Palaeontologica Polonica* 49(3): 365–392.

Sanders, W. J. and E. R. Miller. 2002. New proboscideans from the Early Miocene of Wadi Moghara, Egypt. *Journal of Vertebrate Paleontology* 22(2): 388–404.

Sarwar, M. 1977. Taxonomy and distribution of the Siwalik Proboscidea. *Bulletin of the Department of Zoology, University of the Punjab* 10:1–172.

Savage, R. J. G. 1971. Review of the fossil mammals of Libya. In *Symposium on the Geology of Libya*,edited by C. Gray, 217–225. Tripoli: Faculty of Science, University of Libya.

Semprebon, G. M., W. J. Sanders, A. Lister, M. E. Morgan, T. E. Cerling, G. Rivals, U. Göhlich, and J. M. Fahlke. 2015. Dietary reconstruction of fossil proboscideans from the Siwalik Series of Pakistan using enamel microwear and carbon isotopes. *Journal of Vertebrate Paleontology, Progam and Abstracts* 2015:211–212.

Shoshani, J. 1993. Elephants: The super keystone species. *Swara* 16:25–29.

Shoshani J. 1996. Para- or monophyly of the gomphotheres and their position within Proboscidea. In *The Proboscidea: Evolution and Palaeoecology of Elephants and Their Relatives*, edited by J. Shoshani and P. Tassy, 149–177. Oxford: Oxford University Press.

Shoshani, J., E. M. Golenberg, and H. Yang. 1998. Elephantidae phylogeny: Morphological versus molecular results. *Acta Theriologica* suppl. 5:89–122.

Shoshani, J., R. C. Walter, M. Abraha, S. Berhe, P. Tassy, W. J. Sanders, G. H. Marchant, Y. Libsekal, T. Ghirmai, and D. Zinner. 2006. A proboscidean from the late Oligocene of Eritrea; A "missing link" between early Elephantiformes and Elephantimorpha, and biogeographic implications. *Proceedings of the National Academy of Sciences, USA* 103(46): 17296–17301.

Tassy, P. 1983a. Les Elephantoidea Miocènes du Plateau du Potwar, Groupe de Siwalik, Pakistan. Ire partie: Introduction, cadre chronologique et géographique, mammutidés, amébélodontidés. *Annales de Paléontologie* 69(2): 99–136.

Tassy, P. 1983b. Les Elephantoidea Miocènes du Plateau du Potwar, Groupe de Siwalik, Pakistan. IIe partie: Choerolophodontes et gomphotheres. *Annales de Paléontologie* 69(3): 235–298.

Tassy, P. 1983c. Les Elephantoidea Miocènes du Plateau du Potwar, Groupe de Siwalik, Pakistan. IIIe partie: Stégodontidés, éléphantoïdes indéterminés. Restes postcrâniens. Conclusions. *Annales de Paléontologie* 69(4): 317–354.

Tassy, P. 1986. *Nouveaux Elephantoidea (Mammalia) dans le Miocène du Kenya; essai de réévaluation systématique*. Cahiers de Paléontologie—Travaux de paléontologie est-africaine. Paris: Editions du CNRS.

Tassy, P. 1989. The "Proboscidean Datum Event": How many proboscideans and how many events?. In *European Neogene Mammal Chronology*, edited by E. H. Linday, V. Fahlbusch, and P. Mein, 237–252. New York: Plenum Press.

Tassy, P. 1990. Phylogénie et classification des Proboscidea (Mammalia): Historique et actualité. *Annales de Paléontologie* 76:159–224.

Tassy, P. 1994. Les proboscidiens fossiles (Mammalia) du Rift occidental, Ouganda. In *Geology and Palaeobiology of the Albertine Rift Valley, Uganda-Zaire*, Vol. 2: *Paléobiologie/Palaeobiology*, edited by B. Senut, M. Pickford, and D. Hadoto, 215–255. Orléans: C.I.F.E.G., Occasional Publication 29 (1994).

Tassy, P. 1996. Who is who among the Proboscidea?. In *The Proboscidea: Evolution and Palaeoecology of Elephants and Their Relatives*, edited by J. Shoshani and P. Tassy, 39–48. Oxford: Oxford University Press.

Tassy, P. 2013. L'anatomie cranio-mandibulaire de *Gomphotherium angustidens* (Cuvier, 1817) (Proboscidea, Mammalia): Données issues du gisement d'en Péjouan (Miocène Moyen du Gers, France). *Geodiversitas* 35(2): 377–445.

Tassy, P. 2014. L'odontologie de *Gomphotherium angustidens* (CUVIER, 1817) (Proboscidea, Mammalia): Données issues du gisement d'En Péjouan (Miocène Moyen du Gers, France). *Geodiversitas* 36(1): 35–115.

Tassy, P. 2016. Proboscidea. Late Miocene mammal locality of Küçükçekmece, European Turkey. *Geodiversitas* 38(2): 261–271. https://doi .org/10.5252/g2016n2a7.

Tassy, P. 2019. Remarks on the cranium of *Eozygodon morotoensis* (Proboscidea, Mammalia) from the Early Miocene of Africa, and the question of the monophyly of Elephantimorpha. *Revue de Paléobiologie* 37(2): 593–607.

Tassy, P., P. Anupandhanat, L. Ginsburg, P. Mein, B. Ratanastien, and V. Sutteethorn. 1992. A new *Stegolophodon* (Proboscidea, Mammalia) in the Miocene of northern Thailand. *Géobios* 25:511–523.

Tassy, P., and P. Darlu. 1987. Les Elephantidae: Nouveau regard sur les analyses de parcimonie. *Geobios* 20(4): 487–494.

Tsujikawa, H. 2005. The updated Late Miocene large mammal fauna from Samburu Hills, northern Kenya. *African Study Monographs* suppl. 32:1–50.

Von Koenigswald, W. 2016. The diversity of mastication patterns in Neogene and Quaternary proboscideans. *Palaeontographica Abt. A Palaeontology-Stratigraphy* 307(1-6): 1–41.

Wang, S., and T. Deng. 2011 The first *Choerolophodon* (Proboscidea, Gomphotheriidae) skull from China. *Science China. Earth Sciences* 54:1326–1337.

Wang, S., T. Deng, T. Tang, G. Xie, Y. Zhang, and D. Wang. 2015 Evolution of *Protanancus* (Proboscidea, Mammalia) in East Asia. *Journal of Vertebrate Paleontology* 35(1): e881830. https://doi.org/10.1080 /02724634.2014.881830.

Wang, S., J. Duangkrayom, and X.-W. Yang 2015 Occurrence of the *Gomphotherium angustidens* group in China, based on a revision of *Gomphotherium connexum* (Hopwood, 1935) and *Gomphotherium shensiensis* Chang and Zhai, 1978: Continental correlation of *Gomphotherium* species across the Palearctic. *Paläontologische Zeitschrift* 89:1073–1086.

Wight, A. W. R. 1980. Paleogene vertebrate fauna and regressive sediments of Dur at Talhah, southern Sirt Basin, Libya. *The Geology of Libya*, edited by M. J. Salem and M. T. Busrewil, 1:309–325. London: Academic Press.

16

EQUIDAE FROM THE POTWAR PLATEAU, PAKISTAN

RAYMOND L. BERNOR, MICHÈLE E. MORGAN, SHERRY NELSON, AND GINA M. SEMPREBON

INTRODUCTION

Hussain (1971) revised Siwalik hipparionine horses based on fossil materials in Utrecht (Geological State Museum, Utrecht [GIU]), Munich (Bavarian State Museum, Munich [BSM]), London (Natural History Museum, London [NHM]), New York (American Museum of Natural History, New York [AMNH]), and New Haven, Connecticut (Yale University Peabody Museum [YPM]). Much of the previous work on Siwalik equids was intertwined with the developing understanding of long stratigraphic sections in the Potwar Plateau and northwestern India. The first studies of fossil equids from the Siwaliks were made by Owen (1846) based on material sent by Cautley to the NHM. Falconer and Cautley (1849) recognized two equid genera, *Hippotherium* and *Equus*. Hussain (1971, 10–12) has documented the long sequence of studies on Siwalik hipparions between 1849 (Falconer and Cautley) and 1968 (Forsten) that finally resulted in a seemingly endless parade of taxonomic hyperbole, which concluded with the recognition of two species (Matthew 1929; Colbert 1935; Gromova 1952): *"Hipparion" theobaldi* and *"Hipparion" antelopinum*. Nanda (1979) recognized these two equids from the Upper Siwalik Subgroup, Ambala.

Modern revisions of Siwalik hipparions include MacFadden and Woodburne (1982), Bernor and Hussain (1985), and Wolf, Bernor, and Hussain (2013). All of these authors took into account the historic Geological Survey of India (GSI) collections in Kolkata, which include several skulls of Indian Subcontinent hipparions. MacFadden and Woodburne (1982) recognized five hipparion taxa from the Potwar Plateau, Pakistan: *Hipparion antelopinum*, *"Hipparion" feddeni*, *Cormohipparion theobaldi*, *Cormohipparion* cf. *nagriensis* and *"Hipparion"* sp. Bernor and Hussain (1985) recognized a *"Cormohipparion"* (*Sivalhippus*) group with at least three species, *"C." (S.) theobaldi, "C." (S.) perimensis* and *"C." (S.)* sp. and a ?*Hipparion* group with a single species, ?*Hipparion antelopinum*. Wolf, Bernor, and Hussain (2013) utilized the Harvard-Geological Survey of Pakistan (Harvard-GSP) collections from the Potwar Plateau along with data from MacFadden and Woodburne (1982) and Bernor and Hussain (1985) to develop our current understanding of Pot-

war Plateau hipparion species diversity, stratigraphic ranges, and biogeography by integrating cranial, dental, and postcranial morphology into their systematics. As a result, we follow Wolf, Bernor, and Hussain (2013) in recognizing seven species of Potwar Plateau hipparion: *Cormohipparion* sp., *Cormohipparion* small sp., *Sivalhippus nagriensis*, *Sivalhippus perimensis*, *Sivalhippus theobaldi*, *Sivalhippus anwari*, and *Cremohipparion antelopinum*. The Potwar Plateau records the immigration and short-interval persistence of North American *Cormohipparion* species, endemic evolution of *Sivalhippus*, the immigration of *Cremohipparion* most plausibly from the west, and the biogeographic extension of *Sivalhippus nagriensis* and *Sivalhippus perimensis* into China and of *Sivalhippus theobaldi* and *Sivalhippus perimensis* into Africa (Wolf, Bernor, and Hussain. 2013; Bernor et al. 2010; Sun et al. 2018). The following summarizes the evolutionary history and paleobiology of Siwalik hipparions.

Note that Hipparion or "Hipparion" is an informal name for three-toed horses generally referred to as "hipparionines." *Hipparion* refers formally to a specific clade of hipparion as defined by Bernor et al. (2016). Here we utilize formal generic ranks of hipparion: *Cormohipparion, Hipparion* s.s., *Cremohipparion, Baryhipparion, Sivalhippus, Plesiohipparion, Proboscidipparion, Eurygnathohipparion*. See Bernor et al. (1996, 2016), Bernor and Sun (2015), and Bernor and Sen (2017) for updated definitions of these taxa.

POTWAR PLATEAU HIPPARION SPECIES AND THEIR CHRONOSTRATIGRAPHIC RANGES

Cormohipparion sp., 10.8–10.4 Ma.

The oldest hipparion specimens from the Potwar Plateau are YGSP 20151 and YGSP 20152, right M3 and left P2, respectively, dated to 10.8 Ma (survey block GB1). YGSP 20151 is an early wear M3, while YGSP 20152 (left P2) is virtually unworn and has a crown height of 44.2 mm. *Cormohipparion* sp. ranges in age from 10.8 to 10.4 Ma (Fig. 16.1). No cranial or postcranial

Age (Ma)

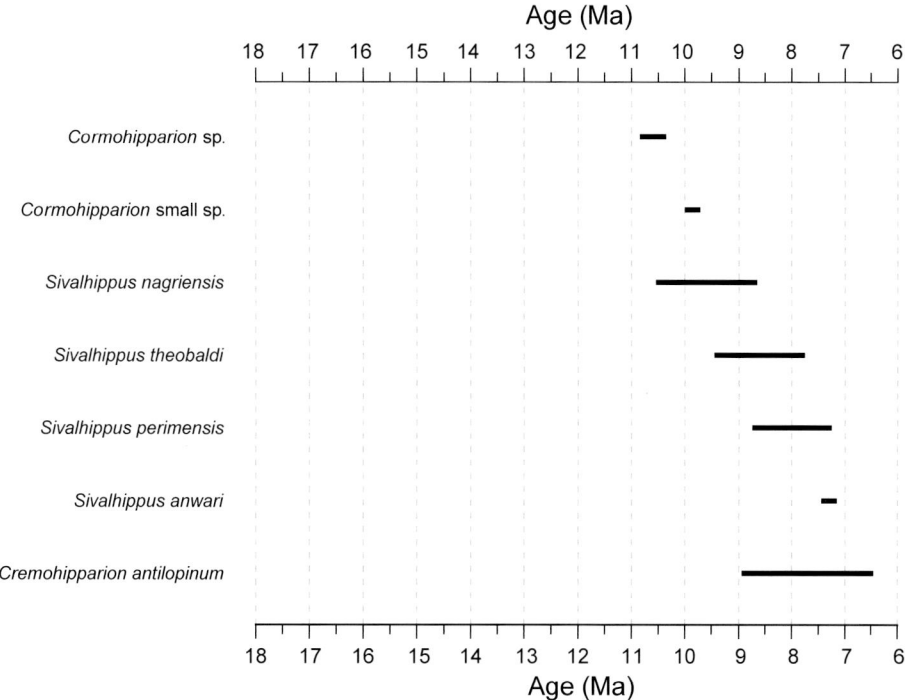

Figure 16.1. Biostratigraphic ranges of Siwalik hipparion taxa in the Potwar Plateau, Pakistan.

fragments are known. The greatest recorded crown height for Siwalik *Cormohipparion* sp. is YGSP 16620 (Locality Y572 dated to 10.4 Ma), a slightly worn left M3 with a crown height of 45.0 mm. Mandibular cheek teeth of *Cormohipparion* sp. are rare; have moderate to simple occlusal ornamentation, V-shaped linguaflexids, and primitive round metaconids and metastylids; and are relatively small like their maxillary counterparts. The relatively low crown height for Siwalik *Cormohipparion* sp. is similar to first-occurring *Cormohipparion sinapensis* from Sinap, Turkey (Bernor et al. 2003). Potwar Plateau *Cormohipparion* sp. differs from the local *Sivalhippus* species in that it is small and has relatively modest plication amplitude of the pre- and postfossettes (re: D 478 right P3 and P4 in middle wear, labeled as YGSP 478/76-25 in Wolf, Bernor, and Hussain, 2013, 26). To compare more closely with first occurring hipparion in Turkey and Central Europe (Bernor et al. 1988, 1997, 2003, 2017; Bernor, Boaz, and Rook 2012; Bernor, Cirilli, and Mittman 2022;) and Africa (Bernor and White 2009), more complete material would be needed, including crania, mandibles, and a more extensive series of cheek teeth and complete postcrania. For now, we remain conservative in recognizing this taxon as *Cormohipparion* sp. and distinct from *Sivalhippus nagriensis*, with which it overlaps in range.

Cormohipparion small sp.—10.0–9.8 Ma.

There are at least two specimens of this small and very enigmatic taxon: a distal metatarsal III (of the central digit; MTIII), YGSP 13556 (Locality Y357, 10.0 Ma; Wolf, Bernor, and Hussain 2013, Plate 4, Fig. 1) and a left P4, YGSP 16381 (survey block KL20, 9.8 Ma; Wolf, Bernor, and Hussain 2013, Plate 13, Fig. 8). The MTIII is the smallest one in the entire YGSP collection. The P4 is likewise very small with the following salient features:

moderately complex plication patterns of the fossettes; pli caballin is double; protocone is oval; hypoglyph is very deeply incised with crown height only 39.0 mm. There are two damaged but very small second central phalanges from Locality Y311 (10.1 Ma), YGSP 20271, and YGSP 11450 that may represent this taxon. Wolf, Bernor, and Hussain (2013) have remarked that very small hipparions such as Siwalik *Cormohipparion* small sp. are not known to occur elsewhere in Eurasia until the latest Miocene (MN13, 6.7 Ma). Perhaps, this species was derived directly from *Cormohipparion* sp. undergoing body size reduction.

Sivalhippus nagriensis, 10.5–9.0 Ma.

Wolf, Bernor, and Hussain (2013) reported *Sivalhippus nagriensis* from Units 1–2 of their informal Siwalik equid biostratigraphic zonation. *Sivalhippus nagriensis* overlaps chronologically with *Cormohipparion* sp. between 10.5 and 10.4 Ma. and continues until 9.0 Ma. The most complete specimen of *S. nagriensis* is YGSP 12507, an adult partial left maxilla with a large, dorsoventrally extensive preorbital fossa (POF) and P4 through M3 from Locality Y330 dated to 9.7 Ma (MacFadden and Woodburne1982, Fig. 13; Wolf, Bernor, and Hussain 2013, Plate 8, Fig. 5). The cheek teeth of *S. nagriensis* are quadrangular and have complexly plicated fossettes; pli caballin is double or multiple; protocone is oval and slightly flattened lingually; hypoglyph is very deeply incised. The hypodigm of *Sivalhippus nagriensis* supports the following diagnostic characters: medium-sized hipparion, POF large, dorsoventrally and posteriorly deep; P2 anterostyle elongate, maximum crown height 60–65 mm; maxillary cheek tooth fossettes highly ornamented with multiple pli caballins; oval protocones with slight lingual flattening; hypoglyphs deeply incised. Mandibular cheek teeth may primitively have deeply extending ectoflexids on p2 and p3; linguaflexids are V- to U- shaped; lack ectostylids on adult cheek teeth, but have them on juvenile cheek teeth; have rounded metaconids while metastylids may be squared and show some pointing lingually; pli caballinids vary from being an open loop to a closed ring. Calcanea, metapodial IIIs, and first phalanx IIIs (of the central digit, 1PHIII) plot within the Höwenegg 95% confidence ellipse indicating similar size. Wolf, Bernor, and Hussain. (2013, 31) emphasized that most of the earlier Siwalik material attributed to *Sivalhippus nagriensis,* including YGSP 12507, YGSP 46718, and YGSP 13803, could confidently be related to the holotype. Autapomorphic characters supporting the taxon *Sivalhippus nagriensis* include increased crown height, consistently lingually flattened protocone, double or triple pli caballins in maxillary molars, and large "blocky" morphology of the cheek teeth.

Sivalhippus theobaldi, 9.4–7.8 Ma.

The type specimen of *Sivalhippus theobaldi* is GSI C153, a juvenile right maxillary fragment with dP1 through 4 initially described by Lydekker (1877) from Keypar, Punjab. This specimen is very large and preserves a POF that extended ventrally deep on the face and cheek teeth that are in a very early stage-of-wear, not allowing characterization of occlusal features. The most complete cranial specimen of *Sivalhippus theobaldi* is AMNH 98728, a very large maxillary fragment with a huge, dorsoventrally extensive, medially deep POF with well delineated peripheral and anterior rim. AMNH 98728 has no specific stratigraphic provenance in that it was discovered by Barnum Brown's collecting team who found it serving as a door stop at the Dhok Pathan rest house (Morris Skinner, pers. comm.). That being said, most of the deposits around the Dhok Pathan resthouse are ~8 Ma. (John C. Barry, pers. comm. 2019). Crown height is not great due to the great dorso-ventral extension of the POF.

Sivalhippus theobaldi is relatively rare in the Potwar Plateau sequence, but readily identifiable by its large size and relatively low crown height: a left P3, YGSP 5089 (Locality Y207, 9.4 Ma), has a crown height of 63.4 mm (Wolf, Bernor, and Hussain 2013, Plate 11, Fig. 1). The greatest crown height recorded for *S. theobaldi* is also the youngest specimen, a right P4, YGSP 52633, which was estimated to be less than 67 mm in height (Locality Y960, 7.8 Ma; Wolf, Bernor, and Hussain 2013, Plate 11, Fig. 8). The largest cheek tooth, YGSP 5114, right M1 (Locality Y211, 9.3 Ma), has a crown height of only 58.1 mm. Crown height was certainly constrained by the dorsoventrally extensive POF. Salient morphological features of the maxillary cheek tooth occlusal surface include fossettes with complex plications, protocone oval, pli caballins weakly double, hypoglyphs very deeply incised.

Mandibular cheek teeth are difficult to discriminate from those of *S. nagriensis,* but are believed to have the following salient features: anterostylid is elongate and mesially pointed; pre- and postflexids have mostly simple margins; metaconids and metastylids are small, closely juxtaposed with some angularity; linguaflexids are mostly V-shaped; ectoflexid is moder-

ately deeply incised with a prominent pli (found also in African *Eurygnathohippus*), entoconid is pointed mesially.

Metatarsal IIIs are massive and represented by YGSP 17774 (Locality Y541, 8.0 Ma). Particularly striking in this specimen are the very elevated values for mid-shaft width and the proximal and distal width and depth dimensions (Wolf, Bernor, and Hussain 2013, Plate 5, Figs. 1, 8b). While later occurring *Sivalhippus anwari* is also very large, it has more slender metapodials overall than *S. theobaldi.* The oldest postcranial material identified for *S. theobaldi* is YGSP 6757 (Locality 221, 9.4 Ma; Wolf, Bernor, and Hussain 2013, Plate 4, Fig. 4). Wolf, Bernor, and Hussain (2013) concluded that *Sivalhippus theobaldi* was most likely derived from *S. nagriensis.* Relative to *S. nagriensis,* *S. theobaldi* had increased skull size and massivity of the distal limb elements, but remained conservative in facial and cheek tooth characters.

Sivalhippus perimensis, 8.7–7.3 Ma.

The type specimen of *Sivalhippus perimensis* is GSI C349, an adult skull with left P2 through M3 derived from the "Nagri Zone" of Perim (Piram) Island, India. The best specimen of *S. perimensis* is AMNH 19761, a complete male skull collected from near the Dhok Pathan rest house. *Sivalhippus perimensis* is derived in having a POF that is placed high and far anterior to the orbit, associated with an increase in maxillary cheek tooth crown height to 70 mm. P2 anterostyle is elongate; cheek teeth are complexly to very complexly plicated; pli caballins are usually double or complex; protocones are flattened lingually and rounded buccally; hypoglyphs are moderately to deeply incised. MacFadden and Woodburne (1982) and Bernor and Hussain (1985) referred a number of GSI specimens to *S. perimensis,* and Wolf, Bernor, and Hussain (2013) referred additional YGSP dental material to *S. perimensis.*

Wolf, Bernor, and Hussain (2013) found that mandibular cheek teeth in the Harvard-GSP collection between 8.7 and 7.8 Ma conform to the morphology and dimensions of a left mandible fragment with p3 through m3, AMNH 19855, collected four miles west of the Dhok Pathan rest house (~8.0 Ma). Mandibular cheek teeth have the following salient features: premolars have rounded to angular metaconid; metastylid is distally squared; linguaflexids have deep U-shape; preflexid and postflexid have simple borders; ectoflexid is deep, terminating short of the metaconid-metastylid valley; protoconid is circular; molars may have a pointed metastylid; linguaflexids have a broader U-shape.

Postcrania in the AMNH collection have been attributed to *S. perimensis* (Wolf, Bernor, and Hussain 2013). *Sivalhippus perimensis* postcrania are more massive than the Höwenegg *Hippotherium primigenium* sample, but are not as massive as *S. theobaldi* and not as large as *S. anwari. Sivalhippus perimensis* postcrania are most abundant between 8.5 and 7.5 Ma. The youngest postcrania of *S. perimensis* occur at 7.4 Ma, while the hardly worn P4 or M1 (YGSP 50131) that is best referred to *S. perimensis* is 7.3 Ma. Wolf, Bernor, and Hussain (2013) have argued that *Sivalhippus perimensis* likely evolved from *S. nagriensis* with the following character complexes being involved: reduction of POF dorso-ventral height and length, increased cheek tooth crown height, and increased body mass with more massive distal postcranial elements.

Sivalhippus anwari, 7.4–7.2 Ma.

The type specimen of *Sivalhippus anwari* is YGSP 50388 and is a late juvenile right maxillary cheek tooth series with P2 through M3 and partial left cheek tooth series with P4 through M3 (Wolf, Bernor, and Hussain 2013). Plausibly associated with the holotype are an incisor and a number of postcranial specimens including femur and tibia fragments, astragalus and calcaneum, proximal MTII, MTIII and MTIV in association with other MTIII shaft fragments and a 1PHIII.

Sivalhippus anwari is a large hipparionine horse with very high crowned cheek teeth, which achieve a maximum height of nearly 85 mm. P2 has a very elongate anterostyle; molars have complex fossette plications; pli caballins are double to complex; protocone is lingually flattened and buccally rounded; hypoglyph is deeply incised. Wolf, Bernor, and Hussain (2013) have referred a mandibular dentition, YGSP 16803 (from Locality Y581 dated to 7.2 Ma), which expresses a mosaic of primitive and advanced characters, to *Sivalhippus anwari*. The deeply incised premolar ectoflexids are plesiomorphic, commonly disappearing in Old World clades after the Early Vallesian (Bernor et al. 1997, 2017). The pointed metastylids and broad U-shaped linguaflexids are advanced features found in African *Eurygnathohippus* and Eurasian *Plesiohipparion* and *Proboscidipparion* (Bernor et al. 2010; Bernor and Sun 2015). The elongate P2 is a character found in *Proboscidipparion* (Bernor and Sun 2015).

Postcranials are large with robust MTIIIs being elongate. Metacarpal IIIs (of the central digit; MCIIIs) are proximally deep and very wide; 1PHIIIs are elongate and robustly built. The extremely high crown height of *S. anwari* likely predicts the absence of a POF. Wolf, Bernor, and Hussain (2013) have suggested that *S. anwari* occurs between 7.4 and 7.2 Ma. and was likely derived from *S. perimensis*.

Cremohipparion antelopinum, 8.9–6.5 Ma

The type specimen of *Cremohipparion antelopinum* is BMNH 2647, a right subadult maxilla fragment with P2 through M3. A medium sized juvenile left maxilla fragment, BMNH 2646 with dP2 through 4 and M1, and P2 exposed in its crypt is likewise referred to *Cr. antelopinum*. BMNH M16170 (MacFadden and Woodburne 1982, Fig. 5) is a poorly preserved adult maxillary fragment with P through M3 and left P3 through M3 with a crushed POF located high on the face. Bernor and Hussain (1985, Fig. 14) referred a skull fragment AMNH 19492 to *"Cormohipparion"* (*Sivalhippus*) sp., noting that the P2 through M3 compared closely in their size and morphology to the topotypic BMNH series of *Cr. antelopinum*. Colbert (1935) had originally referred AMNH 19492 to *Hipparion antelopinum*. MacFadden and Woodburne (1982, Fig. 17) assigned this specimen to *"Hipparion"* sp. Wolf, Bernor, and Hussain (2013) noted that while the POF is relatively shallow, it is dorso-ventrally extensive and has a short preorbital bar (POB) as is typical in species of

Cremohipparion. While the cheek teeth are more complex than the type specimen of *Cr. antelopinum*, AMNH 19492 was ultimately referred to *Cr. antelopinum* by Wolf, Bernor, and Hussain (2013). In addition, Hussain (1971:32) referred to a nearly complete but distorted subadult skull (BSM H689): "All the incisors and canines on both sides are present. Amongst the cheek teeth dP1 is present while M3 had not yet erupted. The right cheek teeth series shows unworn permanent P2, dP3, dP4, M1 and just worn M2. The left cheek teeth series contains a part of dP2, dP3, dP4, M1 and just worn M2." MacFadden and Woodburne (1982), Bernor and Hussain (1985), and Wolf, Bernor, and Hussain (2013) recognized this attribution.

Wolf, Bernor, and Hussain (2013) referred a number of smaller maxillary cheek teeth in the YGSP collection to *Cremohipparion antelopinum*. These ranged in age from 8.9 Ma to 6.5 Ma. Salient features of the maxillary cheek teeth are that medium sized, P2 anterostyle is short; cheek teeth have relatively simple to moderate fossette plication frequency; pli caballins are usually single to, at the most, double; protocone is rounded-oval; hypoglyph is moderately to deeply incised.

BMNH 2652 is a right mandible with p2 through m3 referred to *Hipparion antelopinum* by Falconer and Cautley (1849) and Lydekker (1886). MacFadden and Woodburne (1982) noted that the teeth of this specimen are relatively simple with no developed pli caballinids or stylids (no protostylid or ectostylid). The Harvard-GSP collection has some poorly preserved mandibular teeth that, based on their size, have been referred to *Cremohipparion antelopinum* (Wolf, Bernor, and Hussain 2013).

The postcrania of *Cr. antelopinum* are strongly divergent from Siwalik *Sivalhippus* spp. Metapodial IIIs are remarkable for having very slender midshaft dimensions like *Cormohipparion sinapensis* and other members of the *Cremohipparion* clade. MTIIIs from the BMNH (M16681, 17865) and the AMNH (19669) exhibit the elongate slender morphology of *Cr. antelopinum*.

Wolf, Bernor, and Hussain (2013) have referred Harvard-GSP material to *Cr. antelopinum* ranging in age from 8.9 to 6.5 Ma, with the most abundant material occurring between 8.7 and 7.8 Ma. Wolf, Bernor, and Hussain (2013) stated that the AMNH and BMNH material referred in their work likely ranged from 8.5 to 7.8 Ma. *Cremohipparion antelopinum* is a lineage distinct from the *Sivalhippus* lineage; the former likely immigrated into the Siwalik province from the west (Ukraine, Iran, Greece and likely Turkey). Work currently underway by Jukar and Bernor has established that not only the type specimen, but other material of *Cremohipparion antelopinum* is found in Tatrot Pliocene horizons of Northwest India.

BIOGEOGRAPHY

Hipparionine horses first immigrated into Eurasia and Africa near the base of the late Miocene. Berggren and Van Couvering (1974) established the concept of a *"Hipparion"* Datum with the assertion that a North American *"Hipparion"* made an instantaneous range extension across Eurasia and Africa. Van Couvering and Miller (1971) proposed that this "event" corresponded to

Lippolt, Gentner, and Wimmenauer's (1963) date of 12.5 Ma on volcanic sediments from the Höwenegg (Hegau, SW Germany) with abundant skeletal remains of *Hippotherium primigenium* (Bernor et al.1997). This date came into conflict with Opdyke et al.'s (1979) much younger age of 9.5 Ma for the first occurring *"Hipparion"* in the Potwar Plateau, Pakistan. Both dates have since been revised. Höwenegg is now interpreted as being 10.3 Ma (Swisher 1996), while current magnetostratigraphic correlation to the Gradstein et al. (2012) geomagnetic polarity timescale (GPTS) supports an age of 10.8 Ma for the earliest equids in the Potwar Plateau. Dates from the Valles Penedes, Spain, support a *"Hipparion"* Datum of 11.1 Ma., whereas in the Spanish Daroca section and in Turkey a *"Hipparion"* Datum of 10.8 Ma has been found. Woodburne and Swisher (1995) concluded in their review of the *"Hipparion"* Datum that a correlation of the *"Hipparion"* Datum with GPTS C5n.2n was the "consensus" Old World *"Hipparion"* Datum.

Woodburne (2009) and Bernor et al. (2017) studied what are arguably the oldest and most primitive *"Hipparion"* assemblages of the Old World, the Pannonian C assemblages from Atzelsdorf, Gaiselberg ,and Mariathal, Vienna Basin with an age between 11.4 Ma. and 11.0 Ma. Bernor et al. (2017) defined those characters of the maxillary and mandibular dentition that were primitive, corresponded to North American *Cormohipparion* sp. of Woodburne (2009) and proposed the taxonomic content of first appearing *"Hipparion"* in Eurasia and Africa. Bernor et al. (2017) presented data that support the early divergence of *Hippotherium* in central and western Europe and the occurrence of *Cormohipparion* in the oldest stratigraphic levels of the Potwar Plateau, Pakistan, Sinap, Turkey, and Bou Hanifia, Algeria. North American *Cormohipparion* sp. was the source of the Old World *"Hipparion"* Datum, and the *Cormohipparion* sp. from the Punchbowl Formation, California, exhibits the skull and cheek-tooth morphology most similar to the Pannonian C *Hippotherium* sp. (Woodburne 2009; Bernor et al. 2017). We follow Sun et al. (2018) in advocating the renaming of the *"Hipparion"* Datum as the *Cormohipparion* Datum based on these new data. This convention was followed by Bernor et al. (2021) and by Bernor, Cirilli, and Mittman (2022).

Cormohipparion sp. is the first occurring *"Hipparion"* in the Potwar Plateau (see Fig. 16.1), ranging from 10.8 to 10.4 Ma. (Wolf, Bernor, and Hussain 2013). It is the most plausible ancestor for the genus *Sivalhippus*. The *Sivalhippus* clade would appear to have originated on the Indian subcontinent and evolved there between 10.4 and 7.2 Ma. *Sivalhippus nagriensis* persisted between 10.4 and 9.3 Ma. and apparently diverged into two separate species, *Sivalhippus theobaldi* (9.4–7.8 Ma), a very large form with a huge POF and very robust limbs , and *Sivalhippus perimensis*, a large form with a small dorsally placed POF and robust limbs. Bernor et al. (2010) noted a very close morphological resemblance between *Sivalhippus theobaldi* and the Ugandan taxon *"Hipparion" macrodon*, referring it to *Sivalhippus macrodon* (Kakara Fm., Toro, Uganda; Eisenmann 1994). Bernor and Harris (2003) and Bernor et al. (2010) found that *Sivalhippus perimensis* showed a striking resemblance to Lotha-

gam Lower Nawata, Kenya,*"Hipparion" turkanense* in cranial and postcranial morphology; Wolf, Bernor, and Hussain (2013) referred the Lothagam form to *Sivalhippus turkanensis*. Bernor, Kaiser, and Wolf (2008) noted that a large hipparion 1PHIII from Sahabi, Libya, also conformed well with *Sivalhippus turkanensis*. Sun et al. (2018) identified two species of *Sivalhippus* from the Late Miocene of China: *S. platyodus* and *S. ptychodus*. In their study, Sun et al. (2018) determined that *S. platyodus* is likely derived from *S. nagriensis* and that *S. ptychodus* is closely related to *S. perimensis*. The China *Sivalhippus* species likely first appeared between 9 and 7 Ma. There is no evidence that the latest occurring Potwar Plateau *Sivalhippus*, *S. anwari*, occurred outside the Indian subcontinent.

Cremohipparion includes several species from Greece (*Cr. nikosi*, *Cr. matthewi*, *Cr. mediterraneum*, *Cr. proboscideum*), Italy and North Africa (*Cr. periafricanum*), Iran and Ukraine (*Cr. moldavicum* and *Cr. matthewi*), China (*Cr. forstenae* and *Cr. licenti*) and the Indian subcontinent (*Cr. antelopinum*; Bernor et al.1996; Wolf, Bernor, and Hussain 2013). Siwalik *Cremohipparion* is relatively rare and not well represented by critical skull material. Wolf, Bernor, and Hussain (2013) estimated that the earliest Potwar Plateau *Cremohipparion* occur at 8.9 Ma. The most likely geographic source for Siwalik *Cremohipparion* is from the west, from the Subparatethyan Province (Bernor 1983, 1984; Ratoi et al. 2022). The small form *Cormohipparion* small sp. is rare, enigmatic, and may be derived from local *Cormohipparion*.

PALEODIET

Mesowear

Wolf et al. (2012) and Wolf, Bernor, and Hussain (2013) studied dental mesowear on equid paracones (M2 preferred) using the cusp height and shape methods developed by Kaiser and Solounias (2003) and the 0–7 point scale of Mihlbachler et al. (2011). A total of 83 teeth were scored. These teeth are from sites dating between 10.8 and 7.2 Ma. Siwalik equid mesowear scores indicate a mixed-feeding diet and trended toward increased grazing and/or abrasion in geologically younger specimens. Additional equid teeth were analyzed subsequently, bringing the sample size to 97 with 75 identified to species (Table 16.1). The pattern remains the same with 97% of teeth scored as high and 93% as round. The two species with Milhbachler scores >3.2, *Sivalhippus perimensis* and *Sivalhippus anwari*, have first occurrences at 8.5 Ma and 7.4 Ma, respectively. All other species have average scores ranging from 2.8 to 3.0.

Microwear

The enamel microwear features of 32 equid molars (29 identified to species) were examined by Gina Semprebon using a light stereomicroscope and 35× magnification following the cleansing, molding, casting, and examination regime developed by Solounias and Semprebon (2002) and the screening protocol for taphonomic defects of King, Andrews, and Boz (1999). The number of pits (rounded features with approximately similar lengths and widths) and scratches (elongated features

Table 16.1. Equid Body Size Estimates, Mesowear, and Microwear

Taxa	Range in Potwar Plateau Siwaliks (Ma)	Body-size Estimate in kg (n)	Average Mesowear Score (n)	Microwear Average Scratches (n)	Microwear Average Pits (n)	Microwear Scratch/pit Ratio (n)	Microwear Mean Scratch Texture Score (n)
Cormohipparion sp.	10.8–10.4	120–200					
Cormohipparion small sp.	10.0–9.8	112 ± 6 (2)					
Sivalhippus nagriensis	10.5–8.7	247 ± 37 (34)	2.8 (45)	14.6 (24)	25.9 (24)	0.55 (24)	1.4 (24)
Sivalhippus theobaldi	9.4–7.8	440 (1)	3.0 (5)	19.5 (1)	23.5 (1)	0.83 (1)	2 (1)
Sivalhippus perimensis	8.7–7.3	274 ± 52 (11)	3.4 (16)	13.8 (3)	32.7 (3)	0.42 (3)	1.7 (3)
Sivalhippus anwari	7.4–7.2	427 ± 72 (3)	3.3 (2)*				
Cremohipparion antelopinum	8.9–6.5	239 ± 71 (2)	2.6 (7)	10 (1)	36.5 (1)	0.27 (1)	1 (1)

Note: *includes specimens attributed to *S.* cf. *anwari*.

with straight, parallel sides) were assessed following methods described in Solounias and Semprebon (2002). In Figure 16.2, scratch and pit values for taxa are compared to Gaussian confidence ellipses ($p = 0.95$) on the centroid for values of extant leaf-browsers and grazers from the comparative extant database (Solounias and Semprebon 2002). The largest samples obtained were of *Sivalhippus nagriensis*, which clearly shows a wide dispersion of scratch and pit values indicative of a mixed-feeding browse/grass dietary regime. *Sivalhippus perimensis* individuals are dispersed both within and outside of the browsing morphospace; thus, this species apparently had a varied diet (i.e., mixed feeding), while the single individual of *Sivalhippus theobaldi* falls just outside the grazing morphospace. The single individual of *Cremohipparion antelopinum* in our sample is typical of extant browsers.

Scratch textures were qualitatively scored as being either predominantly fine, coarse, or a mixture of fine and coarse types of textures per tooth surface following the criteria described in Solounias and Semprebon (2002) and Semprebon et al. (2004) to differentiate these textures. A scratch texture score was obtained by assigning a score of 0 to teeth with predominantly fine scratches per tooth surface, a 1 to those with a mixture of fine and coarse types of textures, and a 2 to those with predominantly coarse scratches per tooth surface. Individual scratch width scores for a sample were then averaged to obtain the mean scratch texture score for that taxon (Rivals and Solounias 2007). Average scratch and pit values for each taxon were used to calculate scratch/pit ratios.

Table 16.1 summarizes mean scratch texture scores and scratch/pit ratios by species. Although numbers are small, the scratch/pit ratio is lower in species appearing 8.6 Ma and younger than in earlier appearing species. Scratch texture scores correlate positively with mesowear scores, which indicates the likely exposure to more abrasive elements through time (i.e., coarser diet and/or exposure to more exogenous grit on dietary items).

Isotopes

More than 100 equid teeth have been sampled for stable carbon and oxygen isotopic analysis. These teeth span 4 million years (10.5–6.5 Ma) and include specimens identified to species

as well as more fragmentary fossils identifiable only as hipparion. First, we consider equid stable carbon isotope values throughout the Siwalik record (Figure 16.3a). Estimated average ingested stable carbon isotope values of equid diets range from −29 to −17.2‰, with values trending higher through geologic time. Here, an enrichment factor of 14‰ is added to equid $\delta^{13}C$ values following Cerling et al. (2009), and an additional atmospheric CO_2 $\delta^{13}C$ correction factor of 0 to 2.5‰ is added following Passey et al. (2009). Over several decades, Cerling and colleagues have amassed a large dataset of stable carbon isotope values for extant African large mammals (e.g., Cerling, Harris, and Passey, 2003; Cerling et al., 2015). In Chapter 5 (this volume), Cerling et al. use four isotopically defined dietary categories (see Figure 5.5): C_3-hyperforager (no measurable foraging on C_4 plants), predominantly C_3–foragers (<25% C_4 plants), C_3-C_4 mixed feeders (25–75% C_4 plants), and C_4–grazers (>75% C_4 plants). When we apply these isotopically defined categories to the Siwalik equids, we find that prior to 9.4 Ma some equids fall into the C_3–hyperforager category, which suggests feeding in forests (equid mesowear indicates mixed C_3 feeding). The first equid C_3–C_4 mixed feeders, meaning at least 25% of the diet from C_4 grasses, appear by 9.7 Ma although most equids remained predominantly browsers at that time. (One of us, Michèle Morgan, interprets the data to indicate the first detection of C_4 plant consumption by equids at 10.4 Ma, in the <25% C_4 plants diet category.) By 8.9 Ma, equids were consistently feeding on C_4 grasses in addition to C_3 plants. Through enamel apatite isotope values alone, we cannot distinguish whether the C_3 plant component included C_3 grasses or only browse. Some individuals placed in the predominantly C_3–foraging category until 7.4 Ma, when all sampled equids have >25% C_4 grass in their diets. Co-occurring species show nearly complete overlap with respect to $\delta^{13}C$ values. As can be determined isotopically, there were no dietary niche differences between sympatric species.

Given that equids yield carbon isotope values consistent with both predominantly browse and mixed feeding between 9.7–7.4 Ma, we next consider intra-tooth sampling of individuals to examine whether C_4 grasses may have been a seasonal component of diets. Intra-tooth sampling consisted of taking 10–23 samples from a single tooth, running the entire height

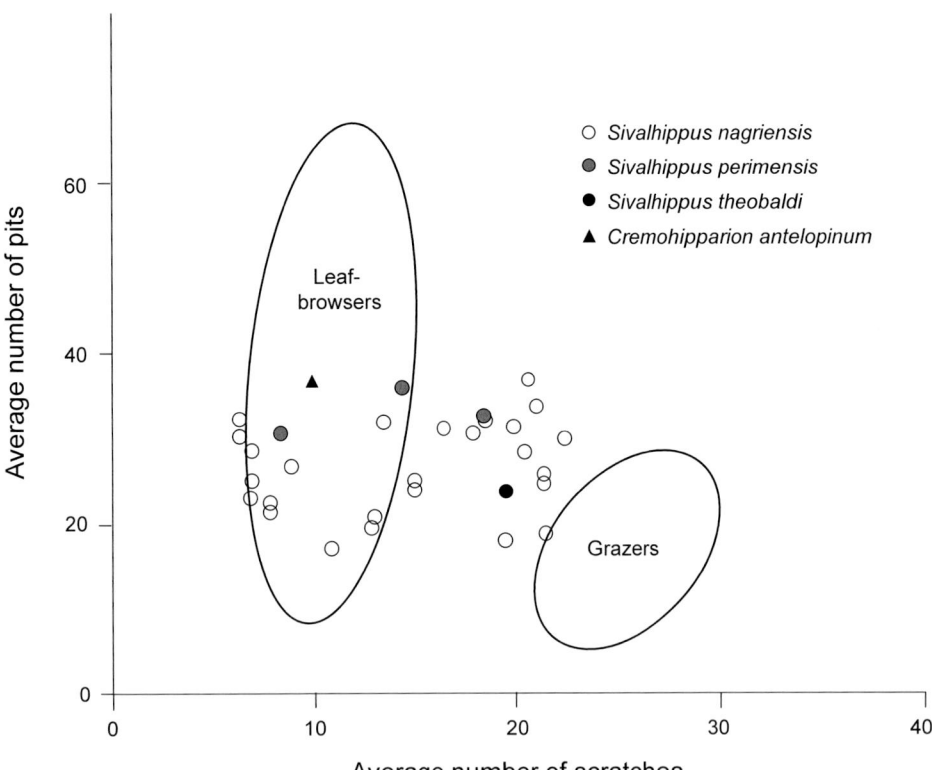

○ *Sivalhippus nagriensis*
● *Sivalhippus perimensis*
● *Sivalhippus theobaldi*
▲ *Cremohipparion antelopinum*

Figure 16.2. Microwear patterns of Siwalik hipparion taxa. Gaussian confidence ellipses ($p = 0.95$) on the centroid for extant leaf-browsers and grazers database (Solounias and Semprebon 2002).

of the crown from root to occlusal surface. Because hypsodont teeth take a year or more to develop, such intra-tooth sampling can detect intra-annual seasonal oscillations that the animal experienced during tooth growth (Bryant et al.1996; Fricke and O'Neil 1996; Sharp and Cerling 1998; Gadbury et al. 2000; Fox and Fisher 2001; Nelson 2005). We sampled 22 individuals for intra-tooth seasonal variability between 10.2 and 6.5 Ma (Figure 16.4). Eight teeth between 10.2–9.1 Ma yield diets of primarily browse year-round. Of the 14 teeth dated between 8.7–6.5 Ma, 3 yield diets of primarily browse year-round, 2 yield diets that are mostly mixed C_3-C_4 with some values that fall into the primarily browse category, and the remaining 9 teeth have at least 75% of samples in the mixed-feeder category. These results suggest that for the majority of individuals, once C_4 grasses became a component of their diet, they ingested them year-round.

Stable oxygen isotope data for Siwalik equids consistently display a considerable degree of variation consistent with obligate drinkers ingesting water from a variety of contexts, as well as from forage. Between 10 and 7 Ma, the oxygen isotope values shift towards more enriched values (see Fig. 16.3b) while the range remains fairly constant at about 8‰, similar to the annual variation in $\delta^{18}O$ values of precipitation in the region today. This isotopic shift supports gradual regional change in the average annual stable oxygen isotopic composition of rainfall in the Potwar Plateau as well as in the snowmelt derived from the western Himalayas, which contributes to Indus area catchments today and presumably also did so in the Late Miocene.

LOCOMOTION/SUBSTRATE USE

Hipparionine equids were terrestrial cursorial quadrupeds. We have noted differences in body size and degree of robustness among the Siwalik hipparionine taxa based on available postcranial material, and we infer that this morphological variation is associated with behavioral interspecies differences. The *Sivalhippus nagriensis* had a postcranial morphology similar to the Höwenegg population of *Hippotherium primigenium*, which has been assessed as being adapted to running and to springing and leaping in a subtropical forest-woodland context (Bernor et al. 1988, 1997). *Sivalhippus perimensis* was more heavily built, and *S. theobaldi* even more heavily built; the implication is that increased weight meant less springing ability. *Sivalhippus anwari* was a larger form with elongate but robustly built distal foot elements, but was overall similar to *S. perimensis* and likely capable of open country running. *Cremohipparion antelopinum* had elongated slender distal limb elements and was adapted for more sustained, open-country running.

BODY SIZE

Bernor, Wolf, and colleagues collected standardized measurements of postcranials following Eisenmann et al. (1988) and Bernor et al. (1997) where M is the abbreviation for the measurement number. Three measurements, 1PHIII length (measurement M1), 1PHIII proximal depth (M5) and MCIII distal minimum height (M13, distal minimal depth of lateral condyle) were used preferentially to estimate body mass as

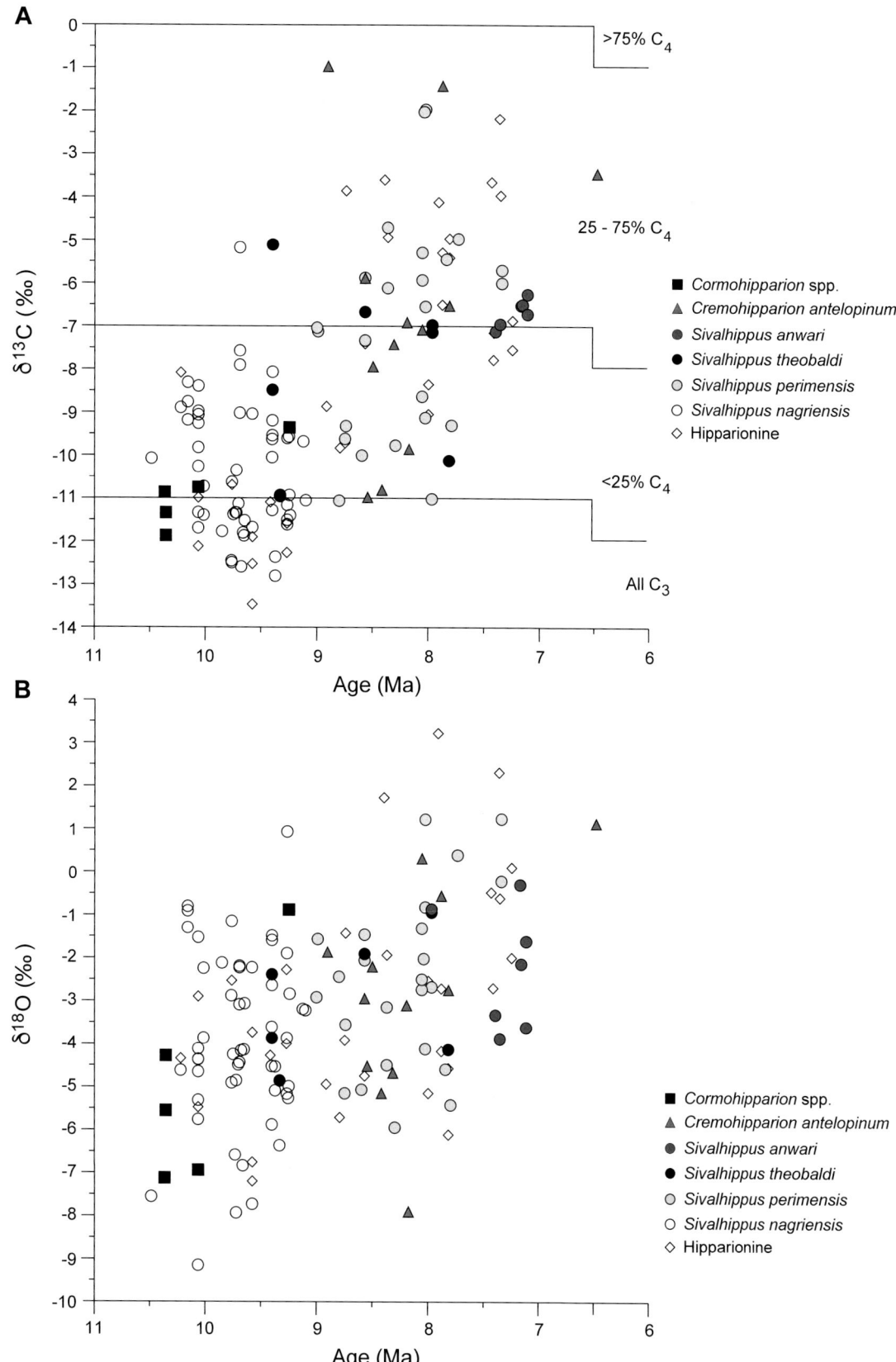

A

>75% C$_4$

25 - 75% C$_4$

■ *Cormohipparion* spp.
▲ *Cremohipparion antelopinum*
● *Sivalhippus anwari*
● *Sivalhippus theobaldi*
○ *Sivalhippus perimensis*
○ *Sivalhippus nagriensis*
◇ Hipparionine

<25% C$_4$

All C$_3$

B

■ *Cormohipparion* spp.
▲ *Cremohipparion antelopinum*
● *Sivalhippus anwari*
● *Sivalhippus theobaldi*
○ *Sivalhippus perimensis*
○ *Sivalhippus nagriensis*
◇ Hipparionine

Figure 16.3. A. Siwalik equid stable carbon isotope values from tooth enamel indicate increasing foraging on C$_4$ grasses through geologic time, beginning at 9.7 Ma.
B. Siwalik equid stable oxygen isotope values from tooth enamel are typical of obligate drinkers and indicate a slight shift to ingestion of heavier water (more positive δ^{18}O values) through time.

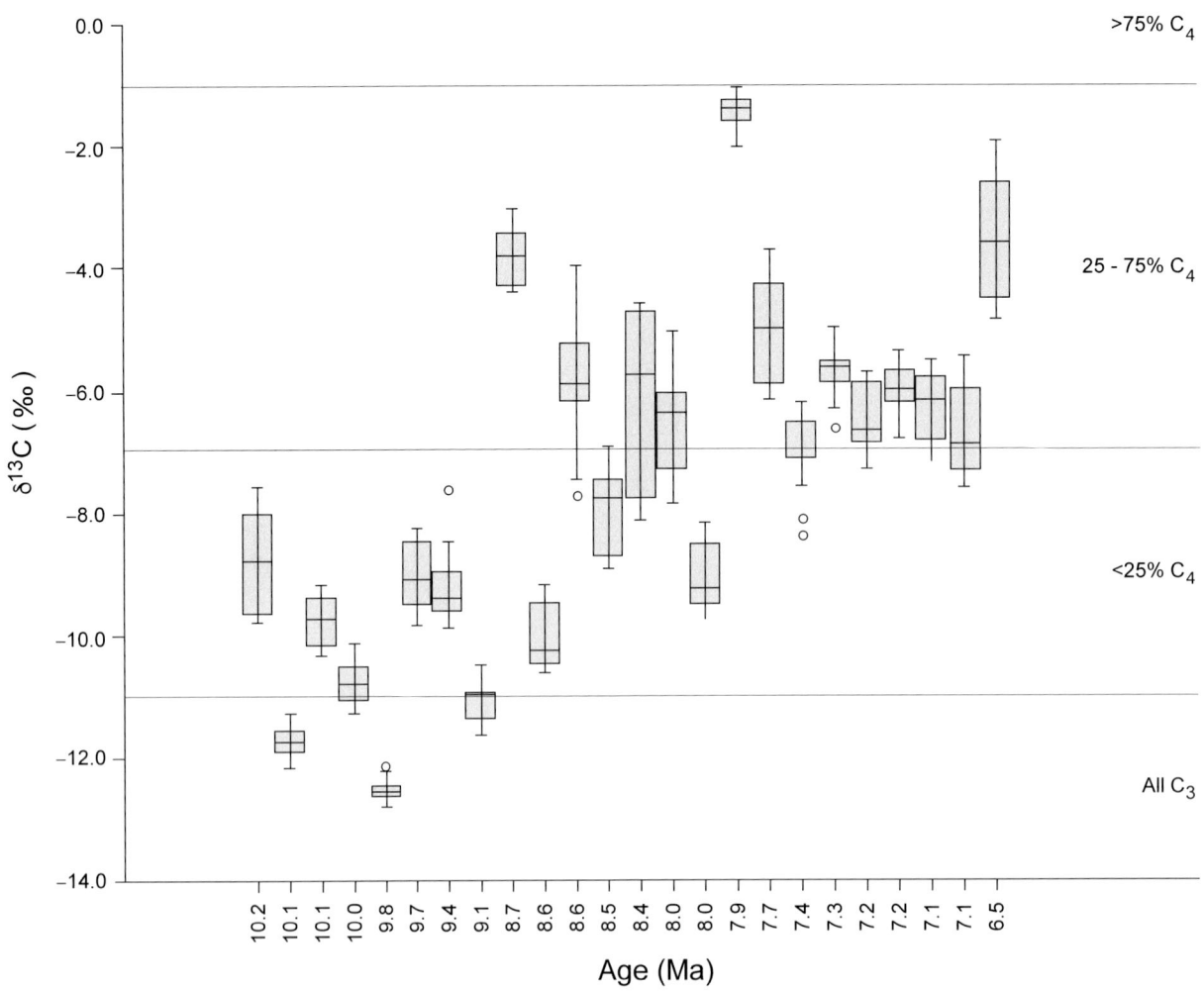

Figure 16.4. Siwalik equid intra-tooth sampling of stable carbon isotopes through geologic time. Each individual tooth is represented by 10–23 samples taken along the length of the crown from occlusal surface to root. For each tooth, the middle line represents the median value. The bottom of the box is the 25th percentile, while the top of the box is the 75th percentile. T-bars extend to 1.5 times the height of the box or to the minimum and maximum values. Points are outliers.

recommended by Alberdi et al. (1995). In addition, MTIII measurements M12 (distal maximal depth of mid-sagittal keel) and M13 (distal minimal depth of lateral condyle) were included to increase sample size. Regression equations used to estimate body mass for these measurements are from Eisenmann et al. (1988).

Body size estimates for Siwalik equids range from 110 kg to 500 kg ($n = 69$) with 90% falling between 140 and 350 kg. While there are no measured postcranial estimates for the first equid recorded in the Siwaliks, *Cormohipparion* sp., teeth and available postcranial elements are generally larger than specimens attributed to *Cormohipparion* small sp. and smaller than specimens of *Sivalhippus*. Measured postcranial elements for the small equid taxon, *Cormohipparion* small sp., yield size estimates of 110 to 120 kg (see Table 16.1). The most commonly preserved taxon, *Sivalhippus nagriensis*, generated size estimates between 175 kg and 350 kg and overlapped in size with *S. perimensis*, *Cremohipparion antelopinum*, and most specimens not identified to the species level. This size range is comparable to that of extant feral equid species except for the larger *Equus grevyi* (350–450 kg; Kingdon 1979; Rubenstein 2010; Faith et al. 2013). Two Siwalik equid species overlap in size with *E. grevyi*: *S. theobaldi* yields an estimate of 440 kg ($n = 1$), with largest likely exceeding 500 kg, and *S. anwari* is estimated at 330–500 kg ($n = 3$).

DIMORPHISM

Body size dimorphism is not pronounced in hipparionines, and has not been documented in the Siwalik species. Sexual dimorphism has been noted in similarly aged Höewenegg hipparionines, with males having large canines and females having small canines (and occasionally fetuses *in situs utero* providing unambiguous identification of sex; Bernor et al. 1997; Bernor, Cirilli, and Mittman 2022). The available sample of Siwalik hipparionine fossils is too limited to assess sexual dimorphism within species.

HABITAT PREFERENCE

Siwalik equids exploited a diversity of habitats present on the Potwar Plateau during the Late Miocene but likely required frequent access to fresh water in ponds or streams. They consumed a range of C_3 and C_4 vegetation, which implies foraging in both forested and more open habitats with diets ranging from pure browse to pure graze. For those hipparionine species with good sample sizes, there are no indications of dietary specialization, at least as can be inferred isotopically and via microwear analyses, although there is a trend toward a greater percentage of C_4 forage in the diet of equids through time. Dental wear analyses are consistent with this interpretation as features indicate increased abrasion in younger equid specimens (<8.5 Ma) as compared to older specimens (>8.5 Ma). The observed variation in $\delta^{13}C$ values from intra-tooth sampling does not match the expected pattern if there were a consistent annual seasonal shift in diet between C_3 and C_4 forage. Instead, intra-tooth sampling data suggest that equids that consumed C_4 forage did so year-round. The oxygen isotope data support categorization of Siwalik hipparionines as obligate drinkers.

CONCLUSION

Hipparionine horses first occurred in the Siwaliks at 10.8 Ma. Once they appeared, they were ubiquitous on the landscape, at least as reflected in the fossil record. (Thus, between 10.8 and 7.8 Ma, equids have been found at 75% of all localities in the Potwar Plateau.) *Cormohipparion* sp. was the initial species and persisted in the local record until 10.4 Ma (Wolf, Bernor, and Hussain 2013; Bernor et al. 2017). The Potwar Plateau hipparionine sequence documents the regional evolutionary radiation of the *Sivalhippus* clade, including *S. nagriensis* (10.5–9.0 Ma), *S. theobaldi* (9.4–7.8 Ma), *S. perimensis* (8.7–7.3 Ma), and *S. anwari* (7.4–7.2 Ma). Also documented is the apparent immigration of the elongate, slender-limbed form *Cremohipparion* into the region circa 8.9 Ma., perhaps continuing until 6.5 Ma or less. *Cormohipparion* small sp. was a rare and enigmatic small form with a brief duration, 10.0–9.8 Ma, and may have been an endemic species that evolved directly from *Cormohipparion* sp. Hipparionines were not sexually dimorphic in body mass (Bernor et al. 1997), were cursorial, and likely inhabited a range of habitats requiring frequent access to fresh water via ponds and streams. Paleodietary studies of Siwalik hipparion suggest that these equids were generalists and consumed a variety of graze and browse throughout the late Miocene. Stable carbon isotope analysis detects incorporation of C_4 grasses into hipparion diets by 9.7 Ma, or earlier depending on interpretation. In the Siwaliks, *Sivalhippus perimensis* spanned the temporal interval during which C_4 grasses became more dominant on the floodplains; stable carbon isotope values for this taxon range from entirely C_3 to entirely C_4 diets. Once C_4 grasses became part of the hipparion diet, they were apparently incorporated year-round, based on serial isotopic profiles of equid teeth. Mesowear and microwear studies are consistent with a mixed-feeding diet and increasingly more abrasive forage.

The *Sivalhippus* lineage extended its range into Africa and China in the Late Miocene. Bernor et al. (2010) and Wolf, Bernor, and Hussain (2013) argued that *S. theobaldi* likely extended its range into eastern Africa (Uganda, *S. macrodon*) ~9 Ma and that *S. perimensis* extended its range into northern and eastern Africa ~9–7 Ma. (*Sivalhippus perimensis*; Bernor and Harris 2003; Bernor et al. 2008). Sun et al. (2018) likewise identified two taxa of *Sivalhippus*—*S. platyodus* and *S. ptychodus*—which they argue were plausibly derived from *S. nagriensis* and *S. perimensis*, respectively. These Chinese taxa likely first occurred in the 9–7 my interval. *Cremohipparion antelopinum* most likely derived from a *Cremohipparion* species that immigrated from the west, where the genus was abundantly represented by multiple species.

ACKNOWLEDGMENTS

We appreciate the opportunity to contribute to this volume: our collective work has been reevaluated and refined over the past several decades. We thank the Geological Survey of Pakistan staff for their support, all who participated in fieldwork to build the hipparion horse collection, and John Barry who developed the hipparion biostratigraphy and geochronologic correlation which is the backbone of our study. John Barry was also instrumental in facilitating transfer of the hipparion collections to R. L. Bernor for study. We also wish to acknowledge the important systematic work conducted in the early 1980s by Bruce MacFadden and Michael O. Woodburne, and to thank Taseer Hussain who funded Bernor's NSF sponsored postdoctoral research fellowship at Howard University to study Siwalik hipparion collections in Kolkata, Munich, and New York. We thank Dominik Wolf for his earnest work on the Harvard-GSP Siwalik hipparion collections for this PhD dissertation at Howard University and the ensuing monograph, and we thank Boyang Sun for his comparative work on Siwalik and Chinese hipparions and Omar Cirilli for his statistical evaluation of Siwalik hipparions in comparison to other Eurasian and African hipparions. Finally, we acknowledge the more than 50-year support and encouragement of David Pilbeam.

References

Alberdi, M. T., J. L. Prado, and E. Ortiz-Jaureguizar. 1995. Patterns of body size changes in fossil and living Equini (Perissodactyla). *Biological Journal of the Linnean Society* 54(4): 349–370.

Berggren, W. A., and J. A. Van Couvering. 1974. The late Neogene: Biostratigraphy, geochronology, and paleoclimatology of the last 15 million years in marine and continental sequences. *Palaeogeography, Palaeoclimatology, Palaeoecology* 16(1–2): 1–216.

Bernor, R. L. 1983. Geochronology and zoogeographic relationships of Miocene Hominoidea. In *New Interpretations of Ape and Human Ancestry*, edited by R. L. Ciochon and R. Corrucini, 21–64. New York: Plenum Press.

Bernor, R. L. 1984. A zoogeographic theater and biochronologic play: The time-biofacies phenomena of Eurasian and African Miocene mammal provinces. *Paléobiologie continentale* 14(2): 121–142.

Bernor, R. L., M. Armour-Chelu, H. Gilbert, T. M. Kaiser, and E. Schulz. 2010. Equidae. In *Cenozoic Mammals of Africa*, edited by L. Werdelin and W. J. Sanders, 685–721. Berkeley: University of California Press.

Bernor, R. L, M. M. Ataabadi, K. Meshida, and D. Wolf. 2016. The Maragheh hipparions, Late Miocene of Azarbaijan, Iran. *Palaeobiodiversity and Palaeoenvironments* 96(3): 453–488.

Bernor, R. L., N. T. Boaz, and L. Rook. 2012. *Eurygnathohippus feibeli* (Perissodactyla: Mammalia) from the Late Miocene of Sahabi (Libya) and its evolutionary and biogeographic significance. *Bolletino della Societa Paleontologica Italiana* 51(1): 39–48.

Bernor, R. L., O. Cirilli, and H.-W. Mittmann. 2022. Höwenegg *Hipptherium primigenium*: Geological context, cranial and postcranial morphology paleoecological and biogeographic importance. *Historical Biology* 35(8): 1376–1390.

Bernor, R. L., U. B. Göhlich, M. Harzhauser, and G. M. Semprebon. 2017. The Pannonian C hipparions from the Vienna Basin. *Palaeogeography, Palaeoclimatology, Palaeoecology* 476:28–41.

Bernor, R. L., and J. M. Harris. 2003. Systematics and evolutionary biology of the Late Miocene and Early Pliocene hipparionine equids from Lothagam, Kenya. In *Lothagam: The Dawn of Humanity in Eastern Africa*, edited by M. G. Leakey and J. M. Harris, 387–438. New York: Columbia University Press.

Bernor, R. L., and S. T. Hussain. 1985. An assessment of the systematic, phylogenetic and biogeographic relationships of Siwalik hipparionine horses. *Journal of Vertebrate Paleontology* 5(1): 32–87.

Bernor, R. L., T. M. Kaiser, and D. Wolf. 2008. Revisiting Sahabi equid species diversity, biogeographic patterns, and dietary preferences. *Garyounis Scientific Bulletin* Special Issue(5): 159–167.

Bernor, R. L., F. Kaya, A. Kakkinen, J. Saarinnen, and M. Fortelius 2021. Old World hipparionine evolution, biogeography, climatology and ecology. *Earth-Science Reviews* 221(October): 103784. https://doi.org/10.1016/j.earscirev.2021.103784.

Bernor, R. L., G. D. Koufos, M. O. Woodburne, and M. Fortelius. 1996. The evolutionary history and biochronology of European and Southwest Asian Late Miocene and Pliocene hipparionine horses. In *The Evolution of Western Eurasian Neogene Mammal Faunas*, edited by R. L. Bernor, V. Fahlbusch, and H.-W. Mittmann, 307–338. New York: Columbia University Press.

Bernor, R. L., J. Kovar-Eder, D. Lipscomb, F. Rögl, S. Sen, and H. Tobien. 1988. Systematic, stratigraphic, and paleoenvironmental contexts of first-appearing Hipparion in the Vienna Basin, Austria. *Journal of Vertebrate Paleontology* 8(4): 427–452.

Bernor, R. L., R. S. Scott, M. Fortelius, J. Kappelman, and S. Sen. 2003. Systematics and evolution of the Late Miocene hipparions from Sinap, Turkey. In *The Geology and Paleontology of the Miocene Sinap Formation, Turkey*, edited by M. Fortelius, J. Kappelman, S. Sen and R. L. Bernor, 220–281. New York: Columbia University Press.

Bernor, R. L., and S. Sen. 2017. The Early Pliocene Plesiohipparion and Proboscidipparion (Equidae, Hipparionini) from Çalta, Turkey (Ruscinian Age, c. 4.0 Ma). *Geodiversitas* 39(2): 285–314.

Bernor, R. L., and B. Sun. 2015. Morphology through ontogeny of Chinese *Proboscidipparion* and *Plesiohipparion* and observations on their Eurasian and African relatives. *Vertebrata PalAsiatica* 53(1): 73–92.

Bernor, R. L., H. Tobien, L. A. Hayek, and H. W. Mittmann. 1997. The Höwenegg Hipparionine horses: Systematics, stratigraphy, taphonomy and paleoenvironmental context. *Andrias* 10:1–230.

Bernor, R. L., and T. D. White. 2009. Systematics and biogeography of "*Cormohipparion*" *africanum*, early Vallesian (MN 9, ca. 10.5 Ma) of Bou Hanifia, Algeria. *Papers on Geology, Vertebrate Paleontology,*

and Biostratigraphy in Honor of Michael O. Woodburne. Bulletin, Museum of Northern Arizona 65:635–658.

Bryant, J. D., Philip N. F., W. J. Showers, and B. J. Genna. 1996. Biologic and climatic signals in the oxygen isotopic composition of Eocene-Oligocene equid enamel phosphate. *Palaeogeography, Palaeoclimatology, Palaeoecology* 126(1–2): 75–89.

Cerling, T. E., S. A. Andanje, S. A. Blumenthal, F. H. Brown, K. L. Chritz, J. M. Harris, J. A. Hart, et al. 2015. Dietary changes of large herbivores in the Turkana Basin, Kenya from 4 to 1 Ma. *Proceedings of the National Academy of Sciences*, 112(37): 11467–11472.

Cerling, T. E., J. M. Harris, and B. H. Passey. 2003. Diets of East African Bovidae based on stable isotope analysis. *Journal of Mammalogy* 84(2): 456–470.

Cerling, T. E., G. Wittemyer, J. R. Ehleringer, C. H. Remien, and I. Douglas-Hamilton. 2009. History of animals using isotope records (HAIR): A 6-year dietary history of one family of African elephants. *Proceedings of the National Academy of Sciences* 106(20): 8093–8100.

Colbert, E. H. 1935. Siwalik mammals in the American Museum of Natural History. *Transactions of the American Philosophical Society* 26:1–401.

Eisenmann, V. 1994. Equidae of the Albertine Rift Valley, Uganda Zaire. *Geology and Paleobiology of the Albertine Rift Valley, Uganda-Zaire* 2:289–307.

Eisenmann, V., M. T. Alberdi, C. de Giuli, and U. Staesche. 1988. Methodology. In *Studying Fossil Horses: Collected Papers after the New York International Hipparion Conference, 1981*, edited by Mike Woodburne and Paul Sondaar, 1:1–71. Leiden: Brill Archive.

Faith, J. T., C. A. Tryon, D. J. Peppe, and D. L. Fox. 2013. The fossil history of Grévy's zebra *(Equus grevyi)* in equatorial East Africa. *Journal of Biogeography* 40(2): 359–369.

Falconer, H., and P. T. Cautley. 1849. Fauna Antiqua Sivalensis, being the fossil zoology of the Sewalik Hills, in the north of India. In *Fauna Antiqua Part 9, Equidae, Camelidae and Sivatherium*. London: Smith, Elder and Co.

Forsten, Ann. 1968. Revision of the palearctic Hipparion. *Acta Zoologica Fennica* 119:1–134.

Fox, D. L, and D. C. Fisher. 2001. Stable isotope ecology of a Late Miocene population of *Gomphotherium productus* (Mammalia, Proboscidea) from Port of Entry Pit, Oklahoma, USA. *Palaios* 16(3): 279–293.

Fricke, H. C., and J. R. O'Neil. 1996. Inter-and intra-tooth variation in the oxygen isotope composition of mammalian tooth enamel phosphate: Implications for palaeoclimatological and palaeobiological research. *Palaeogeography, Palaeoclimatology, Palaeoecology* 126(1–2): 91–99.

Gadbury, C., L. Todd, A. H. Jahren, and R. Amundson. 2000. Spatial and temporal variations in the isotopic composition of bison tooth enamel from the Early Holocene Hudson–Meng Bone Bed, Nebraska. *Palaeogeography, Palaeoclimatology, Palaeoecology* 157(1–2): 79–93.

Gradstein, F. M, J. G. Ogg, M. Schmitz, and G. Ogg. 2012. *The Geologic Time Scale 2012*. Amsterdam: Elsevier.

Gromova, V. I. 1952. Le genre Hipparion. *Bureau de Recherches Géologiques et Minières, C.E.D.P* 12:1–288.

Hussain, S. T. 1971. Revision of Hipparion (Equidae, Mammalia) from the Siwalik Hills of Pakistan and India: Mit 19 Tabellen. *Abhandlungen der Bayerischen Akademie der wissenschaften, Mathematisch-Naturwissenschaftliche Abteilung* 147:1–68.

Kaiser, T. M., and N. Solounias. 2003. Extending the tooth mesowear method to extinct and extant equids. *Geodiversitas* 25(2): 321–345.

King, T., P. Andrews, and B. Boz. 1999. Effect of taphonomic processes on dental microwear. *American Journal of Physical Anthropology* 108(3): 359–373.

Kingdon, J. 1979. *East African Mammals: Large Mammals*. Chicago: University of Chicago Press.

Lippolt, H. J., W. Gentner, and W. Wimmenauer. 1963. Altersbestimmungen nach der Kalium-Argon-Methode an tertiären Eruptivgesteinen Südwestdeutschlands. *Jahreshefte des geologischen Landesamts Baden-Württemberg* 6:507–538.

Lydekker, R. 1877. Notices of new and rare mammals from the Siwaliks. *Records of the Geological Survey of India* 10:76–83.

Lydekker, R. 1886. *Catalogue of the Fossil Mammalia in the British Museum (Natural History): Part III. Containing the Order Ungulata, Suborders Perissodactyla, Toxodontia, Condylartha, and Amblypoda.* London: Taylor and Francis.

MacFadden, B. J., and M. O. Woodburne. 1982. Systematics of the Neogene Siwalik hipparions (Mammalia, Equidae) based on cranial and dental morphology. *Journal of Vertebrate Paleontology* 2(2): 185–218.

Matthew, W. D. 1929. Critical observations upon Siwalik mammals. *Bulletin of the American Museum of Natural History* 61:427–560.

Mihlbachler, M. C., F. Rivals, N. Solounias, and G. M. Semprebon. 2011. Dietary change and evolution of horses in North America. *Science* 331(6021): 1178–1181.

Nanda, A. C. 1979. Fossil equids from the Upper Siwalik subgroup of Ambala, Haryana. *Himalayan Geology* 8:149–177.

Nelson, S. V. 2005. Paleoseasonality inferred from equid teeth and intra-tooth isotopic variability. *Palaeogeography, Palaeoclimatology, Palaeoecology* 222(1–2): 122–144.

Opdyke, N. D., E. Lindsay, G. D. Johnson, N. N. Johnson, R. A. K. Tahirkheli, and M. A. Mirza. 1979. Magnetic polarity stratigraphy and vertebrate paleontology of the Upper Siwalik Subgroup of northern Pakistan. *Palaeogeography, Palaeoclimatology, Palaeoecology* 27:1–34.

Owen, R. 1846. *A History of British Fossil Mammals, and Birds.* 560 pp. London: John Van Voorst.

Passey, B. H., L. K. Ayliffe, A. Kaakinen, Z. Zhang, J. T. Eronen, Y. Zhu, L. Zhou, T. E. Cerling, and M. Fortelius. 2009. Strengthened East Asian summer monsoons during a period of high-latitude warmth? Isotopic evidence from Mio-Pliocene fossil mammals and soil carbonates from northern China. *Earth and Planetary Science Letters* 277(3–4): 443–452.

Ratoi, B. G., B. S. Haiduc, G. Semprebon, P. Tibuleac, and R. L. Bernor. 2022. The Turolian Hipparions from Cioburciu site (Republic of Moldova): Systematics and paleodiet. *Rivista Italiana di Paleontologia E Stratigrafia* 128(2): 411–425.

Rivals, F., and N. Solounias. 2007. Differences in tooth microwear of populations of caribou (*Rangifer tarandus*, Ruminantia, Mammalia) and implications to ecology, migration, glaciations and dental evolution. *Journal of Mammalian Evolution* 14(3): 182.

Rubenstein, D. I. 2010. Ecology, social behavior, and conservation in zebras. *Advances in the Study of Behavior* 42:231–258.

Semprebon, G. M., L. R. Godfrey, N. Solounias, M. R. Sutherland, and W. L. Jungers. 2004. Can low-magnification stereomicroscopy reveal diet? *Journal of Human Evolution* 47(3): 115–144.

Sharp, Z. D., and T. E. Cerling. 1998. Fossil isotope records of seasonal climate and ecology: Straight from the horse's mouth. *Geology* 26(3): 219–222.

Solounias, N., and G. Semprebon. 2002. Advances in the reconstruction of ungulate ecomorphology with application to early fossil equids. *American Museum Novitates* 3366:1–49.

Sun, B., X. Zhang, Y. Liu, and R. L. Bernor. 2018. *Sivalhippus ptychodus* and *Sivalhippus platyodus* (Perissodactyla, Mammalia) from the Late Miocene of China. *Rivista Italiana di Paleontologia e Stratigrafia* 124(1): 1–22.

Swisher, C. C. 1996. New $^{40}AR/^{39}Ar$ dates and their contribution toward a revised chronology for the late Miocene Nonmarine of Europe and West Asia. In *The Evolution of Western Eurasian Neogene Mammal Faunas*, edited by R. L. Bernor, V. Fahlbusch, and H.-W. Mittmann, 271–289. New York: Columbia University Press.

Van Couvering, J. A., and J. A. Miller. 1971. Late Miocene marine and nonmarine time scale in Europe. *Nature* 230(5296): 559.

Wolf, D., R. L. Bernor, and S. T. Hussain. 2013. A systematic, biostratigraphic, and paleobiogeographic reevaluation of the Siwalik hipparionine horse assemblage from the Potwar Plateau, northern Pakistan. *Palaeontographica* 300(2013): 1–115.

Wolf, D., G. M. Semprebon, and R. L. Bernor. 2012. New observations on the paleodiet of the late Miocene Höwenegg (Hegau, Germany) *Hippotherium primigenium* (Mammalia, Equidae) *Bollettino della Societa Paleontologica Italiana* 51(3): 185–191.

Woodburne, M. O. 2009. The early Vallesian vertebrates of Atzelsdorf (Late Miocene, Austria). 9. *Hippotherium* (Mammalia, Equidae). *Annalen des Naturhistorischen Museums in Wien* 111(A): 585–604.

Woodburne, M. O., and C. C. Swisher III. 1995. Land mammal high-resolution geochronology, intercontinental overland dispersals, sea level, climate and vicariance. *Geochronology Times Scales and Global Stratigraphic Correlation. SEPM Special Publication* 54:335–364.

SIWALIK CHALICOTHERIIDAE

SUSANNE COTE AND MARGERY C. COOMBS

INTRODUCTION

The Chalicotherioidea (*sensu* Coombs 1998) is a group of extinct ungulates best known for having claws instead of hooves. The dentition of chalicotheres unequivocally indicates an herbivorous diet and perissodactyl affinities. The claws, on the other hand, were originally compared with those of pangolins and sloths. Only in 1877 did Filhol recognize that the claws and teeth belonged to a single taxon, a perissodactyl (Filhol 1877).

The Chalicotherioidea first appear in the fossil record at the beginning of the Eocene in China and Mongolia (Radinsky 1964; Bai 2008). In the Middle Eocene, two Asian genera reached North America (Coombs 1998). A chalicotherioid is also documented from the Middle Eocene of Myanmar (Remy et al. 2005).

The Chalicotheriidae first appears in the Oligocene and includes two subfamilies, the Chalicotheriinae and the Schizotheriinae. Oligocene chalicotheriids, most placed in the genus *Schizotherium*, are known from localities across Asia and Europe (Coombs 1978a). During this time, no chalicotheres are known from North America or Africa.

The Paleogene–Neogene boundary was important in chalicothere evolution, with the earliest representatives of the subfamily Chalicotheriinae appearing in Asia at this time. Postcranial elements of the Chalicotheriinae are distinctively modified and thus readily identified, while the dentition, at least in the earliest representatives, is little changed. Members of the subfamily Schizotheriinae have elongated molars at this time, and most species have fused the proximal and middle phalanges of digit II of the manus to form a bone called the duplex. Schizotheriinae crossed the Bering Land Bridge to North America around 23 Ma and rapidly spread east and south (Coombs et al. 2001; Wood and Ridgwell 2015). The Chalicotheriinae appear in eastern Africa by at least 20 Ma (Coombs and Cote 2010).

Chalicotheres reached their broadest geographic distribution during the Miocene, when most of the known occurrences from South Asia occur. Yet chalicothere remains are rarely numerous anywhere, except in a few unusual assemblages. Chalicothere fossils from South Asia include a variety of dental and postcranial elements that must be interpreted in the context of more complete material found in other regions.

By the end of the Miocene, the Chalicotheriidae were extinct in North America and Europe. The schizotheriine genus *Ancylotherium* persisted as late as the Early Pleistocene in both Africa (Butler 1965; Coombs and Cote 2010) and China (Li and Deng 2003; Li and Xue 2004). The Chalicotheriinae, represented primarily by the genus *Nestoritherium*, continued in Asia, but nowhere else, into the Early Pleistocene (Butler 1965; Qiu 2002; Li and Deng 2003).

Both the Chalicotheriinae and the Schizotheriinae first appeared in South Asia in the latest Oligocene to earliest Miocene and are found in the Chitarwata Formation of the Bugti Hills (Antoine et al. 2013) and at Zinda Pir (Lindsay et al. 2005). In the Middle Siwaliks (Middle and Late Miocene), chalicothere remains have historically been attributed to *Chalicotherium salinum* (Butler 1965; Pickford 1982; Khan et al. 2009). In the Upper Siwaliks, the subfamily Chalicotheriinae is represented by the species *Nestoritherium sivalense*, which is the last known chalicothere in South Asia.

TAXA FROM THE MIDDLE SIWALIKS

At least five taxa are present in the Neogene of South Asia, but not all occur in the Potwar Plateau.

Anisodon salinus, 14.1–7.4 Ma (Fig. 17.1)

The holotype of *A. salinus* (subfamily Chalicotheriinae) is NHML M12239, an isolated M3 first described by Forster Cooper (1922).

Over 100 specimens were collected by the Harvard-GSP project, approximately 60 of which are presently curated at the Harvard University Paleoanthropology Laboratory, with the remainder in Pakistan. Additional material includes 11 specimens collected by B. Brown in 1922, now in the American Museum of Natural History (AMNH), of which 6 were figured by Colbert (1935); 3 specimens at the Yale Peabody

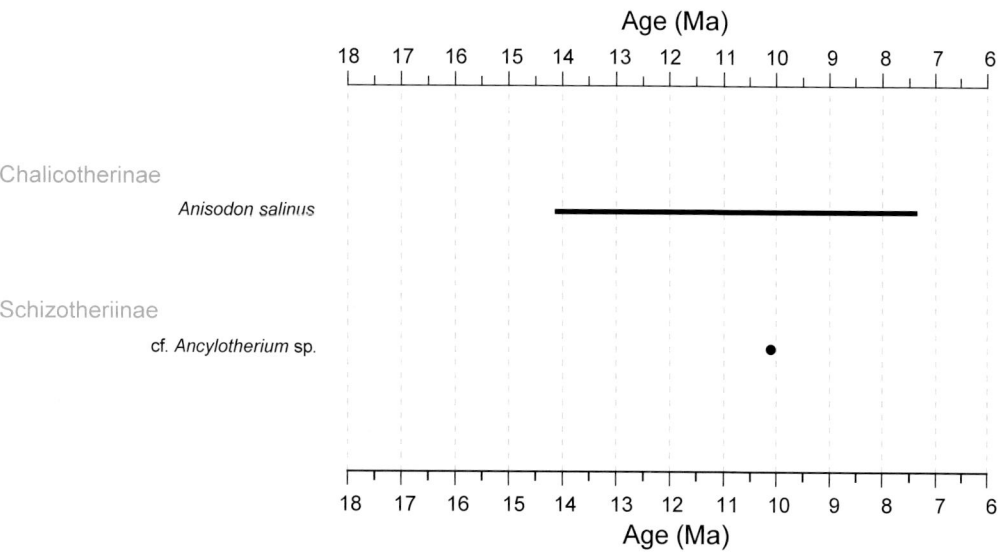

Museum (YPM), collected in the 1930s from near both Chinji and Dhok Pathan villages (Pickford 1982); material in the Geological Survey of India Museum in Kolkata (some of which was described by Pickford 1982); and material at Punjab University in Lahore (PUPC) (Khan et al. 2009). Of the three specimens described by Khan et al. (2009), however, only PUPC 07/56 belongs to *A. salinus*; PUPC 07/80 and 02/154, respectively a P2 and juvenile lower jaw, belong to rhinocerotids.

This classic Siwalik chalicotheriine has been referred to *Macrotherium* (Forster Cooper 1922; Colbert 1935; de Bonis et al. 1995) and *Chalicotherium* (Butler 1965). Following a major revision of the Chalicotheriinae by Anquetin, Antoine, and Tassy (2007), the Siwalik species has been referred to *Anisodon* by Semprebon, Sise, and Coombs (2011), Chen et al. (2012, 2016), Fahlke, Coombs, and Semprebon (2013), and Xue et al. (2014).

The holotype of *A. salinus* came from the area near Chinji village (Middle Miocene exposures), as did much of the referred AMNH material described by Colbert (1935), although some specimens come from the younger Dhok Pathan Formation (Pickford 1982; Flynn et al. 1995; Barry et al. 2002). Jaws and teeth of *A. salinus* are the most diagnostic elements, readily compared with those of *Anisodon grande*, the earliest known representative of the Chalicotheriinae in Europe (Heissig 1999). *A. grande* is also the best known of the Chalicotheriinae, with nearly complete skeletal material of multiple individuals available from Sansan, France (Anquetin, Antoine, and Tassy 2007) and Neudorf, Slovakia (Zapfe 1979).

Morphology of the Siwalik chalicotheriine material supports its inclusion in the anisodont clade of Anquetin, Antoine, and Tassy (2007), corresponding to the Tribe Anisodontini of Coombs and Göhlich (2019). Key characters that differentiate anisodonts from other chalicotheriines include brevirostry, loss of the retromolar space, having the external wall of the metacone and the metastyle oriented mesiodistally on M3, retaining a wide postfossette on M3, and having a P2 that is wider than long (Anquetin, Antoine, and Tassy 2007). The holotype

specimen of *A. salinus* (NHML M12239, M3) has a wide postfossette and mesiodistally oriented metastyle, as do all other M3s from the Middle Siwaliks (Fig. 17.2a, b). No P2s of *A. salinus* are preserved, so the proportions of this tooth cannot be observed (Table 17.1). Anquetin, Antoine, and Tassy (2007) included a weak or absent "metastylid" among the synapomorphies defining the anisodont clade (the "metastylid" is probably a twinned metaconid; Hooker 1994). As noted by Fahlke, Coombs, and Semprebon (2013), this character can be troublesome because *Chalicotherium goldfussi* from the Dinotheriensande localities of Eppelsheim, Germany, has a weak "metastylid," while in at least two anisodonts (*Nestoritherium wuduense, N. fuguense*) it is well developed (Xue et al. 2014). "Metastylid" development can also vary somewhat within a species. Nonetheless, the "metastylid" is absent or weak in all known lower molars of *A. salinus* (Fig. 17.2c).

Mandibles from the Middle Siwaliks are also consistent with membership in the anisodont clade. For example, YGSP 6006 (Locality Y337, 9.7 Ma; Fig. 17.2c, d, e; Pickford 1982, Figs. 10–11) has the following anisodont characters: a coronoid process and condyle that ascend abruptly distal to the tooth row, an expanded mandibular angle, a mandibular corpus that increases in height distally, and no retromolar space.

YGSP 6006 and almost all the other known mandibles are from juveniles or young adults. It is thus difficult to assess the adult condition of the jaw symphysis, which in known specimens extends only to the distal edge of p2, in contrast to the notably longer symphysis (to the p3–p4 boundary) in derived Late Miocene anisodonts. However, YPM 19686 (Locality L82, 7.4 Ma), a robust lower jaw of a mature individual with two worn molars, also has a short symphysis that ends approximately opposite the alveoli for p2. Therefore, it appears that Middle Siwalik anisodonts lacked the notably longer symphysis (to the p3–p4 boundary) seen in derived anisodonts such as *Anisodon macedonicus* and *Nestoritherium wuduense* (Anquetin, Antoine, and Tassy 2007). In addition to the shorter symphysis,

Table 17.1. Measurements (in mm) for Important Dental Specimens of *Anisodon salinus*

Both adult and deciduous teeth are included. Measurements are maximum length and width; estimated measurements are denoted by "e."
MD = mesiodistal; BL = buccolingual, which is equivalent to talonid width in lower premolars and molars.

Specimen Number	P3 MD	P3 BL	P4 MD	P4 BL	M1 MD	M1 BL	M2 MD	M2 BL	M3 MD	M3 BL	p2 MD	p2 BL	dp3 MD	dp3 BL	dp4 MD	dp4 BL	m1 MD	m1 BL	m2 MD	m2 BL
YGSP 27715	—	—	18.4	20.8	—	—	—	—	—	—	—	—	—	—	—	—	—	—	—	—
YGSP 4256	—	—	19.5	25.9	—	—	—	—	—	—	—	—	—	—	—	—	—	—	—	—
YGSP 19802[1]	15.8	19.6e	—	—	—	—	—	—	—	—	—	—	—	—	—	—	—	—	—	—
YGSP 23046	—	—	—	—	—	—	—	—	—	—	—	—	—	—	17.8	10.5	—	—	—	—
YGSP 9665	—	—	—	—	—	—	—	—	—	—	—	—	—	—	20.0e	10.1	—	—	—	—
YGSP 48738	—	—	—	—	27.6	27.3	—	—	—	—	—	—	—	—	—	—	—	—	—	—
YGSP 47310	—	—	—	—	—	—	—	—	44.6	41.8	—	—	—	—	—	—	—	—	—	—
YGSP 6006	—	—	—	—	—	—	—	—	—	—	11.8	10.0	—	—	—	—	24.8	14.7	30.0 e	17.3
YGSP 30972	—	—	18.2	20.5	29.2	25.9	—	34.0e	—	—	—	—	—	—	—	—	—	—	—	—
YPM 20096	—	—	—	—	—	—	38.1	—	—	—	—	—	—	—	—	—	—	—	—	—
YPM 19288[2]	—	—	—	—	—	—	—	—	—	—	—	—	—	—	—	—	—	—	29.5 e	14.9
YPM 19686	—	—	—	—	—	—	—	—	—	—	—	—	—	—	—	—	21.9	13.6	28.0 e	17.9e
AMNH 19437	—	—	—	—	—	—	—	—	—	—	—	—	—	—	—	—	—	—	31.3	16.1
AMNH 29834	—	—	—	—	—	—	37.0e	32.7	—	—	—	—	—	—	—	—	—	—	—	—
AMNH 19647	—	—	—	—	—	—	38.5e	38.0e	—	—	—	—	—	—	—	—	—	—	—	—
AMNH 19577	—	—	—	—	—	—	—	—	—	—	—	—	16.0	8.3	19.0	12.4	27.2	17.2	—	—
YGSP 3700[3]	—	—	—	—	—	—	39.5e	36.5e	—	—	—	—	—	—	—	—	—	—	—	—
YGSP 51644[3]	—	—	—	—	—	—	34.0e	36.0e	—	—	—	—	—	—	—	—	—	—	—	—

[1] This tooth is probably P3, but is very worn.

[2] This tooth is either m2 or m3, which are generally similar in size.

[3] These upper molars are likely M2, but could possibly be M3, which are similar in size.

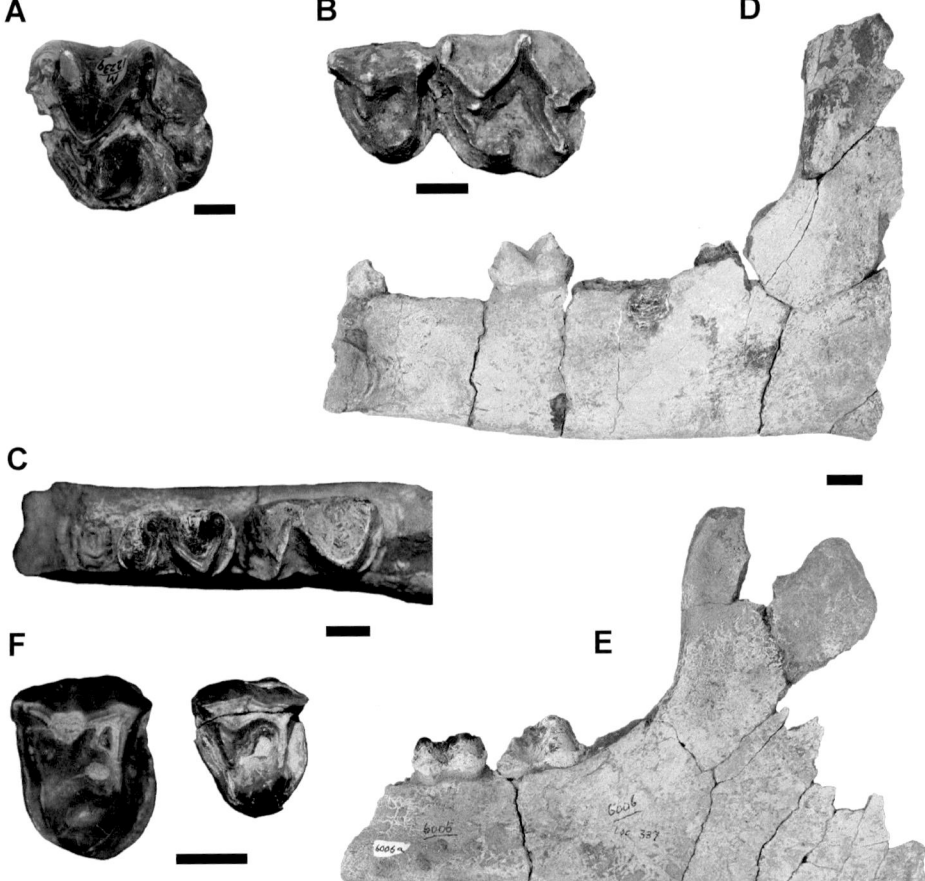

Figure 17.2. Representative specimens of *Anisodon salinus*. A. NHML M12239, left M3 holotype in occlusal view (image courtesy of J. Anquetin); B. YGSP 30972, left maxilla with P4 and M1 in occlusal view; C.–E. YGSP 6006: left mandible in occlusal view with m1 and m2 (C), right mandible in lingual view with p2, m1, and partially erupted m3 (D), and left mandible in buccal view with m1 and m2 (E); F. YGSP 4256 (left) and YGSP 27715 (right), two P4s with distinct occlusal morphologies. All scale bars 1 cm.

A. *salinus* does not show a strong trend toward brevirostry, as occurs in derived anisodonts that have notably shortened the premolar row. In general, the jaw and tooth morphology suggests that *A. salinus*, like *A. grande*, had a relatively plesiomorphic morphology among the anisodonts. Other authors have also noted similarities between *A. salinus* and *A. grande* (de Bonis et al. 1995; Semprebon, Sise, and Coombs 2011).

In contrast, Chen et al. (2012) linked *A. salinus* phylogenetically with the Late Miocene species *A. macedonicus*, best known from Dytiko 3, Macedonia, Greece (MN 13; de Bonis et al. 1995). Later, Chen et al. (2016) suggested that *A. yuanmouensis*, a Late Miocene species from southern China and Myanmar (see Chavasseau et al. 2010), was closely related to *A. salinus*. The mandible of *A. yuanmouensis*, like that of *A. salinus*, retains a relatively short symphysis into the Late Miocene (extending approximately to the p2–p3 boundary).

cf. *Ancylotherium* sp. (~10.1 Ma)

YGSP 17639, a worn and broken m3 from Locality Y596, is here referred to cf. *Ancylotherium* sp. (subfamily Schizotheriinae). This molar (Fig. 17.3) is much larger than teeth of *Anisodon salinus* and has a well-developed "metastylid" (talonid width = 28.0 mm). The identification of the tooth as m3 is based on its size and the condition of its distal and distolabial cingulum; despite its substantial wear, it had no distal neighbor in the tooth row. The trigonid is damaged but was clearly narrower than the talonid (see Fig. 17.3). A comparison of YGSP 17639 with the m3 of other large Miocene Schizotheriinae shows many similarities in shape and proportions. Worn lower molars of Miocene schizotheriines are similar and can be difficult to assign at lower taxonomic levels. In terms of size, location, and time, *Ancylotherium* is probably the best candidate genus. *Ancylotherium* was broadly distributed across Asia (China, Afghanistan, Iran, Turkey) and into southeastern Europe (and Africa) during the Late Miocene as part of the Pikermian biome (Solounias et al. 1999; Solounias et al. 2013).

OTHER CHALICOTHERES IN PAKISTAN

"Chalicotherium" pilgrimi (Forster Cooper 1920)

The holotype of *"C." pilgrimi* is NHML M12166, associated M1 and M2. The species includes limited dental and postcranial material from the upper member of the Chitarwata Formation

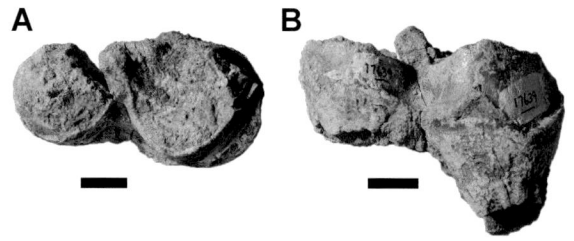

Figure 17.3. cf. *Ancylotherium* sp. YGSP 17639, right m3 in occlusal (A) and buccal (B) views. All scale bars 1 cm. Images by Walter Coombs.

(earliest Miocene) of the Bugti Hills (Forster Cooper 1920; Butler 1965; Antoine et al. 2013).

The original placement of the holotype molars within *Schizotherium* relied on the relatively primitive nature of the teeth (Forster Cooper 1920). The material from Bugti, however, also includes a derived metatarsal II and two proximal phalanges that clearly belong to the subfamily Chalicotheriinae; thus Butler (1965) referred the species to *Chalicotherium*. Anquetin, Antoine, and Tassy (2007) substantially redefined the genus *Chalicotherium*, but *"C." pilgrimi* was not included in this review. It is unlikely that *"C." pilgrimi* actually belongs to *Chalicotherium* as redefined, but no alternate generic name has so far been proposed. *"C." pilgrimi* could also profitably be compared with early Miocene chalicotheriine material from eastern Africa (Coombs and Cote 2010). There is no evidence of a close phylogenetic relationship between *"C." pilgrimi* and the later chalicotheriines of the Siwalik Potwar Plateau.

Phyllotillon naricus (Pilgrim 1908)

The schizotheriine genus *Phyllotillon* was created by Pilgrim (1910). *P. naricus* is known from somewhat better material than *"C." pilgrimi*, including several maxillae and mandibles with cheek teeth (Pilgrim 1912), although some of the teeth are badly worn. Additional cranial and postcranial material is lacking, except for several fused proximal and middle phalanges (duplexes) of digit II of the manus (Forster Cooper 1920). Pilgrim never formally designated a type. The five specimens figured by Pilgrim (1912, Plate 12) might be considered syntypes.

The Chitarwata Formation in the Bugti Hills spans the Oligocene–Miocene boundary (Antoine et al. 2013). *P. naricus* is best known from the earliest Miocene localities (equivalent to European MN 1–2) of the upper member but has also been documented in the lower Bugti Member, with the upper boundary dated at 22.6 ± 2.9 Ma (Antoine et al. 2013; Métais et al. 2009). *P. naricus* is thus one of the earliest members of the Schizotheriinae to have the fused duplex bone that characterizes all members of the subfamily except for *Schizotherium* and *Borissiakia* (Coombs 1989). The limited material of *P. naricus* makes it difficult to define the genus and species and differentiate it from contemporaries elsewhere. After the earliest Miocene, there is no further evidence of *P. naricus* in South Asia. Schizotheriinae became rare in this region but flourished elsewhere.

Additional Oligocene and Miocene Material

Chalicotheres have also been reported from the Zinda Pir Dome in the Sulaiman Range of central Pakistan (Raza et al. 2002; Lindsay et al. 2005). Lindsay et al. (2005, Table 1) listed seven chalicotheriid specimens from five localities spread throughout the Chitarwata and Vihowa formations. We were able to confirm that four of these are chalicotheriid. The oldest is a small proximal phalanx (Z 2064 from Locality Z151) and likely represents the widespread Oligocene genus *Schizotherium*, which Antoine et al. (2013) also listed as occurring at Bugti. A second proximal phalanx from Locality Z139 (Z 2042) is from higher in the Chitarwata Formation and dates to the

Oligocene–Miocene boundary (Lindsay et al. 2005; Antoine et al. 2013). This larger specimen is probably from the manus of a chalicotheriine and thus represents one of the earliest representatives of the subfamily Chalicotheriinae.

A proximal phalanx (Z 2444) and right metatarsal IV (Z 166) are known from the Early Miocene section of the Vihowa Formation at Zinda Pir (Locality Z116, ~19 Ma; Lindsay et al. 2005). Both are referable to the Chalicotheriinae. The metatarsal (Z 166) is especially interesting. While no metatarsal IV of *Anisodon salinus* is available for comparison, Z 166 can be compared with metatarsal IV of *Anisodon grande* from Neudorf, Slovakia (Zapfe 1979), and undescribed chalicotheriine specimens (KNM-RU 3469 and KNM-RU 3468) from Rusinga Island, Kenya. The size and shape of Z 166 are more similar to the Kenyan material; the generic assignment is at present indeterminate. Antoine et al. (2013) listed *Anisodon* sp. from the middle part of the Vihowa Formation in the Bugti Hills (level W; Middle Miocene). Their identification was based on "a few postcranial and dental specimens" that showed similarities to counterparts in *A. grande* from Sansan.

Chalicotheres have also been reported from the Manchar Beds of Sind Province in southern Pakistan. Lydekker (1876) figured a middle phalanx, Geological Society of India D99, as a pangolin, *Manis sindiense,* later identified as a chalicothere (Pilgrim 1910; Colbert 1935). Raza and colleagues (1984) reported an undescribed specimen of an indeterminate chalicotheriid in the basal part of the Manchar Formation, thought to be roughly 15 or 16 Ma, but this specimen is more likely rhinocerotid (J. Barry, pers. comm.).

Nestoritherium sivalense (Falconer and Cautley 1843)

The holotype of *N. sivalense* includes left (NHML M15367, P3–M3) and right (NHML M15366, P4–M3) maxillae, probably from a single palate. Additional material includes a left mandible with p4–m3 (NHML M36374) and the cast (NHML M2710) of an anterior skull and jaw figured by Falconer and Cautley (1845), de Blainville (1849), and Falconer (1868) but subsequently lost (Holland and Peterson 1914; Colbert 1935; Anquetin, Antoine, and Tassy 2007). Catalogue information for all these specimens lists them as "Upper Siwaliks" of the Siwalik Hills (India), and they are thus Plio-Pleistocene in age.

N. sivalense is the type species of the genus *Nestoritherium* Kaup, 1859. The species was included in the phylogenetic analysis by Anquetin, Antoine, and Tassy (2007), who positioned it as a derived member of the anisodont clade; these authors applied the generic name *Anisodon* to all members of this lineage. Subsequently, Chen et al. (2012) retained the name *Nestoritherium* for certain derived Asian anisodonts, including *N. sivalense*. The anisodont clade survived into the Early Pleistocene of China, where it has been found at a number of localities (e.g., Qiu 2002; Li and Deng 2003). Because Chen et al. (2012) and Xue et al. (2014) linked *N. sivalense* with species found in China, it is more likely that *Nestoritherium* was an immigrant to South Asia than that it evolved in situ from *Anisodon salinus*. *Nestoritherium* has also been reported in the Pleisto-

cene of Java (von Koenigswald 1940; Hooijer 1964), but the species-level identification of this material is unclear.

PALEOBIOLOGY OF MIDDLE SIWALIK CHALICOTHERES

Stratigraphic Range and Diversity

Anisodon salinus had an unusually long stratigraphic range among Siwalik mammals. *Anisodon salinus* is known from Siwalik localities dating from 14.1 Ma (Locality Y478) to 7.4 Ma (Locality L82), a span of almost 7 million years.

This stratigraphic range is based on dental material that is clearly identifiable as belonging to the Chalicotheriinae. The youngest specimen (YPM 19686, 7.4 Ma) is almost one million years younger than the next oldest material of *A. salinus* (Locality Y172, 8.3 Ma), but the Harvard-GSP project confidently attributes this specimen to locality L82. Two phalanges and a metapodial from Localities Y682 (15.2 Ma) and Y689 (15.5 Ma) are currently in Pakistan and not included in this study but are unlikely to have been misidentified. On that basis, the first appearance datum (FAD) for *A. salinus* may be 15.5 Ma, but because we cannot verify this, we report a date of 14.1 Ma.

Size variation is common in chalicotheres and is usually attributable to sexual dimorphism (Coombs 1975). It is possible that the specimens attributed to *A. salinus* are too variable in size for a single species, but no element is represented by a large enough sample to evaluate this idea rigorously. It is also difficult to discern any particular trends in size. Morphologically, there is little material that could diagnose additional taxa, but a few elements are suggestive. Pickford (1982) noted some morphological variation among the preserved lunars and pointed out that phalanges from the Chinji Formation tended to be smaller than those from the Dhok Pathan Formation. These differences are not consistent among all specimens, and as noted by Pickford (1982), phalanges cannot always be assigned to a specific digit, which is another source of size variation.

A few postcranial specimens are of special interest. YGSP 46946, a right calcaneum (Locality Y496, 12.4 Ma), is large with a proportionally short, robust tuber calcis and a large sustentacular area. In contrast, AMNH 19436, a smaller calcaneum figured by Colbert (1935), is more similar to calcanea of *A. grande*, figured by Zapfe (1979). YGSP 16439, a right metatarsal II (Survey block RK1, 10.4 Ma) is unusually short for its width, even in the context of the typically short metatarsal II seen in other Chalicotheriinae, such as *A. grande* (Zapfe 1979). Unfortunately, this is the only metatarsal II available in the Siwalik collections. YGSP 20065, a metatarsal III (Locality Y634, 12.3 Ma), does not show any unusual proportions. Postcranial elements in Schizotheriinae can be important diagnostic elements (Coombs 1978b), as is also the case in rhinocerotids (Antoine et al. 2010), but they have been little studied in Chalicotheriinae, except by Zapfe (1979). The differences are thus difficult to evaluate. Nevertheless, specimens YGSP 46946 and YGSP 20065 present the possibility that additional species exist in the sample attributed to *A. salinus*.

Especially compelling are differences in the morphology of the P4 (see Fig. 17.2f). Two specimens that are roughly 13 million years old (YGSP 27715 and YGSP 30972) are very similar to each other and may represent the usual morphology for *A. salinus*. A younger specimen (YGSP 4256, Locality Y182, 9.2 Ma) is wider (buccolingually) than long. It also has an enclosed fossette (formed by a small crista and crochet emanating from the ectoloph and metaloph, respectively) and a shorter protoloph with a very small paraconule. This may be the best evidence available for a second taxon lurking within the *A. salinus* hypodigm, but we see no comparable differences in other teeth.

Ecomorphology and Diet

Several authors have hypothesized that the unusual body plan of chalicotheres, with their short weight-bearing hindlimbs and long forelimbs, was an adaptation for bipedal browsing (Schaub 1943; Zapfe 1979; Coombs 1983). Chalicotheres are thought to have stood on their hind limbs while browsing, using their clawed forelimbs as hooks to stabilize themselves and grab vegetation. Schizotheriinae had a body shape somewhat like that of living okapis (Giraffidae), with long forelimbs and neck, while Chalicotheriinae had a body shape more like that of a gorilla (Zapfe 1979; Tassy 1978). As reconstructed by Zapfe (1979), *Anisodon grande* had very long arms with the front claws held flexed during walking; the neck was short, as were the distal hindlimbs. Although we do not have enough of the skeleton of *A. salinus* to confidently reconstruct its proportions, they were probably similar to those of *A. grande*.

Semprebon, Sise, and Coombs (2011) included a small sample of *A. salinus* teeth from the AMNH collection in their study of chalicothere dental microwear. Their results showed that all Chalicotheriinae were browsers, a finding consistent with the brachydont dentition of the group. Higher frequencies of large pits and puncture-like pits than in modern leaf-browsers may indicate a reliance on fruits, nuts, or robust seeds and perhaps bark and twigs, in addition to leaves. *A. salinus* was classified as a probable fruit-browser in Semprebon, Sise, and Coombs's (2011) discriminant function analysis, although it showed fewer large pits and coarse scratches than *A. grande* and *Chalicotherium goldfussi* from Europe, suggesting less obdurate fruit or seed consumption.

Stable carbon and oxygen isotope data for *A. salinus* indicate an entirely C_3-based diet sourced from high in the canopy (Nelson 2003, 2007; Badgley et al. 2008; Chapter 25, this volume). Carbon isotope values show no change through time, but all of the available chalicothere isotope data predate the transition to C_4 ecosystems beginning at ~ 8.5–8.0 Ma (Badgley et al. 2008). Many other browsing and frugivorous taxa disappeared regionally at that time, while others incorporated more C_4 into their diets. There are no data to suggest how late-surviving *A. salinus* might have responded to this large change in vegetation, but it does appear to have survived the transition to more C_4-dominated ecosystems for a time.

Habitat Preferences

Most authors consider chalicotheres to be indicators of the presence of trees or shrubs, including closed-canopy forests, open woodlands, or riparian forests near watercourses in otherwise drier areas, depending on the taxon and region. Chalicotheres are virtually never found in deposits representing open grassland environments. Middle to Late Miocene Chalicotheriinae have been typically associated with closed, moist, subtropical environments (Coombs 1989; Coombs and Cote 2010). They often co-occur with other forest-dwelling taxa, such as tragulids, deinotheres, and hominoids. In addition to the Siwalik record, Eurasian Chalicotheriinae co-occur with hominoids at Çandir and Paşalar, Turkey (Geraads, Begun, and Güleç 2003), Neudorf, Slovakia (Begun, Güleç, and Geraads 2003), the Dinotheriensande localities at Eppelsheim, Germany (Franzen et al. 2003), Lufeng, China (Chen et al. 2016), and the Irawaddy Formation, Myanmar (Chavasseau et al. 2010).

Body Size

The average body mass of *A. salinus* was approximately that of a small horse, which is moderate for chalicotheres, and smaller than the average for the better-known species *A. grande*. Flynn et al. (1995) used a weight estimate of 250–1,000 kg. Chalicotheriidae display sexual dimorphism in size (Coombs 1975), and the size variation seen in specimens of *A. salinus* from similar-aged faunal assemblages suggests that this species was no exception. Sexual dimorphism indicates some degree of competition among males and a degree of social activity at least during the mating season.

Biogeographic Relationships

During the Middle to Late Miocene, the predominant Siwalik chalicotheres were anisodont chalicotheriines referable to *Anisodon salinus*. In the Middle Miocene, *Anisodon grande*, which is also a relatively basal member of the anisodont clade, appears in Europe and Anatolian Turkey during MN 5-6 (Coombs and Göhlich 2019); more derived *Anisodon*, such as *A. macedonicus*, occurs in much of the same region in the late Miocene (Fahlke, Coombs, and Semprebon 2013). In the late Miocene, *Anisodon yuanmouensis* occurs in southwestern China and Myanmar. Chen et al. (2016) observed differences between *A. salinus* and *A. yuanmouensis* but also reported several similarities, many of them plesiomorphic relative to characters of *Nestoritherium* from northern China. They interpreted *A. yuanmouensis* and *A. salinus* as part of a geographic continuum of conservative anisodont chalicotheriines living in subtropical, forested environments, south and east of the Himalayas, into the Late Miocene.

The occurrence of a schizotheriine chalicothere at Locality Y596 is of some biogeographic interest. The molar from Y596 is the only known record of an *Ancylotherium*-like schizotheriine from South Asia, and potentially one of the oldest representatives of the genus found anywhere. The schizotheriine from La Grive St. Alban, France (MN 7/8) is older and represents

Ancylotherium (Fahlke and Coombs 2009), not *Metaschizotherium* as previously thought. Notably, the oldest *Ancylotherium* from Africa is also 10 Ma (Nakali, Kenya; Handa et al. 2021) suggesting a major range expansion for derived schizotheriines at that time. The best-known member of the genus, *A. pentelicum*, is characteristic of the Pikermian biome, and occurs as far east as Maragheh, Iran (Bernor 1986) and Molayan, Afghanistan (Brunet, Heintz, and Battail 1984). This is the geographically closest occurrence of the genus to the Siwaliks, though all occurrences of *A. pentelicum* are younger. Members of the Pikermian biome are not common in the Siwaliks, and they are often considered to represent different biogeographic provinces. Siwalik rodents, artiodactyls, and carnivores do show evidence for at least some exchange of Pikermian and Siwalik taxa (Flynn et al. 2014, 2024; Chapter 13, this volume) but most of these are younger occurrences. A 10.1 Ma cf. *Ancylotherium* molar in the Siwaliks raises interesting questions about the geographic origins of the genus and demonstrates a broader distribution of schizotheriine chalicotheres in the Late Miocene.

CONCLUSIONS

Chalicotheres occur sporadically in South Asia in deposits dating from the latest Oligocene to the Early Pleistocene. Both the Chalicotheriinae and Schizotheriinae are present, although schizotheriines occur only rarely in this region after the Early Miocene. We document for the first time the presence of a schizotheriine in the Middle Siwalik sequence of the Potwar Plateau, based on a single tooth. This find suggests the immigration of a species from the Pikermian biome into the Siwaliks around 10 Ma.

The best documented chalicothere from the Middle Siwaliks is *Anisodon salinus*. Previously placed in the genus *Chalicotherium*, *A. salinus* is clearly an anisodont chalicotheriine based on the upper molar morphology of its holotype. Mandibular and molar characters used in recent phylogenetic analyses to define the anisodont clade further confirm the attribution of the Siwalik chalicothere to *Anisodon* (Anquetin, Antoine, and Tassy 2007; Fahlke, Coombs, and Semprebon 2013). *A. salinus*, as currently recognized, is a long-lived species known from over 100 specimens spanning 14.1 to 7.4 Ma. The absence of unusual variation or a trend in body size appears consistent with a single, sexually dimorphic taxon. However, some specimens raise the possibility that additional species might have been present. For example, we note two different morphs of the P4. The remaining dental and postcranial material is too fragmentary to sort into more than one taxon. Future recovery of more diagnostic material may reveal that some material now referred to *A. salinus* represents additional species.

Based on earlier occurrences at Bugti, Zinda Pir, and Manchar, it is clear that chalicotheriids have been present in South Asia since at least the Late Oligocene and persisted throughout the Early Miocene. While tantalizing, the known material is presently inadequate to clarify the origins of *A. salinus*. *Nestoritherium sivalense* is the youngest chalicothere from South Asia.

It also belongs to the anisodont clade but appears to be more closely related to species of *Nestoritherium* from China than to *A. salinus*. It is more likely an immigrant from central or eastern Asia than to have evolved in situ from *A. salinus*. Future collecting in the region is needed to clarify the species diversity and phylogenetic relationships of the Chalicotheriidae.

ACKNOWLEDGMENTS

We thank the editors for asking us to contribute to this volume and John C. Barry for helpful discussions about the material during our visit to Harvard. Sayyed Ghyour Abbas provided photos that clarified the morphology of specimens figured by Khan et al. (2009), and Pierre-Olivier Antoine helped us to identify some doubtful chalicothere specimens as rhinocerotids. Shaokun Chen shared his knowledge of Chalicotheriinae from China. Madison Bradley assisted with preparation of figures. We also thank Jin Meng and Judy Galkin (AMNH), Pip Brewer (NHM London), and Christopher Norris and Daniel Brinkman (YPM) for access to specimens. Funding was provided by the University of Calgary and the Natural Sciences and Engineering Research Council of Canada.

References

Anquetin, J., P.-O. Antoine, and P. Tassy. 2007. Middle Miocene Chalicotheriinae (Mammalia, Perissodactyla) from France, with a discussion on chalicotheriine phylogeny. *Zoological Journal of the Linnean Society* 151(3): 577–608.

Antoine, P.-O., K. F. Downing, J.-Y. Crochet, F. Duranthon, L. J. Flynn, L. Marivaux, G. Métais, A. R. Rajpar, and G. Roohi. 2010. A revision of *Aceratherium blanfordi* Lydekker, 1884 (Mammalia: Rhinocerotidae), from the early Miocene of Pakistan: Postcranials as a key. *Zoological Journal of the Linnean Society* 160(1): 139–194.

Antoine, P. -O., G. Métais, M. J. Orliac, J.-Y. Crochet, L. J. Flynn, L. Marivaux, A. R. Rajpar, G. Roohi, and J.-L. Welcomme. 2013. Mammalian Neogene biostratigraphy of the Sulaiman Province, Pakistan. In *Fossil Mammals of Asia: Neogene Biostratigraphy and Chronology*, edited by X. Wang, L. J. Flynn, and M. Fortelius, 400–422. New York: Columbia University Press.

Badgley, C., J. C. Barry, M. E. Morgan, S. V. Nelson, A. K. B., T. E. Cerling, and D. Pilbeam. 2008. Ecological changes in Miocene mammalian record show impact of prolonged climatic forcing. *Proceedings of the National Academy of Sciences* 105(34): 12145–149.

Bai, B. 2008. A review on Chinese Eocene chalicotheres *Eomoropus* and *Grangeria*. In *Proceedings of the 11th Annual Meeting of the Chinese Society of Vertebrate Paleontology*, edited by W. Dong, 19–30. Beijing: China Ocean Press.

Barry, J. C., M. E. Morgan, L. J. Flynn, D. Pilbeam, A. K. Behrensmeyer, S. M. Raza, I. A. Khan, C. Badgley, J. Hicks, and J. Kelley. 2002. Faunal and environmental change in the Late Miocene Siwaliks of northern Pakistan. *Paleobiology Memoirs* 3(S2): 1–71.

Begun, D. R., E. Güleç, and D. Geraads. 2003. Dispersal patterns of Eurasian hominoids: Implications from Turkey. In *Distribution and Migration of Tertiary Mammals in Eurasia, a Volume in Honour of Hans de Bruijn*, edited by J. W. F. Reumer and W. Wessels. *Deinsia, Annual of the Natural History Museum, Rotterdam* 10(1): 23–39.

Bernor, R. 1986. Mammalian biostratigraphy, geochronology, and zoogeographic relationships of the Late Miocene Maragheh Fauna, Iran. *Journal of Vertebrate Paleontology* 6(1): 76–95.

Brunet, M., E. Heintz, and B. Battail. 1984. Molayan (Afghanistan) and the Khaur Siwaliks of Pakistan: An example of biogeographic isolation of late Miocene mammalian faunas. *Geologie en Mijnbouw* 63(1): 31–38.

Butler, P. M. 1965. East African Miocene and Pleistocene chalicotheres. Fossil mammals of Africa No. 18. *Bulletin of the British Museum (Natural History) Geology* 10:165–237.

Chavasseau, O., Y. Chaimanee, P. Coster, E.-G. Emonet, A. N. Soe, A. A. Kyaw, A. Maung, M. Rugbumrung, H.la Shwe, and J.-J. Jaeger. 2010. First record of a chalicothere from the Miocene of Myanmar. *Acta Palaeontologica Polonica* 55(1): 13–22.

Chen, S.-K., T. Deng, W. He, and S.-Q. Chen. 2012. A new species of Chalicotheriinae (Perissodactyla, Mammalia) from the Late Miocene in the Linxia Basin of Gansu, China. *Vertebrata PalAsiatica* 50(1): 53–73.

Chen, S.-K., L.-B. Pang, T. Deng, and G.-Q. Qi. 2016. Chalicotheriidae (Mammalia, Perissodactyla) from the *Lufengpithecus* locality of Lufeng, Yunnan Province, China. *Historical Biology* 28(1–2): 270–279.

Colbert, E. H. 1935. Siwalik mammals in the American Museum of Natural History. *Transactions of the American Philosophical Society* 26:1–198.

Coombs, M. C. 1975. Sexual dimorphism in chalicotheres (Mammalia, Perissodactyla). *Systematic Biology* 24(1): 55–62.

Coombs, M. C. 1978a. Additional *Schizotherium* material from China, and a review of *Schizotherium* dentitions (Perissodactyla, Chalicotheriidae). *American Museum Novitates* 2647:1–18.

Coombs, M. C. 1978b. Reevaluation of Early Miocene North American *Moropus* (Perissodactyla, Chalicotheriidae, Schizotheriinae). *Bulletin of the Carnegie Museum of Natural History* 4:1–62.

Coombs, M.C. 1983. Large mammalian clawed herbivores: A comparative study. *Transactions of the American Philosophical Society* 73(7):1–96.

Coombs, M. C. 1989. Interrelationships and diversity in the Chalicotheriidae. In *The Evolution of Perissodactyls*, edited by D. R. Prothero and R. M. Schoch, 438–457. New York: Oxford University Press.

Coombs, M. C. 1998. Chalicotherioidea. In *Evolution of Tertiary Mammals of North America*, Vol. 1: *Terrestrial Carnivores, Ungulates, and Ungulate-Like Mammals*, edited by C. M. Janis, K. M. Scott, and L. L. Jacobs, 511–524. Cambridge: Cambridge University Press.

Coombs, M. C., and S. Cote. 2010. Chalicotheriidae. In *Cenozoic Mammals of Africa*, edited by L. Werdelin and W. J. Sanders, 659–667. Berkeley: University of California Press.

Coombs, M. C., and U. B. Göhlich. 2019. *Anisodon* (Perissodactyla, Chalicotheriinae) from the Middle Miocene locality Gračanica (Bugojno Basin, Gornji Vakuf, Bosnia and Herzegovina). *Palaeobiodiversity and Palaeoenvironments* 100(2): 363–372.

Coombs, M. C., R. M. Hunt Jr., E. Stepleton, L. B. Albright III, and T. J. Fremd. 2001. Stratigraphy, biochronology, biogeography, and taxonomy of Early Miocene small chalicotheres in North America. *Journal of Vertebrate Paleontology* 21(3): 607–620.

de Blainville, H. M. D. 1849. Des Anoplotheriums. In *Ostéographie ou description iconographique comparée du squelette et du système dentaire des Mammifères récents et fossils pour servir de base à la zoologie et à la géologie. Atlas— Tome Quatrième: Quaternates—Maldentés.* Paris: J. B. Ballière et Fils.

de Bonis, L., G. Bouvrain, G. Koufos, and P. Tassy. 1995. Un crane de chalicothere (Mammalia, Perissodactyla) du Miocène Superieur de Macedoine (Grèce): Remarques sur la phylogénie des chalicotheriinae. *Palaeovertebrata* 24(102): 135–176.

Fahlke, J. M., and M. C. Coombs. 2009. Dentition and first postcranial description of *Metaschizotherium fraasi* Koenigswald, 1932 (Perissodactyla: Chalicotheriidae) and its occurrence on a karstic plateau—new insights into schizotheriine morphology, relationships, and ecology. *Palaeontographica Abteilung A* 290(1–3): 65–129.

Fahlke, J. M., M. C. Coombs, and G. M. Semprebon. 2013. *Anisodon* sp. (Mammalia, Perissodactyla, Chalicotheriidae) from the Turolian of Dorn-Dürkheim 1 (Rheinhesssen, Germany): Morphology, phylogeny, and palaeoecology of the latest chalicothere in central Europe. *Palaeobiodiversity and Palaeoenvironments* 93:151–170.

Falconer, H. 1868. On *Chalicotherium sivalense*. In *Palaeontological Memoirs and Notes of the late Hugh Falconer*, edited by C. Murchison, 1:208–226, plate 17. London: Robert Hardwicke.

Falconer, H., and P. T. Cautley. 1843. On some fossil remains of *Anoplotherium* and giraffe, from the Siwalik Hills, in the North of India. *Proceedings of the Geological Society of London* 4:235–249.

Falconer, H., and P. T. Cautley. 1845. *Fauna antiqua sivalensis. Being the fossil zoology of the Sewalik Hills, in the north of India. Illustrations—Part I. Proboscidea.* London: Smith, Elder, and Company.

Filhol, H. 1877. Recherches sur les Phosphorites du Quercy: Étude des fossils qu'on rencontre et spécialement des mammifères. *Annales des Sciences Géologiques* 8:1–340.

Flynn, L. J., J. C. Barry, M. E. Morgan, D. Pilbeam, L. L. Jacobs, and E. H. Lindsay. 1995. Neogene Siwalik mammalian lineages: Species longevities, rates of change, and modes of speciation. *Palaeogeography, Palaeoclimatology, Palaeoecology* 115(1–4): 249–264.

Flynn, L. J., M. E. Morgan, D. Pilbeam, and J. C. Barry. 2014. "Endemism" relative to space, time, and taxonomic level. *Annales Zoologici Fennici* 51(1–2): 245–258.

Flynn, L. J., S. M. Raza, M. E. Morgan, J. C. Barry, and D. Pilbeam. 2024. The Potwar Siwaliks: An impressive Neogene record of terrestrial rocks and fossils. In *Geology's Significant Sites and their Contributions to Geoheritage*, edited by R. M. Clary, E. J. Pyle, and W. M. Andrews, 1–12. London: Geological Society Special Publication 543.

Forster Cooper, C. 1920. Chalicotheroidea from Baluchistan. *Proceedings of the Zoological Society, London* 1920:357–366.

Forster Cooper, C. 1922. LV—*Macrotherium salinum* sp. n., a new chalicothere from India. *Annals and Magazine of Natural History* 10(59): 542–544.

Franzen, J. L., O. Fejfar, G. Storch, and V. Wilde. 2003. Eppelsheim 2000— New discoveries at a classic locality. In *Distribution and Migration of Tertiary Mammals in Eurasia, a Volume in Honour of Hans de Bruijn*, edited by J. W. F. Reumer and W. Wessels. *Deinsia, Annual of the Natural History Museum, Rotterdam* 10:217–234.

Geraads, D., D. R. Begun, and E. Güleç. 2003. The middle Miocene hominoid site of Çandir, Turkey: General paleoecological conclusions from the mammalian fauna. *Courier Forschungs Institut Senckenberg* 240:241–250.

Handa, N., M. Nakatsukasa, Y. Kunimatsu, T. Tsubamoto, and H. Nakaya. 2021. The Chalicotheriidae (Mammalian, Perissodactyla) from the Upper Miocene Nakali Formation, Kenya. *Historical Biology* 33(12): 3522–3529.

Heissig, K. 1999. Family Chalicotheriidae. In *The Miocene Land Mammals of Europe*, edited by G. E. Rössner and K. Heissig, 189–192. Munich: Verlag Dr. Friedrich Pfeil.

Holland, W. J., and O. A. Peterson. 1914. The osteology of the Chalicotheroidea with special reference to a mounted skeleton of *Moropus elatus* Marsh, now installed in the Carnegie Museum. *Memoirs of the Carnegie Museum* 3:189–406.

Hooijer, D. A. 1964. New records of mammals from the Middle Pleistocene of Sangiran, central Java. *Zoologische Mededelingen Rijksmuseum Naturlijke Historie Leiden* 40:73–88.

Hooker, J. J. 1994. The beginning of the equoid radiation. *Zoological Journal of the Linnean Society* 112(1–2): 29–63

Kaup, J. J. 1859. *Beiträge zur näheren kenntniss der Urweltlichen Säugethiere. Viertes Heft.* Darmstadt: Eduard Zernin.

Khan, M. A., M. Iqbal, M. Akhtar, and M. Hassan. 2009. Chalicotheres in the Siwaliks of Pakistan. *Pakistan Journal of Zoology* 41(6): 429–435.

Li, X., and K. Deng. 2003. Early Pleistocene chalicothere fossils from Huangjiawan, Zhen'an, Shaanxi, China. *Vertebrata PalAsiatica* 41(4): 332–336.

Li, X., and X.-X. Xue. 2004. Early Pleistocene mammalian fauna of Huangjiawan, Zhen'an, Shaanxi, and its companioned plants. *Acta Palaeontologica Sinica* 43(3): 388–399.

Lindsay, E. H., L. J. Flynn, I. U. Cheema, J. C. Barry, K. Downing, A. R. Rajpar, and S. M. Raza. 2005. Will Downs and the Zinda Pir Dome. *Palaeontologia Electronica* 8(1): 19A.

Lydekker, R. 1876. Molar teeth and other remains of Mammalia. *Palaeontologia Indica* ser. 10(1): 19–87, plates 4–10.

Métais, G., P.-O. Antoine, S. H. Baqri, J.-Y. Crochet, D. de Franceschi, L. Marivaux, and J.-L. Welcomme. 2009. Lithofacies, depositional environments, regional biostratigraphy and age of the Chitarwata Formation in the Bugti Hills, Balochistan, Pakistan. *Journal of Asian Earth Sciences* 34(2): 154–167.

Nelson, S. V. 2003. *The Extinction of Sivapithecus: Faunal and Environmental Changes Surrounding the Disappearance of a Miocene Hominoid in the Siwaliks of Pakistan.* Boston: Brill.

Nelson, S. V. 2007. Isotopic reconstructions of habitat change surrounding the extinction of *Sivapithecus*, a Miocene hominoid, in the Siwalik Group of Pakistan. *Palaeogeography, Palaeoclimatology, Palaeoecology* 243(1–2): 204–222.

Pickford, M. 1982. Miocene Chalicotheriidae of the Potwar Plateau, Pakistan. *Tertiary Research* 4:13–29.

Pilgrim, G. E. 1908. The Tertiary and post-Tertiary freshwater deposits of Baluchistan and Sind with notices of new vertebrates. *Records of the Geological Survey of India, Calcutta* 37:139–166.

Pilgrim, G. E.. 1910. Notices of new mammalian genera and species from the Tertiaries of India. *Records of the Geological Survey of India, Calcutta* 40:63–71.

Pilgrim, G. E. 1912. The vertebrate fauna of the Gaj Series in the Bugti Hills and the Punjab. *Memoirs of the Geological Survey of India, Calcutta* 4:1–83.

Qiu, Z.-X. 2002. *Hesperotherium*—A new genus of the last chalicotheres. *Vertebrata PalAsiatica* 40(4): 317–325.

Radinsky, L. B. 1964. *Paleomoropus*, a new Early Eocene chalicothere (Mammalia, Perissodactyla), and a revision of Eocene chalicotheres. *American Museum Novitates* 2179:1–28.

Raza, S. M., J. C. Barry, G. E. Meyer, and L. Martin. 1984. Preliminary report on the geology and vertebrate fauna of the Miocene Manchar Formation, Sind, Pakistan. *Journal of Vertebrate Paleontology* 4(4): 584–599.

Raza, S. M., I. U. Cheema, W. R. Downs, A. R. Rajpar, and S. C. Ward. 2002. Miocene stratigraphy and mammal fauna from the Sulaiman Range, Southwestern Himalayas, Pakistan. *Palaeogeography, Palaeoclimatology, Palaeoecology* 186(3–4): 185–197.

Remy, J.-A., J.-J. Jaeger, Y. Chaimanee, U. A. N. Soe, L. Marivaux, J. Sudre, S. T. Tun, B. Marandat, and E. Dewaele. 2005. A new chalicothere from the Pondaung Formation (late Middle Eocene of Myanmar). *Comptes Rendus Palevol* 4(4): 341–349.

Schaub, S. 1943. Die Vorderextremität von *Ancylotherium pentelicum* Gaudry und Lartet. *Schweizerischen Palaeontologischen Abhandlungen* 64:1–36.

Semprebon, G. M., P. J. Sise, and M. C. Coombs. 2011. Potential bark and fruit browsing as revealed by stereomicrowear analysis of the peculiar clawed herbivores known as chalicotheres (Perissodactyla, Chalicotheroidea). *Journal of Mammalian Evolution* 18:33–55.

Solounias, N., J. M. Plavcan, J. Quade, and L. Witmer. 1999. The paleoecology of the Pikermian biome and the savanna myth. In *Evolution of the Neogene Terrestrial Ecosystems in Europe*, edited by J. Agustí, L. Rook, and P. Andrews, 427–444. Cambridge: Cambridge University Press.

Solounias, N., G. M. Semprebon, M. C. Mihlbachler, and F. Rivals. 2013. Paleodietary comparisons of ungulates between the Late Miocene of China, and Pikermi and Samos in Greece. In *Fossil Mammals of Asia: Neogene Biostratigraphy and Chronology*, edited by X. Wang, L. J. Flynn, and M. Fortelius, 676–692. New York: Columbia University Press.

Tassy, P. 1978. *Chalicotherium*: Le "cheval-gorille." *La Recherche* 9(87):283–285.

von Koenigswald, G. H. R. 1940. Neue *Pithecanthropus*-Funde 1936–1938. *Wetenschappelijke Mededeelingen Dienst Mijnbouw Nederlandsch-Indie* 28:1–232.

Wood, A. R., and N. M. Ridgwell. 2015. The first Central American chalicothere (Mammalia, Perissodactyla) and the paleobiogeographic implications for small-bodied chalicotheres. *Journal of Vertebrate Paleontology* 35(3): e923893.

Xue, X.-X., T. Deng, M. C. Coombs, and Y.-X. Zhang. 2014. New chalicothere materials from the Late Miocene of Fugu, Shaanxi, China. *Vertebrata PalAsiatica* 52(4): 401–426.

Zapfe, H. 1979. *Chalicotherium grande* (Blainv.) aus der miozänen Spaltenfüllung von Neudorf an der March (Devinská Nová Ves), Tschechoslowakei. *Neue Denkschriften Naturhistorisches Museums Wien* 2:1–282.

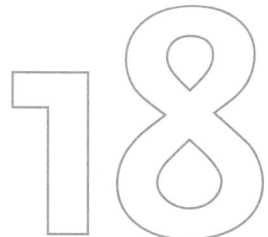

RHINOCEROTIDS FROM THE SIWALIK RECORD

PIERRE-OLIVIER ANTOINE

INTRODUCTION

This chapter provides a synthetic survey of the rhinocerotid record in the Siwalik faunal sequence (taxa and assemblages, taxonomic diversity, body weight estimates, and ecological inferences).

Rhinocerotidae have by far the longest temporal range among South Asian perissodactyls: the earliest representatives are recorded in the late Middle Eocene of Myanmar and Vietnam (Holroyd et al. 2006; Böhme et al. 2014), and their record is then virtually uninterrupted, culminating in three Recent Asian species (Laurie, Lang, and Groves 1983; Antoine 2012). By contrast, other Rhinocerotoidea, such as Hyracodontidae (including giant rhinos, better known as indricotheres or baluchitheres) and Amynodontidae (hippo-like amphibious forms) co-occurred with rhinocerotids in South Asia, but are restricted to the Paleogene period (Antoine et al. 2004, 2013). Rhinocerotids were particularly diverse and abundant throughout the Neogene in South Asia (e.g., Antoine and Welcomme 2000; Antoine 2012; Antoine et al., in press). Several comprehensive studies concerning the Siwalik Rhinocerotidae were published in the nineteenth century (Falconer and Cautley 1846–1849; Lydekker 1881, 1884, 1886) and early twentieth (Pilgrim 1908, 1910, 1912, 1913; Matthew 1929; Forster Cooper 1934; Colbert 1934, 1935). Even so—albeit with a few notable exceptions (Heissig 1972; Guérin 1979; Khan et al. 2011, 2014)—alpha-taxonomy of the Siwalik rhinocerotids has long been underinvestigated, despite intense fieldwork in the Potwar Plateau from the 1970s to the 2000s. In nearby areas, such as the Sulaiman Province of Pakistan, rhinocerotids were closely investigated in recent decades, based on field collections from the Zinda Pir and the Bugti Hills (Antoine and Welcomme 2000; Antoine 2002; Raza et al. 2002; Antoine, Duranthon, and Welcomme 2003; Antoine et al. 2003, 2010, 2013).

MATERIALS AND METHODS

The current study is based mainly on specimens from the Lower and Middle Siwaliks of the Potwar Plateau, collected by the Geological Survey of Pakistan-Yale, later Harvard-GSP joint expeditions since 1973, and presently curated in the Harvard University Paleoanthropology Laboratory, Cambridge (HUPL; ~720 specimens), in the Collections de Paléontologie, Muséum National d'Histoire Naturelle in Paris (MNHN; 59 specimens), and in the Palaeontology Department of the Pakistan Museum of Natural History in Islamabad (PMNH; 1973–1980 field seasons; ~50 specimens). Siwalik specimens described by Matthew (1929), Forster Cooper (1934), Colbert (1934, 1935), and Heissig (1972) and curated in the American Museum of Natural History (AMNH, New York), the Natural History Museum (NHM, London), and the Bayerische Staatssammlung für Paläontologie und Geologie (BSP, Munich), were also studied. Material stored in the Indian Museum in Kolkata was not studied.

Comparative material consists primarily of specimens from Neogene deposits of nearby areas in Pakistan (Bugti Hills: MNHN and Muséum d'Histoire naturelle in Toulouse, France; Zinda Pir: HUPL). Undescribed Thai material stored in the Rajabhat Institute (Nakhon Ratchasima) and the Sahat Sakhan Dinosaur Centre (Sahat Sakhan, Kalasin) was also useful for comparison. Stratigraphy and taxonomy of other South Asian rhinocerotid occurrences were cautiously surveyed from the literature. Figure 18.1 illustrates the main rhinocerotid-yielding localities and areas, while Figure 18.2 and Table 18.2 report the stratigraphic ranges of documented taxa in the Siwalik faunal sequence.

Body weights were estimated using both dental dimensions (area of m1: Legendre 1989; length of M1, M2, and M3: Fortelius and Kappelman 1993) and postcranial dimensions (astragalus: Tsubamoto 2014) or were extrapolated from those of present-day rhinoceroses (Guérin 1980: *Rhinoceros* aff. *sondaicus* and *Rhinoceros* aff. *sivalensis*). Raw estimates appear in Table 18.1, and mean values for each documented taxon are reported in Table 18.2. The available material did not allow for characterizing sexual dimorphism or intraspecific body-weight changes through time.

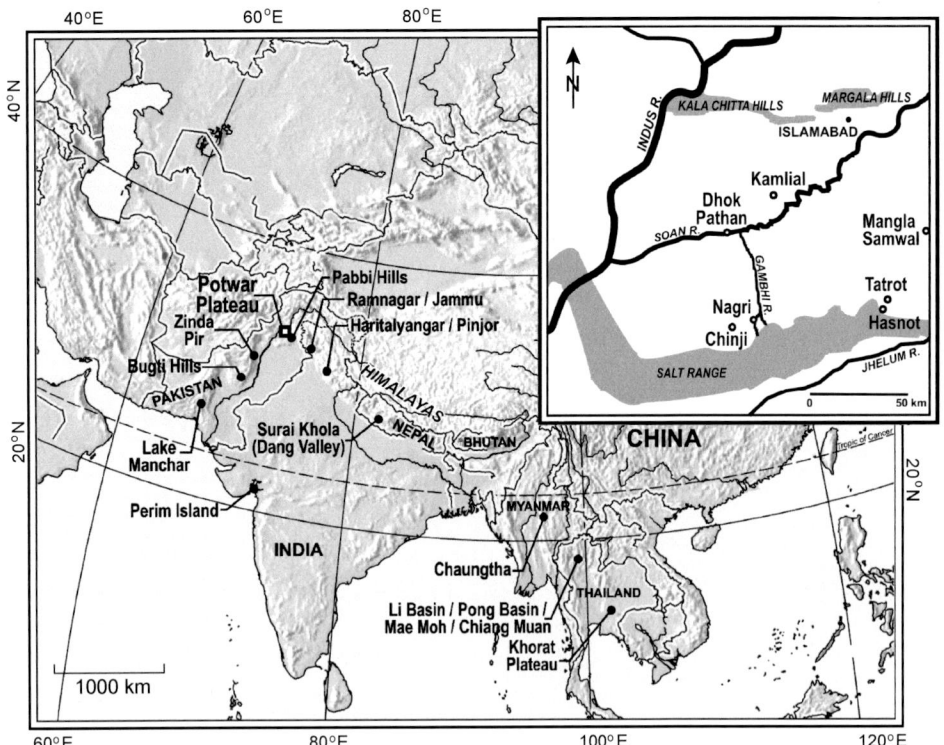

Figure 18.1. Map of the main rhinocerotid-yielding areas discussed in the text (Pakistan, India, Nepal, Bhutan, Thailand, and Myanmar), with special emphasis on the Potwar Plateau (detailed at top right); after West et al. (1978), Pilbeam et al. (1979), Nanda (2002), Chavasseau et al. (2006), and Antoine et al. (2013).

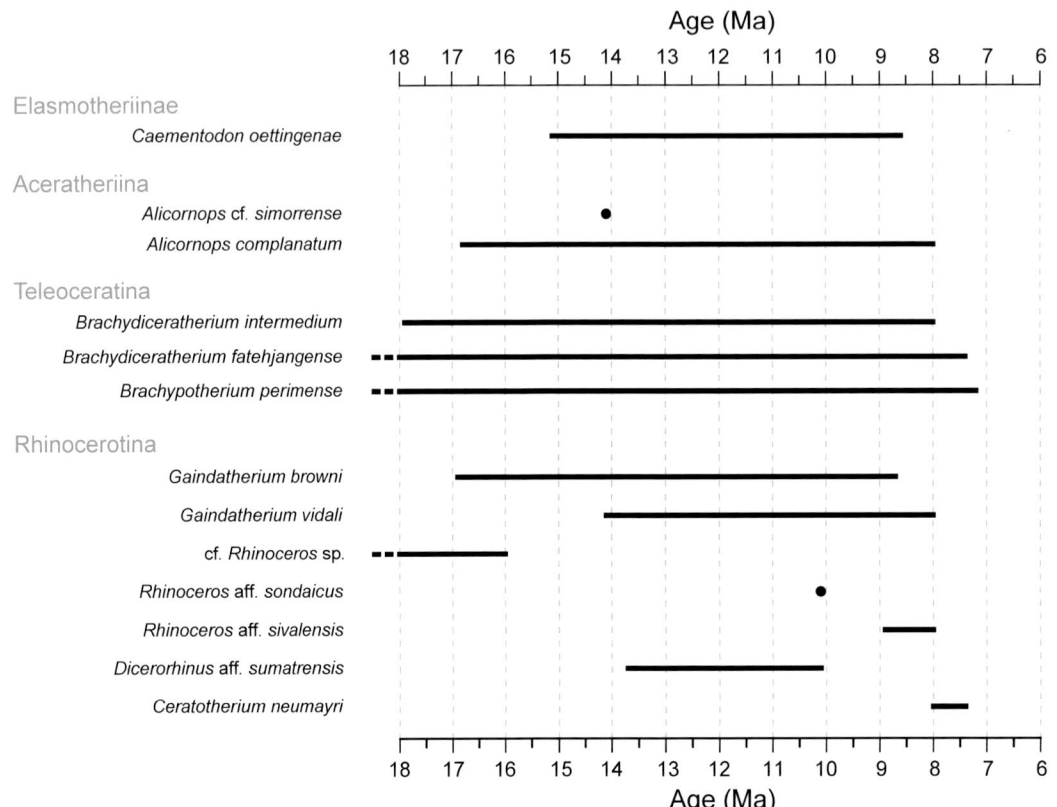

Figure 18.2. Stratigraphic ranges of rhinocerotids in the Potwar Plateau Siwalik sequence.

Table 18.1. Body Weight Estimates for Rhinocerotid Taxa Recognized in the Siwalik Faunal Sequence of the Potwar Plateau and in Miocene Deposits of Sulaiman Province

Taxon	Body Weight (kg)	Reference	Specimen	Mean Value (kg) and Standard Deviation [SD]
Bugtirhinus	334	Pers. obs.	m1 NHM w/n (min)	—
praecursor	412	Pers. obs.	m1 NHM M15345 (max)	—
	460	Pers. obs.	M3 Pak 762	477
	517	Pers. obs.	M1 NHM M15361	[94]
	571	Pers. obs.	M2 NHM M15361	—
	571	Antoine and Welcomme (2000)	Astragalus Pak 788	—
Caementodon	368	Pers. obs.	Astragalus 1956 II 385 (min)	—
oettingenae	571	Pers. obs.	Astragalus YGSP 40848 (max)	—
	597	Pers. obs.	M3 CH 018	—
	658	Pers. obs.	m1 YGSP 47182 from Y76	812
	796	Pers. obs.	Extrapolated from m2 1956 II 371	[389]
	1,185	Pers. obs.	M1 1956 II 364	—
	1,511	Pers. obs.	M2 1956 II 364	—
Protaceratherium	130	Ginsburg et al. (1992)	M3 Hang Mon (Vietnam)	153
sp.	175	Pers. obs.	Extrapolated from cuboid Pak 1876	[32]
Plesiaceratherium	459	Pers. obs.	M2 Pak 1042	—
naricum	488	Pers. obs.	Astragalus Pak 1142 (min)	—
	525	Pers. obs.	M3 Pak 1042	554
	600	Pers. obs.	Astragalus Pak 1683 (max)	[97]
	699	Pers. obs.	Extrapolated from m2 Pak 1661	—
Pleuroceros	694	Pers. obs.	Astragalus Pak 1139 (min)	—
blanfordi	815	Pers. obs.	Astragalus Pak 1138 (max)	—
	922	Pers. obs.	M3 Pak 1013 (min)	—
	1,254	Pers. obs.	M3 Pak 918 (max)	—
	1,267	Pers. obs.	m1 Pak 1038 (min)	1,525
	1,398	Pers. obs.	m1 Pak 1038 (max)	[725]
	1,762	Pers. obs.	M1 Pak 1061 (min)	—
	1,916	Pers. obs.	M2 Pak 1019 (min)	—
	2,144	Pers. obs.	M2 Pak 760 (max)	—
	3,081	Pers. obs.	M1 Pak 1012a (max)	—
Mesaceratherium	871	Pers. obs.	Astragalus Pak 1136 (min)	—
welcommi	1,341	Pers. obs.	m1 Pak 1023 (min)	—
	1,406	Pers. obs.	Astragalus Pak 1134 (max)	—
	1,412	Pers. obs.	M3 Pak 1051 (min)	1,615
	1,479	Pers. obs.	M3 Pak 1032b (max)	[503]
	1,753	Pers. obs.	m1 Pak 1054 (max)	—
	2,265	Pers. obs.	M2 Pak 1032a (min)	—
	2,390	Pers. obs.	M2 Pak 2203 (max)	—
Alicornops	542	Cerdeño and Sanchez (2000)	Astragalus of *A. simorrense* (min)	—
cf. simorrense	728	Cerdeño and Sanchez (2000)	m1 of *A. simorrense* (min)	875
	909	Cerdeño and Sanchez (2000)	Astragalus of *A. simorrense* (max)	[333]
	1,321	Cerdeño and Sanchez (2000)	m1 of *A. simorrense* (max)	—
Alicornops	515	Heissig (1972)	Astragalus NG 349	—
complanatum	787	Heissig (1972)	m1 1956 II 393 (min)	821
	1,162	Heissig (1972)	m1 1956 II 392 (max)	[325]
Brachypotherium	1,859	Pers. obs.	Astragalus Pak 1680 (min)	1,922
gajense	1,985	Pers. obs.	Astragalus Pak 1872 (max)	[89]
Prosantorhinus	662	Pers. obs.	Astragalus Pak 2126	—
shahbazi	882	Pers. obs.	extrapolated from m3 Pak 880	1,223
	1,465	Pers. obs.	M2 Pak 1030	[556]
	1,884	Pers. obs.	M1 Pak 1029	—
Brachydiceratherium	1,118	Pers. obs.	Astragalus YGSP 17757	1,905
intermedium	2,693	Pers. obs.	extrapolated, m2 YGSP 24184f	[1,113]

(continued)

Table 18.1. continued

Taxon	Body Weight (kg)	Reference	Specimen	Mean Value (kg) and Standard Deviation [SD]
Brachydiceratherium fatehjangense	1,224	Heissig (1972)	M3 CHJ 9 (min)	—
	1,458	Pers. obs.	Astragalus Pak 183 (min)	—
	1,512	Pers. obs.	Astragalus CHC 10 (max)	—
	1,730	Pers. obs.	M3 Pak 1012c (max)	1,999
	1,972	Pers. obs.	M2 Pak 2207	[633]
	2,433	Pers. obs.	M1 Pak 878	—
	2,724	Pers. obs.	m1 Pak 1069 (min)	—
	2,943	Pers. obs.	estimated from m1 Pak 1446 (max)	—
Brachypotherium perimense	1,534	Heissig (1972)	M1 1956 II 439 (min)	—
	1,619	Heissig (1972)	M3 CHO 6 (min)	—
	2,028	Heissig (1972)	M2 1956 II 438 (min)	—
	3,291	Pers. obs.	M3 CCZ RH 22 (Thailand) (max)	—
	3,747	Pers. obs.	Astragalus YGSP 21109 (min)	3,712
	4,263	Heissig (1972)	M1 1956 II 448 (max)	[1,581]
	4,464	Pers. obs.	Astragalus from Thailand (max)	—
	4,662	Heissig (1972)	M2 1956 II 460 (max)	—
	5,413	Pers. obs.	m1 YGSP 50554 (min)	—
	6,097	Pers. obs.	m1 from Thailand (max)	—
Gaindatherium cf. *browni*	726	Pers. obs.	m1 Pak 1659	810
	894	Pers. obs.	extrapolated from p4 Pak 1660	[119]
Gaindatherium browni	627	Heissig (1972)	M1 1956 II 241 (min)	—
	635	Heissig (1972)	M3 1956 II 241 (min)	—
	716	Heissig (1972)	M3 1956 II 241 (max)	—
	841	Khan et al. (2014)	m1 PUPC 11/101 (min)	—
	871	Heissig (1972)	Astragalus CHD 4 (min)	1,069
	882	Heissig (1972)	M2 1956 II 241 (min)	[449]
	1,257	Heissig (1972)	Astragalus 1956 II 331 (max)	—
	1,419	Khan et al. (2014)	M2 PUPC 09/58	—
	1,431	Heissig (1972)	m1 1956 II 247 (max)	—
	2,013	Heissig (1972)	M1 CHK 6 (max)	—
Gaindatherium vidali	492	Heissig (1972)	M3 NG 351	—
	627	Heissig (1972)	M1 NG 350	—
	909	Pers. obs.	Astragalus YGSP 47077 (min)	829
	953	Heissig (1972)	m1 1956 II 260	[268]
	1,163	Heissig (1972)	Astragalus CHD 3 (max)	—
cf. *Rhinoceros* sp.	1,209	Pers. obs.	Astragalus Pak 1458 (min)	—
	1,257	Pers. obs.	Astragalus YGSP 47266 (max)	1,738
	1,831	Pers. obs.	extrapolated from m1 Pak 171A	[673]
	2,654	Pers. obs.	M2 YGSP 17538	—
Dicerorhinus aff. *sumatrensis*	803	Pers. obs.	M3 YGSP 28100	—
	1,248	Pers. obs.	M1 YGSP 28100 (min)	1,232
	1,645	Pers. obs.	M1 YGSP 28100 (max)	[421]
Ceratotherium neumayri	1,619	Pers. obs.	M3 YGSP 17993	—
	2,469	Antoine and Saraç (2005)	Astragalus AK5-634 (min)	2,487
	3,373	Antoine and Saraç (2005)	Astragalus AK5-423 (max)	[877]

Notes: Binomials in bold font refer to species found on the Potwar Plateau. Estimates were calculated based primarily on the length of M1, M2, or M3 (Fortelius and Kappelman 1993), the area of m1 (Legendre 1989), and the trochlear width of the astragalus (Tsubamoto 2014), with minimum-maximum ranges provided whenever possible.

SYSTEMATIC OVERVIEW OF SIWALIK RHINOCEROTIDS

Most rhinocerotid groups known in the Old World are represented in Neogene deposits of Siwaliks and adjacent areas (see Table 18.2). The relationships among Siwalik rhinocerotids recognized at the species level, as shown in Figure 18.3, are mapped on formal phylogenetic trees based on Antoine, Duranthon, and Welcomme (2003), Antoine et al. (2010), and Becker, Antoine, and Marident (2013). Elasmotheriinae are represented by only one species of Elasmotheriina; in sharp contrast, Rhinocerotinae are quite diverse throughout the studied interval, with at least 12 species referred to Aceratheriina, Teleoceratina, and Rhinocerotina (see Fig. 18.3).

The **Elasmotheriina**, from the Miocene–Pleistocene of Eurasia and the Miocene of Africa, were considered the sister

Table 18.2. Rhinocerotid Species Recognized in the Siwalik Faunal Sequence

Group	Species	Loc. FLO	FLO (Ma)	LLO (Ma)	Loc. LLO	N spec.	N Loc.	Locomotion	Diet	BW (kg)
Elasmotheriina	Caementodon oettingenae	Y833	15.1	8.6	Y444	48	30	Cursorial	Mixed feeder	812
Aceratheriina	Alicornops cf. simorrense	Y478	14.1	14.1	Y478	1	1	Cursorial	Browser	875
Aceratheriina	Alicornops complanatum	Y802	16.8	8.0	Y547	33	8	Cursorial	Browser	821
Aceratheriina	Alicornops sp.	Y802	16.8	8.1	Y606	16	11	Cursorial	Browser	—
Teleoceratina	Brachydiceratherium intermedium	Y721	17.9	8.0	Y541	44	10	Cursorial	Browser	1905
Teleoceratina	Brachydiceratherium fatehjangense	Y747	17.9	7.4	Y382	44	29	Graviportal	Browser	1999
Teleoceratina	Brachypotherium perimense	Y739	18.0	7.2	KL15	180	62	Graviportal	Browser	3712
Rhinocerotina	Gaindatherium browni	Y692*	16.9	8.7	Y544	74	25	Cursorial	Browser	1069
Rhinocerotina	Gaindatherium vidali	Y501	14.1	8.0	Y1007	16	12	Cursorial	Browser	829
Rhinocerotina	Gaindatherium sp.	Y747	17.9	8.4	Y28	58	18	Cursorial	Browser	—
Rhinocerotina	cf. Rhinoceros sp.	Y744	16.7	16.0	Y592	9	2	Cursorial	Mixed feeder	1738
Rhinocerotina	Rhinoceros aff. sondaicus	Y311	10.1	10.1	Y311	1	1	Cursorial	Browser	~1500
Rhinocerotina	Rhinoceros aff. sivalensis	Y980	8.9	8.0	KL11	34	4	Cursorial	Mixed feeder	~2000
Rhinocerotina	Dicerorhinus aff. sumatrensis	Y641	13.7	10.1	Y311	10	5	Cursorial	Browser	1232
Rhinocerotina	Ceratotherium neumayri	Y399	8.0	7.4	Y907	6	3	Cursorial	Grazer	2487

Notes: BW = body weight estimate, in kg; FLO = first local occurrence; LO = last local occurrence; Loc. = locality; N = number; pc = postcranial; spec. = specimen. Locomotor types are inferred through appendicular skeletal features. Diet (browser or mixed feeder) is hypothesized on the basis of tooth-crown height, amount of cement present, and morphological similarity with modern analogs. Body weight range estimates result from the equations of Legendre (1989) and Fortelius and Kappelman (1993).

*The age of locality Y692 is not well constrained. Current estimates range from 16.5 to 17.3 Ma; the midpoint is used here.

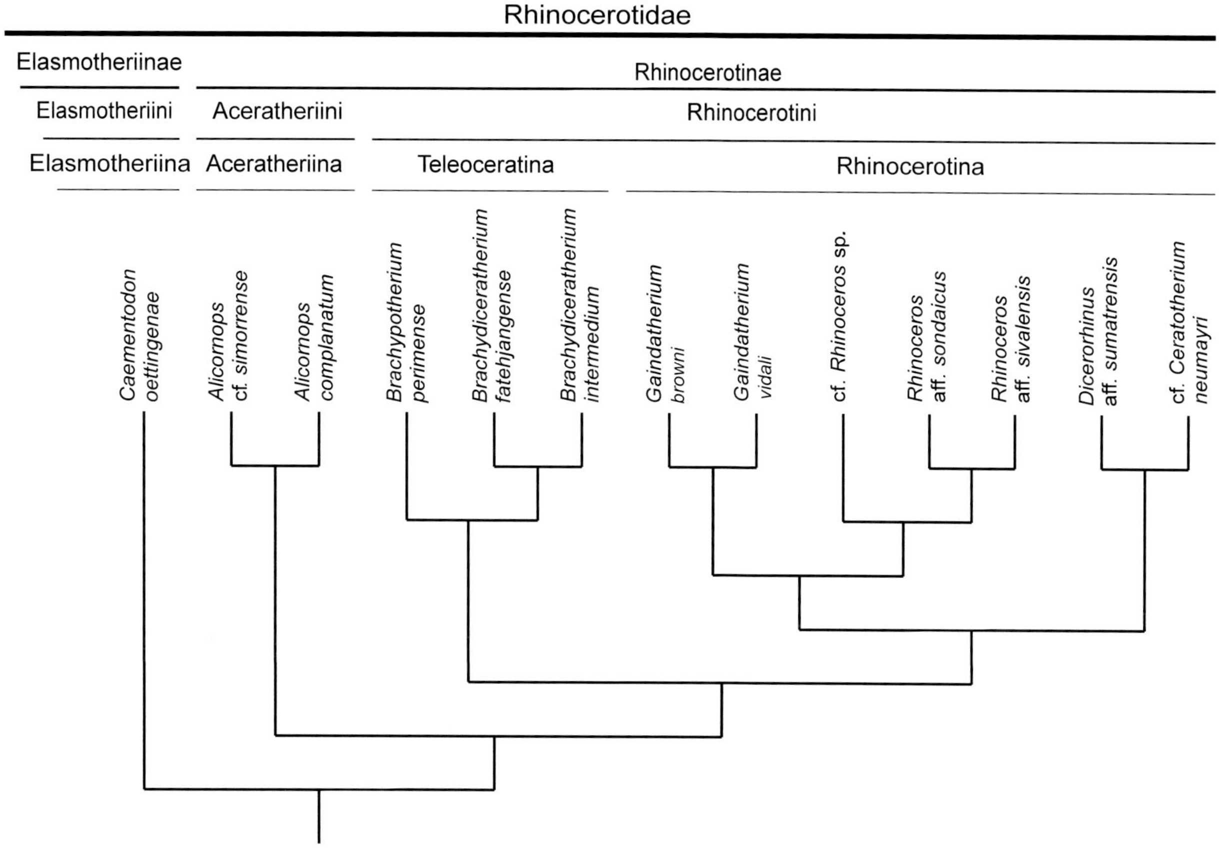

Figure 18.3. Informal phylogenetic relationships of rhinocerotids from the Siwalik faunal sequence of the Potwar Plateau. Phylogenetic framework follows Antoine et al. (2010), Becker et al. (2013), and Sizov, Klementiev, and Antoine (2024).

group of the Rhinocerotina until phylogenetic analyses demonstrated that this clade was more closely related to North American Diceratheriini and Menoceratina, within Elasmotheriinae, with an early differentiation from Rhinocerotinae (see Fig. 18.3; Antoine 2002; Antoine, Duranthon, and Welcomme 2003). Early elasmotheriines were tapir-sized, slender-limbed rhinos; some of them had small nasal horns, while the latest representatives of this clade were mammoth-sized, with a huge frontal horn and ever-growing teeth (e.g., Pleistocene *Elasmotherium*; Antoine 2002).

The only elasmotheriine recognized in the Siwalik faunal sequence is *Caementodon oettingenae* Heissig, 1972, an early-diverging taxon thus far recorded in the Chinji and Nagri Formations of the Potwar Plateau (Heissig 1972; Antoine 2002). New data later extend its stratigraphic range to the lower Dhok Pathan Formation (8.6 Ma; see Table 18.2) and earlier to the middle Kamlial Formation (15.1 Ma). Around 50 cranial, dental, and postcranial remains are recognized, originating from 30 localities (body weight ~810 kg; Figs. 18.4a, b, 18.5a). The earliest elasmotheriine, *Bugtirhinus praecursor* Antoine and Welcomme, 2000, is smaller (body weight ~480 kg) and restricted to the lowermost Miocene deposits of Sulaiman Province (Fig. 18.6; Antoine et al. 2013), which attests to the importance of the Indian subcontinent in the early evolution of this clade. A close relative of this taxon was recently described from coeval deposits from Vietnam (*Bugtirhinus* sp.; Prieto et al. 2018).

Among Aceratheriini Rhinocerotinae, **Aceratheriina** are hornless extinct rhinos, widespread in the Oligocene–Miocene of North America and Eurasia, as well as in the Miocene of Africa (e.g., Geraads 2010). Most of them are slender-limbed, tetradactyl, and brachydont, even though a small group known as chilotheres developed short-limbs and high-crowned teeth (e.g., Antoine, Duranthon, and Welcomme 2003).

At least two taxa from the Lower and Middle Siwaliks document the genus *Alicornops* Ginsburg and Guérin, 1979, the representatives of which lack the enlarged hippo-like symphysis typical of more derived Asian genera such as *Chilotherium* Ringström, 1924, and *Acerorhinus* Kretzoi, 1942. "*Aceratherium* sp." from the Chinji Formation (Heissig 1972) and "*Alicornops* sp., cf. *A. simorrense*" from the upper Nagri Formation/lower Dhok Pathan Formation (Guérin 1979) probably belong to the same taxon, which can be provisionally designated as *Alicornops* cf. *simorrense* (body weight ~880 kg; see Fig. 18.6). This taxon is documented by a single metatarsal from Locality Y478 (14.1 Ma) in the collections surveyed here. This occurrence is coeval with the maximal abundance of *A. simorrense* in Europe (Heissig 2012). The other species is *Alicornops complanatum* (Heissig 1972), which is much better represented and occurs throughout the Dhok Pathan Formation in the Potwar Plateau (Colbert 1935; Heissig 1972; new data) and in coeval deposits of the Bugti Hills (Antoine, Duranthon, and Welcomme 2003). The specimens examined span a 16.8–8.0 Ma interval and consist of 33 cranio-mandibular, dental, and postcranial remains from eight localities (mainly Y450 and Y545; Figs. 18.4c, 18.5b). Body weight approximates 820 kg. The upper Dhok Pathan

specimens described by Heissig (1972) and referable to *A. complanatum* would be among the last occurrences of aceratheriines throughout Eurasia, but their numeric age is not constrained.

Sixteen mandibular, dental, and postcranial specimens were assigned to *Alicornops* sp., for want of a better alternative. They first occur at 16.8 Ma (i.e., earlier than the first appearance datum [FAD] of *Alicornops* in Europe, estimated at ~15.0 Ma; Antoine, Duranthon, and Welcomme 2003) and are likely to document another representative of *Alicornops* in the Siwalik faunal sequence.

Teleoceratina are hippo-like extinct Rhinocerotini, closely related to Rhinocerotina (see Fig. 18.3). Most of them had hornless skulls, barrel-shaped bodies, and shortened limbs adapted to swamps and riverbanks. Teleoceratines span the Late Oligocene–latest Miocene interval in Eurasia, the Miocene epoch in Africa, and the Early Miocene–Early Pliocene in North and Central America (Prothero 2005; Sizov, Klementiev, and Antoine 2024). Most teleoceratines are interpreted as browsers based on both dental morphology and isotopic studies (MacFadden 1998). Available isotopic data on Siwalik teleoceratines indicate a C_3 plant diet, consistent with this interpretation (see Chapter 25, this volume). All three teleoceratine species formally recognized in the Siwalik sequence are long-lived (~10 myr) and widely co-occurred throughout their ranges.

After long controversies regarding its generic assignment, the puzzling "*Rhinoceros sivalensis intermedius* Lydekker, 1884," from the Chinji and Nagri Formations, was referred to as "*Chilotherium intermedium intermedium*" by Heissig (1972), who thus considered it a short-limbed aceratheriine. A later formal phylogenetic analysis instead allocated its holotype to the Teleocerotina (Antoine, Duranthon, and Welcomme 2003, Fig. 4). Among the 42 undescribed specimens unambiguously documenting this taxon in the Siwalik faunal sequence between 17.9 and 8.0 Ma (10 localities), 27 dental and postcranial remains found in association at Locality Y678 (same individual) confirm its assignment to teleoceratines (Figs. 18.4d, 18.5c, d). These specimens further point to the early-diverging genus *Diaceratherium*, very similar to specimens of the Early Miocene *D. aginense* from western Europe, based on postcranial traits (slender zeugopods and autopods; e.g., Répelin 1917), and testifying to the survival of long-limbed teleoceratines in South Asia long after their supposed demise elsewhere in Eurasia. A recent phylogenetic analysis of Eurasian teleoceratines confirms the assignment of both species to *Brachydiceratherium* Lavocat, 1951 (hereafter *Bd.*), whereas the genus *Diaceratherium* should be restricted to its type species, *D. tomerdingense* Dietrich, 1931 (Sizov, Klementiev, and Antoine 2024). I therefore consider the binomen *Brachydiceratherium intermedium* (Lydekker, 1884) to be appropriate here. Body weight is estimated at ~1,900 kg. A close ally of *D. intermedium* showing similar postcranial features occurs in the late Early Miocene Tagay assemblage of Olkhon Island, in Siberia (Sizov, Klementiev, and Antoine 2024). The Pliocene occurrence of "*C. intermedium*," as mentioned by

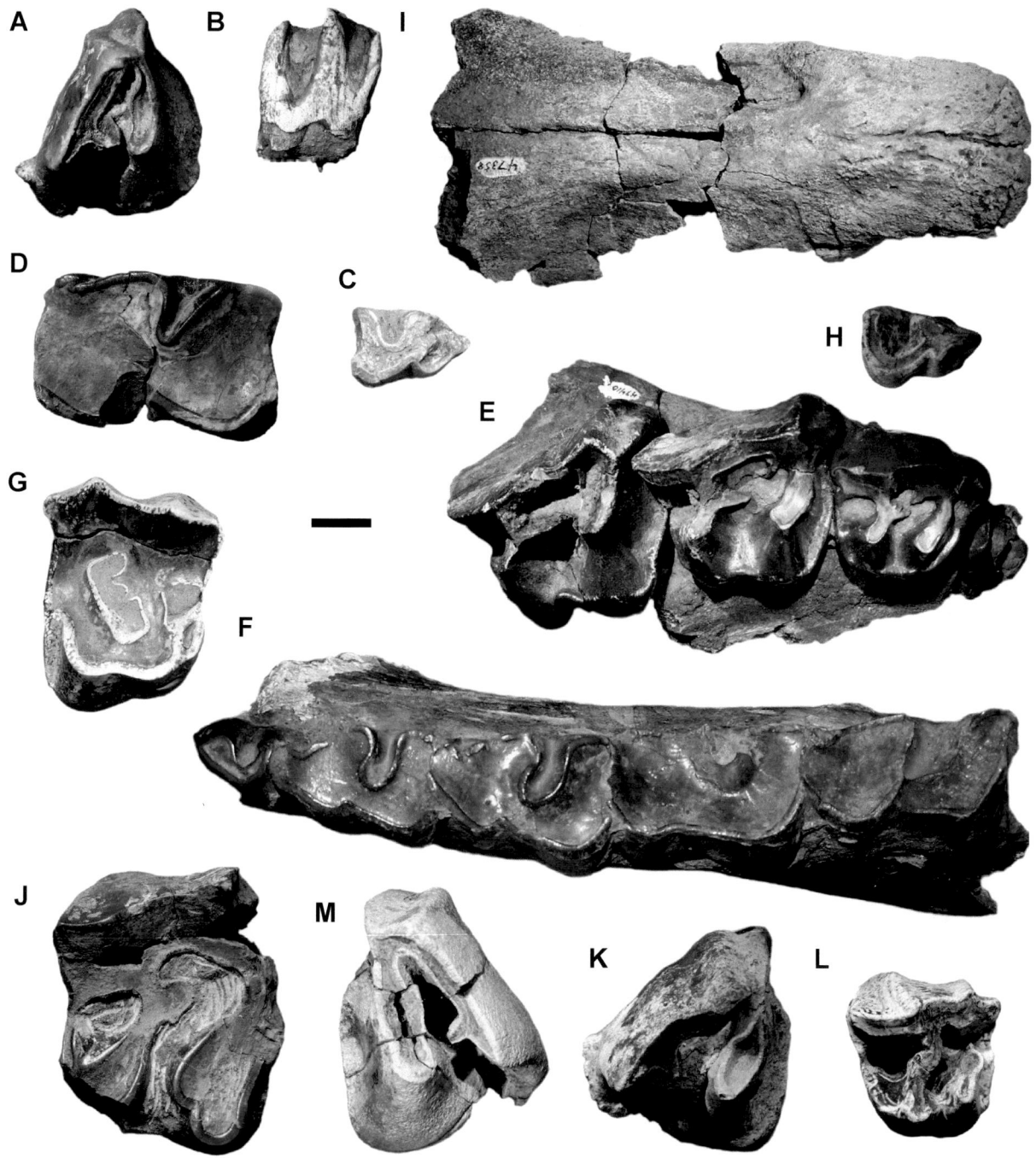

Figure 18.4. Rhinocerotid fossils from the Siwalik record of the Potwar Plateau: cranio-dental remains. A,, B. *Caementodon oettingenae* Heissig, 1972: right M3 (YGSP 50690, Loc. Y942), in occlusal view (A); left m1 (YGSP 47182, Loc. Y76), in lingual view (B). Valleys are filled by cement. C. *Alicornops complanatum* (Heissig 1972): right p2, in occlusal view (YGSP 17960, Loc. Y504). D. *Brachydiceratherium intermedium* (Lydekker 1884): left m2, in occlusal view (YGSP 24184f, from Loc. Y678). E. *Brachydiceratherium fatehjangense* (Pilgrim 1910): juvenile right maxilla with D1–4, in occlusal view (YGSP 47140, Loc. Y496). F. *Brachypotherium perimense* (Falconer and Cautley 1847): left jaw with p2-m2, in occlusal view (YGSP 20609d, Loc. Y311). G. *Gaindatherium browni* Colbert, 1934: left P4, in occlusal view (YGSP 26657, Loc. Y695). H. *Gaindatherium vidali* Heissig, 1972: right p2, in occlusal view (Y51741, Loc. Y1007). I. *Gaindatherium* sp.: nasal bones, in dorsal view (YGSP 47358, Loc. Y647). J. cf. *Rhinoceros* sp.: right M1, in occlusal view (YGSP 17538, Loc. Y592). K. *Rhinoceros* aff. *sivalensis*: right M3, in occlusal view (YGSP 28225, Loc. Y744). L. *Dicerorhinus* aff. *sumatrensis*: right P3, in occlusal view (YGSP 28099, Loc. Y705 locality). M. *Ceratotherium neumayri* (Osborn 1900): left M3, in occlusal view (YGSP 17993, survey block KL08). Scale bars = 15 mm (A–E, G, H, J–M) and 20 mm (F, I).

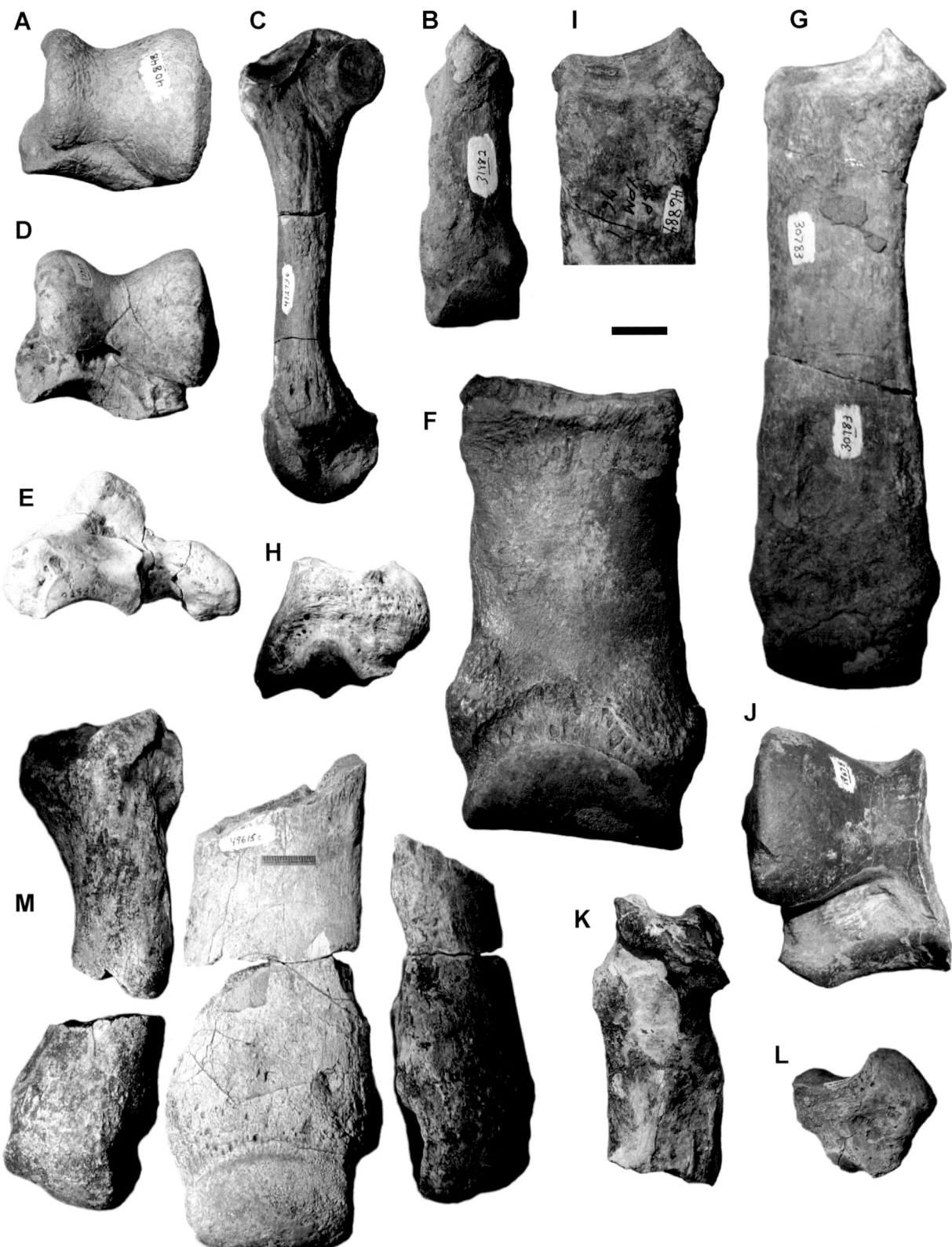

Figure 18.5. Rhinocerotid fossils from the Siwalik record of the Potwar Plateau: postcranial remains. A. *Caementodon oettingenae* Heissig, 1972: left astragalus, in anterior view (YGSP 40848, Loc. Y76). B. *Alicornops complanatum* (Heissig 1972): right Mt2, in anterior view (YGSP 31482, Loc. Y545). C., D. *Brachydiceratherium intermedium* (Lydekker 1884): right Mc4 (C), in medial view (YGSP 41273c, Loc. Y678); left astragalus (D), in anterior view (YGSP 17757, Loc. Y545). E. *Brachydiceratherium fatehjangense* (Pilgrim 1910): right magnum in medial view (YGSP 32536, Loc. Y738). F. *Brachypotherium perimense* (Falconer and Cautley 1847): right third metatarsal, in anterior view (YGSP 22160, Loc. Y735). G., H. *Gaindatherium browni* Colbert, 1934: left Mc3 (G), in anterior view (YGSP 30783, Loc. Y76); right scaphoid (H), in medial view (Y50918, Loc. Y76); to be compared with the scaphoid of *Dicerorhinus* aff. *sumatrensis*. I. *Gaindatherium vidali* Heissig, 1972: left Mc3, proximal fragment (YGSP 46884, Loc. Y76), in anterior view; to be compared with the Mc3 referred to as *G. browni*. J. Cf. *Rhinoceros* sp.: right broken astragalus, in anterior view (YGSP 19678, Loc. Y592). K. *Rhinoceros* aff. *sivalensis*: proximal fragment of a right Mc2, in posterior view (Y52781, Loc. Y980); note the presence of a large trapezium-facet. L. *Dicerorhinus* aff. *sumatrensis*: right incomplete scaphoid, in medial view (Y51445, Loc. Y76). M. *Ceratotherium neumayri* (Osborn 1900): right metacarpus (YGSP 49615, Loc. Y399), in anterior view. Scale bar = 20 mm.

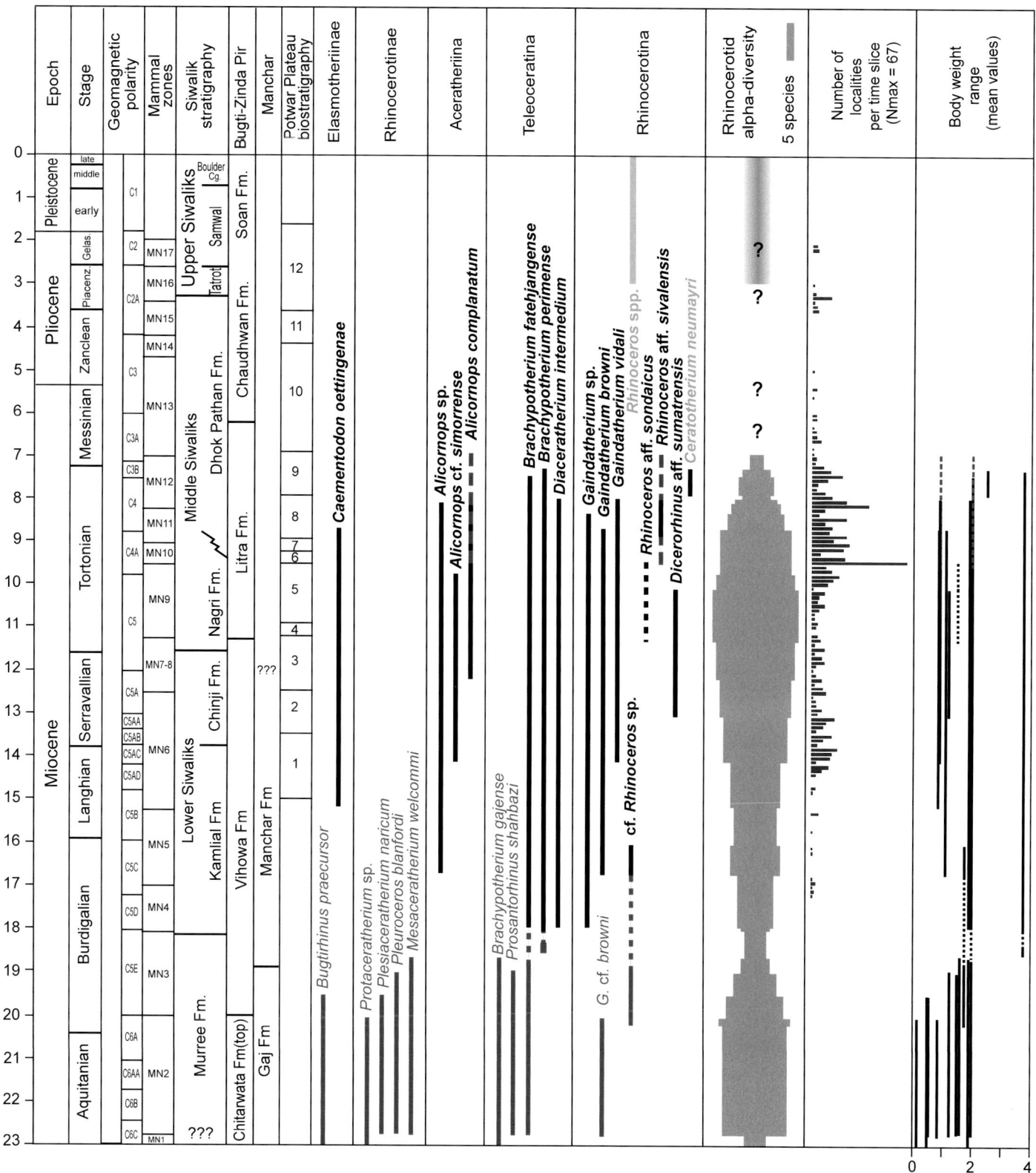

Figure 18.6. Stratigraphic ranges, alpha-diversity, and body weight range for the Rhinocerotidae documented in the Neogene of the Potwar Plateau (bold-type and black) and adjacent areas from Pakistan (Sulaiman Province; in red and dark grey), India, and Nepal (blue). Based on Antoine et al. (2013), Patnaik (2013), and original data. Biostratigraphy of the Potwar Plateau was revised after Pilbeam et al. (1979), Barry et al. (2002, 2013), and Wang et al. (2013). Number of localities per time slice and mean body weight estimates are from Barry et al. (2013) and Table 18.2, respectively.

Patnaik (2013) in the Siwaliks of India, is not reliable. Two larger teleoceratines are recorded in the Siwalik faunal sequence. These similarly long-ranged (>10 myr) and widely co-occurring species were the most abundant Miocene rhinos in the Potwar Plateau. *Brachydiceratherium fatehjangense* (Pilgrim 1910) has a continuous record in the Siwalik sequence, is documented by 44 cranio-dental and postcranial remains from 29 localities (Figs. 18.4e, 18.5e), and spans the Kamlial–Dhok Pathan interval (17.9–7.4 Ma). Its body weight was ~2,000 kg. Only three specimens (6.8%) belong to juvenile individuals. The first appearance datum of this species dates back to the earliest Miocene in central Pakistan (see Fig. 18.6; Antoine et al. 2013). In the Siwalik sequence of Jammu Sub-Himalaya, India, *Bd. fatehjangense* occurs in lower Murree sediments (early Miocene; Patnaik 2013); Chavasseau et al. (2006) report *Bd. fatehjangense* in the Middle Miocene Chaungtha locality of central Myanmar (Irrawaddy Formation; see Fig. 18.1).

Brachypotherium perimense (Falconer and Cautley 1847) is the largest and heaviest teleoceratine ever known (mean body weight of ~3,710 kg with some individuals [over]estimated at 6,100 kg; see Tables 18.1, 18.2) with extremely short carpals and tarsals (Figs. 18.4f, 18.5f). This species was coeval with *Bd. fatehjangense*, but occurred more widely in the Siwalik sequence. It is recognized from 180 cranio-mandibular, dental, and postcranial remains from 62 localities (18.0–7.2 Ma). Juveniles, subadults, and adults are represented by 20 (11.1%), 3 (1.7%), and 154 specimens (87.2%), respectively. Within the rhinocerotid record from the Potwar Plateau, *B. perimense* follows the closest age distribution with respect to natural populations of recent rhinoceroses (Chitwan Park; Laurie, Lang, and Groves 1983). *Brachypotherium perimense* was originally described by Falconer (1845) from the Late Miocene of Perim Island, in Gujarat, western India (see Fig. 18.1; today "Piram Island"; Chauhan 2008). This huge species is also documented in the Early–Late Miocene interval in the Sulaiman Province (see Fig. 18.5; Antoine et al. 2013), in Middle Miocene deposits of the Dang Valley in Nepal (Surai Khola; Corvinus and Rimal 2001), in the Siwaliks of India (13.2–9.8 Ma; Patnaik 2013) and in upper Miocene localities of Central Thailand (Chaimanee et al. 2004).

Rhinocerotina are medium- to large-size rhinos with modern representatives living either in riversides and moist forests (Asian rhinos; Antoine 2012) or savannahs (African rhinos; Guérin 1980). All five living rhino species belong to this clade; three live in South Asia, of which two have only a nasal horn: the large *Rhinoceros unicornis* (Indian rhino) and the lesser *R. sondaicus* (nearly extinct Javan rhino). The third, *Dicerorhinus sumatrensis* (Sumatran rhino), also has a frontal horn, as do African rhinos (*Diceros bicornis* and *Ceratotherium simum*, black and white rhino, respectively). The fossil record of Rhinocerotina is so far restricted to the Old World, and this clade is particularly well represented in the Siwalik faunal sequence (see Fig. 18.3). Among Rhinocerotidae, only Rhinocerotina have a well-constrained record after the Miocene–Pliocene transition in the Siwalik record. Limited stable isotope data suggest that

Miocene rhinocerotines foraged on C_3 plants (see Chapter 25, this volume).

Gaindatherium Colbert, 1934, is a group of medium-size, one-horned rhinocerotines, closely related to *Lartetotherium* Lartet, 1837, from the late Early–early Late Miocene of Europe (Antoine, Duranthon, and Welcomme 2003; Antoine et al. 2010). At least two representatives of *Gaindatherium* occur throughout the Miocene epoch in the Potwar Plateau: *Gaindatherium browni* Colbert, 1934, is documented through 75 dental and postcranial remains (Figs. 18.4g, 18.5g, h), with a temporal range from at least 16.5–8.7 Ma (25 localities). (The age of the oldest locality is not well constrained; see Table 18.2.) Most specimens represent adults (72/75 = 96%). Estimated body weight is ~1,070 kg. The sample studied by Heissig (1972) was restricted to the Middle Miocene Chinji Formation, while Patnaik (2013) provides a 12.7–8.85 Ma range for this species in the Siwaliks of India. Compared to *G. browni*, its sister species *G. vidali* Heissig, 1972, is slightly smaller (body weight ~830 kg; Figs. 18.4h, 18.5i) with more slender proportions in the appendicular skeleton (corresponding to Rhinocerotini type 2 and type 1 of Heissig [1972], respectively). In good agreement with Heissig (1972), *G. vidali* is also less abundant than *G. browni*, with 16 dental and postcranial remains from 12 localities (ranging from 14.1 to 8.0 Ma). The two species have concurrent ranges, and they co-occur in four localities (Y76, Y311, Y502, and Y647; see Fig. 18.6). *Gaindatherium vidali* was originally reported only from the early Late Miocene Nagri Formation from the Potwar Plateau (Heissig 1972), but Khan et al. (2011) have described some remains from the Middle Miocene Chinji Formation. Due to considerable morphological and metric overlap between the two species, 58 dental and postcranial specimens were assigned only to *Gaindatherium* sp. (see Fig. 18.4I), with a first local occurrence at 17.9 Ma (Y747). *Gaindatherium* cf. *browni*, recorded in ~21 Ma deposits of the Chitarwata Formation in the Bugti Hills, provides the FAD for the genus (see Fig. 18.6; Antoine et al. 2013). Its body weight approximates 810 kg.

Close allies or putative representatives of the living genera *Rhinoceros* and *Dicerorhinus* have also been recognized in the Siwalik sequence. Ten dental and postcranial remains from two localities of the Kamlial Formation (Y744 and Y592; Figs. 18.4Jj 18.5j) attest to the presence of a large one-horned rhinocerotine during the late Early Miocene in the Potwar Plateau (16.7–16.0 Ma). This taxon is most likely conspecific with "cf. *Rhinoceros* sp." as recognized in the 19 Ma Assemblage B of the Sulaiman Province (see Fig. 18.6; Antoine et al. 2013). Its body weight is estimated at ~1,740 kg (see Table 18.2). A small isolated DP1 from the Y311 locality (YGSP 10466; 10.1 Ma) documents an early relative of the lesser one-horned rhino, namely *Rhinoceros* aff. *sondaicus*, as described in coeval deposits of the Nagri Formation by Heissig (1972). The size of this tooth indicates a body weight of ~1,500 kg. A larger representative of *Rhinoceros* is recorded in four localities in the interval between 8.9 and 8.0 Ma through 34 dental and postcranial specimens (30 of which belong to the same individual; YGSP 49617 at Locality Y897). Their general dimensions yield body weight estimates of approximately 2,000 kg (Figs. 18.4k, 18.5k). These remains strongly resemble those assigned to *Rhinoceros* aff. *sivalensis* by Heissig (1972) in the Dhok Pathan Formation and in coeval deposits from the Bugti Hills (~8.5 Ma; Antoine et al. 2013). *Rhinoceros sivalensis* Falconer and Cautley, 1847, occurs in Early Pleistocene deposits of Pakistan and India (see Patnaik 2013), and a latest Miocene–Pliocene ghost lineage might be inferred for this taxon in the Himalayan foothills (Raza et al. 1984).

A palate, three isolated upper premolars, and six slender postcranial bones of small to medium dimensions were attributed to a close ally of the two-horned Sumatran rhino, *Dicerorhinus* aff. *sumatrensis* (Figs. 18.4l, 18.5l). The fossils come from five localities in the Chinji and Nagri Formations, with a temporal range of 13.7–10.1 Ma (see Fig. 18.6). Body weight is estimated at ~1,230 kg. The same taxon was previously described in the Chinji Formation (Heissig 1972). The remains from the middle and upper Chinji Formation, referred to as "*Didermocerus* cf. *abeli*" by Heissig (1972), probably document *Dicerorhinus* aff. *sumatrensis* instead of "*Aceratherium abeli* Forster-Cooper, 1934," an enigmatic rhinocerotid restricted to Oligocene layers from the Bugti Hills (Antoine et al. 2013, 417).

Another putative two-horned rhinocerotine was documented in upper Miocene deposits (8.0–7.4 Ma), based on an isolated M3 (Fig. 18.4m), and five postcranial bones of large dimensions (four bones belong to the same right hand; see Fig. 18.5M), were assigned to *Ceratotherium neumayri* (Osborn 1900). This species was widespread throughout the eastern Mediterranean in the Late Miocene, with maximum abundance around 9–8 Ma (e.g., Antoine et al. 2012). The body weight of *C. neumayri* was around 2,500 kg (see Table 18.1). It is recognized for the first time in the Siwalik faunal sequence in this summary.

Other rhinocerotine taxa from the Upper Siwaliks of Pakistan and India (Late Pliocene–early Middle Pleistocene), such as *Rhinoceros palaeindicus* Falconer and Cautley, 1847 (junior synonym of *R. unicornis*; e.g., Laurie, Lang, and Groves 1983; Antoine 2012), and *Rhinoceros platyrhinus* Falconer and Cautley, 1847 (for a review, see Pandolfi and Maiorino 2016), have not been recognized in material currently available. Until recently, all three extant Asian rhinos were present in the Indian Himalayan foothills (e.g., Antoine 2012). *R. unicornis* was widespread in the Late Pleistocene of peninsular India (Chauhan 2008). It no longer occurs in Pakistan, but it is still living in India, Nepal, and Bhutan (Antoine 2012).

RHINOCEROTID DIVERSITY AND BODY WEIGHT CHANGES THROUGHOUT THE SIWALIK SEQUENCE

When considered as part of one paleobiogeographic province, the Potwar Plateau (Kamlial Formation–Middle Dhok Pathan Formation; Barry et al. 2002, 2013) and Sulaiman Province (Barry et al. 2005; Antoine et al. 2013) together provide an exceptional overview of rhinocerotid diversity and evolutionary dynamics for the Neogene interval in South Asia. Regarding

the subsequent interval (latest Miocene–?Early Pliocene; Upper Dhok Pathan Formation), no chronostratigraphic constraint is available for rhinocerotid remains described by Heissig (1972). Moreover, a reliable taxonomic revision of Upper Siwalik rhinocerotines is badly needed (these taxa are gathered under the name "*Rhinoceros* spp."; see Fig. 18.6). Hence, the last 7 million years of Siwalik rhinocerotids are not discussed further here.

Regardless of the number of suprageneric groups considered, there are two intervals of high species diversity (see Fig. 18.6). They are not obviously correlated with the number of localities documenting a given time slice in the Potwar Plateau (see Fig. 18.6, right columns), although they might be linked to ecological factors (such as climate and food supply), as discussed below.

After a short interval with only three co-occurring species (23.0–22.7 Ma), there is a diversity peak in the Early Miocene (Aquitanian and earliest Burdigalian times, 22.7–19.5 Ma), with 8–10 coeval species recorded in the Bugti Hills. This interval coincides with warm, moist climatic conditions and the presence of extensive tropical forests in South Asia. Such an ecosystem could support species-rich browser-dominated herbivore guilds as documented in Sulaiman Province (e.g., Antoine et al. 2013) and supported by isotopic analyses of tooth enamel (e.g., Martin, Bentaleb, and Antoine 2011). During this interval, body weights are evenly distributed and balanced within a 150–2,000 kg domain. Four estimated species weights are less than 1,000 kg (the tapir-sized *Protaceratherium* sp.: 150 kg; *Bugtirhinus praecursor*: ~480 kg; *Plesiaceratherium naricum*: ~550 kg; *Gaindatherium* cf. *browni*: ~810 kg; see Table 18.1), whereas *Pleuroceros blanfordi* (~1,530 kg), *Mesaceratherium welcommi* (~1,620 kg), and teleoceratines are much heavier (*Prosantorhinus shahbazi*: ~1,220 kg; *Brachypotherium gajense*: ~1,920 kg; *Brachydiceratherium fatehjangense*: ~2,000 kg).

Rhinocerotid species diversity then decreases gradually, with the possible presence of three species between 18.7–18.0 Ma, hypothesized due to (1) a gap between the top of Assemblage B in the Sulaiman Province (Antoine et al. 2013) and the base of the Kamlial Formation in the Potwar Plateau (Wang et al. 2013) and (2) the relative paucity of Kamlial localities (Barry et al. 2013). In the species-poor interval from 19.0–16.7 Ma, a marked loss of small body sizes and shift toward heavier body weights is recorded, with a range of 1,740–3,710 kg. The lighter rhinocerotid is cf. *Rhinoceros* sp. (~1,740 kg; see Table 18.1). Once again, teleoceratines comprise the heavier component of the fauna, with *Brachydiceratherium intermedium* (~1,900 kg; "replacing" *B. gajense* in the assemblage), *Bd. fatehjangense* (~2,000 kg), and the first appearance of the heavy *B. perimense* (~3,710 kg).

The next stage (16.7–7.0 Ma) has species-rich assemblages, with the greatest body weight range (810–3,710 kg). A second diversity peak following that of the Early Miocene occurs in the early Late Miocene (11–10 Ma, lower Nagri Formation), with up to 12 co-occurring species, despite low numbers of localities per time slice (see Fig. 18.6; Barry et al. 2013). These rhino assemblages are species-rich until 8.7 Ma (around eight species at the Nagri–Dhok Pathan transition; see Fig. 18.6), after which diversity decreases again to only two species at ~7 Ma (middle Dhok Pathan). This decline occurs in spite of high numbers of localities per time slice (see Fig. 18.6; Barry et al. 2013; Wang et al. 2013). This last interval coincides with the replacement of wet monsoonal forest by dry woodland then by savannah (Badgley et al. 2008). A salient feature of the 17.9–7.0 Ma interval, aside from the persistence of heavy, water-dependent taxa, is the co-occurrence of four rhinocerotids having similar body weights between 800–900 kg (*Caementodon oettingenae*: ~810 kg; *Alicornops complanatum*: ~820 kg; *A.* cf. *simorrense*: ~880 kg; *Gaindatherium vidali*: ~830 kg), probably with different feeding habits. For the rhinocerotines, body weights were non-overlapping over a wide range (*Gaindatherium browni*: ~1,070 kg; *Dicerorhinus* aff. *sumatrensis*: ~1,230 kg; *Rhinoceros* aff. *sondaicus*: ~1,500 kg; cf. *Rhinoceros* sp.: ~1,740 kg; *R.* aff. *sivalensis*: ~2,000 kg; *Ceratotherium neumayri*: ~2,490 kg; see Table 18.1).

In summary, throughout the sequence, taxonomic diversity is particularly unbalanced among Rhinocerotidae, with only 2 species of Elasmotheriinae but over 20 species recognized for Rhinocerotinae (see Fig. 18.4, Fig. 18.5). Most rhinocerotid species have long stratigraphic ranges in the Siwalik faunal sequence (see Fig. 18.6), with the notable exception of cf. *Rhinoceros* sp., *Rhinoceros* aff. *sondaicus* (if distinct from *R. sondaicus*), and *Ceratotherium neumayri*. The elasmotheriines *Bugtirhinus praecursor* (23.0–19.5 Ma) and *Caementodon oettingenae* (15.1–8.6 Ma) did not overlap. In addition, the ~4.4 myr-long gap (no elasmotheriine in the Kamlial Formation) implies that *C. oettingenae* is an early Middle Miocene immigrant from another Eurasian region (other representatives of *Caementodon* are known in the late Early–early Middle Miocene of both the Caucasus and China; see Antoine 2002). Early-diverging Rhinocerotinae (*Protaceratherium* sp., *Plesiaceratherium naricum*, *Pleuroceros blanfordi*, and *Mesaceratherium welcommi*) comprise almost half of species richness during the earliest interval (23–19 Ma; see Fig. 18.6). They co-occur with *Bugtirhinus praecursor*, the earliest representatives of Teleoceratina (either short-ranging *Brachypotherium gajense* and *Prosantorhinus shahbazi* or long-ranging *Bd. fatehjangense*), and the first Rhinocerotina (*Gaindatherium* cf. *browni* and cf. *Rhinoceros* sp.). Their gradual disappearance is the main component of the diversity decrease following the Early Miocene peak. A major turnover in rhinocerotids occurs then, with the first local occurrence of the teleoceratines *Brachypotherium perimense* and *Brachydiceratherium intermedium* (~18 Ma), followed by the appearance of the rhinocerotine *Gaindatherium browni* (16.9 Ma) and the aceratheriine *Alicornops* sp. (16.8 Ma) in the middle Kamlial Formation. Rhinocerotid assemblages are then stable, with only the first occurrence of *Caementodon oettingenae* at 15.1 Ma and the disappearance of cf. *Rhinoceros* sp. (12.4 Ma) as noticeable events. Afterwards, diversity increases to reach a peak between 11 and 10 Ma, with representatives of all suprageneric groups among Rhinocerotidae (Elasmotheriina: *Caementodon oettingenae*; Aceratheriina: *Alicornops* cf. *simorrense* and *A. complanatum*; Teleoceratina: *Brachydiceratherium intermedium*, *Bd. fatehjangense*, and *Brachypotherium perimense*; Rhinocerotina: *Gaindatherium*

browni, *G. vidali*, *Rhinoceros* aff. *sondaicus*, *R.* aff. *sivalensis*, and *Dicerorhinus* aff. *sumatrensis*). In younger intervals, species richness decreases gradually until the last rhinocerotid record surveyed here (7.2 Ma). The immigrant *Ceratotherium neumayri* is briefly present in the 8.0–7.4 Ma interval and may be from the eastern Mediterranean, where it is a conspicuous element of Tortonian mammalian faunas (e.g., Antoine et al. 2012). The persistence of *Rhinoceros* aff. *sivalensis*, *Alicornops complanatum*, and *Brachypotherium perimense* in deposits from the Upper Dhok Pathan Formation (latest Miocene–Early Pliocene) was described by Heissig (1972). They probably document the last occurrences of both Aceratheriina and Teleoceratina on a global scale, but their age is not well constrained chronostratigraphically. Although a few localities document this stratigraphic interval (see Fig. 18.6; Barry et al. 2013), rhinocerotid remains are not present in the available sample (presently curated at HUPL or MNHN).

CONCLUSION

The analysis of ~2,000 cranio-mandibular, dental, and postcranial specimens from Miocene deposits of Pakistan (~800 from the Potwar Plateau and 1,200 from Sulaiman Province) allows the recognition of more than 20 rhinocerotid species. The major identified turnover between 18.7–18.0 Ma consists of the demise of early-diverging representatives of Rhinocerotinae and the onset of assemblages including two or three small aceratheriines, two or three large teleoceratines, and up to eight small to large rhinocerotines (from 18–7 Ma). In the Potwar Plateau, rhinocerotid species richness increased regularly from the earliest Kamlial record to reach a peak at 11–10 Ma, with at least 9 coeval species (mostly browsers). One-horned and two-horned rhinocerotines and their sister taxa (hippo-like teleoceratines) dominated the megaherbivore guilds throughout the 18–7 Ma interval, under humid, then drier environmental conditions. Notably, this study confirms the teleoceratine affinities of "*Rhinoceros sivalensis intermedius* Lydekker, 1884," here designated as *Brachydiceratherium intermedium* (Lydekker 1884), and includes the first recognition of the large two-horned rhinocerotine *Ceratotherium neumayri* in South Asia. A thorough taxonomic revision of rhinocerotid remains originating from younger intervals of the Upper Siwaliks would be critically important, especially for better constraining the final demise of stem-rhinocerotids and the emergence of modern Asian rhinoceroses.

ACKNOWLEDGMENTS

I am deeply indebted to all the collectors in the field (Potwar Plateau, Bugti Hills, and Zinda Pir) and colleagues in charge of paleontological collections I visited for this study (Harvard University Paleoanthropology Laboratory, Cambridge; American Museum of Natural History, New York; Muséum National d'Histoire Naturelle, Paris; Muséum de Toulouse; Université de Montpellier; Université Lyon 1; Natural History Museum, London; Bayerische Staatssammlung für Paläontologie und

Geologie, Munich; Pakistan Museum of Natural History, Islamabad). I would like to warmly thank Michèle E. Morgan, Catherine Badgley, and David Pilbeam for inviting this contribution, John C. Barry and Larry J. Flynn for providing critical information regarding the Siwalik faunal sequence, and all of them for their inestimable kindness.

References

Antoine P.-O. 2002. Les Elasmotheriina (Mammalia, Perissodactyla): Phylogénie et paléobiogéographie. *Mémoires du Muséum National d'Histoire Naturelle, Paris, Série C, Sciences de la Terre* 188: 359 pp.

Antoine, P.-O. 2012. Pleistocene and Holocene rhinocerotids (Mammalia, Perissodactyla) from the Indochinese Peninsula. *Comptes Rendus Palevol* 11(2–3):159–168.

Antoine, P.-O., D. Becker, L. Pandolfi, and D. Geraads. In press. Evolution and fossil record of Old World Rhinocerotidae. In *Rhinos of the World: Ecology, Conservation and Management*, edited by M. Melletti, D. Balfour, and B. Talukdar. Springer Nature.

Antoine, P.-O., K. F. Downing, J.-Y. Crochet, F. Duranthon, L. J. Flynn, L. Marivaux, G. Métais, A. R. Rajpar, and G. Roohi. 2010. A revision of *Aceratherium blanfordi* Lydekker, 1884 (Mammalia: Rhinocerotidae) from the early Miocene of Pakistan: postcranials as a key. *Zoological Journal of the Linnean Society* 160:139–194.

Antoine, P.-O., S. Ducrocq, L. Marivaux, Y. Chaimanee, J.-Y. Crochet, J.-J. Jaeger, and J. L. Welcomme. 2003. Early rhinocerotids (Mammalia, Perissodactyla) from South Asia and a review of the Holarctic Paleogene rhinocerotid record. *Canadian Journal of Earth Sciences* 40: 365–374.

Antoine, P.-O., F. Duranthon, and J.-L. Welcomme. 2003. *Alicornops* (Mammalia, Rhinocerotidae) dans le Miocène supérieur des Collines Bugti (Balouchistan, Pakistan): implications phylogénétiques. *Geodiversitas* 25:575–603.

Antoine, P.-O., G. Métais, M. J. Orliac, J.-Y. Crochet, L. J. Flynn, L. Marivaux, A. R. Rajpar, G. Roohi, and J.-L. Welcomme. 2013. Mammalian Neogene biostratigraphy of the Sulaiman Province, Pakistan. In *Fossil Mammals of Asia: Neogene Biostratigraphy and Chronology*, edited by X.-M. Wang, L. J. Flynn, and M. Fortelius, 400–422. New York: Columbia University Press.

Antoine, P.-O., S. M. I. Shah, I. U. Cheema, J.-Y. Crochet, D. de Franceschi, L. Marivaux, G. Métais, and J.-L. Welcomme. 2004. New remains of the baluchithere *Paraceratherium bugtiense* (Pilgrim, 1910) from the Late/latest Oligocene of the Bugti Hills, Balochistan, Pakistan. *Journal of Asian Earth Sciences* 24:71–77.

Antoine P.-O., and J.-L. Welcomme. 2000. A new rhinoceros from the Bugti Hills, Baluchistan, Pakistan: the earliest elasmotheriine. *Palaeontology* 43, 795–816.

Badgley, C., J. C. Barry, M. E. Morgan, S. V. Nelson, A.K. Behrensmeyer, T. E. Cerling, and D. Pilbeam, D. 2008. Ecological changes in Miocene mammalian record show impact of prolonged climatic forcing. *PNAS* 105:12145–12149.

Barry, J. C., M. E. Morgan, L. J. Flynn, D. Pilbeam, A. K. Behrensmeyer, S. M. Raza, I. A. Khan, C. Badgley, J. Hicks, and J. Kelley. 2002. Faunal and environmental change in the Late Miocene Siwaliks of Northern Pakistan. *Paleobiology* 28(3): 1–71.

Barry, J. C., A. K. Behrensmeyer, C. E. Badgley, L. J. Flynn, H. Peltonen, I. U. Cheema, D. Pilbeam, D., et al. 2013. The Neogene Siwaliks of the Potwar Plateau, Pakistan. In *Fossil Mammals of Asia: Neogene Biostratigraphy and Chronology*, , edited by X.-M. Wang, L. J. Flynn, and M. Fortelius, 373–399. New York:Columbia University Press.

Barry, J.C., S. Cote, L. MacLatchy, E. H. Lindsay, R. Kityo, and A. R. Rajpar. 2005. Oligocene and early Miocene ruminants (Mammalia, Artiodactyla) from Pakistan and Uganda. *Palaeontologia Electronica* 8(1): 1–29.

Becker, D., P.-O. Antoine, and O. Maridet. 2013. A new genus of Rhinocerotidae (Mammalia, Perissodactyla) from the Oligocene of Europe. *Journal of Systematic Palaeontology* 11·947–972.

Behrensmeyer, A. K., J. Quade, T. E. Cerling, J. Kappelman, I. A. Khan, P. Copeland, L. Roe, et al. 2007. The structure and rate of late Miocene expansion of C_4 plants: Evidence from lateral variation in stable isotopes in paleosols of the Siwalik Group, northern Pakistan. *Geological Society of America Bulletin* 119(11–12): 1486–1505.

Böhme, M., M. Aigstorfer, P.-O. Antoine, E. Appel, P. Havlik, N. V. Hung, G. Métais, et al. 2014. Na Duong (Northern Vietnam)—An exceptional window into Eocene ecosystems from South-East Asia. *Zitteliana A* 53:120–167.

Chaimanee, Y., V. Suteethorn, P. Jintasakul, C. Vidthayanon, B. Marandat, and J.-J. Jaeger. 2004. A new orang-utan relative from the Late Miocene of Thailand. *Nature* 427:439–41.

Chauhan, P. R. 2008. Large mammal fossil occurrences and associated archaeological evidence in Pleistocene contexts of peninsular India and Sri Lanka. *Quaternary International*, 192, 20–42.

Chavasseau, O., Y. Chaimanee, S. T. Tun, A. N. Soe, J. C. Barry, B. Marandat, J. Sudre, L. Marivaux, S. Ducrocq, and J.-J. Jaeger. 2006. Chaungtha, a new Middle Miocene mammal locality from the Irrawaddy Formation, Myanmar. *Journal of Asian Earth Sciences* 28:354–362.

Colbert, E. H., and B. Brown. 1934. A new rhinoceros from the Siwalik Beds of India. *American Museum Novitates* no. 749.

Colbert, E. H. 1935. Siwalik mammals in the American Museum of Natural History. *Transactions of the American Philosophical Society* (New Series) 26:1–401.

Corvinus, G., and L. N. Rimal. 2001. Biostratigraphy and geology of the Neogene Siwalik Group of the Surai Khola and Rato Khola areas in Nepal. *Palaeogeography, Palaeoclimatology, Palaeoecology* 165:251–279.

Falconer, H., 1845. Description of some fossil remains of Dinotherium, Giraffe, and other mammalia, from the Gulf of Cambay, western coast of India, chiefly from the collection presented by Captain Fulljames, of the Bombay Engineers, to the Museum of the Geological Society. *Quarterly Journal of the Geological Society* 1:356–372.

Falconer, H., and P. T. Cautley. 1846–1849. *Fauna antiqua sivalensis*, being the fossil zoology of the Sewalik Hills, in the north of India. Edited by H. Falconer. London: Smith Elder.

Forster Cooper, C. 1934. XIII. The extinct rhinoceroses of Baluchistan. *Philosophical Transactions of the Royal Society of London, Series B* 223:569–616.

Fortelius, M., and J. Kappelman. 1993. The largest land mammal ever imagined. *Zoological Journal of the Linnean Society* 108:85–101.

Geraads, D. 2010. Rhinocerotidae. In *Cenozoic mammals of Africa*, edited by L. Werdelin and W. J. Sanders, 669–683. Berkeley: University of California Press.

Guérin, C. 1979. Rhinocerotidae In *Miocene Sediments and Faunas of Pakistan*, Postilla Number 179, edited by D. R. Pilbeam, A. K. Behrensmeyer, J. C. Barry, and S. M. Ibrahim Shah, 36. New Haven, CT: Peabody Museum of Natural History Yale University.

Guérin, C. 1980. *Les Rhinocéros (Mammalia, Perissodactyla) du Miocène terminal au Pléistocène supérieur en Europe occidentale. Comparaison avec les espèces actuelles.* Documents du Laboratoire de Géologie de l'Université de Lyon, Sciences de la Terre no. 79:1–421.

Heissig, K. 1972. Paläontologische und geologische Untersuchungen im Tertiär von Pakistan. 5. Rhinocerotidae (Mamm.) aus den unteren und mittleren Siwalik-Schichten. *Abhandlungen der Bayerischen Akademie der*

Wissenschaften, Mathematisch-Naturwissenschaftliche Klasse, Neue Folge 152:1–112.

Holroyd, P.A., T. Tsubamoto, N. Egi, R. L. Ciochon, M. Takai, S. Turra Tun, C. Sein, and G. F. Gunnell. 2006. A rhinocerotid perissodactyl from the late Middle Eocene Pondaung Formation, Myanmar. *Journal of Vertebrate Paleontology* 26:491–494.

Khan, A. M., E. Cerdeño, M. Akhtar, M. A. Khan, A. Iqbal, and M. Mubashir. 2014. New fossils of *Gaindatherium* (Rhinocerotidae, Mammalia) from the Middle Miocene of Pakistan. *Turkish Journal of Earth Sciences* 23:452–461.

Khan, A. M., E. Cerdeño, M. A. Khan, M. Akhtar, and M. Ali. 2011. *Chilotherium intermedium* (Rhinocerotidae, Mammalia) from the Siwaliks of Pakistan: Systematic implications. *Pakistan Journal of Zoology* 43(4):651–663.

Laurie, W. A., E. M. Lang, and C. P. Groves. 1983. *Rhinoceros unicornis*. *Mammalian Species* 211:1–6.

Legendre, S. 1989. Les communautés de mammifères du Paléogène (Eocène supérieur et Oligocène) d'Europe occidentale: structures, milieux et évolution. *Münchner Geowissenschaftliche Abhandlungen (Reihe A)* 16:1–110.

Lydekker, R. 1881. Siwalik Rhinocerotidae. *Memoirs of the Geologcal Survey of India—Palaeontologica Indica* series 10(2): 1–62.

Lydekker, R. 1884. Additional Siwalik Perissodactyla and Proboscidea. *Memoirs of the Geological Survey of India—Palaeontologia Indica* series 10(3):1–34.

Lydekker, R. 1886. *Catalogue of the Fossil Mammalia in the British Museum (Natural History).* London: Taylor and Francis.

MacFadden, B. J. 1998. Tale of two rhinos: Isotopic ecology, paleodiet, and niche differentiation of Aphelops and Teleoceras from the Florida Neogene. *Paleobiology* 24:274–286

Martin, C., I. Bentaleb, I., and P.-O. Antoine. 2011. Pakistan mammal tooth stable isotopes show paleoclimatic and paleoenvironmental changes since the early Oligocene. *Palaeogeography, Palaeoclimatology, Palaeoecology* 311:19–29.

Matthew, W. D. 1929. Critical observations upon Siwalik mammals. *Bulletin of the American Museum of Natural History* 56:437–560.

Nanda, A. C. 2002. Upper Siwalik mammalian faunas of India and associated events. *Journal of Asian Earth Sciences* 21:47–58.

Osborn, H. F. 1900. Phylogeny of Rhinoceroses of Europe. *Memoirs of the American Museum of Natural History* 13:229–267.

Pandolfi, L., and L. Maiorino. 2016. Reassessment of the largest Pleistocene rhinocerotine Rhinoceros platyrhinus (Mammalia, Rhinocerotidae) from the Upper Siwaliks (Siwalik Hills, India). *Journal of Vertebrate Paleontology* 36(2): e1071266.

Patnaik, R. 2013. Indian Neogene Siwalik mammalian biostratigraphy: An overview. In *Fossil Mammals of Asia: Neogene Biostratigraphy and Chronology*, edited by X.-M. Wang, L. J. Flynn, and M. Fortelius, 423–444. New York: Columbia University Press.

Pilbeam, D. R., A. K. Behrensmeyer, J. C. Barry, and S. M. I. Shah, eds. 1979. Miocene sediments and Faunas of Pakistan. *Postilla* 179:1–45.

Pilgrim, G. E. 1908. The Tertiary and post-Tertiary freshwater deposits of Baluchistan and Sind, with notes on new vertebrates. *Records of the Geological Survey of India* 37:139–66.

Pilgrim, G. E. 1910. Notice on new mammal genera and species from the Tertiaries of India. *Records of the Geological Survey of India, Calcutta* 15:63–71.

Pilgrim, G. E. 1912. The vertebrate fauna of the Gaj Series in the Bugti Hills and the Punjab. *Paleontologia Indica* n.s. 4:1–83.

Pilgrim, G. E. 1913. The correlation of the Siwaliks with mammal horizons of Europe. *Records of the Geological Survey of India* 43:264–326.

Prieto, J., P.-O. Antoine, M. Böhme, J. van der Made, G. Métais, L. T. Phuc, Q. T. Quan, et al. 2018. Biochronological and paleobiogeographical significance of the earliest Miocene mammal fauna from Northern Vietnam. *Palaeobiodiversity and Palaeoenvironments* 98:287–313. https://doi.org/10.1007/s12549-017-0295-y.

Prothero, D. R. 2005. *The Evolution of North American Rhinoceroses.* Cambridge: Cambridge University Press.

Raza, S. M., I. U. Cheema, W. R. Downs, A. R. Rajpar, and S. C. Ward. 2002. Miocene stratigraphy and mammal fauna from the Sulaiman Range, Southwestern Himalayas, Pakistan. *Palaeogeography, Palaeoclimatology, Palaeoecology* 186:185–197.

Répelin, J. 1917. Études paléontologiques dans le sud-ouest de la France (Mammifères). Les rhinocérotidés de l'Aquitanien supérieur de l'Agenais (Laugnac). *Annales du Muséum d'Histoire naturelle de Marseille* 16:1–47.

Sizov, A., A. Klementiev, and P.-O. Antoine. 2024. An early Miocene skeleton of *Brachydiceratherium* Lavocat, 1951 (Mammalia, Perissodactyla), from the Baikal area, Russia, and a revised phylogeny of Eurasian teleoceratines. *PCI Paleontology.* https://doi.org/10.1101/2022.07.06.498987.

Tsubamoto, T. 2014. Estimating body mass from the astragalus in mammals. *Acta Palaeontologica Polonica* 59:259–265.

Wang, X.-M., L. J. Flynn, and M. Fortelius. 2013. Introduction: Toward the establishment of a continental Asian biostratigraphic and geochronologic framework. In *Fossil mammals of Asia: Neogene Biostratigraphy and Chronology*, edited by X.-M. Wang, L. J. Flynn, and M. Fortelius, 1–25. New York: Columbia University Press.

SIWALIK SUIDAE AND PALAEOCHOERIDAE

OLIVIER CHAVASSEAU

INTRODUCTION

The Potwar Plateau of Northern Pakistan, which is part of the Neogene Siwalik Group of Indo-Pakistan, has a nearly continuous sedimentary sequence ranging from the Early Miocene to the Pleistocene. A series of field campaigns in this region, mostly in the twentieth century, have documented an exceptionally rich mammal fauna from hundreds of localities, making the Siwaliks of Pakistan a critical record of Neogene mammalian evolution. Coupled biostratigraphic (e.g., Barry et al. 1982) and magnetostratigraphic studies (e.g., Johnson et al. 1982; Tauxe and Opdyke 1982) have established a precise relative and absolute chronology of the Potwar Plateau localities (see Chapter 2). This exceptional fossil record combined with a well constrained chronology have allowed detailed studies of faunal and environmental changes in the Neogene of Pakistan (e.g., Barry et al. 1995, 2002, 2013; Flynn et al. 1995, 2013; Badgley et al. 2016). Such investigations are of particular interest in the Siwalik record, which documents an essential piece of the evolutionary puzzle of Asian hominoids (e.g., Pilbeam et al. 1977).

This chapter summarizes the contribution of the Suidae to the faunal, environmental, and biogeographical changes of the Potwar Plateau sequence. This group is notably relevant to the understanding of *Sivapithecus* evolution since fossil suids and hominoids have often been similar in terms of habitat, diet, and size. The chapter begins with a synthesis of suoid diversity in the Siwaliks, including an ecomorphological analysis of their dietary and habitat preferences. It then addresses faunal change among the Siwalik suoids, their potential relationships with environmental or climatic change, and the evolution of this group (biogeographical affinities, main dispersal events, origin of species).

THE SUIDAE

The Suidae (wild boars, hogs) are part of the superorder Cetartiodactyla (Montgelard et al. 1997) and superfamily Suoidea, in which they are the sister group of the Tayassuidae (peccaries) among extant representatives of the group (Gentry and Hooker 1988). Suidae are characterized by elongate, low, dorsally flat, or weakly concave skulls presenting a high braincase (Vaughan, Ryan, and Czaplewski 2013) and a long snout terminating with a cartilaginous disk used for rooting (although some fossil suids likely did not possess this structure according to Orliac [2007]); dental formula I1–3/i3 C1/c1 P2–4/p2–4 M3/m3; thick mandibular corpus and lower tooth row presenting an anteriolabial to posterolingual orientation relative to the corpus long axis (Ducrocq et al. 1998); splayed outward, sexually dimorphic, and often ever-growing lower canines with triangular section; splayed outward and/or upward, sexually dimorphic, and often ever-growing upper canines with three enamel bands (van der Made 1997); broadening of the maxilla and dentary bones at the level of the canines; M1/m1–M2/m2 with four main cusps and several other cusps/conules (Fig. 19.1) that can become enlarged in some lineages; M3/m3 more elongate with a fifth main cusp (hypoconulid/distocone) (see Fig. 19.1) that can be associated with several additional distal cusps in some representatives; paraconule fused to mesial cingulum on upper molars (van der Made 1997); upper molars with unfused lingual roots and four-rooted M1–M2 (van der Made 1997); paraxonic members with four digits on hand and foot; digits II–V smaller than digits III–IV; unfused metapodials with dorsally expanded distal keels on central metapodials and dorsally expanded gulleys on associated first phalanges (van der Made 1997; Lihoreau et al. 2007); astragalus with marked angulation between proximal and distal trochlea.

The Suidae emerged in the Eocene of Asia. The most ancient representatives attributed to this family dating from the latest Middle to Late Eocene of southern China (*Eocenchoerus*; Liu 2001, 2003) and Late Eocene of Thailand (*Siamochoerus*; Ducrocq et al. 1998). Attribution of these Late Eocene taxa to the Suidae has, however, been debated, and they are often considered to be stem suoids, rather than true suids, with close relationships to other Eocene taxa from China and Thailand such as *Odoichoerus*, *Egatochoerus*, and *Huaxiachoerus*. (See phylogenetic and systematic assessments of Liu [2003], except for *Eocenchoerus*, Harris and Liu [2007], and Orliac, Antoine, and

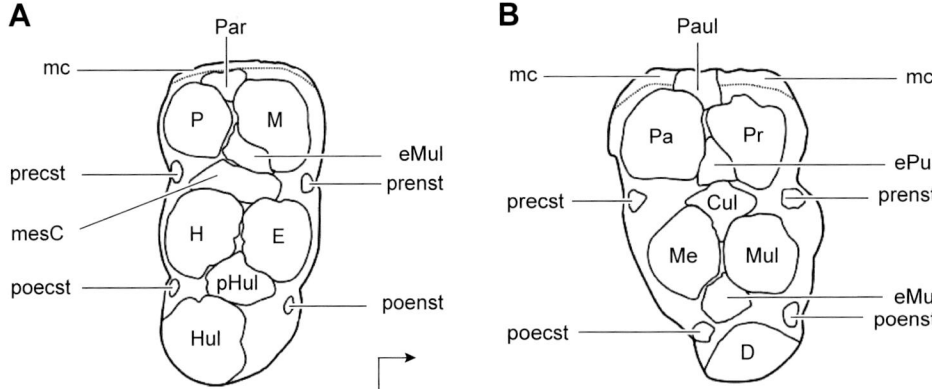

Figure 19.1. Suoid dental nomenclature (after Boisserie et al. 2010; Orliac et al. 2011, 2015). A. Left m3 (occlusal view); B: right M3 (occlusal view). Abbreviations: Cul = centroconule, D = distocone, eMul = endometaconule/-id, E = entoconid, ePul = endoprotoconule, H = hypoconid, Hul = hypoconulid, mc = mesial cingulum/-id, Me = metacone, M = metaconid, Mul = metaconule, mesC = mesoconulide, pHul = prehypoconulide, P = protoconid, Pa = paracone, Par = paraconid, Paul = paraconule, poecst = postectostyle/-id, poenst = postendostyle/-id, Pr = protocone, precst = preectostyle/-id, prenst = preendostyle/-id. The arrow indicates the mesiolingual side of the teeth.

Ducrocq [2010].) According to Orliac, Antoine, and Ducrocq (2010), who revised the phylogeny of suoids, the unambiguous Suidae are divided into four main subfamilies: Hyotheriinae, Listriodontinae, Tetraconodontinae, and Suinae.

The subfamily Hyotheriinae, which appeared in the earliest Miocene, has often been considered as an ancestral suid group from which other subfamilies evolved (e.g., Pickford 1993). This phylogenetic assumption has been based on the mostly conservative cranial and dental morphology presented by the Hyotheriines: many of them possess low-crowned simple molars (low complexity of molar cusp and groove pattern) and premolars (simple trenchant premolars with very limited or absent molarization) and have short skulls and short jaws with reduced or absent postcanine diastema. The monophyly of the group has been questioned by Liu (2003) but reaffirmed by Orliac, Antoine, and Ducrocq (2010), who pointed out three unambiguous synapormorphies on the anterior dentition supporting a clade Hyotheriinae. The latter authors also challenged the basal position of the Hyotheriinae, having obtained a topology in which the Listriodontinae represent the most basal offshoot among the Suidae. The Hyotheriinae are known in Eurasia and Africa from the earliest Miocene to the Middle Miocene.

The Listriodontinae are Eurasian and African suids that developed bunolophodont or lophodont molars as well as enlarged incisors. The oldest representatives of this subfamily are known in the Early Miocene of the Zinda Pir Dome, the Bugti Hills, and the Manchar Formation of Pakistan (Orliac et al. 2009; Métais et al. 2016). The listriodontines diversified in the Early and Middle Miocene and are known in Asia and Europe until the early Late Miocene (~10 Ma). According to Orliac, Antoine, and Ducrocq (2010), the Listriodontinae comprise the tribes Listriodontini and Kubanochoerini. The latter group, with both African and Asian taxa, was considered to be a distinct subfamily and reconstructed as the sister group of the Listriodontinae in the phylogenetic analysis of Liu (2003). The

Kubanochoerini (limited to the genus *Kubanochoerus*, following Orliac, Antoine, and Ducrocq [2010]), which appear in the early Miocene of Africa and are last recorded in the late Middle Miocene of Asia, are characterized by a frontal horn in males.

The Tetraconodontinae are Eurasian and African suids possessing inflated posterior premolars. The subfamily has been extensively used for biochronology in the Mio-Pliocene of Africa, where it diversified and evolved rapidly following a general trend involving premolar reduction, third molar lengthening, achievement of molar hypsodonty, and size increase. In contrast, Asian Tetraconodontinae remained low-crowned and evolved toward premolar gigantism. The origin of this subfamily remains enigmatic. Their fossil record extends from the Middle Miocene to the Early Pleistocene. Phylogenetic analysis (Liu 2003; Orliac, Antoine, and Ducrocq 2010) place the Tetraconodontinae as sister group of the Suinae.

Suinae are the subfamily that includes all extant suids. Suinae are recorded from the late Middle Miocene to early Late Miocene and became the dominant suid subfamily during the Late Miocene. They originated and diversified in Eurasia and migrated to Africa during the Pliocene (Harris and Cerling 2002).

In addition, the Palaeochoerinae are Oligocene to Miocene Eurasian suoids. Their unity and phylogenetic position have been long debated, and this group has been placed alternately within the Suidae (e.g,. Pickford 1988; Liu 2003) and the Tayassuidae (e.g., Pickford 1993; van der Made and Han 1994), as well as in their own family, the Palaeochoeridae. Root fusions on molars and the posterior position of the paraconule are notable synapomorphies of the group (van der Made 1997). The concept of Palaeochoeridae is adopted here, although phylogenetic analyses of Orliac, Antoine, and Ducrocq (2010) have not systematically demonstrated the monophyly of this group. The Siwalik suoid, *"Schizochoerus" gandakasensis* (=*Yunnanochoerus gandakasensis* according to van der Made and Han [1994]), which has fused lingual roots on upper molars, is considered here as a species within the Palaeochoeridae and, although not a suid, is included in this study. The binomial name *Yunnanochoerus gandakasensis* will be used hereafter instead of the name *Schizochoerus gandakensensis* proposed by Pickford (1978), van der Made and Han (1994) having shown substantial differences in premolar morphology and proportions between the Siwalik material and the European material of *Schizochoerus*. This choice is reinforced by the revision of the phylogenetic position of *Schizochoerus*, now considered to be within Suidae by Orliac, Antoine, and Ducrocq (2010).

The first descriptions of suid fossils from the Siwalik record were by Falconer and Cautley (1847), followed by Lydekker (1884). Pilgrim (1926) provided a more complete description of the fauna and attempted to reconstruct their phylogenetic relationships. Colbert (1935a, 1935b) recognized significantly greater intraspecific variation and simplified the taxonomy of previous authors, proposing an improved, synthetic vision of the Siwalik suid relationships. Pickford (1988) revised the fauna, recognizing 14 species in 5 subfamilies (Listriodontinae, Hyotheriinae, Kubanochoerinae, Suinae, and Tetraconodontinae). The presence of kubanochoeres (here considered as a tribe of Listriodontinae) in Pakistan and India, as proposed by Pickford (1988), has been highly debated since the rare *Libycochoerus fategadensis*, based on a single fossil of poor stratigraphic provenance, was interpreted by Fortelius, van der Made, and Bernor (1996) as a Listriodontini. Pickford also recognized the presence of a Kubanochoerini in the Early Miocene of the Bugti Hills (Pakistan), with the rare species *Libycochoerus affinis* (Pilgrim 1908) known by a single maxilla fragment. Fortelius, van der Made, and Bernor (1996), van der Made (1996), and Orliac et al. (2009) have all interpreted this species as a representative of the Listriodontini. In addition, Orliac et al. (2009) have recognized the validity of *Listriodon guptai* (Early Miocene of Pakistan, notably present in the Lower Siwaliks), a species synonymized by Pickford (1988) with *L. pentapotamiae*. Thus, besides the debated presence of Kubanochoerini in the Siwaliks of Pakistan, the taxonomic framework adopted by Pickford (1988) may have significantly underestimated the diversity of the Listriodontini in the Siwaliks of Pakistan, this tribe being represented by the sole *L. pentapotamiae* in the latter work. Furthermore, at least some of the material allocated to *Hyotherium pilgrimi* by Pickford (1988) is doubtfully assigned to the Hyotheriinae. Specimens from the Geological Survey of India collections GSI B682 and GSI B686 have laterally expanded P3's with large and buccally bulged paracones and distinct protocones that differ markedly from those of *Hyotherium soemmeringi*, *H. meissneri*, and *H. shanwangense*, which display a weak paracone and a postero-lingual cingulum instead of a protocone on the P3. The P3 of these specimens resembles those of tetraconodontines with modestly-enlarged premolars and inflated paracones such as *Parachleuastochoerus*. The lower jaw of the American Museum of Natural History specimen AMNH 19423, also attributed to *H. pilgrimi* by Pickford (1988), possesses a p4 larger and longer than m1, which is not diagnostic of *Hyotherium* but closely matches *Parachleuastochoerus* (p4 as long as m1 but slightly narrower in *P. crusafonti*

[see Pickford 1981]; p4 longer than and as broad as m1 in *P. sinensis* [see Pickford and Liu 2001]). Several fossils from the Chinji Formation cluster with GSI B682 and GSI B686. This is the case of the Geological Survey of Pakistan specimen YGSP 8712, a maxilla fragment with P3–P4 attributed by Pickford (1988) to *Conohyus indicus*, whose teeth are smaller than those of *C. sindiensis* and consequently too small to match the lower premolars of the rest of the hypodigm, which are larger than those of *C. sindiensis* (Pickford and Gupta 2001). Until a more complete revision of this species is performed, *"Hyotherium" pilgrimi* is regarded here as a Tetraconodontinae belonging to the genus *Parachleuastochoerus*. Hence, I consider that there is presently no sufficient evidence supporting the presence of the subfamily Hyotheriinae in the fossil record of the Potwar Plateau.

Figure 19.2 displays the temporal distribution of the Siwalik Suidae with ranges of species documented between 18 and 6 Ma in the Potwar Plateau (Fig. 19.2a) and including suids from Haritalyangar, India (Fig 19.2b). Most of the species recognized by Pickford (1988) are retained in Fig. 19.2b. A group of unnamed species noted by Barry et al. (2002) is also included since they illuminate the species-level diversity of the Potwar Plateau suids. Seventeen taxa of suids and two taxa of palaeochoerids have been formally identified in this area. In the best documented time interval (~14–7 Ma), the number of co-occurring suoids peaks at seven species. For the latest Miocene/Pliocene period, this number is between two and four species, probably because of a poorer fossil record in the corresponding part of the Potwar Plateau. As available, the average body weight estimates of these taxa, based on m1 surface area, are included in Table 19.1, along with first and last occurrence dates.

DIET AND HABITAT OF THE SIWALIK SUIDAE

Diet of Living Suids

The living members of *Sus* and *Potamochoerus* are omnivorous and have low-crowned and bunodont teeth. Eurasian *Sus scrofa* is a particularly opportunistic omnivore with a diet comprised of roots, plants, insects, birds, and mammals (e.g., Schley and Roper 2003). *Sus scrofa* lives in a wide variety of habitats but is dependent on vegetation for shelter. African *Potamochoerus porcus* feeds mainly on roots, fruit, and insects (Breytenbach and Skinner 1982) and inhabits a wide range of habitats, particularly wooded areas that provide shelter. Southeast Asian *Babyrousa babyrussa* inhabits mainly tropical forest, has bunodont and low-crowned dentition, and is omnivorous with a preference for fruits and leaves (Leus 1994). The African forest hog *Hylochoerus meinertzhageni* has brachydont teeth with a tendency to lophodonty and is a forest-dwelling grazer that feeds mainly on C_3 herbs, grasses, and sedges (Cooke 1985; Kingdon 1997;). Stable carbon isotope analyses of tooth enamel place this suid in the C_3-hyperbrowser category (Harris and Cerling 2002). Sub-Saharan *Phacochoerus aethiopicus*, a grazer that feeds primarily on C_4 grasses (Poaceae) and sedges

A

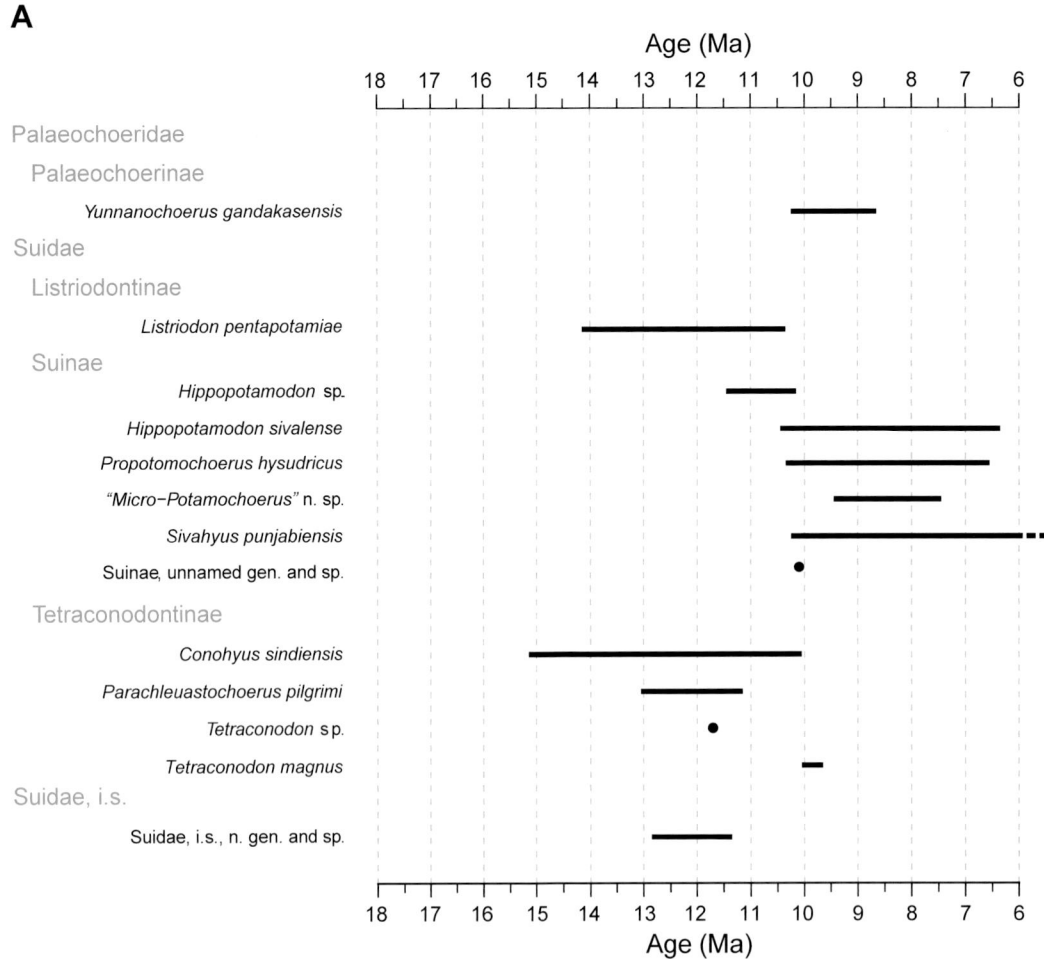

Figure 19.2. Temporal ranges of the suoids of the Siwaliks of Pakistan (A) and including India (B). The rare palaeochoerid *Pecarichoerus orientalis* Colbert, 1933 from the Chinji Formation, *Listriodon guptai* from the Kamlial Formation, and *Sus* are not represented. The succession and duration of the sedimentary formations represented are those of the Potwar Plateau according to Barry et al. (2002). The average body weight of the taxa was determined by the surface area of their m1 according to the formula of Legendre (1989): ln $(W) = 3.5346 + 1.5416 * \ln (S)$ with S the m1 area (in mm^2) and W the body weight (in g). The surface areas were approximated from linear measurements (mesiodistal length × buccolingual breadth) acquired in GSP, AMNH, and NHM collections and completed by measurements given by Pickford (1988). The body weight of *Yunnanochoerus gandakansensis* is only estimated; the m1 measurements for two individuals are extrapolated from roots/alveoli.

(Cyperaceae; Kingdon 1997; Harris and Cerling 2002; Treydte et al. 2006), is characterized by hypsodont molars.

This brief overview of the diet and habitat of the common extant forms shows that among suids, omnivory is associated with bunodont to brachydont teeth, folivory with lophodont-brachydont teeth, and grazing with hypsodont teeth.

Fossil Suids

The tooth morphology of fossil suids can thus provide general inferences about their diet. Of course, the relationships between tooth morphology and diet are more complex than those stated above. For example, hypsodont browsers were found in equids (MacFadden et al. 1999) as well as in antilocaprids (Fortelius and Solounias 2000), which demonstrates that high-crowned taxa are not systematically C$_4$ grazers or mixed C$_3$/C$_4$ feeders. Conversely, brachydont species can be C$_4$ con-

sumers, as is notably the case with the Late Miocene African suid *Nyanzachoerus* (Harris and Cerling 2002).

Because tooth morphology can provide ambiguous interpretations of diet beyond general dietary categories, dental microwear and isotopic data of Siwalik suids and, in some cases, of phylogenetically close taxa from other geographic areas can be used to further characterize the diet of fossil suids. Microwear analyses of several Siwalik mammals, including some of the most abundant suids, were performed by Morgan (1994) and Nelson (2003). These two studies used different methodologies and are not directly comparable: Morgan (1994) counted features on scanning electron microscope images taken at 500×, while Nelson (2003) counted features on casts magnified at 35× with a stereomicroscope following the methodology of Solounias and Semprebon (2002). Because dietary inferences obtained by these two methodologies are

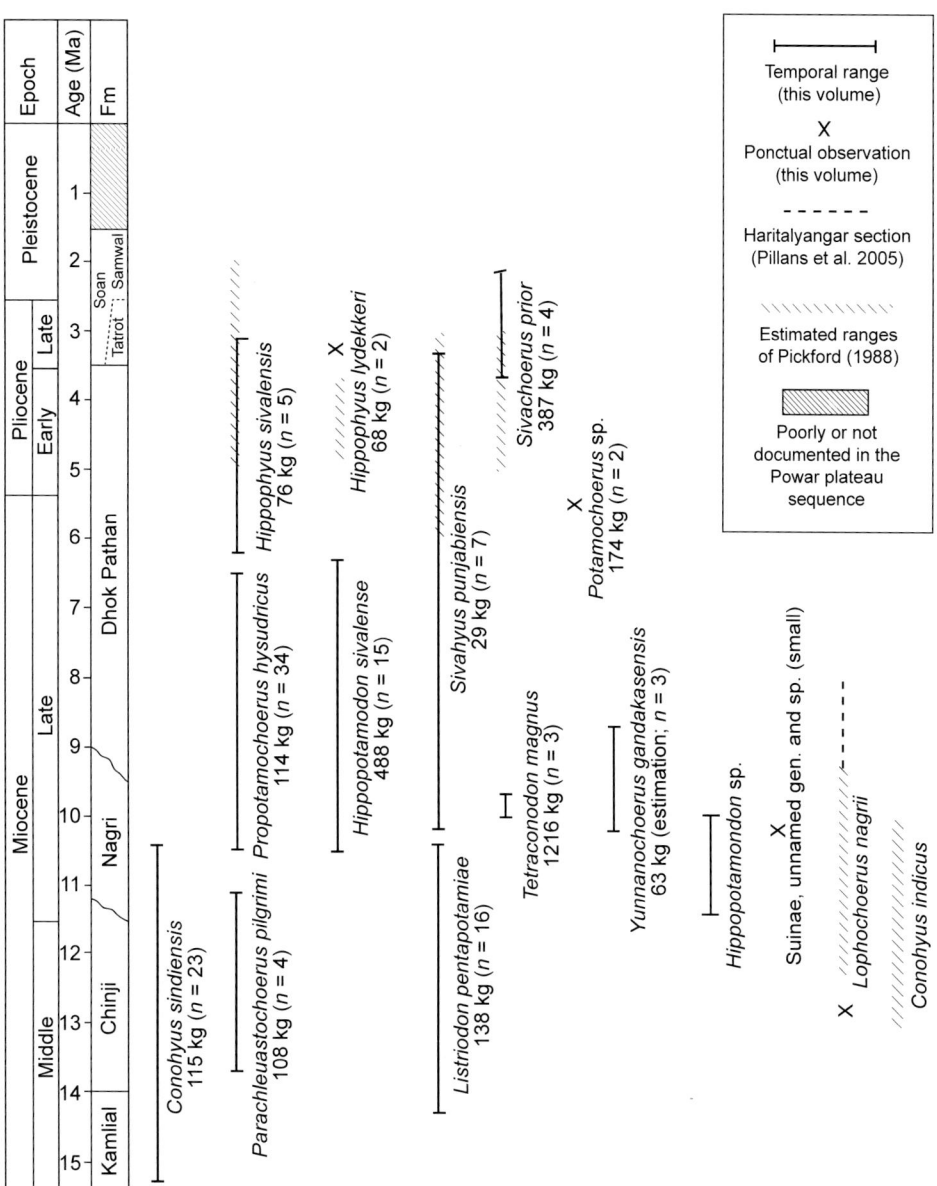

Figure 19.2. (Continued)

sometimes very different, I rely primarily on isotopic data to complement the tooth morphology analysis. The suid isotopic data (enamel $\delta^{18}O$, $\delta^{13}C$) discussed below originate from the dataset of Badgley et al. (2008). (Note that all of the isotopic values cited hereafter were measured on the carbonate group of enamel and are expressed in the PDB (Pee Dee Belemnite) standard.

Bunodont and Low-Crowned Suids

A majority of Siwaliks suids falls within the bunodont/low-crowned category, which suggests an omnivorous diet. This is the case for members of the subfamily Tetraconodontinae,

represented in the Neogene of Pakistan mainly by *Conohyus sindiensis*, *C. indicus*, *Sivachoerus prior*, and *Tetraconodon magnus*. The rare tetraconodont *Lophochoerus nagrii*, possibly a small version of *Conohyus* (judging by its moderately inflated p4 with a large protoconid), was described only from the Siwalik beds of India (see Pilgrim 1926; Pickford 1988; van der Made and Han 1994). These taxa exhibit various degrees of hypertrophy of the posterior premolars, which greatly influences their dental proportions: P3/p3 and P4/p4 are large in *Sivachoerus* and gigantic in *Tetraconodon* relative to *Conohyus*. Posterior premolars of the Siwalik tetraconodontines have a simple organization: p3 and p4 are dominated by a large, high

Table 19.1. Body Weight Estimates of Suoid Species

Subfamily	Species	Locality of FO	FO (Ma)	LO (Ma)	Locality of LO	Average Body Weight (kg)	Number of Specimens for Body Weight Estimate
Listriodontinae	*Listriodon pentapotamiae*	Y655	14.1*	10.4	Y395	138	16
Tetraconodontinae	*Conohyus sindiensis*	Y682	15.1*	10.1**	Y728	115	23
Tetraconodontinae	*Parachleuastochoerus pilgrimi*	Y822	13.0	11.2	Y455	108	4
Tetraconodontinae	*Tetraconodon magnus*	Y493	10.0	9.7	Y329	1,216	3
Tetraconodontinae	*Sivachoerus prior*	D15	3.6	2.1	D67	387	4
Suinae	*Hippopotamodon sivalense*	Y454 Y395	10.4	6.4	Y930	488	15
Suinae	*Propotomochoerus hysudricus*	D27	10.3	6.6	Y453	114	34
Suinae	*Sivahyus punjabiensis*	L94	10.2	3.3	D7	29	7
Suinae	*Hippohyus sivalensis*	Y860	4.8	3.2	L105	76	5
Suinae	*Hippohyus lydekkeri*	Tatrot Formation	3.3 (3.0–3.6)	3.3 (3.0–3.6)	Tatrot Formation	68	2
Suinae	*Potamochoerus* n. sp.	Y553	5.5	5.5	Y553	174	2
Palaeochoerinae	*Yunnanochoerus gandakasensis*	Y258	10.2	8.7	Y445	63	3

*Older specimens may represent this species. ** Younger specimens may represent this species. FO = First Occurrence. LO = Last Occurrence.

protoconid, while P3 has a large protocone. Finally, P4 is bicuspid in *Conohyus* with only protocone and paracone, faintly tricuspid in *Tetraconodon*, and tricuspid in *Sivachoerus* with a distinct metacone. The morphology and the relative size of the tetraconodontine premolars indicate that they probably had significant bite forces, particularly for *Conohyus sindiensis*, *C. indicus*, and *Tetraconodon magnus*, in which strongly inclined buccal walls are associated with reduced areas of the P3 and P4. The striking crown morphology of tetraconodonts is accompanied by modifications of the root pattern, necessary to accommodate the forces generated in occlusion: the p3 of *Conohyus* possesses a large posterior root often partially subdivided into two. *Tetraconodon* and *Sivachoerus*, which display larger posterior premolars than *Conohyus*, have a three-rooted p3 (see Lydekker, 1884, Plate 11 for *Sivachoerus*). The peculiar posterior premolars of tetraconodonts have been considered to represent an adaptation to a hard food–based diet (e.g., Cooke 1985; Pickford 1988, 48). The various degrees of hypertrophy of the posterior premolars possibly represent different stages of adaptation to the crushing of hard food or a refinement of feeding strategy based on hard food. Nevertheless, tooth morphology alone cannot fully predict diet of these suids.

The microwear analysis of Morgan (1994) excluded a root-dominated diet for *Conohyus* since it shows many fewer pits than *Potamochoerus porcus*, a frequent rooter. *C. sindiensis* has depleted δ^{13}C values (mean = −11.5‰; n = 6; Fig. 19.3), which implies that it fed in predominantly closed habitat and was thus probably a forest browser. In comparison, Quade et al. (1995) report that *Conohyus simorrensis* from Paşalar (Middle Miocene, Turkey) had δ^{13}C values bracketed between −9.0 and −10.1‰ (n = 3), which suggests that it may have fed in more open habitats than *C. sindiensis*. The difference between the isotopic values of *Conohyus simorrensis* and *C. sindiensis* despite similar occlusal morphology illustrates the limited value of tooth morphology alone. Nelson (2003, 2007) established that the very

large-bodied *Tetraconodon magnus* was, like *Conohyus sindiensis*, a forest-dweller (δ^{13}C range: −12.8‰ to −11.1‰; see Fig. 19.3).

According to Nelson (2003), *Tetraconodon* had enriched δ^{18}O values compared with other suids (−3.4‰, n = 1; range of −9.2‰ to −5.8‰ for other suids of the same age) and even all other taxa (mean value = −5‰). She proposed three hypotheses to explain the enriched oxygen values of this suid: first, a low dependence on drinking rainfall water may explain the high δ^{18}O values, water dependence being considered as a cause of important variation in δ^{18}O values of living suids (Harris and Cerling 2002); second, foraging on fruits fallen from the upper part of the canopy, as these fruits have enriched δ^{18}O values relative to ground-level growing vegetation due to the limited canopy effect on oxygen fractionation; and third, consumption of drinking water from open habitats, pools rather than flowing water, as more evaporative contexts have higher δ^{18}O values. The latter hypothesis is not favored here because contemporaneous suoids with low δ^{13}C values in the range of *Tetraconodon* (−13‰ to −11‰) have systematically lower δ^{18}O values (see Nelson, 2003, 49–50). Interestingly, another taxon contemporaneous with *Tetraconodon* and combining low δ^{13}C and high δ^{18}O values is *Sivapithecus* (mean δ^{13}C = −13‰, mean δ^{18}O = −2.6‰). Microwear and isotope data have suggested that *Sivapithecus* was feeding on fruits from the upper canopy and was likely including hard objects such as hard fruits in its diet (Nelson 2003). Hence, *Tetraconodon* was perhaps also feeding on hard objects from the upper canopy similar to *Sivapithecus*, a hypothesis that is in agreement with its extremely hypertrophied premolars. The available data on *Tetraconodon* do not preclude the hypothesis that this suid was, perhaps like *Sivapithecus* (Nelson 2003), getting most of its water from food.

Most of the Suinae represented in the Siwalik fossil record are bunodont and low-crowned. In contrast to tetraconodonts, they have simple, sharp-cusped premolars. Occlusal morphology of these Suinae became more complex through the sequence, probably indicating different feeding strategies than those of tetraconodonts. For the dominant Late Miocene Suinae *Hippopotamodon sivalense* and *Propotamochoerus hysudricus,* Nelson (2003) inferences from dental microwear are that these species were omnivorous with a diet mostly based on fruit and roots, close to that of the extant African suid *Potamochoerus. Hippopotamodon sivalense* and *Propotamochoerus hysudricus* display a wider range of $\delta^{13}C$ values than the late Middle and early Late Miocene suids (see Fig. 19.3). From 10.4 to 8.0 Ma, a majority of the isotopic data from mammalian tooth enamel samples show lower $\delta^{13}C$ values (−13.0‰ to −10.0‰), which signifies that *Hippopotamodon* and *Propotamochoerus* were living, like their earlier Miocene allies, mainly in closed habitats. Nevertheless, two $\delta^{13}C$ values of *Hippopotamodon* at ~9.3 Ma are more positive (−8.6‰ and −9.4‰), which suggests that some *Hippopotamodon* occasionally exploited more open habitats. A similar interpretation is proposed for *Propotamochoerus* with a $\delta^{13}C$ value at 8.9 Ma of −7.8 ‰. This value is more positive than the threshold value of −8‰ over which the consumption of C_4 plants is considered as significant (e.g., Cerling et al. 1997) and may represent the first record of the incorporation of C_4 plants in the diet of this suid. After 8 Ma, the $\delta^{13}C$ values for both species became more positive, and four individuals passed the −8‰ threshold (maximum $\delta^{13}C$ value for *H. sivalense* = −6.2‰; maximum $\delta^{13}C$ value for *P. hysudricus* =

−6.7‰). These isotopic data indicate that the habitats and diets of *Hippopotamodon* and *Propotamochoerus* probably ranged from woodlands to more open vegetation and that C_4 plants were plausibly part of their diet. After 8 Ma, the dietary and habitat range of these two suids may have expanded as evidenced by an increase in the range of $\delta^{13}C$ values.

The Suinae *Hippohyus* and *Sivahyus*, sometimes placed in the tribe Hippohyini, deviate from the typical bunodont morphology of suines with cheek teeth displaying complex patterns due to accentuated, long, and mostly mesiodistally oriented crests on the main cusps, which are separated by deep mesiodistal furrows (Fig. 19.4). As a result, when slightly worn, the cusps appear somewhat "H"-shaped. This pattern is accentuated on the metaconid by incorporation with little wear of the endometaconulid. Additional wear also reinforces the "H" pattern of the hypoconid owing to fusion with the mesoconulid and the hypoconulid on m1–m2 and with the mesoconulid and the prehypoconulid on m3. On upper molars wear rapidly creates fusion between the paraconule and the protocone and between the centroconule and the metaconule. These fusions form extended enamel ridges oriented mesiodistally or slightly obliquely.

The molars show very deep mesiodistal valleys between the main cusps of lobes 1–2 while transverse valleys are blocked by the mesoconulid/prehypoconulid and centroconule/endometaconule cusps. The Hippohyini show increased premolar

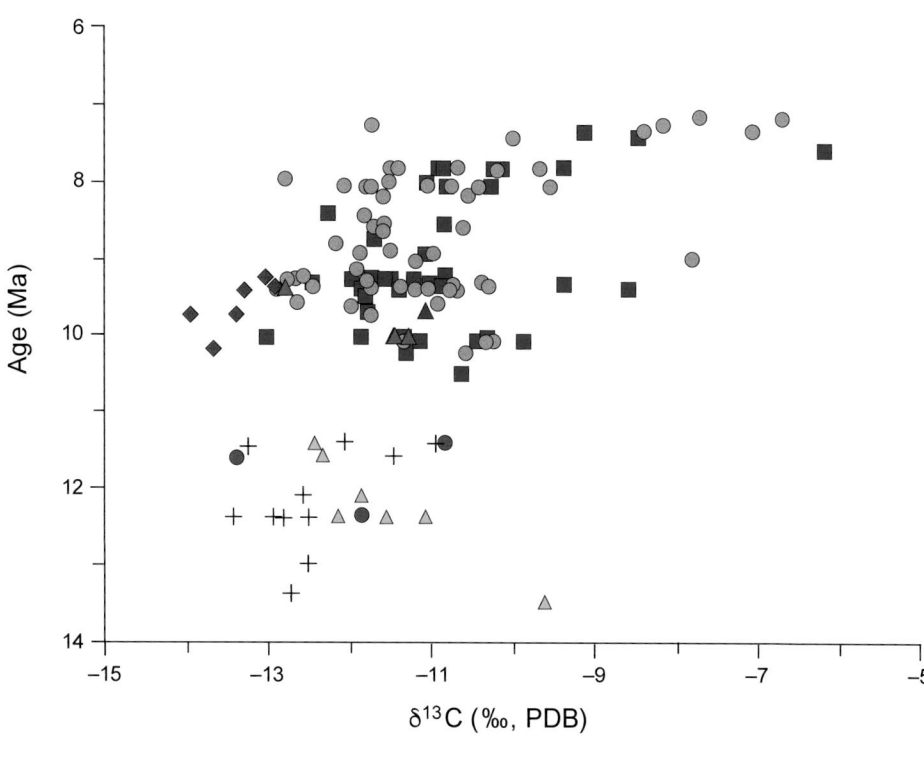

Figure 19.3. Enamel carbonate $\delta^{13}C$ (‰, PDB) record of the main Siwalik suids through time. Values from Morgan, Kingston, and Marino (1994), Nelson (2003), and Badgley et al. (2008).

● *Parachleuastochoerus pilgrimi*
+ *Listriodon pentapotamiae*
■ *Hippopotamodon sivalense*
● *Propotamochoerus hysudricus*
◆ *Yunnanochoerus gandakasensis*
△ *Conohyus sindiensis*
▲ *Tetraconodon magnus*

and molar crown heights (Pickford 1988). Maximum hypso-donty indices measured in a sample of M3s and m3s of Hip-pohyini from Harvard-GSP project specimens currently on loan to Harvard, National History Museum, London (NHM) and AMNH collections are slightly greater than 100 in *Hippohyus*. In comparison with the hypsodonty values of the Neogene African suids, those of *Hippohuys* correspond well to those of *Nyan-zachoerus jaegeri* (Fig. 19.5), a slightly hypsodont tetraconodon-tine from the Pliocene of East Africa (Harris and Cerling 2002), but are much lower than the markedly hypsodont tetracon-odontine *Notochoerus* or the suine *Metridiochoerus*, whose hypso-donty indices are frequently greater than 150 and can reach 250 (Cooke 2007). The evolutionary trend presented by the Hip-pohyini (increase in molar crown height, deepening of cusp grooves and valleys, M3/m3 lengthening with extra distal cusps, and fourth lobe frequently achieved on m3) somewhat re-sembles features found within African suid lineages that be-came progressively adapted to C$_4$ grazing during the Pliocene (*Nyanzachoerus-Notochoerus*, *Metridiochoerus*, *Kolpochoerus*).

Among the Hippohyini, *Hippohyus* clearly shows more special-ized dentition than *Sivahyus*; the latter retains a more plesiomor-phic dentition closer to that of the suine *Propotamochoerus*, for example. In *Hippohyus*, tooth crowns are higher (see Fig. 19.5), and chewing likely had a more important lateral component than in *Sivahyus*, judging by the more accentuated mesiodistal shape of the main cusps as well as the paraconule and the cen-troconule on upper molars. The grooves on cusps and in the valleys between the cusps are also much deeper in *Hippohyus* and are associated with more accentuated mesiodistal crests than in *Sivahyus*. The shallower grooves on the molars of *Siv-ahyus* are especially noticeable on the lingual cusps of upper molars (Lewis collection specimen YPM 13812) and on the protoconid and the mesial side of the metaconid on lower mo-lars (YPM 19149, YPM 19150), which show poorly expressed grooves. The paraconule/paraconid and centroconule/meso-conulid are also comparatively lower in *Sivahyus* and conse-quently fuse with the main cusps at a more advanced stage of wear. Pickford (1988) noted that metrical variation within *Hippohyus* is greater than that observed in a single species of suid. According to this author, *Hippohyus sivalensis* can be dis-tinguished from *H. lydekkeri* in having higher molar crowns and larger molar size (especially M3/m3). However, the ob-served size overlap between these species is greater than that figured by Pickford (1988), and the hypsodonty indices found for these species are nearly identical in the small sample avail-able; this finding indicates that species separation within *Hip-pohyus* needs further revision.

   As noted by Pickford (1988), the modification of Hippo-hyine dental morphology from the primitive bunodont organization of Suinae probably reflects a change in diet: the complex molar pattern and large shearing surface generated by deep grooves associated with elongated and mesio-distally oriented crests attest to an important lateral component of mastication movements. As compared with other Siwalik su-ines (Pickford 1988), the more vertically implanted incisors,

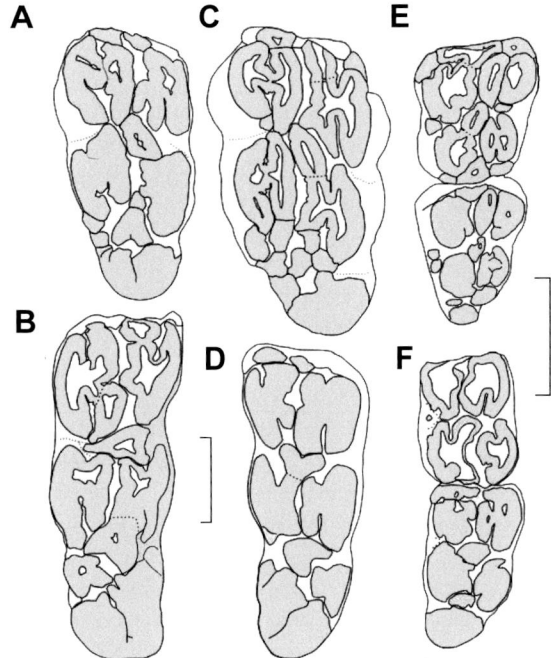

Figure 19.4. Molar morphologies of the Hippohyini *Hippohyus* and *Sivahyus*. A., B.: *Hippohyus sivalensis*. Right M3 DP 22 (A) and right m3 L=YPM 19089 (B). C., D.: *Hippohyus lydekkeri*. Right M3 YGSP 52635 (C) and left m3 M 12772 (D). E., F.: *Sivahyus punjabiensis*. Left M2-M3 YPM 13812 (E) and left m2-m3 YPM 19149 (F). Scale bars = 1 cm. All drawings in occlusal views.

which are associated with increased crown height on premo-lars and molars, imply that the Hippohyini may have been mixed-feeders with potentially a C$_4$ plant component in their diet and that some specialized *Hippohyus* may have been C$_4$ grazers. This hypothesis is corroborated by the fact that *Hip-pohyus* has comparable hypsodonty indices with *Nyanzachoe-rus jaegeri* (see Fig. 19.5), which is isotopically documented as a C$_4$ grazer (Harris and Cerling 2002). *Sivahyus* has slightly lower crown height than *Hippohyus* (see Fig. 19.5), the former being more similar to *Nyanzachoerus kanamensis*, a brachyodont browser that incorporated C$_4$ plants in its diet (Harris and Cer-ling 2002). The noted differences in dental specialization among the Hippohyini probably imply different feeding strate-gies, *Sivahyus* being more specialized for browsing and *Hip-pohyus* more specialized for grazing, perhaps incorporating C$_4$ grasses. Neither microwear nor isotopic data are available for *Hippohyus* and *Sivahyus*.

The Lophodont Suoids

Listriodon pentapotamiae differs radically from the other Siwalik suids in having bilophodont molars (Fig. 19.6). On lower mo-lars, the metalophid and the hypolophid are slightly convex posteriorly and placed on the posterior face of the cusp. The protoloph and metaloph show position and concavity that are opposite to those of the lophids. Lower molars show low but distinct accessory cusps, which are linked to the hypoconid and,

A

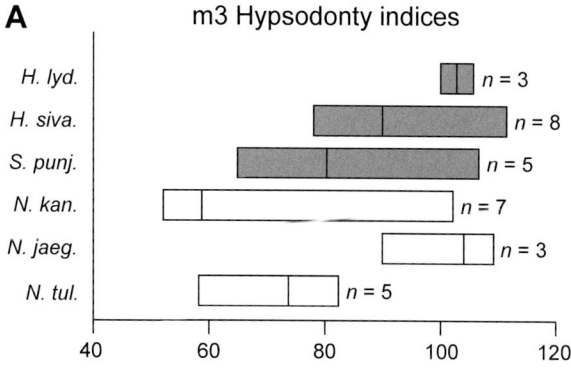

B

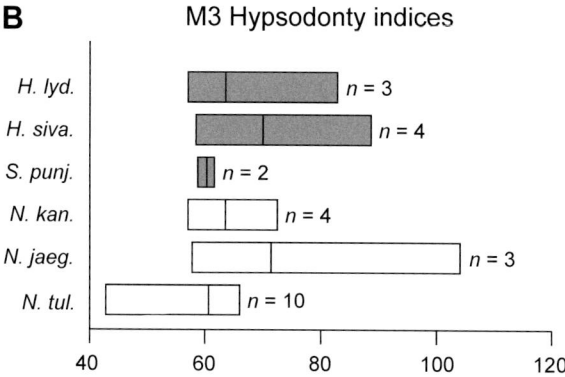

Figure 19.5. Hypsodonty indices, (crown height) : (buccolingual breadth) × 100, showing minimum, maximum and median values of the Hippohyini *Hippohyus lydekkeri* (*H. lyd.*), *H. sivalensis* (*H. siva.*), and *Sivahyus punjabiensis* (*S. punj.*) compared with those of African tetraconodontines *Nyanzachoerus tulotos* (*N. tul.*), *N. kanamensis* (*N. kan.*), and *N. jaegeri* (*N. jaeg.*). A. m3; B. M3. *N. tulotos* and *N. kanamensis* were brachydont browsers incorporating C4 plants in their diet. *N. jaegeri* was a slightly hypsodont C4-grazer (Harris and Cerling 2002). Data for Hippohyini: own measurements from NHM, AMNH, GSP, and YPM collections. Data for *Nyanzachoerus* from Cooke and Ewer (1972).

on m3, to the hypoconulid by oblique crests. Wear facets are visible on both sides of these crests, the more lingually oriented being predominant. Upper molars have weaker or absent accessory cusps and oblique crests in the valley. In some specimens, a main postero-buccal facet is discernible. *L. pentapotamiae* also possesses a metaloph on P4.

The brachydont and bilophodont molars of *L. pentapotamiae* indicate that this species was probably folivorous. This interpretation finds support in the highly modified incisors and symphysis of this taxon: the incisors are very broad, and the lower incisors are extremely procumbent. The cutting function of the incisors was probably fulfilled mainly by i1–i2 and I1, which occlude together and accumulate most of the wear. A wear facet appears on i3 on some specimens. Contrary to the assertion of Pickford (1988, 27), this facet is not restricted to females: AMNH 19395 (Colbert 1935b, 234), a male specimen judging by the hypsodont canine of triangular section (Pick-

ford 1988:27), presents a distinct subvertical wear facet on both i3s. The individual was old, as indicated by the heavily worn pairs of central incisors. This pattern suggests that the development of the facet on i3s is related to individual age rather than to sex. The symphysis of *L. pentapotamiae* displays several derived traits. It is broad from the incisors to the canines, a feature probably linked to great enlargement of the lower incisors. The symphysis is remarkably long (~8–11 cm in adults) and reaches posteriorly to the anterior border of p3. It is strongly deflected ventrally to sub-horizontal. The long and broad morphology of the anterior portion of the dentary and the broad incisors somewhat resembles those of ruminants and is likely to be related to the prehension of food (van der Made 1996). Some authors have interpreted the broad snout and incisors of listriodontines as an adaptation for grazing (van der Made 1996 and references herein). This interpretation is in contradiction with the molar morphology and some of the microwear analyses available for *Listriodon*, which suggest a folivorous diet for this suid (see below).

The transverse structure of the loph(id)s and the predominant location of wear on these loph(id)s do not indicate that *Listriodon* had a mostly antero-posterior component of mastication. On the contrary, the presence of oblique crests associated with buccally and lingually directed facets on molars shows that chewing in *L. pentapotamiae* probably involved important transverse movement. This interpretation is consistent with the study of Hunter and Fortelius (1994), who conclude that the wear facets of the lophodont *L. splendens*, a listriodontine with a molar morphology close to that of *L. pentapotamiae*, indicate important transverse movements during mastication. According to these authors, the degree of lophodonty in listriodontine suids is likely positively associated with increased transverse movements during mastication.

L. pentapotamiae has deep insertions of masticatory muscles (see Fig. 19.6c, d), a condition that may be correlated with the great length of its mandible. The angular region of the mandible is massive and circular in shape. Its ventral border, markedly curved externally, features a large masseteric fossa. The angular region is likely sexually dimorphic as in some other Cetartiodactyla such as hippopotamids—thus, more expanded in putative males (with hypsodont lower canines of triangular or subtriangular cross-section) but less so in the female YGSP 4527 (brachydont canine of ovoid section; see Pickford 1988). On the lingual side, a prominence situated at the boundary of the alveolar and angular regions separates the deep insertion for the medial pterygoid muscle and the insertion for the digastric muscle. This structure is common in listriodontines (Orliac 2006). The insertion for the digastric muscle, used to pull down and retract the mandible in extant suids (Barone 1989), is large and placed along a horizontal band near the ventral edge of the jaw. Beneath the lingual prominence, the lingual border of the dentary often shows a marked indentation known as the facial vascular notch. This notch, which accommodates the facial vein and artery, is also well-marked in other listriodontines and is comparatively deeper than in extant suids (see Orliac 2006; Pickford and Morales 2003). The snout of *Listriodon*

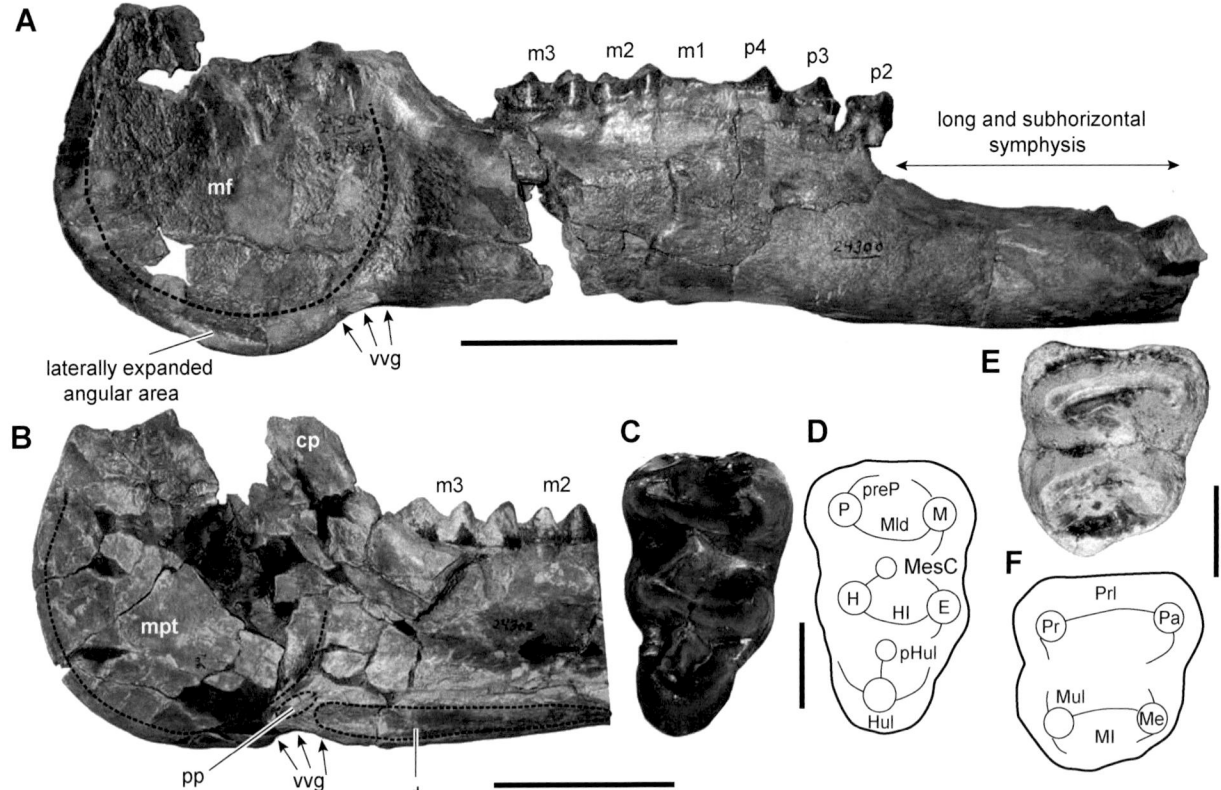

Figure 19.6. Listriodon pentapotamiae. A. Buccal view of the mandible YGSP 24300. B. lingual view of the posterior part of the left mandible YGSP 24302; C: right m3 AM 93819 in occlusal view (mirror mirage); D. schematic representation of a left m3 in occlusal view; E. left M3 M 31869 in occlusal view; F. schematic representation of a left M3 in occlusal view. Dashed lines denote accessory/variable crests. Abbreviations : cp = coronoid process, d = digastric muscle insertion, Hl = hypolophid, mf = masseteric fossa, Ml = metaloph, mpt = medial pterygoid insertion, Mld = metalophid, pp = procumbent process, preP = preprotocristid, Prl = protolophe., vvg = ventral vascular groove. See Fig. 19.1 for other abbreviations. Scale bars = 5cm for A and B and 1 cm for C and E.

pentapotamiae shows weak insertions for the rhinarium muscles; this points to a reduced snout disc and limited rooting ability (van der Made 1996) in comparison with suines. Orliac (2007) has noticed that the listriodontines shared weak rhinarium muscles and has hypothesized the absence of a snout disc in some of them.

Morgan (1994) proposed from her microwear analysis that *Listriodon pentapotamiae* was a forest-dwelling grazer. This interpretation is in contradiction with the morphology of its molars, which suggest a folivorous diet since bilophodont molars are often associated with a browsing habit in mammals. This alternate point of view finds support in the study of two Middle Miocene representatives of *Listriodon* from Turkey, which interprets the lophodonty of these suids as a specialization for folivory (Hunter and Fortelius 1994). Based on dental morphology, these authors have inferred a folivorous diet for *Listriodon* cf. *splendens*, a species that is morphologically very close to *L. pentapotamiae*. Their microwear analysis of this species has revealed that the large facets of the loph(id)s, although mostly pitted, present mesio-lingual and mesio-buccal striae, which suggest both antero-posterior and transverse movements during mastication. Considering their overall morphological similar-

ity, the masticatory mechanisms of *L. pentapotamiae* and *L.* cf. *splendens* were likely similar, as were their diets. With a narrow range of $\delta^{13}C$ values from −13.5‰ and −11.0‰ (Morgan, Kingston, and Marino 1994; Fig. 19.3; see Chapter 25, this volume) *Listriodon pentapotamiae* was probably a primarily closed-habitat C_3 forager).

The palaeochoerid *Yunnanochoerus gandakasensis* is another lophodont species recorded from the early Late Miocene of the Siwaliks of Pakistan. This species, originally placed in the genus *Taucanamo* (Pickford 1976), was transferred to *Schizochoerus* because of its lophodont molars. Later, van der Made and Han (1994) and then van der Made (1997) placed this species within the genus *Yunnanochoerus* based on its elongated lower premolars. *Yunnanochoerus gandakasensis* can be distinguished from *Y. lufengensis*, the type species of the genus, by a larger posterior lobe on m3, a weaker development of the posthypocristid and prehypoconulid crista on m3, and weaker constrictions between lobes 1 and 2 on molars. The lophodont molars of *Y. gandakasensis* (Fig. 19.7) are convergent with those of *Listriodon* and represent probable adaptations to folivory. Apart from the root fusions, the molars of *Yunnanochoerus* are differentiated from those of *Listriodon* in having notably more slender

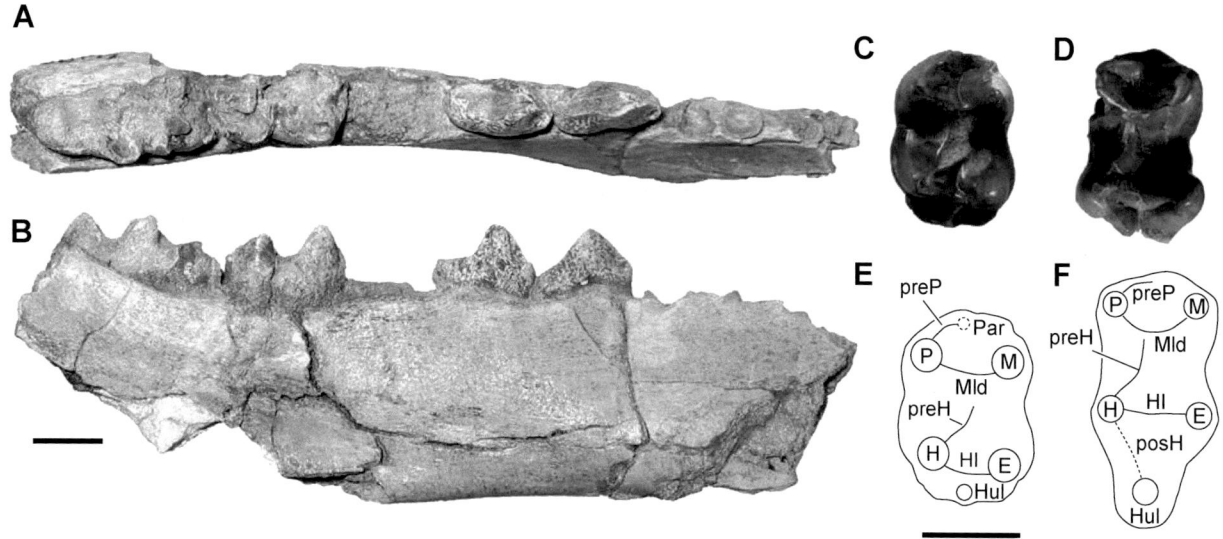

Figure 19.7. Yunnanochoerus gandakasensis. A., B. Holotype YGSP 4192 in occlusal (A) and lingual (B) views. C. YGSP 13056, right m1 or m2 in occlusal view. D. YGSP 4630, left m3 fragment in occlusal view. E. Schematic interpretation of YGSP 13056. F. schematic representation of a left m3. Abbreviations: posH = posthypocristid and prehypoconulid crista, preH = prehypocristid. See Fig. 19.1 and 19.6 for other abbreviations. Dashed crests and cusps are variably expressed in the sample. Scale bars = 1 cm.

tubercules, more marked constrictions between the lobes, a cristid obliqua (prehypocristid) that often connects the metalophid, and a posthypocristid sometimes present on m3. Whether or not the slight differences in the crest development of molars imply subtle differences in diet and mastication of these two species is unclear. The trenchant premolars with high protoconid and low mesial cingulum and hypoconid displayed by the type specimen of *Yunnanochoerus gandakasensis* also differ markedly from the p4 of *Listriodon,* which is more molariform and possesses a metalophid and additional shearing crest on the talonid. Although the more molariform p4 of *Listriodon* can be interpreted as being inherited phylogenetically (tendency of premolar molarization observed in the Listriodontinae; see van der Made 1996), its lophodont pattern reflects mastication with increased shearing function on premolars in comparison with the predominantly cutting function expected for the premolars of *Yunnanochoerus.*

Stable isotope analysis of *Yunnanochoerus gandakasensis* tooth enamel has yielded consistently low values ($\delta^{13}C = -13.0$ ‰; $n = 2$ and $\delta^{18}O = -5.1$ ‰ and -9.2‰) compared to other Siwalik taxa (Nelson, 2003, 2007; see Chapter 25, this volume). The isotopic data, together with the dental morphology, suggest that *Yunnanochoerus,* like *Listriodon,* was most probably a browser and inhabited quite closed and wet habitats.

Patterns of Change in Habitat and Diet throughout the Siwalik Sequence

The Middle Miocene suids from Pakistan, *Listriodon pentapotamiae, Conohyus sindiensis* and *Parachleuastochoerus pilgrimi* (formerly *Hyotherium pilgrimi*), were all closed-habitat C_3-browsers, as inferred from dental morphology, microwear, and stable isotopes. The forest environment inferred for the suid fossil

record is in agreement with the interpretation based on isotopic analyses of other Middle Miocene taxa (Morgan, Kingston, and Marino 1994). Despite Late Miocene faunal turnover, the early Late Miocene was not marked by significant changes in diet and habitat of the Suidae. The palaeochoerid *Yunnanochoerus gandakasensis* and the tetraconodont *Tetraconodon magnus* were both associated with forest habitats, *Yunnanochoerus* being presumably folivorous. Most of the isotopic data from *Hippopotamodon* and *Propotamochoerus* indicate that these suids were living in habitats similar to Middle Miocene members of the group. However, the somewhat enriched $\delta^{13}C$ values of these later suines indicate some exploitation of more open habitats with incorporation of some C_4 plants in their diet. This change in habitat and diet observable at the end of the ranges of *Hippopotamodon sivalense* and *Propotamochoerus hysudricus* was likely driven by the rapid expansion of C_4 plants recorded between 8.7 and 6.8 Ma, indicating a progressive opening of the landscape (Morgan, Kingston, and Marino 1994; Cerling et al. 1997; Nelson 2003, 2005, 2007; Chapters 24 and 25, this volume).

The dental traits of the Pliocene *Hippohyus* and *Sivahyus* suggest a C_4-based diet and thus imply a more open habitat for these suids. Although isotopic and microwear data are not available for Upper Siwalik suids, it is likely that the Late Pliocene–Pleistocene taxa were living in more open habitats than the Miocene suids and probably consuming some C_4 plants.

FAUNAL EVENTS AND CHANGES

The most reliable recent reference concerning the chronology of the Siwalik mammals is Barry et al. (2002); Chapter 2, this volume. Additional data, from the Pliocene period, have been extracted from the database managed by John Barry (Barry, pers. comm.). Unless noted otherwise, all of the first appearance data (FAD) and last appearance data (LAD) cited here come from these sources.

Lower Miocene and lower Middle Miocene suid remains of the Siwaliks are poorly represented. Thus, the analysis of faunal change begins in the latest part of the Middle Miocene, marked by the FAD of *Listriodon pentapotamiae* and *Conohyus sindiensis* in the Potwar Plateau region. The most recently available age calibrations for these inferred FADs are 15.2 Ma for *Conohyus sindiensis* and 14.3 Ma for *Listriodon pentapotamiae*. *Listriodon pentapotamiae* and *Conohyus sindiensis* both have last occurrences in the Potwar at 10.4 Ma (although there is a single upper premolar from Locality Y578, poorly dated to between 10.1 and 9.6 Ma, which has been identified as *C. sindiensis* by J. Barry). The last occurrences of these two species are nearly contemporaneous with the appearance of the long-duration species *Hippopotamodon sivalense* and *Propotamochoerus hysudricus* at 10.5 Ma, which triggered the domination of the Suinae subfamily in the late Neogene.

The faunal turnover in suids between ~11 and ~10 Ma, also noted in other families of artiodactyls and rodents (Barry et al. 2002), suggests significant reorganization of components of the mammalian community (as detailed in Chapter 25, this volume). Barry et al. (2002) proposed that the faunal change was caused mostly by species interactions because it could not be correlated with any detectable local environmental or climatic change. This hypothesis finds support in the fact that the diet and habitat of the immigrant suids did not differ significantly from those of their predecessors. At the same time, this interpretation cannot explain the disappearance of *Listriodon* around 11–10 Ma throughout Eurasia as well as the extinction of *Conohyus simorrensis* in Europe and *C. sindiensis* in Pakistan around 10 Ma (van der Made 1989–1990). The 11–10 Ma faunal turnover in Pakistan partially resembles the situation in Europe, where the Suinae *Microstonyx* and *Propotamochoerus* became the dominant suids after the extinction of *Conohyus* and *Listriodon* in Mammal Neogene zone MN9 deposits (~10 Ma). In China, the Suinae also became the dominant subfamily of suids (*Microstonyx* and *Propotamochoerus*) after the extinction of *Listriodon* (Liu 2003). Thus, it is possible that the 11–10 Ma turnover among suid communities in Pakistan reflects species interactions caused by the evolutionary success of the suines relative to other subfamilies of suids.

The early Late Miocene of the Siwalik sequence was marked by extinctions and appearances of species with long temporal ranges. In addition, several short-duration species such as *Yunnanochoerus gandakasensis*, *Tetraconodon magnus*, an unnamed small suine from Locality Y311, and a species of *Hippopotamodon* (see below for more details about this species) appeared. Note that *Yunnanochoerus gandakasensis* appeared at the end of the range of *Listriodon pentapotamiae* or just after its extinction and might thus have occupied a similar niche to *L. pentapotamiae*. In western Europe, a similar situation is posited as *Schizochoerus vallesiensis* and *Listriodon splendens* show comparable ranges (Agustí 1999). It is possible that competition drove the replacement of *Listriodon* by *Yunnanochoerus* among lophodont suoids, but considering that *Yunnanochoerus* is not a commonly found taxon in the fauna, this replacement may reflect a global niche reorganization among suoids (e.g., domination of the suines in the fauna after 10.5 Ma and replacement of *Listriodon* and *Conohyus*) rather than direct competition between *Listriodon* and *Yunnanochoerus*.

As noted by Morgan et al. (1995) and Chapter 25 (this volume), the early Late Miocene witnessed changes in suid body size. The more abundant suids of the 14–10 Ma interval all have average body weights close to those of extant suids, with values around 100 kg (see Fig. 19.1), as inferred from dental measurements (Morgan et al. 1995). In contrast, the suids present during the 10–9 Ma interval display a much wider range of body weights owing to the appearance of large species: during this interval, the large *Hippopotamodon sivalense* (~500 kg) and the gigantic *Tetraconodon magnus* (more than 1,000 kg) coexisted with the small suoid *Yunnanochoerus gandakasensis* (~60 kg).

The forest-habitat suoids *Tetraconodon magnus* and *Yunnanochoerus gandakasensis* have inferred LADs in the Potwar Plateau of 9.4 Ma and 8.1 Ma, respectively. The extinction of *T. magnus* seems too early to be correlated with the forest fragmentation that began after 9 Ma (e.g. Morgan, Kingston, and Marino 1994; Barry et al. 2002; Nelson 2003, 2005, 2007), which may have contributed to the extinction of both *Sivapithecus* (LAD at 8.4 Ma) and *Y. gandakasensis*.

Another period of faunal change occurred during the latest Miocene, beginning around 7.0 Ma. This period was notably marked by last appearances of forest- dependent and frugivorous large mammals (Barry et al. 2002; Nelson 2003), including *Hippopotamodon sivalense*, which had been a common member of the fauna since its appearance at 10.5 Ma. The rapid increase of open habitats between 8 and 6 Ma revealed by isotopic analysis of paleosols (Quade and Cerling 1995) and enamel of large mammals (Morgan, Kingston, and Marino 1994; Nelson 2003, 2005) is a probable cause of the local extinction of *H. sivalense*: this species shows an abrupt isotopic shift toward more enriched $\delta^{13}C$ values, which is presumably a response to changes in vegetation. The local extinction of *Propotamochoerus hysudricus* at 6.5 Ma is probably for the same reasons as the extinction of *Hippopotamodon sivalense*. The Potwar Plateau Siwalik suoid fossil record documents a sequence of local extinctions in which forest-dependent taxa such as *Yunnanochoerus* disappeared first, followed by the more generalist and ubiquitous suids *Propotamochoerus* and *Hippopotamodon*.

The Pliocene fauna of the Siwaliks is mostly comprised of *Hippohyus*, *Sivahyus* and *Sivachoerus*. A rare species referable to *Potamochoerus*, based on its symmetrical p3–p4 with a strong protoconid, simple talon of m3, and lateral bulge of the mandible, is present around the Mio–Pliocene transition (see Fig. 19.2b). These suids are too rare to provide accurate FADs and LADs. Material collected by the Harvard-GSP field project provides a minimum estimate of their ranges, which differs sometimes markedly from the estimates of Pickford (1988; see Fig. 19.2b). The presence of the putative open-habitat *Hippohyus* and *Sivahyus* in the Pliocene reinforces the idea that the Late Miocene faunal change was linked to environmental change. The Upper Siwaliks document other less common taxa, including a potential representative of the extant genus *Sus*.

First Appearance of the Suinae in the Siwaliks

While it is classically reported that Suinae appear around 10 Ma in the fossil record (Pickford and Liu 2001), Liu (2003) noted that the Middle Miocene Chinese suid *Miochoerus* represented a possible early member of the subfamily. Van der Made, Belinchon and Montoya (1998) have suggested the possibility of a record of the Suinae as early as the Early Miocene. In Thailand, Suinae have been documented from lignite seams (Pickford et al. 2004) dated at 12.4–12.2 Ma (Coster et al. 2010). In the Siwalik sequence, *Hippopotamodon sivalense* and *Propotamochoerus hysudricus* are preceded by a rare suine that appeared more than one million years earlier. This species, referred to as *Hippopotamodon* sp. in Figure 19.2a and "?Hippopotamodon Y450 unnamed species" in Barry et al. (2002), has a temporal distribution estimated at 11.4–10.2 Ma. This material may correspond to the Suinae *Propotamochoerus* sp. reported by van der Made, Belinchon, and Montoya (1998) from the Chinji Formation. *Hippopotamodon* sp. is best represented by a lower jaw with m2–m3 (YGSP 14284): the teeth display a well-defined groove that allows an attribution to Suinae. The m3 has a centrally-placed, elongated, and wide hypoconulid, which recalls the condition in *Hippopotamodon sivalense* and *Microstonyx major*. The mesoconulid and prehypoconulid are extremely large. The m3 (length ~38 mm) is smaller than that of *Hippopotamodon sivalense* and close in size and morphology to the material from the Nagri Formation published by van der Made and Hussain (1989) as *Microstonyx major*. I agree with Liu, Kostopoulos, and Fortelius (2004) in considering that the latter fossils are closer to *Hippopotamodon sivalense* by having larger premolars than *Microstonyx major*. It is therefore plausible that all the "smaller" material represents a single species of *Hippopotamodon* distinct from *H. sivalense*.

The Range of *Sus* in the Siwaliks

In the early study of Siwalik Suidae by Lydekker (1884), a large number of the taxa were placed in the genus *Sus*. Subsequent revisions of the fauna (Pilgrim 1926; Colbert 1935b, Pickford 1988) led to considerable change, with most species being transferred to other genera or synonymised. According to Pickford (1988), *Sus* is found in the Plio–Pleistocene deposits and may range back to the latest Miocene. The presence of *Sus* in the Harvard-GSP project collections is questionable. Fossils attributed by others to this genus do not contain derived features unique to *Sus* and may instead belong to representatives of the Hippohyini.

Although fossils collected prior to the 1970s provide clear evidence of the presence of *Sus* in the Siwaliks of Pakistan, these specimens lack precise geographic and stratigraphic provenance. NHM fossils are usually labeled "Siwalik Hills," while most of the AMNH fossils are labeled "Pinjor," "Tatrot," or "Hasnot" zones. These zones are currently recognized as Late Pliocene–Pleistocene sequences (e.g., Dennell, Coard, and Turner 2006). A partial skull and a maxilla fragment currently

curated at Harvard together with a few specimens at the AMNH collected by de Terra and Teilhard de Chardin (1936) come from Locality 103 (also named DP 54) situated in the Haro River valley, in the northern Potwar Plateau. This site has a maximum age estimate of ~1.2 Ma (Saunders and Dawson 1998). *Sus* was thus definitely present in the Siwaliks during the Lower Pleistocene.

The fossil record of *Sus* in the Siwaliks is scarce and suffers from the poor provenance of early collections. A presence of *Sus* by the Late Pliocene–Pleistocene is reasonable, but there is no strong support for a Lower Pliocene or latest Miocene record of the genus in the Siwaliks.

BIOGEOGRAPHY, DISPERSIONS, AND ORIGIN OF THE SPECIES

Middle Miocene

There is no satisfying candidate for the ancestor of *Conohyus sindiensis*. This problem is intimately related to the obscure origins of the Tetraconodontinae.

The origins of *Listriodon pentapotamiae* are also uncertain. *Listriodon pentapotamiae* is hypothesized to have evolved locally from *L. guptai* (e.g., Pickford and Morales 2003), a bunolophodont species defined from scarce remains from the early Miocene and early Middle Miocene of Bhaghothoro (Sind) and also present in the Siwaliks.

Faunal comparisons between the Siwaliks and Thailand (Ducrocq et al. 1994) reveal several genus-level similarities during the Middle Miocene. Species-level similarities between the Siwaliks and Thailand or Myanmar have also been proposed for non-suoid large mammals such as rhinocerotids and proboscideans (e.g., Chavasseau et al. 2006, 2009). These faunal similarities support biogeographic connections within South Asia during the Middle Miocene. *Conohyus thailandicus*, a suid discovered in Thailand and Myanmar, and *C. sindiensis* are very close morphologically, differentiated notably by their dental proportions (Ducrocq et al. 1997; Chavasseau et al. 2006). They share lower premolars with more expressed buccal bulges and second lobe buccal crown expansions than the European species *C. simorrensis* and *C. steinheimensis*, which indicates that they are likely closely related species. However, there is probably some regional endemism of Southeast Asia during this period. For instance, the association of *Listriodon* and *Conohyus*, very common in the Siwaliks, has never been documented in Southeast Asia, where the only possible *Listriodon*-bearing locality currently known is in Myanmar (Bender 1983).

The Middle Miocene fauna of Paşalar (Turkey) differs from that of the Siwalik Chinji Formation: Paşalar yielded a *Conohyus* species closer to the European species *C. simorrensis* and a palaeochoerid *Taucanamo inonuensis* unique to Anatolia (Fortelius and Bernor 1990). Moreover, the presence of two listriodontine lineages, with one fully lophodont form close or identical to the European species *Listriodon splendens* and another more bunodont species (Fortelius and Bernor 1990), is not observed in the Chinji Formation of Pakistan. Bernor and

Tobien (1990) argued that the mammalian assemblage of Paşalar showed that Anatolia exhibited a significant degree of endemism in the Middle Miocene.

Northern China has a suoid fauna which differs from that of the Siwaliks. It includes *Kubanochoerus gigas, Hyotherium shanwangense,* and *Sinapriculus linquensis* (Pearson 1928; Liu, Fortelius, and Pickford 2002). The suids of southern China show affinities with those of northern Thailand at the end of the Middle Miocene (Pickford et al. 2004).

These comparisons show that during the Middle Miocene, the Siwaliks had no obvious faunal affinities with Anatolia and northern China. The faunal resemblance with localities from Thailand and Myanmar suggests that these areas might have been part of a single South Asian biogeographical province during the Middle Miocene.

Late Miocene

The gigantic suid *Tetraconodon* appeared in the Siwaliks at 10.0 Ma, after the local extinction of *Conohyus sindiensis* (although there is a poorly dated specimen of *C. sindiensis* that could be younger than 10.0 Ma). The highly inflated premolars displayed by *Tetraconodon* make it unlikely to be a descendant of *Conohyus sindiensis.* Pickford (1988) and Pickford and Gupta (2001) proposed that *Conohyus indicus,* a late Middle Miocene–early Late Miocene species with slightly larger premolars than *C. sindiensis,* was a potential ancestor of *Tetraconodon.* The scarcity of the record of *Conohyus indicus* together with the large difference in premolar size between *C. indicus* and *Tetraconodon* make the relationship unclear. An alternative hypothesis is that *Tetraconodon* originated in Myanmar, where the genus is represented by two species, *T. minor* and *T. intermedius* (van der Made 1999; Htike et al. 2005), and subsequently migrated to the Indian subcontinent during the early Late Miocene. *T. minor* is interpreted as the most primitive species owing to its small premolars (for the genus); consequently, this species would be expected to predate or be contemporaneous with the more derived form of *Tetraconodon* in the Siwaliks. However, a Late Miocene age is generally asserted for the fossil assemblages that comprise *T. minor.* The poorly constrained chronology of the Miocene deposits of Myanmar as well as the limited knowledge of the intraspecific variation of the *Tetraconodon* species make this difficult to resolve.

The exclusive presence of *Tetraconodon* in Pakistan, India, and Myanmar demonstrates that biogeographic connections across southern Asia were maintained in the Late Miocene. The Late Miocene beds of Thailand have yielded *Hippopotamodon* cf. *sivalense* and *Propotamochoerus* cf. *hysudricus* (Chaimanee et al. 2004, 2006). *Propotamochoerus hysudricus* and *Hippopotamodon sivalense* were also identified in Myanmar (Htike et al. 2006; Sein, van der Made, and Rössner 2009; Jaeger et al. 2011). These discoveries indicate that the geographical distribution of *Propotamochoerus hysudricus* and *Hippopotamodon sivalense* probably encompassed southern Asia. Analyses of the Late Miocene large mammal fauna of Myanmar has reaffirmed a strong faunal resemblance to the Siwaliks of Pakistan (Chavasseau et al. 2013, 2018).

The Late Miocene fauna of northern China includes *Microstonyx* and *Chleuastochoerus,* the latter suid being unknown outside of China. The Late Miocene fauna of southern China is notably composed of the palaeochoerid *Yunnanochoerus* (van der Made and Han 1994) and the suids *Chleuastochoerus* and *Molarochoerus,* the latter being a dentally specialized suine (Pickford, Liu, and Pan 2004). The genera *Chleuastochoerus* and *Molarochoerus* are not found in southern Asia; thus, there is a significant dissimilarity between the suoids of southern Asia and southern China although common species have been identified (e.g., *Propotamochoerus wui* in Myanmar; Sein, van der Made, and Rössner 2009). Despite the presence of *Yunnanochoerus* in the Siwaliks, the Chinese suoids of the Late Miocene are mostly dissimilar to those of the Siwalik Late Miocene, which, in turn, suggests that Pakistan and China were in different biogeographical provinces. The information given by the micromammals is, however, somewhat different from that provided by the suoids: Flynn and Qi (1982) demonstrated that some rhizomyid rodents from Lufeng, China, were similar to taxa found in the Siwaliks.

Pliocene and Pleistocene

The Pliocene suids of Asia are poorly known. *Sivachoerus prior* is a member of the Pliocene fauna of the Siwaliks and has an observed range of 3.6–2.1 Ma in Pakistan. The Tetraconodontinae are represented in the early Late Miocene by three rare species: *Tetraconodon magnus, Conohyus indicus,* and *Lophochoerus nagrii.* It is interesting to note that *Lophochoerus,* described from the Haritalyangar area (India; ~10–8 Ma in Pillans et al. 2005), represents the latest Miocene record of a tetraconodont in the Siwaliks until the appearance of *Sivachoerus prior* in the Pliocene. *S. prior* is problematic because it be cannot be satisfyingly linked to any Miocene species in the Siwaliks. Moreover, hypothesizing a longer range for this species (owing to its rarity) would not bridge the temporal gap of the tetraconodont record. It is thus probable that the *Sivachoerus* lineage evolved elsewhere during the Late Miocene and then migrated to South Asia in the Pliocene. Kullmer (1999) proposed derivation from the African *Nyanzachoerus,* but a southeast Asian origin of *Sivachoerus* is also possible because *Sivachoerus* is also known from the latest Miocene and Pliocene of Myanmar (Takai et al. 2006; Thein et al. 2011).

The genera *Hippohyus* and *Sivahyus* are the other main suid taxa in the Pliocene of Pakistan. Colbert (1935a) hypothesized that they represented a single lineage phylogenetically close to those of *Sus* and *Phacochoerus.* Pickford (1988) considered the pattern of Hippohyini to be derived from that of the primitive Suinae *Propotamochoerus.* This hypothesis suffers from the lack of intermediate fossils. There are two possibilities regarding the evolution of Hippohyini. First, the Hippohyini evolved in another area and migrated to Pakistan and India in the latest Miocene–Early Pliocene. Limiting this interpretation is the absence of described Hippohyini or affiliated suid outside the Siwaliks of Indo-Pakistan. Second, the Hippohyini evolved in situ. This alternative finds modest support in the poorer fossil record of the Potwar Plateau after 7 Ma (Barry et al. 2002) and in the possible presence of conservative forms of Hippohyini in the Pliocene.

The poor knowledge of the Pliocene suids of Asia renders biogeographical analysis difficult. The presence of *Sivachoerus* in both Pakistan and Myanmar shows that there were at least minimal biogeographical links within southern Asia during the Pliocene. That the two highly specialized Suinae, *Sivahyus* and *Hippohyus*, have never been described outside of the Siwaliks of Pakistan and India possibly reflects their endemism.

CONCLUSION

The Neogene Siwalik sequence has yielded a minimum of 19 species of suoids and is thereby one of the richest faunas of the Old World. Contemporaneous or co-occurring species diversity is typically three or four species, and species richness peaks at seven around 10 Ma (see Fig. 19.2). In comparison, the Mio–Pliocene Sinap Formation of Turkey has yielded 10 suoid species with a maximum of 4 co-occurring species (van der Made 2003). According to Liu (2003), 15 suoid taxa are known in northern China from the Early Miocene to the Pliocene, and 5 suoids have been documented in the Late Miocene of southern China. With at least 30 species recognized, the Mio–Pliocene European fossil record has a greater species diversity than that of the Siwaliks (van der Made 1989–1990), but this diversity is aggregated over a much larger geographical area and covers the entire Miocene and Pliocene periods.

The suoids, like the overall large and small mammal record, show two main episodes of faunal change, between 11 and 10 Ma and between 7 and 6 Ma. Although climatic factors may be implicated, the first faunal turnover was probably also influenced by biotic interactions. This turnover significantly reorganized the suoid fauna, as shown by the notable disappearance of listriodontines, as well as the appearance of *Hippopotamodon sivalense* and *Propotamochoerus hysudricus*, which shows the increasing dominance of the suine subfamily. Body weight structure changed markedly with the emergence of larger species.

The second faunal turnover is probably primarily due to the climatic and vegetation changes that affected the Potwar Plateau during the Late Miocene: *P. hysudricus* and *Hippopotamodon sivalense* went extinct not long after an isotopic shift in their diet, suggesting a change in foraging behavior driven by expansion of C_4-dominated habitats. They were replaced by suids better adapted to open landscapes (*Potamochoerus*, *Sus*), including grazing taxa (*Hippohyus*, *Sivahyus*). More forest-dependent taxa such as *Yunnanochoerus* disappeared near the beginning of the C_4 expansion, about 0.2 My before the ape *Sivapithecus* disappeared.

Although the Siwalik suoids show genus-level similarities to European suids, the Siwalik suoids are most similar to those found in the faunas of Southeast Asia; this reflects a significant degree of provinciality across South Asia during the Neogene. There are a number of unresolved issues concerning the Siwalik suoids that await additional discovery of fossil specimens and their study:

Except for the most common taxa, species are still poorly known, especially in the Early and early Middle Miocene and in the Plio-Pleistocene.

Relationships between the primitive genera and the post-Miocene forms of Suinae are unclear.

In this exceptional Siwalik fossil record from 13 to 7 Ma, there is no demonstrated example of phyletic speciation and little documented evidence of phyletic evolution. The clearest example of phyletic evolution is *Hippopotamodon sivalense*, which shows an increase in size during its temporal range (e.g., Pickford 1988).

ACKNOWLEDGMENTS

Many thanks to David Pilbeam for inviting me to contribute to this volume. I thank Michèle E. Morgan and Catherine Badgley for their excellent editorial work, John C. Barry and Lawrence J. Flynn for discussions about Siwalik faunas. Thanks to Andy Currant (NHM) and Judy Galkin (AMNH) for access to fossil suoid collections.

References

Agustí, J. 1999. A critical re-evaluation of the Miocene mammal units in Western Europe: dispersal events and problems of correlation. In *Hominoid Evolution and Climatic Change in Europe*. Vol. 1: *The Evolution of Neogene Terrestrial Ecosystems in Europe*, edited by J. Agustí, L. Rook and P. Andrews, 84–112. Cambridge: Cambridge University Press.

Badgley, C., J. C. Barry, M. E. Morgan, S. V. Nelson, A. K. Behrensmeyer, T. E. Cerling, and D. R. Pilbeam. 2008. Ecological changes in Miocene mammalian record show impact of prolonged climatic forcing. *Proceedings of the National Academy of Sciences of the United States of America* 105(34): 12145–12149.

Badgley, C., M. S. Domingo, J. C. Barry, M. E. Morgan, L. J. Flynn, and D. Pilbeam. 2016. Continental gateways and the dynamics of mammalian faunas. *Comptes Rendus Palevol* 15(7): 763–779.

Barone, R. 1989. *Anatomie comparée des mammifères domestiques*. Tome 2, *Arthrologie et myologie*. Paris: Vigot.

Barry, J. C., A. K. Behrensmeyer, C. E. Badgley, L. J. Flynn, H. Peltonen, I. U. Cheema, D. Pilbeam, et al. 2013. The Neogene Siwaliks of the Potwar Plateau, Pakistan. In *Fossil Mammals of Asia: Neogene Biostratigraphy and Chronology*, edited by X.-M. Wang, L. J. Flynn and M. Fortelius, 373–399. New York: Columbia University Press.

Barry, J. C., E. H. Lindsay, and L. L. Jacobs. 1982. A biostratigraphic zonation of the Middle and Upper Siwaliks of the Potwar Plateau of Northern Pakistan. *Palaeogeography, Palaeoclimatology, Palaeoecology* 37(1): 95–130.

Barry, J. C., M. E. Morgan, L. J. Flynn, D. Pilbeam, A. K. Behrensmeyer, S. M. Raza, I. A. Khan, C. Badgley, J. Hicks, and J. Kelley. 2002. Faunal and environmental change in the Late Miocene Siwaliks of northern Pakistan. *Paleobiology* 28(2): 1–71.

Barry, J. C., M. E. Morgan, L. J. Flynn, D. Pilbeam, L. L. Jacobs, E. H. Lindsay, S. M. Raza, and N. Solounias. 1995. Patterns of faunal turnover and diversity in the Neogene Siwaliks of Northern Pakistan. *Palaeogeography, Palaeoclimatology, Palaeoecology* 115(1–4): 209–226.

Bender, F. 1983. *Geology of Burma*. Vol. 16. Berlin: Gebrüder Borntraeger.

Bernor, R. L., and H. Tobien. 1990. The mammalian geochronology and biogeography of Pasalar (Middle Miocene, Turkey). *Journal of Human Evolution* 19(4–5): 551–568.

Breytenbach, C. J., and J. D. Skinner. 1982. Diet, feeding and habitat utilization by bushpigs *Potamochoerus porcus* Linnaeus. *South African Journal of Wildlife Research* 12(1): 1–7.

Cerling, T. E., J. M. Harris, B. J. MacFadden, M. G. Leakey, J. Quade, V. Eisenmann, and J. R. Ehleringer. 1997. Global vegetation change through the Miocene/Pliocene boundary. *Nature* 389(6647): 153–158.

Chaimanee, Y., V. Suteethorn, P. Jintasakul, C. Vidthayanon, B. Marandat, and J.-J. Jaeger. 2004. A new orang-utan relative from the Late Miocene of Thailand. *Nature* 427(6973): 439–441.

Chaimanee, Y., C. Yamee, P. Tian, K. Khaowiset, B. Marandat, P. Tafforeau, C. Nemoz, and J.-J. Jaeger. 2006. *Khoratpithecus piriyai*, a Late Miocene hominoid of Thailand. *American Journal of Physical Anthropology* 131(3): 311–323.

Chavasseau, O., A. A. Khyaw, Y. Chaimanee, P. Coster, E.-G. Emonet, A. N. Soe, M. Rugbumrung, S. T. Tun, and J.-J. Jaeger. 2013. Advances in the biostratigraphy of the Irrawaddy Formation (Myanmar). In *Fossil Mammals of Asia: Neogene Biostratigraphy and Chronology*, edited by X.-M. Wang, L. J. Flynn and M. Fortelius, 461–474. New York: Columbia University Press.

Chavasseau, O., Y. Chaimanee, S. T. Tun, A. N. Soe, J. C. Barry, B. Marandat, J. Sudre, L. Marivaux, S. Ducrocq, and J.-J. Jaeger. 2006. Chaungtha, a new Middle Miocene mammal locality from the Irrawaddy Formation, Myanmar. *Journal of Asian Earth Sciences* 28(4–6): 354–362.

Chavasseau, O., Y. Chaimanee, C. Yamee, P. Tian, M. Rugbumrung, B. Marandat, and J.-J. Jaeger. 2009. New Proboscideans (Mammalia) from the Middle Miocene of Thailand. *Zoological Journal of the Linnean Society* 155(3): 703–721.

Chavasseau, O., C. Sein, H. Bocherens, F. Bibi, Y. Chaimanee, J. Maugoust, and J.-J. Jaeger. 2018. The *Khoratpithecus*-bearing fauna from the Central Basin of Myanmar: faunal composition, biochronology and biogeographic affinities. 5th International Palaeontological Congress, Paris, July 9–13.

Colbert, E. H. 1935a. Distributional and phylogenetic studies on Indian fossil mammals. IV, The phylogeny of the Indian Suidae and the origin of the Hippopotamidae. *American Museum Novitates* 799:1–24.

Colbert, E. H. 1935b. Siwalik Mammals in the American Museum of Natural History. *Transactions of the American Philosophical Society* n.s. 26(9): 1–401.

Cooke, H. B. S. 1985. Plio-Pleistocene Suidae in relation to African hominid deposits. In *L'environnements des hominidés au Plio-Pléistocène*, edited by Y. Coppens, 101–117. Paris: Masson.

Cooke, H. B. S. 2007. Stratigraphic variation in Suidae from the Shungura Formation and some coeval deposits. In *Hominin Environments in the East African Pliocene: An Assessment of the Faunal Evidence*, edited by R. Bobe, Z. Alemseged and A. K. Behrensmeyer, 107–127. Dordrecht: Springer.

Cooke, H. B. S., and R. F. Ewer. 1972. Fossil Suidae from Kanapoi and Lothagam, Northwestern Kenya. *Bulletin of the Museum of Comparative Zoology* 143(3): 150–236.

Coster, P., M. Benammi, Y. Chaimanee, O. Chavasseau, E.-G. Emonet, and J.-J. Jaeger. 2010. A complete magnetic-polarity stratigraphy of the Miocene continental deposits of Mae Moh Basin, northern Thailand, and a reassessment of the age of hominoid-bearing localities in northern Thailand. *Geological Society of America Bulletin* 122(7–8): 1180–91.

Dennell, R., R. Coard, and A. Turner. 2006. The biostratigraphy and magnetic polarity zonation of the Pabbi Hills, northern Pakistan: An Upper Siwalik (Pinjor Stage) Upper Pliocene-Lower Pleistocene fluvial sequence. *Palaeogeography, Palaeoclimatology, Palaeoecology* 234(2–4): 168–185.

De Terra, H., and P. Teilhard de Chardin. 1936. Observations on the Upper Siwalik Formation and later Pleistocene deposits in India. *Proceedings of the American Philosophical Society* 76(6): 791–822.

Ducrocq, S., Y. Chaimanee, V. Suteethorn, and J.-J. Jaeger. 1994. Age and paleoenvironment of Miocene mammalian faunas from Thailand. *Palaeogeography, Palaeoclimatology, Palaeoecology* 108 1–2): 149–163.

Ducrocq, S., Y. Chaimanee, V. Suteethorn, and J.-J. Jaeger. 1997. A new species of *Conohyus* (Suidae, Mammalia) from the Miocene of northern Thailand. *Neues Jahrbuch für Geologie und Paläontologie, Monatshefte* H(6): 348–360.

Ducrocq, S., Y. Chaimanee, V. Suteethorn, and J. Jaeger. 1998. The earliest known pig from the Upper Eocene of Thailand. *Palaeontology* 41(1): 147–156.

Falconer, H., and P. T. Cautley. 1847. *Fauna Antiqua Sivalensis, Being the Fossil Zoology of the Sewalik Hills, in the North of India. Part 8: Suidae and Rhinocerotidae.* London: Smith, Elder, and Co.

Flynn, L. J., J. C. Barry, M. E. Morgan, D. Pilbeam, L. L. Jacobs, and E. H. Lindsay. 1995. Neogene Siwalik mammalian lineages: Species longevities, rates of change, and modes of speciation. *Palaeogeography, Palaeoclimatology, Palaeoecology* 115(1–4): 249–264.

Flynn, L. J., E. H. Lindsay, D. Pilbeam, S. M. Raza, M. E. Morgan, J. C. Barry, et al. Opdyke. 2013. The Siwaliks and Neogene evolutionary biology in South Asia. In *Fossil Mammals of Asia: Neogene biostratigraphy and chronology*, edited by X.-M. Wang, L. J. Flynn and M. Fortelius, 353–372. New York: Columbia University Press.

Flynn, L. J., and G.-Q. Qi. 1982. Age of the Lufeng, China, hominoid locality. *Nature* 298(5876): 745–747.

Fortelius, M., and R. L. Bernor. 1990. A provisional systematic assessment of the Miocene Suoidea from Pasalar, Turkey. *Journal of Human Evolution* 19(4–5): 509–528.

Fortelius, M., J. van der Made, and R. L. Bernor. 1996. A new listriodont suid, *Bunolistriodon meidamon* sp. nov. from the Middle Miocene of Anatolia. *Journal of Vertebrate Paleontology* 16(1): 149–164.

Fortelius, M., and N. Solounias. 2000. Functional characterization of ungulate molars using the abrasion-attrition wear gradient: A new method for reconstructing paleodiets. *American Museum Novitates* 2000(3301): 1–36.

Gentry, A. W., and J. J. Hooker. 1988. The phylogeny of the Artiodactyla. In *The Phylogeny and Classification of the Tetrapods*, edited by Michael J. Benton, 235–272. Systematics Association Special Volume 35B. Oxford: Clarendon Press.

Harris, J. M., and T. E. Cerling. 2002. Dietary adaptations of extant and Neogene African suids. *Journal of Zoology* 256(1): 45–54.

Harris, J. M., and L. Liu. 2007. Superfamily Suoidea. In *The Evolution of Artiodactyls*, edited by D. R. Prothero and S. Foss, 130–150. Baltimore: John Hopkins University Press.

Htike, T., T. Tsubamoto, M. Takai, M. Natori, N. Egi, M. Maung, and C. Sein. 2005. A revision of *Tetraconodon* (Mammalia, Artiodactyla, Suidae) from the Miocene of Myanmar and description of new species. *Paleontological Research* 9(3): 243–253.

Htike, T., T. Tsubamoto, M. Takai, N. Egi, Z. M. M. Thein, M. Maung, and C. Sein. 2006. Discovery of *Propotamochoerus* (Artiodactyla, Suidae) from the Neogene of Myanmar. *Asian Paleoprimatology* 4:173–185.

Hunter, J. P., and M. Fortelius. 1994. Comparative occlusal morphology, facet development and microwear in two sympatric species of *Listriodon* (Mammalia: Suidae) from the Middle Miocene of Western Anatolia. *Journal of Vertebrate Paleontology* 14(1): 105–126.

Jaeger, J.-J., A. N. Soe, O. Chavasseau, P. Coster, E.-G. Emonet, F. Guy, R. Lebrun, et al. 2011. First hominoid from the Late Miocene of the

Irrawaddy Formation (Myanmar). *PLOS ONE* 6 (4):e17065. https://doi.org/10.1371/journal.pone.0017065.

Johnson, N. M., N. D. Opdyke, G. D. Johnson, E. H. Lindsay, and R. A. K. Tahirkheli. 1982. Magnetic polarity stratigraphy and ages of Siwalik Group rocks of the Potwar Plateau, Pakistan. *Palaeogeography, Palaeoclimatology, Palaeoecology* 37(1): 17–42.

Kingdon, J. 1997. *The Kingdon Field Guide to African Mammals.* San Diego, CA: Academic Press.

Kullmer, O. 1999. Evolution of African Plio-Pleistocene suids (Artiodactyla: Suidae) based on tooth pattern analysis. *Kaupia* 9:1–34.

Legendre, S. 1989. Les communautés de mammifères du Paléogène (Eocène supérieur et Oligocène) d'Europe occidentale: Structures, milieux et évolution. *Münchner Geowissenschaftliche Abhandlungen Reihe A: Geologie und Paläontologie* 16:1–110.

Leus, K. Y. G. 1994. Foraging, behavior, food selection and diet digestion of *Babyrousa babyrussa* (Suidae, Mammalia). Unpublished PhD diss., University of Edinburgh.

Lihoreau, F., J. C. Barry, C. Blondel, Y. Chaimanee, J.-J. Jaeger, and M. Brunet. 2007. Anatomical revision of the genus *Merycopotamus* (Artiodactyla; Anthracotheriidae): Its significance for Late Miocene mammal dispersal in Asia. *Palaeontology* 50(2): 503–524.

Liu, L. 2001. Eocene suoids (Artiodactyla, Mammalia) from Bose and Yongle basins, China, and the classification and evolution of the Paleogene suoids. *Vertebrata palasiatica* 39(2): 116–128.

Liu, L. 2003. Chinese fossil Suoidea. Systematics, evolution and paleoecology. PhD diss., University of Helsinki.

Liu, L., M. Fortelius, and M. Pickford. 2002. New fossil Suidae from Shanwang, Shandong, China. *Journal of Vertebrate Paleontology* 22(1): 152–163.

Liu, L., D. S. Kostopoulos, and M. Fortelius. 2004. Late Miocene *Microstonyx* remains (Suidae, Mammalia) from Northern China. *Geobios* 37(1): 49–64.

Lydekker, R. 1884. Indian Tertiary and post-Tertiary vertebrata. Siwalik and Narbada Bunodont Suina. *Memoirs of the Geological Survey of India* 10(3): 35–104.

MacFadden, B. J., and M. O. Woodburne. 1982. Systematics of the Neogene Siwalik hipparions (Mammalia, Equidae) based on cranial and dental morphology. *Journal of Vertebrate Paleontology* 2(2): 185–218.

Métais, G., R. A. Lashari, M. Pickford, and M. A. Warar. 2016. New material of *Listriodon guptai* Pilgrim, 1926 (Mammalia, Suidae) from the basal Manchar Formation, Sindh, Pakistan: Biochronological and paleobiogeographic implications. *Paleontological Research* 20(3): 226–232.

Montgelard, C., F. M. Catzeflis, and E. Douzery. 1997. Phylogenetic relationships of artiodactyls and cetaceans as deduced from the comparison of cytochrome b and 12s rRNA mitochondrial sequences. *Molecular Biology and Evolution* 14(5): 550–559.

Morgan, M. E. 1994. Paleoecology of Siwalik Miocene hominoid communities: Stable carbon isotope, dental microwear, and body size analyses. Unpublished PhD diss., Harvard University.

Morgan, M. E., C. Badgley, G. F. Gunnell, P. D. Gingerich, J. W. Kappelman, and M. C. Mass. 1995. Comparative paleoecology of Paleogene and Neogene mammalian faunas: Body-size structure. *Palaeogeography, Palaeoclimatology, Palaeoecology* 115(1–4): 287–317.

Morgan, M. E., J. D. Kingston, and B. D. Marino. 1994. Carbon isotopic evidence for the emergence of C_4 plants in the Neogene from Pakistan and Kenya. *Nature* 367(6459): 162–165.

Nelson, S. V. 2003. *The Extinction of Sivapithecus. Faunal and Environmental Changes Surrounding the Disappearance of a Miocene Hominoid in the Siwaliks of Pakistan.* Boston: Brill.

Nelson, S. V. 2005. Paleoseasonality inferred from equid teeth and intra-tooth isotopic variability. *Palaeogeography, Palaeoclimatology, Palaeoecology* 222(1–2): 122–144.

Nelson, S. V. 2007. Isotopic reconstructions of habitat change surrounding the extinction of *Sivapithecus*, a Miocene hominoid, in the Siwalik Group of Pakistan. *Palaeogeography, Palaeoclimatology, Palaeoecology* 243(1–2): 204–222.

Orliac, M. J. 2006. *Eurolistriodon tenarezensis*, sp. nov., from Montréal-du-Gers (France): Implications for the systematics of the European Listriodontinae (Suidae, Mammalia). *Journal of Vertebrate Paleontology* 26(4): 967–980.

Orliac, M. J. 2007. Does the snout disc make the pig? *Journal of Morphology* 268(12): 1113–1113.

Orliac, M. J., P.-O. Antoine, and S. Ducrocq. 2010. Phylogenetic relationships of the Suidae (Mammalia, Cetartiodactyla): New insights on the relationships within Suoidea. *Zoologica Scripta* 39(4): 315–330.

Orliac, M. J., P.-O. Antoine, G. Métais, L. Marivaux, J.-Y. Crochet, J.-L. Welcomme, S. R. H. Baqri, and G. Roohi. 2009. *Listriodon guptai* Pilgrim, 1926 (Mammalia, Suidae) from the early Miocene of the Bugti Hills, Balochistan, Pakistan: New insights into early Listriodontinae evolution and biogeography. *Naturwissenschaften* 96(8): 911–920.

Orliac, M., F. Guy, Y. Chaimanee, J.-J. Jaeger, and S. Ducrocq. 2011. New remains of *Egatochoerus jaegeri* (Mammalia, Suoidea) from the Late Eocene of peninsular Thailand. *Palaeontology* 54(6): 1323–1335.

Orliac, M. J., L. Karadenizli, P.-O. Antoine, and S. Sen. 2015. Small hyotheriine suids (Mammalia, Artiodactyla) from the late early Miocene of Turkey and a short overview of early Miocene small suoids in the Old World. *Palaeontologia Electronica* 18(2): 30A.

Pearson, H. S. 1928. Chinese Fossil Suidae. *Palaeontologia Sinica, Series C* 5(5): 1–75.

Pickford, M. 1976. A new species of *Taucanamo* (Mammalia: Artiodactyla) from the Siwaliks of the Potwar Plateau, Pakistan. *Pakistan Journal of Zoology* 8(1): 13–20.

Pickford, M. 1978. The taxonomic status and distribution of *Schizochoerus* (Mammalia, Tayassuidae). *Tertiary Research* 2(1): 29–38.

Pickford, M. 1981. *Parachleuastochoerus* (Mammalia, Suidae). *Estudios geologicos* 37(3): 313–320.

Pickford, M. 1988. Revision of the Miocene Suidae of the Indian Subcontinent. *Münchner Geowissenschaftliche Abhandlungen Reihe A: Geologie und Paläontologie* 12:1–92.

Pickford, M. 1993. Old World suoid systematics, phylogeny, biogeography and biostratigraphy. *Paleontologia i evolucio* 26–27:237–269.

Pickford, M., and S. S. Gupta. 2001. New specimen of *Conohyus indicus* (Lydekker, 1884) (Mammalia: Suidae) from the base of the Late Miocene, Jammu, India. *Annales de Paléontologie* 87(4): 271–281.

Pickford, M., J.-H. Liu, and Y.-R. Pan. 2004. Systematics and functional morphology of *Molarochoerus yuanmouensis* (Suidae, Mammalia) from the Late Miocene of Yunnan, China. *Comptes Rendus Palevol* 3(8): 691–704.

Pickford, M., and L. Liu. 2001. Revision of the Miocene Suidae of Xiaolongtan (Kaiyuan), China. *Bollettino della Societa Paleontologica Italiana* 40(2): 275–283.

Pickford, M., and J. Morales. 2003. New Listriodontinae (Mammalia, Suidae) from Europe and a review of listriodont evolution, biostratigraphy and biogeography. *Geodiversitas* 25(2): 347–404.

Pickford, M., H. Nakaya, Y. Kunimatsu, H. Saegusa, A. Fukuchi, and B. Ratanasthien. 2004. Age and taxonomic status of the Chiang Muan (Thailand) hominoids. *Comptes Rendus Palevol* 3(1): 65–75.

Pilbeam, D., G. E. Meyer, C. Badgley, M. D. Rose, M. Pickford, A. K. Behrensmeyer, and I. S. M. Shah. 1977. New hominoid primates from Siwaliks of Pakistan and their bearing on hominoid evolution. *Nature* 270(5639): 689–695.

Pilgrim, G. E. 1908. The Tertiary and post-Tertiary fresh-water deposits of Baluchistan and Sind, with notices of new vertebrates. *Records of the Geological Survey of India* 37:139–166.

Pilgrim, G. E. 1926. The fossil Suidae of India. *Memoirs of the Geological Survey of India* 8(4): 1–68.

Pillans, B., M. Williams, D. Cameron, R. Patnaik, J. Hogarth, A. Sahni, J. C. Sharma, F. Williams, and R. L. Bernor. 2005. Revised correlation of the Haritalyangar magnetostratigraphy, Indian Siwaliks: Implication for the age of the Miocene hominoids *Indopithecus* and *Sivapithecus*, with a note on a new hominid tooth. *Journal of Human Evolution* 48(5): 507–515.

Quade, J., and T. E. Cerling. 1995. Expansion of C_4 grasses in the Late Miocene of Northern Pakistan: Evidence from stable isotopes in paleosols. *Palaeogeography, Palaeoclimatology, Palaeoecology* 115(1–4): 91–116.

Quade, J., T. E. Cerling, P. Andrews, and B. Alpagut. 1995. Paleodietary reconstruction of Miocene faunas from Pasalar, Turkey, using stable carbon and oxygen isotopes of fossil tooth enamel. *Journal of Human Evolution* 28(4): 373–384.

Saunders, J. J., and B. K. Dawson. 1998. Bone damage patterns produced by extinct hyena, *Pachycrocuta brevirostris* (Mammalia: Carnivora), at the Haro River Quarry, northwestern Pakistan. *National Science Museum Monographs* 14:215–242.

Schley, L., and T. J. Roper. 2003. Diet of wild boar *Sus scrofa* in western Europe, with particular reference to consumption of agricultural crops. *Mammal Review* 33(1): 43–56.

Sein, C., J. van der Made, and G. Rössner. 2009. New Material of *Propotamochoerus* (Suidae, Mammalia) from the Irrawaddy Formation, Myanmar. *Neues Jahrbuch für Geologie und Paläontologie Abhandlungen* 251(1): 17–31.

Solounias, N., and G. Semprebon. 2002. Advances in the reconstruction of ungulates ecomorphology with application to early fossil equids. *American Museum Novitates* 2002(3366): 1–49.

Takai, M., H. Saegusa, T. Htike, and Z. M. M. Thein. 2006. Neogene mammalian fauna in Myanmar. *Asian Paleoprimatology* 4:143–72.

Tauxe, L., and N. D. Opdyke. 1982. A time framework based on magnetostratigraphy for the Siwalik sediments of the Khaur area, northern Pakistan. *Palaeogeography, Palaeoclimatology, Palaeoecology* 37(1): 43–61.

Thein, Z. M. M., M. Takai, H. Uno, J. G. Wynn, N. Egi, T. Tsubamoto, T. Htike, et al. 2011. Stable isotope analysis of the tooth enamel of Chaingzauk mammalian fauna (late Neogene, Myanmar) and its implication to paleoenvironment and paleogeography. *Palaeogeography, Palaeoclimatology, Palaeoecology* 300(1–4): 11–22.

Treydte, A. C., S. M. Bernasconi, M. Kreuzer, and P. J. Edwards. 2006. Diet of the common warthog (*Phacochoerus africanus*) on former cattle grounds in a Tanzanian savanna. *Journal of Mammalogy* 87(5): 889–898.

Van der Made, J. 1989–1990. A range-chart for European Suidae and Tayassuidae. *Paleontologia i evolucio* 23:99–104.

Van der Made, J. 1996. Listriodontinae (Suidae, Mammalia), their evolution, systematics and distribution in time and space. *Contributions to Tertiary and Quaternary Geology* 33(1–4): 3–254.

Van der Made, J. 1997. Systematics and stratigraphy of the genera *Taucanamo* and *Schizochoerus* and a classification of the Palaeochoeridae (Suoidea, Mammalia). *Proceedings of the Koninklijke Nederlandse Akademie van Wettenschappen* 100(1–2): 127–139.

Van der Made, J. 1999. Biometrical trends in the Tetraconodontinae, a subfamily of pigs. *Transactions of the Royal Society of Edinburgh: Earth Sciences* 89(3): 199–225.

Van der Made, J. 2003. Suoidea (Artiodactyla). In *Geology and Paleontology of the Miocene Sinap Formation, Turkey*, edited by M. Fortelius, J. Kappelman, S. Sen and R. L. Bernor, 308–27. New York: Columbia University Press.

Van der Made, J., M. Belinchon, and P. Montoya. 1998. Suoidea (Mammalia) from the Lower Miocene locality of Bunol, Valencia, Spain. *Geobios* 31(1): 99–112.

Van der Made, J., and D. Han. 1994. Suoidea from the Upper Miocene hominoid locality of Lufeng, Yunnan Province, China. *Proceedings of the Koninklijke Nederlandse Akademie van Wettenschappen* 97(1): 27–82.

Van der Made, J., and S. T. Hussain. 1989. 'Microstonyx' major (Suidae, Artiodactyla) from the type area of the Nagri formation, Siwalik group, Pakistan. *Estudios geologicos* 45(5–6): 409–416.

Vaughan, T. A., J. M. Ryan, and N. J. Czaplewski. 2013. *Mammalogy*, 6th ed. Philadelphia: Saunders College Publishing.

20

SIWALIK HIPPOPOTAMOIDEA

FABRICE LIHOREAU AND JEAN-RENAUD BOISSERIE

INTRODUCTION

Since the 2000s, systematic and phylogenetic studies have strongly linked two cetartiodactyl families, the Anthracotheriidae and the Hippopotamidae, as sister taxa in the superfamily Hippopotamoidea (*sensu* Gentry and Hooker 1988) within Cetancodonta (Boisserie, Lihoreau, and Brunet 2005; Spaulding et al. 2009; Boisserie et al. 2010; Orliac et al. 2010; Lihoreau et al. 2015; Gomes Rodrigues et al. 2021). Consequently, the group "Anthracotheriidae" is paraphyletic. The Hippopotamoidea are characterized by diachronic acquisition of several independent morphological features linked to a semiaquatic lifestyle (Houssaye et al. 2021; Orliac et al. 2023). This adaptation favored the abundance of well-preserved fossils (Boisserie et al. 2011; Lihoreau et al. 2014). The semiaquatic, large-herbivore niche enhances the value of this superfamily for reconstructing Cenozoic environments and paleogeography (Boisserie et al. 2011).

The oldest representative of Hippopotamoidea *sensu stricto* (i.e., the clade "Anthracotheriidae" + Hippopotamidae) is a small bunodont anthracothere, *Siamotherium*, from the late Middle Eocene (~40 Ma) of southeastern Asia (Suteethorn et al. 1988; Ducrocq et al. 2000; Soe et al. 2017). It exhibited numerous dental anomalies in an early illustration of the morphological plasticity of the group (Ducrocq et al. 1995). Anthracotheres dispersed as early as the Late Eocene to North America and Europe (Kron and Manning 1998; Grandi and Bona 2017; Lihoreau and Ducrocq 2007) and somewhat later to Africa (Holroyd et al. 2010; Lihoreau, Hautier and Mahboubi 2015). They diversified rapidly with different patterns in each continental region (Lihoreau and Ducrocq 2007). Such large-scale dispersal events occurred many times during the history of the superfamily and facilitated its survival during most of the Cenozoic (Lihoreau and Ducrocq 2007). The anthracotheres experienced diachronic extinctions on each landmass. Their latest representatives on the Indian subcontinent and in southeastern Asia occurred around 2 Ma (Hooijer 1952; Steensma and Hussain 1992; Dennell 2005).

The earliest Hippopotamidae, *Morotochoerus* and *Kulutherium*, are known from the Early Miocene of eastern Africa (Orliac et al. 2010; Tsubamoto, Kunimatsu, and Nakatsukasa

2015). They most likely evolved from an earlier African lineage of anthracotheres, the best-known representative being *Epirigenys*, an Oligocene bothriodontine (Boisserie and Lihoreau 2006; Orliac et al. 2010; Lihoreau et al. 2015). The evolutionary history of Hippopotamidae occurred mostly in Africa, but was also marked by several dispersals to Eurasia at the end of the Miocene and during the Early Pleistocene, with South Asian populations surviving until the Late Pleistocene (e.g., Badam 2007). The two living representatives of the Hippopotamoidea— *Hippopotamus* and *Choeropsis*—occur, respectively, in sub-Saharan freshwater rivers and lakes and in West African rainforests north of the Gulf of Guinea.

The Siwalik fossil record has played a critical role in the history of Hippopotamoidea because hypothesized phylogenetic relationships between anthracotheres and hippopotamuses were based on morphological resemblances between Siwalik taxa (Falconer and Cautley 1847; Lydekker 1876; Colbert 1935). Initially, a high number of hippopotamoid species was recognized in Siwalik deposits, which led to confusing taxonomy (Falconer and Cautley 1836; Lydekker 1876, 1877a, 1877b, 1878, 1884a, 1885; Pilgrim 1910, 1913, 1917). Colbert (1935) made the first attempt to clarify this situation but lacked precise stratigraphic placement of the many localities. More recently, field programs organized by David Pilbeam and the Geological Survey of Pakistan (initially Yale-GSP, then Harvard-GSP, with all recovered fossils having YGSP prefix) in the Potwar Plateau resulted in the recovery of many new specimens with accurate geographic and temporal provenience (Pilbeam et al. 1979; Barry, Behrensmeyer, and Monaghan 1980; Johnson et al. 1985; Tauxe and Badgley 1988; Barry et al. 2002). In particular, anatomical study of this well-dated Miocene anthracothere material has made possible new understandings of the variation within and relationships among these taxa and thereby reinterpretation of their status (Lihoreau et al. 2004a, 2004b, 2007). Furthermore, the YGSP material documents hippopotamoid evolution within one region over millions of years in response to regional environmental change (Barry et al. 2002; Badgley et al. 2008). We present a synthesis of current knowledge on Siwalik hippopotamoids

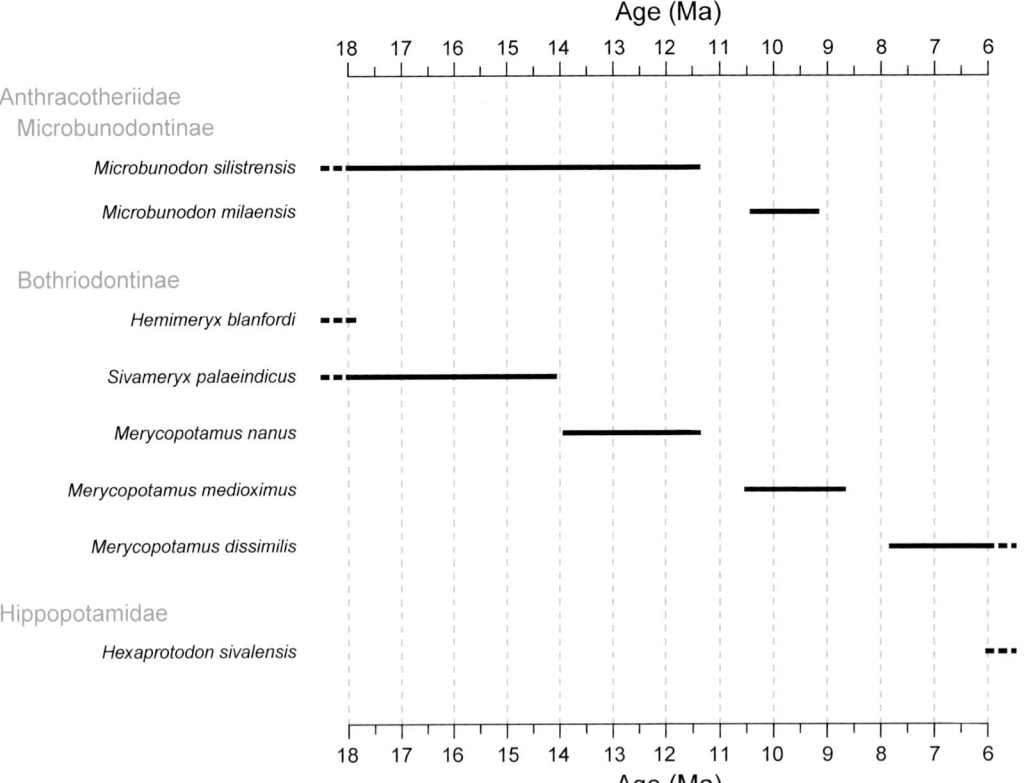

Figure 20.1. Simple diagram with species ranges of Siwalik taxa.

(Fig. 20.1) and a paleoecological interpretation of each group in order to highlight important transitions during the Neogene and their global relevance.

METHODS

Here we review all hippopotamoid species found in the Potwar Plateau by describing their main morphological characteristics. These descriptions are based on our direct observations of YGSP material, as well as other Siwalik material, notably from the collections of the Natural History Museum in London (NHM). Tooth descriptions follow the dental nomenclature in Boisserie et al. (2010). Phylogenetic placements for each taxon are drawn from the published literature. We also include the inferred paleobiology for each species.

The only extant hippopotamoids are two African species. A brief overview of hippopotamid biological traits is helpful for ecological interpretations of fossil hippopotamoids. Hippopotamid morphology is dominated by adaptations to (1) semiaquatic life; (2) social interactions, notably fights linked to population control in the context of high-density social groups; and (3) low-quality food accessible around water bodies. These adaptations include prolonged- to ever-growing incisors and canines, high-crowned molars with elongated crests, robustness of the symphyseal region of the mandible, osteomuscular complexes controlling gape (developed angular process, elevated nuchal crest, robust zygomatic apophyses), and sensory cranial superstructures (elevated orbits, enlarged nasal cavity). These features are particularly marked in the common hippo,

Hippopotamus amphibius, but are little expressed in the other living species, *Choeropsis liberiensis.* Premature paleoecological interpretations of fossil morphologies have been based on these extant interspecific differences. Species sharing plesiomorphic traits with *C. liberiensis* (e.g., low orbits, low-crowned cheek teeth, short muzzle) have been interpreted as less water-dependent and reliant on low-quality food from relatively closed-canopy forest. These interpretations should be considered with caution, because they do not account for intertwined dietary, habitat, and behavior-related adaptations of cranial morphology in fossil hippopotamids. Furthermore, many "archaic: traits of *C. liberiensis,* whose fossil lineage is currently unknown, could be related to its singular habitat of tropical rainforests, which are rarely sampled in the fossil record. Recent isotopic and microwear analyses support the semiaquatic habits and low-quality diet of fossil hippopotamids (Boisserie et al. 2005b; Clementz, Holroyd, and Koch 2008; Harris et al. 2008; Boisserie and Merceron 2011; Uno et al. 2011).

Since extant hippopotamids appear to be derived within Hippopotamoidea, they cannot readily be used to interpret the ecology of fossil hippopotamoids. In the present chapter, our inferences are in part based on alternative comparative data. We discuss dental microwear analysis of *Microbunodon, Merycopotamus,* and *Hexaprotodon sivalensis,* using methods and comparative datasets established by Merceron et al. (2005), as well as Merceron, Viriot, and Blondel (2004), Boisserie et al. (2005b), and Boisserie and Merceron (2011).

In addition, we analyzed stable isotopes (δ^{13}C and δ^{18}O, using data provided by Michèle Morgan; see Chapters 5 and 25, this volume) of tooth enamel of *Microbunodon* and *Merycopotamus* species for the time interval from 13.7 Ma to 6.3 Ma, allowing estimates of the contribution of C_3 versus C_4 plants to the diet. We follow diet classes as in Uno et al. (2011), with $\delta^{13}C \leq 8\text{‰}$ corresponding to C_3-dominated diets and C_3 from closed forest for $\delta^{13}C \leq 12\text{‰}$ following Badgley et al. (2008), $\delta^{13}C$ values between -8‰ to -2‰ corresponding to mixed C_3/C_4 diets and $\delta^{13}C \geq -2\text{‰}$ corresponding to C_4-dominated diets. In addition, enamel geochemistry may identify putative semiaquatic habits (noted $\delta^{18}O = \text{mean} \pm SD\text{‰}$). Clementz, Holroyd, and Koch (2008) have highlighted the correlation between $\delta^{18}O$ values of the extant semiaquatic hippopotamus and associated terrestrial herbivorous mammals (noted as $t\delta^{18}O$). Following this correlation, we predict the mean value of a semiaquatic mammal based on data from associated large herbivorous mammals (noted as $e\delta^{18}O$) in the Siwalik sequence and compare this result to the mean $\delta^{18}O$ values of Siwalik anthracoteres. In order to have enough specimens to make these comparisons, we use time intervals of 0.2 to 0.4 million years instead of localities (noted as $\delta^{18}O_{time\ interval}$).

Functional interpretations of anthracothere species are based on comparative studies of extant ungulate anatomy by Herring (1975), Solounias and Dawson-Saunders (1988), and Janis (1995). Body-size estimates are based on talus measurements using regressions of Martinez and Sudre (1995).

SPECIES DESCRIPTIONS AND PALEOECOLOGY

Within Microbunodontinae, the genus *Microbunodon*, following the diagnosis of Lihoreau and Ducrocq (2007), is a small anthracothere with buno-selenodont molars, upper canines transversely compressed with continuous growth in males and a small lower canine in contact with I3. *Microbunodon silistrensis* (Pentland, 1828; Figs. 20.2a, b, c, d, h) is known from the Late Oligocene of Zinda Pir to the early Late Miocene (last occurrence at 11.4 Ma). Probable specimens of *Microbunodon* cf. *silistrensis* occur in the Early Oligocene (Z108 and C2; Lindsay et al. 2005; Métais et al. 2009; Antoine et al. 2013). This species occurs in localities of Bengal and Ramnagar, India; the Potwar Plateau, Sind, Zinda Pir Dome, and Bugti Hills, Pakistan; and Thanbinkan, Myanmar (Lihoreau et al. 2004b; Lindsay et al. 2005; Tsubamoto et al. 2012). According to Lihoreau et al. (2004b), the teeth are equivalent in size to those of the type species *M. minimum*, with a smaller style on molars and with the premolar row longer than the molar row. On the mandible this species displays a keeled genio-hyoideus muscle attachment that extends anteriorly in ventral view.

Until the beginning of the twenty-first century, representatives of *Microbunodon* were considered endemic to the Late Oligocene of Europe, but, numerous fossils from the Potwar Plateau support the presence of this genus on the Indian subcontinent (Lihoreau et al. 2004b). Indeed, the generic attribution of *M. silistrensis* has been discussed since Pomel (1848). A close relationship between European and Asian anthracotheres was first proposed by Pilgrim (1910, 1912, 1913) but the scarce and incomplete material known at the beginning of the twenti-

eth century led to rejection of this hypothesis (Forster Cooper 1924; Matthew 1929; Colbert 1935; Cabard 1976; Raza et al. 1984; Pickford 1987; Pickford and Rogers 1987). The new material from the Potwar Plateau, notably the diagnostic anterior-tooth morphology, supports assignment to Microbunodontinae (Lihoreau and Ducrocq 2007), which includes the genera *Microbunodon*, at least some *Anthracokeryx* (Scherler, Lihoreau, and Becker. 2018), and *Geniokeryx* (Ducrocq 2020). The species *M. silistrensis* displays a long temporal range (Early Oligocene to Late Miocene), even if revision is necessary, notably for the Bugti material. A specimen from the Eocene Naduo Formation in China has been interpreted as *Microbunodon* sp. (Tsubamoto 2010), which could anchor the origin of the genus in the Late Eocene of southeastern Asia.

Despite few postcranial remains, one talus (YGSP 17377 from Locality Y500 dated to 12.1 Ma) provides the basis for an estimate of the body size of *M. silistrensis* as ~25 kg, slightly larger than the estimate for European *M. minimum* (mean of 21 ± 5 kg; Lihoreau 2003). *Microbunodon silistrensis* has more slender limbs than the European species and exhibits a narrow muzzle, a shallow mandible with an undeveloped gonial angle, and a concave nuchal border. Moreover, *Microbunodon* has a blade-like canine in males (Brunet 1968; Tsubamoto 2010); such was never recovered in the Potwar Plateau but has been inferred from mandibular morphology (Lihoreau et al. 2004b). Following Kingdon (1997), extant cetartiodactyls that exhibit such canines (moschids and tragulids) are nocturnal forest-dwellers with territorial and solitary habits or forming small groups.

We analyzed dental microwear of seven specimens of Siwalik *Microbunodon* M2 (Table 20.1) from Pakistan. The proportion of pits for each specimen compared to all microwear features is close to or greater than 50%, which is similar to those in mixed-feeding species (Merceron, Viriot, and Blondel 2004). Although this small, chronologically heterogeneous sample cannot be used for reliable inference of diet, it is possible to exclude grazing and exclusively leaf-browsing habits. Larger samples of *Microbunodon minimum* (18 specimens) from the Late Oligocene of La Milloque, France, have microwear patterns similar to those of the roe deer, a frugivorous browser (Lihoreau 2003). *M. silistrensis* from Pakistan does not differ markedly from the European species, *M. minimum*, but the greater number of scratches in *M. silistrensis* may indicate a diet that included more grass, perhaps seasonally. Only three specimens of *M. silistrensis* from localities dated between 12.3 and 11.5 Ma have been isotopically sampled; they all display $\delta^{13}C$ values of a C_3 diet but not a primarily frugivorous diet ($\delta^{13}C_{YGSP\ 18928} = -10.7\text{‰}$, $\delta^{13}C_{YGSP\ 21063} = -10.6\text{‰}$, $\delta^{13}C_{YGSP\ 19831} = -10.9\text{‰}$).

This small anthracothere could be a closed-habitat dweller, mixed feeder, and primarily solitary with territorial behavior similar to that of the extant water chevrotain.

Microbunodon milaensis Lihoreau et al. 2004b (Fig. 20.2e, f, g) is known from the Late Miocene (10.4 Ma) to Late Pliocene (Lihoreau et al. 2004b; Takai et al. 2016) in localities from the Nagri Formation and Irrawaddy sediments. In the Potwar Plateau, this species occurs between 10.4 Ma (Y572) and 9.2 Ma (Y317). In terms of geography, *M. milaensis* is known from the

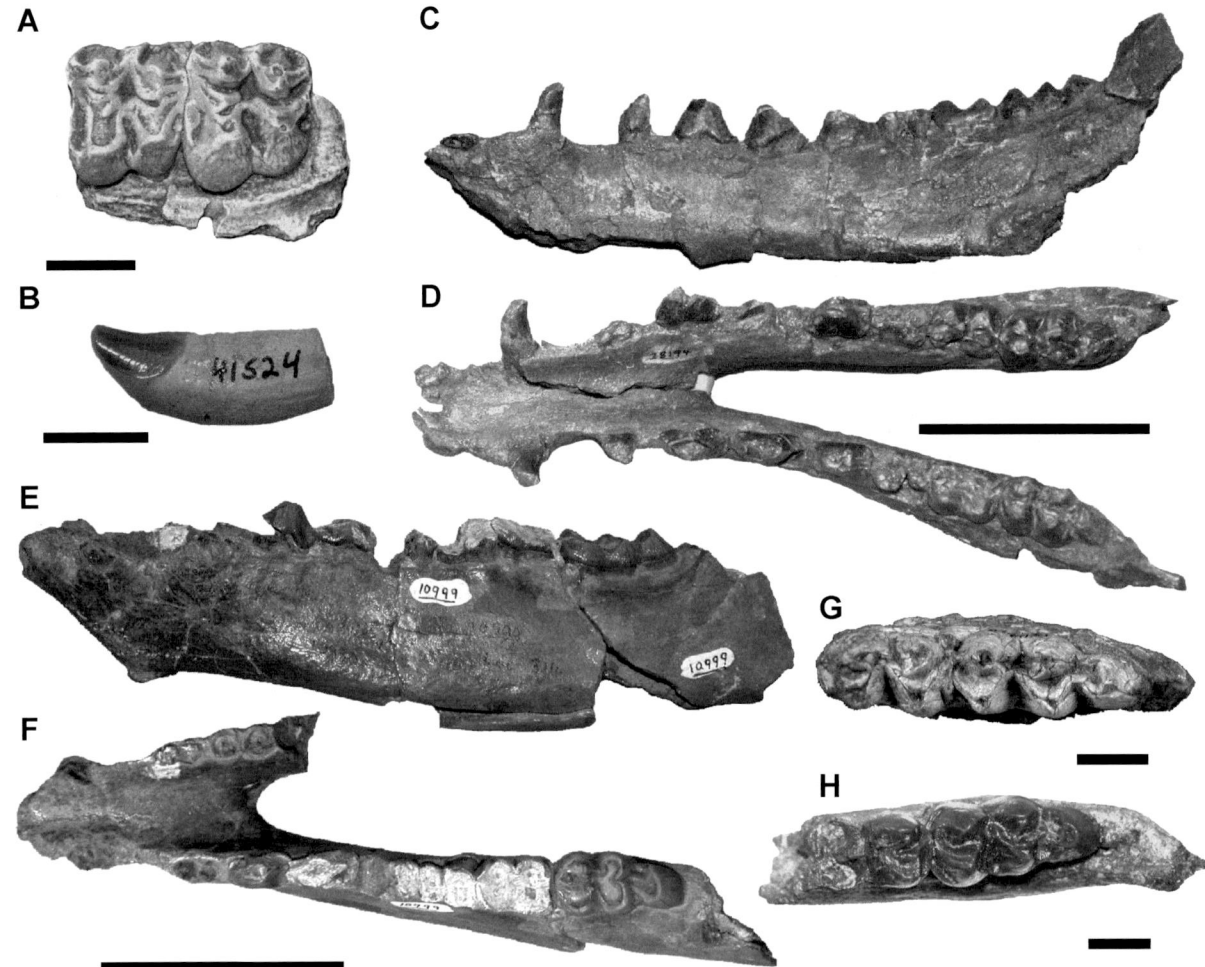

Figure 20.2. Dental and mandibular remains of Microbunodontinae from Potwar plateau. A.–D., H. *Microbunodon silistrensis*: A. left M2-M3 (YGSP 24297) in occlusal view; B. right lower canine (YGSP 41524) in lingual view; C., D. mandible (YGSP 28194) in labial and occlusal views; H. left m2-m3 (YGSP 21063) in occlusal view. E.–G. *Microbunodon milaensis*: E., F. mandible with left p3-m3 (YGSP 10999, type) in labial and occlusal views; G. left m2-m3 (YGSP 12448, paratype) in occlusal view. Scale bars for A., B., G.–H. = 1 cm; scale bars for C.,D., E., F. = 5 cm.

Potwar Plateau, Pakistan and from Yenangyaung, Chaingzauk, and Gwebin, Myanmar (Lihoreau et al. 2004b; Tsubamoto et al. 2012; Chavasseau et al. 2013).

According to Lihoreau et al. (2004b), *M. milaensis* is larger than other species of the genus, shows a biradiculate p1 and, lacks a diastema between p1 and p2. The morphology of the symphysis is close to that of *M. silistrensis*.

All specimens attributed to *M. milaensis* are from the Potwar Plateau or Irrawady sediments, except for two m3s at the NHM whose origins are unknown. Recent discoveries of *M. milaensis* in Myanmar document a broad geographical range (Tsubamoto et al. 2012; Chavasseau et al. 2013), suggesting that the species survived until the late Pliocene in this area (Takai et al. 2016); its last occurrence in the Potwar Plateau was at 9.2 Ma.

The two Pakistani species share similar symphyseal morphology, particularly the anterior position of the muscular insertions, the ventral expansion of the genial crista, and the small genial fossae (Lihoreau et al. 2004b). This condition does not exist in the European *Microbunodon minimum* and could be a synapomorphy of the Pakistani species. *Microbunodon milaensis* is a larger, more derived species than *M. silistrensis* and replaces the latter in the Potwar Plateau, suggesting a possible ancestor-descendant relationship.

Although fossil material is limited for this species, we can determine that it is larger than *M. silistrensis*. Two specimens, for which we studied microwear, are consistent with microwear results of *M. silistrensis* (see Table 20.1). Mean δ^{13}C values of -12.3 ± 0.3‰ from four individuals, dated between 10.4 Ma and 9.2 Ma, indicate a C_3-dominated diet under a closed forest canopy (see also Lihoreau 2003). It appears that this genus had stable dietary ecology since the Late Oligocene.

The Merycopotamini are selenodont bothriodontines with enlarged lower canines and cuspidate premolars (Lihoreau et al. 2017, Gernelle et al. 2023). This tribe is well represented in the Potwar Plateau by three genera.

The genus *Sivameryx* is characterized, following Pickford (1991) and Holroyd et al. (2010), by its quasi-pentacuspidate upper molars with loop-like parastyles and mesostyles and a long postcanine diastema with a flange-like protuberance leaning laterally. The type species, *Sivameryx palaeindicus* (Lydekker, 1877a; Fig. 20.3a), occurs possibly from the Early Oligocene of the Zinda Pir Dome (Z108, Z125; Lindsay et al. 2005; Métais et al. 2009; Antoine et al. 2013), but is recognized with certainty from the latest Oligocene to the earliest Miocene (Z151, Z129; Lihoreau 2003; Lindsay et al. 2005; Antoine et al. 2013) to the Middle Miocene (14.1 Ma) from localities in the Chitarwata, Vihowa, Kamlial, and Manchar Formations. In the Potwar Plateau, the species occurs in localities dated from 17.9 Ma (Y747) to 14.1 Ma (Y478). Geographically, the species occurs in Sind, the Zinda Pir Dome, the Bugti Hills, and the Potwar Plateau of Pakistan, as well as in Bengal, Jammu, and Kashmir, India. It is also probably present in Africa as discussed below.

Sivameryx palaeindicus is a medium-sized merycopotamine, larger than *Sivameryx moneyi* (Moghra, Egypt). It has two accessory cusps on the postparacrista of upper premolars and only one on the preprotocristid of the lower premolars; a low angle between the ventral border of the symphysis and the lower tooth row (17.9°–21.0°); and sexually dimorphic canines. The absence of significant morphological and biometrical differences between *S. palaeindicus* (see Fig. 20.3a) and *S. africanus* (Andrews 1914; Fig. 20.3b) led Lihoreau and Ducrocq (2007) to infer a single taxon. Indeed, the first diagnosis of Lydekker (1877b) for *Sivameryx palaeindicus* focused on upper-molar morphology, notably the reduced paraconule. Such characters led to an increase in the number of Asian bothriodontine species based on the relative development of the paraconule. Many other species have also been created based on mandibular material despite the lack of diagnostic characters (Pilgrim 1908). The discovery at Karungu, Kenya, of an isolated lower molar led to the creation of *S. africanus* (Andrews 1914) despite its strong resemblance to known *S. palaeindicus* (MacInnes 1951; Black 1978). Revisions of the Neogene African and Pakistani anthracothere material by Pickford (1987, 1991) led the author to propose differential diagnoses based on molar-row length. At that time, no upper-molar row was known in Africa, and no lower-molar row was known in Asia. Review of material from the Indian subcontinent, together with newly described material from Sind and the Potwar Plateau, highlighted the morphological and biometrical similarities (Lihoreau 2003; Lihoreau and Ducrocq 2007; Holroyd et al. 2010; Lihoreau et al. 2017). An edentulous cranial fragment (YGSP 21064) from locality Y855 (Bugti area) might belong to *S. palaeindicus*, as only one anthracothere species has been identified from dentitions found at this locality (cf. *Sivameryx palaeindicus* in Orliac et al. 2023). This specimen is compatible in size with *S. palaeindicus*. Comparisons of this specimen with a recently described skull from the Early Miocene of Kalodirr, Kenya, attributed to *Sivameryx africanus* (Rowan, Adrian, and Grossman 2015), will require revision of this genus.

In cladistic analyses, *Sivameryx palaeindicus* is linked to an *Elomeryx*-like lineage (Lihoreau and Ducrocq 2007). The same result is obtained with maximum-likelihood analyses on a morphological matrix (Lihoreau et al. 2015), although this result is not well supported due to the poor fossil record for early bothriodontines (Lihoreau et al. 2017). The discovery of representatives of *Elomeryx* in Asia, and particularly on the Indian subcontinent, supports this relationship (Ducrocq and Lihoreau 2006; Métais et al. 2009). During the Oligocene, the Indian subcontinent was a center of diversification for bothriodontines, particularly Merycopotamini (Lihoreau et al. 2017, Gernelle et al. 2023). Numerous anthracothere remains, such as *Telmatodon* and *Parabrachyodus* (Pickford 1987) from the Bugti Hills, recently revised and included within Merycopotamini, have help to resolve phylogenetic relationships between *Elomeryx* and this tribe (Gernelle et al. 2023). *Sivameryx palaeindicus* can be considered as the origin of the *Merycopotamus-Libycosaurus* lineage that appears on the Indian subcontinent during the Middle Miocene, around 14 Ma.

Body-mass estimates based on four tali from the early Middle Miocene of the Potwar Plateau (YGSP 24001, YGSP 31150, and YGSP 41909) and the latest Oligocene–earliest Miocene of the Zinda Pir dome (Z 2065) indicate a mean weight of 116.1 ± 21.3 kg. Because of scarce and poorly preserved material, a thorough dietary analysis of this species is not yet possible. Even so, its brachydont teeth, shallow mandible, and selenodont molars with a tendency toward tetracuspidy on upper molars suggest browsing or frugivorous habits. The recent investigation of the cranial fragment (YGSP 21064) indicates a tegmen tympani inflation interpreted as the first steps toward an adaptation of underwater hearing that will reach a higher degree in *Merycopotamus* (Orliac et al. 2023).

The genus *Hemimeryx* is represented by the type and only known species, *Hemimeryx blanfordi* Lydekker, 1883, and is known from the Late Oligocene of the Bugti Hills to the Middle Miocene of the Potwar Plateau, with fossils from the Chitarwata, Manchar, Vihowa, and Kamlial Formations. The species is present at locality Y687 (17.9 Ma) in the Potwar Plateau. The geographic distribution includes Lundo Chur (J1 level) south of the Zin range (Lihoreau et al. 2017), Dera Bugti, Sind, and Potwar Plateau, Pakistan. According to Lihoreau et al. (2016), *Hemimeryx blanfordi* is a medium-sized merycopotamine with tetracuspidate upper molars, the absence of postprotocrista on P4, four lower premolars with accessory cuspids on the preprotocristid, and a monozonal schmelzmuster on molar enamel.

This rare species is known only from 12 specimens (isolated molars, one mandible, and possibly a phalanx). Its possible presence in the Potwar Plateau is due to the discovery of a first central phalanx of a large anthracothere (Fig. 20.3c, d, e, f) in the Kamlial Formation ("gigantic anthracotheres" in Barry 1985; Barry and Flynn 1990). The phalanx belongs to an anthracothere due to the presence of a lateral and medial ligament foveae (Fig. 20.3e; unlike in suids), the symmetry of the distal articular facet (Fig. 20.3c, d; unlike in ruminants), and the abrupt delimitation of the articular facet in plantar-palmar view (Fig. 20.3d). The specimen is similar to *H. blanfordi* (Lihoreau 2003), as it belongs to a large species of anthracothere, approximately the size of a small *Brachyodus*. It is less dorsoplantary flattened (Fig. 20.3e, f) than phalanges from other

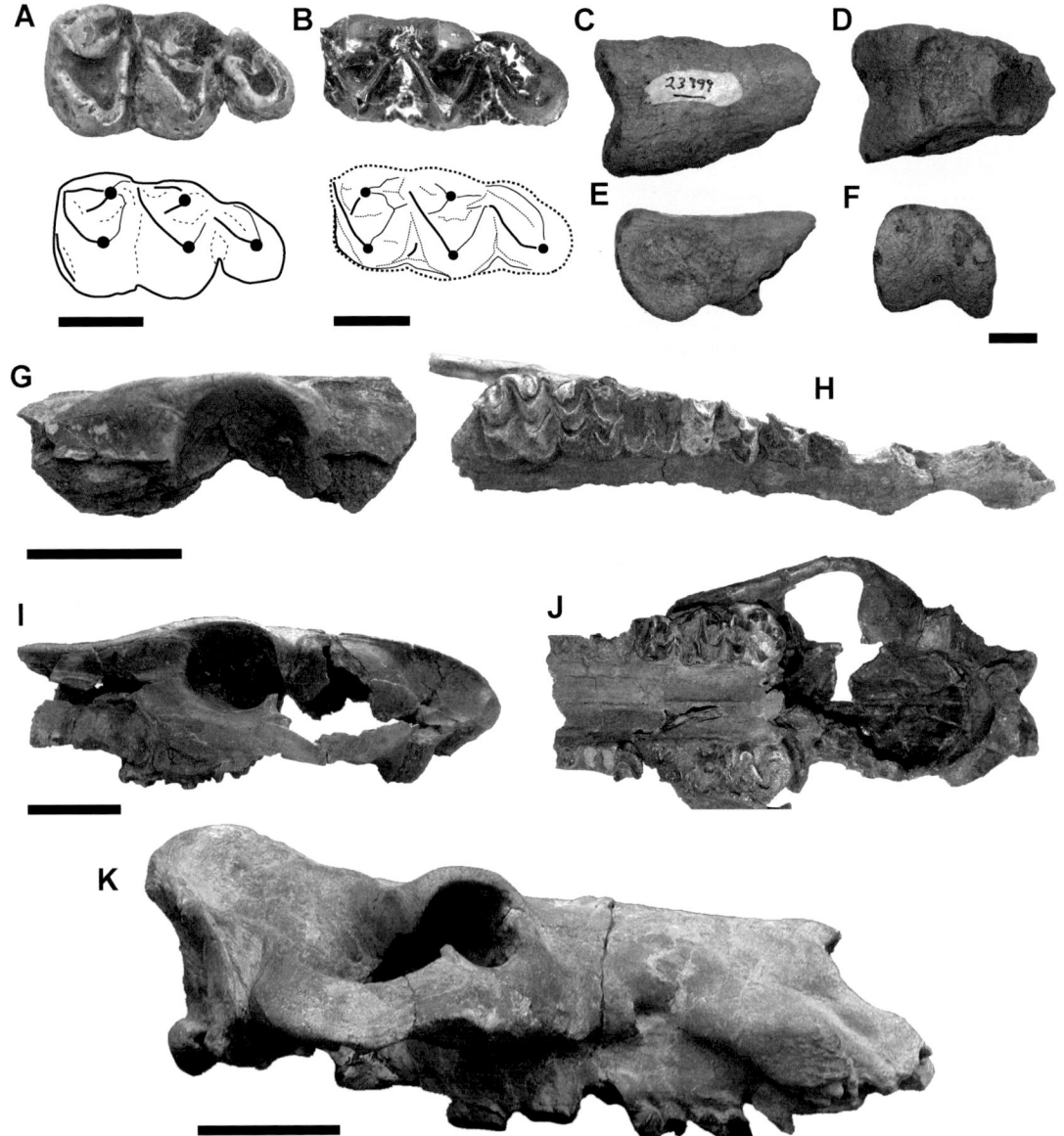

Figure 20.3. Bothriodontines and a hippopotamid. A. *Sivameryx palaeindicus*, left m3 (YGSP 40812, Kamlial Formation, Potwar Plateau) in occlusal view (photography and drawing). B. *Sivameryx africanus*, left m3 (NHM, no number, Gebel Zelten, Libya) in occlusal view (photography and drawing). C.–F. cf. *Hemimeryx blanfordi*, first central phalanx (YGSP 23999, Kamlial Formation., Potwar Plateau) in dorsal (C), plantar (D), lateral (E), and distal (F) views. G.–H. *Merycopotamus nanus*, fragments of skull (YGSP 47189, Chinji Formation, Potwar Plateau) in lateral and palatal views. I.–J. *Merycopotamus medioximus*, skull (YGSP 13310, Nagri Formation, Potwar Plateau) in lateral and palatal views. K. *Hexaprotodon sivalensis*, skull (2269, lectotype NHM, Siwalik Hills, India) in lateral view. Scale bars = 1 cm (A.–F.), 5 cm (G.–J.) and 10 cm (K.).

known anthracotheres from this area (e.g., *Merycopotamus nanus*). More fossils are needed to determine its presence with certainty in the Potwar Plateau, but this species is known in levels of equivalent age in Sind and Bugti (Raza et al. 1984; Métais et al. 2009; Lihoreau et al. 2017).

A recent revision of Merycopotamini proposed *Hemimeryx* as the sister taxon of *Sivameryx* and sister group to the *Merycopotamus-Libycosaurus* lineage (Lihoreau et al. 2017). Tetracuspidy appears in the Late Oligocene and might be convergent with that of *Merycopotamus* if we consider the latter derived from *Sivameryx*.

Little can be inferred about the paleobiology of *H. blanfordi*. Based on the area of the only known first lower molar and the regression formulas proposed in Damuth and MacFadden (1990), body mass estimates range from 356 kg (using the non-selenodont regression) to 483 kg (using the all-ungulate regression). The cheek teeth are brachydont and selenodont, and upper molars are tetracuspidate with large wear facets on the cuspid flank (as in derived Merycopotamini). This morphology is compatible with browsing dietary habits. It is noteworthy that the mandible is long and shallow, with a long diastema between c and p1, similar to that of *Giraffa*. Recent investigations of enamel microstructure reveal abrasive radial enamel rather than crack-resistant decussated enamel, which suggests feeding on tough leathery leaves (Lihoreau et al. 2017).

The genus *Merycopotamus* is characterized, following Lihoreau et al. (2007), by selenodont tetracuspidate upper molars with a loop-like or divided mesostyle, a bent symphysis with the concave ventral border, and the absence of a premetacristid on lower molars.

Merycopotamus nanus Falconer, 1868 (Fig. 20.3g, h), is known from the Middle Miocene of India, Nepal, and Pakistan, in localities of the Kamlial, Chinji, and possibly Manchar Formations. In the Potwar Plateau, localities range in age from 13.9 Ma (Y694) to 11.4 Ma (Y76). This is the smallest species of *Merycopotamus*. It has a dorsal margin of the orbit at the level of the cranial roof, a shallow mandible with a weakly marked vascular impression, weak sexual dimorphism of the canines, and a continuous loop-like mesostyle on upper molars (Lihoreau et al. 2007).

A revision of the genus (Lihoreau et al. 2007) proposed *M. pusillus* Lydekker, 1885, as a junior synonym of *Merycopotamus nanus* Falconer, 1868. Therefore, despite the common use in the literature of *M. pusillus*, this name must be abandoned for *M. nanus*.

Merycopotamus is prevalent in the Middle to Late Miocene and Pliocene of the Indian subcontinent and Southeast Asia. It is clearly derived from a lineage of Early Miocene selenodont anthracotheres of the Indian subcontinent, sister to *Sivameryx-Hemimeryx* (Lihoreau et al. 2017). A succession of three species implies anagenesis within a long-lived lineage. *Merycopotamus nanus* is the earliest representative of this lineage, which is derived from an early merycopotamine probably from the Indian subcontinent; it remains endemic to this area.

Numerous tali have been recovered from localities where confident species determinations based on dentitions are possible. Therefore, estimates of body mass are available for each species of *Merycopotamus* (see Lihoreau et al. 2007). Eleven specimens of *Merycopotamus nanus* indicate a body mass of 125 ± 15 kg. Sexual dimorphism is low in this species, and aquatic morphological specializations are not apparent (see Fig. 20.3g), in contrast to the youngest species in this genus (Fig. 20.3i).

Microwear patterns of 18 M2 specimens of *M. nanus* (see Table 20.1) indicate a mean value of 28.8 (SD = 13.6) pits, thereby excluding strict grazing as a dietary habit (Merceron et al. 2004). The mean value of 28.7 (SD = 6.6) for scratches also excludes a traditional browsing and frugivorous diet (Merceron et al. 2007). Microwear patterns for six *M. nanus* specimens resemble those of a generalist feeder such as *Cervus elaphus* (Merceron et al. 2007). The mean $\delta^{13}C$ value of 26 *M. nanus* specimens from localities dated between 13.7 Ma and 11.4 Ma is -10.6 ± 1‰, indicating a C_3-dominated diet. These results are congruent with brachydont tooth morphology, a shallow mandible, and an arched lower incisor row, probably indicating selective browsing habits (Solounias and Dawson-Saunders 1988; Janis 1995).

Merycopotamus nanus is common in major channel deposits (MC-U2 facies) and rare in floodplain deposits (FP-P facies;

Barry et al. 2013, Chapter 4, this volume). This taphonomic distribution is similar to that of *Listriodon pentapotamiae* and *Conohyus sindiensis* and reflects large fossil accumulations near major channels. The $\delta^{18}O_{SMOW}$ values of this species are significantly lower ($\delta^{18}O_{13.7-13.4} = 22.6 \pm 1.9$‰, $n=3$, t test: $p < 0.03$; $\delta^{18}O_{13.1-12.8} = 22.3 \pm 1.4$‰, $n=6$, t test: $p < 0.02$) than those of other ungulates (Bovidae, Suidae, and Giraffidae) from the same time interval ($t\delta^{18}O_{13.7-13.4} = 25.9 \pm 2.2$‰, $n=17$; $t\delta^{18}O_{13.1-12.8} = 25.1 \pm 2$‰; $n=10$). Moreover, these values are within the range of estimated aquatic values of this time interval (following Clementz, Holroyd, and Koch 2008). These features suggest that *Merycopotamus nanus* is adapted to aquatic environments and may have at least semiaquatic habits. This is congruent with its petrosal and inner ear morphologies (Orliac et al. 2023).

Merycopotamus medioximus (Lihoreau, et al. 2004a; Fig. 20.3i, j) occurs in Late Miocene localities of the Nagri, Dhok Pathan, and Irrawady formations. In the Potwar Plateau, it is known from 10.5 Ma (Y 454) to 8.7 Ma (Y408). In addition to the Potwar Plateau, the species is known from Thailand, Iraq, and Myanmar.

This large species of *Merycopotamus* has orbits that are slightly elevated above the cranial roof, an anterior main palatine foramen (opening at P1-P2), upper molars with a mesostyle notch but not fully divided, transverse broadening of metapodials, and a tetradactyl manus.

Merycopotamus medioximus represents a temporal and morphological transition between *M. nanus* and *M. dissimilis* and is the first *Merycopotamus* species to disperse outside the Indian subcontinent, being found in Iraq, Thailand (Lihoreau et al. 2007), and Myanmar (Chavasseau et al. 2013). One specimen from Thailand suggests that the contemporaneous *Merycopotamus tachangensis* (Hanta et al. 2008) differs from other *Merycopotamus* species by its posterior main palatine foramen, mesostyle morphology, and lack of contact between the nasal and the frontal. Some proposed characters are clearly present in the *M. medioximus* holotype, such as the lobe-like nasofrontal suture, the supraorbital foramen morphology, and the posterior position of the maxilopalatine suture. The creation of a new species, *M. tachangensis*, seems premature when considering the great variation present in the material from the Potwar Plateau, notably in mesostyle morphology. Many *Merycopotamus* specimens from the Potwar Plateau appear to be transitional between two species and therefore have not been attributed. These include *Merycopotamus* sp. that lie between *M. nanus* and *M. medioximus* from localities dating between 10.8 Ma (Y779) and 10.5 Ma (Y 259) and some *Merycopotamus* specimens that lie between *M. medioximus* and *M. dissimilis* from localities dating between 8.7 Ma (Y981) and 7.8 Ma (Y917).

Merycopotamus medioximus is derived relative to *M. nanus* and represents the sister taxon of *M. dissimilis*. We consider it a transitional species within the anagenetic lineage of South Asian *Merycopotamus*. As the earliest *Merycopotamus* found outside the Indian subcontinent, it is possibly ancestral to the African *Libycosaurus*. Indeed, *Libycosaurus* resembles *Merycopotamus* closely and may be the sister group of all *Merycopotamus* species (Lihoreau et al. 2006) or possibly included within the *Merycopotamus* clade (Lihoreau et al. 2017, 2021).

Nine astragalar specimens of *M. medioximus* suggest a body mass around 260 ± 67 kg. The microwear features of a majority of *M. medioximus* individuals indicate a grazing diet ($n = 8$; see Table 20.1), while three specimens indicate a browsing diet. The mean values of pits and scratches (mean pits = 25, SD = 15.9; mean scratches = 27, SD = 8.1) are close to those of *M. nanus* but with higher standard deviations. Unlike *M. nanus*, two distributions are clearly identifiable. Specimens from the same locality (Y311) have microwear features of both grazers and browsers. The mean δ^{13}C value from 21 individuals dating between 10.2 Ma and 8.7 Ma is -11.3 ± 1‰ and reflects a C_3-dominated diet similar to that of *M. nanus*. *Merycopotamus medioximus* is clearly a mixed feeder. It differs from *M. nanus* in being recovered from floodplain channel deposits, which suggests different habitat preferences (Barry et al. 2013, Chapter 4, this volume).

Some anatomical characters indicate adaptation to a semi-aquatic lifestyle, notably the elevation of the orbits above the cranial roof (see Fig. 20.3i) known in other aquatic Merycopotamini and Hippopotaminae. In addition, there is an increase of body mass, enlargement of metapodials, and increase in sexual dimorphism. The $\delta^{18}O_{SMOW}$ values of this species do not differ significantly but mean values ($\delta^{18}O_{9.4-9.2} = 24.2 \pm 1.3$‰, $n = 9$, t test: $p < 0.1$; $\delta^{18}O_{9-8.7} = 24.4 \pm 1.2$‰, $n = 4$, t test: $p = 0.1$) are lower than those of other contemporaneous ungulates (Bovidae, Suidae, and Equidae) ($t\delta^{18}O_{9.4-9.2} = 25.7 \pm 2.5$‰, $n = 75$; $t\delta^{18}O_{9-8.7} = 26.0 \pm 2.5$‰, $n = 29$). Unlike in *Merycopotamus nanus*, all the mean δ^{18}O values obtained are slightly higher than those of estimated aquatic values of the same time interval ($e\delta^{18}O_{9.4-9.2} = 23$‰, $e\delta^{18}O_{9-8.7} = 23.3$‰), requiring cautious inference concerning aquatic adaptations in contrast to indicators from skull morphology. In the same way, microanatomical morphology of the humerus and femur of *M. medioximus* is not so different from that of a terrestrial artiodactyl (Cooper et al. 2016; Houssaye et al. 2021).

Merycopotamus dissimilis (Falconer and Cautley, 1836) occurs from Late Miocene to Early Pleistocene localities from the Dhok Pathan, Tatrot, and Pinjor Formations and the Pabbi Hills sequence. It is known in the Potwar Plateau from 7.8 Ma (Y948) to 3.3 Ma (L103). Fossils of this species occur near Chandigarh, India; in the Potwar Plateau, Mangla-Samwal, and Pabbi Hills, Pakistan; at Mingun and Yenanyoung, Irrawaddy River, Myanmar; in Surai Khola, Nepal; and in Cijulang, Java Island, Indonesia.

Merycopotamus dissimilis is similar in size to *M. medioximus* but displays more sexual size variation. Following Lihoreau et al. (2007), this species is notably recognized by its elevated orbits, its main palatine foramen aperture anterior to P1, the fully divided mesostyle, a very deep mandibular corpus and symphysis morphology, hypsoarhizodont lower canines, and a simple hypoconulid on M3.

This species displays strong body-size dimorphism that initially suggested the existence of two species (Lydekker 1884a; Lihoreau et al. 2007). The systematics became confusing, as the name designating the small variety (putative female speci-

mens) was already used for the holotype of *M. nanus* (Falconer 1868).

This species is probably the descendant of a *Merycopotamus medioximus*–like ancestor and originated in the Indian subcontinent or Southeast Asia. It is the most derived species of the *Merycopotamus* lineage as well as the youngest representative of anthracotheres. Unfortunately, the species is poorly sampled, and most of the attributed material is not dated. Further discrimination between Miocene, Pliocene, and Pleistocene specimens is possible.

A body mass estimate of 220 ± 94 kg based on five tali of *M. dissimilis* is similar to size estimates for *M. medioximus* but with bimodal variation. Indeed, differences between males and females are more pronounced than in other *Merycopotamus*; this parallels the evolutionary trends observed in *Libycosaurus* in Africa (Lihoreau et al. 2014). Notably, the lower canines become hypselodont with enamel along the entire tooth, as in *Hippopotamus*, and the sagittal crest is more developed. This size dimorphism together with the development of larger, hypselodont canines may indicate frequent social interactions in this species. We also note a pronounced ventral notch on the mandible. This character, together with the lowering of the mandibular condyle relative to the tooth row, and the development of a mandibular ventral process for the insertion of a strong digastric muscle, can be interpreted as adaptations for greater gape (Herring 1972; Lihoreau 2003). This morphology, which occurs in extant hippopotamids and tayassuids, is related to intraspecific competition and social hierarchy in the former and defense against predators and maintenance of herd organization in the latter (Herring 1975).

Microwear of seven specimens of *M. dissimilis* sampled between 7.8 Ma and 5.8 Ma (see Table 20.1) show numerous scratches (mean of 32.1, SD = 3) and few pits (mean of 19.8, SD = 7.1). This small percentage of pits is characteristic of grazing (Merceron et al. 2007). The mean δ^{13}C value of 7.8 ± 3.1‰, obtained from 13 individuals dated between 8.0 Ma and 6.3 Ma, corresponds to mixed C_3/C_4 diets. Over shorter intervals, the mean δ^{13}C values vary from a C_3-dominated diet, as in *M. medioximus*, before 7.8 Ma ($\delta^{13}C_{8.0-7.8} = -11.2 \pm 1.1$‰, $n = 5$) to mixed C_3/C_4 diets ($\delta^{13}C_{7.4-7.0} = -6.9 \pm 1.4$‰, $n = 4$) with a tendency toward a C_4-dominated diet after 6.5 Ma ($\delta^{13}C_{6.5-6.3} = -4.6 \pm 0.7$‰, $n = 4$). This dietary shift between 8 and 7 Ma coincides with the appearance of the morphological characteristics of *M. dissimilis*: deepening of the mandible and enlargement of the rostral part of the mandible with a straight incisor row, in contrast to the narrow mandible with curved incisor row in *M. nanus*. The change to a straight incisor row suggests that *M. dissimilis* become less selective when browsing, although enlargement of the canine might affect the morphology of the mandibular symphysis. The development of the maxillary tuberosity on the zygomatic arch indicates stronger insertion of the zygo-mandibular muscle, responsible for pronounced lip extension in grazers. A deep mandibular corpus with a large, convex gonial angle has been considered a grazing adaptation (Solounias and Dawson-Saunders 1988; Janis 1995). In addition, the opening of the mesostyle on the

Table 20.1. Dental Microwear Results for M²/ of *Hexaprotodon sivalensis, Microbunodon silistrensis, Microbunodon milaensis, Merycopotamus nanus, Merycopotamus medioximus,* and *Merycopotamus dissimilis*

Specimen	Loc	Ns	Np	Specimen	Loc	Ns	Np	Specimen	Loc	Ns	Np
Hexaprotodon sivalensis				*Merycopotamus medioximus*				*Merycopotamus nanus*			
15935	—	34	110	NG111	Y311	13	24	YGSP 32849	Y806	27	57
15938	—	37	75	YGSP 11457	Y311	25	27	YGSP 4428	Y189	33	30
16365	—	50	43	YGSP 13523	Y239	17	67	YGSP 4427	Y189	33	14
16380	—	9	96	YGSP 52174	Y311	37	8	YGSP 16011	Y381	11	33
17068	—	25	38	YGSP 32739	CH10	29	16	YGSP 16019	Y381	19	30
17419	—	43	83	AMNH 9917		19	37	YGSP 19674	Y496	29	16
M44682	—	73	73	YGSP 49562	Y891	30	15	YPM 20083	—	26	28
M489	—	65	59	YGSP 4597	Y193	25	14	YGSP 25524	Y705	33	15
no number	—	44	77	YGSP 10631	Y311	30	26	YGSP 25479	—	27	42
no number	—	43	34	YGSP 14318	Y450	38	21	YGSP 47189	Y735	25	30
36824	—	45	28	YGSP 14969	Y408	34	20	YGSP 30572	Y515	35	28
Microbunodon silistrensis				*Merycopotamus dissimilis*				YGSP 45817	Y502	26	12
YGSP 18928	Y496	34	59	YGSP 50101	Y910	37	30	YGSP 28124	Y705	39	46
YGSP 21063	Y494	26	21	YGSP 50107	Y910	31	17	YGSP 45729	Y496	32	23
YGSP 41301	Y738	34	23	YGSP 7506	Y584	29	32	YGSP 47146	Y647	29	39
S99	Sind	25	45	YGSP 16418	Y551	33	18	YGSP 20525	Y494	30	11
Z263	Dala	25	44	YGSP 50337	Y930	31	18	YGSP 23549	Y675	37	16
Microbunodon milaensis				Y50102	Y910	28	13	YGSP 20399	Y640	25	49
YGSP 5174	Y211	26	25	YGSP 50867	Y948	35	14	—	—	—	—
YGSP 9791	Y269	27	25	—		—	—	—	—	—	—

Notes: Specimens with prefix YGSP are from localities of the Harvard-GSP collection from the Potwar Plateau. Specimens of *Hexaprotodon sivalensis* are from unknown localities in the Siwalik Hills of India and are curated at the Natural History Museum (NHM), London (M number for old collections). Some other numbers have specified letter(s): S= Sind, Z = Zinda Pir, NG= Nagri. Ns = number of scratches; Np = number of pits.

upper molars allows greater chewing motion and thus might also be linked to grazing. It is important to note that the morphological traits that characterize this species are present slightly before the shift to a C_4-dominated diet, which suggests that grazing preceded the adoption of a full C_4 diet.

The orbits are more elevated above the cranial roof than in *M. medioximus,* and the metapodials are wider; these suggest greater adaptation to a semiaquatic lifestyle. The $\delta^{18}O_{SMOW}$ values of this species are significantly lower ($\delta^{18}O_{8.0–7.8} = 24.1 \pm 1.1‰$, $n = 5$, t-test: $p < 0.01$; $\delta^{18}O_{6.5–6.3} = 24.7 \pm 0.9‰$, $n = 4$, t-test: $p < 0.02$) or only lower ($\delta^{18}O_{7.4–7.1} = 26.0 \pm 1.8‰$, $n = 4$, t test: $p = 0.34$) than those of terrestrial ungulates (Bovidae, Suidae, and Equidae) for the same time interval ($t\delta^{18}O_{8.0–7.8} = 27.4 \pm 2.5‰$, $n = 74$; $t\delta^{18}O_{7.4–7.1} = 27.7 \pm 3.4‰$, $n = 28$; $t\delta^{18}O_{6.5–6.3} = 31.06 \pm 3.7‰$, $n = 5$). Two mean values obtained are equivalent or slightly lower than estimated aquatic values of the same time interval ($e\delta^{18}O_{8.0–7.8} = 24.6‰$, $e\delta^{18}O_{6.5–6.3} = 28.1‰$), and one mean is only slightly higher ($e\delta^{18}O_{7.4–7.1} = 25.1‰$). These results support the interpretation of semiaquatic habits in this species. Microanatomy of stylopod of *M. dissimilis* rather resembles that of an aquatic mammal (Cooper et al. 2016, Houssaye et al. 2021). Overall, the morphology and inferred behavior, diet, and aquatic adaptations of *M. dissimilis* are similar to those of small individuals of extant *Hippopotamus amphibius.*

Within Hippopotamidae, the genus *Hexaprotodon* is notably defined by its hexaprotodonty, the i2 usually being the smallest; the robustness and height of the mandibular symphysis; and an elevated sagittal crest on a transversally compressed braincase. Some constant features of this genus appear to be primitive, such as the lachrymal contact with the frontal, not the nasal (from Boisserie 2007).

Hexaprotodon was named by Falconer and Cautley (1836) as a subgenus for hippopotamid remains found in the Siwalik Hills. Following Coryndon (1977), *Hexaprotodon* served to designate most species not displaying the derived morphology of the genus *Hippopotamus.* A phylogenetic revision of the family Hippopotamidae (Boisserie 2005) proposed a restricted definition of *Hexaprotodon* on the basis of synapomorphic characters, notably those related to the morphology of the mandibular symphysis. This revision implied that (1) *Hexaprotodon* is extinct; (2) most of its representatives are found in southern and southeastern Asia; and (3) its origin is likely African (Boisserie et al. 2005a).

Hexaprotodon includes two species from Africa (Chad, Ethiopia), possibly five species from southern Asia (India, Pakistan, Nepal, Myanmar), and possibly four species from Sri Lanka and Indonesia. In the Siwalik Hills, three species were recognized: the Mio-Pliocene *Hex. sivalensis,* the Plio-Pleistocene (?) *Hex. iravaticus* (Falconer and Cautley 1847), and the Pleistocene *Hex. duboisi* (Hooijer 1950). The hippopotamid material recovered by the H-GSP group can be attributed to *Hex. sivalensis.*

The type species, *Hexaprotodon sivalensis* Falconer & Cautley, 1836 (Fig. 20.2k), occurs in late Miocene to late Pliocene localities, roughly between 6 Ma and 2 Ma (see Barry et al. 2002) and potentially as young as 1.7 Ma (Dennell, Coard, and Turner 2006).

In Pakistan, fossils are known from Jalalpur, Rohtas, and Hasnot (Tatrot, Kotal Kund) in the Potwar Plateau and in the Mangla-Samwal Anticline (Hussain et al. 1992). The species also occurs in the Chandigarh area, India (Nanda 1980), and in the Surai Khola Siwalik sequence of Nepal (Corvinus and Rimal 2001).

Hexaprotodon sivalensis is notably characterized by its moderate to large size, its low orbits, a long mandibular symphysis relative to its width, i2 two-thirds to about equal in size to i1, and i3 slightly smaller than i1 (adapted from Hooijer 1950).

In 1950, Hooijer conducted an extensive study of morphological variation within *Hip. amphibius* and rejected subspecies divisions based on the high degree of intrapopulation variation. He also documented morphological variation in Asian fossil species. Those species were based on relatively marked differences, but Hooijer (1950) noted morphological intergradation between several of them, with characters being especially variable in *Hip. amphibius*. He considered that observed differences could not justify specific differentiation and therefore defined seven to eight subspecies within *Hex. sivalensis* and retained *Hex. iravaticus* as a different species.

Here, we use subspecies as geographical divisions of species, as proposed by Mayr (1964). The distribution of Hooijer's subspecies is not congruent with this concept: for example, *Hex. sivalensis namadicus* and *Hex. sivalensis palaeindicus* would have been contemporaneous in the Narmada Hills of central India (Hooijer 1950). This use of "subspecies" is difficult to apply to the fossil record without clearly defined and precisely controlled geographic and stratigraphic contexts. Most subspecies defined by Hooijer (1950) are based on morphotypes exhibiting substantial differences. At least some of them constitute lineages within chronospecies. This is notably the case for *Hex. sivalensis sivalensis* and *Hex. sivalensis palaeindicus* (Lydekker 1884b). We prefer to use specific rank for *Hex. sivalensis sivalensis sensu* Hooijer (1950), as well as for the other Asian taxa until further data lead to different interpretations (but see de Visser 2008).

The precise origin of Hippopotamidae within the anthracotheres was only recently established. The earliest hippopotamids are African (Pickford 1983; Boisserie et al. 2010; Orliac et al. 2010). Different phylogenetic scenarios have been proposed (Boisserie and Lihoreau 2006; Boisserie et al. 2011): (1) a late emergence from Middle Miocene Asian bothriodontines, (2) an early emergence from Early Miocene Asian bothriodontines, or (3) a deeply nested African emergence from a *Bothriogenys*-like stock. Archaic bothriodontine remains from the Oligocene of Kenya identified as *Epirigenys lokonensis*, are a plausible transitional form between Bothriodontinae and Hippopotamidae (Lihoreau et al. 2015), thus supporting the third scenario.

The most complete cladistic analysis of the Hippopotamidae thus far (Boisserie 2005, Figs. 6, 7, 10) placed *Hex. sivalensis* in a clade with *Hex. palaeindicus* and *Hex. bruneti*, the latter being an African basal Pleistocene representative of *Hexaprotodon*. In this analysis, the clade *Hexaprotodon* was found to be the sister group of a clade including the genus *Hippopotamus* as well as various Plio–Pleistocene forms of African species endemic to the Turkana and Afar depressions. The clade encompassing *Hippopotamus* and *Hexaprotodon* (bearing most of the Plio–Pleistocene diversity of the family) has for a sister group the genus *Archaeopotamus*, a primitive form known principally from eastern Africa and Arabia (Boisserie 2005; Boisserie and Bibi 2022).

These results (e.g., see Boisserie 2007, Fig. 8.4) imply that *Hexaprotodon* probably emerged in Africa at the end of the Miocene from a basal stock of hippopotamines close to *Archaeopotamus* and was characterized by a derived dentition relative to kenyapotamines (Early to Late Miocene in Africa), but with plesiomorphic symphyseal morphology (Boisserie et al. 2017a, 2017b). The earliest known representative is *Hexaprotodon garyam* from central Africa, dated to ~7 Ma (Boisserie et al. 2005a). It is morphologically close to *Hex. sivalensis*, except for its plesiomorphic postcanine dentition, so is potentially a good model for the basal stock of the Asian hippopotamines. Finally, *Hex. sivalensis* is likely not closely related to *Choeropsis liberiensis*, the extant Liberian hippopotamus, as initially thought by various authors (see Boisserie and Eltringham 2013).

The relationships of *Hex. sivalensis* with the Late Miocene southern European taxa, *Hex.? pantanelli*, *Hex.? siculus*, and *Hex.? crusafonti*, are unclear, as all are known only from fragmentary remains (but see recent work by Martino et al. 2020, 2021). Most other Asian *Hexaprotodon* and the African Pliocene representative, *Hex. bruneti* (Boisserie and White 2004), exhibit derived conditions compared to *Hex. sivalensis*, notably reduced i2 and enlarged i3, elevated orbits, and higher-crowned molars. This view is also supported by a phylogenetic analysis based prominently on Asian material (Thaung and Takai 2016).

Hex. iravaticus, in contrast, differs from *Hex. sivalensis* mainly by its smaller size and more elongated symphysis. At the same time, the lectotype of this species (no. 14771, NHM) calls for caution regarding the elongated symphysis, as its erupting P2 indicates a juvenile specimen. In the living species, specimens with a non-erupted P2 display morphology characterized by poorly developed canine processes and consequently a relatively narrow rostrum. A clarification of the diagnosis and distribution of *Hex. iravaticus* is needed in order to understand its exact relationships with *Hex. sivalensis*. An important contribution by Htike (2012) described new material from the Pliocene of the Irrawady Basin, Myanmar, which demonstrates the presence of a small hippopotamid showing a marked size reduction from *Hex. sivalensis*. As proposed by Boisserie (2005), we support designating all of the distinctively small, adult specimens from Myanmar as *Hexaprotodon* sp.; they clearly represent a species that differs from *Hex. sivalensis*.

Based on astragalus measurements (five specimens at the NHM), *Hexaprotodon* is by far the largest fossil hippopotamoid reported in this chapter, with an estimated average weight of 1,226 kg (range 1,142 kg–1,339 kg). This estimate is lower than the average estimates for *Hip. amphibius* (1,546 kg for males, 1,385 kg for females; Klingel 2013). This difference between the Siwalik hippopotamid and its extant relative is consistent with craniodental dimensions. What is not fully understood, given the available sample, is the degree of sexual size dimorphism. The mandibular material of *Hex. sivalensis* displays a relatively small range of morphological variation that is usually attributable to sex differences in hippopotamines (notably canine dimensions, spread of canine processes, incisor dimensions, depth of ramus). This difference suggests that population structure and social interactions possibly differed from those of *Hip. amphibius*, a gregarious species in which aggressive males live peripheral to groups of females and calves. *Choeropsis liberiensis*, although rarely observed, is solitary or found in very small groups including a male, a female, and a calf (Robinson 2013). The dimorphism in canines and incisors in this species was described as relatively strong (Weston 1997), so *C. liberiensis* is not necessarily a good model for *Hex. sivalensis*. Given current evidence, it is too speculative to infer the social organization of *Hex. sivalensis*.

Other important ecological traits of hippopotamids are their semiaquatic habits, which are well-documented for the extant species and inferred for several fossil species, including *Hexaprotodon* from the Late Neogene of southern Asia. Cranially, *Hex. sivalensis* displays moderately elevated orbits, a trait linked in various animals to life at the interface between water and air (e.g., Boisserie et al. 2011). It is not possible to infer the degree of dependence on water from orbit elevation, because some species displayed little specialization in this regard (e.g., *Archaeopotamus harvardi*) and yet were probably semiaquatic according to isotopic data (Cerling, Harris, and Leakey 2003). In any case, later forms from the Indian subcontinent display a higher degree of orbit elevation, which suggests greater specialization of primitive semiaquatic habits (Boisserie et al. 2011, Fig. 5). Semiaquatic habits are also suggested by another line of evidence: long bone microanatomy observed on transverse sections (Houssaye et al. 2021) indicated that *Hex. garyam*, potentially close to the basal stock for *Hex. sivalensis*, was a semiaquatic species, although less specialized than *Hip. amphibius* in this ecology.

Stable isotope data provide support for dependence of hippopotamids on water bodies, based on the comparison of oxygen isotopes of dental enamel to those of contemporaneous faunal elements with strong terrestrial habits. Several such studies have been conducted for African fossil hippopotamids (e.g., Franz-Odendaal, Lee-Thorp, and Anusuya 2002; Cerling, Harris, and Leakey 2003; Brachert et al. 2010; Lihoreau et al. 2014), but few data are published for southern Asia. Zin-Maung-Maung-Thein et al. (2011) reported isotopic results for 16 specimens from Myanmar, most likely lower Pliocene in age, and these results suggest semiaquatic habits for material attributed to at least two species, *Hex. sivalensis* and *Hex. iravaticus*. More recently, Patnaik (2015) published a similar result based on a single specimen for Early Pleistocene material from the Indian Siwaliks. Altogether, the morphological and isotopic evidence is suggestive of semiaquatic habits for *Hex. sivalensis*, similar to *Hip. amphibius*.

Little can be inferred about diet from hippopotamid craniomandibular morphology, which is marked by the ever-growing

rostral dentition that is related to social interactions. Stable carbon isotope data are again few. The studies of Zin-Maung-Maung-Thein et al. (2011) and Patnaik (2015) indicate a high proportion of C_4 plants in the diet of southern Asian fossil hippopotamids, dominated by grass consumption.

As for anthracotheres, we compared the microwear results obtained for the extant *Hip. amphibius* ($N = 40$) and a sample of 11 specimens for *Hex. sivalensis* from the Siwaliks (NHM collections; see Table 20.1). Both scratches and pits are particularly abundant in *Hip. amphibius* compared to values seen in other ungulates (Boisserie et al. 2005b). Common hippopotamuses mainly rely on fresh leaves of grasses with relatively few developing phytoliths, allegedly responsible for abundant fine scratches. The high number of pits can be ascribed to ingestion of large amounts of grit from plucking plants out of the ground rather than cutting leaves (Boisserie et al. 2005b). *Hexaprotodon sivalensis* is similar to *Hip. amphibius* in the number of scratches but has significantly more pits (see Table 20.1; Mann-Whitney U test: $P < 0.01$; see Boisserie et al. [2005b] for details on *Hip. amphibius*). Both species are similar in retaining many scars but few large scars. The small fossil sample calls for caution in interpreting these results. *Hex. sivalensis* may have consumed larger amounts of non-grass plants or larger amounts of grit than did *Hip. amphibius*. On both fossil and living species, the fineness of scratches is remarkable when compared to that of extant ungulates feeding on C_4 grasses (see, for example, Merceron et al. 2004; Merceron, Viriot, and Blondel 2004). In conclusion, *Hex. sivalensis* may have had a more variable diet than that of *Hip. amphibius*.

BIOGEOGRAPHIC EVENTS AND CLIMATIC CHANGES

Early Miocene: Diversification of bothriodontines

Whereas anthracothere diversity decreased on other continents during the Early Miocene (Lihoreau and Ducrocq 2007), an important diversification is recorded on the Indian subcontinent at the beginning of the Neogene (Antoine et al. 2013; see Fig. 20.3). Those anthracotheres originated from Paleogene ancestors present on the Indian subcontinent (Métais et al. 2009; Lihoreau et al. 2017). There, the Merycopotamini appeared at the end of Oligocene (Lihoreau et al. 2017) and evolved into two species, the large *Hemimeryx* and the smaller *Sivameryx*. Both are characterized by selenodont tetracuspidate or sub-tetracuspidate molars, pustulate premolar crests, and secondary radial enamel, which might correspond to a dietary specialization to herbaceous vegetation (Lihoreau et al. 2017). The African *Afromeryx* probably emerged from this diversification as well (Pickford 1991), along with the Asian genera *Gonotelma* and *Telmatodon* (Pickford 1987), but this evolutionary history is still poorly known. This diversification led to the predominance of early merycopotamines at the beginning of the Miocene, following the extinction of the Anthracotheriinae (Lihoreau and Ducrocq 2007; Antoine et al. 2013). The glaciation event at the Oligocene-Miocene boundary around 24Ma

(Mi 1) and the acceleration of Tibetan Plateau uplift (Zachos et al. 2001; Fig. 20.4) might have transformed the environment by creating wet habitats. Anthracotheriines did not display aquatic adaptations (Tütken and Absolon, 2015) unlike bothriodontines, and these adaptations might have enabled Miocene bothriodontines to thrive. *Microbunodon* persisted during the Early Miocene without important change since the Oligocene, but the Early Miocene fossil remains of this genus require revision, which may lead to different interpretation of the evolution of this small, supposedly terrestrial anthracothere.

Miocene: Dispersals and Ecological Specialization

The first dispersal of anthracotheres occurred around 18 Ma, with the expansion of the geographic range of *Sivameryx* (documented from the Bengal area to Lake Victoria, Kenya, and Sahara, Libya, via the Levantine corridor [Grossman et al. 2019]). Some authors have named this event the "Proboscidean Datum Event" (Madden and Van Couvering 1976; Antoine et al. 2003), marking a major exchange between African and Asian mammals following closure of the Tethys (see Fig. 20.4). The genus *Afromeryx* appeared in Africa at the same time as *Sivameryx* and also seems to have had an Asian origin (Pickford 1987, 1991). A second dispersal event from Asia to Africa took place around 10–11 Ma with the arrival of *Libycosaurus* in eastern and northern Africa (Lihoreau, Hautier, and Mahmoubi. 2015, Lihoreau et al. 2021), the genus being more closely related to the Asian *Merycopotamus* than to other African selenodont bothriodontines (Lihoreau and Ducrocq 2007; Lihoreau et al. 2017, 2021). When this genus arrived in Africa, the earlier African anthracotheres had been extinct for at least 2 million years— with the exception of their offshoot, the Hippopotamidae.

Before the time of this second dispersal, the genus *Merycopotamus* is present in the Potwar Plateau and is interpreted as a semiaquatic generalist herbivore (see Fig. 20.4). Restricted to the Indian subcontinent initially, perhaps blocked from dispersal by mountain barriers (such as the Zagros and Himalayas; Lihoreau et al. 2007), this genus is later recognized in sediments around 11–10 Ma from Pakistan, Iraq, and Thailand, with the species *M. medioximus*, a distribution documenting connection between different Asian areas at the time of a major faunal turnover in Siwalik mammal communities (Barry et al. 2002; Lihoreau et al. 2007, Chapter 25, this volume). The semiaquatic habits of this species suggest possible dispersal routes via various palaeodrainage systems, connections possibly triggered by a major marine regression at ~11-10 Ma (Lihoreau et al. 2007, 2021; see Fig. 20.4). The youngest species of the genus, *M. dissimilis*, was subsequently restricted to the Indian subcontinent, the Irrawady Basin (Myanmar), and Java (Indonesia).

The Potwar Plateau offers a rare continuous record of an anthracothere lineage in a single region over more than 10 million years (see Fig. 20.1). During the Middle and Late Miocene, the genus *Merycopotamus* shows adaptations to changing environments (see Fig. 20.4). It appears that all species of *Merycopotamus* show preference for aquatic environments. However, greater adaptation to such environments, notably the position of sense organs on the cranium, the doubling of estimated body weight, and the widening of autopods (metapodials and phalanx), occurs at the transition between *M. nanus* and *M. medioximus* (between 11.5 Ma and 10.0 Ma; see Fig. 20.4). This transition also corresponds to changes in environments of fossil preservation, from accumulations in large channels to those of floodplain deposits (Barry et al. 2013; see Fig. 20.4). *Merycopotamus medioximus* appears to be associated with floodplain habitats that become dominant during the late Miocene (Barry et al. 2002, Chapter 4, this volume) and indicate a variety of wetland and well-drained habitats.

Between *M. medioximus* and *M. dissimilis*, an increase in body-size dimorphism, an increase in the canine size of males, and adaptations for gaping ability all signify an increase in social behavior, as inferred for *Libycosaurus* in Africa at approximately the same time (between 10 Ma and 7 Ma; Lihoreau et al. 2014). These changes are interpreted as reflecting increasing intraspecific competition, perhaps exacerbated by increasing seasonal aridity that may have restricted preferred habitats. Evolution of the masticatory complex in this genus suggests a dietary change from a browsing to a grazing diet, changes also indicated in microwear patterns and stable-isotope data (shift from a C_3-dominated diet to increased consumption of C_4 plants). The transition from *M. medioximus* to the more gregarious grazing *M. dissimilis* occurs within the context of increasing aridity (Zhisheng et al. 2001; Barry et al. 2002; see Fig. 20.4; Chapter 24, this volume), as proposed for African *Libycosaurus*, and probably linked to more marked seasonality.

Terminal Miocene: First Occurrence of Hippopotamids

The earliest representative of *Hexaprotodon* occurs in the Late Miocene of Chad (Boisserie et al. 2005a), between 7.4 Ma and 6.5 Ma (Vignaud et al. 2002; Lebatard et al. 2008). *Hexaprotodon* was then documented in the Siwaliks at Locality Y551 reported as dated to 5.9 Ma in Barry et al. (2002), recalibrated to 6.0 Ma using Gradstein et al. (2012); a poorly constrained locality could place the first occurrence as early as 6.2 Ma. A few specimens known from Hasnot could be dated to ~7 Ma (John Barry, pers. comm.). Unfortunately, most of them do not retain much morphological information in relation to their state of preservation, except a distal putative P3 (AMNH 19732A) that differs from *Hex. sivalensis* in retaining a stronger (plesiomorphic) protocone and a smaller size.

The dispersal route may not have included eastern Africa and the Arabian Peninsula, because between 7 Ma and 6 Ma there are no records of *Hexaprotodon* there; rather, there is evidence of a more primitive genus of hippopotamines, *Archaeopotamus* (e.g., Weston and Boisserie 2010). An alternative dispersal route involves the Mediterranean Basin. Although some hippopotamid remains have been documented in the Upper Miocene of southern Europe, they are too poorly known to establish a clear relationship with *Hexaprotodon*.

The spread of *Hexaprotodon* to Asia could be contemporaneous with the Messinian salinity crisis starting at 5.96 Ma (Krijgsman et al. 1999; see Fig. 20.4) and with mammalian exchanges between Africa and the Iberian Peninsula (Agustí,

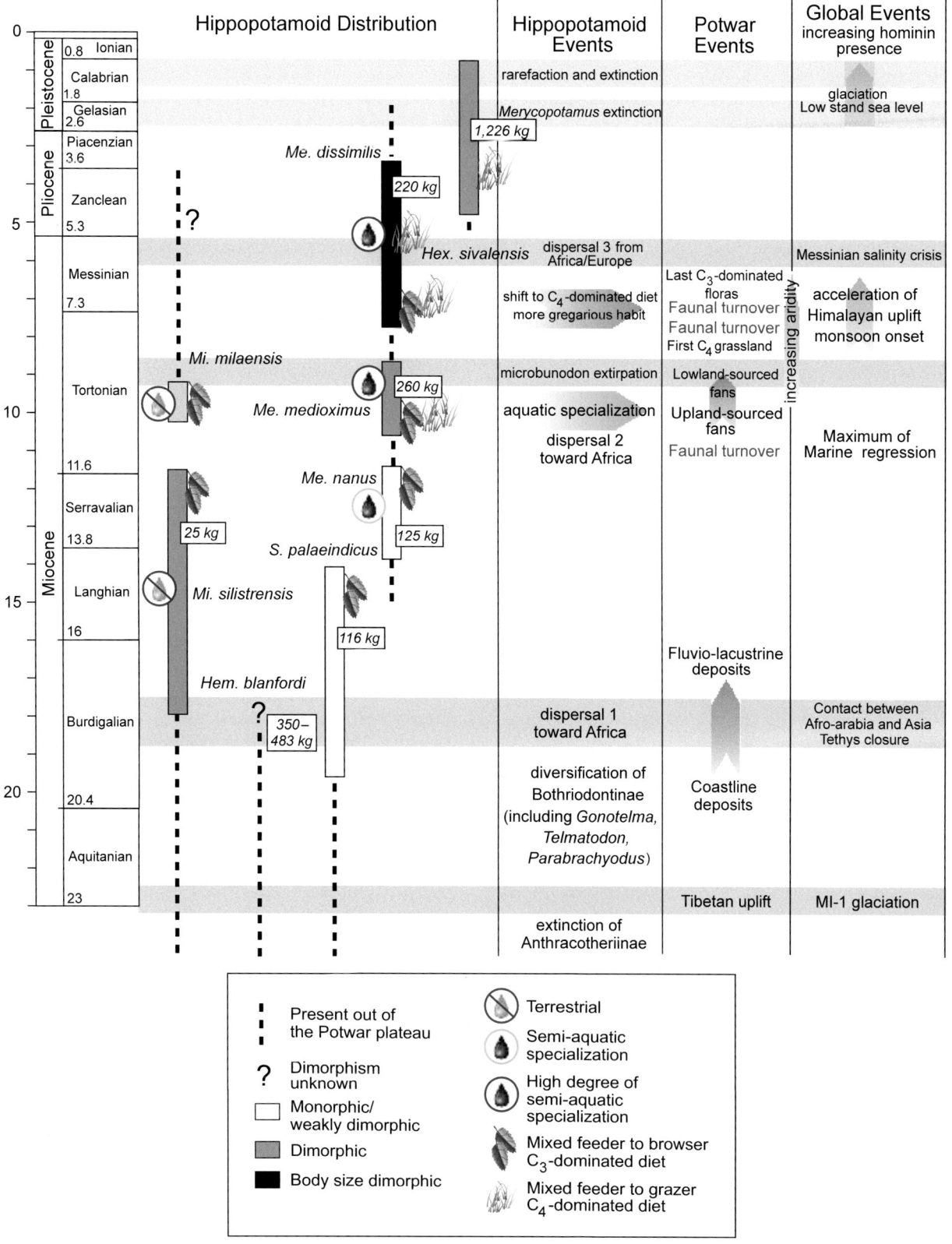

Figure 20.4. Synthesis of Neogene hippopotamoid evolution on the Potwar Plateau, with proposed correlations to local and global events.

Garcés, and Krijgsman 2006). There is no conclusive evidence of *Hexaprotodon* being able to cross extensive areas of sea water. The abundance of *Hexaprotodon* on the Sunda Shelf during the Pleistocene (e.g., Hooijer 1950) is probably due to land connections with mainland Asia during low sea-level stands. No hippopotamids are known in Wallacea. This distribution is quite different from that of *Hippopotamus*, which colonized oceanic islands such as Madagascar, Cyprus, and Crete. In these insular settings, hippopotamids underwent marked body-size dwarfing, as well as dental and postcranial modifications. If Late Miocene hippopotamids lacked the ability to disperse in the marine realm, the Messinian crisis may have represented an opportunity for them to spread to Eurasia. Clarification of dispersal routes would require improving the record of Late Miocene hippopotamids from the Mediterranean area.

The appearance of *Hexaprotodon* in the Indian subcontinent represents a geographic expansion of the large herbivore semiaquatic niche, documented by two contemporaneous species of large and very large size (~220 kg and ~1,200 kg for *Merycopotamus* and *Hexaprotodon*, respectively). Both were probable grazers, taking advantage of the spread of grasses during and after the Late Miocene (see Fig. 20.4). Competition between these species may have been low because of their marked size differences.

Extinction of Hippopotamoids in Asia

Microbunodon thrived on the Indian subcontinent during the Neogene even as it became extinct in Europe at the end of the Oligocene (Lihoreau et al. 2004b). In the Potwar Plateau, *Microbunodon* persists until 9.2 Ma, and its local extinction occurs around the time of last occurrences of lorisoids and tetraconodontine suids, as well as some tragulids and boselaphine bovids (Barry et al. 2002), and somewhat before the last occurrence of the hominoid *Sivapithecus* (Nelson 2007; Chapter 12, this volume). Forested habitats are inferred for all of these taxa; their local extinction appears linked to fragmentation of mesic forests and the spread of floodplain grasslands (Badgley et al. 2008; see Fig. 20.4). Among hippopotamoids, only *Merycopotamus* adapted locally to these environmental changes. The genus *Microbunodon* apparently survived until the Late Pliocene in the Irrawaddy Basin (Takai et al. 2016), which suggests that southeastern Asia served as a refuge for *M. milaensis*.

Merycopotamus is known south of the Himalayas during the entire Pliocene but few remains are described or well-dated, thus hindering morphological comparisons with the Late Miocene specimens from the Potwar Plateau. The last occurrence of *Merycopotamus* on the Potwar Plateau is at 3.3 Ma (Barry et al. 2002), but younger maxillary remains were described from Mangla-Samwal (2.6–2.4 Ma; Steensma and Hussain 1992), and possible postcranials from the Pabbi Hills were attributed to this taxon (2.2–1.9 Ma; Dennell, Coard, and Turner 2006). *Merycopotamus* persisted until the beginning of the Pleistocene and thus coexisted with hippopotamids for at least 5 million years (see Fig. 20.4).

Barry et al. (2002) estimated that *Hexaprotodon sivalensis* survived until 2.2 Ma in the Siwaliks, while Dennell, Coard, and Turner (2006) indicated that it may have been present in the region until 1.7 Ma. *Hexaprotodon* persisted elsewhere in Asia (India, Java, and possibly Sri Lanka) during most of the Pleistocene (Storm 2001; Badam 2007; Indriati et al. 2011).

The extinction of Siwalik *Hexaprotodon* as a water-dependent megaherbivore was proposed by Jablonski (2004) as a marker of environmental change in Asia on the grounds of the rarity of specimens after 3.5 Ma, as noted by Barry et al. (2002) for the Siwaliks, and after 2.4 Ma according to Hussain et al. (1992) for the Mangla-Samwal Anticline Range. For Jablonski (2004), this rarity and eventual local extinctions were related to changes in the monsoon regime, with longer dry seasons creating unfavorable conditions for year-round maintenance of large, perennial water bodies needed by large semiaquatic mammals.

The correlation with climatic change is not fully supported by the gradual reduction in specimens of *Hexaprotodon* and *Merycopotamus* in the Siwaliks. Other explanations such as the tectonic dynamics of the Himalayas as well as biotic interactions should be taken into account. The rise of the Himalayas resulted in local changes of hydrological dynamics that were likely to impact large semiaquatic animals. The increased presence of hominins, notably *Homo sapiens* during the Late Pleistocene, could be another factor in the final extinction of *Hexaprotodon*.

CONCLUSIONS

The Siwalik record of the Potwar Plateau has played a major role in studies of Neogene hippopotamoid evolution, both as the original diversification center for Merycopotamini and as a refuge for microbunodontines, which vanished elsewhere at the end of the Paleogene. The exceptional quality of the Potwar Siwalik record documents the adaptations of the genus *Merycopotamus* that echo the evolutionary history of *Libycosaurus* in Africa. A correlation between semiaquatic specialization and aridification on both continents during the same time interval is plausible. With regard to hippopotamoid evolution, the Indian subcontinent acted as a relatively isolated area, Himalayan uplift providing barriers to dispersal exchanges with the rest of the Old World, and triggering rapid evolution during any rapid environmental shifts.

Many questions remain unresolved. The first concerns the origin of the genus *Merycopotamus* and requires a thorough survey of anthracothere material from the Kamlial Formation to document the transition from early Merycopotamini to *Merycopotamus nanus*. The timing and mode of acquisition of semiaquatic habits in this tribe are also unclear, since early *Merycopotamus* appears already adapted to this lifestyle.

The paleoecology of Microbunodontinae is less well known than for the European Paleogene *M. minimum* and requires further study. Comparative analysis of faunas and environments at the Oligocene–Miocene transition in Europe and the Indian subcontinent might contribute to explaining the different timing of extinction in each region and clarify the role of South Asia as a refuge.

Still poorly understood are the conditions that favored the arrival of hippopotamids in South Asia, the ecological consequences of this arrival, and details of their regional evolution and diversification.

Finally, the Pleistocene extinction of hippopotamoids (*Merycopotamus* and *Hexaprotodon*) needs further study, particularly the relative contributions of climatic shifts and biotic interactions.

ACKNOWLEDGMENTS

We would like to thank David Pilbeam, John Barry, and Larry Flynn for entrusting us with the study of the Siwalik Hippopotamoidea and for welcoming us at Harvard University. We are indebted to the curators of the Siwalik mammal collections at the AMNH (Denny Diveley, Judith Galkin), the Yale Peabody Museum (Daniel Brinkman, Vanessa Rhue), and the NHM in London (Jerry Hooker, Pip Brewer). We would like to thank Michel Brunet for suggesting that we pursue this work on hippopotamoids that he initiated decades ago. We are very grateful to Michèle Morgan, Catherine Badgley, and David Pilbeam for inviting us to participate in this book project.

References

Agustí, J., M. Garcés, and W. Krijgsman. 2006. Evidence for African–Iberian exchanges during the Messinian in the Spanish mammalian record. *Palaeogeography, Palaeoclimatology, Palaeoecology* 238(1–4): 5–14. https://doi.org/10.1016/j.palaeo.2006.03.013.

Andrews, C. W. 1914. On the Lower Miocene vertebrates from British East Africa, collected by Dr. Felix Oswald. *Quarterly Journal of the Geological Society, London* 70(1–4): 163–186.

Antoine, P.-O., G. Métais, M. J. Orliac, J.-Y. Crochet, L. J. Flynn, L. Marivaux, A. R. Rajpar, G. Roohi, and J.-L. Welcomme. 2013. Mammalian Neogene biostratigraphy of the Sulaiman Province, Pakistan. In *Fossil Mammals of Asia: Neogene Biostratigraphy and Chronology*, edited by X.-M. Wang, L. J. Flynn and M. Fortelius, 400–442. New York: Columbia University Press.

Antoine, P.-O., J.-L. Welcomme, L. Marivaux, I. Baloch, M. Benammi, and P. Tassy. 2003. First record of Paleogene Elephantoidea (Mammalia, Proboscidea) from the Bugti Hills of Pakistan. *Journal of Vertebrate Paleontology* 23(4): 977–980.

Badam, G. L. 2007. *The Central Narmada Valley—A Study in Quaternary Palaeontology and Allied Aspects*. New Delhi: Indira Gandhi Rashtriya Manav Sangrahalaya, Bhopal and D. K. Printworld (P) Ltd.

Badgley, C., J. C. Barry, M. E. Morgan, S. V. Nelson, A. K. Behrensmeyer, T. E. Cerling, and D. Pilbeam. 2008. Ecological changes in Miocene mammalian record show impact of prolonged climatic forcing. *Proceedings of the National Academy of Sciences* 105(34): 12145–12149.

Barry, J. C. 1985. Les variations des faunes du Miocene Moyen et Supérieur des Formation des Siwalik du Pakistan. *L'Anthropologie* 89(3): 267–269.

Barry, J. C., A. K. Behrensmeyer, C. E. Badgley, L. J. Flynn, H. Peltonen, I. U. Cheema, D. Pilbeam, et al. 2013. The Neogene Siwaliks of the Potwar Plateau, Pakistan. In *Fossil Mammals of Asia: Neogene Biostratigraphy and Chronology*, edited by X.-M. Wang, L. J. Flynn and M. Fortelius, 373–399. New York: Columbia University Press.

Barry, J. C., A. K. Behrensmeyer, and M. Monaghan. 1980. A geologic and biostratigraphic framework for Miocene sediments near Khaur Village, Northern Pakistan. *Postilla* 183:1–19.

Barry, J. C., and L. J. Flynn. 1990. Key biostratigraphic events in the Siwaliks sequence. In *European Neogene Mammal Chronology*, edited by E. H. Lindsay, V. Fahlbusch and P Mein, 557–571. New York: Plenum Press.

Barry, J. C., M. E. Morgan, L. J. Flynn, D. Pilbeam, A. K. Behrensmeyer, S. M. Raza, I. A. Khan, C. Badgley, J. Hicks, and J. Kelley. 2002. Faunal and environmental change in the Late Miocene Siwaliks of Northern Pakistan. *Paleobiology Memoirs 28* (Memoir 3 suppl. to no. 2): 1–72.

Black, C. C. 1978. Anthracotheriidae. In *Evolution of African Mammals*, edited by V. J. C Maglio and H. B. S. Cooke, 423–434. Cambridge, MA: Harvard University Press.

Boisserie, J.-R. 2005. The phylogeny and taxonomy of Hippopotamidae (Mammalia: Artiodactyla): A review based on morphology and cladistic analysis. *Zoological Journal of the Linnean Society* 143(1): 1–26.

Boisserie, J.-R.. 2007. Family Hippopotamidae. In *The Evolution of Artiodactyls*, edited by D. R. Prothero and S. E. Foss, 106–119. Baltimore, MD: Johns Hopkins University Press.

Boisserie, J.-R., and F. Bibi. 2022. Hippopotamidae from the Baynunah Formation. In *Sands of Time: Ancient Life in the Late Miocene of Abu Dhabi, United Arab Emirates*, edited by F. Bibi, B. Kraatz, M. J. Beech, and A. Hill, 243–260. Cham: Springer International Publishing.

Boisserie, J.-R., and S. K. Eltringham. 2013. Genus *Choeropsis*—Pigmy hippopotamus. In *Mammals of Africa*, edited by J. Kingdon and M. Hoffmann, 78–79. London: Bloomsbury.

Boisserie, J.-R., R. E. Fisher, F. Lihoreau, and E. M. Weston. 2011. Evolving between land and water: Key questions on the emergence and history of the Hippopotamidae (Hippopotamoidea, Cetancodonta, Cetartiodactyla). *Biological Reviews* 86(3): 601–625. https://doi.org/10.1111/j.1469-185X.2010.00162.x.

Boisserie, J.-R., C. Kiarie, F. Lihoreau, and I. Nengo. 2017a. Middle Miocene *Kenyapotamus* (Cetartiodactyla, Hippopotamidae) from Napudet, Turkana Basin, Kenya. *Journal of Vertebrate Paleontology* 37(1): e1272055. https://doi.org/10.1080/02724634.2017.1272055.

Boisserie, J.-R., and F. Lihoreau. 2006. Emergence of Hippopotamidae: New scenarios. *Comptes Rendus Palevol* 5(5): 749–756. https://doi.org/10.1016/j.crpv.2005.11.004.

Boisserie, J.-R., F. Lihoreau, and M. Brunet. 2005. The position of Hippopotamidae within Cetartiodactyla. *Proceedings of the National Academy of Sciences* 102(5): 1537–1541.

Boisserie, J.-R., A. Likius, P. Vignaud, and M. Brunet. 2005a. A new Late Miocene hippopotamid from Toros-Menalla, Chad. *Journal of Vertebrate Paleontology* 25(3): 665–673.

Boisserie, J.-R., and G. Merceron. 2011. Correlating the success of Hippopotaminae with the C_4 grass expansion in Africa: Relationship and diet of Early Pliocene hippopotamids from Langebaanweg, South Africa. *Palaeogeography, Palaeoclimatology, Palaeoecology* 308(3–4): 350–361. https://doi.org/10.1016/j.palaeo.2011.05.040.

Boisserie, J.-R., F. Lihoreau, M. Orliac, R. E. Fisher, E. M. Weston, and S. Ducrocq. 2010. Morphology and phylogenetic relationships of the earliest known hippopotamids (Cetartiodactyla, Hippopotamidae, Kenyapotaminae): *Kenyapotamus* and hippopotamid origins. *Zoological Journal of the Linnean Society* 158(2): 325–366. https://doi.org/10.1111/j.1096-3642.2009.00548.x.

Boisserie, J.-R., G. Suwa, B. Asfaw, F. Lihoreau, R. L. Bernor, S. Katoh, and Y. Beyene. 2017b. Basal Hippopotamines from the Upper Miocene of Chorora, Ethiopia. *Journal of Vertebrate Paleontology* 37(3): e1297718. https://doi.org/10.1080/02724634.2017.1297718.

Boisserie, J.-R., A. Zazzo, G. Merceron, C. Blondel, P. Vignaud, A. Likius, H. T. Mackaye, and M. Brunet. 2005b. Diets of modern and Late Miocene hippopotamids: Evidence from carbon isotope composition and micro-wear of tooth enamel. *Palaeogeography, Palaeoclimatology,*

Palaeoecology 221(1–2): 153–174. https://doi.org/10.1016/j.palaeo.2005.02.010.

Boisserie, J.-R., and T. White. 2004. A new species of Pliocene Hippopotamidae from the Middle Awash, Ethiopia. *Journal of Vertebrate Paleontology* 24(2): 464–473.

Brachert, T. C., G. B. Brügmann, D. F. Mertz, O. Kullmer, F. Schrenk, D. E. Jacob, I. Ssemmanda, and H. Taubald. 2010. Stable isotope variation in tooth enamel from Neogene hippopotamids: Monitor of meso and global climate and rift dynamics on the Albertine Rift, Uganda. *International Journal of Earth Sciences* 99:1663–1675.

Brunet, M. 1968. Découverte d'un crane d'Anthracotheriidae, *Microbunodon minimum* (Cuvier), à La Milloque (Lot et Garonne). *Comptes Rendus de l'Académie des Sciences de Paris* 267(8): 835–838.

Cabard, P. 1976. Monographie du Genre *Microbunodon* Deperet, 1908 (Mammalia, Artiodactyla, Anthracotheriidae) de l'Oligocène Supérieur d'Europe de l'ouest. Thèse de 3ème cycle. Poitiers: Université de Poitiers.

Cerling, T. E., J. M. Harris, and M. G. Leakey. 2003. Isotope paleoecology of the Nawata and Nachukui Formations at Lothagam, Turkana Basin, Kenya. In *Lothagam: The Dawn of Humanity in Eastern Africa*, edited by M. G. Leakey and J. M. Harris, 605–624. New York: Columbia University Press.

Chavasseau, O., A. A. Khyaw, Y. Chaimanee, P. Coster, E.-G. Emonet, A. N. Soe, M. Rugbumrung, S. T. Tun, and J.-J. Jaeger. 2013. Advances in the biochronology and biostratigraphy of the continental Neogene of Myanmar. In *Fossil Mammals of Asia: Neogene Biostratigraphy and Chronology*, edited by X.-M. Wang, L. J. Flynn and M. Fortelius, 461–474. New York: Columbia University Press.

Clementz, M. T., P. A. Holroyd, and P. L. Koch. 2008. Identifying aquatic habits of herbivorous mammals through stable isotope analysis. *PALAIOS* 23(9): 574–585. https://doi.org/10.2110/palo.2007.p07-054r.

Colbert, E. H. 1935. Siwalik mammals in the American Museum of Natural History. *American Philosophical Society* 26:1–401.

Cooper, L. N., M. T. Clementz, S. Usip, S. Bajpai, S. T. Hussain, and T. L. Hieronymus. 2016. Aquatic habits of cetacean ancestors: integrating bone microanatomy and stable isotopes. *Integrative and Comparative Biology* 56(6): 1370–1384. Corvinus, G., and L. N. Rimal. 2001. Biostratigraphy and geology of the Neogene Siwalik Group of the Surai Khola and Rato Khola areas in Nepal. *Paleogeography, Palaeoclimatology, Palaeoecology* 165:251–279.

Coryndon, S. C. 1977. The taxonomy and nomenclature of the Hippopotamidae (Mammalia, Artiodactyla) and a description of two new fossil species. *Proceedings of the Koninklijke Nederlandse Akademie van Wetenschappen* 80(2): 61–88.

Damuth, J., and B. J. MacFadden. 1990. *Body Size in Mammalian Paleobiology: Estimation and Biological Implications.* Cambridge: Cambridge University Press.

Dennell, R., R. Coard, and A. Turner. 2006. The biostratigraphy and magnetic polarity zonation of the Pabbi Hills, Northern Pakistan: An Upper Siwalik (Pinjor stage) Upper Pliocene–Lower Pleistocene Fluvial Sequence. *Palaeogeography, Palaeoclimatology, Palaeoecology* 234(2–4): 168–185. https://doi.org/10.1016/j.palaeo.2005.10.008.

Dennell, R. W. 2005. Early Pleistocene hippopotamid extinctions, monsoonal climates, and river system histories in South and South West Asia: Comment on Jablonski (2004) "The hippo's tale: How the anatomy and physiology of Late Neogene hexaprotodon shed light on Late Neogene environmental change." *Quaternary International* 126–128(January): 283–287. https://doi.org/10.1016/j.quaint.2004.05.001.

De Visser, J. A. 2008. The extinct genus *Hexaprotodon* Falconer & Cautley, 1836 (Mammalia, Artiodactyla, Hippopotamidae) in Asia: Paleoecol-

ogy and taxonomy. Unpublished PhD diss., Utrecht: Universiteit Utrecht.

Ducrocq, S. 2020. Taxonomic revision of *Anthracokeryx thailandicus* Ducrocq, 1999 (Anthracotheriidae, Microbunodontinae) from the Upper Eocene of Thailand. *Vertebrata PalAsiatica* 58(4): 293–304. https://doi.org/10.19615/j.cnki.1000-3118.200618.

Ducrocq, S., Y. Chaimanee, V. Sutheethorn and J.-J. Jaeger. 1995. Dental anomalies in Upper Eocene Anthracotheriidae: A possible case of inbreeding. *Lethaia* 28:355–360.

Ducrocq, S., and F. Lihoreau. 2006. The occurrence of Bothriodontines (Artiodactyla, Mammalia) in the Paleogene of Asia with special reference to *Elomeryx*: Paleobiogeographical implications. *Journal of Asian Earth Sciences* 27(6): 885–91. https://doi.org/10.1016/j.jseaes.2005.09.004.

Ducrocq, S., A. N. Soe, A. K. Aung, M. Benammi, B. B., Y. Chaimanee, T.T., T. Thein, and J.-J. Jaeger. 2000. A new anthracotheriid artiodactyl from Myanmar, and the relative ages of the Eocene anthropoïd primate-bearing localities of Thailand (Krabi) and Myanmar (Pondaung). *Journal of Vertebrate Paleontology* 20(4): 755–760.

Falconer, H. 1868. *Palaeontological Memoirs.* London: R. Hardwicke.

Falconer, H., and P. T. Cautley. 1836. Note on the fossil *Hippopotamus* of the Siwalik Hills. *Asiatic Researches* 19:39–53.

Falconer, H., and P. T. Cautley. 1847. *Fauna Antiqua Sivalensis, Atlas.* London.

Forster Cooper, C. 1924. The Anthracotheriidae of the Dera Bugti Deposits in Baluchistan. *Memoirs of the Geological Survey of India* 8(2): 1–56.

Franz-Odendaal, T. A., J. A. Lee-Thorp, and C. Anusuya. 2002. New evidence for the lack of C_4 grassland expansions during the Early Pliocene at Langebaanweg, South Africa. *Paleobiology* 28(3): 378–388.

Gentry, A. W., and J. J. Hooker. 1988. The phylogeny of the Artiodactyla. In *The Phylogeny and Classification of the Tetrapods*, Vol. 2: *Mammals*, edited by M. J. Benton, 235–272. Oxford: Clarendon Press.

Gernelle, K., F. Lihoreau, J.-R. Boisserie, L. Marivaux, G. Métais and P.-O. Antoine. 2023. New material of *Parabrachyodus hyopotamoides* from Samane Nala, Bugti Hills (Pakistan) and the origin of Merycopotamini (Mammalia: Hippopotamoidea). *Zoological Journal of the Linnean Society* 198(1): 278–309.

Gomes Rodrigues, H., F. Lihoreau, M. Orliac and J.-R Boisserie. 2021. Characters from the deciduous dentition and its interest for phylogenetic reconstruction in Hippopotamoidea (Cetartiodactyla: Mammalia). *Zoological Journal of the Linnean Society* 193(2): 413–431.

Gradstein, F., J. G. Ogg, M. Schmitz, and G. Ogg. 2012. *The Geologic Time Scale.* Amsterdam: Elsevier, 1176.

Grandi, F., and F. Bona. 2017. *Prominatherium dalmatinum* from the Late Eocene of Grancona (Vicenza, NE Italy). The oldest terrestrial mammal of the Italian peninsula. *Comptes Rendus Palevol* 16(7): 738–745. https://doi.org/10.1016/j.crpv.2017.04.002.

Grossman, A., R. Calvo, R. López-Antoñanzas, F. Knoll, G. Hartman and R. Rabinovich. 2019. First record of *Sivameryx* (Cetartiodactyla, Anthracotheriidae) from the Lower Miocene of Israel highlights the importance of the Levantine Corridor as a dispersal route between Eurasia and Africa. *Journal of Vertebrate Paleontology* 39(2): e1599901. https://doi.org/10.1080/02724634.2019.1599901.

Hanta, R., B. Rathanastien, Y. Kunimatsu, H. Saegusa, H. Nakaya, S. Nagaoka, and P. Jintasakul. 2008. A new species of Bothriodontinae, *Merycopotamus thachangensis* (Cetartiodactyla, Anthracotheriidae) from the Late Miocene of Nakhon Ratchasima, northeastern Thailand. *Journal of Vertebrate Paleontology* 28(4): 1182–1188.

Harris, J. M., T. E. Cerling, M/ G. Leakey, and B. H. Passey. 2008. Stable isotope ecology of fossil hippopotamids from the Lake Turkana Basin of East Africa. *Journal of Zoology* 275(3): 323–331. https://doi.org/10.1111/j.1469-7998.2008.00444.x.

Herring, S. W. 1972. The role of canine morphology in the evolutionary divergence of pigs and peccaries. *Journal of Mammalogy* 53(3): 500–512.

Herring, S. W. 1975. Adaptations for gape in the *Hippopotamus* and its relatives. *Forma et Functio* 8:85–100.

Holroyd, P. A., F. Lihoreau, G. F. Gunnell, and E. R. Miller. 2010. Anthracotheriidae. In *Cenozoic Mammals of Africa*, edited by L. Werdelin and W. J. Sanders, 843–851. Berkeley: University of California Press.

Hooijer, D. A. 1950. The fossil Hippopotamidae of Asia, with notes on the recent species. *Zoologische Verhandelingen* 8(1): 1–123.

Hooijer, D. A. 1952. Fossil mammals and the Plio-Pleistocene boundary in Java. *Proceedings of the Koninklijke Nederlanse Akademie von Wetenschappen* 55(4): 395–398.

Houssaye, A., F. Martin, J.-R. Boisserie and F. Lihoreau. 2021. Paleoecological inferences from long bone microanatomical specializations in Hippopotamoidea (Mammalia, Artiodactyla). *Journal of Mammalian Evolution* 28(3): 847–870.

Hussain, S. T., G. D. van der Bergh, K. J. Steensma, J. A. de Visser, J. de Vos, M. Arif, J. A. Van Dam, P. Y. Sondaar, and S. B. Malik. 1992. Biostratigraphy of the Plio-Pleistocene continental sediments (Upper Siwaliks) of the Mangla-Samwal Anticline, Azad Kashmir, Pakistan. *Proceedings of the Koninklijke Nederlandse Akademie van Wetenschappen* 95(1): 65–80.

Indriati, E., C. C. Swisher, C. Lepre, R. L. Quinn, R. A. Suriyanto, A. T. Hascaryo, R. Grün, et al. 2011. The age of the 20 meter Solo River Terrace, Java, Indonesia, and the survival of *Homo erectus* in Asia, edited by Fred H. Smith. *PLOS ONE* 6(6): e21562. https://doi.org/10.1371/journal.pone.0021562.

Jablonski, N. G. 2004. The hippo's tale: How the anatomy and physiology of Late Neogene *Hexaprotodon* shed light on Late Neogene environmental change. *Quaternary International* 117(1): 119–123. https://doi.org/10.1016/S1040-6182(03)00121-6.

Janis, C. M. 1995. Correlation between craniodental morphology and feeding behaviour in ungulates : Reciprocal illumination between living and fossil taxa. In *Functionnal Morphology in Vertebrate Paleontology*, edited by J. Thomason, 76–98. Cambridge: Cambridge University Press.

Johnson, N. M., J. Stix, L. Tauxe, P. F. Cerveny, and R. A. K. Tahirkheli. 1985. Paleomagnetic chronology, fluvial processes, and tectonic implications of the Siwalik deposits near Chinji Village, Pakistan. *Journal of Geology* 93(1): 27–40.

Kingdon, J. 1997. *The Kingdon Field Guide to African Mammals*. Academic press.

Klingel, H. 2013. *Hippopotamus Amphibius*—Common hippopotamus. In *Mammals of Africa*, edited by J. Kingdon and M. Hoffmann, 68–78. London: Bloomsbury.

Krijgsman, W., F. J. Hilgen, I. Raffi, F. J. Sierro, and D. S. Wilson. 1999. Chronology, causes and progression of the Messinian salinity crisis. *Nature* 400(6745): 652–655. https://doi.org/10.1038/23231.

Kron, D. G., and E. Manning. 1998. Anthracotheriidae. In *Evolution of Tertiary Mammals of North America*, edited by C. M. Janis, K. M. Scott, and L. L. Jacobs, 1:381–388. Cambridge: Cambridge University Press.

Lebatard, A.-E., D. L. Bourlès, P. Duringer, M. Jolivet, R. Braucher, J. Carcaillet, M. Schuster, et al. 2008. Cosmogenic nuclide dating of *Sahelanthropus tchadensis* and *Australopithecus bahrelghazali*: Mio-Pliocene hominids from Chad. *Proceedings of the National Academy of Sciences* 105(9): 3226–3231.

Lihoreau, F. 2003. Systématique et paléoécologie des Anthracotheriidae [Artiodactyla ; Suiformes] du Mio-Pliocène de l'Ancien Monde : Im-

plications paléobiogéographiques. PhD diss., Université de Poitiers, 395 pp.

Lihoreau, F., J. Barry, C. Blondel, and M. Brunet. 2004a. A new species of Anthracotheriidae, *Merycopotamus medioximus* nov. sp. from the Late Miocene of the Potwar Plateau, Pakistan. *Comptes Rendus Palevol* 3(8): 653–662. https://doi.org/10.1016/j.crpv.2004.07.001.

Lihoreau, F., C. Blondel, J. Barry, and M. Brunet. 2004b. A new species of the genus *Microbunodon* (Anthracotheriidae, Artiodactyla) from the Miocene of Pakistan: Genus revision, phylogenetic relationships and palaeobiogeography. *Zoologica Scripta* 33(2): 97–115.

Lihoreau, F., J. Barry, C. Blondel, Y. Chaimanee, J.-J. Jaeger, and M. Brunet. 2007. Anatomical revision of the genus *Merycopotamus* (Artiodactyla; Anthracotheriidae): Its significance for Late Miocene mammal dispersal in Asia. *Palaeontology* 50(2): 503–524.

Lihoreau, F., L. Alloing-Séguier, P.-O. Antoine, J.-R. Boisserie, L. Marivaux, G. Métais, and J.-L. Welcomme. 2017. Enamel microstructure defines a major paleogene hippopotamoid clade: The Merycopotamini (Cetartiodactyla, Hippopotamoidea). *Historical Biology* 29(7): 947–957.

Lihoreau, F., J.-R. Boisserie, C. Blondel, L. Jacques, A. Likius, H. T. Mackaye, P. Vignaud, and M. Brunet. 2014. Description and palaeobiology of a new species of *Libycosaurus* (Cetartiodactyla, Anthracotheriidae) from the Late Miocene of Toros-Menalla, Northern Chad. *Journal of Systematic Palaeontology* 12(7): 761–798. https://doi.org/10.1080/14772019.2013.838609.

Lihoreau, F., J.-R. Boisserie, F. K. Manthi, and S. Ducrocq. 2015. Hippos stem from the longest sequence of terrestrial cetartiodactyl evolution in Africa. *Nature Communications* 6 (6264). https://doi.org/10.1038/ncomms7264.

Lihoreau, F., J.-R. Boisserie, L. Viriot, Y. Coppens, A. Likius, H. T. Mackaye, P. Tafforeau, P. Vignaud, and M. Brunet. 2006. Anthracothere dental anatomy reveals a Late Miocene Chado-Libyan bioprovince. *Proceedings of the National Academy of Sciences* 103(23): 8763–67.

Lihoreau, F., and S. Ducrocq. 2007. Family Anthracotheriidae. In *The Evolution of Artiodactyls*, edited by D. R. Prothero and S. C. Foss, 89–105. Baltimore, MD: John Hopkins University Press

Lihoreau, F., Lionel H., and M. Mahboubi. 2015. The new Algerian locality of Bir El Ater 3: Validity of *Libycosaurus algeriensis* (Mammalia, Hippopotamoidea) and the age of the Nementcha Formation. *Palaeovertebrata* 39(2)-e1: 1–11.

Lihoreau, F., Essid El Mabrouk, Hayet Khayati Ammar, L. Marivaux, Wissem Marzougui, Rodolphe Tabuce, Rim Temani, Monique Vianey-Liaud and Gilles Merzeraud 2021. The *Libycosaurus* (Hippopotamoidea, Artiodactyla) intercontinental dispersal event at the early Late Miocene revealed by new fossil remains from Kasserine area, Tunisia. *Historical Biology* 33(2): 146–158. https://doi.org/10.1080/08912963.2019.1596088.

Lindsay, E. H., L. J. Flynn, I. U. Cheema, J. C. Barry, K. F. Downing, A. R. Rajpar, and S. M. Raza. 2005. Will Downs and the Zinda Pir Dome. *Palaeontologia Electronica* 8(1): 1–19.

Lydekker, R. 1876. Notes on the osteology of *Merycopotamus dissimilis*. *Records of the Geological Survey of India* 9:144–153.

Lydekker, R. 1877a. Note on the Genera *Choeromeryx* and *Rhagatherium*. *Records of the Geological Survey of India* 10:225.

Lydekker, R. 1877b. Notices of new or rare mammals from the Siwaliks. *Records of the Geological Survey of India* 10: 76–83.

Lydekker, R. 1878. Notices of Siwalik mammals. *Records of the Geological Survey of India* 11:64–104.

Lydekker, R. 1883. Siwalik selenodont Suina. *Palaeontologia Indica* 10 11): 143–177.

Lydekker, R. 1884a. Note on a new species of *Merycopotamus*. *Geological Magazine* 1(7): 545–547.

Lydekker, R. 1884b. Siwalik and Narbada bunodont Suina. *Memoirs of the Geological Survey of India Paleontologia Indica* 10(3): 35–104.

Lydekker, R. 1885. Note on a third species of *Merycopotamus. Records of the Geological Survey of India* 18:145–146.

MacInnes, D. G. 1951. Miocene Anthracotheriidae from East Africa. *Fossil Mammals of Africa* 4:1–24.

Madden, C. T., and J. A. van Couvering. 1976. The proboscidean datum event: Early Miocene migration from Africa. *Geological Society of America Abstracts with Programs*, 992–993.

Martinez, J.-N., and J. Sudre. 1995. The astragalus of Paleogene artiodactyls : Comparative morphology, variability and prediction of body mass. *Lethaia* 28:197–209.

Martino, R., J. Pignatti, L. Rook, and L. Pandolfi. 2020. Systematic revision of hippopotamid remains from the Casino basin, Tuscany, Italy. *Fossilia* 2020:29–31.

Martino, R., J. Pignatti, L. Rook, and L. Pandolfi. 2021. Hippopotamid dispersal across the Mediterranean in the latest Miocene: A re-evaluation of the Gravitelli record from Sicily, Italy. *Acta Palaeontologica Polonica* 66(3): S67–S78.

Matthew, W. D. 1929. Critical observation upon Siwalik Mammals. *Bulletin of the American Museum of Natural History* 56(7): 437–560.

Mayr, E. 1964. *Animal Species and Evolution*. Cambridge, MA: The Belknap Press of Harvard University Press.

Merceron, G., C. Blondel, M. Brunet, S. Sen, N. Solounias, L. Viriot, and E. Heintz. 2004. The Late Miocene paleoenvironment of Afghanistan as inferred from dental microwear in artiodactyls. *Palaeogeography, Palaeoclimatology, Palaeoecology* 207(1–2): 143–163. https://doi.org/10.1016/j.palaeo.2004.02.008.

Merceron, G., C. Blondel, L. Viriot, G. D. Koufos, and L. de Bonis. 2007. Dental microwear analysis of bovids from the Vallesian (Late Miocene) of Axios Valley in Greece: Reconstruction of the habitat of *Ouranopithecus macedoniensis* (Primates, Hominoidea). *Geodiversitas* 29(3): 421–433.

Merceron, G., L. de Bonis, L. Viriot, and C. Blondel. 2005. Dental microwear of fossil bovids from Northern Greece: Paleoenvironmental conditions in the eastern Mediterranean during the Messinian. *Palaeogeography, Palaeoclimatology, Palaeoecology* 217(3–4): 173–185. https://doi.org/10.1016/j.palaeo.2004.11.019.

Merceron, G., L. Viriot, and C. Blondel. 2004. Tooth microwear pattern in roe deer (*Capreolus Capreolus* L.) from Chizé (western France) and relation to food composition. *Small Ruminant Research* 53:125–132.

Métais, G., P.-O. Antoine, S. R. Hassan Baqri, J.-Y. Crochet, D. de Franceschi, L. Marivaux, and J.-L. Welcomme. 2009. Lithofacies, depositional environments, regional biostratigraphy and age of the Chitarwata Formation in the Bugti Hills, Balochistan, Pakistan. *Journal of Asian Earth Sciences* 34(2): 154–167. https://doi.org/10.1016/j.jseaes.2008.04.006.

Nanda, A. C. 1980. *Hexaprotodon Sivalensis* Falconer and Cautley from the Upper Siwalik Subgroup, Ambala, Harayana. *Recent Researches in Geology* 5:161–174.

Nelson, S. V. 2007. Isotopic reconstructions of habitat change surrounding the extinction of *Sivapithecus*, a Miocene hominoid, in the Siwalik Group of Pakistan. *Palaeogeography, Palaeoclimatology, Palaeoecology* 243(1–2): 204–222. https://doi.org/10.1016/j.palaeo.2006.07.017.

Orliac, M., J.-R. Boisserie, L. MacLatchy, and F. Lihoreau. 2010. Early Miocene hippopotamids (Cetartiodactyla) constrain the phylogenetic and spatiotemporal settings of hippopotamid origin. *Proceedings of the National Academy of Sciences* 107(26): 11871–11876. https://doi.org/10.1073/pnas.1001373107.

Orliac, M., M. Mourlam, J.-R. Boisserie, L. Costeur, and F. Lihoreau. 2023. Evolution of semiaquatic habits in hippos and their extinct relatives:

Insights from the ear region. *Zoological Journal of the Linnean Society* 198(4): 1092–1105.

Patnaik, R. 2015. Diet and habitat changes among Siwalik herbivorous mammals in response to Neogene and Quaternary climate changes: An appraisal in the light of new data. *Quaternary International* 371:232–243.

Pentland, J. B. 1828. Description of fossil remains of some animals from the north-east border of Bengal. *Transactions Of the Geological Society of London* 2(2): 393–394.

Pickford, M. 1983. On the origins of Hippopotamidae together with descriptions of two new species, a new genus and a new subfamily from the Miocene of Kenya. *Geobios* 16(2): 193–217.

Pickford, M. 1987. Révision des Suiformes (Artiodactyla, Mammalia) de Bugti (Pakistan). *Annales de Paléontologie* 73:289–350.

Pickford, M. 1991. Revision of the Neogene Anthracotheriidae of Africa. In *The Geology of Libya*, edited by M. J. Salem and M. T. Busrewil, 4:1491–1525. New York: Academic Press.

Pickford, M., and D. Rogers. 1987. Révision des Suiformes de Bugti, Pakistan. *Compte Rendu de l'Académie des Sciences de Paris* 305(2): 634–646.

Pilbeam, D., A. K. Behrensmeyer, J. C. Barry, and S. M. Shah. 1979. Miocene sediments and faunas of Pakistan. *Postilla* 179:1–45.

Pilgrim, G. E. 1908. The Tertiary and post-Tertiary freshwater deposits of Baluchistan and Sind with notices of new vertebrates. *Records of the Geological Survey of India* 37: 139–169.

Pilgrim, G. E. 1910. Notices of new mammalian genera and species from the Tertiaries of India. *Records of Geological Survey of India* 40:63–71.

Pilgrim, G. E. 1912. The Vertebrate fauna of the Gaj Series in the Bugti Hills and the Punjab. *Memoirs of the Geological Survey of India Paleontologia Indica* 4:1–30.

Pilgrim, G. E. 1913. The correlation of the Siwaliks with the mammal horizon of Europe. *Records of the Geological Survey of India* 43(4): 264–326.

Pilgrim, G. E. 1917. Note on some recent mammal collections from the basal beds of the Siwalik. *Records of Geological Survey of India* 48: 98–101.

Pomel, A. 1848. Recherches sur les caractères et les rapports entre deux des divers genres vivants et fossiles des mammifères ongulés. *Compte Rendu Hebdomadaire des Séances de l'Académie des Sciences* 24:686–688.

Raza, S. M., J. C. Barry, G. E. Meyer, and L. Martin. 1984. Preliminary report on the geology and vertebrate fauna of the Miocene Manchar Formation, Sind, Pakistan. *Journal of Vertebrate Paleontology* 4(4): 584–599.

Robinson, P. T. 2013. *Choeropsis liberiensis*—Pigmy hippopotamus. In *Mammals of Africa*, edited by J. Kingdon and M. Hoffmann, 80–83. London: Bloomsbury.

Rowan, J., B. Adrian, and A. Grossman. 2015. The first skull of *Sivameryx africanus* (Anthracotheriidae, Bothriodontinae) from the Early Miocene of East Africa. *Journal of Vertebrate Paleontology* 35(3). https://doi.org/10.1080/02724634.2014.928305.

Scherler, L., F. Lihoreau,, and D. Becker. 2019. To split or not to split *Anthracotherium*? A phylogeny of Anthracotheriinae (Cetartiodactyla: Hippopotamoidea) and its palaeobiogeographical implications. *Zoological Journal of the Linnean Society* 185(2): 487–510.

Soe, A. N., O. Chavasseau, Y. Chaimanee, C. Sein, J.-J. Jaeger, X. Valentin, and S. Ducrocq. 2017. New remains of *Siamotherium pondaungensis* (Cetartiodactyla, Hippopotamoidea) from the Eocene of Pondaung, Myanmar: Paleoecologic and phylogenetic implications *Journal of Vertebrate Paleontology* 37(1). https://doi.org/10.1080/02724634.2017.1270290.

Solounias, N., and B. Dawson-Saunders. 1988. Dietary adaptations and paleoecology of the Late Miocene ruminants from Pikermi and Samos in Greece. *Paleogeography, Palaeoclimatology, Palaeoecology* 65(3–4): 149–172.

Spaulding, M., M. A. O'Leary, and J. Gatesy. 2009. Relationships of Cetacea (Artiodactyla) among mammals: Increased taxon sampling alters interpretations of key fossils and character evolution, edited by Andrew Allen Farke. *PLOS ONE* 4(9): e7062. https://doi.org/10.1371/journal.pone.0007062.

Steensma, K. J., and S. T. Hussain. 1992. *Merycopotamus dissimilis* (Artiodactyla, Mammalia) from the Upper Siwalik Subgroup and its affinities with Asian and African forms. *Proceedings of the Koninklijke Nederlanse Akademie von Wetenschappen* 95(1): 97–108.

Storm, P. 2001. The evolution of humans in Australasia from an environmental perspective. *Palaeogeography, Palaeoclimatology, Palaeoecology* 171(3–4): 363–383. https://doi.org/10.1016/S0031-0182(01)00254-1.

Suteethorn, V., E. Buffetaut, R. Helmcke-Invagat, J.-J. Jaeger, and Y. Jongkanjansoontorn. 1988. Oldest known Tertiary mammals from South-East Asia: Middle Eocene primate and anthracotheres from Thailand. *Neues Jahrbuch für Geologie und Paläontologie, Monatshefte* 9:563–570.

Takai, M., Y. Nishioka, T. Htike, M. Maung, K. Khaing, Zin-Maung-Maung-Thein, T. Tsubamoto, and N. Egi. 2016. Late Pliocene *Semnopithecus* fossils from Central Myanmar: Rethinking of the evolutionary history of cercopithecid monkeys in Southeast Asia. *Historical Biology* 28(1–2): 172–188. https://doi.org/10.1080/08912963.2015.1018018.

Tauxe, L., and C. Badgley. 1988. Stratigraphy and remanence acquisition of a palaeomagnetic reversal in alluvial Siwalik Rocks of Pakistan. *Sedimentology* 35: 697–715.

Thaung, H. 2012. Review on the taxonomic status of *Hexaprotodon Iravaticus* (Mammalia, Artiodactyla, Hippopotamidae) from the Neogene of Myanmar. *Shwebo University Research Journal* 3(1): 94–110.

Thaung, H., and M. Takai. 2016. Reevaluation of the phylogeny and taxonomy of the Asian fossil hippopotamuses. *Universities Research Journal* 8(1): 171–177.

Tsubamoto, T. 2010. Recognition of *Microbunodon* (Artiodactyla, Anthracotheriidae) from the Eocene of China. *Paleontological Research* 14(2): 161–165. https://doi.org/10.2517/1342-8144-14.2.161.

Tsubamoto, T., Y. Kunimatsu, and M. Nakatsukasa. 2015. A lower molar of a primitive, large hippopotamus from the Lower Miocene of Kenya.

Paleontological Research 19(4): 321–327. https://doi.org/10.2517/2015PR015.

Tsubamoto, T., H. Thaung, Zin-Maung-Maung-Thein, N. Egi, Y. Nishioka, Maung-Maung, and M. Takai. 2012. New data on the Neogene Anthracotheres (Mammalia, Artiodactyla) from Central Myanmar. *Journal of Vertebrate Paleontology* 32(4): 956–964. doi.org/10.1080/02724634.2012.670176.

Tütken, T., and J. Absolon. 2015. Late Oligocene ambient temperatures reconstructed by stable isotope analysis of terrestrial and aquatic vertebrate fossils of Enspel, Germany. *Palaeobiodiversity and Palaeoenvironments* 95(1): 17–31.

Uno, K. T., T. E. Cerling, J. M. Harris, Y. Kunimatsu, M. G. Leakey, M. Nakatsukasa, and H. Nakaya. 2011. Late Miocene to Pliocene carbon isotope record of differential diet change among East African herbivores. *Proceedings of the National Academy of Sciences* 108(16): 6509–6514. https://doi.org/10.1073/pnas.1018435108.

Vignaud, P., P. Duringer, H. T. Mackaye, A. Likius, C. Blondel, J.-R. Boisserie, L. de Bonis, et al. 2002. Geology and palaeontology of the Upper Miocene Toros-Menalla Hominid Locality, Chad. *Nature* 418(6894): 152–155.

Weston, E. M. 1997. A biometrical analysis of evolutionary change within the Hippopotamidae. Unpublished PhD diss., Cambridge University.

Weston, E. M., and J.-R. Boisserie. 2010. Hippopotamidae. In *Cenozoic Mammals of Africa*, edited by L. Werdelin and W. Sanders, 853–871. Berkeley: University of California Press.

Zachos, J., M. Pagani, L. Sloan, E. Thomas, and K. Billups. 2001. Trends, rhythms, and aberrations in global climate 65 Ma to Present. *Science* 292(5517): 686–693. https://doi.org/10.1126/science.1059412.

Zhisheng, A., J. E. Kutzbach, W. L. Prell, and S. C. Porter. 2001. Evolution of Asian monsoons and phased uplift of the Himalaya-Tibetan Plateau since Late Miocene times. *Nature* 411(6833): 62–66.

Zin-Maung-Maung-Thein, M. Takai, H. Uno, J. G. Wynn, N. Egi, T. Tsubamoto, T. Htike, et al. 2011. Stable isotope analysis of the tooth enamel of Chaingzauk mammalian fauna (Late Neogene, Myanmar) and its implication to paleoenvironment and paleogeography. *Palaeogeography, Palaeoclimatology, Palaeoecology* 300(1–4): 11–22. https://doi.org/10.1016/j.palaeo.2010.11.016.

SIWALIK TRAGULIDAE

JOHN C. BARRY

INTRODUCTION

Although the number and status of the various tragulid species are uncertain, they are an important component of Neogene Siwalik mammal faunas on the Indian subcontinent. Fossil collections that included tragulids were discovered early in the exploration of the Siwaliks, apparently being among Falconer's materials, which were recovered as early as 1842 (Lydekker 1876). The Natural History Museum in London and the Geological Survey of India Museum in Kolkata notably hold the holotypes for most of the named species, but other important collections include those of the Panjab University in Chandigarh, the Wadia Institute of Himalayan Geology in Dehra Dun, the American Museum of Natural History, the Yale Peabody Museum, the Bavarian State Collection of Palaeontology and Geology, and the Geological Institute, State University of Utrecht. More recently, large collections of Siwalik fossils have been made by collaborative projects including Harvard University, the Geological Survey of Pakistan, the University of Arizona, and the Pakistan Museum of Natural History. While the provenance of the earlier collections is often in question, for the most part the more recent collections are from well-dated stratigraphic sections and can be placed within a reliable chronostratigraphic framework (Raza et al. 1984, 2002; Lindsay et al. 2005; Antoine et al. 2013; Barry et al. 2013; Flynn et al. 2013; Chapter 2, this volume), making them especially valuable.

While the closely related lophiomerycids and perhaps tragulids themselves (Antoine et al. 2013) were present in Oligocene faunas of the subcontinent (Barry et al. 2005; Métais, Welcomme, and Ducrocq 2009), the oldest securely identified Siwalik tragulids are from the upper Chitarwata Formation in southwestern Punjab. They are approximately 22.1 Ma. Tragulids persist in modern faunas from India and Sri Lanka (various species of *Moschiola*) and extinct species are present as late as 3.3 Ma in the Tatrot Formation of Pakistan and perhaps even as late as 1.5 Ma in the Pinjor Formation of India (Gaur, Vasishat, and Sankhyan 1980).

In contrast to extant tragulids, Miocene tragulids are species rich with as many as 20 species, 4 or 5 of which may co-occur (Fig. 21.1, Table 21.1). This is an old pattern, as there are at least three and perhaps four species belonging to three genera present by 18.7 Ma. This level of richness compares favorably to the five to seven co-occurring Siwalik Miocene bovids. Siwalik Early and Middle Miocene tragulids are also relatively common, although they become scarce in the latest Miocene and Pliocene. In addition, the fossil species differ from extant species in having a much wider range of body sizes. The smallest fossil species is estimated to be just over 1 kg, while at least four species are over 25 kg, with the largest estimated to be nearly 76 kg. Thus, the small fossil species overlap with the smallest species of extant *Tragulus*, while the larger species are very much larger than individuals of the African *Hyemoschus aquaticus* (ca. 7 to 16 kg; Kingdon 1997) and approach medium size bovids and cervids.

MATERIALS

The collections used in this study include recently discovered specimens in the Geological Survey of Pakistan (GSP) and Pakistan Museum of Natural History (PMNH) collections, as well as fossils in the Yale Peabody Museum (YPM) and American Museum of Natural History (AMNH). In total there are more than 3,700 specimens that consist mostly of isolated teeth, jaws, and disarticulated skeletal elements, nearly all of which are fragmentary. There is very little cranial material and very few associated teeth and skeletal elements belonging to one individual. As with other Siwalik species, the fragmentary, disarticulated preservation makes it difficult to link dental and postcranial morphologies to characterize the species.

In this chapter I use an index of crown height (hypsodonty) derived from the height of the metaconid divided by the trigonid width as measured on unworn or lightly worn lower second and third molars. This index is distinct from indices using a measure of tooth length instead of width (Fortelius et al. 2002; Guzmán-Sandoval and Rössner, 2019). Although the width has a large measurement error, the index has the advantage of being applicable to third molars, which are more

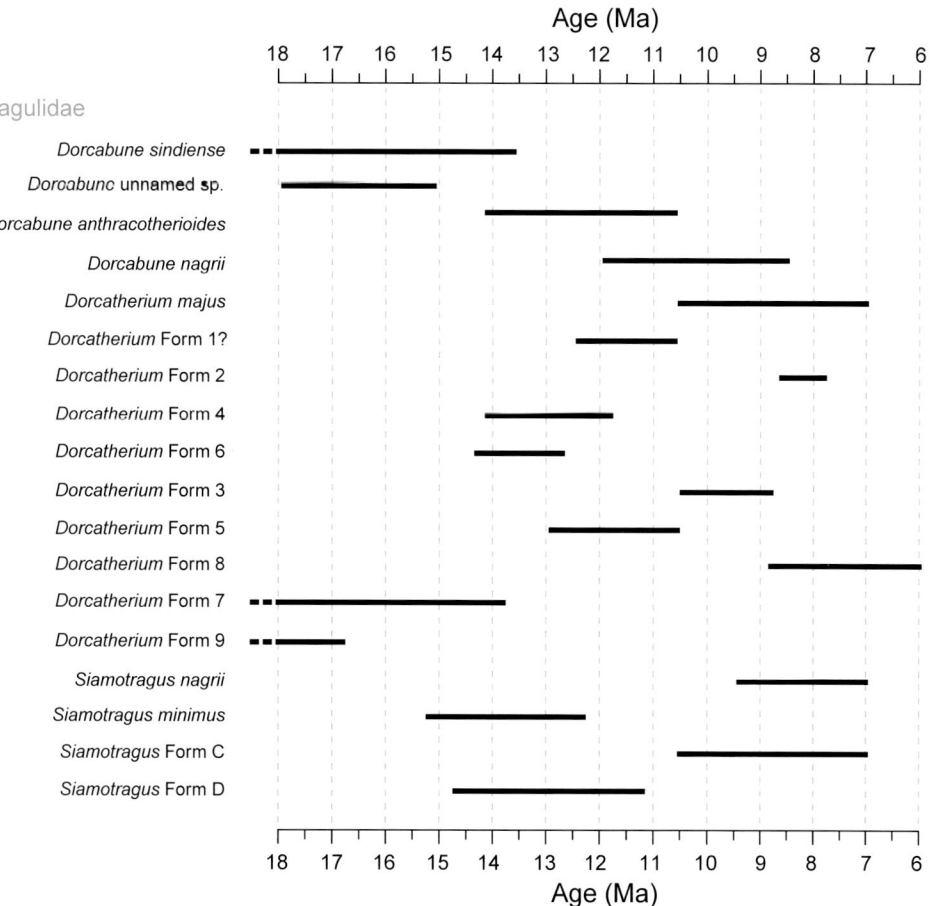

common in collections, and it allows comparisons to the findings of Damuth and Janis (2011) for extant species. Molar and premolar terminology follows Janis and Scott (1987), Gentry, Rössner, and Heizmann (1999), and Bärmann and Rössner (2011) although I use "mesoconid" in place of "protoconid" or "mesolabial conid" when discussing lower premolars. Other modifications are noted in the text.

The term "Siwaliks" refers to the Neogene fluvial deposits found along the southern margin of the Himalayas in Pakistan, India, and Nepal (Flynn et al. 2013; Patnaik 2013). Siwalik sediments are particularly well exposed on the Potwar Plateau in northern Pakistan, but significant exposures of the Vihowa and Chitarwata Formations are also known in southwestern Punjab and Baluchistan, as well as of the Manchar Formation in Sind. Although only sparsely fossiliferous, the Vihowa, Chitarwata, and Manchar Formations are significant because they contain fossils of earliest Miocene and latest Oligocene age (Khan et al. 1984; Raza et al. 1984, 2002; Lindsay et al. 2005; Antoine et al. 2013), while the Potwar Plateau localities are of Early Miocene through Pleistocene age. Siwalik fossils are found in many different depositional environments, with the richest localities typically being in small abandoned channels and channel margins of the fluvial systems (Chapters 2 and 4, this volume). In both northern and southwestern Punjab and

in Baluchistan exposures are extensive, with more than 1,400 individual fossil sites now known, a large number of which can be placed on stratigraphic sections and their ages estimated (Chapter 3, this volume).

SPECIES ACCOUNTS

Three genera have customarily been recognized among Siwalik tragulids: (1) *Dorcabune* with six nominal species; (2) *Dorcatherium* with three; and (3) *Siamotragulus* with three. To these *Vishnumeryx daviesi* (Lydekker 1886) should be added and perhaps *Tragulus sivalensis*, also described by Lydekker (1882, 1884). However, the status of several named species is uncertain and, because their types are inadequate and of unknown provenance and age, systematic revision is difficult. The problem is further complicated because while the genera differ morphologically, species within a genus are more similar and can be distinguished easily only on the basis of size and relative proportions. Accordingly, pairs of contemporaneous species are most readily separated, while it is more difficult to distinguish between time successive species, especially those not separated by long temporal gaps in the record. As a result, the stratigraphic ranges of some species are poorly known, while relationships with tragulids in Eurasia and Africa are obscure.

Table 21.1. Siwalik Tragulid Taxa, Range, and Estimated Body Mass

Taxon, Author, Date	Holotype / Collection[1]	Age (Ma)[2]	Est. Body Mass (kg)[3]
Tragulidae Milne-Edwards, 1864			
***Dorcabune* Pilgrim, 1910**			
Dorcabune anthracotheroides Pilgrim, 1910	GSI B 580 (Lectotype)	14.1–10.6	43–77
Dorcabune nagrii Pilgrim, 1915	GSI B 590 (Lectotype)	11.9–8.5	30–35
Dorcabune sindiense Pilgrim, 1915	GSI B 598 (Lectotype)	18.7–13.6	18–21
Dorcabune unnamed species	YGSP	17.9–15.1	34–40
***Dorcatherium* Kaup, 1833**			
Group A			
Dorcatherium majus Lydekker, 1876	GSI B 197 (Lectotype)	10.5–7.0	18–43
Dorcatherium Form 1?	YGSP	12.4–10.6	14–30.5
Dorcatherium Form 2	YGSP	8.6–7.8	73
Dorcatherium Form 4	YGSP	14.1–11.8	9.2–19.7
Dorcatherium Form 6	YGSP	14.3–12.7	6.3–11.5
Group B			
Dorcatherium Form 3	YGSP	10.5–8.8	3.8–9.0
Dorcatherium Form 5	YGSP	12.9–10.5	4.6–8.5
Dorcatherium Form 8	YGSP	8.8–6.0	4.6–7.6
Group unknown			
Dorcatherium Form 7	YGSP	22.1–13.8	4.5–8.1
Dorcatherium Form 9	YGSP	18.7–16.8	10–13
***Siamotragulus* Thomas, Ginsburg, Hintong, and Suteethorn, 1990**			
Siamotragulus nagrii (Prasad, 1968)	GSI 18079	9.4–7.0	1.7–2.4
Siamotragulus minimus (West, 1980)	H-GSP 1983	15.2–12.3	0.8–1.1
Siamotragulus bugtiensis Ginsburg, Morales, and Soria, 2001	PAK 2485	22.1–ca. 18.1	2.9–4.8
Siamotragulus Form C	YGSP	10.5–7.0	1.2–1.5
Siamotragulus Form D	YGSP	14.7–11.2	2.0–3.8
***Tragulus* Brisson, 1762**			
Tragulus sivalensis Lydekker, 1882	GSI B 360	Late Miocene	ca. 2
***Vishnumeryx* Pilgrim, 1939**			
Vishnumeryx daviesi (Lydekker, 1886)	NHML M 3492	Pliocene? (3.5–3.3)	ca. 10–16

[1] Institutional Abbreviations:

    H-GSP: Geological Survey of Pakistan, Islamabad, collections made by Howard University-GSP Project

    YGSP: Geological Survey of Pakistan, Islamabad, collections made by Yale/Harvard University-GSP Project

    AMNH: American Musuem of Natural History, New York

    GSI: Geological Survey of India, Kolkata

    NHML: Natural History Museum London

    PMNH: Pakistan Museum of Natural History, Islamabad

    PAK: Collections made by a joint project of the University of Montpellier 2, the Muséum National d'Histoire Naturelle in Paris, and the University of Balochistan

[2] Observed first and last occurrences.

[3] Mass estimates are based on regressions developed by Morgan (1994) using the distal width of the astragalus.

Dorcabune Pilgrim, 1910

Dorcabune has previously been thought to be confined to Asia and has not been extensively discussed in the literature. For that reason I discuss it at some length here. The teeth and some skeletal elements of the genus are distinctive, while the species differ in size so that with only a few exceptions the fossils can be sorted into species. Three or four species of *Dorcabune* are present in the new collections, including the type species *Dorcabune anthracotheroides*, as well as *Dorcabune nagrii* and *Dorcabune sindiense*. In addition to these three taxa, there is possibly a fourth species, as a few fossils from the Kamlial Formation are too large for *Dorcabune sindiense*, but too small for *Dorcabune anthracotheroides*. There is, moreover, a suggestion of strong sexual dimorphism in the species, adding to the complexity of recognizing species.

The species of *Dorcabune* most obviously differ from other tragulids in their low crowned, bulbous cusps, but they are distinctive in other features as well. In the upper molars the post-protocrista has two limbs, one posteriorly directed and often flanked by a distinct lingual furrow, and the other a labially directed "neocrista" (Gentry, Rössner, and Heizmann 1999). Together these two crests form a distinctive cross-shaped wear figure. In addition the mesostyle is large and bulbous, the lingual and labial faces of both the paracone and especially the metacone are inflated, the postmetacrista is short and not inflected labially, the metastyle is weak or even absent, and the premetaconulecrista may develop a separate central cuspule.

In the mandible p1 is present and has a fused double root. The p2 is tricuspid, with a tall and trenchant premesoconulid, a tall mesoconid with a faint posterior cleft, and a posterolabial conid with a posteriorly directed posterolabial cristid and posterolingual cristid flanking a shallow posterolingual cleft that widens posteriorly. The p3 is also tricuspid, with a trenchant premesoconulid, a mesoconid with a deep posterior cleft flanked by salient lingual and labial postmesoconid cristids (the lingual of which is aligned with the preposterolabial cristid), and a posterolabial conid with a short posterolingual cristid and a longer posterolabial cristid that is separated from a small transverse stylid. The p4 is relatively long (about 90% the length of p3), with the mesoconid having a deep posterior cleft bounded by a lingual postmesocristid that almost closes the posterior cleft and a labial postmesocristid with two segments, the more distal of which is transversely oriented. The lower molars are wide relative to their length. The labial faces of the metaconid and entoconid are inflated, the anterior face of the metaconid is rounded with a weak premetacristid, the anterior segment of the preprotocristid is transverse, the anterior face of the entoconid is flattened with weak preentocristid and in some cases a weak secondary lingual fold, while the posterior face of the entoconid is round and the postentocristid very weak or absent. An associated distal tibia, astragalus, cuboid-navicular complex, and intact metatarsal of Dorcabune anthracotheroides show that in this adult individual the distal tibia and fibula were fused. The ecto-mesocuneiform and cuboid-navicular form a single unit, while the astragalus has a well-marked facet for the medial malleolus of the tibia, a deep pit for a posterior astragalofibular ligament, a shallow lateral notch between the proximal and distal trochleae, and a subdued lateral prominence for the superior calcaneal articulation. The latter two features give the lateral face of the astragalus a bovid-like aspect and allow the astragalus of Dorcabune to be recognized. Metatarsals III and IV are fused, but retain separate medullary cavities and a rectangular cross section midshaft. The lateral metatarsals II and V were not fused to metatarsals III and IV. A proximal third metacarpal lacks the interlocking mechanism described by Morales, Sánchez, and Quiralte (2012) for Dorcatherium naui.

Dorcabune anthracotheroides Pilgrim, 1910
(Includes *Db. hyaemoschoides* Pilgrim, 1915)

This is the largest species of *Dorcabune* and the largest tragulid in the Siwaliks (Figs. 21.2 and 21.3, Table 21.1). The wide range of measurements of the astragalus suggests it may have had a high degree of sexual dimorphism. Farooq et al. (2007b) and Khan et al. (2012) have referred approximately 8 Ma specimens from the Hasnot area to this species. It is difficult to be certain, but features of the illustrated upper molars are of *Dorcatherium majus*, while the metrics of the referred lower molars are close to those of *Db. nagrii* and *Dt. majus*.

Dorcabune nagrii Pilgrim, 1915
(Includes *Db. latidens* Pilgrim, 1915)

Smaller than *Dorcabune anthracotheroides* and overlapping with larger species of *Dorcatherium*, this species has a more extensive chronostratigraphic range than previously recorded (Figs. 21.2 and 21.3, Table 21.1).

Dorcabune sindiense Pilgrim, 1915
(Includes *Db. welcommi* Ginsburg, Morales, and Soria, 2001)

The smallest and oldest species, *Dorcabune sindiense* has molars that are lower crowned than those of *Db. nagrii*. *Db. sindiense* is present in the Vihowa and Chitarwata Formations, the uppermost Gaj Formation, and the lower third of the Manchar Formation. On the Potwar Plateau specimens have been found in the Kamlial and lowest Chinji Formations. Other than for its slightly smaller size, *Dorcabune welcommi*, which was based on a mandible from the Vihowa Formation near Dera Bugti, cannot be differentiated from *Dorcabune sindiense*.

The oldest specimen in the YGSP collection is approximately 18.7 Ma, but Antoine et al. (2013, Fig. 16.4B) suggest a first appearance closer to 20 Ma. (Note that in their figure Chron C6 is mislabeled as C5E.)

Dorcabune unnamed species?

Although clearly belonging to *Dorcabune*, specimens from Early and Middle Miocene sites intermediate in size between *Db. sindiense* and *Db. anthracotheroides* (Figs. 21.2 and 21.3) may belong to an unrecognized species. If different, this species would be broadly contemporaneous with *Db. sindiense* (Table 21.1).

Dorcatherium Kaup, 1833

There is a general consensus that *Dorcatherium* needs revision with a sharper definition of the genus that is not compromised by inclusion of species belonging to other lineages (Rössner 2007; Sánchez et al. 2010; Rössner and Heissig 2013; Aiglstorfer, Rössner, and Böhme 2014, Mennecart et al. 2018; Roussiakis 2022; Sanchez et al. 2022). Among Siwalik tragulids four selenodont species have historically been referred to *Dorcatherium*, although as discussed in the following, *Dorcatherium nagrii* and *Dorcatherium minimus* and perhaps *Afrotragulus akhari* are best referred to *Siamotragulus*. This leaves *Dorcatherium majus* and *Dorcatherium minus* as potential species of *Dorcatherium*. However, these two plus *Dorcatherium dehmi* (Guzmán-Sandoval and Rössner, 2019) and two species assigned to *Afrotragulus* by Sanchez et al. (2022) belong to a complex assemblage that could include as many as nine distinct species, as can be seen in a plot of the medial length of the astragalus against time with separation into clusters that are likely to be different species (Fig. 21.3). Generally, while the most visible distinctions between the putative *Dorcatherium* species reflect differences in size and proportions, examination of dental material to some extent supports these clusters, although I do not discuss dental characters here. For consistency, I use the same designations for these clusters (referred to as "Forms") as in Barry (2014).

The species of the Siwalik *Dorcatherium* complex partially overlap those of *Dorcabune* in size but possess molars that are more selenodont and higher crowned. Other distinctive features, all of which are variable and affected by wear, include the following: the post-protocrista of the upper molars has a single, posteriorly directed limb, which in some species curves toward the center of the tooth; the mesostyles are less bulbous

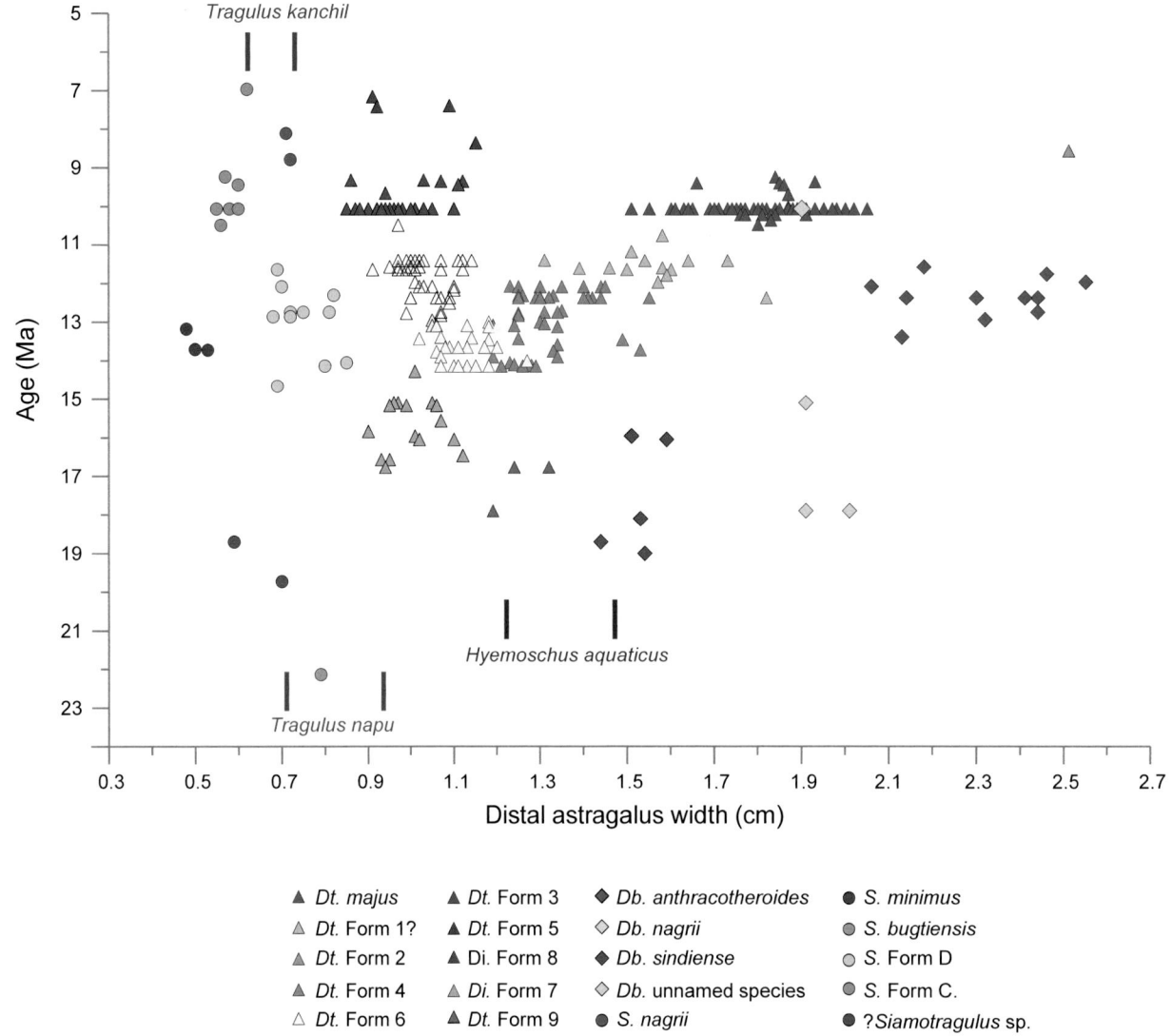

Figure 21.2. Plot of the distal width of the astragalus for 20 nominal species against time. For comparison the minimum and maximum values of small samples of three extant species are shown as well.

than in *Dorcabune* and the lingual and labial faces of both the paracone and especially the metacone are not as inflated and have weaker ribs, while the postmetacrista is slightly inflected labially.

In the mandible p1 is present with a single root. The p2 is bicuspid and shorter than p3, with a large blade-like premesocristid and a posterolabial conid with a lingually directed posterolingual cristid and posteriorly directed posterolabial cristid. The two cristids together flank a shallow posterolingual cleft that widens posteriorly. The p3 is tricuspid, with a trenchant premesoconulid, a mesoconid with in some species a posterior cleft flanked by salient labial and lingual postmesoconid cristids, and a posterolabial conid with a short posterolingual cristid and a longer curved posterolabial cristid separated from a small transverse stylid. The p4 is also relatively long (about 90% the length of p3), with the posterior of the mesoconid having a deep posterior cleft bounded by lingual and labial cris-

tids. Both the p3 and p4 are highly variable in morphological details within species. In contrast to *Dorcabune*, in these species it is the labial not the lingual postmesoconid cristid of the p3 that is aligned with the preposterolabial cristid.

The lower molars are narrow relative to their length. The labial faces of the metaconid and entoconid are less inflated than in *Dorcabune*, while the metaconid has a strong, trenchant premetacristid, and the entoconid has trenchant pre- and postentocristids.

Postcranially, the distal tibia and fibula are fused as are the ecto-mesocuneiform and cuboid-navicular. The astragalus has a distinct notch between the lateral sides of the offset proximal and distal trochleae and a large prominence for the superior lateral calcaneal articulation. Metatarsals III and IV are fused along much of their lengths, but retain separate medullary cavities and a rectangular cross section at midshaft. The proximal portion of metatarsal II is frequently fused to the third,

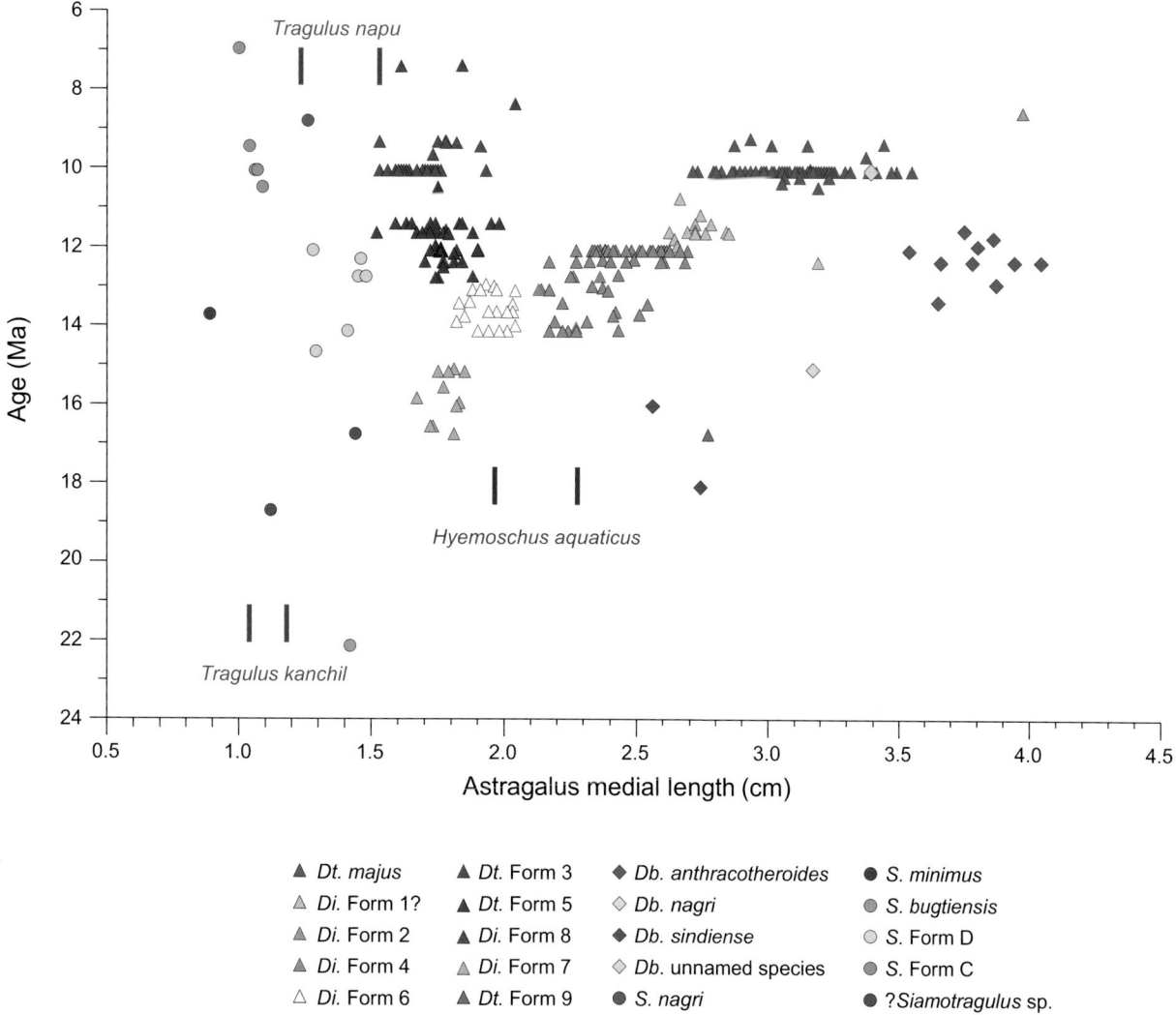

Figure 21.3. Plot of the medial length of the astragalus against time for the same species as in *Figure 21.2*. Also has for comparison minimum and maximum values for three extant species.

but metatarsal V is only rarely fused to the fourth. Metacarpals III and IV have the interlocking mechanism described by Morales, Sánchez, and Quiralte (2012) for *Dorcatherium naui*.

Differences in skeletal proportions suggest that taxa of the Siwalik *Dorcatherium* complex fall into two groups. One group, exemplified by Form 5, has short and stouter limb elements and the other, exemplified by Form 6, relatively long and slender limbs. This is most easily seen in the dimensions of intact metatarsals (Barry 2014, Fig. 1 and Textfigure 6), but is likely to be expressed in other limb elements as well. While the astragalar dimensions give good separation for the Late Miocene, the Middle Miocene clusters are not as distinct from one another (Figs. 21.2 and 21.3).

It is difficult to determine the stratigraphic ranges of several of the time successive species of *Dorcatherium*. This is because although samples are metrically distinct, in the absence of characteristic morphologies individual specimens can be difficult to place, especially those that occur temporally between

clusters. An example is the specimen of *Dt.* Form 7 at 1.01 cm and 14.3 Ma on Figure 21.2.

There are two commonly used names available for Siwalik species, *Dorcatherium majus* Lydekker, 1876, and *Dorcatherium minus* Lydekker, 1876. Because of its large size, it is relatively unambiguous as to what material should be referred to *Dt. majus*. In the case of *Dt. minus*, however, it is difficult to determine to which of several forms the name should be applied. The single, small maxilla with M2 and M3 on which Lydekker based his description of *Dt. minus* (#1301, later renumbered B 195) was acquired by W. Theobald between 1873 and 1876 in the vicinity of the village of Hasnot (or Asnot) (Lydekker 1885). Later Pilgrim (1910a, 1910b) also listed the species as coming from "Chenji," and Colbert (1935) subsequently referred 18 additional specimens to Lydekker's species, all but two of which were from the Chinji Formation (~ca. 14 to 11 Ma). Since then, other authors have followed in referring material from the Middle Miocene as well as the Late Miocene to *Dt. minus* (e.g. Barry 1995; Khan and Akhtar 2005, 2013; Khan et al. 2005, 2006, 2012, 2017; Farooq et al. 2007a; Iqbal et al. 2011; Batool et al. 2014; Nanda 2015; Samiullah et al. 2015; Sehgal 2015).

While there are exposures of the Chinji Formation six to seven kilometers to the west of Hasnot, they are sparsely fossiliferous compared to the very fossiliferous Late Miocene exposures between the villages of Padhri and Hasnot and extending several kilometers to the south. These younger exposures range in age between 8.0 and 6.4 Ma and in the nineteenth century were probably more accessible to Theobald and others than the Chinji Formation exposures to the west. I believe these Late Miocene exposures are most likely the source of Lydekker's type. Among the forms discussed in the following, five (Forms 3, 5, 6, 7, and 8) are of an appropriate size for Dt. minus. However, three (Forms 5, 6, and 7) are restricted to horizons older than 10.5 Ma and on those grounds can be ruled out, assuming that Lydekker's holotype was Late Miocene. The remaining two forms (Forms 3 and 8) are also of an appropriate size and known from Late Miocene localities near Hasnot and thus likely candidates. At present, however, it is uncertain which, if either, of these two forms corresponds to Dt. minus.

Group A—slender limb forms.

There are five forms in this group, all of which are larger than species of Group B (Fig. 21.2). There is some variation in premolar structure that may separate the species.

Dorcatherium majus Lydekker, 1876

Listed as Dorcatherium Form 1 in Barry (2014), this is a very large and ubiquitous species (Table 21.1), known from nearly 17% of the localities in its stratigraphic range.

The holotype came from near Khushalgarh, just west of the Indus along the Kohat road. Its age is not known, but other species cited as from the area (Lydekker 1885) indicate there is probably a full range of Middle and Late Miocene age sediments present.

Dorcatherium Form 1?

This was referred to as "Dorcatherium cf. majus" in Barry et al. (2002). It is intermediate in size between Dt. majus and Dt. Form 4 (Table 21.1, Figs. 21.2 and 21.3). Although the referred specimens are all older than Dt. majus, they need further analysis to determine if this is truly a separate species. It is larger than Dt. Form 4, which, taken together with the temporal overlap (Table 21.1), indicates it is a species separate from Dt. Form 4. Like Dt. majus it had a long, slender metatarsal.

Dorcatherium Form 2

This is a gigantic species, much larger than Dt. majus and perhaps even larger than Dorcabune anthracotheroides (Table 21.1, Figs. 21.2 and 21.3). It is currently known for certain only from associated postcranials from a single 8.6 Ma site. A proximal metatarsal places this species in Group A.

Three high crowned Dt. majus third molars that are younger than 8 Ma may belong to this species (Fig. 21.4a). However, the lengths of these three molars are only 2.28, 2.34, and 2.60 cm and within the range of third molars of other Dt. majus (Fig. 21.4b). The ratio of average third molar length to average astragalus distal width in Dt. majus gives a predicted third molar length greater than 3.3 cm for the Dt. Form 2 astragalus.

Dorcatherium Form 4

This is the most common Dorcatherium species in the Middle Miocene Siwaliks, comprising about 45% of all Dorcatherium specimens between 14.3 and 10.6 Ma. Intact slender metatarsals ally it with the larger Dt. Form 1? and Dt. majus.

This form has been confused with Dt. minus, but the fossils in the AMHH collections identified as Dt. minus by Colbert (1935) belong to four different forms (Table 21.2), and this is probably true of other collections. Two (Forms 5 and 6) are small and of an appropriate size for Lydekker's holotype maxilla, but much of Colbert's material represents a species larger and probably older than Lydekker's Dt. minus holotype. This includes eight specimens that I place in Dt. Form 4 (Table 21.2). Five of those eight were referred to Dt. dehmi, Dt. naui, or Dt. minus by Guzmán-Sandoval and Rössner (2019) (Table 21.2). There are in addition 13 other specimens in the AMNH collections not discussed by Colbert (1935), which I place in Dt. Form 4. Guzmán-Sandoval and Rössner (2019) referred two of these to Dt. minus (Table 21.2).

Guzmán-Sandoval and Rössner (2019) attributed 20 Siwalik specimens in the Bavarian State Collection of Palaeontology and Geology and AMNH collections to the European species Dt. naui, which they differentiated from other Siwalik species on the basis of size, crown height, and the morphology of the dp4. All of the specimens apparently are from the Chinji Formation. With m3 lengths between 1.75 and 2.01 cm, the Siwalik specimens attributed to Dt. naui are much the same size as Dt. Form 4, but larger than either Dt. Form 5 or Dt. Form 6. (Fig. 21.5a). Crown heights in all these forms show a wide range of values, with no consistent pattern, although by comparison to some Dt. majus they are low crowned (Figs. 21.4a and 21.5b). A more detailed comparison of the Dt. Form 4 and the European material of Dt. naui is needed to resolve the issue, but it is possible and perhaps even likely that Dt. Form 4 is the same species as Dt. naui.

Dorcatherium Form 6

This is the smallest of the Group A Dorcatherium species. It overlaps slightly with Dt. Form 4 in the width of the distal astragalus (Fig. 21.2) and body mass (Table 21.1), but the medial length of the astragalus (Fig. 21.3) gives good separation, even for specimens that overlap in their distal dimensions. Two intact slender metatarsals ally this form with the larger members of Group A.

Dt. Form 6 includes three AMNH specimens identified as Dt. minus by Colbert (1935). These were referred to Dt. naui, Dt. minus, and Dt. dehmi by Guzmán-Sandoval and Rössner (2019). Six additional AMNH specimens not mentioned by Colbert (1935) belong to this form, two of which were identified as Dt. dehmi and Dt. naui by the same authors (Table 21.2).

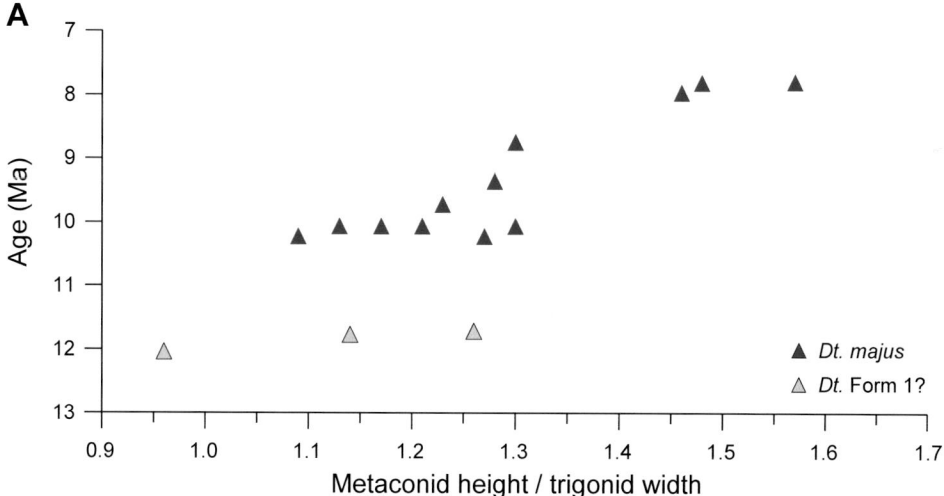

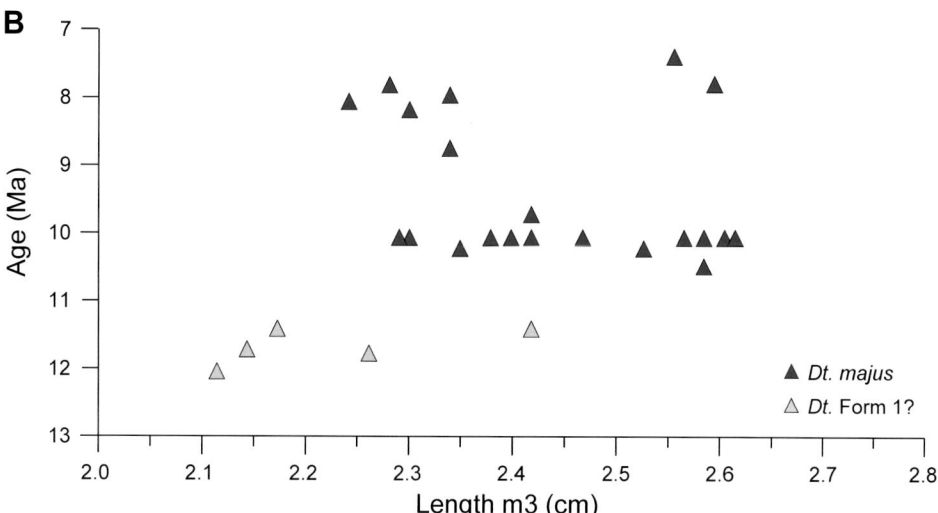

Figure 21.4. Plots of A. index of hypsodonty (metaconid height/ trigonid width) against time for unworn or lightly worn lower third molars of *Dt. majus* and *Dt.* Form 1; B. length of lower third molars against time of the same species.

Group B—stout limb forms

This group has three members, all smaller than those of Group A. Collectively Group B forms overlap in time with Group A, although *Dt.* Form 8 persists to a much younger age (Table 21.1)

Dorcatherium Form 3

This is a common tragulid in the Late Miocene and could correspond to Lydekker's *Dorcatherium minus.* (But see the following discussion of *Dt.* Form 8.) In Barry et al. (2002) it was referred to as "*Dorcatherium* Y311 species." A complete short, stout metatarsal allies it with the Group B forms.

This and the older *Dt.* Form 5 are the most difficult members of the Siwalik *Dorcatherium* complex to separate. The mean of the distal width of the astragalus of *Dt.* Form 3 is only about 6% smaller than the mean of *Dt.* Form 5 (16% when expressed as body mass), while the individual measurements show considerable overlap (Figs. 21.2 and 21.3). Nevertheless, the distribution of the values of the two groups is quite differ-

ent (Fig. 21.6) and a Mann-Whitney U Test indicates the two samples are unlikely to have been drawn from a single population (Z-Score: −4.72884, with a $p < 0.00001$), strengthening a claim for two separate species.

Dorcatherium Form 5

In Barry et al. (2002) this was referred to as "*Dorcatherium* Y259 species" and it was common in the Middle Miocene. *Dt.* Form 5 is well separated from the contemporaneous *Dt.* Form 4 (Figs. 21.2 and 21.3), but has a mean distal astragalar width only about 8% smaller (20% in terms of body mass) than the older *Dt.* Form 6. However, an intact 12.4 Ma metatarsal indicates it is in Group B, not Group A.

Three AMNH specimens identified as *Dt. minus* by Colbert (1935) and three identified as *Dt. dehmi* by Guzmán-Sandoval and Rössner (2019) are included in this form (Table 21.2). With m3 lengths from 1.39 to 1.56 cm (Guzmán-Sandoval and Rössner 2019, Table 7), the material attributed to *Dt. dehmi* is approximately the same size as this form (Fig. 21.5a), but the holotype

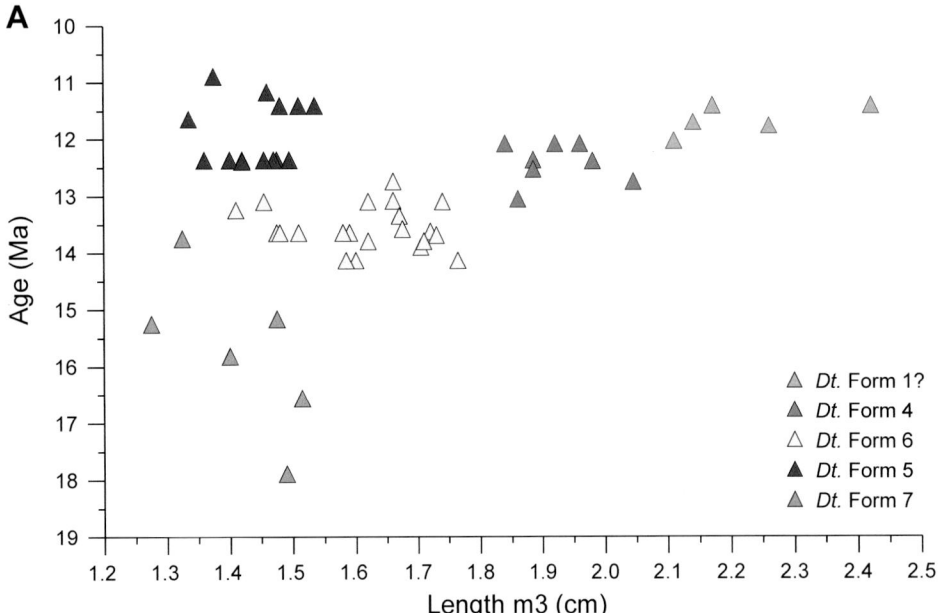

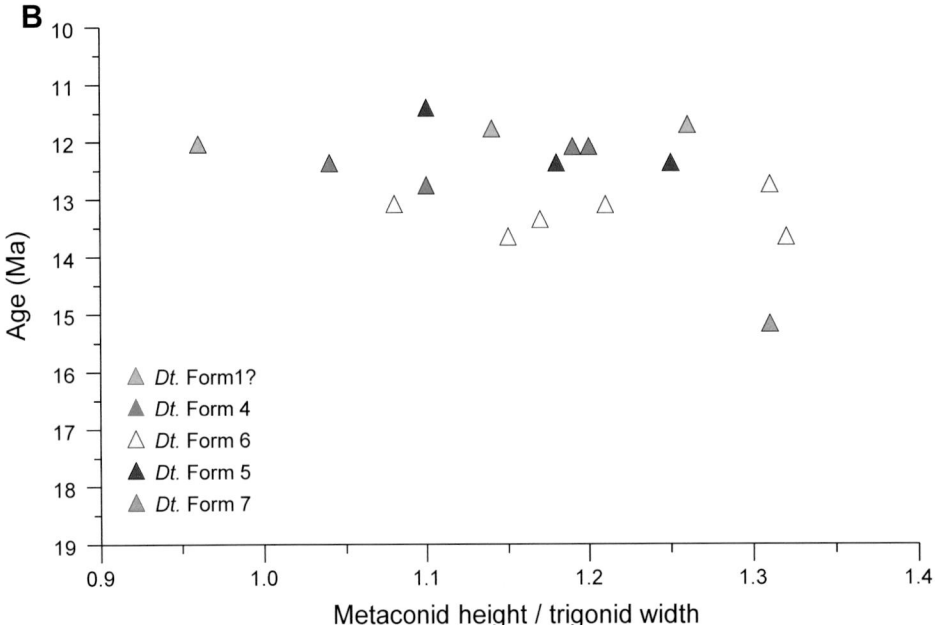

Figure 21.5. Plots of A. length of lower third molar against time for five species of *Dorcatherium*; B. index of hypsodonty (metaconid height/ trigonid width) against time for unworn or lightly worn lower third molars of the same five species.

of *Dt. dehmi* is much younger. (See the following discussion of *Dt.* Form 8.)

Dorcatherium Form 8

This form is also common in the Late Miocene. In Barry et al. (2002) it was referred to as *"Dorcatherium* Y373 species." Its placement in Group B is based on a relatively short and stout metacarpal IV and an intact radius.

The lower molars of *Dt.* Form 8 are consistently larger than those of *Dt.* Form 3 and also higher crowned as shown by sparse data (Fig. 21.7a, b). The m3 dimensions are distinct from those of *Dt.* Form 3, but measurements on the astragalus (Figs.

21.2 and 21.3) suggest a more complex situation, with some small specimens co-occurring with larger dentitions. A more complete analysis of the postcranial material may clarify the issues. One AMNH specimen of this form (AMNH 29887) was attributed to *Dt. dehmi* by Guzmán-Sandoval and Rössner (2019). The specimen is from the Hasnot area and is approximately 7.4 to 7.3 Ma (Table 21.2).

As noted, if the holotype of *Dt. minus* (GSI B 195) is as young as I propose, then there are only two forms recognized here that are of both an appropriate age and size to potentially be the same as Lydekker's species. The holotype of *Dt. minus* is a maxilla with two molars, thought to be M2 and M3 (Lydekker 1876;

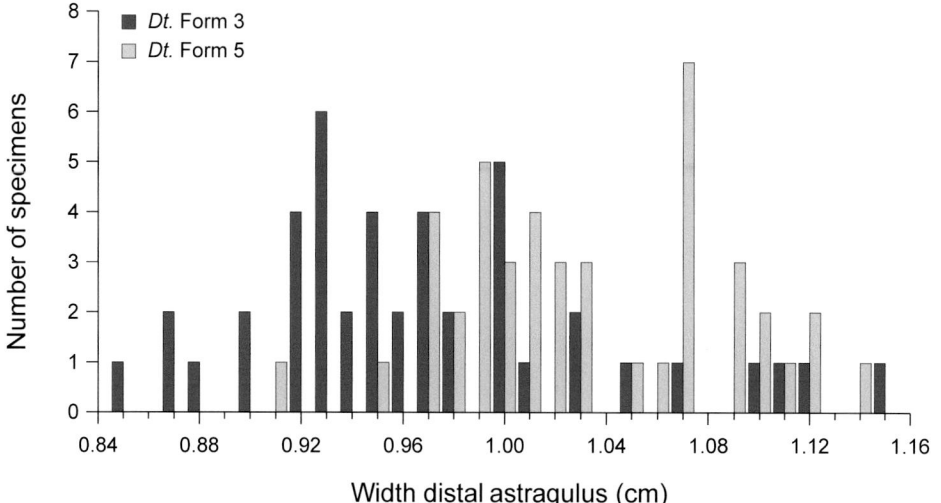

Figure 21.6. Histogram of the frequency of values for width of distal astragalus for *Dt.* Form 3 and *Dt.* Form 5.

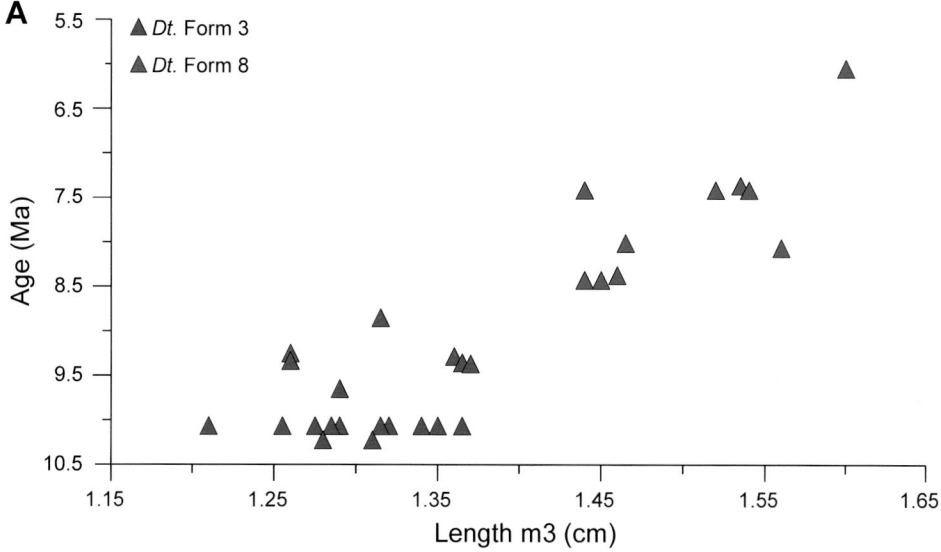

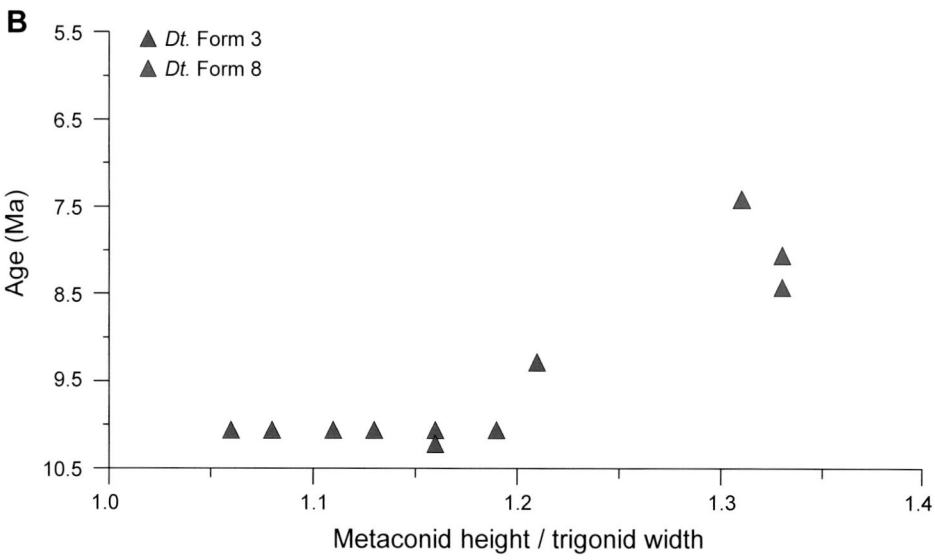

Figure 21.7. Plots of: A. length of lower third molar against time for *Dt.* Form 3 and *Dt.* Form 8; B. index of hypsodonty (metaconid height/ trigonid width) against time for unworn or lightly worn lower third molars of the same two species.

Pilgrim 1915). Although the forms discussed here are recognized on the basis of postcranials and not dentitions, measurements of molars assigned to *Dt.* Form 8 (Fig. 21.8a) show that the M3 of the B 195 holotype of *Dt. minus* is in the range of *Dt.* Form 8 and much larger than those of *Dt.* Form 3. Consequently, I think it is most likely that *Dt.* Form 8 is the same species as *Dt. minus*. Such an equivalence would restrict the occurrence of *Dt. minus* to the Late Miocene and possibly to less than 9 Ma.

Among the recognized Late Miocene forms, the dentition of *Dt.* Form 8 is characterized not only by greater crown height, but also by its size. The holotype maxilla of *Dt. dehmi* (SNSB-BSPG 1956 II 2615) was said to come from Parlewala 3 (Guzmán-Sandoval and Rössner 2019), a site in the Dhok Pathan Formation "about 5 km WSW" of the Dhok Pathan Resthouse (letter from R. Dehm to D. Pilbeam, September 11, 1989). The Parlewala locality is in Geomagnetic Polarity Chron 4n.2n (Barry et al. 2002) and between 8.1 and 7.8 Ma, which is within the time range of *Dt.* Form 8. Its age and size (Fig. 21.8b) thus suggest that *Dt.* Form 8 and the holotype of *Dt. dehmi* are the same species. If so, then the holotype of *Dt. dehmi* is also likely to be the same species as *Dt. minus*. Other AMNH specimens

attributed to *Dt. dehmi*, however, are older and likely to represent at least three other forms (Table 21.2).

Sanchez et al. (2022) recognized a new species of tragulid (*Afrotragulus moralesi*), based on a single m3 from a locality near Hasnot that was thought to be approximately 6.5 Ma. Although the body weight estimate is much greater than that estimated for *Dt.* Form 8, the length of the holotype (1.54 cm) plus its presumed age suggest it is the same species as *Dt.* Form 8 (Fig. 21.7a). If so, then it is also likely to be the same species as *Dt. minus*, which might therefore belong to *Afrotragulus*.

Group unknown.

Two forms cannot be placed in either Group A or Group B, as neither complete metatarsals nor other critical limb elements are known for them. Found only in the poorly fossiliferous Kamlial Formation, both are older than any of the Group A or Group B forms.

Dorcatherium Form 7

Although a very poorly known form, given its age it is of considerable interest. There are no intact metatarsals or other limb

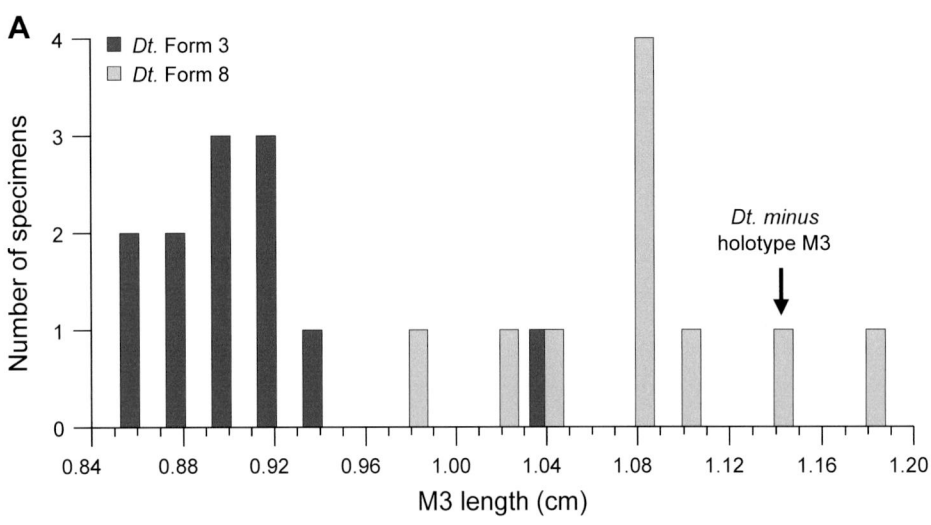

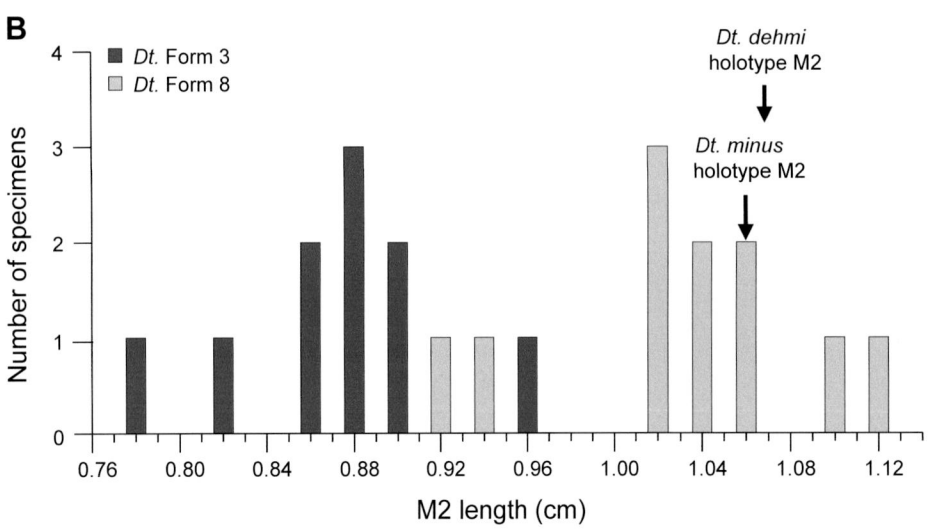

Figure 21.8. Histograms of A. length of upper third molar of *Dt.* Form 3 and *Dt.* Form 8; B. length of upper second molar of the same two species. In both diagrams the lengths of the holotypes of *Dt. minus* (GSI D 195) and *Dt. dehmi* (SNSB-BSPG 1956 II 2615) are shown by arrows. The respective values of the holotypes are B 195 M2 (1.06 cm), M3 (1.14 cm), and SNSB-BSPG 1956 II 2615 M2 (1.07 cm).

Table 21.2. American Museum of Natural History Fossil Tragulid Specimens

Specimen Number	This Study	Colbert (1935)	Guzmán-Sandoval and Rössner (2019)	Element	Locality	Estimated Age (Ma)
AMNH 19302	*Dt. majus*	*Dt. majus*	*Dt. majus*	Upper molar	?	?
AMNH 19304	*Dt. majus*	*Dt. majus*	*Dt. majus*	M1–M2	?	?
AMNH 19369	*Dt. majus*	*Dt. majus*	*Dt. majus*	m2–m3	?	?
AMNH 19520	*Dt. majus*	*Dt. majus*	*Dt. majus*	m1–m2	15	ca. 8–7
AMNH 19524	*Dt. majus*	*Dt. majus*	*Dt. majus*	p4-m2	115	ca. 9.4–8.3
AMNH 19939	*Dt. majus*	*Dt. majus*	*Dt. majus*	m3	9–13	ca. 11.2–10.8
AMNH 19623-2	*Dt.* Form 1?			Lower molar	144	?
AMNH 19307	*Dt.* Form 4	*Dt. minus*	*Dt. minus*	m2–m3	?	?
AMNH 19309	*Dt.* Form 4	*Dt. minus*		Lower molar	?	?
AMNH 19313	*Dt.* Form 4	*Dt. minus*	*Dt. naui*	M1–M3	?	?
AMNH 19367	*Dt.* Form 4	*Dt. minus*	*Dt. dehmi*	m1–m2	51	13.8–13.7
AMNH 19374	*Dt.?*Form 4	*Dt. minus*		Lower molar	?	?
AMNH 19517	*Dt.* Form 4	*Dt. minus*	*Dt. naui*	dP4–M1	108	ca. 11.2–10.8
AMNH 19594-1	*Dt.* Form 4			Lower molar	52	14.1–13.9
AMNH 19594-2	*Dt.* Form 4			m2	52	14.1–13.9
AMNH 19609	*Dt.* Form 4	*Dt. minus*	*Dt. naui*	p3–m3	56	13.2–12.9
AMNH 19623-3	*Dt.* Form 4			Upper molar	?	?
AMNH 19644	*Dt.* Form 4			M1–M2	?	?
AMNH 19833	*Dt.?*Form 4	*Dt. minus*		Lower molar	159	?
AMNH 29953	*Dt.* Form 4			m3	108	ca. 11.2–10.8
AMNH 32587-1	*Dt.* Form 4			dp3	?	?
AMNH 39301	*Dt.?*Form 4			Upper molar	?	?
AMNH 39302	*Dt.* Form 4		*Dt. minus*	m1–m2	124	ca. 13.7–12.9
AMNH 39303-1	*Dt.* Form 4			Upper molar	52	14.1–13.9
AMNH 39303-2	*Dt.* Form 4		*Dt. minus*	dP3–M1	52	14.1–13.9
AMNH 39303-3	*Dt.* Form 4			dP2–M1	52	14.1–13.9
AMNH 39329	*Dt.* Form 4			Lower molar	126	13.9–13.7
AMNH 39516	*Dt.* Form 4			m3	?	?
AMNH 19365	*Dt.* Form 6	*Dt. minus*	*Dt. naui*	m2–m3	?	?
AMNH 19366	*Dt.* Form 6	*Dt. minus*	*Dt. minus*	m2–m3	?	?
AMNH 19368	*Dt.* Form 6	*Dt. minus*	*Dt. dehmi*	m1–m3	58	14.2–14.1
AMNH 19611	*Dt.* Form 6			Lower molar	?	?
AMNH 29970	*Dt.* Form 6			p4–m1	51	13.8–13.7
AMNH 32587-2	*Dt.* Form 6			m3	?	?
AMNH 32587-3	*Dt.* Form 6			Lower molar	?	?
AMNH 32588	*Dt.* Form 6		*Dt. dehmi*	m3	?	?
AMNH 32742	*Dt.* Form 6		*Dt. naui*	m2–m3	?	?
AMNH 19305	*Dt.* Form 5	*Dt. minus*		m1–m3	?	?
AMNH 19306	*Dt.* Form 5	*Dt.* sp	*Dt. dehmi*	M1–M3	59	11.3–11.0
AMNH 19623-1	*Dt.* Form 5			Lower molar	?	?
AMNH 20043	*Dt.* Form 5		*Dt. dehmi*	m3	59	11.3–11.0
AMNH 29855	*Dt.* Form 5	*Dt. minus*	*Dt. dehmi*	M2–M3	?	?
AMNH 29856	*Dt.* Form 5	*Dt. minus*	*Dt. minus*	M1–M3	124	ca. 13.7–12.9
AMNH 29887	*Dt.* Form 8	*Dt.* sp	*Dt. dehmi*	dp4–m1	113	ca. 7.4–7.3
AMNH 32589	*Dt.* Form 8			M2–M3	135	6.8–6.5
AMNH 19310	*Dt.* Form 7	*Dt. minus*	*Dt. dehmi*	p4–m3	51	13.8–13.7
AMNH 29854	*Dt.* Form 7	*Dt.* sp	*Dt. guntianum*	m2–m3	?	?
AMNH 19613	*S.* Form D	*Dt.* sp	*Dt. nagri*	m1–m3	123	12.0–11.8
AMNH 39511	*V. daviesi*			P2–M3	?	?

elements that would place it in either Group A or B. It is present at the base of the Kamlial Formation on the Potwar Plateau (ca.17.9 Ma), while a tragulid of similar size has been found in the Manchars of Sind and in the Vihowa and Chitarwata Formations. The oldest of these occurrences is about 22.1 Ma. *Dt.* Form 7 is slightly smaller than *Dt.* Form 5, making it the smallest member of the Siwalik Middle Miocene *Dorcatherium* complex. Together with *Siamotragulus bugtiensis*, the oldest *Dt.* Form

7 specimens co-occur with the primitive pecoran *Bugtimeryx pilgrimi* Ginsburg, Morales, and Soria, 2001.

One specimen of *Dt.* Form 7 in the AMNH was referred by Guzmán-Sandoval and Rössner (2019) to *Dt. dehmi* and a second to the European species *Dt. guntianum* (Table 21.2). With m3 lengths between 1.27 and 1.45 cm (Guzmán-Sandoval and Rössner 2019, Table 8), the Siwalik specimens attributed to *Dt. guntianum* are the same size as those of *Dt.* Form 7 (Fig. 21.5a).

In western Europe *Dt. guntianum* occurs between about 17.5 and 13.3 Ma (Aiglstorfer, Rössner, and Böhme 2014). This overlaps with the range of *Dt.* Form 7, and I think it is likely that *Dt.* Form 7 is the same species as *Dt. guntianum*.

Sanchez et al (2022) based a new species of *Afrotragulus* (*Af. megalomilos*) on the AMNH specimen (AMNH 19319) Guzmán-Sandoval and Rössner (2019) had referred to *Dt. dehmi*. The estimated age (13.8–13.7 Ma) and size of the m3 (1.39 cm in Guzmán-Sandoval and Rössner [2019], 1.467 cm in Sanchez et al. [2022], and 1.33 cm from my measurements on the original) suggest it is the same species as *Dt.* Form 7 (Fig. 21.5a where AMNH 19319 is the youngest specimen).

Dorcatherium Form 9

A very small number of postcranials and fragments of molars from the Kamlial Formation establish the presence of a second Early Miocene selenodont tragulid. It is larger than *Dt.* Form 7 and nearly the size of *Dt.* Form 4 (Fig. 21.2). There is, however, a substantial temporal gap between Forms 9 and 4 and for that reason I have separated them. A 17.9 million year old locality at the base of the Kamlial Formation with both Forms 7 and 9 documents the oldest instance of two co-occurring selenodont *Dorcatherium* species in the Siwaliks. There are so few specimens in the collections that its stratigraphic range is likely to be more extensive than presently known.

Siamotragulus Thomas, Ginsburg, Hintong, and Suteethorn, 1990

Two very small taxa among described Siwalik species were initially attributed to *Dorcatherium* as *Dt. nagrii* from the Late Miocene and *Dt. minimus* from the Middle Miocene. Both species are represented in the YGSP collections, along with *Siamotragulus bugtiensis* from the older Chitarwata Formation and, tentatively, two additional undescribed forms (referred to here as *Siamotragulus* Form C and *Siamotragulus* Form D). The three named species and the two undescribed forms can be attributed to *Siamotragulus* on the basis of dental and postcranial attributes. Distinctive features include (1) a very slender and compressed p3 with a small, barely deflected parastylid, distinct posterolabial conid, and posteriorly directed posterolabial cristid; (2) a slender p4 with long anterior cristid, small parastylid, and a narrow posterior cleft bounded by lingual and lateral postcristids; (3) lower molars with intact and salient "M" structure and upper molars with highly variable cingulum; (4) slender, elongated limbs with in some cases the distal tibia and fibular malleolus fused; (5) metatarsals III and IV slender and fully fused with a single medullary cavity; and (6) metacarpals III and IV fused, but with separate medullary cavities.

The coexistence on the Potwar Plateau of pairs of a smaller and slightly larger species of *Siamotragulus* is similar to the coexistence of *Tragulus napu* with smaller species of *Tragulus* (*T. javanicus* or *T. kanchil*) over a wide geographic area in southeast Asia and the islands of Indonesia (Meijaard and Groves 2004).

Siamotragulus nagrii (Prasad, 1968)

This is a large species of *Siamotragulus* (Fig. 21.1, Table 21.1). Mandibles and associated hind limb elements of one individual combine a slender and simple p4 with fully fused metatarsals III and IV. The lower p3s of other individuals are very slender, with a large posterolabial conid and a posteriorly directed posterolabial cristid. This species is restricted to the Late Miocene.

Siamotragulus minimus (West, 1980)

West (1980) based this species on an upper M3, but the hypodigm included a tiny astragalus. The species is smaller than extant species of *Tragulus* such as *T. javanicus* and *T. kanchil* and is the smallest of the Siwalik *Siamotragulus* species (Fig. 21.1, Table 21.1). The oldest remains of this species from the Potwar are 15.2 Ma, and it appears to be restricted to the Middle Miocene.

Sanchez et al (2022) described a very small species of *Afrotragulus* (*Af. akhtari*) from the Middle Miocene Chinji Formation. The species is based on a mandible with m2 and m3, two additional mandible fragments with m2 and m3, and an isolated M2. The mandible of *Siamotragulus minimus* is not yet known, but the *Af. akhtari* M2 is the size of the M2 on a 13.1 Ma maxilla referred to *S. minimus*.

Siamotragulus bugtiensis Ginsburg, Morales, and Soria, 2001

Ginsburg, Morales, and Soria (2001) described dental and postcranial remains of *Siamotragulus bugtiensis* from the upper member of the Chitarwata Formation near Dera Bugti. It is a large species for the genus (Fig. 21.2, Table 21.1). Specimens from the Chitarwata and Vihowa Formations at Zinda Pir can be assigned to this species as well. The species is so far known only from the Early Miocene

Siamotragulus Form D

The older of the unnamed *Siamotragulus* forms, *S.* Form D overlaps in time with *S. minimus,* but is larger, being approximately the size of *Tragulus napu*. A well-preserved proximal metacarpal complex shows that at least this species had fused MC III and MC IV, with separate medullary cavity. It is the size of *S. nagrii*, but separated from it by a nearly 2 million year gap (Table 21.1).

Siamotragulus Form C

This form is only slightly larger than *Siamotragulus minimus*. The two smallest species of *Siamotragulus* both have long stratigraphic ranges (Table 21.1), but do not overlap as neither is found in the highly fossiliferous localities around 11.4 Ma. For this reason I consider them to be potentially separate species.

Additional very small tragulids are known from older levels of the Kamlial, Vihowa, and Chitarwata Formations (from 16.6 to ca. 19.7 Ma). They are the size of Middle Miocene *Siamotragulus* (Fig. 21.2 as ?*Siamotragulus* sp.), and, taken together with *Siamotragulus bugtiensis*, they suggest that a complex of very small species was present in the Early Miocene. The available material, however, does not include skeletal or dental elements that would allow it to be unequivocally referred to *Siamotragulus*.

Tragulus Brisson, 1762
Tragulus sivalensis Lydekker, 1882

This species is especially problematic. It is known from only the holotype, an upper molar which is not particularly distinctive and is of uncertain age. Lydekker's specimen was said to come from near Hasnot (Lydekker 1884, 1885) and is therefore probably Late Miocene. Lydekker (1882, 1884) and subsequent authors did not publish measurements, but assuming the figure in Lydekker (1884) is accurate, approximate dimensions are length = 0.9 mm and width = 0.8 mm. This is within the range of values for upper molars of *S. nagrii*, which is known from the same area and time period and has a similar morphology.

Vishnumeryx Pilgrim, 1939
Vishnumeryx daviesi (Lydekker, 1886)

Both Lydekker (1886) and Pilgrim (1939) considered this species to be a bovid and in the case of Lydekker a species of the small four-horned antelope, *Tetracerus*. Pilgrim (1939) noted both the holotype and a referred mandible (M 16535) were from the Cautley collection and therefore collected before 1842. The locality is given as only "Upper Siwalik." The mandible is a small bovid, but the holotype maxilla and another maxilla in the AMNH (AMNH FM 39511) have dental characters of tragulids, including: (1) P3 long with bladed metastyle and large, isolated, crescent-shaped parastyle; (2) P4 triangular and weakly asymmetric with bladed metastyle and anterior crest of protocone not connected to the parastyle; and (3) upper molars with weak lingual cingula. In addition, three specimens in the YPM and one in the PMNH can tentatively be assigned to this species. The four are from localities near Tatrot and are 3.3 to 3.5 Ma.

The tragulid affinities of *Vishnumeryx daviesi* are clear, but whether *Vishnumeryx* should be recognized as a separate genus depends to some extent on revision of *Dorcatherium*. Nonetheless, the species is of special interest on two counts: first, its late occurrence in the Siwaliks (but see Gaur, Vasishat, and Sankhyan [1980], who cite a Pinjor occurrence for *Dorcatherium* sp. at ~1.5 Ma), and second, on the relatively high degree of hypsodonty. While not impressive by bovid standards, *Vishnumeryx daviesi* is still remarkably high crowned for a tragulid. Unfortunately, there are no known mandibular molars suitable to estimate a hypsodonty index. The species is approximately the same size as *Hyemoschus aquaticus*.

DISCUSSION

Siwalik tragulids have generally been thought to have long temporal ranges, as have European and African species (Geraads 2010; Aiglstorfer, Rössner, and Böhme 2014). Nevertheless, with the exceptions of *Dorcabune sindiense* and *Dorcatherium* Form 7,

the species I recognize have much shorter durations, being between 0.8 and 4.6 million years with an average duration just over 3 million years (Table 21.1). Setting aside *Dorcabune sindiense* and *Dorcatherium* Form 7, which are from the poorly fossiliferous Kamlial Formation, there is no relationship between the age and duration of a species. In both number of specimens and number of species, the tragulid assemblages are dominated by *Dorcatherium*, which, allowing for taphonomic influences, possibly reflects the importance of those species in the living communities.

Within the three multi-species Siwalik genera, there are temporally successive as well as contemporaneous congeneric species. Whether any of the sets of successive species form ancestor-descendant pairs is difficult to determine because of the homogeneous morphology of the genera. Furthermore, given the presence of related species outside the Indian subcontinent, such an analysis will have to be part of a more comprehensive study of the individual genera or the family as a whole.

Some nine species of Miocene tragulids have been recognized in Europe (Rössner 2007) and six in Africa (Geraads 2010, Sanchez et al. 2014, Musalizi et al 2023), although there are questions about the validity of some species (Aiglstorfer, Rössner, and Böhme 2014). While the relationships of Siwalik tragulids with African and European species have been partially investigated (Guzmán-Sandoval and Rössner 2019), answering the question as to whether any Siwalik species is the same as an African or European species is complicated, because the species have few morphological differences and there is considerable intraspecific variation. Moreover, the characterization of several Siwalik species has been compromised by inclusion of specimens belonging to other species.

Dorcabune is frequently cited as restricted to southern Asia, but it is now known from China (Han 1986) and Vietnam (Prieto et al. 2017), extending the geographic range and the temporal range to between 7.1 and 6.3 Ma (Dong and Qi 2013; dates adjusted to timescale of Gradstein et al. 2012) and even into the Early Pleistocene (Han 1974, 1986). *Dorcabune liuchengense* Han, 1974, and *Dorcabune progressus* (Yan, 1978) are both the size of the Late Miocene *Dorcabune nagrii*.

European tragulids are customarily placed in *Dorcatherium*, with species characterized as either selenodont or bunoselenodont (Rössner 2007). The latter include *Dorcatherium crassum* (Lartet, 1851) and *Dorcatherium vindebonense* von Meyer, 1846, from the Early and Middle Miocene and *Dorcatherium peneckei* (Hofmann, 1893) from the Middle Miocene. These three bunoselenodont species have similarities to *Dorcabune* (for example, see Mottl 1961, Plates 4–7; Fahlbusch 1985, Plates 1 and 2) and are potentially species of *Dorcabune*, although they need closer study. *Dorcatherium peneckei* is about the size of Late Miocene *Dorcabune nagrii* and the unnamed Middle Miocene species, although some specimens approach *Dorcabune anthracotheroides* in size. *Dorcatherium crassum* and the slightly larger *Dorcatherium vindebonense* are much the same size as *Dorcabune sindiense*.

In addition van der Made (1997) and Koufus (2021) have cited a possible *Dorcabune anthracotheroides* from Crete and Greece. The astragalus from Crete, however, lacks the characteristic morphology of *Dorcabune* (Barry 2014) and metrically is comparable to specimens of *Dorcatherium majus*. The dental material might belong to a large *Dorcabune*, although the lower molars of the various Siwalik species do not have strong labial cingulids.

Until recently, *Dorcabune* has not been recognized in Africa (Geraads 2010), although dentitions of *Dorcatherium chappuisi* Arambourg, 1933 from Rusinga (~18–16.6 Ma) and perhaps Fort Ternan (~14–13.4 Ma), which I have examined, have distinctive characters of *Dorcabune*, as do some specimens of *Dorcatherium iririensis* Pickford, 2002; a conclusion supported by inclusion of both species in *Dorcabune* by Musalizi et al. (2023).

In Europe there are two common selenodont species of *Dorcatherium* (*Dt. naui* Kaup and Scholl, 1834, and *Dt. guntianum* von Meyer, 1846), as well as two rarer species (*Dt. jourdani* Depéret, 1887, and *Dt. puyhauberti* Arambourg and Piveteau, 1929). *Dt. naui* and *Dt. guntianum* have stratigraphic durations of about 3.5 and 4.2 million years, respectively, with the Early and Middle Miocene *Dt. guntianum* being succeeded by the Middle and Late Miocene *Dt. naui* (Aiglstorfer, Rössner, and Böhme 2014). The Late Miocene occurrences of *Dt. jourdani* and *Dt. puyhauberti* are more sparse and their durations are shorter; there being only six occurrences of *Dt. jourdani* and four of *Dt. puyhauberti*. Dental measurements show *Dt. guntianum* is about the same size as *Dorcatherium* Form 7 (Alba et al. 2011, Fig. 3; Aiglstorfer, Rössner, and Böhme 2014, Fig. 3), which is also about the same age. *Dt. naui* and *Dt. puyhauberti* are both about the size of the Siwalik *Dorcatherium* Form 4, which, with a first appearance at 14.1 Ma, is older than either European species.

In Africa one or two species attributed to *Dorcatherium* are currently recognized: *Dt. pigotti* Arambourg, 1933 and perhaps *Dt. libiensis* Hamilton, 1973. (The latter is considered a junior synonym of *Dt. pigotti* by Pickford [2001] and Geraads [2010]). The ages of these species are respectively about 19.5 Ma, and 16.5 Ma (Geraads 2010). *Dt. pigotti* is the smallest with considerable size variation (Whitworth 1958; Pickford 2002) and is the size of the smaller Siwalik species, including Forms 3, 5, 6, and 7, which collectively encompass its temporal range.

In China only one *Dorcatherium* species has been recognized, *Dt. orientale* Qiu and Gu, 1991. With an age from about 18.5 to ca. 15.5 Ma (Qiu et al. 2013), it appears to be smaller than both the early Middle Miocene Siwalik *Dorcatherium* Form 7 and the much younger *Dt.* Form 3.

Siamotragulus is based on *S. sanyathanai* Thomas. Ginsburg, Hintong, and Suteethorn, 1990, record a Middle Miocene species from Thailand; a second, smaller Lower Miocene species, *S. haripounchai* Mein and Ginsburg, 1997, is also known from Thailand. *Siamotragulus* has been recognized in East Africa as the well-known Lower Miocene *S. songhorensis* (Sánchez et al. 2014), but it has not been reported from Europe. *Siamotragulus* is very similar to the late Miocene *Yunnanotherium simplex* Han, 1986, known from southern China (Qiu and Gu 1991).

The species of *Siamotragulus* are all close in size and difficult to distinguish morphologically. *Siamotragulus bugtiensis* is noted as differing from *S. sanyathanai* in being slightly smaller, with lower crowned molars and an unfused cuboid-navicular and ecto-mesocuneiform complex—a feature that varies within *S. sanyathanai* and perhaps other species (Ginsburg, Morales, and Soria 2001; Sánchez et al. 2014). Dental measurements of *S. bugtiensis* indicate it is about 20% larger than the East African *S. songhorensis* (Whitworth 1958, Tables VII and VIII; Ginsburg, Morales, and Soria 2001, Table 1). *S. nagrii* and *S.* Form D are both approximately the same size as *S. songhorensis*, while *S. minimus* and *S.* Form C are both even smaller. *S. bugtiensis* and *S. songhorensis* are both the oldest known species on their respective continents, appearing by 22 Ma.

Evolutionary Events

Even with their first Early Miocene appearance in southern Asia, tragulids are diverse, having three distinctive genera. Subsequently additional species appear, including at least seven species by 14.1 Ma (Fig. 21.1). The sudden appearance of multiple species of *Dorcatherium* and *Siamotragulus* between 15 and 14 Ma, however, is most likely an artifact of preservation because the number of sites and fossils increases dramatically between the Kamlial and Chinji Formations (Chapter 3 this volume, Fig. 3.3a). Between 14 and 8 Ma there are consistently five or six contemporaneous species present (Fig. 21.1).

Prior to 9.5 Ma tragulids are common in the fossil collections, comprising from 30% to 65% of non-giraffoid ruminants. After 9.5 Ma the number of specimens (but not number of species) gradually declines to just 8% at the end of the Miocene; when finding a tragulid is a rare prize. Although preservation biases due to size need to be considered, these numbers may approximate relative abundance in the living communities and mark changes in ecological structure.

The decline in tragulid abundance throughout the Late Miocene Siwaliks corresponds to significant environmental change (Badgley et al. 2008). The Siwaliks were deposited on low-relief floodplains presumably providing abundant cover and fruit (Chapter 2, this volume) essential for the survival of extant tragulids. Isotopic analyses of paleosols show vegetation was dominated by C_3 plants until around 8 Ma, after which C_3 plants became less common with increased C_4 grasslands (Chapters 5 and 24, this volume), which could account for the decline in the relative abundance of tragulids.

The distal width of the astragalus provides a measure of size, from which equivalent body masses can be calculated (Morgan 1994). Figure 21.2 shows measurements of distal width for 380 specimens of 20 of the putative species. Species of *Dorcabune* achieve large size before species of *Dorcatherium*, but within both there is a trend toward increasing size, which becomes most apparent after 14 Ma with the appearance of new species of the slender-limbed Group A forms of *Dorcatherium*.

Extant tragulids are all small, with median masses ranging from less than 2 kg to 12 kg. They are territorial, highly selective feeders with a diverse diet, prefer to be near water, and need cover to hide from predators. The Siwalik species have a much larger size range than extant species, with a third of the species having median values over 20 kg (Table 21.1).

Sexual Dimorphism
Extant tragulids are not strongly dimorphic in size, but males have enlarged canines and in *Tragulus* dermal pelvic shields. Isolated, saber-like, and presumably male fossil canines are known, but difficult to assign to species. One large canine is associated with the postcranials of the giant *Dt.* Form 2, and another very large, nearly complete specimen from the upper Chinji Formation probably belongs to *Dorcabune anthracotheroides*. Other specimens have been tentatively assigned to *Dt. majus*, *Dt.* Form 5, and *Dt.* Form 6 and it seems likely that all the species had dimorphic canines.

Apart from the dorsal dermal shield itself, the innominates of mature male *Tragulus* have a bony bridge arising from the superior edges of the ilium and ischium that connects to the lateral margins of the sacrum and first caudal vertebra. This structure is absent in female and immature male individuals. In addition, in males the sacrum is solidly fused to the ilium at the sacroiliac joint. While innominate fragments are known for most Siwalik tragulids, very few preserve the critical anatomy, other than for the sacroiliac joint. In none is there an indication that a dermal shield might have been present.

Habitats and Ecomorphology
Locomotion and Substrate Preference
All Siwalik tragulids were terrestrial, but specific analyses of their habitat preferences have not been done. Studies on the ecomorphology of coexisting Middle and Late Miocene bovids

(Chapter 23, this volume; Scott, Kappelman, and Kelley 1999), suggest the predominance of dense cover and soft substrates, which would have been suitable for tragulids, especially the smaller species.

Diet
Diet can be assessed in Siwalik tragulids using isotopic composition of the enamel, microwear on the teeth, and degree of hypsodonty. $^{13}C/^{12}C$ ratios in enamel in particular have been used to infer the isotopic composition of an animal's diet, distinguishing diets based on C_3 plants, which would be predominantly fruit and browse, from diets with significant amounts of C_4 grasses (Chapter 5, this volume; Nelson 2003). In contrast to other techniques, the isotopic composition of enamel is established when the tooth forms. It therefore may not reflect diet over the animal's lifetime, but only in the first year or two. Also, because isotopic composition of atmospheric CO_2 varied over time, interpretation of the isotopic composition of tooth enamel needs to take such changes over time into consideration (Chapter 5, this volume; Cerling et al. 2010; Tipple, Meyers, and Pagani 2010). For the interval between 6.5 and 12.5 Ma, $\delta^{13}C$ values more negative than −7‰ can be taken to indicate a diet composed predominantly of C_3 plants. Such isotopic values are typical of frugivores and browsers. Values more positive than −7‰ indicate a diet that might have included a minor component of C_4 grass (Chapter 5, this volume).

The $\delta^{13}C$ data for tragulids are sparse, with only 26 sampled individuals (Fig. 21.9). Reported values range from −11.8 to −7.3‰, with only one sample more positive than −8.5‰. It is from a 7.4 Ma *Dorcatherium majus* individual having a large enough positive deviation to suggest it possibly had a minor

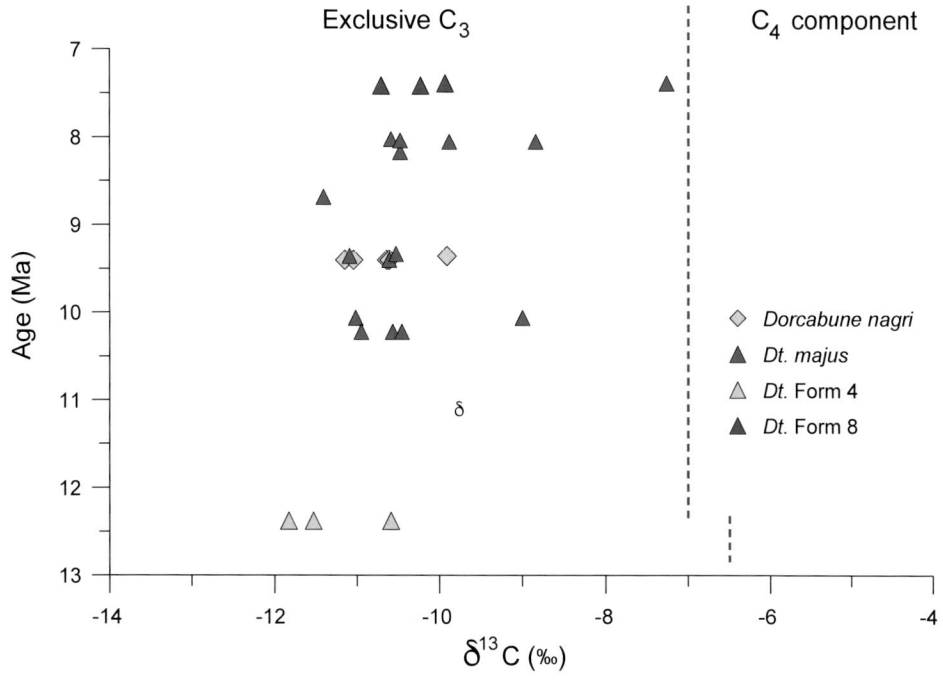

Figure 21.9. $\delta^{13}C$ values against age of sample. The vertical blue line marks the boundary between exclusively C_3 diets and those that may have had a C_4 component.

component of C_4 plants in its diet or fed on plants from a more open or dryer setting. (The specimen is the youngest *Dt. majus* on Figure 21.4b.) The diets of all other individuals are presumed composed exclusively of C_3 plants which might have been either browse or fruit, or both.

Microscopic wear on tooth enamel has also been used to infer an animal's diet and distinguish grazing from browsing and, with less success, fruit-dominated from browse-dominated diets. ("Browse" includes not just leaves, but also buds and young shoots.) Unlike isotopic ratios, microwear records diet shortly before the animal died and again may not reflect diet over the animal's lifetime. Considerable microwear data were collected from Siwalik tragulids by Nelson (2003), with data from 85 individuals representing 4 species (Fig. 21.10). The values fall within the ranges of browsers and frugivore/browsers, as established by Solounias and Semprebon (2002). From this data Nelson (2003) concluded that the tragulids took browse and fruit, with the heavy wear seen in some individuals a result of feeding on hard fruits and seeds. The larger species *Dorcabune nagrii*, *Dorcatherium majus*, and *Dt.* Form 3 generally took more fruit than browse, while the two smaller species of *Siamotragulus* may have consumed a greater proportion of browse (Nelson 2003).

Hypsodonty is an adaptation to resist tooth wear, whether from an abrasive diet, the volume of food, or a longer life. Hypsodonty reflects not just actual use during an individual's life, but probably also the dentition's potential use or even a species'

evolutionary history (Damuth and Janis 2011). As an adaptation for an abrasive diet, hypsodonty most likely reflects design for rarely consumed fallback foods, not necessarily the frequently eaten preferred foods, and may therefore not be a reliable indicator of actual diet.

An index of hypsodonty (metaconid height/trigonid width) for 46 unworn or lightly worn lower third molars and 4 lower second molars is plotted against time on Figure 21.11, along with a line delineating feeding categories (Damuth and Janis 2011). Although the measure of width used here is slightly greater than that of Damuth and Janis (2011), the results are very close and internally consistent. Most specimens are brachydont, with only one *Dt. majus* being classified as mesodont. After about 8.5 Ma, however, there is an apparent shift toward increased hypsodonty. This is apparent in *Dt. majus*, *Dt.* Form 8, and *Siamotragulus nagrii*, three species of very different sizes. The difference in hypsodonty between *Dt.* Form 8 and *Dt.* Form 3 has already been noted (Fig. 21.7b). There are no known lower molars of *Vishnumeryx daviesi* or any unworn upper molars, but the appearance of a more hypsodont Pliocene tragulid would be a continuation of trends seen in the latest Miocene. Increased hypsodonty in bovids appears at about the same time (Chapter 23, this volume).

The isotopic, microwear, and hypsodonty data are consistent, which indicates that the various tragulid species had diets composed largely of fruit and browse derived from C_3 plants. There are, however, indications after 8.5 Ma of a slight shift

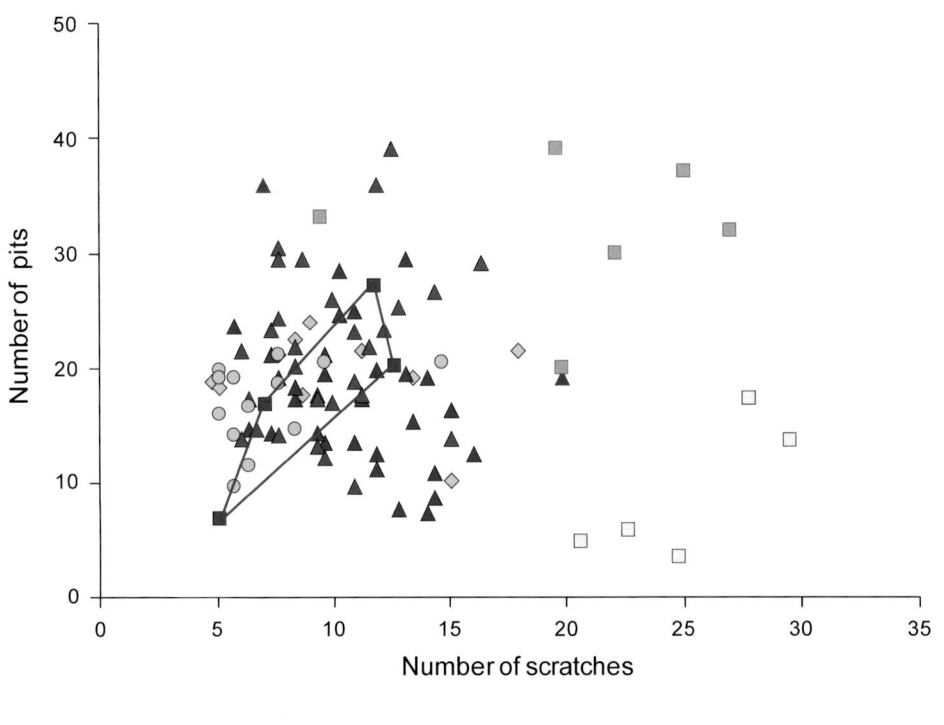

Figure 21.10. Dental microwear with number of scratches versus number of pits for four fossil tragulid species (data from Nelson [2003]) and extant grazers, browsers, and frugivore/browsers (data from Solounias and Semprebon [2002]). Quadrilateral is the region occupied by extant browsing bovids.

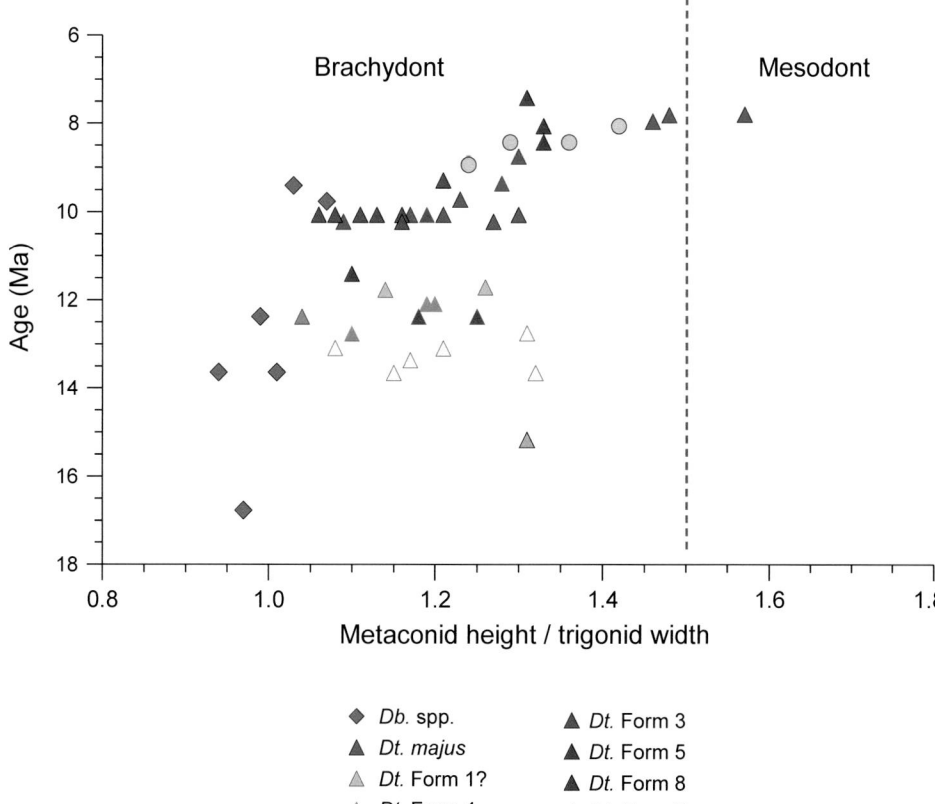

Figure 21.11. Index of hypsodonty (metaconid height/trigonid width) versus age for ten tragulid species. The data include 47 unworn or lightly worn lower third molars. The vertical red line delineates crown height categories related to feeding (Damuth and Janis 2011).

Legend:

◆ *Db.* spp.   ▲ *Dt.* Form 3

▲ *Dt. majus*   ▲ *Dt.* Form 5

▲ *Dt.* Form 1?   ▲ *Dt.* Form 8

▲ *Dt.* Form 4   △ *Dt.* Form 7

△ *Dt.* Form 6   ◉ *S. nagrii*

toward more abrasive diets. Locomotor adaptations of bovids suggest the predominance of relatively bushy and dense undergrowth before 11.4 Ma, with increasing openness in the Late Miocene. The decrease in the relative abundance of tragulids can be explained by such changes.

CONCLUSIONS

With as many as 20 species, Siwalik tragulids have greater species richness than previously thought. Of the 20 putative species, more than half have yet to be described and named (Table 21.1). Species of *Dorcatherium* (some of which might belong to *Afrotragulus*) predominate in the collections, both in number of species (nine) and number of specimens. But three other genera are also known: *Dorcabune* with perhaps four species, *Siamotragulus* with five, and *Vishnumeryx* with one. Little is known about the relationships of any of the Siwalik tragulids either with one another or with species from Africa, Europe, or northern Asia. In the Siwalik *Dorcatherium* there are two groups of species, one characterized by long, slender limbs and the other by shorter, stouter limbs. In both groups of *Dorcatherium* and in *Siamotragulus* there are pairs of earlier and later species that could represent ancestor and descendant species, but this has not yet been demonstrated. *Dorcatherium* is present in Europe and Africa, while *Siamotragulus* is known in Africa, but not Europe. *Dorcabune* has usually been consid-

ered to be strictly Asian, but it appears to be represented by bunoselenodont species in Europe and Africa.

The appearance between 15 and 14 Ma of four additional species is most likely an artifact of preservation, but the turnover of species between 11.2 and 10.4 Ma is not, although whether it was one or a series of events is not clear (Chapter 25, this volume). The Late Miocene turnover involved six or seven species with both disappearances and appearances, most of which occurred between 10.6 and 10.5 Ma. Bovids show similar changes, as do other artiodactyl groups.

The Siwalik tragulids are assumed to have been terrestrial and, as with extant species, probably inhabitants of areas of dense cover with close access to water. The $\delta^{13}C$ and microwear data indicate fruit-dominated and browsing diets based on C_3 vegetation, which is consistent with their brachydont dentitions. There is, however, a latest Miocene trend toward slightly more hypsodont dentitions.

ACKNOWLEDGMENTS

I thank the editors David Pilbeam, Catherine Badgley, and Michèle Morgan for inviting me to participate. The research on Siwalik tragulids is a result of a long collaboration with the Geological Survey of Pakistan and I appreciate the efforts of the Directors General and members of the survey who have over the years supported the project. Financial support has included grants from the National Science Foundation, the Smithsonian Foreign Currency Program, the Wenner-Gren Foundation for Anthropological Research, and the American School for Prehistoric Research. I would also like to acknowledge the abiding support of Francine Berkowitz of the Smithsonian Office of International Relations and the enthusiasm of many colleagues.

References

Aiglstorfer, M., G. E. Rössner, and M. Böhme. 2014. *Dorcatherium naui* and pecoran ruminants from the late Middle Miocene Gratkorn locality (Austria). *Palaeobiodiversity and Palaeoenvironments* 94:83–123.

Alba, D. M., S. Moyà-Solà, J. M. Robles, I. Casanovas-Vilar, C. Rotgers, R. Carmona, and J. Galindo. 2011. Middle Miocene tragulid remains from Abocador de Can Mata: The earliest record of *Dorcatherium naui* from Western Europe. *Geobios* 44:135–150.

Antoine, P.-O., G. Métais, M. J. Orliac, J.-Y. Crochet, L. J. Flynn, L. Marivaux, A. R. Rajpar, G. Roohi, and J.-L. Welcomme. 2013. Mammalian Neogene biostratigraphy of the Sulaiman Province, Pakista. In *Fossil Mammals of Asia: Neogene Biostratigraphy and Chronology*, edited by X.-M. Wang, L. J. Flynn and M. Fortelius, 400–442. New York: Columbia University Press.

Badgley C, J. C. Barry, M. E. Morgan, S. V. Nelson, A. K. Behrensmeyer, T. E. Cerling, and D. Pilbeam. 2008. Ecological changes in Miocene mammalian record show impact of prolonged climatic forcing. *Proceedings of the National Academy of Sciences* 105(34): 12145–12149.

Bärmann, E. V., and G. E. Rössner. 2011. Dental nomenclature in Ruminantia: Towards a standard terminological framework. *Mammalian Biology* 76:762–768.

Barry, J. C. 1995. Faunal turnover and diversity in the Terrestrial Neogene of Pakisan. In *Paleoclimate and Evolution, with Emphasis on Human Origins*, edited by E. S. Vrba, G. H. Denton, T. C. Partridge, and L. H. Burckle, 115–134. New Haven, CT: Yale University Press.

Barry, J. C. 2014. Fossil tragulids of the Siwalik Formations of southern Asia. *Zitteliana B* 32:53–61.

Barry J. C., A. K. Behrensmeyer, C. Badgley, L. J. Flynn, H. Peltonen, I. U. Cheema, D. Pilbeam, et al. 2013. The Neogene Siwaliks of the Potwar Plateau, Pakistan. In *Fossil Mammals of Asia: Neogene Biostratigraphy and Chronology*, edited by X.-M. Wang, L. J. Flynn and M. Fortelius, 373–399. New York: Columbia University Press.

Barry, J. C., S. Cote, L. MacLatchy, E. H. Lindsay, R. Kityo, and A. R. Rajpar. 2005. Oligocene and Early Miocene ruminants (Mammalia, Artiodactyla) from Pakistan and Uganda. *Palaeontologia Electronica* 8(1): 22A.

Barry, J. C., M. E. Morgan, L. J. Flynn, D. Pilbeam, A. K. Behrensmeyer, S. M. Raza, I. A. Khan, C. Badgley, J. Hicks, and J. Kelley. 2002. Faunal and environmental change in the Late Miocene Siwaliks of northern Pakistan. *Paleobiology Memoirs* 3:1–72.

Batool, A., M. A. Khan, M. Akhtar, and N. A. Qureshi. 2014. New remains of tragulids (Mammalia, Tragulidae) from Dhok Pathan Formation of Hasnot (Late Miocene), Pakistan. *Pakistan Journal of Zoology* 46(5): 1323–1336.

Cerling, T. E., J. M. Harris, M. G. Leakey, B. H. Passey, and N. E. Levin. 2010. Stable carbon and oxygen isotopes in East African mammals: Modern and fossil. In *Cenozoic Mammals of Africa*, edited by L. Werdelin and W. J. Sanders, 941–952. Berkeley: University of California Press.

Colbert, E. H. 1935. Siwalik mammals in the American Museum of Natural History. *Transactions of the American Philosophical Society* n.s. 26:1–401.

Damuth, J., and C. M. Janis. 2011. On the relationship between hypsodonty and feeding ecology in ungulate mammals, and its utility in palaeoecology. *Biological Reviews* 86:733–758.

Dong, W., and G.-Q. Qi. 2013. Hominoid-Producing Localities and Biostratigraphy in Yunnan. In *Fossil Mammals of Asia: Neogene Biostratigraphy and Chronology*, edited by X.-M. Wang, L. J. Flynn and M. Fortelius, 293–313. New York: Columbia University Press.

Fahlbusch, V. 1985. Säugetierreste (*Dorcatherium, Steneofiber*) aus der Miozänen Braunkohle von Wackersdorf/Oberpfalz. *Mitteilungen Bayerischen Staatssammlung für Paläontologie und Historische Geologie* 25:81–94.

Farooq, U., M. A. Khan, M. Akhtar, A. M. Khan, and Z. Ali. 2007a. *Dorcatherium minus* from the Siwaliks, Pakistan. *Journal of Animal and Plant Sciences* 17(3–4): 86–89.

Farooq, U., M. A. Khan, M. Akhtar, and A. M. Khan. 2007b. *Dorcabune anthracotheriides* (Ariodactyla: Ruminantia: Tragulidae) from Hasnot, the Middle Siwaliks, Pakistan. *Pakistan Journal of Zoology* 39(6): 353–360.

Flynn, L. J., E. H. Lindsay, D. Pilbeam, S. M. Raza, M. E. Morgan, J. C. Barry, C. E. Badgley, et al. 2013. The Siwaliks and Neogene Evolutionary Biology in South Asia. In *Fossil Mammals of Asia: Neogene Biostratigraphy and Chronology*, edited by X.-M. Wang, L. J. Flynn and M. Fortelius, 353–372. New York: Columbia University Press.

Fortelius, M., J. Eronen, J. Jernvall, L. Liu, D. Pushkina, J. Rinne, A. Tesakov, I. Vislobokova, Z. Zhang, and L. Zhou. 2002. Fossil mammals resolve regional patterns of Eurasian climate change over 20 million years. *Evolutionary Ecology Research* 4:1005–1016.

Gaur, R., R. N. Vasishat, and A. K. Sankhyan. 1980. The first record of *Dorcatherium* (Tragulidae) from the Pinjor Formation, East of Chandigarh. *Bulletin Indian Geological Association* 1:67–70.

Gentry, A. W., G. E. Rössner, and E. P. J. Heizmann. 1999. Suborder Ruminantia. In *The Miocene Land Mammals of Europe*, edited by G. E. Rössner, and K. Heissig, 225–258. Munich: Verlag Pfeil.

Geraads, D. 2010. Tragulidae. In *Cenozoic Mammals of Africa*, edited by L. Werdelin and W. J. Sanders, 723–729. Berkeley: University of California Press.

Ginsburg, L., J. Morales, and D. Soria. 2001. Les Ruminantia (Artiodactyla, Mammalia) du Miocène des Bugti (Balouchistan, Pakistan). *Estudios Geologicos* 57(3–4):155–170.

Gradstein, F. M., J. G. Ogg, M. D. Schmitz, and G. M. Ogg. 2012. *The Geological Time Scale 2012*. Amsterdam: Elsevier.

Guzmán-Sandoval, J. A., and G. E. Rössner. 2019. Miocene chevrotains (Mammalia, Artiodactyla, Tragulidae) from Pakistan. *Historical Biology* 33(6): 743–776. https://doi.org/ 10.1080/08912963.2019.1661405.

Han, D. 1974. First discovery of *Dorcabune* in China. *Vertebrata Palasiatica* 12:217–221.

Han, D. 1986. Fossils of Tragulidae from Lufeng, Yunnan. *Acta Anthropologica Sinica* 5:68–78.

Iqbal, M., M. A. Khan, M. Atiq, T. Ikram, and M. Akhtar. 2011. *Dorcatherium minus* (Tragulidae-Artiodactyla-Mammalia) from the Nagri type area of the Nagri Formation, Middle Siwaliks, northern Pakistan: New collection. *Yerbilimleri* 32(1): 59–68.

Janis, C. M., and K. M. Scott. 1987. The interrelationships of higher ruminant families with special emphasis on the members of the Cervoidea. *American Museum Novitates* 2893:1–85.

Khan, M. A., S. G. Abbas, M. A. Babar, S. Kiran, A. Riaz, and M. Akhtar. 2017. *Dorcatherium* (Mammalia: Tragulidae) from Lower Siwaliks of Dhok Bun Amir Khatoon, Punjab, Pakistan. *Pakistan Journal of Zoology* 49(3): 883–888.

Khan, M. A., and M. Akhtar. 2005. Discovery of a deciduous premolar of *Dorcatherium minus* from the Middle Siwaliks of Pakistan. *Punjab University Journal of Zoology* 20(2): 181–188.

Khan, M. A., and M. Akhtar. 2013. Tragulidae (Artiodactyla, Ruminantia) from the Middle Miocene Chinji Formation of Pakistan. *Turkish Journal of Earth Sciences* 22(2): 339–353.

Khan, M. A., M. Akhtar, G. Iliopoulos, and Hina. 2012. Tragulids (Artiodactyla, Ruminantia, Tragulidae) from the Middle Siwaliks of Hasnot (Late Miocene), Pakistan. *Rivista Italiana di Paleontologia e Stratigrafia* 118(2): 325–341.

Khan M. A., A. Ghaffar, U. Farooq, and M. Akhtar. 2006. Ruminant fauna from the Tertiary Hills (Neogene) of the Siwaliks of Pakistan. *Journal of Applied Sciences* 6:131–137.

Khan, M. J., S .T. Hussain, M. Arif, and H. Shaheed. 1984. Preliminary paleomagnetic investigations of the Manchar Formation, Gaj River Section, Kirthar Range, Pakistan. *Geological Bulletin University of Peshawar* 17:145–152.

Kingdon, J. 1997. *The Kingdon Field Guide to African Mammals*. London: Academic Press.

Koufos, G. D. 2021. New tragulid remains from the early/middle Miocene and a revision of their occurrence in Greece. *Historical Biology* 33(10): 2371–2386. https://doi.org/10.1080/08912963.2020.1795650.

Lindsay, E. H., L. J. Flynn, I. U. Cheema, J. C. Barry, K. F. Downing, A. R. Rajpar, and S. M. Raza. 2005. Will Downs and the Zinda Pir Dome. *Palaeontologia Electronica* 8(1): 19A.

Lydekker, R. 1876. Indian Tertiary and Post-Tertiary Vertebrata. Descriptions of the molar teeth and other remains of Mammalia. *Memoirs of the Geological Survey of India, Palaeontologia Indica* Series 10, I, part 2:19–87.

Lydekker, R. 1882. Note on some Siwalik and Jamna mammals. *Records of the Geological Survey of India* 15:28–33.

Lydekker, R. 1884. Indian Tertiary and Post-Tertiary Vertebrata. Rodents, ruminants and synopsis of Mammalia. *Memoirs of the Geological Survey of India, Palaeontologia Indica* Series 10, III, part 3:105–134.

Lydekker, R. 1885. *Catalogue of the remains of Siwalik vertebrata contained in the Geological Department the Indian Museum, Calcutta. Part I. Mammalia.* Calcutta, India. 116 pp.

Lydekker, R. 1886. Siwalik Mammalia—Supplement I, addendum. *Memoirs of the Geological Survey of India, Palaeontologia Indica* Series 10, IV, part 1:19–21.Meijaard, E., and C. P. Groves. 2004. A taxonomic revision of the *Tragulus* mouse-deer (Artiodactyla). *Zoological Journal of the Linnean Society* 140(1):63–102.

Mennecart, B., Adrien de Perthuis, G. E. Rössner, J. A. Guzmánd, Aude de Perthuis, and L. Costeura. 2018. The first French tragulid skull (Mammalia, Ruminantia, Tragulidae) and associated tragulid remains from the Middle Miocene of Contres (Loir-et-Cher, France), *Comptes Rendus Palevol* 17(3): 189–200.

Métais, G., J.-L. Welcomme, and S. Ducrocq. 2009. Lophiomerycid ruminants from the Oligocene of the Bugti Hills (Balochistan, Pakistan). *Journal of Vertebrate Paleontology* 29(1):1–12.

Morales, J., I. M. Sánchez, and V. Quiralte. 2012. Les Tragulidae (Artiodactyla) de Sansan. In *Mammifères de Sansan*, edited by S. Peigné and S. Sen. *Mémoires du Muséum national d'Histoire naturelle*, 203:225–247.

Morgan, M. E. 1994. Paleoecology of Siwalik Miocene hominoid communities: Stable carbon isotope, dental microwear, and body size analyses. Unpublished PhD diss., Harvard University.

Musalizi, S., J. Schnyder, L. Segalen and G. E. Rössner. 2023 Early and Middle Miocene Tragulidae of the Napak Region (Uganda) including the oldest African tragulids: Taxonomic revision, stratigraphical background, and biochronological framework, *Historical Biology*, 35(12): 2456–2503. https://doi.org/10.1080/08912963.2022.2144285

Mottl, M. 1961. Die Dorcatherien (Zwerghirsche) der Steiermark. *Mitteilungen des Museums für Bergbau, Geologie, Technick, Landesmuseum "Joanneum" Graz* 22:21–71.

Nanda, A. C. 2015. *Siwalik Mammalian Faunas of the Himalayan Foothills, with Reference to Biochronology, Linkages and Migration*. Dehradun: Wadia Institue of Himalayan Geology.

Nelson, S. V. 2003. *The Extinction of Sivapithecus: Faunal and Environmental Changes Surrounding the Disappearance of a Miocene Hominoid in the Siwaliks of Pakistan.* American School of Prehistoric Research Monographs, vol. 1. Boston: Brill Academic Publishers.

Patnaik, R. 2013. Indian Neogene Siwalik Mammalian Biostratigraphy: An Overview. In *Fossil Mammals of Asia: Neogene Biostratigraphy and Chronology*, edited by X.-M. Wang, L. J. Flynn and M. Fortelius, 423–444. New York: Columbia University Press.

Pickford, M. 2001. Africa's smallest ruminant: A new tragulid from the Miocene of Kenya and the biostratigraphy of East African Tragulidae. *Geobios* 34(4): 437–447.

Pickford, M. 2002. Ruminants from the Early Miocene of Napak, Uganda. *Annales de Paléontologie* 88:85–113

Pilgrim, G. E. 1910a. Notices of new mammalian genera and species from the Tertiaries of India. *Records of the Geological Survey of India* 40:63–71.

Pilgrim, G. E. 1910b. Preliminary note on a revised classificaton of the Tertiary freshwater deposits of India. *Records of the Geological Survey of India* 40:185–205.

Pilgrim, G. E. 1915. The dentition of the tragulid genus *Dorcabune*. *Records of the Geological Survey of India* 45:226–238.

Pilgrim, G. E. 1939. The fossil Bovidae of India. *Memoirs of the Geological Survey of India, Palaeontologia Indica* New Series, 26(1): 1–356.

Prieto, J., P.-O. Antoine, J. van der Made, G. Métais, L. T. Phuc, Q. T. Quan, S. Schneider, et al. 2017. Biochronological and palaeobiogeographical significance of the earliest Miocene mammal fauna from Northern Vietnam. *Palaeobiodiversity and Palaeoenvironments* 98(2): 287–313. https://doi.org/10.1007/s12549-017-0295-y.

Qiu, Z.-x. and Y.-m. Gu. 1991. The Aragonian vertebrate fauna of Xlacaowan, Jiangsu-8. *Dorcatherlum* (Tragulidae, Artiodactyla). *Vertebrata Palasiatica* 29:21–37.

Qiu, Z.-X., Z.-D. Qiu, T. Deng, C.-K. Li, Z.-Q. Zhang, B.-Y. Wang, and X.-M. Wang. 2013. Neogene land mammal stages/ages of China: Toward the goal to establish an Asian land mammal stage/age scheme. In *Fossil Mammals of Asia: Neogene Biostratigraphy and Chronology*, edited by X.-M. Wang, L. J. Flynn and M. Fortelius, 29–90. New York: Columbia University Press.

Raza, S. M, J. C. Barry, G. E. Meyer, and L. Martin. 1984. Preliminary report on the geology and vertebrate fauna of the Miocene Manchar Formation, Sind, Pakistan. *Journal of Vertebrate Paleontology* 4(4):584–599.

Raza, S. M., I. U. Cheema, W. R. Downs, A. R. Rajpar, and S. C. Ward. 2002. Miocene stratigraphy and mammal fauna from the Sulaiman Range, Southwestern Himalayas, Pakistan. *Palaeogeography, Palaeoclimatology, Palaeoecology* 186(3–4): 185–197.

Rössner, G. E. 2007. Family Tragulidae. In *The Evolution of Artiodactyls*, edited by D. R. Prothero and S. E. Foss, 213–220. Baltimore, MD: Johns Hopkins University Press.

Rössner, G. E., and K. Heissig. 2013. New records of *Dorcatherium guntianum* (Tragulidae), stratigraphical framework, and diphyletic origin of Miocene European tragulids. *Swiss Journal of Geosciences* 106:335–347.

Roussiakis, S. 2022. The fossil record of Tragulids (Mammalia: Artiodactyla: Tragulidae) in Greece. In *Fossil Vertebrates of Greece*, Vol. 2, edited by E. Vlachos, 335–350. Cham: Springer.

Samiullah, K., F. Jabee, R. Yasin, S. Yaqub, F. Jalal, S. Ahmad, B. Rasool, K. Feroz, and M. Akhtar. 2015. *Dorcatherium* from the Dhok Bun Ameer Khatoon, Chinji Formation, Lower Siwalik Hills of

Pakistan. *Journal of Biodiversity and Environmental Sciences* 6(3): 408–422.

Sánchez, I. M., V. Quiralte, J. Morales, and M. Pickford. 2010. A new genus of tragulid ruminant from the early Miocene of Kenya. *Acta Palaeontologica Polonica* 55(2): 177–187.

Sánchez, I. M., V. Quiralte, M. Ríos, J. Morales, and M. Pickford. 2014. First African record of the Miocene Asian mouse-deer *Siamotragulus* (Mammalia, Ruminantia, Tragulidae): Implications for the phylogeny and evolutionary history of the advanced selenodont tragulids. *Journal of Systematic Palaeontology* 13(7): 543–556. https://doi.org/10.1080/14772019.2014.930526.

Sánchez, I. M., S. G. Abbas, M. A. Khan, M. A. Babar, V. Quiralte, and D. DeMiguel. 2022. The first Asian record of the mouse-deer *Afrotragulus* (Ruminantia, Tragulidae) reassess its evolutionary history and offers insights on the influence of body size on *Afrotragulus* diversification. *Historical Biology* 34(8): 1544–1559. https://doi.org/10.1080/08912963.2022.2050719.

Scott, R. S., J. W. Kappelman, and J. Kelley. 1999. The paleoenvironment of *Sivapithecus parvada*. *Journal of Human Evolution* 36(3): 245–274.

Sehgal, R. K. 2015. Mammalian faunas from the Siwalik sediments exposed around Nurpur, District Kangra (H.P.): Age and palaeobiogeographic implications. *Himalayan Geology* 36(1): 9–22.

Solounias, N., and G. M. Semprebon. 2002. Advances in the reconstruction of ungulate ecomorphology and application to early fossil equids. *American Museum Novitates* 3366:1–49.

Tipple, B. J., S. R. Meyers, and M. Pagani. 2010. Carbon isotope ratio of Cenozoic CO_2: A comparative evaluation of available geochemical proxies. *Paleoceanography* 25(3): PA3202. https://doi.org/10.1029/2009PA001851.

van der Made, J. 1997. Pre-Pleistocene land mammals from Crete. In *Pleistocene and Holocene Fauna of Crete and Its First Settlers*, edited by D. S. Reese, 69–79. Madison, WI: Monographs in World Archaeology 28.

West, R. M. 1980. A minute new species of *Dorcatherium* (Tragulidae, Mammalia) from the Chinji Formation near Daud Khel, Mianwali District, Pakistan. *Milwaukee Public Museum, Contributions in Biology and Geology* 33:1–6.

Whitworth, T. 1958. Miocene ruminants of East Africa. *Fossil Mammals of Africa* 15:1–50.

22

SIWALIK GIRAFFOIDEA

NIKOS SOLOUNIAS AND MELINDA DANOWITZ

INTRODUCTION

The term "Giraffoidea" refers to a clade of ruminant artiodactyls that includes four families of living and extinct lineages. The family Giraffidae includes all ruminants that possess a canine with an extra lobe and ossicones. The species are all large with short lumbar vertebrae and a downward-tilted pelvis. The metapodials are notably elongated with a flat back. The cheek teeth have crenulated enamel. The closest family is the Palaeomerycidae. Members of this family differ from Giraffidae in having an occipital horn, an arched back, smaller body size with a more variable size range (small and large species), and noncrenulated enamel in cheek teeth. Two closely related families are the Antilocapridae (here including Climacocerinae) and the Lagomerycidae. These are collectively grouped in the Giraffoidea and differ from ruminants in the Bovoidea and Cervoidea.

Siwalik fossils collected by members of the Harvard-Geological Survey of Pakistan (Harvard-GSP) research project include more than 2,000 giraffoid specimens from the Potwar Plateau, most of which belong to the Giraffidae. Important comparative fossil collections are in the Indian Museum, Kolkata (IMC), the Natural History Museum in London (NHM), the Paläontologisches Museum in Munich, the Yale Peabody Museum (YPM), and the American Museum of Natural History (AMNH) in New York. We have visited all of these collections except that held by the museum in Kolkata.

The critical early publications that described Siwalik giraffid taxa were based on fragmentary, mostly dental material (Bettington 1846 Falconer 1868; Falconer and Cautley 1868; Lydekker 1876, 1883; Pilgrim 1910b, 1911; Matthew 1929; and Colbert 1935). The new giraffoid specimens are also fragmentary but some are more complete than in the older collections. The new material includes cranial and dental specimens with a few braincases, along with many ossicones, maxillae, and mandibles, as well as numerous postcranials. The discovery of complete metapodials is significant both for species identification and functional morphology.

It is not possible to assign postcranial material to known species, when, for example, the published types are dental or fragmentary skull material. The same is true if the type is an upper dentition and the new find is a lower dentition. Comparisons with material from other collections beyond those of South Asia have been essential. In particular, instrumental in our determinations have been specimens from Gebel Zelten, Libya; Moruorot Hill and Fort Ternan, Kenya; Samos and Pikermi, Greece; several localities in Shanxi and Gansu Provinces of China; as well as collections in France, Spain, Iran, Iraq, and Turkey. Two further issues are noteworthy: many giraffids possess similar dentitions, and many co-occurring species are of similar size. An interesting feature of both the Siwalik species and those from other localities is that they cluster into two sizes. The smaller size is about that of *Okapia*, the extant okapi, the larger about that of *Giraffa*, the extant giraffe, or *Bubalus*, the Asian buffalo. Prior to our study, most Siwalik giraffid fossils were sorted into two categories: a smaller group and a larger group (identified as *Giraffokeryx punjabiensis* and *Bramatherium megacephalum*, respectively). Many new genera and species have recently been identified. The presence of two size clusters in Siwalik Giraffidae differs from that of other giraffoids, bovids, and cervids, in which a wider spectrum of body sizes is present. One convenient way to document trends in body size over time is by measuring the numerous astragali (in particular, the width of the astragalar head).

Ossicones are important taxonomically as they differ among species. Previous research has not utilized ossicones in detail. This is because secondary growth, an external layer of bone on the ossicones of *Giraffa*, gives the impression that ossicones are variable and hence systematically unreliable (Spinage 1968), but this is not the case. The ossicones of extinct species, unlike those of *Giraffa*, have either organized secondary bone growth or none. The ossicones are highly specific in form and resemble the horn cores of bovids. In the present study, we have relied on ossicone morphology for the determination of many taxa. Metapodials and astragali are also important; hence, our diagnoses of species from astragali and metapodials can be connected to published descriptions based on dentitions and skulls (Ríos, Danowitz and Solounias 2016; Solounias and Danowitz 2016a).

SYSTEMATIC APPROACH AND GENERAL PATTERNS

For our systematic study, we used fossils from other regions of the world for comparison. Table 22.1 lists the current determination of identified species from the Siwalik record. (We are preparing a separate monograph on Giraffoidea, in which new taxa will be formally named, many key specimens will be figured, and a cladistic analysis will be provided.) The classification presented here is based on Hamilton (1978), Solounias (2007), and Ríos, Sánchez, and Morales (2017). Many taxa in this study are new, but with few exceptions, at the family or subfamily level the classification does not differ substantially from that of Hamilton (1978). It does however entail some notable changes. One involves placement of *Triceromeryx* and *Zarafa* in the Palaeomerycidae. Another is that *Climacoceras* is classified along with Merycodontinae of North America within the Antilocapridae. This assignment supports the hypothesis that giraffids are close to *Antilocapra*, a relationship that is currently supported by genetic data (Hassanin and Douzery 2003). *Orangemeryx* and *Prolibytherium* are placed within the Lagomerycidae; they are larger in size, but the horn morphology is similar in these two species. *Okapia* had no fossil record until this study, and the recognition of a new genus close to *Okapia* has interesting implications. The recognition of several species of the Sivatheriinae expands this clade in the Miocene. *Decennatherium* and a new closely related genus are early members of the Sivatheriinae (see Table 22.1).

Figure 22.1 illustrates the current temporal distribution of identified taxa, which range in age from 18.5 to 7.1 Ma in the Siwalik record. A general body-mass estimate is provided, with size equivalents of modern species in Table 22.1. This review reveals a substantial diversity of taxa previously unknown from the Siwalik record. The giraffoids include seven Palaeomerycidae, two Antilocapridae, and three Lagomerycidae. These species are new for the Siwaliks. The number of non-giraffid species is comparable to that of other Miocene localities in Central Europe and North and Central Africa. Their abundance in terms of number of specimens in the Siwalik record, however, is low.

The Palaeomerycidae, Climacocerinae (Antilocapridae), Lagomerycinae (Lagomerycidae), and Giraffidae are phylogenetically close. Based on fossil data, we estimate the origins of these families to be about 18.5 to 19.5 Ma, when they presumably evolved from more archaic non-horned ruminants. The Siwalik fossil record supports this estimate. Their ancestors would be found in the Late Oligocene or earliest Miocene. We consider that the Gelocidae are the most plausible ancestors of Giraffidae and perhaps Palaeomerycidae. *Prodremotherium elongatum* from the Phosphorite Beds of Quercy, France, resembles a small giraffid (Janis and Scott 1987; Danowitz and Solounias 2015). The Siwalik record shares many giraffoid species with other regions of Eurasia and Africa. The richness of taxa suggests that the Indian subcontinent was a primary region for the origin and diversification of these taxa.

Different kinds of head ornaments characterize the giraffoid families. Palaeomerycidae and Giraffidae are the only fami-

lies with ossicones, and they share this derived character. Palaeomerycidae and Dromomerycidae share occipital horns, but this is due to convergence. Lagomerycidae, Dromomerycidae, and Antilocapridae (Climacocerinae) share the feature of long horns as expansions of the frontal bones. Janis and Scott (1987) have provided a complex interpretation of how these taxa are related to Cervidae and other ruminant groups.

The detailed morphology of giraffid ossicones distinguishes every species. The same is true for the morphology of the metapodials. The number of species is high in the Siwalik record, which suggests that the Indian subcontinent was critical in the evolution and diversification of this family. The non-giraffid giraffoids are rare as fossils, whereas giraffids are common. Possibly, the former inhabited uplands and were occasional members of the floodplain ecosystem, while the more abundant Giraffidae were more persistent inhabitants of the floodplains. The two groups possibly differed ecologically.

SYSTEMATIC DETERMINATIONS

Order Pecora Linnaeus, 1758

GIRAFFOIDEA SIMPSON, 1931
Family Palaeomerycidae Lydekker, 1883

The seven species of Palaeomerycidae documented in the Siwalik record range in size from very small (duiker size) to large (okapi size) and comparable in terms of size range to the Bovidae.

Subfamily Palaeomerycinae Lydekker, 1883

Palaeomerycinae, n. gen. and sp.
This new genus and species is distinguished by the shape of the ossicone and the presence of a deep pit for the nuchal ligament (YGSP 30933 from Locality Y691 dated to 13.1 Ma). The specimen is a complete braincase with supraorbital ossicones. The bullae are large, round, and bulbous, and the deep nuchal-ligament pits signify differences in the mode of use of the neck and head in fighting. The anterior curvature of the orbital ossicones is atypical for many giraffoid taxa. The position of the ossicone is known as it was found in this anterior orientation next to the skull. The diversity and geographic range of Palaeomerycidae are expanded by the recognition of this new taxon.

Palaeomerycinae, gen. and sp. indet.
A large-bodied taxon, the size of *Bramatherium*, is known only from an upper molar (see *Sivatherium giganteum* in Lydekker [1883], Plate 21, Fig. 1). The large molar possesses an elevated rib on the metacone unlike molars of giraffids. It also has the characteristic palaeomerycid metaconule fold. It is not represented in the Harvard-GSP Siwalik collection. This exceptionally large palaeomerycid is very rare. The specimen is from the "Siwaliks of Punjab" (Lydekker 1883) with no specific locality information.

Tauromeryx Astibia, Morales, and Moya Sola 1998
Type species: *Tauromeryx turiasonensis* Astibia, Morales, and Moya Sola 1998

Table 22.1. Siwalik Giraffoids

Giraffoidea	Age Range	Body Size Estimate	Modern Analogue for Size Estimate
Family Palaeomerycidae			
Palaeomerycinae n. gen. and sp.	13.1 Ma	~200–350 kg	*Okapia johnstoni*
Palaeomerycinae gen. and sp. indet.	Unknown[1]	Very large	
Tauromeryx n. sp.	12.0 Ma	~200–350 kg	*Okapia johnstoni*
cf. *Tauromeryx* sp.	10.1 Ma	13–30 kg	*Gazella thomsoni*
Triceromeryx n. sp.	12.1 Ma	50–100 kg	*Tragelaphus imberbis*
cf. *Triceromeryx* n. sp.	11.6–10.1 Ma	50–100 kg	*Tragelaphus imberbis*
cf. *Orygotherium* sp.	7.2 Ma	~14 kg	*Cephalophus natalensis*
Family Antilocapridae			
Climacocerinae			
Climacocerinae, n. gen. and sp.	17.9 Ma	25–60 kg	*Rupicapra rupicapra*
Climacoceras n. sp.	11.4–10.0	150–200 kg	*Oryx beisa*
Family Lagomerycidae			
Lagomerycinae			
Orangemeryx n. sp.	13.8 Ma	45–70 kg	*Odocoileus virginianus*
cf. *Orangemeryx* n. sp.	14.1–12.7 Ma	40–75 kg	*Aepyceros melampus*
Prolibytherium fusus	LO 18.8 Ma	40–75 kg	*Aepyceros melampus*
Family Giraffidae			
Progiraffinae			
Progiraffa exigua	18.1–9.4 Ma	50–100 kg	*Tragelaphus imberbis*
Giraffokerycinae			
Giraffokeryx primaevus	Chinji Fm.	?	
Giraffokeryx punjabiensis	14.1–11.4 Ma	~200–350 kg	*Okapia johnstoni*
Bohlininae			
Bohlininae n. gen. and sp.	13.8–11.4 Ma	~500–900 kg	*Giraffa camelopardalis* (smaller)
Honanotherium sp.	14.0–11.4 Ma	~500–900 kg	*Giraffa camelopardalis* (smaller)
Giraffinae			
Giraffa punjabiensis	9.4–8.1 Ma	800–1,200 kg	*Giraffa camelopardalis*
Okapiinae			
Ua pilbeami	14.1–8.1 Ma	~200–350 kg	*Okapia johnstoni*
Okapiinae, n. gen. and sp.	9.6–9.4 Ma	~200–350 kg	*Okapia johnstoni*
Sivatheriinae			
Decennatherium asiaticum	13.0–10.1 Ma	~200–700 kg	*Alces alces*
Libytherium sp. 1	10.1 Ma	~600–1,200 kg	*Bubalus bubalis*
Libytherium sp. 2	9.6–9.4 Ma	~600–1,200 kg	*Bubalus bubalis*
Bramatherium megacephalum	10.2–8.0? Ma	~600–1,200 kg	*Bubalus bubalis*
Bramatherium n. sp.	13.7 Ma	120–315 kg	*Tragelaphus strepsiceros*
Helladotherium grande	10.1–8.8 Ma	800–1,200 kg	*Giraffa camelopardalis*
Samotheriinae			
Injanatheriium hazimi	12.4–7.5 Ma	~500–900 kg	*Giraffa camelopardalis* (small)

[1] Lydekker, 1883.

Tauromeryx n. sp.

Two bilateral occipital horn-base fragments (YGSP 42343 from Locality Y502 dated to 12.0 Ma) and an astragalus are the basis for recognizing a new palaeomerycid genus and species. The occipital horn fragments are oval in cross-section and possess a robust core with a smooth surface. The occipital horns are directed dorso-laterally and cranially and are completely separated at the base. The occipital surface is broad, notably flat, and deeply concave under the occipital horns. There are no marks for the nuchal ligament. Two ridges are present on the lateral aspect of the occipital crest. The astragalus is similar to that of *Palaeomeryx eminens* from Steinheim, Germany.

This taxon is similar to other palaeomerycids, such as *Triceromeryx pachecoi* (Villalta, Crusafont, and Lavocat 1946), *Ampelomeryx ginsburgi* (Duranthon et al. 1995), *Xenokeryx amidalae* (Sánchez et al. 2015), and *Tauromeryx turiasonensis* (Astibia, Morales, and Moya Sola 1998) in possessing occipital horns. Unlike other palaeomerycids, the occipital horns are subdivided all the way to their base to form two distinct shafts; in previously named taxa, the occipital horns are fused at the base. Hence, the Siwalik specimen may be plesiomorphic.

cf. *Tauromeryx* sp.

A single ossicone (YGSP 21816 from Locality Y311 dated to 10.1 Ma) is similar to *Tauromeryx*. It possesses forward curvature and

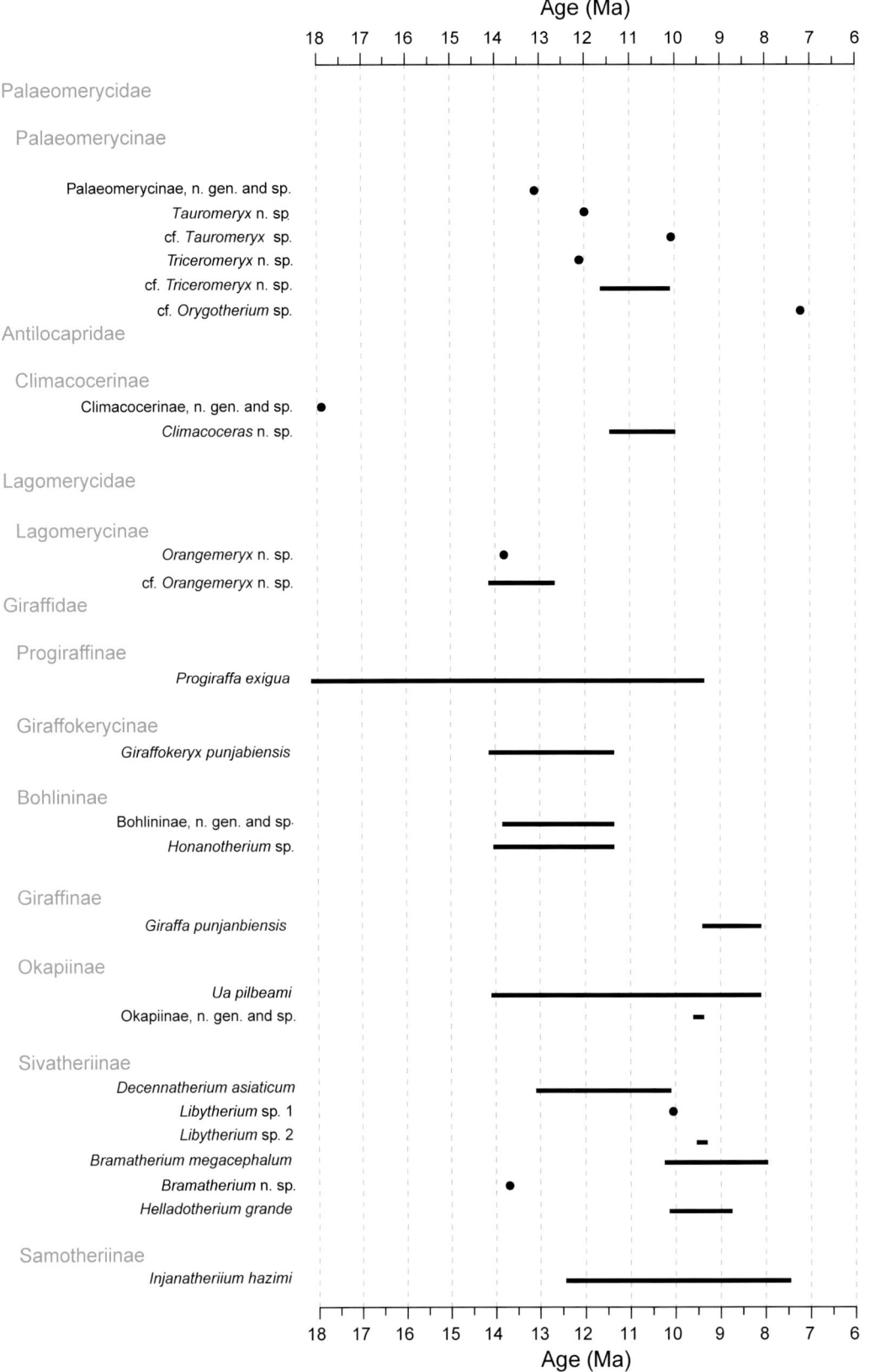

Figure 22.1. Biostrati-graphic range of Giraffoidea in the Siwalik record of Pakistan.

longitudinal grooving. The surface is smooth in both taxa; it resembles a gazelle horn core. Details of the surface, however, and a cylindrical cross-section make it more like a giraffoid, most similar to *Tauromeryx*.

Triceromeryx Villalta, Crusafont Pairo, and Lavocat, 1946
Type species: *Triceromeryx pachecoi* Villalta, Crusafont-Pairo, and Lavocat, 1946

Triceromeryx n. sp.
This taxon is recognized from two ossicone fragments (YGSP 46990 and YGSP 47194, both from Locality Y500 dated to 12.1 Ma). Long surficial deep grooves are separated by thin well-defined, continuous but irregular ridges. The surface between the ridges is pitted and grooved with oval depressions. The crests of the ridges are covered with bumpy irregular growths. Distally the apical end is rounded. In cross-section, the ossicone cortex is notably thick. The two skull fragments reveal a thick calvaria without sinuses. This is an unusual braincase for a ruminant. The inner surface of the calvaria shows a broad, bony, rounded transverse ridge between the cerebellum and cerebrum. An occipital horn is tentatively assigned to this taxon. The horn is of similar size to the braincase. The occipital horn commences as a single elongated shaft. The distal part is not preserved. The surface is grooved, unlike in *Triceromeryx pachecoi*. Both taxa share a strong anterior thickening at the base of the occipital horn.

 Triceromeryx pachecoi, the type species, has smooth, rounded, irregular ridges on the surface of its supraorbital ossicones. It has a forked occipital horn (Crusafont-Pairó 1952; known only from Spain). In the new species, the ossicone ridges are sharper and more continuous. The calvaria is notably thick with the broad transverse ridge, which is also known in *Nyanzameryx*, *Samotherium*, and *Okapia*. The paracone and metacone of the molars are separate, as in bovids, and they have the metacononule fold, as in palaeomerycids. The lower p4 is complex in its cuspids.

cf. *Triceromeryx* n. sp.
Two ossicones (YGSP 21897 from Locality Y504 dated to 11.6 Ma and YGSP 23715 from Locality Y311 dated to 10.1 Ma) are the basis for recognizing another new taxon. This species is similar to *Triceromeryx pachecoi*. An apical ossicone has a unique surface with irregular oval pitting over the entire surface, the pits being transverse, oblique, and vertical. The apex terminates in a rounded end. There are irregular growths as in *Triceromeryx pachecoi*. Some consider this specimen heavily altered by post-mortem weathering, but no other Siwalik specimen among thousands has this type of weathering. Both specimens, although from different localities, have similar surficial morphology.

cf. *Orygotherium* sp.
A single ossicone YGSP 50593 (from Locality Y581 dated at 7.2 Ma) represents a notably small-bodied taxon, estimated as the size of a small duiker. The ossicone surface is grooved and

strongly inclined posteriorly. There is a sinus basal to the ossicone and a large, restricted frontal sinus. The specimen has no pedicle, and its position on the skull is unlike that of Bovidae. It is directed posteriorly. Because of its small size, we have tentatively related it to *Orygotherium* (Vislobokova 2004). *Orygotherium* is known only from dentitions, so our assessment is based on size.

Family Antilocapridae Gray, 1866 (including Merycodontidae Matthew, 1904; Climacoceridae Hamilton, 1978)

Subfamily Climacocerinae (new this study)
Climacoceras MacInnes, 1936
Type species: *Climacoceras africanus* MacInnes, 1936

Climacocerinae, n. gen. and sp.
Two horn fragments (YGSP 31790 and YGSP 31796 from Locality Y747 dated to 17.9 Ma) have tubular straight central shafts that are equally thick at the proximal and distal ends. There is no internal canal, and the horns display a gentle curvature of the long axis. The horn surface has straight, longitudinal, extremely fine grooving. The size is comparable to specimens of *Climacoceras africanus* (MacInnes 1936).

 These fragments resemble a frontlet, two horns attached to a piece of frontal (NHM M26690), from Gebel Zelten, Libya, which was described as a female of *Prolibytherium magnieri* (Sánchez et al. 2010). There is a general similarity between the male skulls of *Prolibytherium* and the presumably female frontlet, which involves a long medial thickening (along the axis) in the palmate male horn in the same orientation as that of the horn shaft of the frontlet. In our estimation, the females of *Prolibytherium* were most likely hornless, as are most other basal ruminants. In addition, the frontlet has no branches, a condition that differs from that of the presumed males of *Prolibytherium*. This Libyan frontlet represents a new genus and species of Climacocerinae, systematically close to *Prolibytherium* and *Climacoceras* but differing from both in having no branching of the horns. The long axis of the horn curves the same way in all three taxa.

 In Maboko Island, Kenya, Miocene fossil horns of *Climacoceras* are abundant and often occur as fragments, including small fragmented apices (age about 17 Ma). In the Siwalik record, no small apices were recovered. The Siwalik ruminant sample comprises thousands of specimens; the fact that we did not recover any apical tips suggests that the Siwalik taxon may not have had branches. It is reasonable therefore to infer that this rare taxon from the Siwaliks was similar to the Gebel Zelten frontlet (NHM M26690), which also lacks branches. The Siwalik specimens are placed in the same taxon with the Gebel Zelten specimen rather than the Maboko taxon.

 The present study places Climacocerinae, a new subfamily, in the Antilocapridae. Species of Climacocerinae are the size of a reindeer (90–180 kg). Climacocerinae superficially resemble Cervidae because some species have branching horns. The horn base, however, is not detachable, and there is no cervid pedicle or burr. This condition is evident from *Climacoceras gen-*

tryi, the Gebel Zelten frontlet, and specimens described by Thomas (1984). The base and the horn are one continuous structure with the frontal, as is the horn of *Antilocapra* and the Merycodontinae. It is worth noting that Merycodontinae and some of the Antilocaprinae possess branching horns.

The presence of Climacocerinae in the Siwalik record in the Potwar Plateau, with an age range of 17.9–10.0 Ma, is the first record outside of North Africa (the Gebel Zelten material is dated at about 16 Ma). The Siwalik specimens, although fragmentary, are pristine, and the fine grooving on the surface presents new information about this species. *Nyanzameryx* from Nyakach, Kenya, confirms the fine grooving (Thomas 1984). In contrast, other taxa, such as Antilocaprinae and *Orangemeryx*, have smooth horn surfaces. In all Antilocapridae, the horns are simple extensions of the frontal. We propose that the Siwalik specimens are conspecific with the taxon from Gebel Zelten (NHM M26690).

Climacoceras n. sp.

A large horn fragment (YGSP 47197 from Locality Y76 dated at 11.4 Ma) representing the shaft base is larger than in Climacocerinae, n. gen. and sp. The fragment has a thin cortex. In cross-section, the elongated rod fibers of the horn compose most of the external cortex. On the surface, the ridges appear as fine lineations. The inner cortex consists of disorganized trabeculae that create an in-folded inner surface, similar to specimens figured in MacInness (1936, Fig. 3a, b). The medulla attaches tightly to the in-folded surface. The medullary trabeculae are long and wavy and directed latero-medially. An additional distal horn fragment (YGSP 13977 from Locality Y251 dated at 10.0 Ma) has extremely fine grooving. They terminate on pointed apices which is characteristic.

Family Lagomerycidae Pilgrim, 1941

Subfamily Lagomerycinae, this study
Lagomeryx Roger, 1904 (cited in Simpson 1945)
Orangemeryx Morales, Soria, and Pickford, 1999
Type species: *Orangemeryx hendeyi* Morales, Soria, and Pickford 1999

Orangemeryx n. sp.

We recognize a new species for a partial skull with horns (YGSP 23698 from Locality Y666 dated to 13.8 Ma). The horn base above the frontal is lateromedially compressed in cross-section with a median keel and lateral keels running along the shaft. The horn surface is very smooth without sutures at the base and is therefore an extension of the frontal. Internally, the horn has a small central canal. The frontals are highly pneumatized and elevated in the intercornual area.

cf. *Orangemeryx* n. sp.

A new species is recognized from a partial skull (YGSP 31810 from Locality Y750 dated to 12.7 Ma) and a partial horn (YGSP 30608 from Locality Y758 dated to 14.1 Ma). The horn is situated above the edge of the orbit and inclined posteriorly; the shaft is straight. Small sinuses are present. There are two laterally positioned keels on the horn. The surface has many short discontinuous grooves, while others are more continuous. The horn is antero-posteriorly compressed with a concave area on the posterior aspect. The base of the horn is quadrangular in cross-section. The cortex is very thin and the trabecular medullary bone is uniform and disorganized.

The first new species is similar to *Orangemeryx hendeyi* from Arrisdrift in Namibia (early Middle Miocene about 17.0–17.5 Ma; Morales, Soria and Pickford1999). The *Orangemeryx* taxa appear to have horns that are extensions of the frontal bone, as in Lagomerycinae and the North American Dromomerycidae. It is likely that these Siwalik specimens belong to Lagomerycidae and not to Palaeomerycidae, which possess supraorbital ossicones. The second new species is problematic as the horn is grooved. In that respect, it differs from *Orangemeryx* and is more similar to *Climacoceras*. It differs from Palaeomerycidae as the base of the horn is narrow and the horn is an extension of the frontal. In *Palaeomeryx* and Giraffidae, the base of the ossicones is always broad and is a separate ossification from the frontal.

Prolibytherium Arambourg, 1961

Type species: *Prolibytherium magnieri* Arambourg, 1961
Prolibytherium fusus Danowitz, Domalski, and Solounias, 2016

A braincase from the Zinda Pir sequence (Z162 from Locality Z124 dated to 18.8 Ma) is the basis for recognizing this species. The occipital condyles are fused at the midline to create a single articular surface. This trait is unique for the genus. The bulla is bulbous. The posterior basioccipital tuberosities are large. The occipital has the characteristic hourglass appearance of Giraffidae. There are two massive occipital ridges in the mastoid area. This taxon appears to have had an occipital horn emanating on the dorsal side but caudally. The occipital edge of the skull is broken so some anatomical details are uncertain.

In the type species, *Prolibytherium magnieri*, originally known from Gebel Zelten, Libya (Early Middle Miocene, about 16.5 Ma; Hamilton 1973), several horn specimens appear to be extensions of the frontal. The horns show venation impressions as in Cervidae, but the cortex is very thin. The Siwalik specimen (*Prolibytherium fusus*) is a well-preserved braincase, but the horns are broken off and missing. Their outline shows that they were positioned posteriorly as in the Gebel Zelten species. This identification establishes a zoogeographic connection between the Siwalik faunal province and Gebel Zelten. The Zinda Pir locality is older than Gebel Zelten (18.8 Ma versus 16.5 Ma). Several calvariae in the Libyan specimen show exquisite details of the horns, which seem to be extensions of the frontal. The single upper dentition is heavily worn but shows that the enamel surface is not crenulated. The mandibular rami are thick and resemble those of *Antilocapra*. If there are dentitions present in the Siwalik record, they may be mistaken for bovids. In contrast, the enamel surfaces of teeth of Giraffidae, including *Progiraffa*, are crenulated. Thus, *Progiraffa* and all Giraffidae are dentally distinct from *Prolibytherium*.

The systematic position of *Prolibytherium* has been debated. It has been assigned to Palaeomerycidae based on the "horns"

(Janis and Scott 1987; Prothero and Liter 2007; Solounias 2007). Other studies suggest a closer relationship to *Climacoceras* based on similarities of the female *Prolibytherium* horns to those of *Climacoceras* and on the shared orientation of the cranial appendages in relation to the skull (Sánchez et al. 2010; Danowitz, Domalski and Solounias 2016). We consider *Prolibytherium* to be more closely related to *Climacoceras* based on the branching of the horn and absence of the occipital horn. The horns branch in *Climacoceras*, *Orangemeryx*, and *Prolibytherium* but not in Dromomerycidae or in Climacocerinae new genus and new species.

Family Giraffidae Gray, 1821

Fossils of the Giraffidae are notably common in the Siwalik record, with over 2,000 specimens recorded.

Subfamily Progiraffinae
Progiraffa Pilgrim, 1908
Progiraffa exigua Pilgrim, 1908

Progiraffa has been studied extensively (Barry et al. 2005). *Canthumeryx sirtensis* Hamilton 1973 is a junior synonym. However, *Zarafa zelteni* Hamilton, 1973, is a palaeomerycid (this study). *Progiraffa exigua* is a primitive giraffid. The lower premolars are plesiomorphic with long and with open lingual cuspids (Hamilton 1978). Similar specimens from the Siwaliks were identified by Pilgrim (1911; the type) and Matthew (1929). The Siwalik sample includes dentitions, astragalus, and the distal metapodial from localities ranging in age from 18.1 to 9.4 Ma. Although no ossicone has been identified in the Siwaliks, Moruorot Hill, or Gebel Zelten, this species probably did possess ossicones, as do all giraffids and their sister taxon the Palaeomerycidae.

Progiraffa exigua is one of the smaller giraffids. The teeth are low-crowned. The enamel walls of the m3 bulge as in extreme brachydont taxa. In the upper molars, the mesostyle and parastyle are strong, and the anterior rib is well developed. There are elongated basal pillars.

The metatarsal is long and slender. The proximal end of the metatarsal head flares outward. There is a notably flat plantar surface with a minimal central trough. The medial and lateral ridges are weakly developed. The astragalus is plesiomorphic in having a large head relative to the trochlea.

Subfamily Giraffokerycinae Solounias, 2007
Giraffokeryx primaevus (Churcher 1970)

This species is known from a dentary (AMNH 19323) from the Chinji Formation. The specific age is not known. This specimen is similar to the Fort Ternan taxon (AMNH 19849). The lower p4 is not complex and has a single simple stylid extending from the mesolingual conid. Overall, the stylids are simple.

Giraffokeryx punjabiensis Pilgrim, 1910a
Although the original specimens, and hence the type, are a mixture of fragmentary dentitions (Pilgrim 1911), most researchers consider a skull at the AMNH (19475) found 60 m above the level of the Chinji Rest House and 6 km west of that loca-

tion, Punjab (Colbert 1935), to be the definitive specimen. The original dentitions could belong to the same taxon as the skull. There are two pairs of ossicones: the anterior pair positioned anterior to the orbit and the left and right ossicones fused at the base of the midline; the posterior pair is positioned well caudal to the orbits in the center of the temporalis line and fossa. The cranial and caudal ossicones are similar and cannot be easily distinguished as fragments detached from the skull. This similarity is a characteristic of this genus. The caudal ossicone surface has long grooves, and there are anterior and posterior keels, creating a triangular shape in cross-section. The ossicone cortex consists of two thin layers. The medulla is composed of a simple trabecular structure. The apex of the ossicone terminates in a nonexpanded, rounded end (no knob). The caudal side of the ossicone base is a separate conglomeration of bone units that appear distinct in trabecular structure. They form a basal bulge with a surface of disorganized trabeculae. The frontal boss is distinctive in possessing a broad base medially and no visible base laterally. Thus, the boss is exposed on the medial attachment of the ossicone with the frontal. Two large supraorbital foramina are connected with flumina. Numerous smaller supraorbital foramina are positioned anteriorly. The supraorbital sinuses are very small, and the frontal sinuses are absent. The occipital crest projects strongly backward. In dorsal view, the occipital forms a flat edge. The bulla is compressed. The posterior basioccipital tuberosities are small and separate from the occipital condyles. The nasolacrimal canal is single and small at the cranial orbital edge. In the molars, the mesostyle blends into the metacone. The lingual basal pillars are small. The lower p4 is complex and has two stylids extending from the mesolingual conid (Pilgrim 1911, Plate 2, Fig. 1). Overall, the stylids are complex. The second cervical vertebra has a bifid spinous process. In the third cervical vertebra, the spinous process is positioned cranially. The cranial bulge is intermediate between rounded and flattened. The ventral tubercle is positioned cranially, and the anterior arch is uninterrupted. The metatarsal is slender and long, similar to the metatarsal of *Giraffokeryx primaevus* from Fort Ternan, Kenya (figured by Churcher 1970). There is no significant flaring of the distal shaft toward the condyles. *G. punjabiensis* was about the size of the living okapi.

Both species occur in the Siwalik record, *G. primaveus* being much rarer than *G. punjabiensis*. *G. primaevus* is plesiomorphic dentally but the postcranials of the two species are similar (Harris, Solounias, and Geraads 2010; Ríos, Danowitz and Solounias 2016). This contrast suggests evolutionary change in diet but not in locomotion. *G. primaevus* is known from Fort Ternan (13.7 Ma; Pickford et al. 2006), which is about the same age as the specimen from the type Chinji Formation (~14.0–11.5 Ma). The genus is also known from Paşalar, Turkey (13.7–14.9 Ma; Alpagut, Andrews, and Marin 1990) as *Giraffokeryx* aff. *punjabiensis* (Gentry 1987, 2003).

These are plesiomorphic giraffids (Solounias 2007; Ríos, Danowitz and Solounias 2016), which possess an open nasolacrimal canal similar to that of *Zarafa*, whereas later giraffids have a closed canal. *Giraffokeryx punjabiensis* shares similarities with the sivatheres, *Sivatherium* and *Bramatherium*, including secondary shortening of the cervical vertebrae and the presence of two pairs of ossicones (Geraads and Güleç 1999; Danowitz, Domalski and Solounias 2015). Although further research and phylogenetic analyses are needed, we feel that *Giraffokeryx* represents a plausible ancestor for the sivatheres, a group whose earliest occurrence is in the Siwalik record.

Subfamily Bohlininae Solounias, 2007
Bohlinia Matthew, 1929

Bohlinia attica (Gaudry and Lartet 1856) is from Pikermi, Greece (Kostopoulos, Koliadimou and Koufos 1996; Likius, Vignaud, and Brunet 2007; Marra et al. 2011; Parizad et al. 2019). The upper P3 is circular in occlusal view, and the parastyle and metastyle curve inward toward the mesostyle. There are no frontal sinuses. The ossicones are massive at the base and do not terminate in knobs. The neck is long. The astragalus is less squared than in *Giraffa*. The medial epicondyle of the femur is small. The metapodials are long with a deep ventral trough.

Bohlininae, n. gen. and sp.
Before the description of the material, we present the relevant taxa. Some of these taxa also need amendments in morphology and determination.

The new taxon is based on a metatarsal, radius, metacarpals, astragalus, tibia, and an atlas from several localities ranging in age from 13.8 to 11.4 Ma. The atlas is long as in *Giraffa*. Metacarpal YGSP 47187 from Locality Y76 dated to 11.4 Ma is the most slender and longest of any ruminant—thus unique among ruminants. We also refer a maxilla from Tung Gur, Mongolia, (11.8–13.0 Ma;) to this new genus and species. This species resembles *Bohlinia* in the metapodials with a deep trough and sharp ridges. In appearance, this new taxon is similar to the long-legged dibatag (*Ammodorcas*) but was tall with a long neck, like the gerenuk (*Litocranius*). It was the size of a small giraffe.

Similar species are found in the Chinji Formation (AMNH collection), in Tung Gur, Mongolia (AMNH collection), and in Macedonia, Greece. Presently, it is not possible to relate these species more closely to the new genus as the material is fragmentary and sampled from different ages and localities over huge geographic distances.

Honanotherium Bohlin, 1926

Honanotherium schlosseri, Bohlin, 1926, is the type species. *Honanotherium bernori* is a smaller species from Maragheh, described by Solounias and Danowitz (2016b).

Honanotherium sp.
A lower p3 (YGSP 19852 from Locality Y76 dated to 11.4 Ma) should be assigned to *Honanotherium*, a species smaller than *H. schlosseri*. Also, a lower p3 (YGSP 39136 from Locality Y44 dated to 14.0 Ma) and a fragmentary maxilla (YGSP 27662 from Locality Y507 dated to 11.7 Ma) are assigned to *Honanotherium*. The lower p3 is primitive, and the premolars are bulbous in occlusal view.

The Siwalik record of *Honanotherium* currently consists of a few teeth. Colbert (1935) identified the mandible from Bhandar on the Potwar Plateau as *Giraffa punjabiensis*, but it is more similar to *Honanotherium schlosseri*. Other upper molars in the collection also resemble those of *Honanotherium*. The metapodial of *Honanotherium* has a deep trough as in *Bohlinia* but is shorter. The dentition of *Honanotherium* is plesiomorphic to those of *Bohlinia* and *Giraffa*.

Vishnutherium Lydekker, 1876
Vishnutherium iravadicum Lydekker, 1876, Plate 9, Figures 1 and 2

This taxon was originally known from a fragmentary mandibular dentition from the Siwaliks of Myanmar not Pakistan. This species has lower molars with notably large, distinct stylids and cingulids. The upper molars have well developed buccal styles. The mesostyle is double. It is a rare taxon with interesting biogeographic connections. The molars of the figured type have as a key characteristic extremely well-developed lingual stylids.

Vishnutherium priscillum (*Giraffa priscilla* Matthew, 1929)
This is a smaller species with similar morphology to *V. iravadicum*. It is known from an upper molar figured by Pilgrim (1911). A second molar was identified at the AMNH (824 Locality Y49—Hasnot, Punjab). (We do not know the age of these specimens and the figure from Pilgrim is the only documented record of this species.)

Overdeveloped styles and stylids occur in *Bohlinia* specimens from Pikermi. We place *Vishnutherium* in the Bohlininae based on the large distinct styles of the upper molars, the stylids of the lower molars, and the deep trough of the metapodials.

Subfamily Giraffinae Gray, 1821
Giraffa Brisson, 1756
Giraffa camelopardalis Linnaeus, 1758

Giraffa camelopardalis is the type species of this genus. Recently the living species, *Giraffa camelopardalis*, has been divided into four species on the basis of genetic differences (Fennessy et al. 2016). The four species are *Giraffa camelopardalis*, *G. reticulata*, *G. tippelskirchi*, and *G. giraffa*. The morphology of these recently recognized species has not been well studied.

Giraffa punjabiensis Pilgrim, 1910a

This species is based on dentitions in the Indian Museum (Kolkata), described but not figured in Pilgrim (1910a, 1911). The dentition is large and similar in size to that of the modern giraffe. The new material includes an ossicone, an astragalus, and a fragment of a metatarsal (Danowitz, Barry, and Solounias 2017). The ossicone is similar to that of the extant giraffe. It is small and curves posteriorly with a small apical knob. There is little secondary bone growth on the distal surface. The underside is concave and suggests the presence of a small frontal boss. The astragalus is squared and resembles that of *Bohlinia attica* more than that of *G. camelopardalis*. The metatarsal has no trough as in *G. camelopardalis* (Danowitz, Barry, and Solounias 2017).

This species is the oldest known member of *Giraffa* (8.1–9.4 Ma), with a probable origin in the Indian subcontinent. It coexisted with Bohlininae new genus and species and *Honanotherium* sp. This species is of comparable size to the modern giraffe. The new ossicone specimen must have belonged to a young individual, as it is still detached. This species may be ancestral to the younger *Giraffa sivalensis*. Eventually, a species of *Giraffa* dispersed into Africa.

Giraffa sivalensis Falconer and Cautley, 1843

This Pleistocene species is recognized from fossils in the Natural History Museum (London). The sample includes dentitions, a third cervical vertebra, a partial humerus, a mid-shaft metatarsal, proximal and mid-shaft metacarpals, a proximal phalanx, and a middle phalanx.

Giraffa sivalensis is within the size range of the modern giraffe. The dentition and the cervical vertebra are morphologically similar to those of the modern giraffe (Danowitz et al. 2015). The distal humerus differs in the epicondyles. The metatarsal and the metacarpal, however, differ clearly from those of the modern giraffe in having a deep trough. The length of the metapodials cannot be determined due to their fragmentary nature. The premolars have tall basal pillars, and the molars have wavy central cavities and lack basal pillars. The upper third molar has a small hypocone. The third cervical vertebra is elongate and slender; it has a cranio-caudally long but dorso-ventrally low spinous process that is positioned caudally on the dorsal lamina. There are elongated cranial articular processes that are not connected to the dorsal lamina. A thin ventral ridge extends the length of the ventral vertebral surface. The humerus has a tall minor tubercle and a rounded, confined head. The crest on the shaft is long and distinct. The medial epicondyle is large. There is a significant V-shaped notch on the condyle. The metacarpal fragments indicate that it is an elongated limb bone, wide proximally and distally constricted centrally. The central trough is notably deep. The proximal phalanx is gracile and elongated, with a broad lateral palmar eminence. The shaft is broad at the base and the head and thinner centrally.

This species is the youngest *Giraffa* in the Siwalik sequence, with *Giraffa* becoming extinct during the Pleistocene in the Indian Subcontinent.

Subfamily Okapiinae Bohlin, 1926
Okapia Lankester, 1901

The living species, *Okapia johnstoni* Lankester, 1901, is described in Lankester (1902, 1907), Colbert (1938), Hamilton (1978), Fraipont (2007), and Danowitz and Solounias (2015).

Ua pilbeami Solounias, Smith, and Ríos Ibáñez, 2022
A new taxon is represented by a partial skull with both ossicones and orbit (YGSP 28483 from Locality Y758 dated to between 13.9 and 14.1 Ma) (Solounias, Smith, and Ríos Ibáñez 2022). The ossicones are long, rounded, and curve gently inward, forming a U in anterior view. They show slight torsion, and the apex terminates in a small expanded secondary growth

with an undercut groove. There is a small canal inside. Long ridges occur on the surface. The anterior frontals are strongly pneumatized. Additional ossicone specimens from localities dated from 13.1 to 8.1 Ma are similar to the ones described. These features are similar to morphologies of *Okapia*. This taxon is characterized by upper premolar teeth described in Matthew (1929, Fig. 44), in which the anterior style (parastyle) is completely separated from the antero-labial cone even in late wear. The parastyle is positioned buccally at the level of the mesostyle. These premolar features are encountered only in *Okapia*. The metapodials assigned to this taxon are similar to those of the living okapi but longer.

Okapiinae, n. gen. and sp.
A second taxon, Okapiinae, new genus and new species, is a larger-sized species similar to *Ua pilbeami*. The ossicone apices possess complex expanded knobs (YGSP 9733 from Locality Y265 dated to 9.6 Ma; YGSP 53273 from Locality Y309 dated to 9.4 Ma). A maxilla in Matthew (1929, Fig. 45) also has okapi-like premolars. We propose to link these fossils with the previous material as a large species of the same genus.

The recognition of two species (smaller and larger), based on the specialized morphology of the upper premolars P2 and P3, the expanded sinus of the frontals, and the apical ossicone growth, as similar to *Okapia* is novel. It is a recognition of a Miocene taxon in the Okapiinae. It is probable that the modern species, *Okapia johnstoni*, arose from an ancestor like these from the Indian subcontinent and dispersed to Africa. The living species occurs today in the mountainous rainforests of northeastern Zaire. *Afrikanokeryx leakeyi*, known from the Baringo Formation of Kenya is another Okapiinae. It has ossicone canals and large bullae (Harris, Solounias, and Geraads 2010).

Subfamily Sivatheriinae Bonaparte, 1850
Decennatherium Crusafont-Pairó, 1952
Decennatherium asiaticum

The type species is *Decennatherium pachecoi* Crusafont-Pairó, 1952; see Morales (1985) and Ríos, Sánchez and Morales (2016) and Ríos, Danowitz, and Solounias (2019). The Siwalik metapodial material was published as *Decennatherium asiaticum* (Ríos, Danowitz, and Solounias 2019). These metapodials are similar to those of *Decennatherium* spp. from Spain, which range in age from 10 to 8 Ma.

New material assigned to *Decennatherium* includes a skull (YGSP 47357 from Locality Y647 dated at 12.4 Ma) and ossicones (including YGSP 6392 from Locality Y311, 10.1 Ma, and YGSP 47192 from Locality Y514, 13.0 Ma). The ossicones are not fused at the base, unlike in *Decennatherium pachecoi* and *D. rex* known from Spain. The skull is similar to that of *Bramatherium permimense* (Colbert 1935; Ríos, Danowitz, and Solounias 2019) in the orbits, sinuses, and the large tall ossicone bosses. This species is also similar to *Bramatherium perimense* in the location of the anterior ossicones behind the orbits on huge bosses, the extent of the frontal sinuses, and the protruding orbits. The ossicones, however, are not fused at the base. The species is also similar to *B. perimense* in the shortened braincase and basioccipital morphology. It differs from *Giraffokeryx*, which has two ossicone pairs, the anterior pair being fused at the base, the posterior pair being more compressed and curving backward instead of outward and inward. In *Giraffokeryx*, the ossicones are deeply grooved. Both taxa have a bulge at the base of the ossicone. In *D. asiaticum*, the ossicones are positioned dorsally and slightly posterior to the orbit, whereas in *Giraffokeryx*, the anterior pair lies anterior to the orbit, and the posterior pair lies well posterior to the orbit. The frontal sinuses are more developed, as in *Bramatherium*, whereas in *Giraffokeryx*, they are absent.

Libytherium Pomel, 1892

The genus *Libytherium* was erected by Pomel (1892) for giraffid-like dental material from Pliocene deposits of Oran, Algeria. He not only erected a new genus but also a new species, *Libytherium maurusium*, assigning a mandibular fragment with a partial p3, p4–m3 as the holotype (Pomel 1892m Plate 5, Figs. 1, 2; Plate 6, Figs. 1, 2). An ossicone was added later from the same locality. We find this genus to be distinct from *Sivatherium giganteum*. Most authors have used the genus *Sivatherium* for all the *Libytherium* and *Sivatherium* material. The main difference is that in *Sivatherium giganteum*, the metacarpals are notably shorter than the metatarsals, a unique feature. In *Sivatherium* the posterior ossicones have no canals and are composed of hundreds of internal fine placations. External to the placations are flat ribbons of bone material. In *Libytherium* the metapodials are sub-equal in length, and the ossicones possess canals and are not composed of fine placations. There are no ribbons of bone on the surface, which is grooved or smooth depending on the species.

Several ossicones in the Siwaliks are assigned to *Libytherium*.

Libytherium sp. 1
Libytherium sp. 1 is represented by ossicones (e.g., YGSP 46530 from Locality Y311 dated to 10.1 Ma). Several dentitions and postcranial elements, including a radius, cubonavicular, calcaneum, metatarsal, and distal phalanx may be referred to this taxon.

Libytherium sp. 2
Libytherium sp. 2 is recognized by both smaller ossicones (YGSP 14823 from Y391 dated to 8.3 Ma and YGSP 51540 from Locality Y998 dated to 8.1 Ma) and massive ossicones with deep grooves on all sides (YGSP 7024) as well as a compressed lower p4 (YGSP 14387 from Locality Y452 dated to 7.3 Ma). The ossicone surface has thick ridges that twist around the shaft, and discrete lumps of secondary bone growth are linearly distributed on the outer edge. The ossicones are straight, and their morphology is similar to that of specimens that are recognized as *Sivatherium maurusium* from Langebaanweg, South Africa, and from Libya.

Libytherium most likely evolved from *Decennatherium*, which in turn evolved from *Giraffokeryx*. The best known so far is species 1. It possessed a short snout, and the metapodials were long as in *Bramatherium* and *Decennatherium*. The diagnostic features are the curved ossicone with an abrupt para-apex and bumps and the short mandibular symphysis. The twisted ossicone resembles specimens from Langebaanweg (Singer and Boné 1960).

Sivatherium Falconer and Cautley, 1835
Sivatherium giganteum Falconer and Cautley, 1835

Sivatherium giganteum Falconer and Cautley, 1835 is the type for *Sivatherium*. This Pleistocene species is recognized in the Siwaliks from a skull with four ossicones, additional ossicone specimens, dentitions, cervical vertebrae, metacarpal, astragalus, and metatarsal. This species is the largest giraffid known (a little larger than *Bos gaurus* and *Bubalus arnee*). The ossicones have fine internal placations and no sinuses. The metacarpals are notably shorter than the metatarsals. The females are hornless and the frontal sinuses are small.

Bramatherium Falconer, 1845

Bramatherium perimense is a Late Miocene taxon from Perim Island. Bettington (1846) figured and described specimens but Falconer (1845, 1868) named the genus and species (Matthew 1929).

The type is a maxilla at the NHM (M 48933). The best specimens are the referred skull from Perim Island (AMNH 1936 cast of skull preserved in the Royal Society of Surgeons) and a skull and complete mandible with incisor arcade and metapodials from the Dhok Pathan Formation (AMNH 19771; Colbert 1935). This is a large-bodied species.

Bramatherium megacephalum
A skull and complete neck at YPM was described by Lewis (1939). The Harvard-GSP collection includes ossicones, dentitions, cervical vertebra, metacarpal, proximal phalanx, tibia, astragalus, and metatarsal. This is a large species (about the size of *Bos gaurus*).

The new collections include ossicones (YGSP 10473, YGSP 20614, and YGSP 30862 from Locality Y311, 10.1 Ma; YGSP 53273 from Locality Y309, 9.4 Ma; and Y46693 from RH001, 8.0 Ma), a maxilla (YGSP 7459 from Locality Y212, 9.4 Ma), a metacarpal (YGSP 16818c from Locality Y450, 10.2 Ma), and a mandible (YGSP 17499 from Locality Y311). These ossicone apices have expanded knobs, unlike other ossicones of *Bramatherium megacephalum*. The posterior ossicones possess two internal canals, a larger and a smaller canal separated by a thin plate of bone. The mandibular diastema is short. The astragalus is not as squared as in the other species. The metatarsal has a much better developed medial ridge than lateral ridge.

Bramatherium n. sp.
This new species, recognized from YGSP 20597 (from Locality Y640 dated to 13.7 Ma), is notably small-bodied but otherwise similar to other *Bramatherium*. The central canal of the ossicone is small. The ossicone is straight and grooved with a small

apex that does not form a knob. The metacarpal trough is medium in depth. The ridges on the ventral surface are uneven; the medial ridge protrudes more than the lateral ridge as in *B. megacephalum*.

Three skulls of *Bramatherium* are known from the Siwaliks and one from Turkey. It is interesting that the posterior ossicones possess two canals, which suggests that these ossicones arose by the fusion of two ossicones. It is possible that the anterior ossicones also had two canals. This observation raises the possibility that the ancestor of *Bramatherium* may have possessed three pairs of ossicones. The similarity of the anterior and posterior pairs of ossicones is a special feature of this genus. In this regard, *Bramatherium* is similar to *Giraffokeryx*, in which the four ossicones are very similar in structure.

From *Giraffokeryx*, two lineages arose. (1) In the *Decennatherium* and *Libytherium* lineage the ossicones are very similar in shape. (2) *Decennatherium asiaticum* forms a lineage with species of *Bramatherium*. The ossicones of *D. asiaticum* possess a huge boss, a shared feature with *Bramatherium*. Since the ossicones of *D. asiaticum* are not fused and possess a basal caudal bulge, this taxon may be ancestral to *Bramatherium*. *Bramatherium* is also found at Kavakder, Turkey (Geraads and Güleç 1999).

Helladotherium Gaudry, 1860
Helladotherium grande Pilgrim, 1911

Helladotherium grande (Pilgrim 1911, Plate 3, Fig. 1a; Lydekker 1883, Plate 16, Fig. 7—Dhok Pathan, specific locality unknown) is represented by upper molars and a crushed skull. The lingual surface of the upper molars is strongly inclined. The Harvard-GSP collection has teeth from four localities (e.g., YGSP Y11981 from Locality Y311, 10.1 Ma).

The specimen, initially described by Lydekker, was identified as *Vishnutherium* (Lydekker 1883). It does not have the enlarged styles of *Vishnutherium*. We agree with Pilgrim (1911) as to its assignment to *Helladotherium*. It is a large giraffid the size of *Libytherium* sp. 1. It differs from *Vishnutherium iravadicum* in the absence of the continuous lingual cingulum at the base of the molars and in having normal-sized buccal styles. *Helladotherium* is also known from Pikermi, Samos and localities in Turkey.

Subfamily Samotheriinae Hamilton, 1978

Injanatherium Heintz, Brunet, and Sen, 1981, is from Injana Province, Bakhtiari Formation of Iraq. For a review see Heintz, Brunet, and Sen (1981).

Injanatherium hazimi Heintz, Brunet, and Sen, 1981
We place *Injanatherium hazimi*, the sole species, in the Samotheriinae because of a new reconstruction of the holotype skull. This taxon is very similar to *Samotherium boissieri*. It differs in the bulla and in having a shorter ossicone with a flat apex and a thinner orbital rim. It also has smaller sinuses at the base of the ossicone. The ossicone has no central canal and a notably smooth surface. The supraorbital foramen is without flumina and is situated anterior to the ossicone. In *Samotherium*, there

are flumina and the supraorbital foramen is positioned medially to the ossicone. The metapodials of *Injanatherium* are long and slender with a deep trough and sharp ventral ridges. In *Samotherium*, they are short and wide with a shallow trough and dull ridges. *Injanatherium hazimi* may be ancestral to *Samotherium*. *Injanatherium arabicum* Morales, Soria, and Thomas, 1987, is a *Giraffokeryx* (see above). The Harvard-GSP collection has more than a dozen specimens that can be assigned to *Injanatherium hazimi* (e.g., ossicone YGSP 14106 from Locality Y378 dated to 7.5 Ma).

DIETARY HABITS

Paleodietary inferences are based on mesowear and microwear analyses. We used a mesowear approach that combines features on the labial-most and lingual enamel bands of the paracone and metacone (Mihlbachler et al. 2011; Danowitz et al. 2016). For the mesowear analysis, some of the fossil teeth have not been identified at the species level, especially for the smaller-sized Giraffoidea. Based on mesowear data for 165 teeth, most of the dentitions that we examined are interpreted as browsers. The only taxon that suggests slightly mixed feeding is *Bramatherium megacephalum*. The Pleistocene *Sivatherium giganteum* was a grazer. The browsing pattern persists for all specimens through the Siwalik sequence of the Potwar Plateau.

We have microwear data for 114 teeth, documented with the traditional light-microscope microwear method (Solounias and Semprebon 2002) and a new transect method. Two sets of data were generated: one in which the species were identified, the other in which the species were not. For both datasets, we examined changes in microwear features over time. All specimens were interpreted as browsing. There were no significant changes in microwear features throughout the Potwar Siwalik record. The persistent browsing diets over time in the Siwalik record imply availability of browse for large ruminants through the sequence. It was notable that the teeth assigned to *Libytherium* sp. 1 had microwear indicative of extreme browsing. This is interesting as the Pleistocene descendant, *Sivatherium giganteum*, is considered a grazer by mesowear analysis. Data for mesowear and microwear can be obtained by request from the authors.

The microwear data for Siwalik species indicate greater reliance on browsing (broad-leaved angiosperms) than do the giraffids of the Pikermian Biome, which include species of mixed feeders (broad-leaved vegetation and grasses) (Solounias, Rivals, and Semprebon 2010).Comparison of the two faunas implies a more open-canopy habitat in the Pikermian Biome and more forested vegetation in the Siwaliks. *Bramatherium megacephalum* is the only Siwalik species that is slightly more similar to giraffids of the Pikermian Biome. It is notable that *Bramatherium* is also found in Turkey (Geraads and Güleç 1999) and on Samos, Greece. Another interesting pattern is that the only giraffoids absent from the Siwalik record are the Palaeotraginae/Samotheriinae, which are common in the Pikermian Biome and more grazing than all other giraffids (e.g., Solounias 1981; Danowitz et al. 2016).

BODY-SIZE TRENDS

Most giraffoid type specimens are ossicones, rarely skulls. Critical Siwalik specimens include ossicones, braincases, frontal fragments, and a few dentitions. A few more complete skulls are available in other collections (e.g., for *Giraffokeryx*, *Bramatherium*, and *Sivatherium*). For some body-size estimates, we used comparable specimens of extant species (see Table 22.1). It was not possible to use one morphological variable for all body-size estimates. The pedicle of cervids was used as an estimate of size for giraffoid ossicones. Metapodials are also available, but in giraffids, they are large in relation to body weight, a specialized trait for the family; hence estimates based on cervids or bovids would be inaccurate. We used metapodials of *Bramatherium*, *Decennatherium*, *Libytherium*, *Okapia*, and *Giraffa* for size estimates. For these taxa, partial skeletons are known. Two types of size estimates were produced. When possible, the body size of individual species was inferred from the following variables: the length of the braincase from occipital condyles to orbit; ossicone length and width at the base; the length of the upper M2; the length of P2 to M3; and the astragalar width (Fig. 22.2).

Body-mass estimates are also based on regressions developed by Morgan (1994) using the distal width of the astragalus, but without species identification. Although we have many astragali, most species of giraffoids are not known from that element. In addition, many species have very similar body-size dimensions, which complicates identifications. Giraffid astragali are useful for documenting the size range of contemporaneous species and trends over time. Giraffid astragali can be separated from bovid astragali, and most contemporaneous Bovidae are smaller (see Fig. 25.7, this volume). The smallest giraffid known in the Siwalik record is the size of the living okapi. This size for a Miocene ruminant is notably large. The largest astragali are within the size range of extant *Giraffa*.

There are generally two sizes of astragali in Siwalik giraffids: an earlier smaller and a later larger size, changing between ~11 and 10 Ma with no appreciable intermediates. The giraffids begin as large ruminants, the earlier smaller astragali indicating weights of 150 kg and increasing to 300 kg around 11 Ma. After 10 Ma, taxa begin at 600 kg and reach as much as 1,250 kg by around 9 Ma. The Palaeomerycidae, Climacocerinae, and Lagomerycidae overlap in size with the Bovidae and, unlike the Giraffidae, are not bimodal in body size.

CONCLUSION

This review reveals a substantial diversity of ruminant taxa previously unknown from the Siwalik record. The giraffoid taxa range in age from 18.5 to 7.1 Ma in the Potwar Plateau Siwalik record. A few species extend into the Plio-Pleistocene. The distribution of astragali shows a hiatus with few specimens between 11 and 10 Ma and few specimens prior to 14.1 Ma. The giraffoids include seven species of Palaeomerycidae, three of Lagomerycidae, and two of Antilocapridae. These 12 species are new recognitions for the Siwaliks. The number of Siwalik Giraffidae documented in the Potwar Plateau region is 19 taxa. A rough set of size estimates is provided with modern size equivalents.

In terms of feeding ecology, all of the dentitions examined are interpreted as browsing. The only taxon that suggests slightly mixed feeding is *Bramatherium megacephalum*. The Pleistocene *Sivatherium giganteum* from the NHM collection was a

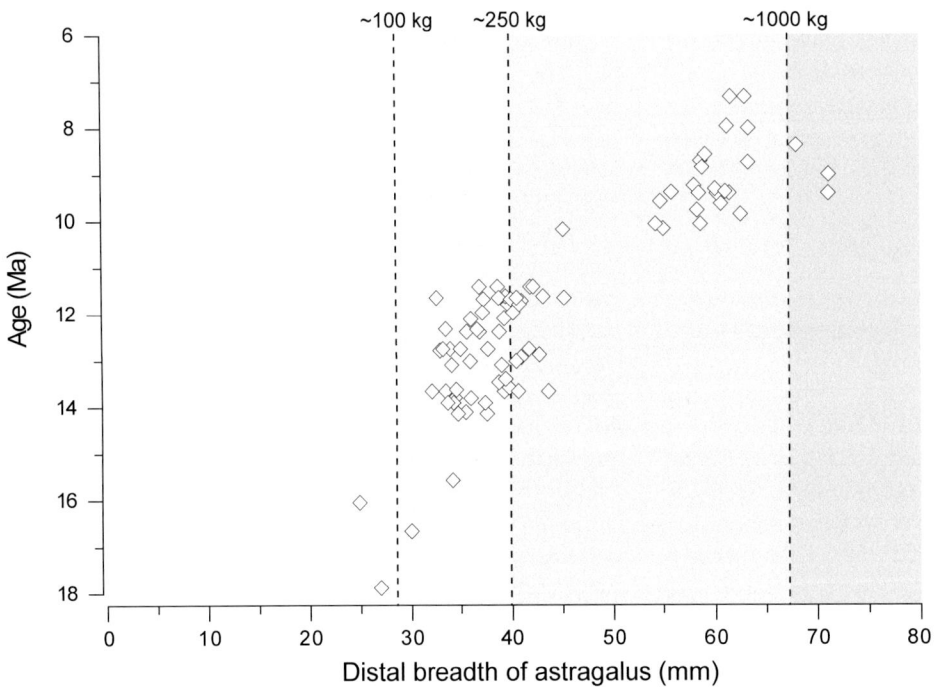

Figure 22.2. Astragalar measurements of Siwalik giraffoids and estimated body mass after Morgan (1994).

grazer. *Libytherium* sp. 1 was a browser. There is no detectable change over time in the browsing habit. The notable change from medium-sized to larger species over time does not involve a change in diet. The question then is, Why did the giraffids get bigger? It is possible that other species were changing in the Siwalik region and that the giraffids became larger to avoid foraging competition.

ACKNOWLEDGMENTS

We especially thank John Barry for numerous interactions and his help with so many issues over the years. We also thank him for putting up with us working around his desk where the fossils are. We also thank Larry Flynn, Michèle Morgan, Catherine Badgley, David Pilbeam and the Department of Human and Evolutionary Biology at Harvard University. We thank the departments of Mammalogy and Paleontology of the AMNH for access to specimens. We thank Jerry Hooker and Pip Brewer at the NHM London. We thank Sevket Sen and Christine Argot and MNHN de Paris. We thank Gertrud Rössner for access to collections at the Paläontologishes Museum in München. We also thank Daniel Brinkman and Chris Norris at YPM. Funds were covered by Nikos Solounias and by the Anatomy Department at NYIT-COM.

References

Alpagut, B., P. Andrews, and Martin, L. 1990. New hominoid specimens from the Middle Miocene site at Paşalar, Turkey. *Journal of Human Evolution* 19(4–5): 397–422.

Arambourg, C. 1961. *Prolibytherium magnieri*, un Velléricorne nouveau du Burdigalien de Libye. *Comptes Rendus Sommaires de la Société Géologique de France* 3:61–62.

Astibia, H., J Morales, and S. Moya Sola. 1998. *Tauromeryx*, a new genus of Palaeomerycidae (Artiodactyla, Mammalia) from the Miocene of Tarazona de Aragón (Ebro Basin, Aragón, Spain). *Bulletin de la Société Géologique de France* 169(4): 471–477.

Barry, J. C., S. Cote, L. MacLatchy, E. H. Lindsay, R. Kityo, and A. R. Rajpar. 2005. Oligocene and Early Miocene ruminants (Mammalia, Artiodactyla) from Pakistan and Uganda. *Palaeontologia Electronica* 8(1): 1–29.

Bettington, A. 1846. Art. XVIII. Memorandum on certain fossils, more particularly a new ruminant found at the Island of Perim, in the Gulf of Cambay. *Journal of the Royal Asiatic Society* 8(15): 340–348.

Bohlin, B. 1926. Die Familie Giraffidae. *Palaeontologica Sinica* ser. C 1:1–178.

Brisson, M. J. 1756. Regnum animale in classes IX. distributum, sive, synopsis methodica. Paris: Jean-Baptiste Claude Bauche.

Churcher, C. S. 1970. Two new Upper Miocene giraffids from Fort Ternan, Kenya, East Africa: *Palaeotragus primaevus* n. sp. and *Samotherium africanum* n. sp. *Fossil Vertebrates of Africa* 2:1–106.

Colbert, E. H. 1935. Siwalik mammals in the American Museum of Natural History. *Transactions American Philosophical Society* 26:1–401.

Colbert, E. H. 1938. The relationships of the okapi. *Journal of Mammalogy* 19(1): 47–64.

Crusafont Pairó, M. 1952. Los Jiráfidos fósiles de España. Dip. Prov. Barcelona. *Consejo Superior de Investigaciones Científicas, Memorias Y Comunicaciones del Instituto Geologico* 8:1–239.

Danowitz, M., J. C. Barry, and N. Solounias. 2017. The earliest ossicone and post-cranial record of *Giraffa*. *PLOS ONE* 12: e0185139.

Danowitz, M., R. Domalski, and N. Solounias. 2015. The cervical anatomy of *Samotherium*, an intermediate necked Giraffid. *Royal Society Open Science* 2(11): 150521.

Danowitz, M., R. Domalski, and N. Solounias. 2016. A new species of *Prolibytherium* (Ruminant, Mammalia) from Pakistan, and the functional implications of an atypical atlanto-occipital morphology. *Journal of Mammalian Evolution* 23:201–207.

Danowitz, M., S. Hou, M. Mihlbachler, V. Hastings, and N. Solounias. 2016. A combined-mesowear analysis of Late Miocene giraffids from north Chinese and Greek localities of the Pikermian Biome. *Palaeogeography, Palaeoclimatology, Palaeoecology* 449:194–204.

Danowitz, M., and N. Solounias. 2015. The cervical osteology of *Okapia johnstoni* and *Giraffa camelopardalis*. *PLOS ONE* 10(8): e0136552.

Danowitz, M., A. Vasilyev, V. Kortlandt, and N. Solounias, N. 2015. Fossil evidence and stages of elongation of the *Giraffa camelopardalis* neck. *Royal Society Open Science* 2(10): 150393.

Duranthon, F., S. Moyà Sola, H. Astibia, and M. Köhler. 1995. *Ampelomeryx ginsburgi* nov. gen., nov. sp. (Artiodactyla, Cervoidea) et la famille des Palaeomerycidae. *Comptes-Rendus de l'Académie des Sciences-Paris. Série 2. Sciences de la terre et des planetes* 321(4): 339–346.

Falconer, H. 1845. Description of some fossil remains of *Deinotherium*, giraffe, and other mammals. *Quarterly Journal of the Geological Society.* 1(1): 356–371.

Falconer, H. 1868. *Palaeontological Memoirs and Notes of the late Hugh Falconer, A.M., M.D. Fauna antiqua sivalensis*, Vol. 1., edited by C. Murchison. London: R. Hardwicke.

Falconer, H., and P. T. Cautley. 1868. On some fossil remains of *Anoplotherium* and giraffe, from the Siwalik Hills. In *Palaeontological Memoirs and Notes of the Late Hugh Falconer, A.M., M.D.*, Fauna antiqua sivalensis, Vol. 1., edited by C. Murchison, 190–207. London: R. Hardwicke.

Fennessy, J., T. Bidon, F. Reuss, V. Kumar, P. Elkan, M. A. Nilsson, M. Vamberger, U. Fritz, and A. Janke. 2016. Multi-locus analyses reveal four giraffe species instead of one. *Current Biology* 26(18): 2543–2549.

Fraipont, J. 1907. Okapia. Contributions à la faune du Congo, Vol.1 Annales du Museé du Congo, 118 pp. Brussels:Imprimerie Veuve Mònnom.

Gaudry, A. 1860. Sur les Antilopes trouvées a Pikermi (Grèce). *Bulletin de la Société Geologique de France* 18:388–399.

Gaudry, A. 1860. Résultats des no fouilles exécutées en Grèce sous les auspices de l'Académie à Pikermi (Grèce). *Comptes Rendus de l'Académie des Sciences* 51:802–804.

Gentry, A. W. 1987. Ruminants from the Miocene of Saudi Arabia. *Bulletin of the British Museum (Natural History) Geology* 41(4): 433–439.

Gentry, A. W. 2003. Ruminantia (Artiodactyla). In *Geology and Paleontology of the Miocene Sinap Formation, Turkey*, edited by M. Fortelius, J. Kappelman, S. Sen, and R. L. Bernor, 332–379. New York: Columbia University Press.

Geraads, D., and E. Güleç. 1999. A *Bramatherium* skull (Giraffidae, Mammalia) from the late Miocene of Kavakdere (Central Turkey). Biogeographic and phylogenetic implications. *Bulletin of the Mineral Research and Exploration* 121:51–56.

Gray, J. E. 1821. On the natural arrangement of the vertebrose animals. *London Medical Repository* 15(1): 296–310.

Gray, J. E. 1866a. XLIII.–Notes on the pronghorn buck (*Antilocapra*) and its position in the system. *Annals and Magazine of Natural History* 18(106): 323–326.

Gray, J. E. 1866b. LXIII.–Additional note on the Antilocapridae. *Annals and Magazine of Natural History* 18(106): 468–469.

Hamilton, W. R. 1973. The Lower Miocene ruminants of Gebel Zelten, Libya. *Bulletin of the British Museum (Natural History) Geology* 24(3):73–150.

Hamilton, W. R. 1978. Fossil giraffes from the Miocene of Africa and a revision of the phylogeny of the Giraffoidea. *Philosophical Transactions of the Royal Society of London. Series B* 283(996): 165–229.

Hassanin, A., and E. J. Douzery. 2003. Molecular and morphological phylogenies of Ruminantia and the alternative position of the Moschidae. *Systematic Biology* 52(2): 206–228.

Heintz, E., M. Brunet, and S. Sen. 1981. Un nouveau giraffidé du Miocène supérieur d'Irak: *Injanatherium hazimi* ng, n. sp. *Comptes Rendus de l'Académie des* Sciences *Série II Sciences de la Vie* 292:423–426.

Harris, J. M., N. Solounias, and D. Geraads. 2010. Giraffoidea. In *Cenozoic Mammals of Africa*, edited by L. Werdelin and W. J. Sanders, 797–811. Berkeley: University of California Press.

Janis, C. M., and K. M. Scott. 1987. The interrelationships of higher ruminant families with special emphasis on the members of Cervoidea. *American Museum Novitates* 2893:1–85.

Kostopoulos, D. S., Koliadimou, K. K. and G. D. Koufos. 1996. *The giraffids (Mammalia, Artiodactyla) from the late Miocene mammalian localities of Nikiti (Macedonia, Greece). Palaeontographica Abteilung* A239(1–3): 61–88.

Lankester, E. R. 1902. On *Okapia*, a new genus of Giraffidae, from Central Africa. *The Transactions of the Zoological Society of London* 16(6): 279–314.

Lankester, E. R. 1907. On the existence of rudimentary antlers in the okapi. *Proceedings of the Zoological Society of London* 77(1): 126–135.

Lewis, G. E. 1939. A new *Bramatherium* skull. *American Journal of Science* 237(4): 275–280.

Likius, A., P. Vignaud, and M. Brunet. 2007. Une nouvelle espèce du genre *Bohlinia* (Mammalia, Giraffidae) du Miocène Supérieur de Toros-Menalla, Tchad. *Comptes Rendus Palevol* 6(3): 211–220.

Linnaeus C. (1758). *Systema naturae per regna tria naturae, secundum classes, ordines, genera, species, cum characteribus, differentiis, synonymis, locis. Vol. 1: Regnum animale.* Holmiae: Impensis Laurentii Salvii.

Lydekker, R. 1876. Indian Tertiary and post-Tertiary Vertebrata. Descriptions of the molar teeth and other remains of Mammalia. *Memoirs of the Geological Survey of India, Palaeontologica Indica* 1(2): 19–83.

Lydekker, R. 1883. Siwalik Camelopardalidae. *Memoirs of the Geological Survey of India, Palaeontologica Indica* 2(4): 116–141.

MacInnes, D. G. 1936. A new genus of fossil deer from the Miocene of Africa. *Zoological Journal of the Linnean Society* 39(267): 521–530.

Marra, A. C., N. Solounias, G. Carone, and L. Rook. 2011. Palaeogeographic significance of the giraffid remains (Mammalia, Artiodactyla) from Cessaniti (Late Miocene, Southern Italy). *Geobios* 44(2–3): 189–197.

Matthew, W. D. 1904. A complete skeleton of the *Merycodus*. *Bulletin of the American Museum of Natural History* 20:101–130.

Matthew, W. D. 1929. Critical observations upon Siwalik mammals: (exclusive of Proboscidea). *Bulletin of the American Museum of Natural History* 56:437–560.

Mihlbachler, M. C., F. Rivals, N. Solounias, and G. M. Semprebon. 2011. Dietary change and evolution of horses in North America. *Science* 331(6021): 1178–1181.

Morales, J. 1985. Nuevos datos sobre "*Decennatherium pachecoi*" (Crusafont, 1952) (Giraffidae, Mammalia): Descripción del cráneo de Matillas. *Coloquios de la Catedra de Paleontologia Madrid* 40:51–58.

Morales, J., D. Soria, and M. Pickford. 1999. New stem giraffoid ruminants from the early and middle Miocene of Namibia. *Geodiversitas* 21(2): 229–253.

Morales, J., D. Soria, and H. Thomas. 1987. Les Giraffidae (Artiodactyla, Mammalia) d'Al Jadidah du Miocène Moyen de la Formation Hofuf (province du Hasa, Arabie Saoudite). *Geobios* 20(4): 441–467.

Morgan, M. E. 1994. Paleoecology of Siwalik Miocene hominoid communities: Stable carbon isotope, dental microwear, and body size analyses. PhD diss., Harvard University.

Parizad, E., M. M. Ataabadi, M. Mashkour, and N. Solounias. 2019. First giraffid skulls (*Bohlinia attica*) from the Late Miocene Maragheh fauna, Northwest Iran. *Geobios* 53:23–34.

Pickford, M., Y. Sawada, R. Tayama, Y. K. Matsuda, T. Itaya, H. Hyodo, and B. Senut. 2006. Refinement of the age of the Middle Miocene Fort Ternan Beds, Western Kenya, and its implications for Old World biochronology. *Comptes Rendus. Géoscience* 338(8): 545–555.

Pilgrim, G. E. 1908. The Tertiary and post-Tertiary freshwater deposits of Baluchistan and Sind, with notices of new vertebrates. *Records of the Geological Survey of India* 37(2): 139–166.

Pilgrim, G. E. 1910a. Preliminary note on a revised classification of the Tertiary freshwater deposits of India. *Records of the Geological Survey of India* 40(3): 185–205.

Pilgrim, G. E. 1910b. Notices of new mammalian genera and species from the Tertiaries of India. *Records of the Geological Survey of India* 40(1): 63–71.

Pilgrim, G. E. 1911. The fossil Giraffidae of India. *Palaeontologia Indica* 4(1): 1–29.

Pilgrim, G. E. 1941. X.–The relationship of certain variant fossil types of "horn" to those of the living Pecora. *The Annals and Magazine of Natural History* 7(38): 172–184.

Pomel, A., 1892. Sur le *Libytherium maurusium*, grand ruminant du terrain Pliocène de l'Algérie. *Comptes rendus hebdomadiares des séances de l'Académie des Sciences* 115:100–102.

Prothero, D. R., and M. R. Liter. 2007. Family Palaeomerycidae. In *The Evolution of Artiodactyls*, edited by D.R. Prothero and S. E. Foss, 241–248. Baltimore, MD: Johns Hopkins University Press.

Ríos, M., M. Danowitz, and N. Solounias. 2016. First comprehensive morphological analysis on the metapodials of Giraffidae. *Palaeontologia Electronica* 19(3): 52A, 1–39.

Ríos, M., M. Danowitz, and N. Solounias. 2019. First identification of *Decennatherium* Crusafont, 1952 (Mammalia, Ruminantia, Pecora) in the Siwaliks of Pakistan. *Geobios* 57:97–110.

Ríos, M., I. M. Sánchez, and J. Morales. 2016. Comparative anatomy, phylogeny, and systematics of the Miocene giraffid *Decennatherium pachecoi* Crusafont, 1952 (Mammalia, Ruminantia, Pecora): State of the art. *Journal of Vertebrate Paleontology* 36(5): e1187624.

Ríos, M., I. M. Sánchez, and J. Morales. 2017. A new giraffid (Mammalia, Ruminantia, Pecora) from the late Miocene of Spain, and the evolution of the sivathere-samothere lineage. *PLOS ONE* 12(11): e0185378.

Sánchez, I. M., J. L. Cantalapiedra, M. Ríos, V. Quiralte, and J. Morales. 2015. Systematics and evolution of the Miocene three-horned palaeomerycid ruminants (Mammalia, Cetartiodactyla). *PLOS ONE* 10: e0143034.

Sánchez, I. M., V. Quiralte, J. Morales, B. Azanza, and M. Pickford. 2010. Sexual dimorphism of the frontal appendages of the Early Miocene African pecoran *Prolibytherium* Arambourg, 1961 (Mammalia, Ruminantia). *Journal of Vertebrate Paleontology* 30(4): 1306–1310.

Simpson, G. G. 1931. A new classification of mammals. *Bulletin of the American Museum of Natural History* 59:258–294.

Simpson, G. G. 1945. The principles of classification and a classification of mammals. *Bulletin of the American Museum of Natural History* 85:1–130.

Singer, R., and E. Boné. 1960. Modern giraffes and the fossil giraffids of Africa. *Annals of the South African Museum* 45:375–548.

Solounias, N. 1981. Mammalian fossils of Samos and Pikermi. Part 2. Resurrection of a classic Turolian fauna. *Annals of the Carnegie Museum* 50:231–269.

Solounias, N. 2007. Family Giraffidae. In *The Evolution of Artiodactyls*, edited by D. R. Prothero and S. E. Foss, 257–277. Baltimore, MD: Johns Hopkins University Press.

Solounias, N., and G. M. Semprebon. 2002. Advances in the reconstruction of ungulate ecomorphology and application to early fossil equids. *American Museum Novitates* 3366:1–49.

Solounias, N., and M. Danowitz. 2016a. Astragalar morphology of selected Giraffidae. *PLOS ONE* 11(3): e0151310. https://doi.org/ 10.1371/journal.pone.0151310.

Solounias, N., and M. Danowitz. 2016b. The Giraffidae of Maragheh and the identification of a new species of *Honanotherium*. *Paleobiodiversity and Paleoenvironments* 96:489–506.

Solounias, N. and M. Ríos Ibáñez. 2023. Giraffoids from the Siwaliks of Pakistan. bioRxiv 2023.10.06.561267; doi: https://doi.org/10.1101/2023.10.06.561267.

Solounias, N., F. Rivals, and G. M. Semprebon. 2010. Dietary interpretation and paleoecology of herbivores from Pikermi and Samos (late Miocene of Greece). *Paleobiology* 36(1): 113–136.

Solounias, N., S. Smith, and M. Ríos Ibáñez. 2022. *Ua pilbeami*: A new taxon of Giraffidae (Mammalia) from the Chinji Formation of Pakistan with phylogenetic proximity to Okapia. *Bollettino della Societa Paleontologica Italiana* 61(3): 319–326.

Spinage, C. A. 1968. Horns and other bony structures of the skull of the giraffe and their functional significance. *East African Wildlife Journal* 6:53–61.

Thomas, H. 1984. Les Giraffoidea et les Bovidae Miocènes de la Formation Nyakach (Rift Nyanza, Kenya) *Palaeontographica Abteilung* A 183(1–3): 64–89.

Villalta J. F., M. Crusafont, and R. Lavocat. 1946. Primer Hallazgo en Europa de rumiantes fósiles tricornios. *Publicaciones del Museo del Sabadell. Communicaciones científicas. Palaeontología* 1946: 1–4.

Vislobokova, I. 2004. New species of *Orygotherium* (Palaeomerycidae, Ruminantia) from the Early and Late Miocene of Eurasia. *Annalen des Naturhistorischen Museums Wien. Serie A.* 106:371–385.

SIWALIK BOVIDAE

ALAN W. GENTRY, NIKOS SOLOUNIAS, JOHN C. BARRY,
AND S. MAHMOOD RAZA

INTRODUCTION

Bovids make up an important part of the Neogene Siwalik mammal faunas of the Indian subcontinent. They also have a long history of study, with fossil bovid limb bones from Tibet being sold to Captain W. S. Webb in 1819 (Pilgrim 1939). Discovery continued through the nineteenth and twentieth centuries, and consequently there are now large collections in museums in India and Pakistan. These include, among others, the Geological Survey of India in Kolkata (GSI), Panjab University in Chandigarh, and the Wadia Institute in Dehra Dun, as well as the Pakistan Museum of Natural History and the Geological Survey of Pakistan (GSP), both in Islamabad. In Europe and North America there are collections in the Natural History Museum in London, the Bavarian State Museum for Palaeontology and Historical Geology, and the Geological Institute, State University of Utrecht, in addition to those in the American Museum of Natural History (AMNH) and Yale Peabody Museum. Beginning in 1973, collections of Siwalik fossils have been made by collaborative projects involving Harvard University, Howard University, the GSP, and the Pakistan Museum of Natural History, as well as by the University of Peshawar and a consortium including Dartmouth College and the University of Arizona. Unlike earlier collections, for the most part this recently obtained material is from well-dated stratigraphic sections and can be placed within a secure chronostratigraphic framework (Raza et al. 1984; Lindsay et al. 2005; Antoine et al. 2013; Barry et al. 2013; Flynn et al. 2013; Chapters 2 and 3, this volume). These more recent collections, together with the older collections, contain a large and diverse assemblage of extinct bovids, documenting aspects of morphology but leaving much unknown about phylogeny and paleobiology. The present chapter is a summary of our current understanding, based on an ongoing systematic study.

Bovidae are a family of ruminants characterized by horns formed as postorbital outgrowths of the frontals in which horny sheaths cover bony cores and in which neither sheath nor core is branched or seasonally shed (Janis and Scott 1987; Gentry 2010). Males always have horns, which assume a variety of shapes and presumably act primarily for species recognition. Females in most species either have smaller horns or lack them. Horns may not have been present in the earliest hypsodontines, here treated as a distinct bovid subfamily, and may have evolved in hypsodontines separately from other bovids (Bibi et al. 2009; Gentry 2010). Bovids have dental and postcranial features not shared with giraffoids, and these together with consistent differences in size (Chapter 22, this volume) allow bovid skeletal and dental elements to be recognized.

The oldest horned bovids in southern Asia are from the Vihowa Formation (Ginsburg, Morales, and Soria 2001; Métais et al. 2003; Barry et al. 2005) and are at least 18.1 Ma (Lindsay et al. 2005; recalibrated for the Gradstein et al. [2012] timescale). Apart from a single hypsodontine species and ambiguous postcranials, bovids are unknown in the older Chitarwata Formation—at least not as species that have horns (Ginsburg, Morales, and Soria 2001; Barry et al. 2005; Antoine et al. 2013). Throughout the Middle and Late Miocene there are frequently five to seven co-occurring species, making bovids the most species rich artiodactyl family in the Siwaliks. With the exception of Late Miocene equids, Siwalik bovids are also the most abundant of the larger mammals throughout the Miocene (Barry et al. 2002). Furthermore, they show a considerable range in body mass, ranging from around 6 kg to over 350 kg, compared to a range of approximately 3.5 to over 900 kg for extant African bovids (Kingdon 1997).

Materials

Including specimens from the Sulaimans in Baluchistan and Manchars in Sind, the post-1973 collections comprise over 10,000 bovid fossils. Most remains are isolated teeth, jaws, and postcranials, and as is typical of Siwalik fossils, most are disarticulated and fragmentary. Although there are partial skulls in the collections, cranial material mostly consists of isolated, broken horn cores. There are few associated teeth and skeletal elements that clearly belong to a single individual (and only one instance of associated skeletal, dental, and cranial material). The fragmentary, disarticulated fossils make it difficult to link dental, cranial, and postcranial morphologies and thus to fully characterize species.

Bovid species are most easily diagnosed and identified by their horn cores. But while horn cores exhibit morphological differences between species, albeit with considerable variation, dentitions and skeletal remains are much more uniform, making it difficult to assign them to species. This is unfortunate because horn cores and skulls are rare and the use of postcranials or teeth would in many cases extend stratigraphic ranges, more accurately documenting first and last occurrences. Nevertheless, there are differences in size that can be used to separate some species; a strategy that works best for the Late Miocene species, but is more problematic for those that are older.

We recognize 42 bovid species (or likely species) from the Late Oligocene and Miocene deposits of Pakistan, 24 of which have previously been named (Table 23.1). Of these, 14 were created by G. Pilgrim (1910, 1937, 1939), with most holotypes from collections in the Geological Survey of India in Kolkata and the American Museum of Natural History in New York. The Harvard-GSP project collection contains specimens of all but two or possibly three of the previously recognized species: *Ruticeros pugio*, *Pachyportax giganteus*, and perhaps *Selenoportax vexillarius* if it differs from *Selenoportax* aff. *vexillarius*. Although 3 of the 18 unnamed taxa in Table 23.1 are known only from teeth (?Hypsodontinae sp., Bovidae sp. 1, and ?Antilopinae, small spp.), the others with cranial material are well enough known to define and name. (These are in varying pre-publication stages.)

Occurrence and Age of the Fossils

The ages and depositional environments of the fluvial Siwalik deposits are discussed in Chapters 2 and 4 of this volume. As noted, Siwalik sediments are found along the southern margin of the Himalayas (Corvinus and Rimal 2001; Flynn et al 2013; Patnaik 2013; Nanda 2015), but are particularly well exposed in northern Pakistan, where most of the collections under study originate. Significant exposures of fluvial and fluviodeltaic deposits are also known in southwestern Punjab, Baluchistan, and Sind. They are important because they contain fossils of earliest Miocene and latest Oligocene age (Raza et al. 1984; Lindsay et al. 2005; Antoine 2013), while the northern localities are of Early Miocene through Pleistocene age. Throughout Pakistan post-1973 collecting has produced more than 1,400 individual fossil sites, a large number of which have ages estimated to within 100,000 years or better (Chapter 3, this volume; Lindsay et al. 2005; Barry et al. 2013). However, the holotypes and referred specimens of most named species were collected in the late nineteenth or early twentieth centuries and specific localities and ages are not known for them. Most critically this includes the types of *?(Sivaceros) vedicus*, *Miotragocerus punjabicus*, *Ruticeros pugio*, *Caprotragoides potwaricus*, *Nisidorcas planicornis*, *Kobus porrecticornis*, and although discovered more recently *Pachyportax giganteus*. As noted in Chapter 3 of this volume (Fig. 3.3), the distribution of sites over time is uneven, with very few sites before 14 or after 7 Ma, while between 7 and 14 Ma some intervals have very few sites

and others many. Critically, this uneven distribution biases the stratigraphic ranges of the species. Finally, while most localities have only a few bovid fossils, some have many, making it possible to analyze variation within a single species.

SPECIES ACCOUNTS

The 42 species represent as many as 27 genera, some of which are endemic to the Siwaliks while others are known elsewhere. Boselaphini predominate with 17 species, but Hypsodontinae, Bovini, Antilopini, Reduncini, and Caprini are also known. The species, their temporal ranges (based on first and last occurrences), and estimated body mass are listed in Table 23.1. Their established temporal ranges are shown in Figure 23.1. In this section we note some of the essential distinguishing characteristics of the taxa.

?Family Bovidae Gray, 1821
Subfamily Hypsodontinae Köhler, 1987
Palaeohypsodontus Trofimov, 1958
Palaeohypsodontus zinensis Métais, Antoine, Marivaux, Welcomme, and Ducrocq, 2003

A large species of *Palaeohypsodontus*, with very hypsodont lower molars. This species was probably hornless. The holotype mandible and some postcranials of *Palaeohypsodontus zinensis* came from the Late Oligocene Chitarwata Formation at Dera Bugti (Métais et al. 2003). Barry et al. (2005) added two younger teeth from the Vihowa Formation and a damaged mandible from a much older locality in the Chitarwata Formation.

Hypsodontus Sokolov, 1949
Hypsodontus sokolovi (Gabunia, 1973)

This is a small species of *Hypsodontus* with moderately long horn cores having small diameters relative to length and without keels. The horn cores are nearly straight with little or no divergence. Teeth are extremely hypsodont.

Hypsodontus pronaticornis Köhler, 1987

In this large species of *Hypsodontus* the horn cores are moderately long, curving outwards and forwards, with homonymous torsion and sinuses in the frontals and horn pedicels. The molars are very hypsodont.

?Hypsodontinae sp.

A single mandible with m3 from the Kamlial Formation (15.2 Ma) is larger than *Palaeohypsodontus zinensis* and *Hypsodontus sokolovi*. Unidentified pecoran postcranial material from the base of the Kamlial Formation (17.9 Ma) probably also belongs to this species. Hypsodontinae are the most likely group to have reached such a size and degree of hypsodonty at this period.

Family Bovidae Gray, 1821
Subfamily indet.
Bovidae sp. 1

A larger species than either the Vihowa Formation boselaphine discussed next or *Eotragus noyei*, this species is represented

Table 23.1. Siwalik Bovid Taxa, Range, and Estimated Body Mass

Taxon, Author, Date	Collection / Holotype[1]	Age (Ma)[2]	Estimated Mass (kg)[3]
?Bovidae Gray, 1821			
Hypsodontinae Köhler, 1987			
Palaeohypsodontus zinensis Métais et al., 2003	ISEM DBJ2—A1	(25.5–19.7)–18.2	Class VI–VII?
Hypsodontus sokolovi (Gabunia, 1973)	P.I.N. 428/10	14.1–12.1	45–50
Hypsodontus pronaticornis Köhler, 1987	not numbered	13.7–12.8	160?
?Hypsodontinae sp.	GSP	?17.9–15.2	34–40
Bovidae Gray, 1821			
Bovidae sp. 1	GSP	16.0–15.1	15–18
Bovinae Gray, 1821			
Boselaphini Knottnerus-Meyer, 1907			
Unnamed genus and sp.	GSP	(18.3–17.9)	9–12
Eotragus noyei Solounias, Barry, Bernor, Lindsay, and Raza., 1995	GSP, YGSP 31616	17.9–?16.8	9–12
Elachistoceras khauristanensis Thomas, 1977	GSP, YGSP 4262	14.1?–7.5	5–12
Strepsiportax unnamed sp.	GSP	14.1–12.4	15–30?
Strepsiportax gluten Pilgrim, 1937	AMNH 19746	12.4–(11.1–10.8)	30–50?
Sivaceros unnamed sp.	GSP	14.1–13.4	15–30?
Sivaceros gradiens Pilgrim, 1937	AMNH 19448	13.0–(11.3–11.0)	20–40?
?(*Sivaceros*) *vedicus* Pilgrim, 1939	GSI B 796	9.8	20–40?
Sivoreas eremita Pilgrim, 1939	GSI B 578	14.1–12.1	20–40?
Helicoportax praecox Pilgrim, 1937	AMNH 19476	13.4–(11.2–10.8)	30–50?
Selenoportax vexillarius Pilgrim, 1937	AMNH 19748	(11.2–10.8)	Class VII–VIII
Selenoportax aff. *vexillarius*	GSP	10.6–10.0	80–150
Miotragocerus pilgrimi (Kretzoi, 1941)	GSI B 631	10.6–8.8	30–85
Miotragocerus punjabicus (Pilgrim, 1910)	GSI B 486	8.7–6.4	60–100?
Tragoportax salmontanus Pilgrim, 1937	AMNH 19467	8.4–7.2	60–100?
?Boselaphini, unnamed genus and sp,	GSP	9.4?–7.3	Class VII–VIII
Ruticeros pugio Pilgrim, 1939	GSI B 579	?	Class VI–VII?
?Bovini Gray, 1821			
Pachyportax falconeri (Lydekker, 1886)	NHML 37262	9.8–8.9	90–195
Pachyportax dhokpathanensis Pilgrim, 1937	GSI B 488 and B 489	9.0–7.2	150–355
Pachyportax giganteus Akhter, 1995	PMNH 87/323	?	Class IX?
?Bovinae			
Unnamed genus and sp..	GSP	9.6–9.0	Class VIII?
Antilopinae Gray, 1821			
Unnamed genus and sp.	GSP	14.1–13.0	15–20?
Antilopini Gray, 1821			
?*Gazella* sp.	GSP	13.2–11.7	10–15?
Gazella lydekkeri Pilgrim, 1937	AMNH 19663	10.2–6.3	10–20?
Prostrepsiceros vinayaki (Pilgrim, 1939)	GSI B 799	9.4?–7.9	Class VI–VII
?*Prostrepsiceros* large sp.	GSP	8.2	Class VII?
Nisidorcas planicornis (Pilgrim, 1939)	NHML 37264	9.0	Class VI–VII?
?*Protragelaphus* sp.	GSP	10.8	Class VII?
Large, unnamed genus and sp.	GSP	9.4–8.8	Class VIII?
Reduncini Knottnerus-Meyer, 1907			
Kobus sp. 1	GSP	9.4–8.2	18–32?
Kobus porrecticornis (Lydekker, 1878)	GSI B 229	8.1–7.7	Class VII
Kobus sp. 2	GSP	7.4–(7.2–6.6)	Class VII
Antilopini or Caprini			
Tethytragus sp.	GSP	13.4?–12.3?	30–50?
?*Caprotragoides potwaricus* Pilgrim, 1939	GSI B 630	10.2	Class VII?
?*Dorcadoryx* sp.	GSP	9.7–9.4	Class VII?
Caprini Gray, 1821			
Protoryx aff. *solignaci* (Robinson, 1972)	SGT T 3657	11.3	Class VII
?Antilopinae			
Small spp.	GSP	8.0–6.4	Class VI?

[1]Institutional Abbreviations:

    ISEM: Institut des Sciences de l'Evolution, Université de Montpellier II, Montpellier

    P.I.N.: Institute of Paleonology, Academy of Sciences, Moscow

    GSP: Geological Survey of Pakistan, Islamabad

    AMNH: American Musuem of Natural History, New York

    GSI: Geological Survey of India, Kolkata

    NHML: Natural History Museum London

    PMNH: Pakistan Museum of Natural History, Islamabad

    SGT: Service Géologique de Tunisie

[2]Observed first and last occurrences. Values in parenthesis are ranges for a single endpoint.

[3]Mass estimates are based on regressions developed by Morgan (1994) using the distal width of the astragalus. Class Ranges: VI = 8–20 kg, VII = 20–100 kg, VIII = 100–250 kg, IX = 250–800 kg.

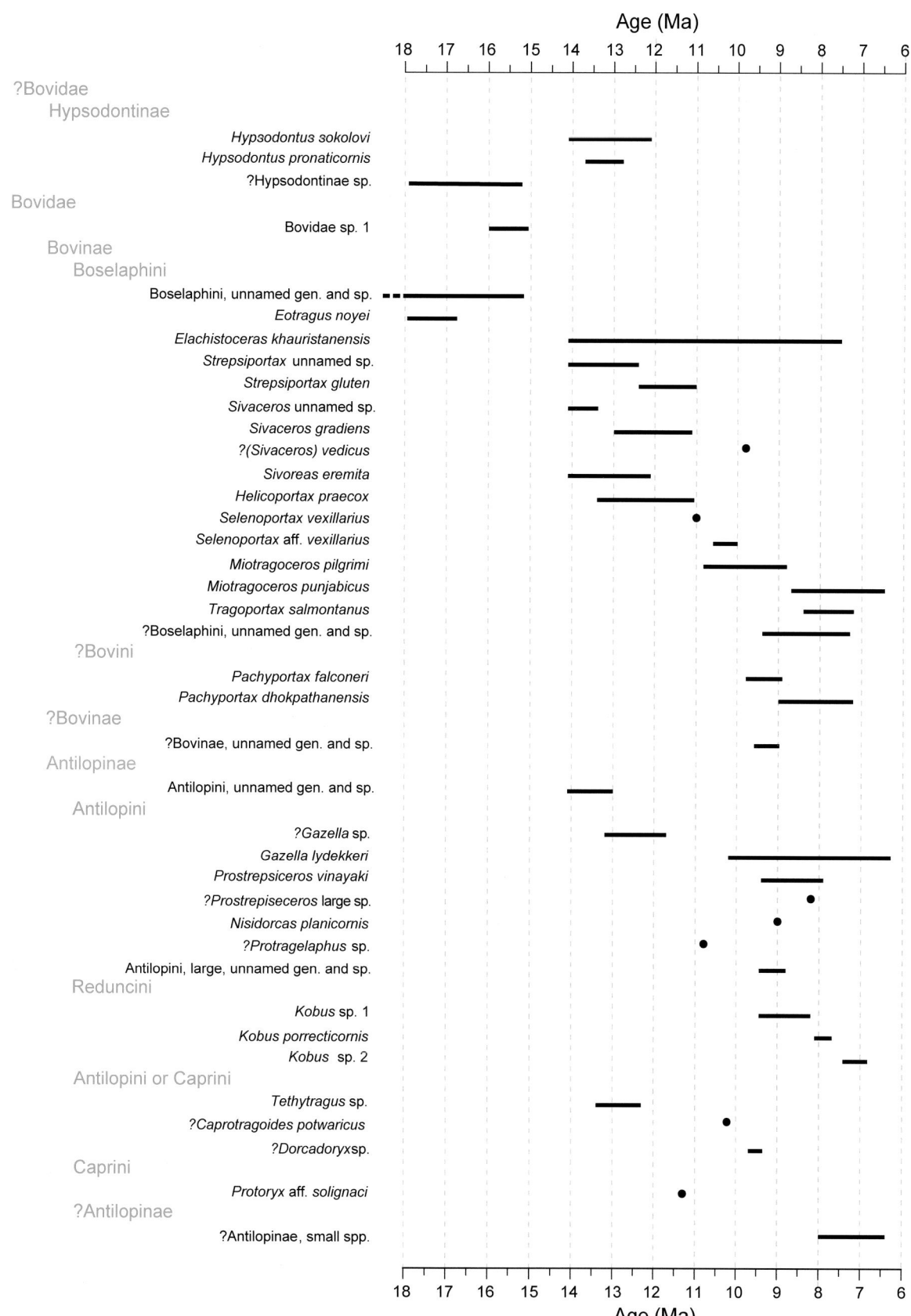

Figure 23.1. Observed stratigraphic ranges determined from first and last occurrences.

by fragmentary tooth remains, possibly two horn core bases, and some postcranials. All the specimens are from the Kamlial Formation and indicate that by 16 Ma there was a larger bovid coexisting with a hypsodontine and a smaller boselaphine.

Subfamily Bovinae Gray, 1821
Tribe Boselaphini Knottnerus-Meyer, 1907
Unnamed genus and sp.

As noted in Solounias et al. (1995), there is a small bovid present in the late Early Miocene of Zinda Pir. Known from compressed, keeled horn core fragments, an associated cranium, and teeth, it is distinctively different from *Eotragus*. It demonstrates the existence of two divergent boselaphine lineages by the Early Miocene.

Eotragus Pilgrim, 1939
Eotragus noyei Solounias, Barry, Bernor, Lindsay, and Raza, 1995

This small *Eotragus* may be the same species as *Eotragus minus* Ginsburg, Morales, and Soria, 2001. The holotype and referred material are from the base of the Kamlial Formation (17.9 Ma), while the correlations of Antoine et al. (2013) place the *Eotragus minus* holotype near 19 Ma. Additional, older remains also referred to *Eotragus minus* are known from the Upper Member of the Chitarwata Formation between 21.9 and 20.4 Ma (Antoine et al. 2013).

Elachistoceras Thomas, 1977
Elachistoceras khauristanensis Thomas, 1977

A small species, with short, straight horn cores of small dimensions and an irregularly oval cross section. The horns are inclined backwards with slight anterior curvature. They were inserted on the cranial vault and not laterally on the dorsal orbital rim as in Neotragini. The teeth are brachyodont and smaller than in *Eotragus noyei*.

Fossils of small bovids are found through the Middle and Late Miocene Siwaliks, and separation of a small Antilopinae from a tiny boselaphine is difficult. The oldest horn core of *E. khauristanensis* is 13.1 Ma and differs somewhat from Thomas's Late Miocene material, suggesting it might be a separate, earlier species.

Strepsiportax Pilgrim, 1937

Thenius (1956) synonymized *Strepsiportax* with the Middle Miocene European *Protragocerus*, which was accepted by Gentry (1970) and Thomas (1984). We now believe, however, that the type species, *Protragocerus chantrei*, is too poorly known to make useful comparisons with other genera.

Strepsiportax unnamed species

A *Strepsiportax* with short, mediolaterally compressed horn cores having a variably developed posterolateral keel. The horn cores diverge little and have a slight backward curvature with only poor or no demarcation. There are sinuses in parts of the frontals that just reach into the front of the pedicel. Females were hornless.

Strepsiportax gluten Pilgrim, 1937

A *Strepsiportax* differing from the preceding species in having longer horn cores with a more definite posterolateral keel and the degree of divergence lessening more strongly distally. The horn cores have more torsion and a straighter course in sideview. Sinuses are present in the front part of the pedicels and rise to a centrally positioned peak within the pedicel. *Strepsiportax gluten* includes *Strepsiportax chinjiensis* Pilgrim, 1937. Females are presumed hornless.

The holotype of *Strepsiportax gluten* and a referred braincase (AMNH 29745) are between 11.2 and 10.8 Ma, which makes them the youngest known specimens of the species.

Sivaceros Pilgrim, 1937
Sivaceros unnamed species

A small species of *Sivaceros* with short horn cores that are inserted fairly widely apart with an ill-defined anterior keel descending to an anterior position lateral to the supraorbital pit. The demarcation has a stepped appearance, and the horn cores diverge in anterior view, with the degree of divergence lessening distally. The cranium is high and narrow, and the cranial roof bends down sharply at bregma.

Sivaceros gradiens Pilgrim, 1937

A *Sivaceros* differing from the preceding species with horn cores being longer with a stronger expression of the keels. The horn cores are less compressed, and the maximum transverse width lies more posteriorly. The anterior keel descends to an anteromedial insertion with the supraorbital pit lying lateral to this line of descent. The horn cores are almost parallel in anterior view and the insertions are closer together, with an incipient intercornual ridge. Sinuses in the frontals extend into the pedicels, with their highest point reaching posteriorly into the horn core proper.

We estimate the age of the holotype as being between 11.3 and 11.0 Ma, which makes it the youngest known *Sivaceros gradiens*.

?(Sivaceros) vedicus Pilgrim, 1939

A small species with compressed horn cores of moderate length having salient anterior and posterolateral keels. The maximum transverse thickness is close to the anteroposterior center. The horns diverge slightly, but there is no alteration in the degree of divergence distally. The horn cores are inclined backwards, lack a demarcation, and are inserted close together. There is no torsion, but the anterior keel turns anteromedially as it descends. There are large sinuses extending up into the horn core.

This is a very late occurrence of *Sivaceros*. The locality for Pilgrim's holotype is stated to be "Nila," a village on the south side of the Soan River surrounded by outcrops of the Dhok Pathan Formation that are around 7.8 Ma. However, on the north side of the river there are highly fossiliferous outcrops that are as old as 9.4 Ma. It is possible that Pilgrim's holotype came from the older outcrops.

Sivoreas Pilgrim, 1939
Sivoreas eremita Pilgrim, 1939

A small to moderate size boselaphine with a low and wide cranium. The horn cores are moderately long without demarcations. Their salient anterior and posterolateral keels are tightly twisted around a straight axis, with the anterior keel descending to an anteromedial insertion extending on to the pedicel. The horn cores are divergent, with the degree of divergence remaining constant from base to tip. The horn core insertions are wide apart and there are no sinuses in the pedicels, but some are present in the frontals below and lateral to supraorbital pits. The premolar row is short relative to the molar row with P2 shorter than P3.

The key feature of *Sivoreas* is the tight torsion of the horn cores. Made and Hussain (1994) founded a second species, *Sivoreas sondaari*, believing their material was stratigraphically older with weaker torsion than *Sivoreas eremita*. We do not find change in torsion over time in our sample, much of which came from the same 14 Ma level as the *Sivoreas sondaari* holotype.

Helicoportax Pilgrim, 1937
Helicoportax praecox Pilgrim, 1937

A small to moderate size boselaphine, with moderately long, straight horn cores having a strong anterior keel and a posteromedial keel. There is usually a slight or even a marked torsion of the keels around an otherwise straight axis. The horn cores are compressed but less so than in *Strepsiportax* or *Sivaceros*. They have fairly strong divergence, but little or no distal diminution in the degree of divergence. The pedicels grow forward and medially, but the frontals are not raised between horn bases. The large supraorbital pits lie lateral to the line of descent of the pedicels. There are no sinuses in the pedicels. This species includes *Helicoportax tragelaphoides* Pilgrim, 1937.

We estimate the age of the holotype as between 11.1 and 10.8 Ma, and as a result the species may have temporally overlapped with *Selenoportax vexillarius* and hipparionine horses.

Selenoportax Pilgrim, 1937

With *Selenoportax* we pass from the Middle to the Late Miocene. We accept previous opinions (Pilgrim 1937, 1939; Thomas 1984) that *Selenoportax* is related to *Helicoportax*, but contra Gentry, Solounias, and Barry (2014), we now recognize only the type species.

Selenoportax vexillarius Pilgrim, 1937

A moderate to large size boselaphine with a low and wide skull and horn cores that are long with strong anterior and posteromedial keels. A posterolateral keel is also present, but less pronounced. There is only slight mediolateral compression, although the medial surface is flattened. The anterior keel descends to a medial insertion with the horn cores inserted widely apart. Divergence of the horn cores is marked but distally diminishes greatly so that they have a crescentic appearance in anterior view (and thus the reference to the moon it the generic name). There is only slight torsion of the keels

around the axis with no apparent demarcations. Sinuses are restricted to the frontals, not reaching into the pedicels. Supraorbital pits are large (exceptionally so for Boselaphini) and lie lateral to the line of descent of the pedicels. The frontals are not raised between horn core bases. The surface behind the horn core insertions is smooth.

The holotype skull is the only known specimen. We estimate it to be between 11.2 and 10.8 Ma. As noted, it may have temporally overlapped with *Helicoportax praecox*.

Selenoportax aff. *vexillarius*

The YGSP material is from a less gracile and possibly larger animal than the holotype of *Selenoportax vexillarius* but is otherwise morphologically similar. The species is very common and marks the appearance of a much larger bovid than had been present before (Fig. 23.2). Because it is the only species in its size class throughout most of its stratigraphic range, we refer to it not only horn cores, but also teeth and postcranials.

Miotragocerus Stromer, 1928

Miotragocerus is a characteristic and common Late Miocene boselaphine. In our usage the genus includes species formerly in *Tragocerus* and *Tragoportax*. There are two time-successive *Miotragocerus* species in the Siwaliks, *Miotragocerus pilgrimi* and *Miotragocerus punjabicus*. The teeth of *Miotragocerus punjabicus* and the closely related *Tragoportax salmontanus* acquire a degree of occlusal complexity, unlike Late Miocene boselaphines elsewhere but similar to the larger *Pachyportax*, a likely Bovini.

Miotragocerus pilgrimi (Kretzoi, 1941)

A small to moderate size *Miotragocerus*. The horn cores have their greatest transverse diameter lying posteriorly, and slight divergence with the degree of divergence lessening distally. There are one or more demarcations with slight torsion of the anterior keel. Large sinuses frequently rise higher in the posterior parts of the pedicel than they do more anteriorly. Females were hornless.

After 9.9 Ma horn cores become slightly larger on average (about 7% for the anteroposterior diameter) and slightly more compressed. The oldest certainly identified horn cores are 10.2 Ma, but skeletal elements suggest the species may have been present as early as 10.5 or even 10.6 Ma. It is very common, and because throughout its stratigraphic range it is the only species in its size class, we refer teeth and postcranials to it.

Miotragocerus pilgrimi has similarities to species of *Sivaceros* that suggest a phylogenetic relationship, although not a direct ancestor-descendant one. Instead, *Miotragocerus pilgrimi* appears to be part of an earliest Late Miocene radiation of *Miotragocerus* species in Europe.

Miotragocerus punjabicus (Pilgrim, 1910)

A *Miotragocerus* differing from the preceding species in the horn cores being larger and the sinuses more extensive within the frontals. It also has a prominent intercornual ridge between closely inserted horn cores and a very large, deep preorbital fossae. If the holotype of *Tragocerus brown* is characteristic of

Miotragocerus punjabicus, it would have had long, backwardly-curved horn cores. Crania are high and narrow, and the roof is nearly horizontal in profile throughout its length. Females apparently were hornless, as skulls without horns cores have been found at the same localities as skulls with them. If correctly assigned to this species, they suggest that there was not marked sexual dimorphism in size. It is a very common species.

Miotragocerus punjabicus includes *Tragocerus browni* Pilgrim, 1937 and *Tragoportax islami* Pilgrim, 1939. The holotype of *Tragocerus browni* is one of the oldest specimens, but we have not been able to assign it an age with confidence. It is between 8.6 and 8.4 Ma.

Tragoportax Pilgrim, 1937

We retain only the type species for this genus, but it might be insufficiently distinct to be separated from *Miotragocerus*.

Tragoportax salmontanus Pilgrim, 1937

A species characterized by short horn cores with the maximum transverse thickness just behind the anteroposterior center. The horn cores are inclined backwards, with only slight or no backward curvature. The degree of divergence diminishes distally with slight torsion of the anterior keel and usually a demarcation. The preorbital fossa is large, and the horn core insertions close, with raised frontals between the bases and a rugose surface behind. The cranium is low and wide, and the back of the cranial roof curves downward.

This is a slightly smaller species than *Miotragocerus punjabicus* but overlapping in size. It was probably less common than *Miotragocerus punjabicus*.

?Boselaphini, unnamed genus and species

This species has short and rather small horn cores, no keels, less compression than in the two *Miotragocerus* species, strong backward inclination, insertions close, and inflated pedicels containing extensive sinuses.

The species has a long period of sympatry with both *Miotragocerus punjabicus* and *Tragoportax salmontanus* and, being about the same size, can only be distinguished by its horn cores. It was a rare species.

Ruticeros pugio Pilgrim, 1939

Ruticeros pugio is based on a left horn core from an unknown locality, but presumably one in the Dhok Pathan Formation (Pilgrim 1939, caption for Plate IV, Fig. 2). It is about the size of a Miocene gazelle. Two Late Miocene horn cores in London agree with the holotype in the accentuated longitudinal grooves and ridges. It is not represented in the Geological Survey of Pakistan collections.

?Tribe Bovini Gray, 1821
Pachyportax Pilgrim, 1937

Pachyportax is much larger than *Miotragocerus* and slightly larger than *Selenoportax*. The horn cores are moderately long and slightly compressed mediolaterally, with a strong anterior keel, a posterolateral edge or keel, and a more rounded posterome-

dial edge. The anterior keel descends to an anterior insertion. The horns are inclined backwards and inserted widely apart, with moderate to strong divergence. Crescentic curvature in anterior view is not pronounced. There are sinuses in the frontals and in *Pachyportax dhokpathanensis* in the base of the horn core. Frontals are not raised between horn bases, while the preorbital fossa is shallow and poorly delimited. Molar teeth over time increasingly acquire large basal pillars, a more complicated outline of the central fossettes, more prominent styles and stylids, and stronger labial ribs.

Lydekker (1876) based the type species (*Pachyportax latidens*) on a lower and an upper molar. The upper molar was chosen by Pilgrim (1937) as lectotype. Bibi (2007a) has argued that the lectotype tooth of *Pachyportax latidens* should be treated as belonging to a *nomen dubium*. *Pachyportax latidens* has to remain the type species. Consequently we propose that the name *Pachyportax dhokpathanensis* should be used as a full species. We therefore accept four species of *Pachyportax*: the type species, *Pachyportax latidens* containing only the lectotype tooth, *Pachyportax falconeri*, *Pachyportax dhokpathanensis*, and *Pachyportax giganteus* of Akhtar (1995), which is not represented in our material. Some of the tooth characters of *Pachyportax* are suggestive of Bovini (Bibi 2007a). The morphological limits of its species are uncertain.

Pachyportax falconeri (Lydekker, 1886)

Larger than *Selenoportax* aff. *vexillarius*, this species has a moderately wide and low skull and horn cores with appreciable curvature in anterodorsal view. The medial surface of the horn cores is more rounded in cross-section than the lateral surface. Frontal sinuses are of restricted extent. The cranial roof may be slightly sloping. This is a common species.

Pachyportax dhokpathanensis Pilgrim, 1937

A species larger than *Pachyportax falconeri*. Its skull is moderately wide and high, while the maximum transverse diameter of the horn core lies centrally rather than posteriorly. The horn cores sometimes have flattened lateral surfaces, and are more inclined backwards than in *Selenoportax* aff. *vexillarius*. Supraorbital pits are small. The braincase has a rugose dorsal surface.

Pachyportax falconeri and *Pachyportax dhokpathanensis* may overlap after 9.0 Ma for some unknown length of time. Massive horn cores with the characteristic form of *Pachyportax dhokpathanensis* appear just after 8.9 Ma, while the youngest well-dated horn core of *Pachyportax falconeri* is only 100,000 years older. Very large dental and skeletal remains that may belong to *Pachyportax dhokpathanensis* are also found at 9.0 Ma, while small specimens persist until much later. *Pachyportax dhokpathanensis* is a common species, known from many specimens and localities.

Pachyportax giganteus Akhtar, 1995

This species was based on a cranium from the Dhok Pathan Formation (Akhtar 1995). It is larger than the *Pachyportax dhokpathanensis* holotype (Akhtar 1995, Table 1). The horn cores are short with strong anterior and posterolateral keels. The maximum mediolateral width is positioned centrally along the anteroposterior axis of the cross-section mainly because of a well-defined medial keel. The horn core has a demarcation and in frontal view shows a limited degree of smooth curvature as the degree of divergence diminishes from base to tip. The frontals have a weakly to moderately rugose surface behind the horn core bases, and the posterior part of the cranial roof turns downward as it approaches the occiput.

?Bovinae, unnamed genus and species

This is a mysterious species known from only a few specimens. It has long horn cores, a sharp medially inserted "anterior" keel, and a posterolateral keel. The horn cores are barely compressed, and between the keels the posteromedial surface is flat, while the anterolateral surface shows a considerable bulge. Divergence and backward inclination are moderate and constant, and the insertions are wide apart. The horn cores are straight without torsion, spiraling, or backward curvature. Sinuses are visible where the front of the pedicels merge with the frontals.

The keels on these horn cores are narrow ridges on the body of the horn core rather than intersections between surfaces in different planes. In this they are like the Tragelaphini, but they lack torsion or spiraling. The insertion position above the back of the orbits is more posterior than in boselaphines.

Subfamily Antilopinae Gray, 1821

Current classifications of extant Antilopinae include all bovids other than the Bovinae. Siwalik Antilopinae include three tribes: Antilopini, Reduncini, and Caprini. The Reduncini and Antilopini may be closely related, and Siwalik fossils bear on this problem. One early non-boselaphine species is a member of the Antilopinae, but cannot be assigned to a tribe.

Unnamed genus and species

A small species, with horn cores that are small and short to moderately long, without keels, and not compressed. There are strong longitudinal ridges on the posterolateral corner or in the middle of the lateral side, and the maximum transverse diameter is at the middle of the anteroposterior diameter. The horn cores are slightly divergent in anterior view with a barely perceptible lessening degree of divergence at the tips. The horn cores are moderately inclined and straight in side-view or have a very slightly concave profile along the front edge. Their insertions are wide apart, and the frontals are not raised between the horn core bases. There are no sinuses, but a large supraorbital foramen or a small supraorbital pit is present, although not developing a *Gazella*-like triangular depression.

The youngest horn cores of this species are 13.0 Ma, but lower dentitions that may belong to it instead of the following *?Gazella* sp. would extend its stratigraphic range to 11.7 Ma.

Tribe Antilopini Gray, 1821

Living Antilopini comprise the gazelles, several other mainly African genera, and the spiral-horned *Antilope* Pallas, 1766. *Gazella* and several extinct spiral-horned antilopines are found in the Miocene of the Siwaliks.

Gazella Blainville, 1816

The 14 or so extant species of *Gazella* are often split into many more species, with the sub-Saharan African ones placed in *Nanger* and *Eudorcas*. They are not very distinctive, and identifying Miocene fossils as definitely *Gazella* is difficult. The following Siwalik species of doubtful generic identity is from the Middle Miocene.

?Gazella sp.

The main difference between the little-compressed horn cores of this species and those of the preceding unnamed form lies in their being very slightly curved backwards along the posterior edge and more strongly so along the anterior edge. The horn cores also have a less irregular cross-section and some flattening of a more definitely developed lateral surface.

Gazella lydekkeri Pilgrim, 1937

This is a small gazelle with long horn cores that are elliptical in cross-section with the lateral surface flatter than the medial and slightly to moderately divergent. There are no pedicel sinuses, but narrow supraorbital pits are present, while the cranial roof shows relatively little posterior down turning. The females were hornless.

The teeth are those of a generalized bovid having smooth enamel and a low degree of hypsodonty. The premolars relative to molars are longer than in some extinct and most extant gazelles.

Prostrepsiceros Major, 1891

Prostrepsiceros is one among a number of Late Miocene spiral-horned antilopines that overlapped with and exceeded the size of contemporaneous gazelles. The genus is mainly known from after 9 Ma in southeast Europe and southwest Asia. Four lineages succeed *Prostrepsiceros vallesiensis* of the earlier Late Miocene, with the Siwalik *Prostrepsiceros vinayaki* being one.

Prostrepsiceros vinayaki (Pilgrim, 1939)

This species has small horn cores with anticlockwise torsion and weak spiraling. The horn cores are compressed, with the lateral surface flatter than the medial, a strong anterior keel, and a weaker posterior one. The maximum thickness is central or posterior and the supraorbital pit is large.

As with *?(Sivaceros) vedicus*, Pilgrim's holotype is said to have come from "Nila," but could instead have come from older, highly fossiliferous outcrops to the north of the village as did several of our specimens. The ages of species of *Prostrepsiceros* elsewhere are poorly constrained. At 9.4 Ma, the first occurrence of this species in the Siwaliks is approximately 600,000 years younger than occurrences at Sinap in Turkey (Gentry 2003; Kappelman et al. 2003).

?Prostrepsiceros large sp.

We have a single horn core of this species, the anterior surface of which is damaged. It has a posterolateral keel showing anticlockwise torsion, a more rounded anterior edge, a flattened lateral surface, and divergence decreasing upwards with slight backward curvature. There is a frontal sinus immediately lateral to the pedicel. The horn core is about the same size as that of *?Protragelaphus* sp., but torsion is weaker, and its lateral surface is noticeably flatter.

Nisidorcas Bouvrain, 1979

Although the type is a Siwalik specimen, *Nisidorcas planicornis* is best known from the Late Miocene of southeast Europe and southwest Asia (Bouvrain 1979; Köhler 1987; Kostopoulos 2006, 2016).

Nisidorcas planicornis (Pilgrim, 1939)

A spiral horned Antilopini slightly larger than Late Miocene *Gazella*, this species has moderately long horn cores with the anterior keel blunt or absent, the posterolateral one stronger, and the lateral surface flatter than the medial. The horn core is slightly compressed with moderate divergence, weak anticlockwise spiraling, and a large supraorbital pit.

The holotype was collected in the 1830s, probably from Piram Island in the Gulf of Khambhat in India. It was the only known specimen until Bouvrain (1979) described specimens from Greece, Turkey, and Macedonia. In our collections a specimen from the Potwar Plateau has an age of 9.0 Ma, an early record for *N. planicornis* and probably older than the material from Turkey or Greece.

Protragelaphus Dames, 1883

Protragelaphus skouzesi, the type species, is known from localities that are between about 9.9 and 7.1 Ma.

?Protragelaphus sp.

This species is known from a single horn core with a strong anterior and weaker posterior keel that are tightly twisted around a straight axis. Torsion is anticlockwise. The horn is compressed and has a large anterior sinus that extends more than half way up the pedicel and is separated by a bony partition from a smaller posterior one.

This is the oldest spiral-horned antilopine in the collection, and at 10.8 Ma it is older than species of *Protragelaphus* elsewhere. The lack of open spiraling is similar to *Ouzocerus*, *Protragelaphus*, and *Palaeoreas* among western Late Miocene antilopines, although species of these genera have stronger posterior keels. The Siwalik specimen is slightly larger and more compressed than *Protragelaphus skouzesi*.

Large, unnamed genus and species

This is a mysterious and large species known from just three specimens. The robust, keeled horn cores have anticlockwise torsion around a straight axis. The horn cores are inserted close together, are not mediolaterally compressed, diverge moderately, and were probably long. A somewhat flattened basal posterior surface becomes distally a prominent and wide groove flanked by two keels.

This is not a boselaphine species, but the only large non-boselaphine Miocene antelope with anticlockwise torsion,

keeled horn cores, and tight spiraling is *Criotherium argalioides* (Major, 1890) from Samos and Turkey. *C. argalioides* shows no resemblance to our species, which is an endemic, very large, spiral-horned member of the Antilopinae.

Tribe Reduncini Knottnerus-Meyer, 1907

Present-day reduncines are the African *Redunca* and *Kobus*. They live near water, and their distinctive teeth are moderately hypsodont with enamel-infolding, implying a grazing habit. Females are hornless. Fossil reduncines have long been known in southern Asia from the Pinjor back to the Miocene.

Kobus A. Smith, 1840

The principal Miocene species in the Siwaliks is *Kobus porrecticornis*, to which we add two species, an earlier *Kobus* sp. 1 and a later *Kobus* sp. 2.

Kobus sp. 1

The first reduncine to appear in the Siwaliks is a gazelle-sized species at 9.4 Ma, over 2 million years before unambiguous reduncines appear in Africa. Its horn cores have no unambiguous reduncine characters, but do show small differences from contemporaneous *Gazella* including broader and shorter pedicels. Unfortunately no horn cores are sufficiently well preserved to link these characters with a greater backward inclination than in *Gazella*.

Kobus porrecticornis (Lydekker, 1878)

Horn cores of this species were probably quite long, with the front surface rounded. A posterolateral keel is often present. The horn cores have slight mediolateral compression, with the lateral surface flatter than the medial and the maximum transverse thickness central or slightly posterior. Transverse ridges are poorly developed, but the horn cores curve backwards with longitudinal grooving, strong divergence, inclined insertions, and low pedicels. Supraorbital fossa are quite deep and sinuses are absent from the pedicel.

Kobus porrecticornis was thought to be an antilopine by Pilgrim (1939), who founded *Dorcadoxa* for it. Subsequently Gentry (1970) suggested it was a reduncine, and Thomas (1980) placed *Dorcadoxa* in *Kobus*. This species is very like some of the early African reduncines, but nearly a million years older than definite African reduncines.

Kobus sp. 2

While its molars are only slightly bigger than those of *Kobus porrecticornis*, the horn cores of this species are substantially larger.

?Antilopinae, small spp.

A number of unidentified teeth of small size occur in the Late Miocene. They are not species of *Gazella* or *Kobus* nor are they like the teeth discussed below under *Caprotragoides*. All are larger and more hypsodont than teeth of the boselaphine *Elachistoceras*. They appear to belong to two or more species.

Tribe Antilopini or Caprini
Tethytragus Azanza and Morales, 1994

Tethytragus is a Middle to Late Miocene bovid of Europe and Turkey, first known from about 16.4 Ma.

Tethytragus sp.

Horn cores from four localities as well as teeth from another three belong to this rare species. The horn cores are backwardly curved, more compressed than those of other *Tethytragus* species, with the lateral surface probably flatter than the medial one. The horn cores are without keels or transverse ridges and are weakly divergent. There are sinuses in the frontals anteromedially to small supraorbital pits, but probably not within the pedicels. The oldest and youngest specimens tentatively attributed to this species are teeth.

Caprotragoides Thenius, 1979

This genus was founded for the species "Hippotraginae genus indet. (cf. *Tragoreas*) *potwaricus*" Pilgrim, 1939. It is based on a specimen thought to have come from the "Nagri stage of the Middle Siwaliks" (Pilgrim 1939, 87). Thomas (1984) identified another specimen from Ramnagar, India, but the locality data are deficient.

?Caprotragoides potwaricus Pilgrim, 1939

A single horn core and an upper molar possibly represent Pilgrim's species. The horn core tapers only slightly and is long. There are no keels or transverse ridges, but faint longitudinal grooves are present. It is moderately compressed, with medial and lateral sides flattened. The maximum transverse thickness lies posteriorly, and the horn core appears to have slight and constant divergence. It is backwardly curved, without torsion. The pedicel is short and without sinuses. The upper molar is brachyodont.

Both YGSP specimens may be approximately the same age as Pilgrim's holotype, but younger than the Ramnagar specimen mentioned by Thomas (1984). The related East African *Gentrytragus thomasi* Azanza and Morales, 1994, from Fort Ternan would be much older at ca. 13.7 Ma, but *Gentrytragus gentryi* (Thomas, 1981) from Member E of the Ngorora Formation would be approximately contemporaneous (Thomas 1981; Gentry 2010; Werdelin 2010). Critically, *Gentrytragus* has hypsodont molars, while the Siwalik molar is brachyodont.

Dorcadoryx Teilhard de Chardin and Trassaert, 1938

Several species of *Dorcadoryx* are known from China from around 8 to 4 Ma. The Siwalik fossils are much older.

?Dorcadoryx sp.

The YGSP collections have four horn cores of this species, including a frontlet. Their distinctive features are the tall front edge of the pedicel and the oblique seating of the long axis of the cross-section, which results in an inward rotation of the horn cores. Distally there are anterior and posterior keels that

are less apparent at the base of the horn core. The divergence of the horn cores is slight and constant.

The frontlet is not that of *?(Sivaceros) vedicus*, differing as it does in the less rapid tapering of its longer horn cores, presence of keels only distally, rounded medial surface, and absence of sinuses in the base of the horn core and pedicel. Compared with *?Caprotragoides potwaricus*, *?Dorcadoryx* sp. is smaller, the maximum transverse thickness of the horn core lies more anteriorly, the anterior edge of the pedicel is taller, and the backward curvature is very slight.

Tribe Caprini Gray, 1821

The Caprini, often treated as a subfamily, contains nearly all Eurasian bovids that are not Boselaphini, Antilopini, or Reduncini. They are a genetically and geographically coherent group (Gatesy et al.1997; Hassanin et al. 2012), but difficult to characterize morphologically.

The three fossil genera *Palaeoryx*, *Protoryx*, and *Pachytragus* were originally found at the Turolian localities of Pikermi, Samos, and Maragheh. Robinson (1972) described the first pre-Turolian member of the group from the Beglia Formation of Tunisia as *Pachytragus solignaci*. We treat *Pachytragus* as congeneric with *Protoryx*.

Protoryx Major, 1891

There is much confusion about what species should be included in this genus. Moreover, Robinson's species *Protoryx solignaci* partially predates hipparionine equids as does material from Turkey and maybe Kenya (Nakaya et al. 1984; Köhler 1987; Gentry 2003). Otherwise species of *Protoryx* are Turolian and Eurasian.

Protoryx aff. *solignaci* (Robinson, 1972)

The single incomplete skull is a unique and unexpected find. Similar in size to other *Protoryx*, the skull has several features of *Protoryx solignaci*: strong compression of the horn cores, a weaker distinction between the top of the pedicel and the horn core base, little elevation of the midfrontals between the horn bases, the braincase widening posteriorly, and a flat occipital surface. The frontal sinus is not appreciably developed.

Other taxa

There are additional specimens in the YGSP collections that do not fit with any of the species listed above but are too fragmentary to justify adding further taxa. There may be as many as eight more taxa in the Late Miocene, including from the poorly known interval after 7 Ma. Notable among these is a 6.8 Ma horn core that has a combination of reduncine and hippotragine features.

DISCUSSION

Summary of Species Changes

It is reasonable to think of the first bovids as similar to *Eotragus* in Eurasia or *Namacerus* in Africa. This expectation, however, is complicated by the Hypsodontinae, an older group apparently present in central Asia in the Early Oligocene (Huang 1985; Wang 1992; Meng and McKenna 1998; Dmitrieva 2002; Tsubamoto, Takai, and Egi 2004), in southern Asia before the end of the Oligocene (Métais et al. 2003), and elsewhere in Europe and Africa by the Early Miocene. Hypsodontines may have evolved separately from other bovids, perhaps as an Old World branch from the same stem as Antilocapridae (Janis and Manning 1998; but see Kostopoulos 2014). Early hypsodontines were apparently hornless and only evolved horns about when *Eotragus* appeared (between about 22.0–18.0 Ma). The very limited Siwalik occurrences include two species of *Hypsodontus*, both disappearing before the end of the Middle Miocene.

Fossil Boselaphini have been known since the middle of the nineteenth century, when they were discovered in Middle and Late Miocene European sites, and in the Siwaliks they are by far the most important bovid group. It is very likely that Boselaphini arose from an *Eotragus*-like ancestor, and species of *Eotragus* are reported from ca. 17.9 Ma in the Kamlial Formation and ca. 19 Ma in the Vihowa Formation (Solounias et al. 1995; Ginsburg, Morales, and Soria 2001; Antoine et al. 2013). In addition, teeth and postcranials from the uppermost Chitarwata Formation have been identified as bovids (Antoine et al. 2013). At between 20 and 22 Ma these would be among the oldest known bovids other than for hypsodontines.

As noted in Solounias et al. (1995), there is a second small boselaphine present in the Vihowa Formation of Zinda Pir (ca. 18.1 Ma) that is different from contemporaneous species of

Eotragus. With compressed, keeled horn cores this species demonstrates the existence of two divergent boselaphine lineages by 18 Ma.

Subsequently, the Siwaliks have what is probably the best Middle through Late Miocene record of diverse and abundant boselaphines. Between 17 and 14.1 Ma the record is sparse, containing mostly dental and skeletal remains with only two horn cores. The material indicates that in addition to a hypsodontine at least two boselaphines were present, one the size of *Eotragus noyei* and one somewhat larger. The two horn cores are too incomplete to identify, but are unlike either older or younger horn cores.

Beginning at 14.1 Ma there is an increase in boselaphine diversity, with three species in the range of 15 to 40 kg appearing together with a smaller species. These species belong to four lineages, with a fifth added at 13.4 Ma with the appearance of *Helicoportax praecox*. Throughout the rest of the Middle Miocene there are up to five boselaphine species present (Fig. 23.1).

Between 11.2 and 10.5 Ma there is complete turnover in the boselaphines and other bovids, with *Miotragocerus pilgrimi* and *Selenoportax* aff. *vexillarius* becoming the most common species. Significantly, these and succeeding Late Miocene species are larger than Middle Miocene boselaphines (Fig. 23.2), and with the appearance of *Pachyportax* (ca. 9.8 Ma) Siwalik bovids reach the size of the greater kudu *Tragelaphus strepsiceros* (Jarman 1974; Kingdon 1997). However, very small species persist (Fig. 23.2) producing an overall range of sizes similar to that of extant bovids.

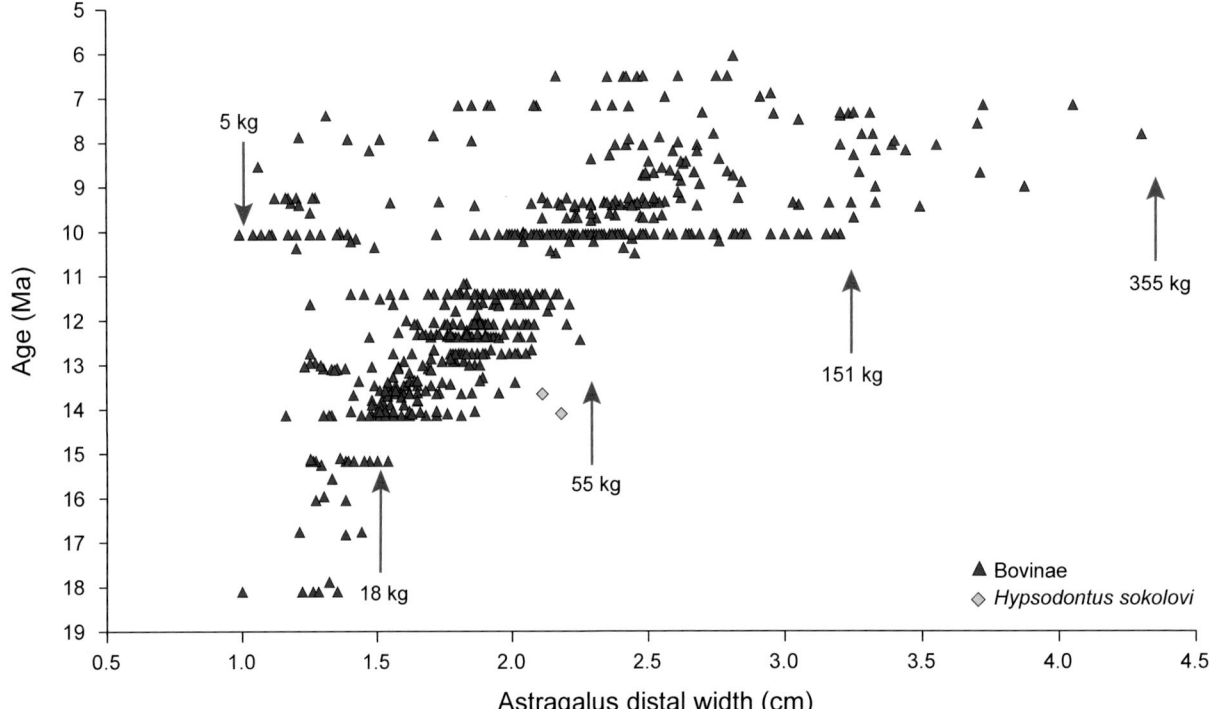

Figure 23.2. Measurements of the distal width (cm) of the astragalus for 593 bovinae and 2 hypsodontinae plotted against geologic age. Arrows indicate equivalent weights, based on regressions of Morgan (1994).

Miotragocerus pilgrimi was succeeded by *Miotragocerus punjabicus* sometime between 8.8 and 8.7 Ma, with the change in both horn cores and teeth appearing to have been gradual. In addition, two other species of similar size appear, one a new genus and species and the other *Tragoportax salmontanus*. Similarly, *Selenoportax* is replaced by *Pachyportax*, which in addition to an increase in size evolves bovine-like teeth (Bibi 2007a).

Apart from *Eotragus*, the oldest boselaphines outside the Siwaliks are in East Africa (Fort Ternan, 13.8–13.7 Ma) and Russia (Belometchetskaja, perhaps 16.4–14.2 Ma). In both locations they coexisted with hypsodontines, as did *Strepsiportax*, *Sivaceros*, and *Sivoreas* in the Siwaliks. *Kipsigicerus labidotus* from Fort Ternan is coeval with the older, unnamed Siwalik species of *Strepsiportax* but morphologically closer to the younger *Strepsiportax gluten*. Less is known of the Belometchetskaja *Paratragocerus*, but it might be closer to *Sivaceros* and *Miotragocerus*. *Sivoreas* may also have been present in East Africa, but does not appear there until about 12.0 Ma. Two boselaphines appeared in Europe late in the Middle Miocene: *Austroportax* with resemblances to *Strepsiportax* and *Miotragocerus* possibly related to *Sivaceros*.

An unresolved question is, then, whether a species of *Sivaceros* gave rise to *Miotragocerus*, which appears in Europe in the earliest Late Miocene. *Sivaceros gradiens* is an unlikely ancestor for *Miotragocerus pilgrimi*, which is presumably an immigrant. Also of interest is the relationship between *Helicoportax* and *Selenoportax*. Although it is unlikely that *Selenoportax* descends from *Helicoportax*, it is significant that several characteristic features of *Helicoportax praecox* are found in *Selenoportax vexillarius* and *Selenoportax* aff. *vexillarius*. Potential ancestors of *Selenoportax* are not known outside the Siwaliks.

Other bovid groups were present in the Siwaliks, and much of their importance is their bearing on the origins of the Antilopini, Reduncini, and Caprini. A small, undescribed species that appears at 14.1 Ma is the earliest likely member of the Antilopinae. It is succeeded by an indeterminate species of *Gazella* (within the Antilopini), and then at 10.2 Ma by *Gazella lydekkeri*. This remains until about 8.0 Ma one of the most commonly fossilized Siwalik species.

Too little is known of the Siwalik spiral-horned Antilopini to judge how they relate to those elsewhere. In the Late Miocene there is a diverse assemblage of at least five species, including *Prostrepsiceros vinayaki*, ?*Prostrepsiceros* large sp., *Nisidorcas planicornis*, and ?*Protragelaphus* sp., as well as a large unnamed species. The spiral horned species are represented by only a few specimens and unsurprisingly have shorter temporal ranges than contemporary boselaphines. Also, in contrast to boselaphines, with the exception of the unnamed species, Siwalik spiral-horned Antilopini have closely related species elsewhere in Eurasia. While the ages of those related species are generally poorly constrained, in the case of ?*Protragelaphus* sp. and perhaps *Nisidorcas planicornis* the Siwalik occurrences are the oldest known.

Siwalik Reduncini have long been known from the Pliocene Tatrot and younger formations and possibly from the younger parts of the Dhok Pathan Formation (Pilgrim 1939). The YGSP/Harvard collection establishes that a reduncine first appeared at 9.4 Ma, which is older than African reduncines by 1 to 2 million years (Bibi et al. 2009; Gentry 2010).

Neither Hippotragini nor Alcelaphini have been recorded from the Miocene of the Siwaliks, although both are known from the Late Pliocene Pinjor Formation (Pilgrim 1939). African Hippotragini appear at about 7.0 Ma (Geraads et al. 2008), while Thomas (1980) has identified African alcelaphine teeth of a similar age.

Fossils of Caprini are rare in the Siwaliks. Three taxa (*Tethytragus* sp. at 13.4–12.3 Ma, ?*Caprotragoides potwaricus* at 10.2 Ma, and ?*Dorcadoryx* sp. at 9.7–9.4 Ma) could belong to either Antilopini or Caprini, but their more precise affinities are uncertain. The single 11.3 Ma *Protoryx* aff. *solignaci* cranium precedes the appearance of hipparionine horses in the Siwaliks and agrees best with *Protoryx solignaci* also found in transitional Middle to Late Miocene deposits in Tunisia and Turkey. The relationships of this species with the younger Turolian species of *Protoryx*, *Palaeoryx*, and *Pachytragus* are not resolved.

Evolutionary Events

Throughout the Miocene, Siwalik bovid assemblages are dominated by boselaphines, which by 14 Ma have become species rich and abundant. In the Middle Miocene antilopines and perhaps caprines are also present, so that, together with the hypsodontines, there are at least ten bovid species at 13 Ma. Fewer co-existing species are known in the Late Miocene (Fig. 23.1), although bovids continue to be very abundant. Significantly, shortly before 9 Ma reduncines and the large, bovine-like *Pachyportax* appear, appearances that are earlier than elsewhere in Africa or Eurasia.

The dominance of Siwalik boselaphines contrasts to antilopine-dominated assemblages elsewhere during the Late Miocene. Further, although tragulids and giraffoids are diverse and abundant throughout the Miocene, both in the Siwaliks and elsewhere (Chapters 21 and 22, this volume), cervids are not known in the Siwaliks for certain until the Late Pliocene (but see Ghaffar, Khan, and Akhtar 2010; Ghaffar, Akhtar, and Nayye 2011).

The abrupt appearance at 14.1 Ma of a number of species is most likely an artifact of preservation because the numbers of fossils and sites increase dramatically between the Kamlial and Chinji Formations (Chapters 2 and 3, this volume). Although the known fossil record from the Kamlial Formation suggests low diversity among bovids, the number of Kamlial Formation fossils is so few that our collection is likely to be missing species that were present. Similarly, the marked decrease in the number of bovids after 7 Ma is a result of poor preservation.

The distal width of the astragalus provides a measure of size from which equivalent weights can be calculated based on regressions developed by Michèle Morgan (1994). Figure 23.2 shows the measurements for 595 individuals, including 2 likely hypsodontines. Arrows indicate equivalent weights at five points. Weights based on regressions of femoral and metapodial dimensions (Table 23.2; Kappelman 1986; Scott 2004) are compatible with this data.

Table 23.2. Habitat Preferences of Siwalik Bovids Based on Discriminate Function Analysis of Femoral and Metapodial Dimensions

Species	Forest	Heavy Cover	Light Cover	Mountain	Estimated Body Mass (kg)[1]
Indeterminant	7	1	3	1	16–46
?*Gazella* sp.		1			16
Selenoportax aff. *vexillarius*	1	2	3	1	83–142
Miotragocerus pilgrimi	7	2	8		38–64

[1] Estimated body mass and habitat preferences from Kappelman (1986, 1988), Scott et al. (1999), and Scott (2004).

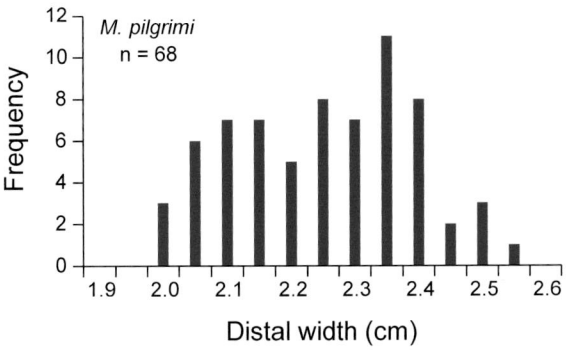

Figure 23.3. Histogram of measurements of the distal width (cm) of the astraglus for 68 presumed adult *Miotragocerus pilgrimi*.

Throughout the Miocene Siwalik bovids increase in size, with a noticeably large increase between 11.4 and 10.5 Ma (Fig. 23.2). At that time (except for *Hypsodontus pronaticornis*, which on the basis of teeth we estimate had a body mass near 160 kg) individuals over 55 kg appear for the first time. This body mass increase, which involves primarily the boselaphines, occurs at the same time as a major turnover of species. After this abrupt increase, the slower general trend toward larger size continues. Tragulids and giraffoids are similar (Chapters 21 and 22, this volume), although tragulids become increasingly less abundant in the Late Miocene. The apparent loss of large bovids and a more restricted size range after 7 Ma are probably a result of a sparse fossil record. A weaker, but also abrupt, increase is apparent between 15.2 and 14.1 Ma; an increase that also involves an expansion of the range of sizes and is possibly accompanied by a turnover of species.

Sexual Dimorphism

Most extant bovids are sexually dimorphic, with females having smaller or no horns, while in those species over 15 to 20 kg females are usually smaller in body mass. Horns are absent in extant female reduncines and boselaphines, but present in extant female bovines and many Antilopini. Hornless skulls are known for *Strepsiportax* unnamed species, *Miotragocerus pilgrimi*, *Selenoportax* aff. *vexillarius*, *Gazella lydekkeri*, and perhaps *Nisidorcas planicornis*. Pilgrim (1937, 751, 1939, 178) also attributed hornless skulls to *Helicoportax*, but these could just as well be *Strepsiportax*. Hornless female skulls of either *Miotragocerus punjabicus* or *Tragoportax salmontanus* are also known.

Assessing size dimorphism is not possible for most Siwalik species because of limited sample sizes. However, a sample of measurements on adult astragali of *Miotragocerus pilgrimi* (Fig. 23.3) from one site is large enough (n = 68) to suggest an underlying bimodal distribution. The tabulated values of Figure 23.3 are the measurements, but the equivalent weights range from 37.7 to 75.8 kg. If the values between 2.3 and 2.4 cm are taken to be typical males and those between 2.10 and 2.15 cm typical females, then the dimorphic ratio (m/f) based on the equivalent body mass values is approximately 1.3. By comparison to extant species of the same body size (Jarman

1983; Gagnon and Chew 2000), this is a modest degree of size dimorphism.

Habitats and Ecomorphology

Extant bovids display strong linkages between size and diet, behavior, sexual dimorphism, and habitat and we assume these same relationships apply to the extinct species of the Siwaliks. Small species, those less than around 20 kg, are often territorial, non-dimorphic, highly selective feeders with a diverse diet and need cover to hide from predators. Larger species in contrast may avoid dense cover, have a less selective and lower quality diet, congregate in larger herds, and may be very sexually dimorphic in size (Jarman 1974; Estes 1993; Kingdon 1997). Antilopini are something of an exception, with even small species relying on speed and agility to escape predators, and thus can inhabit more open habitats. The persistence of small species throughout the Late Miocene may indicate that there was always some dense cover available, but the appearance of ever larger species does not necessarily entail the appearance of increasingly more open habitats.

All Siwalik bovids were terrestrial, but more specific analyses of their habitat preferences have been inconclusive. Kappelman (1986, 1988), Scott, Kappelman, and Kelley (1999), and Scott (2004) used discriminate functions of femoral and metapodial dimensions to determine habitat preferences using small samples. Recognized ancient habitat types included forest, heavy cover (encompassing bush, woodland, and swamp or proximal to water), light cover (light bush, tall grass, and hilly areas), plains (edges, ecotones, and open and arid country), and mountain (rolling hills and mountains). Thirteen Middle and earliest Late Miocene specimens were analyzed by Kappelman (1986, 1988) and Scott, Kappelman, and Kelley (1999), only one of which could be identified to species level. Of the 12 indeterminates, 7 were classified as indicating forest, 1 as heavy cover, 3 as light cover, and 1 as mountain. Although to some extent inconclusive, the predominance of forest and heavy cover suggests the presence between 12.4 and 11.4 Ma of relatively closed habitats with soft substrates, although there may also have been areas with light bush or even grass. A small metatarsal likely to be ?*Gazella* sp. was classified as heavy cover with a high probability, while the metacarpal classified as mountain was regarded by Scott (2004, 64) as anomalous. However, *Tethytragus* sp, a likely caprine, is known from the same time interval.

Twenty-four additional specimens from a single Late Miocene locality can be identified to species: 7, which include both femora and metapodials, are referred to *Selenoportax* aff. *vexillarius*, while 17, also both femora and metapodials, are *Miotragocerus pilgrimi* (Scott, Kappelman, and Kelley 1999; Scott 2004). The results for both species appear inconsistent, with individuals within each species grouped as forest, heavy cover, or light cover (Table 23.2). (One *Selenoportax* aff. *vexillarius* metatarsal was classified as mountain with a low probably and considered anomalous.) Scott (2004) suggested that these results might indicate the presence of more species then was recognized on the basis of horn cores.

Diet

Diet has been assessed in Siwalik bovids using a variety of approaches, including isotopic composition, aspects of tooth wear, and degree of hypsodonty. Specifically, $^{12}C/^{13}C$ ratios in enamel have been used to infer the isotopic composition of an animal's diet and thus to distinguish predominantly browsing from grazing diets (Chapters 5 and 25, this volume; Nelson 2003; Bibi 2007b). In contrast to other techniques, enamel isotopic composition is determined when the tooth forms and represents a brief interval in the animal's life, perhaps only the first one or two years, as well as part of the prenatal period. It may therefore not reflect diet over the animal's lifetime.

After discounting the effects of industrialization on atmospheric $\delta^{13}C$ composition, $\delta^{13}C$ values more negative than −8‰ result from a diet composed predominantly of C_3 plants, which are typically trees and shrubs, as well as grasses growing in the cooler, wetter conditions of forests and dense woodlands. Such isotopic values would be expected in both browsers and frugivores, although in some individuals there may have been a minor component of C_4 vegetation. Values more positive than −8‰ indicate a mixed diet that may have included plants using the C_4 photosynthetic pathway, such as grasses growing in open conditions under water stress. Values more positive than −1‰ reflect a diet dominated by grazing on C_4 grasses (Chapter 5, this volume; Uno et al. 2011).

These modern thresholds are applicable to about 6.5 Ma, but for older samples they must be modified (Chapter 5, this volume) because there was a shift in the $\delta^{13}C$ composition of the atmosphere. For the interval between 6.5 and 12.5 Ma, $\delta^{13}C$ values more negative than −−7‰ can be taken to indicate a diet composed predominantly of C_3 plants, while values more positive than −7‰ indicate a diet that might have included a component of C_4 grass. Another change in atmospheric $\delta^{13}C$ composition at about 12.5 Ma necessitates a further 0.5‰ modification.

Another qualification is the effect of body mass and digestive physiology on isotopic enrichment between diet and tooth enamel. Taxa discussed in Chapter 5 of this volume are restricted to those more than 100 kg, which only *Pachyportax*

dhokpathanensis and the larger individuals of *Pachyportax falconeri* and *Selenoportax* aff. *vexillarius* reach (Table 23.1). In the following discussion we include all bovid taxa that have been sampled regardless of size.

The $\delta^{13}C$ values for 161 samples taken from 155 individuals between 14.2 and 5.7 Ma range from −14.1 to 3.3‰ (Fig. 23.4a). Of the 161 samples, 114 (71%) are less than −10‰, 36 (22%) between −10 and −7‰, 6 (4%) between −7 and 0‰, and 5 (3%) more positive then 0‰. Between 14.2 and 11.2 Ma few of the sampled specimens can be identified to species, but all the $\delta^{13}C$ values are less than −7.0, indicating diets dominated by or exclusively of C_3 plants. This includes samples from both high-crowned species of *Hypsodontus*, which might be expected to have been grazing to some extent. There is in addition a suggestion of a weak trend toward more negative values in the Middle and earliest Late Miocene, even after the sample values have been adjusted for the 0.5 $\delta^{13}C$ shift at 12.5 Ma. This can be seen when the 47 samples between 14.2 and 11.2 Ma are divided into 5 equal intervals and sample medians are calculated for each (Fig. 23.4b).

In the Late Miocene more of the sampled specimens can be identified to species. *Elachistoceras khauristanensis*, *Miotragocerus pilgrimi*, *Selenoportax* aff. *vexillarius*, *Pachyportax falconeri*, and the species of *Kobus* probably had exclusively C_3 diets, although they show considerable variation (Table 23.3). At about 8.2 Ma, however, there is shift toward more positive values with a higher percentage of individuals between −10 and −7‰, including species of *Kobus* and *Miotragocerus punjabicus* and *Pachyportax dhokpathanensis*. Subsequently, in the latter two species this trend culminates with individuals having a minor C_4 component to their diets (Fig. 23.4c, d; Table 23.3). For instance in *Pachyportax dhokpathanensis* mean $\delta^{13}C$ values shift at 8.2 Ma from −11.4‰ (n = 9) to −9.6‰ (n = 18) and then to −7.5‰ (n = 5) after 7.5 Ma. The older *Selenoportax* aff. *vexillarius* and *Pachyportax falconeri* had mean $\delta^{13}C$ values of −11.9‰ (n = 7) and −11.1‰ (n = 10) respectively, which are similar to those of pre–8.2 Ma *Pachyportax dhokpathanensis*.

Microscopic wear on tooth enamel has also been used to infer diet and distinguish between several general categories of diet: predominantly grazing from browsing, but also with less success fruit-dominated from browse-dominated diets or those many species with "mixed" diets (Chapter 24 this volume; Gagnon and Chew 2000; Solounias and Semprebon 2002; Solounias, Rivals, and Semprebon 2010). In contrast to isotopic studies, because teeth wear rapidly, microwear records diet shortly before the animal died and therefore may not reflect diet over the animal's lifetime.

Nelson (2003) collected microwear data on a small sample of Late Miocene species, including *Miotragocerus pilgrimi*, *Kobus porrecticornis* and *Kobus* sp. 2, *Pachyportax dhokpathanensis*, *Elachistoceras khauristanensis*, *Gazella lydekkeri*, and medium-sized unidentified boselaphines. She concluded that most species took browse and fruit and after about 8 Ma maybe some C_4 grass. That is, they had mixed diets. The one exception was *Kobus* sp. 2, with many fine scratches Nelson (2003) thought indicated C_4 grazing.

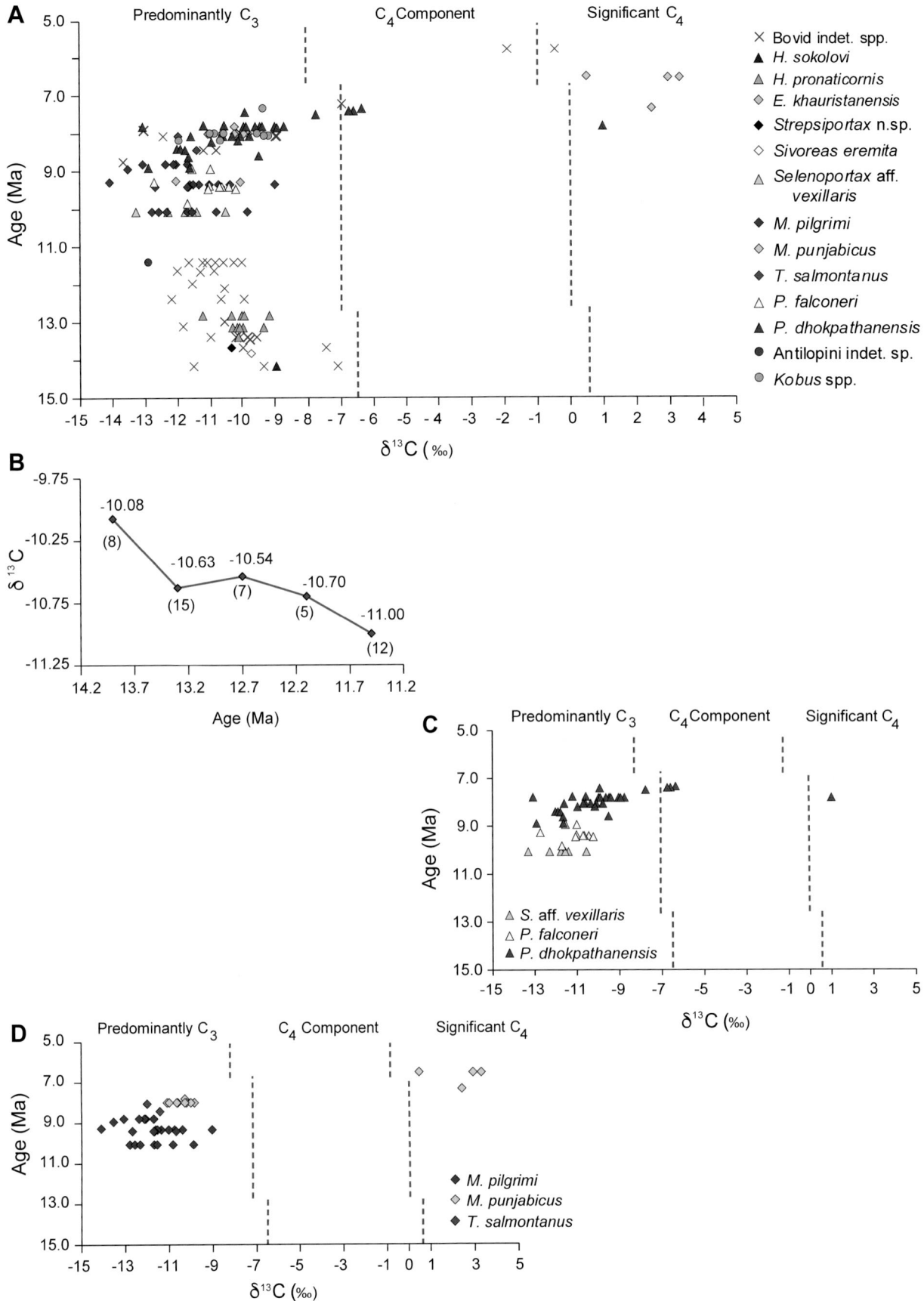

Figure 23.4. A. One hundred and sixty-one δ¹³C sample values versus age of sample. The vertical blue and red lines mark the boundaries between exclusively C₃ diets, those that may have a C₄ component, and those that probably had a significant C₄ component; B. Sample medians for five equal length intervals between 14.2 and 11.2 Ma showing a shift to more negative δ¹³C values; C. Forty-nine δ¹³C sample values for *Selenoportax* aff. *vexillarius*, *Pachyportax falconeri*, and *Pachyportax dhokpathanensis* showing a shift to more positive values over time; D. Forty δ¹³C sample values for *Miotragocerus pilgrimi*, *Miotragocerus punjabicus*, and *Tragoportax salmontanus* showing an abrupt shift to more positive values over time.

Table 23.3. Stable Carbon Isotope Values of Siwalik Bovids

Species	N	δ¹³C ‰ Mean	δ¹³C ‰ Median	Range ‰	δ¹³C ‰ Minimum	δ¹³C ‰ Maximum
Hypsodontus sokolovi	1	−9.0	−9.0			
Hypsodontus pronaticornis	14/3[1]	−10.0	−10.1	2.1	−11.3	−9.2
Strepsiportax unnamed species	1	10.4				
Sivoreas eremita	1	−9.8				
Elachistoceras khauristanensi	2	−11.1		1.9	−12.0	−10.1
Miotragocerus pilgrimi	25/23[1]	−11.7	−11.7	5.1	−14.1	−9.0
Miotragocerus punjabicus						
pre-7.5 Ma	9/7[1]	−10.4	−10.3	1.2	−11.1	−9.8
post-7.5 Ma	3	2.9	2.9	0.9	2.4	3.3
Tragoportax salmontanus	2	−11.7		0.6	−12.0	−11.4
Selenoportax aff. *vexillarius*	7	−11.9	−11.8	2.8	−13.3	−10.6
Pachyportax falconeri	10	−11.1	−11.0	2.5	−12.7	−10.2
Pachyportax dhokpathanensis						
pre-8.0 Ma	15	−11.1	−11.0	3.4	−12.9	−9.5
8.0–7.5 Ma	12	−9.1	−9.6	14.1	−13.1	1.0
post-7.5 Ma	5	−7.5	−6.8	3.6	−9.9	−6.4
Kobus sp. 1	2/1	−11.3		1.3	−12.0	−10.7
Kobus porrecticornis	6/4[1]	−10.1	−10.1	1.8	−11.0	−9.2
Kobus sp. 2	1	−9.4				

Note: Table does not include 37 samples, mostly of *Pachyportax dhokpathensis*, with imprecise age estimates.

[1] Number samples / number of individuals.

Nelson's counts of scratches and pits, plus unpublished data collected by G. M. Semprebon, place most individuals (Fig. 23.5) within the region occupied by browsers, mixed feeders, and fruit-dominated browsers (Solounias and Semprebon 2002). In contrast, extant bovid grazers are characterized by large numbers of scratches and few pits (Solounias and Hayek 1993), and several individuals lie near the grazer region (Fig. 23.5). Two are 6.5 Ma individuals of *Miotragocerus punjabicus* with positive δ¹³C values, while a third 8.0 Ma *Kobus porrecticornis* has a δ¹³C value of −10.9‰, suggesting it grazed on C₃ grasses.

Six specimens of *Pachyportax dhokpathanensis* have both microwear and δ¹³C values, two of which are of interest. The first has a δ¹³C value of −10.7‰, while the microwear suggests grazing. The other, whose very low number of scratches suggests browsing, has a δ¹³C value of −6.6‰. Although the sample size is very small, these data, together with the highly skewed variation seen in the number of scratches (5 to 22, with a median of 8.5), suggest mixed feeding on a seasonal or geographic basis in *Pachyportax dhokpathanensis*.

Mesowear is the wear of the tooth crown as a whole, as seen in the relative height and bluntness of cusps, and has been used to differentiate between diet categories (Fortelius and Solounias 2000). Mesowear is thought to reflect use of the tooth throughout the life of the animal and is characterized as a gradient from attrition-dominated (typical of browsers) to abrasion-dominated (typical of grazers), with browse-dominated and graze-dominated mixed feeders as intermediates.

Our scoring of data collected by Miriam Belmaker on GSP specimens (Belmaker et al. 2007) uses a five point system (0–4) that combines scores for cusp height and shape, with the higher scores indicating a more abrasive diet. Among extant taxa, more than 95% of browsers score less than 2, while nearly a fourth of the scores for grazers are 3 or higher (Fig. 23.6a; data for extant species from Fortelius and Solounias [2000] and Blondel et al. [2010] converted and normalized for sample size.) Mixed feeders are intermediate, but most similar to browsers.

Overall the Siwalik species exhibit the full range of scores from 0 to 4, but the sample includes a mix of species and feeding types of different ages. A subset of specimens between 14.1 and 11.0 Ma (Fig. 23.6b) may include as many as ten species. Although a small sample, all the teeth score less than 2, and the sample as a whole is most similar to browsers and mixed feeders. A larger Late Miocene sample also contains many low-scoring teeth, but also a small number with scores of 2 or more. A preliminary study (Morgan et al. 2015) focused on two presumed Late Miocene lineages: *Selenoportax* aff. *vexillarius-Pachyportax falconeri-Pachyportax dhokpathanensis* and *Miotragocerus pilgrim-Miotragocerus punjabicus*. In both lineages the teeth acquired increasing occlusal complexity over time, a trend not present elsewhere in Late Miocene boselaphines, and mesowear also suggests the younger species had a more abrasive diet, with median scores changing from 0 to 1 (Fig. 23.6c, d).

For specimens used in multiple studies, mesowear is consistent with microwear and isotopic signals. A low number of scratches and evidence for C₃-dominated diets combine with scores indicating less tooth abrasion.

Hypsodonty is an adaptation related to tooth wear and as such reflects not only the use of a tooth throughout the life of the animal, but also its potential use or even a species' evolutionary history (Damuth and Janis 2011). Hypsodonty is usually explained as an adaptation for grazing or an otherwise

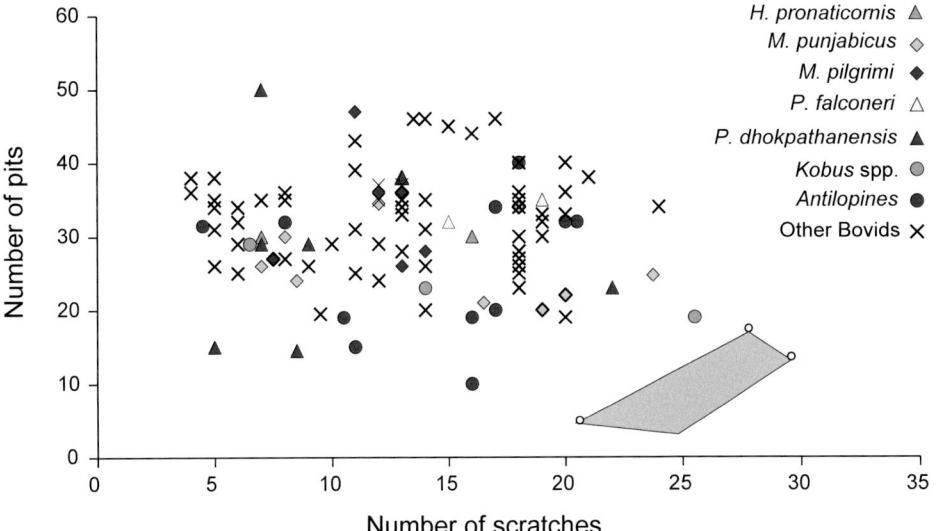

H. pronaticornis △
M. punjabicus ◇
M. pilgrimi ◆
P. falconeri △
P. dhokpathanensis ▲
Kobus spp. ●
Antilopines ●
Other Bovids ✕

Figure 23.5. Microwear data (scratches versus pits) for 109 individuals from seven taxa plus indeterminate bovids. Shaded quadrilateral is the region occupied by extant grazing bovids. Extant bovid data from Solounias and Semprebon 2002.

abrasive diet, or perhaps longer life. However, as an adaptation for an abrasive diet, hypsodonty most likely reflects design for rarely consumed fall-back foods and not necessarily the food most frequently eaten. It may therefore not be a reliable indicator of diet.

An index of hypsodonty (metaconid height / molar width) for 52 unworn or lightly worn lower third molars is plotted against time on Figure 23.7 with the classification of Damuth and Janis (2011) for extant species. Most sampled molars are mesodont, although after 8.2 Ma molars of *Kobus* and *Pachyportax dhokpathanensis* are more hypsodont, as are a few of *Miotragocerus punjabicus*. The appearance of higher crowned teeth in these three taxa is consistent with other indicators of diet. At 14.1 Ma *Hypsodontus sokolovi* is an important exception, for while very hypsodont, microwear and isotopic data indicate a nonabrasive, C_3-dominated diet. *Hypsodontus pronaticornis* is also high crowned, but there are no intact, unworn lower molars in the collections.

CONCLUSIONS

We recognize 42 bovid species (or probable species) from the Late Oligocene and Miocene deposits of Pakistan. Twenty-seven are species that have previously been named (Table 23.1), while another 15 have cranial material suitable for defining and naming a species. Of the named species, the GSP collection contains specimens of all but 2 or possibly 3: *Ruticeros pugio*, *Pachyportax giganteus*, and perhaps *Selenoportax vexillarius*.

Boselaphini predominate in the collections, both in terms of number of species (17) and number of specimens (~75%), but Hypsodontinae, Bovini, Antilopini, Reduncini, and Caprini are also known. Most of the boselaphine genera are endemic to the Siwaliks of the subcontinent (*Elachistoceras, Strepsiportax, Sivaceros, Helicoportax, Selenoportax, Tragoportax, Ruticeros,* and most likely *Sivoreas*), as are also the hypsodontine *Palaeohypsodontus* and the bovini *Pachyportax*. Significantly, the few Re-

duncini and Caprini taxa (*Kobus, Tethytragus, Caprotragoides, Dorcadoryx,* and *Protoryx*) are cosmopolitan in their distributions, as are *Hypsodontus, Eotragus, Miotragocerus, Gazella, Prostrepsiceros, Nisidorcas, Protragelaphus,* and *Dorcadoryx.*

Little is known about the phylogeny of any of the Siwalik bovids. Among the boselaphines there are several earlier and later species pairs within genera that may represent ancestor and descendent species, but a phylogenetic analysis has not yet been undertaken. In contrast to the boselaphines, the Siwalik spiral-horned Antilopini have closely related species in Eurasia, but too little is known to judge how they are related. The ages of the Eurasian species are poorly constrained, but in the case of *?Protragelaphus* sp. and perhaps *Nisidorcas planicornis* the Siwalik occurrences are the oldest known. A 14.1 Ma small undescribed species is the earliest likely member of the Antilopinae.

Although Siwalik Pliocene Reduncini were long known, the post-1973 GSP collections establish the presence of a reduncine as early as 8.0 or 9.0 Ma, which is 1 to 2 million years older than African reduncines. Caprini are very rare in the Siwaliks, but the single *Protoryx* aff. *solignaci* occurrence just precedes the appearance of hipparionine horses as does *Protoryx solignaci* in Tunisia and Turkey. Neither Hippotragini nor Alcelaphini have been recorded from the Miocene Siwaliks, although both are known from younger horizons. Both tribes appear in Africa at about 7.0 Ma.

The appearance at 14.1 Ma of a number of bovid species is most likely an artifact of preservation, but the complete turnover of species and expansion in size between 11.2 and 10.5 Ma is not, although how rapidly and exactly when within that interval it took place are not clear (Figs. 23.1 and 23.2). The brief presence in this interval of *Selenoportax vexillarius, ?Protragelaphus* sp., and *Protoryx* aff. *solignaci* suggests a complex, perhaps multistep event.

Locomotor adaptations suggest the predominance of relatively closed habitats between 12.4 and 11.4 Ma, although there

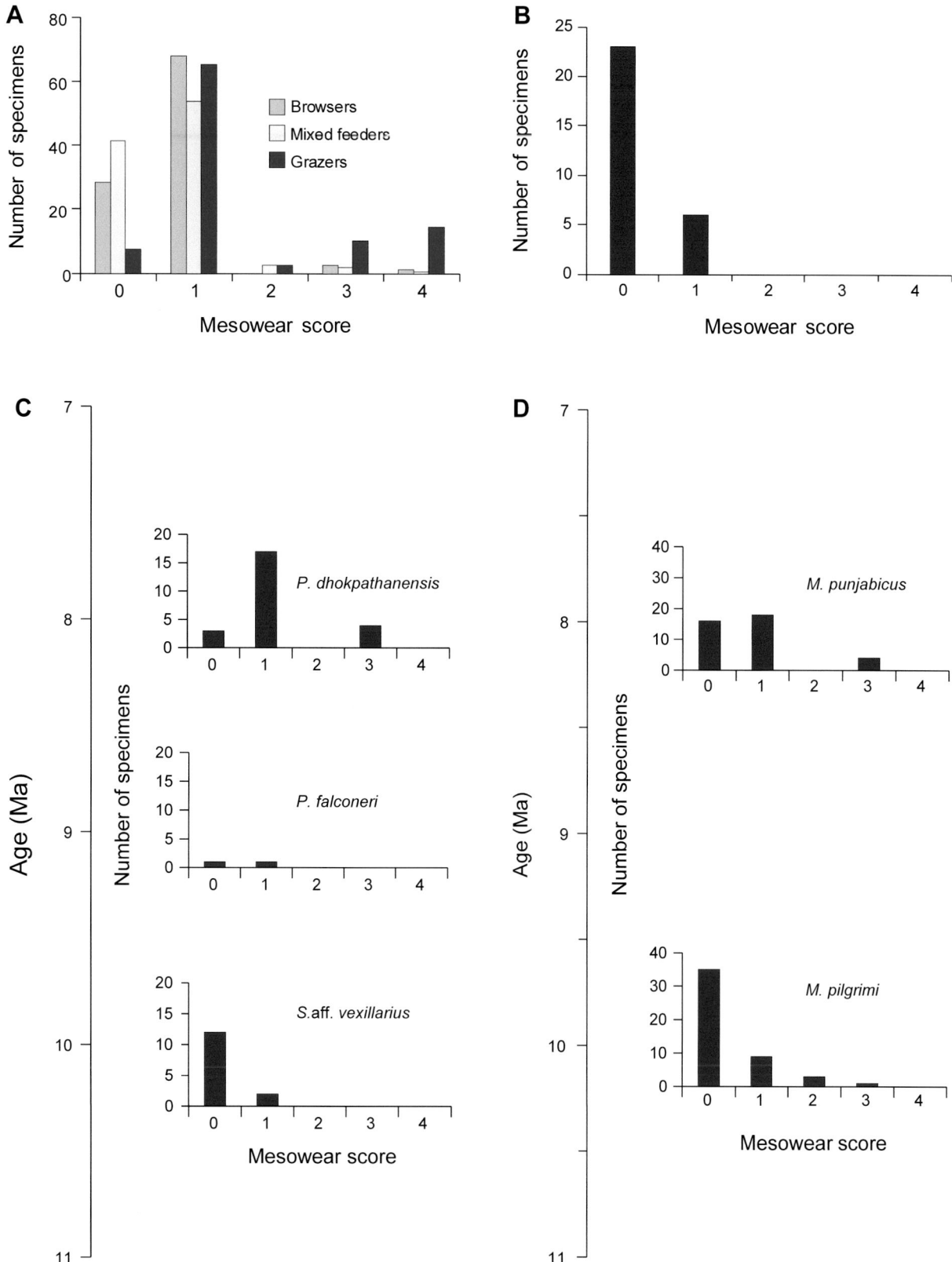

Figure 23.6. A. Mesowear scores for extant browsers, mixed feeders, and grazers; B. Mesowear scores for 29 Middle Miocene Siwalik bovids; C. and D. Mesowear scores versus specimen age for two presumed lineages.

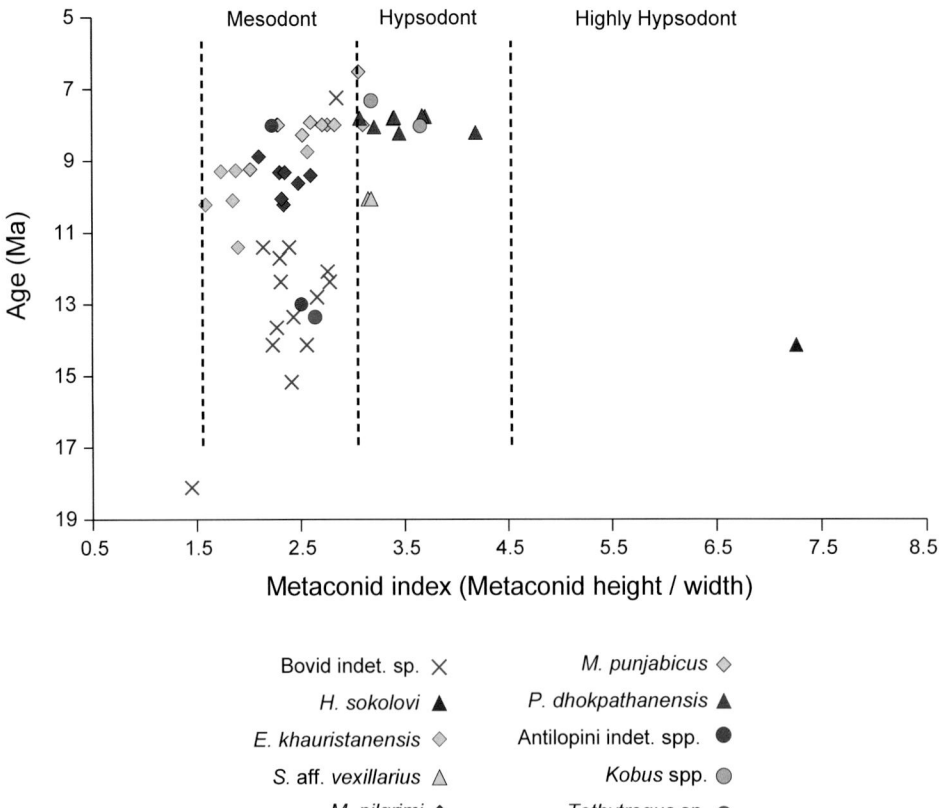

Figure 23.7. Hypsodont scores versus specimen age. The index is the height of the metaconid divided by width of the trigonid at the level of the premetacristid. Vertical lines separate categories of Damuth and Janis (2011).

Bovid indet. sp. ✕		*M. punjabicus* ◇
H. sokolovi ▲		*P. dhokpathanensis* ▲
E. khauristanensis ◇		Antilopini indet. spp. ●
S. aff. *vexillarius* △		*Kobus* spp. ●
M. pilgrimi ◆		*Tethytragus* sp. ●

may have been areas with light bush or even grass. The δ^{13}C and dental data support this conclusion. Subsequently the isotopic, dental, and postcranial data are more complicated. While the results are ambiguous, analysis of 10.1 Ma postcranials suggests the presence of more light cover compared to 11.4 Ma. However, data on diet over the longer period between 10.5 and 8.0 Ma indicate the continuing presence of an assemblage of species with diets based on C_3 vegetation.

After 8.0 and perhaps as early as 8.5 Ma there is clear evidence for some species shifting their diets, with microwear, mesowear, hypsodonty, and stable isotope data for *Kobus* spp., *Pachyportax dhokpathanensis*, and medium size boselaphines variously indicating the presence of C_4 plants or substantial grazing. After 7.5 Ma, some taxa possibly had diets dominated by C_4 vegetation.

ACKNOWLEDGMENTS

We thank David Pilbeam, Catherine Badgley, and Michèle Morgan for inviting us to participate in this undertaking and the Department of Human Evolutionary Biology, Harvard University, for the provision of facilities. This research is a result of a long collaboration with the Geological Survey of Pakistan, and we are grateful to the Directors General and members of the Survey who have over the years supported our project. Financial support has come through a number of grants from the National Science Foundation, the Smithsonian Foreign Currency Program, the Wenner-Gren Foundation for Anthropological Research, and the American School for Prehistoric Research. We also acknowledge the support of Francine Berkowitz of the Smithsonian Office of International Relations.

References

Akhtar, M. 1995. *Pachyportax giganteus*, new species (Mammalia: Artiodactyla: Bovidae) from the Dhok Pathan, District Chakwal, Punjab, Pakistan. *Pakistan Journal of Zoology* 27:337–340.

Antoine, P.-O., G. Métais, M. J. Orliac, J.-Y. Crochet, L. J. Flynn, L. Marivaux, A. R. Rajpar, G. Roohi, and J.-L. Welcomme. 2013. Mammalian Neogene biostratigraphy of the Sulaiman Province, Pakistan. In *Fossil Mammals of Asia: Neogene Biostratigraphy and Chronology*, edited by X.-M. Wang, L. J. Flynn and M. Fortelius, 400–442. New York: Columbia University Press.

Barry J. C., A. K. Behrensmeyer, C. Badgley, L. J. Flynn, H. Peltonen, I. U. Cheema, D. Pilbeam, et al. 2013. The Neogene Siwaliks of the Potwar Plateau, Pakistan. In *Fossil Mammals of Asia: Neogene Biostratigraphy and Chronology*, edited by X.-M. Wang, L. J. Flynn and M. Fortelius, 373–399. New York: Columbia University Press.

Barry, J. C., S. Cote, L. MacLatchy, E. H. Lindsay, R. Kityo, and A. R. Rajpar. 2005. Oligocene and Early Miocene ruminants (Mammalia, Artiodactyla) from Pakistan and Uganda. *Palaeontologia Electronica* 8(1): 22A.

Barry, J. C., M. E. Morgan, L. J. Flynn, D. Pilbeam, A. K. Behrensmeyer, S. M. Raza, I. A. Khan, C. Badgley, J. Hicks, and J. Kelley. 2002. Faunal and environmental change in the Late Miocene Siwaliks of northern Pakistan. *Paleobiology Memoirs* 3(S2): 1–72.

Belmaker, M., S. Nelson, M. Morgan, J. Barry, and C. Badgley. 2007. Mesowear analysis of ungulates in the Middle to Late Miocene of the

Siwaliks, Pakistan: Dietary and paleoenvironmental implications. *Society of Vertebrate Paleontology Meeting Program and Abstracts* 2007:46A.

Blondel, C., G. Merceron, L. Andossa, M. Taisso, P. Vignaud, and M. Brunet. 2010. Dental mesowear analysis of the late Miocene Bovidae from Toros-Menalla (Chad) and early hominid habitats in Central Africa. *Palaeogeography, Palaeoclimatology, Palaeoecology* 292(1–2): 184–191.

Bibi, F. 2007a. Origin, Paleoecology, and Paleobiogeography of Early Bovini. *Palaeogeography, Palaeoclimatology, Palaeoecology* 248(1–2): 60–72.

Bibi, F. 2007b. Dietary niche partitioning among fossil bovids in late Miocene C_3 habitats: Consilience of functional morphology and stable isotope analysis. *Palaeogeography, Palaeoclimatology, Palaeoecology* 253(3–4): 529–538.

Bibi, F., M. Bukhsianidze, A. W. Gentry, D. Geraads, D. S. Kostopoulos, and E.S.Vrba. 2009. The fossil record and evolution of Bovidae: State of the field. *Palaeontologia Electronica* 12(3): 10A.

Bouvrain, G. 1979. Au nouveau genre de Bovidè de la fin du Miocéne. *Bulletin de la Société Géologique de France* S7-XXI(4): 507–511.

Corvinus, G., and L. N. Rimal. 2001. Biostratigraphy and geology of Neogene Siwalik Group of Surai Khola and Rato Khola areas in Nepal. *Palaeogeography, Palaeoclimatology, Palaeoecology* 165(3–4): 251–279.

Damuth, J. and C. M. Janis. 2011. On the relationship between hypsodonty and feeding ecology in ungulate mammals, and its utility in palaeoecology. *Biological Reviews* 86(3): 733–758.

Dmitrieva, E. L. 2002. On the early evolution of bovids. *Paleontological Journal* 36(2): 204–206.

Estes, R. D. 1993. *The Safari Companion: A Guide to Watching African Mammals*. White River Junction, VT: Chelsea Green Publishing Company.

Flynn, L. J., E. H. Lindsay, D. Pilbeam, S. M. Raza, M. E. Morgan, J. C. Barry, C. E. Badgley, et al. 2013. The Siwaliks and Neogene evolutionary biology in South Asia. In *Fossil Mammals of Asia: Neogene Biostratigraphy and Chronology*, edited by X.-M. Wang, L. J. Flynn and M. Fortelius, 353–372. New York: Columbia University Press.

Fortelius, M., and N. Solounias. 2000. Functional characterization of ungulate molars using the abrasion-attrition wear gradient: A new method for reconstructing paleodiets. *American Museum Novitates* 3301:1–36.

Gagnon, M., and A. Chew. 2000. Dietary preferences in extant African Bovidae. *Journal of Mammalogy* 81(2): 490–511.

Gatesy, J., G. Amato, E. Vrba, G. Schaller, and R. DeSalle. 1997. A cladistic analysis of mitochondrial ribosomal DNA from the Bovidae. *Molecular Phylogenetics and Evolution* 7(3): 303–319.

Gentry, A. W. 1970. The Bovidae (Mammalia) of the Fort Ternan fossil fauna. In *Fossil Vertebrates of Africa*, edited by L. S. B. Leakey and R. J. G. Savage, 2:243–324. London: Academic Press.

Gentry, A. W. 2003. Ruminantia (Artiodactyla). In *Geology and Paleontology of the Miocene Sinap Formation, Turkey*, edited by M. Fortelius, J. Kappelman, S. Sen, and R. L. Bernor (eds.), 332–379. New York: Columbia University Press.

Gentry, A. W. 2010. Bovidae. In *Cenozoic Mammals of Africa*, edited by L. Werdelin and W. J. Sanders, 741–796. Berkeley: University of California Press.

Gentry, A. W., N. Solounias, and J. C. Barry. 2014. Stability in higher level taxonomy of Miocene bovid faunas of the Siwaliks. *Annales Zoologici Fennici* 51:49–56.

Geraads, D., C. Blondel, A. Likius, H. T. Mackaye, P. Vignaud, and M. Brunet. 2008. New Hippotragini (Bovidae, Mammalia) from the Late Miocene of Toros-Menalla (Chad). *Journal of Vertebrate Paleontology* 28(1): 231–242.

Ghaffar, A., M. Akhtar, and A. Q. Nayye. 2011. Evidences of early Pliocene fossil remains of tribe Cervini (Mammalia, Artiodactyla, Cervidae) from the Siwaliks of Pakistan. *Journal of Animal and Plant Sciences* 21(4): 830–835.

Ghaffar, A., M. A. Khan, and M. Akhtar. 2010. Early Pliocene Cervids (Artiodactyla-Mammalia) from the Siwaliks of Pakistan. *Yerbilimleri* 31:217–231.

Ginsburg, L., J. Morales, and D. Soria. 2001. Les Ruminantia (Artiodactyla, Mammalia) du Miocène des Bugti (Balouchistan, Pakistan). *Estudios Geologicos* 57(3–4): 155–170.

Gradstein, F. M., J. G. Ogg, M. D. Schmitz, G. M. Ogg. 2012. *The Geological Time Scale 2012*. Amsterdam: Elsevier.

Hassanin, A., F. Delsuc, A. Ropiquet, C. Hammer, B. J. van Vuuren, C. Matthee, M. Ruiz-Garcia, et al. 2012. Pattern and timing of diversification of Cetartiodactyla (Mammalia, Laurasiatheria), as revealed by a comprehensive analysis of mitochondrial genomes. *Comptes Rendus Biologies* 335(1): 32–50.

Huang, X. 1985. Fossil bovids from the Middle Oligocene of Ulantalal, Nei Mongol, *Vertebrata Palasiatica* 23:152–160.

Janis, C. M., and E. Manning. 1998. Antilocapridae. In *Evolution of Tertiary Mammals of North America*, edited by C. M. Janis, K. M. Scott, and L. L. Jacobs,1: 491–507. Cambridge: Cambridge University Press.

Janis, C. M., and K. M. Scott. 1987. The interrelationships of higher ruminant families with special emphasis on the members of the Cervoidea. *American Museum Novitates* 2893:1–85.

Jarman, P. 1974. The social organisation of antelope in relation to their ecology. *Behaviour* 48:215–267.

Jarman, P. 1983. Mating system and sexual dimorphism in large, terrestrial, mammalian herbivores. *Biological Reviews* 58(4):485–520.

Kappelman, J. W. 1986. The paleoecology and chronology of the Middle Miocene hominoids from the Chinji Formation of Pakistan. Unpublished PhD diss., Harvard University.

Kappelman, J. W. 1988. Morphology and locomotor adaptations of the bovid femur in relation to habitat. *Journal of Morphology* 198(1): 119–130.

Kappelman, J., A. Duncan, M. Feseha, J.-P. Lunkka, D. Ekart, F. McDowell, T.M. Ryan, and C. C. Swisher. 2003. Chronology. In *Geology and Paleontology of the Miocene Sinap Formation, Turkey*, edited by M. Fortelius, J. Kappelman, S. Sen, and R. L. Bernor, 41–68. New York: Columbia University Press.

Kingdon, J. 1997. *The Kingdon Field Guide to African Mammals*. New York: Academic Press.

Köhler, M. 1987. Boviden des turkischen Miozäns (Känozoikum und Braunkohlen der Türkei 28). *Paleontologia i Evolucio* 21:133–246.

Kostopoulos, D. S. 2006. The late Miocene vertebrate locality of Perivolaki, Thessaly, Greece. 9. Cervidae and Bovidae. *Palaeontographica A* 276(1–6): 151–183.

Kostopoulos, D. S. 2014. Taxonomic re-assessment and phylogenetic relationships of Miocene homonymously spiral-horned antelopes. *Acta Palaeontologica Polonica* 59(1): 9–29.

Kostopoulos, D. S. 2016. Palaeontology of the Upper Miocene vertebrate localities of Nikiti (Chalkidiki Peninsula, Macedonia, Greece). Artiodactyla. *Geobios* 49:119–134.

Lindsay, E. H., L. J. Flynn, I. U. Cheema, J. C. Barry, K. F. Downing, A. R. Rajpar, and S. M. Raza. 2005. Will Downs and the Zinda Pir Dome. *Palaeontologia Electronica* 8(1): 19A.

Lydekker, R. 1876. Descriptions of the molar teeth and other remains of Mammalia. *Memoirs of the Geological Survey of India, Palaeontologia Indica* 10(2): 19–87.

Made, J., van der, and S. T. Hussain. 1994. Horn cores of *Sivoreas* (Bovidae) from the Miocene of Pakistan and utility of their torsion as a taxonomic tool. *Geobios* 27(1): 103–111.

Meng, J. and M. C. McKenna. 1998. Faunal turnovers of Palaeogene mammals from the Mongolian Plateau. *Nature* 394:364–367.

Métais, G., P.-O. Antoine, L. Marivaux, J.-L. Welcomme, and S. Ducrocq. 2003. New artiodactyl ruminant mammal from the Late Oligocene of Pakistan. *Acta Palaeontologica Polonica* 48(3): 375–382.

Morgan, M. E. 1994. Paleoecology of Siwalik Miocene hominoid communities: Stable carbon isotope, dental microwear, and body size analyses. Unpublished PhD diss., Harvard University.

Morgan, M. E, L. Spencer, E. Scott, M. S. Domingo, C. Badgley, J. C. Barry, L. J. Flynn, and D. Pilbeam. 2015. Detecting differences in forage in C_3-dominated ecosystems: Mesowear and stable isotope analyses of Miocene bovids from northern Pakistan. *Society of Vertebrate Paleontology Meeting Program and Abstracts*, 2015:184.

Nakaya, H., M. Pickford, Y. Nakano, and H. Ishida. 1984. The Late Miocene large mammal fauna from the Namurungule Formation, Samburu Hills, northern Kenya. *African Study Monographs Kyoto*, Supplementary Issue 2:87–131.

Nanda, A. C. 2015. *Siwalik Mammalian Faunas of the Himalayan Foothills, with Reference to Biochronology, Linkages and Migration.* Dehradun, India: Wadia Institute of Himalayan Geology.

Nelson, S. V. 2003. *The Extinction of Sivapithecus: Faunal and Environmental Changes Surrounding the Disappearance of a Miocene Hominoid in the Siwaliks of Pakistan.* American School of Prehistoric Research Monographs, vol. 1. Boston: Brill Academic Publishers.

Patnaik, R. 2013. Indian Neogene Siwalik Mammalian Biostratigraphy: An Overview. In *Fossil Mammals of Asia: Neogene Biostratigraphy and Chronology*, edited by X.-M. Wang, L. J. Flynn and M. Fortelius, 423–444. New York: Columbia University Press.

Pilgrim, G. E. 1910. Notices of new mammalian genera and species from the Tertiaries of India. *Records of the Geological Survey of India* 40:63–71.

Pilgrim, G. E. 1937. Siwalik antelopes and oxen in the American Museum of Natural History. *Bulletin of the American Museum of Natural History* 72(7): 729–874.

Pilgrim, G. E. 1939. The fossil Bovidae of India. *Memoirs of the Geological Survey of India, Palaeontologia Indica* n.s. 26:1–356.

Raza S. M, J. C. Barry, G. E. Meyer, and L. Martin. 1984. Preliminary report on the geology and vertebrate fauna of the Miocene Manchar Formation, Sind, Pakistan. *Journal of Vertebrate Paleontology* 4(4): 584–599.

Robinson, P. 1972. *Pachytragus solignaci*, a new species of caprine bovid from the late Miocene Beglia Formation of Tunisia. *Notes du Service Géologique de Tunisie* 37:73–94.

Scott, R. S. 2004. The comparative paleoecology of Late Miocene Eurasian hominoids. Unpublished PhD diss., University of Texas at Austin.

Scott, R. S., J. W. Kappelman, and J. Kelley. 1999. The paleoenvironment of *Sivapithecus parvada*. *Journal of Human Evolution* 36(3): 245–274.

Solounias, N., J. C. Barry, R. L. Bernor, E. H. Lindsay, and S. M. Raza. 1995. The oldest bovid from the Siwaliks, Pakistan. *Journal of Vertebrate Paleontology* 15(4): 806–814.

Solounias, N. and L-A.C. Hayek. 1993. New methods of tooth microwear analysis and application to dietary determination of two extinct antelopes. *Journal of Zoology (London)* 229(3): 421–445.

Solounias, N., F. Rivals, and G. M. Semprebon. 2010. Dietary interpretation and paleoecology of herbivores from Pikermi and Samos (late Miocene of Greece). *Paleobiology* 36(1): 113–136.

Solounias, N., and G. M. Semprebon. 2002. Advances in the reconstruction of ungulate ecomorphology and application to early fossil equids. *American Museum Novitates* 3366:1–49.

Thenius, E. 1956. Zur Entwicklung des Knochenzapfens von *Protragocerus* Depéret aus dem Miozän. *Geologie Berlin* 5:308–318.

Thomas, H. 1980. Les bovidés du Miocène Supérieur des couches de Mpesida et de la Formation de Lukeino (district de Baringo, Kenya). In *Proceedings of the 8th Pan-African Congress of Prehistory, Nairobi, 1977*, edited by R. E. F. Leakey and B. A. Ogot, 82–91. Nairobi.

Thomas, H. 1981. Les Bovidés miocènes de la Formation de Ngorora du Bassin de Baringo (Kenya). *Proceedings Koninklijke Nederlandse Akademie van Wetenschappen B*, 84(3): 335–409.

Thomas, H. 1984. Les Bovidés anté-hipparions des Siwaliks inférieurs (plateau du Potwar, Pakistan). *Mémoires de la Société Géologique de France Paris (1924)* n.s. 145:1–68.

Tsubamoto, T., M. Takai, and N. Egi. 2004. Quantitative analyses of biogeography and faunal evolution of Middle to Late Eocene mammals in East Asia. *Journal of Vertebrate Paleontology* 24(3): 657–667.

Uno, K. T., T. E. Cerling, J. M. Harris, Y. Kunimatsu, M. G. Leakey, M. Nakatsukasa, and H. Nakaya. 2011. Late Miocene to Pliocene carbon isotope record of differential diet change among East African herbivores. *Proceedings of the National Academy of Sciences* 108(16): 6509–6514.

Wang, B. 1992. The Chinese Oligocene: A preliminary review of mammalian localities and local faunas. In *Eocene–Oligocene Climatic and Biotic Evolution*, edited by D. R. Prothero and W. A. Berggren, 529–547. Princeton, NJ: Princeton University Press.

Werdelin, L. 2010. Chronology of Neogene mammal localities. In *Cenozoic Mammals of Africa*, edited by L. Werdelin and W. J. Sanders, 27–43. Berkeley: University of California Press.

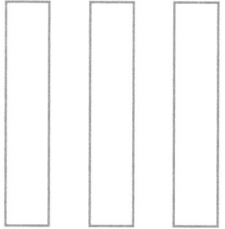

PREAMBLE TO SYNTHETIC CHAPTERS

CATHERINE BADGLEY, MICHÈLE E. MORGAN, AND
DAVID PILBEAM

The following four chapters integrate data from the foregoing work to present paleoenvironmental reconstructions and studies of mammalian community structure and faunal dynamics. Chapter 24 draws together global, regional, and local geological and biological evidence as the basis for a paleoecological synthesis of the Potwar Plateau between 18 and 6 Ma. Central to this chapter are four detailed, evidence-based artistic reconstructions by Julius Csotonyi that provide realistic views of Siwalik ecosystems at different times. Three depict the faunas and floras of single fossil localities, and the fourth is an aerial view of the Late Miocene sub-Himalayan alluvial plain. Chapters 25 and 26 focus on different aspects of faunal structure and change. Chapter 25 characterizes functional attributes of six mammalian paleocommunities between 14.1 and 7.4 Ma, each representing a 100-kyr time interval, using first and last occurrence data for species. Chapter 26 analyzes

rates of faunal change and uses confidence intervals to estimate taxonomic richness through the entire time span between ~18 to ~6 Ma. While the approaches of Chapters 25 and 26 differ, the big-picture findings are similar. Both chapters utilize the dataset for mammalian taxa presented in Table III.1. Chapters 24, 25, and 26 figure or refer to the large data set of stable carbon and oxygen isotope data from Siwalik herbivores to which many have contributed—included primary data are compiled in Table III.2. The book's final chapter reflects on the value of long fossil records for exploring questions relating to evolution of biotas and ecosystems at varying timescales. It offers research highlights from the Harvard-GSP project and suggests ideas and directions for future work. As acknowledged in Chapter 1, this book is a product of the efforts of hundreds of people who have contributed to the success of field and laboratory work over many decades of collaboration.

Table III.1. Species Table for Taxa Documented between 18 and 6 Ma in the Potwar Plateau Siwaliks

Order[1]	Family	Name[2]	Observed First Occurrence[3]	Observed Last Occurrence	Species Origination[4]	Size Class[5]
Lipotyphla	Erinaceidae	Galerix wesselsae	18.800	14.039	I	1
Lipotyphla	Erinaceidae	Galerix rutlandae	14.300	11.600	I	1
Lipotyphla	Erinaceidae	Schizogalerix sp.	11.410	10.222	I	2
Lipotyphla	Erinaceidae	Erinaceinae, gen. B and sp. indet.	13.085	13.085	I	1
Lipotyphla	Erinaceidae	Erinaceinae, gen. C and sp. indet.	7.338	7.338	I	1
Lipotyphla	Soricidae	cf. Crocidura sp.	14.039	6.517	E	1
Lipotyphla	Soricidae	Suncus honeyi	10.492	6.517	I	1
Lipotyphla	Soricidae	Beremendiini, gen. and sp. indet.	8.749	8.749	I	1
Scandentia	Tupaiidae	cf. Dendrogale sp. 1	18.700	10.492	I	1
Scandentia	Tupaiidae	cf. Dendrogale sp. 2	18.700	13.050	I	1
Scandentia	Tupaiidae	cf. Dendrogale sp. 3	14.045	7.338	E	1
Scandentia	Tupaiidae	cf. Tupaia sp.	12.779	11.230	E	1
Scandentia	Tupaiidae	Tupaiidae, n. gen. and sp.	11.644	8.703	I	3
Scandentia	Tupaiidae	Tupaiidae, gen. and sp. indet. (tiny)	8.703	8.703	I	1
Lagomorpha	Ochotonidae	Alloptox sp.	17.892	17.892	I	2
Lagomorpha	Leporidae	Alilepus elongatus	7.394	6.517	I	4
Rodentia	Diatomyidae	Diatomys chitaparwalensis	17.892	17.892	I	1

(continued)

Table III.1. continued

Order[1]	Family	Name[2]	Observed First Occurrence[3]	Observed Last Occurrence	Species Origination[4]	Size Class[5]
Rodentia	Diatomyidae	*Willmus maximus*	11.183	11.183	I	2
Rodentia	Ctenodactylidae	*Sayimys minor*	21.400	15.962	I	1
Rodentia	Ctenodactylidae	*Sayimys intermedius*	18.750	15.100	E	1
Rodentia	Ctenodactylidae	*Sayimys* cf. *S. giganteus*	16.763	16.465	E	2
Rodentia	Ctenodactylidae	*Sayimys sivalensis*	15.169	12.749	E	1
Rodentia	Ctenodactylidae	*Sayimys chinjiensis*	12.377	10.063	E	1
Rodentia	Ctenodactylidae	*Sayimys* sp. B	12.377	11.183	E	1
Rodentia	Ctenodactylidae	*Sayimys perplexus*	7.394	7.394	E	1
Rodentia	Thryonomyidae	*Kochalia geespei*	16.045	13.391	E	1
Rodentia	Thryonomyidae	*Kochalia* n. sp.	17.892	16.763	I	1
Rodentia	Thryonomyidae	*Paraulacodus indicus*	13.085	12.749	I	3
Rodentia	Hystricidae	*Hystrix primigenia*	7.999	3.300	I	7.1
Rodentia	Gliridae	*Myomimus sumbalenwalicus*	13.800	12.327	I	1
Rodentia	Gliridae	*Myomimus* n. sp. A	11.644	10.063	E	1
Rodentia	Gliridae	*Myomimus* n. sp. B	9.400	7.966	E	1
Rodentia	Sciuridae	*Aliveria* sp.	16.763	15.962	I	2
Rodentia	Sciuridae	cf. *Ratufa* n. sp.	16.763	13.589	I	3
Rodentia	Sciuridae	*Ratufa sylva*	10.492	9.400	E	3
Rodentia	Sciuridae	*Tamiops* large sp.	14.279	8.749	E	1
Rodentia	Sciuridae	*Tamiops* small sp.	16.763	10.063	E	1
Rodentia	Sciuridae	*Tamiops* giant sp.	7.966	7.966	E	2
Rodentia	Sciuridae	*Callosciurus* sp.	14.279	9.240	I	2
Rodentia	Sciuridae	*Dremomys* sp.	14.279	7.394	I	2
Rodentia	Sciuridae	*Oriensciurus* n. sp.	18.700	17.878	I	3
Rodentia	Sciuridae	*Heteroxerus* sp.	14.279	13.183	I	3
Rodentia	Sciuridae	*Atlantoxerus* sp.	14.279	13.183	I	2
Rodentia	Sciuridae	*Sciurotamias* Kamlial sp.	16.045	12.756	I	2
Rodentia	Sciuridae	*Sciurotamias* sp. A (small)	14.279	11.410	E	2
Rodentia	Sciuridae	*Sciurotamias* sp. B (large)	11.183	7.157	E	2
Rodentia	Sciuridae	*Tamias* "pre-urialis"	14.279	13.589	I	1
Rodentia	Sciuridae	*Tamias urialis*	13.183	11.183	E	1
Rodentia	Sciuridae	*Tamias* sp., cf. *T. gilaharee*	10.492	7.963	I	2
Rodentia	Anomaluridae	Anomaluridae, n. gen. and sp. indet.	16.763	16.763	E	1
Rodentia	Spalacidae	*Prokanisamys arifi*	19.730	17.878	E	1
Rodentia	Spalacidae	*Prokanisamys major*	18.690	15.103	E	1
Rodentia	Spalacidae	*Prokanisamys* n. sp. (small)	16.763	13.085	E	1
Rodentia	Spalacidae	*Prokanisamys* n. sp.	14.279	12.327	E	1
Rodentia	Spalacidae	*Kanisamys indicus*	16.763	11.400	E	1
Rodentia	Spalacidae	*Kanisamys potwarensis*	14.279	13.391	I	2
Rodentia	Spalacidae	*Kanisamys nagrii*	11.644	9.692	E	2
Rodentia	Spalacidae	*Kanisamys sivalensis*	9.401	7.337	E	2
Rodentia	Spalacidae	*Protachyoryctes tatroti*	7.793	7.096	E	3
Rodentia	Spalacidae	*Eicooryctes kaulialensis*	7.815	7.256	E	2
Rodentia	Spalacidae	*Rhizomyides punjabiensis*	11.230	10.100	I	2
Rodentia	Spalacidae	*Rhizomyides* sp. indet.	8.216	8.216	E	2
Rodentia	Spalacidae	*Rhizomyides sivalensis*	7.256	6.517	E	3
Rodentia	Spalacidae	*Miorhizomys nagrii*	9.706	9.240	I	3
Rodentia	Spalacidae	*Miorhizomys tetracharax*	9.377	7.929	E	4
Rodentia	Spalacidae	*Miorhizomys* cf. *M. pilgrimi*	9.446	7.394	E	5
Rodentia	Spalacidae	*Miorhizomys choristos*	8.425	6.967	E	4
Rodentia	Spalacidae	*Miorhizomys micrus*	9.333	9.333	E	2
Rodentia	Spalacidae	*Miorhizomys blacki*	9.258	9.258	E	3
Rodentia	Cricetidae	*Democricetodon* sp. A	17.892	13.400	E	1
Rodentia	Cricetidae	*Democricetodon* sp. B, C	16.045	10.492	E	1
Rodentia	Cricetidae	*Democricetodon* sp. E	14.279	10.100	E	1
Rodentia	Cricetidae	*Democricetodon kohatensis*	17.892	10.063	E	1
Rodentia	Cricetidae	*Democricetodon fejfari*	13.800	8.703	E	1
Rodentia	Cricetidae	*Megacricetodon aguilari*	17.892	12.756	I	1
Rodentia	Cricetidae	*Megacricetodon mythikos*	17.892	12.749	E	1
Rodentia	Cricetidae	*Megacricetodon sivalensis*	16.800	12.749	E	1
Rodentia	Cricetidae	*Megacricetodon daamsi*	15.169	12.756	E	1

Table III.1. continued

Order[1]	Family	Name[2]	Observed First Occurrence[3]	Observed Last Occurrence	Species Origination[4]	Size Class[5]
Rodentia	Cricetidae	Punjabemys mikros	16.045	12.749	E	1
Rodentia	Cricetidae	Punjabemys leptos	16.045	12.300	E	1
Rodentia	Cricetidae	Punjabemys downsi	16.045	13.200	E	1
Rodentia	Gerbillinae	Myocricetodon sivalensis	17.892	12.779	E	1
Rodentia	Gerbillinae	Myocricetodon large sp.	16.763	12.749	E	1
Rodentia	Gerbillinae	Dakkamys barryi	15.169	12.749	I	1
Rodentia	Gerbillinae	Dakkamys n. sp.	14.279	11.404	E	1
Rodentia	Gerbillinae	Dakkamys asiaticus	13.183	10.492	E	1
Rodentia	Gerbillinae	Paradakkamys chinjiensis	12.377	10.492	E	1
Rodentia	Gerbillinae	Mellalomys lavocati	16.763	12.377	E	1
Rodentia	Gerbillinae	Mellalomys perplexus	17.892	10.492	I	1
Rodentia	Gerbillinae	Abudhabia pakistanensis	8.703	8.703	I	1
Rodentia	Murinae	Potwarmus primitivus	17.892	12.300	E	1
Rodentia	Murinae	Potwarmus minimus	17.892	14.039	E	1
Rodentia	Muridae	Antemus chinjiensis	13.800	12.749	E	1
Rodentia	Murinae	"post-Antemus" n. sp.	12.377	12.327	E	1
Rodentia	Murinae	"pre-Progonomys" n. sp. X	11.644	11.404	E	1
Rodentia	Murinae	Murinae, gen. D and n. sp.	11.644	10.200	E	1
Rodentia	Murinae	Murinae, gen. C and n. sp.	11.183	11.183	E	1
Rodentia	Muridae	Progonomys morganae	11.644	10.063	E	1
Rodentia	Muridae	Progonomys hussaini	10.500	10.063	E	1
Rodentia	Muridae	Karnimata fejfari	10.500	10.063	E	1
Rodentia	Muridae	"Karnimata–Progonomys" n. sp.	11.200	11.200	E	1
Rodentia	Muridae	Progonomys debruijni	9.692	8.174	E	1
Rodentia	Muridae	Progonomys n. sp.	7.966	7.157	E	1
Rodentia	Muridae	Parapodemus badgleyae	10.492	10.063	E	1
Rodentia	Muridae	Parapodemus hariensis	9.269	7.966	E	1
Rodentia	Muridae	Karnimata darwini	9.700	8.000	E	1
Rodentia	Muridae	Karnimata huxleyi	7.840	6.517	E	1
Rodentia	Muridae	Karnimata n. sp. (large)	9.240	7.966	E	1
Rodentia	Muridae	Parapelomys cf. P. robertsi	7.800	7.337	E	1
Rodentia	Muridae	Parapelomys robertsi	7.200	6.517	E	1
Rodentia	Muridae	Mus absconditus	7.966	7.157	I	1
Rodentia	Muridae	Mus auctor	6.517	6.517	E	1
Rodentia	Muridae	Murinae, gen. A and n. sp.	6.517	6.517	I	1
Primates	Lorisidae	Nycticeboides simpsoni	9.240	8.929	E	3
Primates	Lorisidae	Microloris pilbeami	10.492	9.400	I	1
Primates	Lorisidae	Nycticeboides sp.	15.169	15.169	I	2
Primates	Lorisidae	cf. Nycticeboides sp.	9.200	9.200	E	2
Primates	Lorisidae	Lorisidae, gen. and sp. indet. (small)	12.779	12.779	I	1
Primates	Lorisidae	Lorisidae, gen. and sp. indet. (large)	16.045	15.169	I	4
Primates	Sivaladapidae	Sivaladapis palaeindicus	15.496	11.644	E	5
Primates	Sivaladapidae	Indraloris kamlialensis	15.169	15.169	E	4
Primates	Catarrhini, i.s.	cf. Dionysopithecus sp.	16.000	16.000	I	5
Primates	Catarrhini, i.s.	Catarrhini, gen. and sp. indet.	12.088	12.088	I	5
Primates	Hominoidea, i.s.	Sivapithecus indicus	12.749	11.404	I	7.1
Primates	Hominoidea, i.s.	Sivapithecus parvada	10.063	10.063	E	7.2
Primates	Hominoidea, i.s.	Sivapithecus sivalensis	10.228	8.534	E	7.1
Primates	Cercopithecidae	Mesopithecus sivalensis	7.691	6.400	I	6
Creodonta	Hyaenodontidae	Dissopsalis carnifex	17.000	10.015	. . .	7.2
Creodonta	Hyaenodontidae	Hyainailouros sulzeri	19.000	11.400	. . .	9
Creodonta	Hyaenodontidae	(?)Metapterodon unnamed sp.	13.680	10.063	. . .	6
Carnivora	Carnivora, i.s.	Family incerta sedis, unnamed gen. and sp.	10.063	10.063	. . .	5
Carnivora	Arctoidea	Arctoidea, gen. and sp. indet.	11.550	11.552	. . .	7.2
Carnivora	Amphicyonidae	Amphicyon palaeindicus	17.892	11.616	. . .	8
Carnivora	Amphicyonidae	Amphicyon lydekkeri	12.749	7.878	. . .	9
Carnivora	Amphicyonidae	Amphicyon giganteus	19.000	17.892	. . .	9
Carnivora	Amphicyonidae	cf. Agnotherium antiquum	15.818	10.063	. . .	8
Carnivora	Ursidae	Indarctos punjabiensis	8.056	7.766	. . .	8
Carnivora	Mellivorinae	Eomellivora necrophila	10.063	9.356	. . .	7.1

(continued)

Table III.1. continued

Order[1]	Family	Name[2]	Observed First Occurrence[3]	Observed Last Occurrence	Species Origination[4]	Size Class[5]
Carnivora	Mustelinae	*Cernictis lydekkeri*	14.000	12.088	. . .	5
Carnivora	Mustelinae	*Plesiogulo crassa*	6.660	6.660	. . .	7.2
Carnivora	Mustelinae	cf. *Anatolictis* sp. A	9.446	6.517	. . .	3
Carnivora	Mustelinae	cf. *Anatolictis* sp. B	14.138	14.138	. . .	3
Carnivora	Mustelinae	cf. *Anatolictis* sp. C	12.290	12.290	. . .	4
Carnivora	Mellivorinae	cf. *Hoplictis anatolicus*	9.662	9.240	. . .	6
Carnivora	Mustelinae	*Circamustela* sp.	9.329	7.394	. . .	4
Carnivora	Lutrinae	*Enhydriodon falconeri*	6.427	6.427	. . .	6
Carnivora	Lutrinae	*Sivaonyx bathygnathus*	9.415	6.436	. . .	6
Carnivora	Lutrinae	*Vishnuonyx chinjiensis*	13.652	11.572	. . .	5
Carnivora	Lutrinae	*Vishnuonyx* large sp..	11.616	10.063	. . .	6
Carnivora	Lutrinae	cf. *Torolutra* sp.	7.394	7.157	. . .	5
Carnivora	Herpestidae	*Urva* small sp.	13.589	6.967	. . .	3
Carnivora	Herpestidae	*Urva* medium sp.	15.103	6.967	. . .	4
Carnivora	Herpestidae	*Urva* large sp.	13.050	7.158	. . .	4
Carnivora	Viverridae	Viverrid gen. et sp. indet (large)	17.996	12.130	. . .	7.1
Carnivora	Viverridae	*Vishnuictis salmontanus*[6]	10.063	7.414	. . .	5
Carnivora	Viverridae	*Vishnuictis chinjiensis*	14.300	10.492	. . .	5
Carnivora	Viverridae	*Vishnuictis hasnoti*	10.492	6.497	. . .	6
Carnivora	Viverridae	*Mioparadoxurus meini*	10.063	7.304	. . .	5
Carnivora	Viverridae	Paradoxurinae gen. et sp. indet.	12.088	9.240	. . .	6
Carnivora	Percrocutidae	*Percrocuta carnifex*	14.019	9.633	. . .	7.2
Carnivora	Percrocutidae	*Percrocuta grandis*	9.428	6.334	. . .	7.2
Carnivora	Hyaenidae	*Protictitherium* sp.	9.676	9.676	. . .	6
Carnivora	Hyaenidae	"*Thalassictis*" *proava*	13.617	10.222	. . .	6
Carnivora	Hyaenidae	cf. "*Thalassictis*" *montadai*[6]	14.000	11.500	. . .	7.1
Carnivora	Hyaenidae	*Lepthyaena sivalensis*	11.971	8.174	. . .	7.1
Carnivora	Hyaenidae	(?)*Lepthyaena pilgrimi*	9.248	7.326	. . .	7.1
Carnivora	Hyaenidae	*Lycyaena macrostoma*	8.739	6.427	. . .	7.2
Carnivora	Hyaenidae	*Adcrocuta eximia*	8.488	8.056	. . .	7.2
Carnivora	Barbourofelidae	(?)*Sansanosmilus serratus*	15.169	11.620	. . .	7.1
Carnivora	Barbourofelidae	(?)*Barbourofelis piveteaui*	11.410	8.569	. . .	7.2
Carnivora	Felidae	*Paramachaerodus sivalensis*	10.063	7.010	. . .	7.2
Carnivora	Felidae	*Sivasmilus copei*	14.138	11.410	. . .	7.1
Carnivora	Felidae	cf. *Miomachairodus* sp.[6]	11.410	7.966	. . .	8
Carnivora	Felidae	(?)*Amphimachairodus giganteus*	9.415	4.657	. . .	8
Carnivora	Felidae	cf. *Lokotunjailurus* sp.	8.056	8.056	. . .	7.2
Carnivora	Felidae	*Sivaelurus chinjiensis*	13.797	12.377	. . .	7.2
Carnivora	Felidae	Felidae, indet. small spp.	13.901	7.809	. . .	5
Carnivora	Felidae	Felidae, indet. medium spp.	21.950	6.517	. . .	7.1
Carnivora	Felidae	Felidae, indet. large spp.	13.751	7.394	. . .	7.1
Tubulidentata	Orycteropodidae	*Amphiorycteropus* n. sp.	14.147	12.000	I	7.1
Tubulidentata	Orycteropodidae	*Amphiorycteropus browni*	12.669	9.415	E	7.1
Tubulidentata	Orycteropodidae	*Amphiorycteropus* sp. indet.	8.000	8.000	E	7.1
Tubulidentata	Orycteropodidae	*Orycteropus* sp.	10.159	8.193	I	7.1
Pholidota	Manidae	cf. *Manis* sp.	9.269	7.096	I	5
Proboscidea	Deinotheriidae	*Prodeinotherium pentapotamiae*	17.892	11.671	I	10.3
Proboscidea	Deinotheriidae	*Deinotherium indicum*	13.100	8.056	E	10.3
Proboscidea	Mammutidae	*Zygolophodon metachinjiensis*	13.838	12.377	I	10.3
Proboscidea	Gomphotheriidae	*Choerolophodon corrugatus*	14.138	7.365	I	10.3
Proboscidea	Gomphotheriidae	*Protanancus chinjiensis*	13.997	6.517	I	10.3
Proboscidea	Gomphotheriidae	cf. '*Konobelodon*' or '*Torynobelodon*' sp.	8.670	8.670	I	10.3
Proboscidea	Gomphotheriidae	*Gomphotherium browni*	13.901	11.404	E	10.3
Proboscidea	Gomphotheriidae	*Paratetralophodon hasnotensis*	8.703	7.414	I	10.3
Proboscidea	Gomphotheriidae	*Anancus perimensis*	8.891	8.891	E	10.3
Proboscidea	Stegodontidae	*Stegolophodon* sp., cf. *S. progressus*	13.652	7.300	I	10.3
Proboscidea	Stegodontidae	*Stegolophodon stegodontoides*	9.401	7.258	E	10.3
Proboscidea	Stegodontidae	*Stegolophodon maluvalensis*	7.414	7.414	E	10.3
Perissodactyla	Equidae	*Cormohipparion* sp.	10.810	10.356	I	8
Perissodactyla	Equidae	*Cormohipparion* small sp.	10.042	9.837	E	8
Perissodactyla	Equidae	*Sivalhippus nagriensis*	10.486	8.749	E	9
Perissodactyla	Equidae	*Sivalhippus theobaldi*	9.401	7.812	E	9

Table III.1. continued

Order[1]	Family	Name[2]	Observed First Occurrence[3]	Observed Last Occurrence	Species Origination[4]	Size Class[5]
Perissodactyla	Equidae	*Sivalhippus perimensis*	8.749	7.338	E	9
Perissodactyla	Equidae	*Sivalhippus anwari*	7.394	7.157	E	9
Perissodactyla	Equidae	*Cremohipparion antilopinum*	8.916	6.480	I	8
Perissodactyla	Chalicotheriidae	*Anisodon salinus*	14.138	7.365	E	9
Perissodactyla	Chalicotheriidae	cf. *Ancylotherium* sp.	10.106	10.106	I	9
Perissodactyla	Rhinocerotidae	*Caementodon oettingenae*	15.129	8.591	I	10.1
Perissodactyla	Rhinocerotidae	*Alicornops* cf. *simorrense*	14.138	14.138	E	10.1
Perissodactyla	Rhinocerotidae	*Alicornops complanatum*	16.763	7.966	I	10.1
Perissodactyla	Rhinocerotidae	*Brachydiceratherium intermedium*	17.878	7.966	I	10.2
Perissodactyla	Rhinocerotidae	*Brachydiceratherium fatehjangense*	>22	7.396	I	10.2
Perissodactyla	Rhinocerotidae	*Brachypotherium perimense*	>22	7.247	E	10.3
Perissodactyla	Rhinocerotidae	*Gaindatherium browni*	16.900	8.731	E	10.1
Perissodactyla	Rhinocerotidae	*Gaindatherium vidali*	14.118	7.966	E	10.1
Perissodactyla	Rhinocerotidae	cf. *Rhinoceros* sp.	>20	16.045	I	10.2
Perissodactyla	Rhinocerotidae	*Rhinoceros* aff. *sondaicus*	10.063	10.063	E	10.2
Perissodactyla	Rhinocerotidae	*Rhinoceros* aff. *sivalensis*	8.891	7.961	E	10.1
Perissodactyla	Rhinocerotidae	*Dicerorhinus* aff. *sumatrensis*	13.652	10.063	E	10.1
Perissodactyla	Rhinocerotidae	*Ceratotherium neumayri*	7.963	7.356	I	10.2
Artiodactyla	Sanitheriidae	*Sanitherium schlagintweiti*	17.965	14.118	I	7.1
Artiodactyla	Palaeochoeridae	*Yunnanochoerus gandakasensis*	10.159	8.703	I	7.2
Artiodactyla	Suidae	*Listriodon pentapotamiae*	14.138	10.372	E	8
Artiodactyla	Suidae	*Hippotamodon* n. sp.	11.410	10.222	I	8
Artiodactyla	Suidae	*Hippopotamodon sivalense*	10.372	6.436	E	9
Artiodactyla	Suidae	*Propotamochoerus hysudricus*	10.348	6.631	I	8
Artiodactyla	Suidae	*Sivahyus punjabiensis*	10.228	3.282	I	7.1
Artiodactyla	Suidae	Suidae, i.s., n. gen. and sp.	12.756	11.410	I	7.2
Artiodactyla	Suidae	Suinae, unnamed gen. and sp.	10.063	10.063	I	7.1
Artiodactyla	Suidae	"*Micro-Potamochoerus*" n. sp.	9.356	7.471	I	7.1
Artiodactyla	Suidae	*Conohyus sindiensis*	15.090	10.105	E	8
Artiodactyla	Suidae	*Parachleuastochoerus pilgrimi*	13.664	11.153	I	8
Artiodactyla	Suidae	*Tetraconodon magnus*	10.015	9.692	E	10.1
Artiodactyla	Anthracotheriidae	*Microbunodon silistrensis*	28.400	11.400	I	7.1
Artiodactyla	Anthracotheriidae	*Microbunodon milaensis*	10.366	9.248	E	7.1
Artiodactyla	Anthracotheriidae	*Hemimeryx blanfordi*	23.900	17.878	I	9
Artiodactyla	Anthracotheriidae	*Sivameryx palaeindicus*	24.100	14.138	I	8
Artiodactyla	Anthracotheriidae	*Merycopotamus nanus*[7]	15.129	11.404	E	8
Artiodactyla	Anthracotheriidae	*Merycopotamus medioximus*	10.486	8.685	E	9
Artiodactyla	Anthracotheriidae	*Merycopotamus dissimilis*	7.814	3.300	E	9
Artiodactyla	Hippopotamidae	*Hexaprotodon sivalensis*	6.049	3.499	E	10.1
Artiodactyla	Tragulidae	*Dorcabune sindiense*	18.690	13.617	E	6
Artiodactyla	Tragulidae	*Dorcabune* unnamed sp.	17.892	15.103	E	7.1
Artiodactyla	Tragulidae	*Dorcabune anthracotheroides*	14.138	10.581	E	7.2
Artiodactyla	Tragulidae	*Dorcabune nagrii*	11.856	8.544	E	7.1
Artiodactyla	Tragulidae	*Dorcatherium majus*	10.492	7.006	E	7.1
Artiodactyla	Tragulidae	*Dorcatherium* Form 1?	12.434	10.567	E	7.1
Artiodactyla	Tragulidae	*Dorcatherium* Form 2	8.569	7.809	E	7.2
Artiodactyla	Tragulidae	*Dorcatherium* Form 4	14.138	11.792	E	6
Artiodactyla	Tragulidae	*Dorcatherium* Form 6	14.279	12.749	E	6
Artiodactyla	Tragulidae	*Dorcatherium* Form 3	10.548	8.794	E	5
Artiodactyla	Tragulidae	*Dorcatherium* Form 5	12.928	10.486	E	5
Artiodactyla	Tragulidae	*Dorcatherium* Form 8	8.794	5.979	E	5
Artiodactyla	Tragulidae	*Dorcatherium* Form 7	22.140	13.752	E	5
Artiodactyla	Tragulidae	*Dorcatherium* Form 9	18.734	16.763	E	6
Artiodactyla	Tragulidae	*Siamotragulus nagrii*	9.415	6.967	E	4
Artiodactyla	Tragulidae	*Siamotragulus minimus*	15.169	12.327	E	4
Artiodactyla	Tragulidae	*Siamotragulus* Form C	10.492	6.967	E	4
Artiodactyla	Tragulidae	*Siamotragulus* Form D	14.664	11.183	E	5
Artiodactyla	Palaeomerycidae	Palaeomerycidae, n. gen. and sp.	13.085	13.085	I	9
Artiodactyla	Palaeomerycidae	*Tauromeryx* n. sp.	11.971	11.971	I	9
Artiodactyla	Palaeomerycidae	cf. *Tauromeryx* sp.	10.063	10.063	I	7.1
Artiodactyla	Palaeomerycidae	*Triceromeryx* n. sp	12.088	12.088	I	7.2

(continued)

Table III.1. continued

Order[1]	Family	Name[2]	Observed First Occurrence[3]	Observed Last Occurrence	Species Origination[4]	Size Class[5]
Artiodactyla	Palaeomerycidae	cf. *Tricomeryx* n. sp.	11.644	10.063	E	7.2
Artiodactyla	Palaeomerycidae	cf. *Orygotherium* sp.	7.157	7.157	I	6
Artiodactyla	Antilocapridae	Climococerinae, n. gen. and sp.	17.892	17.892	I	7.1
Artiodactyla	Antilocapridae	*Climacoceras* n. sp.	11.410	10.015	I	8
Artiodactyla	Antilocapridae	*Orangemeryx* n. sp.	13.797	13.797	E	7.2
Artiodactyla	Antilocapridae	cf. *Orangemeryx* n. sp.	14.051	12.749	I	7.2
Artiodactyla	Giraffidae	*Progiraffa exigua*	18.100	9.415	I	7.2
Artiodactyla	Giraffidae	*Giraffokeryx punjabiensis*	14.147	11.400	I	9
Artiodactyla	Giraffidae	Bohlininae n. gen. and n. sp.	13.800	11.400	I	9
Artiodactyla	Giraffidae	*Honanotherium* sp.	14.019	11.410	I	9
Artiodactyla	Giraffidae	*Giraffa punjabiensis*	9.415	8.056	E	10.1
Artiodactyla	Giraffidae	*Ua pilbeami*	14.051	8.056	I	9
Artiodactyla	Giraffidae	Okapiinae n. gen. and n. sp.	9.599	9.401	I	9
Artiodactyla	Giraffidae	*Decennatherium asiaticum*	13.047	10.063	E	9
Artiodactyla	Giraffidae	*Libytherium* sp. 1	10.063	10.063	E	10.1
Artiodactyla	Giraffidae	*Libytherium* sp. 2	9.377	7.328	E	10.1
Artiodactyla	Giraffidae	*Bramatherium megacephalum*	10.222	7.960	E	10.1
Artiodactyla	Giraffidae	*Bramatherium* n. sp.	13.652	13.652	I	8
Artiodactyla	Giraffidae	*Helladotherium grande*	10.063	8.751	E	10.1
Artiodactyla	Giraffidae	*Injanatherium hazimi*	12.377	7.471	I	9
Artiodactyla	Bovidae	*Hypsodontus sokolovi*	14.138	12.147	E	7.1
Artiodactyla	Bovidae	*Hypsodontus pronaticornis*	13.652	12.800	E	8
Artiodactyla	Bovidae	?Hypsodontinae sp.	17.900	15.200	I	7.1
Artiodactyla	Bovidae	Bovidae sp. 1	16.045	15.103	I	6
Artiodactyla	Bovidae	*Eotragus noyei*	17.892	16.800	I	6
Artiodactyla	Bovidae	*Elachistoceras khauristanensis*	14.138	7.471	I	6
Artiodactyla	Bovidae	*Strepsiportax* unnamed sp.	14.138	12.371	I	7.1
Artiodactyla	Bovidae	*Strepsiportax gluten*	12.377	11.081	E	7.1
Artiodactyla	Bovidae	*Sivaceros* unnamed. sp.	14.138	12.749	I	7.1
Artiodactyla	Bovidae	*Sivaceros gradiens*	12.900	11.304	E	7.1
Artiodactyla	Bovidae	?(*Sivaceros*) *vedicus*	9.836	9.836	E	7.1
Artiodactyla	Bovidae	*Sivoreas eremita*	14.138	12.088	I	7.1
Artiodactyla	Bovidae	*Helicoportax praecox*	13.400	11.200	I	7.1
Artiodactyla	Bovidae	*Selenoportax vexillarius*	~11	~11	I	8
Artiodactyla	Bovidae	*Selenoportax* aff. *vexillarius*	10.567	10.015	E	8
Artiodactyla	Bovidae	*Miotragocerus pilgrimi*	10.632	8.792	I	7.2
Artiodactyla	Bovidae	*Miotragocerus punjabicus*	8.670	6.350	E	7.2
Artiodactyla	Bovidae	*Tragoportax salmontanus*	8.425	7.157	I	7.2
Artiodactyla	Bovidae	?Boselaphini, unnamed gen. and sp.	9.401	7.328	I	7.2
Artiodactyla	Bovidae	*Pachyportax falconeri*	9.837	8.850	E	8
Artiodactyla	Bovidae	*Pachyportax dhokpathanensis*	8.996	7.157	E	9
Artiodactyla	Bovidae	?Bovinae, unnamed gen. and sp.	9.568	7.210	I	8
Artiodactyla	Bovidae	Antilopini, unnamed gen. and sp.	14.138	13.007	I	6
Artiodactyla	Bovidae	?*Gazella* sp.	13.200	11.674	I	6
Artiodactyla	Bovidae	*Gazella lydekkeri*	10.222	6.257	E	6
Artiodactyla	Bovidae	*Prostrepsiceros vinayaki*	9.391	7.929	I	7.1
Artiodactyla	Bovidae	?*Prostrepsiceros* large sp.	8.174	8.174	E	7.2
Artiodactyla	Bovidae	*Nisidorcas planicornis*	9.048	9.048	I	7.1
Artiodactyla	Bovidae	?*Protragelaphus* sp.	10.759	10.759	I	7.1
Artiodactyla	Bovidae	Antilopini, large unnamed gen. and sp.	9.367	8.794	I	8
Artiodactyla	Bovidae	*Kobus* sp. 1	9.400	8.200	I	7.1
Artiodactyla	Bovidae	*Kobus porrecticornis*	8.056	7.735	E	7.1
Artiodactyla	Bovidae	*Kobus* sp. 2	7.416	7.218	E	7.2
Artiodactyla	Bovidae	*Tethytragus* sp.	13.357	12.327	I	7.1
Artiodactyla	Bovidae	?*Caprotragoides potwaricus*	10.222	10.159	I	7
Artiodactyla	Bovidae	?*Dorcadoryx* sp.	9.729	9.367	I	7.1
Artiodactyla	Bovidae	*Protoryx* aff. *solignaci*	11.321	11.321	I	7.1
Artiodactyla	Pecora, i.s.	"Genus A" small sp.	17.892	17.892	I	6?

[1]Follows order of chapters. Due to late changes, two rodent taxa in Table 11.1, *Sayimys baskini* and Pteromyinae, gen. and sp. indet., and one bovid in Table 23.1, Boselaphini, unnamed gen. and sp., are not included in this table or in Chapter 25 and 26 analyses. | [2]Informal names do not conform to single standard due to prior usage in publications and presentations. | [3]Observed first and last occurrence ages of taxa reported to three decimal places as this level used for basis of confidence interval estimates in Chapter 26. Observed first and last occurrences are rounded to nearest 100-kyr interval in Chapter 25. | [4]Inferred origin of taxa, endemic origination (E) and immigrant (I). See Chapter 25 for details. | [5]There are ten size classes with additional subdivisions as follows: 1: <100 g; 2: 100–249 g; 3: 250–999 g; 4: 1.0–2.9 kg; 5: 3.0–7.9 kg; 6: 8–19 kg; 7: 20–99 kg (7.1: 20–49 kg; 7.2: 50–99 kg); 8: 100–249 kg; 9: 250–800 kg; 10: >800 kg (10.1: >800–1500 kg; 10.2: 1500–3000 kg; 10.3: >3000 kg). | [6]Observed range differs slightly from that in chapter 13 due to recent work. First and last occurrence ages of taxa for this table reflect understanding in May 2023. | [7]First occurrence date includes anthracothere fossils between 15.1 and 14.0 Ma that are likely this taxon (and used in Chapter 26 analyses).

Table III.2. Stable Carbon and Oxygen Isotope Data from Tooth Enamel Apatite of Siwalik Mammals[1]

Specimen number[2]	Family or Order[3]	Name[4]	Tooth sampled	Locality[5]	Age (Ma)[6]	δ^{13}C (V-PDB)[7]	δ^{18}O (V-PDB)[7]
YGSP 23094	Hominoidea, i.s.	*Sivapithecus indicus*	left p or m	Y663	11.6	−11.9	−5.7
YGSP 30622	Hominoidea, i.s.	*Sivapithecus indicus*	m	Y750	12.7	−11.7	−1.3
YGSP 13811	Hominoidea, i.s.	*Sivapithecus sivalensis*	left m	Y224	9.4	−12.7	−2.7
YGSP 13933	Hominoidea, i.s.	*Sivapithecus sivalensis*	molar	Y260	9.3	−13.1	−1.9
YGSP 47021	Hominoidea, i.s.	*Sivapithecus sivalensis*	right p or m	Y182	9.2	−13.9	−6.0
YGSP 47023	Hominoidea, i.s.	*Sivapithecus sivalensis*	left M	Y182	9.2	−12.8	−1.7
YGSP 47024	Hominoidea, i.s.	*Sivapithecus sivalensis*	left M2	Y317	9.2	−12.4	−0.9
YGSP 49567	Hominoidea, i.s.	*Sivapithecus sivalensis*	p or m	Y891	8.7	−12.8	−4.1
YGSP 26856a	Deinotheriidae	*Deinotherium indicum*	M2	Y311	10.1	−11.9	−6.6
YGSP 38932	Deinotheriidae	*Deinotherium indicum*	molar	Y251	10.0	−12.2	−3.5
YGSP 49579	Deinotheriidae	*Deinotherium indicum*	molar	Y494	12.4	−13.3	−7.9
YGSP 50685a	Deinotheriidae	*Deinotherium indicum*	m2	Y450	10.2	−11.4	−6.0
YGSP 51237	Deinotheriidae	*Deinotherium indicum*	molar	Y311	10.1	−9.6	−8.5
YGSP 52493	Deinotheriidae	*Deinotherium indicum*	molar	Y260	9.3	−11.7	−1.9
YGSP 17458	Deinotheriidae	*Prodeinotherium pentapotamiae*	M2	Y590	18.0	−11.0	−7.7
YGSP 30896	Deinotheriidae	*Prodeinotherium pentapotamiae*	molar	Y591	16.0	−12.1	−8.8
YGSP 41592	Deinotheriidae	*Prodeinotherium pentapotamiae*	left m1	Y716	12.9	−11.1	−7.4
YGSP 42024	Deinotheriidae	*Prodeinotherium pentapotamiae*	left m3	Y506	14.0	−11.6	−6.0
YGSP 20517	Deinotheriidae	deinothere	molar	Y494	12.4	−12.4	−5.4
YGSP 41885	Deinotheriidae	deinothere	molar	Y478	14.1	−12.7	−7.3
YGSP 42256	Deinotheriidae	deinothere	premolar	Y496	12.4	−12.3	−6.1
YGSP 45761	Deinotheriidae	deinothere	molar	Y507	11.7	−12.2	−6.6
YGSP 45920	Deinotheriidae	deinothere	molar	Y500	12.1	−12.8	−7.5
YGSP 46958	Deinotheriidae	deinothere	molar	Y504	11.6	−10.5	−8.5
YGSP 47138b	Gomphotheriidae	*Choerolophodon corrugatus*	right m2	Y878	14.1	−10.2	1.0
YGSP 42246	Gomphotheriidae	*Choerolophodon corrugatus*	M3	Y513	13.1	−10.3	−6.7
YGSP 50028	Gomphotheriidae	*Choerolophodon corrugatus*	M3	Y917	7.8	−9.3	−7.6
YGSP 52744	Gomphotheriidae	*Choerolophodon corrugatus*	molar	Y980	8.9	−10.3	−5.3
YGSP 52892	Gomphotheriidae	*Paratetralophodon hasnotensis*	right m3	Y943	7.4	−10.3	−9.6
YGSP 11022	Gomphotheriidae	gomphothere	right dP2	Y313	9.4	−11.6	−6.5
YGSP 20338	Gomphotheriidae	gomphothere	molar	Y640	13.7	−8.8	−4.6
YGSP 26705	Gomphotheriidae	gomphothere	m	Y716	12.9	−12.4	−6.8
YGSP 41488	Gomphotheriidae	gomphothere	m1	Y311	10.1	−11.5	−6.0
YGSP 45862	Gomphotheriidae	gomphothere	molar	Y76	11.4	−11.3	−5.5
YGSP 46707	Gomphotheriidae	gomphothere	molar	Y865	6.9	−7.5	−3.9
YGSP 46962	Gomphotheriidae	gomphothere	right m3	Y76	11.4	−9.6	−8.4
YGSP 47216	Gomphotheriidae	gomphothere	right m3	Y694	13.9	−10.6	−7.2
YGSP 49794	Gomphotheriidae	gomphothere	molar	Y913	6.4	−5.7	−5.4
YGSP 50193	Gomphotheriidae	gomphothere	molar	Y906	7.8	−11.3	−7.9
YGSP 50325	Gomphotheriidae	gomphothere	molar	Y935	7.4	−10.5	−6.8
YGSP 50758	Gomphotheriidae	gomphothere	right m3	D13	6.5	−7.0	−7.2
YGSP 51559	Gomphotheriidae	gomphothere	molar	KL11	8.0	−11.0	−7.0
YGSP 51635	Gomphotheriidae	gomphothere	molar	Y406	8.9	−10.1	−3.0
YGSP 51692	Gomphotheriidae	gomphothere	molar	Y545	8.1	−11.0	−3.5
YGSP 52552	Gomphotheriidae	gomphothere	molar	Y960	7.8	−9.8	−4.9
YGSP 52745	Gomphotheriidae	gomphothere	molar	Y980	8.9	−10.7	−5.8
YGSP 50236a	Stegodontidae	stegodontid	right m3	Y926	7.4	−9.1	−8.6
YGSP 19726	Stegodontidae	stegodontid	m	Y547	8.0	−11.0	−6.0
YGSP 31122	Stegodontidae	stegodontid	left dP3	Y767	12.4	−11.8	−4.6
YGSP 41134	Stegodontidae	stegodontid	dp or dP	Y647	12.4	−11.7	−7.4
YGSP 41489	Stegodontidae	stegodontid	molar	Y311	10.1	−10.0	−8.7
YGSP 49854	Stegodontidae	stegodontid	molar	Y899	8.2	−8.6	−5.4
YGSP 50039	Stegodontidae	stegodontid	M3	Y581	7.2	−7.6	−8.0
YGSP 50185	Stegodontidae	stegodontid	M	Y934	7.4	−7.7	−6.7
YGSP 50197	Stegodontidae	stegodontid	molar	Y933	7.3	−9.3	−8.5
YGSP 50236c	Stegodontidae	stegodontid	molar	Y926	7.4	−11.7	−9.2
YGSP 22462	Elephantoidea	elephantoid	molar	Y515	12.1	−9.9	−9.6
YGSP 24298	Elephantoidea	elephantoid	molar	Y688	11.5	−12.6	−7.1
YGSP 26831	Elephantoidea	elephantoid	molar	Y690	13.1	−9.1	−2.0
YGSP 30650	Elephantoidea	elephantoid	molar	Y690	13.1	−11.4	−4.3
YGSP 38931	Elephantoidea	elephantoid	molar	Y440	7.5	−8.2	−1.8

(continued)

Table III.2. continued

Specimen number[2]	Family or Order[3]	Name[4]	Tooth sampled	Locality[5]	Age (Ma)[6]	δ13C (V-PDB)[7]	δ18O (V-PDB)[7]
YGSP 38933	Elephantoidea	elephantoid	molar	Y258	10.2	−11.7	−5.8
YGSP 38935	Elephantoidea	elephantoid	molar	Y865	6.9	−6.6	−5.6
YGSP 38936	Elephantoidea	elephantoid	molar	Y860	4.8	0.6	−2.8
YGSP 41508	Elephantoidea	elephantoid	molar	Y842	13.4	−10.5	−2.7
YGSP 41582	Elephantoidea	elephantoid	M	Y76	11.4	−9.9	−5.8
YGSP 41606	Elephantoidea	elephantoid	molar	Y848	13.4	−10.8	−5.1
YGSP 42205	Elephantoidea	elephantoid	molar	Y507	11.7	−10.6	−6.8
YGSP 49892	Elephantoidea	elephantoid	molar	Y908	7.0	−6.0	−2.5
YGSP 49924	Elephantoidea	elephantoid	molar	D28	6.7	−2.8	−5.4
YGSP 49964	Elephantoidea	elephantoid	M	Y918	7.8	−11.0	−8.0
YGSP 50152	Elephantoidea	elephantoid	molar	Y910	7.2	−10.1	−5.3
YGSP 50315	Elephantoidea	elephantoid	molar	Y939	7.2	−2.1	−6.7
YGSP 50324a	Elephantoidea	elephantoid	m	Y940	7.4	−3.9	−4.4
YGSP 50361	Elephantoidea	elephantoid	molar	Y941	7.4	−7.2	−4.7
YGSP 50379	Elephantoidea	elephantoid	molar	Y943	7.4	−9.3	−5.2
YGSP 50774	Elephantoidea	elephantoid	molar	Y927	7.4	−6.1	−3.2
YGSP 50859	Elephantoidea	elephantoid	molar	Y951	7.8	−10.2	−5.8
YGSP 51576	Elephantoidea	elephantoid	molar	Y539	8.1	−11.3	−5.9
YGSP 51625	Elephantoidea	elephantoid	molar	Y993	8.0	−10.2	−4.8
YGSP 51774	Elephantoidea	elephantoid	molar	Y1000	8.1	−12.0	−1.0
YGSP 51782	Elephantoidea	elephantoid	molar	Y547	8.0	−12.2	−6.6
YGSP 52492	Elephantoidea	elephantoid	molar	KL01	9.3	−10.8	−0.4
YGSP 20593	Proboscidea	proboscidean	molar	Y640	13.7	−11.7	−4.0
YGSP 3052	Proboscidea	proboscidean	molar	Y113	8.4	−10.3	−7.6
YGSP 41941	Proboscidea	proboscidean	premolar	Y514	13.0	−10.9	−5.1
YGSP 49753	Proboscidea	proboscidean	molar	Y907	7.4	−2.2	−2.9
YGSP 50903	Proboscidea	proboscidean	molar	Y950	7.8	−11.9	−6.7
YGSP 51749	Proboscidea	proboscidean	molar	Y599	8.1	−11.2	−3.4
YGSP 5290	Proboscidea	proboscidean	molar	Y211	9.3	−12.0	−5.8
YGSP 19770	Perissodactyla	perissodactyl	right M	Y240	9.3	−11.0	−3.3
YGSP 45921	Perissodactyla	perissodactyl	M	Y500	12.1	−11.4	−5.3
YGSP 19802	Chalicotheriidae	Anisodon salinus	left P	Y182	9.2	−11.1	−3.8
YGSP 27715	Chalicotheriidae	Anisodon salinus	right P4	Y698	13.0	−10.5	−6.4
YGSP 30540	Chalicotheriidae	Anisodon salinus	right m	Y311	10.1	−11.9	−0.8
YGSP 31100	Chalicotheriidae	Anisodon salinus	left p1	Y767	12.4	−9.9	−5.3
YGSP 41898	Chalicotheriidae	Anisodon salinus	right m	Y478	14.1	−11.5	−3.4
YGSP 42442a	Chalicotheriidae	Anisodon salinus	left M	Y491	13.8	−9.6	−4.5
YGSP 45836	Chalicotheriidae	Anisodon salinus	m	Y504	11.6	−9.8	−4.3
YGSP 47129	Chalicotheriidae	Anisodon salinus	left M	Y535	8.4	−10.9	−2.4
YGSP 47310	Chalicotheriidae	Anisodon salinus	left M	Y750	12.7	−10.1	−4.2
YGSP 51644	Chalicotheriidae	Anisodon salinus	left M	Y262	9.6	−11.8	−3.9
YGSP 6006a	Chalicotheriidae	Anisodon salinus	left m1	Y337	9.7	−11.7	−1.0
YGSP 9665	Chalicotheriidae	Anisodon salinus	left m	Y76	11.4	−11.8	−1.0
YGSP 16620	Equidae	Cormohipparion sp.	right M3	Y572	10.4	−10.9	−7.1
YGSP 27728	Equidae	Cormohipparion sp.	right m2	Y729	10.4	−11.9	−4.3
YGSP 16806a	Equidae	Sivalhippus anwari	right P3	Y373	7.2	−6.5	−0.3
YGSP 49997	Equidae	Sivalhippus anwari	right P4	Y457	7.4	−7.1	−3.3
YGSP 50338ah	Equidae	Sivalhippus anwari	right P2	Y927	7.4	−7.0	−3.9
YGSP 50603	Equidae	Sivalhippus anwari	left P3	Y581	7.2	−6.5	−2.1
YGSP 50803b	Equidae	Sivalhippus anwari	right p4	Y925	7.1	−6.5	−2.4
YGSP 10112	Equidae	Sivalhippus nagriensis	left p4	Y309	9.4	−9.6	−4.5
YGSP 10640	Equidae	Sivalhippus nagriensis	right m2	Y311	10.1	−10.3	−4.4
YGSP 11188	Equidae	Sivalhippus nagriensis	right m3	Y182	9.2	−11.4	−2.8
YGSP 11730	Equidae	Sivalhippus nagriensis	left M3	Y317	9.2	−9.4	−0.9
YGSP 11881	Equidae	Sivalhippus nagriensis	left M3	Y311	10.1	−11.7	−5.3
YGSP 12506	Equidae	Sivalhippus nagriensis	left M1	Y330	9.7	−7.6	−2.2
YGSP 12604	Equidae	Sivalhippus nagriensis	right P3	Y261	9.7	−11.8	−6.8
YGSP 12947	Equidae	Sivalhippus nagriensis	right M3	Y251	10.0	−10.7	−2.2
YGSP 13042	Equidae	Sivalhippus nagriensis	left P3	Y250	9.7	−11.3	−4.9
YGSP 13175	Equidae	Sivalhippus nagriensis	left p4	Y311	10.1	−11.3	−4.7
YGSP 13703	Equidae	Sivalhippus nagriensis	right M2	KL01	9.3	−11.6	−3.9

Table III.2. continued

Specimen number[2]	Family or Order[3]	Name[4]	Tooth sampled	Locality[5]	Age (Ma)[6]	δ^{13}C (V-PDB)[7]	δ^{18}O (V-PDB)[7]
YGSP 13859	Equidae	*Sivalhippus nagriensis*	left p4	Y362	9.7	−7.9	−3.1
YGSP 15475	Equidae	*Sivalhippus nagriensis*	right P2	Y448	9.1	−9.7	−3.2
YGSP 15595	Equidae	*Sivalhippus nagriensis*	left P2	KL06	9.7	−11.9	−4.1
YGSP 16007	Equidae	*Sivalhippus nagriensis*	right m1	KL04	9.8	−12.5	−2.9
YGSP 16742	Equidae	*Sivalhippus nagriensis*	left P4	Y579	9.7	−11.4	−4.2
YGSP 17109	Equidae	*Sivalhippus nagriensis*	left dP	Y262	9.6	−12.2	−8.1
YGSP 18047	Equidae	*Sivalhippus nagriensis*	right P4	Y450	10.2	−8.9	−4.6
YGSP 19735	Equidae	*Sivalhippus nagriensis*	right m1	KL05	9.6	−9.0	−2.2
YGSP 19810	Equidae	*Sivalhippus nagriensis*	right M3	Y312	9.4	−9.2	−1.6
YGSP 19883	Equidae	*Sivalhippus nagriensis*	left M1	Y311	10.1	−10.8	−6.9
YGSP 22096	Equidae	*Sivalhippus nagriensis*	right M3	Y311	10.1	−9.1	−9.2
YGSP 38956	Equidae	*Sivalhippus nagriensis*	right p4	Y875	9.8	−10.6	−1.2
YGSP 38959	Equidae	*Sivalhippus nagriensis*	left p3	Y251	10.0	−11.4	−3.9
YGSP 46248	Equidae	*Sivalhippus nagriensis*	left m3	Y418	9.3	−10.1	−5.6
YGSP 46327	Equidae	*Sivalhippus nagriensis*	right M3	Y869	9.7	−11.1	−4.5
YGSP 46510a	Equidae	*Sivalhippus nagriensis*	right M1	Y258	10.2	−9.2	−0.9
YGSP 46669	Equidae	*Sivalhippus nagriensis*	left M3	Y875	9.8	−12.5	−4.9
YGSP 49761	Equidae	*Sivalhippus nagriensis*	right p2	Y904	10.5	−10.1	−7.6
YGSP 50509	Equidae	*Sivalhippus nagriensis*	left m3	Y311	10.1	−9.7	−5.5
YGSP 5097	Equidae	*Sivalhippus nagriensis*	right P4	Y210	9.1	−11.1	−3.2
YGSP 51362	Equidae	*Sivalhippus nagriensis*	right M3	Y311	10.1	−9.3	−5.8
YGSP 51647	Equidae	*Sivalhippus nagriensis*	left P3	Y262	9.6	−11.7	−7.7
YGSP 51838	Equidae	*Sivalhippus nagriensis*	right P4	Y329	9.7	−9.0	−2.2
YGSP 51860	Equidae	*Sivalhippus nagriensis*	left m1	KL01	9.3	−11.2	−1.9
YGSP 51863	Equidae	*Sivalhippus nagriensis*	left m1	KL01	9.3	−9.6	0.9
YGSP 52038	Equidae	*Sivalhippus nagriensis*	left P3	KL01	9.3	−11.5	−5.2
YGSP 52055	Equidae	*Sivalhippus nagriensis*	left M1	KL03	9.4	−12.8	−4.5
YGSP 52056	Equidae	*Sivalhippus nagriensis*	right p2	KL03	9.4	−12.4	−5.1
YGSP 5270	Equidae	*Sivalhippus nagriensis*	right p4	Y211	9.3	−11.0	−6.4
YGSP 53633	Equidae	*Sivalhippus nagriensis*	left P3	Y311	10.1	−9.8	−1.5
YGSP 54562	Equidae	*Sivalhippus nagriensis*	right M3	Y312	9.4	−11.3	−5.9
YGSP 54563	Equidae	*Sivalhippus nagriensis*	right M3	Y317	9.2	−10.9	−5.0
YGSP 54564	Equidae	*Sivalhippus nagriensis*	right M1	Y225	9.4	−10.1	−3.6
YGSP 54575	Equidae	*Sivalhippus nagriensis*	left P2	Y337	9.7	−11.3	−6.6
YGSP 5733	Equidae	*Sivalhippus nagriensis*	left M1	Y330	9.7	−5.2	−4.4
YGSP 8683	Equidae	*Sivalhippus nagriensis*	left P3	Y246	9.8	−11.8	−2.1
YGSP 8829	Equidae	*Sivalhippus nagriensis*	left M3	Y250	9.7	−10.4	−7.9
YGSP 9488	Equidae	*Sivalhippus nagriensis*	right M3	Y258	10.2	−8.8	−0.8
YGSP 9846	Equidae	*Sivalhippus nagriensis*	left M3	Y258	10.2	−8.3	−1.3
YGSP 9989	Equidae	*Sivalhippus nagriensis*	left M2	Y274	9.6	−11.5	−3.1
YGSP 14801	Equidae	*Sivalhippus perimensis*	left m1	Y168	8.3	−9.8	−5.9
YGSP 15293	Equidae	*Sivalhippus perimensis*	left P3	Y435	7.7	−5.0	0.4
YGSP 15610	Equidae	*Sivalhippus perimensis*	left M1	Y457	8.6	−10.0	−5.1
YGSP 16238	Equidae	*Sivalhippus perimensis*	left m3	Y540	8.0	−2.0	1.2
YGSP 16960	Equidae	*Sivalhippus perimensis*	left M1	Y547	8.0	−11.0	−2.7
YGSP 18379	Equidae	*Sivalhippus perimensis*	left P4	Y608	8.6	−5.9	−1.5
YGSP 18418	Equidae	*Sivalhippus perimensis*	right M1	Y606	8.1	−9.5	−3.1
YGSP 19724	Equidae	*Sivalhippus perimensis*	right p4	Y434	7.8	−9.3	−5.4
YGSP 19743	Equidae	*Sivalhippus perimensis*	left P4	Y604	8.6	−7.3	−2.1
YGSP 3050	Equidae	*Sivalhippus perimensis*	left P4	Y113	8.4	−6.1	−4.5
YGSP 49548f	Equidae	*Sivalhippus perimensis*	left M2	Y890	8.0	−6.5	−0.8
YGSP 49667	Equidae	*Sivalhippus perimensis*	left p4	Y891	8.7	−9.6	−5.2
YGSP 49837	Equidae	*Sivalhippus perimensis*	left p2	Y906	7.8	−5.4	−4.6
YGSP 51839	Equidae	*Sivalhippus perimensis*	left m1	Y890	8.0	−9.1	−4.1
YGSP 54565	Equidae	*Sivalhippus perimensis*	left M1	Y158	8.7	−9.3	−3.6
YGSP 54566	Equidae	*Sivalhippus perimensis*	left P2	Y158	8.7	−9.3	−3.6
YGSP 54567	Equidae	*Sivalhippus perimensis*	right P3	Y149	9.0	−7.1	−1.6
YGSP 54568	Equidae	*Sivalhippus perimensis*	left P3	Y33	8.0	−2.0	−2.0
YGSP 54573	Equidae	*Sivalhippus perimensis*	right P4	Y205	9.0	−7.0	−2.9
YGSP 598a	Equidae	*Sivalhippus perimensis*	right p4	Y28	8.4	−4.7	−3.2
YGSP 17772	Equidae	*Sivalhippus theobaldi*	right P4	Y541	8.0	−8.8	−3.7

(*continued*)

Table III.2. continued

Specimen number[2]	Family or Order[3]	Name[4]	Tooth sampled	Locality[5]	Age (Ma)[6]	δ^{13}C (V-PDB)[7]	δ^{18}O (V-PDB)[7]
YGSP 17789	Equidae	*Sivalhippus theobaldi*	right P2	Y541	8.0	−7.1	−0.9
YGSP 18378	Equidae	*Sivalhippus theobaldi*	left M2	Y608	8.6	−6.2	−1.4
YGSP 47081	Equidae	*Sivalhippus theobaldi*	right M1 or M2	Y312	9.4	−6.8	−3.1
YGSP 50016	Equidae	*Sivalhippus theobaldi*	left P2	Y917	7.8	−10.4	−4.0
YGSP 5114	Equidae	*Sivalhippus theobaldi*	right P4	Y211	9.3	−10.9	−4.9
YGSP 13636	Equidae	*Sivalhippus* sp.	right M2	Y350	9.4	−11.1	−4.3
YGSP 14053	Equidae	*Sivalhippus* sp.	left P4	Y370	7.2	−6.9	−2.0
YGSP 14054	Equidae	*Sivalhippus* sp.	left m2	Y370	7.2	−7.6	0.1
YGSP 17083	Equidae	*Sivalhippus* sp.	left P2	Y262	9.6	−11.9	−6.8
YGSP 18046	Equidae	*Sivalhippus* sp.	right P3	Y450	10.2	−8.1	−4.3
YGSP 18377	Equidae	*Sivalhippus* sp.	left P4	Y608	8.6	−7.4	−4.7
YGSP 3176	Equidae	*Sivalhippus* sp.	right M1	Y17	8.0	−9.0	−2.5
YGSP 3699a	Equidae	*Sivalhippus* sp.	left P3	Y158	8.7	−3.9	−1.4
YGSP 4197	Equidae	*Sivalhippus* sp.	molar	Y176	8.8	−9.8	−5.7
YGSP 50017	Equidae	*Sivalhippus* sp.	right M2	Y917	7.8	−5.0	−4.6
YGSP 50130	Equidae	*Sivalhippus* sp.	right M	Y932	7.3	−5.8	0.6
YGSP 50322	Equidae	*Sivalhippus* sp.	left M3	Y940	7.4	−4.0	−0.6
YGSP 50394	Equidae	*Sivalhippus* sp.	right p or m	Y944	7.4	−7.8	−2.7
YGSP 15403	Equidae	*Cremohipparion antelopinum*	left P3	Y443	8.5	−7.9	−2.2
YGSP 15542	Equidae	*Cremohipparion antelopinum*	left M3	KL12	8.5	−11.0	−4.5
YGSP 19742	Equidae	*Cremohipparion antelopinum*	right m1	Y604	8.6	−5.9	−2.9
YGSP 3326	Equidae	*Cremohipparion antelopinum*	left P3	Y141	8.9	−1.0	−1.9
YGSP 49535a	Equidae	*Cremohipparion antelopinum*	left M3	Y542	8.1	−7.1	0.3
YGSP 49784	Equidae	*Cremohipparion antelopinum*	right m1	Y912	6.5	−3.4	1.1
YGSP 50870	Equidae	*Cremohipparion antelopinum*	left P3	Y948	7.8	−6.6	−2.8
YGSP 54570	Equidae	*Cremohipparion antelopinum*	right M1	Y24	8.2	−9.9	−7.9
YGSP 54571	Equidae	*Cremohipparion antelopinum*	left M3	Y7	8.3	−7.4	−4.7
YGSP 54572	Equidae	*Cremohipparion antelopinum*	left M3	Y176	8.2	−6.9	−3.1
YGSP 54576	Equidae	*Cremohipparion antelopinum*	right P4	Y166	8.8	−10.9	−3.7
YGSP 579b	Equidae	*Cremohipparion antelopinum*	right M2	Y29	7.9		
YGSP 637a	Equidae	*Cremohipparion antelopinum*	right P4	Y30	8.4	−10.8	−5.2
YGSP 15069	Equidae	hipparionine	left m	Y419	7.4	−2.2	2.3
YGSP 15097	Equidae	hipparionine	right M3	Y174	8.4	−3.6	1.7
YGSP 15462	Equidae	hipparionine	right m3	KL14	8.9	−8.9	−4.9
YGSP 27729	Equidae	hipparionine	left M3	Y729	10.4	−11.4	−6.3
YGSP 328	Equidae	hipparionine	right dP	Y17	8.0	−8.4	−5.1
YGSP 38915a	Equidae	hipparionine	left dP	Y544	8.7	−8.8	−5.7
YGSP 38917	Equidae	hipparionine	M	Y262	9.6	−12.5	−3.7
YGSP 38921	Equidae	hipparionine	right M	Y262	9.6	−13.5	−7.2
YGSP 38930	Equidae	hipparionine	M	Y311	10.1	−12.1	−5.5
YGSP 38943	Equidae	hipparionine	M	Y876	7.4	−3.7	−0.5
YGSP 38944	Equidae	hipparionine	M	Y1020	7.4		
YGSP 38949	Equidae	hipparionine	M	Y260	9.3	−11.6	−2.3
YGSP 38957	Equidae	hipparionine	right M	Y875	9.8	−10.7	−2.5
YGSP 38958	Equidae	hipparionine	M	Y260	9.3	−12.3	−4.0
YGSP 49565	Equidae	hipparionine	right P4	Y891	8.7	−9.7	−3.9
YGSP 50871	Equidae	hipparionine	right m2	Y948	7.8	−5.4	−6.1
YGSP 53637	Equidae	hipparionine	left m	Y311	10.1	−11.0	−2.9
YGSP 54569	Equidae	hipparionine	right M1	Y112	7.9	−5.3	−2.7
YGSP 54574	Equidae	hipparionine	right P4	Y15	7.9	−6.5	−4.2
YGSP 663	Equidae	hipparionine	m3 or M3	Y32	8.4	−4.9	−1.9
YGSP 16366	Rhinocerotidae	*Alicornops complanatum*	left P4	Y542	8.1	−11.4	−4.5
YGSP 31469	Rhinocerotidae	*Alicornops complanatum*	left P4	Y545	8.1	−12.6	−3.6
YGSP 49585d	Rhinocerotidae	*Alicornops complanatum*	left P3	Y604	8.6	−12.0	−4.2
YGSP 23303	Rhinocerotidae	*Brachydiceratherium fatehjangense*	right m3	Y682	15.2	−11.2	−7.6
YGSP 24184f	Rhinocerotidae	*Brachydiceratherium intermedium*	left m2	Y678	15.1	−11.2	−5.7
YGSP 17113	Rhinocerotidae	*Brachypotherium perimense*	right m3	Y262	9.6	−12.5	−7.3
YGSP 17115	Rhinocerotidae	*Brachypotherium perimense*	right M	Y262	9.6	−12.1	−8.7
YGSP 24069	Rhinocerotidae	*Brachypotherium perimense*	right P2	Y647	12.4	−12.1	−8.5
YGSP 24282	Rhinocerotidae	*Brachypotherium perimense*	right m3	Y648	11.7	−11.4	−9.0
YGSP 26656	Rhinocerotidae	*Brachypotherium perimense*	right M2	Y695	12.4	−12.8	−8.8

Table III.2. continued

Specimen number[2]	Family or Order[3]	Name[4]	Tooth sampled	Locality[5]	Age (Ma)[6]	δ¹³C (V-PDB)[7]	δ¹⁸O (V-PDB)[7]
YGSP 30569	Rhinocerotidae	*Brachypotherium perimense*	right m	Y515	12.1	−12.2	−7.6
YGSP 31073	Rhinocerotidae	*Brachypotherium perimense*	left m3	Y634	12.3	−12.6	−10.2
YGSP 31108	Rhinocerotidae	*Brachypotherium perimense*	left M2	Y767	12.4	−12.2	−7.7
YGSP 31577	Rhinocerotidae	*Brachypotherium perimense*	right p4	Y775	12.3	−12.7	−9.3
YGSP 31577	Rhinocerotidae	*Brachypotherium perimense*	right p4	Y775	13.1	−13.3	−8.6
YGSP 3165	Rhinocerotidae	*Brachypotherium perimense*	left m	Y20	8.0	−11.9	−5.9
YGSP 32571	Rhinocerotidae	*Brachypotherium perimense*	left M1	Y60	12.9	−11.6	−6.4
YGSP 41082	Rhinocerotidae	*Brachypotherium perimense*	left m	Y728	10.4	−12.8	−9.7
YGSP 41124	Rhinocerotidae	*Brachypotherium perimense*	left dP2	Y496	12.4	−11.4	−8.5
YGSP 41135	Rhinocerotidae	*Brachypotherium perimense*	right M2	Y647	12.4	−11.8	−8.0
YGSP 41350	Rhinocerotidae	*Brachypotherium perimense*	right m2	Y706	12.3	−11.9	−8.5
YGSP 41427	Rhinocerotidae	*Brachypotherium perimense*	left M1	Y695	12.4	−11.7	−8.5
YGSP 46328	Rhinocerotidae	*Brachypotherium perimense*	left p2	Y869	9.7	−12.5	−3.9
YGSP 49521	Rhinocerotidae	*Brachypotherium perimense*	left M3	KL11	8.0	−11.9	−7.0
YGSP 50515	Rhinocerotidae	*Brachypotherium perimense*	right M1	Y311	10.1	−12.0	−9.2
YGSP 51649	Rhinocerotidae	*Brachypotherium perimense*	left M3	Y262	9.6	−13.1	−7.5
YGSP 52190	Rhinocerotidae	*Brachypotherium perimense*	right M3	Y504	11.6	−11.2	−7.8
YGSP 52778	Rhinocerotidae	*Brachypotherium perimense*	right P2	Y980	8.9	−11.0	−3.7
YGSP 52842	Rhinocerotidae	*Brachypotherium perimense*	left M1	Y981	8.7	−11.9	−5.6
YGSP 19966	Rhinocerotidae	*Gaindatherium browni*	left M3	Y350	9.4	−12.4	−4.7
YGSP 24067a	Rhinocerotidae	*Gaindatherium browni*	left p3	Y647	13.7	−10.6	−3.9
YGSP 26657	Rhinocerotidae	*Gaindatherium browni*	left P4	Y695	12.4	−12.1	−6.3
YGSP 52177	Rhinocerotidae	*Gaindatherium browni*	left P3	Y311	10.1	−11.5	−5.0
YGSP 52195	Rhinocerotidae	*Gaindatherium browni*	right p3	Y311	10.1	−9.7	−5.7
YGSP 51741	Rhinocerotidae	*Gaindatherium vidali*	right p2	Y1007	8.0	−10.4	−4.0
YGSP 22459	Rhinocerotidae	*Gaindatherium* sp.	left dP1	Y515	12.1	−12.5	−9.1
YGSP 28007	Rhinocerotidae	*Gaindatherium* sp.	left P3	Y740	15.8	−12.6	−9.2
YGSP 32366a	Rhinocerotidae	*Gaindatherium* sp.	left P3	Y76	11.4	−11.1	−4.7
YGSP 40704	Rhinocerotidae	*Gaindatherium* sp.	left p	Y76	11.4	−10.8	−7.4
YGSP 41587	Rhinocerotidae	*Gaindatherium* sp.	right P2	Y716	12.9	−11.2	−5.7
YGSP 52166	Rhinocerotidae	*Gaindatherium* sp.	right dP	Y28	8.4	−11.7	−5.9
YGSP 51572	Rhinocerotidae	*Rhinoceros aff. sivalensis*	right P3	KL11	8.0	−11.4	−3.3
YGSP 28099	Rhinocerotidae	*Dicerorhinus aff. sumatrensis*	right P3	Y705	12.4	−11.1	−8.1
YGSP 38925	Rhinocerotidae	rhinoceratine	molar	Y604	8.6	−12.3	
YGSP 17993	Rhinocerotidae	*Ceratotherium neumayri*	left M3	KL08	7.8	−12.2	−5.8
YGSP 17464	Rhinocerotidae	rhinocerotid	right M	Y590	18.0	−11.7	−5.7
YGSP 18440	Rhinocerotidae	rhinocerotid	left m	Y606	8.1	−11.7	−4.4
YGSP 26588	Rhinocerotidae	rhinocerotid	right m	Y681	14.7	−11.0	−5.9
YGSP 38916	Rhinocerotidae	rhinocerotid	molar	Y547	8.0	−11.8	−2.3
YGSP 40750	Rhinocerotidae	rhinocerotid	molar	Y828	13.4	−11.8	−6.8
YGSP 41577	Rhinocerotidae	rhinocerotid	molar	Y847	13.4	−10.5	−6.7
YGSP 45760	Rhinocerotidae	rhinocerotid	left M	Y507	11.7	−11.6	−3.4
YGSP 49920	Rhinocerotidae	rhinocerotid	M	D28	6.7	−1.6	−5.6
YGSP 50459	Rhinocerotidae	rhinocerotid	M	Y947	7.8	−11.0	−6.9
YGSP 50514	Rhinocerotidae	rhinocerotid	left m	Y311	10.1	−10.8	−6.7
YGSP 50597	Rhinocerotidae	rhinocerotid	right m	Y581	7.2	−4.2	−2.7
YGSP 50873	Rhinocerotidae	rhinocerotid	M	Y948	7.8	−11.7	−5.1
YGSP 50902	Rhinocerotidae	rhinocerotid	M	Y950	7.8	−8.8	0.7
YGSP 51694	Rhinocerotidae	rhinocerotid	premolar	Y545	8.1	−10.9	−3.3
YGSP 51734	Rhinocerotidae	rhinocerotid	M	Y889	8.0	−13.3	−4.3
YGSP 51776	Rhinocerotidae	rhinocerotid	m	Y1000	8.1	−8.3	−3.4
YGSP 51796	Rhinocerotidae	rhinocerotid	left m	Y448	9.1	−12.7	−2.1
YGSP 13056	Palaeochoeridae	*Yunnanochoerus gandakasensis*	left m	Y250	9.7	−13.4	−4.7
YGSP 14499	Palaeochoeridae	*Yunnanochoerus gandakasensis*	right m2	Y269	9.4	−13.3	−9.0
YGSP 4630	Palaeochoeridae	*Yunnanochoerus gandakasensis*	left m	Y182	9.2	−13.0	−5.1
YGSP 46525	Palaeochoeridae	*Yunnanochoerus gandakasensis*	left m	Y258	10.2	−13.7	−3.0
YGSP 5474	Palaeochoeridae	*Yunnanochoerus gandakasensis*	left m2	Y310	9.4	−12.9	−8.0
YGSP 5912	Palaeochoeridae	*Yunnanochoerus gandakasensis*	right m	Y337	9.7	−13.9	−8.9
YGSP 23073	Suidae	*Listriodon pentapotamiae*	left P4	Y496	12.4	−12.8	−3.0
YGSP 23639a	Suidae	*Listriodon pentapotamiae*	left m2	Y661	13.4	−12.7	−6.8

(continued)

Table III.2. continued

Specimen number[2]	Family or Order[3]	Name[4]	Tooth sampled	Locality[5]	Age (Ma)[6]	δ^{13}C (V-PDB)[7]	δ^{18}O (V-PDB)[7]
YGSP 26389j	Suidae	*Listriodon pentapotamiae*	left M3	Y699	13.0	−12.5	
YGSP 27030	Suidae	*Listriodon pentapotamiae*	right m3	Y61	11.6	−11.5	−4.7
YGSP 28132	Suidae	*Listriodon pentapotamiae*	m	Y705	12.4	−12.9	−7.1
YGSP 30674a	Suidae	*Listriodon pentapotamiae*	right M2	Y705	12.4	−12.5	−7.3
YGSP 31239	Suidae	*Listriodon pentapotamiae*	left m3	Y773	11.5	−13.2	−6.6
YGSP 31623	Suidae	*Listriodon pentapotamiae*	right m3	Y772	12.4	−13.4	−6.9
YGSP 32572	Suidae	*Listriodon pentapotamiae*	right m3	Y500	12.1	−12.6	−4.5
YGSP 46145	Suidae	*Listriodon pentapotamiae*	molar	Y76	11.4	−11.0	−4.9
YGSP 9663	Suidae	*Listriodon pentapotamiae*	right m2	Y76	11.4	−12.1	0.0
YGSP 12708b	Suidae	*Hippopotamodon sivalense*	left P4	Y260	9.3	−11.7	−4.5
YGSP 12861	Suidae	*Hippopotamodon sivalense*	right m1	Y309	9.4	−11.9	−8.1
YGSP 13524	Suidae	*Hippopotamodon sivalense*	left m3	Y239	9.3	−11.1	−6.4
YGSP 13972	Suidae	*Hippopotamodon sivalense*	right m3	Y251	10.0	−11.4	−5.7
YGSP 14112	Suidae	*Hippopotamodon sivalense*	left M3	Y260	9.3	−11.5	−4.6
YGSP 16049	Suidae	*Hippopotamodon sivalense*	left M3	Y251	10.0	−11.9	−6.6
YGSP 16085	Suidae	*Hippopotamodon sivalense*	right m1	Y182	9.2	−10.8	−3.3
YGSP 16099	Suidae	*Hippopotamodon sivalense*	left m1	Y310	9.4	−10.9	−6.2
YGSP 17937	Suidae	*Hippopotamodon sivalense*	left m2	KL12	8.5	−10.8	−4.0
YGSP 18276	Suidae	*Hippopotamodon sivalense*	molar	Y454	10.5	−10.6	−6.0
YGSP 23714	Suidae	*Hippopotamodon sivalense*	left P2	Y311	10.1	−11.2	−7.6
YGSP 27915	Suidae	*Hippopotamodon sivalense*	left m3	Y260	9.3	−11.2	−5.7
YGSP 31502	Suidae	*Hippopotamodon sivalense*	right P4	Y406	8.9	−11.1	−4.3
YGSP 330	Suidae	*Hippopotamodon sivalense*	right m1	Y17	8.0	−11.1	−4.1
YGSP 38971c	Suidae	*Hippopotamodon sivalense*	right M1	Y327	9.3	−11.7	−9.0
YGSP 4086	Suidae	*Hippopotamodon sivalense*	right M1	Y174	8.4	−12.3	−6.9
YGSP 41401	Suidae	*Hippopotamodon sivalense*	right M1	Y311	10.1	−9.9	−5.1
YGSP 46308	Suidae	*Hippopotamodon sivalense*	left m1	Y873	7.6	−6.2	−4.8
YGSP 46568	Suidae	*Hippopotamodon sivalense*	left M3	Y251	10.0	−13.0	−6.5
YGSP 47176	Suidae	*Hippopotamodon sivalense*	m	Y159	9.4	−11.4	−6.6
YGSP 49514	Suidae	*Hippopotamodon sivalense*	right dP3	Y599	8.1	−10.3	−0.9
YGSP 49561	Suidae	*Hippopotamodon sivalense*	right m	Y891	8.7	−11.7	−6.4
YGSP 4984	Suidae	*Hippopotamodon sivalense*	left m2	Y159	9.4	−8.6	−4.7
YGSP 50019	Suidae	*Hippopotamodon sivalense*	right m3	Y917	7.8	−10.2	−4.7
YGSP 50134g	Suidae	*Hippopotamodon sivalense*	right m3	Y932	7.3	−9.1	−5.6
YGSP 50349	Suidae	*Hippopotamodon sivalense*	left m3	Y930	6.3	−7.0	−5.3
YGSP 50419	Suidae	*Hippopotamodon sivalense*	left M2	Y946	7.8	−10.1	−5.5
YGSP 50455	Suidae	*Hippopotamodon sivalense*	left p4	Y947	7.8	−9.4	−8.1
YGSP 50891	Suidae	*Hippopotamodon sivalense*	right P4	Y950	7.8	−10.9	−4.9
YGSP 51745	Suidae	*Hippopotamodon sivalense*	left dP4	Y599	8.1	−10.8	3.5
YGSP 51963	Suidae	*Hippopotamodon sivalense*	right m3	Y1005	9.5	−11.8	−7.6
YGSP 52040	Suidae	*Hippopotamodon sivalense*	left P4	KL01	9.3	−12.0	−6.0
YGSP 5212	Suidae	*Hippopotamodon sivalense*	left m1	Y211	9.3	−12.5	−5.6
YGSP 52401	Suidae	*Hippopotamodon sivalense*	right M	Y311	10.1	−10.4	−4.8
YGSP 52626	Suidae	*Hippopotamodon sivalense*	molar	Y960	7.8	−10.9	−5.0
YGSP 5276	Suidae	*Hippopotamodon sivalense*	left M1	Y211	9.3	−9.4	−7.0
YGSP 52894	Suidae	*Hippopotamodon sivalense*	M	Y943	7.4	−8.5	−4.8
YGSP 53009	Suidae	*Hippopotamodon sivalense*	left M	Y450	10.2	−11.3	−6.8
YGSP 8972	Suidae	*Hippopotamodon sivalense*	left m3	Y251	10.0	−10.3	−5.5
YGSP 10224	Suidae	*Propotomochoerus hysudricus*	left m3	Y309	9.4	−11.8	−9.2
YGSP 10225	Suidae	*Propotomochoerus hysudricus*	right m3	Y309	9.4	−11.2	−6.5
YGSP 12201	Suidae	*Propotomochoerus hysudricus*	right m	Y311	10.1	−11.3	−6.4
YGSP 13678	Suidae	*Propotomochoerus hysudricus*	left P4	Y367	9.0	−11.2	−5.2
YGSP 13855	Suidae	*Propotomochoerus hysudricus*	left m3	Y226	9.4	−11.4	−7.8
YGSP 14044	Suidae	*Propotomochoerus hysudricus*	left M3	Y262	9.6	−12.6	−6.9
YGSP 14045	Suidae	*Propotomochoerus hysudricus*	left m3	Y370	7.2	−8.2	−3.0
YGSP 14105	Suidae	*Propotomochoerus hysudricus*	right m3	Y377	7.3	−11.7	−3.3
YGSP 14502	Suidae	*Propotomochoerus hysudricus*	right P4 or M1	Y269	9.4	−11.0	−6.9
YGSP 14972	Suidae	*Propotomochoerus hysudricus*	left m3	KL05	9.6	−10.9	−7.9
YGSP 15739a	Suidae	*Propotomochoerus hysudricus*	left m	Y309	9.4	−10.8	−4.2
YGSP 16132	Suidae	*Propotomochoerus hysudricus*	left m3	Y310	9.4	−10.7	−5.6
YGSP 16969	Suidae	*Propotomochoerus hysudricus*	left p4	Y545	8.1	−9.5	−1.5

Table III.2. continued

Specimen number[2]	Family or Order[3]	Name[4]	Tooth sampled	Locality[5]	Age (Ma)[6]	$\delta^{13}C$ (V-PDB)[7]	$\delta^{18}O$ (V-PDB)[7]
YGSP 17578	Suidae	*Propotomochoerus hysudricus*	right M3	Y311	10.1	−10.4	−7.2
YGSP 17777	Suidae	*Propotomochoerus hysudricus*	left m1	Y541	8 0	−11.5	−5.1
YGSP 17928	Suidae	*Propotomochoerus hysudricus*	right P4	Y603	8.5	−11.6	−3.0
YGSP 17968a	Suidae	*Propotomochoerus hysudricus*	left M2	Y604	8.6	−11.7	−5.1
YGSP 18124	Suidae	*Propotomochoerus hysudricus*	left m3	Y452	7.3	−8.4	−3.7
YGSP 18132	Suidae	*Propotomochoerus hysudricus*	right M3	Y452	7.3	−7.1	−2.5
YGSP 19733	Suidae	*Propotomochoerus hysudricus*	right P4	Y547	8.0	−12.8	−7.3
YGSP 19765	Suidae	*Propotomochoerus hysudricus*	right M2	Y166	8.8	−12.2	−2.8
YGSP 19971	Suidae	*Propotomochoerus hysudricus*	left P4	Y350	9.4	−10.7	−6.0
YGSP 22080	Suidae	*Propotomochoerus hysudricus*	right M3	Y311	10.1	−10.3	−8.8
YGSP 31474	Suidae	*Propotomochoerus hysudricus*	left M3	Y545	8.1	−11.0	−2.7
YGSP 351	Suidae	*Propotomochoerus hysudricus*	left M2	Y24	8.2	−10.6	−5.0
YGSP 37006	Suidae	*Propotomochoerus hysudricus*	right M3	Y227	9.4	−12.4	−7.4
YGSP 40100a	Suidae	*Propotomochoerus hysudricus*	left m	Y403	9.3	−10.3	−5.0
YGSP 41368b	Suidae	*Propotomochoerus hysudricus*	right m1	Y269	9.4	−12.9	−9.2
YGSP 4554	Suidae	*Propotomochoerus hysudricus*	right M3	Y182	9.2	−12.6	−5.4
YGSP 457	Suidae	*Propotomochoerus hysudricus*	right m2	Y24	8.2	−11.6	−5.6
YGSP 4599	Suidae	*Propotomochoerus hysudricus*	left m3	Y193	9.0	−7.8	−7.0
YGSP 46666	Suidae	*Propotomochoerus hysudricus*	left m1	Y875	9.8	−11.7	−8.6
YGSP 47127	Suidae	*Propotomochoerus hysudricus*	??	Y535	8.4	−11.8	−4.1
YGSP 4813	Suidae	*Propotomochoerus hysudricus*	left M3	Y203	8.7	−11.6	−5.2
YGSP 49515	Suidae	*Propotomochoerus hysudricus*	left m3	Y599	8.1	−10.4	−4.3
YGSP 49540	Suidae	*Propotomochoerus hysudricus*	left M	Y542	8.1	−12.1	−3.9
YGSP 49816	Suidae	*Propotomochoerus hysudricus*	right M2	Y916	7.2	−7.7	1.3
YGSP 49840a	Suidae	*Propotomochoerus hysudricus*	right m2	Y906	7.8	−11.5	−7.3
YGSP 49844	Suidae	*Propotomochoerus hysudricus*	right M3	Y906	7.8	−10.2	−6.8
YGSP 4988	Suidae	*Propotomochoerus hysudricus*	right m2	Y159	9.4	−11.0	−9.4
YGSP 50027	Suidae	*Propotomochoerus hysudricus*	right m3	Y917	7.8	−9.7	−7.3
YGSP 50374	Suidae	*Propotomochoerus hysudricus*	right m2	Y943	7.4	−10.0	−7.1
YGSP 50423	Suidae	*Propotomochoerus hysudricus*	m3	Y946	7.8	−10.7	−4.4
YGSP 50574	Suidae	*Propotomochoerus hysudricus*	left P4	Y581	7.2	−6.7	−4.6
YGSP 50975	Suidae	*Propotomochoerus hysudricus*	left M3	Y450	10.2	−10.6	−8.2
YGSP 51554	Suidae	*Propotomochoerus hysudricus*	left m2	Y542	8.1	−11.8	−4.0
YGSP 51575	Suidae	*Propotomochoerus hysudricus*	right m3	Y539	8.1	−10.8	−2.2
YGSP 51700	Suidae	*Propotomochoerus hysudricus*	left m3	Y545	8.1	−11.8	−6.0
YGSP 51710	Suidae	*Propotomochoerus hysudricus*	right m	Y1008	9.0	−11.0	−4.5
YGSP 51793	Suidae	*Propotomochoerus hysudricus*	M3	Y448	9.1	−11.9	−8.4
YGSP 51849	Suidae	*Propotomochoerus hysudricus*	left m2	Y269	9.4	−11.7	−7.6
YGSP 51918	Suidae	*Propotomochoerus hysudricus*	right m3	Y444	8.6	−10.6	−5.3
YGSP 52049	Suidae	*Propotomochoerus hysudricus*	right m3	KL14	8.9	−11.9	−4.3
YGSP 52623	Suidae	*Propotomochoerus hysudricus*	left M2	Y960	7.8	−11.4	−5.5
YGSP 52762	Suidae	*Propotomochoerus hysudricus*	right m3	Y980	8.9	−11.5	−7.6
YGSP 6137	Suidae	*Propotomochoerus hysudricus*	right P4	Y221	9.4	−10.3	−4.7
YGSP 6226	Suidae	*Propotomochoerus hysudricus*	right m2	Y228	9.6	−12.0	−5.8
YGSP 6753	Suidae	*Propotomochoerus hysudricus*	left M3	Y227	9.4	−11.0	−5.7
YGSP 7654	Suidae	*Propotomochoerus hysudricus*	right P4	Y182	9.2	−11.9	−5.3
YGSP 9355	Suidae	*Propotomochoerus hysudricus*	left m3	Y260	9.3	−12.7	−6.9
YGSP 9975	Suidae	*Propotomochoerus hysudricus*	left m2	Y260	9.3	−12.7	−6.9
YGSP 26980	Suidae	*Conohyus sindiensis*	m	Y500	12.1	−11.9	−10.9
YGSP 28126	Suidae	*Conohyus sindiensis*	right M3	Y705	12.4	−11.5	−9.0
YGSP 30673	Suidae	*Conohyus sindiensis*	left m3	Y705	12.4	−12.1	−7.9
YGSP 31406	Suidae	*Conohyus sindiensis*	right m3	Y61	11.6	−12.3	−6.8
YGSP 45724	Suidae	*Conohyus sindiensis*	left m3	Y496	12.4	−11.1	−5.8
YGSP 757	Suidae	*Conohyus sindiensis*	left M3	Y38	13.5	−9.6	−6.8
YGSP 9659	Suidae	*Conohyus sindiensis*	left M3	Y76	11.4	−12.4	−6.7
YGSP 25528	Suidae	*Parachleuastochoerus pilgrimi*	right M3	Y705	12.4	−11.8	−7.6
YGSP 45866	Suidae	*Parachleuastochoerus pilgrimi*	right M3	Y76	11.4	−10.8	−6.5
YGSP 46192	Suidae	*Parachleuastochoerus pilgrimi*	right M	Y83	11.6	−13.4	−8.2
YGSP 12463	Suidae	*Tetraconodon magnus*	P3	Y329	9.7	−11.1	−3.0
YGSP 13971	Suidae	*Tetraconodon magnus*	P4	Y251	10.0	−11.3	−7.4
YGSP 17069	Suidae	*Tetraconodon magnus*	right P	Y315	9.4	−12.8	−3.4

(continued)

Table III.2. continued

Specimen number[2]	Family or Order[3]	Name[4]	Tooth sampled	Locality[5]	Age (Ma)[6]	δ¹³C (V-PDB)[7]	δ¹⁸O (V-PDB)[7]
YGSP 8971a	Suidae	*Tetraconodon magnus*	right M2	Y251	10.0	−11.5	−5.3
YGSP 27877	Suidae	*Tetraconodon* sp.	right p3	Y735	11.7	−12.1	−5.8
YGSP 18928	Anthracotheriidae	*Microbunodon silistrensis*	right m	Y496	12.4	−10.7	−7.1
YGSP 19831	Anthracotheriidae	*Microbunodon silistrensis*	right P3	Y495	11.6	−10.9	−3.8
YGSP 21063	Anthracotheriidae	*Microbunodon silistrensis*	left m2	Y494	12.4	−10.6	−5.1
YGSP 11797	Anthracotheriidae	*Microbunodon milaensis*	left p	Y317	9.2	−12.1	−6.9
YGSP 16612	Anthracotheriidae	*Microbunodon milaensis*	left m1	Y572	10.4	−13.1	−7.9
YGSP 5124	Anthracotheriidae	*Microbunodon milaensis*	right p2	Y211	9.3	−11.6	−7.2
YGSP 9799	Anthracotheriidae	*Microbunodon milaensis*	left p	Y269	9.4	−12.3	−5.8
YGSP 15070a	Anthracotheriidae	*Merycopotamus dissimilis*	left M	Y419	7.4	−5.7	−5.1
YGSP 16813	Anthracotheriidae	*Merycopotamus dissimilis*	right p4	Y373	7.2	−8.1	−6.1
YGSP 18157	Anthracotheriidae	*Merycopotamus dissimilis*	left m3	D13	6.5	−3.8	−5.2
YGSP 49859	Anthracotheriidae	*Merycopotamus dissimilis*	right M	Y903	6.4	−4.2	−5.6
YGSP 50041	Anthracotheriidae	*Merycopotamus dissimilis*	left P	Y920	7.0	−5.6	−2.1
YGSP 50101	Anthracotheriidae	*Merycopotamus dissimilis*	left M	Y910	7.2	−8.1	−5.6
YGSP 50337	Anthracotheriidae	*Merycopotamus dissimilis*	left m2	Y930	6.3	−4.9	−6.3
YGSP 50721	Anthracotheriidae	*Merycopotamus dissimilis*	right p4	D13	6.5	−5.5	−7.1
YGSP 50868	Anthracotheriidae	*Merycopotamus dissimilis*	left P	Y948	7.8	−10.1	−7.9
YGSP 50894	Anthracotheriidae	*Merycopotamus dissimilis*	left p	Y950	7.8	−11.1	−5.9
YGSP 52631	Anthracotheriidae	*Merycopotamus dissimilis*	right M3	Y960	7.8	−10.1	−6.9
YGSP 10846	Anthracotheriidae	*Merycopotamus medioximus*	left M	Y310	9.4	−10.2	−8.0
YGSP 10886	Anthracotheriidae	*Merycopotamus medioximus*	left m3	Y312	9.4	−11.8	−3.0
YGSP 13523b	Anthracotheriidae	*Merycopotamus medioximus*	right p3	Y239	9.3	−11.6	−7.5
YGSP 14318	Anthracotheriidae	*Merycopotamus medioximus*	right M2 or M3	Y450	10.2	−10.2	−7.0
YGSP 14515	Anthracotheriidae	*Merycopotamus medioximus*	right M3	Y269	9.4	−12.9	−6.4
YGSP 14969	Anthracotheriidae	*Merycopotamus medioximus*	left m1	Y406	8.7	−12.0	−4.7
YGSP 15572	Anthracotheriidae	*Merycopotamus medioximus*	right M2	Y285	9.8	−12.8	−4.3
YGSP 26878	Anthracotheriidae	*Merycopotamus medioximus*	left m3	Y311	10.1	−11.0	−5.2
YGSP 4596	Anthracotheriidae	*Merycopotamus medioximus*	right m3	Y193	9.0	−11.1	−6.7
YGSP 4597c	Anthracotheriidae	*Merycopotamus medioximus*	right m3	Y193	9.0	−10.0	−7.2
YGSP 47066	Anthracotheriidae	*Merycopotamus medioximus*	right M1	Y317	9.2	−8.8	−5.6
YGSP 49562	Anthracotheriidae	*Merycopotamus medioximus*	left m3	Y891	8.7	−11.4	−6.9
YGSP 50490	Anthracotheriidae	*Merycopotamus medioximus*	left p	Y311	10.1	−10.7	−6.9
YGSP 5122	Anthracotheriidae	*Merycopotamus medioximus*	left m3	Y211	9.3	−11.5	−6.4
YGSP 52174	Anthracotheriidae	*Merycopotamus medioximus*	right m	Y311	10.1	−11.9	−6.0
YGSP 5528	Anthracotheriidae	*Merycopotamus medioximus*	left M	Y310	9.4	−11.4	−7.8
YGSP 6169	Anthracotheriidae	*Merycopotamus medioximus*	left M2	Y224	9.4	−10.5	−6.6
YGSP 6225	Anthracotheriidae	*Merycopotamus medioximus*	right m3	Y228	9.6	−12.1	−7.6
YGSP 6320	Anthracotheriidae	*Merycopotamus medioximus*	right M	Y311	10.1	−11.4	−7.7
YGSP 6746	Anthracotheriidae	*Merycopotamus medioximus*	left M2	Y227	9.4	−11.6	−6.2
YGSP 9003	Anthracotheriidae	*Merycopotamus medioximus*	left m1	Y251	10.0	−12.1	−7.7
YGSP 20525	Anthracotheriidae	*Merycopotamus nanus*	left M2	Y494	12.4	−10.9	−8.1
YGSP 21070	Anthracotheriidae	*Merycopotamus nanus*	right p4	Y502	12.0	−11.9	−5.1
YGSP 21077	Anthracotheriidae	*Merycopotamus nanus*	right M	Y496	12.4	−10.4	−7.3
YGSP 21081	Anthracotheriidae	*Merycopotamus nanus*	right m	Y495	11.6	−10.8	−6.4
YGSP 21084	Anthracotheriidae	*Merycopotamus nanus*	left M	Y495	11.6	−10.9	−6.5
YGSP 23246	Anthracotheriidae	*Merycopotamus nanus*	M	Y663	11.6	−10.4	−5.0
YGSP 23549	Anthracotheriidae	*Merycopotamus nanus*	right m1	Y675	12.7	−10.4	−9.7
YGSP 26311b	Anthracotheriidae	*Merycopotamus nanus*	right m3	Y713	12.9	−8.7	−11.0
YGSP 26769	Anthracotheriidae	*Merycopotamus nanus*	right P3	Y714	12.8	−8.6	−7.4
YGSP 27785	Anthracotheriidae	*Merycopotamus nanus*	right p4	Y504	11.6	−10.9	−8.4
YGSP 27786	Anthracotheriidae	*Merycopotamus nanus*	right p4	Y504	11.6	−11.0	−8.2
YGSP 28124	Anthracotheriidae	*Merycopotamus nanus*	right m2	Y705	12.4	−11.4	−7.6
YGSP 28283	Anthracotheriidae	*Merycopotamus nanus*	P4	Y500	12.1	−11.3	−8.8
YGSP 30500	Anthracotheriidae	*Merycopotamus nanus*	right m3	Y750	12.7	−9.6	−7.2
YGSP 30572	Anthracotheriidae	*Merycopotamus nanus*	left M2	Y515	12.1	−10.8	−7.9
YGSP 30989	Anthracotheriidae	*Merycopotamus nanus*	right M	Y714	12.8	−9.7	−8.4
YGSP 31217	Anthracotheriidae	*Merycopotamus nanus*	left m	Y691	13.1	−9.6	−8.1
YGSP 31258	Anthracotheriidae	*Merycopotamus nanus*	right M	Y650	13.1	−10.8	−8.0
YGSP 41028	Anthracotheriidae	*Merycopotamus nanus*	m	Y711	13.1	−9.4	−7.2
YGSP 41571	Anthracotheriidae	*Merycopotamus nanus*	right M	Y847	13.4	−9.8	−9.6

Table III.2. continued

Specimen number[2]	Family or Order[3]	Name[4]	Tooth sampled	Locality[5]	Age (Ma)[6]	$\delta^{13}C$ (V-PDB)[7]	$\delta^{18}O$ (V-PDB)[7]
YGSP 45679	Anthracotheriidae	*Merycopotamus nanus*	left m3	Y495	11.6	−11.4	−7.6
YGSP 45817	Anthracotheriidae	*Merycopotamus nanus*	right m	Y502	12.0	−11.2	7.0
YGSP 46589	Anthracotheriidae	*Merycopotamus nanus*	right M3	Y76	11.4	−10.6	−8.4
YGSP 47146	Anthracotheriidae	*Merycopotamus nanus*	right m3	Y647	12.4	−11.9	−8.6
YGSP 52196	Anthracotheriidae	*Merycopotamus nanus*	left M	Y38	13.5	−10.5	−6.0
YGSP 31899	Anthracotheriidae	*Merycopotamus* sp.	right m3	Y779	10.8	−10.2	−6.0
YGSP 16239	Anthracotheriidae	*Merycopotamus* sp.	right m3	Y540	8.0	−12.0	−5.2
YGSP 49962	Anthracotheriidae	*Merycopotamus* sp.	left P	Y906	7.8	−12.7	−7.1
YGSP 10220	Tragulidae	*Dorcabune nagrii*	left m3	Y309	9.4	−11.2	−8.2
YGSP 11155	Tragulidae	*Dorcabune nagrii*	left M	Y310	9.4	−9.9	−1.6
YGSP 27044	Tragulidae	*Dorcabune nagrii*	right m3	Y224	9.4	−11.0	−6.5
YGSP 50921	Tragulidae	*Dorcabune nagrii*	left P4	Y227	9.4	−10.6	−8.3
YGSP 6857	Tragulidae	*Dorcabune nagrii*	left M	Y227	9.4	−10.6	−2.4
YGSP 18979	Tragulidae	*Dorcatherium* Form 4	right m	Y496	12.4	−10.6	−4.1
YGSP 19068	Tragulidae	*Dorcatherium* Form 4	right m3	Y496	12.4	−11.5	−8.0
YGSP 27675	Tragulidae	*Dorcatherium* Form 5	left m	Y496	12.4	−11.8	−8.2
YGSP 50210	Tragulidae	*Dorcatherium* Form 8	left m3	Y935	7.4	−10.2	−1.7
YGSP 50403	Tragulidae	*Dorcatherium* Form 8	right m3	Y941	7.4	−9.9	−5.6
YGSP 50997	Tragulidae	*Dorcatherium* Form 8	left m3	Y935	7.4	−10.7	−3.1
YGSP 10965	Tragulidae	*Dorcatherium majus*	left m3	Y314	9.3	−10.5	−3.6
YGSP 15126	Tragulidae	*Dorcatherium majus*	right m3	Y450	10.2	−10.9	−6.6
YGSP 16236	Tragulidae	*Dorcatherium majus*	left p or m	Y540	8.0	−10.6	−1.7
YGSP 16270	Tragulidae	*Dorcatherium majus*	left m	Y539	8.1	−8.8	−2.5
YGSP 16845	Tragulidae	*Dorcatherium majus*	right m2	Y450	10.2	−10.5	−5.5
YGSP 47203	Tragulidae	*Dorcatherium majus*	left m3	Y311	10.1	−9.3	−7.1
YGSP 49530	Tragulidae	*Dorcatherium majus*	right m2 or m3	Y888	8.1	−9.8	−2.4
YGSP 499	Tragulidae	*Dorcatherium majus*	right m	Y24	8.2	−10.5	−3.8
YGSP 50639	Tragulidae	*Dorcatherium majus*	right m1 or m2	Y450	10.2	−10.6	−3.9
YGSP 52556	Tragulidae	*Dorcatherium majus*	right m	Y941	7.4	−7.3	−1.9
YGSP 52833	Tragulidae	*Dorcatherium majus*	right p or m	Y981	8.7	−11.4	−3.5
YGSP 5567	Tragulidae	*Dorcatherium majus*	right m1 or m2	Y310	9.4	−11.1	−5.6
YGSP 7251	Tragulidae	*Dorcatherium majus*	right m	Y311	10.1	−11.0	−4.8
YGSP 27733	Giraffoidea	giraffid	left m	Y729	10.4	−12.8	−8.5
YGSP 10099a	Giraffoidea	giraffid	left P4	Y309	9.4	−11.0	−0.2
YGSP 11295	Giraffoidea	giraffid	left m3	Y311	10.1	−10.4	−5.7
YGSP 11298	Giraffoidea	giraffid	molar	Y311	10.1	−12.0	−5.9
YGSP 1159	Giraffoidea	giraffid	right M	Y76	11.4	−11.4	−5.9
YGSP 11678a	Giraffoidea	giraffid	right m3	Y317	9.2	−10.3	−3.3
YGSP 12176	Giraffoidea	giraffid	molar	Y311	10.1	−12.1	−5.8
YGSP 12360	Giraffoidea	giraffid	left M	Y324	8.9	−12.5	−2.1
YGSP 12399	Giraffoidea	giraffid	right P3	Y325	8.8	−11.8	−4.1
YGSP 14171	Giraffoidea	giraffid	right p4	Y455	11.2	−11.7	−2.0
YGSP 14317	Giraffoidea	giraffid	right m1	Y450	10.2	−12.6	−7.4
YGSP 14387	Giraffoidea	giraffid	left p4	Y452	7.3	−10.2	−0.1
YGSP 14395	Giraffoidea	giraffid	right m2 or m3	Y452	7.3	−11.1	2.4
YGSP 15063	Giraffoidea	giraffid	right m3	Y406	8.9	−11.7	0.0
YGSP 15390	Giraffoidea	giraffid	left P2	Y337	9.7	−12.8	−5.6
YGSP 16058	Giraffoidea	giraffid	left P	Y251	10.0	−11.8	−2.6
YGSP 16147	Giraffoidea	giraffid	right P4	Y309	9.4	−11.4	−4.2
YGSP 16249	Giraffoidea	giraffid	molar	Y544	8.7	−11.7	−1.0
YGSP 16759	Giraffoidea	giraffid	left p or m	KL03	9.4	−11.2	−2.2
YGSP 17164	Giraffoidea	giraffid	right m	Y496	12.4	−12.0	−5.1
YGSP 18340	Giraffoidea	giraffid	right M	Y609	8.8	−11.8	0.0
YGSP 18385	Giraffoidea	giraffid	p4	Y605	8.1	−11.5	−0.9
YGSP 18412	Giraffoidea	giraffid	M	Y605	8.1	−10.9	−0.9
YGSP 19756	Giraffoidea	giraffid	left M	Y604	8.6	−12.3	−0.9
YGSP 19846	Giraffoidea	giraffid	left m	Y76	11.4	−12.8	−3.1
YGSP 19860	Giraffoidea	giraffid	left m3	Y76	11.4	−10.0	0.8
YGSP 2025	Giraffoidea	giraffid	left M	Y83	11.6	−12.0	−2.9

(continued)

Table III.2. continued

Specimen number[2]	Family or Order[3]	Name[4]	Tooth sampled	Locality[5]	Age (Ma)[6]	δ13C (V-PDB)[7]	δ18O (V-PDB)[7]
YGSP 20380	Giraffoidea	giraffid	left M	Y640	13.7	−10.7	0.4
YGSP 21903	Giraffoidea	giraffid	left m	Y504	11.6	−11.3	−5.7
YGSP 21905	Giraffoidea	giraffid	left P4	Y504	11.6	−11.6	−5.6
YGSP 21926	Giraffoidea	giraffid	left m3	Y504	11.6	−11.7	−4.9
YGSP 23645	Giraffoidea	giraffid	left M1	Y661	13.4	−11.8	−1.5
YGSP 23646	Giraffoidea	giraffid	left M3	Y661	13.4	−10.6	−4.0
YGSP 23652	Giraffoidea	giraffid	left m	Y661	13.4	−11.2	−4.1
YGSP 23827	Giraffoidea	giraffid	left M	Y640	13.7	−10.9	−6.4
YGSP 24335	Giraffoidea	giraffid	right M3	Y60	12.9	−11.6	−5.9
YGSP 26648	Giraffoidea	giraffid	right M	Y695	12.4	−13.2	−2.1
YGSP 26874	Giraffoidea	giraffid	right P4	Y311	10.1	−11.5	−4.2
YGSP 26940	Giraffoidea	giraffid	left m	Y499	12.1	−11.6	−1.6
YGSP 27674	Giraffoidea	giraffid	left m3	Y495	11.6	−9.7	0.0
YGSP 27860	Giraffoidea	giraffid	right m1 or m2	Y647	12.4	−11.6	−4.9
YGSP 27873	Giraffoidea	giraffid	right P	Y735	11.7	−11.5	−6.8
YGSP 2798c	Giraffoidea	giraffid	M	Y19	8.0	−11.1	−0.3
YGSP 30988a	Giraffoidea	giraffid	left m	Y710	13.0	−11.8	−5.5
YGSP 31102	Giraffoidea	giraffid	right m	Y767	12.4	−11.5	−3.2
YGSP 31214	Giraffoidea	giraffid	left m3	Y691	13.1	−9.4	−2.1
YGSP 31292	Giraffoidea	giraffid	right P2	Y311	10.1	−9.9	−6.8
YGSP 31642	Giraffoidea	giraffid	right M	Y772	12.4	−11.4	−7.7
YGSP 32110	Giraffoidea	giraffid	molar	Y804	11.1	−14.0	−6.5
YGSP 3211a	Giraffoidea	giraffid	right M	Y130	8.0	−11.5	2.4
YGSP 3548	Giraffoidea	giraffid	molar	Y163	9.4	−14.5	−2.1
YGSP 38918	Giraffoidea	giraffid	P	Y699	13.0	−11.1	−3.5
YGSP 38920	Giraffoidea	giraffid	molar	Y547	8.0	−10.6	−5.5
YGSP 38922	Giraffoidea	giraffid	right M	Y504	11.6	−11.3	−2.2
YGSP 38945	Giraffoidea	giraffid	left m	Y258	10.2	−13.2	−1.5
YGSP 38947	Giraffoidea	giraffid	left m2 or m3	Y661	13.4	−11.3	−6.8
YGSP 38948a	Giraffoidea	giraffid	right m1	Y311	10.1	−13.0	0.8
YGSP 40145	Giraffoidea	giraffid	molar	Y403	9.3	−10.2	−2.0
YGSP 40148	Giraffoidea	giraffid	M	Y413	9.4	−12.5	−2.2
YGSP 41348	Giraffoidea	giraffid	right m	Y834	11.7	−11.0	−5.1
YGSP 41384	Giraffoidea	giraffid	right M3	Y698	13.0	−10.2	−6.5
YGSP 41826	Giraffoidea	giraffid	right m	Y478	14.1	−12.0	−0.4
YGSP 4564	Giraffoidea	giraffid	M	Y182	9.2	−12.0	−5.7
YGSP 4602	Giraffoidea	giraffid	left p4	Y193	9.0	−11.7	−6.3
YGSP 46621	Giraffoidea	giraffid	left m3	Y496	12.4	−10.8	−0.4
YGSP 4671	Giraffoidea	giraffid	left P	Y182	9.2	−12.3	−1.5
YGSP 46853	Giraffoidea	giraffid	left m	Y496	12.4	−11.2	−7.1
YGSP 47141	Giraffoidea	giraffid	right m	Y647	12.4	−12.3	−5.7
YGSP 4920	Giraffoidea	giraffid	left p4	Y196	9.0	−11.5	−4.4
YGSP 49576	Giraffoidea	giraffid	left m3	Y406	8.9	−12.3	2.2
YGSP 50504	Giraffoidea	giraffid	right p4	Y311	10.1	−11.1	−3.2
YGSP 50839	Giraffoidea	giraffid	left m1 or m2	Y452	7.3	−11.8	2.4
YGSP 5183	Giraffoidea	giraffid	left m3	Y211	9.3	−12.0	−4.8
YGSP 5232	Giraffoidea	giraffid	left M	Y211	9.3	−11.0	−3.6
YGSP 5694	Giraffoidea	giraffid	right M	Y17	8.0	−14.2	1.3
YGSP 6002b	Giraffoidea	giraffid	left M	Y337	9.7	−12.1	−4.6
YGSP 8787	Giraffoidea	giraffid	left P3	Y236	9.6	−12.6	−1.3
YGSP 8888	Giraffoidea	giraffid	right M	Y252	10.2	−11.8	−0.8
YGSP 8955	Giraffoidea	giraffid	left m	Y251	10.0	−13.0	0.0
YGSP 32872	Bovidae	Hypsodontus pronaticornis	left m3	Y795	12.8	−10.0	−5.8
YGSP 33121	Bovidae	Hypsodontus pronaticornis	right dP4	Y479	13.4	−10.2	−4.6
YGSP 40789	Bovidae	Hypsodontus pronaticornis	right M	Y824	13.1	−10.1	−7.8
YGSP 52444	Bovidae	Hypsodontus sokolovi	right m3	Y478	14.1	−9.0	−1.9
YGSP 40402	Bovidae	Elachistoceras khauristanensis	right m	Y182	9.2	−12.0	−8.5
YGSP 9888	Bovidae	Elachistoceras khauristanensis	right m3	Y260	9.3	−10.1	−4.1
YGSP 20575	Bovidae	Strepsiportax unnamed sp.	left m3	Y640	13.7	−10.4	−4.6

Table III.2. continued

Specimen number[2]	Family or Order[3]	Name[4]	Tooth sampled	Locality[5]	Age (Ma)[6]	δ^{13}C (V-PDB)[7]	δ^{18}O (V-PDB)[7]
YGSP 1431	Bovidae	*Sivoreas eremita*	right P or M	Y69	13.8	−9.8	−6.0
YGSP 11222	Bovidac	*Selenoportax aff. vexillarius*	left M	Y311	10.1	−10.5	−3.2
YGSP 30830	Bovidae	*Selenoportax aff. vexillarius*	left M	Y311	10.1	−11.8	−3.3
YGSP 38929	Bovidae	*Selenoportax aff. vexillarius*	right P4	Y311	10.1	−12.3	−6.5
YGSP 38964	Bovidae	*Selenoportax aff. vexillarius*	right m1 or m2	Y311	10.1	−13.3	−7.9
YGSP 38965	Bovidae	*Selenoportax aff. vexillarius*	right m	Y311	10.1	−12.3	−2.0
YGSP 53566	Bovidae	*Selenoportax aff. vexillarius*	left M3	Y311	10.1	−11.4	−5.8
YGSP 53567	Bovidae	*Selenoportax aff. vexillarius*	right M	Y311	10.1	−11.6	−3.1
YGSP 10610	Bovidae	*Miotragocerus pilgrimi*	left m	Y311	10.1	−9.9	−8.8
YGSP 11593	Bovidae	*Miotragocerus pilgrimi*	right m3	Y312	9.4	−11.2	−1.8
YGSP 12368	Bovidae	*Miotragocerus pilgrimi*	right m	Y325	8.8	−13.1	−7.3
YGSP 12380	Bovidae	*Miotragocerus pilgrimi*	left m2	Y325	8.8	−12.1	−5.5
YGSP 12385	Bovidae	*Miotragocerus pilgrimi*	right M3	Y325	8.8	−11.7	−2.4
YGSP 12402	Bovidae	*Miotragocerus pilgrimi*	left m3	Y325	8.8	−12.1	−7.0
YGSP 12406	Bovidae	*Miotragocerus pilgrimi*	left m3	Y325	8.8	−12.4	−5.6
YGSP 12973	Bovidae	*Miotragocerus pilgrimi*	right m3	Y311	10.1	−11.7	−8.0
YGSP 14103	Bovidae	*Miotragocerus pilgrimi*	right M3	Y311	10.1	−11.5	−2.5
YGSP 15118	Bovidae	*Miotragocerus pilgrimi*	right m3	Y227	9.4	−11.7	0.3
YGSP 38923	Bovidae	*Miotragocerus pilgrimi*	left m	Y406	8.9	−13.5	−6.5
YGSP 38927	Bovidae	*Miotragocerus pilgrimi*	left M	Y311	10.1	−12.8	−3.1
YGSP 38952	Bovidae	*Miotragocerus pilgrimi*	left M	Y260	9.3	−14.1	−7.3
YGSP 38963	Bovidae	*Miotragocerus pilgrimi*	left m3	Y311	10.1	−12.6	−4.8
YGSP 5105	Bovidae	*Miotragocerus pilgrimi*	right m	Y211	9.3	−10.8	−6.6
YGSP 5110	Bovidae	*Miotragocerus pilgrimi*	left m3	Y211	9.3	−9.0	−5.6
YGSP 5127	Bovidae	*Miotragocerus pilgrimi*	right m3	Y211	9.3	−11.6	−5.1
YGSP 5214	Bovidae	*Miotragocerus pilgrimi*	right M1	Y211	9.3	−10.4	−5.3
YGSP 5279	Bovidae	*Miotragocerus pilgrimi*	left m	Y211	9.3	−11.4	−7.9
YGSP 5281	Bovidae	*Miotragocerus pilgrimi*	right M	Y211	9.3	−11.0	−8.0
YGSP 53574	Bovidae	*Miotragocerus pilgrimi*	right m3	Y311	10.1	−10.8	−9.8
YGSP 53599	Bovidae	*Miotragocerus pilgrimi*	right M	Y311	10.1	−12.3	−8.9
YGSP 7310	Bovidae	*Miotragocerus pilgrimi*	right dP4	Y227	9.4	−12.7	0.5
YGSP 15261	Bovidae	*Miotragocerus punjabicus*	left m2	Y17	8.0	−10.6	−1.9
YGSP 18144	Bovidae	*Miotragocerus punjabicus*	left m3	D13	6.5	2.9	3.5
YGSP 18149	Bovidae	*Miotragocerus punjabicus*	right M	D13	6.5	3.3	2.7
YGSP 18210	Bovidae	*Miotragocerus punjabicus*	left m3	Y453	6.5	0.5	−1.3
YGSP 21985	Bovidae	*Miotragocerus punjabicus*	right M3	Y17	8.0	−10.2	−2.7
YGSP 263a	Bovidae	*Miotragocerus punjabicus*	left m3	Y17	8.0	−10.7	−2.5
YGSP 421	Bovidae	*Miotragocerus punjabicus*	right m2	Y17	8.0	−10.6	1.7
YGSP 50418	Bovidae	*Miotragocerus punjabicus*	left M2	Y946	7.8	−10.3	−4.8
YGSP 51111	Bovidae	*Miotragocerus punjabicus*	left p4	Y452	7.3	2.4	−3.3
YGSP 9203	Bovidae	*Miotragocerus punjabicus*	right m2	Y17	8.0	−10.5	−3.3
YGSP 9208	Bovidae	*Miotragocerus punjabicus*	left P or M	Y17	8.0	−10.3	0.1
YGSP 16232a	Bovidae	*Tragoportax salmontanus*	left m3	Y535	8.4	−11.4	−2.7
YGSP 17007	Bovidae	*Tragoportax salmontanus*	left m3	Y545	8.1	−12.0	−1.9
YGSP 1129	Bovidae	boselaphine	left M3	Y76	11.4	−10.3	−4.4
YGSP 1156	Bovidae	boselaphine	left m3	Y76	11.4	−11.7	−7.0
YGSP 17011a	Bovidae	boselaphine	right m2	Y545	8.1	−12.4	−6.5
YGSP 23594	Bovidae	boselaphine	right m3	Y647	12.4	−10.7	−2.5
YGSP 38914	Bovidae	boselaphine	right m	Y545	8.1	−9.0	−0.3
YGSP 38962	Bovidae	boselaphine	right M	Y860	4.8	2.1	−1.8
YGSP 40732	Bovidae	boselaphine	left m3	Y828	13.4	−9.6	−7.8
YGSP 40857	Bovidae	boselaphine	right M	Y502	12.0	−11.6	−5.0
YGSP 41133	Bovidae	boselaphine	left M2	Y647	12.4	−10.0	−3.3
YGSP 41387	Bovidae	boselaphine	left m3	Y661	13.4	−10.2	−5.1
YGSP 41752	Bovidae	boselaphine	left m3	Y478	14.1	−9.4	−6.6
YGSP 50082	Bovidae	boselaphine	right M	Y923	7.2	−7.0	−3.7
YGSP 50942	Bovidae	boselaphine	right M3	Y953	7.9	−10.3	−3.5
YGSP 51687	Bovidae	boselaphine	left M3	Y545	8.1	−9.6	0.7
YGSP 103	Bovidae	*Pachyportax dhokpathanensis*	left m1	Y12	8.2	−10.2	−6.9

(continued)

Table III.2. continued

Specimen number[2]	Family or Order[3]	Name[4]	Tooth sampled	Locality[5]	Age (Ma)[6]	δ¹³C (V-PDB)[7]	δ¹⁸O (V-PDB)[7]
YGSP 14920	Bovidae	*Pachyportax dhokpathanensis*	right m2	Y400	7.8	−13.1	−3.8
YGSP 15190	Bovidae	*Pachyportax dhokpathanensis*	left m	Y14	8.2	−10.2	−6.1
YGSP 15281	Bovidae	*Pachyportax dhokpathanensis*	right M2	KL08	7.8	−11.2	−6.8
YGSP 15338	Bovidae	*Pachyportax dhokpathanensis*	right m3	KL08	7.8	−10.6	−4.5
YGSP 15411	Bovidae	*Pachyportax dhokpathanensis*	P or M	Y444	8.6	−9.5	−2.3
YGSP 17014	Bovidae	*Pachyportax dhokpathanensis*	right m1	Y545	8.1	−11.6	−4.2
YGSP 17017	Bovidae	*Pachyportax dhokpathanensis*	right M	Y545	8.1	−10.4	−3.8
YGSP 17886	Bovidae	*Pachyportax dhokpathanensis*	right M3	Y600	8.1	−10.1	−1.9
YGSP 31402	Bovidae	*Pachyportax dhokpathanensis*	left m2	Y599	8.1	−9.8	−2.4
YGSP 38946	Bovidae	*Pachyportax dhokpathanensis*	left m	Y440	7.5	−7.8	−4.0
YGSP 38951	Bovidae	*Pachyportax dhokpathanensis*	M	KL11	8.0	−10.4	−3.1
YGSP 4087	Bovidae	*Pachyportax dhokpathanensis*	right M	Y174	8.4	−12.0	−3.5
YGSP 4088	Bovidae	*Pachyportax dhokpathanensis*	left m3	Y174	8.4	−11.9	−7.4
YGSP 4785	Bovidae	*Pachyportax dhokpathanensis*	right m	Y201	8.6	−11.7	−4.1
YGSP 49828	Bovidae	*Pachyportax dhokpathanensis*	left m3	Y898	7.8	−9.0	−3.5
YGSP 49993	Bovidae	*Pachyportax dhokpathanensis*	right m2	Y457	7.4	−6.8	−3.1
YGSP 50399	Bovidae	*Pachyportax dhokpathanensis*	right p3	Y941	7.4	−6.6	−2.5
YGSP 50417	Bovidae	*Pachyportax dhokpathanensis*	right m3	Y946	7.8	−9.1	−3.1
YGSP 50865	Bovidae	*Pachyportax dhokpathanensis*	left m3	Y948	7.8	−9.4	−5.5
YGSP 50866	Bovidae	*Pachyportax dhokpathanensis*	left m3	Y948	7.8	−9.9	−3.5
YGSP 50886	Bovidae	*Pachyportax dhokpathanensis*	right M	Y950	7.8	−8.8	−5.6
YGSP 51747	Bovidae	*Pachyportax dhokpathanensis*	right M	Y599	8.1	−10.7	−4.1
YGSP 51812	Bovidae	*Pachyportax dhokpathanensis*	left M3	KL11	8.0	−10.6	−5.7
YGSP 52369	Bovidae	*Pachyportax dhokpathanensis*	left m	Y33	8.0	−10.6	−4.7
YGSP 52573	Bovidae	*Pachyportax dhokpathanensis*	right M	Y965	7.3	−6.4	−3.1
YGSP 52608	Bovidae	*Pachyportax dhokpathanensis*	left m2	Y960	7.8	−10.0	−2.4
YGSP 52670	Bovidae	*Pachyportax dhokpathanensis*	M	Y961	7.8	1.0	−1.5
YGSP 52724	Bovidae	*Pachyportax dhokpathanensis*	right m	Y943	7.4	−9.9	−5.3
YGSP 52787	Bovidae	*Pachyportax dhokpathanensis*	right M	Y980	8.9	−11.7	−5.3
YGSP 52788	Bovidae	*Pachyportax dhokpathanensis*	left M	Y980	8.9	−12.9	−6.2
YGSP 53015	Bovidae	*Pachyportax dhokpathanensis*	right m2	Y950	7.8	−9.5	−5.2
YGSP 54577	Bovidae	*Pachyportax dhokpathanensis*	right m	Y947	7.8	−9.6	−4.5
YGSP 5836	Bovidae	*Pachyportax dhokpathanensis*	right m3	Y214	8.2	−11.0	−4.5
YGSP 10296	Bovidae	*Pachyportax falconeri*	left P3	Y309	9.4	−10.7	−1.9
YGSP 11045	Bovidae	*Pachyportax falconeri*	left m3	Y312	9.4	−10.5	−2.0
YGSP 12337	Bovidae	*Pachyportax falconeri*	left M	Y324	8.9	−11.3	−7.6
YGSP 13126	Bovidae	*Pachyportax falconeri*	left m2	Y286	9.8	−11.7	−7.8
YGSP 3608	Bovidae	*Pachyportax falconeri*	left M	Y162	9.4	−10.7	−3.6
YGSP 38950	Bovidae	*Pachyportax falconeri*	left M	Y260	9.3	−12.7	−0.7
YGSP 6761	Bovidae	*Pachyportax falconeri*	left M	Y221	9.4	−11.0	−2.6
YGSP 38926	Bovidae	antilopine	left m3	Y76	11.4	−12.9	−3.2
YGSP 17887	Bovidae	*Kobus porrecticornis*	left M	Y600	8.1	−9.2	−4.4
YGSP 2817	Bovidae	*Kobus porrecticornis*	right m3	Y17	8.0	−10.2	−1.0
YGSP 49550	Bovidae	*Kobus porrecticornis*	left p or m	Y606	8.1	−9.3	−3.2
YGSP 9207	Bovidae	*Kobus porrecticornis*	left m3	Y17	8.0	−10.8	−2.1
YGSP 461	Bovidae	*Kobus* sp. 1	left m3	Y24	8.2	−11.3	−5.6
YGSP 50824	Bovidae	*Kobus* sp. 1	right m2 or m3	Y452	7.3	−9.4	1.9
YGSP 121a	Bovidae	bovid	right M	Y11	7.9	−13.0	0.1
YGSP 1316	Bovidae	bovid	right M	Y41	13.6	−10.0	−2.8
YGSP 20387	Bovidae	bovid	left m3	Y640	13.7	−7.5	−3.5
YGSP 23061	Bovidae	bovid	left m3	Y76	11.4	−11.1	−10.5
YGSP 23663	Bovidae	bovid	right m3	Y661	13.4	−10.1	−5.0
YGSP 25120	Bovidae	bovid	right M3	Y76	11.4	−10.3	−2.0
YGSP 26798	Bovidae	bovid	left m3	Y690	13.1	−11.8	−8.5
YGSP 31077	Bovidae	bovid	left m3	Y58	13.0	−10.6	−6.6
YGSP 37023	Bovidae	bovid	right m3	Y495	11.6	−10.9	−4.5
YGSP 38924	Bovidae	bovid	molar	Y496	12.4	−12.2	
YGSP 38928	Bovidae	bovid	m3	Y388	8.7	−13.7	−5.7
YGSP 38938	Bovidae	bovid	left m	Y861	4.7	−1.9	−2.0

Table III.2. continued

Specimen number[2]	Family or Order[3]	Name[4]	Tooth sampled	Locality[5]	Age (Ma)[6]	$\delta^{13}C$ (V-PDB)[7]	$\delta^{18}O$ (V-PDB)[7]
YGSP 38939	Bovidae	bovid	left M	Y863	5.8	−0.5	−2.9
YGSP 38940	Bovidae	bovid	left m3	Y860	4.8	0.8	0.6
YGSP 38953	Bovidae	bovid	right M	RH08	5.8	−1.9	
YGSP 38960	Bovidae	bovid	left M	Y861	4.7	−6.5	3.9
YGSP 38961	Bovidae	bovid	right M	Y860	4.8	2.2	−1.7
YGSP 40730	Bovidae	bovid	left m3	Y828	13.4	−11.0	−5.6
YGSP 40837	Bovidae	bovid	right M	Y504	11.6	−11.3	−3.0
YGSP 41388	Bovidae	bovid	right m3	Y661	13.4	−9.9	−7.0
YGSP 41570	Bovidae	bovid	left m	Y847	13.4	−9.8	−5.4
YGSP 47123	Bovidae	bovid	right M3	Y535	8.4	−10.8	−2.8
YGSP 47125	Bovidae	bovid	left M	Y535	8.4	−11.2	−3.5
YGSP 47179	Bovidae	bovid	left m3	Y76	11.4	−10.8	−3.7
YGSP 47387	Bovidae	bovid	left m3	Y883	12.1	−10.6	−8.9
YGSP 51277	Bovidae	bovid	right M	Y83	11.6	−12.0	−11.0
YGSP 51515	Bovidae	bovid	right M3	Y76	11.4	−10.1	−7.8
YGSP 52170	Bovidae	bovid	right m3	Y76	11.4	−10.6	−4.0
YGSP 52450	Bovidae	bovid	right m3	Y478	14.1	−7.1	−4.9
YGSP 52487	Bovidae	bovid	right m1 or m2	Y76	11.4	−11.2	−8.2
YGSP 52491	Bovidae	bovid	right m3	Y478	14.1	−11.5	−6.2

[1] Data verified and compiled (n=742 specimens) by Michèle Morgan. Ninety percent of enamel samples were analyzed at the Stable Isotope Ratio Facility of Environmental Research at the University of Utah under the direction of Thure Cerling, with remaining samples analyzed in the Department of Earth and Planetary Sciences at Harvard University, except for three that were analyzed at the Woods Hole Oceanographic Institute. Most tooth enamel samples were collected by Michèle Morgan and Sherry Nelson between 1992 and 2011, with additional samples collected by M. Soledad Domingo in 2013–2014. For specimens that were sampled more than once, the average value is included. Table includes previously unpublished data as well as data published in Morgan, Kingston, and Marino 1994; Nelson 2005, 2007; Badgley et al. 2008; and Morgan et al. 2009. The contributions of many people to the collecting, identification, and analysis of these fossils are gratefully acknowledged.

[2] Specimen number in the Harvard-GSP database

[3] Higher level taxon attribution

[4] Most resolved taxon name

[5] Locality number in the Harvard-GSP database

[6] Estimated midpoint age of locality using Gradstein et al. (2012) GPTS.

[7] Stable carbon and oxygen isotope values reported relative to the Vienna Pee Dee Belemnite standard. In this table values are not adjusted for changes in $\delta^{13}C$ of atmospheric CO_2 (see Chapter 5, this volume).

24

RECONSTRUCTING SIWALIK MIOCENE PALEOECOLOGY FROM ROCKS AND FAUNAS

ANNA K. BEHRENSMEYER AND MICHÈLE E. MORGAN

INTRODUCTION

This chapter focuses on geological and biological evidence for Siwalik ecosystems and how this evidence (Fig. 24.1) is synthesized into reconstructions of physical environments and vegetation in the Potwar Plateau during the interval between 18 and 6 Ma. Starting with the large-scale physical setting and reviewing global climatic indicators, we next summarize characteristics of the sub-Himalayan alluvial landscape and then incorporate paleoecological evidence from soils, invertebrates, and vertebrates, with additional data on Miocene flora from the eastern-central Siwaliks, where there are well-preserved fossil plant records. These data together with information from modern South Asian ecosystems provide the basis for detailed paleoenvironmental reconstructions of three representative localities during the Middle and Late Miocene (Y496, Y182, and Y581), which are artistically rendered by Julius Csotonyi. We use these portrayals of the ancient habitats and the science behind them to illustrate how different types of evidence can be combined in reconstructing Siwalik paleoecology. The sequence of locality-scale reconstructions also provides a framework for describing major ecological changes over time in the western Siwaliks.

PHYSICAL SETTING OF SIWALIKS IN THE MIOCENE OF SOUTHERN ASIA

Sedimentary rocks known as "the Siwaliks," which stretch along the southern border of the Himalayas for more than 2,000 km, resulted from the filling of a vast, elongate basin formed along the boundary between the Indian subcontinent and southern Asia (Fig. 2.1, this volume; Fig. 24.2). At the western edge of the Himalayas today, several mountain ranges (e.g., Pir Panjal range) extend southward to the Arabian Sea; these began to form with the closure of the Tethys Sea in the Late Eocene. In the Miocene (see Fig. 24.2), the Potwar Plateau region was bounded to the north and west by mountain ranges but was bordered by low elevations toward the south and east. Although they were not as high as today's mountains, many of which exceed 5,000 m in height, these Miocene moun-

tains would have presented barriers to movements of plants and animals, potentially affecting their patterns of immigration and emigration. The rate and timing of Miocene uplift of the Himalayas are the subject of continuing debate (e.g., Hu et al. 2016) although high sediment loads deposited in the Bengal Fan off the eastern shore of India from about 13.5 Ma to the end of the Miocene indicate high rates of sediment erosion associated with large-scale uplift (Pickering et al. 2020; Polissar et al. 2021), and similarly there is evidence for massive sediment input to the Arabian Sea by a precursor of the Indus River (Feakins et al. 2020).

GLOBAL CLIMATE AND OCEAN CIRCULATION BETWEEN 18 AND 6 MA

Although a global climatic cooling trend characterizes the Middle and Late Miocene, several warming periods occurred within the 18–6 Ma time span of this study (Holbourn et al. 2013, 2018, 2022; Lübbers et al. 2019; Westerhold et al. 2020). Uplift of the Himalayas and mountain ranges westward into Europe affected regional and global circulation patterns and created topographic barriers for plants and animals (see Fig. 24.2). Periods of lower sea level, associated with cooler temperatures and expansion of Antarctic ice, also opened new dispersal routes for biota by creating intercontinental land bridges and expanding continental margins (Kominz et al. 2008; Miller et al. 2011; Lübbers et al. 2019; Holbourn et al. 2022). Miocene tectonic events profoundly affected ocean circulation patterns. The dominant westward circum-equatorial current of the Late Paleogene was associated with warm oceans and continents. During the Miocene the circum-equatorial current weakened and broke up, with increased narrowing of both the Panama Seaway between North and South America and the Tethys Seaway between Arabia and Eurasia (von der Heydt and Dijkstra 2006) and a reduction of deep-water flow through the Indonesian Gateway connecting the tropical Pacific and Indian Oceans (Kuhnt et al. 2004). These tectonic changes altered ocean currents and the dynamics of heat transport from the

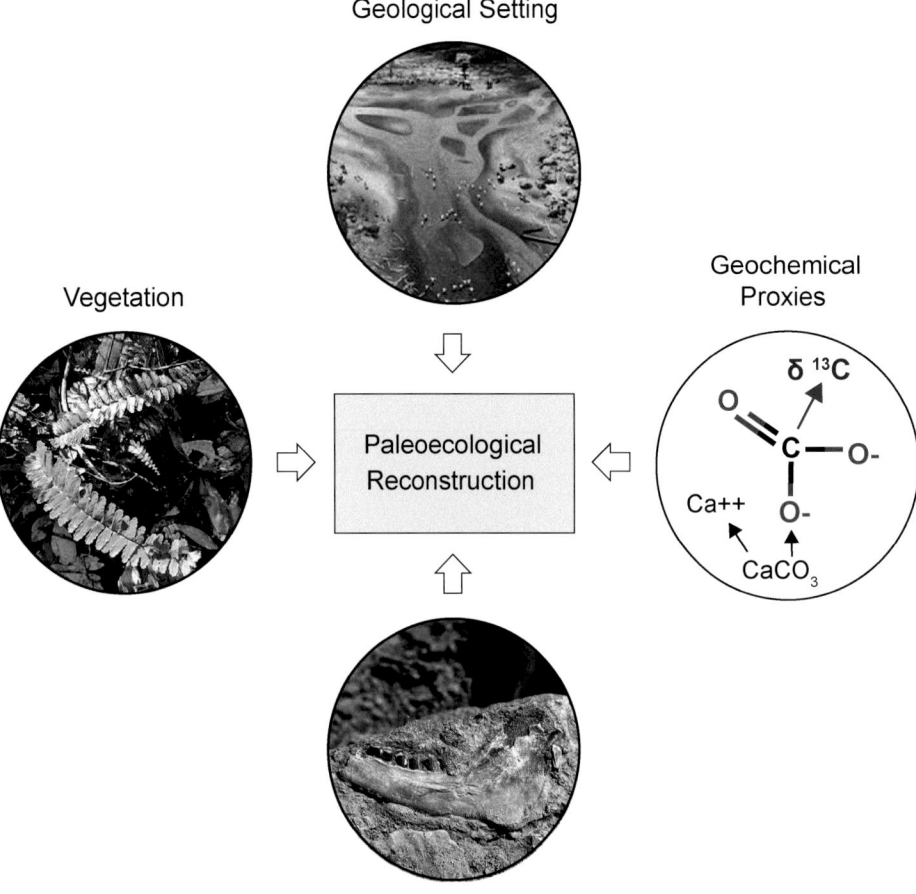

Geological Setting

Vegetation

Geochemical Proxies

Paleoecological Reconstruction

Fossils and Taphonomy

Figure 24.1. Schematic representation of the major categories of evidence that we use in reconstructing Siwalik Miocene paleoecology. See also Table 24.1.

tropics to higher latitudes. A strengthened Antarctic circumpolar current further thermally isolated Antarctica and led to the development of a massive East Antarctic ice sheet that has persisted since the late Middle Miocene. By the end of the Miocene, ocean circulation patterns and temperatures were broadly similar to those observed today (Lyle, Dadey, and Farrell 1995; Fig. 24.3).

Against the backdrop of a general Miocene cooling trend, there were several punctuated warming and cooling episodes that likely affected conditions in the Potwar Plateau region between 18 and 6 Ma (Figs. 24.2, 24.3). Milankovitch cycles and attendant effects on insolation also would have caused shorter-term variations (e.g., over 10^{4-5} years) in regional temperature and rainfall, leading to increases and decreases in the rates of weathering, erosion, and transport of sediment from the rising mountains (Filippelli 1997; Sarin 2001; Laskar et al. 2004; Lupker et al. 2013).

The first of these global climatic episodes is the Middle Miocene Climatic Optimum (MCO), a prolonged period of warmer temperatures between ~17 and 14.7 Ma (Fig. 24.3a). The onset of the MCO is documented by an abrupt decrease in stable carbon and oxygen isotope values of benthic foraminifera from several Integrated Ocean Drilling Program Sites (Diester-Haass et al. 2009; Holbourn et al. 2015). These isoto-

pic shifts are linked to a major perturbation of the global carbon cycle (Holbourn et al. 2015) and increased atmospheric CO_2 levels (Kürschner, Kvaček, and Dilcher 2008) that may be connected to increased volcanic activity (Hodell and Woodruff 1994). Persisting for more than 2 million years, the MCO ends with a return to the Miocene cooling trend starting around 14.6 Ma. Factors contributing to the end of the MCO include a shift to shorter periods of Milankovitch-driven climate variability (e.g., from eccentricity to obliquity cycles and increased marine carbonate accumulation [Holbourn et al. 2014]).

A series of stepped cooling events between 13.9 and 9.0 Ma are documented in the oxygen isotope records of benthic foraminifera from the South China Sea (Holbourn et al. 2013). At 13.9 Ma, there were markedly lower sea surface temperatures (estimates up to 6° C lower) and increased isolation of Antarctica and expansion of the East Antarctic ice sheet (Shevenell, Kennett, and Lea 2004; Holbourn et al. 2013). A second major cooling event occurred at 13.1 Ma, also associated with an increase in ice volume on Antarctica (Holbourn et al. 2013) and coincident with the onset of an equatorial marine carbonate decline, which has been attributed to significant change in continental contributions of calcium and carbonate to oceans associated with mountain-building (Lubbers et al. 2019; Steinthorsdottir et al. 2021). Less pronounced cooling events

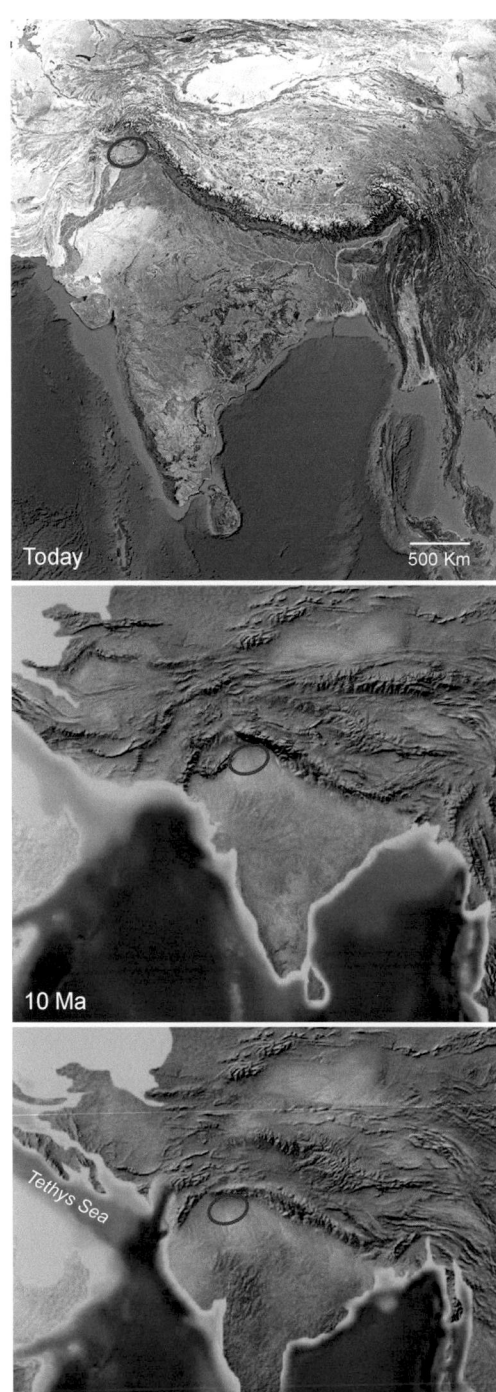

Figure 24.2. Paleogeographic maps of southern and central Asia across 20 million years, showing the seaways and mountain ranges to the west and north of the Potwar Plateau region (red ovals) that helped to shape the evolutionary and ecological history in our fossil record. Land connections and barriers affected faunal movements, as did coastal corridors during times of low sea level. (10 Ma and 20 Ma paleomap credit: Ron Blakey, © 2016 Colorado Plateau Geosystems Inc. Tethys Sea added to map by authors.)

occurred during the Late Miocene at 10.6, 9.9, and 9.0 Ma (Holbourn et al. 2013). Within this pattern of stepped cooling, an intense brief warming period between 10.8 and 10.7 Ma, within a single high-amplitude 100-kyr eccentricity cycle, is recorded in both the South China Sea (Holbourn et al. 2013) and the Indian Ocean (Lubbers et al. 2019).

These climate episodes likely drove fluctuations in sea level, with potentially dramatic consequences for expanding and contracting the areas of continental margins that created and submerged bridges between land areas. Emergent connections between Africa, Europe, southwest Asia, and the Indian subcontinent would have provided routes for dispersal of organisms across and between continents. The presence of interior high mountains and plateaus during the Late Paleogene into the Neogene, from central Europe east to the Indian subcontinent (Fig. 24.2), may have concentrated favorable habitats and dispersal corridors for many species close to coastal margins. During the Miocene, million-year fluctuations in sea-level were typically 5–20 m (Kominz et al. 2008). The rapid cooling event and East Antarctic ice build-up at 13.9 Ma may have lowered sea level by as much as 50 m (Kominz et al. 2016).

The Miocene was a transformative time for terrestrial ecosystems. To what degree did the global-scale climate and ocean circulation changes described above control regional climate patterns, and how did the physiographic effects of the rising Himalaya affect the western part of the sub-Himalayan foreland basin (i.e., our study area)? It is likely that these different scales of earth-system processes contributed to the events and trends recorded in the Siwalik sedimentary and paleontological record. We evaluate ideas about different scales of environmental cause and effect in the following sections to build an overall synthesis of Siwalik paleoecology and how it changed over time.

THE SUB-HIMALAYAN ALLUVIAL PLAIN

Tectonic Setting

The physical landscape sets the stage for the life that colonizes and then sustains itself through ecological processes developed on this landscape. Research on the fluvial lithofacies and architecture by many geologists and sedimentologists over decades of research in Pakistan provides a rich source of information on the complex river systems that flowed into the sub-Himalayan foreland basin and deposited the Siwalik Group. Based on documentation of lateral and vertical patterns in the sediments described in Chapter 2 of this volume, we can say quite a lot about the successive landscapes represented by the four Siwalik formations between 18 and 6 Ma.

The early phase of "docking" of the northward-drifting Indian subcontinent against Asia in the Late Eocene closed the eastern end of the Tethys Sea and initiated uplift of the Himalayas (Leinders et al. 1999; Ali and Aitchison 2008; Hu et al. 2016). There is evidence from the Chitarwata Formation in western Pakistan of marine regression, also represented in an unconformity on Eocene marine rocks (see Fig. 2.2, this

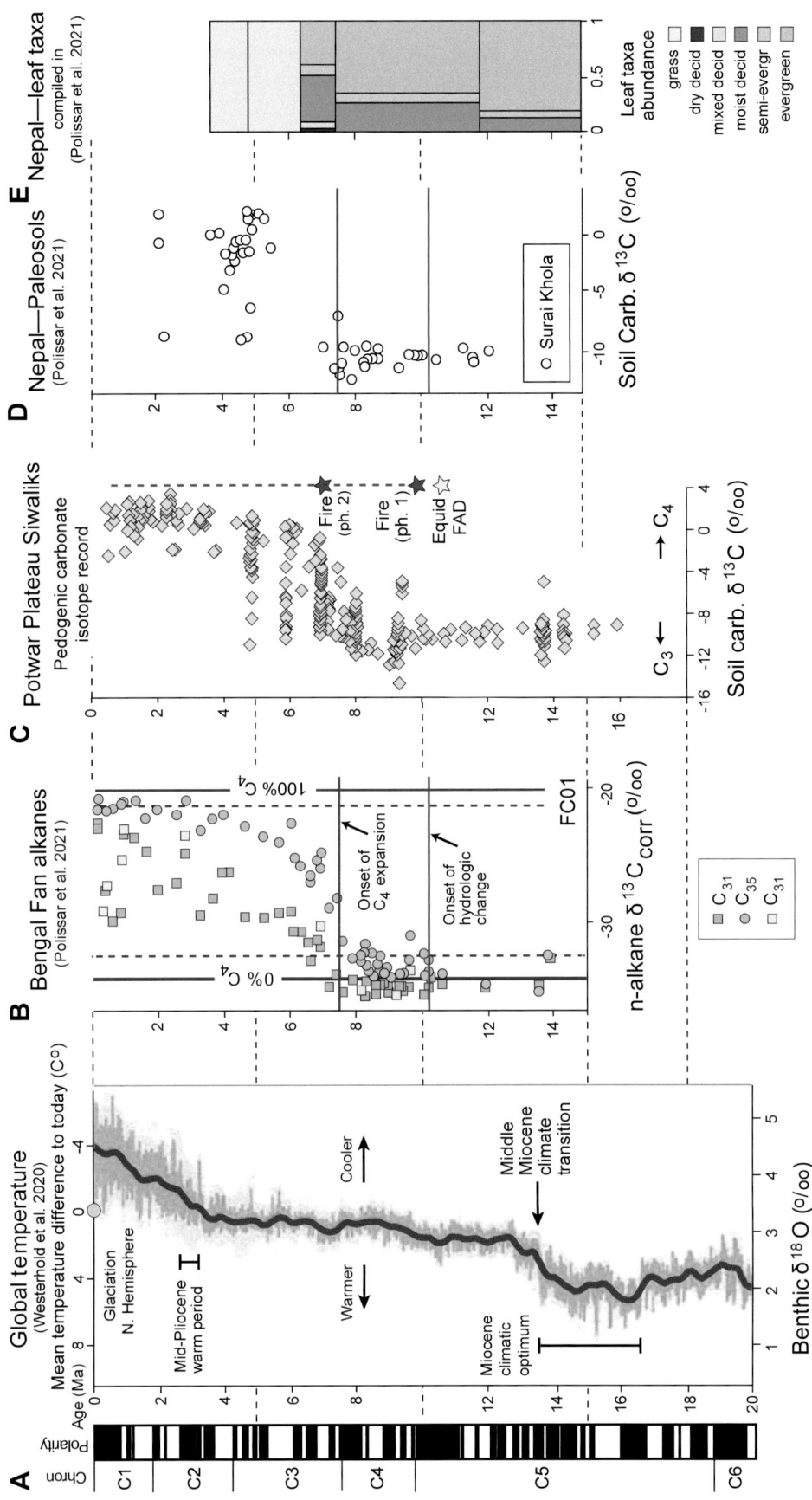

Figure 24.3. Comparison of different climate and vegetation proxies for the time interval coinciding with the Potwar Plateau Siwalik record. A. Global temperature curve based on δ18O records from benthic foraminifera, with lower axis values reversed to reflect warmer temperatures at times of lower δ18O values (adapted from Westerhold et al. 2020). Blue curve shows resampled data smoothed over 20-kyr intervals and red curve data smoothed over 1- myr intervals to reflect different rhythms and trends in Earth's carbon cycle and temperature. Green dot on upper scale indicates the average oxygen isotope composition of the last 10 kyr. B. Environmental trends based on δ13C isotope values in different plant wax (biomarker) n-alkanes from Bengal Fan (IODP Site 1457); FC01 (lower scale bar) indicates n-alkane C31 data from Freeman and Colarusso 2001). The onset of hydrological change toward increased seasonal evaporation is based on δ2H and other proxy records, not shown here; see Polissar et al. 2021, Fig. 5). C. Compiled Potwar Siwalik δ13C data for pedogenic carbonate; see also Fig. 24.7 and Chapter 5, this volume. D. Pedogenic carbonate δ13C data for the Nepal Surai Khola sequence, showing later C3–C4 transition 1,000 km east of the Potwar Plateau (Polissar et al. 2021). (x-axis in B, C, and D in ppm). E. Compiled data showing vegetation change based on leaf taxa preserved in the Surai Khola sequence (see references in Polissar et al. 2021).

volume), that the future Potwar Plateau was temporarily emergent and undergoing erosion during this time (Downing and Lindsay 2005). As the Indian plate was being subducted under Asia, a trough formed along the boundary, becoming the actively subsiding basin that provided accommodation space for sediments eroding from the rising mountains (see Fig. 24.2; also Fig. 2.1, this volume). The balance of sustained subsidence and sediment input maintained this basin as slightly "over-filled," forming a vast subaerial plain with multiple river systems that persisted for millions of years. Based on dominantly southeastern paleocurrent directions (Sheikh 1984; Khan 1993; Willis 1993a, 1993b) and evidence from drill cores in the Bay of Bengal (Polissar et al. 2021), the rivers recorded in our Potwar Siwalik sequence connected to the ancestral Ganges River farther east. However, deep-sea drill cores also provide evidence for a Miocene mountain-sourced drainage flowing south (i.e., the ancestral Indus River), which deposited sediments in the Arabian Sea (Clift et al. 2002; Feakins et al. 2020). This suggests that there were two different trunk rivers, one farther west that formed the ancestral Indus and another connecting eastward to the ancestral Ganges. In any case, the uplift of the Potwar Plateau starting around 3.5 Ma (Fig. 2.3, this volume; Raynolds 1981; Raynolds and Johnson 1985; Raynolds et al. 2022) caused a shift in current direction such that drainages in the upper Dhok Pathan and Soan Formations subsequently flowed southwest and connected with the Indus into the Indian Ocean. The paleochannels preserved in the Potwar Siwalik Group are interpreted as major tributaries to this trunk river because, although large (up to 1 km or more across), they are not on the scale of the Indus or Ganges (Willis 1993a, 1993b; Zaleha 1997).

Depositional Setting

We can visualize the Siwalik foreland basin as a vast, flat alluvial plain with large, braided channels and smaller meandering channels interspersed with broad expanses of floodplain covered with vegetation and scattered bodies of open water or wetlands (Fig. 24.4). The nature of the vegetation is discussed further below, but undoubtedly there were areas with varying tree, bush, thicket, and grass cover forming a mosaic of habitats. Although generally flat, the alluvial plain would have had several meters of local relief adjacent to channels, with slightly higher areas created by crevasse-splay deposits and levees, interspersed with lower areas occupied by active and abandoned channels and floodplain swamps and ponds. The terrain thus would have been characterized by subtle topographic highs and lows that controlled the depth of the water table and therefore the deployment of different vegetation types (Figs. 2.8, 2.10, this volume; Figs. 24.4, 24.5, 24.9–14).

Water availability was an important control on both the physical and biological features of the Siwalik ecosystem. Many lines of evidence point to considerable seasonal variation in the water table as well as periodic major floods that eroded parts of the floodplain to cause avulsions, when an established channel was abandoned and water flow relocated on the floodplain to form a new one. Significant amounts of the finer-grained floodplain sediments were deposited rapidly enough to retain primary bedding structures, without evidence for rooting or other land surface features (Chapters 2 and 4, this volume). Many of these may have been temporary (e.g., seasonal) lakes, as there is little evidence of persistent or deep lacustrine paleoenvironments in the Siwalik floodplain deposits. The presence of crocodilians, fish, and other aquatic animals and algal stromatolites in abandoned channel deposits provides evidence for some bodies of water that were present for decades to (possibly) millennia. The well-developed paleosols, which are often clearly accretionary (i.e., built vertically over time), attest to long periods of stability of subaerial land surfaces. Floods added small increments of sediment during the wet seasons, while intervening dry seasons allowed soil organisms and plants to become active after the flood waters receded. The near absence of preserved carbon in the form of plant remains or other (un-mineralized) organic materials indicates a relatively high level of oxidation within the soils. This is also consistent with large seasonal fluctuations in the water table as well as relatively high soil temperatures expected for the region's subtropical paleolatitude of ~29 degrees (Tauxe and Opdyke 1982).

This fluvial depositional setting resulted in a mosaic of both stable and changing ecological conditions for the plants and animals inhabiting the alluvial plain (see Fig. 24.4), which was sustained over millions of years across the extensive sub-Himalayan region. The successive formations—Kamlial, Chinji, Nagri, and Dhok Pathan—represent large-scale shifts in fluvial conditions within the window provided by the Potwar Plateau outcrops. These formations are also recognized in Siwalik exposures farther east in India, and differences across this wide area likely reflect variable subsidence, changing positions of the major rivers in the foreland basin, fluctuations in sediment input, and movements along the main boundary fault (Willis 1993a, 1993b; see Fig. 2.3, this volume). All along the ~2,000-km rising mountain front, however, the alluvial landscape of the Potwar region would have looked much the same from 18 to 6 Ma, perhaps even up to ~2.0 Ma.

The Potwar Plateau outcrops record two different colors of channel sands, blue-gray and buff-brown, which represent two different river systems draining different parts of the rising Himalaya Mountains (see Fig. 24.4). With the help of paleomagnetic isochrons (Chapter 2, this volume) and differences in the mineralogical composition of the sandstones, we can show that these were contemporaneous and represent mountain- versus foothill-sourced drainage systems (Behrensmeyer and Tauxe 1982; see Fig. 24.4). The blue-gray system was dominated by relatively unweathered sands derived from the high mountain ranges, while the buff sands and associated floodplain sediments derived from the more highly weathered foothills. Both filled up the sub-Himalayan basin and provided plants and animals with different substrates, channel flow patterns, and floodplain soils over distances of tens to hundreds of kilometers. Stable isotope research shows that this scale of lateral environmental variability is reflected in pedogenic carbonates and dietary signals recorded

Figure 24.4. The western Siwalik sub-Himalayan alluvial plain ecosystem at ~9 Ma (similar in age to Locality Y182, Fig. 24.11). The reconstruction is based on scientific evidence discussed in this book and illustrated by Julius Csotonyi. The view is toward the north, with a large, mountain-sourced river entering from the upper left and a smaller, foothill-sourced tributary entering from the upper right. The scene occurs after the end of the rainy season and shows ponding, sediment deposition, and debris from recent flooding, with smoke from a wildfire visible in a drier area on the upper right. Low hills are visible in the near distance and higher mountains in the far distance. A full-size image of this reconstruction is available at the end of the chapter.

by the Siwalik mammals (Chapter 5, this volume; Behrensmeyer et al. 2007; Morgan et al. 2009).

Landscape and Soils

Stable floodplain surfaces between the river channels were subject to soil formation, with biological, chemical, and physical processes typical of alluvial soils modifying the sediment structure and mobilizing nutrients that supported biological productivity. These processes were superimposed on varying mixtures of sands, silts, and clays laid down by annual flooding events. Original bedding structures were overprinted with features typical of soils—massive (unbedded) texture, ped structures, mottling, root and burrow traces, and vertical zonation in elements such as calcium and iron caused by water percolation through the soil (Retallack 1991; Willis 1993a; Behrensmeyer, Willis, and Quade 1995; Zaleha 1997; Retallack 1998). Many of these features involved the actions of micro- and macro-organisms and provide indirect evidence for their important roles in the cycles of life that characterized the floodplain ecosystems.

Pedogenesis occurred over wide range of time intervals, from years to millennia, as is evident from different stages of soil development recorded in the Siwalik strata (Retallack 1991; Cerling and Quade 1993). Some paleosols can be traced for many kilometers along the outcrops, and these represent the most stable, emergent parts of the alluvial plain (Behrensmeyer 1987; Behrensmeyer, Willis, and Quade 1995; see Fig. 2.8, this volume) before grading into wetter areas or channel-margin environments. The sharp upper contacts of many of the paleosols, with minimal evidence of erosion, attest to rapid burial as seasonal flooding and channel avulsion spread new sediment over the slowly subsiding land surfaces. Lateral studies of individual paleosols also show that some of these eventually

converged with under- or overlying soils. This convergence demonstrates that the floodplains were constructed of thin, extensive lobes of fine-grained sediment, which, along with the coarser channel deposits, built up thousands of meters of the Siwalik rock sequence (Willis 1993a; Willis and Behrensmeyer 1994; Behrensmeyer, Willis, and Quade 1995).

The stable soil-forming areas would have occurred on elevated areas of the floodplain, such as surfaces of crevasse-splay lobes, along alluvial ridges near active and abandoned channels, and areas distant from major channels that were not affected by seasonal flooding. These land surfaces supported a patchwork of habitats, some of which were likely linear and roughly parallel to channels that traversed the floodplain (see Fig. 24.4). Their distribution would also have been controlled by the frequency of channel avulsion, which in turn depended on the balance of sediment input and subsidence of the alluvial plain. The slower average rates of sediment accumulation in the Kamlial and Chinji Formations (~14 cm/kyr and ~22 cm/kyr; Table 2.1 and Fig. 2.9, this volume) are associated with more laterally extensive paleosols than in the Dhok Pathan Formation, where deposits of foothill-sourced, small- to medium-scale rivers built much of the regional alluvial plain, and accumulation rates approached 40–60 cm/kyr. Channel avulsion drove fluvial deposition in all Potwar Siwalik formations, but the Kamlial and Chinji supported larger abandoned channels and broader stable land surfaces with well-established vegetation. The Nagri Formation is composed mostly of multistorey sandstones but has large-scale, soil-capped abandoned-channel fills and floodplain deposits that continue the pattern of laterally extensive land surfaces (Willis 1993a; Chapters 2 and 4, this volume). At the scale of ~1 to 5 km, the older alluvial environments were more spatially ho-

Figure 24.5. Key to vertebrate taxa in Fig. 24.4: 1. Amphicyonid, 2. *Sivapithecus sivalensis*, 3. *Deinotherium indicum*, 4. *Caementodon oettingenae*, 5. Machairodontine, 6. *Dorcatherium majus*, 7. *Miotragocerus pilgrimi*, 8. *Bramatherium megacephalum*, 9. Flamingos, 10. *Alicornops complanatum*, 11. *Varanus*, 12. *Stegalophodon*, 13. *Sivalhippus nagriensis*, 14. *Hippopotamodon sivalense*, 15. Catfish, 16. *Crocodylus*, 17. Vulture, 18. Tortoise, 19. *Pachyportax falconeri*, 20. *Kobus*, 21. Ostrich, 22. *Gazella lydekkeri*, 23. *Gavialis*, 24. *Brachypotherium perimense*. See also Table 24.3.

mogeneous than those of the Dhok Pathan Formation. Avulsion of the smaller buff-system channels left behind many abandoned "shoe-string" sand bodies, and these linear channel segments were bordered by raised levees and crevasse-splay deposits that would have been stable for ~10^2–10^3 years, supporting a complex patchwork of disturbed (i.e., low-canopy grass, bush) to late-successional vegetation (i.e., mature forests and woodlands).

Although most of the Miocene paleosols of the Potwar Plateau have soil carbonate, (Chapter 5, this volume; Quade and Cerling 1995), there are numerous examples of well-developed soils lacking carbonate throughout the sequence. These are documented in lateral panels as well as long stratigraphic sections (Khan 1993; Willis 1993a, 1993b; Zaleha 1997; Behrensmeyer et al. 2007) and testify to different environmental conditions that prevented calcium carbonate ($CaCO_3$) nodule formation. Many of these paleosols have iron-manganese (Fe-Mn) nodules formed by seasonal leaching and precipitation of iron hydroxide minerals. The occurrence of such minerals suggests periodic acidic and reducing conditions that either dissolved soil carbonate or prevented its precipitation. Such paleosols could represent climatic conditions with greater rainfall and shorter seasonal intervals of soil drying or wetter areas with vegetation that buffered the land surface from desiccation. These paleosols could also reflect unrecognized landscape variation and climate cycles (e.g., Milankovitch orbital cycles) in the overall trend toward drier and more seasonal conditions through the Miocene Siwalik sequence (see Fig. 24.3).

Pedogenic carbonate precipitation is controlled by transpiration by plants growing on the surface, combined with periods of drying (Cerling et al. 1989; Cerling and Quade 1993). Carbonate-bearing paleosols often have well-defined rhizoliths

that testify to precipitation associated with living or decaying plant roots. Some of the Siwalik rhizoliths are over 1 m in length and 2–4 cm diameter—sizes that indicate the presence of fairly large trees or shrubs and a seasonally-deep water table. Smaller root traces (e.g., mm in diameter) are very common in the paleosols and usually have diffuse gray halos rather than distinct carbonate rhizoliths. Varying patterns of mottling and rhizolith formation may indicate different chemical conditions associated with roots of different sizes, with small roots more often associated with reducing decay processes (Retallack 1990). The frequent co-occurrence of Fe-Mn and $CaCO_3$ nodules, which require different conditions for precipitation, provides evidence for changing water availability during the formation of many Siwalik paleosols, from initially water-saturated and reducing (Fe-Mn) to later drying as the floodplain aggraded ($CaCO_3$), or vice versa.

Geochemical Proxies

Geochemical analysis shows that considerable iron oxidation took place during the formation of the Siwalik soils. The ratio of goethite (G) and hematite (H) preserved in paleosols provides evidence for original soil moisture, with higher ratios indicating more moisture at the time of pedogenesis (see Table 24.1). A pilot study by Abradjevitch et al. (2009) correlated the ratio of G/H with $\delta^{18}O$ (in carbonate) in paleosol samples at three different stratigraphic levels (Chinji Formation at 13.5 Ma, Dhok Pathan Formation at 9 Ma and at 6.8 Ma). Consistent results show that soil samples with higher G/H also have lower $\delta^{18}O$, an indication of lower levels of evaporation during soil formation, which can also reflect soil temperature gradient. In soil profiles in the Rohtas sequence, samples farther below the top of the soil (Abradjevitch et al. 2009) have

Table 24.1. Proxies and Other Evidence Used for Reconstructing Siwalik Paleoecology

Paleoecological Element	Proxies, Evidence	Information Source	Geographic Scale	~Temporal Scale (yrs)	Paleoecological Inferences
Climate	Alkenones, diatom organics, boron isotopes	Deep sea drill cores	Global	10^5 to 10^6	Carbon dioxide levels
	Cyclostratigraphy	Climate modeling	Global	10^5 to 10^6	Cycles in climate variables
	Oxygen isotopes in foraminifera	Deep sea drill cores	Global/continental	10^5 to 10^6	Changes in Temperature and Precipitation over time
	Paleogeography; marine seismic and paleomagnetic data	Deep sea drill cores; tectonics	Global/continental	10^5 to 10^6	Ocean circulation patterns
	Paleogeography	Orogeny; oceanography	Global/continental	10^6	Atmospheric circulation patterns
	Variation in $\delta^{18}O$ in shells, teeth	Invertebrates, mammals	Regional	10^1	Seasonality
	Variation in $\delta^{18}O$ in pedogenic carbonate	Paleosol carbonates	Regional	10^2 to 10^3	Seasonality
	Biomarker PAHs, retene	Paleosols	Regional	10^2 to 10^3	Change in length of dry season over time
Physical Environments	Structural geology, stratigraphy, sedimentology	Geological mapping and analysis	Continental	10^6	Tectonic setting
	Preserved stratigraphic records	Geological mapping and analysis	Continental	10^5 to 10^6	Land connections and barriers
	Lithofacies, paleogeography	Sedimentology, stratigraphy	Local, regional	10^4 to 10^5	Physiography of alluvial plain
	Channel sizes and flow patterns	Field documentation	Local, regional	10^4 to 10^5	Drainage systems and source areas
	Paleosol morphology	Field documentation	Local, regional	10^3 to 10^4	Land surface topography, lateral variability
	G/H ratios (Goethite/Hematite)	Analysis of paleosol Fe (iron-bearing chemicals)	Local, regional	10^3 to 10^4	Soil moisture
	Lithofacies patterns—lateral and vertical	Sedimentology, stratigraphy	Local	10^2 to 10^3	Variation in habitats, topography
Vegetation	Stable isotopes in plant wax biomarkers, PAHs	Bengal and Indus fan drill cores	Global/continental	10^5 to 10^6	C_3–C_4 vegetation, fire
	Nepalese fossil plant record, modern South Asian flora	Macro and micro plant fossils; modern flora	Subcontinental	10^5	Vegetation types, change over time
	Carbon and oxygen isotopic signals	Pedogenic carbonate, biomarkers	Regional	10^2 to 10^3	C_3–C_4 variation in space and time
	Biomarker $\delta^{13}C$, PAHs	Paleosols, sediments	Regional	10^2 to 10^3	C_3–C_4 variation in space and time
	Rhyzoliths, root traces	Paleosols, sediments	Local	10^1 to 10^2	Vegetation—root patterns and sizes
Species Paleobiology	Functional morphology	Anatomy, taxonomic group	Individual	10^5	Likely habitat, locomotion, diet
	Body size estimates	Anatomy	Individual	10^5	Population dynamics, range, metabolic constraints on forage quality/quantity
	Diet: microwear analysis of dentition	SEM, image analysis	Individual	10^1	Diet (herbivores); browser, grazer, mixed-feeder
	Diet: mesowear analysis of dentition	Fossils	Individual	10^1	Diet (herbivores); abrasiveness of forage
	Diet: isotopic analysis of enamel	Fossils	Individual	10^1	Diet: C_3/C_4 forage; water sources
	Diet: hypsodonty	Anatomy	Individual	10^5	Grazing versus browsing adaptation
Faunal Community	Co-occurring taxa, taphonomy at fossil localities	Faunal assemblages	Local	10^2 to 10^4	Composition of faunal and floral community
	Co-occurring taxa within stratigraphic intervals	Faunal assemblages	Western alluvial plain	10^4 to 10^5	Composition of Siwalik metacommunity
	Modern analogues	South Asian ecosystems	Local	10^5 to 10^6	Invertebrates, physiognomy of vegetation

higher G/H ratios. These results also suggest that soil temperatures varied laterally, which could reflect the amount of vegetation cover and shade (Behrensmeyer et al. 2007).

The Siwalik paleosols also preserved biomolecules (biomarkers) from the vegetation that grew on the original soils. The $\delta^{13}C$ in different hydrocarbon molecules (*n*-alkanes; Freeman and Colarusso 2001) showed good agreement with the shift from C_3 to C_4 vegetation prior to ~6 Ma recorded by Siwalik pedogenic carbonate and dental enamel (Quade and Cerling 1995; Nelson 2007; Badgley et al. 2008; Morgan et al. 2009) and also provided a correlation to a similar shift in alkane biomarkers preserved in Bengal Fan drill-core sediments (Tauxe and Feakins 2020; Karp et al. 2021; see Fig. 24.3). Further work on paleosol biomarkers (polycyclic aromatic hydrocarbons, or PAHs) reveals that changes in the frequency of natural fires was part of a transformation of South Asian Miocene floodplain ecosystems (Karp, Behrensmeyer, and Freeman 2018; details below).

EVIDENCE FOR VEGETATION

The Siwalik floral record in the Potwar Plateau region is very limited. Occasional leaf impressions and chunks of fossil wood occur but are too fragmentary to offer significant taxonomic information and ecological context (B. F. Jacobs, pers. comm., 2020). Several efforts to recover pollen and phytoliths have been unsuccessful (Barry 1995; Jacobs, Kingston, and Jacobs 1999; V. Korasidis, pers. comm., 2022). We therefore rely on proxies from isotopic analyses of soil carbonates, biomarkers, and pedogenic features, as well as on floral records from Siwalik sequences in India and Nepal, and biomarker and microfloral remains from Indus and Bengal Fan cores, to reconstruct vegetation patterns and infer temperature and precipitation trends through time. Although vertebrate functional anatomy, dental morphology, and isotopes also can be linked to vegetation, we focus below on the suite of independent proxies listed above to compare with paleoecological interpretations based on faunas.

One thousand km east of the Potwar Plateau, the Surai Khola section of Nepal preserves pollen and plant macrofossils in the Middle and Late Miocene (Figs. 24.3, 24.6; Hoorn, Ohja, and Quade 2000; Prasad 2008; Srivastava et al. 2018; Polissar et al. 2021). Between about 11.5 and 8 Ma, pollen samples indicate predominantly subtropical to temperate broad-leaved forests in the Himalayan foothills and floodplain, with conifers growing at higher altitudes. Grasses, comprising less than a third of the average pollen count, are interpreted as open patches within the forests and woodlands, not as expanses of open grassland (Hoorn, Ohja, and Quade 2000). After 8 Ma, tropical C_3 trees and shrubs are still present in stable areas on the paleo-Gangetic floodplain, but grasslands occupy the more frequently disturbed and poorly drained floodplain regions and become dominant in the latest Miocene (Hoorn, Ohja, and Quade 2000; Fig. 24.3d, e). Soil carbonates from this section record the first evidence of C_4 plants at 7 Ma (Quade and Cerling 1995), but the pollen record suggests that C_3 grasses were relatively common elements of the flora prior to that time.

Two rich paleofloras from the Surai Khola section, one from the late Middle Miocene (between 13 and 11 Ma) and the other from the Late Miocene (younger than 9 Ma) document a shift in the species composition of the flora from wet evergreen taxa (more than half of identified species) to more deciduous taxa (Figs. 24.3, 24.6; Srivastava et al. 2018; Polissar et al. 2021; Adhikari, Srivastava, and Paudayal 2024). Dry deciduous species and a higher frequency of deciduous species are present in the latest Miocene and Pliocene (Prasad and Awasthi 1996; Corvinus and Rimal 2001; Polissar et al. 2021). These floristic changes are consistent with an increase in mean annual temperature of 2–3 degrees Celsius, more seasonal precipitation, and greater fire frequency (Srivastava et al. 2018; Karp et al. 2021). Despite increased seasonality in precipitation caused by reduced winter rainfall sourced from the west, estimated mean annual

Miocene plant record in Surai Khola, Nepal, ~ 1000 km east of the Potwar Plateau

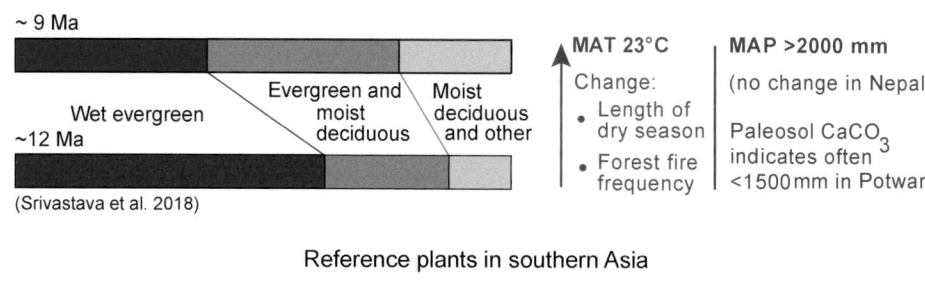

~ 9 Ma

Wet evergreen
~12 Ma

Evergreen and moist deciduous

Moist deciduous and other

(Srivastava et al. 2018)

MAT 23°C
Change:
- Length of dry season
- Forest fire frequency

MAP >2000 mm
(no change in Nepal)

Paleosol $CaCO_3$ indicates often <1500 mm in Potwar

Reference plants in southern Asia

Ficus   *Calamus*   *Shorea*   *Asplenium*

Figure 24.6. Summary of evidence from the plant record (pollen and macroplants) from the Miocene Surai Khola deposits in Nepal, ~1,000 km east of the Potwar Plateau (Srivastava et al. 2018), with images of selected modern plant genera recorded by the Nepalese fossils. The Surai Khola area was wetter and less seasonal over the same time interval when the western sub-Himalayan ecosystems were transitioning to drier conditions (Tauxe and Feakins 2020), and the preserved plant record in this area provides a reference for the vegetation in the Potwar Siwaliks prior to this transition.

precipitation (MAP) in the eastern Siwaliks remained above 2,000 mm throughout the Late Miocene (Srivastava et al. 2018). An increase in the length of the dry season, rather than reduced mean annual precipitation, likely lowered the frequency of evergreen species. Today, these species do not occur in areas with dry seasons regularly exceeding four months (Morley 2000; Eiserhardt, Couvreur, and Baker 2017).

The high frequency of soil carbonate nodules in the western Siwaliks and their absence in the eastern-central Siwaliks are consistent with a pronounced rainfall gradient across the Himalayan foreland basin beginning by 14 Ma, if not earlier (Vögeli et al. 2017). Bulk soil organic carbon values diverged between the eastern-central and western Siwaliks around 7 Ma, a time when C_4 grasses became prevalent in the Potwar region as west-sourced winter precipitation decreased (Fig. 24.3c). Simultaneously, the wetter eastern region continued to support mostly C_3 vegetation (Vögeli et al. 2017). Paleosol carbonate data suggest that the vegetation shift began in the western Siwaliks as early as 7.8 Ma but did not reach the eastern-central Siwaliks (Nepal) until at least 7 Ma, and possibly as late as 6 Ma (Tauxe and Feakins 2020).

Soil carbonate nodules generally indicate mean annual precipitation of less than 1,000–1,250 mm (Quade and Cerling 1995; Retallack 2005). The absence of soil carbonate in many Potwar Plateau paleosols suggests prolonged periods of higher and less seasonal rainfall, perhaps in the range of 2,000 mm, as estimated for the eastern-central Siwaliks, until about 8 Ma. The trend toward more positive values after 8 Ma in the paleosol stable oxygen isotope record (Figs. 24.3, 24.7) is consistent with decreasing water availability on the landscape and longer dry seasons (Quade and Cerling 1995). Variation in the presence of carbonates and other pedogenic features could also reflect precessional cycles, as yet an untested idea. A transition from C_4 to mixed C_3–C_4 vegetation through a single 2 m accretionary paleosol profile in the Rohtas section, dated at 4.9 Ma (Behrensmeyer et al. 2007, Fig. 9b), suggests that some soils aggraded at a rate appropriate for recording portions of precession cycles, although other factors such as plant succession could be involved. This example also attests to the continuing presence of C_3 vegetation into the Pliocene in the western Siwaliks.

Soil biomarkers, particularly polycyclic aromatic hydrocarbons (PAHs), indicate the frequency of fires on the landscape. An increase in fire frequency occurred between 11.5 and 10 Ma (Karp, Behrensmeyer, and Freeman 2018), well before the establishment of C_4 grasslands on the floodplain in the latest Miocene (Quade. Cerling. and Bowman 1989). This occurred in two distinct phases (Fig. 24.3c): the first involved burning of conifers and possibly C_3 grasslands, and the second coincided with burning during the transition to C_4 grasslands between 8 and 6 Ma (Karp, Behrensmeyer, and Freeman 2018). After 10 Ma, the presence of conifers in the drainage basin, as measured by concentration of the diagnostic organic molecule retene, decreases. Karp, Behrensmeyer, and Freeman (2018) posit an accompanying increase in moist deciduous plant species including grasses, largely C_3 grasses, although C_4 grasses may also

have been present. High molecular weight n-alkanes from Siwalik paleosols suggest an increased presence of C_4 plants as early as 9 Ma (Freeman and Colarusso 2001), while carbon-isotopic values of mammalian tooth enamel from the region indicate C_4 foraging in equids by 10 Ma (Morgan, Kingston, and Marino 1994). Variation in rainfall patterns, on multiple temporal scales, likely would have kept the C_3–C_4 composition of floristic elements on the landscape in flux throughout the 3-million-year isotope transition interval.

Biomarkers in Bengal and Indus Fan cores also document an increase in fire frequency and presence of C_4 grasses in sub-Himalayan regions during the Late Miocene (Feakins et al. 2020; Tauxe and Feakins 2020; Karp et al. 2021; Polissar et al. 2021). In the Bengal Fan, plant waxes show greater seasonality after 10 Ma (Polissar et al. 2021), while analyses of PAHs and plant waxes indicate marked increase in fire frequency and C_4 vegetation at 7.4 Ma (Karp et al. 2021) (Fig. 24.3c). Fire is interpreted as a contributing factor in opening and maintaining C_4 grasslands in South Asia (Karp et al. 2021). Geochemical analyses of several organic and inorganic markers from an Indus Fan core indicate that the C_4 transition in peninsular India began at about 7.2 Ma (Tauxe and Feakins 2020), with the earliest traces occurring between 9.0 and 8.7 Ma (Feakins et al. 2020).

FAUNAL EVIDENCE

The rich mammalian fossil record, together with less common but critical non-mammal vertebrates and rare invertebrates, provides important clues about the landscape and vegetation. Ecomorphological indicators (traits) based on species anatomy and phylogenetic relationships (see Table 24.1) are the basis for inferences about diet, locomotion, and shelter, which allow us to identify key attributes of the Miocene ecosystems that must have been present to sustain these animals. For example, at large spatiotemporal scale, a high diversity of browsing megaherbivore taxa (i.e., species greater than 800 kg in size) requires abundant forage (e.g., trees, shrubs, and forbs). Species richness within feeding guilds is also related to habitat scale and productivity as well as to the dietary or other specializations of species occupying the habitat, which shape the trophic relationships of the faunal paleocommunity at any given time. Broad characteristics of the Siwalik mammalian fauna and their trends through time are addressed in detail in Chapters 25 and 26 of this volume; here we summarize key environmental proxies based on taxa in our fossil assemblages (see Table 24.1, with time-slice and locality-based examples in Tables 24.2 and 24.3).

In the absence of direct floral evidence, we focus on reconstructing the diets of the most abundant large-mammal group, the ungulates, as evidence for Siwalik vegetation. Differences in the relative proportions of trees, shrubs, and grasses on the landscape and the contribution of cool-season C_3 grasses and warm-season C_4 grasses and sedges are recorded in ungulate tooth and jaw morphology, tooth-wear patterns, and the stable isotope composition of their dental enamel (Chapter 5, this volume). Dietary analyses also inform us about species biology

Table 24.2. Environmental Context, Specimen Data, and Faunal Information for Three Featured Artistic Reconstructions of Siwalik Paleocommunities by Julius Csotonyi

	Y496	Y182	Y581	
Formation	Chinji	Dhok Pathan	Dhok Pathan	
Age	12.4 Ma	9.2 Ma	7.2 Ma	
Depositional setting	Large abandoned channel	Floodplain channel	Small tributary to larger channel	
Season	~2 months after end of rainy season	Near end of rainy season	Middle of dry season	
Time of day	Early afternoon	Early morning	Mid-morning	
# Catalogued fossils	1071	955	148	
# Mammal taxa	60	50	20	
Small	22	17	5	
Large	30	28	13	
Mega	8	5	2	
# Reptile taxa	6	7	4	
# Avian taxa	1	1	2	
# Amphibian taxa		1		
# Fish taxa		1	1	
# Invertebrate taxa	1	3		

Featured Taxa				Common Name
Fauna				
Scandentia	**Tupaiid**			Treeshrew
Primates	*Sivapithecus indicus*	*Sivapithecus sivalensis*		Ape
			Mesopithecus sivalensis	Monkey
		Nycticeboides simpsoni		Loris
Rodentia	*Tamiops* **large species**	*Ratufa sylva*		Squirrel
		Miorhizomys nagrii		Bamboo rat
			Hystrix primigenia	Porcupine
Lagomorpha			*Alilepus elongatus*	Rabbit
Creodonta	*Hyainailouros sulzeri*			Hyaenodont
Carnivora	*Vishnuonyx chinjiensis*			Otter
		Anatolictis	*Anatolictis*	Mongoose
		Vishnuictis chinjiensis		Civet
		Percrocuta grandis	**Percrocuta grandis**	Hyaena-like carnivore
		Machairodontine		Sabertooth cat
Proboscidea	*Zygolophodon metachinjiensis*	**Deinotherium indicum**	**Stegolophodon**	Proboscidean
Tubulidentata		*Orycteropus*		Aardvark
Artiodactyla	**Listriodon pentapotamiae**	*Hippopotamodon sivalense*	**Propotamochoerus hysudricus**	Suid
			Merycopotamus dissimilis	Anthracothere
	Dorcabune anthracotheroides	*Dorcatherium majus*	**Dorcatherium**	Tragulid
		Giraffa	*Giraffa*	Giraffe
		Bramatherium megacephalum		Giraffid
	Elaschistoceras khauristanensis		**Tragoportax salmontanus**	Bovid
		Pachyportax falconeri	*Pachyportax dhokpathenensis*	Bovid
		Gazella lydekkeri	**Kobus**	Bovid
Perissodactyla	*Anisodon salinus*			Chalicothere
		Sivalhippus nagriensis	*Sivalhippus perimensis*	Equid
			Sivalhippus anwari	Equid
	Brachypotherium perimense		*Brachypotherium perimense*	Rhinocerotid
	Gaindatherium browni			Rhinocerotid
Reptiles		**Varanid** (*Varanus*)	*Varanus*	Lizard
		Scincidae (skink)	Agamid (cf. *Calotes*)	Lizard
	Python	*Python*	*Python*	Snake
	Acrochordus dehmi	**Acrochordus dehmi**	*Bungarus*	Snake
		Daboia	**Colubroid** (*Sivaophis*)	Snake
	Chitra	Geochelone		Turtle
	Nilssonia	*Nilssonia*		Turtle
	Batagur	*Geochlemys*		Turtle
			cf. *Indotestudo*	Tortoise
	Crocodylia (*Rhamphosuchus*)	**Crocodylia** (*Crocodylus*)	**Crocodylia** (*Crocodylus*)	Crocodilian
	Crocodylia (*Gavialis*)	**Crocodylia** (*Gavialis*)		Crocodilian

(continued)

Table 24.2. continued

Featured Taxa				Common Name
Birds	**Ciconid** (stork)	Ciconid (stork)	**Tadornini** (shelduck)	
		Pelecanus		Pelican
	Alcedinidae (kingfisher)		**Otidae** (bustard)	
Amphibian	Anuran (tree frog)			
Fish	***Chaca pachyptreygia***			Catfish
Invertebrates	*Pila*			Snail
		Bivalvia		
	Ants	Termites	Dung beetles	
	Spider	Beetle		
	Butterfly	Wasp		
	Beetle			
	Millipede			

Notes: Taxa shown in bold are documented by fossils at the locality and other taxa represent fossil occurrences in similar-age localities.; italics/formal name = identified fossil specimens. Some taxa (e.g., invertebrates) are based on modern analogues in southern Asian ecosystems (see text; Figs. 24.9–24.14).

and trends within lineages, topics addressed in faunal chapters and synthesized in Chapter 25 of this volume.

Teeth play prominent roles in both the procurement and initial processing of food for most vertebrates. Tooth morphology and wear patterns, combined with jaw structure, reveal texture and toughness of food, as well as abrasiveness that may be caused by adhering sediment particles rather than the food itself (Zliobaite et al. 2018; Fortelius et al. 2019). For example, dental mesowear studies of Siwalik equids and bovids suggest increased abrasion in specimens younger than 10 Ma as compared to those older than 10 Ma (Morgan et al. 2015). Sources of increased abrasion include phytoliths in grasses and grit adhering to forage that would be expected in drier, more seasonal conditions.

Isotopic analyses of herbivore enamel indicate relative contributions of C_3 and C_4 plants to an individual's diet and can also identify obligate drinkers (Levin et al. 2006). In the earlier part of our record, 18–13 Ma, Siwalik herbivores were feeding on C_3 vegetation, which likely included C_3 grasses (monocots) as well as trees and shrubs (dicots; Fig. 24.7). Most genera that foraged on C_3 plants show isotopic variation in carbon on the order of 4–5‰ per mil over a million years, an amount comparable to the variation observed within serially sampled individuals over a period of a year (see Chapter 25, this volume). This individual/species similarity in isotopic variation suggests forage preferences that remained relatively stable for very long periods of time. Observed isotopic shifts through time in Siwalik herbivores may partly reflect external influences on plant isotopic values such as atmospheric carbon dioxide levels and water stress, in addition to changing forage preferences (Badgley et al. 2008; see Fig. 24.7). After 10 Ma, equids expanded their isotopic niche breadth by adding C_4 grasses, as did some other groups after 9 Ma (see Fig. 24.7).

Species-level isotopic data are presented in Chapter 25 of this volume and in some faunal chapters; here we divide taxa into four groups, smaller mammals (species < 100 kg), large mammals (species 100–800 kg), megaherbivores (species > 800 kg), and equids (see Fig. 24.7). Vegetation predictions follow a model developed by Magill, Ashley, and Freeman (2013) that estimates percentage of C_4 vegetation and C_3 woody and non-woody vegetation from soil organic matter and leaf lipid biomarkers. Prior to 9.8 Ma, paleosol carbonates indicate mainly woodland, while sampled herbivores record isotopic values indicating a mixture of forest and woodland vegetation. Forest and woodland vegetation continue to dominate until 8 Ma, although wooded-grassland habitats are indicated after 9.5 Ma by equids and by paleosol carbonates at 8 Ma (Polissar et al. 2021). At 7 Ma and thereafter, there are no isotopic indicators of forest vegetation in soils with pedogenic carbonate, and woodland and wooded-grassland vegetation dominated, with some open grassland vegetation. Most herbivores foraged on a mixture of C_3 and C_4 vegetation (Badgley et al. 2008; Chapter 25, this volume).

Non-mammalian vertebrates provide additional information about the paleoecology of the Siwaliks, although sample sizes are relatively small. Bird fossils for the entire collection number only 141 specimens, plus 34 eggshell fragments, but diversity is relatively high (Chapter 9, this volume). Larger taxa are most common, and a range of diets is represented. The avifauna is dominated by water-associated herbivores, omnivores, carnivores, and piscivores, as would be expected based on higher probability of preservation in aquatic settings. Land herbivores include a ratite (ostrich—based on eggshells), a second large, likely flightless taxon, and a bustard likely associated with open habitats. The only large birds of prey preserved in our record include a vulture and an eagle. Between 12.8 and 9.3 Ma, the avifauna includes pheasants and flamingos, which indicates both forested and saline aquatic habitats in a heterogeneous mosaic with some open areas as well. Ostrich appears in the record at ~9.4 Ma, and the loss of forest-adapted birds is consistent with the general Siwalik trend toward open, drier environments. In spite of this, however, Dhok Pathan Formation localities include wetland and aquatic-adapted species, which testifies to the continued availability of mesic habitats in an increasingly C_4-dominated landscape (as shown below in Fig. 24.13 for Locality Y581). The Siwalik herpetological fossil record represents both aquatic and terrestrial environments and

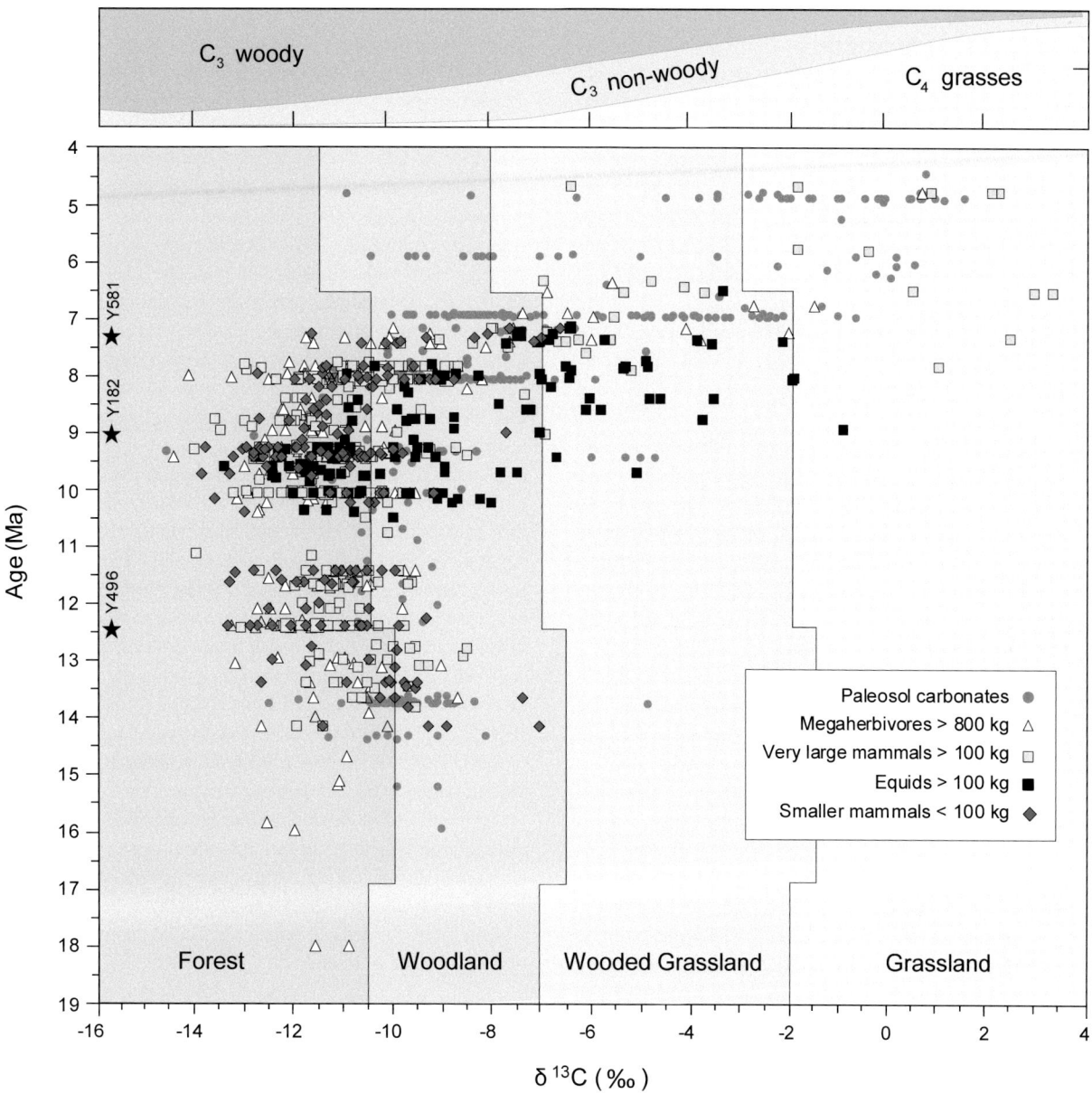

Figure 24.7. Compiled $\delta^{13}C$ data for Potwar Siwalik paleosol carbonates and mammalian enamel (see Table III.2) showing ranges of variation for successive time intervals and overall change through time across the C_3–C_4 transition between 10.0 and 6.5 Ma. Equids were the earliest group to show a transition to C_4 diet (Badgley et al. 2008). Vertical lines mark shifts in atmospheric carbon dioxide values (as discussed in Chapter 5, this volume; Magill, Ashley, and Freeman 2013). Stars show the stratigraphic positions of the three localities used for detailed paleoecological reconstructions. Note that $\delta^{13}C$ in pedogenic carbonates indicates that woodland persisted after 7 Ma, but very large and megaherbivore enamel $\delta^{13}C$ data record only wooded grassland and grassland.

provides important evidence for the paleoecology of the faunal communities. Frog remains were recovered along with other small vertebrates in screen-washing operations but have not yet been studied. Crocodilians, turtles, and squamates (lizards and snakes) are all well represented (Chapter 8, this volume). The persistence of both crocodilians and aquatic turtles throughout the study interval (18–6 Ma) supports the continuing presence of permanent aquatic ecosystems. The loss of large, piscivorous crocodilians (e.g., >8 m in length) and snakes in the Nagri and Dhok Pathan Formations is interpreted as evi-

dence for a decrease in the scale and permanence of water bodies (channels, lakes) in the Siwalik fluvial systems. The reptile record documents a trend toward greater taxonomic diversity among snakes, as well as increasing occurrences of large testudinids and large *Varanus* lizards higher in the Siwalik sequence. These observations are consistent with the ecosystem shift to more heterogeneous, open, and seasonal environments (Chapters 2, 5, 25, this volume), which would have provided increased opportunities for grazing in terrestrial testudinids and basking for squamates and turtles.

Siwalik fish fossils (Chapter 6, this volume) provide valuable information about environments and freshwater fishes of southern Asia, including the first Neogene records of the freshwater ray family Dasyatidae and the catfish family Chacidae. These two families occur in modern Ganges-Brahmaputra drainage systems but not in the modern Indus, supporting the eastward flow of Siwalik fluvial systems during the Miocene, and a separate, south-directed drainage system for the ancestral Indus (Chapter 2, this volume). The fossil record includes large fish individuals (up to 1 meter length), which indicate large bodies of water. The abundance of two obligate air-breathing taxa, *Clarias* and *Channa* (Graham 1997), which encase themselves in mud during dry periods, suggests that ephemeral aquatic environments such as abandoned channels may have enhanced preservation of these taxa.

The invertebrate record in our Potwar Siwalik sequence consists of a limited sample of fossil mollusks (gastropods, bivalves) collected in depositional settings that also include vertebrate remains. The "ear-shaped" opercula of the genus *Pila* (apple snails) occur at numerous localities throughout the study interval and are particularly common between 11.7 and 9.1 Ma (see Fig. 6.1, this volume). Strong growth banding in the well-preserved calcitic opercula, which reach up to 4 cm in length, suggests that they record seasonality in water availability. Modern species of this semiaquatic snail in southern Asia are known to aestivate during the dry season (Prashad 1925), and Gurung (Chapter 6, this volume) regards the occurrence of this genus as evidence of aquatic habitats subject to seasonal drought. Four opercula from the 9.8 Ma time interval (survey block KL20; see Chapter 3, this volume) were sampled for $\delta^{18}O$ isotope fluctuations to test for variable growth due to wet-dry seasonal cycles (N. Loughlin, pers. comm., 2014). Preliminary evidence supports marked seasonality in the Miocene habitats occupied by *Pila*, although there is considerable variability among individuals. Bivalves in the family Unionidae occur throughout the sequence (14–6 Ma) and indicate perennial freshwater aquatic habitats (river channels, ponds; Chapter 6, this volume). The larger taxa and individuals (e.g., *Pseudodon*) occur in the Chinji Formation but are rare later on, which suggests more favorable, perhaps less seasonal aquatic habitats for unionids prior to 11 Ma.

PALEOECOLOGICAL SYNTHESIS

Bringing together the many lines of evidence presented above enables detailed reconstructions of the ecology of the sub-Himalayan alluvial plain between 18 and 6 Ma. As examples of this synthesis, we focus on three "snapshot" reconstructions based on individual fossil localities dated at 12.4, 9.2, and 7.8 Ma, respectively. Produced in collaboration with artist Julius Csotonyi, these portrayals of places and times are evidence-based, realistic views of landscapes with plants and animals (Figs. 24.7–24.14). The taphonomy of the fossil assemblages at these localities indicates that accumulations of skeletal remains resulted from attritional deaths with limited transport by fluvial processes. Each assemblage is a time-averaged faunal sample accumulated over intervals of ~10^3 years by a variety of taphonomic processes (Chapter 4, this volume). Key species of small, large and mega mammals, birds, non-avian reptiles, and fishes were identified for each reconstruction, and most of these taxa are represented by fossils found at the locality (Table 24.2). In the absence of preserved plant remains, we used fossil plant records from the Nepalese Miocene along with present-day South Asian plant taxa to reconstruct the vegetation (Srivastava et al. 2018). Stable isotopes and biomarkers through the Siwalik sequence indicate the likely proportion of C_3 versus C_4 vegetation, seasonality, and the wildfire in the ecosystem for each of the three localities (see Figs. 24.3, 24.7, Tables 24.1–3). Reconstructions also include arthropods, fungi, and other organisms that were not directly recorded but would be expected in such habitats, based on modern analogues.

The sequence of reconstructions spans 5 million years and illustrates the transformation of the Siwalik landscape from predominantly closed forest and wooded to open and grassy vegetation (Fig. 24.8). This transformation is recorded by the C_3–C_4 stable isotope transition and by changes in the vertebrate fauna. Each scene includes 20 to 30 species, and all major Siwalik mammalian clades and many other vertebrate groups are visible in at least one of the three scenes. Within each time interval, other contemporaneous fossil localities represent slightly different samples of the ecological communities. The paleoecological information from all localities within each time slice documents the different habitats and fossil-preserving environments of the Siwalik alluvial plain. When examined in chronological order, these time slices represent the major changes in Siwalik paleoecology over our 12-million-year study interval (see Chapters 25–26, this volume).

Our goals for science-based artistic reconstructions of the three target localities are to

- feature diversity of species present in Siwaliks, including representatives from all faunal chapters in this book— families and orders now extinct as well as those present today
- convey ecological change during the Miocene, from mostly forest and woodland to mostly woodland and grassland
- feature ecologically representative settings on the alluvial plain during different seasons and times of day
- take into account taphonomic processes that accumulated and modified the preserved faunal remains
- use artistic perspective to provide views of both large and small organisms
- create realistic scenes that show connections and interactions among the organisms, the vegetation, and the physical habitat of the local community.

12.4 Ma Y496 Reconstruction

Locality Y496 is a rich fossil locality dated to 12.4 Ma (see Table 24.2; Figs. 24.8, 24.9, 24.10; see also Fig. 4.3a, this volume). The fluvial strata associated with the fossils represent a large, abandoned channel that filled slowly with periodic in-

Table 24.3. Environmental Context and Faunal Information for Aerial View Reconstruction of the Siwalik Alluvial Plain Paleocommunity at about 9.2 Ma by Julius Csotonyi

Formation	Dhok Pathan
Age	9.2 Ma
Depositional setting	View of Siwalik alluvial plain with foothills and mountains to the north
Season	End of the rainy season
Time of day	Late morning

Featured taxa

Fauna	Taxon	Common Name
Primates	*Sivapithecus sivalensis*	Ape
Carnivora	Machairodontine	Sabertooth cat
	Amphicyonid	"Bear dog"
Proboscidea	*Stegalophodon*	Proboscidean
	Deinotherium indicum	Proboscidean
Artiodactyla	*Hippopotamodon sivalense*	Suid
	Dorcatherium majus	Tragulid
	Bramatherium megacephalum	Giraffid
	Miotragocerus pilgrimi	Bovid
	Gazella lydekkeri	Bovid
	Pachyportax falconeri	Bovid
	Kobus	Bovid
Perissodactyla	*Sivalhippus nagriensis*	Horse
	Alicornops complanatum	Rhinocerotid
	Caementodon oettingenae	Rhinocerotid
	Brachypotherium perimense	Rhinocerotid
Reptiles	*Varanus*	Lizard
	Tortoise	
	Crocodylia (*Crocodylus*)	Crocodilian
	Crocodylia (*Gavialis*)	Crocodilian
Birds	Ostrich	
	Vulture	
	Flamingo	
Fish	Catfish	

Note: List of taxa based on those occurring at Y182 and similar-aged localities (see Table 24.2 and Figs. 24.4–24.5).

fluxes of coarse- to fine-grained sediment (Chapter 4, this volume). Skeletal remains accumulated over time in a depression left by this channel, where occasional flooding and temporary flow helped bury and preserve the remains. The abandoned channel likely retained ponded water during the dry season, and associated lush vegetation also served as an attractor for herbivores and their predators. Two other fossil localities at the same stratigraphic level occur within several hundred meters of Y496 and contribute to our understanding of the fauna and habitats of that time.

Time of day and season: The scene depicts an early afternoon shortly after the end of the rainy season, with water levels still high in river channels and ponds in abandoned channels.

Vegetation: Some plants are flowering, and the evergreen vegetation is lush and green. Dense trees and shrubs are consistent with isotopic data indicating forest and woodland vegetation during this time. Our choice of specific plants includes *Shorea*, *Ficus*, *Calamus*, and *Asplenium*, as well as leaf types, epiphytes, and fungi and is based on modern ecosystems in southeast Asia.

Fauna: We included vertebrates that broadly resemble living species as well as members of extinct groups such as creodonts, chalicotheres, and mammutids (see Table 24.2). Twelve mammal species are depicted, representing 12 of 60 species documented at the locality. Only 2 of the 22 small mammal species are featured in the scene—both arboreal. The forest floor and open patches would have been inhabited by cricetids and murids. Twelve bovid and tragulid species occur at Y496; the scene features the smallest bovid (<10 kg) and the largest tragulid (>50 kg). The three megaherbivores, two rhinocerotids, and one proboscidean represent only 25% of megaherbivore species richness at 12.4 Ma. The scene also shows a variety of invertebrates that would be expected in a moist, densely vegetated environment.

Overall Visualization: The illustration provides an oblique downward view from the lower canopy toward a small stream with dense thickets and trees, which opens in the distance into a sunny glade where a mammutid proboscidean browses. Water is flowing slowly in the channel segment on the right. Most of the scene is a shaded, damp woodland floor with a smaller abandoned channel pond on the left and overhanging bushes

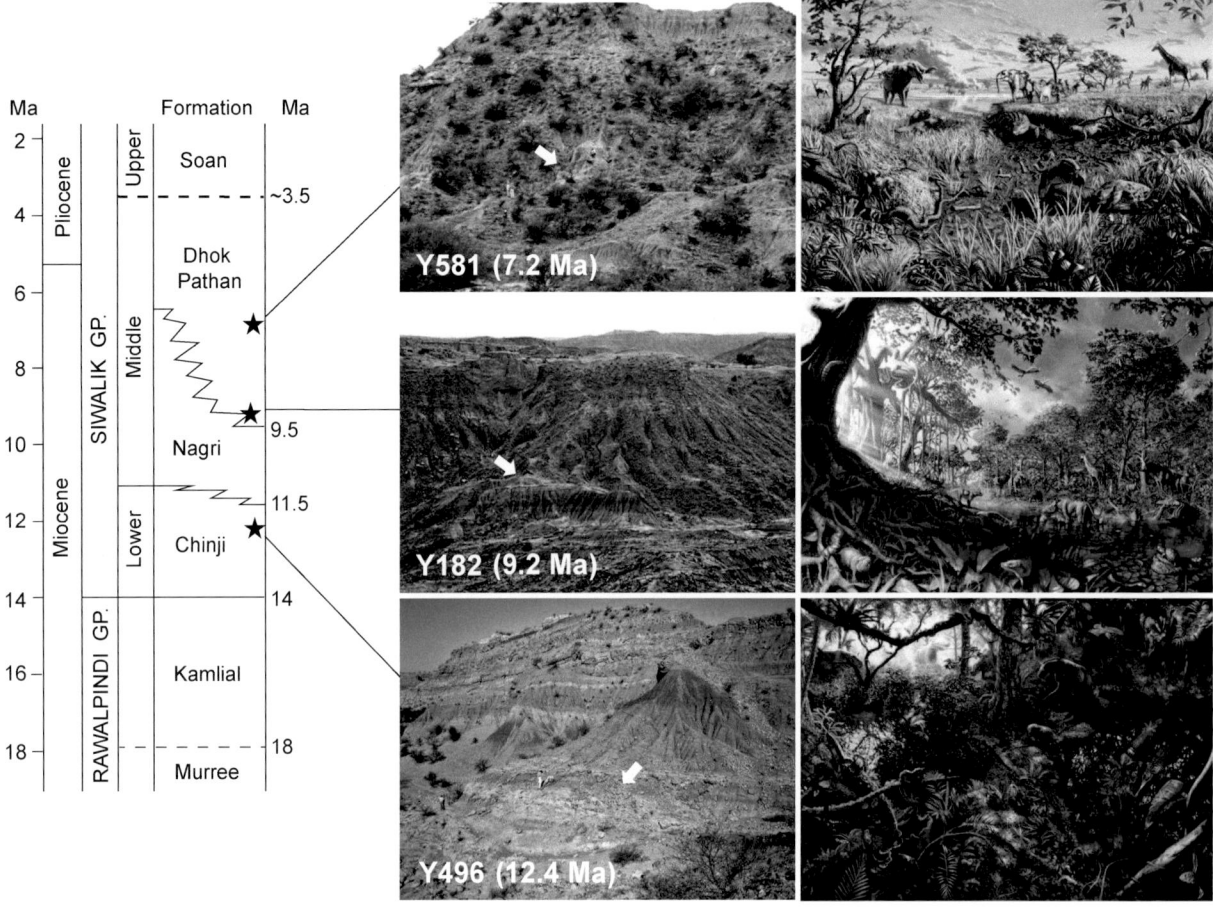

Figure 24.8. Images of the outcrop areas of the three localities used for paleoecological reconstructions next to their paleoecological reconstructions by Julius Csotonyi. White arrows indicate the main concentrations of surface fossils. Stratigraphic column shows where these localities occur in the succession of Siwalik formations. See text and Table 24.2 for details, also Figs. 24.9–24.14.

Figure 24.9. Artistic reconstruction by Julius Csotonyi of Siwalik paleoecology at Chinji Formation Locality Y496 (12.4 Ma). Time of Day: early afternoon; Time of Year: a month or two after the end of the rainy season. A full-size image of this reconstruction is available at the end of the chapter.

or trees. Some wetter areas are trampled by animals, and the ground shows accumulations of plant debris left by the recent floods.

Detail: Central in the foreground is the litter-covered, muddy edge of a partly abandoned channel with trails made by large animals coming to drink. A low bush and dappled shade hide a large creodont interested in two large tragulids near the water. A small bovid off the trail is alert. A chalicothere and two rhinocerotids are visible in the middle distance, browsing on low bushes and young trees. Near the chalicothere, a suid drinks, and a large crocodile lurks in the water. Several kinds of turtles are in the water or on logs, and large snakes are partially submerged near the bank. An otter eats a catfish on the bank. In the canopy foreground, a tree squirrel is carrying a red flower, and a treeshrew is eating a millipede. Higher up a python curls over a branch. Visible in branches above the water are a female and juvenile ape. A kingfisher is in flight, while a stork perches on a branch. The leaves show damage from insects and fungi, and broken branches provide evidence of herbivore impact on the vegetation.

9.2 Ma Y182 Reconstruction

Locality Y182 is dated to 9.2 Ma and is similar in fossil productivity to Y496 (see Table 24.2, Figs. 24.8, 24.11- 24. 12; see also Fig. 4.3b, this volume). The depositional setting is a crevasse splay near a floodplain channel that would have overtopped its banks during periods of heavy rain. The fossil-bearing deposits contain reworked soil nodules indicating that such flooding caused local erosion of channel banks and floodplain sediments, concentrating these nodules and any buried or surface bones into discrete, coarse-grained layers. Skeletal remains also accumulated over time in the depression formed by a channel,

where seasonal flooding helped bury and preserve fragmentary bones and teeth. Many other fossil localities at the same stratigraphic level contribute to our understanding of the fauna and habitats at that time.

Time of day and season: This scene shows an early, misty morning near the end of the rainy season.

Vegetation: Compared with Y496, the environment of Y182 had more open-canopy deciduous vegetation resulting from longer dry seasons and more frequent flood disturbance. Our choice of plant species is based on modern analogues for similar habitats and includes a variety of leaf types, epiphytes, fruits, and aquatic vegetation. Specific genera include *Ficus*, *Dillenia*, *Borassus*, *Dendrocalamus*, and *Potamogeton*.

Fauna: The animals depicted for this time and place include those that broadly resemble living species as well as members of extinct groups such as saber-tooth cats, deinotheres, and sivatheres (see Table 24.2). The 17 mammals in this morning scene represent 12 of 50 species at the locality plus 5 from time-equivalent localities. Small mammals are particularly abundant at Y182 (especially murines, not depicted); two arboreal species (one primate and one rodent) and one burrowing rodent are represented in the scene. The five large artiodactyl and perissodactyl species include the earliest species attributed to Bovini and equids. Two types of megaherbivore giraffoids and several deinotheres browse on the western (right-hand) bank. Also featured are the aardvark *Orycteropus*, the ape *Sivapithecus*, and several carnivores.

Overall Visualization: This is a ground-level perspective from the edge of the stream bank looking south across the shallow

Figure 24.10. Key to taxa depicted in Fig. 24.9: 1. *Brachypotherium perimense*, 2. Stork, 3. *Zygolophodon metachinjiensis*, 4. *Python*, 5. *Gaindatherium browni*, 6. *Hyainailouros sulzeri*, 7. *Tamiops* (large species), 8. Spider, 9. Tupaiid eating a millipede, 10. Butterfly, 11. Beetle, 12. Tree frog, 13. *Elaschistoceras khauristanensis*, 14. *Dorcabune anthracotheroides*, 15. *Listriodon pentapotamiae*, 16. *Anisodon salinus*, 17. Kingfisher, 18. *Sivapithecus indicus*, 19. *Gavialis*, 20. *Chitra*, 21. *Batagur*, 22. *Rhamphosuchus*, 23. *Acrochordus dehmi*, 24. *Nilssonia*, 25. *Pila*, 26. *Vishnuonyx chinjiensis*, 27. *Chaca pachyptreygia*, 28. Ants. See also Table 24.2 and text.

Figure 24.11. Artistic reconstruction by Julius Csotonyi of Siwalik paleoecology at Dhok Pathan Formation Locality Y182 (9.2 Ma). Time of Day: early morning (note aardvark and other crepuscular/nocturnal species); low slanting light with morning mist; Time of Year: soon after the end of the rainy season. A full-size image of this reconstruction is available at the end of the chapter.

Figure 24.12. Key to taxa depicted in Fig. 24.11: 1. *Varanus*, 2. Python, 3. *Ratufa sylva*, 4. Ciconid (stork), 5. Machairodontine, 6. *Orycteropus*, 7. *Miorhizomys nagrii*, 8. *Daboia*, 9. skink, 10. *Pelacanus*, 11. *Percrocuta grandis*, 12. *Vishnuictis chinjiensis*, 13. *Anatolictis*, 14. *Deinotherium indicum*, 15. Geochelone, 16. *Hippopotamodon sivalense*, 17. *Sivalhippus nagriensis*, 18. *Giraffa*, 19. *Pachyportax falconeri*, 20. *Bramatherium megacephalum*, 21. *Sivapithecus sivalensis*, 22. *Nycticeboides simpsoni*, 23. *Nilssonia*, 24. *Gavialis*, 25. *Gazella lydekkeri*, 26. *Crocodylus*, 27. *Acrochordus dehmi*, 28. *Geochlemys*, 29. *Dorcatherium majus*. Bivalves, termites, and wasp not labeled. See also Table 24.2 and text.

channel, which is flowing slowly away from the viewer. The near bank has exposed roots and overhanging branches from large trees. Animals are approaching to drink or are foraging on lush vegetation growing along the opposite, more distant bank.

Detail: In the left foreground, an aardvark burrow and termite activity are sheltered in a tangle of exposed *Ficus* roots, along with decomposing fruit and leaf litter. Nearby, a burrowing rodent (rhizomyine) emerges into the early morning light. A

mongoose and civet are alert to activity near the massive *Ficus*, while a daboia (snake) is resting under a tree root. Climbing the tree is a varanid lizard, and a flying squirrel and a loris perch high above. Farther away, a python is coiled in a tree, and larger carnivores—a saber-toothed cat and a hyaenid—look across the channel. On the opposite bank, a crocodile is eating a fish; in the water are turtles and a snake, along with a large suid and her young. Across the channel, giraffoids, various bovids of different sizes, tragulids, equids, and deinotheres are foraging or moving through the streamside vegetation, and a small group of apes is foraging in a tree. Large storks fly northward along the channel.

7.2 Ma Y581 Reconstruction

Fossil vertebrates are less common in the later part of our study interval. Dated at 7.2 Ma, Y581 is one of the richer localities around this time and has both small and large mammals in the sample of 148 catalogued fossils (see Table 24.2, Figs. 24.8, 24.13–24.14). The skeletal remains are associated with small floodplain channels and associated soils, and these accumulated

over time as sediments carried by seasonal flooding helped to bury and preserve the remains.

Time of day and season: The scene shows mid-morning in the middle of the dry season.

Vegetation: Plant life is abundant and mostly green, with grass just turning yellow in drier places. Some deciduous trees and shrubs are present along the channel but grasses dominate the floodplain. Named plant genera include *Saccharum*

Figure 24.13. Artistic reconstruction by Julius Csotonyi of Siwalik paleoecology at Dhok Pathan Formation Locality Y581 (7.2 Ma). Time of Day: mid-morning with full sunlight; Time of Year: middle of the dry season; note grass fire smoldering in the distance. A full-size image of this reconstruction is available at the end of the chapter.

Figure 24.14. Key to taxa depicted in Fig. 24.13 : 1. *Kobus*, 2. *Hystrix primigenia*, 3. *Percrocuta grandis*, 4. *Dorcatherium*, 5. *Mesopithecus sivalensis*, 6. *Sivaophis*, 7. *Bungarus*, 8. *Python*, 9. Agamid (cf. *Calotes*), 10. *Alilepus elongatus*, 11. *Propotamochoerus hysudricus*, 12. *Anatolictis*, 13. *Merycopotamus dissimilis*, 14. *Stegolophodon*, 15. *Brachypotherium perimense*, 16. *Crocodylus*, 17. *Varanus*, 18. *Tadornini* (shelduck), 19. cf. *Indotestudo*, 20. Otidae (bustard), 21. *Sivalhippus anwari*, 22. *Sivalhippus perimensis*, 23. *Pachyportax dhokpathenensis*, 24. *Tragoportax salmontanus*, 25. *Giraffa*. Dung beetles not labeled. See also Table 24.2.

(broomsedge) and *Potamogeton* (pondweed); other plants are based on modern grassland habitats in southern Asia.

Fauna: The 16 mammals represent 10 of 20 species recorded in the relatively small sample from Y581 plus 6 others from time-equivalent localities (see Table 24.2). These include species in living genera as well as members of extinct groups such as anthracotheres, hipparionine equids, and stegolophodontid proboscideans. Rodents recorded at Y581 are dominated by small murines, but the scene includes only the porcupine *Hystrix*. Large mammals include reduncine, bovine and boselaphine bovids, and small and large hipparionines. Also featured are the lagomorph, *Alilepus* and the colubine monkey, *Mesopithecus*.

Visualization: We are viewing this scene from standing height, looking northward along a muddy tributary channel in the foreground toward a larger channel and more distant grassy alluvial plain. Trees are present but sparse, and grasses cover most of the area. A fire is burning in the distance, indicating the drier seasonal conditions of this time period.

Detail: Two hyaenas are feeding on a dead suid (*Propotamochoerus*). A large crocodile is approaching, also interested in the carcass, as are a varanid lizard and mongoose. Closer to the viewer, a hare, python, and small lizard are camouflaged in foliage, while two birds and a tortoise are more visible. On the left (west) side of the small channel, a monkey with a juvenile perched on her back is alert to a nearby venomous snake, while a small tragulid shelters in the shade of the fallen tree and another snake rises from trampled grass. Trampled bones, tracks, and dung are visible on the mud. Slightly farther away, two anthracotheres are wallowing in a muddy pool. A porcupine rests in a small tree with a watchful hyaena underneath and a reduncine bovid and elephantoid are farther away on the western edge of the channel. On the eastern side of the channel is another elephantoid along with two species of equid, two species of bovid, and a giraffe. In the distance, across the larger channel, are two large rhinocerotids, and behind them is smoke from the grass fire.

SUMMARY AND CONCLUSIONS

The Siwalik geological, geochemical, and paleontological record provides a wealth of information for science-based reconstructions of the Miocene ecology of the western sub-Himalayan alluvial plain. The different types of evidence (see Fig. 24.1, Table 24.1) can be combined into realistic representations of plants and animals in their habitats at a given point in time. We have a high level of confidence in our paleoecological and taphonomic interpretations of localities and time-intervals (i.e., ~10^2–10^3 years) due to the care we have taken to integrate many types of evidence to understand local, time-averaged records. By assembling these into a history of ecological change through the Potwar Siwalik record (Chapters 25 and 26, this volume), we can then consider how observed patterns in this record correspond to continental- to global-scale forces, which are possible drivers of ecological change (see Fig. 24.3). This must be done with caution, in part because correlation in time does not necessarily mean causation, but it is also highly probable that Siwalik ecological history was affected by regional to global forces over the past 20 million years, particularly during our 18–6-million-year study interval.

The global climate record, as documented by changes in $\delta^{18}O$ in benthic foraminifera, shows a warm interval between 17 and 14.7 Ma (Miocene Climatic Optimum). This corresponds to the Kamlial Formation faunal record, which indicates forested habitats (C_3 vegetation) in the western sub-Himalayan foreland basin, and geological evidence for an extensive alluvial plain with a relatively high water table and low seasonality. There is no obvious temporal correlation between increased faunal change in the lower part of our record and marked changes in global climate (Chapters 25–26, this volume). After 14 Ma, global temperatures cooled but were still relatively warm (see Fig. 24.3), and forested habitats continued through the Chinji and lower Nagri Formations up to about 10 Ma, with localized patches of C_3 and C_4 grass. Major ecological changes began in the Potwar Siwaliks around 10 Ma, with the increase in fire frequency, the appearance of equids and decline in forest-adapted vertebrates, changes in tooth microwear and dental isotopes indicating C_4 diets, and $\delta^{13}C$ in the Indus Fan core biomarker record indicating increased C_4 vegetation on land. The combined evidence points to increasing seasonality in rainfall that resulted in longer dry seasons and more open alluvial landscapes resulting from fire and the adaptive advantage of warm-season growth in C_4 plants. Correlations between deep sea drill core and on-land records of a major transition from dominant C_3 to dominant C_4 vegetation are clear (see Fig. 24.3), and the Potwar Siwalik record shows that this transition occurred between 8.5 and 6 Ma in our study area. The C_3–C_4 transition broadly correlates with a decline in global CO_2 levels, cooling global temperatures, and changes in the hydrology of the rising Himalayas, although the relationships among these large-scale drivers of environmental change continue to be debated.

In this chapter, we have focused on synthesizing multiple lines of scientific evidence as the basis for reconstructing Siwalik paleoecology, which also provided the foundation for a collaboration between the science and artist Julius Csotonyi to create realistic visualizations of extinct but once vibrant plant and animal communities. Three locality-based ground views and one aerial view were selected as examples of this science-art collaboration, representing Siwalik ecology at 12.4 Ma (Y496), 9.2 Ma (Y182) and 7.2 Ma (Y581). The temporal sequence of the reconstructions captures the profound change in Siwalik ecosystems through the C_3–C_4 transition (woodland and forest to subtropical grassland) in southern Asia and the accompanying decline in vertebrate biodiversity. The following chapters draw on all the information presented above to synthesize detailed analyses of faunal persistence, change, and evolution though the Potwar Siwalik record.

ACKNOWLEDGMENTS

The goal of this chapter was to bring together the many different types of evidence about Potwar Siwalik paleoecology that have been a hallmark of our collective research and to consider how this evidence might relate to broader-scale environmental processes. We have done our best to fairly represent a large body of work resulting from long collaborations with the extended Siwalik team. The collaboration with artist Julius Csotonyi (via virtual discussions during the pandemic lockdown) brought to life the Siwalik plant and animal communities and was a wonderful experience. We are very grateful to him for his contributions to this chapter and to the book overall. We thank Bonnie Fine-Jacobs for essential advice on plant taxa to include in the visual reconstructions. Work on seasonality in *Pila* opercula was contributed by Nora Loughlin, and Vera Korasidis kindly reexamined sediment samples for pollen. Finally, although the two authors of this chapter take responsibility for its content, we sincerely thank Catherine Badgley, John Barry, Larry Flynn and David Pilbeam for their helpful suggestions and encouragement—contributions that also are evident in many other chapters in this volume.

References

Abrajevitch, A., K. Kodama, A. K. Behrensmeyer, and C. Badgley. 2009. Magnetic Signature of moisture availability in ancient subtropical soils: A pilot rock magnetic study of Neogene paleosols in Pakistan. American Geophysical Union, Fall Meeting 2009, abstract id. GP43B-0853.

Adhikari, P., G. Srivastava, and K. N. Paudayal. 2024. An overview of the Middle Miocene to Early Pleistocene flora of the Siwalik Sediments in Nepal. In *Flora and Vegetation of Nepal: Plant and Vegetation*, vol 19, edited by M. B. Rokaya and S. R. Sigdel, 89–111. Cham: Springer.

Ali, J. R., and J. C. Aitchison. 2008. Gondwana to Asia: Plate tectonics, paleogeography and the biological connectivity of the Indian subcontinent from the Middle Jurassic through latest Eocene (166–35 Ma). *Earth-Science Reviews* 88(3–4): 145–166.

Badgley, C., J. C. Barry, M. E. Morgan, S. V. Nelson, A. K. Behrensmeyer, T. E. Cerling, and D. Pilbeam. 2008. Ecological changes in Miocene mammalian record show impact of prolonged climatic forcing. *Proceedings of the U.S. National Academy of Sciences* 105(34): 12145–12149.

Barry, J. C. 1995. Faunal turnover and diversity in the terrestrial Neogene of Pakistan. In *Paleoclimate and Evolution, with Emphasis on Human Origins*, edited by E. S. Vrba, G. H. Denton, T. C. Partridge, and L. H. Burkle, 115–134. New Haven, CT: Yale University Press.

Behrensmeyer, A. K. 1987. Miocene fluvial facies and vertebrate taphonomy in northern Pakistan. In *Recent Developments in Fluvial Sedimentology*, edited by F. G. Ethridge, R. M. Flores, and M. D. Harvey, 169–176. Society for Economic Paleontology and Mineralogy Special Publication 39, Tulsa, OK.

Behrensmeyer, A. K., and L. Tauxe. 1982. Isochronous fluvial systems in Miocene deposits of northern Pakistan. *Sedimentology* 29(3): 331–352.

Behrensmeyer, A. K., B. J. Willis and J. Quade. 1995. Floodplains and paleosols of Pakistan Neogene and Wyoming Paleogene deposits: A comparative study. *Palaeogeography, Palaeoclimatology, Palaeoecology* 115(1–4): 37–60.

Behrensmeyer, A. K., J. Quade, T. E. Cerling, J. Kappelman, I. A. Khan, P. Copeland, L. Roe, et al. 2007. The structure and rate of late Miocene expansion of C_4 plants: Evidence from lateral variation in stable isotopes in paleosols of the Siwalik Group, northern Pakistan. *Geological Society of America Bulletin* 119(11–12): 1486–1505.

Cerling, T. E., and J. Quade. 1993. Stable carbon and oxygen isotopes in soil carbonates. *Washington DC American Geophysical Union Geophysical Monograph Series* 78:217–231.

Cerling, T. E., J. Quade, Y. Wang, and J. R. Bowman. 1989. Carbon isotopes in soils and palaeosols as ecology and palaeoecology indicators. *Nature* 341(6238): 138–139.

Clift, P., C. Gaedicke, R. Edwards, J. Il Lee, P. Hildebrand, S. Amjad, R. S. White, and H. U. Schlüter. 2002. The stratigraphic evolution of the Indus Fan and the history of sedimentation in the Arabian Sea. *Marine Geophysical Researches* 23:223–245.

Corvinus, G., and L. N. Rimal. 2001. Biostratigraphy and geology of the Neogene Siwalik Group of the Surai Khola and Rato Khola areas in Nepal. *Palaeogeography, Palaeoclimatology, Palaeoecology* 165(3–4): 251–279.

Diester-Haass, L., K. Billups, D. R. Gröcke, L. François, V. Lefebvre, V., and K. C. Emeis. 2009. Mid-Miocene paleoproductivity in the Atlantic Ocean and implications for the global carbon cycle. *Paleoceanography* 24(1):1–19.

Downing, K. F. and Lindsay, E. H., 2005. Relationship of Chitarwata Formation paleodrainage and paleoenvironments to Himalayan tectonics and Indus River paleogeography. *Palaeontologia Electronica* 8(1): 1–12.

Eiserhardt, W. L., T. L. Couvreur, and W. J. Baker. 2017. Plant phylogeny as a window on the evolution of hyperdiversity in the tropical rainforest biome. *New Phytologist* 214(4): 1408–1422.

Feakins, S. J., H. M. Liddy, L. Tauxe, V. Galy, X. Feng, J. E. Tierney, Y. Miao, and S. Warny. 2020. Miocene C_4 grassland expansion as recorded by the Indus Fan. *Paleoceanography and Paleoclimatology* 35(6): e2020PA003856.

Filippelli, G. M. 1997. Intensification of the Asian monsoon and a chemical weathering event in the Late Miocene–Early Pliocene: Implications for late Neogene climate change. *Geology* 25(1): 27–30.

Freeman K. H., and L. A. Colarusso. 2001. Molecular and isotopic records of $_{C4}$ grassland expansion in the late Miocene. *Geochimica et Cosmochimica Acta* 65(9): 1439–1454.

Fortelius, M., F. Bibi, H. Tang, I. Žliobaitė, J. T. Eronen, and F. Kaya. 2019. The nature of the Old World savannah palaeobiome. *Nature Ecology & Evolution* 3(4): 504.

Graham, J. B., ed. 1997. *Air-Breathing Fishes: Evolution, Diversity, and Adaptation*. Cambridge, MA: Academic Press.

Hodell, D. A., and F. Woodruff. 1994. Variations in the strontium isotopic ratio of seawater during the Miocene: Stratigraphic and geochemical implications. *Paleoceanography* 9(3): 405–426.

Holbourn, A. E., W. Kuhnt, S. C. Clemens, K. G. D. Kochhann, J. Jöhnck, J. Lübbers, and N. Andersen. 2018. Late Miocene climate cooling and intensification of southeast Asian winter monsoon. *Nature Communications* 9: art. 1584. https://doi.org/10.1038/s41467-018-03950-1.

Holbourn, A., W. Kuhnt, S. Clemens, W. Prell, and N. Andersen. 2013. Middle to late Miocene stepwise climate cooling: Evidence from a high-resolution deep water isotope curve spanning 8 million years. *Paleoceanography* 28(4): 688–699.

Holbourn, A., W. Kuhnt, K. G. Kochhann, N. Andersen, and K. J. Sebastian Meier. 2015. Global perturbation of the carbon cycle at the onset of the Miocene Climatic Optimum. *Geology* 43(2): 123–126.

Holbourn, A., W. Kuhnt, M. Lyle, L. Schneider, O. Romero, and N. Andersen. 2014. Middle Miocene climate cooling linked to intensification of eastern equatorial Pacific upwelling. *Geology* 42(1): 19–22.

Holbourn, A., W. Kuhnt, K.G. Kochhann, K.M. Matsuzaki, and N. Andersen. 2022. Middle Miocene climate–carbon cycle dynamics: Keys for understanding future trends on a warmer Earth? *Geological Society of America Special Paper* 556:1–19.

Hoorn, C., T. Ohja, and J. Quade. 2000. Palynological evidence for vegetation development and climatic change in the sub-Himalayan Zone (Neogene, Central Nepal). *Palaeogeography, Palaeoclimatology, Palaeoecology* 163(3–4): 133–161.

Hu, X., E. Garzanti, J. Wang, W. Huang, W. An, and A. Webb. 2016. The timing of India-Asia collision onset—Facts, theories, controversies. *Earth-Science Reviews* 160:264–299.

Jacobs, B. F., J. D. Kingston, and L. L. Jacobs. 1999. The origin of grass-dominated ecosystems. *Annals of the Missouri Botanical Garden* 86(2): 590–643.

Karp, A. T., A. K. Behrensmeyer, and K. H. Freeman. 2018. Grassland fire ecology has roots in the Late Miocene. *Proceedings of the National Academy of Sciences* 115(48): 12130–12135.

Karp, A. T., K. T. Uno, P. J. Polissar, and K. H. Freeman. 2021. Late Miocene C_4 grassland fire feedbacks on the Indian subcontinent. *Paleoceanography and Paleoclimatology* 36(4): e2020PA004106.

Khan, I. A. 1993. Evolution of Miocene-Pliocene fluvial paleoenvironments in eastern Potwar Plateau, northern Pakistan. Unpublished PhD diss., Binghamton University.

Kominz, M. A., J. V. Browning, K. G. Miller, P. J. Sugarman, S. Mizintseva, and C. R. Scotese. 2008. Late Cretaceous to Miocene sea-level estimates from the New Jersey and Delaware coastal plain coreholes: An error analysis. *Basin Research* 20(2): 211–226.

Kominz, M. A., K. G. Miller, J. V. Browning, M. E. Katz, and G. S. Mountain. 2016. Miocene relative sea level on the New Jersey shallow continental shelf and coastal plain derived from one-dimensional backstripping: A case for both eustasy and epeirogeny. *Geosphere* 12(5): 1437–1456.

Kuhnt, W., A. Holbourn, R. Hall, M. Zuvela, and R. Käse. 2004. Neogene history of the Indonesian throughflow. *Continent-Ocean Interactions within East Asian Marginal Seas. Geophysical Monograph* 149:299–320.

Kürschner, W. M., Z. Kvaček, and D. L. Dilcher. 2008. The impact of Miocene atmospheric carbon dioxide fluctuations on climate and the evolution of terrestrial ecosystems. *Proceedings of the National Academy of Sciences* 105(2): 449–453.

Laskar, J., P. Robutel, F. Joutel, M. Gastineau, A. C. M. Correia, and B. Levrard. 2004. A long-term numerical solution for the insolation quantities of the Earth. *Astronomy & Astrophysics* 428(1): 261–285.

Leinders, J. J. M., M. Arif, H. de Bruijn, H., S. T. Hussain, and W. Wessels. 1999. Tertiary continental deposits of northwestern Pakistan and remarks on the collision between the Indian and Asian plates. *Deinsea* 7(1): 199–214.

Levin, N. E., T. E. Cerling, B. H. Passey, J. M. Harris, and J. R. Ehleringer. 2006. A stable isotope aridity index for terrestrial environments. *Proceedings of the National Academy of Sciences* 103(30): 11201–11205.

Lübbers, J., W. Kuhnt, A. E. Holbourn, C. T. Bolton, E. Gray, Y. Usui, K. G. Kochhann, S. Beil, and N. Andersen. 2019. The Middle to Late Miocene "Carbonate Crash" in the equatorial Indian Ocean. *Paleoceanography and Paleoclimatology* 34(5): 813–832.

Lupker, M., C. France-Lanord, V. Galy, J. Lavé, and H. Kudrass. 2013. Increasing chemical weathering in the Himalayan system since the Last Glacial Maximum. *Earth and Planetary Science Letters* 365:243–252.

Lyle, M., K. A. Dadey, and J. W. Farrell. 1995. The Late Miocene (11–8 Ma) Eastern Pacific carbonate crash: Evidence for reorganization of deep-water circulation by the closure of the Panama Gateway. In *Proceedings of the Ocean Drilling Program, Scientific Results*, Vol. 138, edited by N. G. Pisias, L. A. Mayer, T. R. Janecek, A. Palmer-Jelson, and T. H. van Andel, 821–838.

Magill, C. R., G. M. Ashley, and K. H. Freeman. 2013. Ecosystem variability and early human habitats in eastern Africa. *Proceedings of the National Academy of Sciences* 110(4): 1167–1174.

Miller, K. G., G. S. Mountain, J. D. Wright, and J. V. Browning. 2011. A 180-million-year record of sea level and ice volume variations from continental margin and deep-sea isotopic records. *Oceanography* 24(2): 40–53.

Morgan, M. E., A. K. Behrensmeyer, C. Badgley, J. C. Barry, S. Nelson, S. and D. Pilbeam. 2009. Lateral trends in carbon isotope ratios reveal a Miocene vegetation gradient in the Siwaliks of Pakistan. *Geology* 37(2): 103–106.

Morgan, M. E., J. D. Kingston, and B. D. Marino. 1994. Carbon isotopic evidence for the emergence of C_4 plants in the Neogene from Pakistan and Kenya. *Nature* 367(6459): 162–165.

Morgan, M. E., L. Spencer, E. Scott, M. S. Domingo, C. Badgley, J. C. Barry, L. J. Flynn, and D. Pilbeam. 2015. Detecting differences in forage in C_3-dominated ecosystems: Mesowear and stable isotope analyses of Miocene bovids from northern Pakistan. *Journal of Vertebrate Paleontology, Program and Abstracts*, 184.

Morley, R. 2000. *Origin and Evolution of Tropical Rain Forests*. Chichester, UK: John Wiley & Sons.

Nelson, S. V. 2007. Isotopic reconstructions of habitat change surrounding the extinction of Sivapithecus, a Miocene hominoid, in the Siwalik Group of Pakistan. *Palaeogeography, Palaeoclimatology, Palaeoecology* 243(1–2): 204–222.

Pickering, K. T., H. Pouderoux, L. C. McNeill, J. Backman, F. Chemale, S. Kutterolf, K. L. Milliken, et al. 2020. Sedimentology, stratigraphy and architecture of the Nicobar Fan (Bengal–Nicobar Fan System), Indian Ocean: Results from International Ocean Discovery Program Expedition 362. *Sedimentology* 67(5): 2248–2281.

Polissar, P. J., K. T. Uno, S. R. Phelps, A. T. Karp, K. H. Freeman, and J. L. Pensky. 2021. Hydrologic changes drove the Late Miocene expansion of C_4 grasslands on the northern Indian subcontinent. *Paleoceanography and Paleoclimatology* 36(4): 1–22. 10.1029/2020PA004108.

Prashad, B. 1925. Anatomy of the common Indian apple-snail, *Pila globosa*. *Zoological Survey of India* 25(5): 137–142.

Prasad, M. 2008. Angiospermous fossil leaves form the Siwalik foreland basins and their palaeoclimatic implications. *Palaeobotanist* 57(1–3): 177–215.

Prasad, M., and N. Awasthi. 1996. Contribution to the Siwalik flora from Surai Khola sequence, western Nepal, and its palaeoecological and phytogeographical implications. *Palaeobotanist* 43(3): 1–42.

Quade, J., and T. E. Cerling. 1995. Expansion of C_4 grasses in the Late Miocene of northern Pakistan: Evidence from stable isotopes in paleosols. *Palaeogeography, Palaeoclimatology, Palaeoecology* 115(1–4): 91–116.

Quade, J., T. E. Cerling, and J. R. Bowman. 1989. Development of Asian monsoon revealed by marked ecological shift during the latest Miocene in northern Pakistan. *Nature* 342(6246): 163–166.

Raynolds, R. G. H. 1981. Did the ancestral Indus flow into the Ganges drainage? *Geological Bulletin of the University of Peshawar* 14:141–150.

Raynolds, R. G., and G. D. Johnson. 1985. Rate of Neogene depositional and deformational processes, north-west Himalayan foredeep margin, Pakistan. *Geological Society, London, Memoirs* 10(1): 297–311.

Raynolds, R. G., G. D. Johnson, C. D. Frost, H. M. Keller, M. G. McMurtry, and C. F. Visser. 2022. Magnetic polarity maps and time maps in the eastern Potwar Plateau: Applications of magnetic polarity stratigraphy. *Journal of Himalayan Earth Science* 55(2): 1–20.

Retallack, G. J. 1990. *Soils of the Past: An Introduction to Paleopedology*. Boston: Unwin Hyman.

Retallack, G. J. 1991. Untangling the effects of burial alteration and ancient soil formation. *Annual Review of Earth and Planetary Sciences* 19(1): 183–206.

Retallack, G. J. 1998. Core concepts of paleopedology. *Quaternary International* 51–52:203–212.

Retallack, G. J. 2005. Pedogenic carbonate proxies for amount and seasonality of precipitation in paleosols. *Geology* 33(4): 333–336.

Sarin, M. M. 2001. Biogeochemistry of Himalayan rivers as an agent of climate change. *Current Science* 81(11): 1446–1450.

Sheikh, K. A. 1984. Use of magnetic reversal time lines to reconstruct the Miocene Landscape near Chinji Village, Pakistan. Unpublished MS thesis, Dartmouth College.

Shevenell, A. E., J. P. Kennett, and D. W. Lea. 2004. Middle Miocene southern ocean cooling and Antarctic cryosphere expansion. *Science* 305(5691): 1766–1770.

Srivastava, G., K. N. Paudayal, T. Utescher, and R. C. Mehrotra. 2018. Miocene vegetation shift and climate change: Evidence from the Siwalik of Nepal. *Global and Planetary Change* 161:108–120.

Steinthorsdottir, M., H. K. Coxall, A. M. de Boer, M. Huber, N. Barbolini, C. D. Bradshaw, N. J. Burls, et al. 2021. The Miocene: The future of the past. *Paleoceanography and Paleoclimatology* 36(4): e2020PA004037.

Tauxe, L., and S. J. Feakins. 2020. A reassessment of the chronostratigraphy of Late Miocene C_3–C_4 transitions. *Paleoceanography and Paleoclimatology* 35(7): e2020PA003857.

Tauxe, L., and N. D. Opdyke. 1982, A time framework based on magnetostratigraphy for the Siwalik sediments of the Khaur area, northern Pakistan. *Palaeogeography, Palaeoclimatology, Palaeoecology* 37(1): 43–61.

Vögeli, N., Y. Najman, P. van der Beek, P. Huyghe, P. M. Wynn, G. Govin, I. van der Veen, and D. Sachse. 2017. Lateral variations in vegetation in the Himalaya since the Miocene and implications for climate evolution. *Earth and Planetary Science Letters* 471:1–9. https://doi.org/10.1016/j.epsl.2017.04.037.

Von der Heydt, A., and H. A. Dijkstra. 2006. Effect of ocean gateways on the global ocean circulation in the Late Oligocene and Early Miocene. *Paleoceanography* 21(1): 1–18.

Westerhold, T., N. Marwan, A. J. Drury, D. Liebrand, C. Agnini, E. Anagnostou, J. S. Barnet, et al. 2020. An astronomically dated record of Earth's climate and its predictability over the last 66 million years. *Science* 369(6509): 1383–1387.

Willis, B. 1993a. Ancient river systems in the Himalayan foredeep, Chinji Village area, northern Pakistan. *Sedimentary Geology* 88(1–2): 1–76.

Willis, B. 1993b. Evolution of Miocene fluvial systems in the Himalayan foredeep through a two kilometer-thick succession in northern Pakistan. *Sedimentary Geology* 88(1–2): 77–121.

Willis, B. J., and A. K. Behrensmeyer. 1994. Architecture of Miocene overbank deposits in northern Pakistan. *Journal of Sedimentary Research* 64(1b): 60–67.

Zaleha, M. J. 1997. Siwalik paleosols (Miocene, northern Pakistan): Genesis and controls on their formation. *Journal of Sedimentary Research* 67(5): 821–839.

Zliobaite, I., H. Tang, J. Saarinen, M. Fortelius, J. Rinne, and J. Rannikko. 2018. Dental ecometrics of tropical Africa: Linking vegetation types and communities of large plant-eating mammals. *Evolutionary Ecology Research* 19(2): 127–147.

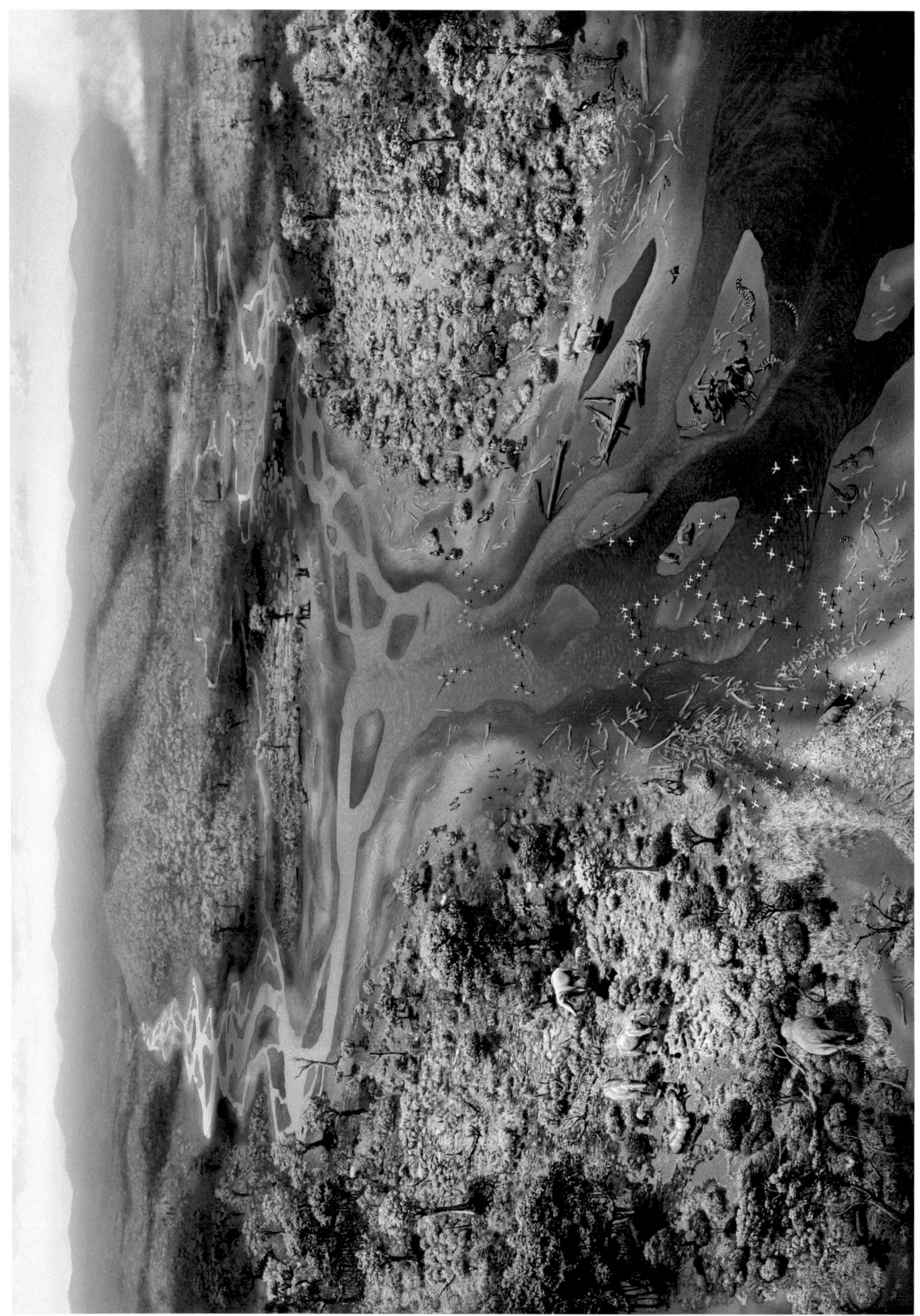

Figure 24.4.

Figure 24.9.

Figure 24.11.

Figure 24.13.

25 SIWALIK MAMMALIAN COMMUNITY STRUCTURE AND PATTERNS OF FAUNAL CHANGE

MICHÈLE E. MORGAN, LAWRENCE J. FLYNN, AND DAVID PILBEAM

INTRODUCTION

The collision of the Indian and Asian plates during the Cenozoic left the Indian subcontinent surrounded by significant physical barriers to the north, west, and east, with the northern barrier effectively impermeable to faunal exchanges, the others variably so. Beginning in the Oligocene, a South Asian paleobioprovince developed that, at least for the resident vertebrates, became increasingly endemic despite ongoing arrivals of immigrants from outside the bioprovince (see Fig. 2.1). Most of these vertebrates are recovered as fossils from exposures in the Siwalik Hills, a belt up to 2,000 m in elevation extending across almost the entire northern Indian subcontinent. Hence the informal term "Siwaliks" is commonly used to refer to both the rocks and the fossil fauna and flora preserved within them.

Starting in 1973, a collaborative research project with the Geological Survey of Pakistan (now referred to as the Harvard–GSP project) (see Chapter 1, this volume) has included collection, documentation, and analysis of fossils from the Siwaliks exposed in the Potwar Plateau in northern Pakistan, where fluvial sediments span the interval between the Early Miocene and the Pleistocene (see Part I chapters). For geopolitical reasons our fieldwork ceased there around 2000, and since then efforts have focused on documentation and analysis. In 2002, using then current systematic analyses, we reported species lineages through the Late Miocene sequence of the Potwar Plateau between 11 and 6 Ma and were able to place paleomagnetically calibrated localities into "bins" as brief as 100 kyr (Barry et al. 2002). Expanding on earlier analyses (e.g., Barry et al. 1995), we also examined Late Miocene faunal change for relationships with inferred habitat change. Badgley et al. (2008) looked more closely at faunal turnover between ~8 and 6 Ma and related this to the well-documented change from C_3 plant-dominated to C_4 grass-dominated floodplain habitats (Quade, Cerling, and Bowman 1989; Behrensmeyer et al. 2007). A decade after Barry et al. (2002), analysis was extended through the entire Miocene Potwar Siwalik sequence with foci on both sedimentary contexts of fossil localities and the potential effects of immigrants on faunal turnover (Barry et al. 2013; Flynn et al. 2013, 2014; Badgley et al. 2015). Half of all fossil localities are now resolved in age to within 100 kyr

and another quarter to within 200 kyr (see Chapters 2 and 3 of this volume for details).

As reviewed in Chapters 2–5 of this volume, several decades of Siwalik research saw increasing focus on systematics, sedimentology, chronostratigraphy, and isotope paleoecology, as well as on reconstructing evolving landscapes and habitats across the plateau between 18 and 6 Ma. Chapters 6–23 of this volume document current systematic studies, predominantly on mammals, wherever possible to the species level and within as broad a biogeographic framework as feasible. These faunal studies make possible more refined assessments of biostratigraphic ranges for the paleoprovince, including faunal immigrations and emigrations (criteria for recognizing immigrants and endemic originations are discussed later in this chapter). Immigration and emigration patterns provide insight into the timing and ecological nature of gateways in paleoprovince boundaries (Badgley et al. 2015).

We acknowledge that collecting areas on the Potwar Plateau represent a small window on the South Asian bioprovince but believe that the recovered fauna samples a regional paleocommunity that is reasonably representative of the sub-Himalayan metacommunity. Faunal change is constant; it varies markedly in tempo, but adjacent well-sampled 100-kyr intervals always record some first or last occurrences of species. Here we focus in particular on a 7-million-year analytical window in the Middle to Late Miocene and characterize 6 mammalian paleocommunities, each of which represents 100-kyr time slices, at 14.1, 12.7, 11.4, 10.1, 8.7, and 7.4 Ma, respectively. Throughout, we maintain a focus on indicators of climate and habitat change and possible impacts on faunal change at both lineage and community levels.

Our new analyses show that mammalian species richness declines over time, with the decline driven almost entirely by reduction in number of small-mammal species. Throughout this 7-myr interval, the mega-mammals, primarily proboscideans and rhinocerotids, are remarkably species-rich, peaking at 18 species within a 100-kyr bin (and with as many as 9 species recorded from a single locality), and generally have long species durations, or residence times, in the Potwar Plateau. We analyze both species richness and species abundance within groups; abundant taxa have large numbers of specimens compared to taphonomically equivalent taxa (Chapter 4, this volume). Tragulid ruminants, for example, are both species-rich and abundant. Within rodents, murines are abundant shortly after their first occurrence, but the species-rich sciurids were never abundant. Within large mammals, the greatest faunal change occurs between 11.4 and 10.1 Ma and is coincident with a marked expansion in the body-size structure of large herbivores. Overall, immigrant species are recorded in the Potwar fauna for somewhat shorter durations (residence times) than species originating within the paleobioprovince.

These and other aspects of faunal turnover and phyletic evolution can be viewed against current inferences of habitats and climates (Chapters 5, 24, this volume) and as contributing to the ongoing debate about the relative macroevolutionary roles of biotic and abiotic processes affecting species survival and extinction.

APPROACH

Our analyses focus on the mammalian faunal groups for which we have the most complete and consistent data over the time period of interest—thus, the non-volant, non-carnivoran, herbivorous and omnivorous taxa. Throughout this chapter, unless specifically noted, all references to the Siwalik mammalian fauna in the text, tables, and figures exclude bats and carnivorans. Bats are poorly represented in the Siwalik fossil record due to taphonomic factors (see Chapter 10, this volume). Carnivorans (carnivores and creodonts) are briefly mentioned in descriptions of the six well-sampled mammalian communities, but their fragmentary nature makes species-level identification challenging and renders species residence times uncertain (see Chapter 13, this volume). Although their high species richness but relative rarity is notable, including them in our analyses complicates evaluation of their presence in the successive paleocommunities and their impact as predators on other mammalian species. (See Chapter 26 of this volume, which does include them.)

The functional attributes of species in a modern community are a major source of information about available habitats (e.g., forest, grassland, pond). One of the biggest analytic challenges we face is the non-analogic nature of most Siwalik mammalian species and hence of the paleocommunities; few genera are represented in extant faunas. We employ ecomorphological approaches (see Chapter 24 and Table 24.1 for additional details), and isotopic data, to characterize significant paleobiological features of mammalian species: body mass, diet, foraging behavior, and shelter habitats. Faith et al. (2018) and Faith, Rowan and Du (2019) use East African examples to demonstrate that until even as recently as 700 kyr ago, all paleocommunities were functionally non-analogic, even hundreds of thousands of years after they had become taxonomically modern. Nevertheless, despite the non-analogic characteristics, in reconstructing our paleocommunities we rely on basic principles derived from the dynamics of extant communities. In contrast to the mammals, the avian paleofaunas include many genera and possibly a few species that are extant (see Chapter 9 of this volume).

The entire fossil assemblage from each 100-kyr time slice is treated as a paleocommunity in which we recognize structural differences from modern ecological communities. While recognizing that 100-kyr is a geologically brief interval (and for terrestrial Miocene biostratigraphic and paleontological studies close to the best that can be achieved), it is still an order of magnitude, or more, longer than the span for "neo-communities" (10 kyr or less; Bennett 1990, 2004).

Body Size

We sort mammal species into three size groups: small (< 1 kg); large (1–800 kg), and mega (> 800 kg, with at least some individuals of mega-sized species exceeding 1,000 kg; Owen-Smith 1988). Within these three size groups are further subdivisions that are used for some analyses and are noted in Table III.1 of this volume. The small-mammal fossil record is recovered mostly from concentrations of isolated teeth that we infer to represent raptor or occasional carnivore accumulations (see Chapters 4 and 10 of this volume for details). These accumulated in swales or were transported short distances along small streams before final burial. Derived from lenses of sediments within fossil localities, the small-mammal record likely represents shorter periods of time-averaging than other fossil localities (i.e., years to tens of years rather than hundreds to thousands of years; Badgley, Downs, and Flynn 1998; Chapter 4, this volume). Highly arboreal taxa such as lorises and bats, which are volant, are very underrepresented. With these exceptions, we believe that well-sampled, small-mammal assemblages approximate the standing species richness of the small-mammal community in the Potwar region. In contrast, the large- and mega-mammal records consist primarily of disarticulated craniodental and postcranial specimens from species that died near ponds or waterways, some having tooth marks and other evidence of mammalian and crocodilian predators (see Chapter 4 of this volume). Most carnivoran fossils are also derived from channel-related depositional contexts, but a few fossil localities associated with floodplain paleosols are identified as dens of carnivores. Thus, the Siwalik fossil record of small mammals is largely one of discrete concentrations separated by gaps in space and time (Fig. 25.1a), while that of large mammals and mega mammals is more evenly distributed throughout the record (Fig. 25.1b).

Figure 25.1 notes first occurrences of species. High numbers of newly occurring species do not always map to highly fossiliferous 100-kyr intervals, nor do low numbers of newly occurring species always map to less well-sampled 100-kyr intervals.

Species First Occurrences: Immigrants and Endemics

To examine the tempo and mode of species first occurrences, we categorize newly occurring species as either immigrants or endemic originations, while recognizing that these assignments are hypotheses of the actual nature of species originations. For us, endemic originations record the initial occurrence of new species within genera already documented in the Siwalik

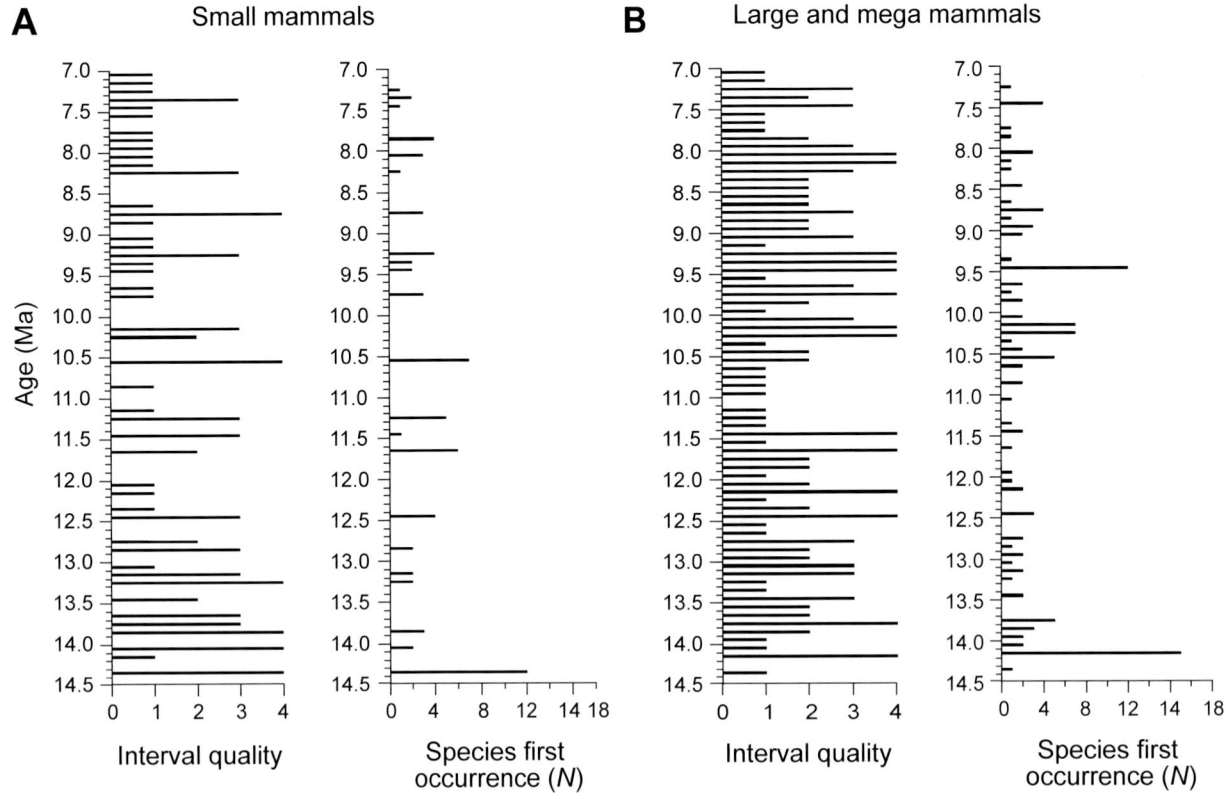

Figure 25.1. Interval quality and number of species first occurrences for small mammals (A) and large mammals plus megaherbivores (B). Interval quality (0–4) reflects number of specimens: 0 (0); 1 (<100); 2 (100–249); 3 (250–499); 4 (>500). High numbers of observed first occurrences do not always correspond to data quality. (Species first occurrence data from Table III.1.)

metacommunity. Immigration is invoked for cases in which a predecessor is not recorded; most cases involve the arrival of new families or genera. Occasionally, faunal specialists have proposed the origination of new genera within the sub-Himalayan bioprovince or instances of the immigration of congeneric species (see Part II chapters, particularly Chapters 11, 18, and 23, of this volume). They may also infer likely region of origination relative to the Siwaliks (see chapters in Part II and Table III.1, of this volume). Taxa are described as coming from the west when older relatives occur in Africa, western Asia, or Europe and from the east if older relatives are in South China or Southeast Asia. Identifying emigrations or range-extensions is more challenging, since these are dependent on both sound alpha taxonomy and relevant well-dated fossil sequences outside the South Asian bioprovince. We note a few well-supported range-extensions of species (discussed in Chapters 11, 16, 21, and 23). Immigration rates provide some insight into the fluctuating permeability of the bioprovince boundaries, while endemic origination rates to some extent reflect selection on local faunas in response to biotic factors, with or without changing environments.

Residence Time

To approximate species residence times (RT) in the Potwar sequence, we use biostratigraphic ranges between first and last documented occurrences for small-, large- and mega-mammal

species as presented in Chapters 10–12 and 14–23 of this volume. We are aware that Potwar RT is not equivalent to the actual species duration within the Potwar or the larger sub-Himalayan paleobioprovince (see Fig. 2.1 of this volume for regional map). Indeed this is our reason for choosing RT over species range or other terms. We use first/last occurrence (see Table III.1) rather than appearance/disappearance in text and tables to highlight these as the currently available primary data. We have chosen not to estimate confidence intervals, or theoretical extensions of temporal ranges, because we have concerns about the feasibility of assigning priors in statistical analyses (but see Chapter 26 of this volume for an elegant analysis including confidence intervals). We acknowledge the varying quality of our fossil record and recognize that there are many cases where "missing" taxa would have been periodically truly absent from the Potwar Plateau region while remaining members of the larger paleobioprovince metacommunity. Hence, we count range-through taxa as members of the paleocommunity and in summaries of species richness.

Frequency Index

On a few occasions we use a frequency index (FI) to document the frequency at which a taxon occurs at different fossil localities across a defined interval of time, expressed as a percentage. Taxa with a high FI are found at many fossil localities. Frequency can be independent of abundance as detailed in later examples.

Faunal Change Indices

As measures of faunal change to summarize data and highlight trends we use several simple metrics based on Simpson's discussion of faunal resemblance (Simpson 1960).

The persistence index (PI) is the percentage of species present at the beginning of any time interval of interest that is also present at the end of that time interval. Hence, $PI = (C_{N1N2}/N1) * 100$, where N1 is the number of taxa present at the beginning of the interval, N2 is the number of species present at the end of the interval, and C is the number of taxa in common between N1 and N2. The PI is most easily interpreted when the numbers of species at the beginning and end of the interval are similar. A high index reflects the relative stability of the earlier faunal community, while a low index reflects change in the original faunal community through declining diversity and/or first occurrence of new species.

The new species index (NSI) is the percentage of newly occurring taxa during the time interval of interest that are present at the end of that time interval, relative to the number of taxa present at the end of the time interval of interest; hence $NSI = (N2 - C/N2) * 100$. A high index reflects the addition of many new faunal elements via immigration or endemic origination, while a low index reflects low novel change in community composition.

Adding the PI and NSI together records increases (>100) or decreases (<100) in species richness of taxa during the time interval of interest.

Some of our faunal metrics utilize a 5-point 500,000-year running average. Such time-averaging smooths out variation due to fluctuating sampling intensity and other biases without obscuring signals of faunal trends.

Description of Siwalik Faunas and Faunal Change

Key features of Siwalik mammalian paleocommunities are summarized in Tables 25.1 and 25.2, with additional details presented in the following section. For each time slice examined, these tables include broad taxonomic characterizations of the small-, large-, and mega-mammal species, estimates of mammal size structure, and numbers of new species grouped as immigrants and endemic originations (again, range-throughs are included). The tables also include summary descriptions of paleocommunities, non-mammalian faunal elements, and ecomorphological indicators. Table 25.1 focuses on four Siwalik time slices within the interval from 17.9 to 15.2 Ma, while Table 25.2 summarizes the six paleocommunities noted earlier at 14.1, 12.7, 11.4, 10.1, 8.7 and 7.4 Ma. Our primary reason for beginning with the earlier Kamlial Formation fauna is that many Early Miocene species persist well into the Middle Miocene (some even into the Late Miocene) and provide essential context for later faunas.

Guiding questions discussed in detail later are:

- How do species richness and the pace of faunal change vary through time and across the three major mammalian size groupings?
- How do abundance and richness of taxa vary through time?

- How does community structure vary through time as assessed by proxies such as body size and stable isotope values of fossil enamel apatite?
- Does permeability of the bioprovince boundaries vary through time?
- Are there detectable effects of immigrants on endemic origination rates?
- Do residence times vary, and do they differ across size categories or between immigrants and endemics?
- What can the Siwalik Miocene mammal sequence tell us about the tempo and mode of species originations, extinctions, or range extensions and contractions?

KAMLIAL FORMATION PALEOBIOLOGY

The Kamlial Formation, deposited between ~18 and ~14 Ma, is less fossiliferous and was less intensively prospected during our fieldwork than the overlying Chinji Formation (see Chapter 2 of this volume). Kamlial paleosols frequently have nodular carbonates at depths greater than 1.5 m, and using modern soil analogues, annual rainfall between 18 and 14 Ma in this region is estimated at 1,000–1,250 mm (Quade and Cerling 1995). Limited stable carbon isotope data available from paleosol and fossil tooth enamel carbonate samples between 17.9 and 15.2 Ma indicate locally heterogeneous C_3 habitats, as summarized above, with abundant forest and some moist deciduous and non-woody vegetation (see Chapter 5 of this volume for background and details). Six megaherbivore teeth provide pure C_3 vegetation stable carbon isotope values, consistent with this interpretation (see Fig. 24.7). A serially sampled deinothere molar dated to 17.9 Ma indicates variation in isotopic composition of forage, likely linked to seasonal variation in diet.

Several productive fossil horizons about 1 million years apart through the Kamlial Formation sample the fauna across the Miocene Climate Optimum (MCO), ~17 to 14.7 Ma (Holbourn et al. 2015). Characteristics of four paleocommunities are summarized in Table 25.1, and species lists for small, large, and mega mammals are presented in Figure 25.S21 in the "Supplementary Figures" section in this chapter. The first good samples of micromammals dated to 17.9 Ma record 19 small-mammal species, the 15 rodents representing mostly groups that persist until later. One notable component is a species of the archaic Diatomyidae, a family afterward absent locally until ~11 Ma. *Kochalia* sp., a primitive cane rat, represents an Early Miocene immigrant lineage from Africa. Further documenting connections with Africa, *Prokanisamys*, a primitive rhizomyine recorded in Libya (Wessels 2009), is an early Middle Miocene emigrant from southern Asia to northern Africa. Rodent species richness increases to 26 species at younger localities, and the taxonomic structure remains stable over several million years at the generic level (Chapter 11, this volume).

The large-mammal fauna between 17.9 and 15.2 Ma is dominated by artiodactyls, with 10 to 13 species present across 6 families. Within these families, progiraffids are at first the most

Table 25.1. Principal Characteristics of Paleocommunities at Selected Time Slices, 17.9–15.2 Ma

Time slice	Formation	Mammalian Community Composition	Small Mammals (SM)	Large Mammals (LM)	Mega Mammals (MM)	Carnivorans	Mammal Size Structure	New Faunal Elements: Immigrants (since previous time slice)	New Faunal Elements: Endemic Origination (since previous time slice)	Community Characterization	Other Faunal Elements	Ecomorphological Indicators
17.9 Ma	Kamlial	37 species: 19 SM; 13 LM; 5 MM	15 rodents, 1 hedgehog, 2 treeshrews, 1 ochotomid	13 artiodactyls	4 rhinocerotids, 1 proboscidean	At least 5 species from 4 families: 1 creodont (Hyaenodontidae) and 4 carnivores (Amphicyonidae, Viverridae, Felidae)	Majority of species <1 kg; Median size SM: <100 g; LG: 20–49 kg; MM: 1,500–3,000 kg	SM: *Diatomys, Kochalia, Alloptox, Megacricetodon, Mellalomys;* LM: climacocerine: hypsodontine; *Eotragus,* i.s.; Pecora, i.s.; MM: *Prodeinotherium*	SM: *Sayimys, Democricetodon* (2), *Megacricetodon, Myocricetodon, Potwarmus* (2); LM: *Dorcabune;* MM: *Brachydiacratherium*	Last occurrence of Diatomyidae until ~11 Ma; rodents are primarily predecessors of species present after 14 Ma; progiraffids are most abundant artiodactyls; rhinocerotids are more species-rich than proboscideans, a trend for the next 10 myr	...	Presence of humid forest—treeshrews, arboreal sciurids; serially sampled deinothere indicates seasonal variation in diet/isotopic composition of forage
16.8 Ma	Kamlial	42 species: 25 SM; 10 LM; 7 MM	22 rodents, 1 hedgehog, 2 treeshrews	10 artiodactyls	6 rhinocerotids, 1 proboscidean	At least 5 species from 4 families: 2 creodonts and 3 carnivores	Majority of species <1 kg; median size SM: <100 g; LG: 20–49 kg; MM: 1,500–3,000 kg	SM: *Aliveria,* cf. *Ratufa;* MM: *Alicornops*	SM: *Sayimys, Tamiops,* anomalurid, *Prokanisamys, Kanisamys, Megacricetodon, Myocricetodon, Mellalomys;* MM: *Gaindatherium*	Rodents increase in species richness; tragulids are most abundant artiodactyls.	Avian fauna includes humid-climate, open-country species.	Presence of humid forest—treeshrews, arboreal sciurids, catarrhine primates
16.0 Ma	Kamlial	46 species: 28 SM; 11 LM; 7 MM	25 rodents, 1 hedgehog, 2 treeshrews	9 artiodactyls, 2 primates	6 rhinocerotids, 1 proboscidean	At least 5 species from 4 families: 2 creodonts and 3 carnivores	Majority of species <1 kg; median size SM: <100 g; LG: 20–49 kg; MM: 1,500–3,000 kg	SM: *Sciurotamias;* LM: lorisid, cf. *Dionysopithecus,* bovid	SM: *Kochalia, Democricetodon, Punjabemys* (3)	Cricetids and gerbillids are species-rich; tragulids are most abundant artiodactyls.	Avian fauna includes open-country species.	Presence of humid forest—treeshrews, arboreal sciurids, catarrhine primates
15.2 Ma	Kamlial	49 species: 30 SM; 13 LM; 6 MM	26 rodents, 1 hedgehog, 1 primate, 2 treeshrews	10 artiodactyls, 3 primates	5 rhinocerotids, 1 proboscidean	At least 7 species from 5 families: 2 creodonts and 5 carnivores; Barbourofelidae documented in Siwaliks	Majority of species <1 kg; median size SM: <100 g; LG: 8–19 kg; MM: 1,500–3,000 kg	SM: *Dakkamys, Nycticeboides*	SM: *Sayimys, Megacricetodon;* LM: *Sivaladapis, Indraloris, Siamotragus*	Cricetids and gerbillids are species rich; tragulids are most abundant artiodactyls.	...	Presence of humid forest—treeshrews, arboreal sciurids, sivaladapids

Table 25.2. Principal Characteristics of Paleocommunities at Selected Time Slices, 14.1–7.4 Ma

Time Slice (Ma)	Formation	Mammalian Community Composition	Small Mammals (SM)	Large Mammals (LM)	Mega Mammals (MM)	Carnivorans	Mammal Size Structure	New Faunal Elements: Immigrants (since previous time slice)	New faunal Elements: Endemic Origination (since previous time slice)	Community Characterization	Other Faunal Elements	Ecomorphological Indicators
14.1	Chinji	74 species: 39 SM; 25 LM; 10 MM	35 rodents, 2 hedgehogs, 2 treeshrews	22 artiodactyls, chalicothere, aardvark, sivaladapid	8 rhinocerotids; 2 proboscideans	At least 11 species from 7 families: 2 creodonts, 9 carnivores; newly appearing families: Mustelidae, Herpestidae	Majority of species <1 kg; median size SM: <100 g; LM 20–49 kg; MM: >1,500 kg	2 families and 18 genera since 15.2 Ma; SM: 4 rodents (2 rhizomyines); 5 genera sciurids; spalacid LM: Orycteropodidae; 5 genera bovids; lagomerycine; okapine; *Giraffokeryx*; MM: *Gomphotheriidae*; *Caementodon*	14 species compared to 15.1 Ma; SM: 4 rodents (2 rhizomyines); LM: 7 artiodactyls; *Listriodon*, *Conohyus*, 4 tragulids, *Hypsodontus*, *Anisodon*; MM: *Alicornops*, *Gaindatherium*	Archaic: sanitheres; SM: sciurids peak in species richness but uncommon; cricetids abundant and species-rich; LM: immigrant giraffids triple body size of largest artiodactyls; bovids most abundant artiodactyls; MM: rhinocerotids have high species richness; three proboscidean families present; creodont, *Hyainailouros sulzeri*, largest carnivoran at >400 kg; some amphicyonid and felid species likely exceed 250 kg.	Avian fauna includes wetland and water-body dependent species; open-country to forest-dependent species; molluscs indicate shallow slow-moving water and stable lake and river freshwater environments.	Presence of humid forest—treeshrews, arboreal sciurids, sivaladapid; bovid isotopic and microwear data consistent with C_3 browsing, mixed-feeding, grazing; megaherbivore species richness indicates abundant sub-canopy browse.
12.7	Chinji	79 species: 31 SM; 33 LM; 15 MM	26 rodents, shrew, hedgehog, 3 treeshrews	28 artiodactyls, chalicothere, 2 aardvarks, ape, sivaladapid	8 rhinocerotids; 7 proboscideans	At least 22 species from 10 families: 2 creodonts and 20 carnivores; newly appearing families: Hyaenidae, Percrocutidae	Majority of species are >8 kg; median size SM: <100 g; LM 20–49 kg; MM: >3,000 kg	4 families and 13 genera; SM: Gliridae, *Paralaucodus*; LM: Hominidae; new suid genus; 2 genera bohlinines; 3 genera bovids; MM: Mammutidae, Stegodontidae, *Protanancus*, *Dicerorhinus*	16 species; SM: soricid, 2 tupaiids, *Dakkamys*, *Tamias*, *Democricetodon*, *Dakkamys*, *Antemus*; LM: *Merycopotamus*, *Dorcatherium*, *Sivaceros*, *Hypsodontus*, *Amphiorycteropus*, *Decennatherium*; MM: *Deinotherium*; *Gomphotherium*	Highest mammal species richness; SM: 13 cricetid and gerbilline rodents, *Antemus*, abundant hedgehogs are >10% SM specimens; LM: bovids >40% artiodactyl fossils; most bovids <50 kg; MM: immigrant *Dicerorhinus* is extant genus	Avian fauna includes wetland and water-body-dependent species; open-country to forest-dependent species; molluscs indicate shallow slow-moving water and stable lake and river freshwater environments.	Presence of humid forest—treeshrews, sivaladapids, arboreal sciurids, a galericine, *Sivapithecus* (immigrant from west); bovid isotopic and dental wear data consistent with C_3 browsing, mixed-feeding, and grazing (microwear) and moderately abrasive diet (mesowear); megaherbivore species richness indicates abundant sub-canopy browse.

(continued)

Table 25.2. continued

Time Slice	Formation	Mammalian Community Composition	Small Mammals (SM)	Large Mammals (LM)	Mega Mammals (MM)	Carnivorans	Mammal Size Structure	New Faunal Elements: Immigrants (since previous time slice)	New faunal Elements: Endemic Origination (since previous time slice)	Community Characterization	Other Faunal Elements	Ecomorphological Indicators
11.4	Chinji	69 species: 28 SM; 28 LM; 13 MM	22 rodents, shrew, hedgehog, 4 treeshrews	25 artiodactyls, ape, chalicothere, aardvark	8 rhinocerotids; 5 proboscideans	At least 21 species from 8 families; 2 creodonts and 19 carnivores; last occurrence of very large hyaenodontine creodont	Majority of species are >20 kg; median size SM: <100 g; LM: 20–49 kg; MM: 1,500–3,000 kg	6 genera; SM: *Schizogalerix*, tupaiid; LM: *Hippopotamodon*, climacocerine, samotheriine, palaeomerycid	11 species; SM: 2 *Sayimys*, *Myomimus*, *Kanisamys*, *Paradakkamys*; 3 murines LM 1 bovid, 2 tragulids	Mammal species richness markedly lower than at 12.7 Ma; SM: abundant murine "Pre-Progonomys" accounting for 50% rodents; LM: bovids and tragulids >65% artiodactyl fossils, most bovids <50 kg; MM: no new species relative to 12.7 Ma	Avian fauna includes wetland and water-body-dependent species, open-country to forest-dependent species; alkaline or saline water bodies; molluscs indicate shallow slow-moving water and stable lake and river freshwater environments.	Presence of humid forest—treeshrews, arboreal sciurids, galericines, *Sivapithecus*; murine dominance and decreased SM species richness suggest increased competition or less potential niche space for SMs; bovid isotopic and dental wear data consistent with C_3 browsing, mixed-feeding, and grazing (microwear) and moderately abrasive diet (mesowear);megaherbivore species richness indicates abundant sub-canopy browse.
10.1	Nagri	69 species: 23 SM; 30 LM; 16 MM	18 rodents, lorisid, 2 treeshrews, 2 insectivores	23 artiodactyls, 2 chalicotheres, equid, 2 aardvarks, 2 apes	3 giraffids; 9 rhinocerotids; 4 proboscideans	At least 22 species from 9 families; 2 creodonts and 20 carnivores	Majority of species are >20 kg; median size SM: <100 g; LM 20–49 kg; MM: 1,500 kg	2 families and 12 genera; SM: *Suncus*, immigrant *Tamias*, *Rhizomyides*, *Microloris*; LM: Equidae, Palaeochoeridae, *Orycteropus*, 3 suid genera, paleomerycid; *Miotragocerus*; c.f. *Ancylotherium*	20 species; SM: 3 murines, 2 squirrels *Sciurotamias*, *Ratufa*; LM: 2 apes, 2 anthracotheres, 1 suid, 3 tragulids, 2 bovids, 1 equid; MM: *Rhinoceros*; 3 okapiines	Archaic elements: last occurrence creodonts at 10 Ma; LM species turnover with expanded body size structure; SM: murines comprise >70% SM specimens, murine *Progonomys* and *Karnimata* lineages distinct and abundant; LM: excepting giraffoids, few species persist from 11.4 Ma; most LM fossils from species >50 kg; equids present at majority of fossil localities; MM: giraffids now megaherbivores; extant genus *Rhinoceros* present	Avian fauna includes wetland and water-body-dependent species, open-country to forest-dependent species; alkaline or saline water bodies; molluscs indicate shallow slow-moving water with some periodic drought and stable lake and river freshwater environments.	Presence of lorisid, ape, treeshrew, and insectivores indicates forest vegetation; continued murine dominance and decreased SM species richness may indicate reduced overall habitat productivity, change in predator preferences, or increased competition among SMs; equids consuming C_4 grass; bovid isotopic and dental wear data indicate more graze and grit in diet; megaherbivore species richness still indicates abundant sub-canopy browse but perhaps reduced relative to earlier time slices; giraffids now megaherbivores.

	Site	Total species	Insectivores/ small mammals	Artiodactyls & others	Giraffids/rhinos/proboscideans	Carnivores	Body size	Taxa (1)	Taxa (2)	Abundance/ ecology	Birds/molluscs	Interpretation
8.7	Dhok Pathan	67 species: 18 SM; 31 LM; 18 MM	12 rodents, 3 shrews, 3 treeshrews	21 artiodactyls, 4 equids, chalicothere, ape, aardvark, 2 rodents, pangolin	3 giraffids, 8 rhinocerotids, 7 proboscideans	At least 21 species from 8 families; creodonts are no longer present.	Majority of species are >20 kg; Median size SM: <100 g; LM 20–49 kg; MM: 1,500–3,000 kg	1 family, 12 genera; SM: gerbil *Abudhabia*; unnamed genus Soricidae, unnamed tupaiid; LM: Manidae, unnamed suid genus "*Micro-Potamochoerus*," 4 bovid genera including oldest reduncine *Kobus*, *Cremohipparion*, MM: *Paratetralophodon*, immigrant ambelodontine	18 species: SM: *Myomimus*, 4 murines, *Kanisamys*; LM: 2 tachyoryctines (*Miorhizomys*), 2 tragulids, 2 bovids including oldest bovine, 2 equids; MM: *Stegolophodon*, *Rhinoceros*; *Giraffa* at 9.4 Ma; *Libytherium*	Most abundant taxonomic group shifts across all size groups; SM: rhizomyines abundant, taterilline gerbil present; LM: equids abundant and species-rich; 2 burrowing rhizomyines >1 kg, bovines and reduncines present, tragulids uncommon and species-rich; MM: proboscidean fossils more common than rhinocerotid fossils, absence of small primates.	Avian fauna includes wetland and water-body-dependent species, open-country to forest-dependent species; molluscs indicate large rivers and shallow slow-moving water.	Taterilline gerbil *Abudhabia* suggests some open-canopy habitat; bovid isotopic and dental wear data indicate more graze and grit in diet; bovine and reduncine bovids suggest more open vegetation; megaherbivore species richness still indicates abundant sub-canopy browse but perhaps reduced relative to earlier time slices; fewer rhinocerotids.
7.4	Dhok Pathan	48 species: 13 SM; 25 LM; 10 MM	10 rodents, 2 shrews, treeshrew	15 artiodactyls, 3 equids, chalicothere, monkey, 3 rodents, leporid, pangolin	1 giraffid, 3 rhinocerotids, 6 proboscideans	Poorly represented, but at least 17 species from 6 families; Ursidae has appeared in the Siwalks	Majority of species are >20 kg; median size SM: <100 g; LM 50–99 kg; MM: >3,000 kg	3 families, 6 genera; SM: *Mus*, LM: Leporidae, Colobidae, Hystricidae, *Tragoportax*; MM: *Ceratotherium*	10 species SM: 3 murines, 2 tachyoryctines *Eicooryctes* and *Protachyoryctes*; *Sayimys* reappears; LM: *Kobus*, *Merycopotamus*, *Sivalhippus*; MM: *Stegolophodon*	Least species rich mammalian fauna; absence of glirids, cricetids, hominoids, palaeochoerids, aardvarks, elasmothere, and acerathine rhinocerotid; SM: last documented treeshrew; LM: hypsodont rhizomyines present, immigrant porcupine, leporid, and colobine monkey; last documented *Anisodon*; MM: proboscideans more species-rich than rhinocerotids, immigrant grazer *Ceratotherium*	Avian fauna includes wetland and water-body-dependent species, open-country to forest-dependent species; molluscs indicate large rivers and shallow slow-moving water.	Forested vegetation persists as indicated by monkey and treeshrew; anthracothere shows hippo-like adaptations to semiaquatic lifestyle and graze diet; greater consumption of C_4 grasses including equids, bovids, suids, anthracothere, and rhinocerotids (occlusal morphology, dental wear, and isotopic data).

frequently documented artiodactyls, but tragulids are most common at 16.8, 16.0, and 15.2 Ma; this shift may reflect changes in tragulid species' original relative abundance and could be linked to the MCO. Although chalicotheres are recorded earlier in the Indian subcontinent, the few chalicothere fossils recovered between 17.9 and 15.2 Ma, likely from *Anisodon*, are too fragmentary to be attributed to species (Chapter 17, this volume).

Carnivorans include creodonts and carnivores. A minimum of ten species from five families are documented during the time of the Kamlial Formation, with five species persisting through most of the Middle Miocene. The largest carnivoran is the hyaenodontine creodont, *Hyainailouros sulzeri*, a species with a body mass upward of 400 kg. It has a minimum residence time of 7.6 myr (~19 to 11.4 Ma) in this region. Carnivores include amphicyonids, herpestids, viverrids, barbourofelids, and felids. The smallest carnivores are the herpestids; the extant mongoose genus *Urva*, first documented at 15.1 Ma, is distributed throughout southern Asia, Africa, and in southern Europe. The largest carnivores, at least two perhaps time-successive species of amphicyonids, are present and range in size from 150 to 300 kg. The smaller *Amphicyon palaeindicus* is last encountered at 11.6 Ma (see Chapter 13 of this volume for details).

Within megaherbivores, rhinocerotids are more diverse than proboscideans. At 17.9 Ma four rhinocerotids from two subfamilies are present; three of these persist, and by 15.1 Ma two additional species representing two more subfamilies appear. Only one recognized proboscidean, *Prodeinotherium pentapotamiae*, is recorded between 17.9 and 15.2 Ma.

Between 17.9 and 15.2 Ma, the Potwar mammalian species richness increased from 37 to 49 species, with many archaic faunal elements such as sanitheres, progiraffids, thryonomyids, and anomalurids still present. The monotonic increase in number of mammalian species over this nearly 3 myr period may reflect under-sampling at 17.9 Ma, although the high persistence indices between the better-sampled 100 kyr bins, along with presence of nearly all species at each well-sampled 100 kyr bin, suggest that we have not missed sampling the more common Kamlial species (Table 25.3). By 14.1 Ma, there are at least 74 mammal species in the region, and the fossil record documents both the arrival of immigrants and the regional endemic originations during and immediately after the MCO (see below).

POTWAR DEPOSITIONAL ENVIRONMENTS AND HABITATS IN MIDDLE AND LATE MIOCENE POST–KAMLIAL FORMATION PALEOCOMMUNITIES

Chinji Formation deposits overlie those of the Kamlial Formation, and for the following 3 million years between 14 and 11 Ma, the Potwar Plateau was dominated by large fluvial belts 1–2 km wide with numerous smaller paleochannels (Willis and Behrensmeyer 1995). Many of these were stable for periods of 1–10 kyr, and across the floodplains there were seasonal ponds and swamps as well as topographically higher and drier land surfaces associated with abandoned channels, levees, and crevasse splays (Chapter 4, this volume).

Stable carbon isotope data from paleosols are sparse but indicative of C_3 vegetation throughout the Chinji Formation (Quade, Cerling and Bowman 1989; see also Chapter 5, this volume). About a third of studied Chinji paleosols lack carbonates; these soils may have formed during periods of higher annual rainfall (>1,250 mm average annual rainfall) and/or less variable moisture availability throughout the year and/or interannually (Quade and Cerling 1995). Neither paleosol nor enamel stable carbon isotope data indicate predominantly closed-canopy wet evergreen forests, but rather support a heterogeneous habitat mosaic with dense forest along stable river channels and wetlands and woodlands with varying degrees of open canopy and grass patches. It is likely that non-woody C_3 plants (grasses, forbs, legumes) comprised on occasion a significant fraction of the vegetation (see Fig. 24.7 and Chapters 5 and 24 of this volume). Given the variety of factors affecting interpretation of stable carbon isotope values in the −8 to −10 ‰ range (see Polissar et al. 2021), it is currently unclear whether C_4 grasses were entirely absent or were minor components of the non-woody Chinji Formation vegetation. Recent recalibration of macroplant and pollen assemblages from the Surai Khola sequence in Nepal (Polissar et al. 2021) places the shift at around 12 Ma from a predominantly evergreen habitat (over 60% wet evergreen) to one with <40% wet evergreen vegetation and an expanded contribution of moist deciduous plants (Srivastava et al. 2018) including C_3 grasses (Hoorn, Ohja, and Quade 2000). This suggests that while still predominantly forested, habitats in Nepal were becoming more open with an increasingly non-woody understory, and this is also likely for the Potwar Plateau as discussed in Chapter 24 (this volume).

The sandstone-dominated Nagri Formation replaces the mudstone-dominated Chinji Formation time-transgressively between approximately 11.2 and 11.0 Ma in the Potwar Plateau (see Fig 2.2). The primary change involves the incursion of a much larger river system into the region with channel belts typically more than 5 km wide. (Willis 1993; Chapters 2, 4, this volume). Many streams are interpreted as having strongly seasonal flow with deposition occurring during major flood episodes (Willis and Behrensmeyer 1995). Paleosol carbonate isotope values are similar to those found in the Chinji Formation, indicating mostly C_3 vegetation including non-woody plants. By around 10.5 Ma, soil biomarkers suggest increased fire frequency and reduced abundance of conifers in the region; these data are interpreted as supporting greater presence of C_3 grasses and dry angiosperm forests in the region (Karp, Behrensmeyer, and Freeman 2018; Karp et al. 2021).

Between 9.5 and 9.0 Ma, the Nagri Formation and Dhok Pathan Formation depositional systems coexisted on the Potwar Plateau before the larger-scale (Nagri) fluvial system shifted away from the region. In the Dhok Pathan Formation, frequent well-drained soils and smaller networks of vegetated channels crossing the alluvial plain are consistent with decreased habitat heterogeneity (Barry et al. 2002), and likely also with reduced presence of mature tree canopy vegetation. Karp et al. (2021) note that between 9 and 6 Ma, oxygen isotope data from paleosol carbonates and tooth enamel, augmented by leaf and pol-

Table 25.3. Species Counts of Mammals at 11 Early to Late Miocene Time Slices, 17.9–6.5 Ma and Faunal Indices Reflecting Change between Time Slices

Age (Ma)	Species (N)	Indices	Small Mammals	Large Mammals	Mega Mammals	All Mammals
17.9	N at 17.9 Ma (N documented prior to 17.9 Ma)		19 (8)	13 (8)	5 (3)	37 (19)
		Persistence	78.9	76.9	100.0	81.1
		New species	40.0	0.0	28.6	28.6
	Documented only between 17.8 and 16.9 Ma		0	0	0	0
16.8	N at 16.8 Ma (N continuing from 17.9 Ma)		25 (15)	10 (10)	7 (5)	42 (30)
		Persistence	88.0	80.0	100.0	88.1
		New species	21.4	27.3	0.0	19.6
	Documented only between 16.7 and 16.1 Ma		0	0	0	0
16.0	N at 16.0 Ma (N continuing from 16.8 Ma)		28 (22)	11 (8)	7 (7)	46 (37)
		Persistence	92.9	90.0	85.7	91.3
		New species	13.3	23.1	0.0	14.3
	Documented only between 15.9 and 15.3 Ma		0	0	0	0
15.2	N at 15.2 Ma (N continuing from 16.0 Ma)		30 (26)	13 (10)	6 (6)	49 (42)
		Persistence	90.0	61.5	100.0	83.7
		New species	30.8	68.0	40.0	44.6
	Documented only between 15.1 and 14.2 Ma		0	0	0	0
14.1	N at 14.1 Ma (N continuing from 15.2 Ma)		39 (27)	25 (8)	10 (6)	74 (41)
		Persistence	56.4	80.0	90.0	69.9
		New species	29.0	39.4	40.0	35.4
	Documented only between 14.0 and 12.8 Ma		2	3	0	5
12.7	N at 12.7 Ma (N continuing from 14.1 Ma)		31 (22)	33 (20)	15 (9)	79 (51)
		Persistence	58.1	63.6	86.7	65.8
		New species	35.7	25.0	0.0	24.6
	Documented only between 12.6 and 11.5 Ma		1	4	0	5
11.4	N at 11.4 Ma (N continuing from 12.7 Ma)		28 (18)	28 (21)	13 (13)	69 (52)
		Persistence	50.0	39.3	92.3	53.6
		New species	39.1	63.3	25.0	46.4
	Documented only between 11.3 and 10.2 Ma		3	5	2	7
10.1	N at 10.1 Ma (N continuing from 11.4 Ma)		23 (14)	30 (11)	16 (12)	69 (37)
		Persistence	39.1	53.3	75.0	53.6
		New species	50.0	46.7	33.3	43.9
	Documented only between 10.0 and 8.8 Ma		5	7	2	14
8.7	N at 8.7 Ma (N continuing from 10.1 Ma)		18 (9)	31 (16)	18 (12)	67 (37)
		Persistence	33.3	53.3	44.4	44.8
		New species	53.8	36.0	20.0	37.5
	Documented only between 8.6 and 7.5 Ma		2	4	0	6
7.4	N at 7.4 Ma (N continuing from 8.7 Ma)		13 (6)	25 (16)	10 (8)	48 (30)
		Persistence	23.1	36.0	10.0	27.1
		New species	57.1	0.0	0.0	23.5
	Documented only between 7.3 and 6.6 Ma		1	1	0	2
6.5	N at 6.5 Ma (N continuing from 7.4 Ma)		7 (3)	9 (9)	1 (1)	17 (13)

Note: For small mammals, large mammals, mega mammals, and all mammals, the number of species counted at each time slice is noted and followed by a number in parentheses indicating the number of species continuing from the previous interval. For example, at the 6.5 Ma time slice, 7 small mammal species are counted; 3 of them were also counted in the 7.4 Ma time slice and 4 are "new," that is, not documented at 7.4 Ma.

len data from Bengal Fan cores, signal increasing regional aridity, including seasonal aridity. While woodland, thicket, and open-canopy forests likely still predominated, stable carbon isotope paleosol and enamel data indicate presence of wooded grassland with shrubs and grasses, including up to 25% C_4 grasses, by at least 8.2 Ma (Feakins et al. 2020).

By 7.5 Ma, more open and less spatially diverse habitats characterized the floodplains (Willis and Behrensmeyer 1995; Barry et al. 2002; Badgley et al. 2008). Soil biomarkers indicate

increased fire frequency in the Potwar region (Karp, Behrensmeyer, and Freeman 2018; Feakins et al. 2020; Karp et al. 2021), and paleosols suggest decreased precipitation and lower water tables on floodplains. There is a particularly marked increase in C_4 vegetation between 7.5 and 7.2 Ma (Feakins et al. 2020; Tauxe and Feakins 2020), corresponding in time to a climate shift also documented in benthic and planktic oxygen and carbon isotopic drill core data in South China Sea cores (Holbourn et al. 2018). Detailed lateral studies of paleosols through a Late

Miocene Potwar section at Rohtas show varying proportions of C_3 and C_4 plants on the landscape, trending from 25% C_4 at 8 Ma toward nearly 100% C_4 floodplain vegetation by 4 Ma (Behrensmeyer et al. 2007).

In addition to the mammalian fauna, other taxa including birds, non-avian reptiles, amphibians, fish, and molluscs provide useful information about local Potwar habitats. The molluscan and fish records (Chapters 6 and 7, this volume) document rivers and streams of varying size, lakes and ponds, and swampy areas (see Fig. 4.12, this volume). Species indicate warm and humid environments, and there is also evidence for marked dry seasons. The non-avian reptile fauna (Chapter 8, this volume) reflects similar environments and habitats, with evidence for increasing seasonality and the addition of more openness and heterogeneity after ~11 Ma. The avian fauna described in Chapter 9 of this volume includes many genera that are extant as well as some taxa close to extant species, which can be used to infer past preferred feeding behaviors and habitats. Unsurprisingly, birds dependent on wetlands and water bodies such as cormorants and ducks are found throughout the sequence, as are open-country taxa including storks and bustards. Species dependent on forested habitats, such as pheasants and peafowl, are not documented after 9.4 Ma.

In summary, complementary lines of evidence point to fluvial paleoenvironments and vegetational habitats that were heterogeneous throughout the Potwar sequence, but less three-dimensionally diverse and increasingly open after 11 Ma.

POST-KAMLIAL PALEOCOMMUNITIES: TIME-SLICE RECONSTRUCTIONS

The 14.1 Ma Paleocommunity

This paleocommunity lies at the base of the Chinji Formation. At 14.1 Ma there are 74 mammalian species, including 39 small mammals, 25 large mammals, and 10 megaherbivores (Table 25.2; Figure 25.S2 in the "Supplementary Figures" section in this chapter). (As noted, species counts exclude bats and carnivorans.) Of the 74 mammal species, 37 have been present in the Potwar region since at least 15.2 Ma, with 18 of these co-existing since at least 17.9 Ma.

The most striking difference between the 14.1 Ma mammalian paleocommunity and that present at 15.2 Ma is the net increase in species richness of 25 species, from 49 to 74. The high persistence index (PI, 84%) of the fauna present at 15.2 Ma (Table 25.3) indicates little effect of newly appearing species on survivability of species already present in the region. First occurrences are nearly evenly divided among immigrants and endemic originations (Table 25.S2). This pattern of faunal change—namely, persistence of existing species with addition of new species—resulted in a species-rich paleocommunity on the western Siwalik alluvial plain.

The 14.1 Ma bin itself does not include a small-mammal locality, but there are good small-mammal localities dated to ~14.3 and 14.0. Twenty-four rodent species continue from at least 15.2 Ma. Sciurids peak in species richness with ten species, seven of

them immigrants, including both ground squirrels and tree squirrels. As proposed by Flynn and Wessels (2013), the immigrant squirrels indicate dispersal from both the east and the west following the MCO. While cricetids are both species-rich and abundant at 14.0 Ma (40% of recovered rodent specimens; Fig. 25.2a), sciurids, with just 5% of specimens, remain species-rich yet uncommon. The very low numbers of recovered sciurid fossils suggest low abundance, but predator bias and taphonomic factors may strongly influence the sciurid fossil record.

The 25 large-mammal species include 9 immigrant genera with first occurrences in the Potwar sequence between ~ 15 and 14.1 Ma. Five of these newly appearing genera are bovids, all under 50 kg (see Chapter 23 of this volume); bovids comprise >40% of artiodactyl specimens. Giraffids are the largest artiodactyls, weighing up to 300 kg (Fig. 25.2b).

The 10 megaherbivores include eight rhinocerotids from four subfamilies and two proboscideans from two families (Fig. 25.2c).

The 12.7 Ma Paleocommunity

Potwar mammalian species richness peaks between 13.8 and 12.8 Ma, reaching 84 species at 13.7 and 12.8 Ma. At 12.7 Ma, 79 species are documented, including 31 small mammals, 33 large mammals, and 15 megaherbivores (see Table 25.2; Table 25.S2). The high species richness indicates varied and productive habitats supporting diverse mammals across all three size classes. (As one of many non-analog examples, this is both greater mammalian species richness and a different mammalian size distribution pattern than observed today in most parts of the Indian subcontinent [Roberts 1977].)

Between 14.1 and 12.7 Ma four families first occur: Mammutidae, Stegodontidae, Hominidae, and Gliridae (all plausibly arriving from the west). With an overall PI of 70 (Table 25.3), the Potwar fauna shows more turnover than during the previous interval.

Among the small mammals the diverse muroids include 13 cricetid and gerbilline species as well as the murine *Antemus chinjiensis* (see Fig. 25.2a). A significant contributor to the small-mammal fauna is the hedgehog *Galerix rutlandae*, represented by up to 10% of small-mammal specimens in localities dated to around 12.7 Ma. The 12.7 Ma bin is the last occurrence for seven small-mammal species, including three megacricetodontines.

Of the 33 large-mammal species, 28 are artiodactyls. Apes immigrated from the west; *Sivapithecus* has a first occurrence at 12.7 Ma.

The 15 megaherbivore species include 7 proboscideans and 8 rhinocerotids (see Fig. 25.2c), a number unprecedented in the Recent or even late Neogene fossil record. Up to nine megaherbivore species have been documented within individual fossil localities, which represent time-constrained samples of faunal communities on the order of hundreds to thousands of years (Chapter 4, this volume).

The 11.4 Ma Paleocommunity

With 69 species, the earliest Late Miocene 11.4 Ma mammalian fauna is markedly less species-rich than that at 12.7 Ma (see

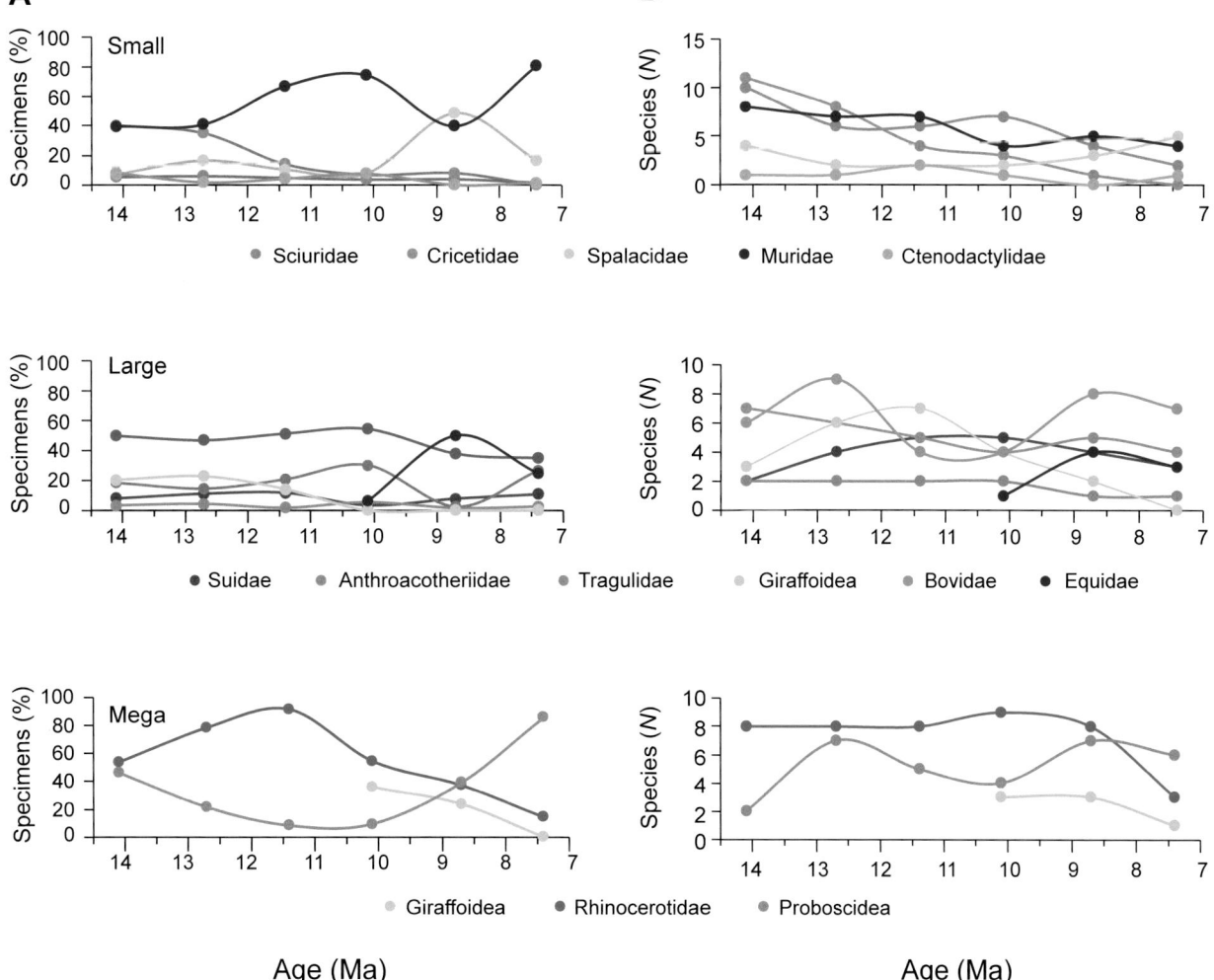

Figure 25.2. Percentage of specimens (A) and number of species (B) for selected small, large, and mega mammal families and orders at six 100-kyr time slices from 14.1 to 7.4 Ma. Number of species and their relative abundance of specimens vary markedly. Equids are not documented in the Siwaliks prior to 10.8 Ma. Giraffoids are plotted as both large and megamammals. (Included mammal specimen numbers at time slices range from 67–646 [small], 267–3404 [large], and 14–543 [mega]. Species number derived from Table III.1.)

Table 25.2), especially within families Cricetidae and Bovidae. Of the 28 small-mammal, 28 large-mammal, and 13 megaherbivore species, 75% were also present at 12.7 Ma (see Table 25.3). Sampling is dense through this interval, minimizing likely taphonomic or other sampling variables that may bias species richness.

As compared to the 12.7 Ma mammalian paleocommunity, the reduced number of small-mammal species, together with the dominance of murines (more than 60% of the small-mammal assemblage), suggests strong interspecies competition (direct or indirect) perhaps amplified by substantial change in habitat structure that resulted in fewer discrete potential niches for small mammals. Although there is no evidence for major global climatic change between 12.7 and 11.4 Ma (Holbourn et al. 2013; Lübbers et al. 2019), regional indicators (Prasad 2008; Srivastava et al. 2018; Polissar et al. 2021) signal an increase in the deciduous component of Potwar forests and

thereby imply that habitats may have become spatially or temporally patchier, with more C_3 shrub and grass understory (see Chapter 24, this volume).

Aside from bovids, the large-herbivore community is relatively similar in species composition to that present at 12.7 Ma. Persisting species include suids *Listriodon pentapotamiae* and *Conohyus sindiensis*, giraffid *Giraffokeryx punjabiensis*, and tragulids *Dorcabune anthracotheroides* and *Siamotragus* Form D. Despite the decrease in bovid species richness from ten to four species (see Table 25.S2 and Table III.1 for details), bovids remain the most abundant artiodactyl group followed by tragulids and giraffids.

The 13 megaherbivore species (8 rhinocerotids and 5 proboscideans) include no new species; the relative stability of the megaherbivore community implies continued abundance of browse <2 meters in height (Dutoit 1990; Dinerstein 1992).

The 10.1 Ma Paleocommunity

The 10.1 Ma mammalian fauna has 69 species (23 small mammals, 30 large mammals, and 16 megaherbivores), similar in number to the 11.4 Ma paleocommunity; however, with a PI of 54 (see Table 25.3), it is a very different fauna, especially with regard to the large herbivores (see Table 25.2). There has been nearly complete faunal turnover among artiodactyls, excepting giraffoids. This faunal change coincides with the first occurrence of equids (*Cormohipparion* at 10.8 Ma) and palaeochoerids, an extinct suoid group (see Chapters 16, 19, this volume). Details of this faunal change at 100-kyr intervals, with numbers of newly appearing immigrant and endemic species and last occurrences, are shown in Figure 25.3.

Of the 17 new artiodactyl species present at 10.1 Ma, 6 are probable immigrants from the west with first occurrences between 10.3 and 10.1 Ma, while 11 are likely originations endemic to the Siwalik bioprovince, with first occurrences between 10.6 and 10.1 Ma. Accompanying the large-mammal taxonomic changes is a marked increase in the range of body size of ungulate taxa, which signals a new guild structure of browsers and grazers discussed later this chapter.

Equid species present at 10.1 Ma include C_4 grasses in their forage (stable isotope data are discussed below). Other local and regional vegetation and climate indicators (Chapter 24, this volume) suggest that by 10.5 Ma, the C_3 habitats, still domi-nated by trees, were increasingly open, with occasional periods of greater seasonal dryness (Karp, Behrensmeyer, and Freeman 2018; Karp et al. 2021).

The small-mammal fauna shows an abundance of murines (the four murine species comprise 73% of small-mammal specimens at 10.1 Ma; no other family accounts for more than 8% [see Fig. 25.2a]). The decrease in small-mammal species richness from that at 11.4 is due to reductions primarily in cricetines and gerbillines. The continuing trend of decreasing small-mammal species richness from the peak at 14 Ma and change in rodent community structure may indicate some or all the following: reduced overall habitat productivity and/or complexity; a shift in prey preferences of predators responsible for accumulating the preserved remains; strong competition among the small mammals not detectable in a fossil record.

The 16 megaherbivore species now include giraffids (see Fig. 25.2c). Increased giraffid size along with neck elongation suggest selection for foraging at higher browse levels than other ungulate browsers (Danowitz et al. 2015), which is possibly linked to increased patchiness and therefore expanded ecotones for C_3 vegetation. The proboscidean and rhinocerotid species composition remains stable, with 14 species continuing from 11.4 Ma. Rhinocerotids are at maximum richness, with nine species documented.

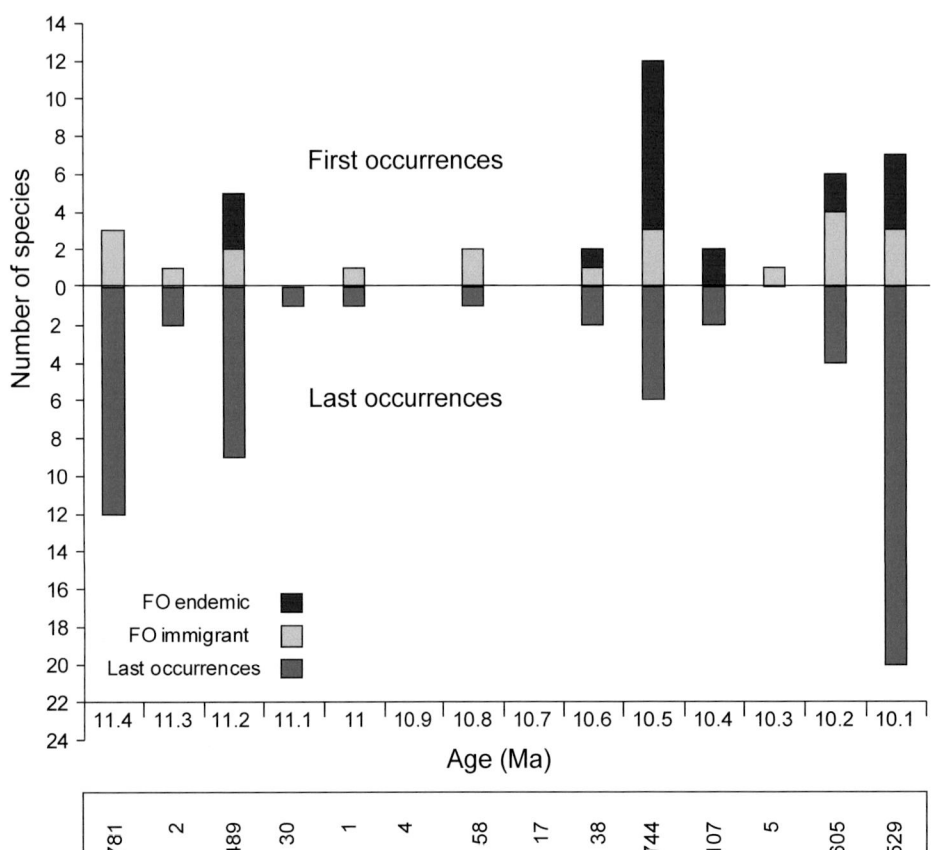

Figure 25.3. First occurrences and last occurrences (blue, below line) for species from 11.4 to 10.1 Ma, highlighting mode of origination (immigrant, green, and endemic, purple). Box with number of catalogued mammal specimens per 100-kyr-interval details the varying quality of the fossil record. (Species first occurrence data from Table III.1.)

We note that our reconstruction of the 10.1 Ma paleocommunity is heavily dependent on one fossil-rich, complex locality at 10.1 Ma, Y311, which is unique in the number of documented mammalian (and bird) species. Three large-mammal species are known only from Locality Y311: the ape *Sivapithecus parvada*, at 40–80 kg double the size of other Siwalik *Sivapithecus* species, the chalicothere cf. *Ancyclotherium* sp., and a small suid representing an unnamed genus. Among mega mammals, the genus *Rhinoceros* appears briefly at 10.1 Ma and then is not recorded again until 8.9 Ma. In comparison with other localities dating between 10.2 and 10.0 Ma, Y311 has an unusually high concentration of tragulid fossils and fewer equid, bovid, and suid specimens; this implies habitat differences, possibly denser vegetation associated with standing water provided by the large abandoned Y311 channel (see Chapter 4 of this volume). The fauna includes a creodont; the Siwaliks record the last occurrence of the order, and a minimum of 12 species of Carnivora. (See Chapter 13 of this volume for details.)

The 8.7 Ma Paleocommunity

The 8.7 Ma faunal community is characterized by high relative abundance of spalacids, equids, and proboscideans, which represents a shift in each size group from earlier dominance of murids, bovids and rhinocerotids, respectively. Relative to the richer 10.1 Ma fauna, the 8.7 Ma mammalian fauna has slightly fewer species, including 18 small mammals, 29 large mammals, and 18 megaherbivores, and a PI of 45 (see Tables 25.2, 25.3; Table 25.S2).

Subterranean bamboo rats (rhizomyines) are first recorded in the Siwaliks at 9.7 Ma, and at 8.7 Ma there are two burrowing rhizomyine spalacids that exceed 1 kg in size. The smaller high-crowned rhizomyine *Kanisamys sivalensis* is abundant. The increase in rhizomyine species richness and relative abundance may indicate greater availability and exploitation of underground resources, such as tubers or grass rhizomes. Together, spalacids and murines comprise more than 85% of identified rodent specimens at 8.7 Ma. Another notable change within the rodent fauna is the decrease in species richness of sciurids from seven at 10.1 Ma to four at 8.7 Ma (including absence of ratufines).

Bovid tribes Bovini and Reduncini (including extant genus *Kobus*) are present, as are Boselaphini and Antilopini. Tragulids continue to be species-rich but decrease in both range of species' body size and relative abundance. Within equids, *Cremohipparion*, characteristic of the Pikermian Biome, has immigrated from the west (Ataabadi et al. 2013), while derived *Sivalhippus* species have dispersed both eastward and westward from the Siwalik bioprovince (see Chapter 16 of this volume).

Giraffa, interpreted as originating within the Indian subcontinent (Chapter 22, this volume), is present in addition to other megaherbivore giraffids. Proboscidean species richness has increased to seven species, adding three species, two immigrants and one endemic origination. Observed change in both proboscidean species composition and increased abundance, relative to rhinocerotids (see Fig. 25.2c), may indicate a reduction in dense browse such as thickets preferred by browsing rhinocerotids (Owen-Smith 2021).

Global cooling is documented in the marine benthic foraminiferal $\delta^{18}O$ record (Holbourn et al. 2013; Lübbers et al. 2019). Local and regional enamel and paleosol isotopes record both a more stressed C_3 vegetation and a minor C_4 component along with a probable decrease in average annual precipitation compared to that at 10.1 Ma (Quade and Cerling 1995; Karp, Behrensmeyer, and Freeman 2018; Feakins et al. 2020; Karp et al. 2021; Polissar et al. 2021 [see Chapters 5 and 24, this volume]).

The 7.4 Ma Paleocommunity

The 48 species, comprising 13 small mammals, 25 large mammals and 10 megaherbivores, document further decline in species richness especially among non-ungulate large mammals and rhinocerotids (see Tables 25.2, 25.3; Table 25.S2). Colobine monkeys have appeared in the region, but apes are not recorded (last occurrence at 8.5 Ma). Other faunal elements that are no longer present include palaeochoerids, aardvarks, glirids, and cricetids. New faunal elements include leporids and extant rodent genera *Mus* and *Hystrix* (both likely immigrants from the west). The 7.4 Ma faunal community is characterized within each size group by high relative abundance of murids, bovids, and proboscideans.

Rhinocerotids are represented by three species (two browsers and one grazer); this relatively sudden reduction in species number, compared to >8 species throughout the previous 8 myr, is very probably associated with a decrease in the overall proportion of consistently productive browse in the region. After 8 Ma, C_4 grasses are an increasing component of the regional and local floodplain vegetation—up to 50% by 7.4 Ma (Feakins et al. 2020; Polissar et al. 2021)—and are regularly consumed by many ungulates (Badgley et al. 2008) and proboscideans. Forested vegetation persists, as evidenced by uncommon colobine monkeys and tree shrews, but given evidence for widespread C_4 grasses, perhaps less continuously and in smaller forested areas and gallery forests along perennial drainages.

The Latest Miocene Fauna

The upper Dhok Pathan Formation is not very fossiliferous in the Potwar Plateau. A rich locality at 6.5 Ma and sparse younger fossil data document continued decreasing mammal species richness (see Table 25.3). Hippopotamidae have a first occurrence at 6.0 Ma, joining the mega-mammal group along with giraffids, rhinocerotids, and proboscideans. The only recognized clearly tree-dependent mammal species is the uncommon colobine monkey, *Mesopithecus*.

PATTERNS OF CHANGE

The 18 to 6 Ma fossil record in the Potwar Plateau offers a temporally long but geographically narrow window on a very extensive bioprovince. Within the 12-myr span, we have focused on the best-sampled Middle–Late Miocene interval between 14.1 and 7.4 Ma. Other temporally overlapping South Asian faunal sequences record similar species or genera to those found in the Potwar (Patnaik 2013). Having summarized highlights of the six successive mammalian paleocommunities at 14.1, 12.7, 11.4, 10.1, 8.7, and 7.4 Ma, we now return to the questions posed earlier in the chapter to guide our analysis of faunal change. Here we explore how well the Siwalik fossil record can address these questions and what conclusions we can draw regarding possible determinants of lineage and community evolution. Throughout, we assume that our compiled faunal records are representative of broad evolutionary and ecological patterns characterizing the original biome of the Siwalik alluvial plain and except where noted are not significantly affected by collecting or taphonomic biases (Chapters 3 and 4, this volume).

How Do Species Richness and the Pace of Faunal Change Vary through Time and across the Three Major Mammalian Size Groupings?

Mammalian species richness declines from Middle to Late Miocene from a high of 84 species at 13.7 and 12.8 Ma to 48 species at 7.4 Ma. More than half of this decrease in species richness occurs within the small mammals (i.e., those less than 1 kg in size). In preceding sections and in Table 25.2, small-, large- and mega-mammal species richness is summarized for the six well-sampled 100-kyr bins that we interpret as paleocommunities and use as baseline metrics for assessing faunal turnover. Figure 25.4 shows the 500 kyr running averages for small-, large-, and mega-mammal species richness over the nearly 7-myr interval. Small-mammal species richness declines sharply between 12.8 and 12.2 Ma and again between 10.5 and 10.0 Ma (Flynn et al. 2023). A comparison of small-mammal localities from two 3-million-year intervals (14.0–11.0 Ma and 9.5–6.5 Ma) illustrates that decreased species richness is not related to sample size (Figure 25.5). After 14.0 Ma large mammals average 25–32 species, except between 11.4 and 10.3 Ma, when there is temporarily lower richness. Mega-mammal species richness is relatively stable from 13.5 to 8.8 Ma and then declines gradually.

Faunal change over time is measured using the PI and NSI to describe both continuity of taxa and change in species richness. Table 25.3 presents PI and NSI for the five intervals between 14.1 and 7.4 Ma; three of the intervals are 1.3-myr long while two are 1.4-myr long.

Patterns of faunal change rarely align for the three mammal size groups. Presented as a 0.5 myr running average in Figure 25.6, species change is low for megaherbivores (PI greater than 95% and NSI less than 5% after an initial peak at 13.9 Ma) until after 8.5 Ma. More taxonomic change is observed for large herbivores than for megaherbivores. The interval between 11.4 and 10.1 Ma includes both low PI and high NSI; these reflect low faunal continuity but relatively stable species richness. For small mammals, excepting the two-million-year period between 12.3 and 10.3 Ma, similar or lower levels of faunal continuity are recorded as compared to large mammals. Small-mammal peaks in NSI at 11.4; 9.2–9.4 Ma are asynchronous to intervals of lowest faunal continuity.

Change in both Siwalik species richness and faunal composition differs following two significant global isotopic shifts in the marine record at 13.9 and 13.1 Ma (discussed in Chapter 24 of this volume). These two isotopic excursions, particularly the

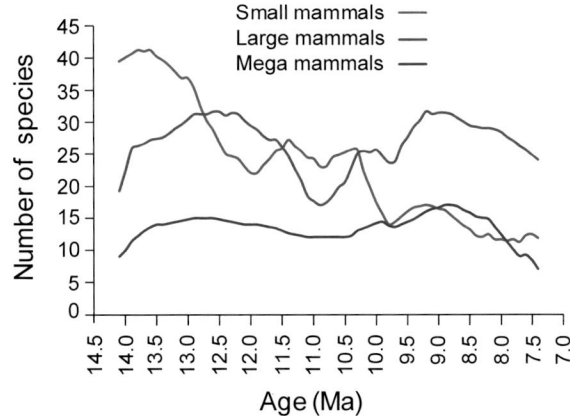

Figure 25.4. Species richness expressed as 500-kyr running averages for small, large, and mega mammals from 14.1 to 7.4 Ma. Note stepped decline in species richness for small mammals, fluctuating species richness with trough around 11.0 Ma for large mammals, and generally lower and more stable species richness for mega mammals. (Data derived from Table III.1.)

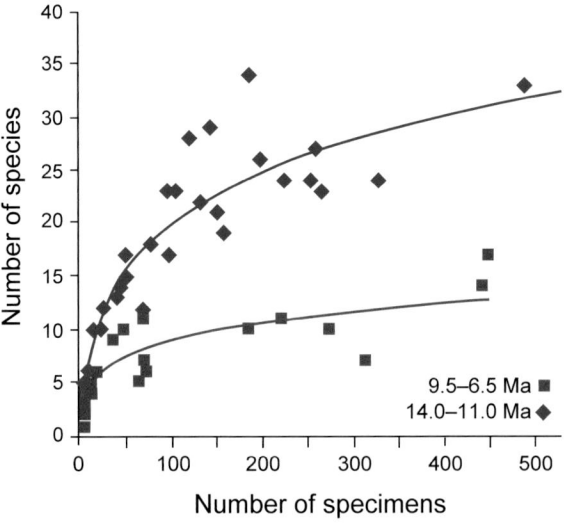

Figure 25.5. Comparison between two 3-myr intervals of small-mammal species richness per number of identified specimens shows reduced species richness in localities younger than 9.5 Ma as compared to localities older than 11.0 Ma. Difference in species richness apparent with fewer than 100 specimens/locality. Regression lines show clear distinction between the younger group of localities (9.5–6.5 Ma: $n = 22$ localities; $y = 150.71x^{0.02} - 150.57$) and the older group of localities (14.0–11.0 Ma: $n = 25$ localities; $y = 26.85x^{0.13} - 29.50$). (X-axis truncated at 500 specimens; 2 localities have >500 specimens with greatest number being 875. Regression analysis performed with PAST 4.13.)

first, are linked to significant increases in East Antarctic ice volume (Holbourn et al. 2005, 2007, 2013; Lübbers et al. 2019). We observe that the Potwar fauna at 13.7 Ma is very similar to that present at 14.1 Ma with a high PI across small-, large-, and mega-mammal species (see Fig. 25.6), which indicates little impact of the marine isotopic shift on species persistence in the

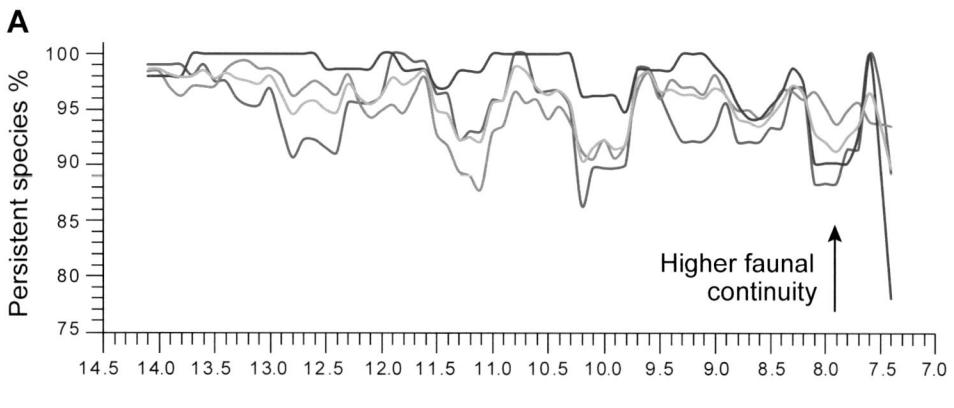

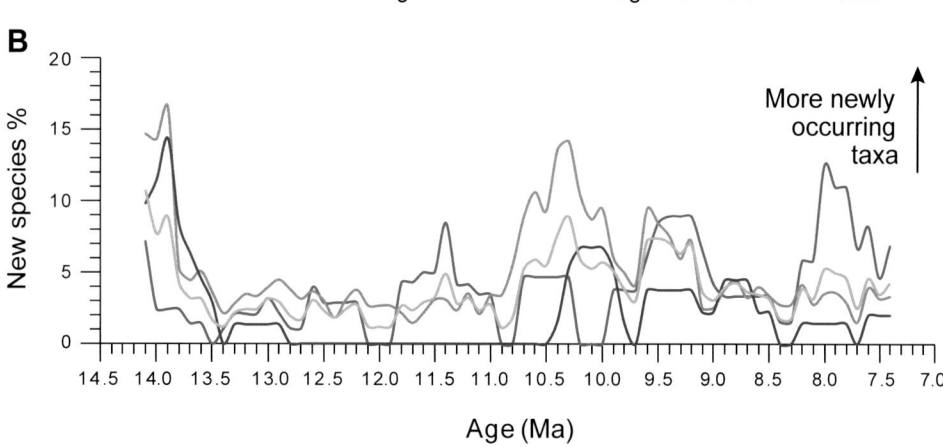

Figure 25.6. Persistence Index (A) and New Species Index (B) expressed as 500 kyr running averages for small, large, mega, and all species from 14.1 to 7.4 Ma. Higher percentages reflect both greater faunal continuity (A) and occurrence of new taxa (B). (Data derived from Table III.1.)

Potwar. However, mammalian species richness increases from 74 to 84 species. At 13.7 Ma, small-mammal diversity peaks with 42 species, and there is little change in the artiodactyl or overall large-herbivore community from species present at 14.1 Ma. More change is seen among proboscideans, with representatives of two families, Mammutidae and Stegodontidae, appearing between 13.9 and 13.7 Ma.

Between 13.2 and 12.7 Ma, the overall PI is lower than for the 14.1–13.7 interval, driven by last occurrences of small mammals (see Fig. 25.4). Rodents decline in species richness from 34 to 26 species, and last occurrences are seen among ground squirrels, small rhizomyines, cricetids, and gerbillines. Four artiodactyls appear between 13.2 and 12.7 Ma, including immigrant genera of bovids and suids. Megaherbivores add one proboscidean species.

How Do Abundance and Richness of Taxa Vary through Time?

Relative abundance and species richness data are analyzed and presented at the family level, except for Proboscidea (see Fig. 25.2). Counts of species richness include range-throughs and derive from first and last occurrence data (see Table 25.S2 and Table III.1). Most small-mammal specimens are identified to the species level, while many large- and mega-mammal postcranial and fragmentary dental specimens are identified to the family level.

We assume some level of positive correlation between relative specimen abundance and original biomass of each taxonomic group (while acknowledging taphonomic factors; see Chapter 4 of this volume). The microvertebrate fossil assemblages are interpreted as accumulations by predators such as raptorial birds and small mammalian carnivores. An elegant study of live-capture and raptor pellet analyses in the western United States (Terry 2010) supports the proposition that the species composition of the Siwalik small-mammal fossil assemblages generally reflects original prey richness and abundance—albeit with some taphonomic filtering based on predator preference and access (e.g., diurnal versus nocturnal activity cycles) along with fossilization and recovery factors. An analogous long-term study of non-carnivoran large- and mega-mammal skeletal elements recovered at Amboseli, Kenya, showed fidelity of relative abundance and richness to modern census data for common species (Western and Behrensmeyer 2009; Miller et al. 2014).

Species with high relative abundance within a given time interval generally have a high FI—that is, they are documented at many localities. There are also many species documented at low relative abundance present at a high percentage of localities. For example, among 33 rodent species documented across a group of five localities dated to 13.6–13.7 Ma (Table 25.4), ten species are recorded at all five localities (100% FI). Two of these, *Punjabemys downsi* and *Antemus chinjiensis*, are very abundant, totaling 39.5% of all specimens across the localities, while two, *Democricetodon fejfari* and *Megacricetodon mythikos*, are minimally abundant, totaling just 1.6% of all specimens. Of five rodent families, sciurids, cricetids, spalacids, murids, and ctenodactylids (see Fig. 25.2a), murids are first or second in species richness in all but the oldest paleocommunity and are usually the most abundant. In contrast, while squirrels are also first or second in species richness in most of the paleocommunities, they are always low in abundance.

For large mammals we survey five artiodactyl families, bovids, tragulids, giraffids, suids, and anthracotheres, and track equids following their first occurrence at 10.8 Ma. Of these ungulates (see Fig. 25.2b), bovids always have high species richness and are nearly always the most abundant group (>40% of specimens). Equids become very abundant by 8.7 Ma. Tragulids maintain relatively high species richness in most paleocommunities but vary considerably in relative abundance, perhaps because of their affinity for well-watered major channel depositional settings (Clauss and Rössner 2014).

Megaherbivores include rhinocerotids, proboscideans, and, beginning at 10.1 Ma, giraffids and are still high in species richness at the end of this 7-myr interval. Proboscidean relative abundance and species richness exceed those of rhinocerotids at 7.4. Ma (see Fig. 25.2c). This switch tracks habitat change between 8.7 and 7.4 Ma; C_4 grasses become more abundant, and megaherbivore species include more grazers and mixed-feeders than selective browsers. This enduring species richness among megaherbivores is striking.

How Does Community Structure Vary through Time as Assessed by Proxies Such as Body Size and Stable Isotope Values of Fossil Enamel Apatite?

Community structure can also be assessed using various ecomorphological attributes, such as body-size distribution and dietary indicators, including tooth wear and stable isotope analysis of enamel. Figure 25.7 summarizes shifts in the distribution of species body size across the six paleocommunities for small, large, and mega mammals. While species of most

Table 25.4. Rodent Species Occurrences and Frequency Indices at Lower Chinji Microsites

Rodent Species	Locality Y59 (13.7 Ma) Specimens (N)	Locality Y640 (13.7 Ma) Specimens (N)	Locality Y641 (13.7 Ma) Specimens (N)	Locality Y651 (13.6 Ma) Specimens (N)	Locality Y430 (13.6 Ma) Specimens (N)	Frequency Index (across 5 localities)	Percentage of Total Specimens Identified to Species (across 5 localities)
Kochalia geespei	. . .	. . .	. . .	. . .	2	20	0.3
Sayimys sivalensis	1	. . .	. . .	3	4	60	1.2
Myomimus sumbalenwalicus	6	2	1	. . .	1	80	1.5
Callosciurus sp.	. . .	. . .	. . .	1	. . .	20	0.1
Dremomys sp.	. . .	. . .	. . .	. . .	3	20	0.4
Tamias "pre-urialis"	15	7	3	8	5	100	5.6
Tamiops large sp.	2	. . .	1	. . .	. . .	40	0.4
Tamiops small sp.	1	. . .	. . .	. . .	. . .	20	0.1
Sciurotamias sp. A (small)	1	. . .	2	. . .	1	60	0.6
cf. *Ratufa* n. sp.	. . .	. . .	. . .	1	. . .	20	0.1
Prokanisamys n. sp.	3	3	1	. . .	. . .	60	1.0
Prokanisamys n. sp. (small)	. . .	. . .	. . .	. . .	1	20	0.1
Kanisamys potwarensis	2	. . .	1	17	23	80	6.4
Kanisamys indicus	22	12	8	31	19	100	13.6
Democricetodon fejfari	1	1	1	1	1	100	0.7
Democricetodon sp. E	. . .	. . .	2	5	16	60	3.4
Democricetodon sp. B-C	12	1	4	. . .	1	80	2.7
Democricetodon kohatensis	7	1	3	11	8	100	4.4
Democricetodon sp. A	2	. . .	. . .	. . .	. . .	20	0.3
Megacricetodon daamsi	14	. . .	. . .	7	. . .	40	3.1
Megacricetodon mythikos	1	2	1	1	1	100	0.9
Megacricetodon sivalensis	9	7	9	9	1	100	5.2
Megacricetodon aguilari	6	4	8	9	3	100	4.4
Punjabemys leptos	3	1	2	2	9	100	2.5
Punjabemys mikros	1	1	2	. . .	2	80	0.9
Punjabemys downsi	11	7	2	16	93	100	19.1
Myocricetodon sivalensis	1	. . .	. . .	. . .	. . .	20	0.1
Myocricetodon large sp.	1	. . .	. . .	. . .	1	40	0.3
Dakkamys barryi	1	. . .	. . .	. . .	1	40	0.3
Mellalomys lavocati	. . .	. . .	. . .	. . .	2	20	0.3
Mellalomys perplexus	. . .	1	. . .	. . .	. . .	20	0.1
Potwarmus primitivus	1	. . .	1	. . .	. . .	40	0.3
Antemus chinjiensis	31	12	26	23	39	100	19.4

families fall within one of the three primary size groups, rhizomyine spalacids and giraffids add much larger-sized species after 10 Ma, each spanning two size groups. Between 14.1 and 11.4 Ma most small-mammal species are less than 100 g in size. After 11.4 Ma there is a greater number of larger-sized species, and the specimen size distribution shows an abundance of larger-sized spalacids at 8.7 Ma. Within large mammals, the relative proportion of herbivore species in the 20–100 kg size range remains relatively constant between 14.1 and 8.7, although across the six paleocommunities the relative abundance of 20–100 kg large-mammal specimens shifts markedly. Whereas from 14.1 to 11.4 Ma, 60–70% of catalogued specimens are from species less than 50 kg, this drops to 30–40% from 10.1 to 7.4 Ma. At 10.1 Ma species in the 50–99 kg group are most abundant. After 10.1 Ma more than 60% of large-herbivore specimens are from species in the 250–800 kg range.

For mega mammals, the size distribution of species is relatively stable, but with a sharp increase in relative number of species >3,000 kg between 8.7 and 7.4 Ma.

Of note is the extraordinary number of contemporaneous megaherbivore species documented in the Potwar region: at least 10 and as many as 18 species between 14 and 8 Ma (see Fig. 25.4, Table 25.2). There are no living or recent communities with this degree of species richness to provide explanatory clues (Faith et al. 2018; Faith, Rowan and Du 2019). Clearly, Siwalik Miocene habitats were highly productive and communities structured in ways to allow many megaherbivores to coexist over millions of years. Average megaherbivore species size increases after 11.4 Ma (Fig. 25.7) with the addition of more proboscidean species and larger rhinocerotids.

Inferences about diet are based on tooth anatomy (crown height, occlusal morphology, enamel thickness), microwear and

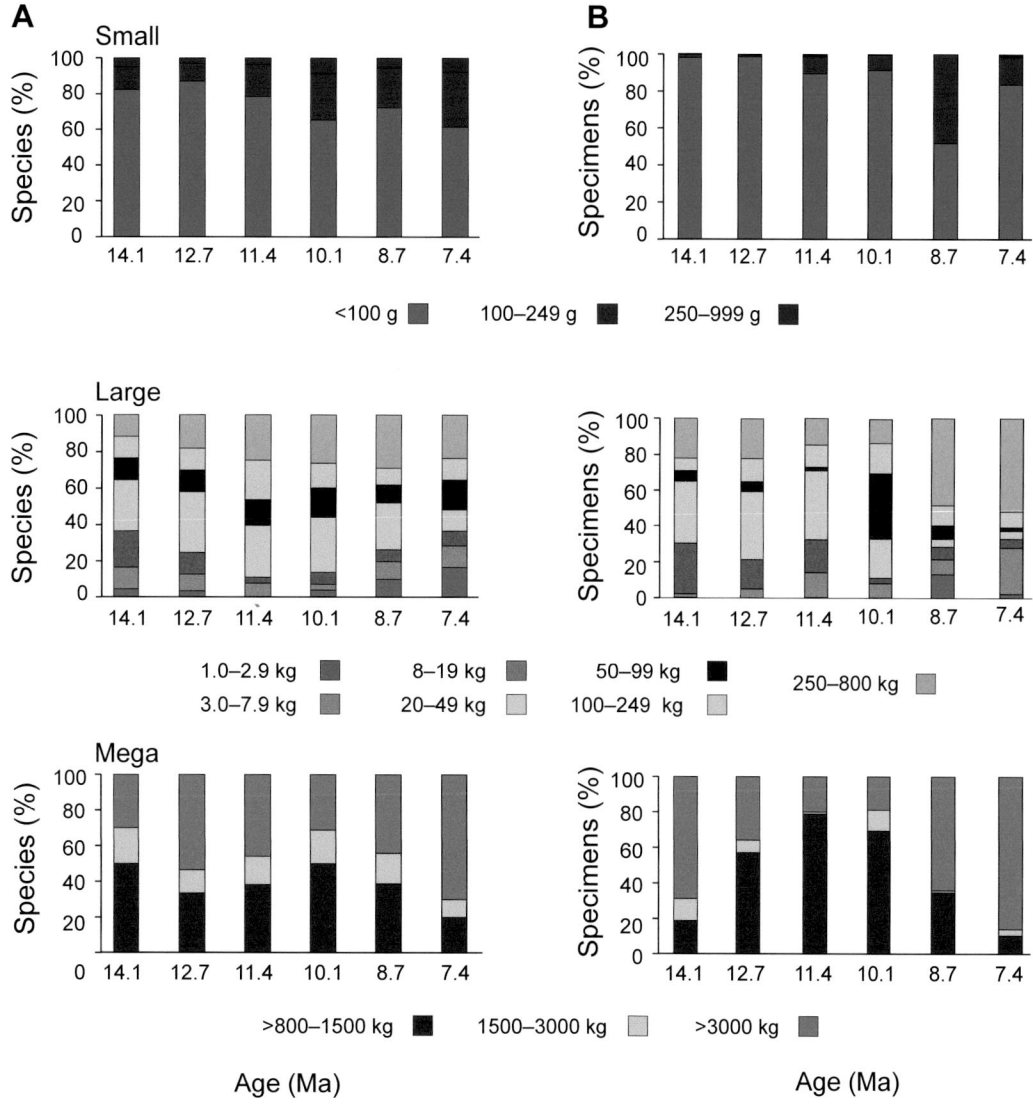

Figure 25.7. Size distribution with percentages of species (A) and specimens (B) in 13 size subdivisions. Distributions are compared for the six 100-kyr time slices, 14.1–7.4 Ma. While generally similar over time, the relative abundances in different sized groups do not always parallel species number. (Species data are derived from Table III.1; included specimen counts per 100-kyr time slice range from 448 (at 7.4 Ma) to 4373 (at 10.1 Ma).

mesowear, and enamel stable isotopic values. Herbivores were predominantly browsers and mixed-feeders throughout much of the Potwar sequence, with increasingly frequent signs of some grazing after 10 Ma (see Chapters 15–23 for details on particular taxa). Most megaherbivores were primarily browsers prior to 10 Ma, after which there is more evidence for mixed-feeding (browse and graze, including C_3 grasses). Stable carbon and oxygen isotope data from more than 70 large- and mega-mammal teeth representing four orders (Primates, Artiodactyla, Perissodactyla, and Proboscidea) provide evidence about the isotopic ecology of species and the types of forage available to herbivores (see Table III.2). The overall pattern between 18 and 6 Ma is summarized in Chapters 5 and 24 of this volume. In brief, the earliest indicator of C_4 grasses in the diet of equids occurs at around 10 Ma, and increased C_4 foraging by species from three orders occurs between 8 and 6 Ma. (Mouse data also show C_4 foraging after 8.0 Ma [Kimura et al. 2013].) The oxygen isotope record demonstrates a slow and consistent shift over millions of years towards more positive values, reflecting ever more ^{18}O-enriched average precipitation and water sources in the region (Quade and Cerling 1995).

We characterize the isotopic niche space of 25 species, plus older and younger giraffids, in Table 25.5. Included are apes, ruminant artiodactyls (bovids, tragulids and giraffids), non-ruminant artiodactyls (suids, palaeochoerids, and anthracotheres), perissodactyls (equids, chalicotheres and rhinocero-

tids), and deinotheres. For each taxon, mean and median stable carbon and oxygen isotope values are given, along with the lowest and highest stable isotope values and the age range of teeth sampled; each taxon is represented by at least five specimens. Sampling is densest between 13.5 and 7.5 Ma but spans the six focal paleocommunities as shown in Figure 25.8.

Two striking findings emerge from these data. First, there are no exclusive C_4 grazers; nearly half of the species sampled consumed some C_4 grass, but none specialized on C_4 grass (Fig. 25.8 and Table 25.5). This is in stark contrast to the foraging strategies of many large and megaherbivore species inhabiting savannah ecosystems in eastern and southern Africa today, the only regions supporting ungulate diversity somewhat comparable to that documented in the Siwalik Miocene record (Said 2003; Cotterill 2004; Owen-Smith 2021). Average $\delta^{13}C$ values of Siwalik species range from a low of −14.4‰ (palaeochoerid *Yunnanochoerus gandakasensis* samples range in age from 10.2 to 9.2 Ma) to a high of −6.8 ‰ (bovid *Miotragocerus punjabicus* samples range in age from 8.0 to 6.5 Ma). The Late Miocene isotopic record of East African mammalian herbivores records median values greater than −8‰ for all families except deinotheres by 9.6 Ma (Uno et al. 2011). In the Siwaliks, only two species, the equid *Sivalhippus anwari* and the anthracothere *Merycopotamus dissimilis*, have median $\delta^{13}C$ values higher than −8‰. While many species may have regularly foraged on browse, C_3 grasses were likely a significant presence

Table 25.5. Siwalik Fossil Enamel Stable Carbon and Oxygen Isotope Summary Statistics for 25 Species and Two Groups of Giraffids

Family	Species	N Specimens*	Age Range of Specimens (Ma)	$\delta^{13}C$ Median Value (‰)	$\delta^{13}C$ Mean ± SD (‰)	Range $\delta^{13}C$ Values (‰)	$\delta^{18}O$ Median Value (‰)	$\delta^{18}O$ Mean ± SD (‰)	Range $\delta^{18}O$ values (‰
Hominoidea	*Sivapithecus sivalensis*	6	(9.4, 8.8)	−13.8	−13.9±0.5	(−14.9, −13.4)	−2.3	−2.9±1.9	(−6.0, −0.9
Bovidae	*Selenoportax aff. vexillarius*	7	(10.1, 10.1)	−12.8	−12.9±0.8	(−14.3, −11.5)	−3.3	−4.5±2.2	(−7.9, −2.0
Bovidae	*Pachyportax falconeri*	7	(9.8, 8.9)	−12.0	−12.1±0.9	(−13.7, −11.5)	−2.6	−3.8±2.8	(−7.8, −0.7
Bovidae	*Pachyportax dhokpathanensis*	40	(8.9, 7.3)	−11.3	−11.0±2.4	(−15.1, 0.0)	−4.1	−4.2±1.9	(−8.3, 0.2)
Bovidae	*Miotragocerus pilgrimi*	23	(10.1, 8.8)	−12.7	−12.2±2.5	(−14.5, −1.9)	−5.6	−5.5±2.8	(−9.8, 0.5)
Bovidae	*Miotragocerus punjabicus*	11	(8.0, 6.5)	−11.3	−6.8±6.5	(−11.7, 2.3)	−1.9	−1.1±2.7	(−1.3, 3.5)
Giraffidae	(11 Ma and older)	34	(14.1, 11.1)	−12.6	−12.6±0.9	(−15.0, −10.7)	−4.5	−3.9±2.4	(−7.7, 0.8)
Giraffidae	(10.3 Ma and younger)	45	(10.4, 7.3)	−12.8	−12.8±1.0	(−15.5, −10.9)	−2.1	−2.5±2.8	(−8.5, 2.4)
Tragulidae	*Dorcabune nagrii*	5	(9.4, 9.4)	−11.6	−11.7±0.5	(−12.2, −10.9)	−2.4	−5.4±3.2	(−8.3, −1.6
Tragulidae	*Dorcatherium majus*	13	(10.2, 7.4)	−11.5	−11.2±1.1	(−12.4, −8.3)	−3.8	−4.1±1.7	(−7.1, −1.7
Suidae	*Listriodon pentapotamiae*	10	(13.4, 11.4)	−13.8	−13.6±0.8	(−14.4, −12.0)	−5.8	−5.2±2.3	(−7.3, 0.0)
Suidae	*Conohyus sindiensis*	7	(13.5, 11.4)	−12.9	−12.6±0.8	(−13.4, −11.1)	−6.8	−7.7±1.7	(−10.9, −5.8
Suidae	*Hippopotamodon sivalense*	39	(10.4, 6.4)	−11.9	−11.7±1.5	(−14.0, −7.0)	−5.6	−5.4±2.1	(−9.0, 3.5)
Suidae	*Propotomochoerus hysudricus*	62	(10.2, 7.2)	−12.3	−12.0±1.3	(−13.9, −7.7)	−5.5	−5.6±2.1	(−9.4, 1.3)
Palaeochoeridae	*Yunnanochoerus gandakasensis*	6	(10.2, 9.2)	−14.3	−14.4±0.4	(−14.9, −13.9)	−6.6	−6.5±2.5	(−9.0, −3.0
Anthracotheriidae	*Merycopotamus nanus*	25	(13.5, 11.4)	−11.8	−11.7±0.7	(−12.9, −10.1)	−7.9	−7.7±1.3	(−11.0, −5.4
Anthracotheriidae	*Merycopotamus medioximus*	21	(10.2, 8.7)	−12.4	−12.3±1.0	(−13.9, −9.8)	−6.7	−6.4±1.3	(−6.4, −3.0
Anthracotheriidae	*Merycopotamus dissimilis*	11	(7.8, 6.4)	−6.7	−7.8±2.8	(−12.1, −4.2)	−5.9	−5.8±1.5	(−7.9, −2.1
Equidae	*Sivalhippus nagriensis*	51	(10.2, 9.1)	−11.8	−11.4±1.5	(−13.8, −6.2)	−4.4	−4.4±2.2	(−9.2, 0.9)
Equidae	*Sivalhippus perimensis*	20	(9.0, 7.7)	−8.1	−8.2±2.6	(−12.0, −3.0)	3.1	−3.2±1.9	(−5.9, 1.2)
Equidae	*Sivalhippus theobaldi*	6	(9.4, 7.8)	−8.5	−9.4±2.0	(−11.9, −7.2)	−3.4	−3.0±1.5	(−4.9, −0.9
Equidae	*Sivalhippus anwari*	5	(7.4, 7.1)	−7.5	−7.7±0.3	(−8.1, −7.5)	−2.4	−2.4±1.4	(−3.9, −0.3
Equidae	*Cremohipparion antelopinum*	12	(8.9, 6.5)	−8.3	−8.4±3.1	(−12.0, −2.0)	−3.0	−3.1±2.4	(−7.9, 1.1)
Chalicotheriidae	*Anisodon salinus*	12	(14.1, 8.4)	−12.1	−12.0±0.8	(−13.0, −10.8)	−3.9	−3.4±1.8	(−6.4, −0.8
Rhinocerotidae	*Brachypotherium perimense*	24	(13.1, 8.0)	−13.1	−13.1±0.6	(−14.8, −12.0)	−8.3	−7.7±1.7	(−10.2, −3.7
Rhinocerotidae	*Gaindatherium browni*	5	(13.7, 9.4)	−12.5	−12.4±1.1	(−13.4, −10.7)	−5.0	−5.1±0.9	(−6.3, −3.9
Deinotheriidae	*Deinotherium indicum*	6	(12.4, 9.3)	−12.8	−12.7±1.2	(−14.3, −10.6)	−6.3	−5.7±2.6	(−8.5, −1.9

*A minimum of five unassociated teeth were sampled for each taxon.

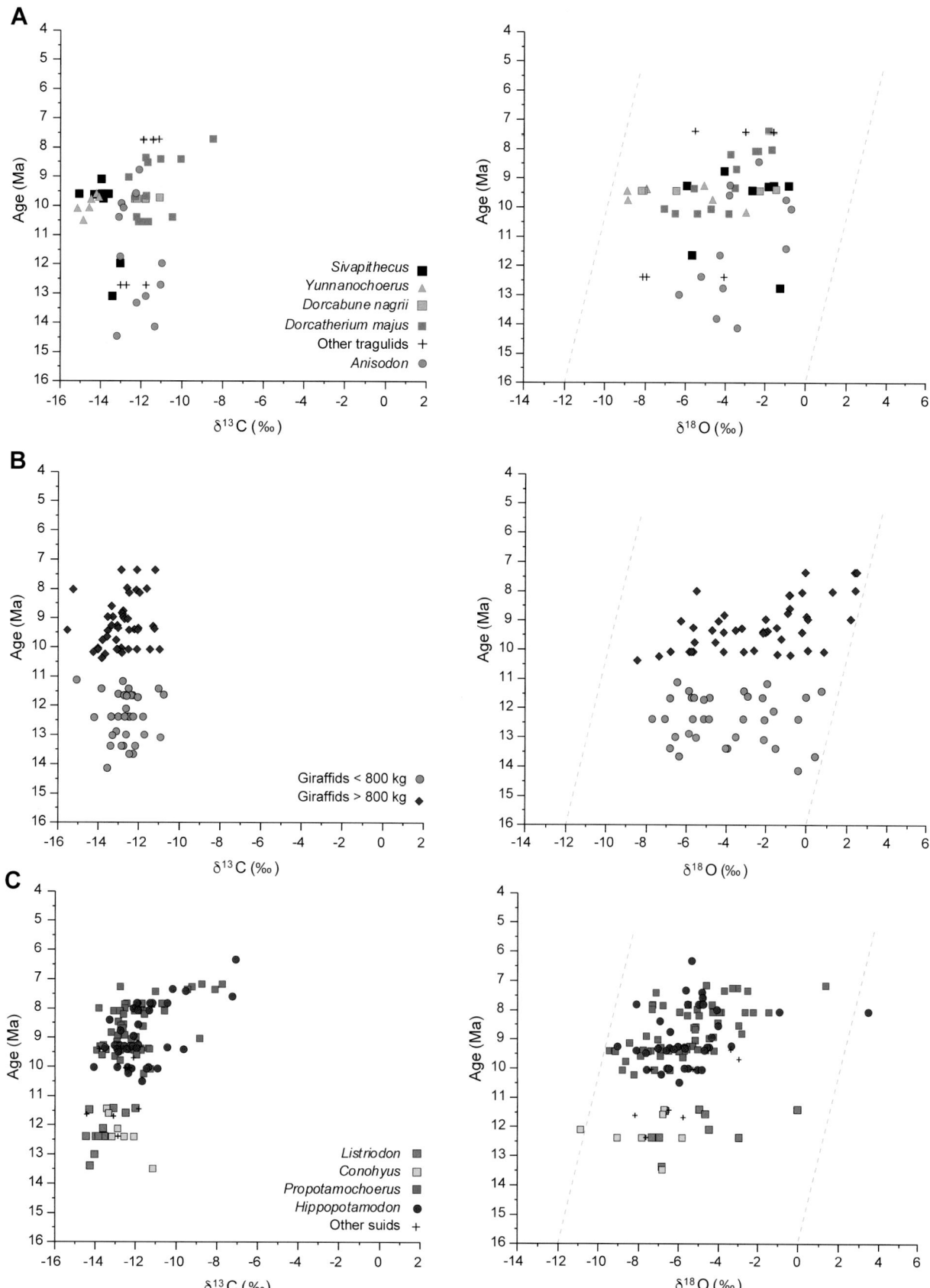

Figure 25.8. Stable carbon and oxygen isotope data of selected Siwalik mammals: A. hominoids, palaeochoerids, tragulids and chalicotheres; B. giraffoids; C. suids; D. anthracotheriids; E. bovids; F. equids; G. rhinocerotids; H. proboscideans. For oxygen isotope data dashed lines encompass most values and highlight change through time. (Data derived from from Table III.2.)

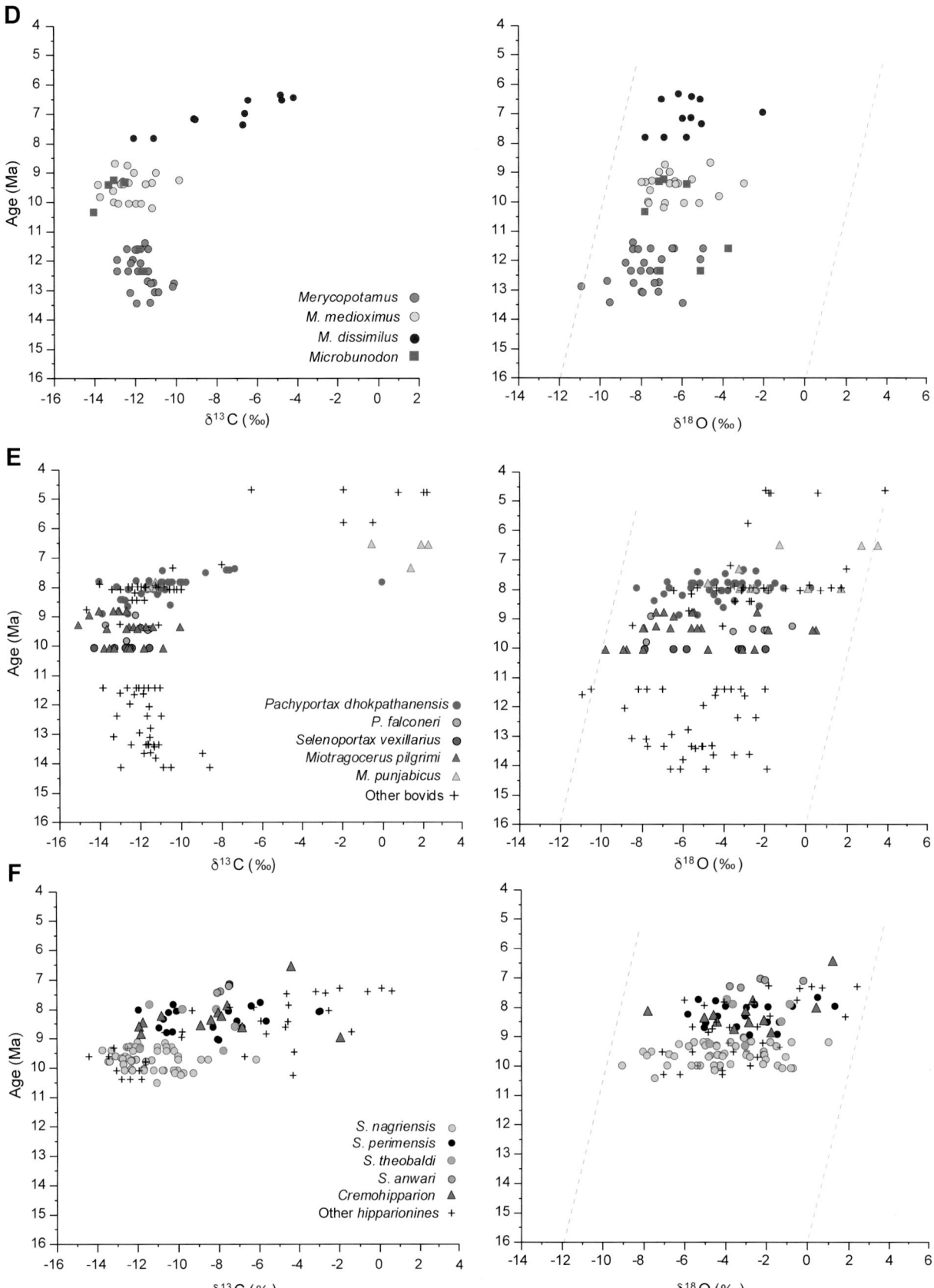

Figure 25.8. (Continued)

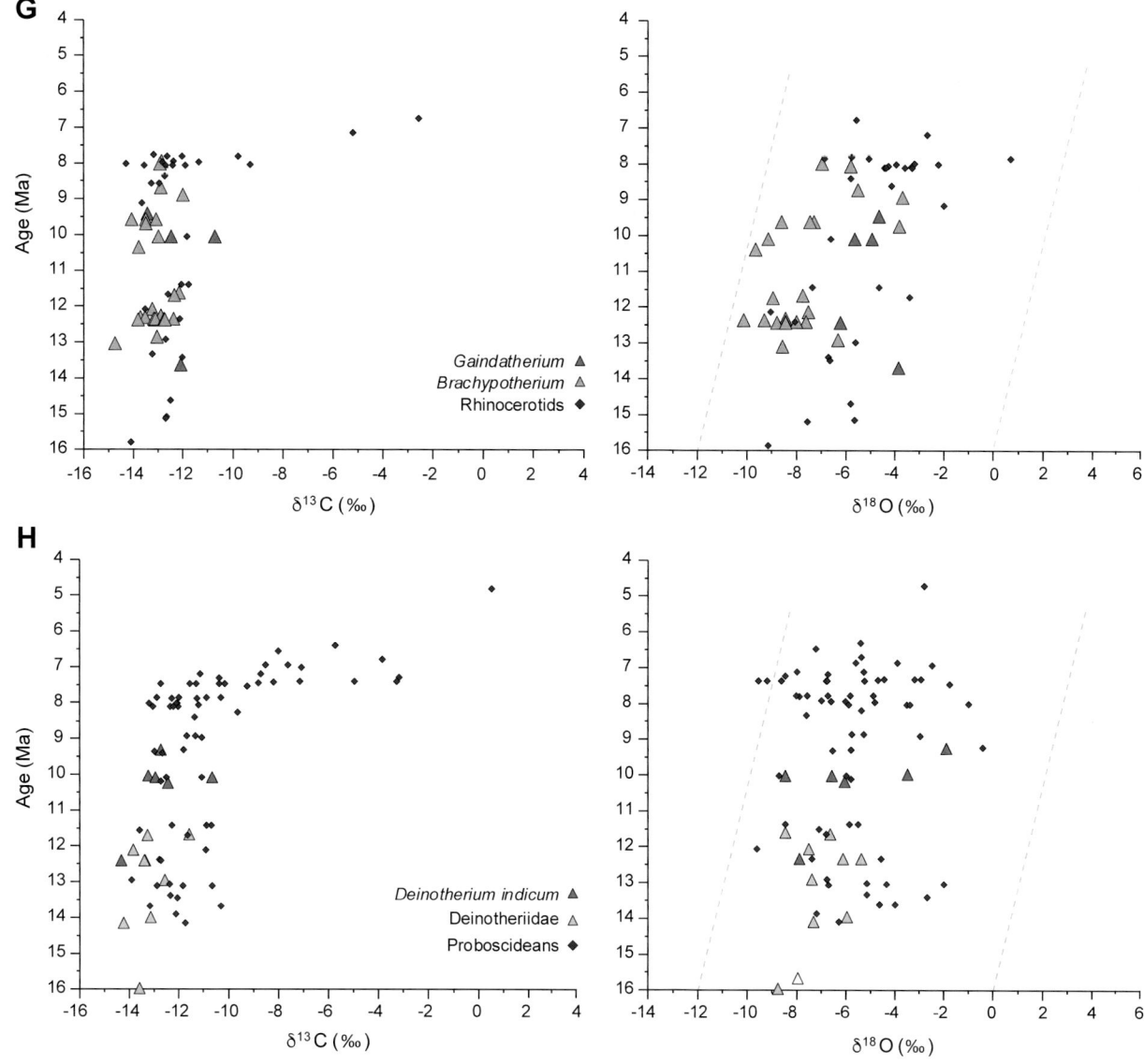

Figure 25.8. (Continued)

in the region, some 15–35% of groundcover, well into the Late Miocene. Estimates for abundance of C_3 grasses derive from dental wear studies of Siwalik taxa, pollen records from Nepal, and models based on soil carbonates from modern East African ecosystems (Hoorn, Ohja, and Quade 2000; Prasad 2008; Magill, Ashley, and Freeman 2013; Srivastava et al. 2018).

The second striking finding is that most of the 14 species with isotopic signatures indicating exclusive C_3 diets have a 2–3‰ range in $\delta^{13}C$ values (see Table 25.5). This is about half of the total range of $\delta^{13}C$ values observed in any 100-kyr interval between 14 and 11 Ma (see Fig. 25.8) and is comparable to the recorded range in many extant East African mammal species (Cerling et al. 2015). The 2–3‰ range in $\delta^{13}C$ values within Siwalik species reflects species diet breadth, plant tissue variation, and change over millions of years, all of which suggest a high degree of stability.

Equids, identified as early consumers of C_4 vegetation (Morgan, Kingston, and Marino 1994) and generally having high-crowned teeth suitable for serial sampling, have been a focus for stable isotope analysis (e.g., Bryant et al. 1996; Nelson, 2005; Chapter 16, this volume). In the Siwaliks there is perhaps surprisingly little isotopic evidence for niche differentiation among the five equid species sampled (see Table 25.5). The oldest *Sivalhippus* species, *S. nagriensis*, has the lowest mean $\delta^{13}C$ value (−11.4 ‰), while younger species, *Sivalhippus perimensis* and *Cremohipparion antilopinum*, include values indicating an exclusively C_4 diet. All species have values in both the C_3 and mixed C_3–C_4 dietary range except for the youngest-appearing *S. anwari*, which has only mixed C_3–C_4 isotopic values (see Fig. 25.8f). Dental microwear of equids is consistent with a mixed C_3–C_4 diet including some browse; mesowear suggests a trend toward more abrasive forage through time (see

Chapter 16 of this volume). More than 20 equid premolars and molars have been serially sampled (see Chapter 16 of this volume). The average intra-tooth variation in δ¹³C is 1.9 ± 0.8‰ with a minimum of 0.6‰ and a maximum of 3.6‰. Enamel profiles of four equid teeth dated from 10.1 to 7.4 Ma are detailed in Figure 25.9. Each data point sampled about 2 mm of crown height and, taking into consideration the timing and shape of enamel deposition, represents several weeks to a month of carbon incorporation into enamel (Bendrey et al. 2015). Intra-tooth variation within equids is not very different from the total isotopic range of most C₃-foraging species. As noted in Chapter 16 of this volume, individual animals did not consume mostly C₄ grass for several months and then switch to mostly C₃ graze or browse for several months. Instead, on an annual scale there was modest variation in the isotopic composition of an animal's diet. Observed differences between individuals through the section may reflect the flux of C₃ and C₄ grasses in the region on precessional and longer-

frequency Milankovitch scales—cooler and wetter periods favoring more forested and woodland habitats with predominantly C₃ grasses and warmer and drier periods favoring more woodland to wooded-grassland habitats with predominantly C₄ grasses. Compared to equids, *Brachypotherium perimense*, the largest rhinocerotid and inferred C₃-browser, exhibits a narrow range of both intra-tooth and inter-tooth variation in teeth dated to between 8.0 and 12.3 Ma, which suggests species fidelity to a limited range of available forage as assessed isotopically (Fig. 25.9).

Both the first relatively abundant large bovid greater than 100 kg, *Selenoportax* aff. *vexillarius*, and the first bovine, *Pachyportax falconeri*, are C₃ foragers (see Table 25.5, Fig. 25.8e); large relatively high-crowned ruminants might be expected to exploit available C₄ grasses. The larger bovine *P. dhokpathanensis* and two boselaphine *Miotragocerus* species were mixed C₃–C₄ foragers and include some 100% C₄ grass isotopic sample values. Dental wear studies support browsing and mixed feeding diets

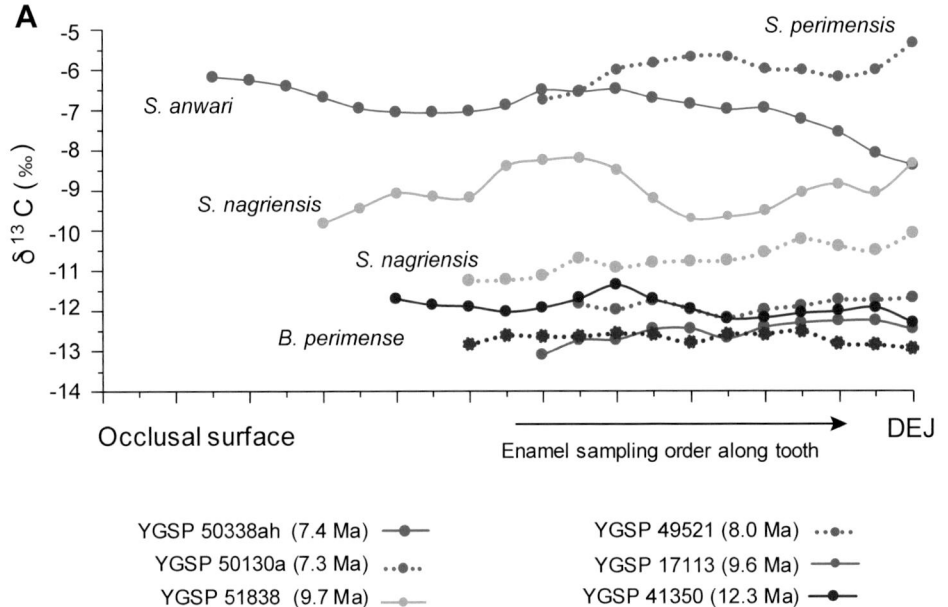

Figure 25.9. Serial sampling of four equid (*Sivalhippus* spp.) and four rhinocerotid (*Brachypotherium perimense*) teeth for stable carbon (A) and oxygen (B) isotopes with specimen number and locality age noted. Alignment at right edge of figure is at youngest enamel values for each tooth, the dentine enamel junction (DEJ). In all teeth equids have more positive values than rhinocerotids. Tooth position as follows: YGSP 50338ah (right P2), YGSP 50130a (left M3), YGSP 51838 (right P4), YGSP 12947 (right M3), YGSP 41350 (right m2), YGSP 49521 (left M3), YGSP 17113 (right m3), YGSP 31577 (right m2).

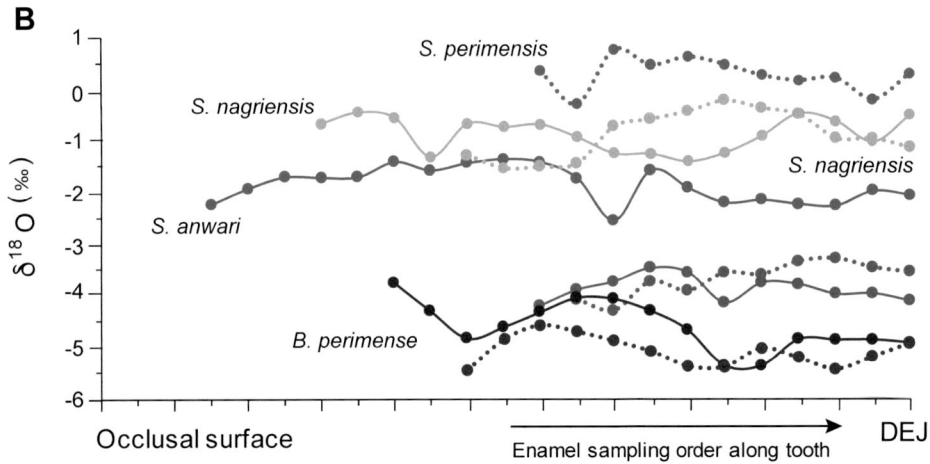

(see Chapter 23 of this volume). In addition to bovids and equids, the most hippo-like anthracothere *Merycopotamus dissimilis*, the suines *Hippopotamodon sivalense* and *Propotamochoerus hysudricus*, and some rhinocerotids and proboscideans consumed some C_4 vegetation (see Table 25.5, Fig. 25.8c–g; Chapters 15, 18–20, this volume).

The stable oxygen isotope data from fossil herbivore enamel are interpreted within the context of the observed increase of $\delta^{18}O$ values over millions of years recorded in both paleosol carbonates and tooth enamel (see Figs. 5.4, 25.8, Chapter 5, this volume). Taxonomic differences, for example between giraffids and anthracotheres, were noted in a study that focused on mammals sampled over 30 km within a 200-kyr (9.2–9.4 Ma) interval (Morgan et al. 2009). The large anthracothere lineage *Merycopotamus*, along with the suid *Conohyus sindiensis* and the rhinocerotids *Gaindatherium browni* and *Brachypotherium perimense*, consistently have lower $\delta^{18}O$ values than other species; they are relatively depleted in ^{18}O, which suggests obligate drinking from water sources such as small streams, not evaporating ponds (see Table 25.5, Fig. 25.8c,d,g). Taxa with higher than average $\delta^{18}O$ values include the ape *Sivapithecus*, all giraffids, the chalicothere *Anisodon*, and the suid *Listriodon pentapotamiae* (see Fig. 25.8a–c). All but the suid are canopy and/or above 2 m browse-feeders; their higher $\delta^{18}O$ values are consistent with deriving significant water from leaves and fruit. *Sivapithecus* has the most restricted range in $\delta^{18}O$ within this group. The other species with high $\delta^{18}O$ values is the Late Miocene bovid *Miotragocerus punjabicus*, interpreted as an obligate drinker; this species also has the highest average $\delta^{13}C$ value (see Fig. 25.8e). Equids, also interpreted as obligate drinkers, record a wide range of oxygen isotope values (see Fig. 25.8f).

Intra-tooth profiles of equid teeth provide additional insights into the behavior of these hipparionines. The mean $\delta^{18}O$ range within a single tooth is 2.5 ± 0.7 ‰ with a maximum of 3.9‰ (from a tooth dated to 8.0 Ma), and a minimum of 1.2‰ (from a tooth dated to 7.2 Ma). Three of the four teeth shown in Figure 25.9 have sharp inflection points near lowest $\delta^{18}O$ values suggesting a relatively modest but rapid change in $\delta^{18}O$ values of water sources and plants, perhaps associated with onset of monsoonal rains (Nelson 2005). As for carbon isotopes, observed variation in oxygen isotopes does not support a seasonal behavioral change in pattern or source of water consumption. Rather, the differences between animals reflect regional-scale change in the oxygen isotopic composition of water in fluvial systems from precipitation and snowmelt as well as in consumed vegetation. The four *Brachypotherium perimense* teeth show a generally similar pattern with a narrower range of variation compared to equids (see Fig. 25.9).

What can the isotopic record contribute regarding the habitat structure for these herbivores between 14.1 and 7.4 Ma? At the family and subfamily level, some taxa have a consistent C_3 diet derived primarily from wooded and forested habitats. These taxa—primates, tragulids, palaeochoerids, giraffids, chalicotheres, and deinotheres—include frugivorous and folivorous species foraging at ground-level and/or from leaves and fruit

two or more meters above ground level. Some obtained a high percentage of their water from evaporative tissues of plants while others drank from non-evaporative water bodies. We have evidence for behavioral differences between the most common suid species present between 14.1 and 10.4 Ma, with *Listriodon pentapotamiae* having higher $\delta^{18}O$ values than *Conohyus sindiensis*. Later common suids, such as *Hippopotamodon sivalense* and *Propotamochoerus hysudricus*, are similar isotopically (see Fig. 25.8c). From their first records in the Potwar region, equids exploited a wide range of forage as mixed-feeders and were the first documented consumers of C_4 grasses. This behavioral change is not reflected until after 8 Ma in co-occurring mixed-feeding species, including some bovids, suids, anthracotheres, rhinocerotids, and proboscideans. If some species were mostly grazers, they were consuming both C_3 as well as C_4 grasses, at least until after 7 Ma. While woodland habitats were regularly present in the region, after 8 Ma forest habitats shrank and more wooded-grasslands spread (see Fig. 24.7). These changes are reflected in loss of arboreally dependent species and addition of species able to forage in more open areas with C_4 grasses (Nelson 2007; Badgley et al. 2008).

Does Permeability of the Bioprovince Boundaries Vary through Time?

Surrounded by mountains to the west and particularly the north, mountains and large rivers to the east, and the Indian Ocean to the south, the Indian subcontinent has presented significant physical barriers to potential immigrants since its docking with the Asian landmass (e.g., Reynolds, Copley, and Hussain 2015). We can use immigrant mammalian taxa to assess varying permeability of the bioprovince boundaries (Badgley et al. 2015). We do not know where many immigrant taxa originated, but for those with a sufficient fossil record, most arrived from the west, from what is now Afghanistan, Iran, and Anatolia, and farther west from Europe and Afro-Arabia. Lack of sufficient information about the paleontological record and its calibration in bioprovinces adjacent to the Indian subcontinent limit our ability to assess emigration from the Siwalik bioprovince.

The pattern of Siwalik mammal immigration rates between 14.1 and 7.4 Ma does reveal variation in bioprovince boundary permeability (see also Chapter 26, this volume). For mammals overall, immigration rates average 11.0 species per million years. There are no documented first occurrences of immigrant species between 10.1 and 9.7 Ma and 8.4 and 8.0 Ma; between 12.7 and 12.1 Ma, there is only one immigrant (the giraffid *Injanatherium* at 12.4 Ma.) These data do not suggest impermeable boundaries for prolonged periods of time (i.e., for more than 300 kyr).

A smoothed curve of 500 kyr running averages for immigrants calculated over the 14.1 to 7.4 Ma interval shows the highest rates of immigration, exceeding or nearing 2 species per 100 kyr, occurring between 14.1 and 13.8 Ma, 10.4 and 10.2 Ma, and 9.6 and 9.4 Ma (Fig. 25.10). Long intervals with low immigration rates (≤ 0.8 species per 100 kyr) occur between 12.8 and 11.6 Ma and after 8.5 Ma and suggest intervals of lower permeability reflecting physical and/or biological barriers.

For both large and small mammals, the interval with the largest number of immigrant species lies between 11.4 and 10.1 Ma, especially after 10.7 Ma, and includes species from 12 different families. The 13 large-mammal immigrants include first occurrences of equids and paleochoerids and five bovids from multiple tribes. The five small-mammal immigrants include the large shrew *Suncus*, a diatomyid, a tree squirrel, a tachyoryctine, and a loris. The highest immigration rate interval (10.7–10.1 Ma) follows a period of low sea level (Holbourn et al. 2013; Lübbers et al. 2019), but our record shows that newly arriving immigrant species are not restricted to such events. The 10.1-8.7 Ma interval has the second highest number of immigrants across 10 families, with 11 large-mammal species, 4 small-mammal species, and 2 megaherbivores.

Are There Detectable Effects of Immigrants on Endemic Origination Rates?

Between 14.1 and 7.4 Ma, there are no intervals in the Potwar sequence longer than 200 kyr without at least one new species occurrence—immigrant or endemic. Endemic origination rates are mostly low between 14.0 and 10.8 and higher between 10.7 and 8.7 Ma. A moderate rate of endemic species originations between 12.6 and 12.2 Ma (six species across five families) coincides with an interval of low immigration rates and is one of relative global, regional, and local climate stability (Holbourn et al. 2013; Lübbers et al. 2019; Polissar et al. 2021; Fig. 25.10).

The first occurrence rates of immigrant and endemically originating species generally track each other across time, sug-

gesting no negative interactions. There is a possible exception between 12.6 and 12.2 Ma, when there is a moderate endemic origination rate and a low immigration rate, probably reflecting barriers restricting immigration during that interval, and another between 8.2 and 7.6 Ma, when immigration rates are low but endemic origination rates are relatively high.

A total of 178 newly appearing species is recorded during our interval of study, the number appearing via immigration (78 species, 44%) being fewer than that originating endemically (100 species). With 197 last occurrences, overall species richness declines over our 6.7-myr study interval.

Do Residence Times Vary through Time, and Do They Differ across Size Categories or between Immigrants and Endemics?

Residence times document observed ranges in the Potwar sequence. Although they are probably shorter than those for many species within the wider bioprovince, and shorter still than actual species durations, we treat them as useful estimates of the latter (Liow 2007). Mean residence times for species recorded at one or more 100-kyr bins between 14.1 and 7.4 Ma are shorter for small- and large-mammal species (2.4 and 2.2 myr, respectively) than for mega-mammal species (4.0 myr; Table 25.6). These results are partially in accord with a broad analysis of temperate Neogene taxa suggesting that small mammals have longer species durations than large mammals (Liow et al. 2008). Megamammals have much longer species durations than either small or large mammals in the Siwalik record.

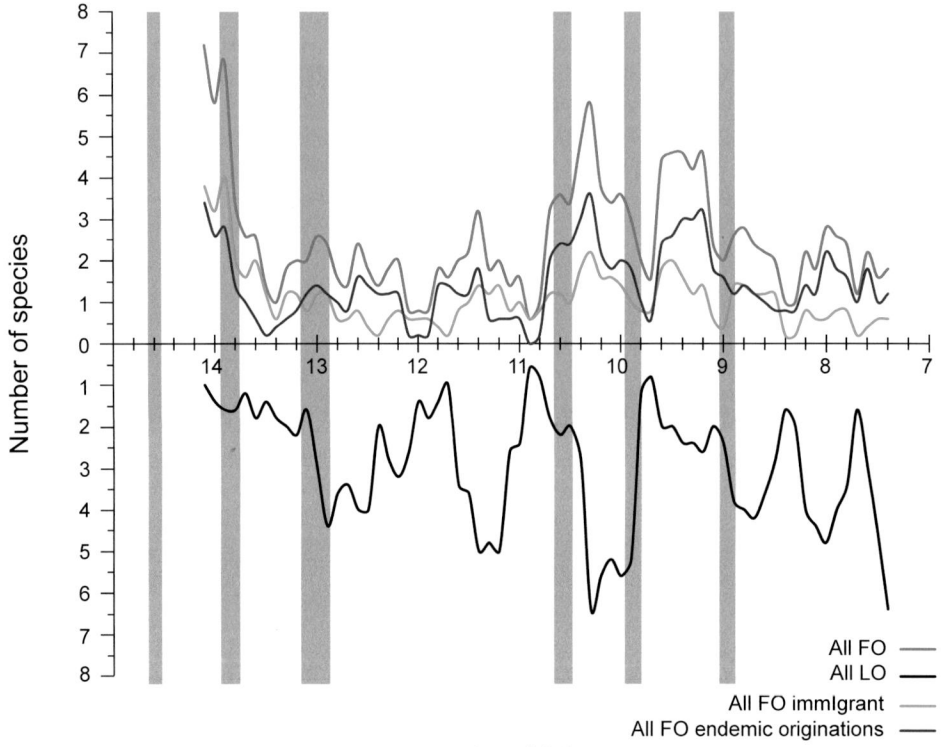

Figure 25.10. First and last occurrences of all mammals expressed as 500 kyr running averages, with first occurrences further separated into immigrant and endemic origination from 14.1 to 7.4 Ma. Periods of global cooling (in the marine realm, noted as gray bands; Holbourn et al. 2013) are not tracked by numbers of species' first and last occurrences. (Data derived from Table III.1.)

Age (Ma)

An earlier study of then-available data (Flynn et al. 1995) reported that small mammal median residence times were shorter than those for large mammals, but this pattern is not supported here. Current data show similar median residence times for small and large mammals (1.8 and 1.7 myr, respectively) with mega mammals having longer median residence times (2.2 myr) for the period between 14.1 and 7.4 Ma (see Table 25.6).

For perspective, Gingerich and Gunnell (1995) surveyed 125 species of Early Eocene (~55 to 52 Ma) mammals and noted a median duration (RT) of 0.76 myr, which is much shorter than that reported here for the Siwaliks. Gingerich and Gunnell (1995) explored factors that might bear on this discrepancy, and Gingerich (1991) noted the interesting possibility that median species duration may have increased over Cenozoic time.

Mean residence times of newly appearing Potwar species decrease over time. Between 15.2 and 14.1 Ma nine newly appearing species have RT longer than 5 myr, compared to three species after 14.1 Ma. Comparison of successive 2.5-myr intervals, 14.1–11.4 Ma and 11.2–8.7 Ma, reveals changing median and mean residence time for small, large, and mega mammals (Table 25.7). Sample sizes are comparable for the earlier and younger cohort in each size group. The reduction in mean residence time is significant for small and mega mammals (Mann Whitney U-test: $U = 402.5$, $p < 0.01$ [small]; $U = 95$, $p < 0.0005$ [mega]).

Rounding RT for small-, large- and mega-mammal species to the nearest 0.5 myr interval (Fig. 25.11) shows that most small and large mammals appearing between 14.1 and 8.7 Ma have an RT of 1.5 myr or less. Mega mammals differ in that most species have an RT of more than 2 myr. There are many taxa with RTs of 0.1 myr—33 of the 45 species in the 0.1–0.5 myr bin; the figures are strongly skewed toward immigrant species for large mammals (13 of 16), evenly divided among

small mammals (6 and 6), and primarily endemic originations for mega mammals (4 of 5). We also note the paucity of taxa with an RT of 0.6–1.0 myr, which suggests that if species survive an interval during which there would have been several Milankovitch-paced cycles of insolation variation (see Chapter 24 of this volume), then they are likely to persist at least another million years.

Endemic originations comprise the majority of new species for small and mega-sized mammals, while large mammals are mostly immigrants (Fig. 25.11). Within large mammals, species with RTs less than the mean and median values are predominantly immigrant taxa, with a higher proportion of longer RTs among endemic originations. Across all size classes, over 70% of the 45 species with RT ≤ 0.5 myr appear after 11.4 Ma. These data hint at changing selection pressures, possibly related to decreased stability of many species-preferred habitats, as forest fragmentation and seasonality increased in the Late Miocene.

There are family-level differences in the median residence time of small- and large- mammal species. Residence times for five small-mammal rodent families and five large-mammal artiodactyl families plus equids are presented in Figure 25.12. Among small mammals, the median RT of murid and spalacid species is less than 2 myr, while for sciurid and cricetid species it is greater than 2 myr (Fig. 25.12a). Among large mammals (Fig. 25.12b), most bovid and equid species have an RT less than 2 myr, while other families have a median RT greater than 2 myr. These differences indicate higher speciation and/or extinction rates for bovids, equids, murids, and spalacids than for tragulids, giraffids, sciurids and cricetids; excepting giraffids, families in the latter group have less observed ecomorphological change in the Potwar than families in the former group.

In summary, average residence times for Potwar herbivores decline over the 7-myr interval of analysis, reflecting increased local extinction relative to origination rates with a consequent decrease in species richness.

What Can the Siwalik Miocene Mammal Sequence Tell Us about the Tempo and Mode of Species Originations and Extinctions and of Range Extensions and Contractions?

Well over a century ago Lyell and Darwin debated whether the origin of species and the rates of change were gradual (Bennett 2004). In a seminal paper Eldredge and Gould (1972) asked the same questions more formally about tempo and mode. Were originations rapid or slow? Was lineage change steadily directional and gradual or stable and minimal?

The Potwar sequence reinforces that of other fossil records (Behrensmeyer et al. 1997; Liow 2007; Casanovas-Vilar et al. 2016): hard-tissue morphological stasis is far and away the norm, and any change is usually minimal and gradual.

The Potwar record also presents some examples of species originations, including what is in geological time both rapid change over intervals as short as or shorter than 100 kyr (following Vrba 1993) and gradual change over millions of years. Still, it is important to note that evolutionary change occurs

Table 25.6. Residence Times for Small-, Large-, and Mega-Mammal Taxa Documented between 14.1 and 7.4 Ma

Mammal Size	N Species	Median (myr)	Mean ± SD (myr)
Small (<1 kg)	96	1.8	2.4 ± 2.2
Large (1–800 kg)	110	1.7	2.2 ± 2.4
Mega (>800 kg)	30	2.15	4.0 ± 4.2

Table 25.7. Comparison of Residence Times for Small-, Large, and Mega-Mammal Taxa across Two Time Intervals

Mammal Size	First Occurrence	N species*	Median (myr)	Mean ± SD (myr)
Small (<1 kg)	14.1–11.4 Ma	22	1.5	2.0 ± 2.0
	11.2–8.7 Ma	26	0.5	1.0 ± 1.1
Large (1–800 kg)	14.1–11.4 Ma	41	1.8	2.1 ± 1.6
	11.2–8.7 Ma	45	1.6	1.6 ± 1.4
Mega (>800 kg)	14.1–11.4 Ma	9	5.0	4.4 ± 2.6
	11.2–8.7 Ma	12	1.1	1.0 ± 0.8

*Documented for an interval of time between 14.1 and 7.4 Ma.

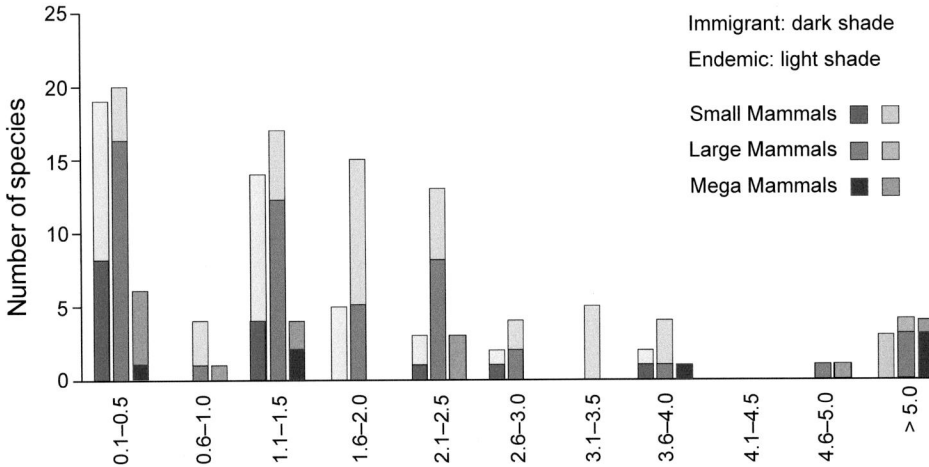

Figure 25.11. Species residence-time duration for taxa with first occurrences from 14.1 to 8.7 Ma. Numbers of species of small, large, and mega mammals are presented separately. Across all size groups the largest peak is in the 0.1–0.5 myr bin, which suggests that many taxa do not become well established in Siwalik paleocommunities. Note the paucity of species from all size groups with residence times in the 0.6–1.0 myr bin from any size group (Data derived from Table III.1.)

on what for paleontologists are much briefer, annual to generation-interval, time scales. As Reznick and Ghalambor (2005, 459) observe, "Our perceptions of the rate of evolution . . . based on the fossil record . . . have played an important historical role in shaping our impressions of what evolution is like [and] are strongly biased . . . because inferences from the fossil record are based on long-term averages that are likely to include long intervals of no change or reversals in the direction of change." While this is undoubtedly true when comparing short- and long-term patterns of change in an evolving lineage, the Siwalik fossil record provides compelling evidence for gradual morphological evolution at the temporal scale of 100 kyr and suggests environmental and/or behavioral "sorting" of genetically controlled traits.

Early murines provide an example of major morphological change, interpreted as anagenic. *Potwarmus* at 14.0 Ma was replaced by *Antemus* no more than 200 kyr later at 13.8 Ma; this change is marked morphologically by connection of a new anterolingual cusp on M1, enhancing chevron shearing crests. Another example of anagenesis is a long Early and Middle Miocene succession of gundi (Ctenodactylidae) species in the genus *Sayimys*; Baskin (1996) documented 8 myr of anagenic change within the genus from >18 to 10.1 Ma. Kimura et al. (2013) detailed the Late Miocene non-splitting evolution from 10.5 to 8.0 Ma within the murine lineages *Progonomys* and *Karnimata*. The origin of the *Progonomys* and *Karnimata* lineages themselves from the informally named "pre-*Progonomys* sp. X" (see Chapter 11 of this volume for details) appears to be a case of "gradual" cladogenesis, as inferred from abundant and well-calibrated samples during the interval of change (Kimura, Flynn, and Jacobs 2015, 2021; Flynn, Kimura, and Jacobs 2020; Chapter 11, this volume). The potential adaptive benefits of these documented morphological changes are presently unknown.

Two examples of endemic origination involve the succession of the boselaphine *Miotragocerus pilgrimi* by the larger *M. punjabicus* between 8.8 and 8.7 Ma, with observed change in both dentition and horn core morphology, and the replacement of the first bovine, *Pachyportax falconeri*, documented in the Siwaliks between 9.4 Ma and 9.0 Ma, by *P. dhokpathanensis* (80%

larger in body size), first occurring at 8.9 Ma and persisting until 7.2 Ma (see Chapter 23 of this volume).

The record also documents instances in which descendant lineages overlap with their ancestors. For example, the first large suine, *Hippopotamodon* sp. (a distinct but currently unnamed species), is recorded between 11.4 and 10.2 Ma and overlaps for 300 kyr with the species *H. sivalense* (recorded between 10.4 and 6.4 Ma), which is 80% larger on average. The rhizomyine *Kanisamys indicus*, documented from 16.8 to 11.4 Ma, overlaps for 200 kyr with its high-crowned derivative *K. nagrii*, present between 11.6 and 9.7 Ma (Flynn 1986). These new species may have originated outside the local "Potwar area" but within the wider Siwalik bioprovince.

Bamboo rats provide an example of an adaptive radiation. Rhizomyinae originated in the Oligocene. It is likely that they were always low-vegetation feeders and facultative burrowers, as inferred from a specimen of the non-fossorial *Kanisamys* preserved in its burrow at 8.7 Ma (Flynn 2009); otherwise, the genus displays no obvious burrowing adaptations. *Miorhizomys*, the first genus of the tribe Rhizomyini, appears at 9.7 Ma as a probable immigrant, likely from elsewhere in Southeast Asia. It shows multiple morphological burrowing adaptations in the skull and humerus, and we interpret it as committed to fossorial life, presumably exploiting underground food sources (Flynn 2009). This is a first-time behavior for Siwalik rodents. The genus subsequently radiated, with multiple species established in the small-mammal community (some exceeding 1 kg in size and considered large mammals for analyses), but without apparent effects on other muroids.

Species extinctions are challenging to pinpoint in the fossil record. One example of extinction that is fairly well supported is the last known occurrence of the order Creodonta in the world. Originating in North America during the Paleocene, creodonts were once common in Asia, Europe, and Africa. They were diverse in the Indian subcontinent during the Eocene. The youngest Potwar creodont fossil currently recognized dates to 10.0 Ma (see Chapter 13 of this volume).

Range extensions are easier to recognize when an emigrant species appears in another well-documented region. Within

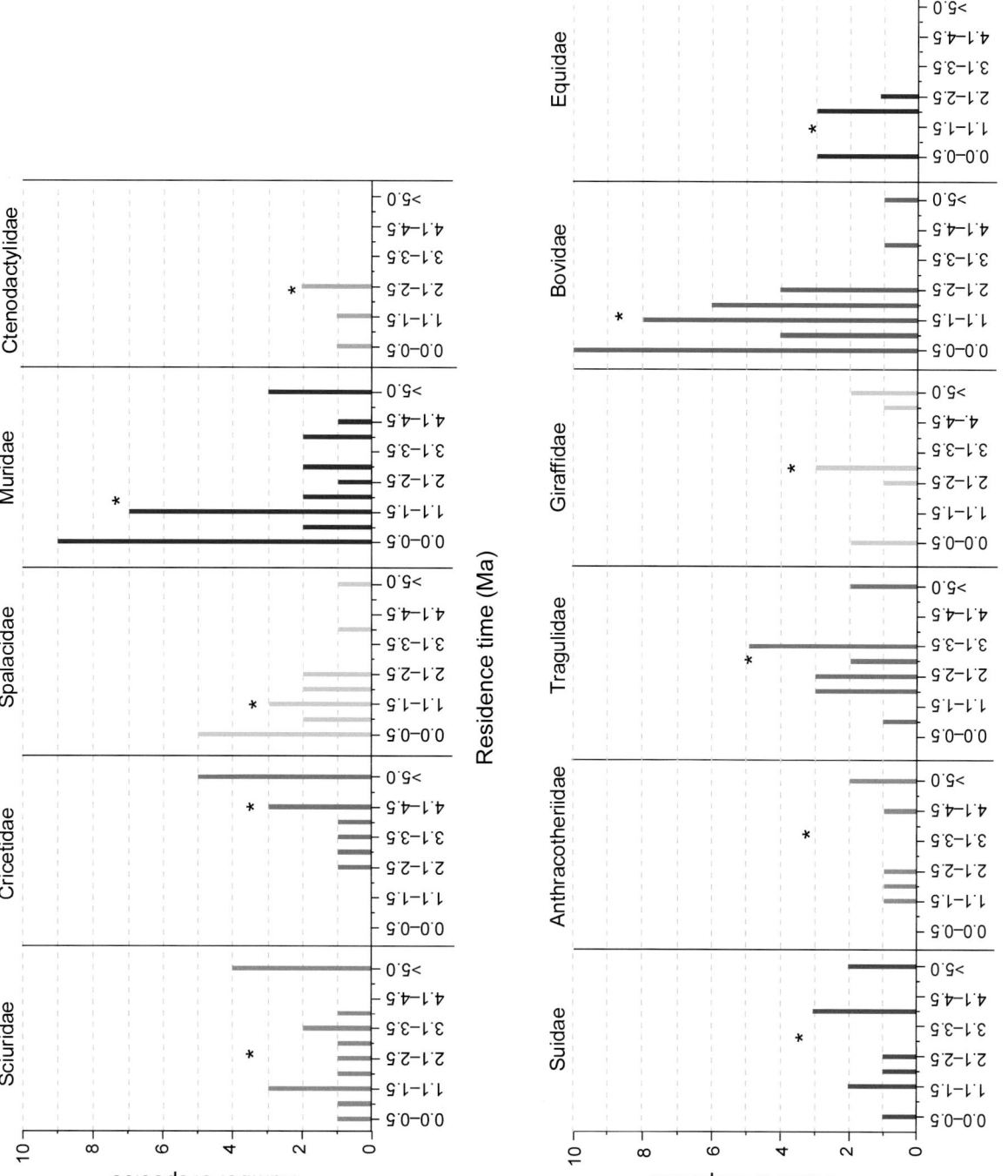

Figure 25.12. Residence time (RT) compared to number of species for selected families of rodents (A) and artiodactyl families plus Equidae (B). Median RT indicated by * for each family. (Data derived from Table III.1.)

small mammals, the endemic murine *Progonomys* emigrated westward from the Siwaliks into the circum-Mediterranean region after 11 Ma (Wessels 2009; Flynn and Wessels 2013; Casanovas-Vilar et al. 2016; see Chapter 11 of this volume for details). Two endemic large-mammal lineages also clearly expanded their geographic range from the Siwaliks. During the Late Miocene the anthracothere *M. medioximus* dispersed both eastward and westward and occurred in Iraq, Thailand, and Myanmar (Lihoreau et al. 2007; Chavasseau et al. 2013; see Chapter 20 of this volume for details). Similarly, species of the equid genus *Sivalhippus* dispersed both eastward into China and westward into Africa around 9 Ma (Bernor et al. 2010; Wolf, Bernor, and Hussain 2013; Sun et al. 2018; see Chapter 16 of this volume for details).

DETERMINANTS OF CHANGE

In Chapter 3 of the *Origin of Species*, Charles Darwin provided a characteristically diplomatic summary of causes of change, recognizing the importance of climate and habitat change while continuing his emphasis on competition:

> Climate plays an important part in determining the average numbers of a species, and periodical seasons of extreme cold or drought, I believe to be the most effective of all checks. I estimated that the winter of 1854–55 destroyed four-fifths of the birds in my own grounds; and this is a tremendous destruction, when we remember that ten per cent is an extraordinarily severe mortality from epidemics with man. The action of climate seems at first sight. . . . quite independent of the struggle for existence but in so far as climate chiefly acts in reducing food, it brings on the most severe struggle between individuals whether of the same or of distinct species, which subsist on the same kind of food. (Darwin 1859, 68)

Over the past half-century, the discussion has been to an extent formalized around two causal poles—the so- called Red Queen (Van Valen 1977) and Court Jester (Barnosky 2001) models, describing respectively biotic (interspecific competition, predation) and abiotic (habitat / climate change) as contrasting factors in evolution. There is consensus that temporal and spatial scales influence the apparent relative importance of biotic and abiotic factors and that neither the Red Queen nor the Court Jester model should be viewed as exclusive (Barnosky 2005; Benton 2009).

Abiotic factors are perhaps more likely to be invoked than biotic interactions as explanations for change due to challenges in interpreting a paleontological record. In a recent literature review, Fraser et al. (2020) recognized this, countering by presenting an impressive range of cases in support of the importance of biotic interactions. We think it highly plausible that biotic drivers played major roles in shaping some of the macroevolutionary and macroecological shifts documented in the Potwar Siwalik record.

Throughout this chapter we have drawn attention to spatial scales—local, regional, global. Of equal importance is temporal scale. Bennett (1990, 11) noted that Darwin's great achievement was the linking of "processes observable today to patterns in the fossil record." This was emphasized by Reznick and Ghalamor (2005): "Evolution may well be concentrated in small, brief events and will only be seen if it is looked for, such as in the context of an experiment or as part of an individual mark-recapture study that is associated with the quantification of individual traits, as in work on Galapagos finches. . . . Evolutionary trends, such as the fossil record, are an epiphenomenon of average long-term trends in these many discrete events (Reznick and Ghalamor 2005, 459; see Grant and Grant 2002).

Stephen Jay Gould drew distinctions between the tiers of "ecological time" and "geological time" (Gould 1985). The processes about which we would like to know, such as microevolution and competition, "processes observable today," act in ecological time. They are driven by natural selection acting in "generation time." This contrasts with the patterns and cadences of macroevolution, which play out over the much longer arc of geological time. Bennett (1990, 2004) added an important third tier, "Milankovitch time." The Milankovitch astronomical cycles, precessional (~23 kyr), obliquity (~41 kyr) and eccentricity (~100 and ~400 kyr), determine regular changes in global and regional insolation patterns (Laskar et al. 2004). Variation in their amplitudes strongly influences the pace, degree, and intensity of critically important seasonality. Much shorter "sub-precessional" cycles are also documented, at least in the Pliocene and Pleistocene (for example, Wilson et al. 2014; Colcord et al. 2018; Lupien et al. 2020). The amplitude of precessional and obliquity cycles, determining the rate and magnitude of annual change in seasonality, would have been modulated by changes in eccentricity.

These regular and predictable fluctuations would promote continual restructuring and reorganization of communities, a pattern well documented for both plants and animals during the very marked climate and habitat variations of Late Pleistocene time (Davis 1983; Graham et al. 1996; Bennett 2004). Populations of plant and animal species have individual habitat preferences and continually track shifts in climates and habitats and thereby alter the competitive landscape for any particular species. Even as recently as a few thousand years ago associations of species (communities) were quite different from those found today; communities are ephemeral on Milankovitch timescales. Such constant shuffling normally obliterates much microevolutionary change since such changes accumulate on short generation-length timescales.

When viewed on a million-year timescale, the Siwalik faunas documented in the Potwar Plateau appear, like other Miocene faunal sequences, relatively stable, reflecting the presence of often long-lived, co-occurring species. Referring to such long-lived species, the paleocommunities can be interpreted as "resilient." We are fortunate to be able to sort and sequence Potwar fossil assemblages into 100 kyr bins, which reveals that there is always at least some faunal change between adjacent bins. Most species recorded during the 14.1–7.4 Ma of our study have short RTs; for example, just over 15% of species (38/236) are recorded in only one 100 kyr bin, while 23% are present for 500 kyr or less. Such "brief" intervals would experience multiple Milankovitch-driven cycles of climate and probable habitat change.

Following Darwin's wise insight, it is important to remind ourselves that climates and habitats are always changing and doing so on multiple scales: annual, generational, centennial, millennial, Milankovitch, geological. No matter how modestly, community composition is constantly changing in response to these habitat shifts, as does the competitive landscape. Species are always subject to competition, facing new habitats or new community members or both. Both biotic and abiotic factors will be interactively involved in all but the most rapid and high mortality events affecting species survival in a region (e.g., disease, bolide impact, major tectonic events, dramatic global climate change; see Reznick and Ricklefs 2009).

What follows are three examples examining this dynamic from the perspective of the Potwar fossil record.

Possible Competition among Muroid Rodents

While competition is a reality of species interactions, it is difficult to identify competition in particular cases. The abundant Siwalik rodent record presents a possible case of interspecific competition. During the late Middle Miocene, between 12.7 and 12.4 Ma, there is a dramatic change in the relative abundance of Murinae, Gerbillinae, and Cricetidae (see Chapter 11 of this volume), with Murinae replacing the other groups. At 12.7 Ma the moderately abundant stem murine *Antemus chinjiensis* (~15% of specimens) coexisted with gerbilline and cricetid species, while by 12.4 Ma "post-*Antemus*" comprised over 65% of murines (Fig. 25.14, p. 463). Relative to *Antemus*, the 12.4 Ma "post-*Antemus*" molars exhibit stronger transverse connections linking all cusps in inclined shearing chevrons, plausibly giving an adaptive advantage by increasing chewing efficiency. This geologically brief 300-kyr interval falls midway through the Chinji Formation, during which time we find no evidence for any significant global climate or regional habitat fluctuations. The rate of overall faunal change is low, as reflected in residence times and origination and extinction rates, yet the murines show significant change in both abundance and morphology. We conclude that interspecific competition remains a likely explanation for increased murine abundance around the *Antemus* to "post-*Antemus*" transition.

Following the first occurrence of "post-*Antemus*," non-murine species became less common, apart from hamsters (*Democricetodon* species), which remained species-rich and moderately abundant. At 10.5 Ma, *Democricetodon* was still common, but less than a half million years later, at 10.1 Ma, it was greatly diminished in abundance, while murines provided almost 75% of specimens (see Fig. 25.2a). By 9.2 Ma both gerbillines and cricetids were gone from the Potwar muroid record (Fig. 25.14), although cricetids briefly reappeared at 8.7 Ma (see Chapter 11, this volume, for details). Despite their high abundance, murine species richness never exceeded five species from 12.7 to 9.2 Ma, in contrast to the high richness of gerbillines and cricetids before 11.6 Ma.

We hypothesize that this reduction and eventual extirpation of gerbils and cricetids was driven by habitat change, which

rendered them less well-adapted, and by interspecific competition with the more successfully adapted murines. Interspecific competition is widespread among small vertebrates, particularly among closely related rodents (Grant 1972), and may reflect rates of reproduction, need for space or resources, or influence of other species. The turnover among muroids, the success of mice over hamsters and gerbils, can therefore be seen as a two-step process, during which the relative effects of biotic and abiotic factors shifted over time.

Faunal Turnover among Ungulates

In the interval between 11.4 and 10.1 Ma, while there is minimal change in species richness, there is marked faunal turnover among ungulates. Sixteen artiodactyl species present at 11.4 Ma are no longer documented at 10.1 Ma, and 17 artiodactyl species recorded at 10.1 Ma were not present at 11.4. Ma, all these having first occurrences between 10.6 and 10.1 Ma (see Table 25. S2 for taxa). Equids appear in the Potwar sequence as immigrants by 10.8 Ma, along with other taxa around that time. This turnover also involves a significant shift in size structure of artiodactyls as reflected in the distribution of their astragalus size (Morgan 1994; Morgan et al. 1995). Giraffids and bothriodontine anthracotheres show change toward larger-sized species, while bovids, tragulids and suids expand their size range by adding larger species (Fig. 25.13, p. 462). As noted previously, habitats were becoming more open through this interval, with more non-woody vegetation including grasses present by at least 10.5 Ma along with increased fire frequency (Hoorn, Ohja, and Quade 2000; Karp, Behrensmeyer, and Freeman 2018; Srivastava et al. 2018; Karp et al. 2021; Polissar et al. 2021).

Around 10.8 and 10.7 Ma, two brief rapid warming episodes were punctuated by marked cooling episodes (Holbourn et al. 2013) and can be plausibly linked with lowered sea levels: this could have expanded gateways into the Siwalik bioprovince. Our ability to probe the timing and causes of the faunal change is constrained by patchy fossil productivity between 11.1 and 10.6 Ma. Until this sparse interval is better sampled elsewhere in the Potwar region, it remains challenging to untangle the role of immigrants or the relative contributions of abiotic and biotic factors to the burst of first and last occurrences. The current record hints at an interval of high faunal change among artiodactyls between 10.6 and 10.4 Ma, with seven endemic originations, one immigrant, and four last occurrences.

When we recognize that multiple macroevolutionary processes can select for larger body size within lineages (Huang et al. 2017), the pattern of size change and relative abundance among Siwalik ungulates, particularly bovids, tragulids, and equids, implicates habitat change as a contributing factor, with bovids and equids expanding their range of body size, increasing in abundance, and consuming grasses, while tragulids (Clauss and Rössner 2014), with a likely affinity to "near-aquatic" habitats and fruit and high-quality browse, decrease in relative abundance and body size. Prior to the arrival of equids at 10.8 Ma, bovids and tragulids comprised 75% of large-herbivore fossil specimens identified to at least family level and were similar in estimated body mass; at 11.4 Ma the four bovid

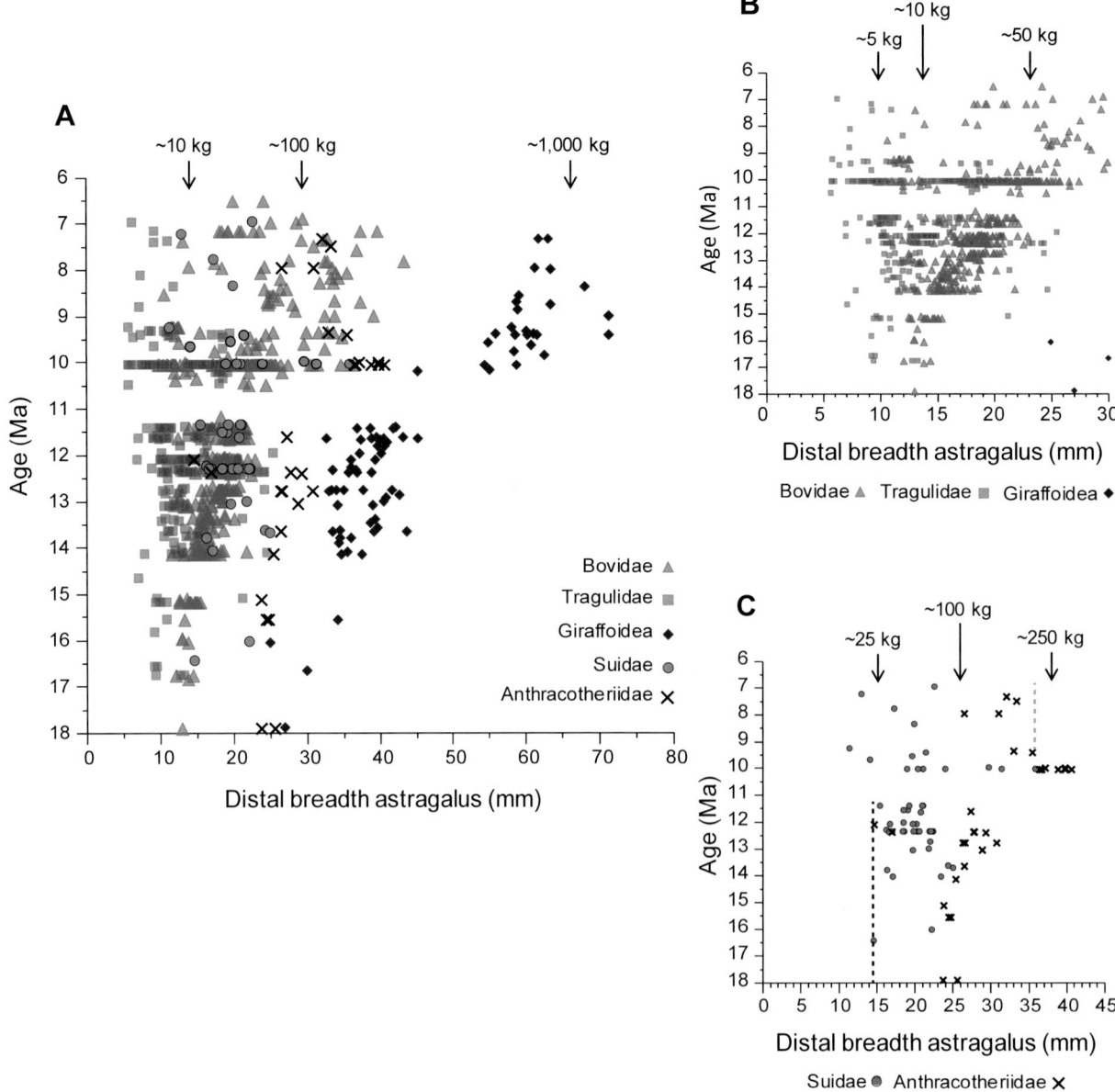

Figure 25.13. Size distribution in four families of artiodactyls as reflected by distal astragalar width (mm) of five families (A) with estimated kg based on regressions derived from extant artiodactyls (Morgan et al. 1995). Details for tragulids and bovids less than 100 kg (B) and for suids and anthracotheres (C) are included. Note stepped increases in maximum size at around 14 and 10 Ma; occupation of 20–30 mm size between 15.5 and 10.5 Ma by anthracotheres; and the large size gap after 10 Ma between the largest bovids and anthracotheres (maximum size around 45 mm), and giraffids. Overlap between bovids and tragulids and relative rarity of tragulids after 9 Ma is detailed in (B). In (C) the observed range of the large suid *Hippopotamodon sivalense* is indicated by an orange line and that of the small anthracothere *Microbunodon silistrensis* by a black line; more astragalar specimens for these species were not available for measurement.

species averaged 30 kg while the five tragulid species averaged 26 kg (see Tables 21.1 and 23.1 for estimated body size range [kg] of species; midpoint values used here to calculate estimated average bovid and tragulid species size). The immigration of equids had little effect on bovid and tragulid relative abundance as assessed at 10.1 Ma, but the average body size of bovid species increased from 30 to 49 kg (*n* = 5 species) while that of tragulids decreased from 25 to 18 kg (*n* = 4 species). The estimated body size of the equid species present at 10.1 Ma is

250 kg (see Table 16.1). By 9.4 Ma, the average mass difference between bovid species (mean = 70 kg; *n* = 9 species) and tragulid species (mean = 15 kg; *n* = 5 species) was 55 kg; the two equid species averaged 325 kg. Despite taxonomic and body-size changes, bovids show the greatest relative abundance of specimens throughout this 2-myr interval. Tragulid species richness remained stable over the interval, but their relative abundance declined while equids commonly outnumbered them in postcranial specimens (Fig. 25.15). Observed change in size and

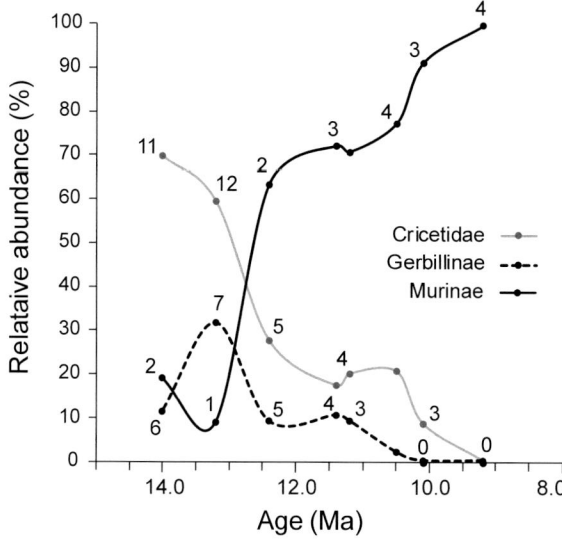

Figure 25.14. Relative abundance of cricetids, gerbillines and murines between 14.1 and 9.2 Ma. Numbers of species at each data point are noted by small numerals above the curve. (Data from Table III.1.) Number of included specimens across eight 100 kyr bins is 2796 (range of included specimens/bin is 167–712).

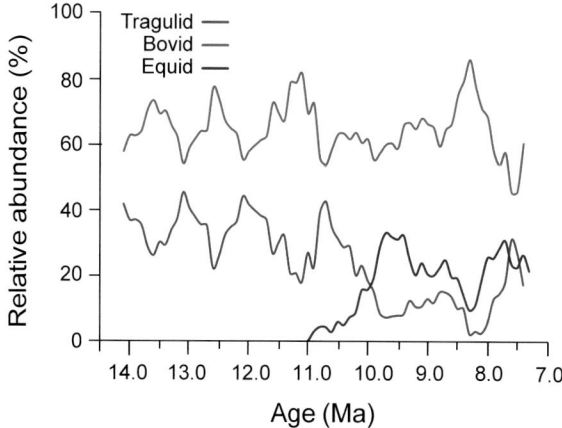

Figure 25.15. Relative abundance of bovids, tragulids, and equids as reflected in numbers of postcranial specimens and expressed as 500 kyr running averages between 14 and 7 Ma. Note both the consistent dominance of bovids, and their increase and proportional decrease in tragulids shortly following the first occurrence of equids. A single locality at 7.4 Ma has abundant tragulids.

abundance of species within these three families is consistent with a shift to more open habitats by at least 10.5 Ma, followed by a steady gradual trend to increasingly open and seasonal habitats over the following two million years, presenting more resource-variant habitats (Andrews, Lord, and Nesbit-Evans 1979; Bhat, Kempes, and Yeakel 2019).

In an elegant analysis of ruminant gut and life-history adaptations, Clauss and Rössner (2014) plausibly argue for their physiological role in explaining observed patterns of evolution among tragulids, bovids, and giraffids. They note that pecoran ruminants have an omasum between reticulum and abomasum, which makes moisture re-absorption from food easier. They propose that the lack of an omasum in tragulids limited their ability to competitively forage as mixed feeders, and specifically "may have made tragulids susceptible to resource competition in the context of increased openness of landscapes during the Neogene" (Clauss and Rössner 2014, 90).

Determining whether interspecific competition is an important factor is more difficult. The first occurrence of immigrant equids by 10.8 Ma falls in a poorly fossiliferous interval (see Fig. 25.1). Once present in the fauna, their pattern of greater abundance than artiodactyls in three of the four depositional environments (see Fig. 4.10b) indicates that equids became a dominant group in the faunal communities after 10.8 Ma. Other immigrants, bovids, tragulids, and giraffids, also appear, including larger taxa. Of interest is the presence between 11.3 and 10.8 Ma of three immigrant single-bin "singletons": *Protoryx* aff. *solignaci*, *Selenoportax vexillarius*, and ?*Protragelopus* sp. (see Table III.1). That they are recorded in a poorly fossiliferous interval suggests they may not have been rare species at the time. But that they are recorded at no other time suggests that they did not successfully become established in the com-

munity. Noting again the likely slow pace of habitat change, biotic interactions are likely significant contributing factors.

Late Miocene Faunal Change

A good case for the role of habitat change as one driver of faunal change involves events during and immediately following the 8.7 to 7.4 Ma interval. We have discussed earlier the evidence that large-scale environmental change began in the Potwar at around 7.8 Ma with a shift from predominantly C_3 vegetation to floodplain vegetation dominated by C_4 grasses (see also Chapters 5 and 24, this volume). What had long been a predominantly browsing large and mega mammal herbivore community in highly productive and heterogeneous habitat types gradually changed over a few million years to a community with a significant number of grazers in less productive and more homogeneous habitat types. A landmark study of paleosol stable isotopes in the Rohtas area of the Potwar allows us to estimate rates of change; the average carbon isotope values shifted from around −10‰ at 8 Ma to ca. −1.2‰ at about 5 Ma, around 3‰ per million years (Behrensmeyer et al. 2007). Along with this gradual shift would have been many marked and briefer shifts, possibly precessionally driven. These are most easily detectable when vegetation fluctuates between mostly C_3 and mostly C_4 plants (for example, for the Potwar, see Behrensmeyer et al. 2007, Fig. 9, or for the Pleistocene of Olduvai, Tanzania, see Colcord et al. 2018, Fig. 3).

Badgley et al. (2008) described the faunal consequences of this habitat change (see also Barry et al. 2002), and Chapter 26 (this volume) presents an insightful analysis of species and functional richness of the mammalian fauna. Here we have presented new details about tempo and mode of change in small, large and mega mammals. The small-mammal fauna became murine- and rhizomyine-dominated, bovids and equids of increasing body size characterized the large mammals, and stegodontine proboscideans became more abundant than rhinocerotids (megaherbivore species richness remained high with 11 species in the region at 7.4 Ma). There were fewer arboreal frugivores and folivores, and more mammals were grazing, with a C_4 grass forage component being isotopically detectable in many mammalian families (see Fig. 25.8).

Clearly, more open-habitat adapted species entered the region and persisted, while those dependent on forests and woodlands became rarer or locally extinct. Although the quality of the fossil record decreases after 7.4 Ma, observed faunal change between 8.7 and 7.4 Ma shows that first and last occurrences are staggered within each feeding guild (frugivores, omnivores, mixed feeders, grazers). There is no evidence for large-scale rapid faunal change.

CONCLUSIONS

The Siwalik Miocene mammal sequence documented in the Potwar Plateau region of northern Pakistan provides a well-calibrated record spanning ~12 million years. It is well-suited for exploring the dynamics of community structure and faunal change at all taxonomic levels, specific and above. We interpret it within a framework of current understanding of Miocene global- to subcontinental-scale tectonic and climatic conditions and how these affect regional and local habitat structure.

In analyzing all mammalian groups except carnivorans and bats, we describe mammalian community structure using species richness, species abundance, and species ecomorphological attributes, as well as patterns of faunal change including species residence times and the tempo and mode of species originations and extinctions. Rather than asking whether biotic or abiotic factors drove evolutionary change, wherever possible we have attempted to assess their fluctuating or joint contributions to faunal turnover. We used the Potwar record for three case studies of such faunal change: late Middle Miocene murine rodent ascendance, notable for the replacement of gerbils and hamsters by true mice; early Late Miocene artiodactyl faunal turnover and change in guild structure; and Late Miocene faunal change during C_4 grassland expansion.

We find that faunal change is asynchronous within and between mammalian size groups. There is marked change in species richness through the record. This is especially the case for small mammals, which show overall species richness decline, with two particular intervals of accelerated decline between 13 and 12 Ma and 10.3 and 9.7 Ma. In contrast, large mammals fluctuate in species richness but show no overall net change. Megaherbivores are the most stable over intervals of several million years, their species richness between 14 and 8 Ma being particularly noteworthy—a minimum of 10 and a maximum of 18 contemporaneous species. (As many as

nine megaherbivore species occur within a single fossil locality.)

There are no concentrated turnover pulses affecting the whole mammal fauna. The differing trajectories and patterns of change across the three size groups show that effects of habitat change on the mammalian fauna varied both through time and across taxa.

Residence times of species are highly variable. Megaherbivores have longer residence times than small or large mammals; more than two-thirds of megaherbivore species appearing prior to 13.5 Ma persist for more than 6 million years. Species with very long residence times are not always abundant in our fossil assemblages; examples of relatively ubiquitous taxa with long residence times yet low relative abundance include the chalicothere *Anisodon* and several calloscuirine squirrels.

All observed shifts in morphology or inferred diet plausibly indicate some type of behavioral change within lineages.

Incumbent species and genera show some selective advantage over immigrant species. The majority of species with residence times of 0.1 myr are immigrants, overall having somewhat shorter average residence times than endemics. Yet some immigrant taxa become very successful, as assessed by both relative abundance and ubiquity on the landscape. Equids are a prime example following their first occurrence by 10.8 Ma.

Equids include C_4 grasses in their dietary resource spectrum 2 million years earlier than other ungulate mammals. Equid consumption of C_4 grasses may possibly have been associated initially with preference for more open areas due to predator avoidance or to hipparionine social structure, rather than food preference or food scarcity.

Using our collective work over the past five decades as a starting point, we hope that fieldwork can continue in the Potwar Plateau and other Siwalik sequences to further develop the fossil record and associated sediments to permit even more fine-grained analyses, not least in that critical 11.4–10.1 Ma interval.

ACKNOWLEDGMENTS

We thank the Geological Survey of Pakistan and our many colleagues for their efforts in the field and laboratories around the world that have provided the foundational and interpretative frameworks for this synthesis. Special thanks to John Barry, database master, and to Catherine Badgley, Kay Behrensmeyer, and Mahmood Raza for frequent helpful discussions. Support for fieldwork and laboratory analyses from the American School of Prehistoric Research, National Science Foundation, Smithsonian Foreign Currency Program, and Wenner-Gren Foundation is gratefully acknowledged.

SUPPLEMENTARY FIGURES

FO immigrant | FO endemic origination | LO | SO immigrant | SO endemic origination | Continuing range

Order (abbr)	Family	Name	Size Group	17.9		16.8		16.0		15.2		14.1
Ins	Erinaceidae	*Galerix wesselsae*	S									
Ins	Erinaceidae	*Galerix rutlandae*	S									
Sca	Tupaiidae	cf. *Dendrogale* sp. 1	S									
Sca	Tupaiidae	cf. *Dendrogale* sp. 2	S									
Lag	Ochotonidae	*Alloptox* sp.	S									
Rod	Diatomyidae	*Diatomys chitaparwalensis*	S									
Rod	Ctenodactylidae	*Sayimys minor*	S									
Rod	Ctenodactylidae	*Sayimys intermedius*	S									
Rod	Ctenodactylidae	*Sayimys* cf. *S. giganteus*	S									
Rod	Ctenodactylidae	*Sayimys sivalensis*	S									
Rod	Thryonomyidae	*Kochalia geespei*	S									
Rod	Thryonomyidae	*Kochalia* n. sp.	S									
Rod	Sciuridae	Aliveria sp.	S									
Rod	Sciuridae	cf. *Ratufa* n. sp.	S									
Rod	Sciuridae	*Tamiops* large sp.	S									
Rod	Sciuridae	*Tamiops* small sp.	S									
Rod	Sciuridae	*Callosciurus* sp.	S									
Rod	Sciuridae	*Dremomys* sp.	S									
Rod	Sciuridae	*Oriensciurus* n. sp.	S									
Rod	Sciuridae	*Heteroxerus* sp.	S									

Figure 25.S1. Visual summary of mammal species recorded at four focus 100-kyr time slices, 17.9, 16.8, 16.0, and 15.2, Ma and faunal change between them. First occurrences of species shaded pink; last occurrences shaded blue, and species documented in only one 100-kyr time slice shaded green. Continuing ranges of species shaded orange. Inferred immigration and endemic origination of species are further color-differentiated with darker shade for immigrant and lighter shade for endemic origination. Small mammals (S), < 1 kg in size; large mammals (L), 1–800 kg in size; and mega mammals (M), > 800 kg. Taxonomic order abbreviations as follows: Art (Artiodactyla); Ins (Insectivora); Lag (Lagomorpha); Per (Perissodactyle); Pri (Primates); Pro (Proboscidea); Rod (Rodentia); Sca (Scandentia); Tub (Tubulidentata).

FO immigrant ■   FO endemic origination ▢   LO ▢   SO immigrant ▢   SO endemic origination ▢   Continuing range ▢

Order (abbr)	Family	Name	Size Group	17.9		16.8		16.0		15.2		14.1
Rod	Sciuridae	*Atlantoxerus* sp.	S									
Rod	Sciuridae	*Sciurotamias* Kamlial sp.	S									
Rod	Sciuridae	*Sciurotamias* sp. A (small)	S									
Rod	Sciuridae	*Tamias* "pre-*urialis*"	S									
Rod	Anomaluridae	Anomaluridae, n. gen. and sp. indet.	S									
Rod	Spalacidae	*Prokanisamys arifi*	S									
Rod	Spalacidae	*Prokanisamys major*	S									
Rod	Spalacidae	*Prokanisamys* "n. sp. small"	S									
Rod	Spalacidae	Prokanisamys n. sp.	S									
Rod	Spalacidae	*Kanisamys indicus*	S									
Rod	Spalacidae	*Kanisamys potwarensis*	S									
Rod	Cricetidae	*Democricetodon* sp. A	S									
Rod	Cricetidae	*Democricetodon* sp. B-C	S									
Rod	Cricetidae	*Democricetodon* sp. E	S									
Rod	Cricetidae	*Democricetodon kohatensis*	S									
Rod	Cricetidae	*Megacricetodon aguilari*	S									
Rod	Cricetidae	*Megacricetodon mythikos*	S									
Rod	Cricetidae	*Megacricetodon sivalensis*	S									
Rod	Cricetidae	*Megacricetodon daamsi*	S									
Rod	Cricetidae	*Punjabemys mikros*	S									
Rod	Cricetidae	*Punjabemys leptos*	S									
Rod	Cricetidae	*Punjabemys downsi*	S									
Rod	Muridae	*Myocricetodon sivalensis*	S									
Rod	Muridae	*Myocricetodon* large sp.	S									
Rod	Muridae	*Dakkamys barryi*	S									

Figure 25.S1. (Continued)

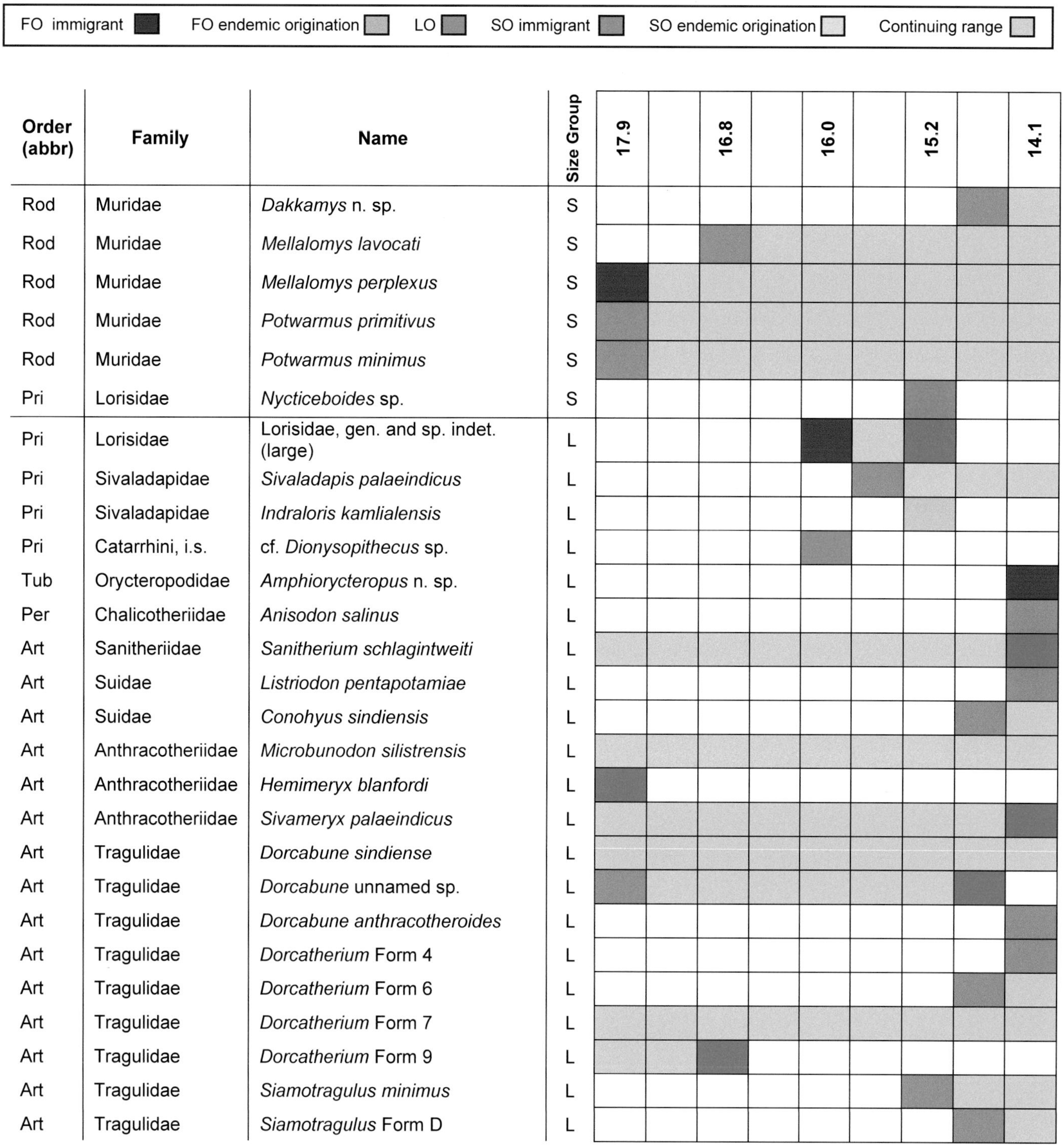

Order (abbr)	Family	Name	Size Group	17.9		16.8		16.0		15.2		14.1
Rod	Muridae	*Dakkamys* n. sp.	S									
Rod	Muridae	*Mellalomys lavocati*	S									
Rod	Muridae	*Mellalomys perplexus*	S									
Rod	Muridae	*Potwarmus primitivus*	S									
Rod	Muridae	*Potwarmus minimus*	S									
Pri	Lorisidae	*Nycticeboides* sp.	S									
Pri	Lorisidae	Lorisidae, gen. and sp. indet. (large)	L									
Pri	Sivaladapidae	*Sivaladapis palaeindicus*	L									
Pri	Sivaladapidae	*Indraloris kamlialensis*	L									
Pri	Catarrhini, i.s.	cf. *Dionysopithecus* sp.	L									
Tub	Orycteropodidae	*Amphiorycteropus* n. sp.	L									
Per	Chalicotheriidae	*Anisodon salinus*	L									
Art	Sanitheriidae	*Sanitherium schlagintweiti*	L									
Art	Suidae	*Listriodon pentapotamiae*	L									
Art	Suidae	*Conohyus sindiensis*	L									
Art	Anthracotheriidae	*Microbunodon silistrensis*	L									
Art	Anthracotheriidae	*Hemimeryx blanfordi*	L									
Art	Anthracotheriidae	*Sivameryx palaeindicus*	L									
Art	Tragulidae	*Dorcabune sindiense*	L									
Art	Tragulidae	*Dorcabune* unnamed sp.	L									
Art	Tragulidae	*Dorcabune anthracotheroides*	L									
Art	Tragulidae	*Dorcatherium* Form 4	L									
Art	Tragulidae	*Dorcatherium* Form 6	L									
Art	Tragulidae	*Dorcatherium* Form 7	L									
Art	Tragulidae	*Dorcatherium* Form 9	L									
Art	Tragulidae	*Siamotragulus minimus*	L									
Art	Tragulidae	*Siamotragulus* Form D	L									

Figure 25.S1. (Continued)

Order (abbr)	Family	Name	Size Group	17.9	16.8	16.0	15.2	14.1
Art	Antilocapridae	Climacocerinae, n. gen. and sp.	L					
Art	Antilocapridae	cf. *Orangemeryx* n. sp.	L					
Art	Giraffidae	Progiraffa exigua	L					
Art	Giraffidae	Giraffokeryx punjabiensis	L					
Art	Giraffidae	*Ua pilbeami*	L					
Art	Bovidae	*Hypsodontus sokolovi*	L					
Art	Bovidae	?Hypsodontinae sp.	L					
Art	Bovidae	Bovidae sp. 1	L					
Art	Bovidae	*Eotragus noyei*	L					
Art	Bovidae	*Elachistoceras khauristanensis*	L					
Art	Bovidae	*Strepsiportax* unnamed sp.	L					
Art	Bovidae	*Sivaceros* unnamed. sp.	L					
Art	Bovidae	*Sivoreas eremita*	L					
Art	Bovidae	Antilopini, unnamed gen. and sp.	L					
Art	Pecora, i.s.	"Genus A" small sp.	L					
Pro	Deinotheriidae	*Prodeinotherium pentapotamiae*	M					
Pro	Gomphotheriidae	*Choerolophodon corrugatus*	M					
Per	Rhinocerotidae	*Caementodon oettingenae*	M					
Per	Rhinocerotidae	*Alicornops* cf. *simorrense*	M					
Per	Rhinocerotidae	*Alicornops complanatum*	M					
Per	Rhinocerotidae	*Brachydiceratherium fatehjangense*	M					
Per	Rhinocerotidae	*Brachypotherium perimense*	M					
Per	Rhinocerotidae	*Brachydiceratherium intermedium*	M					
Per	Rhinocerotidae	*Gaindatherium browni*	M					
Per	Rhinocerotidae	*Gaindatherium vidali*	M					
Per	Rhinocerotidae	cf. *Rhinoceros sp.*	M					

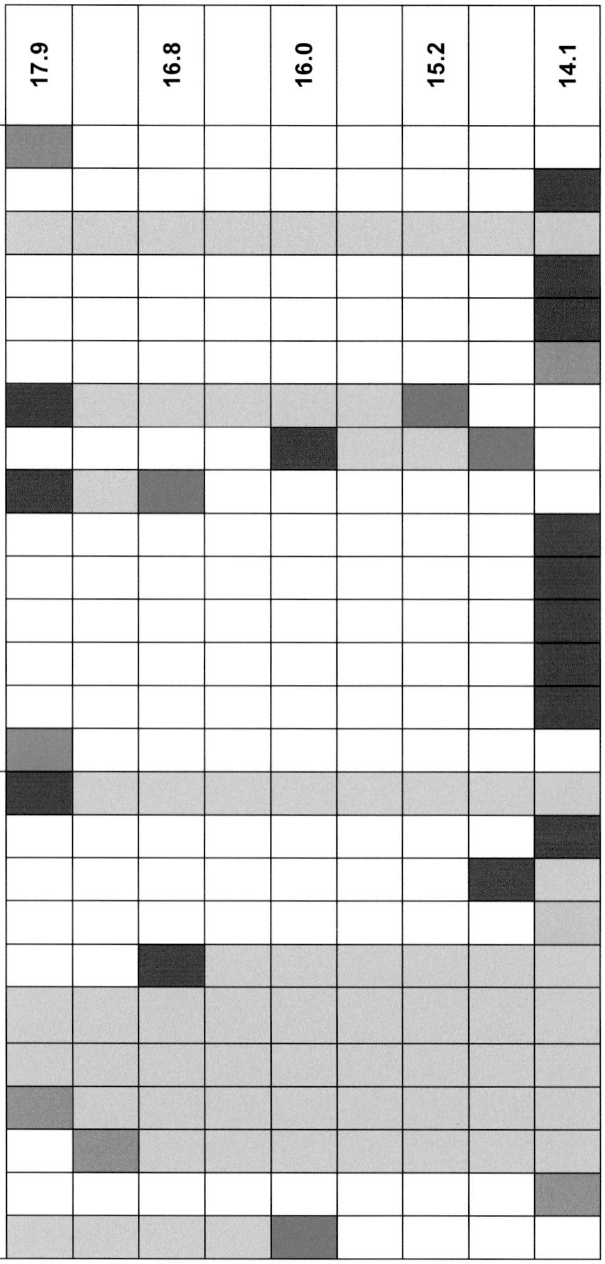

Figure 25.S1. (Continued)

Legend:
- ■ FO immigrant
- ■ FO endemic origination
- ■ LO
- ■ SO immigrant
- ■ SO endemic origination
- ■ Continuing range

Order (abbr)	Family	Name	Size Group	14.1	12.7	11.4	10.1	8.7	7.4	6.5
Ins	Erinaceidae	Galerix wesselsae	S							
Ins	Erinaceidae	Galerix rutlandae	S							
Ins	Erinaceidae	Schizogalerix sp.	S							
Ins	Erinaceidae	Erinaceinae, gen. B and sp. indet.	S							
Ins	Erinaceidae	Erinaceinae, gen. C and sp. indet.	S							
Ins	Soricidae	cf. Crocidura sp.	S							
Ins	Soricidae	Suncus honeyi	S							
Ins	Soricidae	Beremendiini, gen. and sp. indet.	S							
Sca	Tupaiidae	cf. Dendrogale sp. 1	S							
Sca	Tupaiidae	cf. Dendrogale sp. 2	S							
Sca	Tupaiidae	cf. Dendrogale sp. 3	S							
Sca	Tupaiidae	cf. Tupaia sp.	S							
Sca	Tupaiidae	Tupaiidae, n. gen. and sp.	S							
Sca	Tupaiidae	Tupaiidae, gen. and sp. indet. (tiny)	S							
Rod	Diatomyidae	Willmus maximus	S							
Rod	Ctenodactylidae	Sayimys sivalensis	S							
Rod	Ctenodactylidae	Sayimys chinjiensis	S							
Rod	Ctenodactylidae	Sayimys sp. B	S							
Rod	Ctenodactylidae	Sayimys perplexus	S							
Rod	Thryonomyidae	Kochalia geespei	S							
Rod	Thryonomyidae	Paraulacodus indicus	S							
Rod	Gliridae	Myomimus sumbalenwalicus	S							
Rod	Gliridae	Myomimus n. sp. A	S							

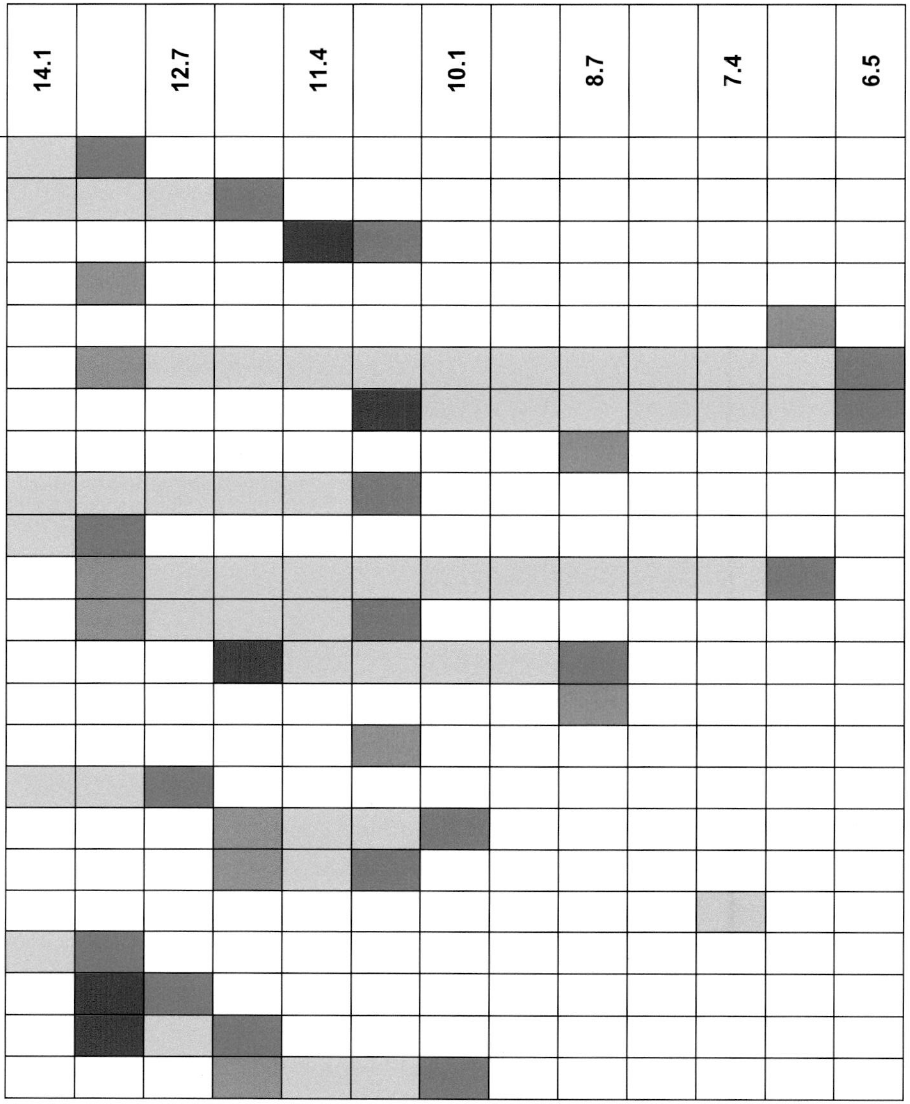

Figure 25.S2. Visual summary of mammal species recorded at six focus 100-kyr time slices, 14.1, 12.7, 11.4, 10.1, 8.7, and 7.4 Ma (plus 6.5 Ma for comparison). Species occurrences (FO, LO, SO) and continuing ranges shaded as in Figure 25.S1. Small mammals (S), < 1 kg in size; large mammals (L), 1–800 kg in size; and mega mammals (M), > 800 kg. Taxonomic order abbreviations as follows: Art (Artiodactyla); Ins (Insectivora); Lag (Lagomorpha); Per (Perissodactyle); Pri (Primates); Pro (Proboscidea); Rod (Rodentia); Sca (Scandentia); Tub (Tubulidentata).

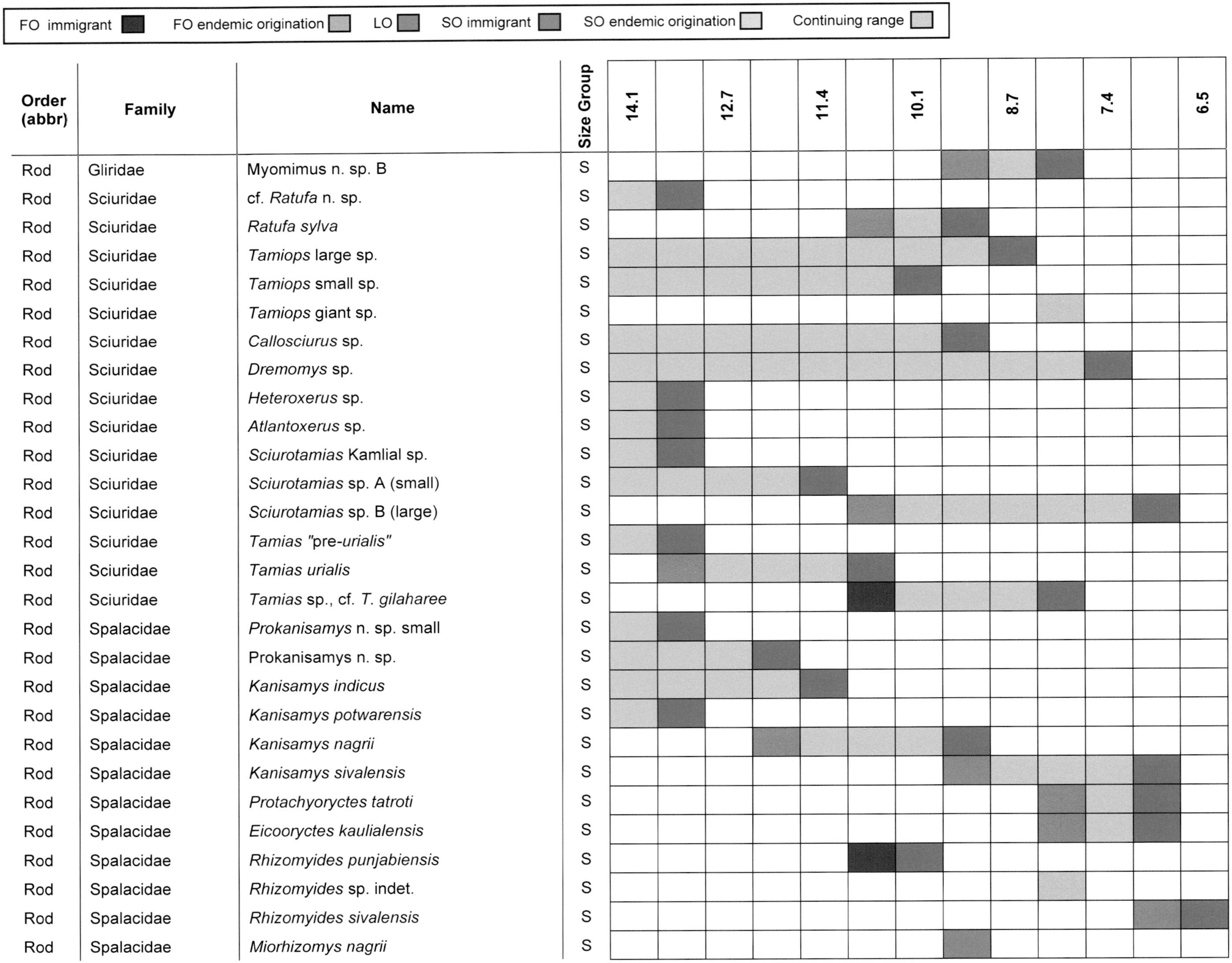

Order (abbr)	Family	Name	Size Group
Rod	Gliridae	Myomimus n. sp. B	S
Rod	Sciuridae	cf. Ratufa n. sp.	S
Rod	Sciuridae	Ratufa sylva	S
Rod	Sciuridae	Tamiops large sp.	S
Rod	Sciuridae	Tamiops small sp.	S
Rod	Sciuridae	Tamiops giant sp.	S
Rod	Sciuridae	Callosciurus sp.	S
Rod	Sciuridae	Dremomys sp.	S
Rod	Sciuridae	Heteroxerus sp.	S
Rod	Sciuridae	Atlantoxerus sp.	S
Rod	Sciuridae	Sciurotamias Kamlial sp.	S
Rod	Sciuridae	Sciurotamias sp. A (small)	S
Rod	Sciuridae	Sciurotamias sp. B (large)	S
Rod	Sciuridae	Tamias "pre-urialis"	S
Rod	Sciuridae	Tamias urialis	S
Rod	Sciuridae	Tamias sp., cf. T. gilaharee	S
Rod	Spalacidae	Prokanisamys n. sp. small	S
Rod	Spalacidae	Prokanisamys n. sp.	S
Rod	Spalacidae	Kanisamys indicus	S
Rod	Spalacidae	Kanisamys potwarensis	S
Rod	Spalacidae	Kanisamys nagrii	S
Rod	Spalacidae	Kanisamys sivalensis	S
Rod	Spalacidae	Protachyoryctes tatroti	S
Rod	Spalacidae	Eicooryctes kaulialensis	S
Rod	Spalacidae	Rhizomyides punjabiensis	S
Rod	Spalacidae	Rhizomyides sp. indet.	S
Rod	Spalacidae	Rhizomyides sivalensis	S
Rod	Spalacidae	Miorhizomys nagrii	S

Time axis (Ma) columns: 14.1, 12.7, 11.4, 10.1, 8.7, 7.4, 6.5

Figure 25.S2. (Continued)

Rod	Spalacidae	*Miorhizomys micrus*
Rod	Spalacidae	*Miorhizomys blacki*
Rod	Cricetidae	*Democricetodon* sp. A
Rod	Cricetidae	*Democricetodon* sp. B-C
Rod	Cricetidae	*Democricetodon* sp. E
Rod	Cricetidae	*Democricetodon kohatensis*
Rod	Cricetidae	*Democricetodon fejfari*
Rod	Cricetidae	*Megacricetodon aguilari*
Rod	Cricetidae	*Megacricetodon mythikos*
Rod	Cricetidae	*Megacricetodon sivalensis*
Rod	Cricetidae	*Megacricetodon daamsi*
Rod	Cricetidae	*Punjabemys mikros*
Rod	Cricetidae	*Punjabemys leptos*
Rod	Cricetidae	*Punjabemys downsi*
Rod	Muridae	*Myocricetodon sivalensis*
Rod	Muridae	*Myocricetodon* large sp.
Rod	Muridae	*Dakkamys barryi*
Rod	Muridae	*Dakkamys* n. sp.
Rod	Muridae	*Dakkamys asiaticus*
Rod	Muridae	*Paradakkamys chinjiensis*
Rod	Muridae	*Mellalomys lavocati*
Rod	Muridae	*Mellalomys perplexus*
Rod	Muridae	*Abudhabia pakistanensis*
Rod	Muridae	*Potwarmus primitivus*
Rod	Muridae	*Potwarmus minimus*
Rod	Muridae	*Antemus chinjiensis*
Rod	Muridae	"post-*Antemus*" n. sp.
Rod	Muridae	"pre-*Progonomys*" n. sp. X

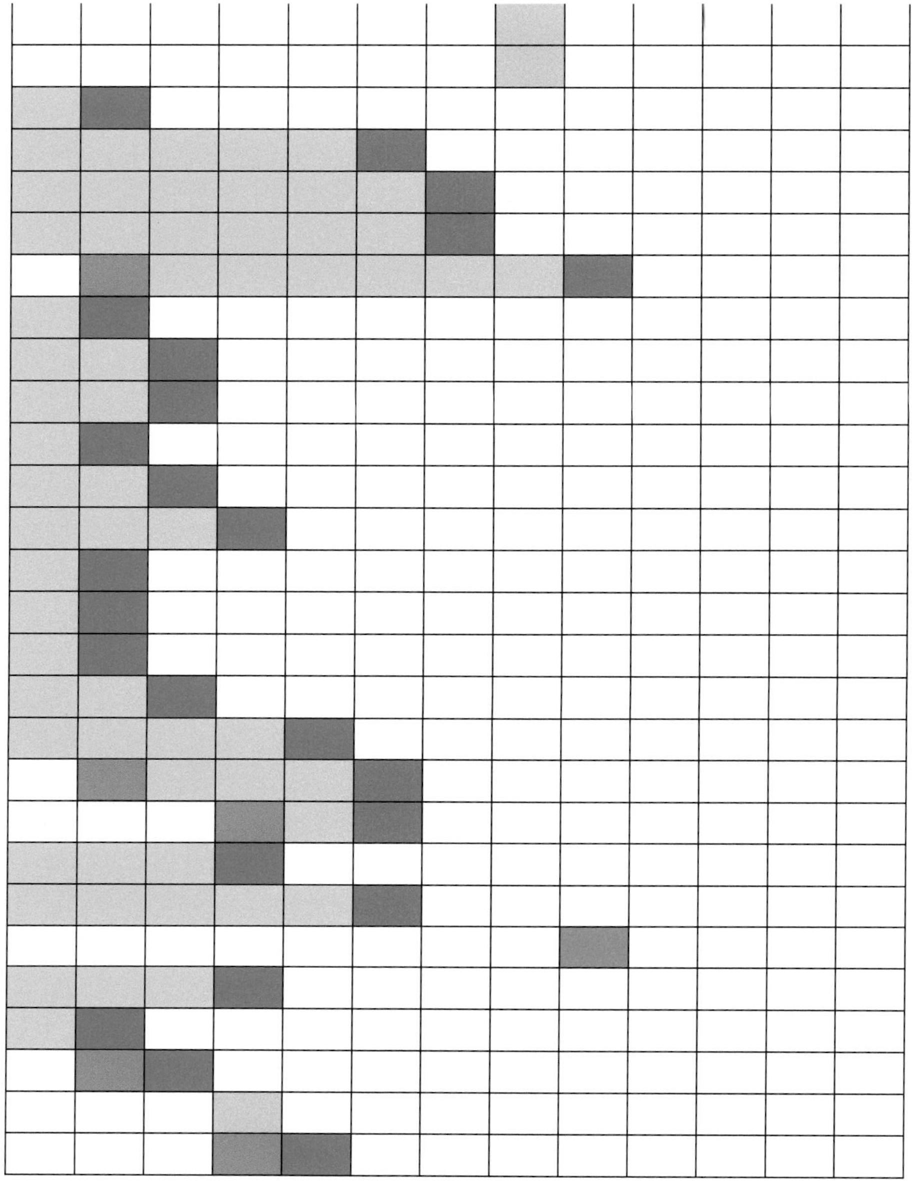

Figure 25.S2. (Continued)

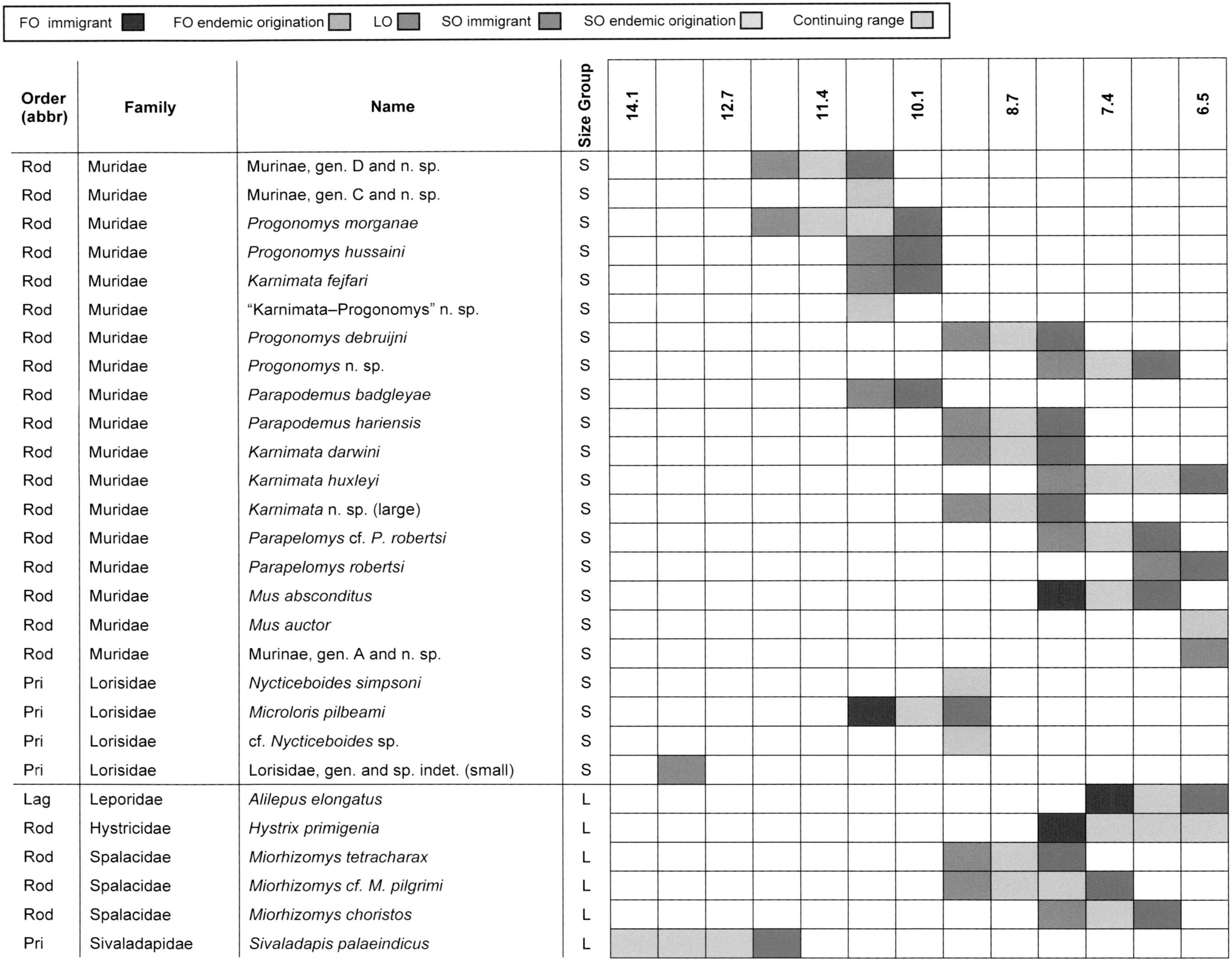
FO immigrant
FO endemic origination
LO
SO immigrant
SO endemic origination
Continuing range
Order (abbr)
Family
Name
Size Group
14.1
12.7
11.4
10.1
8.7
7.4
6.5
Rod Muridae Murinae, gen. D and n. sp. S
Rod Muridae Murinae, gen. C and n. sp. S
Rod Muridae Progonomys morganae S
Rod Muridae Progonomys hussaini S
Rod Muridae Karnimata fejfari S
Rod Muridae "Karnimata–Progonomys" n. sp. S
Rod Muridae Progonomys debruijni S
Rod Muridae Progonomys n. sp. S
Rod Muridae Parapodemus badgleyae S
Rod Muridae Parapodemus hariensis S
Rod Muridae Karnimata darwini S
Rod Muridae Karnimata huxleyi S
Rod Muridae Karnimata n. sp. (large) S
Rod Muridae Parapelomys cf. P. robertsi S
Rod Muridae Parapelomys robertsi S
Rod Muridae Mus absconditus S
Rod Muridae Mus auctor S
Rod Muridae Murinae, gen. A and n. sp. S
Pri Lorisidae Nycticeboides simpsoni S
Pri Lorisidae Microloris pilbeami S
Pri Lorisidae cf. Nycticeboides sp. S
Pri Lorisidae Lorisidae, gen. and sp. indet. (small) S
Lag Leporidae Alilepus elongatus L
Rod Hystricidae Hystrix primigenia L
Rod Spalacidae Miorhizomys tetracharax L
Rod Spalacidae Miorhizomys cf. M. pilgrimi L
Rod Spalacidae Miorhizomys choristos L
Pri Sivaladapidae Sivaladapis palaeindicus L

Figure 25.S2. (Continued)

Pri	Catarrhini, i.s.	Catarrhini, gen. and sp. indet.	L
Pri	Hominidae	*Sivapithecus indicus*	L
Pri	Hominidae	*Sivapithecus parvada*	L
Pri	Hominidae	*Sivapithecus sivalensis*	L
Pri	Cercopithecidae	*Mesopithecus sivalensis*	L
Tub	Orycteropodidae	*Amphiorycteropus* n. sp.	L
Tub	Orycteropodidae	*Amphiorycteropus browni*	L
Tub	Orycteropodidae	*Amphiorycteropus* sp. indet	L
Tub	Orycteropodidae	*Orycteropus* sp.	L
Pho	Manidae	cf. *Manis* sp.	L
Per	Equidae	*Cormohipparion* sp.	L
Per	Equidae	*Cormohipparion* small sp.	L
Per	Equidae	*Sivalhippus nagriensis*	L
Per	Equidae	*Sivalhippus theobaldi*	L
Per	Equidae	*Sivalhippus perimensis*	L
Per	Equidae	*Sivalhippus anwari*	L
Per	Equidae	*Cremohipparion antilopinum*	L
Per	Chalicotheriidae	*Anisodon salinus*	L
Per	Chalicotheriidae	cf. *Ancylotherium* sp.	L
Art	Palaeochoeridae	*Yunnanochoerus gandakasensis*	L
Art	Sanitheriidae	*Sanitherium schlagintweiti*	L
Art	Suidae	*Listriodon pentapotamiae*	L
Art	Suidae	*Hippotamodon* sp.	L
Art	Suidae	*Hippopotamodon sivalense*	L
Art	Suidae	*Propotomochoerus hysudricus*	L
Art	Suidae	*Sivahyus punjabiensis*	L
Art	Suidae	Suidae, i.s., n. gen. and sp.	L
Art	Suidae	Suidae, unnamed gen. and sp.	L

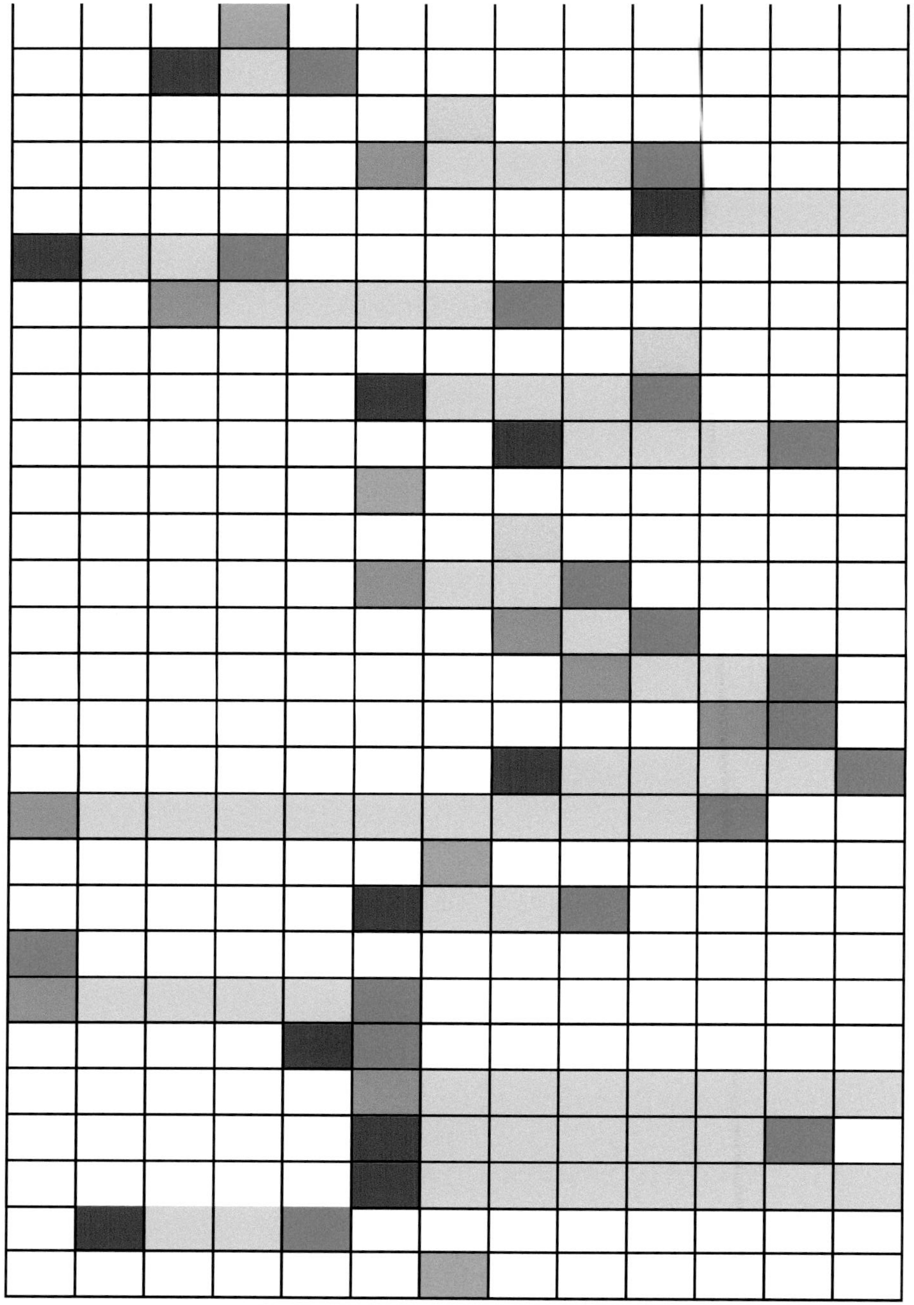

Figure 25.S2. (Continued)

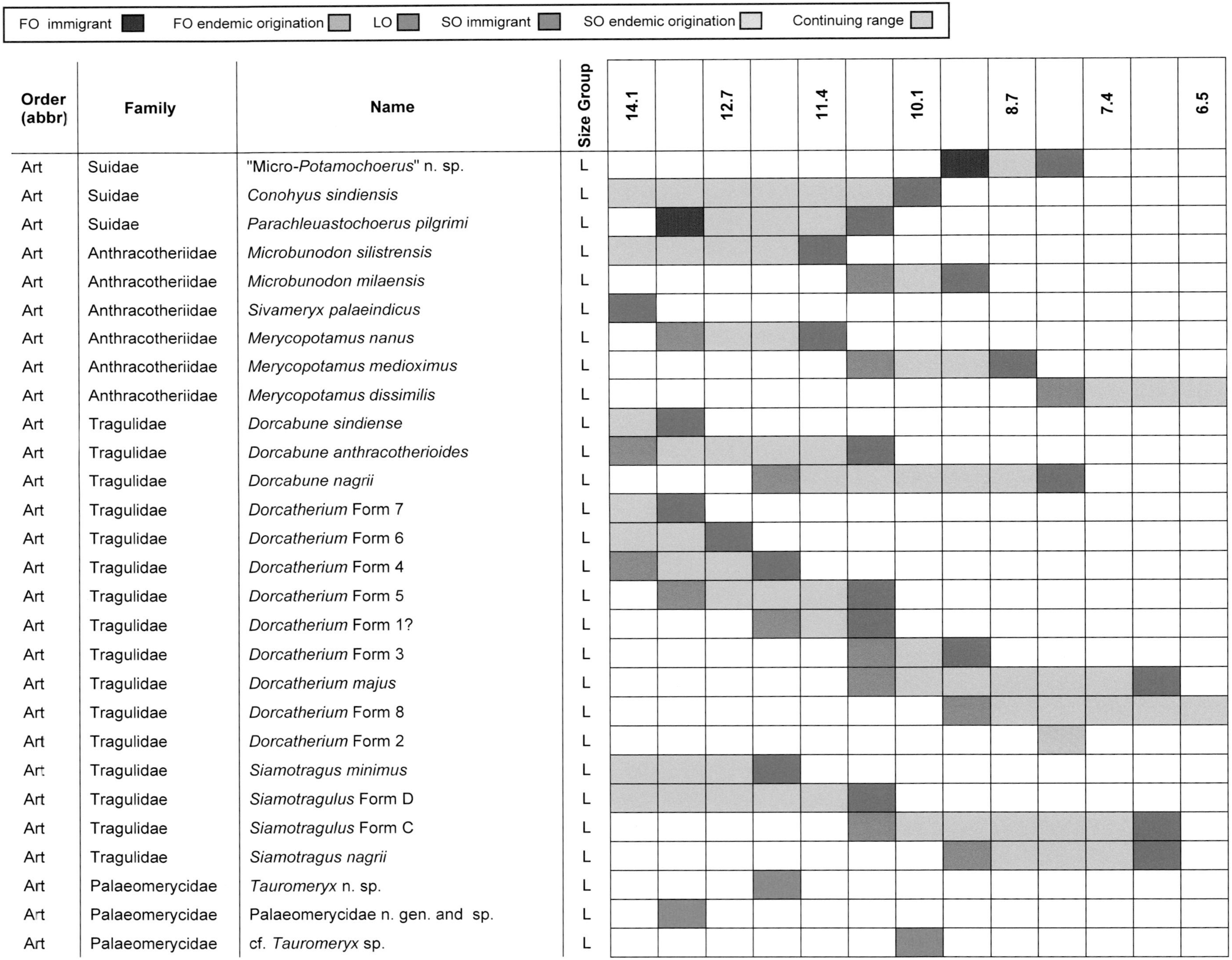

Legend: FO immigrant | FO endemic origination | SO immigrant | LO | SO endemic origination | Continuing range

Order (abbr)	Family	Name	Size Group	14.1	12.7	11.4	10.1	8.7	7.4	6.5
Art	Suidae	"Micro-Potamochoerus" n. sp.	L							
Art	Suidae	Conohyus sindiensis	L							
Art	Suidae	Parachleuastochoerus pilgrimi	L							
Art	Anthracotheriidae	Microbunodon silistrensis	L							
Art	Anthracotheriidae	Microbunodon milaensis	L							
Art	Anthracotheriidae	Sivameryx palaeindicus	L							
Art	Anthracotheriidae	Merycopotamus nanus	L							
Art	Anthracotheriidae	Merycopotamus medioximus	L							
Art	Anthracotheriidae	Merycopotamus dissimilis	L							
Art	Tragulidae	Dorcabune sindiense	L							
Art	Tragulidae	Dorcabune anthracotherioides	L							
Art	Tragulidae	Dorcabune nagrii	L							
Art	Tragulidae	Dorcatherium Form 7	L							
Art	Tragulidae	Dorcatherium Form 6	L							
Art	Tragulidae	Dorcatherium Form 4	L							
Art	Tragulidae	Dorcatherium Form 5	L							
Art	Tragulidae	Dorcatherium Form 1?	L							
Art	Tragulidae	Dorcatherium Form 3	L							
Art	Tragulidae	Dorcatherium majus	L							
Art	Tragulidae	Dorcatherium Form 8	L							
Art	Tragulidae	Dorcatherium Form 2	L							
Art	Tragulidae	Siamotragus minimus	L							
Art	Tragulidae	Siamotragulus Form D	L							
Art	Tragulidae	Siamotragulus Form C	L							
Art	Tragulidae	Siamotragus nagrii	L							
Art	Palaeomerycidae	Tauromeryx n. sp.	L							
Art	Palaeomerycidae	Palaeomerycidae n. gen. and sp.	L							
Art	Palaeomerycidae	cf. Tauromeryx sp.	L							

Figure 25.S2. (Continued)

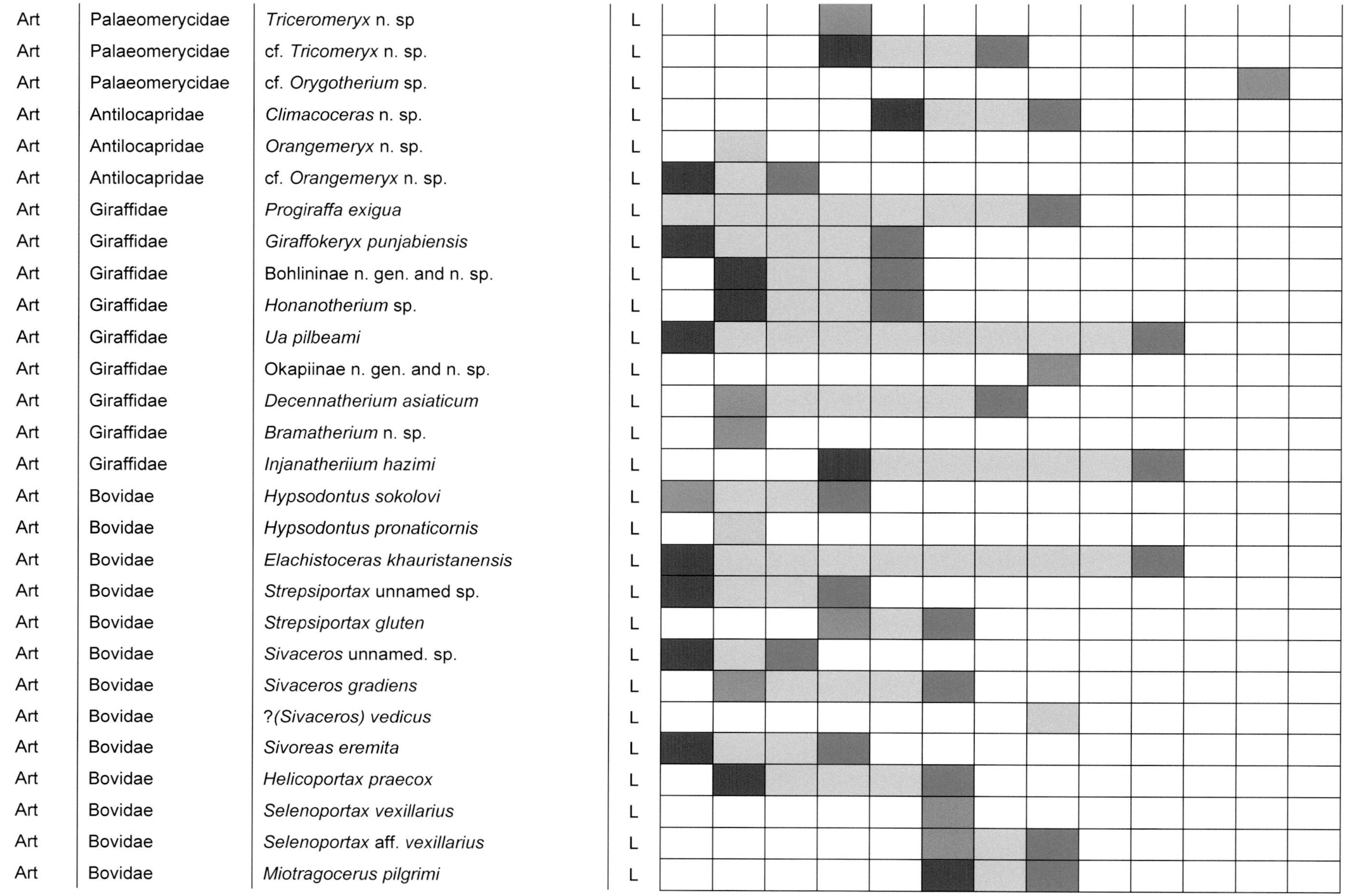

Order	Family	Taxon
Art	Palaeomerycidae	*Triceromeryx* n. sp
Art	Palaeomerycidae	cf. *Tricomeryx* n. sp.
Art	Palaeomerycidae	cf. *Orygotherium* sp.
Art	Antilocapridae	*Climacoceras* n. sp.
Art	Antilocapridae	*Orangemeryx* n. sp.
Art	Antilocapridae	cf. *Orangemeryx* n. sp.
Art	Giraffidae	*Progiraffa exigua*
Art	Giraffidae	*Giraffokeryx punjabiensis*
Art	Giraffidae	Bohlininae n. gen. and n. sp.
Art	Giraffidae	*Honanotherium* sp.
Art	Giraffidae	*Ua pilbeami*
Art	Giraffidae	Okapiinae n. gen. and n. sp.
Art	Giraffidae	*Decennatherium asiaticum*
Art	Giraffidae	*Bramatherium* n. sp.
Art	Giraffidae	*Injanatheriium hazimi*
Art	Bovidae	*Hypsodontus sokolovi*
Art	Bovidae	*Hypsodontus pronaticornis*
Art	Bovidae	*Elachistoceras khauristanensis*
Art	Bovidae	*Strepsiportax unnamed* sp.
Art	Bovidae	*Strepsiportax gluten*
Art	Bovidae	*Sivaceros unnamed.* sp.
Art	Bovidae	*Sivaceros gradiens*
Art	Bovidae	?(*Sivaceros*) *vedicus*
Art	Bovidae	*Sivoreas eremita*
Art	Bovidae	*Helicoportax praecox*
Art	Bovidae	*Selenoportax vexillarius*
Art	Bovidae	*Selenoportax* aff. *vexillarius*
Art	Bovidae	*Miotragocerus pilgrimi*

Figure 25.S2. (Continued)

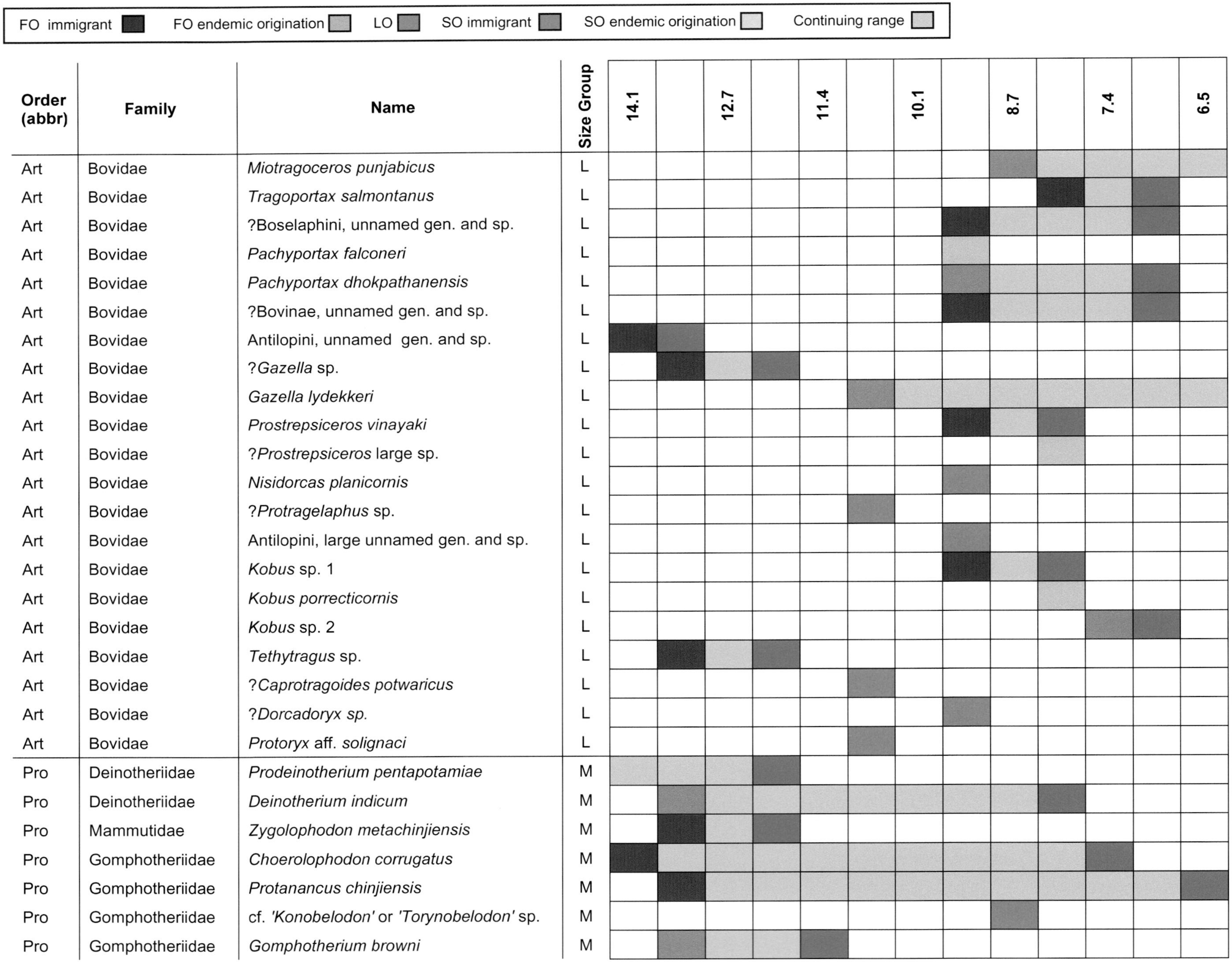

Figure legend:

- FO immigrant
- FO endemic origination
- LO
- SO immigrant
- SO endemic origination
- Continuing range

Time bins (columns): 14.1, 12.7, 11.4, 10.1, 8.7, 7.4, 6.5

Order (abbr)	Family	Name	Size Group
Art	Bovidae	*Miotragoceros punjabicus*	L
Art	Bovidae	*Tragoportax salmontanus*	L
Art	Bovidae	?Boselaphini, unnamed gen. and sp.	L
Art	Bovidae	*Pachyportax falconeri*	L
Art	Bovidae	*Pachyportax dhokpathanensis*	L
Art	Bovidae	?Bovinae, unnamed gen. and sp.	L
Art	Bovidae	Antilopini, unnamed gen. and sp.	L
Art	Bovidae	?*Gazella* sp.	L
Art	Bovidae	*Gazella lydekkeri*	L
Art	Bovidae	*Prostrepsiceros vinayaki*	L
Art	Bovidae	?*Prostrepsiceros* large sp.	L
Art	Bovidae	*Nisidorcas planicornis*	L
Art	Bovidae	?*Protragelaphus* sp.	L
Art	Bovidae	Antilopini, large unnamed gen. and sp.	L
Art	Bovidae	*Kobus* sp. 1	L
Art	Bovidae	*Kobus porrecticornis*	L
Art	Bovidae	*Kobus* sp. 2	L
Art	Bovidae	*Tethytragus* sp.	L
Art	Bovidae	?*Caprotragoides potwaricus*	L
Art	Bovidae	?*Dorcadoryx* sp.	L
Art	Bovidae	*Protoryx* aff. *solignaci*	L
Pro	Deinotheriidae	*Prodeinotherium pentapotamiae*	M
Pro	Deinotheriidae	*Deinotherium indicum*	M
Pro	Mammutidae	*Zygolophodon metachinjiensis*	M
Pro	Gomphotheriidae	*Choerolophodon corrugatus*	M
Pro	Gomphotheriidae	*Protanancus chinjiensis*	M
Pro	Gomphotheriidae	cf. 'Konobelodon' or 'Torynobelodon' sp.	M
Pro	Gomphotheriidae	*Gomphotherium browni*	M

Figure 25.S2. (Continued)

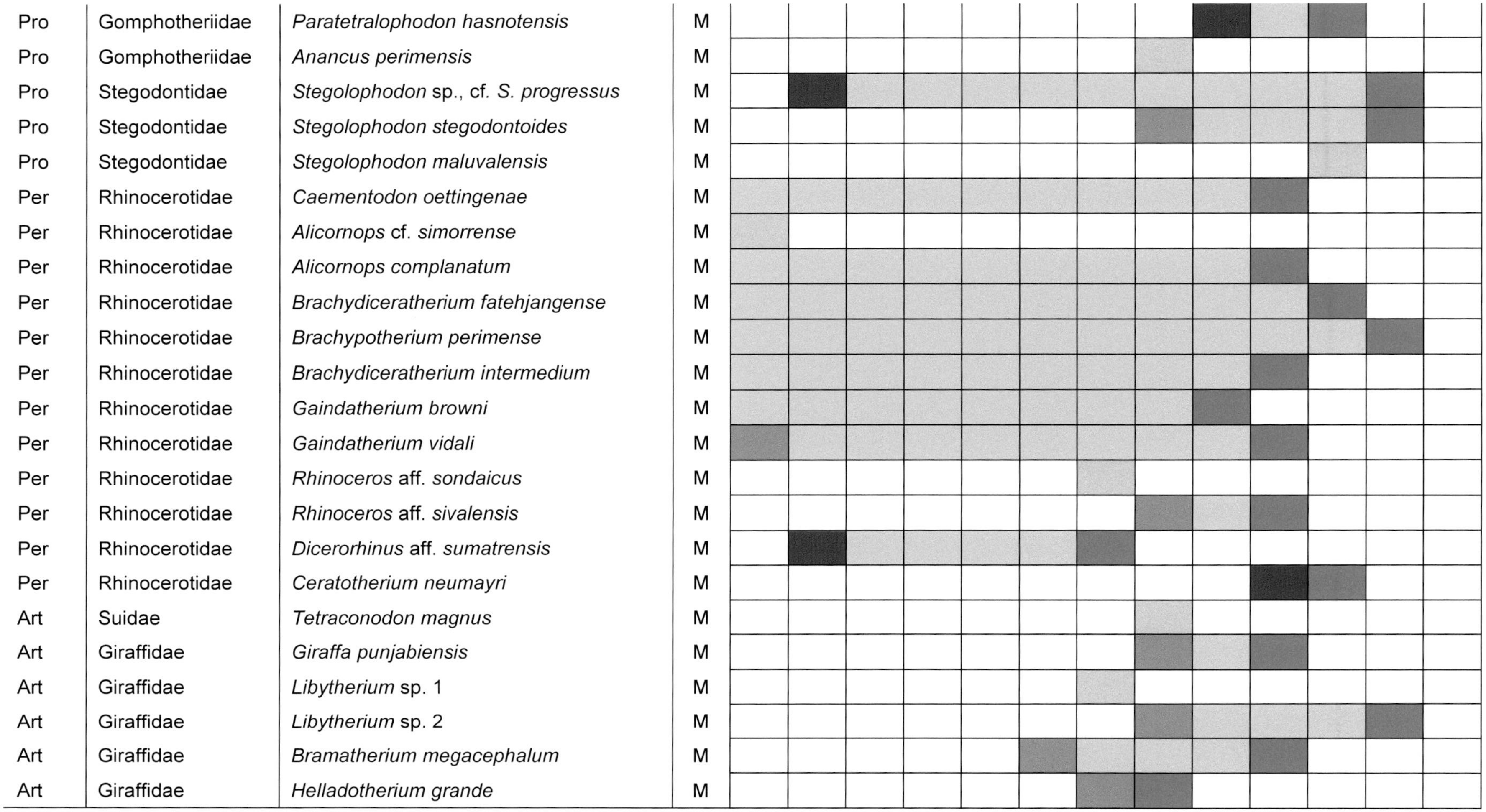

	Family	Species	
Pro	Gomphotheriidae	*Paratetralophodon hasnotensis*	M
Pro	Gomphotheriidae	*Anancus perimensis*	M
Pro	Stegodontidae	*Stegolophodon* sp., cf. *S. progressus*	M
Pro	Stegodontidae	*Stegolophodon stegodontoides*	M
Pro	Stegodontidae	*Stegolophodon maluvalensis*	M
Per	Rhinocerotidae	*Caementodon oettingenae*	M
Per	Rhinocerotidae	*Alicornops* cf. *simorrense*	M
Per	Rhinocerotidae	*Alicornops complanatum*	M
Per	Rhinocerotidae	*Brachydiceratherium fatehjangense*	M
Per	Rhinocerotidae	*Brachypotherium perimense*	M
Per	Rhinocerotidae	*Brachydiceratherium intermedium*	M
Per	Rhinocerotidae	*Gaindatherium browni*	M
Per	Rhinocerotidae	*Gaindatherium vidali*	M
Per	Rhinocerotidae	*Rhinoceros* aff. *sondaicus*	M
Per	Rhinocerotidae	*Rhinoceros* aff. *sivalensis*	M
Per	Rhinocerotidae	*Dicerorhinus* aff. *sumatrensis*	M
Per	Rhinocerotidae	*Ceratotherium neumayri*	M
Art	Suidae	*Tetraconodon magnus*	M
Art	Giraffidae	*Giraffa punjabiensis*	M
Art	Giraffidae	*Libytherium* sp. 1	M
Art	Giraffidae	*Libytherium* sp. 2	M
Art	Giraffidae	*Bramatherium megacephalum*	M
Art	Giraffidae	*Helladotherium grande*	M

Figure 25.S2. (Continued)

References

Andrews, P., J. M. Lord, E. M. and Nesbit-Evans. 1979. Patterns of ecological diversity in fossil and modern faunas. *Biological Journal of the Linnean Society* 11(2): 177–205.

Ataabadi, M. M, R. L. Bernor, D. S. Kostopoulos, D. O. Wolf, Z. A. Orak, G.H. Zare, I. H. Nakaya, ,M.A. Watabe, and M. I. Fortelius. 2013. Recent advances in paleobiological research of the Late Miocene Maragheh fauna, northwest Iran. In *Fossil Mammals of Asia: Neogene Biostratigraphy and Chronology*, edited by X.-M. Wang, L. J. Flynn and M. Fortelius, 546–565. New York: Columbia University Press.

Badgley, C., J. C. Barry, M. E. Morgan, S. V. Nelson, A. K. Behrensmeyer, T. E. Cerling, and D. Pilbeam. 2008. Ecological changes in Miocene mammalian record show impact of prolonged climatic forcing. *Proceedings of the National Academy of Sciences* 105(34): 12145–12149.

Badgley, C., M. S. Domingo, J. C. Barry, M. E. Morgan, L. J. Flynn, and D. Pilbeam. 2015. Continental gateways and the dynamics of mammalian faunas. *Comptes Rendus Palevol* 15(7): 763–779.

Badgley, C., W. R. Downs III, and L. J. Flynn. 1998. Taphonomy of small-mammal fossil assemblages from the Middle Miocene Chinji Formation, Siwalik Group, Pakistan. In *Advances in Vertebrate Paleontology and Geochronology*, edited by Y. Tomida, L. J. Flynn, and L. L. Jacobs, 145–166. National Science Museum of Japan Monographs, No. 14. Tokyo.

Barnosky, A. D. 2001. Distinguishing the effects of the Red Queen and Court Jester on Miocene mammal evolution in the northern Rocky Mountains. *Journal of Vertebrate Paleontology* 21(1): 172–185.

Barnosky, A. D. 2005. Effects of Quaternary climatic change on speciation in mammals. *Journal of Mammalian Evolution* 12(1–2): 247–264.

Barry, J. C., A. K. Behrensmeyer, C. Badgley, L. J. Flynn, H. Peltonen, I. U. Cheema, D. Pilbeam, et al. 2013. The Neogene Siwaliks of the Potwar Plateau and other regions of Pakistan. In *Fossil Mammals of Asia: Neogene Biostratigraphy and Chronology*, edited by X. Wang, L. J. Flynn, and M. Fortelius, 373–399. New York: Columbia University Press.

Barry, J. C., M. E. Morgan, L. J. Flynn, D. Pilbeam, A. K. Behrensmeyer, S. M. Raza, I. A. Khan, C. Badgley, J. Hicks, and Kelley, J., 2002. Faunal and environmental change in the Late Miocene Siwaliks of northern Pakistan. *Paleobiology* 28, no. 2, Supplement (Spring, 2002): 1–71.

Barry, J. C., M. E. Morgan, L. J. Flynn, D. Pilbeam, L. L. Jacobs, E. H. Lindsay, S. M. Raza, and N. Solounias, N. 1995. Patterns of faunal turnover and diversity in the Neogene Siwaliks of northern Pakistan. *Palaeogeography, Palaeoclimatology, Palaeoecology* 115(1–4): 209–226.

Baskin, J. A.1996. Systematic revision of Ctenodactylidae (Mammalia, Rodentia) from the Miocene of Pakistan. *Palaeovertebrata* 25(1): 1–49.

Behrensmeyer, A. K., N. E. Todd, R. Potts, G. E. McBrinn. 1997. Late Pliocene faunal turnover in the Turkana Basin, Kenya and Ethiopia. *Science* 278:1589–1594.

Behrensmeyer, A. K., J. Quade, T. E. Cerling, J. Kappelman, I. A. Khan, P. Copeland, L. Roe, et al. 2007. The structure and rate of late Miocene expansion of C4 plants: Evidence from lateral variation in stable isotopes of the Siwalik Group, northern Pakistan. *GSA Bulletin* 119(11–12): 1486–1505.

Bendrey, R., D. Vella, A. Zazzo, M. Balasse, and S. Lepetz. 2015. Exponentially decreasing tooth growth rate in horse teeth: Implications for isotopic analyses. *Archaeometry* 57(6): 1104–1124.

Bennett, K. D. 1990. Milankovitch cycles and their effects on species in ecological and evolutionary time. *Paleobiology* 16(1): 11–21.

Bennett, K. D. 2004. Continuing the debate on the role of Quaternary environmental change for macroevolution. *Philosophical Transactions of the Royal Society of London B* 359(1442): 295–303.

Benton, M. J. 2009. New take on the Red Queen. *Nature* 463:306–307.

Bernor, R. L., M. Armour-Chelu, H. Gilbert, T. M. Kaiser, and E. Schulz. 2010. Equidae. In *Cenozoic Mammals of Africa*, edited by L. Werdelin and W. J. Sanders, 685–721. Berkeley: University of California Press.

Bhat, U., C. P. Kempes, and J. D. Yeakel. 2009. Scaling the risk landscape drives optimal life-history strategies and the evolution of grazing. *Proceedings of the National Academy of Sciences* 117(3): 1580–1586.

Bryant, J. D., Philip N. F., W. J. Showers, and B. J. Genna. 1996. Biologic and climatic signals in the oxygen isotopic composition of Eocene-Oligocene equid enamel phosphate. *Palaeogeography, Palaeoclimatology, Palaeoecology* 126(1–2): 75–89.

Casanovas-Vilar, I., A. Madern, D. M. Alba, L. Cabrera, I. García-Paredes, L. W. van den Hoek Ostende, D. DeMiguel, D., et al. 2016. The Miocene mammal record of the Vallès-Penedès Basin (Catalonia). *Compte Rendus de l'Académie de l'Science Palevol* 15(7): 791–812.

Cerling, T. E., S. A. Andanje, S.A. Blumenthal, F. H. Brown, K. L. Chritz, J. M. Harris, J. A. Hart, et al. 2015. Dietary changes of large herbivores in the Turkana Basin, Kenya, from 4 to 1 Ma. *Proceedings of the National Academy of Sciences* 112(37): 11467–11472.

Chavasseau, O., A. Aung Khyaw, Y. Chaimanee, P. Coster, E.-G. Emonet, A. Naing Soe, M. Rugbumrung, S. Thura Tun, and J.-J. Jaeger. 2013. Advances in the biostratigraphy of the Irrawaddy Formation (Myanmar). In *Fossil Mammals of Asia. Neogene Biostratigraphy and Chronology*, edited by X. Wang, L. J. Flynn and M. Fortelius, 461–474. New York: Columbia University Press.

Clauss, M., and G. E. Rössner. 2014. Old world ruminant morphophysiology, life history, and fossil record: Exploring key innovations of a diversification sequence. *Annales Zoologici Fennici* 51:80–94.

Colcord, D. E., A. M. Shilling, P. E. Sauer, K. H. Freeman, J. K. Njau, I. G. Stanistreet, H. Stollhofen, K. D. Schick, N. Toth, and S. C. Brassell. 2018. Sub-Milankovitch paleoclimatic and paleoenvironmental variability in East Africa recorded by Pleistocene lacustrine sediments from Olduvai Gorge, Tanzania. *Palaeogeography, Palaeoclimatology, Palaeoecoecology* 495:284–291.

Cotterill, F. B. D. 2004. Mammal Faunal of the Four Corner Area. In *Biodiversity of the Four Corners Area: Technical Reviews*, edited by J. R. Timberlake and S. I. Childes, 2:231–266. Occasional Publications in Biodiversity No. 15. Harare, Zimbabwe: Biodiversity Foundation for Africa, Bulawayo/Zambezi Society.

Danowitz, M., A. Vasilyev, V. Kortlandt, and N. Solounias. 2015. Fossil evidence and stages of elongation of the *Giraffa camelopardalis* neck. *Royal Society Open Science* 2(10): 150393.

Darwin, C. 1859. *On the Origin of Species by Means of Natural Selection, or the Preservation of Favoured Races in the Struggle for Life*, 1st ed. London: John Murray.

Davis, M. B. 1983. Quaternary history of deciduous forests of eastern North America and Europe. *Annals of the Missouri Botanical Garden* 70(3): 550–563.

Dinerstein, E. 1992. Effects of *Rhinoceros unicornis* on riverine forest structure in lowland Nepal. *Ecology* 73(2): 701–704.

Dutoit, J. T. 1990. Feeding-height stratification among African browsing ruminants. *African Journal of Ecology* 28(1): 55–61.

Eldredge, N., and S. J. Gould. 1972. Punctuated equilibria: An alternative to phyletic gradualism. In *Models in Paleobiology*, edited by T. J. M. Schopf, 82–115. San Francisco, CA: Freeman, Cooper and Co.

Faith, J. T., J. Rowan, and A. Du. 2019. Early hominins evolved within nonanalog ecosystems. *Proceedings of the National Academy of Sciences* 116(43): 21478–21483.

Faith, J. T., J. Rowan, A. Du, and P.L. Koch. 2018. Plio-Pleistocene decline of African megaherbivores: No evidence for ancient hominin impacts. *Science* 362(6417): 938–941.

Feakins, S. J., H. M. Liddy, L. Tauxe, V. Galy, X. Feng, J. E. Tierney, Y. Miao, and S. Warny. 2020. Miocene C_4 grassland expansion as recorded by the Indus Fan. *Paleoceanography and Paleoclimatology* 35(6): e2020PA003856.

Flynn, L. J. 1986. Species longevity, stasis, and stairsteps in rhizomyid rodents. In *Vertebrates, Phylogeny*, edited by K. M. Flanagan and J. A. Lillegraven, 273–285. Contributions to Geology, University of Wyoming Special Papers.

Flynn, L. J. 2009. The antiquity of *Rhizomys* and independent acquisition of fossorial traits in subterranean muroids. *Bulletin of the American Museum of Natural History* 331:128–156.

Flynn, L. J., J. C. Barry, M. E. Morgan, D. Pilbeam, L. L. Jacobs, and E. H. Lindsay, E. H. 1995. Neogene Siwalik mammalian lineages: Species longevities, rates of change, and modes of speciation. *Palaeogeography, Palaeoclimatology, Palaeoecology* 115(1–4): 249–264.

Flynn, L. J., Y. Kimura, and L. L. Jacobs. 2020. The murine cradle. In *Biological Consequences of Plate Tectonics: New Perspectives on Post-Gondwana Break-Up—A Tribute to Ashok Sahni*, edited by G. V. R. Prasad and R. Patnaik, 347–362. Vertebrate Paleobiology and Paleoanthropology Series, https://doi.org/10.1007/978-3-030-49753-8_15.

Flynn, L., E. H. Lindsay, D. Pilbeam, S. M. Raza, M. E Morgan, J. C. Barry, C. Badgley, et al. 2013. The Siwaliks and Neogene evolutionary biology in South Asia. In *Fossil Mammals of Asia: Neogene Biostratigraphy and Chronology*, edited by X. Wang, L. J. Flynn, and M. Fortelius, 353–372. New York: Columbia University Press.

Flynn, L J., M.E. Morgan, J. C. Barry, S. M. Raza, I. U. Cheema, and D. Pilbeam. 2023. Siwalik Rodent Assemblages for NOW: Biostratigraphic Resolution in the Neogene of South Asia. In *Evolution of Cenozoic Land Mammal Faunas and Ecosystems: 25 Years of the NOW Database of Fossil Mammals*, edited by I. Casanovas-Vilar, L. W. van den Hoek Ostende, C. M. Janis, and J. Saarinen, 43–58. Vertebrate Paleobiology and Paleoanthropology Series. https://doi.org/10.1007/978-3-031-17491-9.

Flynn, L J., M. E. Morgan, D. Pilbeam, and J. C. Barry. 2014. "Endemism" relative to space, time, and taxonomic level. *Annales Zoologici Fennici* 51:245–258.

Flynn, L .J., and W. Wessels. 2013. Paleobiogeography and South Asian small mammals: Neogene latitudinal faunal variation. In *Fossil Mammals of Asia: Neogene Biostratigraphy and Chronology*, edited by X. Wang, L. J. Flynn, and M. Fortelius, 445–460. New York: Columbia University Press.

Fraser, D., L. C. Soul, A. B. Tóth, M. A. Balk, J. T. Eronen, S. Pineda-Munoz, A. B. Shupinski, et al. 2020. Investigating biotic interactions in deep time. *Trends in Ecology & Evolution* 36(1): 61–75.

Gingerich, P. D. 1991. Fossils and evolution. In *Evolution of Life: Fossils, Molecules, and Culture*, edited by S. Osawa and T. Honjo, 3–20. Tokyo: Springer Verlag.

Gingerich, P. D., and G. F. Gunnell. 1995. Rates of evolution in Paleocene–Eocene mammals of the Clarks Fork Basin, Wyoming, and a comparison with Neogene Siwalik lineages of Pakistan. *Palaeogeography, Palaeoclimatology, Palaeoecology* 115(1–4): 227–247.

Gould, S. J. 1985. The paradox of the first tier: An agenda for paleobiology. *Paleobiology* 11: 2–12.

Graham, R. W., E. L. Lundelius Jr., M. A. Graham, E. K. Schroeder, R. S. Toomey III, E. Anderson, A. D. Barnosky, et al. 1996. Spatial response of mammals to Late Quaternary environmental fluctuations. *Science* 272:1601–1606.

Grant, P., and R. Grant. 2002. Unpredictable evolution in a 30-year study of Darwin's finches. *Science* 296:707–711.

Grant, P. R. 1972. Interspecific competition among rodents. *Annual Review of Ecology and Systematics* 3:79–106.

Holbourn, A. E., W. Kuhnt, S. C. Clemens, K. G. Kochhann, J. Jöhnck, J. Lübbers, and N. Andersen. 2018. Late Miocene climate cooling and intensification of southeast Asian winter monsoon. *Nature Communications* 9(1): 1584.

Holbourn, A., W. Kuhnt, S. Clemens, W. Prell, and N. Andersen. 2013. Middle to late Miocene stepwise climate cooling: Evidence from a high-resolution deep water isotope curve spanning 8 million years. *Paleoceanography* 28(4): 688–699.

Holbourn, A., W. Kuhnt, K. G. Kochhann, N. Andersen, and K. J. Sebastian Meier. 2015. Global perturbation of the carbon cycle at the onset of the Miocene Climatic Optimum. *Geology* 43(2): 123–126.

Holbourn, A., W. Kuhnt, M. Schulz, and H. Erlenkeuser. 2005. Impacts of orbital forcing and atmospheric carbon dioxide on Miocene ice-sheet expansion. *Nature* 438:483–487.

Holbourn, A., W. Kuhnt, M. Schultz, J.-A. Flores, , and N. Andersen. 2007. Orbitally paced climate evolution during the Middle Miocene "Monterey" carbon-isotope excursion. *Earth and Planetary Science Letters* 261(3–4): 534–550.

Hoorn, C., T. Ohja, and J. Quade. 2000. Palynological evidence for vegetation development and climatic change in the sub-Himalayan zone (Neogene, central Nepal). *Palaeogeography, Palaeoclimatology, Palaeoecoecology* 163(3–4): 133–161.

Huang, S., J. T. Eronen, C. M. Janis, J. J. Saarinen, D. Silvestro, and S. A. Fritz. 2017. Mammal body size evolution in North America and Europe over 20 Myr: Similar trends generated by different processes. *Proceedings of the Royal Society B: Biological Sciences* 284:2016.2361.

Karp, A. T., A. K. Behrensmeyer, and K. H. Freeman. 2018. Grassland fire ecology has roots in the Late Miocene. *Proceedings of the National Academy of Sciences* 115(48): 12130–12135.

Karp, A. T., K. T. Uno, P. Polissar, and K. Freeman. 2021. Late Miocene C_4 grassland fire feedbacks on the Indian subcontinent. *Paleoceanography and Paleoclimatology* 36(4): e2020PA004106.

Kimura, Y., L. J. Flynn, and L. L. Jacobs. 2015. A paleontological case study for species delimitation in diverging fossil lineages. *Historical Biology* 28(1–2): 189–198.

Kimura, Y., L. J. Flynn, and L. L. Jacobs. 2021. Tempo and mode: Evidence on a protracted split from a dense fossil record. *Frontiers in Ecology and Evolution* 9:642814.

Kimura, Y., L. L. Jacobs, T. E. Cerling, K. T. Uno, K. M. Ferguson, L. J. Flynn, and R. Patnaik. 2013. Fossil mice and rats show isotopic evidence of niche partitioning and change in dental ecomorphology related to dietary shift in Late Miocene of Pakistan. *PLOS ONE* 8(8): e69308.

Laskar, J., P. Robutel, F. Joutel, M. Gastineau, A. C. M. Correia, and B. Levrard. 2004. A long-term numerical solution for the insolation quantities of the Earth. *Astronomy and Astrophysics* 428:261–285.

Lihoreau, F., J. Barry, C. Blondel, Y. Chaimanee, J.-J. Jaeger, and M. Brunet. 2007. Anatomical revision of the genus *Merycopotamus* (Artiodactyla; Anthracotheriidae): Its significance for Late Miocene mammal dispersal in Asia. *Palaeontology* 50(2): 503–524.

Liow, L. H. 2007. Lineages with long durations are old and morphologically average: An analysis using multiple datasets. *Evolution* 61(4): 885–901.

Liow, L. H., M. Fortelius, E. Bingham, K. Lintulaakso, H. Mannila, L. Flynn, and N. C. Stenseth. 2008. Higher origination and extinction rates in larger mammals. *Proceedings of the National Academy of Sciences* 105(16): 6097–6102.

Lübbers, J., W. Kuhnt, A. E. Holbourn, C. T. Bolton, E. Gray, Y. Usui, K. G. Kochhann, S. Beil, and N. Andersen. 2019. The Middle to Late Miocene "Carbonate Crash" in the equatorial Indian Ocean. *Paleoceanography and Paleoclimatology* 34(5): 813–832.

Lupien, R. L., J. M. Russell, M. Grove, C. C. Beck, C. S. Feibel, and A. S. Cohen. 2020. Abrupt climate change and its influence on hominin evolution during the Early Pleistocene in the Turkana Basin, Kenya. *Quaternary Science Reviews* 245:1–11.

Magill, C. R., G. M. Ashley, and K. H. Freeman. 2013. Ecosystem variability and early human habitats in eastern Africa. *Proceedings of the National Academy of Sciences* 110(4): 1167–1174.

Miller, J. A., A. K. Behrensmeyer, A. Du, S. K. Lyons, D. Patterson, A. Tóth, A. Villasenor, E. Kanga, and D. Reed. 2014. Ecological fidelity of functional traits based on species presence-absence in a modern mammalian bone assemblage (Amboseli, Kenya). *Paleobiology* 40(4): 560–583.

Morgan, M. E. 1994. Paleoecology of Siwalik Miocene hominoid communities: Stable carbon isotope, dental microwear, and body size analyses. PhD diss., Harvard University.

Morgan, M. E., C. Badgley, G. F. Gunnell, P. D. Gingerich, J. W. Kappelman, and M. C. Maas. 1995. Comparative paleoecology of Paleogene and Neogene mammalian faunas: Body-size structure. *Palaeogeography, Palaeoclimatology, Palaeoecology* 115(1–4): 287–317.

Morgan, M. E., A. K. Behrensmeyer, C. Badgley, J. C. Barry, S. Nelson, and D. Pilbeam. 2009. Lateral trends in carbon isotope ratios reveal a Miocene vegetation gradient in the Siwaliks of Pakistan. *Geology* 37(2): 103–106.

Morgan, M. E., J. D. Kingston, and B. D. Marino. 1994. Carbon isotopic evidence for the emergence of C_4 plants in the Neogene from Pakistan and Kenya. *Nature* 367:162–165.

Nelson, S. V. 2005. Paleoseasonality inferred from equid teeth and intratooth isotopic variability. *Palaeogeography, Palaeoclimatology, Palaeoecology* 222:122–144.

Nelson, S. V. 2007. Isotopic reconstructions of habitat change surrounding the extinction of *Sivapithecus*, a Miocene hominoid, in the Siwalik Group of Pakistan. *Palaeogeography, Palaeoclimatology, Palaeoecology* 243(1–2): 204–222.

Owen-Smith, R. N. (1988). *Megaherbivores: The influence of very large body size on ecology*. Cambridge: Cambridge University Press.

Owen-Smith, N. 2021. *Only in Africa: The Ecology of Human Evolution*. Cambridge: Cambridge University Press.

Patnaik, R. 2013. Indian Neogene Siwalik mammalian biostratigraphy: An overview." In *Fossil Mammals of Asia: Neogene Biostratigraphy and Chronology*, edited by X. Wang, L. J. Flynn, and M. Fortelius, 423–444. New York: Columbia University Press.

Polissar, P., K. T. Uno, S. R. Phelps, A. T. Karp, K. H. Freeman, and J. L. Pensky. 2021. Hydrologic changes drove the Late Miocene expansion of C_4 grasslands on the northern Indian subcontinent. *Paleoceanography and Paleoclimatology* 36(4): e2020PA004108.

Prasad, M. 2008. Angiospermous fossil leaves form the Siwalik foreland basins and their palaeoclimatic implications. *Palaeobotanist* 57(1–3): 177–215.

Quade, J., and T. E. Cerling. 1995. Expansion of C_4 grasses in the late Miocene of northern Pakistan: Evidence from stable isotopes in paleosols. *Palaeogeography, Palaeoclimatology, Palaeoecoecology* 115(1–4): 91–116.

Quade, J., T. E. Cerling, and J. R. Bowman. 1989. Development of Asian monsoon revealed by marked ecological shift during the latest Miocene in northern Pakistan. *Nature* 342(6246): 163–166.

Reynolds, K., A. Copley, and E. Hussain. 2015. Evolution and dynamics of a fold-thrust belt: The Sulaiman Range of Pakistan. *Geophysical Journal International* 201(2): 683–710.

Reznick, D. N., and C. K. Ghalambor. 2005. Selection in nature: Experimental manipulations of natural populations. *Integrative and Comparative Biology* 45(3): 456–462.

Reznick, D. N., and R. E. Ricklefs. 2009. Darwin's bridge between microevolution and macroevolution. *Nature* 457(7231): 837–842.

Roberts, T. J. 1977. *Mammals of Pakistan*. London: E. Benn.

Said, M. Y. 2003. Multiscale perspectives of species richness in East Africa. PhD diss., Wageningen University and Research ProQuest Dissertations Publishing, 2003. 28239871.

Simpson, G. G. 1960. Notes on the measurement of faunal resemblance. *American Journal of Science* 258-A:300–311.

Srivastava, G., K. N. Paudayal, T. Utescher, and R. C. Mehrotra. 2018. Miocene vegetation shift and climate change: Evidence from the Siwalik of Nepal. *Global and Planetary Change* 161:108–120.

Sun, B., X. Zhang, Y. Liu, and R. L. Bernor. 2018. *Sivalhippus ptychodus* and *Sivalhippus platyodus* (Perissodactyla, Mammalia) from the Late Miocene of China. *Rivista Italiana di Paleontologia e Stratigrafia* 124(1): 1–22.

Tauxe, L., and S. J. Feakins. 2020. A re-assessment of the chronostratigraphy of late Miocene C_3–C_4 transitions. *Paleoceanography and Paleoclimatology* 35:e2020PA003857.

Terry, R.C. 2010. On raptors and rodents: Testing the ecological fidelity and spatiotemporal resolution of cave death assemblages. *Paleobiology* 36(1): 137–160.

Uno, K. T., T. E. Cerling, J. M. Harris, Y. Kunimatsu, M. G. Leakey, M. Nakatsukasa, and H. Nakaya. 2011. Late Miocene to Pliocene carbon isotope record of differential diet change among East African herbivores. *Proceedings of the National Academy of Sciences* 108(16): 6509–6514.

Van Valen, L. 1977. The Red Queen. *American Naturalist* 111:809–810.

Vrba, E. S. 1993. Turnover-pulses, the Red Queen, and related topics. *American Journal of Science* 293-A:418–452.

Wessels, W. 2009. Miocene rodent evolution and migration: Muroidea from Pakistan, Turkey, and northern Africa. *Geologica Ultraiectina* 307:1–290.

Western, D., and A. K. Behrensmeyer. 2009. Bone assemblages track animal community structure over 40 years in an African savanna ecosystem. *Science* 324(5930): 1061–1064.

Willis, B. J. 1993. Evolution of Miocene fluvial systems in the Himalayan foredeep through a two kilometer-thick succession in northern Pakistan. *Sedimentary Geology* 88(1–2): 77–121.

Willis, B. J., and A. K. Behrensmeyer. 1995. Fluvial systems in the Siwalik Miocene and Wyoming Paleogene. *Palaeogeography, Palaeoclimatology, Palaeoecology* 115(1–4): 13–35.

Wilson, K. E., M. A. Maslin, M. J. Leng, J. D. Kingston, A. L Deino, R. K. Edgar, and A. W. Mackay. 2014. East African lake evidence for Pliocene millennial-scale climate variability. *Geology* 42(11): 955–958.

Wolf, D., R. L. Bernor, and S. T. Hussain. 2013. Evolution, biostratigraphy and geochronology of Siwalik hipparionine horses. *Palaeontographica* 300:1–118.

26 TAXONOMIC AND ECOLOGICAL DYNAMICS OF SIWALIK MAMMALIAN FAUNAS

CATHERINE BADGLEY

INTRODUCTION

The Siwalik record of the Potwar Plateau, Pakistan, preserves a record of terrestrial ecosystem changes over much of the last 18 million years. Over this time, the foreland basin that formed the depositional setting for the Siwalik sequence was deepening, rotating, and moving northward, closer to the Himalayan Mountain front (Opdyke et al. 1982; Chapter 2, this volume). A network of mountain- and foothill-sourced rivers dominated the alluvial plains of the foreland basin (see Fig. 24.4), creating a dynamic landscape of aquatic and terrestrial habitats (Chapters 2, 24, this volume). The alluvial ecosystems persisted over millions of years, with perturbations on time scales ranging from months to millions of years. Yearly to semi-annual flooding would have caused channels to expand over much of the flood basin for weeks to months, followed by retreat of water to confined channel belts, causing a shift in occupancy of animal populations between aquatic and subaerial habitats. Over millions of years, mountain uplift would have altered source areas and, in combination with changes in sea level, widened or closed corridors for animal dispersal in and out of the Indian subcontinent and the Siwalik faunal province. Regional temperatures and the amount and seasonality of precipitation over South Asia would have changed on millenial Milankovitch to million-year time scales (Feakins et al. 2020).

In this chapter, we build upon the synthesis of mammalian faunal change in Chapter 25 in this volume to analyze rates of species origination, extinction, and diversification, as well as changes in taxonomic richness and ecological structure (body size and feeding habit). (I use the term "we" since this chapter draws on information from all earlier chapters and is inherently collaborative.) We use the concept of confidence intervals on stratigraphic ranges of individual species to estimate taxonomic richness from ~18.0 to ~6.0 Ma, a 12 million-year interval of the Miocene that extends over the four formations studied through decades of fieldwork by the Harvard-Geological Survey of Pakistan (Harvard-GSP) research group. This analysis interval represents a sequence of well-exposed outcrops that were the focus of study during the ~25 field seasons of the Harvard-GSP project. We utilize the high temporal resolution

of the Siwalik record (Chapters 2, 3, this volume) to compile species-level occurrences dated to a precision of 100-kyr (thousand year) intervals and conduct most of our analyses at 200-kyr intervals for the sake of statistical assessment of diversification trends (see "Rates of Faunal Change" section below in this chapter). (Eighty percent of Potwar fossil localities have age calibrations of 200 kyr or less, Chapter 3, this volume.) In addition, we document the composition of mammalian faunas in terms of species richness per family and ecological richness (combined size-feeding categories) to assess the timing and magnitude of changes in taxonomic composition and ecological structure over the analysis interval. Building on the biostratigraphic, taxonomic, and paleoecological information of earlier chapters, this chapter (1) compares the richness records of small and large mammals; (2) determines per-capita rates of origination, extinction, and diversification of mammalian faunas; (3) compares the timing and magnitude of change in taxonomic and ecological richness; and (4) evaluates these patterns of faunal change in terms of potential causes and drivers. These objectives are the basis for analyzing ecological and evolutionary processes and require the presence of a fossil record longer than the duration of most species lineages, a high sampling density, high taxonomic resolution, and paleoenvironmental data. The Siwalik record of the Potwar Plateau is exceptional in offering all of these features.

In the next section, we describe the resolution and partitioning of data for these analyses. Then we present a biochronology of residence times based on estimated first and last appearances of species. This dataset is the basis for analyses of diversification and changes in faunal composition over time; methods for these analyses are presented in each section and in Boxes 26.1–26.3.

TEMPORAL AND TAXONOMIC RESOLUTION

All species-level mammalian lineages documented to occur on the Potwar Plateau between 18.1 and 6.1 Ma. comprise the data for these analyses. (These interval boundaries were selected among alternatives for the clearest depiction of peaks of origination and extinction.) Although the distribution of fossil localities and specimens varies markedly through the analysis interval, some of our analytical methods reduce the effects of varying quality of the fossil record. For species that are considered parts of an anagenetic sequence, we combined the species names and their observed durations into a single lineage (e.g., *Sayimys* lineage). We omitted bats (Chiroptera) from analysis, although their original diversity and numbers were likely substantial, because their fossil record is quite sparse and poorly resolved taxonomically (Chapter 10, this volume). We analyzed all other mammalian groups, including carnivores and creodonts (Chapter 13, this volume), which were excluded from the analyses of Chapter 25 of this volume.

We analyzed small mammals and large mammals separately. The distinction is based upon different methods of collection—screen-washing for small mammals and surface collection or excavation for large mammals—as well as substantial differences in life-history traits, ecological interactions, and the spatial scale of community-assembly processes. Small mammals perceive their habitat at a finer grain than do large mammals, and many terrestrial or burrowing species are sensitive to properties of the substrate (Merritt 2010). The short life spans and small home ranges of small mammals enable them to evolve rapidly in response to changes in ecosystem properties (Ernest 2013). In contrast, large mammals have much larger home ranges and longer life spans; few of them have life habits tied to substrates (such as dens or burrows). We used the same criterion as in Chapter 25 of this volume for distinguishing small from large mammals—namely, whether estimated adult body weight was less than or greater than 1.0 kg. (This distinction departs from the body weight of 5 kg that many mammalogists use to separate small from large mammals.) By our criterion, some mammalian orders have both small and large species (e.g., Carnivora, Rodentia). (We did not separate megaherbivores from other large mammals, although they constitute one of the size categories in the analysis of joint size-feeding structure.) This division enables us to compare the timing and pattern of diversification and change in composition for small versus large mammals. The Siwalik small-mammal record includes species in six orders and is dominated by two clades—Lipotyphla (insectivores) and Rodentia (rodents). The large-mammal record include species in ten orders and is dominated by species in Carnivora (phylogenetic carnivores), Artiodactyla (even-toed ungulates), and Perissodactyla (odd-toed ungulates).

For each species, we noted two functional-ecological traits—estimated adult body size and estimated feeding habit—and combined these into joint size-feeding categories. Table 26.1 lists the categories used for each trait. Body size is strongly correlated with metabolic rate, home-range size, and many life-history variables (Peters 1983; Damuth and MacFadden 1990). Adult body weight was estimated from molar or postcranial linear measurements or by direct comparison with a living species of similar size; data come from the faunal chapters of Part II of this volume. Feeding habit indicates the major kinds of plant or animal food resources consumed and

Table 26.1. Body Weight (A) and Feeding Categories (B) Utilized for the Construction of Joint Size-Feeding Categories

A. Body Weight Categories

Size category	Weight Range	Faunal Dataset
Very small	<100 g	Small mammals
Small	100–999 g	Small mammals
Small	1.0–19.9 kg	Large mammals
Medium	20–99 kg	Large mammals
Large	100–249 kg	Large mammals
Very large	250–799 kg	Large mammals
Mega	>800 kg	Large mammals, includes megaherbivores of Chapter 25

B. Feeding Categories for Small and Large Mammals

Feeding category	Dietary Components
Frugivore	Predominantly fruit, nectar
Granivore	Nuts, seeds
Herbivore-browser	Broad-leaf vegetation on ground or in canopy
Herbivore-mixed feeder	Broad-leaf vegetation and grass or sedge
Herbivore-grazer	Predominantly grass or sedge
Rootivore	Underground roots and tubers
Omnivore	Animal, plant, and fungal resources
Invertivore	Insects, other arthropods, worms
Faunivore	Terrestrial or aquatic carnivore

thereby provides information about species interactions, vegetation properties, and climatic conditions (Andrews, Lord, and Nesbit Evans 1979; Badgley and Fox 2000; Fraser et al. 2021). Feeding habit was estimated from a combination of dental morphology, stable-isotopic data ($\delta^{13}C$) for teeth identified to species, and microwear or mesowear data for a subset of ungulate species. Absence of data or poor taxonomic resolution on isotopic or microwear data required us to omit some species from the analysis of functional-ecological traits. Most groups are poorly resolved phylogenetically, so we cannot include quantitative measures of phylogenetic richness in our analysis.

ESTIMATION OF RESIDENCE TIMES

The purpose of determining a confidence interval on the stratigraphic range of Siwalik species is to acknowledge the incompleteness of the fossil record in general, the variation over time in fossil productivity of the Siwalik sequence in the Potwar Plateau, and variation among lineages in their representation in the fossil record (Barry et al. 2002).

The distribution of fossil localities and specimens is uneven through the Potwar Siwalik sequence (see Fig. 3.3A). For the older part of the sequence, the fossil record is sparse, although rich localities occur at fairly regular intervals (nearly every 0.5 million years, myr). Likewise, the youngest portion of the sequence in the areas where we worked has low fossil productivity. In between, from ~14.5 to 7.0 Ma, the fossil record is robust in most 200-kyr time intervals. Fossil localities are less common from the base of the Potwar Siwalik sequence through the Kamlial Formation than in younger formations. Is this sparse fossil record a result of poor preservation, less frequent occupancy of the floodplain ecosystem, or other factors? Taphonomic analysis of Siwalik fossil assemblages (Chapter 4, this volume) indicates that processes of preservation were similar for fossil localities of the Kamlial Formation to those of the overlying richer Chinji, Nagri, and Dhok Pathan Formations. The sedimentary environments of preservation were similar as were taphonomic properties of the fossil assemblages, such as skeletal-element composition and bone-damage patterns. The Kamlial Formation has proportionally more thick sandstones than the Chinji Formation, and Siwalik fossil assemblages are infrequently preserved in this facies (Chapter 4, this volume), and this difference may be a contributing factor. In addition, floodplain channels, which are a productive facies in the Chinji and Dhok Pathan Formations, are less common in the Kamlial Formation (Barry et al. 2013). The taxonomic composition of Kamlial fossil assemblages is distinctive, however, with ten families of small mammals and high richness of rhinocerotids and tragulids (proportionally greater than in younger intervals). These data suggest that fossil preservation and lithology were not the only differences between the record from 18.1 to 14.5 Ma and the record from 14.5 to 7.0 Ma. Rather, the Kamlial Formation documents an ecologically distinctive time interval. For these reasons, we utilized an approach of Marshall (1990, 2010) for estimating confidence intervals and point estimates of actual first and last appearances. An assumption of this method is that fossil preservation is stochastically constant throughout the sequence. This method uses the average gap size between time intervals with fossils of the focal taxon as the basis for calculating a confidence interval.

We calculated 80% confidence intervals for each species with fossils in more than one 100-kyr time interval (see Box 26.1). We selected 80% as a compromise between confidence level and having a large number of lineages for diversification and other analyses; Barry et al. (2002) used this level in their approach to estimating inferred first and last appearances. For analyses here, we used the point estimate of first and last appearance, which is determined as the average gap size (in myr) between all superjacent pairs of fossil localities with the focal taxon, added to the observed first occurrence and subtracted from the observed last occurrence. (If the observed first occurrence of a species was before 18.1 Ma, as happens for some species documented in the Zinda Pir record [Sulaiman region], then we determined the point estimate of estimated first and last appearance for the Potwar record based only on the gap sizes among Potwar localities.) We assumed that species range through the record from the estimated first appearance to the estimated last appearance. Species that occur in a single time interval (singletons) were omitted from some analyses, as noted in the "Rates of Faunal Change" section below. The mean gap size for all small and large mammals is <0.5 myr (see Box 26.1). Species with average gap sizes greater than 1.5 myr were omitted from the diversification analyses, since the uncertainty in their stratigraphic duration is too large to contribute to understanding the dynamics of faunal change over

shorter time intervals (e.g., Domingo et al. 2014). Our analysis interval begins at 18.1 Ma, with the first 0.1-myr time interval extending from 18.100 to 18.001 Ma, and the first 0.2-myr interval extending from 18.100 to 17.901 Ma; subsequent intervals follow the same pattern. While the observed record has no fossil localities older than 18.0 Ma, some estimated first appearances are much older than 18.0 Ma. The analysis interval ends at 6.1 Ma. The 200-kyr interval boundaries facilitated robust bootstrap results for evaluating significant diversification metrics.

Figure 26.1 (p. 485) illustrates four measures of species richness for small and large mammals. Observed species richness is based on the documented first and last occurrences of species, binned in 200-kyr intervals. Estimated species richness is based on the point estimates of first and last appearances with the average gap size extending each species' duration beyond its observed value. Estimated species richness is usually greater than observed richness for both small and large mammals, since the extended residence times elevate richness estimates for most time intervals. However, for some intervals between 18.1 and 15.1 Ma, observed richness is greater; this difference is a consequence of the pruning of the few species with an average gap size greater than 1.5 myr in length from the estimated-richness dataset; most such species occur early in the record. For both small and large mammals, the difference between including and excluding singleton species is minimal for both observed and estimated richness curves (Fig. 26.1). Singletons occur across the sequence, typically one per 200-kyr interval. A few intervals with exceptionally large numbers of fossil specimens have multiple singletons; one such interval that stands out for large mammals occurs at 10.1 Ma, the interval with the richest locality in the Potwar Siwalik sequence (Y311).

RATES OF FAUNAL CHANGE

In order to determine rates of faunal change over time, we evaluated per-capita rates of origination, extinction, and net diversification from 18.1 to 6.1 Ma. This approach scales the number of originations and extinctions per time interval against the number of species present before, during, and after each time interval. We utilized the approach developed by Foote (2000) and employed in an earlier analysis of Siwalik mammal diversification (Badgley et al. 2015). The current analysis uses an expanded faunal dataset and shorter time intervals.

The observed and estimated residence times of small mammals (Fig. 26.2, p. 486) and large mammals (Fig. 26.3, p. 487) are arrayed by the age of observed first occurrence. For small mammals, the pattern of observed residence times shows pulses of first occurrence, especially between 18.1 and 14.1 Ma; the estimated residence times form a more staggered pattern. Likewise, for large mammals, several time intervals show pulses of first occurrence throughout the sequence, with the estimated residence times showing a staggered pattern. These charts include all species of small (100) and large (194) mammals. Pruning species with gap sizes greater than 1.5 myr from each dataset reduced small mammals to 96 species-level lineages and

BOX 26.1

CONFIDENCE INTERVALS AND ESTIMATION OF FIRST AND LAST APPEARANCES

We utilize the method described by Marshall (1990, 2010) as "classical confidence intervals." This approach assumes that the potential for recovering fossils is constant throughout the stratigraphic sequence being analyzed. For a chosen confidence level (C, e.g., 95%, 50%), H as the number of fossil horizons in which the focal taxon is found, and an observed range (R), the formula for calculating a range extension, r, is

$$r = R[(1 - C)^{-1/(H-1)} - 1].$$

For example, if the observed range is 4.0 myr, the confidence level is 80% (0.80), and there are 5 fossil horizons for species X, then

$$r = 4.0\,[(1 - 0.80)^{-1/(5-1)} - 1],$$
$$= 4.0\,[0.495] = 1.98 \text{ myr.}$$

This value of r is quite large in relation to the observed duration of Species X.
If the number of fossil horizons is ten, then

$$r = 4.0\,[0.196] = 0.78 \text{ myr.}$$

The point estimate for the actual first appearance of species X is the average gap size added to the observed first occurrence (FO). The average gap size is simply the observed range divided by the number of gaps (H, the number of fossil horizons in which species X occurs –1). For our analysis, the term "horizon" refers to the number of time intervals (of 100,000 years) in which the species occurs. For some taxa, there may be many more localities documented than time intervals, but since we need an estimate in myr, we use the smallest time interval for which most of our localities can be calibrated. (Many localities can be confidently placed within 100-kyr time intervals; most localities can be confidently placed within 200-kyr time intervals; see Chapter 3, this volume, for explanation.)

For species X with an observed first occurrence at 10.000 Ma and an observed last occurrence at 6.000 Ma, the observed range R is 4.000 myr. If the number of time intervals with documented localities for species X is five, then the average gap size is

$$\text{mean gap size} = R/H - 1 = 4.000/5 - 1 = 1.000 \text{ myr.}$$

If the number of time intervals is 10, then

$$\text{mean gap size} = 4.000/10 - 1 = 0.444 \text{ myr.}$$

For the first instance (5 time intervals), the estimated first appearance is 10.000 + 1.000 Ma = 11.000 Ma. The estimated last appearance is 6.000 − 1.000 Ma = 5.000 Ma.

In the Siwalik record of the Potwar Plateau, the average gap size among small mammals (100 species lineages) is 0.427 myr. For large mammals (194 species lineages), the average gap size is 0.447 myr. (For all mammals combined, the value is 0.440 myr.) We use the estimated first and last appearances based on each species' average gap size in the various analyses presented in this chapter and not the confidence intervals per se.

The assumptions of this method deserve consideration in light of the uneven distribution of fossil localities through the sequence. There are notably fewer localities present between 18.0 and 14.5 Ma and after 7.0 Ma than in between. For reasons presented in the main text, we consider that the potential for fossil preservation is stochastically similar throughout the sequence since fossil localities from the four formations represented in the sequence have generally similar taphonomic attributes (Chapter 4, this volume). Fossil localities are found in similar depositional environments in all four formations, so the taphonomic history is consistent across formations. However, the fossil productivity does vary among formations and is correlated with differences in macrofacies. Fewer floodplain channel localities occur in the Kamlial and Nagri Formations than in the Chinji and Dhok Pathan Formations (Barry et al. 2013); the lower representation of this productive facies is one major reason for the difference in the frequency of fossil localities.

For species arising and disappearing between 14.5 and 7.0 Ma, the assumptions of this method are reasonably well met. For those species with observed first occurrences before 14.5 Ma, the calculated average gap sizes are greater than the overall means for small and large mammals. Confidence intervals for these species are likely underestimated.

A future innovation would be to develop a function for the changing fossil-recovery potential throughout the sequence and to adjust the calculation of confidence intervals with that function. Marshall (2010) discusses potential approaches. Wang et al. (2016) present a Bayesian approach with a changing fossil-recovery potential through the sequence.

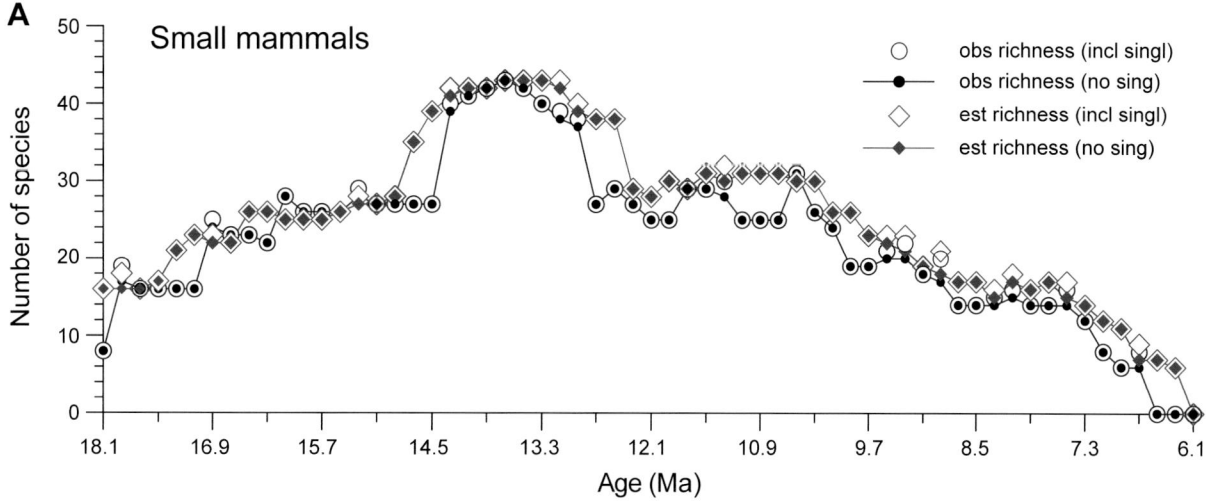

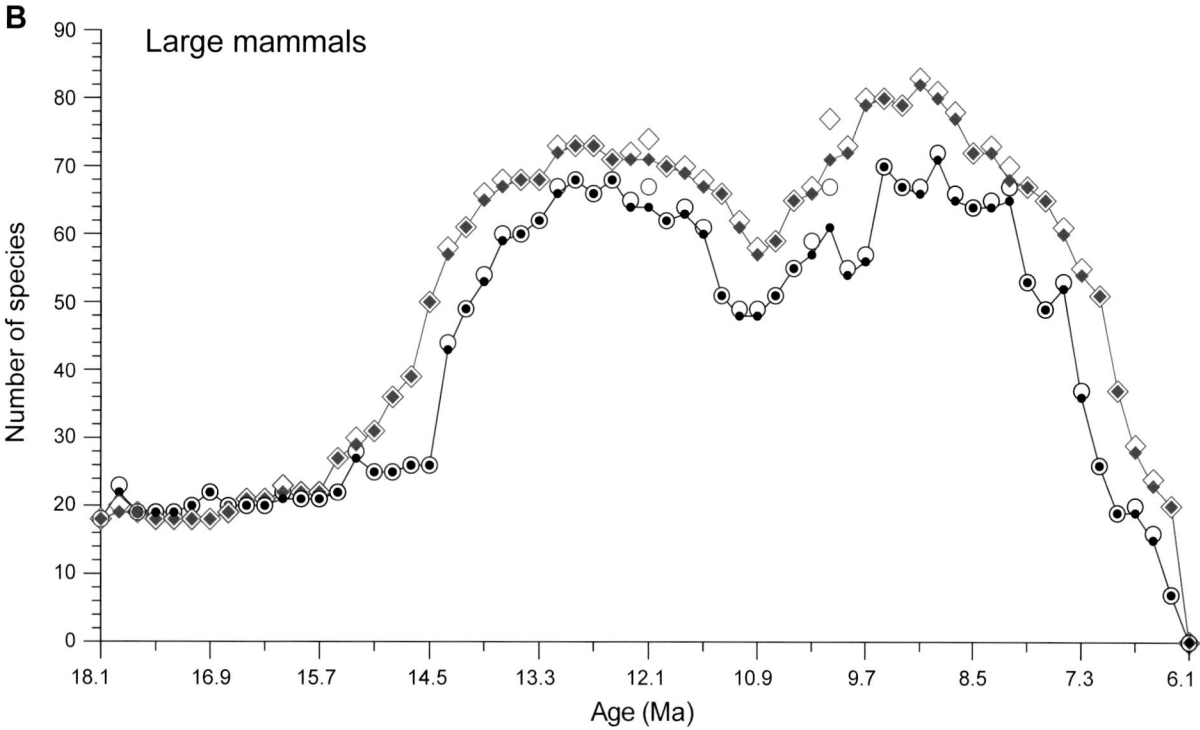

Figure 26.1. Four measures of species richness for (A) small mammals (with estimated body weight <1 kg) and (B) large mammals (with estimated body weight ≥1 kg). Data for small mammals are derived from Figure 26.2 and for large mammals from Figure 26.3 and compiled for 200-kyr time intervals. The range-through assumption is in effect for all measures of richness. Species that occur in a single time interval are considered singletons (even if fossils occur at more than one locality in that interval). Observed richness including singletons (open circles) is the number of species whose documented stratigraphic range begins, ends, or passes through the time interval. Observed richness, no singletons (solid circles), is the number of species that occur in more than one time interval and whose documented stratigraphic range begins, ends, or passes through the time interval. Estimated richness including singletons (open diamonds) is the number of species whose estimated range begins, ends, or passes through the time interval. See main text for explanation of the estimated first and last appearances. Estimated richness, no singletons (solid diamonds), is the number of species that occur in more than one time interval and whose estimated range begins, ends, or passes through the interval. Although singletons occur in many intervals, their omission does not distort the overall trends. Estimated richness is usually greater than observed species richness. We omitted species whose mean gap size between time intervals was >1.5 myr; such large mean gap sizes occur mainly in the oldest 4 myr of the sequence. Omission of these species results in observed species richness being greater than estimated species richness for several time intervals in the record of large mammals (B). Estimated richness was the basis for analyses of diversification and changes in faunal composition.

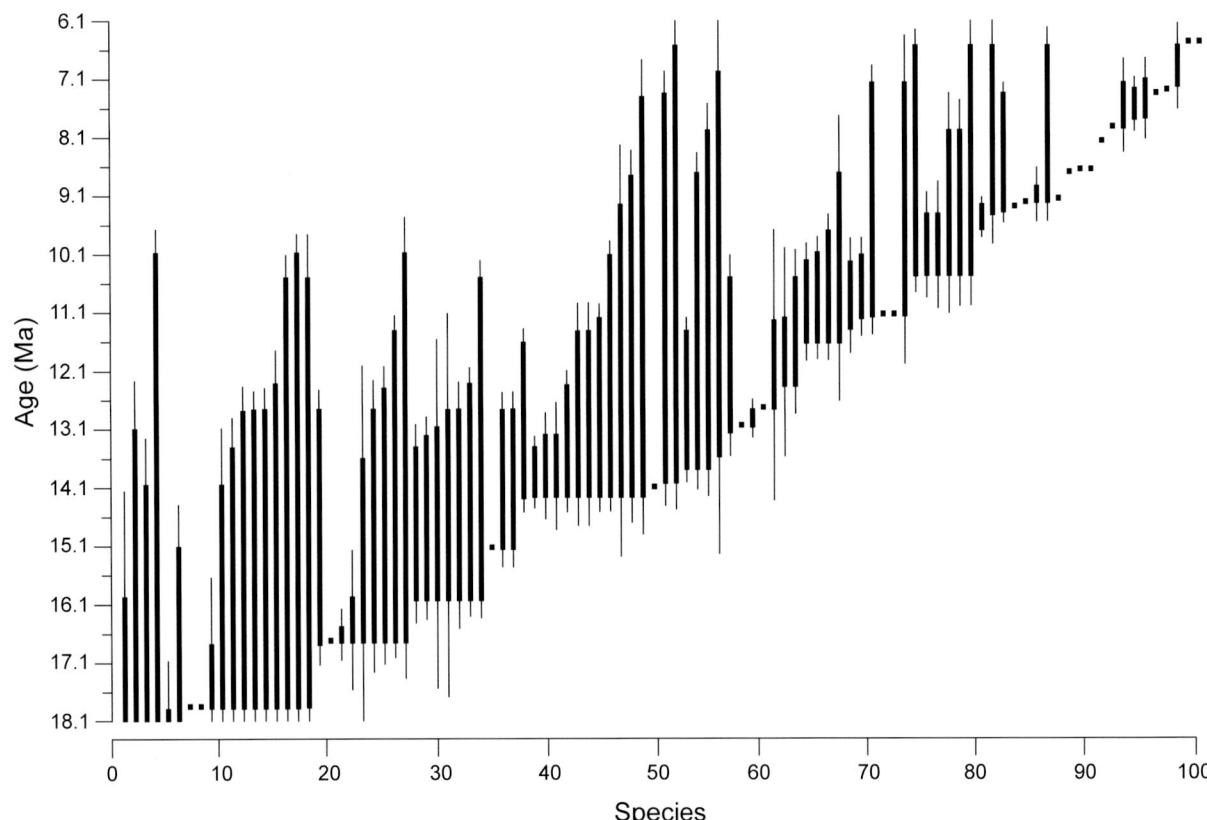

Figure 26.2. Observed and estimated stratigraphic ranges (residence times) for 100 species-level lineages of small mammal. Heavy lines denote the observed range; thin lines depict the estimated first or last appearance based on mean gap size (see text for explanation). Singletons are plotted as small squares that occur in a single time interval. The order of species is based on the age of observed first occurrence. See Table 26.S1 for list of species, observed first and last occurrences, and estimated first and last appearances.

large mammals to 186 lineages. Omission of singletons resulted in 75 small-mammal species and 155 large-mammal species in the diversification analyses. Tables 26.S1 and 26.S2 present the full list of small and large mammals, respectively, with observed and estimated first and last occurrence, estimated body-size and inferred feeding category, along with notes about which species were omitted from particular analyses.

From the estimated residence times, we determined for each 200-kyr interval, (*i*), the following variables:

$O(i)$—the number of species first occurring in an interval either from immigration or endemic speciation,

$E(i)$—the number of species last occurring in an interval either from emigration or global extinction,

$p(i)$—the per-capita origination rate,

$q(i)$—the per-capita extinction rate,

$d(i)$—the per-capita net diversification rate ($p(i) - q(i)$).

We describe the calculation of these variables in Box 26.2. The distinction between immigrant and endemic for the first occurrence of a species follows the practice in Chapter 25 of this volume. If the species has a plausible ancestor in the Siwalik fauna—determined through phylogenetic analysis for some groups or by the presence of an older congener for others, then the first occurrence is considered an endemic origination. Otherwise, it is considered an immigrant.

In order to assess which rates of origination, extinction, and diversification are large enough to stand out from a background of low-level faunal turnover, we bootstrapped the estimated residence times to generate confidence intervals for $p(i)$, $q(i)$, and $d(i)$. We resampled the dataset of estimated residence times in sets of 96 for small mammals and 186 for large mammals 1,000 times and generated means and 2-sigma confidence intervals for $p(i)$, $q(i)$, and $d(i)$. A rate was considered statistically significant if the confidence interval for that time interval did not contain 0. This bootstrap procedure is a variant of the approach described by Foote (2000) and utilized in earlier diversification analyses of fossil mammals (e.g., Badgley and Finarelli 2013; Domingo et al. 2014). Figure 26.4 (p. 489) shows the diversification results for small mammals and Figure 26.5 (p. 491) the results for large mammals. Since estimated species richness and rates of origination and extinction are sensitive to edge effects (whereby fewer species can be sampled at the boundary of a sequence [Foote 2000]), we truncated the horizontal time axis in Figs. 26.4b, c, d and 26.5b, c, d to 6.5 Ma.

Small-Mammal Richness

Species richness of small mammals rose through the first 3.5 million years of the sequence, remained at a broad peak for about 2 million years, then declined unevenly through the younger half of the sequence (Fig. 26.4a). From 16 species

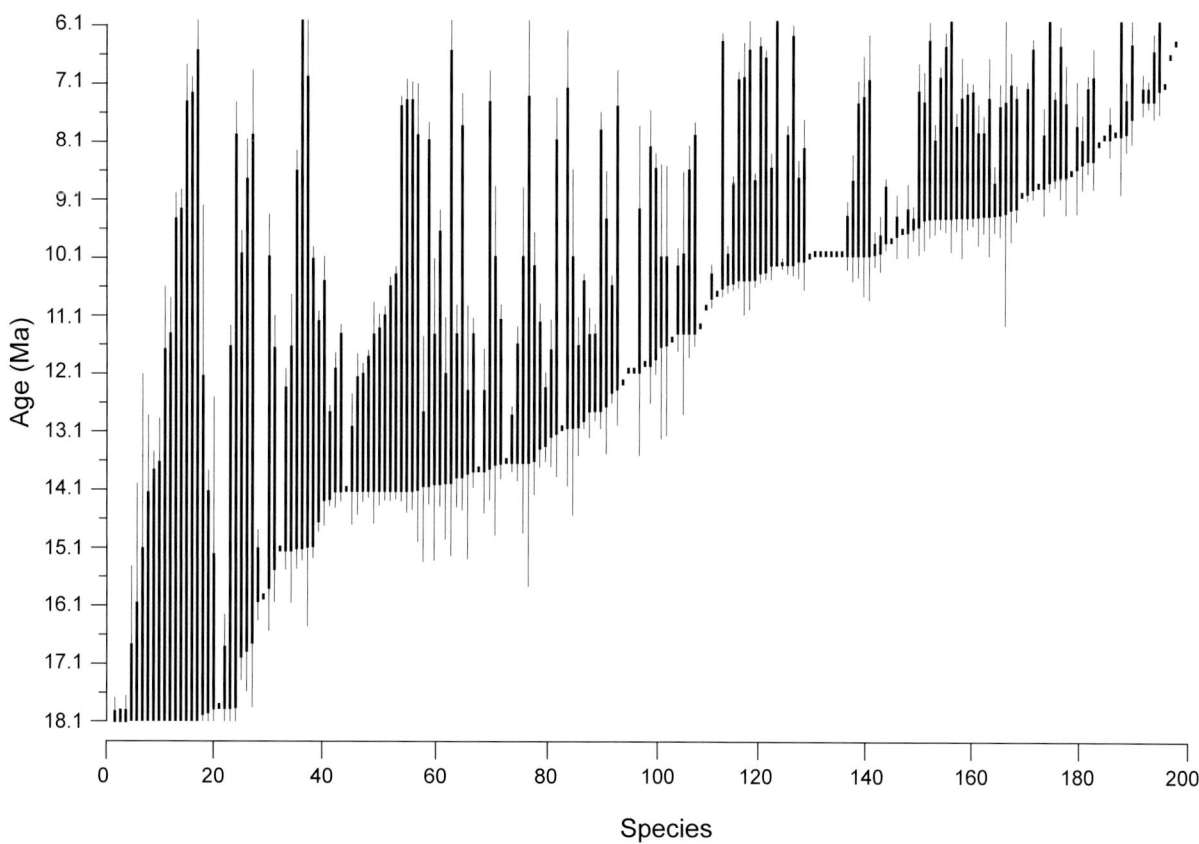

Figure 26.3. Observed and estimated stratigraphic ranges (residence times) for 194 species of large mammal. Heavy lines denote the observed range; thin lines depict the estimated first or last appearance based on mean gap size (see text for explanation). Singletons are plotted as small squares that occur in a single time interval. The order of species is based on the age of observed first occurrence. See Table 26.S2 for list of species, observed first and last occurrences, and estimated first and last appearances.

(estimated richness) at 18.1 Ma, richness increased in 3 steps to a peak of 44 species at 13.7 Ma. An interval of high richness with over 40 species per interval extends from 14.3 to 13.1 Ma. A large drop occurred between 12.5 and 12.3 Ma; then richness fluctuated from 30–32 species per interval until 10.1 Ma, when a long-term decline continued through the next 4 myr. This pattern confirms that documented in Flynn et al. (1998).

The peak of species richness occurs in the well-sampled Chinji Formation. In contrast, the well-sampled late Miocene interval from ~9.7 to 7.5 Ma in the Dhok Pathan Formation documents declining species richness. The first and last 1 million years have fewer fossil localities than at other times, so species richness is under-sampled at the beginning and end of this 12-myr record. However, rich screen-washed sites are present at approximately 0.5-myr intervals throughout the sequence. Thus, despite variation in fossil productivity, these changes in species richness preserve a real trend.

The number of estimated first occurrences (O_i) shows four notable pulses in the first half of the sequence, while six notable pulses of last occurrences (E_i) occur from the middle to the end of the sequence (Fig. 26.4b). Three significant per-capita rates of origination, $p(i)$, occur at 14.7, 14.5, and 11.9 Ma (Fig. 26.4c). Two of these significant rates occur during the

interval of accumulation of richness up to its peak value. The third at 11.9 Ma occurs at the beginning of steady species richness between 11.9 and 10.3 Ma. The significant rates that span 14.7–14.5 Ma include five immigrants and seven endemic originations; a significant diversification rate, $d(i)$, occurs at 14.5 Ma as well (Fig 26.4d). The significant origination rate at 11.9 Ma includes one immigrant and four endemic originations. Among the small-mammal lineages for which an assessment of origin is available, 11 are endemic holdovers from the early Miocene Zinda Pir record, 34 lineages (35%) are immigrants to the Siwalik faunal province, and 52 (54%) are endemic originations (see Chapter 25, this volume). Five time intervals show pulses of estimated last occurrences (E_i); three of these intervals have significant per-capita rates of extinction, $q(i)$ (Fig. 26.4b, c). The earliest pulse occurs at 13.1 Ma with four last occurrences. The largest pulse at 12.5 Ma has nine last occurrences. Two pulses occur at 10.1 and 9.9 Ma, with four and five last occurrences, respectively. A late pulse at 6.9 Ma has four last occurrences. Significant rates of extinction occur at 13.1, 12.5 and 9.9 Ma; the 12.5 Ma interval corresponds to a significant negative diversification rate as well (Fig. 26.4d). None of the notable peaks in last occurrence or significant extinction rate coincides with notable peaks in first occurrence or significant origination rate, although the extinction rate at 12.5 Ma

BOX 26.2

CALCULATION OF PER-CAPITA DIVERSITY DYNAMICS

We used the approach initially developed by Foote (2000) and reformulated algebraically in other publications (e.g., Badgley and Finarelli 2013; Domingo et al. 2014). The goal is to estimate the rates of origination, extinction, and net diversification (origination–extinction) based on the number of taxa appearing and disappearing scaled to the number of taxa passing through a time interval. Thus, an absolute number of first appearances in a time interval leads to a higher origination rate if the number of taxa passing through is small versus large, and likewise for last appearances and extinction rates.

The data for these analyses come from Table 26.S1 for small mammals and Table 26.S2 for large mammals. For each species, we used the estimated first appearance and estimated last appearance (see Box 26.1) as the time of origination and extinction, respectively. First appearances may arise through immigration or endemic speciation. Last appearances may occur through geographic-range shift or global extinction. We excluded singleton species—those that occur in a single time interval—from the analysis (even though some of them are interesting taxonomically or ecologically), since they are especially sensitive to fluctuations in sampling intensity.

Time intervals of 200 kyr begin at 18.1 Ma for our analyses. Even though the Potwar Siwalik sequence begins at 18.0 Ma with the oldest sediments of the Kamlial Formation, the estimates of first appearance for some species are older than the base of the sequence.

We utilized the following variables for each time interval i:

N_{bt}, the number of species present in $i-1$ to $i+1$ (bt, bottom-top boundary crossers),

N_{Ft}, the number of species first appearing in time interval i (Ft) and continuing into interval $i+1$ (originations, $O(i)$),

N_{bL}, the number of species continuing from time interval $i-1$ and last appearing (bL) in interval i (extinctions, $E(i)$).

We applied a continuity correction for N_{bt}, N_{Ft}, and N_{bL} if the count was 0 by adding 0.5; this avoids taking the logarithm of 0 without distorting the counts for the 3 variables (see Finarelli 2007).

For each time interval i, we estimated per-lineage rates of origination, p_i, and extinction, q_i, as follows:

$$p(i) = \text{Ln}\left[\frac{\left[Nbt(i) + NFt(i)\right]}{Nbt(i)}\right],$$

$$q(i) = \text{Ln}\left[\frac{\left[Nbt(i) + NbL(i)\right]}{Nbt(i)}\right].$$

The per-lineage rate of diversification is

$$d(i) = \frac{\left[NFt(i) - NbL(i)\right]}{Nbt(i)}.$$

This expression is equivalent to

$$d(i) = e^{p(i)} - e^{q(i)}.$$

The results of these calculations are displayed in Figures 26.4 (small mammals) and 26.5 (large mammals).

In order to determine the statistical significance of $p(i)$, $q(i)$, and $d(i)$ for each time interval (i), we bootstrapped the residence times of all species in each dataset, sampling the actual residence times at random with replacement. We generated 1,000 replicate samples of the number of species in each dataset ($n = 95$ for small mammals; $n = 186$ for large mammals) and calculated means and 2-sigma confidence intervals for $p(i)$, $q(i)$, and $d(i)$. If the confidence interval did not contain 0, then we considered the rate for that time interval as statistically significant.

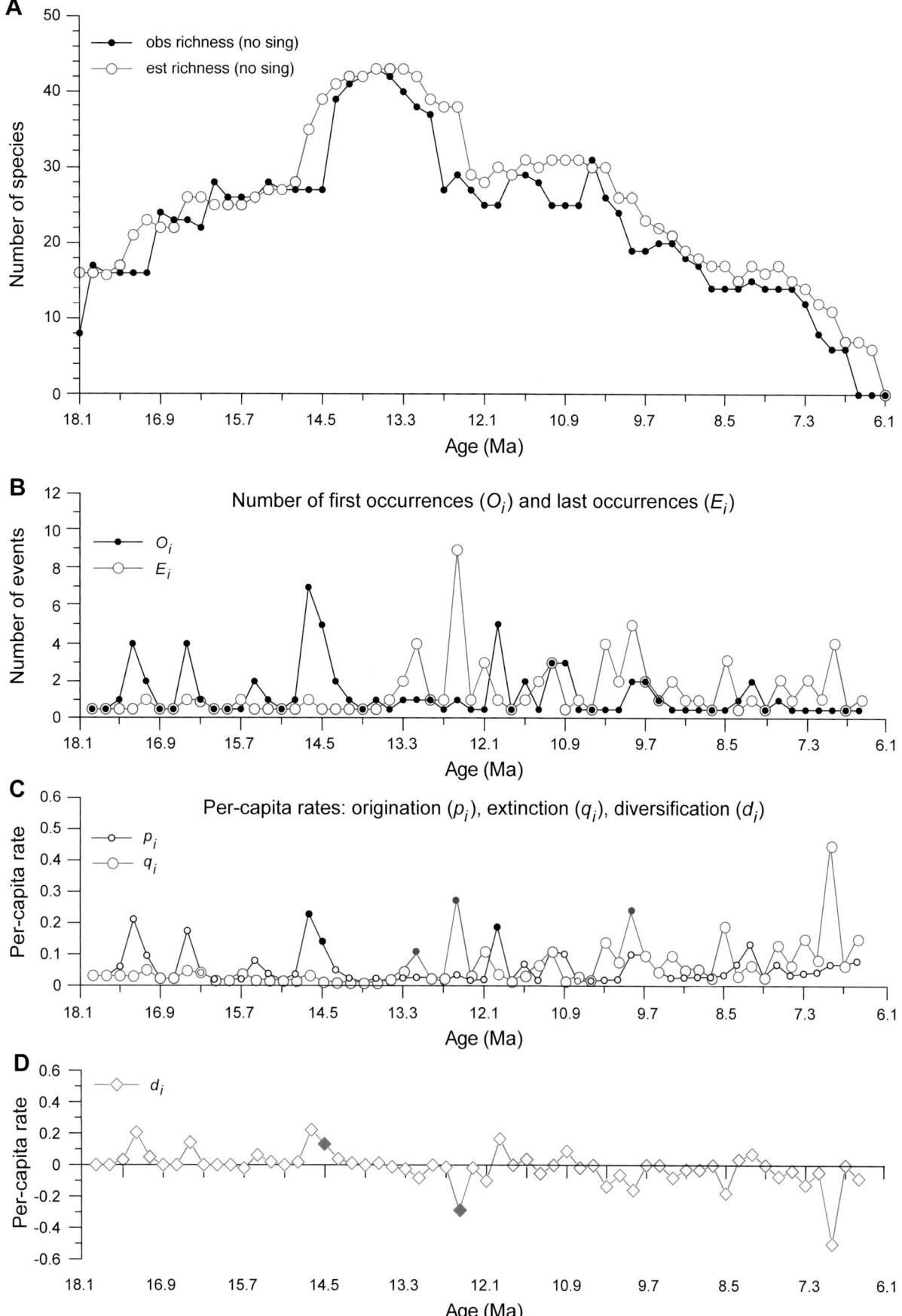

Figure 26.4. Species richness and diversification metrics for large mammals, pruned from 100 to 96 lineages after omitting species with mean gap sizes >1.5 myr and further reduced to 75 species by omission of singletons (no sing). A. Observed and estimated species richness. Data for estimated species richness are the basis for results in B–D. Data are truncated at 6.5 Ma in B–D to reduce edge effects in the estimation of per-capita rates. B. Number of first occurrences (originations, O_i) and last occurrences (extinctions, E_i) in each 200-kyr time interval based on estimated residence times. C. Per-capita rates of origination (p_i) and extinction (q_i) for each time interval. Statistically significant rates are indicated by solid symbols. D. Per-capita diversification rate (origination–extinction) for each time interval. Statistically significant rates are indicated by solid symbols.

and the origination rate at 11.9 Ma contribute to substantial taxonomic turnover over this longer period.

Large-Mammal Richness

The estimated species richness of large mammals hovered around 20 species per interval from 18.1 until 14.9 Ma, rose rapidly to 2 broad peaks that persisted until about 8.0 Ma, then declined rapidly to the end of the record (Fig. 26.5a). From 18 species at 18.1 Ma, species richness ranged from 18–36 species per 200-kyr interval through 15.1 Ma, then rose to 68–74 species per interval between 13.1 and 12.1 Ma to form the first broad peak of richness. Species richness then dropped to 58 species at 10.9 Ma and rebounded to a second peak of 78–83 species at 9.7 Ma. Richness remained above 70 species per interval through 8.3 Ma, then declined steadily to 20 species at 6.3 Ma. As with the small-mammal record, variation in fossil productivity contributed to some of the highs and lows in this pattern but not to the overall trend, for reasons outlined above.

The number of first occurrences (O_i) has three notable pulses, two of which span multiple time intervals (Fig. 26.5b). Each of these pulses is associated with significant per-capita origination rates, $p(i)$ (Fig. 26.5c). The first pulse was at 15.5 Ma. The second began at 14.9 Ma and lasted until 14.3 Ma, with three intervals showing significant origination rates and two intervals with significant positive diversification rates (Fig 26.5d). These pulses are associated with the first increase in species richness. A small pulse occurred at 13.1 Ma and coincided with the first peak in species richness. A cluster of first occurrences between 10.7 and 9.7 Ma have significant origination rates in six of seven consecutive intervals and these are associated with the second peak in species richness.

The number of last occurrences (E_i) shows several pulses that span multiple intervals (Fig. 26.5b). These pulses are all associated with significant per-capita extinction rates (Fig. 26.5c). The double pulse at 11.3 and 11.1 Ma, featuring eight last occurrences, corresponds to the decline in the first broad peak in species richness. Significant negative diversification rates also occurred at 11.3 and 11.1 Ma (Fig. 26.5d). The third pulse includes 8.9 and 8.7 Ma, with four and six last occurrences, respectively, at the initial decline in the second peak of species richness, and a significant extinction rate at 8.7 Ma. The final cluster of last occurrences includes most of the intervals from 7.9 to 6.5 Ma, with four intervals showing six or more last occurrences. Six intervals exhibit significant extinction rates, and four have significant negative diversification rates. None of the significant per-capita origination and extinction rates occurred in the same time intervals, but the interval from 11.3 to 9.1 Ma contains two significant extinction rates followed by six significant origination rates, signifying a protracted interval of faunal turnover.

CHANGE OVER TIME IN FAUNAL COMPOSITION

For both small mammals and large mammals, how much change in faunal composition accompanied the rise and fall in species richness over the Miocene? We analyzed two forms of change in faunal composition: family-level species richness and functional richness (Figs. 26.6–26.9). Family-level richness encompasses ordinal, superfamily, family, and subfamily or tribe categories, depending on the taxonomic group. For small mammals, we combined the two small mustelids and one herpestid as "small Carnivora" in the "Other" category, but we separated Gerbillinae from Murinae for the murids (Fig. 26.6, p. 492). For large mammals, most ungulate groups are represented at the family level. Other groups are represented at the superfamily or ordinal level (e.g., hominoids, suoids, giraffoids, proboscideans, and creodonts; Fig. 26.8, p. 498). Richness was tallied as presence (assuming range-through) of species within 200-kyr intervals for each taxonomic group.

Functional richness refers to joint size-feeding categories (see Table 26.1). We used seven nonoverlapping body-size categories, two for small mammals and five for large mammals. For feeding habit, we used nine feeding categories for small and large mammals, although the occupancy of these categories differs substantially between the two groups (see Tables 26.2, 26.3). Use of joint size-feeding categories enabled us to track change over time in the ecological structure of faunas with one informative variable.

We assessed rates of change in faunal composition by modeling composition as a multinomial distribution with a likelihood approach (Edwards 1992). This assessment has two purposes: one to identify time intervals with substantial change in faunal composition, the other to determine whether significant changes in taxonomic composition were synchronous with significant changes in functional composition. The multinomial distribution expresses the proportional representation of a variable in more than two categories and is the basis for calculating the probability of a sampling outcome when three or more outcomes are possible. The likelihood approach involves comparison of the statistical likelihood of two multinomial distributions, expressed as a ratio. The natural logarithm of the likelihoods enables one ln-likelihood value to be subtracted from the other. For the Siwalik record, the composition of each 200-kyr interval was compared to the composition of the previous time interval using the difference in likelihoods of the multinomial distribution of each interval (see Box 26.3 for details).

Faunal composition consisted of the species-level richness of either taxonomic or ecological categories, as described above. The datasets contain all species with mean gap sizes less than 1.5 myr, including singletons, with estimated first and last appearances providing the residence times. (Groups with few species and few occurrences throughout the 12-myr sequence were lumped into various "other" categories for taxonomic composition. For large mammals, two sets of joint size-feeding categories were grouped as "other omnivores" and "other herbivore grazers.") If the composition changed little from one time interval to the next, then the difference in the natural logarithm of the likelihoods (Delta lnL) was low, below a threshold value of 2.0 for statistically significant change in proportional composition (Edwards 1992). If change in composition from

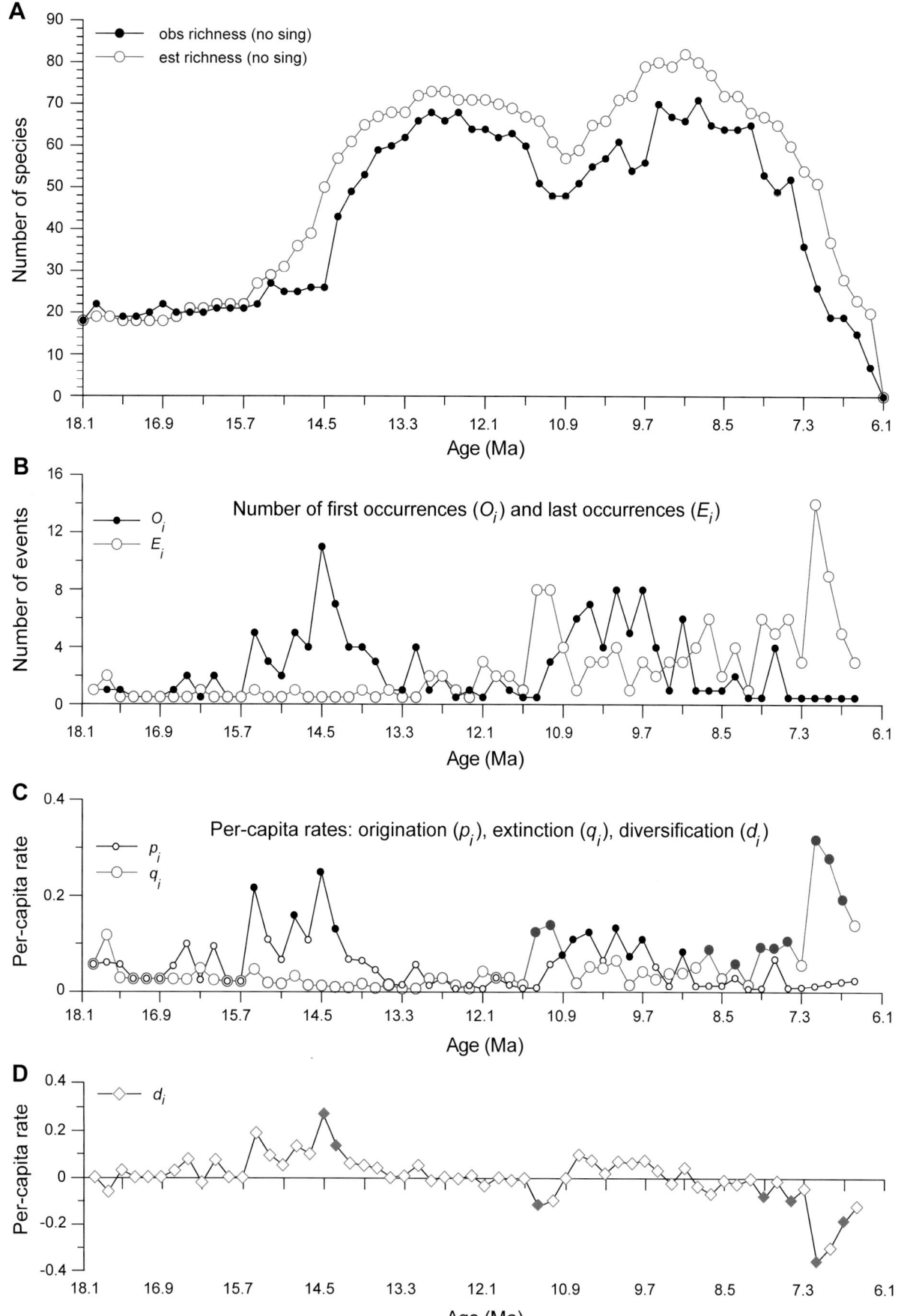

Figure 26.5. Species richness and diversification metrics for large mammals, pruned from 194 to 186 lineages after omitting species with mean gap sizes >1.5 myr and further reduced to 155 species by omission of singletons (no sing). A. Observed and estimated species richness. Data for estimated species richness are the basis for results in B–D. Data are truncated at 6.5 Ma in B–D to reduce edge effects in the estimation of per-capita rates. B. Number of first occurrences (O_i) and last occurrences (E_i) in each 200-kyr time interval based on estimated residence times. C. Per-capita rate of origination (p_i) and extinction (q_i) for each time interval. Statistically significant rates are indicated by solid symbols. D. Per-capita diversification rate (origination–extinction) for each time interval. Statistically significant rates are indicated by solid symbols.

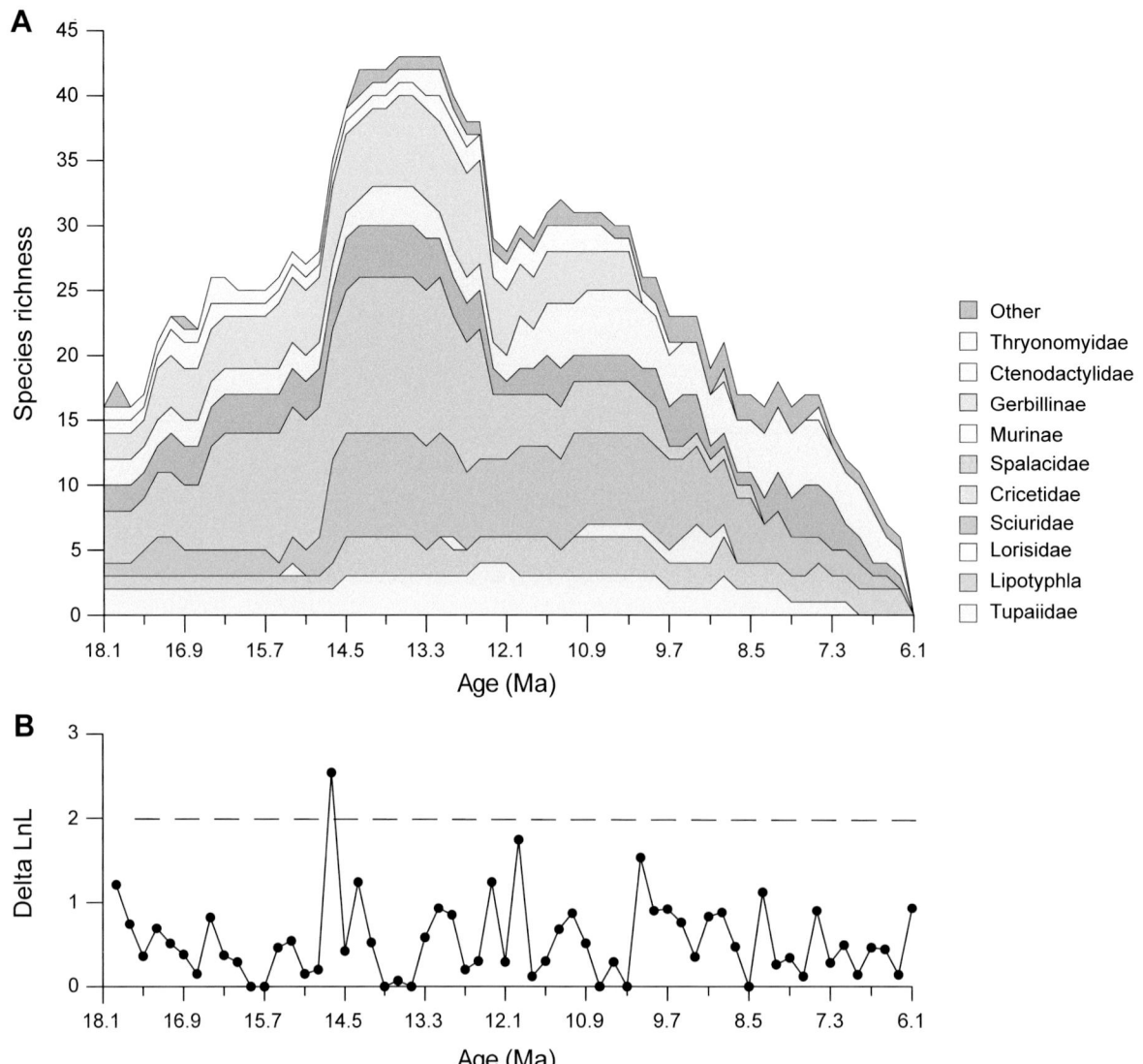

Figure 26.6. Change in taxonomic composition of small mammals over time. A. Species richness of higher taxa (mainly families), based on estimated residence times and compiled for 200-kyr intervals. Singletons were included in this compilation. B. Change in log-likelihood (Delta LnL) for a multinomial model of composition, comparing maximum likelihood estimates based on faunal proportions within one interval with faunal proportions from the prior time interval. If faunal proportions change substantially from the older to younger interval, then the change in log likelihoods is large. The value of 2.0 is a threshold for statistically significant change in faunal proportions. For the small-mammal record, one of the Delta LnL values exceeds that threshold, and two other notable peaks occur. See text for further explanation.

Table 26.2. Joint Size-Feeding Categories Occupied by Siwalik Small Mammals (Estimated Body Weight <1 kg)

Feeding Category	Very Small (<100 g)	Small (100–999 g)
Faunivore	0	2
Invertivore	12	3
Omnivore	5	1
Frugivore	0	0
Granivore	2	7
Herbivore-browser	—	—
Herbivore–mixed feeder	45	12
Herbivore-grazer	—	—
Rootivore	2	9

Note: Each cell contains the number of species in a joint size-feeding group for 100 species. Table 26.S1 lists all the small species, their observed and estimated first and last occurrences, and their estimated size and feeding categories.

one time interval to the next was large, then Delta LnL was near or exceeded 2.0. This approach identifies significant change over relatively short time periods but not over longer time intervals. (In principle, the method could be adjusted to compare the composition of a focal time interval to the composition of intervals 0.5 or 1.0 million years later and thereby recognize significant changes that transpired more slowly.)

For small mammals, change over time in taxonomic composition occurred gradually and largely in step with the major increase and decrease in overall species richness (Fig. 26.6a). Notable changes occurred from the beginning to the end of the sequence. Ctenodacytlids, cricetids, and gerbilline murids were major components of Middle Miocene faunas but were

Table 26.3. Joint Size-Feeding Categories Occupied by Siwalik Large Mammals (Estimated Body Weight ≥1 kg)

Feeding Category	Small (1.0–19.9 kg)	Medium (20–99 kg)	Large (100–249 kg)	Very Large (250–799 kg)	Mega (>800 kg)
Faunivore	14	19	3	4	0
Invertivore	3	4	0	0	0
Omnivore	3	0	5	1	1
Frugivore	6	4	0	0	0
Granivore	0	0	0	0	0
Herbivore-browser	15	13	5	12	25
Herbivore–mixed feeder	3	8	6	5	3
Herbivore-grazer	0	1	0	0	1
Rootivore	3	0	0	0	0

Note: Each cell contains the number of species in a joint size-feeding group for a total of 167 large mammals that could be assessed for both size and feeding habit. Table 26.S2 lists all the large species, their observed and estimated first and last occurrences, and their estimated size and feeding categories.

completely gone by 8.5 Ma. Thryonoymids were present only in the first half of the sequence. The rate of change, as expressed by Delta LnL, exceeds 2.0 only once at 14.7 Ma (Fig. 26.6b). This spike in the rate of change is driven by a large increase in sciurids and modest increase in lipotyphlans and gerbillines. Two additional intervals show values greater than 1.5. One interval occurs as species richness dropped rapidly at 11.9 Ma, when sciurid and gerbilline proportions declined notably. The next interval occurs at 10.1 Ma, when gerbillines dropped out and the number of cricetid species declined rapidly.

A similar overall pattern occurs with change in ecological structure (Fig. 26.7a, p. 497). The composition of size-feeding categories changed notably from the beginning to the end of the sequence, but changes occurred slowly enough that Delta LnL values were well below 2.0 except at 14.7 Ma (Fig. 26.7b), the same interval in which the rate of change in taxonomic composition was significant. The early small-mammal faunas were dominated by very small herbivores, with consistent representation of very small invertivores and omnivores and small granivores. At 14.7 Ma, proportional increases occurred in very small and small herbivores; modest increases also occurred in very small invertivores and granivores. The Late Miocene faunas had a smaller proportion of very small herbivores; small faunivores and small rootivores were persistent components. Very small omnivores and granivores and small invertivores disappeared by 7.5 Ma.

Changes in faunal composition of large mammals were more variable than those of small mammals. With 21 higher taxonomic groups represented in Figure 26.8a (p. 498), it is apparent that many groups were present throughout the entire 12-myr analysis interval. Groups present at the beginning that disappeared before the end include sivaladapids (subsumed under "other primates"), creodonts, and amphicyonids. Groups that appeared at or after 14.5 Ma include hominoids, aardvarks (orycteropodids), chalicotheres (subsumed under "other mammals"), and equids. From 18.1 to 14.5 Ma, large-mammal faunas were dominated by species of tragulids and rhinocerotids. These groups persisted through the entire sequence, but their proportional representation declined as other groups—notably bovids, giraffoids, and suoids, among broadly herbivorous mammals—increased in species richness. Among faunivores, creodonts and amphicyonids were the most species-rich groups from 18.1 to 14.5 Ma. Subsequently, felids + barbourofelids, hyaenids + percrocutids, viverrids, and mustelids increased in richness.

Overall, the large-mammal record shows both gradual and rapid changes in taxonomic composition. The log-likelihood metric of taxonomic change from one time interval to the next has seven pulses that rise above 2.0 (Fig. 26.8b). The earliest pulses occurred at 15.3 and 14.5 Ma, when overall species richness began to increase. Groups contributing to the pulse at 15.3 Ma include proboscideans, other primates, and suoids. Groups contributing to the pulse at 14.5 Ma include proboscideans and bovids, with smaller increases in felids + barbourofelids, hyaenids + percrocutids, orycteropodids, tragulids, and giraffids. A cluster of pulses includes three intervals between 11.1 and 10.1 Ma and coincides with the first decrease and second increase in overall richness. At 11.1 Ma, decreases in giraffids and mustelids influenced this pulse. At 10.7 Ma, increases in proportions of tragulids and decreases in felids + barbourofelids and bovids influenced this pulse. The highest rate of change in the entire sequence at 10.1 Ma involved a notable increase in bovids and modest increases in hominoids, mustelids, and other carnivores as well as a decrease in rhinocerotids. The final two pulses occurred at 7.7 and 6.7 Ma during the long decline in species richness. At 7.7 Ma, declines in giraffids, felids + barbourofelids, and viverrids and increases in equids and other mammals contributed to this pulse. At 6.7 Ma, a decline in tragulids and rhinocerotids and modest declines in large rodents (rhizomyines), felids + barbourofelids, and proboscideans influenced the final pulse.

Ecological structure of large-mammal faunas had the highest magnitudes of change after 11.5 Ma (Fig. 26.9, p. 499). (The cumulative profiles of these two versions of composition differ slightly because about 20 species of known taxonomic affiliation could not be confidently assigned to a feeding category, so the analysis of functional composition contains fewer species than the analysis of taxonomic composition.) Ecological

BOX 26.3

CALCULATION OF LIKELIHOOD FOR THE MULTINOMIAL DISTRIBUTION

In order to assess the significance of change in faunal composition from one time interval to the next, we modeled the distribution of species among taxonomic (or functional) categories as a multinomial distribution and determined the likelihood of that distribution. We then compared the likelihood of the distribution in one time interval to the likelihood of the distribution from the preceding time interval. Essentially, we are asking, How well does the faunal composition of one time interval predict the faunal composition of the next time interval? If the two likelihoods (one, the likelihood of the multinomial based on the distribution for the focal time interval; the other, the likelihood of the multinomial based on the distribution of the preceding interval) are similar, then there has been little faunal change from one interval to the next. If the two likelihoods differ quite a bit, then there has been substantial change in the number of species in one or more taxonomic (or functional) categories from one time interval to the next. The two likelihoods are expressed as a ratio; taking the logarithm of the ratio converts the expression to a sum. If the change in log likelihoods (Delta LnL) exceeds 2.0, then faunal change is considered significantly large between the two time intervals. This approach is explained in Edwards (1992) and utilized in earlier analyses of change in faunal composition (e.g., Badgley and Finarelli 2013; Domingo et al. 2014).

Below, we present a simplified, hypothetical example of the steps in the calculation. Consider a dataset of five families of small mammals with residence times of different lengths in the Siwalik record. We demonstrate the approach for four successive time intervals. The entries in each cell are the number of species present in that time interval for each family.

Table B26.1. Number of Species per Family in Each Time Interval (*i*)

Family	Time Interval (Ma)	18.1	17.9	17.7	17.5
Tupaiidae		2	2	2	2
Lorisidae		0	0	1	1
Sciuridae		1	1	1	2
Cricetidae		1	1	1	4
Ctenodactylidae		1	1	1	0
	Total no. spp.	5	5	6	9

The first step is to eliminate 0s with a continuity correction, since one cannot calculate the logarithm of 0. This is done by adding 0.5 to every cell with a 0 entry. The sum of species for each column is adjusted to reflect this change.

Table B26.2. Count of Species per Family with Continuity Correction to Eliminate Values of 0 (*a_i*)

Family	Time Interval (Ma)	18.1	17.9	17.7	17.5
Tupaiidae		2	2	2	2
Lorisidae		0.5	0.5	1	1
Sciuridae		1	1	1	2
Cricetidae		1	1	1	4
Ctenodactylidae		1	1	1	0.5
	Total no. spp.	5.5	5.5	6	9.5

Next, we calculate the proportional frequency of each family in each time interval; this is the adjusted number of species divided by the total for that column (e.g., 2/5.5 for Tupaiidae for the time interval 18.1 Ma).

Table B26.3. Proportional Frequency (p_i) of Species per Family (adjusted number/total from Table B26.2)

Family	Time Interval (Ma)	18.1	17.9	17.7	17.5
Tupaiidae		0.364	0.364	0.333	0.211
Lorisidae		0.091	0.091	0.167	0.105
Sciuridae		0.182	0.182	0.167	0.211
Cricetidae		0.182	0.182	0.167	0.421
Ctenodactylidae		0.182	0.182	0.167	0.053

The observed multinomial likelihood of the distribution for each time interval (i) is the sum across all families (j) of the count of species, a_j (Table B26.2), times the natural logarithm of the proportional frequency, p_i (Table B26.3). The expression is $LnL(i) = \Sigma a_j * \ln(p_j)$, where a_j is the count of species in family j and p_i is the proportion of species in family j for interval i. For Tupaiidae at 18.1 Ma, the value is $2 \times \ln(0.364) = -2.023$ in Table B26.4 below.

In each cell in Table B26.4, the adjusted count of species in each family, a_j (from Table B26.2), is multiplied by ln of the proportional frequency, p_i (from Table B26.3). All the resulting values are negative, since the logarithm of a value less than 1.0 is negative. The sum is the maximum likelihood estimate for each time interval.

Table B26.4. Calculation of Maximum Likelihood of the Distribution in Each Time Interval

Family	Time Interval (Ma)	18.1	17.9	17.7	17.5
Tupaiidae		−2.023	−2.023	−2.197	−3.116
Lorisidae		−1.199	−1.199	−1.792	−2.251
Sciuridae		−1.705	−1.705	−1.792	−3.116
Cricetidae		−1.705	−1.705	−1.792	−3.460
Ctenodactylidae		−1.705	−1.705	−1.792	−1.472
	Sum (obs ML)	−8.336	−8.336	−9.364	−13.416

Next, we do the same exercise but now multiply the adjusted count of species in each time interval by the natural logarithm of the proportional frequency of the previous time interval (i–1). Our calculations begin in the interval 17.9 Ma, when we compare the observed number of species in each family with the proportional frequency at 18.1 Ma. For Tupaiidae, this calculation is $2 \times \ln(0.364) = -2.203$ in Table B26.5. The sum of each column is the maximum likelihood estimate for three time intervals.

In Table B26.5, the count of species in each family for a given time interval, a_j (from Table B26.2), is multiplied by the proportional frequency of the previous time interval, p_i (from Table B26.3). Again, all the resulting values are negative, since the logarithm of a value less than 1.0 is negative. The sum is the alternate maximum likelihood estimate (alt ML) for each time interval. The first column has no values since there is no previous time interval for comparison.

Table B26.5. Calculation of Alternate Maximum Likelihood

Family	Time Interval (Ma)	18.1	17.9	17.7	17.5
Tupaiidae		—	−2.023	−2.023	−2.197
Lorisidae		—	−1.199	−2.398	−1.792
Sciuridae		—	−1.705	−1.705	−3.584
Cricetidae		—	−1.705	−1.705	−7.167
Ctenodactylidae		—	−1.705	−1.705	−0.896
	Sum (alt ML)	—	−8.336	−9.535	−15.635

Now, we compare the observed maximum likelihood from Table B26.4 with the alternate maximum likelihood from Table B26.5, expressed as the ratio obs ML/alt ML. Since the values are in natural logarithms, this ratio is obs ML– alt ML.

Table B26.6. Change in ln Likelihood (Delta LnL) from One Time Interval to the Next

Time interval (Ma)	18.1	17.9	17.7	17.5
Obs ML	−8.336	−8.336	−9.364	−13.416
Alt ML	—	−8.336	−9.535	−15.635
Delta LnL	—	0	0.171	2.219

The values of Delta LnL indicate that no change in faunal proportions has occurred between 18.1 and 17.9 Ma, that modest change has occurred from 17.9 to 17.7 Ma, and that substantial change has occurred from 17.7 to 17.5 Ma. The original data show a large increase in species of Cricetidae, modest increase in Sciuridae, and a modest decrease in Ctenodactylidae at 17.5 Ma compared to the previous interval.

structure changed significantly, both statistically and biologically, over the sequence. Between 18.1 and 14.5 Ma, small and mega herbivore-browsers were the most species-rich components of the fauna. At 14.7 Ma, species richness increased in most functional groups. Several groups persisted with low species richness from ~14.5 Ma to ~7.3 Ma but disappeared before the end of the sequence. These groups include small invertivores, small omnivores, medium invertivores, medium frugivores, and large herbivore-browsers. Groups that appeared late in the sequence include small rootivores, very large herbivore-mixed feeders, and herbivore-grazers.

The log-likelihood metric rose above 2.0 in three time intervals (Fig. 26.9b). The first such value occurs at 11.1 Ma, marking a moderate decline of species richness. This interval had a decline in small faunivores and very large herbivore-browsers, modest declines in medium and very large faunivores, and modest increases in small omnivores and small herbivore-browsers. The next high value occurs at 10.1 Ma and is followed by two intervals with values of Delta LnL just below 2.0. At 10.1 Ma, small faunivores increased and very large herbivore-browsers decreased compared to the previous time interval, with a modest decrease in large herbivore-browsers and a modest increase in medium frugivores, medium herbivore-browsers, and large faunivores. A final significant pulse occurred at 6.9 Ma, involving a large decrease in megaherbivore-browsers, moderate declines in very large herbivore–mixed feeders and herbivore grazers, and modest declines in small faunivores, small omnivores, medium herbivore-browsers, medium herbivore–mixed feeders, and very large herbivore-browsers.

The three time intervals of significant change in ecological structure all correspond to significant change in taxonomic composition; other pulses of change in taxonomic composi-

tion have corresponding modest changes in ecological structure (with Delta LnL values below 2.0). The late spike of significant change in ecological structure at 6.9 Ma precedes a significant pulse in taxonomic composition at 6.7 Ma (Fig. 26.8b). The offset pulses identify times when notable change in a few taxonomic groups involve small changes across many size-feeding categories. When pulses overlap in both forms of faunal composition, then the major changes in ecological structure are caused by increase or decrease in species richness of a few higher taxa of large mammals, which suggests that there is a phylogenetic signal at these times. A phylogenetic signal implies that the changes in community composition are dominated by eco-evolutionary processes in certain clades. We discuss the biological significance of these patterns in the following section.

EMERGENT PATTERNS

We note several striking aspects of these changes in species richness and faunal composition over time and evaluate potential causes of change.

Richness Patterns for Small and Large Mammals

Despite common features of the record for small mammals and large mammals, there are striking differences. The two records have in common a gradual increase in observed and estimated richness from the base of the Siwalik sequence in the Potwar Plateau to 14.5 Ma (observed richness) and 14.9 Ma (estimated richness), then a rapid increase in richness (Fig. 26.1). Both records also show a sustained decline from about 9 Ma through the end of the analysis interval. Low values by about 7.0 Ma reflect a paucity of fossil localities from 7.1 Ma onward (see Fig. 3.3). Small mammals show an early broad peak

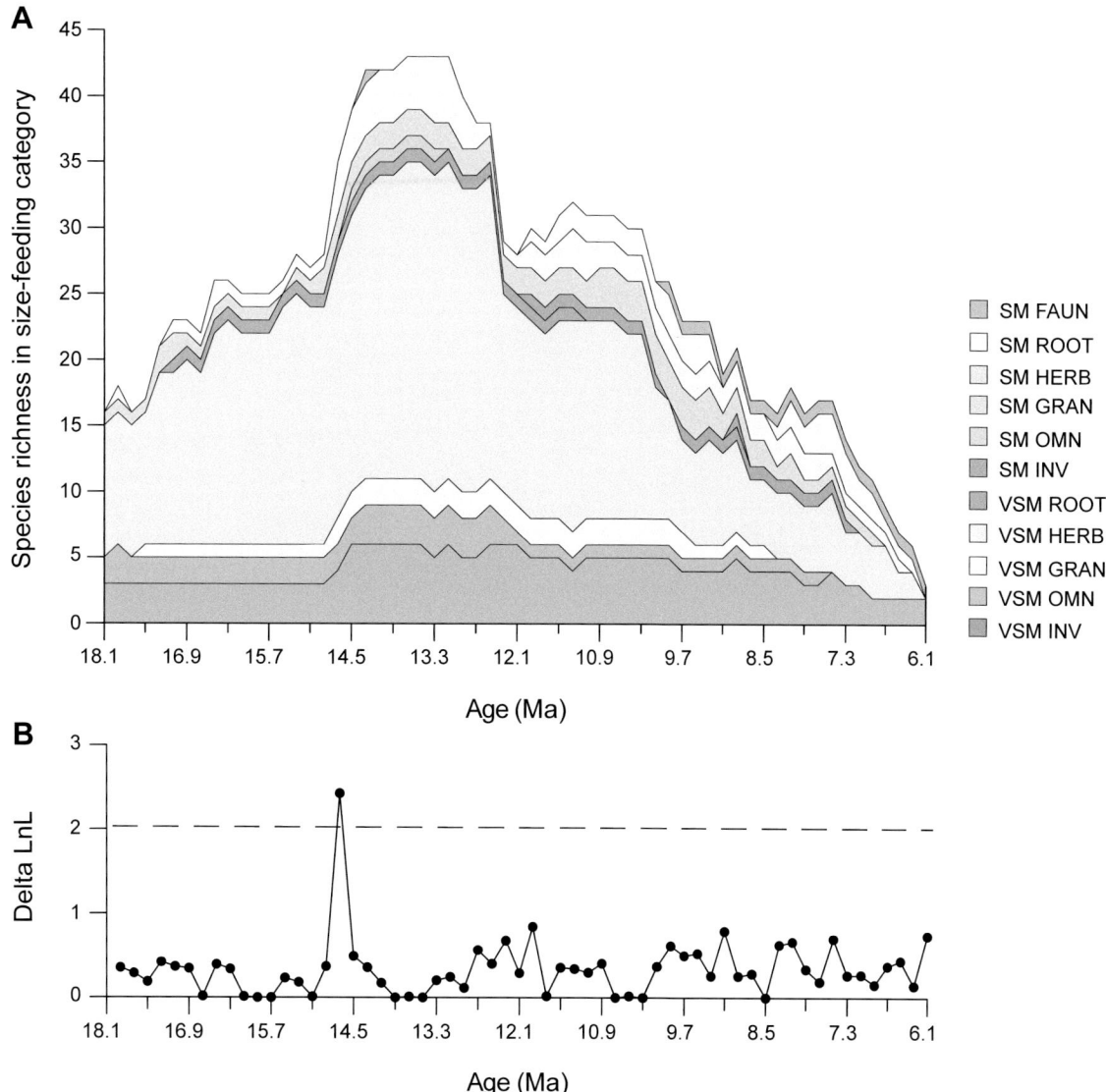

Figure 26.7. Change in ecological structure of small mammals over time. A. Species richness of joint size-feeding categories based on estimated residence times and compiled for 200-kyr intervals; singletons were included. B. Change in log-likelihood (Delta LnL) for a multinomial model of composition, comparing maximum likelihood estimates based on faunal proportions within one interval with faunal proportions from the prior time interval. If faunal proportions change substantially from the older to younger interval, then the change in log likelihoods is large. The value of 2.0 is a threshold for statistically significant change in faunal proportions. For the small-mammal record, one of the Delta LnL values exceeds that threshold. See text for further explanation. Abbreviations: VSM = very small; SM = small; INV = invertivore; OMN = omnivore; GRAN = granivore; HERB = herbivore; ROOT = rootivore; FAUN = faunivore.

in species richness from 14.5 to 12.5 Ma, when ≥40 species occurred per 200-kyr interval. Their long-term decline in species richness began at 12.3 Ma, whereas large-mammal richness was still experiencing its first broad peak from 13.3 to 11.9 Ma. Small-mammal richness continued to decline steadily even over the most richly sampled portion of the Potwar record. Richness declined across taxonomic groups, although faunal proportions changed over the long interval of decreasing richness (Fig. 26.6). These differences suggest that small mammals experienced and reacted to environmental changes differently than large mammals; we return to this topic below.

High Total Mammal Richness

Total mammal richness achieved notably high values, especially between 13.7 and 12.9 Ma, when total richness reached or exceeded 100 species per time interval. The highest value of estimated richness is 116 species at 13.1 Ma; observed richness is also at its highest value with 106 species. This high Middle Miocene value exceeds the species richness of most modern mammalian faunas (and likely underestimates original richness). This richness comes from an area of about 18,000 km² (the sampled area of the Potwar Plateau), although it is likely representative of the fauna of the entire foreland basin extending across South Asia. For comparison, the modern terrestrial

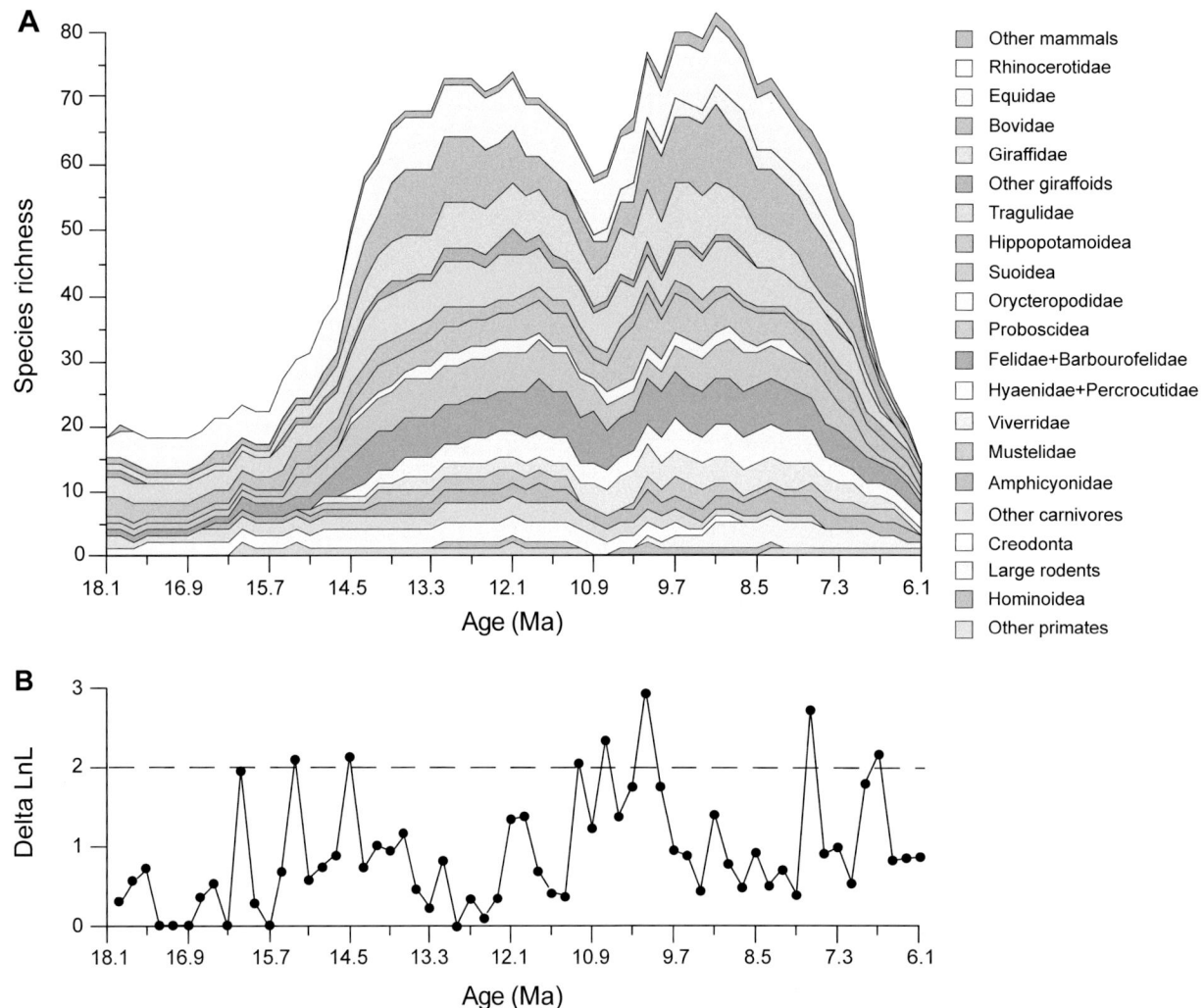

Figure 26.8. Change in taxonomic composition of large mammals over time. A. Species richness of higher taxa (mainly families), based on estimated residence times and compiled for 200-kyr intervals. Singletons were included in this compilation. Faunivorous taxa are colored in reds and oranges, broadly herbivorous taxa in greens, blues, and yellows. B. Change in log-likelihood (Delta LnL) for a multinomial model of composition, comparing maximum likelihood estimates based on faunal proportions within one interval with faunal proportions from the prior time interval. If faunal proportions change substantially from the older to younger interval, then the change in log likelihoods is large. The value of 2.0 is a threshold for statistically significant change in faunal proportions. For the large-mammal record, seven of the Delta LnL values exceed that threshold, and other notable peaks occur. See text for further explanation.

mammal fauna of the Indian subcontinent—excluding bats, introduced species, and species whose ranges occur entirely above 1,000 m in elevation—contains 108 species of small mammals and 118 species of large mammals for a total of 226 species over an area of about 3.2 million km² (Lynx Ediciones 2020). The Indian subcontinent is the physiographic region corresponding to the Indian tectonic plate and includes the foreland basin of the collision zone between the Indian and Asian plates. Thus at peak richness, an area less than 20% of the Indian subcontinent (the foreland basin) contained half the number of species that inhabit the entire peninsula today. This Miocene richness is similar to the richest localities documented today, such as Serengeti National Park, Tanzania (122 species, excluding bats; Misonne and Verschuren 1966; Hendrichs 1970; Kingdon 1974a, 1974b; Magige and Senzota 2006); Garamba

National Park, Democratic Republic of Congo (107 species; Verschuren 1958; Verheyen and Verschuren 1966; Heim de Balsac and Verschuren 1968), and lowland dry forest in central Vietnam (105 species; van Peenen 1970). Many Miocene mammalian faunas have high species richness, especially of ungulates, carnivores, and proboscideans (e.g., Qui and Qui [1995] for China; Janis, Damuth, and Theodor [2000] for North America; Madern and van den Hoek Ostende [2015] for Europe; Faith, Rowan, and Du [2018] for Africa).

Limits to Species Richness

The concept that species richness of ecosystems or clades has an inherent limit such that the rate of addition or subtraction of species decreases or increases, respectively, in proportion to standing richness is known as *diversity dependence* (Rabosky

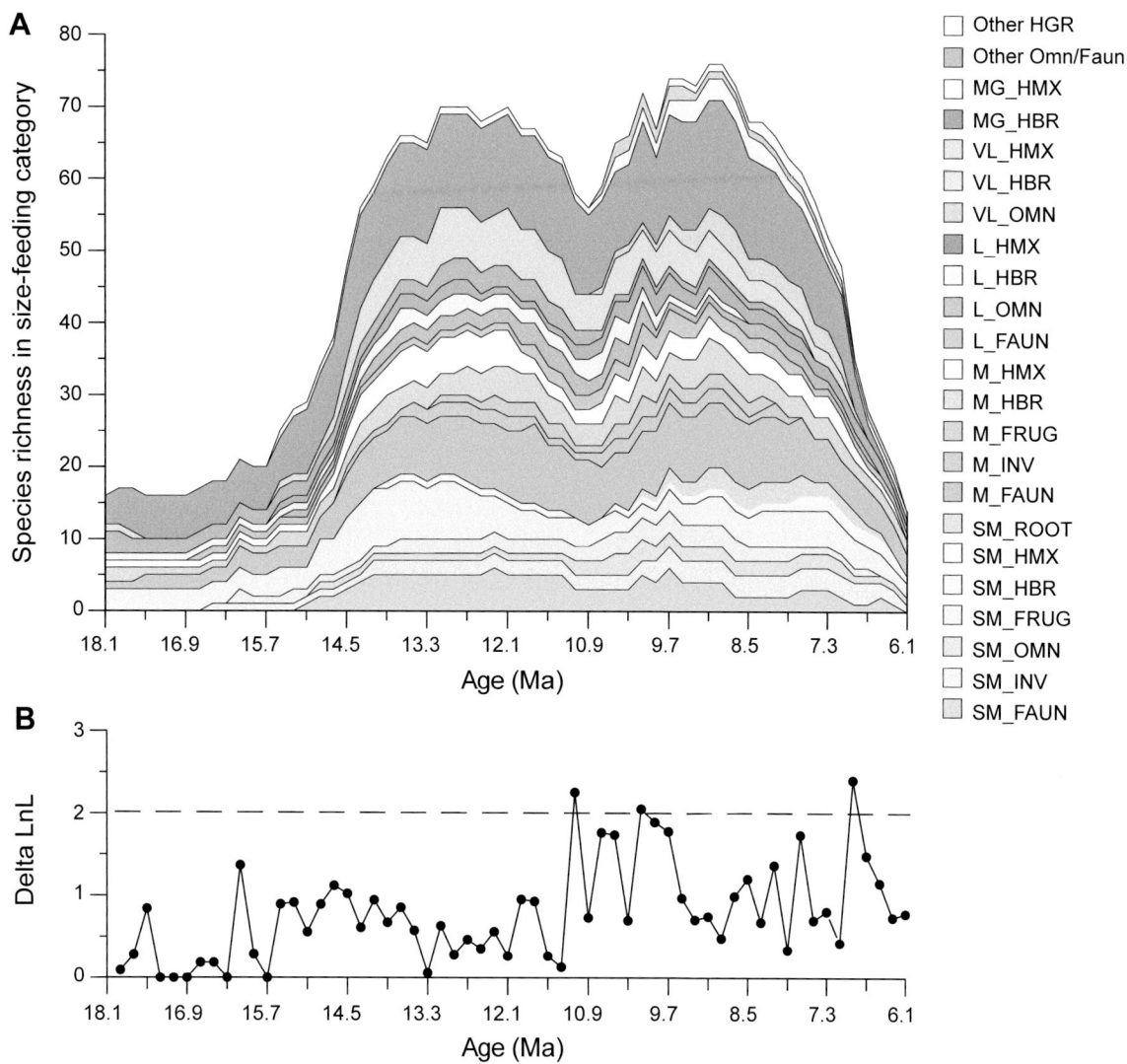

Figure 26.9. Change in ecological structure of large mammals over time. A. Species richness of joint size-feeding categories based on estimated residence times and compiled for 200-kyr intervals. B. Change in log-likelihood (Delta LnL) for a multinomial model of composition, comparing maximum likelihood estimates based on faunal proportions within one interval with faunal proportions from the prior time interval. If faunal proportions change substantially from the older to younger interval, then the change in log likelihoods is large. The value of 2.0 is a threshold for statistically significant change in faunal proportions. For the large-mammal record, three of the Delta LnL values surpass that threshold, and several other notable peaks occur. See text for further explanation. Abbreviations: SM = small; M = medium; L = large; VL = very large; MG = megaherbivore; FAUN = faunivore; INV = invertivore; OMN = omnivore; FRUG = frugivore; HBR = herbivore-browser; HMX = herbivore-mixed feeder; HGR = herbivore-grazer; ROOT = rootivore.

2013). For the Siwalik mammal record, we evaluated whether the number of first occurrences and the per-capita origination rate are correlated with standing richness; likewise for last occurrences and per-capita extinction rate. We analyzed the records of small mammals and large mammals separately and calculated correlation coefficients for all time intervals with ≥12 species present. This practice eliminated five poorly sampled intervals for the small-mammal record and one for the large-mammal record.

If high standing richness has a dampening effect on the number of new species entering the fauna, as either immigrants or endemic originations, then the number of first occurrences

should decline as species richness increases, resulting in a negative correlation. For both small mammals and large mammals, the correlation is positive (Table 26.4). Likewise, the per-capita origination rate should decrease as standing richness increases. For small mammals, the relationship is weakly negative with a non-significant *p*-value. For large mammals, the relationship is also negative, with a marginally non-significant *p*-value (see Table 26.4).

If standing richness has a dampening effect on the addition of new species, then the number of last occurrences and the per-capita rate of extinction should increase as standing richness increases, resulting in a positive correlation. For both small

Table 26.4. Correlation Analysis for Diversification Metrics in Relation to Standing Species Richness for 200-kyr Time Intervals

	Expected Sign (+,−)	Correlation Coefficient, r	p-value (2-tailed)	No. Time Intervals in Analysis
Small mammals (76 species)				
No. of first occurrences	−	0.210	0.125	55
Per-capita origination rate	−	−0.050	0.713	55
No. of last occurrences	+	0.166	0.228	55
Per-capita extinction rate	+	−0.145	0.289	55
Large mammals (186 species)				
No. of first occurrences	−	0.263	0.044	59
Per-capita origination rate	−	−0.252	0.055	59
No. of last occurrences	+	0.052	0.691	59
Per-capita extinction rate	+	−0.204	0.122	59

Note: The analysis is based on estimated residence times and excludes singleton species. Pearson correlation coefficients and associated *p*-values are reported, with significance at $p < 0.05$. Five time intervals were omitted from the small-mammal data, and one time interval was omitted from the large-mammal data, because fewer than 12 species were present in those time intervals. These correlations show weak to negative support for diversity dependence exerting a strong influence on species richness.

mammals and large mammals, there is a weak positive correlation between the number of last occurrences and standing richness, although the *p*-values are non-significant (see Table 26.4). For both small mammals and large mammals, the correlation between per-capita extinction rate and standing richness is negative, with non-significant *p*-values.

Overall, support for diversity dependence is weak to absent (with all correlation coefficients less than 0.3 or greater than −0.3, often with the opposite sign than would support diversity dependence). This finding suggests either that increasing species richness had a positive influence on the addition of new species—through niche creation or the impact of ecosystem engineers (see below)—or that factors other than standing richness controlled the addition to or subtraction of species from mammalian faunas. Other plausible influences include variable opportunities for immigration and emigration due to changes in landscape "gateways" from tectonic or climatic factors, increases or decreases in carrying capacity, or episodes of strong environmental filtering resulting from changes in environmental conditions, such as the increasing seasonal aridity of the Late Miocene (Chapter 24, this volume; Badgley et al. 2008, 2015; Polissar et al. 2021). Changes in fossil preservation because of major changes in facies or outcrop exposure could also affect the apparent addition and deletion of species independently of standing richness.

Diverging Dynamics of Small and Large Mammals

Small mammals and large mammals have substantially different patterns and rates of change in faunal composition. First occurrences were more pulsed in the observed record of species richness than in the estimated record for both groups (Figs. 26.2, 26.3). A few notable time intervals stand out for both the high number of originations and significant per-capita origination rates. For small mammals, three such in-

tervals occur at 14.7, 14.5, and 11.9 Ma (Fig. 26.4b, c). For large mammals, the highest pulses occur between 14.9 and 14.3 Ma and 10.7 and 9.7 Ma (Fig. 26.5b, c). Small mammals and large mammals share pulses of first occurrences in common at 14.7 and 14.5 Ma; the other notable intervals differ for each group.

Last occurrences occurred in notable pulses for both small and large mammals. Intervals with both a high number of last occurrences and significant per-capita extinction rates occur at 13.1, 12.5, and 9.9 Ma for small mammals and at 11.3, 11.1, 8.7, and a cluster of intervals from 7.9 to 6.7 Ma for large mammals. None of the notable pulses in last occurrences for small mammals overlaps with those for large mammals.

Small mammals and large mammals exhibited notable change in family-level taxonomic composition in different time intervals. Their closest pulses occur at 14.7 Ma, when the small-mammal record shows a spike in Delta LnL (Fig 26.6), and at 14.5 Ma, when the large-mammal record shows a spike as well (Fig. 26.8). Both spikes exceed 2.0, rising above the background variation in their respective records. The small-mammal record has no younger spikes that rise above 2.0, whereas the large-mammal record has five younger spikes.

The record of joint size-feeding categories shows asynchronous pulses for small and large mammals. Small mammals have a significant spike at 14.7 Ma (Fig. 26.7), whereas large mammals have the earliest spike at 11.1 Ma (Fig. 26.9). Small mammals have low rates of change in functional composition otherwise. In contrast, large mammals have three pulses in which Delta LnL is above 2.0 at and other values that lie close to this threshold. Thus, the major pulses of change occur asynchronously more than synchronously for small versus large mammals; this contrast suggests either that the causes of change differed or that the responses to causes in common differed between small and large mammals.

DETERMINANTS OF FAUNAL COMPOSITION AND CHANGE

Beyond measuring faunal change, our goal is to infer formative processes. These were likely multidimensional, involving ecological interactions within the Siwalik ecosystem and external processes that affected physical and biotic components of the ecosystem. Criteria for evaluating the influence of factors that affect the appearance and disappearance of species, species richness, and ecological structure of the mammalian fauna include plausibility, mechanism, timing, and evidence that supports or rejects those factors. We note five factors that plausibly exerted a strong influence on mammalian faunas, the first three internal to the ecosystem and the last two external. These factors are not mutually exclusive.

Megaherbivores as Ecosystem Engineers

This concept stems from the study of mammalian megaherbivores (species that weigh >800 kg) in modern ecosystems (Owen-Smith 1988). The feeding and social behavior of elephants, rhinoceroses, and giraffes can cause uprooting of trees, heavy consumption of local vegetation, transport of seeds over long distances, and trampling and bioturbation of the substrate. The cumulative impacts can perpetuate early successional stages of vegetation over the home range of these animals, thereby transforming areas of woodland to shrubland or grassland (Owen-Smith 1988; Hyvarinen et al. 2021). The Siwalik record features up to 18 species of megaherbivore within single 200-kyr time intervals; these animals likely had similar influences on the vegetation and substrates of the Miocene floodplain ecosystems. Their presence likely increased heterogeneity of habitats for other terrestrial animals across the flood basin. Although the Siwalik sequence has no definitive record of local disturbance from megaherbivores, a high degree of bioturbation of floodplain and near-channel sediments is consistent with this influence but could also be caused by other organisms, including diverse aquatic vertebrates. Megaherbivores were present throughout the analysis interval, although the number and proportion of these species dropped rapidly after 7.5 Ma (Fig. 26.9a).

Top-Down Processes

Siwalik mammal faunas had a high number and proportion of faunivorous species, and this proportion increased with overall increase in large-mammal richness (Figs. 26.9a, 26.10). (The richness of mammalian faunivores is remarkable given the sparse quality of their fossil record; Chapter 13, this volume). These predators would have exerted strong top-down pressures on their prey populations and potentially facilitated an increase in the number of prey species. One test of this hypothesis is whether there is a positive correlation between the number of predator species and the number of potential prey species. Figure 26.10 (p. 502) indicates that this relationship for large mammals has a strong positive correlation, with ($r = 0.91$) or without ($r = 0.91$) megaherbivores included as prey species. This result is complicated by the likelihood that other vertebrate

species (e.g., crocodilians) were also significant predators on mammalian herbivores and that prey species would also have included non-mammalian vertebrates (e.g., fishes, snakes, turtles, birds). Taphonomic evidence of the pervasive impact of mammalian faunivores includes the fragmentary original nature of most mammalian fossil remains and the common occurrence of bone-damage resulting from puncturing, chewing, or digestion of skeletal elements (Chapter 4, this volume).

Interspecific Competition

Direct and indirect interspecific competition for food and territory is a plausible process for species with similar ecological attributes, such as those within the same joint size-feeding category (Figs. 26.7, 26.9). Data from stable carbon isotopes, microwear, mesowear, as well as morphometric analysis of jaws or teeth, provide quantitative estimates of similarity of species with regard to feeding habits (e.g., Nelson 2007; Kimura et al. 2013). For species or clades with similar feeding habits, increase in species or populations of one clade while the other clade decreases would support the hypothesis that competition was an important process. The weak support for diversity dependence (see Table 26.4) in both small and large mammals suggests that competition was not the primary control on overall species richness. A plausible case of competition can, however, be made for certain groups, such as the decline of cricetids and gerbillines as murines increased in the small-mammal record (Fig. 26.6a; Chapters 11, 25, this volume). It is plausible that strong top-down pressure from faunivores kept herbivore populations below the size at which interspecific competition was a strong influence.

Properties of the Fluvial System

Large river systems have both stabilizing and disruptive effects on floodplain ecosystems. Annual and long-term fluctuations in water volume change the distribution of water across the floodplain. In rivers of the magnitude of Siwalik channels (smaller channels reconstructed as 80–200 m across and 5–15 m deep, [Willis 1993], on par with the modern Jhelum River, one of the major tributaries to the Indus River), annual to semi-annual floodwaters would have covered much of the floodplain for weeks to months, forcing terrestrial animals to migrate out of the flood basin on a regular basis and enabling aquatic vertebrates to inhabit the floodplain. This regular cycle may have limited the number and kinds of small-mammal species that could persist in the floodplain ecosystem. Burrowing species would be flooded out regularly, and all (non-aquatic) species would have needed to move out in advance of rising floodwaters or survive in tree canopies or ridges on floodplains for weeks (e.g., Sheppe and Osborne 1971; Wijnhoven et al. 2005). Studies of small mammals on modern natural floodplains document considerable mortality in small-mammal populations during the flood season (Anderson et al. 2000; Golet, Hunt, and Koenig 2013).

Receding floodwaters would blanket the landscape with nutrients, stimulating growth in vegetation and leaving ponds behind for drinking water. Avulsion of major channels would

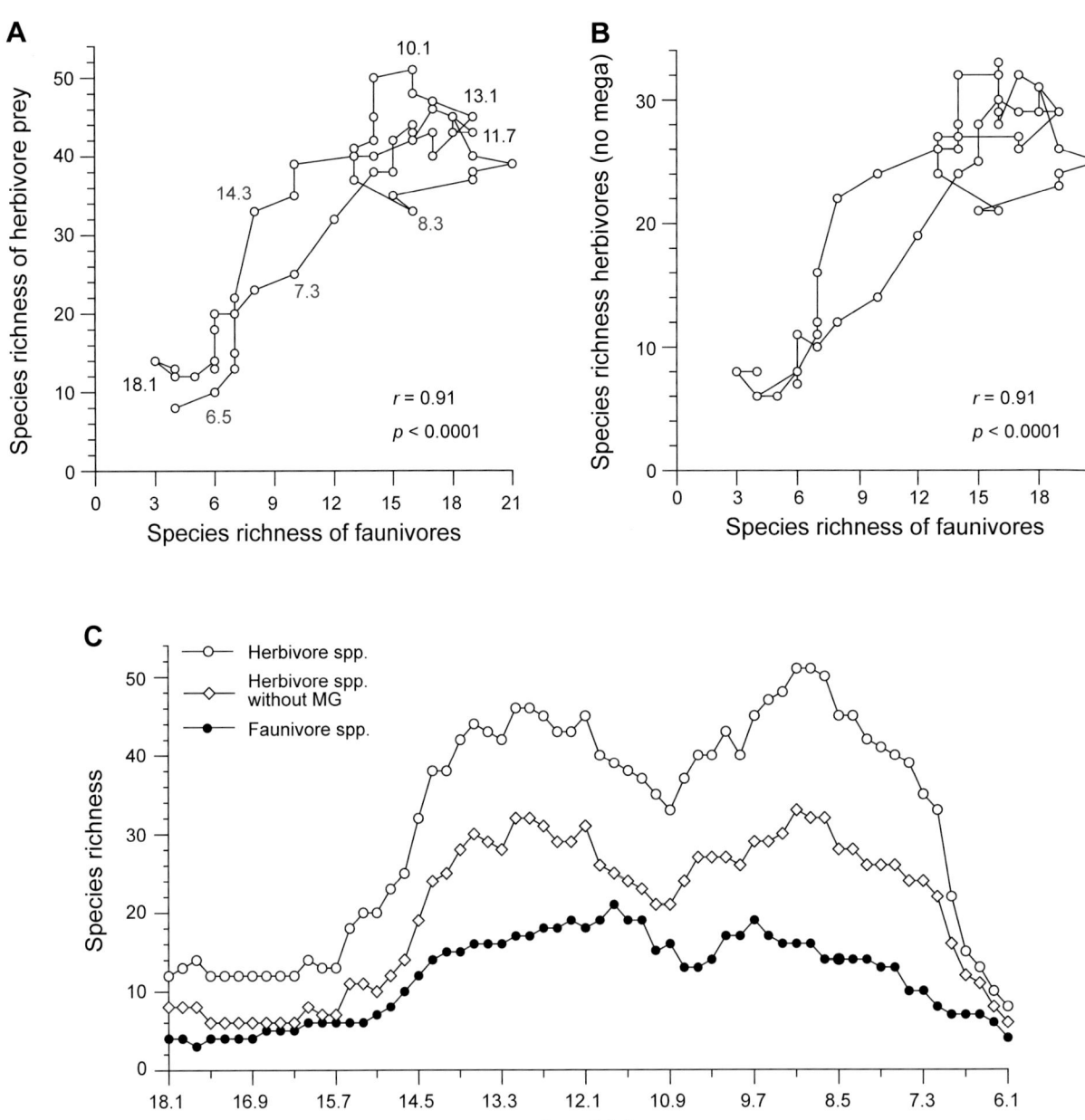

Figure 26.10. Species richness of faunivores and potential herbivore prey for large mammals, with and without megaherbivores (MG). Herbivore prey include all species with feeding categories inferred as frugivore, herbivore-browser, herbivore-mixed feeder, herbivore-grazer, and rootivore across all size categories. A. Scatterplot of faunivore species versus herbivore prey species per time interval. Light line connects time intervals to show changes over time. Species richness of faunivores and herbivore prey is highly correlated (*r* = 0.91, *p* < 0.0001). Selected time intervals are noted; those in red occur during the environmental-stress transitions. B. Same as (A) but omitting megaherbivore species from herbivore prey richness. Data are highly correlated (*r* = 0.91, *p* < 0.0001). C. Species richness over time of faunivores, herbivore prey, and herbivore prey without megaherbivores. Faunivore richness and herbivore richness increase over time in parallel. Herbivore richness including megaherbivores peaks at 9.1 Ma; without megaherbivores, herbivore richness has an early peak at 13.1 Ma and a second peak at 9.1 Ma. Faunivore richness peaks at 11.9 Ma. Herbivore richness begins a steep decline earlier than faunivore richness. The high number of faunivore species over much of the sequence is consistent with the idea that top-down processes contributed to the high number of herbivore species.

leave ponded water and nutrients in the abandoned channels, attracting both terrestrial and aquatic species to favorable habitats. Both annual flooding and longer-term avulsion processes would create areas of disturbed habitat, such that early successional stages of vegetation would recur across the floodplain

(Hughes 1997), thereby maintaining heterogeneous habitats for aquatic and terrestrial vertebrates alike. Alluvial architecture of the Siwalik fluvial system, lateral variation in paleosols, and spatial gradients in stable carbon isotopes of large mammals provide evidence for persistent habitat heterogeneity across the flood basin (Willis 1993; Behrensmeyer et al. 2007; Morgan et al. 2009).

Long-Term Changes in Climate and Vegetation

Several aspects of global and regional environmental change have relevance for Siwalik faunal change. We note these briefly here and take up the subject in the next section. The Miocene Epoch experienced major changes in global temperature, including the Middle Miocene warm interval (Miocene Climatic Optimum, 16.9–14.7 Ma), followed by rapid cooling (Middle Miocene Climatic Transition, 14.7–13.8 Ma) and continued cooling through the Late Miocene (Steinthorsdottir et al. 2021; Clift et al. 2022). Regional expression of these climatic changes in the Potwar Plateau could involve change in macrofacies, paleosol features, vertebrate biotas, and biogeochemistry of sediments, including pollen and biomarkers (e.g., Karp, Behrensmeyer, and Freeman 2018; Feakins et al. 2020).

The following episodes, which are based on recent syntheses of global and regional environmental history (Feakins et al. 2020; Polissar et al. 2021; Steinthorsdottir et al. 2021; Clift et al. 2022), were likely to have been critical environmental influences on the species richness and composition of Siwalik faunas. In the following section, we explain how these changes would have influenced community assembly.

1. The Miocene Climatic Optimum, 16.9–14.7 Ma (Foster, Lear, and Rae 2012; Steinthorsdottir et al. 2021). Increase in temperature would cause geographic-range shifts for many terrestrial species as they tracked their climatic tolerances. Some species may have moved upslope into the Himalayan foothills. This episode may have presented thermoregulatory challenges to mammals occupying warm humid regions (see below).
2. Cooling of the Middle Miocene Climatic Transition, 14.7–13.8 Ma (Holbourn et al. 2013; Steinthorsdottir et al. 2021). Cooling would cause geographic-range shifts, including downslope, for both plants and animals, potentially moving species from montane foothills into the flood basin and stimulating dispersal in low-elevation corridors.
3. Around 10.5 Ma, biomarker evidence for occasional fires on the Potwar Plateau (Karp et al. 2021) and traces of C_4 vegetation in equid diets (Badgley et al. 2008) suggest an increase in seasonality of precipitation. Increase in hematite/goethite ratios in the Indus Fan from 10 to 6 Ma also indicates increasing seasonality of precipitation (Clift et al. 2022).
4. Cooling of 4° C from 10.0 to 5.5 Ma (Feakins et al. 2020; Steinthorsdottir et al. 2021) could contribute to increasing seasonality of precipitation, changes in vegetation composition, and geographic-range shifts of plants and animals.
5. Increase in fire frequency occurred as C_4 monocots became widespread in the Potwar Plateau. Around 7.5 Ma, biomarkers indicate a notable increase in fire frequency and a shift from wood fires to grass fires at 7.4 Ma (Feakins et al. 2020; Karp et al. 2021; Polissar et al. 2021).

Stable isotopes in paleosols indicate a rapid expansion of C_4 vegetation on the landscape at 7.2 Ma (Feakins et al. 2020; Tauxe and Feakins 2020, Karp et al. 2021).

PATTERNS AND PROCESSES OF COMMUNITY ASSEMBLY

The mammalian record of the Potwar Plateau documents a long period of species enrichment followed by a long period of decline. Here we evaluate three aspects of community assembly and disassembly over the record.

Taxonomic Richness and Functional Richness

We examined the relationship between species richness and functional richness. Small mammals show a moderate positive correlation between species richness and the number of size-feeding categories as a measure of functional richness ($r = 0.54$, $p < 0.0001$; Fig. 26.11, p. 504). Species richness increased more rapidly than functional richness. From 18.1 to 16.5 Ma, both rose steeply, then functional richness accumulated slowly while species richness rose rapidly, reaching the highest value (44 species) between 13.7 and 13.1 Ma. The highest number of functional categories (nine) per interval occurred between 11.9 and 8.9 Ma. After 7.5 Ma, both species richness and functional richness fell rapidly toward the end of the record.

Large mammals show a high positive correlation between species richness and the number of functional categories ($r = 0.93$, $p < 0.0001$; Fig. 26.12, p. 505). The temporal relationship differed during the enrichment phase compared to the decline phase. From 18.1 to 10.0 Ma, species richness increased more rapidly than functional richness. The highest species richness (83) occurred at 9.1 Ma, and the highest number of functional categories (23) occurred from 9.5 to 8.7 Ma. From 8.7 to 7.1 Ma, species richness declined more rapidly than functional richness. After 7.1 Ma, functional richness and species richness both declined rapidly. Below we interpret details of these changes in terms of environmental history.

Taxonomic Richness in Relation to the Appearance or Disappearance of Clades

For both small and large mammals, gains and losses of clades would have contributed to changes in phylogenetic richness over time. For small mammals, the enrichment phase of taxonomic and functional richness had a low contribution from the appearance of new clades at the family level. Most higher taxa (11 of 16 families or subfamilies) were present at the beginning of the sequence. Based on estimated first appearances, the appearances of Anomaluridae at 16.8 Ma, Lorisidae at 15.2 Ma, and Soricidae at 14.6 did not add new functional groups (as they were already present), but the appearance of *Urva* small species (the earliest herpestid) at 14.9 Ma added small faunivores to the fauna (see Fig. 26.7). The appearance of a glirid lineage (part of "Other" in Fig. 26.6) at 14.2 Ma likewise did not add a new functional group. Late in the sequence, the last occurrence of eight clades accompanied a drop in richness from 18 to 6 species. Based on estimated last appearances, Lorisidae last

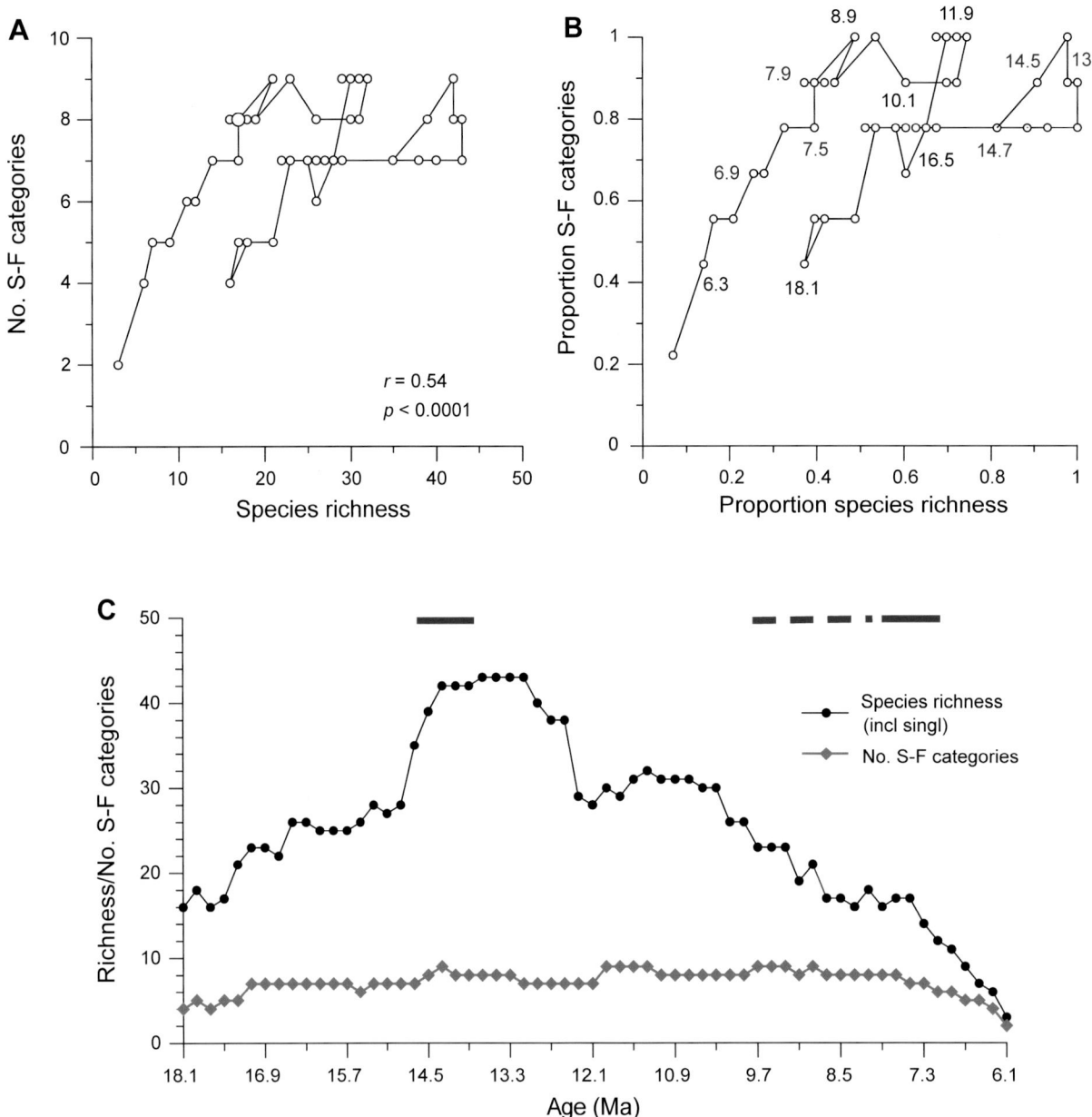

Figure 26.11. Relationship between taxonomic richness and functional richness for small mammals. A. Scatterplot of number of species versus number of joint size-feeding (S-F) categories per time interval. Light line connects successive time intervals. Although the two variables are significantly correlated ($r = 0.54$, $p < 0.0001$), the time trajectories differ between the first half and the second half of the record. B. Data for (A) plotted as proportion of the highest value for each variable. Numbers indicate various time intervals; values in red occur during the early and late environmental stress intervals. C. Estimated species richness (with singletons) and number of size-feeding categories over time. Horizontal red lines indicate the periods of environmental stress transitions. See text for further explanation.

occurred at 8.8 Ma, gerbillines at 8.7 Ma, Cricetidae at 8.4 Ma, Gliridae at 7.5 Ma, Tupaiidae at 7.0 Ma, Sciuridae at 6.5 Ma, Spalacidae at 6.3 Ma, and murines at 6.2 Ma. (The disappearances from 6.5 Ma on likely reflect diminished sampling of the fossil record, since sciurids, murines, and spalacids all occur in Pakistan today and are known from Pliocene or Quaternary fossil assemblages [Chapter 11, this volume].)

For large mammals, the growth phase of taxonomic and functional richness involved few appearances of new orders or

families, although many new genera appeared. At the beginning of the Potwar sequence (by 17.5 Ma) or earlier, 12 of 36 families were present. Immigrant groups that contributed to the rise of species richness or functional richness included Herpestidae at 16.3 Ma, Barbourofelidae at 16.1 Ma, Catarrhini, i.s., and Sivaladapidae at 16.0 Ma (although sivaladapids were present in the older Zinda Pir sequence), Stegodontidae at 15.8 Ma, Gomphotheriidae at 15.2 Ma, Orycteropodidae at 14.4 Ma, Paleomerycidae at 13.2 Ma, and Hominoidea at 12.9 Ma. The

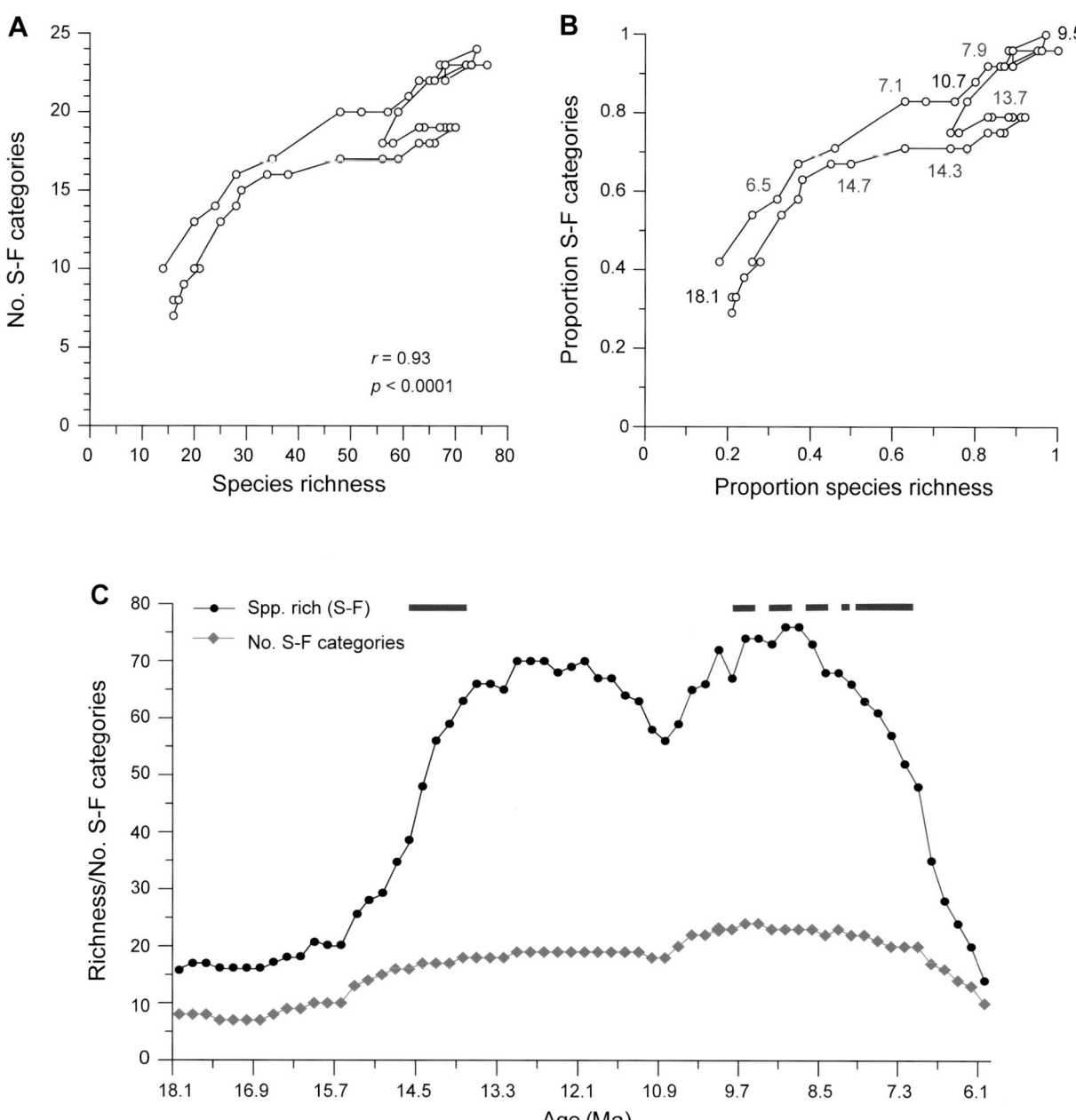

Figure 26.12. Relationship between taxonomic richness and functional richness for large mammals. A. Scatterplot of number of species versus number of joint size-feeding categories (S-F) per time interval. Light line connects successive time intervals. The two variables are significantly correlated (*r* = 0.93, *p* < 0.0001), and the time trajectories are parallel between the first half and the second half of the record. B. Data for (A) plotted as proportion of the highest value for each variable. Numbers indicate various time intervals; values marked in red occur during the early and late environmental stress thresholds. C. Estimated species richness (with singletons) and number of size-feeding categories over time. Horizontal red lines indicate the periods of environmental stress transitions. See text for further explanation.

decline phase from 8.5 to 6.1 Ma involved the loss of Palaeo-choeridae at 8.4 Ma, Hominoidea at 8.3 Ma, Orycteropodidae at 7.7 Ma, and Paleomerycidae at 7.2 Ma. These losses occurred prior to the major decline in fossil localities.

Environmental Stress Thresholds

We recognize environmental stress as a change in ambient conditions great enough to prevent occupancy or to increase mor-

tality for species with particular resource requirements or physiological tolerances (Freedman 2015). Thereby the establishment or removal of an environmental stress leads to a notable change in community composition, including the loss of lineages or functional groups. Based on the significant changes in climate and vegetation noted above, we recognize two significant intervals in environmental stress: a transition from higher to lower stress during the cooling trend of the Middle

Miocene Climatic Transition from 14.7 to 13.9 Ma and a transition from lower to higher stress from 8.1 to 7.0 Ma, with early indicators of this transition starting at 10.5 Ma, during the expansion of C_4 vegetation and increase in fire frequency related to regional cooling and drying.

The Miocene Climatic Optimum featured terrestrial global mean temperatures that are estimated as 5–6° C warmer than those of today (Steinthorsdottir et al. 2021). Estimates of precipitation based on regional studies generally indicate wetter conditions than earlier or later than the Miocene Climatic Optimum and than today (e.g., Hui et al. 2018). Several studies of modern climate change have raised concerns about rising temperatures in areas with high humidity, areas which include subtropical monsoonal regions (e.g., Colwell et al. 2008; Buzan and Huber 2021; Zhang, Held, and Feuglistaler 2021). Hot, humid conditions limit the ability of animals to thermoregulate by evaporative cooling. Exposure to wet-bulb temperatures (the temperatures that can be effectively cooled by evaporation) greater than 35° C for more than 6 hours is lethal in humans. Placental mammals share similar thresholds for moist heat stress and face potentially lethal heat stress in the presence of high heat and humidity, even though they evolved during even higher temperatures of the Late Cretaceous (Huber et al. 2018; Grossnickle, Smith, and Wilson 2019).

South Asia today has some of the highest wet-bulb temperatures on record during the monsoon season, exceeding 28–29° C and resulting in high mortality of people and livestock in recent summers (Buzan and Huber 2021). Would the climatic conditions of the Siwalik faunal province during the Miocene Climatic Optimum have excluded species that could not readily find cool habitats during the warmest times of the year? Small mammals would have been able to utilize microhabitats with cooler temperatures, including burrows and tree hollows, and the fluvial system would have provided cooling ponds for larger animals. An intriguing hypothesis is that the low species and functional richness of the early part of the Potwar Siwalik record may document an interval of inhospitable climatic conditions for many mammals; this hypothesis is hard to test directly so it rests upon plausibility at present. Many species may have moved into the foothills of the mountains where no stratigraphic record was produced. The large peak in small-mammal first appearances at 14.7 Ma was dominated by endemic originations, indicating that closely related species rapidly moved into the Siwalik faunal province during the cooling interval. At this time, species richness and functional richness of both small and large mammals rapidly increased without displacing resident species. The long residence times of lineages during the older portion of the Siwalik record are a consequence of this pattern.

At the beginning of the Potwar sequence (18.1–17.7 Ma), 5 of 11 small-mammal functional groups were present: very small invertivores, very small omnivores, very small herbivores, small granivores, and small herbivores (Fig. 26.7a). Five functional groups appeared sporadically between 17.9 and 14.5 Ma: very small granivores at 17.3 Ma, very small rootivores at 17.0 Ma, small faunivores at 14.9 Ma, and small omnivores at

14.4 Ma. Thus, most (~80%) of the functional richness of small mammals was established by the end of the Miocene Climatic Optimum (Fig. 26.11b), despite the sparser quality of the fossil record during this time. During the first environmental stress threshold, small mammals acquired peak species richness, but the highest functional richness came later. The remaining two functional groups, small rootivores and small invertivores, did not appear until 11.9 and 11.7 Ma, respectively. Up to nine functional categories occurred together in four consecutive time intervals that happened after species richness had begun a notable decline.

For large mammals, 8 of 25 functional groups were present at the beginning of the sequence: small herbivore-browsers, medium faunivores, medium herbivore-browsers, medium herbivore–mixed feeders, large herbivore-browsers, very large faunivores, very large herbivore-browsers, and megaherbivore browsers (Fig. 26.9a). From 17.7 Ma onward, appearances included large faunivores at 16.5 Ma, small invertivores at 16.5 Ma, and small frugivores at 16.0 Ma; megaherbivore–mixed feeders at 15.5 Ma, large herbivore–mixed feeders, and large omnivores at 15.3 Ma; small omnivores at 14.7 Ma; small faunivores at 14.6 Ma; medium invertivores at 14.5 Ma; small herbivore–mixed feeders at 13.7 Ma; and medium frugivores at 12.9 Ma. This enrichment led to the initial build-up of species and functional richness of large mammals. During the cooling interval (14.7–13.8 Ma), species richness increased faster than functional richness (Fig. 26.12a, b). The final functional categories to appear were very large herbivore–mixed feeders at 10.6 Ma, megaherbivore omnivores (one species, part of "other omnivores") at 10.3 Ma, small rootivores at 10.0 Ma, and medium herbivore grazers (one species) at 9.5 Ma.

The second threshold interval from low to high environmental stress had a long buildup, with evidence for occasional fires at 10.5 Ma and early records of C_4 vegetation in Siwalik paleosols and equid diets around 9.0 Ma (Feakins et al. 2020; Karp et al. 2021). Between 7.9 and 7.1 Ma, multiple indicators provide evidence of increasing seasonality of precipitation, general cooling, increased fire frequency, and rapid spread of C_4 vegetation in the fluvial ecosystem. These environmental trends continued through the end of the Miocene to the present day (see Fig. 24.3).

For small mammals, taxonomic and functional richness decreased together during this environmental transition, with functional richness declining more rapidly than species richness (Fig. 26.11a, b). The loss of eight clades, noted above, was accompanied by last occurrences of small invertivores at 8.7 Ma, very small granivores at 8.4 Ma, very small omnivores at 7.5 Ma, very small rootivores at 7.2 Ma, and small granivores at 6.8 Ma. Very small invertivores, very small herbivores, small herbivores, small rootivores, and small faunivores persisted until the end of the sequence.

For large mammals, species richness and functional richness showed a protracted decline, especially from 7.9 Ma on. Species richness declined more rapidly than functional richness (Fig. 26.12b, c). The disappearance of functional categories indicates a sequential loss of species with diets reliant on year-round fruit or browse. The first group to disappear was large herbivore-browsers at 8.6 Ma, then medium frugivores, including hominoids, with last occurrences at 8.3 Ma. Subsequently, medium invertivores last appeared at 7.7 Ma, very large herbivore-browsers at 6.9 Ma, small frugivores at 6.6 Ma, and very large omnivores at 6.3 Ma.

Stable carbon isotopes of a few persisting lineages of medium and large herbivore–mixed feeders recorded a shift from C_3 to C_4 vegetation after ~7.0 Ma (see Fig. 24.6; Nelson 2007; Badgley et al. 2008; Polissar et al. 2021); this change indicates that they had become C_4 grazers. Groups that persisted until the end of the sequence (6.1 Ma) include small omnivores, small herbivore-browsers, small herbivore–mixed feeders, medium faunivores, medium herbivore-browsers, medium herbivore–mixed feeders, large faunivores, large herbivore–mixed feeders, very large omnivores, megaherbivore–browsers, and megaherbivore–mixed feeders, with few species present in each category. As noted earlier, the decline in species richness included loss of several mammalian clades as well.

Could the rise and fall in species richness in this record be an artefact? More specifically, could this pattern be the consequence of substantial edge effects (Foote 2000; Alroy 2010) or the temporal equivalent of the mid-domain effect (Colwell, Rahbek, and Gotelli 2004), whereby taxa are more readily sampled (and therefore more numerous) in the middle rather than at the edges of a temporally or spatially bounded domain? While some edge effects are likely present, we consider that the major trend estimates the actual history for the following reasons. First, longer residence times of both small and large mammals are concentrated in the older third of the sequence, which is the opposite of the expectation from an edge effect. Second, the correspondence in timing between major increases and decreases in species richness and the phases of environmental history suggests a causal relationship for this pattern. Third, changes in taxonomic and functional composition include signals of environmental filtering that cannot be explained by edge effects.

Based on changes in species richness, taxonomic composition, and functional composition, we recognize three phases of mammalian community composition in the Siwalik Potwar record. None of these phases shows stability in richness or composition so they cannot be considered equilibrial communities, such as the Neogene mammalian faunas of the Iberian region (Blanco et al. 2021). Each phase has unique properties, and the transitions between them correspond to major episodes of environmental change.

1. The first phase occurred from 18.1 to 14.7 Ma and included the warm Miocene Climatic Optimum. Both small and large mammals show low but increasing species and functional richness over this period, with low to moderate rates of compositional turnover between adjacent time intervals. Lineages present during this phase included 12 small and 3 large-mammal species that were arboreal. This interval features deeply leached

paleosol with clays indicating wetter conditions than in younger time intervals (Quade and Cerling 1995). The inferred warm, humid climate of this time would have supported forest vegetation over much of the floodplain, similar to vegetation in central and eastern India on similar soils today.

2. The second phase, from 14.5 to 8.9 Ma, began with the cooling of the Middle Miocene Climatic Transition and involved rapid increase in species richness for small and large mammals and major changes in taxonomic or functional composition at 14.7–14.5 Ma. Later changes differed for small and large mammals. Small mammals experienced a step-decline in species richness at 12.5 Ma but without notable change in functional composition. Perhaps certain small-mammal species were harbingers of early environmental changes that would eventually influence most of the fauna.

   For both small and large mammals, species richness and functional richness hovered around high values over several million years (see Figs. 26.1, 26.11, 26.12). Nineteen species of small mammal and seven species of large mammal with arboreal habits lived during much of this phase. The high functional richness is similar to that found in mammalian assemblages from modern tropical forests (e.g., Eisenberg 1981; Emmons, Gautier-Hion, and Dubost 1983). Peaks in functional richness persisted until 8.9 Ma for small mammals and 8.1 Ma for large mammals; then functional richness steadily declined in both records. This second phase ended at 8.7 Ma, as indications of seasonal aridity became more prevalent in sedimentary and faunal records. As noted in Chapter 25 of this volume, an important turnover in ungulates occurred between 11.4 and 10.1 Ma; this interval featured the first appearance of equids, first and last occurrences among lineages of bovids, tragulids, and suoids, and size increase in all artiodactyl families. Paleoenvironmental evidence indicates persistence in tree canopy but reduction in canopy cover and increase in C_3 grass area (Karp, Behrensmeyer, and Freeman; Karp et al. 2021) as early signals of the third phase.

3. The third phase, from 8.7 to 6.1 Ma, involved declining species and functional richness for both small and large mammals. The most rapid rates of decline in species richness occurred from 7.9 to 6.9 Ma for small and large mammals. The disappearance of functional categories followed a sequence consistent with environmental filtering as C_4 vegetation spread rapidly across the ecosystem and fires became more frequent. Eight species of small mammal and four species of large mammal had arboreal habits during this interval. Stable isotopes and biomarkers indicate less canopy cover and increasing areas of C_4 grasslands through this phase; the loss of three-dimensional habitats would have influenced the decline in functional richness of small and large mammals. Paleosols showed evidence of lower and fluctuating water tables, with less ponded water on floodplains

(Quade and Cerling 1995; Behrensmeyer et al. 2007). The entire analysis interval ended with a poorly sampled interval as the faunas continued responding to ecosystem transformation from woodland to monsoonal grassland. This phase signals a regime shift, with fire frequency acting as a positive feedback process in the transition to C_4 grassland and increased aridity.

SUMMARY

Estimates of species residence times preserve the major trends of the observed record for both small and large mammals, while accounting for some of the vagaries of preservation. The estimated record dampens short-term volatility in the observed record and expands the timing of appearances and disappearances over the record, as expected. Singleton species occur throughout both records over the entire analysis interval; their omission does not distort overall trends in species richness.

There are substantial differences in the dynamics of faunal change for small mammals (<1 kg in estimated adult body weight) and those of large mammals (≥1 kg in estimated adult body weight), particularly in the timing of significant episodes of origination and change in taxonomic and functional composition. Both records show significant changes in species richness and composition during the two major transitions in environmental stress, the first being the transition from the warm, humid Miocene Climatic Optimum to rapid cooling, the second being the transition from moderate, humid conditions to seasonally arid, cooler climates with increasing frequency of fire.

For both small and large mammals, species richness showed a sustained period of enrichment with both gradual and rapid increases in first occurrences. Small-mammal richness peaked earlier than large-mammal richness, and the decline from peak richness for small mammals began during the first broad richness peak for large mammals. These differences occurred during a well-sampled part of the Siwalik record and were not likely artefacts of preservation. The time intervals of significant per-capita rates of origination, extinction and diversification are mostly asynchronous for small and large mammals. The only interval of synchronous change occurred at the transition between the first and second phases of community composition, 14.7–14.5 Ma, during the initial cooling after the Miocene Climatic Optimum.

Likelihood analyses of change in taxonomic richness and ecological structure (functional categories) show asynchronous intervals of significant change. For large mammals, herbivore and faunivore richness changed over time in a correlated manner. Faunivore richness was a consistently high proportion of total richness over most of the sequence. Both small mammals and large mammals gained species steadily over the first 5–6 myr of the record, while functional categories increased slowly over the same period. The peak values of species richness and functional richness occurred 2 myr apart for small mammals

but in the same time interval for large mammals. Thus, a persistent feature of the record is the different diversity dynamics of small and large mammals.

Three phases of mammalian community composition occur in the analysis interval, all of them dynamic rather than equilibrial. The first phase, roughly coinciding with the Miocene Climatic Optimum, involved slow increase in species richness and functional categories for both small and large mammals. The second phase began with the cooling period that followed the Miocene Climatic Optimum and involved rapid increase in species richness for small and large mammals and a slower gain of functional richness. This phase includes the peak of species richness and a succeeding lower plateau of richness for small mammals and two broad peaks of species richness for large mammals. For both groups, high species richness and high functional richness characterize this phase, with evidence for extensive canopy cover on the landscape. The third phase occurred as forest canopies shrank and become more open and C_4 vegetation expanded. Declining species richness and loss of functional richness occurred throughout this phase.

The Siwalik record documents several aspects of mammalian faunas that differ from those of the present day—namely, very high peak species richness (especially considering that bats would likely add another 20 or more species), high taxonomic and functional richness of faunivores, and many co-occurring megaherbivore species. This sequence is one of the longest continuous records of mammalian faunal change over an unusually dynamic interval of Cenozoic history. These Miocene faunas have no modern counterparts and thus offer unique insights about faunal responses to environmental changes of the past.

Our analyses provide a foundation for future work that incorporates measures of phylogenetic richness and evaluates the contributions of immigrants versus endemic originations to faunal change using more finely resolved clades than those utilized here. Future fieldwork and phylogenetic, morphometric, and stable-isotopic analyses can refine and expand upon our findings.

ACKNOWLEDGMENTS

We thank the many colleagues whose fieldwork, laboratory analyses, data management, and publications form the foundation for this work. Special thanks to the Geological Survey of Pakistan and the field staff who supported our many field seasons in Pakistan. We acknowledge funding support from the Smithsonian Foreign Currency Program, the National Science Foundation, and many of our home institutions. We also thank John Finarelli for assistance with the bootstrapping analyses, Luke Weaver, and Siwalik collaborators who provided useful comments on early drafts of this chapter.

SUPPLEMENTARY TABLES

Order	Family	Higher Taxon	Species Records Combined into Lineages	Observed Range (Ma) Obs FO	Obs LO	No. Time Intervals (0.1 myr) with Locality	Residence Time (obs, myr), FO–LO	Avg Gap Size, Myr	Est FO, Ma[1]	Est LO, Ma[2]	Feeding Category[3]	Size Category[4]	Size Class Number[5]
Scandentia	Tupaiidae		Tupaiidae, n. gen. and sp.	11.644	8.703	4	2.941	0.980	12.624	7.723	Inv	SM	3
Scandentia	Tupaiidae		cf. *Dendrogale* sp. 1	17.878	10.492	11	7.386	0.739	18.617	9.753	Inv	VSM	1
Scandentia	Tupaiidae		cf. *Dendrogale* sp. 2	16.045	13.050	3	2.995	1.498	17.543	11.553	Inv	VSM	1
Scandentia	Tupaiidae		cf. *Dendrogale* sp. 3	14.045	7.338	19	6.707	0.373	14.418	6.965	Inv	VSM	1
Scandentia	Tupaiidae		cf. *Tupaia* sp.	12.779	11.230	2	1.549	1.549	14.328	9.681	Inv	VSM	1
Scandentia	Tupaiidae		Tupaiidae, gen. and sp. indet. (tiny)	8.703	8.703	1	0.000	0.000	8.703	8.703	Inv	VSM	1
Lipotyphla	Erinaceidae	Erinaceinae	Genus B and sp. indet.	13.085	13.085	1	0.000	0.000	13.085	13.085	Inv	VSM	1
Lipotyphla	Erinaceidae	Erinaceinae	Genus C and sp. indet.	7.338	7.338	1	0.000	0.000	7.338	7.338	Inv	VSM	1
Lipotyphla	Erinaceidae		*Galerix rutlandae*	14.300	11.600	13	2.700	0.225	14.525	11.375	Inv	VSM	1
Lipotyphla	Erinaceidae		*Galerix wesselsae*	18.800	14.039	7	4.761	0.794	19.594	13.246	Inv	VSM	1
Lipotyphla	Erinaceidae		*Schizogalerix* sp.	11.410	10.222	4	1.188	0.396	11.806	9.826	Inv	SM	2
Lipotyphla	Soricidae		Beremendiini, gen. and sp. indet.	8.749	8.749	1	0.000	0.000	8.749	8.749	Inv	VSM	1
Lipotyphla	Soricidae		*Suncus homeyi*	10.492	6.517	9	3.975	0.497	10.989	6.020	Inv	VSM	1
Lipotyphla	Soricidae		cf. *Crocidura* sp.	14.039	6.517	18	7.522	0.442	14.481	6.075	Inv	VSM	1
Primates	Lorisidae		*Nycticeboides simpsoni*	9.240	8.929	2	0.311	0.311	9.551	8.618	Herb	SM	3
Primates	Lorisidae		*Nycticeboides* sp.	15.169	15.169	1	0.000	0.000	15.169	15.169	Herb	SM	2
Primates	Lorisidae		cf. *Nycticeboides* sp.	9.200	9.200	1	0.000	0.000	9.200	9.200	Herb	SM	2
Primates	Lorisidae		*Microloris pilbeami*	10.492	9.400	3	1.092	0.546	11.038	8.854	Herb	VSM	1
Primates	Lorisidae		Lorisidae, gen. and sp. indet (small)	12.779	12.779	1	0.000	0.000	12.779	12.779	Herb	VSM	1
Lagomorpha	Ochotonidae		*Alloptox* sp.	17.892	17.892	1	0.000	0.000	17.892	17.892	Herb	SM	2
Rodentia	Sciuridae		*Oriensciurus* n. sp.	18.700	17.878	1	0.822	0.822	19.522	17.056	Gran/ Omn	SM	3
Rodentia	Sciuridae	Pteromyinae	*Aliveria* sp.	16.763	15.962	2	0.801	0.801	17.564	15.161	Gran/ Omn	SM	2

(continued)

Table 26.S1. continued

Order	Family	Higher Taxon	Species	Records Combined into Lineages	Observed Range (Ma)		No. Time Intervals (0.1 myr) with Locality	Residence Time (obs, myr), FO–LO	Avg Gap Size, Myr	Est FO, Ma[1]	Est LO, Ma[2]	Feeding Category[3]	Size Category[4]	Size Class Number[5]
					Obs FO	Obs LO								
Rodentia	Sciuridae	Ratufinae	cf. *Ratufa* n. sp.		16.763	13.589	3	3.174	1.587	18.350	12.002	Gran/Omn	SM	3
Rodentia	Sciuridae	Ratufinae	*Ratufa sylva*		10.492	9.400	4	1.092	0.364	10.856	9.036	Gran/Omn	SM	3
Rodentia	Sciuridae	Callosciurinae	*Callosciurus* sp.		14.279	9.240	6	5.039	1.008	15.287	8.232	Gran/Omn	SM	2
Rodentia	Sciuridae	Callosciurinae	*Tamiops* large sp.		14.279	8.749	14	5.530	0.425	14.704	8.324	Gran/Omn	VSM	1
Rodentia	Sciuridae	Callosciurinae	*Tamiops* small sp.		16.763	10.063	12	6.700	0.609	17.372	9.454	Gran/Omn	VSM	1
Rodentia	Sciuridae	Callosciurinae	*Tamiops* giant sp.		7.966	7.966	1	0.000	0.000	7.966	7.966	Gran/Omn	SM	2
Rodentia	Sciuridae	Callosciurinae	*Dremomys* sp.		14.279	7.394	12	6.885	0.626	14.905	6.768	Gran/Omn	SM	2
Rodentia	Sciuridae	Marmotinae	*Heteroxerus* sp.		14.279	13.183	3	1.096	0.548	14.827	12.635	Herb/Gran/Omn	SM	3
Rodentia	Sciuridae	Marmotinae	*Atlantoxerus* sp.		14.279	13.183	4	1.096	0.365	14.644	12.818	Herb/Gran/Omn	SM	2
Rodentia	Sciuridae	Marmotinae	*Sciurotamias* Kamlial sp.		16.045	12.756	3	3.289	1.645	17.690	11.112	Herb/Gran	SM	2
Rodentia	Sciuridae	Marmotinae	*Sciurotamias* sp. A (small)		14.279	11.410	7	2.869	0.478	14.757	10.932	Herb/Gran	SM	2
Rodentia	Sciuridae	Marmotinae	*Sciurotamias* sp. B (large)		11.183	7.157	6	4.026	0.805	11.988	6.352	Herb/Gran	SM	2
Rodentia	Sciuridae	Marmotinae	*Tamias* sp., cf. *T. gilaharee*		10.492	7.963	6	2.529	0.506	10.998	7.457	Herb/Gran	SM	2
Rodentia	Sciuridae	Marmotinae	*Tamias urialis* lineage	replaces *T. pre-urialis* and *T. urialis*	14.279	11.183	14	3.096	0.238	14.517	10.945	Herb/Gran	VSM	1
Rodentia	Cricetidae	Cricetinae	*Democricetodon* sp. E		14.279	10.100	19	4.179	0.232	14.511	9.868	Herb/Omn	VSM	1
Rodentia	Cricetidae	Cricetinae	*Democricetodon* sp. A		17.892	13.400	10	4.492	0.499	18.391	12.901	Herb/Omn	VSM	1
Rodentia	Cricetidae	Cricetinae	*Democricetodon kohatensis*		17.892	10.063	26	7.829	0.313	18.205	9.750	Herb/Omn	VSM	1
Rodentia	Cricetidae	Cricetinae	*Democricetodon fejfari*		13.800	8.703	16	5.097	0.340	14.140	8.363	Herb/Omn	VSM	1
Rodentia	Cricetidae	Cricetinae	*Democricetodon* sp. B,–C		16.045	10.492	20	5.553	0.292	16.337	10.200	Herb/Omn	VSM	1

Rodentia	Cricetidae	Megacricetodontinae	Megacricetodon sivalensis		16.800	12.749	13	4.051	0.338	17.138	12.411	Herb/Omn	VSM	1
Rodentia	Cricetidae	Megacricetodontinae	Megacricetodon aguilari		17.892	12.756	17	5.136	0.321	18.213	12.435	Herb/Omn	VSM	1
Rodentia	Cricetidae	Megacricetodontinae	Megacricetodon daamsi		15.169	12.756	9	2.413	0.302	15.471	12.454	Herb/Omn	VSM	1
Rodentia	Cricetidae	Megacricetodontinae	Megacricetodon mythikos		17.892	12.749	15	5.143	0.367	18.259	12.382	Herb/Omn	VSM	1
Rodentia	Cricetidae	Megacricetodontinae	Punjabemys downsi		16.045	13.200	10	2.845	0.316	16.361	12.884	Herb/Omn	VSM	1
Rodentia	Cricetidae	Megacricetodontinae	Punjabemys leptos		16.045	12.300	15	3.745	0.268	16.313	12.033	Herb/Omn	VSM	1
Rodentia	Cricetidae	Megacricetodontinae	Punjabemys mikros		16.045	12.749	8	3.296	0.471	16.516	12.278	Herb/Omn	VSM	1
Rodentia	Spalacidae	Rhizomyinae	Prokanisamys arifi lineage	replaces P. arifi and P. "chinjiarifi"	19.730	13.085	9	6.645	0.831	20.561	12.254	Omn?	VSM	1
Rodentia	Spalacidae	Rhizomyinae	Prokanisamys n. sp.		14.279	12.327	9	1.952	0.244	14.523	12.083	Omn?	VSM	1
Rodentia	Spalacidae	Rhizomyinae	Prokanisamys major		18.690	15.103	6	3.587	0.717	19.407	14.386	Omn?	VSM	1
Rodentia	Spalacidae	Tachyoryctini	Kanisamys sivalensis		9.401	7.337	13	2.064	0.172	9.573	7.165	Root	SM	2
Rodentia	Spalacidae	Tachyoryctini	Kanisamys indicus		16.763	11.400	22	5.363	0.255	17.018	11.145	Root	VSM	1
Rodentia	Spalacidae	Tachyoryctini	Kanisamys nagrii		11.644	9.692	8	1.952	0.279	11.923	9.413	Root	SM	2
Rodentia	Spalacidae	Tachyoryctini	Kanisamys potwarensis		14.279	13.391	6	0.888	0.178	14.457	13.213	Omn?	SM	2
Rodentia	Spalacidae	Tachyoryctini	Protachyoryctes tatroti		7.793	7.096	3	0.697	0.349	8.142	6.748	Root	SM	3
Rodentia	Spalacidae	Tachyoryctini	Eicooryctes kaulialensis		7.815	7.256	4	0.559	0.186	8.001	7.070	Root	SM	2
Rodentia	Spalacidae	Tachyoryctini	Rhizomyides sivalensis		7.256	6.517	3	0.739	0.370	7.626	6.148	Root	SM	3
Rodentia	Spalacidae	Tachyoryctini	Rhizomyides sp. indet.		8.216	8.216	1	0.000	0.000	8.216	8.216	Root	SM	2
Rodentia	Spalacidae	Tachyoryctini	Rhizomyides punjabiensis		11.230	10.100	5	1.130	0.283	11.513	9.818	Root	SM	2
Rodentia	Spalacidae	Rhizomyini	Miorhizomys nagrii		9.706	9.240	5	0.466	0.117	9.823	9.124	Root	SM	3
Rodentia	Spalacidae	Rhizomyini	Miorhizomys micrus		9.333	9.333	1	0.000	0.000	9.333	9.333	Root	SM	2
Rodentia	Spalacidae	Rhizomyini	Miorhizomys blacki		9.258	9.258	1	0.000	0.000	9.258	9.258	Root	SM	3
Rodentia	Muridae	Murinae	Antemus lineage	replaces Antemus chinjiensis, post-Antemus n. sp., pre-Progonomys sp. X	13.800	11.404	12	2.396	0.218	14.018	11.186	Herb/Omn	VSM	1

(continued)

Table 26.S1. continued

Order	Family	Higher Taxon	Species	Records Combined into Lineages	Observed Range (Ma)		No. Time Intervals (0.1 myr) with Locality	Residence Time (obs, myr), FO–LO	Avg Gap Size, Myr	Est FO, Ma[1]	Est LO, Ma[2]	Feeding Category[3]	Size Category[4]	Size Class Number[5]
					Obs FO	Obs LO								
Rodentia	Muridae	Murinae	*Progonomys* lineage	replaces cf. *P. hussaini*, *P. hussaini*, *P. debruijni*, *P.* new sp,	11.200	7.157	15	4.043	0.289	11.489	6.868	Herb/Omn	VSM	1
Rodentia	Muridae	Murinae	*Progonomys morganae*		11.644	10.063	7	1.581	0.264	11.908	9.800	Herb/Omn	VSM	1
Rodentia	Muridae	Murinae	*Karnimata* lineage	replaces *K. fejfari*, *K. darwini*, *K. huxleyi*	10.500	6.517	16	3.983	0.266	10.766	6.251	Herb/Omn	VSM	1
Rodentia	Muridae	Murinae	*Karnimata–Parapelomys* lineage	replaces *Karnimata* large sp., cf. *Parapelomys robertsi*, *P. robertsi*	9.240	6.517	10	2.723	0.303	9.543	6.214	Herb/Omn	VSM	1
Rodentia	Muridae	Murinae	*Parapodemus* lineage	replaces *P. badgleyae* and *P. hariensis*	10.492	7.966	5	2.526	0.632	11.124	7.335	Herb/Omn	VSM	1
Rodentia	Muridae	Murinae	*Mus absconditus*		7.966	7.157	3	0.809	0.405	8.371	6.753	Herb/Omn	VSM	1
Rodentia	Muridae	Murinae	*Mus auctor*		6.517	6.517	1	0.000	0.000	6.517	6.517	Herb/Omn	VSM	1
Rodentia	Muridae	Murinae	"Genus A" and n. sp.		6.517	6.517	1	0.000	0.000	6.517	6.517	Herb/Omn	VSM	1
Rodentia	Murinae	Murinae	"Genus C" and n. sp.		11.183	11.183	1	0.000	0.000	11.183	11.183	Herb/Omn	VSM	1
Rodentia	Murinae	Murinae	"Genus D" and n. sp.		11.644	10.200	6	1.444	0.289	11.933	9.911	Herb/Omn	VSM	1
Rodentia	Murinae	Murinae	*Potwarmus minimus*		17.892	14.039	5	3.853	0.963	18.855	13.076	Herb/Omn	VSM	1
Rodentia	Murinae	Murinae	*Potwarmus primitivus*		17.892	12.300	11	5.592	0.559	18.451	11.741	Herb/Omn	VSM	1
Rodentia	Gerbillinae	Gerbillinae	*Myocricetodon sivalensis*		17.892	12.779	13	5.113	0.426	18.318	12.353	Herb/Omn	VSM	1
Rodentia	Gerbillinae	Gerbillinae	*Myocricetodon* large sp.		16.763	12.749	9	4.014	0.502	17.265	12.247	Herb/Omn	VSM	1
Rodentia	Gerbillinae	Gerbillinae	*Mellalomys perplexus*		17.892	10.492	20	7.400	0.389	18.281	10.103	Herb/Omn	VSM	1
Rodentia	Gerbillinae	Gerbillinae	*Mellalomys lavocati*		16.763	12.377	13	4.386	0.366	17.129	12.012	Herb/Omn	VSM	1
Rodentia	Gerbillinae	Gerbillinae	*Dakkamys* n. sp.		14.279	11.404	7	2.875	0.479	14.758	10.925	Herb/Omn	VSM	1

Order	Family	Subfamily	Species	Notes	Obs FO	Obs LO	N	Gap	Gap	Obs FO + gap[1]	Obs LO − gap[2]	Feeding[3]	Size[4]	Class[5]
Rodentia	Gerbillinae	Gerbillinae	*Dakkamys barryi*		15.169	12.749	9	2.420	0.303	15.472	12.447	Herb/Omn	VSM	1
Rodentia	Gerbillinae	Gerbillinae	*Dakkamys asiaticus*		13.183	10.492	8	2.691	0.384	13.567	10.108	Herb/Omn	VSM	1
Rodentia	Gerbillinae	Gerbillinae	*Paradakkamys chinjiensis*		12.377	10.492	5	1.885	0.471	12.848	10.021	Herb/Omn	VSM	1
Rodentia	Gerbillinae	Gerbillinae	*Abudhabia pakistanensis*		8.703	8.703	1	0.000	0.000	8.703	8.703	Herb/Omn	VSM	1
Rodentia	Ctenodactylidae	Ctenodactylinae	*Sayimys* sp. B		12.377	11.183	2	1.194	1.194	13.571	9.989	Herb	VSM	1
Rodentia	Ctenodactylidae	Ctenodactylinae	*Sayimys perplexus*		7.394	7.394	1	0.000	0.000	7.394	7.394	Herb	VSM	1
Rodentia	Ctenodactylidae	Ctenodactylinae	*Sayimys* lineage	replaces *Sayimys intermedius, S. sivalensis, S. chinjiensis*	18.750	10.063	23	8.687	0.395	19.145	9.668	Herb	VSM	1
Rodentia	Ctenodactylidae	Ctenodactylinae	*Sayimys minor*		21.400	15.962	4	5.438	1.813	23.213	14.149	Herb	VSM	1
Rodentia	Ctenodactylidae	Ctenodactylinae	*Sayimys* cf. *S. giganteus*		16.763	16.465	2	0.298	0.298	17.061	16.167	Herb	SM	2
Rodentia	Diatomyidae		*Diatomys chitaparwalensis*		17.892	17.892	1	0.000	0.000	17.892	17.892	Omn	VSM	1
Rodentia	Diatomyidae		*Willmus maximus*		11.183	11.183	1	0.000	0.000	11.183	11.183	Herb/Omn	SM	2
Rodentia	Gliridae	Myomininae	*Myomimus* lineage	replaces *Myomimus sumbalenwalicus, M.* new sp. A, *M.* new sp. B	13.800	7.966	14	5.834	0.449	14.249	7.517	Omn	VSM	1
Rodentia	Thryonomyidae		*Kochalia* n. sp.		17.892	16.763	2	1.129	1.129	19.021	15.634	Herb	VSM	1
Rodentia	Thryonomyidae		*Kochalia geespei*		16.045	13.391	8	2.654	0.379	16.424	13.012	Herb	VSM	1
Rodentia	Thryonomyidae		*Paraulacodus indicus*		13.085	12.749	3	0.336	0.168	13.253	12.581	Herb	SM	3
Rodentia	Anomaluridae		Anomaluridae, n. gen. and sp. indet.		16.763	16.763	1	0.000	0.000	16.763	16.763	Herb	SM	2
Carnivora	Mustelidae		cf. *Anatolictis* sp. A		9.446	6.517	7	2.929	0.488	9.934	6.029	Carn	SM	3
Carnivora	Mustelidae		cf. *Anatolictis* sp. B		14.138	14.138	1	0.000	0.000	14.138	14.138	Carn	SM	3
Carnivora	Herpestidae		*Urva* small sp.		13.589	6.967	5	6.622	1.656	15.245	5.312	Carn	SM	3

[1] Obs FO + gap size. [2] Obs LO − gap size. [3] Each species was assigned to feeding category of the leading designation, e.g., Gran/Omn assigned to Granivore, etc. [4] SM = small mammal; VSM = very small mammal. [5] Size class number from Chapter 25, this volume, 1: <100 g; 2: 100–249 g; 3: 250–999 g.

Table 26.S2. List of Large Mamals (> 1 kg in adult body weight) with Residence Times in the Analysis Interval

Order	Family	Higher Taxon	Species	Obs Range (Ma) Obs FO	Obs LO	Obs FO Potwar 18.1[1]	Obs LO Potwar 6.0[2]	Comment	No. Time Intervals (0.1 myr) with Locality	Residence Time, (Obs, myr), FO-LO[3]	Avg Gap Size, Myr[4]	Est FO, Ma[5]	Est FO, Ma[6]	Feeding Category[7]	Size Class[8]
OK															
Primates	Sivaladapidae		*Sivaladapis palaeindicus*	15.496	11.644	15.496	11.644	—	8	3.852	0.550	16.046	11.094	Frugivore	5
Primates	Sivaladapidae		*Indraloris kamlialensis*	15.169	15.169	15.169	15.169	—	1	0.000	0.000	15.169	15.169	Frugivore	4
Primates	Catarrhini, i.s.		cf. *Dionysopithecus* sp.	16.000	16.000	16.000	16.000	—	1	0.000	0.000	16.000	16.000	Frugivore	4
Primates	Catarrhini, i.s.		Indet genus indet species	12.088	12.088	12.088	12.088	—	1	0.000	0.000	12.088	12.088	Nd	5
Primates	Hominoidea, i.s.		*Sivapithecus indicus*	12.749	11.404	12.749	11.404	—	9	1.345	0.168	12.917	11.236	Frugivore	7
Primates	Hominoidea, i.s.		*Sivapithecus parvada*	10.063	10.063	10.063	10.063	—	1	0.000	0.000	10.063	10.063	Frugivore	7
Primates	Hominoidea, i.s.		*Sivapithecus sivalensis*	10.228	8.534	10.228	8.534	—	8	1.694	0.242	10.470	8.292	Frugivore	7
Primates	Cercopithecidae	Colobinae	*Mesopithecus sivalensis*	7.691	6.400	7.691	6.400	—	4	1.291	0.430	8.121	5.970	Herbivore-browser	6
Pholidota	Manidae		*Manis* sp.	9.269	7.096	9.269	7.096	—	5	2.173	0.543	9.812	6.553	Invertivore	5
Lagomorpha	Leporidae		*Alilepus elongatus*	7.394	6.517	7.394	6.517	—	4	0.877	0.292	7.686	6.225	Herbivore-mixed feeder	4
Rodentia	Hystricidae		*Hystrix primigenia*	7.999	3.300	7.999	6.000	Count intervals until 6.0 Ma	3	1.999	1.000	8.999	5.001	Herbivore-browser	7
Rodentia	Spalacidae	Rhizomyini	*Miorhizomys tetracharax*	9.377	7.929	9.377	7.929	—	6	1.448	0.290	9.667	7.639	Rootivore	4
Rodentia	Spalacidae	Rhizomyini	*Miorhizomys* cf. *pilgrimi*	9.446	7.394	9.446	7.394	—	5	2.052	0.513	9.959	6.881	Rootivore	5
Rodentia	Spalacidae	Rhizomyini	*Miorhizomys choristos*	8.425	6.967	8.425	6.967	—	4	1.458	0.486	8.911	6.481	Rootivore	4
Creodonta	Hyaenodontidae	Proviverrinae	*Dissopsalis carnifex*	17.000	10.015	17.000	10.015	—	19	6.985	0.388	17.388	9.627	Faunivore	7
Creodonta	Hyaenodontidae	Hyaenodontinae	*Hyainailouros sulzeri*	19.000	11.400	**18.100**	11.400	—	12	6.700	0.609	18.709	10.791	Faunivore	9
Creodonta	Hyaenodontidae	Hyaenodontinae	(?)*Metapterodon* unnamed sp.	13.680	10.063	13.680	10.063	—	4	3.617	1.206	14.886	8.857	Faunivore	6
Carnivora	Carnivora, i.s.		Unnamed gen. and sp.	10.063	10.063	10.063	10.063	—	1	0.000	0.000	10.063	10.063	Faunivore	5

Order	Family	Subfamily	Taxon						N					Diet	
Carnivora	Arctoidea		Indet genus indet species	11.552	11.552	11.552	11.552	—	1	0.000	0.000	11.552	11.552	Faunivore	7
Carnivora	Amphicyonidae	Amphicyoninae	*Amphicyon palaeindicus*	17.892	17.892	11.616	11.616	—	19	6.276	0.349	18.241	11.267	Faunivore	8
Carnivora	Amphicyonidae	Amphicyoninae	*Amphicyon lydekkeri*	12.749	12.749	7.878	7.878	—	17	4.871	0.304	13.053	7.574	Faunivore	9
Carnivora	Amphicyonidae	Amphicyoninae	*Amphicyon giganteus*	19.000	**18.100**	17.892	17.892	Endemic holdover from Early Miocene	1	0.208	0.000	18.100	17.892	Faunivore	9
Carnivora	Amphicyonidae	Amphicyoninae	cf. *Agnotherium antiquum,*	15.818	15.818	10.063	10.063	—	9	5.755	0.719	16.537	9.344	Faunivore	8
Carnivora	Ursidae	Agriotheriinae	*Indarctos punjabiensis*	8.056	8.056	7.766	7.766	—	2	0.290	0.290	8.346	7.476	omnivore	8
Carnivora	Mustelidae	Mustelinae	*Cernictis lydekkeri*	14.000	14.000	12.088	12.088	—	3	1.912	0.956	14.956	11.132	Faunivore	5
Carnivora	Mustelidae	Mustelinae	*Plesiogulo crassa*	6.660	6.660	6.660	6.660	—	1	0.000	0.000	6.660	6.660	Faunivore	7
Carnivora	Mustelidae	Mustelinae	*Circamustela* sp.	9.329	9.329	7.394	7.394	—	2	1.935	1.935	11.264	5.459	Faunivore	4
Carnivora	Mustelidae	Mustelinae	cf. *Anatolictis* sp C	12.290	12.290	12.290	12.290	—	1	0.000	0.000	12.290	12.290	Faunivore	4
Carnivora	Mustelidae	Mellivorinae	cf. *Hoplictis anatolicus*	9.662	9.662	9.240	9.240	—	2	0.422	0.422	10.084	8.818	Faunivore	6
Carnivora	Mustelidae	Mellivorinae	*Eomellivora necrophila*	10.063	10.063	9.356	9.356	—	4	0.707	0.236	10.299	9.120	Faunivore	7
Carnivora	Mustelidae	Lutrinae	*Enhydriodon falconeri*	6.427	6.427	6.427	6.427	—	1	0.000	0.000	6.427	6.427	Faunivore	6
Carnivora	Mustelidae	Lutrinae	*Sivaonyx bathygnathus*	9.415	9.415	6.436	6.436	—	14	2.979	0.229	9.644	6.207	Faunivore	6
Carnivora	Mustelidae	Lutrinae	*Vishnuonyx chinjiensis*	13.652	13.652	11.572	11.572	—	8	2.080	0.297	13.949	11.275	Faunivore	5
Carnivora	Mustelidae	Lutrinae	*Vishnuonyx* large sp.	11.616	11.616	10.063	10.063	—	2	1.553	1.553	13.169	8.510	Faunivore	6
Carnivora	Mustelidae	Lutrinae	cf. *Torolutra* sp.	7.394	7.394	7.157	7.157	—	2	0.237	0.237	7.631	6.920	Faunivore	5
Carnivora	Herpestidae		*Urva* medium sp.	15.103	15.103	6.967	6.967	—	7	8.136	1.356	16.459	5.611	Invertivore	4
Carnivora	Herpestidae		*Urva* large sp.	13.050	13.050	7.158	7.158	—	7	5.892	0.982	14.032	6.176	Invertivore	4
Carnivora	Viverridae		Indet genus large species	17.996	17.996	12.130	12.130	—	3	5.866	2.933	20.929	9.197	Omnivore	7
Carnivora	Viverridae	Viverrinae	*Vishnuictis salmontanus*	10.063	10.063	7.414	7.414	—	8	2.649	0.378	10.441	7.036	Omnivore	5
Carnivora	Viverridae	Viverrinae	*Vishnuictis chinjiensis*	14.299	14.299	10.492	10.492	—	10	3.807	0.423	14.722	10.059	Omnivore	5
Carnivora	Viverridae	Viverrinae	*Vishnuictis hasnoti*	10.492	10.492	6.497	6.497	—	9	3.995	0.499	10.991	5.998	Omnivore	6
Carnivora	Viverridae	Paradoxurinae	*Mioparadoxurus meini*	10.063	10.063	7.304	7.304	—	5	2.759	0.690	10.753	6.614	Frugivore	5
Carnivora	Viverridae	Paradoxurinae	Paradoxurine gen. and sp. indet.	12.088	12.088	9.240	9.240	—	3	2.848	1.424	13.512	7.816	Frugivore	6
Carnivora	Percrocutidae		*Percrocuta carnifex*	14.019	14.019	9.633	9.633	—	13	4.386	0.366	14.385	9.258	Faunivore	7
Carnivora	Percrocutidae		*Percrocuta grandis*	9.428	9.428	6.334	6.334	—	11	3.094	0.309	9.737	6.025	Faunivore	7

(continued)

Table 26.S2. continued

OK

Order	Family	Higher Taxon	Species	Obs Range (Ma) Obs FO	Obs Range (Ma) Obs LO	Obs FO Potwar 18.1[1]	Obs LO Potwar 6.0[2]	Comment	No. Time Intervals (0.1 myr) with Locality	Residence Time, (Obs, myr), F0-LO[3]	Avg Gap Size, Myr[4]	Est FO, Ma[5]	Est FO, Ma[6]	Feeding Category[7]	Size Class[8]
Carnivora	Hyaenidae	Ictitheriinae	"Thalassictis" proava	13.617	10.222	13.617	10.222	—	7	3.395	0.566	14.183	9.656	Faunivore	6
Carnivora	Hyaenidae	Ictitheriinae	Protictitherium sp.	9.676	9.676	9.676	9.676	—	1	0.000	0.000	9.676	9.676	Faunivore	6
Carnivora	Hyaenidae	Ictitheriinae	cf. "Thalassictis" montadai	13.036	11.601	13.036	11.601	—	4	1.435	0.478	13.514	11.123	Faunivore	7
Carnivora	Hyaenidae	Ictitheriinae	Lepthyaena sivalensis	11.971	8.174	11.971	8.174	—	7	3.797	0.633	12.604	7.541	Faunivore	7
Carnivora	Hyaenidae	Ictitheriinae	Lepthyaena (?) pilgrimi	9.248	7.326	9.248	7.326	—	10	1.922	0.214	9.462	7.112	Faunivore	7
Carnivora	Hyaenidae	Hyaeninae	Lycyaena macrostoma	8.739	6.427	8.739	6.427	—	8	2.312	0.330	9.069	6.097	Faunivore	7
Carnivora	Hyaenidae	Hyaeninae	Adcrocuta eximia	8.488	8.056	8.488	8.056	—	2	0.432	0.432	8.920	7.624	Faunivore	7
Carnivora	Barbourofelidae	Barbourofelini	(?) Sansanosmilus serratus	15.169	11.620	15.169	11.620	—	5	3.549	0.887	16.056	10.733	Faunivore	7
Carnivora	Barbourofelidae	Barbourofelini	(?) Barbourofelis piveteaui	11.410	8.569	11.410	8.569	—	8	2.841	0.406	11.816	8.163	Faunivore	7
Carnivora	Felidae	Machairodontinae	Paramachaerodus sivalensis	10.063	7.010	10.063	7.010	—	5	3.053	0.763	10.826	6.247	Faunivore	7
Carnivora	Felidae	Machairodontinae	Sivasmilus copei	14.138	11.410	14.138	11.410	—	6	2.728	0.546	14.684	10.864	Faunivore	7
Carnivora	Felidae	Machairodontinae	cf. Miomachairodus sp.	11.410	7.966	11.410	7.966	—	17	3.444	0.215	11.625	7.751	Faunivore	8
Carnivora	Felidae	Machairodontinae	(?) Amphimachairodus giganteus	9.415	4.657	9.415	**6.000**	Count intervals until 6.0 Ma	7	3.415	0.569	9.984	5.431	Faunivore	8
Carnivora	Felidae	Machairodontinae	cf. Lokotunjailurus sp.	8.056	8.056	8.056	8.056	—	1	0.000	0.000	8.056	8.056	Faunivore	7
Carnivora	Felidae	Felinae	Sivaelurus chinjiensis	13.797	12.377	13.797	12.377	—	3	1.420	0.710	14.507	11.667	Faunivore	7
Carnivora	Felidae		Felidae, indet. small spp.	13.901	7.809	13.901	7.809	Genus and sp indet, <10 kg	12	6.092	0.554	14.455	7.255	Faunivore	5
Carnivora	Felidae		Felidae, indet. medium spp.	21.950	6.517	**18.100**	6.517	Genus and sp indet, 10–30 kg	18	11.583	0.681	18.781	5.836	Faunivore	7
Carnivora	Felidae		Felidae, indet. large spp.	13.751	7.394	13.751	7.394	Genus and sp indet, 30–50 kg	13	6.357	0.530	14.281	6.864	Faunivore	7

Order	Family	Subfamily/Group	Taxon					Endemic holdover from Oligocene						Diet	
Proboscidea	Deinotheriidae		*Prodeinotherium pentapotamiae*	23.000	11.671	**18.100**	11.671		7	6.429	1.072	19.172	10.600	Herbivore-browser	10
Proboscidea	Deinotheriidae	Deinotheriidae	*Deinotherium indicum*	13.148	8.056	13.148	8.056	—	8	5.092	0.727	13.875	7.329	Herbivore-browser	10
Proboscidea	Mammutidae		*Zygolophodon metachinjiensis*	13.838	12.377	13.838	12.377	—	2	1.461	1.461	15.299	10.916	Herbivore-browser	10
Proboscidea	Gomphotheriidae	Choerolophodon-group	*Choerolophodon corrugatus*	14.138	7.365	14.138	7.365	—	20	6.773	0.356	14.494	7.009	Herbivore-browser	10
Proboscidea	Gomphotheriidae	Amebelodontinae	*Protanancus chinjiensis*	13.997	6.517	13.997	6.517	—	7	7.480	1.247	15.244	5.270	Herbivore-browser	10
Proboscidea	Gomphotheriidae	Amebelodontinae	cf. *'Konobelodon'* or *'Torynobelodon'* sp.	8.670	8.670	8.670	8.670	—	1	0.000	0.000	8.670	8.670	Herbivore-browser	10
Proboscidea	Gomphotheriidae	Gomphotheriinae	*Gomphotherium browni*	13.901	11.404	13.901	11.404		6	2.497	0.499	14.400	10.905	Herbivore-browser	10
Proboscidea	Gomphotheriidae	Paratetralophodon-group	*Paratetralophodon hasnotensis*	8.703	7.414	8.703	7.414	—	3	1.289	0.645	9.348	6.773	Herbivore-browser	10
Proboscidea	Gomphotheriidae	Gomphotheriinae	*Anancus perimensis*	8.891	8.891	8.891	8.891	—	1	0.000	0.000	8.891	8.891	Herbivore-browser	10
Proboscidea	Stegodontidae		*Stegolophodon* sp., cf. *S. progressus*	13.652	7.300	13.652	7.300	—	4	6.352	2.117	15.769	5.183	Herbivore-browser	10
Proboscidea	Stegodontidae		*Stegolophodon stegodontoides*	9.401	7.258	9.401	7.258	—	12	2.143	0.195	9.596	7.063	Herbivore-browser	10
Proboscidea	Stegodontidae		*Stegolophodon maluvalensis*	7.414	7.414	7.414	7.414	—	1	0.000	0.000	7.414	7.414	Herbivore-browser	10
Tubulidentata	Orycteropodidae		*Amphiorycteropus* n. sp.	14.147	12.000	14.147	12.000	—	9	2.147	0.268	14.415	11.732	Invertivore	7
Tubulidentata	Orycteropodidae		*Amphiorycteropus browni*	12.669	9.415	12.669	9.415	—	5	3.254	0.814	13.483	8.602	Invertivore	7
Tubulidentata	Orycteropodidae		*Amphiorycteropus* sp. indet	8.000	8.000	8.000	8.000	—	1	0.000	0.000	8.000	8.000	Invertivore	7
Tubulidentata	Orycteropodidae		*Orycteropus* sp.	10.159	8.193	10.159	8.193	—	5	1.966	0.492	10.651	7.702	Invertivore	7
Artiodactyla	Sanitheriidae		*Sanitherium schlagintwetti*	17.965	14.118	17.965	14.118	—	12	3.847	0.350	18.315	13.768	Nd	7
Artiodactyla	Palaeochoeridae	Palaeochoerinae	*Yunnanochoerus gandakasensis*	10.159	8.703	10.159	8.703	—	6	1.456	0.291	10.450	8.412	Herbivore-browser	7
Artiodactyla	Suidae	Listriodontinae	*Listriodon pentapotamiae*	14.138	10.372	14.138	10.372	—	31	3.766	0.126	14.264	10.246	Herbivore-browser	8
Artiodactyla	Suidae	Suinae	*Hippopotamodon* n. sp.	11.410	10.222	11.410	10.222	—	5	1.188	0.297	11.707	9.925	Omnivore	8
Artiodactyla	Suidae	Suinae	*Hippopotamodon sivalense*	10.372	6.436	10.372	6.436	—	26	3.936	0.157	10.529	6.279	Omnivore	9
Artiodactyla	Suidae	Suinae	*Propotamochoerus hysudricus*	10.348	6.631	10.348	6.631	—	32	3.717	0.120	10.468	6.511	Omnivore	8

(continued)

Table 26.S2. continued

Order	Family	Higher Taxon	Species	Obs Range (Ma) Obs FO	Obs Range (Ma) Obs LO	Obs FO Potwar 18.1[1]	Obs LO Potwar 6.0[2]	Comment	No. Time Intervals (0.1 myr) with Locality	Residence Time, (Obs, myr), FO-LO[3]	Avg Gap Size, Myr[4]	Est FO, Ma[5]	Est FO, Ma[6]	Feeding Category[7]	Size Class[8]
OK															
Artiodactyla	Suidae	Suinae	*Sivahyus punjabiensis*	10.228	3.282	10.228	**6.000**	Count intervals until 6.0 Ma	1	4.228	0.000	10.228	6.000	Herbivore-browser	7
Artiodactyla	Suidae	Suinae	Suidae, i.s., n. gen. and sp.	12.756	11.410	12.756	11.410	—	4	1.346	0.449	13.205	10.961	Nd	7
Artiodactyla	Suidae	Suinae	Unnamed gen. and sp.	10.063	10.063	10.063	10.063	—	1	0.000	0.000	10.063	10.063	Nd	7
Artiodactyla	Suidae	Suinae	"Micro-Potamochoerus" n. sp.	9.356	7.471	9.356	7.471	—	6	1.885	0.377	9.733	7.094	Nd	7
Artiodactyla	Suidae	Tetraconodontinae	*Conohyus sindiensis*	15.090	10.105	15.090	10.105	—	26	4.985	0.199	15.289	9.906	Omnivore	8
Artiodactyla	Suidae	Tetraconodontinae	*Parachleuastochoerus pilgrimi*	13.664	11.153	13.664	11.153	—	11	2.511	0.251	13.915	10.902	Omnivore	8
Artiodactyla	Suidae	Tetraconodontinae	*Tetraconodon magnus*	10.015	9.692	10.015	9.692	—	2	0.323	0.323	10.338	9.369	Omnivore	10
Artiodactyla	Anthracotheriidae	Microbunodontinae	*Microbunodon* lineage	28.400	9.248	**18.100**	9.248	Includes *Microbunodon silistrense, M. milaensis*	28	8.852	0.328	18.428	8.920	Herbivore-mixed feeder	7
Artiodactyla	Anthracotheriidae	Bothriodontinae	*Hemimeryx blanfordi*	23.900	17.878	**18.100**	17.878	Endemic holdover from Oligocene	2	0.222	0.222	18.322	17.656	Herbivore-browser	9
Artiodactyla	Anthracotheriidae	Bothriodontinae	*Sivameryx palaeindicus*	24.100	14.138	**18.100**	14.138	—	4	3.962	1.321	19.421	12.817	Herbivore-browser	8
Artiodactyla	Anthracotheriidae	Bothriodontinae	*Merycopotamus* lineage	15.129	3.300	15.129	**6.000**	Includes *M. nanus, M. medioximus, M. dissimilus*	47	9.129	0.198	15.327	5.802	Herbivore-mixed feeder	8

Order	Family	Taxon					Count intervals until 6.0 Ma						Diet	
Artiodactyla	Hippopotamidae	*Hexaprotodon sivalensis*	7.208	3.449	7.208	**6.000**	—	4	1.208	0.403	7.611	5.597	Herbivore–mixed feeder	10
Artiodactyla	Tragulidae	*Dorcabune sindiense*	18.690	13.617	**18.100**	13.617	—	7	4.483	0.747	18.847	12.870	Herbivore-browser	6
Artiodactyla	Tragulidae	*Dorcabune* unnamed sp.	18.500	15.103	**18.100**	15.103	—	2	2.997	2.997	21.097	12.106	Herbivore-browser	7
Artiodactyla	Tragulidae	*Dorcabune anthracotheroides*	14.138	10.581	14.138	10.581	—	24	3.557	0.155	14.293	10.426	Herbivore-browser	7
Artiodactyla	Tragulidae	*Dorcabune nagrii*	11.856	8.544	11.856	8.544	—	14	3.312	0.255	12.111	8.289	Frugivore	7
Artiodactyla	Tragulidae	*Dorcatherium* Form 7	22.140	13.752	**18.100**	13.752	—	15	4.348	0.311	18.411	13.441	Herbivore-browser	5
Artiodactyla	Tragulidae	*Dorcatherium* Form 9	18.734	16.763	**18.100**	16.763	—	2	1.337	1.337	19.437	15.426	Herbivore-browser	6
Artiodactyla	Tragulidae	*Dorcatherium* Form 6	14.279	12.749	14.279	12.749	—	15	1.530	0.109	14.388	12.640	Herbivore-browser	6
Artiodactyla	Tragulidae	*Dorcatherium* Form 4	14.138	11.792	14.138	11.792	—	24	2.346	0.102	14.240	11.690	Herbivore-browser	6
Artiodactyla	Tragulidae	*Dorcatherium* Form 5	12.928	10.486	12.928	10.486	—	21	2.442	0.122	13.050	10.364	Herbivore-browser	5
Artiodactyla	Tragulidae	*Dorcatherium* Form 1?	12.434	10.567	12.434	10.567	—	12	1.867	0.170	12.604	10.397	Herbivore-browser	7
Artiodactyla	Tragulidae	*Dorcatherium* Form 3	10.548	8.794	10.548	8.794	—	17	1.754	0.110	10.658	8.684	Frugivore	5
Artiodactyla	Tragulidae	*Dorcatherium majus*	10.492	7.006	10.492	7.006	—	28	3.486	0.129	10.621	6.877	Herbivore–mixed feeder	7
Artiodactyla	Tragulidae	*Dorcatherium* Form 8	8.794	5.979	8.794	5.979	—	20	2.815	0.148	8.942	5.831	Herbivore-browser	5
Artiodactyla	Tragulidae	*Dorcatherium* Form 2	8.569	7.809	8.569	7.809	—	2	0.760	0.760	9.329	7.049	Herbivore-browser	7
Artiodactyla	Tragulidae	*Siamotragulus minimus*	15.169	12.327	15.169	12.327	—	10	2.842	0.316	15.485	12.011	Herbivore-browser	4
Artiodactyla	Tragulidae	*Siamotragulus* Form D	14.664	11.183	14.664	11.183	—	22	3.481	0.166	14.830	11.017	Herbivore-browser	5
Artiodactyla	Tragulidae	*Siamotragulus* Form C	10.492	6.967	10.492	6.967	—	7	3.525	0.588	11.080	6.380	Herbivore-browser	4
Artiodactyla	Tragulidae	*Siamotragulus nagrii*	9.415	6.967	9.415	6.967	—	15	2.448	0.175	9.590	6.792	Herbivore-browser	4
Artiodactyla	Palaeomerycidae	New genus and n. sp.	11.971	11.971	11.971	11.971	—	1	0.000	0.000	11.971	11.971	Herbivore-browser	9
Artiodactyla	Palaeomerycidae	*Tauromeryx* n. sp.	13.085	13.085	13.085	13.085	—	1	0.000	0.000	13.085	13.085	Herbivore-browser	9
Artiodactyla	Palaeomerycidae	*Tricomeryx* n. sp.	12.088	12.088	12.088	12.088	—	1	0.000	0.000	12.088	12.088	Herbivore-browser	7
Artiodactyla	Palaeomerycidae	cf. *Tricomeryx* n. sp.	11.644	10.063	11.644	10.063	—	2	1.581	1.581	13.225	8.482	Herbivore-browser	7

(continued)

Table 26.S2. continued

Order	Family	Higher Taxon	Species	Obs Range (Ma)		Obs FO Potwar 18.1[1]	Obs LO Potwar 6.0[2]	Comment	No. Time Intervals (0.1 myr) with Locality	Residence Time, (Obs, myr), FO-LO[3]	Avg Gap Size, Myr[4]	Est FO, Ma[5]	Est FO, Ma[6]	Feeding Category[7]	Size Class[8]
				Obs FO	Obs LO										
OK															
Artiodactyla	Palaeomerycidae		cf. *Tauromeryx* sp.	10.063	10.063	10.063	10.063	—	1	0.000	0.000	10.063	10.063	Herbivore-browser	7
Artiodactyla	Palaeomerycidae		cf. *Orygotherium* sp.	7.157	7.157	7.157	7.157	—	1	0.000	0.000	7.157	7.157	Herbivore-browser	6
Artiodactyla	Antilocapridae	Climacocerinae	New genus and sp.	18.500	17.892	**18.100**	17.892	Also from a Zinda Pir Locality	2	0.208	0.208	18.308	17.684	Herbivore-browser	7
Artiodactyla	Antilocapridae	Climacocerinae	*Climacoceras* n. sp.	11.410	10.015	11.410	10.015	—	2	1.395	1.395	12.805	8.620	Herbivore-browser	8
Artiodactyla	Antilocapridae	Lagomerycinae	*Orangemeryx* n. sp.	13.797	13.797	13.797	13.797	—	1	0.000	0.000	13.797	13.797	Herbivore-browser	7
Artiodactyla	Antilocapridae	Lagomerycinae	cf. *Orangemeryx* n. sp.	14.051	12.749	14.051	12.749	—	2	1.302	1.302	15.353	11.447	Herbivore-browser	7
Artiodactyla	Giraffidae	Progiraffinae	*Progiraffa exigua*	18.100	9.415	18.100	9.415	—	21	8.685	0.434	18.534	8.981	Herbivore-browser	7
Artiodactyla	Giraffidae	Giraffokerycinae	*Giraffokeryx punjabiensis*	14.147	11.400	14.147	11.400	—	19	2.747	0.153	14.300	11.247	Herbivore-browser	9
Artiodactyla	Giraffidae	Bohlininae	New genus and n. sp.	13.800	11.400	13.800	11.400	—	10	2.400	0.267	14.067	11.133	Herbivore-browser	9
Artiodactyla	Giraffidae	Bohlininae	*Honanotherium* sp.	14.019	11.410	14.019	11.410	—	3	2.609	1.305	15.324	10.106	Herbivore-browser	9
Artiodactyla	Giraffidae	Giraffinae	*Giraffa punjabiensis*	9.415	8.056	9.415	8.056	—	6	1.359	0.272	9.687	7.784	Herbivore-browser	10
Artiodactyla	Giraffidae	Okapiinae	*Ua pilbeami*	14.051	8.056	14.051	8.056	—	21	5.995	0.300	14.351	7.756	Herbivore-browser	9
Artiodactyla	Giraffidae	Okapiinae	New genus and n. sp.	9.599	9.401	9.599	9.401	—	2	0.198	0.198	9.797	9.203	Herbivore-browser	9
Artiodactyla	Giraffidae	Sivatheriinae	*Decennatherium asiaticum*	13.047	10.063	13.047	10.063	—	3	2.984	1.492	14.539	8.571	Herbivore-browser	9
Artiodactyla	Giraffidae	Sivatheriinae	*Libytherium* sp. 1	10.063	10.063	10.063	10.063	—	1	0.000	0.000	10.063	10.063	Herbivore-browser	10
Artiodactyla	Giraffidae	Sivatheriinae	*Libytherium* sp. 2	9.377	7.328	9.377	7.328	—	4	2.049	0.683	10.060	6.645	Herbivore-browser	10

Artiodactyla	Giraffidae	Sivatheriinae	*Bramatherium megacephalum*	10.222	7.960	—	16	2.262	0.151	10.373	7.809	Herbivore–mixed feeder	10
Artiodactyla	Giraffidae	Sivatheriinae	*Bramatherium* n. sp.	13.652	13.652	—	1	0.000	0.000	13.652	13.652	Herbivore–browser	8
Artiodactyla	Giraffidae	Sivatheriinae	*Helladotherium grande*	10.063	8.751	—	4	1.312	0.437	10.500	8.314	Herbivore–browser	10
Artiodactyla	Giraffidae	Samotheriinae	*Injanatherium hazimi*	12.377	7.471	—	9	4.906	0.613	12.990	6.858	Herbivore–browser	9
Artiodactyla	Bovidae	Hypsodontinae	*Hypsodontus sokolovi*	14.138	12.147	—	6	1.991	0.398	14.536	11.749	Herbivore–mixed feeder	7
Artiodactyla	Bovidae	Hypsodontinae	*Hypsodontus pronaticornis*	13.652	12.800		7	0.852	0.142	13.794	12.658	Herbivore–mixed feeder	8
Artiodactyla	Bovidae	?Hypsodontinae	*?Hypsodontus* sp.	17.900	15.200	—	2	2.700	2.700	20.600	12.500	Nd	7
Artiodactyla	Bovidae	Boselaphini	Indet. genus sp. 1	16.045	15.103	—	4	0.942	0.314	16.359	14.789	Nd	6
Artiodactyla	Bovidae	Boselaphini	*Eotragus noyei*	17.892	16.800	—	3	1.092	0.546	18.438	16.254	Nd	6
Artiodactyla	Bovidae	Boselaphini	*Elachistoceras khauristanensis*	14.138	7.471	—	42	6.667	0.163	14.301	7.308	Herbivore–browser	6
Artiodactyla	Bovidae	Boselaphini	*Strepsiportax* lineage	14.138	11.081	Includes *S.* n. sp. and *S. gluten*	22	3.057	0.146	14.284	10.935	Herbivore–mixed feeder	7
Artiodactyla	Bovidae	Boselaphini	*Sivaceros* lineage	14.138	11.304	Includes *S.* n. sp. and *S. gradiens*	13	2.834	0.236	14.374	11.068	Nd	7
Artiodactyla	Bovidae	Boselaphini	*?(Sivaceros) vedicus*	9.836	9.836	—	1	0.000	0.000	9.836	9.836	Nd	7
Artiodactyla	Bovidae	Boselaphini	*Sivoreas eremita*	14.138	12.088	—	13	2.050	0.171	14.309	11.917	Herbivore–mixed feeder	7
Artiodactyla	Bovidae	Boselaphini	*Helicoportax praecox*	13.400	11.200	—	8	2.200	0.314	13.714	10.886	Nd	7
Artiodactyla	Bovidae	Boselaphini	*Selenoportax vexillarius*	11.000	11.000	—	1	0.000	0.000	11.000	11.000	Nd	8
Artiodactyla	Bovidae	Boselaphini	*Selenoportax* aff. *vexillarius*	10.567	10.015	—	5	0.552	0.138	10.705	9.877	Herbivore–browser	8
Artiodactyla	Bovidae	Boselaphini	*Miotragocerus* lineage	10.632	6.350	Includes *M. pilgrimi* and *M. punjabicus*	32	4.282	0.138	10.770	6.212	Herbivore–mixed feeder	7
Artiodactyla	Bovidae	Boselaphini	*Tragoportax salmontanus*	8.425	7.157	—	7	1.268	0.211	8.636	6.946	Herbivore–mixed feeder	7

(continued)

Table 26.S2. continued

OK

Order	Family	Higher Taxon	Species	Obs Range (Ma) Obs FO	Obs Range (Ma) Obs LO	Obs FO Potwar 18.1[1]	Obs LO Potwar 6.0[2]	Comment	No. Time Intervals (0.1 myr) with Locality	Residence Time, (Obs, myr), FO-LO[3]	Avg Gap Size, Myr[4]	Est FO, Ma[5]	Est FO, Ma[6]	Feeding Category[7]	Size Class[8]
Artiodactyla	Bovidae	Boselaphini	?Boselaphini, unnamed gen. and sp.	9.401	7.328	9.401	7.328	—	4	2.073	0.691	10.092	6.637	Nd	7
Artiodactyla	Bovidae	Boselaphini	Pachyportax falconeri	9.837	8.850	9.837	8.850	—	9	0.987	0.123	9.960	8.727	Herbivore–mixed feeder	8
Artiodactyla	Bovidae	Boselaphini	Pachyportax dhokpathanensis	8.996	7.157	8.996	7.157	—	18	1.839	0.108	9.104	7.049	Herbivore–mixed feeder	9
Artiodactyla	Bovidae	Bovinae?	Bovinae?, genus and species indet.	9.568	7.210	9.568	7.210	—	6	2.358	0.472	10.040	6.738	Nd	8
Artiodactyla	Bovidae	Antilopinae	Unnamed genus and species	14.138	13.007	14.138	13.007	—	3	1.131	0.566	14.704	12.442	Herbivore–browser	6
Artiodactyla	Bovidae	Antilopini	?Gazella sp.	13.200	11.674	13.200	11.674	—	4	1.526	0.509	13.709	11.165	Herbivore–mixed feeder	6
Artiodactyla	Bovidae	Antilopini	Gazella lydekkeri	10.222	6.257	10.222	6.257	—	23	3.965	0.180	10.402	6.077	Herbivore–mixed feeder	6
Artiodactyla	Bovidae	Antilopini	Prostrepsiceros vinayaki	9.391	7.929	9.391	7.929	—	4	1.462	0.487	9.878	7.442	Nd	7
Artiodactyla	Bovidae	Antilopini	?Prostrepsiceros large sp.	8.174	8.174	8.174	8.174	—	1	0.000	0.000	8.174	8.174	Nd	7
Artiodactyla	Bovidae	Antilopini	Nisidorcas planicornis	9.048	9.048	9.048	9.048	—	1	0.000	0.000	9.048	9.048	Nd	7
Artiodactyla	Bovidae	Antilopini	?Protragelaphus sp.	10.759	10.759	10.759	10.759	—	1	0.000	0.000	10.759	10.759	Nd	7
Artiodactyla	Bovidae	Antilopini	Large, unnamed gen. and sp.	9.367	8.794	9.367	8.794	—	3	0.573	0.287	9.654	8.508	Nd	8
Artiodactyla	Bovidae	Reduncinae	Kobus lineage	9.400	7.218	9.400	7.218	Includes K. sp 1, K. porrecticornis and sp. 2	17	2.182	0.136	9.536	7.082	Herbivore–grazer	7
Artiodactyla	Bovidae	Antilopini or Caprini	Tethytragus sp.	13.357	12.327	13.357	12.327	—	5	1.030	0.258	13.615	12.070	Herbivore–mixed feeder	7

Order	Family	Tribe/Subfamily	Taxon						N					Diet	
Artiodactyla	Bovidae	Antilopini or Caprini	*?Caprotragoides potwaricus*	10.222	10.159	10.222	10.159	—	2	0.063	0.063	10.285	10.096	Nd	7
Artiodactyla	Bovidae	Antilopini or Caprini	*?Dorcadoryx* sp.	9.729	9.367	9.729	9.367	—	2	0.362	0.362	10.091	9.005	Nd	7
Artiodactyla	Bovidae	Caprini	*Protoryx* aff. *solignaci*	11.321	11.321	11.321	11.321	—	1	0.000	0.000	11.321	11.321	Nd	7
Artiodactyla	Pecora, i.s.		"Genus A" small sp.	17.892	17.892	17.892	17.892	—	1	0.000	0.000	17.892	17.892	Nd	6?
Perissodactyla	Equidae	Hipparionini	*Cormohipparion* sp.	10.810	10.356	10.810	10.356	—	4	0.454	0.151	10.961	10.205	Herbivore–mixed feeder	8
Perissodactyla	Equidae	Hipparionini	*Cormohipparion* small species	10.042	9.837	10.042	9.837	—	2	0.205	0.205	10.247	9.632	Herbivore–mixed feeder	8
Perissodactyla	Equidae	Hipparionini	*Sivalhippus nagriensis*	10.486	8.749	10.486	8.749	—	16	1.737	0.116	10.602	8.633	Herbivore–mixed feeder	9
Perissodactyla	Equidae	Hipparionini	*Sivalhippus theobaldi*	9.401	7.812	9.401	7.812	—	8	1.589	0.227	9.628	7.535	Herbivore–mixed feeder	9
Perissodactyla	Equidae	Hipparionini	*Sivalhippus perimensis*	8.749	7.338	8.749	7.338	—	12	1.411	0.128	8.877	7.210	Herbivore–mixed feeder	9
Perissodactyla	Equidae	Hipparionini	*Sivalhippus anwari*	7.394	7.157	7.394	7.157	—	3	0.237	0.119	7.513	7.039	Herbivore–mixed feeder	9
Perissodactyla	Equidae	Hipparionini	*Cremohipparion antilopinum*	8.916	6.480	8.916	6.480	—	16	2.436	0.162	9.078	6.318	Herbivore–mixed feeder	8
Perissodactyla	Chalicotheriidae	Chalicotheriinae	*Anisodon salinus*	14.138	7.365	14.138	7.365	—	23	6.773	0.308	14.446	7.057	Herbivore–browser	9
Perissodactyla	Chalicotheriidae	Schizotheriinae	cf. *Ancylotherium* sp.	10.106	10.106	10.106	10.106	—	1	0.000	0.000	10.106	10.106	Herbivore–browser	9
Perissodactyla	Rhinocerotidae	Elasmotheriinae	*Caementodon oettingenae*	15.129	8.591	15.129	8.591	—	20	6.538	0.344	15.473	8.247	Herbivore–mixed feeder	10
Perissodactyla	Rhinocerotidae	Aceratheriina	*Alicornops* cf. *simorrense*	14.138	14.138	14.138	14.138	—	1	0.000	0.000	14.138	14.138	Herbivore–browser	10
Perissodactyla	Rhinocerotidae	Aceratheriina	*Alicornops complanatum*	16.763	7.966	16.763	7.966	—	9	8.797	1.100	17.863	6.866	Herbivore–browser	10
Perissodactyla	Rhinocerotidae	Teleoceratina	*Brachydiceratherium intermedium*	17.878	7.966	17.878	7.966	—	19	9.912	0.551	18.429	7.415	Herbivore–browser	10
Perissodactyla	Rhinocerotidae	Teleoceratina	*Brachydiceratherium fatehjangense*	22.000	7.396	**18.100**	7.396	Endemic holdover from Early Miocene	18	10.704	0.630	18.730	6.766	Herbivore–browser	10

(*continued*)

Table 26.S2. continued

Order	Family	Higher Taxon	Species	Obs Range (Ma) Obs FO	Obs Range (Ma) Obs LO	Obs FO Potwar 18.1[1]	Obs LO Potwar 6.0[2]	Comment	No. Time Intervals (0.1 myr) with Locality	Residence Time, (Obs, myr), FO-LO[3]	Avg Gap Size, Myr[4]	Est FO, Ma[5]	Est FO, Ma[6]	Feeding Category[7]	Size Class[8]
Perissodactyla	Rhinocerotidae	Teleoceratina	*Brachypotherium perimense*	22.000	7.247	**18.100**	7.247	Endemic holdover from Early Miocene	41	10.853	0.271	18.371	6.976	Herbivore-browser	10
Perissodactyla	Rhinocerotidae	Rhinocerotina	*Gaindatherium browni*	16.900	8.731	16.900	8.731	—	13	8.169	0.681	17.581	8.050	Herbivore-browser	10
Perissodactyla	Rhinocerotidae	Rhinocerotina	*Gaindatherium vidali*	14.118	7.966	14.118	7.966	—	8	6.152	0.879	14.997	7.087	Herbivore-browser	10
Perissodactyla	Rhinocerotidae	Rhinocerotina	cf. *Rhinoceros* sp.	20.000	16.045	**18.100**	16.045	Endemic holdover from Early Miocene	2	2.055	2.055	20.155	13.990	Herbivore–mixed feeder	10
Perissodactyla	Rhinocerotidae	Rhinocerotina	*Rhinoceros* aff. *sondaicus*	10.063	10.063	10.063	10.063	—	1	0.000	0.000	10.063	10.063	Herbivore-browser	10
Perissodactyla	Rhinocerotidae	Rhinocerotina	*Rhinoceros* aff. *sivalensis*	8.891	7.961	8.891	7.961	—	3	0.930	0.465	9.356	7.496	Herbivore–mixed feeder	10
Perissodactyla	Rhinocerotidae	Rhinocerotina	*Dicerorhinus* aff. *sumatrensis*	13.652	10.063	13.652	10.063	—	4	3.589	1.196	14.848	8.867	Herbivore-browser	10
Perissodactyla	Rhinocerotidae	Rhinocerotina	*Ceratotherium neumayri*	7.963	7.356	7.963	7.356	—	3	0.607	0.304	8.267	7.053	Herbivore-grazer	10

Note: Boldface numbers indicate residence times that begin earlier than 18.1 Ma or end younger than 6.1 Ma.

[1] Older FO truncated to 18.1 for gap calculation.

[2] Younger LO limited to 6.0 for gap calculation.

[3] FO—LO.

[4] (Residence time / no. time intervals −1) for Potwar record.

[5] Obs FO + gap size.

[6] Obs LO − gap size.

[7] Nd = no data, species excluded from size-feeding categories.

[8] Size class from Chapter 25, this volume, 4: 1.0–2.9 kg, 5: 3.0–7.9 kg, 6: 8–19 kg, 7: 20–99 kg, 8: 100–249 kg, 9: 250–800 kg, 10: >800 kg.

References

Alroy, J. 2010. Fair sampling of taxonomic richness and unbiased estimation of origination and extinction rates. In *Quantitative Methods in Paleobiology*, edited by J. Alroy and G. Hunt, 55–80. The Paleontological Society Papers, Vol. 16.

Anderson, D. C., K. R. Wilson, M. S. Miller, and M. Falck. 2000. Movement patterns of riparian small mammals during predictable floodplain inundation. *Journal of Mammalogy* 81(4): 1087–1099.

Andrews, P., J. M. Lord, and E. M Nesbit Evans. 1979. Patterns of ecological diversity in fossil and modern mammalian faunas. *Biological Journal of the Linnaean Society* 11(2): 177–205.

Badgley, C., J. C. Barry, M. E. Morgan, S. V. Nelson, A. K. Behrensmeyer, T. E. Cerling and D. Pilbeam. 2008. Ecological changes in Miocene mammalian record show impact of prolonged climatic forcing. *Proceedings of the National Academy of Sciences* 105(34): 12145–12149.

Badgley, C., S. M. Domingo, J. C. Barry, M. E. Morgan, L. J. Flynn and D. Pilbeam. 2015. Continental gateways and the dynamics of mammalian faunas. *Comptes Rendus Palevol* 15(7): 763–779.

Badgley, C., and J .A. Finarelli. 2013. Diversity dynamics of mammals in relation to tectonic and climatic history: Comparison of three Neogene records from North America. *Paleobiology* 39(3): 373–399.

Badgley, C., and D. L. Fox. 2000. Ecological biogeography of North American mammals: Species density and ecological structure in relation to environmental gradients. *Journal of Biogeography* 27(6): 1437–1468.

Barry, J. C., A. K. Behrensmeyer, C. E. Badgley, L. J. Flynn, H. Peltonen, I. U. Cheema, D. Pilbeam, et al. 2013. In *Fossil Mammals of Asia: Neogene Biostratigraphy and Chronology*, edited by X. Wang, L. J. Flynn, and M. Fortelius, 373–399. New York: Columbia University Press.

Barry, J. C., M. E. Morgan, L. J. Flynn, D. Pilbeam, A. K. Behrensmeyer, S. M. Raza, I. A. Khan, C. Badgley, J. Hicks, and J. Kelley. 2002. Faunal and environmental change in the Late Miocene Siwaliks of northern Pakistan. *Paleobiology* 28 (2), suppl.

Behrensmeyer, A. K, J. Quade, T. E. Cerling, J. Kappelman, I. A. Khan, P. Copeland, L. Roe, et al. 2007. The structure and rate of Late Miocene expansion of C_4 plants: Evidence from lateral variation in stable isotopes in paleosols of the Siwalik Group, northern Pakistan. *GSA Bulletin* 119(11/12): 1486–1505.

Blanco, F., J. Calatayud, D. M. Martin-Perea, M. S. Domingo, I. Menéndez, J. Müller, M. Hernández-Fernández, and J. L. Cantalapiedra. 2021. Punctuated ecological equilibrium in mammal communities over evolutionary time scales. *Science* 372(6539): 300–303.

Buzan, J. R. and M. Huber. 2021 Moist heat stress on a hotter earth. *Annual Review of Earth and Planetary Sciences* 48:623–655.

Clift, P. D., C. Betzler, S. C. Clemens, B. Christensen, G. P. Eberli, C. France-Lanord, S. Gallagher, et al. 2022. A synthesis of monsoon exploration in the Asian marginal seas. *Scientific Drilling* 31:1–29.

Colwell, R. K., G. Brehm, C. L. Cardelús, A. C. Gilman, and J. T. Longino. 2008. Global warming, elevational range shifts, and lowland biotic attrition in the wet tropics. *Science* 322(5899): 258–261.

Colwell, R. K., C. Rahbek, and N. J. Gotelli. 2004. The mid-domain effect and species richness patterns: What have we learned so far? *American Naturalist* 163(3): E1–E23.

Damuth, J., and B. J. MacFadden. 1990. *Body Size in Mammalian Paleobiology: Estimation and Biological Implications*. New York: Cambridge University Press.

Domingo S. M., C. Badgley, B. Azanza, D. DeMiguel, and M. T. Alberdi. 2014. Diversification of mammals from the Miocene of Spain. *Paleobiology* 40(2): 196–220.

Edwards, A. W. F. 1992. *Likelihood*. Baltimore, MD: Johns Hopkins University Press.

Eisenberg, J. F. 1981. *The Mammalian Radiations: An Analysis of Trends in Evolution, Adaptation, and Behavior*. Chicago: University of Chicago Press.

Elliott, A., and A. Villalta. 2020. *Mammals of South Asia*. Lynx Edicions, Barcelona, Spain.

Emmons, L. H , A. Gautier-Hion, and G. Dubost. 1983. Community structure of the frugivorous-folivorous forest mammals of Gabon. *Journal of Zoology, London* 199(2): 209–222.

Ernest, S. K. M. 2013. Using size distributions to understand the role of body size in mammalian community assembly. In *Animal Body Size*, edited by F. A. Smith and S. K. Lyons, 147–167. Chicago: University of Chicago Press.

Faith, J. T., J. Rowan, and A. Du. 2019. Early hominins evolved within non-analog ecosystems. *Proceedings of the National Academy of Sciences* 116(43): 21478–21483.

Feakins, S. J., H. M. Liddy, L. Tauxe, V. Galy, X. Feng, J. E. Tierney, Y. Miao, and S. Warny. 2020. Miocene C_4 grassland expansion as recorded by the Indus Fan. *Paleoceanography and Paleoclimatology* 35(6):e2020PA003856.

Flynn, L J., W. Downs, M. E. Morgan, J. C. Barry, and D. Pilbeam. 1998. High Miocene species richness in the Siwaliks of Pakistan. In *Advances in Vertebrate Paleontology and Geochronology*, edited by Y. Tomida, L. J. Flynn, and L. L. Jacobs, 167–180. Tokyo: National Science Museum Monograph 14.

Foote, M. 2000. Origination and extinction components of taxonomic diversity: General problems. *Paleobiology* 26(4): 74–102.

Fraser, D. et al. 2021. Biotic interactions in deep time. *Trends in Ecology and Evolution* 36(1): 61–75.

Freedman, B. 2015. Ecological effects of environmental stressors. In *Oxford Research Encyclopedias: Environmental Science*. https://doi.org/10.1093/acrefore/9780199389414.013.1.

Foster, G. L., C. H. Lear, and J. W. B. Rae. 2012. The evolution of pCO₂, ice volume, and climate during the Middle Miocene. *Earth and Planetary Science Letters* 341–344:243–254.

Golet, G. H., J. W. Hunt, and D. Koenig. 2013. Decline and recovery of small mammals after flooding: Implications for pest management and floodplain community dynamics. *River Research and Applications* 29(2): 183–194.

Grossnickle, D. M., S. M. Smith, and G. P. Wilson. 2019. Untangling the multiple ecological radiations of early mammals. *Trends in Ecology and Evolution* 34(10): 936–949.

Heim de Balsac, H., and J. Verschuren. 1968. *Exploration du Parc National de la Garamba, Congo: Mission H. de Saeger (1949–1952)*. Vol. 54, Insectivores. Kinshasa: République démocratique du Congo, Institut des Parcs Nationaux.

Hendrichs, H. 1970. Schätzungen der Huftierbiomasse in der Dornbusch-Savanne nördlich und westlich der Serengetisteppe in Ostafrika nach einem neuen Verfahren und Bemerkungen zur Biomasse der anderen pflanzenfressenden Tierarten. *Säugetierkundliche Mitteilungen* 18:237–255.

Holbourn, A., W. Kuhnt, S. Clemens, W. Prell, and N. Andersen. 2013. Middle to Late Miocene stepwise climate cooling: Evidence from a high-resolution deep-water isotope curve spanning 8 million years. *Paleoceanography* 28(4): 688–699.

Huber, B. T., K. G. MacLeod, D. K. Watkins, and M. F. Coffin. 2018. The rise and fall of the Cretaceous Hot Greenhouse climate. *Global and Planetary Change* 167:1–23.

Hughes, F. M. R. 1997. Floodplain biogeomorphology. *Progress in Physical Geography* 21(4): 501–529.

Hui, Z., J. Zhang, Z. Ma, X. Li, T. Peng, J. Li, and B. Wang. 2018. Global warming and rainfall: Lessons from an analysis of Mid-Miocene climate data. *Palaeogeography, Palaeoclimatology, Palaeoecology* 512:106–117.

Hyvarinen, O., M. Te Beest, E. le Roux, G. Kerley, E. de Groot, R. Vanita, and J. P. G. M. Cromsigt. 2021. Megaherbivore impacts on ecosystem and Earth system functioning: The current state of the science. *Ecography* 44(11): 1579–1594.

Janis, C. M., J. Damuth, and J. M. Theodor. 2000. Miocene ungulates and terrestrial primary productivity: Where have all the browsers gone? *Proceedings of the National Academy of Sciences* 97(14): 7899–7904.

Karp, A. T., A. K. Behrensmeyer, and K. H. Freeman. 2018. Grassland fire ecology has roots in the Late Miocene. *Proceedings of the National Academy of Sciences* 115(48): 12130–12135.

Karp, A. T., K. T. Uno, P. J. Polissar, and K. H. Freeman. 2021. Late Miocene C_4 grassland fire feedbacks on the Indian subcontinent. *Paleoceanography and Paleoclimatology* 36(4): e2020PA004106. https://doi.org/10.1029/2020PA004106.

Kimura, Y., L. L Jacobs, T. E. Cerling, K. T. Uno, K. M. Ferguson, L. J. Flynn, and R. Patnaik. 2013. Fossil mice and rats show isotopic evidence of niche partitioning and change in dental ecomorphology related to dietary shift in Late Miocene of Pakistan. *PLOS ONE* 8(8): e69308.

Kingdon, J. 1974a. *East African Mammals. An Atlas of Evolution in Africa. IIA. Insectivores and Bats*. London: Academic Press.

Kingdon, J. 1974b. *East African Mammals. An Atlas of Evolution in Africa. IIB. Hares and Rodents*. London: Academic Press.

Madern, P. A., and L. W. van den Hoek Ostende. 2015. Going south: Latitudinal change in mammalian biodiversity in Miocene Eurasia. *Palaeogeography, Palaeoclimatology, Palaeoecology* 424:123–131.

Magige, F., and R. Senzota. 2006. Abundance and diversity of rodents at the human-wildlife interface in western Serengeti, Tanzania. *African Journal of Ecology* 44(3): 371–378.

Marshall, C. R. 1990. Confidence intervals on stratigraphic ranges. *Paleobiology* 16(1): 1–10.

Marshall, C. R. 2010. Using confidence intervals to quantify the uncertainty in the end-points of stratigraphic ranges. In *Quantitative Methods in Paleobiology*, edited by J. Alroy and G. Hunt, 291–316. The Paleontological Society.

Merritt, J. F. 2010. *The Biology of Small Mammals*. Baltimore, MD: Johns Hopkins University Press.

Misonne, X., and J. Verschuren. 1966. Les rongeurs et lagomorphes de la région du Parc National du Serengeti (Tanzanie). *Mammalia* 30(4): 517–537.

Morgan, M. E., A. K. Behrensmeyer, C. Badgley, J. C. Barry, S. Nelson, and D. Pilbeam. 2009. Lateral trends in carbon isotope ratios reveal a Miocene vegetation gradient in the Siwaliks of Pakistan. *Geology* 37(2): 103–106.

Nelson, S. V. 2007. Isotopic reconstructions of habitat change surrounding the extinction of *Sivapithecus*, a Miocene hominoid, in the Siwalik Group of Pakistan. *Palaeogeography, Palaeoclimatology, Palaeoecology* 243:204–222.

Opdyke, N. D., N. M. Johnson, G. D. Johnson, E. H. Lindsay, and R. A. K. Tahirkheli. 1982. Paleomagnetism of the Middle Siwalik formations of northern Pakistan and rotation of the Salt Range Decollement. *Palaeogeography, Palaeoclimatology, Palaeoecology* 37:1–15.

Owen-Smith, R. N. 1988. *Megaherbivores*. Cambridge: Cambridge University Press.

Peters, R. H. 1983. *The Ecological Implications of Body Size*. Cambridge: Cambridge University Press.

Polissar, P. J., K. T. Uno, S. R. Phelps, A. T. Karp, K. H. Freeman, and J. L. Pensky. 2021. Hydrologic changes drove the Late Miocene expansion of C_4 grasslands on the northern Indian subcontinent. *Paleoceanography and Paleoclimatology* 36:e2020PA004108.

Quade, J., and T. E. Cerling. 1995. Expansion of C_4 grasses in the Late Miocene of northern Pakistan: Evidence from stable isotopes in paleosols. *Palaeogeography, Palaeoecology* 115(1–4): 91–116.

Qui, Z., and Z. Qui. 1995. Chronological sequence and subdivision of Chinese Neogene mammalian faunas. *Palaeogeography, Palaeoclimatology, Palaeoecology* 116(1–2): 41–70.

Rabosky, D. L. 2013. Diversity-dependence, ecological speciation, and the role of competition in macroevolution. *Annual Review of Ecology, Evolution, and Systematics* 44:481–502.

Sheppe, W., and T. Osborne. 1971. Patterns of use of a floodplain by Zambian mammals. *Ecological Monographs* 41(3): 179–205.

Steinthorsdottir, M., H. K. Coxall, A. M. DeBoer, M. Huber, N. Barbolini, C. D. Bradshaw, N. J. Burls, et al. 2021. The Miocene: The future of the past. *Paleoceanography and Paleoclimatology* 36(4): e2020PA004037.

Tauxe, L., and S. J. Feakins. 2020. A re-assessment of the chronostratigraphy of Late Miocene C_3–C_4 transitions. *Paleoceanography and Paleoclimatology* 35(7):e2020PA003857.

Van Peenen, P. F. D. 1969. *Preliminary Identification Manual for mammals of South Vietnam*. Washington, D.C.: Smithsonian Institution.

Verheyen, W. and J. Verschuren. 1966. Exploration du Parc National de la Garamba, Congo: Mission H. de Saeger (1949–1952). Vol. 50. Rongeurs et lagomorphes. Bruxelles: Institut des Parcs Nationaux du Congo.

Verschuren, J. 1958. Exploration du Parc National de la Garamba, Congo: Mission H. de Saeger (1949–1952). Vol. 9. Ecologie et biologie des grands mammifères (primates, carnivores, ongulés). Bruxelles: Institut des Parcs Nationaux du Congo, 225 pp.

Wijnhoven, S., G. van der Velde, R. S. E. W. Leuven, and A. J. M. Smits. 2005. Flooding ecology of voles, mice and shrews: The importance of geomorphological and vegetational heterogeneity in river floodplains. *Acta Theriologica* 50(4): 453–472.

Willis, B. 1993. Evolution of Miocene fluvial systems in the Himalayan foredeep through a two-kilometer-thick succession in northern Pakistan. *Sedimentary Geology* 88(1–2): 77–121.

Zhang, Y., I. Held, and S. Feuglistaler. 2021. Projections of tropical heat stress constrained by atmospheric dynamics. *Nature Geoscience* 14:133–137.

HIGHLIGHTS OF THE SIWALIK RECORD AND FUTURE RESEARCH OPPORTUNITIES

CATHERINE BADGLEY, MICHÈLE E. MORGAN,
AND DAVID PILBEAM, WITH CONTRIBUTIONS FROM
JOHN C. BARRY, ANNA K. BEHRENSMEYER,
LAWRENCE J. FLYNN, AND S. MAHMOOD RAZA

WHAT CAN WE LEARN FROM STUDYING A LONG FOSSIL RECORD?

A long continuous fossil record offers uniquely valuable insights into ecological and evolutionary properties and processes that are not apparent in present-day ecosystems or in short, isolated records. This statement requires that the record is consistently productive over its duration and that the preserved fossils have taxonomically informative traits. A long record that meets these criteria (such as the Siwalik record of the Potwar Plateau) documents occurrences of lineages, their traits, and assemblages of species over millions of years, providing information about species longevity, rates of trait evolution, the degree of continuity of biotic assemblage composition, and the taxonomic and functional richness of well-preserved clades. With adequate sample sizes, dense sampling through time, and appropriate attention to potential taphonomic biases, microevolutionary change ("adaptive modifications within populations over time" [Reznick and Ricklefs 2009, 837]) or stability can be assessed in relation to the evolutionary fate of particular lineages within a region. Relative abundance can be estimated with potential corrections for probabilities of preservation that vary in relation to body size or habitat. Long records also provide information about community assembly over millennia to millions of years. For example, we can evaluate the relative contributions of immigrant versus endemic originations to species richness. The assembly and disassembly of communities can be assessed in relation to hypotheses of environmental filtering, demographic dominance, or ecological interactions, and the appearance of evolutionary novelties (Swenson 2011). Likewise, the cohesion of biotic composition can be quantified to test whether communities are equilibrial or dynamic (e.g., Blanco et al. 2021). When local or regional environmental data are available, it is possible to assess which aspects of biotic composition appear to track changes in environmental conditions and how lineages vary in persistence during environmental changes. Certain species interactions—such as direct evidence of predation or indirect evidence of resource competition—can be plausibly inferred (Fraser et al. 2021). Assessment of these ecological processes offers opportunities to test hypotheses

about the factors that regulate diversity at the regional ecosystem scale (Benson et al. 2021)—the scale at which population size, geographic range, speciation, and extinction interact. In addition, a long record documents species or genus richness over time, which can be compared with contemporaneous fossil records in nearby basins or on different continents for evidence of regional or global richness gradients or with modern assemblages in broadly similar environments (e.g., humid megathermal environments, mesothermal deserts). At the scale of continents (or ocean basins), multiple long records provide information about the spatial boundaries of distinct biotas in relation to topography, tectonic plate configurations, and climatic gradients. The Siwalik record of northern Pakistan has afforded us the opportunity to investigate many of these properties and processes.

HIGHLIGHTS FROM THE SIWALIK RECORD OF THE POTWAR PLATEAU

Many of the emergent insights from the bounty of the Potwar Siwalik record involve changes over time, although we have also learned details of the spatial ecology and community composition at particular time intervals in our sequence. The ability to anchor fossil localities and specimens in time depends on the detailed chronology that magnetic polarity stratigraphy and the precise stratigraphic position of fossil localities have afforded us. The numerous polarity reversals of the Miocene, including many short-term events, along with multiple stratigraphic reference sections (Chapter 2, this volume) have enabled us to place the majority of fossil localities in stratigraphic position with low measurement error and to estimate the age of most localities to within a particular 100-thousand year (kyr) time interval (Chapter 3, this volume). The commitment to measuring multiple sections and sampling many of them for paleomagnetic data has rewarded us with highly resolved views of lithological, paleoenvironmental, and faunal change, as well as an understanding of the relative persistence of taphonomic processes in the Siwalik depositional system.

Fluvial systems that deposited the Siwalik rock record began to fill basins created by the collision of India with southern Asia more than 25 million years ago and persist up to the present day. Paleorivers of the Potwar Siwaliks flowed eastward into the Ganges-Brahmaputra drainage system until the Late Pliocene, when directions shifted toward the southwest into the Indus River (Burbank, Beck, and Mulder 1996). Deposition of vast quantities of sediment eroded from the rising Himalayas was nearly balanced by steady subsidence of the western foreland basin, with accumulation rates increasing through time from <20 cm/1,000 years to >60 cm/1,000 yr. Our record shows the shifting dominance of mountain- and foothill-sourced river systems over the 14-myr sequence of Kamlial, Chinji, Nagri, and Dhok Pathan Formations (Chapter 2, this volume). Remarkably, in spite of these major changes in lithology and sediment accumulation rates, patterns of fossil preservation were relatively constant, as reflected in the taphonomic features of the vertebrate assemblages (Chapter 4, this volume) and the consistent absence of plant fossils. For the vertebrate record, this means that although the numbers of specimens and localities vary through time (Chapter 3, this volume), the samples that we use to examine long-term evolutionary and ecological patterns (Chapters 25 and 26, this volume) are similar in terms of taphonomic biases. Establishing the history of fossil preservation in our record has strengthened our ability to reconstruct the history of Siwalik biotic change.

Below we summarize highlights from our faunal analyses, primarily of mammalian faunas.

Species Durations (Residence Times)

Observed residence times range from 0.1 myr (a single 100-kyr time interval) to over 10.0 myr, if we assume range-through presence in the Siwalik bioprovince. Estimated residence times are slightly to considerably longer depending on the number of time intervals in which each species occurs (since estimated first and last appearances are based on the mean gap size between occurrences; Chapter 26, this volume). For large mammals (≥1 kg), 56 of 194 species (29%) have observed residence times of less than 1.0 myr; the mean residence time is 2.9 myr. For small mammals (<1 kg), 32 of 100 lineages (32%) have residence times of less than 1.0 myr; the mean residence time is 2.7 myr. Thus, for all mammals, about 30% of species have a geologically brief presence in the Siwalik record and likely also in the original ecosystems. In addition, residence times were greater, on average, for species present prior to 14 Ma compared to younger time intervals, and mean residence times differ among families of small and large mammals (Chapter 25, this volume). The longer mean residence times and notable lineages that persist through the major environmental transitions—and give the impression of faunal resilience during environmental change—are strongly influenced by the minority of species with very long residence times. For example, the giraffid *Progiraffa exigua* has a residence time (based on observed first and last occurrences) of 8.68 myr; the rhino *Brachydiceratherium perimense* has a residence time of 10.85 myr.

Species Richness

The pattern of species richness, determined from overlapping residence times, offers several notable highlights. One is the extraordinary richness of mammal species at the peak of richness 13.1 million years ago (106 species for observed richness, 116 species for estimated richness; Chapters 25 and 26, this volume). These values do not include bats or cryptic species that differ in features other than dental morphology or horn cores. Other Miocene continental records also have high richness of the regional community compared to Late Neogene or modern faunas (e.g., Janis, Damuth, and Theodor 2000, Fortelius et al. 2014). The Siwalik record is noteworthy for species-level documentation of its mammal faunas.

A second highlight is the remarkable number of co-occurring megaherbivores. Certain time intervals have 18 species, and individual fossil localities have up to 8 species of mammals with estimated body weights >800 kg (Chapter 25, this volume). These megaherbivores belong to six families in three orders: Artiodactyla (hippopotamids and giraffids), Perissodactyla (rhinos), and Proboscidea (deinotheres, gomphotheres, elephantoids). Many of these species have very long residence times (Preamble III, Table III.1). Their likely ecological impacts on vegetation growth and productivity through foraging and nutrient and seed dispersal, as well as on substrates through trampling and wallowing, contributed to their roles as ecosystem engineers.

What kind of ecosystem can support such high total species richness and high richness of megaherbivores? The presence of megaherbivore eco-engineers suggests a feedback system operating between vegetation and fauna that may have enhanced plant productivity. Three additional factors come to mind, most efficacious if all were operating. One is high primary productivity in three dimensions, from aquatic vegetation to ground-level herbs and shrubs to tree canopies. Plentiful water over much of the annual cycle would be needed to support this multistory vegetation, as found in humid and mesic forests today. Second, many kinds of faunivores preying on the herbivores would limit population sizes of the many herbivorous species across the body-size spectrum. A third possibility is that some large to mega-sized species were seasonal occupants of the floodplain ecosystem rather than permanent residents. Counter to this possibility, however, a geographic gradient in stable isotopes of carbon among Late Miocene fossil localities indicated habitat fidelity on the scale of tens of km² (Morgan et al. 2009); this study included two megaherbivores (one elephantoid, one rhinocerotid), which suggested that these animals were long-term residents of the floodplain.

Equilibrial versus Dynamic Composition over Time

A persistent theme in the analysis of paleocommunities is the presence of stable community composition over millions of years. This phenomenon has been called *chronofauna* (Olson 1952), *coordinated stasis* (Brett, Ivany, and Schopf 1996), and recently *equilibrial fauna* (Blanco et al. 2021). Such cohesive biotas are documented over millions of years, then show a rapid change in taxonomic or functional composition. Changes like these have been linked to major facies transitions and new ocean gateways that enhance invasions in marine systems (e.g., Harzhauser and Piller 2007; Stigall 2019) and to climatic and tectonic transitions in continental settings (e.g., Figueirido et al. 2019; Blanco et al. 2021).

Siwalik mammalian faunas show a different pattern. Between 18.0 and 6.0 Ma, species richness increased and decreased over millions of years, with intervals in between of high but only brief stable richness (Chapter 26, this volume). Even during the intervals of stable richness, both originations and extinctions continued. Although the dynamics of richness through time differed in substantial ways for small versus large mammals (Chapter 26, this volume), both size groups showed persistent turnover in species composition. Some of the major pulses in originations and extinctions coincided with significant regional environmental transitions. One major turnover in large mammals occurred during an interval without local evidence of changing climate or vegetation but coincided with global cooling and sea-level fall in the marine record (Holbourn et al. 2013) and may represent an interval when new dispersal corridors facilitated an influx of immigrants (Chapter 25, this volume).

Functional versus Taxonomic Composition

The analysis of size and feeding habit of assemblages (Chapters 25, 26, this volume) shows that the functional composition of mammalian faunas had greater persistence than taxonomic composition. Change in functional composition was less volatile than change in taxonomic composition for both small and large mammals (Figs. 26.6–26.9, this volume). This finding is consistent with the analysis of Iberian mammal faunas over the last 21 myr (Blanco et al. 2021) and supports the emphasis on the value of trait-based analyses of paleobiotas. (See "Comparison to Other Continental Sequences" section in this chapter for comparison of the Siwalik and Iberian records.)

The Siwalik record shows a sustained period of taxonomic and functional enrichment, with the greatest increase in species richness occurring at the end of the Miocene Climatic Optimum between 14.7 and 13.9 Ma, an interval of 800 kyr (Figs. 26.11, 26.12, this volume). Functional richness increased more slowly than taxonomic richness for large mammals (Fig. 26.12c, this volume). For small mammals, the number of size-feeding categories in the fauna was highest after the time of peak taxonomic richness, with both peaks occurring much earlier for small mammals than large mammals. For large mammals, the peak of functional richness in the Late Miocene (9.5 Ma) slightly preceded the peak in species richness (9.1 Ma). Also, the size range of ungulates expanded through anagenetic size increase and the appearance of new, larger species (Chapter 25, this volume). From 10.3 to 7.7 Ma, taxonomic richness of small mammals declined dramatically, whereas functional richness persisted; this pattern indicates a thinning of species in functional categories until about 7.1 Ma, when many functional categories disappeared (Fig. 26.11, this volume). For large mammals, species richness began a steep decline at 8.5 Ma,

whereas functional richness remained high until 7.1 Ma and declined steadily thereafter (Fig. 26.12, this volume).

Dynamics of Faunal Change in Small and Large Mammals

The diversity dynamics differed between small mammals and large mammals. We analyzed data for these size groups separately for several reasons. The two datasets were retrieved by different collection processes—screen-washing for small mammals versus surface collection and a few excavations for large mammals. We maintained the same collecting methods throughout the Siwalik sequence, and the depositional environments of fossil assemblages were similar across all four formations studied (Chapter 4, this volume). Small and large mammals differ in life histories, population size, and habitat use. Small mammals generally have higher mass-specific metabolic rates, shorter life spans, higher reproductive rates, and higher population densities than do large mammals (Ernest 2013). Also, the life habits of many small mammals are closely tied to properties of the substrate, including soil texture and composition and canopy height of vegetation (Merritt 2010).

Estimated species richness was initially low for both small and large mammals and remained low from 18.0 to 14.7 Ma. Then richness rose steeply for both groups—a pattern consistent with an increase in fossil productivity, a major change in macrofacies, and a global cooling interval at the end of the Miocene Climatic Optimum (Chapter 26, this volume). From that time onward, the two records of species richness differed. Small mammals had an early peak of species richness between 14.3 and 12.9 Ma, then began to decline—in both large steps and gradual loss of species. In contrast, large mammals had two broad peaks of species richness, the first between 13.3 and 11.9 Ma and the second between 10.1 and 8.7 Ma (Fig. 26.5, this volume), then underwent a steep decline. Small mammals showed more stability in functional composition than large mammals. The major pulses of change in both species richness and functional composition occurred asynchronously more than synchronously. These differences suggest either that causes of faunal change differed in these two broad size categories or that small mammals responded differently than large mammals to common causes.

Causes of Ecosystem Change

The likely causes of ecosystem change, especially in the richness and composition of mammal faunas, were multidimensional and included processes internal to Siwalik ecosystems and others at regional or global scales (Chapters 24–26, this volume). Biotic processes within the Siwalik ecosystems include the effects of megaherbivores as ecosystem engineers, predation pressure from faunivorous vertebrates, competitive interactions among herbivores, and change in vegetation structure and composition. (We focused analysis of competition on herbivores because of the high frequency of specimens but competition was also likely among the rarer carnivores.) Physical processes with both internal and external controls include the fluvial system and regional climate, which were likely the dominant influences on vegetation history. Increased seasonal migration would potentially contribute to increased taxonomic and functional richness by expanding the geographic resource base for the migrating species. Longer dry seasons would have encouraged and facilitated this pattern for mobile species.

Another external physical influence was the fluctuating permeability of mountain barriers around the northwestern, northern, and northeastern margins of South Asia (Indian subcontinent). The main (indirect) evidence for the opening of corridors—either through the mountains or along coastal routes—is the occasional pulses of immigrant lineages with documented ancestors in distant regions. Two examples, based on estimated first appearances, include the pulse of originations between 14.7 and 13.9 Ma, with an influx of immigrants as well as new endemics, and between 11.0 and 10.0 Ma, with both immigrant and endemic first appearances, especially among ungulates. Concurrent size change within lineages as well as first appearances of large species within existing families (Chapter 25, this volume) suggest that changes in vegetation structure toward more open-canopy habitats, as documented in the biomarker record (Karp, Behrensmeyer, and Freeman 2018; Srivastava et al. 2018; Chapter 24, this volume), selected for larger species that can tolerate greater fiber content in their diet than can smaller species (Demment and Van Soest 1985).

The Siwalik record offers more direct evidence of physical rather than biotic drivers of faunal change. Isotopic analyses of mammalian tooth enamel and paleosol carbonates, several forms of biomarker data, and changes in lithofacies are the basis for inferring qualitative and quantitative changes in climate, vegetation, fire frequency, and fluvial systems. While major transitions in faunal richness and composition (of mammals, birds, and reptiles) coincided with episodes of environmental change, the biotic processes mentioned above were likely important contributing factors during times of both faunal change and stability.

DIFFERENT APPROACHES TO ANALYSIS OF FAUNAL CHANGE

Chapters 25 and 26 in this volume take different approaches to the analysis of mammalian faunal change. Here we note the findings in common and the unique insights of each approach. These chapters differ in taxonomic data. Chapter 25 evaluates all mammals except bats and carnivorans and much of the analysis focuses on ungulates and rodents. Three size groups are recognized: small (<1 kg), large (1–800 kg), and mega (>800 kg). Chapter 26 evaluates all mammals except bats and presents separate analyses of small mammals (<1 kg) and large mammals (all species ≥1 kg). Chapter 25 evaluates observed residence times of species and includes some abundance data (proportion of catalogued fossil specimens) in selected analyses. In Chapter 26, residence times are estimated statistically and are somewhat to substantially greater than observed residence times. Finally, Chapter 25 analyzes six time slices of 0.1

myr duration within a well-sampled 7-myr interval. Chapter 26 presents diversity dynamics for a continuous record, compiled in 0.2-myr intervals, over 12 million years that include intervals of low and high fossil productivity.

Despite these differences in datasets and methods, there are noteworthy findings in common. Both analyses recognize an early increase and a late decrease in species richness. The timing of peak richness varies slightly, mainly because of the different presentation of residence times. Both chapters document asynchronous patterns of species richness among major size groups of mammals. Chapter 25 shows that megaherbivore richness is the most stable over millions of years. Many megaherbivore species have long residence times, some longer than 6 myr, with low abundances at all times, the latter partly a demographic consequence of their large body size and possibly a preservation bias against large skeletal elements that remained exposed rather than buried rapidly. Some small mammals also have long residence times in both analyses. Also, both analyses recognize brief intervals with multiple first occurrences of immigrants, suggesting episodes of permeability in montane barriers or expansion of coastal corridors.

The different approaches have unique findings as well. In Chapter 25, the first occurrences of many lineages are distinguished in terms of immigrant or endemic origin. Immigrant lineages have shorter residence times than endemics in the Potwar Plateau, which suggests an incumbency effect (van der Meulen, Peláez-Campomanes, and Levin 2005) based on competitive interactions over habitat use. Some immigrant lineages became very successful in terms of abundance, cladogenesis, or residence time. Several genera of sciurids and hipparionine equids are prime examples. In Chapter 26, functional composition as represented by joint size-feeding categories had greater persistence than taxonomic composition; the number of functional categories increased and decreased over 12 myr more slowly than did species richness.

Both chapters present interpretations of particular episodes of environmental change as the major driver of faunal change, mediated through changes in food and substrate resources. For example, the responses of large mammals to Late Miocene aridification and increased seasonality of moisture are an example of strong environmental filtering (Badgley et al. 2008). Chapter 26 additionally posits that the Middle Miocene increase in taxonomic and functional richness beginning at 14.7 Ma, as well as the concurrent increase in fossil productivity, are a response both to global cooling, with more habitable temperatures for large mammals leading to greater occupancy of the floodplain and to geographic-range shifts within the greater Siwalik bioprovince and from outside South Asia.

COMPARISON TO OTHER CONTINENTAL SEQUENCES

Few fossiliferous sequences are long enough and well enough calibrated chronologically to compare to the Siwalik record. A series of comparisons of the Potwar Siwalik record with the Late Cretaceous to Early Paleogene record of the Bighorn

Basin (northwestern Wyoming, USA), which occurs in a continental basin (originally a foreland basin, now intermontane), revealed the wealth of information documented in each sequence and substantial differences in environments of fossil preservation (Badgley et al. 1995), size and trophic structure of mammalian faunas (Gunnell et al. 1995; Morgan et al. 1995), and overall macroevolutionary patterns (Badgley and Behrensmeyer 1995). The two records represent different time periods—Early and Late Cenozoic—which explains some (but not all) of these differences. Comparisons also highlighted interesting differences in Paleogene and Neogene faunal dynamics and community composition.

The long fossil record of Iberian continental basins offers a useful comparison over the Neogene and with mammalian faunas of broadly similar taxonomic composition (i.e., species from the same orders, families, and genera, as well as a similar body-size range). While nowhere as temporally continuous as the Siwalik record, the shorter sequences of the Iberian record are chronologically well integrated through biostratigraphy, magnetostratigraphy, and ordination analysis (Domingo et al. 2014) and in aggregate span the last 21 myr. The recent analysis of Blanco et al. (2021) compared change over time in taxonomic composition with functional richness. Using network analysis, they identified three "functional faunas" of large mammals with persistent composition in body size, feeding habit, and locomotion despite species-level turnover. An intriguing finding is that the transitions between the successive functional faunas are similar in age and some ecological attributes to the major transitions in large-mammal faunas and ecosystem change in the Siwalik record. The first transition in the Iberian record, between functional faunas 1 and 2, occurred between 15 and 13 Ma and involved an increase in browsing herbivores across the size range of large mammals. Similar changes occurred in the Siwalik record of the Potwar Plateau, where the increase began at 14.7 Ma. The next transition in the Iberian record, between functional faunas 2 and 3, occurred around 9 Ma (10.5–8.5 Ma) and involved replacement of browsers with mixed feeders; stable isotopic analyses of carbon and oxygen indicate a regional increase in seasonality of precipitation (Domingo et al. 2013). The Late Miocene Siwalik record of large mammals shows a sequential loss of frugivores, then browsers, and a modest influx of mixed feeders as well as increasing consumption of C_4 forage in particular resident lineages (Badgley et al. 2008). The functional richness of Siwalik large mammal faunas persisted from 9.5 to 7.1 Ma as taxonomic richness declined. The similarities in ecological dynamics between these two long records invite further comparisons.

RESEMBLANCE TO MODERN ECOSYSTEMS

Siwalik ecosystems have no close modern analogues. The high numbers of megaherbivores, browsing ungulates, and large carnivores are distinctive features of many Miocene mammalian assemblages (e.g., Janis, Damuth, and Theodor 2000; Faith, Rowan, and Du 2019). The persistently high number of co-occurring species with these traits is the basis for our inference

that species interactions were important influences on mammalian species richness and vegetation structure, largely through positive feedback mechanisms. The decline in number and proportion of species in all three functional groups in the Late Miocene and continuing into the Pliocene, as well as the increasing proportion of small to medium-sized ungulates resulted in faunas that were more similar in ecological structure to modern faunas in South Asia, Africa, and North America.

Nonetheless, vestiges of Miocene Siwalik ecosystems are present today. The rich mammalian faunas of southeast Asia, Malaysia, Borneo, and Sumatra have both taxonomic and functional elements similar to the interval of highest species richness in the Late Miocene. Apes, a rhino, the Asian elephant, callosciurine squirrels, forest hogs, multiple felids, and several other arboreal carnivores occur either in present-day or Late Quaternary mammalian faunas of the Malesian biogeographic region (e.g., Bibi and Métais 2016). These ecosystems have multi-canopy forest vegetation; some have masting dipterocarps with their massive seeds dispersed by large mammals and some large lizards (Corlett 1998; Campos-Arceiz and Corlett 2011).

The seasonally arid wooded grasslands of the Late Miocene, with their increasing frequency of fire, have modern counterparts in the monsoonal grasslands of southern Nepal and northeastern India (Assam). Some of these grasslands occupy the floodplains of the eastern Ganges and Brahmaputra Rivers in the present day (Wikramanayake et al. 2002). These regions experience some of the highest values of mean annual precipitation in the world (over 2,500 mm per year) but also have long dry seasons (over 6 months) with little to no rain. The Asian elephant, Asian rhino, boselaphine bovids (e.g., *Tetracercus*), a hyaena, bamboo rats, and native species of *Mus* are some of the taxonomic elements in common with Late Miocene Siwalik mammalian faunas (Bibi and Métais 2016). Riparian forests of these regions support arboreal rodents, carnivores, and primates.

INSIGHTS FOR CONSERVATION

The Siwalik record of environmental and biotic changes offers several insights relevant to conservation and the current extinction crisis. First, the changes in taxonomic and functional diversity of mammals during the environmental transition from the warm, wet conditions of the Middle Miocene to the cooler, drier Late Miocene indicate that present-day ecosystems are considerably altered from their predecessors and not likely to be well suited to the warmer world that is increasingly upon us. Likewise, some of the novel climates of today (Williams, Jackson, and Kutzbach 2007) may have analogues in Middle Miocene values of temperature and precipitation, enabling us to anticipate what kinds of species (in terms of functional traits) are likely to thrive in the novel conditions that are forecast to occur by the end of this century (Barnosky et al. 2017).

In addition, the fossil record demonstrates how species with different traits responded to changes in climate and vegetation of a particular magnitude and duration. For example, in the Siwalik record, small mammals from several clades showed substantial loss of species at the earliest evidence of change in vegetation and increased seasonality of moisture (Chapter 26, this volume; Polissar et al. 2021). Large mammal losses began 2 myr later, and mixed-feeding megaherbivores persisted the longest during the Late Miocene increase in aridity and seasonality.

Another potential insight involves the impact of immigrant species on resident species of similar ecological attributes. In the Siwalik record, immigrant lineages have on average shorter durations than residents, although some immigrants became well established. It is clear from our record, however, that many immigrants failed to become established or did not displace resident lineages (Chapter 25, this volume).

Finally, we can evaluate the range of environmental variability and change that particular ecosystem states can tolerate before they disassemble. The functional traits of species that persist during environmental transitions could inform decisions about managed relocation of vulnerable populations or restoration efforts today (Barnosky et al. 2017).

FUTURE RESEARCH DIRECTIONS

Notwithstanding the accomplishments summarized in this volume, we are keenly aware of many unanswered questions and future research opportunities. One goal of this volume is to provide a foundation for continuing research on the Siwalik record. Below we outline several areas that could add further value to our current state of knowledge.

Fieldwork. Further scouting and collecting in poorly sampled parts of the Potwar sequence would increase the number of fossil localities and specimens in these intervals. For example, increased sampling in the Kamlial Formation would enable testing of the hypothesis that the low documented species richness of both small and large mammals (Chapters 25, 26, this volume) was genuine and not an artefact of smaller sample sizes than in successive formations. Collection of sediment samples for greater coverage and higher density of biomarker analysis would add further details about the composition of vegetation and incidence of fire. Extending our integrated approach of geological, paleontological, and paleoenvironmental documentation to the Upper Siwaliks (Late Miocene to Quaternary) would connect the ecosystem history from the Late Miocene to the present day.

Systematic work. Few clades of Siwalik vertebrates have been the focus of phylogenetic analyses, partly because so much basic taxon identification was needed first and partly because trait data are limited. Where feasible, phylogenetic analyses would illuminate evolutionary relationships among Siwalik lineages and relatives in other regions of the world and contribute to biogeographic analyses of dispersals and geographic range-shifts. Robust phylogenies would facilitate assessment of the immigrant versus endemic origin of new species and enable a test of the hypothesis that the Siwalik bioprovince was

a source or sink for species in particular clades (Fortelius et al. 2014).

Traits. Additional morphometric, functional, and geochemical analyses of fossil specimens would refine our understanding of the ecological attributes of Siwalik species, especially feeding habits and locomotion. These findings would enable assessment of stasis versus anagenetic change in trait properties during different phases of ecosystem history.

Faunal analysis. We have only scratched the surface of potential approaches to analyzing faunal composition and its changes over time. A variety of newer methods would expand insights about species interactions, coherence of community composition, and diversification. Network analysis (e.g., Sidor et al. 2013; Blanco et al. 2021), Bayesian approaches to confidence intervals and estimated first and last appearances (Wang et al. 2016), and the PyRate model of diversification (Silvestro, Salamin, and Schnitzler 2014) are examples.

CONCLUDING THOUGHTS

Reflecting on 50 years of research on this remarkable record, we note several things that made this work possible. The fortunate and timely collaboration with the Geological Survey of Pakistan (and later with the Pakistan Museum of Natural History) provided not only key personnel but also a large support staff and equipment for fieldwork as well as facilities for housing fossil specimens. The composition of the research team included a light and highly motivating leadership style and many early-career scientists of many nationalities working together (Fig. 27.1), studying the Siwalik record in rustic field camps (Fig. 27.2), and providing mutual support in graduate research projects. A dozen PhDs and several postdoctoral fellowships focusing on the Siwalik record were accomplished during the first decade of the Harvard-Geological Survey of Pakistan project. Several individuals who started their research careers in the Potwar Plateau went on to become the leaders of later field seasons, the principal investigators on grant proposals, or those taking initiative for a new line of research. Important elements were the pursuit of common research goals with the ability to change direction, modify questions, and add new goals. For many years, team members met annually for updates and plans for the next phase of research. New methods, such as stable-isotope and biomarker analysis, provided new information that expanded the kinds of questions and interpretations that were possible. Stable funding from several sources, including international programs such as the Smithsonian-administered PL480 Program, as well as the U.S. National Science Foundation, and our own home institutions enabled us to conduct field seasons for nearly three decades and continuing laboratory analyses for another two. We also acknowledge with gratitude those participants who are no longer with us.

The fossil collections themselves deserve a special note of appreciation—they represent the essential "ground truth" on which the accumulated knowledge in this book ultimately rests.

Figure 27.1. Composite photograph of the core research team of the Harvard-Geological Survey of Pakistan project. A: S. Mahmood Raza, S. M. Ibrahim Shah (photo by Dr. Asad Gilani); B: John Barry (photo by C. Badgley); C: Kay Behrensmeyer, Catherine Badgley (photo by Andrew Hill); D (*left to right*): Michèle Morgan, John Wozny, Tom Jorstad, Imran Khan, Larry Flynn, Kay Behrensmeyer, Ev Lindsay, Mohammad Anwar, Will Downs, Jason Head, Khalid Sheikh (photo by C. Badgley); E. Jay Kelley, David Pilbeam, Hod French (photo by A. K. Behrensmeyer); F (*facing camera in center*): Tanya Sher Khan, Jay Kelley, Mahmood Raza, David Pilbeam (photo by A. K. Behrensmeyer).

Figure 27.2. Field camp near the town of Bhilomar, Chinji area, with the screen-washing site in foreground and Salt Range in the distance (photo by A. K. Behrensmeyer).

These collections are the responsibility of the Geological Survey of Pakistan and belong to the people of Pakistan. They represent an internationally recognized, valuable resource to inspire and inform future generations. In addition, fossils need special care and curation to remain secure and useful for understanding how life changes over time. It is our hope that Pakistan's Siwalik collections, as well as the detailed information about them that we have gathered over the past 50 years, can be a lasting scientific legacy for the future.

In closing, we express our deepest gratitude for the opportunity to work in such a remarkable place, on fascinating rocks and fossils, for so many years and with such interesting people. Onward!

References

Badgley, C., J. C. Barry, M. E. Morgan, S. V. Nelson, A. K. Behrensmeyer, T. E. Cerling, and D. Pilbeam. 2008. Ecological changes in Miocene mammalian record show impact of prolonged climatic forcing. *Proceedings of the National Academy of Sciences* 105(34): 12145–12149.

Badgley, C., W. S. Bartels, M. E. Morgan, A. K. Behrensmeyer, and S. M. Raza. 1995. Taphonomy of vertebrate assemblages from the Paleogene of northwest Wyoming and the Neogene of northern Pakistan. *Paleogeography, Paleoclimatology, Paleoecology* 115(1–4): 157–180.

Badgley, C. and A.K. Behrensmeyer. 1995. Preservational, paleoecological, and evolutionary patterns in the Paleogene of Wyoming-Montana and the Neogene of Pakistan. *Paleogeography, Paleoclimatology, Paleoecology* 115(1–4): 318–340.

Barnosky, A. D., E. A. Hadly, P. Gonzalez, J. Head, et al. 2017. Merging paleobiology with conservation biology to guide the future of terrestrial ecosystems. *Science* 355(6325): eaah4787.

Benson, R. B. J., R. Butler, R. A. Close, E. Saupe, and D. L. Rabosky. 2021. Biodiversity across space and time. *Current Biology* 31:R1225–R1236.

Bibi, F., and G. Métais. 2016. Evolutionary history of the large herbivores of South and Southeast Asia (Indomalayan Realm). In *The Ecology of Large Herbivores in South and Southeast Asia*, edited by F. S. Ahrestani and M. Sankaran, 15–88. Dordrecht: Springer Science and Business Media.

Blanco, F., J. Calatayud, D. M. Martin-Perea, M. S. Domingo, I. Menéndez, J. Müller, M. Hernández-Fernández, and J. L. Cantalapiedra. 2021. Punctuated ecological equilibrium in mammal communities over evolutionary time scales. *Science* 372(6539): 300–303.

Brett, C. E., L. C. Ivany, and K. M. Schopf. 1996. Coordinated stasis: An overview. *Paleogeography, Paleoclimatology, Paleoecology* 127(1–4): 1–20.

Burbank D. W., R. A. Beck, and T. Mulder. 1996. The Himalayan foreland basin. In *The Tectonic Evolution of Asia*, edited by A. Yin and T. M. Harrison, 149–188. New York: Cambridge University Press.

Campos-Arceiz, A., and R. T. Corlett. 2011. Big animals in a shrinking world—Studying the role of Asian megafauna as agents of seed dispersal. *Innovation* 10(1): 50–53.

Corlett, R. T. 1998. Frugivory and seed dispersal by vertebrates in the Oriental (Indomalayan) Region. *Biological Reviews* 73:413–448.

Demment, M. W., and P. J. Van Soest. 1985. A nutritional explanation for body-size patterns of ruminant and nonruminant herbivores. *American Naturalist* 125:641–672.

Domingo, L., P. L. Koch, M. Hernández-Fernández, D. L. Fox, M.S. Domingo, and M. T. Alberdi. 2013. Late Neogene and Early Quaternary paleoenvironmental and paleoclimatic conditions in southwestern Europe: Isotopic analyses on mammalian taxa. *PLOS ONE* 8(5): e63739.

Domingo, M. S., C. Badgley, B. Azanza, D. DeMiguel, and M. T. Alberdi. 2014. Diversification of mammals from the Miocene of Spain. *Paleobiology* 40:196–220.

Ernest, S. K. M. 2013. Using size distributions to understand the role of body size in mammalian community assembly. In *Animal Body Size*, edited by F. A. Smith and S. K. Lyons, 147–167. Chicago: University of Chicago Press.

Faith, J. T., J. Rowan, and A. Du. 2019. Early hominins evolved within non-analog ecosystems. *Proceedings of the National Academy of Sciences* 116(43): 21478–21483.

Figueirido, B., P. Palmqvist, J. A. Pérez-Claros, and C. M. Janis. 2019. Sixty-six million years along the road of mammalian ecomorphological specialization. *Proceedings of the National Academy of Sciences* 116(26): 12698–12703.

Fortelius, M., J. T. Eronen, F. Kaya, H. Tang, P. Raia, and K. Puolamäki. 2014. Evolution of Neogene mammals in Eurasia: Environmental forcing and biotic interactions. *Annual Review of Earth and Planetary Sciences* 42:579–604.

Fraser, D., L. C. Soul, A. B. Tóth, M. A. Balk, J. T. Eronen, S. Pineda-Munoz, A. B. Shupinski, et al. 2021. Investigating biotic interactions in deep time. *Trends in Ecology & Evolution* 36(1): 61–75.

Gunnell, G. F., M. E. Morgan, M. C. Maas, and P. D. Gingerich. 1995. Comparative paleoecology of Paleogene and Neogene mammalian faunas: Trophic structure and composition. *Paleogeography, Paleoclimatology, Paleoecology* 115(1–4): 265–286.

Harzhauser, M., and W. E. Piller. 2007. Benchmark data of a changing sea—Paleogeography, paleobiogeography and events in the Central Paratethys during the Miocene. *Paleogeography, Paleoclimatology, Paleoecology* 253(1–2): 8–31.

Holbourn, A., W. Kuhnt, S. Clemens, W. Prell, and N. Andersen. 2013. Middle to Late Miocene stepwise climate cooling: Evidence from a high-resolution deep water isotope curve spanning 8 million years. *Paleoceanography* 28(4): 688–699.

Janis, C. M., J. Damuth, and J. M. Theodor. 2000. Miocene ungulates and terrestrial primary productivity: Where have all the browsers gone? *Proceedings of the National Academy of Sciences* 97(14): 7899–7904.

Karp, A. T., A. K. Behrensmeyer, and K. H. Freeman. 2018. Grassland fire ecology has roots in the Late Miocene. *Proceedings of the National Academy of Sciences* 115 (48): 12130–12135.

Merritt, J. F. 2010. *The Biology of Small Mammals*. Baltimore, MD: Johns Hopkins University Press.

Morgan, M. E., C. Badgley, G. F. Gunnell, P. D. Gingerich, J. Kappelman, and M. C. Maas. 1995. Comparative paleoecology of Paleogene and Neogene mammalian faunas: Body-size structure. *Paleogeography, Paleoclimatology, Paleoecology* 115(1–4): 287–317.

Morgan, M. E., A. K. Behrensmeyer, C. Badgley, J. C. Barry, S. Nelson, and D. Pilbeam. 2009. Lateral trends in carbon isotope ratios reveal a Miocene vegetation gradient in the Siwaliks of Pakistan. *Geology* 37(2): 103–106.

Olson, E. C. 1952. The evolution of a Permian vertebrate chronofauna. *Evolution* 6(2): 181–196.

Polissar, P. J., K. T. Uno, S. R. Phelps, A. T. Karp, K. H. Freeman, and J. L. Pensky. 2021. Hydrologic changes drove the late Miocene expansion of C$_4$ grasslands on the northern Indian subcontinent. *Paleoceanography and Paleoclimatology* 36(4): e2020PA004108.

Reznick, D. N., and R. E. Ricklefs. 2009. Darwin's bridge between micro-evolution and macroevolution. *Nature* 457:837–842.

Sidor, C. A., D. A. Vilhena, K. D. Angielczyk, A. K. Huttenlocker, S. J. Nesbitt, B. R. Peacock, J. S. Steyer, R. M. H. Smith, and L. A. Tsuji. 2013. Provincialization of terrestrial faunas following the end-Permian mass extinction. *Proceedings of the National Academy of Sciences* 110(20): 8129–8133.

Silvestro, D., N. Salamin, and J. Schnitzler. 2014. PyRate: A new program to estimate speciation and extinction rates from incomplete fossil data. *Methods in Ecology and Evolution* 5(10): 1126–1131.

Srivastava, G., K. N. Paudayal, T. Utescher, and R. C. Mehrotra. 2018. Miocene vegetation shift and climate change: evidence from the Siwalik of Nepal. *Global and Planetary Change* 161:108–120.

Stigall, A. L. 2019. The invasion hierarchy: Ecological and evolutionary consequences of invasions in the fossil record. *Annual Review of Ecology, Evolution, and Systematics* 50:355–380.

Swenson, N.G. 2011. The role of evolutionary processes in producing biodiversity patterns and the interrelationships between taxonomic, functional and phylogenetic diversity. *American Journal of Botany* 98(3): 472–480.

Van der Meulen, A. J., P. Peláez-Campomanes, and S. A. Levin. 2005. Age structure, residents, and transients of Miocene rodent communities. *American Naturalist* 165(4): E108–E125.

Wang, S. C., P. J. Everson, H. J. Zhou, D. Park, and D. J. Chudzicki. 2016. Adaptive credible intervals on stratigraphic ranges when recovery potential is unknown. *Paleobiology* 42(2): 240–256.

Wikramanayake E., E. Dinerstein, C. J. Loucks, D. Olson, J. Morrison, J. Lamoreaux, and E. Hamilton-Smith. 2002. *Terrestrial Ecoregions of the Indo-Pacific: A Conservation Assessment*. Washington, DC: Island Press.

Williams, J. W., S. T. Jackson, and J. E. Kutzbach. 2007. Projected distributions of novel and disappearing climates by 2100 A.D. *Proceedings of the National Academy of Sciences* 104(14): 5738–5742.

CONTRIBUTORS

Pierre-Olivier Antoine, Institut des Sciences de l'Évolution, Université de Montpellier, CNRS, IRD, Montpellier, France

Catherine Badgley, Department of Ecology and Evolutionary Biology and Residential College, University of Michigan, Ann Arbor, MI, USA

John C. Barry, Department of Human Evolutionary Biology, Harvard University, Cambridge, MA, USA

Anna K. Behrensmeyer, Department of Paleobiology, National Museum of Natural History, Smithsonian Institution, Washington, DC, USA

Raymond L. Bernor, College of Medicine, Department of Anatomy, Laboratory of Evolutionary Biology, Howard University, Washington DC, USA

Jean-Renaud Boisserie, Laboratoire Paléontologie Évolution Paléoécosystèmes Paléoprimatologie, CNRS, University of Poitiers, Poitiers, France

Thure E. Cerling, Department of Geology and Geophysics, School of Biological Sciences, University of Utah, Salt Lake City, UT, USA

Olivier Chavasseau, Laboratoire Paléontologie Évolution Paléoécosystèmes Paléoprimatologie, CNRS, University of Poitiers, Poitiers, France

Margery C. Coombs, Department of Biology, University of Massachusetts, Amherst, MA, USA

Susanne Cote, Department of Anthropology and Archaeology, University of Calgary, Calgary, Alberta, Canada

Melinda Danowitz, New York Institute of Technology, College of Osteopathic Medicine, Old Westbury, NY, USA

Eric Delson, Department of Vertebrate Paleontology, American Museum of Natural History, New York, NY, USA

Lawrence J. Flynn, Department of Human Evolutionary Biology, Harvard University, Cambridge, MA, USA

Alan W. Gentry, Natural History Museum, London, UK

Damayanti Gurung, Kathmandu, Nepal

Jason J. Head, Department of Zoology and University Museum of Zoology, University of Cambridge, UK

Louis L. Jacobs, Roy M. Huffington Department of Earth Sciences, Southern Methodist University, Dallas, TX, USA

Jay Kelley, Institute of Human Origins and School of Human Evolution and Social Change, Arizona State University, Tempe, AZ, USA

Yuri Kimura, Department of Geology and Paleontology, National Museum of Nature and Science, Tsukuba, Japan

Thomas Lehmann, Messel Research and Mammalogy Department, Senckenberg Research Institute and Natural History Museum Frankfurt, Frankfurt am Main, Germany

Fabrice Lihoreau, Institut des Sciences de l'Évolution, CNRS, Université de Montpellier, CNRS, IRD, Montpellier, France

Everett H. Lindsay, Department of Geosciences, University of Arizona, Tuscon, AZ, USA

Antoine Louchart, École Normale Supérieure de Lyon, Laboratoire de Géologie de Lyon: Terre, Planètes et Environnement, CNRS, UMR 5276, Université Lyon 1, Villeurbanne, France

Michèle E. Morgan, Peabody Museum of Archaeology and Ethnology, Harvard University, Cambridge, MA, USA

Sherry Nelson, Department of Anthropology, University of New Mexico, Albuquerque, NM, USA

David Pilbeam, Department of Human Evolutionary Biology, Harvard University, Cambridge, MA, USA

Jay Quade, Department of Geosciences, University of Arizona, Tuscon, AZ, USA

S. Mahmood Raza, Geological Survey of Pakistan, Islamabad, Pakistan

Lois Roe, Museum of Paleontology, University of Michigan, Ann Arbor, MI, USA

William J. Sanders, Museum of Paleontology, University of Michigan, Ann Arbor, MI, USA

Gina M. Semprebon, School of Health and Natural Sciences, Bay Path University, Longmeadow, MA, USA

Nikos Solounias, Department of Anatomy and Embryology, New York Institute of Technology College of Osteopathic Medicine, Old Westbury, NY, USA

Pascal Tassy, Centre de Recherche en Paléontologie - Paris, Muséum National d'Histoire Naturelle, Paris, France

INDEX

Page numbers followed by *b*, *f*, and *t* indicate boxes, figures, and tables, respectively.

aardvarks, 79–80, 240–42, 244–45, 493; artistic reconstruction, 425–26, 426*f*; Chinji Formation, 65*t*, 437*t*; environmental settings, 68*t*; featured taxa, 419*t*; paleocommunities, 425–26, 426*f*, 437*t*–39*t*; time ranges, 242–44, 244*f*

abiotic factors, 4–5, 9, 433, 460–61, 464

Abudhabia, 183; paleocommunities, 439*t*; species data, 168*t*, 391*t*, 513*t*; time ranges, 174*f*, 471*f*

Accipitridae, 139, 141*t*, 143*f*, 146

Accipitriformes, 139, 141*t*

Aceratheriina, 284, 286, 292; body sizes, 289*f*; paleocommunities, 436*t*; phylogenetic framework, 285*f*; species data, 285*t*, 523*t*; time ranges, 282*f*, 289*f*, 292

Aceratheriini, 285*f*, 286

Aceratherium, 286, 291

Acerorhinus, 286

Acrochordus, 120, 121*f*, 129*f*, 130–31; artistic reconstructions, 424*f*–25*f*, 426*f*; featured taxa, 419*t*

Acrodontans, 118

adaptive radiation, 458

Adcrocuta, 227, 229, 232*t*; species data, 215*t*, 234*t*, 392*t*, 516*t*; time range, 218*f*

Aelurictis sivalensis, 230*t*

Aelurogale sivalensis, 230*t*

Aeluroidea, 226

Aeluropsis annectans, 231

Aepyceros melampus, 355*t*

Afrikanokeryx leakeyi, 361

Afrochoerodon kisumuensis/palaeindicus, 252

Afromeryx, 323–24

Afrotheria, 240

Afrotragulus, 342, 344, 349

Agamidae, 118, 419*t*, 427*f*

age estimates: for fossil localities, 44–47, 46*f*, 47*f*, 50; for fossil remains, 41–44, 50; for fossil sites, 47–49; reliability of, 47–49

Agnotherium, 219–21, 232*t*, 233; species data, 215*t*, 234*t*, 391*t*, 515*t*; time range, 217*f*

Agriotherium, 215*t*, 221, 232*t*

Ailuridae, 215*t*, 221–22

Alcedinidae, 139, 141*t*, 143*f*, 146, 420*t*

Alcelaphini, 379, 384

Alicornops, 285*f*, 287*f*, 288*f*; artistic reconstruction, 413*f*–14*f*; body sizes, 283*t*, 286, 289*f*, 292; featured taxa, 423*t*; paleocommunities, 413*f*–14*f*, 436*t*–37*t*; species data, 285*t*, 393*t*, 523*t*; stable isotope data, 398*t*; time ranges, 289*f*, 468*f*, 477*f*

Alilepus elongatus, 165, 165*f*, 172*f*, 189; artistic reconstruction, 427*f*, 428; featured taxa, 419*t*; species data, 166*t*, 389*t*, 514*t*; time range, 472*f*

Aliveria, 176; paleocommunities, 436*t*; species data, 166*t*, 390*t*, 509*t*; time ranges, 172*f*, 466*f*

Alloptox, 164; paleocommunities, 436*t*; species data, 166*t*, 389*t*, 509*t*; time ranges, 172*f*, 466*f*

alluvial environment: aerial view reconstruction, 413*f*–14*f*, 416*t*, 417–22, 419*t*–20*t*, 423*t*; landscape and soils, 55, 414–15; sub-Himalayan plain, 411–17, 413*f*–14*f*, 481

Alopecocyon, 215*t*, 221–22, 232*t*, 233

Amebelodontinae, 249, 252–53, 255–56, 439*t*, 517*t*

American Museum of Natural History (AMNH), 34, 39–40, 108, 213, 260, 272, 281, 297, 332, 353, 368–69; fossil tragulid specimens, 342, 343*t*; Siwalik Hills Indian Expedition, 117

Ammodorcas, 360

Ampelomeryx ginsburgi, 355

Amphechinus kreuzae, 154

amphibians, 69, 97, 420*t*, 442

Amphictis, 221–22

Amphicyon, 219–20, 221*f*, 232*t*, 440; species data, 215*t*, 234*t*, 391*t*, 515*t*; time ranges, 217*f*

Amphicyonidae, 55, 77, 219–21, 233; artistic reconstruction, 413*f*–14*f*; featured taxa, 423*t*; paleocommunities, 436*t*–37*t*, 440; species data,

215*t*, 391*t*, 515*t*; species richness, 493, 498*f*; time ranges, 217*f*

Amphimachairodus, 230–31, 232*t*; species data, 216*t*, 234*t*, 392*t*, 516*t*; time ranges, 218*f*

Amphiorycteropus, 240–42, 241*f*, 244–45; species data, 392*t*, 517*t*; teeth measurements, 242, 243*t*; time ranges, 244, 244*f*, 467*f*, 473*f*

Amphipelargus, 137, 138*f*, 139, 140*t*, 141–42, 143*f*, 144*t*, 146

Ampullariidae, 100–101, 101*t*, 102*f*, 103

Amynodontidae, 281

Anabantiformes, 109*f*, 109*t*, 112–13

Anancinae, 254–56

Anancus, 254, 256*f*, 392*t*, 477*f*, 517*t*

Anathana, 156, 158–59

Anatidae, 137, 140*t*, 141, 143*f*, 146

Anatinae, 137, 142*t*

Anatolictis, 223, 232*t*; artistic reconstructions, 426*f*, 427*f*; featured taxa, 419*t*; species data, 215*t*, 234*t*, 392*t*, 513*t*, 515*t*; time ranges, 217*f*

Ancylotherium, 272, 275, 275*f*, 277–78; paleocommunities, 438*t*, 445; species data, 393*t*, 523*t*; time ranges, 273*f*, 473*f*

Anderson, Robert, 5

Aneigmachanna, 113

Aneigmachannidae, 112–13

Anepsirhizomys, 167*t*, 181

angler catfish, 111

Anhinga, 138*f*, 139, 143*f*, 145–46; first and last appearances, 144*t*; species data, 141*t*

Anhingidae, 139, 141, 143*f*, 146; localities of occurrence and habitat, 142*t*; species data, 141*t*

Anisodon, 272–78, 464; artistic reconstruction, 424*f*–25*f*; dental specimens, 273, 274*t*; featured taxa, 419*t*; paleocommunities, 424*f*–25*f*, 437*t*, 439*t*, 440; representative specimens, 273, 274*f*; species data, 393*t*, 523*t*; stable isotope data, 396*t*, 450*t*, 455; time ranges, 273*f*, 276, 467*f*, 473*f*

Anisodontini, 273

Ankarapithecus, 207

Anomaluridae, 178, 503–4; faunal change, 466*f*; paleocommunities, 436*t*; species data, 167*t*, 390*t*, 513*t*; time ranges, 173*f*, 466*f*

Anser, 137, 140*t*, 142*t*

Anseriformes, 136–37, 146

Antemus, 183–84, 187*f*, 191, 458, 461; metacommunities, 188, 189*t*; molar morphology, 184–85, 185*f*; paleocommunities, 437*t*, 442; species abundance, 447–48, 448*t*; species data, 168*t*, 391*t*, 511*t*; time ranges, 174*f*, 471*f*

Anthracokeryx, 315

Anthracotheriidae, 55, 80, 108, 313, 315, 317–18; artistic reconstruction, 427*f*, 428; body sizes, 461, 462*f*; Chinji Formation, 65*t*; diet and habitat, 455; diversification, 323–24; ecological specialization, 324; environmental settings, 68*t*, 81–82, 82*f*; extinctions, 325*f*; faunal change, 467*f*; featured taxa, 419*t*; at focus time slices, 474*f*; paleocommunities, 427*f*, 428, 438*t*; residence time, 459*f*; species data, 393*t*, 518*t*; species richness, 443*f*, 448; stable isotope data, 92, 402*t*–3*t*, 450, 450*t*, 452*f*, 455; time ranges, 314*f*, 460

Antilocapra, 354, 358

Antilocapridae, 298, 353–54, 357–58, 364, 378; faunal change, 468*f*; at focus time slices, 475*f*; species data, 355*t*, 394*t*, 520*t*; time ranges, 356*f*

Antilope, 375

Antilopinae, 369, 375–76, 379; species data, 370*t*, 522*t*; stable isotope data, 406*t*; time ranges, 370*t*, 371*f*

Antilopini, 369, 375, 377, 379–80, 384, 445; body sizes, 370*t*; dental microwear, 384*f*; hypsodont scores, 386*f*; species data, 370*t*, 394*t*, 522*t*–23*t*; stable isotope values, 382*f*; time ranges, 370*t*, 371*f*, 468*f*, 476*f*

Anuran, 420*t*

Anwar, Mohammad, 533*f*

apes, 445, 450, 455, 531; artistic reconstruction, 425, 426*f*; featured taxa, 419*t*, 423*t*; paleocommunities, 425, 426*f*, 427, 437*t*–39*t*, 442

Aplodontia, 176

Aplodontiidae, 176

Apodemini, 191

apple snails, 422

aquatic mammals, 68*t*, 81–82, 82*f*

aquatic vertebrates, 80, 84, 108, 501

Aquila, 139, 141, 141*t*, 143*f*

Aragonictis, 223

arboreal mammals, 433, 464, 507, 531; artistic reconstructions, 425, 426*f*; depositional environments, 68*t*, 81–82, 82*f*; paleocommunities, 423, 425, 426*f*, 436*t*–38*t*

Archaeobelodon, 252–53

Archaeopotamus, 322

Archaeopoto, 323

Archosauria, 117

Arctamphicyon, 219

Arctoidea, 219; species data, 215*t*, 234*t*, 391*t*, 515*t*; time ranges, 217*f*

Ardea, 138, 138*f*, 140*t*, 143*f*, 144*t*, 145

Ardeidae, 138, 140*t*, 142*t*, 143*f*, 146

Ardeotis, 137, 140*t*

arthropods, 97

Artiodactyla, 80–81, 153, 278, 461, 481; age distribution, 63*t*, 72; artistic reconstruction, 425,

426*f*; body sizes, 64*t*, 74, 74*f*, 461, 462*f*, 507; depositional environments, 65*t*, 74–75, 75*f*; faunal change, 464, 467*f*–68*f*; featured taxa, 419*t*, 423*t*; at focus time slices, 473*f*–76*f*, 477*f*; Kamlial Formation, 435–40; paleocommunities, 425, 426*f*, 436*t*–39*t*, 442, 444; residence times, 457, 459*f*; skeletal elements, 62*t*, 69, 70*t*, 72, 73*f*; species data, 393*t*–94*t*, 517*t*–23*t*; species richness, 447–48, 528; stable isotope data, 92–93, 93*f*, 450; taphonomic attributes, 62*t*–64*t*

Arvicanthini, 185, 187*f*

arvicolines, 181

Asplenium, 423

Asterratus, 170*f*

Atavocricetodon, 170*f*, 181

Atherurus, 166*t*, 175–76

Atlantoxerus, 177, 177*f*, 189; species data, 166*t*, 390*t*, 510*t*; time ranges, 172*f*, 466*f*, 470*f*

atmospheric CO_2 and $\delta^{13}C$ values, 86–87, 421*f*

attritional mortality, 76–77

Austroportax, 379

avian fauna (avifauna). *See* birds

Babyrousa babyrussa, 297

Badgley, Catherine, 7, 533*f*

Bagarius, 109*t*, 112

Bagridae, 109–11, 109*f*, 109*t*, 111*f*, 113–14, 114*t*

Bagrus, 111*f*

Bakr, Abu, 5

Balaenicipitidae, 138–39, 141*t*, 142*t*, 143*f*, 144*t*, 145–46

baluchimyines, 169

Baluchimys, 170*f*

baluchitheres, 281

bamboo rats, 149–50, 164, 178–81, 179*f*, 180*f*, 187–92, 445, 458, 531; featured taxa, 419*t*

Banda Daud Shah, 176, 179, 182

Barbourofelidae, 229, 440, 504; species data, 216*t*, 392*t*, 516*t*; species richness, 493, 498*f*; time range, 218*f*

Barbourofelini, 516*t*

Barbourofelis, 229, 232*t*, 233; species data, 216*t*, 234*t*, 392*t*, 516*t*; time range, 218*f*

Barbus, 110

Barndt, Jeff, 17

Barry, John, 7, 305–6, 533*f*

Baryhipparion, 260

Batagur, 125, 419*t*, 424*f*–25*f*

bats, 78, 149–50, 153, 156, 158–59, 433, 482

Bavarian State Museum, Munich (BSM), 34, 117, 213, 260, 332, 353, 368

Bayerische Staatssammlung für Paläontologie und Geologie (BSP, Munich), 281

bear dog, 423*t*

Beglia Formation (Tunisia), 377

Behrensmeyer, Kay, 7, 533*f*

Bellamya, 100*t*, 101, 101*t*, 102*f*, 103, 103*f*, 104–5

Bellia (Siebenrockiella) sivalensis, 125

Belmaker, Miriam, 383

Beremendiini, 152*t*, 154*f*, 389*f*, 469*f*, 509*t*

Bighorn Basin (Wyoming, USA) fossil record, 530–31

bioprovince boundaries, 455–56

biostratigraphic surveys, 40, 49–50, 54

biostratigraphy, 16

biotic factors, 4–5, 65, 129, 164, 309, 326–37, 433–34, 460–64, 501, 527–531

birds, 78, 136–48, 138*f*, 152, 420, 501; artistic reconstruction, 427*f*, 428; depositional environments, 81–82, 82*f*; environmental settings, 68*t*; featured taxa, 420*t*, 423*t*; first and last appearances, 144*t*; fossil localities, 141, 142*t*, 445; paleocommunities, 427*f*, 428, 436*t*–39*t*; post–Kamlial Formation, 442; skeletal elements, 69; species data, 140*t*–41*t*; time ranges, 136–39, 142–46, 143*f*

birds of prey, 78

Birket Qarun Formation (Egypt), 111

Bishop, Bill, 7

Bithyniidae, 100

bivalves, 99–104, 101*t*, 102*f*, 103*f*, 422; artistic reconstruction, 426*f*; featured taxa, 420*t*

black rhino, 290

blind mole rats, 178

blue-gray system, 20*f*, 22–23, 24*f*, 413

boars, 295–97

body size(s): categories, 394*t*, 449*f*, 482*t*; changes in lineages, 530; estimates, 415–17, 416*t*; mammalian, 83, 433, 482; range represented by fossil specimens, 73–74; shifts in distribution, 448–49, 449*f*

Bohlinia, 360

Bohlininae, 360; paleocommunities, 437*t*; species data, 355*t*, 394*t*, 520*t*; time ranges, 356*f*, 475*f*

Boidae, 120, 128

bone modification, 61–68, 69*f*

bone-weathering, 71*t*

Boselaphini, 369, 372, 374, 378–81, 384, 386, 458; artistic reconstruction, 427*f*, 428; diet and habitat, 454–55; modern, 531; paleocommunities, 427*f*, 428, 445; species data, 370*t*, 394*t*, 521*t*–22*t*; stable isotope data, 405*t*; time ranges, 371*f*, 476*f*

Bos gaurus, 362

Bothriodontinae, 316, 318*f*, 322–24, 325*f*, 461; species data, 518*t*; species ranges, 314*f*

Bothriogenys, 322

Bovidae, 9, 80–81, 353, 358, 364, 368–88; artistic reconstructions, 427*f*, 428; astragalus widths, 378*f*, 379, 380*f*, 462*f*; body sizes, 370*t*, 380, 380*f*, 461–62, 462*f*; bone-modification features, 69*f*; Chinji Formation, 65*t*, 437*t*; dental microwear and mesowear, 383, 384*f*, 385*f*; diet and habitat, 380–84, 380*t*, 454; ecomorphology, 347, 380–81; environmental settings, 68*t*, 76; faunal change, 463–64, 468*f*; featured taxa, 419*t*, 423*t*; first and last appearances, 507; at focus time slices, 475*f*–76*f*; hypsodont scores, 386*f*; modern, 531; paleocommunities, 423, 426, 427*f*, 428, 436*t*–39*t*, 442–43, 445; relative abundance, 452–53, 463*f*; residence times, 457, 459*f*; sexual dimorphism, 380; species accounts, 369–80, 384; species data, 370*t*, 394*t*, 521*t*–23*t*; species richness, 443*f*, 447–48, 493, 498*f*; stable isotope data, 381, 382*f*, 383, 383*t*, 404*t*–7*t*, 450, 450*t*, 452*f*; time ranges, 369, 370*t*, 371*f*

Bovinae, 372, 375, 378, 380; artistic reconstructions, 425, 426*f*, 428; astragalus widths, 378*f*; diet and habitat, 454–55; paleocommunities, 425, 426*f*, 428, 439*t*; species data, 370*t*, 394*t*, 522*t*; time ranges, 371*f*, 476*f*

Bovini, 369, 374–75, 384, 445; species data, 370*t*; time ranges, 371*f*

Bovoidea, 353

Brachydiceratherium, 285*f,* 287*f,* 288*f,* 290; body sizes, 283*t,* 284*t,* 286, 289*f,* 291; paleocommunities, 436*t;* residence time, 528; species data, 285*t,* 393*t,* 523*t*–24*t;* stable isotope data, 398*t*–99*t;* time ranges, 289*f,* 292, 468*f,* 477*f*

brachydont, 298

Brachypotherium, 285*f,* 287*f,* 288*f,* 290; artistic reconstructions, 413*f*–14*f,* 424*f*–25*f,* 427*f;* body sizes, 283*t,* 284*t,* 289*f,* 290–91; featured taxa, 419*t;* species data, 285*t;* stable isotope data, 450*t,* 454, 454*f,* 455; stratigraphic ranges, 289*f,* 292

Brachyrhizomys, 181

Brachyura, 97

Bramatherium, 353, 361–64; artistic reconstructions, 413*f*–14*f,* 426*f;* featured taxa, 419*t,* 423*t;* species data, 355*t,* 394*t,* 521*t;* time ranges, 356*f,* 475*f,* 477*f*

Branta, 137, 140*t,* 142*t*

brant geese, 137, 139

British Museum of Natural History (BMNH). *See now* Natural History Museum (NHM, London)

Bronchocela, 118, 119*f,* 129*f,* 131

Brotia, 101, 101*t,* 102*f,* 103–4, 103*f,* 104–5

Brown, Barnum, 5, 34, 45, 54

browsers: dental mesowear, 383, 385*f;* dietary components, 482*t;* faunal change, 493–96, 507, 531; functional groups, 506–7; species richness, 501, 502*f*

Bubalus, 353, 355*t,* 362

buff system, 20*f,* 22–23, 24*f,* 413

Bugti Beds (Baluchistan), 169, 192, 199, 252–53

Bugti Hills, Pakistan, 296–97, 315, 317

Bugtimeryx pilgrimi, 343

Bugtimys, 169, 170*f*

Bugtirhinus praecursor, 283*t,* 286, 289*f,* 291–92

Buluk, Kenya, 256

Bungarus, 121, 121*f,* 129*f,* 130–31; artistic reconstruction, 427*f;* featured taxa, 419*t*

bunodont, 299–302

Burmemys, 122

burrowing rodents, 167*t,* 425–26, 426*f*

bustards, 137, 139, 141, 145–46, 420, 442; artistic reconstruction, 427*f;* featured taxa, 420*t;* species data, 140*t*

C₃ grasses, 8, 86–88, 418, 420, 428, 435, 440, 442, 453–55, 463

C₄ grasses, 8, 86–88, 91, 418, 420, 428, 441–42, 445, 453–55, 503, 506–8; earliest indicators, 450; Late Miocene expansion, 463–64

Caementodon, 285*f,* 286, 287*f,* 288*f,* 292; artistic reconstruction, 413*f*–14*f;* body sizes, 283*t,* 289*f;* featured taxa, 423*t;* paleocommunities, 413*f*–14*f,* 437*t;* species data, 285*t,* 393*t,* 523*t;* time range, 289*f,* 468*f,* 477*f*

Calamus, 423

calcium: paleosol profile, 88, 88*f*

calcium carbonate (CaCO₃): fossil enamel, 415; paleosol features, 88, 89*f*

California Academy of Sciences (CAS), 108

Callagur, 125

Calliophis, 121

Callosciurinae, 166*t,* 172*t,* 510*t,* 531

Callosciurus, 177; species abundance, 448*t;* species data, 166*t,* 390*t,* 510*t;* time ranges, 172*f,* 466*f,* 470*f*

Calotes, 118, 419*t,* 427*f*

Campbellpore Basin, 44*t*

cane rats, 149, 164, 166*t,* 173, 187, 189, 192, 435

Canidae, 213, 221

Caprini, 369, 375, 377, 379, 384; species data, 370*t,* 522*t*–23*t;* time ranges, 371*f*

Caprolagus, 165

Caprotragoides, 369, 377, 379, 384; species data, 370*t,* 394*t,* 523*t;* time ranges, 371*f,* 476*f*

carbonates: fossil enamel, 92–93, 93*f,* 265, 267*f,* 268*f,* 300–301, 301*f,* 347–48, 347*f,* 381, 382*f,* 383, 383*t,* 395*t*–407*t,* 413–16, 416*t,* 420, 421*f;* paleosol, 86–91, 89*f,* 90*f,* 90*t,* 412*f,* 414–17, 416*t,* 420, 421*f*

carbon dioxide (CO₂), atmospheric, 86–87, 421*f*

carbon isotopes (δ¹³C): atmospheric, 86–87; as evidence, 414–17, 416*t;* fossil enamel values, 92–93, 93*f,* 395*t*–407*t,* 450–55, 450*t,* 451*f*–53*t,* 454*f;* soil carbonate values, 88–91, 89*f,* 90*f,* 412*f,* 420, 421*f*

Carettochelys insculpta, 122

Carettocheylinae, 122, 123*f,* 128, 129*f,* 131

Carnivora, 78–81, 152–53, 213–39, 245, 278, 433, 481; artistic reconstruction, 425, 426*f;* body sizes, 231, 232*f,* 233; depositional environments, 65*t,* 74–75, 75*f;* environmental settings, 68*t;* featured taxa, 419*t,* 423*t;* first and last occurrences, 233, 236*f;* fossil specimens, 213–14, 214*f,* 218–31; genera present in other regions, 231, 232*t;* paleocommunities, 425–26, 426*f,* 436*t*–39*t,* 440, 445; species data, 214, 215*t,* 218–19, 234*t,* 391*t*–92*t,* 513*t*–16*t;* species durations, 233, 235*f;* species numbers, 231, 232*f,* 233, 235*f;* temporal and taxonomic resolution, 481; time ranges, 217*f,* 233

carnivore damage, 65, 68, 71*t*

carnivores, 433, 481–82, 531; featured taxa, 419*t;* paleocommunities, 436*t*–38*t,* 440; species richness, 493, 498, 498*f*

carp, 110

Castoridae, 163, 178

catfish, 55, 110–14, 413*f*–14*f,* 420*t,* 422, 423*t*

Catla, 110

Cautley, Proby T., 117, 123

Cebus, 205

Central African Museum (Tervuren, Belgium), 108

Cephalogale bugtiensis, 221

Cephalophus natalensis, 355*t*

Ceratotherium, 285*f,* 287*f,* 288*f,* 290–92; body sizes, 284*t,* 289*f;* paleocommunities, 439*t;* species data, 285*t,* 393*t,* 524*t;* stable isotope data, 399*t;* time ranges, 289*f,* 477*f*

Cercopithecidae, 200*f,* 391*t,* 473*f,* 514*t*

Cerling, Thure, 8

Cernictis, 222–23, 232*t;* species data, 215*t,* 234*t,* 392*t,* 515*t;* time range, 217*f*

Cervidae, 9, 353–54, 379

Cervoidea, 353

Ceryle, 138*f,* 139, 141*t,* 143*f,* 144*t,* 145–46

Cetancodonta, 313

Cetartiodactyla, 295

Chaca, 109*t,* 111, 114, 420*t,* 424*f*–25*f*

Chacidae, 109*f,* 109*t,* 110–11, 111*f,* 114, 114*t,* 422

Chalicotheriidae, 272–80, 455, 493; environmental settings, 68*t;* faunal change, 467*f;* featured taxa,

419*t;* at focus time slices, 473*f;* paleocommunities, 423, 437*t*–39*t,* 445; species data, 393*t,* 523*t;* stable isotope data, 396*t,* 450, 450*t,* 451*f,* 455; stratigraphic range, 273*f,* 276–77

Chalicotheriinae, 272–73, 273*f,* 276–77, 523*t*

Chalicotherium, 272–73, 275, 277–78

Channa, 109*t,* 112, 114, 422

channel-lag assemblages, 77

Channidae, 109, 109*f,* 109*t,* 112–13, 113*f,* 114, 114*t*

Channiformes, 112–13

Chaudhwan Formation, 14*f,* 289*f*

Chilgatherium harrisi, 249

Chilotherium, 286

chimpanzees, 203, 204*t*

Chinemys, 125

Chinji Formation, 12–14, 13*f,* 14*f,* 16, 16*f,* 36*f,* 88, 413–14; aardvarks, 240–41; carbonates, 89–90, 89*f,* 90*t;* carnivores and creodonts, 213–14, 216, 218–19, 222–23, 227–28; chalicotheres, 273, 275–76; chronostratigraphy, 43*t*–44*t,* 97*f;* depositional environments, 60*t,* 61*f,* 440; faunal change, 461; fieldwork, 5–8, 534*f;* fishes, 79, 108, 114*t;* fluvial environments, 21–25, 21*f,* 24*f,* 28–29, 60–61; fossil assemblages, 62*t*–64*t,* 68, 73*f,* 74*f,* 435; fossil localities, 22, 38*f,* 54*f,* 56, 60*t,* 61*f,* 79, 108, 483, 484*b;* fossil occurrences, 37, 37*f;* geochemical analysis, 415–17; giraffids, 359–60; habitat, 428; hippopotamids, 318*f,* 319; magnetostratigraphy, 18*f,* 26*t*–28*t;* mass assemblage of fish, 79, 108; molluscs, 100, 100*t,* 101, 102*f,* 104–5, 422; paleocommunities, 81, 419*t,* 424*f,* 437*t*–38*t,* 442; paleosols, 440; primates, 200, 202, 205, 207; proboscideans, 250, 251*f,* 252, 255; reptiles, 117–18, 121–24, 127–31; rhinocerotids, 282*f,* 289*f,* 290–91; sediment accumulation, 414; sedimentation rates, 24, 24*f;* small mammals, 149–50, 153, 155–57, 159, 165, 175–77, 179, 181–83, 187, 487; specimen data, 419*t;* stratigraphic sections, 46*f;* suids, 297, 298*f,* 299*f,* 307; survey blocks, 39*f;* taxonomic composition, 65*t,* 74, 75*f;* tragulids, 337–38, 346–47

chipmunks, 177, 189, 191

Chiroptera, 156, 156*f,* 159, 482

Chitarwata Formation, 14*f,* 15–16; bovids, 368, 372, 378; carnivores and creodonts, 213, 219–22; chalicotheres, 272, 275–76; chronostratigraphy, 19, 41; creodonts, 216; fishes, 109–10; fossil localities, 45; Hippopotamoidea, 317; marine regression, 411, 411*f;* reptile fossils, 119–20, 127; rhinocerotids, 289*f,* 290; small mammal fossils, 150, 155, 157, 169, 179; tragulids, 332–33, 343–44

Chitra, 123, 124*f,* 128, 129*f,* 131; artistic reconstruction, 424*f*–25*f;* featured taxa, 419*t*

Chitta Parwala Nala, 26*t*

Chlamydotis, 137, 138*f,* 140*t*

Chleuastochoerus, 308

Choerolophodon, 251–52, 251*f,* 256; paleocommunities, 437*t;* species data, 392*t,* 517*t;* stable isotope data, 395*t;* time ranges, 256*f,* 468*f,* 476*f*

Choerolophodontinae, 249, 251–52, 255–56

Choeropsis, 313–14, 323

Chopra, S. R. K., 5–6

Chotapohis padhriensis, 122, 129*f*

chronofauna, 528–29

chronostratigraphy, 19–20, 41–46, 97, 97*f;* framework, 41–44, 42*t*–44*t,* 46*f,* 49–50; panel diagrams, 20, 20*f*

Ciconia, 137, 140*t*

Ciconiidae, 137, 141, 143*f,* 146; artistic reconstruction, 426*f;* featured taxa, 420*t;* localities of occurrence and habitat, 142*t;* species data, 140*t*

Ciconiiformes, 137

Circamustela, 223–24, 232*t;* species data, 215*t,* 234*t,* 392*t,* 515*t;* time range, 217*f*

civet, 419*t,* 426

cladogenesis, 458, 503–5, 530

Clarias, 109*t,* 111, 114, 422

Clariidae, 109, 109*f,* 109*t,* 110–11, 111*f,* 113–14, 114*t*

Climacoceras, 354, 357–59; species data, 355*t,* 394*t,* 520*t;* time ranges, 356*f*

Climacoceridae, 357–58

Climacocerinae, 353–54, 357–59, 364; paleocommunities, 436*t,* 438*t;* species data, 355*t,* 394*t,* 520*t;* time ranges, 356*f,* 468*f,* 475*f*

climate change, 83, 323–24, 428, 461; between 18 and 6 Ma, 409–11, 412*f;* cooling events, 410–11, 503; environmental stress thresholds, 505–8; evidence for, 414–17, 416*f;* global cooling, 445, 508; long-time, 503

Colbert, E. H., 5, 231

Colobidae, 439*t*

Colobinae, 514*t*

colobine monkey, 199, 205, 428, 439*t,* 445

Colossochelys atlas, 117, 126, 131

Colubrinae, 122, 129*f*

Colubroidea, 120–22, 128, 130, 130*f,* 131, 419*t*

Columbia University, 6

competition, 460–61, 501

confidence intervals, 483, 484*b*

Conohyus, 299–300, 305–8; body sizes, 299*f,* 300*t;* paleocommunities, 437*t,* 443; species data, 393*t,* 518*t;* stable isotope data, 300–301, 301*f,* 401*t,* 450*t,* 455; time ranges, 297, 298*f*–99*f,* 299*f,* 467*f,* 474*f*

Conohyus thailandicus, 307

Conroy, Glenn, 6

conservation insights, 531–32

coordinated stasis, 528–29

coots, 137

Coppens, Yves, 7

Coraciformes, 139, 141*t*

Cormohipparion, 260–61, 263–64, 268–69; species data, 265*t,* 392*t,* 523*t;* stable isotope data, 267*f,* 396*t;* time ranges, 261*f,* 444, 473*f*

cormorants, 139, 141, 141*t,* 146, 442

Corvidae, 139, 141*t,* 143*f*

Corvinus, Gudrun, 5

Cotter, Gerald, 5

Court Jester hypothesis, 4

Court Jester model, 460

crabs, 97, 233

cranes, 137, 139, 140*t,* 145

Cremnomys, 187

Cremohipparion, 260, 263–64, 266, 268–69; dental mesowear and microwear, 265*t,* 266*f;* paleocommunities, 439*t,* 445; species data, 265, 265*t,* 393*t,* 523*t;* stable isotope data, 267*f,* 398*t,* 450*t,* 453; time ranges, 261*f,* 473*f*

Creodonta, 80, 213–39, 433, 458, 482; body sizes, 231, 232*f;* first and last occurrences, 233, 234*t,* 236*f,* 438*t;* genera in other regions, 231, 232*t;* paleocommunities, 419*t,* 423, 436*t*–39*t,* 440, 445; species data, 214–18, 215*t,* 391*t,* 514*t;* species

durations, 233, 235*f;* species numbers, 231, 232*f,* 233, 235*f;* species richness, 493, 498*f;* specimen numbers, 213–14, 214*f;* time ranges, 217*f,* 233

Cretaceous Maevarano Formation, 79

crevasse splays (CS), 23–24, 56*f,* 57, 60*t;* classification scheme, 59*t;* distribution of fossil animals, 81–82, 82*f;* fossil assemblages, 62*t*–68*t,* 73*f,* 74*f,* 77; fossil localities, 78, 78*t,* 79; fossil preservation, 59–60; taxonomic composition, 65*t*–68*t,* 74–76, 75*f,* 76*f;* unusual fossil concentrations, 78, 83

Cricetidae, 149, 153, 174*f,* 178, 181–82, 184*f;* change over time, 466*f,* 492–93, 492*f,* 494*b*–96*b;* first and last appearances, 447, 504; at focus time slices, 471*f;* metacommunities, 189*t;* paleocommunities, 423, 436*t*–37*t,* 439*t,* 442–43; relative abundance, 461, 463*f;* replacement by murines, 188, 188*f;* residence times, 457, 459*f;* species data, 167*t,* 390*t*–91*t,* 510*t*–11*t;* species richness, 443*f,* 448

Cricetinae, 174*f,* 510*t*

Criotherium argalioides, 376

Crocidura, 155, 155*f,* 159; species data, 152*t,* 389*t,* 509*t;* time ranges, 154*f,* 469*f*

Crocidurinae, 154–55, 155*f*

crocodiles, 55, 79, 108, 117, 426, 427*f,* 428, 433

Crocodilia (or Crocodylia), 126–27, 128*f,* 152, 421, 501; Chinji Formation, 65*t;* depositional environments, 81–82, 82*f;* Dhok Pathan Formation, 77; environmental settings, 68*t,* 74; featured taxa, 419*t,* 423*t;* relative abundance, 129–30, 130*f;* stratigraphic ranges, 129*f*

Crocodilus, 117

Crocodylus, 126–28, 128*f,* 129*f,* 130–31; artistic reconstructions, 413*f*–14*f,* 426*f,* 427*f;* featured taxa, 419*t,* 423*t*

Crocuta talyangari, 227

crows, 139, 141*t*

Cryptomanis, 245–46

Csotonyi, Julius: artistic reconstructions, 409, 413*f*–14*f,* 419*t,* 424*f*–25*f,* 426*f,* 427*f*

Ctenodactylidae, 149, 169, 171–73, 171*f;* change over time, 458, 465*f,* 492–93, 492*f,* 494*b*–96*b;* at focus time slices, 469*f;* residence time, 459*f;* species data, 166*t,* 390*t,* 513*t;* species richness, 443*f,* 448; time ranges, 172*f*

Ctenodactylinae, 166*t,* 171, 173, 513*t*

Ctenodactylus gundi, 172

Ctenohystrica, 169–77, 170*f*–71*f*

Cuoura, 125

Cyclemys, 131

Cynelos, 219

Cyprinidae, 109, 109*f,* 109*t,* 110*f,* 113–14

Cypriniformes, 109*f,* 109*t,* 110

Cyprinus, 109

Daboia, 120, 419*t,* 426, 426*f*

Dada Conglomerate, 14*f,* 41

Dakkamyoides, 183

Dakkamys, 182*f,* 183, 190; paleocommunities, 436*t;* species abundance, 448*t;* species data, 168*t,* 391*t,* 512*t*–13*t;* time ranges, 174*f,* 466*f*–67*f,* 471*f*

darters, 139, 141, 141*t,* 145–46

Dartmouth College, 6, 17, 34, 117, 213, 368

Dartmouth-Peshawar project, 151*f*

Darwin, Charles, 3–4, 460

Dasyatidae, 109–10, 109*f,* 109*t,* 110*f,*

data quality, 47–49

Daud Khel, 13, 27*t,* 149, 154, 157

Decennatherium, 354, 361–64; species data, 355*t,* 394*t,* 520*t;* time ranges, 356*f,* 475*f*

Dehm, Richard, 5

Deinotheriidae, 249–50, 255–56, 455, 528; artistic reconstruction, 425–26, 426*f;* faunal change, 468*f;* at focus time slices, 476*f;* paleocommunities, 425–26, 426*f,* 436*t;* species data, 392*t,* 517*t;* stable isotope data, 395*t,* 435, 450, 450*t;* time range, 256*f*

Deinotherium indicum, 250, 251*f;* artistic reconstructions, 413*f*–14*f,* 426*f;* featured taxa, 419*t,* 423*t;* species data, 392*t,* 517*t;* stable isotope data, 395*t,* 450*t;* time range, 256*f,* 476*f*

Democricetodon, 181–82, 182*f,* 183, 190, 461; paleocommunities, 436*t;* species abundance, 448, 448*t;* species data, 167*t,* 390*t,* 510*t;* time ranges, 170*f,* 174*f,* 466*f,* 471*f*

Democricetodontinae, 188, 188*f*

Dendrogale, 156–59, 157*f;* species data, 152*t,* 389*t,* 509*t;* time ranges, 154*f,* 466*f,* 469*f*

Dendromurinae, 181, 183

Dendromus, 181

dental mesowear and microwear, 8, 383, 416*t,* 417, 428, 437*t*–38*t. See also specific animals*

Deomyinae, 181

depositional environment(s), 35–40, 57, 60*t,* 413–14; artistic reconstruction, 414*f;* classification scheme, 59*t;* distributions across, 61, 61*f,* 68*t,* 81–82, 82*f;* major mammalian groups represented, 65*t;* post–Kamlial Formation paleocommunities, 440–42; taxonomic composition, 74–75, 75*f*

dermopterans, 149

de Terra, Helmut, 5

Dhok Pathan Formation, 5, 12, 13*f,* 14, 14*f,* 16, 16*f,* 19*f,* 88, 413–14; aardvarks, 241–42; aerial view reconstruction, 422, 423*t;* attritional mortality, 76–77; birds, 420; bovids, 372, 374, 379; carnivores and creodonts, 213, 221, 229–30; chalicotheres, 273, 276; chronostratigraphy, 97*f;* creodonts, 218; depositional environments, 60*t,* 61*f,* 440–41; equidae, 262; faunal information, 419*t;* fieldwork, 6–8; fishes, 79, 110, 112, 114*t;* fluvial environments, 22–24, 24*f,* 25–29, 60–61; fluvial transport, 77; fossil assemblages, 62*t,* 63*t,* 65, 68–69, 72–73, 73*f,* 74, 74*f;* fossil localities, 22, 54*f,* 56, 60*t,* 61*f,* 79, 420, 483, 484*b;* fossil occurrences, 37, 37*f;* geochemical analysis, 415–17; giraffids, 362–63; magnetostratigraphy, 18*f,* 26*t*–27*t;* mass accumulations of fish, 79; molluscan fossils, 100, 100*t,* 101, 102*f,* 103–5; paleocommunities, 81, 419*t,* 422, 423*t,* 424*f,* 426*f,* 427*f,* 439*t,* 445; paleosols, 88, 89*f;* panel diagram, 58*f;* pangolins, 246–47; primates, 202, 207; proboscideans, 250–51, 251*f,* 252, 254–56; reptiles, 117, 119–20, 122–24, 126–27, 129–30, 421; rhinocerotids, 282*f,* 286, 289*f,* 290–92; sediment accumulation, 414; small mammals, 149–50, 155–57, 159, 164, 177, 180–81, 186–87, 487; specimen data, 419*t;* suids, 299*f;* taphonomic histories, 76–77; taxonomic composition, 76; tragulids, 342; U-level, 186; unusual fossil concentrations, 78

Dhok Saira, 187

Diaceratherium, 286, 289*f*

Diatomyidae, 169–71, 171*f*, 187, 189–90, 435; faunal change, 465*f*; at focus time slices, 469*f*; paleocommunities, 436*t*; species data, 166*t*, 389*t*–90*t*, 513*t*; time ranges, 172*f*

Diatomys, 169–70, 171*f*, 172*f*, 189; paleocommunities, 436*t*; species data, 166*t*, 389*t*, 513*t*; time range, 170*f*, 466*f*

dibatag, 360

Dicerorhinus, 285*f*, 287*f*, 288*f*, 290–92; body sizes, 284*t*, 289*f*, 292; paleocommunities, 437*t*; species data, 285*t*, 393*t*, 524*t*; stable isotope data, 399*t*; time ranges, 289*f*, 477*f*

Diceros bicornis, 290

Didermocerus cf. *abeli*, 291

Dionysopithecus, 200*f*, 205–7; paleocommunities, 436*t*; species data, 391*t*, 514*t*; time ranges, 467*f*

dipodids, 163

Dipodoidea, 178

Diptychus, 109

dissolution, 72*t*

Dissopsalis, 214, 218, 232*t*, 233; species data, 215*t*, 234*t*, 391*t*, 514*t*; time range, 217*f*

diversification rates, 486–90, 488*b*, 489*f*, 491*f*

diversity dependence, 498–501

documentation, 34–53; digital files, 40, 40*f*; of fossil collection, 39–40, 40*f*; of fossil localities, 37, 38*f*; Google Earth imagery, 38*f*, 40; locality cards, 38*f*, 40, 56; paper documents, 40, 40*f*; Polaroid ground photographs, 38*f*, 40; specimen files, 40, 40*f*; of survey blocks, 37, 39*f*

Dorcabune, 333–35, 345–47, 349; artistic reconstruction, 424*f*–25*f*; astragalus widths, 335, 336*f*–37*f*, 346; body sizes, 334*t*, 335; dental microwear, 348, 348*f*; featured taxa, 419*t*; paleocommunities, 424*f*–25*f*, 436*t*, 443; species data, 393*t*, 519*t*; stable isotope data, 347–48, 347*f*, 403*t*, 450*t*; time ranges, 333*f*, 334*t*, 335, 345, 467*f*, 474*f*

Dorcadoryx, 377, 379, 384; species data, 370*t*, 394*t*, 523*t*; time ranges, 371*f*, 476*f*

Dorcadoxa, 376

Dorcatherium, 333, 335–46, 349; AMNH fossil specimens, 342, 343*t*; artistic reconstructions, 413*f*–14*f*, 426*f*, 427*f*; astragalus widths, 336–38, 336*f*, 337*f*, 340, 341*f*, 346; body sizes, 334*t*, 338–40; dental microwear, 348, 348*f*; featured taxa, 419*t*, 423*t*; hypsodonty, 348, 349*f*; molar morphology, 338, 339*f*, 340–42, 340*f*, 341*f*, 342*f*; species data, 393*t*, 519*t*; stable isotope data, 347–48, 347*f*, 403*t*, 450*t*; time ranges, 333*f*, 334*t*, 337, 343–46, 467*f*, 474*f*

dormice, 149, 164, 166*t*, 176, 176*f*, 189, 191–92

Downs, William R. (Will), 7, 150, 533*f*

Downsimys, 167*t*, 170*f*, 178

Draco, 118

dracoines, 118

Dremomys, 177; species abundance, 448*t*; species data, 166*t*, 390*t*, 510*t*; time ranges, 172*f*, 466*f*, 470*f*

Dromomerycidae, 354, 358–59

ducks, 137, 139, 140*t*, 141, 442

eagles, 139, 141*t*, 420

ecological dynamics, 481–526

ecological time, 460

ecosystem change, 501, 529–30

ecosystems, modern, 531

Egatochoerus, 295

Eicooryctes, 180, 189; paleocommunities, 439*t*; species data, 167*t*, 390*t*, 511*t*; time ranges, 173*f*, 470*f*

Elachistoceras, 372, 381, 384; artistic reconstruction, 424*f*–25*f*; featured taxa, 419*t*; hypsodont scores, 386*f*; species data, 370*t*, 394*t*, 521*t*; stable isotope data, 381, 382*f*, 383*t*, 404*t*; time ranges, 371*f*, 468*f*, 475*f*

Elapoidea, 121, 121*f*, 129*f*

Elasmotheriinae, 284–86, 292; paleocommunities, 439*t*; phylogenetic framework, 285*f*; species data, 285*t*, 523*t*; stratigraphic ranges, 282*f*, 289*f*, 292

Elasmotheriini, 285*f*

Elephantidae, 249, 254

Elephantiformes, 249

elephantimorphs, 250, 255–56

Elephantoidea, 395*t*–96*t*, 427*f*, 428, 528

elephants, 249, 256

Elephas maximus, 249

Elomeryx, 317

emigrants, 433–35

Emys tecta, 117

enamel. *See* Siwalik fossil enamel

endemics, 433–34, 456, 458, 465*f*–68*f*, 530; paleocommunities, 436*t*–39*t*; residence times, 457, 458*f*, 530

Enhydriodon, 224, 225*f*, 232*t*, 233; species data, 215*t*, 234*t*, 392*t*, 515*t*; time range, 217*f*

environmental stress thresholds, 505–8

environmental trends, 409–11, 412*f*, 506

Eocenchoerus, 295

Eomanis, 245

Eomellivora, 222, 232*t*; species data, 215*t*, 234*t*, 391*t*, 515*t*; time range, 217*f*

Eotragus, 369, 372, 377–79, 384; paleocommunities, 436*t*; species data, 370*t*, 394*t*, 521*t*; time ranges, 371*f*, 468*f*

Eozygodon morotoensis, 250

Ephippiorhynchus, 137, 140*t*, 143*f*

Epirigenys, 313, 322

Equidae, 130, 260–71, 428, 460; artistic reconstructions, 425, 426*f*, 427*f*, 428; biogeography, 263–64; body sizes, 83, 265*t*, 266–68, 462; dental microwear and mesowear, 264–65, 265*t*, 266*f*, 453–54; depositional environments, 65*t*, 74–75, 75*f*, 76; diet and habitat, 264–66, 269, 420, 453–55, 464, 503, 506; environmental settings, 68*t*, 74; faunal change, 463–64; featured taxa, 419*t*; first appearance, 493, 507; first occurrence, 444, 461, 463; at focus time slices, 473*f*; fossils, 298; paleocommunities, 425–26, 426*f*, 427*f*, 428, 438*t*–39*t*, 444–45; relative abundance, 452–63, 463*f*; residence times, 457, 459*f*; sexual dimorphism, 268; species data, 392*t*–93*t*, 523*t*; species richness, 443*f*, 448, 462, 498*f*; stable isotope data, 265–66, 267*f*, 268*f*, 396*t*–98*t*, 420, 421*f*, 450, 450*t*, 452*f*, 453–55

equilibrial fauna, 528–29

Equus, 260, 268

Ergilornithidae, 137, 139, 141–42, 143*f*, 146; first and last appearances, 144*t*; localities of occurrence and habitat, 142*t*; species data, 140*t*

Erinaceidae, 153–54, 158; faunal change, 465*f*; at focus time slices, 469*f*; species data, 152*t*, 389*t*, 509*t*; time ranges, 154*f*

Erinaceinae, 154, 154*f*; species data, 152*t*, 389*t*, 509*t*; time ranges, 469*f*

Eritherium azzouzorum, 249

Eritreum melakeghebrekristosi, 253

Eryx, 120

Eudorcas, 375

Eurasian Plate, 3

Euromanis, 245

Eurygnathohippus, 260, 263

Eusciurida, 169, 176–77

Eutamias urialis, 177

extinctions, 207–8, 325*f*, 326–27, 458; per-lineage rate, 486, 488*b*

Falconer, Hugh, 5, 117

Fallomus, 169–70, 170*f*, 189

Fatmi, A. N., 5–6

fauna(s), 3–4; chronofauna, 528–29; data links, 49; equilibrial, 528–29; evidence for, 418–22; fossil locality samples, 22; fossil occurrences, 49; functional, 531; paleoecology reconstruction from, 409–31, 416*t*, 419*t*–20*t*, 423*t*; research directions, 532. *See also* mammalian fauna

faunal change: analytic approaches to, 530; calculation of, 490, 494*b*–96*b*; determinants of, 500–503; patterns of, 9, 244–45, 445–60, 465*f*–68*f*, 496–500; rates of, 483–90; species richness and, 446–47, 446*f*; top-down processes, 501

faunal change indices, 435

faunal zones, 41

faunivores, 496, 503–4, 506–7; dietary components, 482*t*; species richness, 493, 497*f*, 499*f*, 501, 502*f*, 508

feeding habits, 482, 482*t*

Felidae, 77, 213, 229–31, 233, 531; paleocommunities, 436*t*–37*t*, 440; species data, 216*t*, 392*t*, 516*t*; species richness, 493, 498*f*; time ranges, 218*f*

Felinae, 231, 516*t*

Felis concolor, 231–33

Felis lynx, 231

Ferae, 245

Ficus, 423, 426

fieldwork, 1–2, 4–5, 7–8, 532

fires, 427*f*, 428, 503, 506, 508

fishes, 78, 108–16, 233, 422, 501; artistic reconstructions, 426, 426*f*; biogeography, 113–14, 114*t*; biostratigraphic ranges, 109*f*; Chinji Formation, 65*t*, 79, 108; depositional environments, 81–82, 82*f*; diversity, 109–14, 109*t*, 114*t*; environmental settings, 68*t*, 74; featured taxa, 420*t*, 423*t*; mass accumulations, 79, 108; post–Kamlial Formation, 442; relative abundance, 129–30, 130*f*; skeletal elements, 69

flamingos, 137, 139, 141, 145–46, 413*f*–14*f*, 420; featured taxa, 423*t*; species data, 140*t*

floodplain channels (FC), 23, 56*f*, 57–59, 60*t*; classification scheme, 59*t*; distribution of fossil animals, 81–82, 82*f*; fossil assemblages, 62*t*–68*t*, 73*f*, 74*f*, 77; fossil localities, 78, 78*t*, 79–80, 484*b*; fossil preservation, 57–59; taxonomic composition, 65*t*–68*t*, 74–76, 75*f*, 76*f*, 77, 80; unusual fossil concentrations, 78, 83

floodplains (FP), 24–25, 56*f*, 57, 60*t*, 501–2, 507–8; classification scheme, 59*t*; distribution of fossil animals, 81–82, 82*f*; fossil assemblages, 62*t*–68*t*, 73*f*, 74*f*; fossil localities, 78, 78*t*, 79–80; fossil preservation, 60–61; taxonomic composition, 65*t*–68*t*, 74–76, 75*f*, 76*f*, 77, 80; unusual fossil concentrations, 78, 83

flora, Siwalik, 97–98

fluvial paleoenvironment, 20–30, 21*f,* 24*f,* 56–57, 56*f;* depositional setting, 413–14, 414*f;* system properties, 501–3

fluvial transport, 77

flying squirrels, 176–77, 426, 426*f*

Flynn, Larry, 7, 533*f*

folivores, 464

forest hogs, 531

Fortunictis, 231

fossil collection(s), 532–34

fossil record: Bighorn Basin, 530–31; Iberian peninsula, 531; Siwalik. *See* Siwalik fossil record

French, Hod, 533*f*

French Paleontological Mission, 150

frequency index (FI), 434

frogs, 78, 108, 130*f,* 420*t,* 421

frugivores, 464, 493–96, 507, 531; dietary components, 482*t;* species richness, 499*f,* 501, 502*f*

Fulica, 137

functional faunas, 531

functional groups, 506–7, 529–31

functional richness, 503, 504*f,* 529

future research directions, 532

Gabhir Kas section, 18*f,* 26*t–*28*t*

Gaindatherium, 285*f,* 287*f,* 288*f,* 290; artistic reconstruction, 424*f–*25*f;* body sizes, 284*t,* 289*f,* 290–92; featured taxa, 419*t;* paleocommunities, 424*f–*25*f,* 436*t–*37*t;* species data, 285*t,* 393*t,* 524*t;* stable isotope data, 399*t,* 450*t,* 455; time ranges, 289*f,* 292, 468*f,* 477*f*

Gaj Formation, 14*f,* 41, 219, 289*f*

Galericinae (galericines), 153–54, 437*t–*38*t*

Galerix, 153–54, 153*f,* 158–59, 442; first occurrences, 465*f;* metacommunities, 189*t;* species data, 152*t,* 389*t,* 509*t;* time ranges, 154*f,* 465*f,* 469*f*

Galliformes, 136–37, 146

Galloanserae, 136

Gambir Kas, 18*f*

Ganda Kas, 20*f*

Ganga (Ganges) River, 29

Gansophis, 120–21, 129*f*

Garamba National Park, 498

Gaston, Tony, 6

gastropods, 99–100, 101*t,* 102*f–*3*f,* 104–5, 422

Gavialidae, 126

Gavialis, 126–27, 129*f,* 130; artistic reconstructions, 413*f–*14*f,* 424*f–*25*f,* 426*f;* featured taxa, 419*t,* 423*t*

Gazella, 375, 379, 384; artistic reconstructions, 413*f–*14*f,* 426*f;* diet and habitat, 380–81, 380*t;* featured taxa, 419*t,* 423*t;* species data, 355*t,* 370*t,* 394*t,* 522*t;* time ranges, 370*t,* 371*f,* 476*f*

gazelles, 74, 375

Gekkotans, 118, 119*f,* 129*f*

Gelocidae, 354

Geniokeryx, 315

Genotelma, 323

Gentrytragus, 377

Geochelone, 126, 419*t,* 426*f*

geochemical proxies, 415–17

Geoclemys, 124–25, 125*f,* 128, 129*f,* 131, 419*t,* 426*f*

geoemydids, 125, 125*f*

Geological Institute, State University of Utrecht (GIU), 34, 213, 260, 332, 368

Geological Survey of India (GSI), 34, 213, 260, 273, 297, 332, 368–69

Geological Survey of Pakistan (GSP), 34, 117, 126, 126*f,* 150, 213, 272, 281, 313, 368, 432, 532, 534. *See also* Harvard-Geological Survey of Pakistan (Harvard-GSP)

Geomagnetic Polarity Time Scale (GPTS), 47

Geomydidae, 129*f*

gerbillids, 436*t*

Gerbillinae, 181–83, 187*f,* 447, 493, 504; change over time, 490, 492–93, 492*f;* metacommunities, 189*t;* paleocommunities, 437*t,* 442; relative abundance, 461, 463*f;* replacement by murines, 188, 188*f,* 464; species data, 168*t,* 391*t,* 512*t–*13*t;* time ranges, 174*f*

gerbils, 164, 181–83, 182*f,* 187, 190, 192, 461, 464; metacommunities, 189*t;* paleocommunities, 439*t;* proposed relationships, 187*f;* species data, 168*t*

gerenuk, 360

Gill, William, 5

Gingerich, Philip, 6

Giraffa, 318, 353, 360–61, 364; artistic reconstructions, 426*f,* 427*f;* featured taxa, 419*t;* paleocommunities, 426*f,* 427*f,* 439*t,* 445; species data, 355*t,* 394*t,* 520*t;* time ranges, 356*f,* 477*f*

giraffes, 353, 360, 419*t,* 427*f,* 428

Giraffidae, 277, 353–54, 358–60, 364, 461, 463; faunal change, 468*f;* featured taxa, 419*t,* 423*t;* at focus time slices, 475*f,* 477*f;* paleocommunities, 437*t–*39*t,* 442–44; residence times, 457, 459*f;* species data, 355*t,* 394*t,* 520*t–*21*t;* species richness, 448, 493, 498*f,* 528; stable isotope data, 450, 450*t,* 455; time ranges, 356*f*

Giraffinae, 355*f,* 356*f,* 360–61, 520*t*

Giraffoidea, 55, 82, 353–67, 379; artistic reconstruction, 425–26, 426*f;* body sizes, 364, 364*f,* 380, 462*f;* Chinji Formation, 65*t;* environmental settings, 68*t;* paleocommunities, 425–26, 426*f,* 438*t,* 444; species counts, 443*f;* species data, 355*t;* species richness, 493, 498*f;* stable isotope data, 403*t–*4*t,* 451*f;* time ranges, 354, 356*f,* 364

Giraffokerycinae, 355*f,* 356*f,* 359–60, 520*t*

Giraffokeryx, 353, 359–60, 362–64; paleocommunities, 437*t,* 443; species data, 355*t,* 394*t,* 520*t;* time ranges, 356*f,* 468*f,* 475*f*

gleyed soils, 25–28

gliding squirrels, 166*t*

Glires, 150, 153, 164–65, 166*t–*68*t,* 169

Gliridae, 176, 503–4; at focus time slices, 469*f–*70*f;* paleocommunities, 437*t,* 439*t,* 442; species data, 166*t,* 390*t,* 513*t;* time ranges, 172*t*

global cooling, 445, 508. *See also* climate change

glochidia, 104

goethite (G), 415–17, 416*t,* 503

Golunda kelleri, 168*t,* 187

Gomphotheriidae, 249, 251, 255–56, 504, 528; faunal change, 468*f;* at focus time slices, 476*f–*77*f;* species data, 392*t,* 517*t;* stable isotope data, 395*t;* time range, 256*f*

Gomphotheriinae, 253–56, 517*t*

Gomphotherium, 244, 251*f,* 253–54, 256; species data, 392*t,* 517*t;* time ranges, 256*f,* 476*f*

Gonotelma, 325*f*

Google Earth imagery, 38*f,* 40

goose (geese), 137, 139, 140*t*

Gorilla, 202, 204*t*

Gould, Stephen Jay, 460

granivores, 482*t,* 497*f,* 506–7

grasses. *See* C$_3$ grasses, C$_4$ grasses

grass fires, 427*f,* 428, 503, 506, 508

grazers, 464, 506–7; dental mesowear, 383, 385*f;* dietary components, 482*t;* species richness, 499*f,* 501, 502*f*

greater kudu, 378

Griphopithecus alpani, 206

ground squirrels, 166*t–*67*t,* 189, 447

Gruidae, 137, 140*t,* 141, 142*t,* 143*f*

Gruiformes, 137, 140*t,* 146

Grus, 137, 138*f,* 140*t,* 143*f,* 144*t,* 145

gundis, 149, 166*t,* 170*f,* 171–73, 181, 458

gymnures, 153–54

Gyps, 139, 141, 141*t,* 142*t,* 143*f,* 144*t,* 145

habitat change, 461, 463–64

Hadromys, 168*t,* 187

hamsters, 149, 164, 167*t,* 181–82, 461, 464

Hardella, 125

hares, 164, 427*f,* 428

Haritalyangar, 5, 13*f;* fish fossils, 109; primates, 199, 201–2, 205–6; suid fossils, 299*f*

Haro River Quarry, 45, 175, 307

Harvard-Geological Survey of Pakistan (Harvard-GSP), 3–6, 9, 54, 108, 260, 281, 297, 332, 432, 481, 532, 533*f*

Harvard University, 8, 34, 117, 213, 272

Hasnot, 13, 42*t,* 213, 282*f,* 337, 340

Head, Jason, 533*f*

hedgehogs, 149–50, 153–54, 153*f,* 154*f,* 158–59, 192; metacommunities, 189*t;* paleocommunities, 436*t–*38*t,* 442

Helicoportax, 373, 378, 380, 384; species data, 370*t,* 394*t,* 521*t;* time ranges, 371*f,* 475*f*

Helladotherium grande, 363; species data, 355*t,* 394*t,* 521*t;* time range, 356*f,* 477*f*

hematite (H), 415–17, 416*t,* 503

Hemimeryx, 317–19, 323, 325*f;* species data, 393*t,* 518*t;* time ranges, 314*f,* 467*f*

herbivores, 78, 449; diet and habitat, 93, 420, 450, 482*t;* faunal change, 493–96, 507; functional groups, 506–7; prey species, 501, 502*f;* residence times, 457; species richness, 493, 497*f,* 499*f,* 501, 502*f,* 508; stable isotope data, 91–93, 395*t–*407*t.* *See also* megaherbivores

heron, 138–39, 140*t,* 141, 146

Herpestes, 226, 232*t*

Herpestidae, 226, 503–4; change over time, 490, 492*f;* paleocommunities, 437*t,* 440; species data, 215*t,* 392*t,* 513*t,* 515*t;* time ranges, 218*f*

Heterobranchus, 109

Heteroxerus, 177, 177*f,* 189; species data, 166*t,* 390*t,* 510*t;* time ranges, 172*f,* 466*f,* 470*f*

Hexaprotodon, 314, 318*f,* 321–27, 325*f;* dental microwear, 321*t;* species data, 393*t,* 519*t;* time ranges, 314*f*

Hicks, Jason, 17

Hipparion, 206, 260–64, 266, 268–69, 428; dental microwear, 266*f;* stable isotope data, 267*f,* 398*t;* time ranges, 260, 261*f*

Hipparionini, 523*t*

Hippohyus, 308–9; body sizes, 299*f,* 300*t;* dental morphologies, 301–2, 302*f,* 303*f;* time ranges, 299*f,* 306–7, 473*f*

Hippopotamidae, 314*f,* 393*t,* 445, 519*t,* 528

Hippopotamodon, 305, 308; artistic reconstructions, 413f–14f, 426f; body sizes, 299f, 300t, 306, 462f; featured taxa, 419t, 423t; paleocommunities, 413f–14f, 426f, 438t; species data, 393t, 517t; stable isotope data, 300–301, 301f, 400t, 450t, 455; time ranges, 297, 298f–99f, 306–7, 458, 473f

Hippopotamoidea, 313–31, 318f, 325f, 498f

Hippopotamus, 313–14, 322–23, 326

Hippotherium, 260, 262, 264, 266

Hippotragini, 379, 384

Hispanopithecus, 207

Hodsahibia, 170f

hogs, 295–98

Hominidae, 200f, 437t, 442, 473f

Hominoidea, 76, 130, 206, 493, 504–5, 507; paleocommunities, 439t; species data, 391t, 514t; species richness, 493, 498f; stable isotope data, 395t, 450t, 451f

Homo sapiens, 326

Honanotherium, 360; species data, 355t, 394t, 520t; time ranges, 356f, 475f

Hoplictis, 223–24, 232t; species data, 215t, 234t, 392t, 515t; time ranges, 217f

horses, 260, 263–64, 269, 379

Howard University, 6, 34, 213

Huaixiachoerus, 295

Hussain, S. Taseer, 6, 149

Hutchinson, J. Howard, 111

Huxley, Thomas Henry, 9

hyaenas, 427f, 428, 531

Hyaenictitherium, 229

Hyaenidae (hyaenids), 77, 213, 228–29, 233; artistic reconstructions, 426, 426f; paleocommunities, 437t; species data, 215t, 392t, 516t; species richness, 493, 498f; time ranges, 218f

Hyaeninae, 516t

Hyaenodontidae (hyaenodonts), 217–18; Chinji Formation, 65t, 438t; featured taxa, 419t; paleocommunities, 436t, 438t; species data, 215t, 391t, 514t

Hyaenodontinae, 514t

Hyainailouros, 214–16, 218, 232t, 233–34; artistic reconstruction, 424f–25f; featured taxa, 419t; paleocommunities, 424f–25f, 437t, 440; species data, 215t, 234t, 391t, 514t; time range, 217f, 440

Hydrobiidae, 100

Hyemoschus aquaticus, 332, 336f, 337f

Hylobatidae, 206

Hylochoerus meinertzhageni, 297

Hylomys, 153, 159

Hyotheriinae, 296–97

Hyotherium, 297, 305, 308

hypercarnivores, 233

Hypsodontinae, 369, 379, 384; astragalus widths, 378, 378f, 379; paleocommunities, 436t, 439t; species data, 370t, 394t, 521t; time ranges, 370t, 371f, 468f

Hypsodontus, 369, 378, 380–84; astragalus widths, 378, 378f; dental microwear, 384f; hypsodont scores, 384, 386f; paleocommunities, 437t; species data, 370t, 394t, 521t; stable isotope data, 381, 382f, 383t, 404t; time ranges, 371f, 468f, 475f

hypsodonty, 416t; bovid, 383–84, 386f; tragulid, 348, 349f

Hyracodontidae, 281

Hystricidae, 149, 175–76; at focus time slices, 472f; paleocommunities, 439t; species data, 166t, 390t, 514t; time ranges, 172f

Hystrix, 149, 175, 175f, 176, 189, 192; artistic reconstruction, 427f, 428; featured taxa, 419t; metacommunities, 189t; paleocommunities, 42/f, 428, 445; species data, 166t, 390t, 514t; time ranges, 172f, 472f

Iberian peninsula fossil record, 531

Iberictis, 223

ibis, 146

Ictitheriinae, 516t

Ictitherium, 228–29

immigrants, 433–35, 464, 486, 504–5, 530, 532; and endemic origination, 456, 456f; mammal species, 465f–68f; residence times, 457, 458f, 530; at selected time slices, 436t–39t

immigration patterns, 455–56, 456f

Indarctos, 221, 232t; species data, 215t, 234t, 391t, 515t; time range, 217f

Indian Museum Kolkata, 125, 353

Indian Plate, 3

Indian Siwaliks, 139, 199, 201, 205–6, 208, 256, 276

Indian subcontinent, 455; bovids, 368; chalicotheres, 440; creodonts, 458; elasmotheriines, 286; giraffoids, 354, 445; hipparions, 260, 264; hippopotamoids, 313, 315, 317, 319, 323–24, 326; proboscideans, 250; sub-Himalayan alluvial plain, 411–17, 413f–14f; suids, 308; total mammal richness, 498; tragulids, 332

Indo-Malay province, 5–6, 164

Indonaia, 100, 100t, 101, 101t, 102f, 103f, 104–5

Indopithecus, 6, 199, 200f, 204–5, 207

Indoplanorbis, 101, 101t, 102f–3f, 104–5

Indosuncus bhatiai, 155

Indotestudo, 126, 126f, 129f, 131, 419t, 427f

Indraloris, 200f, 201, 391t, 436t, 467f, 514t

indricotheres, 281

Indus River, 29

Injanatheriium, 355t, 356f, 363, 394t, 475f, 521t

Insectivora, 465f, 469f

insectivores, 78, 437t–38t, 481

invertebrates, 68t, 81–82, 82f, 420t, 422–23

invertivores, 482t, 493–96, 497f, 499f, 506–7

isotaphonomic sampling, 8–9, 55, 80, 83

isotopic analysis. *See* stable isotopes

Jacobs, Louis, 7

Jalalpur, 41, 42t

jerboas, 163

Jhelum River, 501

Johnson, Gary, 6, 8, 17

Johnson, Noye, 6, 8, 17

Jorstad, Tom, 533f

Judith River Formation, 79

Kachuga, 125

Kamlial Formation, 12–16, 14f, 16f, 88, 413–14; anthracotheres, 326; bovids, 369, 372, 378–79; carnivorans, 226; chronostratigraphy, 97f; depositional environments, 60t, 61f; faunal record, 428; fieldwork, 5, 7–8, 532; fish fossils, 108, 110, 113, 114t; fluvial environments, 22–24, 24f, 28, 61; fossil assemblages, 62t–64t, 69, 73f, 74f, 435, 483; fossil localities, 22, 54f, 60t, 61f, 79, 483, 484b; fossil occurrences, 37, 37f; future

research directions, 532; hippopotamids, 317, 318f, 319; magnetostratigraphy, 18f, 26t; molluscan fossils, 101; paleobiology, 435–40; paleocommunities, 81, 435, 436t; Pecora *incertae sedis*, 97; primates, 199, 201, 205–7; proboscideans, 250, 251f; reptile fossils, 117, 122–23, 126; rhinocerotids, 282f, 286, 289f, 290–92; sedimentation, 24–25, 24f, 414; small mammal fossils, 150, 155, 157, 171, 174, 176, 182–83; species counts, 440, 441t; stable isotope data, 435; suid fossils, 298f, 299f; tragulids, 343–44, 346

Kanha National Park, 234

Kanisamys, 178–80, 179f, 180f, 458; metacommunities, 189t; paleocommunities, 436t, 438t–39t, 445; species abundance, 448t; species data, 167t, 390t, 511t; time ranges, 173f, 466f, 470f

Kapi ramnagarensis, 206

Kappelman, John, 8, 17

Karewa Intermontane Basin fossils, 109

Karnimata, 184–88, 185f, 187f, 190–91, 458; metacommunities, 189t; paleocommunities, 438t; species data, 168t, 391t, 512t; time ranges, 174f, 472f

Kaulial Kas, 16f, 18f, 27t, 28–29, 205

Kelley, John (Jay), 8, 533f

Kenyapithecus kizili, 206

Khan, Imran, 17, 533f

Khan, Tanya Sher, 533f

Kharinadi Formation (Pasuda), 127

Khaur Area, 6, 13, 13f, 21, 29, 42t–43t

Khoratpithecus, 207

Kilarcola, 182

kingfishers, 139, 141, 141t, 145–46, 420t, 424f–25f

Kipsaramon bonebed, 79

Kipsigicerus labidotus, 379

Kobus, 369, 376; artistic reconstructions, 413f–14f, 427f; dental microwear, 383, 384f; diet and habitat, 381–84, 386; featured taxa, 419t, 423t; hypsodont scores, 384, 386f; paleocommunities, 413f–14f, 427f, 439t, 445; species data, 370t, 394t, 522t; stable isotope data, 381, 382f, 383, 383t, 406t; time ranges, 371f, 476f

Kochalia, 172f, 173–75, 189, 435; paleocommunities, 436t; species abundance, 448t; species data, 166t, 390t, 513t; time ranges, 466f, 469f

Kohat, 149

Kolpochoerus, 302

Konobelodon, 253, 256f, 392t, 476f, 517t

Krishnapithecus krishnaii, 206

Kubanochoerinae, 297

Kubanochoerini, 296–97

Kubanochoerus, 296, 308

kudu, 378

Kuldana Formation (Pakistan), 111

Kulutherium, 313

Kutch, Gujarat (India), 206

Labeo, 109t, 110

Lagomerycidae, 353–54, 355t, 356f, 358–64

Lagomerycinae, 354, 355t, 356f, 358, 437t, 520t

Lagomeryx, 358

Lagomorpha, 92, 150, 153, 163–65; artistic reconstructions, 427f, 428; faunal change, 465f; featured taxa, 419t; at focus time slices, 472f; species data, 166t, 389t, 509t, 514t

Lamellidens, 100–101, 100t, 101t, 102f, 103f, 104–5

Lamont-Doherty Geological Observatory, 34, 213

landfowl, 146

landscape. *See* alluvial environment

Lantanotherium, 159

Laonastes, 169–71

large mammals: analytic approach to, 530; body sizes, 448–49, 449f, 482t; conservation insights for, 532; diversification rates, 490, 491f; extinction rates, 490, 491f; faunal change, 463–64, 467f–68f, 493–96, 498f, 499f, 500, 508, 529; feeding habits, 482t; first and last appearances, 433, 434f, 490, 491f, 504–5, 507, 530; at focus time slices, 472f–76f; fossil localities, 445; fossil occurrences, 37f; fossil record, 433; functional groups, 506–7; functional richness, 529, 531; joint size-feeding categories, 490, 493t, 498f, 499f, 500; origination rates, 490, 491f; paleobiology, 435–40; paleocommunities, 436t–39t, 442–45; prey species, 501, 502f; rarefaction curves, 76f, 81; residence times, 456–57, 457t, 458f, 459f, 483–86, 487f, 506–7, 528; species changes, 446–47, 447f, 460; species counts, 440, 441t, 443f, 498; species data, 514t–24t; species richness, 446–47, 446f, 447–48, 464, 483, 485f, 490, 491f, 493, 496–97, 498f, 499–500, 499f, 500t, 503, 505f, 506–8, 529; stable isotope data, 92–93, 93f, 420; taxonomic richness, 504–5, 505f, 508; temporal and taxonomic resolution, 481

Lartetotherium, 290

Late Miocene, 410–11, 463–64, 500, 503

Laudakia, 118, 119f, 129f

Laurasiatheria, 245

Leiolepis, 118

lemmings, 181

Leporidae (leporids), 164–65; at focus time slices, 472f; metacommunities, 189f; paleocommunities, 439t, 445; species data, 166t, 389t, 514t; time ranges, 172f

Leporinae, 164–65

Lepthyaena, 222, 228, 228f, 229, 232t; species data, 215t, 234t, 392t, 516t; time range, 218f

Leptoptilini, 137, 141

Leptoptilos, 137, 139, 140t, 143f

Leptorhynchus gangeticus, 117

Lewis, G. Edward, 5–6, 34, 45, 54, 205

Libycochoerus, 297

Libycosaurus, 317–20, 324, 326

Libytherium, 362–65; paleocommunities, 439t; species data, 355t, 394t, 520t; time ranges, 356f, 477f

Lindsay, Everett (Ev), 6–8, 17, 533f

Lindsaya, 170f

Lipotyphla, 153–56, 389t, 481, 492f, 493, 509t

Lissemys, 122–23, 123f, 129f, 131

Listriodon, 302–4, 304f, 305–7; artistic reconstruction, 424f–25f; body sizes, 299f, 300t; featured taxa, 419t; paleocommunities, 424f–25f, 437t, 443; species data, 393t, 517t; stable isotope data, 300–301, 301f, 399t–400t, 450t, 455; time ranges, 297, 298f–99f, 467f, 473f

Listriodontinae, 296–97; body sizes, 300t, 306; species data, 517t; temporal ranges, 297, 298f

Listriodontini, 296

lithofacies, 56–57, 56f, 416t

Litocranius, 360

Litra Formation, 14f, 289f

lizards, 78, 117–18, 119f, 421; artistic reconstruction, 426, 427f, 428; depositional environments,

81–82, 82f; environmental settings, 68t; featured taxa, 419t, 423t; relative abundance, 130f

loaches, 110

locality cards, 38f, 40, 56

Lokotunjailurus, 231, 232t; species data, 216t, 234t, 392t, 516t; time range, 218f

Lophibaluchia, 170f

lophiomerycids, 332

Lophochoerus nagrii, 299, 299f, 308

Lophura, 137, 139, 140t, 141, 142t, 143f, 144t

loris, 419t, 426, 433

Lorisidae (lorisids), 199–200, 200f, 201, 503–4; faunal change, 467f, 492f, 494b–95b; paleocommunities, 436t, 438t; species data, 391t, 509t; time ranges, 467f, 472f

Lorisoidea, 200–201, 208

Losodokodon losodokius, 250

Lower Chinji Formation, 16f, 77–78

Lower Siwalik Group, 13–14, 14f; fish fossils, 108, 111–12; molluscan fossils, 100t, 101, 103; proboscideans, 250; rhinocerotids, 281, 286, 289f; small mammal fossils, 150, 171, 182

Lower Vihowa Formation, 16

Loxodonta, 249

Lufeng, Yunnan Province, South China, 181

Lufengpithecus, 207

Lunanosorex, 156

Lutra, 225–26

Lutrinae, 392t, 515t

Lycyaena, 228, 228f, 229, 232t; species data, 215t, 234t, 392t, 516t; time range, 218f

Lydekker, Richard, 5

Lygosominae, 118, 119f, 129f

Lymnaeidae, 100

Macaca sivalensis, 205

Machaerodus sivalensis, 230t

Machairodontinae, 230–31, 413f–14f, 419t, 423t, 426f, 516t

macroevolution, 3–4, 433, 460–61, 531

Macrotherium, 273

Magerictis, 222

magnetostratigraphy, 16–20, 18f, 20f, 26t–28t, 295

Mahâbhârata, 5

major channels (MC), 23, 56f, 57, 60t, 501–2; classification scheme, 59t; distribution of fossil animals, 81–82, 82f; fossil assemblages, 62t–68t, 69–72, 73f–74f, 77; fossil localities, 78, 78t, 79–80; fossil preservation, 57; taphonomic histories, 77; taxonomic composition, 65t–68t, 74–76, 75f–76f, 77, 80; unusual fossil concentrations, 78, 83

Malacothrix, 181

Malayopython, 120

Malaysia, 30, 531

mammal fossils: bone-modification features, 69f; channel-lag assemblages, 77; skeletal elements, 69–72; South Asian, 5–6; stable isotope values, 91–93, 93f, 395t–407t, 450–55, 450t, 451f–53f, 454f; unusual concentrations, 78. *See also* Siwalik fossil record; *specific mammals*

mammalian fauna, 97–98, 433; bioprovince boundaries, 455–56; body sizes, 74, 76, 83, 433; change over time, 490–96, 492f, 494b–96b; community assembly, 503–8; community structure, 9, 80–82, 432–80; composition determinants, 500–503; composition over time, 528–29; determinants of change, 460–64, 500–503; diversity dynamics,

486, 488b; ecological dynamics, 481–526; endemics, 433–34, 436t–39t, 456, 456f, 465f–68f; extinctions, 458, 486, 488b; first and last appearances, 433–34, 444f, 455–56, 456f, 483, 484b; at focus time slices, 469f–77f; functional composition, 529; functional groups, 531; functional richness, 503, 529; immigrants, 433–35, 436t–39t, 455–56, 456f, 465f–68f; interspecific competition, 501; Late Miocene change, 463–64; modern, 531; originations, 456, 456f, 457–58, 486, 488b; paleoecological reconstructions, 423, 423t, 424f–25f, 425, 426f, 427f, 428; paleocommunity reconstructions, 442–45; patterns of change, 445–60, 465f–68f, 496–500; post-Kamlial paleocommunities, 442–45; prey species, 501; rates of change, 483–90; residence times, 456–57, 460; Siwalik, 435, 460–61; species counts, 440, 441t, 444f; species durations, 528; species range extensions and contractions, 458–60; species richness, 446–47, 446f, 447–48, 483, 503, 528, 531; specimen numbers, 66t–67t; taxonomic composition, 65t–68t, 74–76, 75f, 76f, 529; taxonomic dynamics, 481–526; taxonomic richness, 503; temporal and taxonomic resolution, 481–82; total mammal richness, 497–98. *See also* large mammals; mega mammals; small mammals; *specific mammals*

Mammut, 250

Mammutidae, 249–51, 255–56; at focus time slices, 476f; paleocommunities, 423, 437t, 442; species data, 392t, 517t; time range, 256f

Manchar Formation, 14f, 35f; bovids, 368; carnivores, 213, 219–20; chalicotheres, 276, 278; chronostratigraphic framework, 41; creodonts, 213; hippopotamids, 319; Hippopotamoidea, 317; magnetostratigraphy, 20; primates, 206–7; proboscideans, 251f; rhinocerotids, 289f; small mammal fossils, 157, 171–72, 174; suids, 296; tragulids, 333, 343

Mangla-Samwal Anticline, 44t, 282f, 320, 322, 326

Manidae, 244f, 245–47; at focus time slices, 473f; paleocommunities, 439t; species data, 392t, 514t

Manis, 244f, 245–47, 246f, 276; species data, 392t, 514t; time ranges, 473f

Mansar Formation (Jammu/Kashmir), 111

Marmotinae, 166t–67t, 510t

marsupials, 149

Martes, 222

Marymus, 166t, 169, 170f, 189

mastigures, 118

mastodons, 249–51

McRae, Lea, 17

Megachiroptera, 156

Megacricetodon, 182–83, 187f, 190; paleocommunities, 436t; species abundance, 448, 448t; species data, 167t, 390t, 511t; time ranges, 174f, 466f, 471f

Megacricetodontinae, 167t, 174f, 188, 188f, 511t

megaherbivores: artistic reconstructions, 425, 426f; body sizes, 449, 449f, 482t; depositional environments, 76; diet, 450; as ecosystem engineers, 501, 529–30; faunal change, 464, 496, 507; first occurrences, 433, 434f; functional groups, 506–7; paleocommunities, 423, 425, 426f, 437t–39t, 440, 442–45; residence times, 464, 530; species change, 446; species richness, 447–48, 464, 499f, 501, 502f, 507–8, 528, 530; stable isotope data,

420, 421f, 435; temporal and taxonomic resolution, 481. *See also* mega mammals

mega mammals: analytic approach to, 530; body sizes, 448–49, 449f; faunal change, 463–64, 468f, 496; first occurrences, 433, 434f; at focus time slices, 476f–77f; fossil record, 82–83, 433; paleocommunities, 436t–39t, 445; rarefaction curves, 76f, 81; residence times, 456–57, 457t, 458f; species change, 446–47, 447f; species counts, 440, 441t, 443f; species data, 432–33; species richness, 446–47, 446f, 447–48, 499f

Megantereon, 230–31

Melanochelys, 124–25, 125f, 128, 129f, 131

Melanopsidae, 100

Mellalomys, 183, 190; paleocommunities, 436t; species abundance, 448t; species data, 168t, 391t, 512t; time ranges, 174f, 467f, 471f

Mellivorinae, 391t, 392t, 515t

Mellivorodon palaeindicus, 231

Melogale moschata, 223

mergansers, 137

Mergus, 137

Merycodontidae, 357–58

Merycodontinae, 354

Merycopotamus, 314, 316–21, 318f, 323–24, 325f, 455; artistic reconstruction, 427f; dental microwear, 320, 321t; extinction, 325f, 326–27; paleocommunities, 427f, 439t; species data, 393t, 518f; stable isotope data, 402t–3t, 450, 450t, 455; time ranges, 314f, 460, 474f

Mesaceratherium welcommi, 283t, 289f, 291–92

Mesopithecus, 205, 205f, 207, 445; artistic reconstruction, 427f, 428; featured taxa, 419t; species data, 391t, 514t; time ranges, 199, 200f, 208, 473f

Metacrtos, 221

Metailurus, 231

Metapterodon, 216–18, 232t, 233; species data, 215t, 234t, 391t, 514t; time range, 217f

Metaschizotherium, 278

Metridiochoerus, 302

Meyer, Grant, 6

mice, 149, 164, 181, 184–87, 185f, 190–92, 461, 464

Microbunodon, 314–16, 324, 325f, 326; body sizes, 462f; dental and mandibular remains, 315, 316f; dental microwear, 315, 321t; species data, 393t, 518f; stable isotope data, 402t; time ranges, 314f, 467f, 474f

Microbunodontinae, 314f, 315, 316f, 326, 518t

microevolution, 460, 527

Microloris, 200, 391t, 438t, 472f, 509t

Micropithecus, 205

micro-*Potamochoerus,* 298f, 393t, 439t, 474f, 518t

Microstonyx, 306–8

micro-stratigraphy, 22

micro-vertebrates, 77–78, 81

Middle Miocene Climatic Transition, 503, 505–6

Middle Siwalik Group, 13–14, 14f; chalicotheres, 272–78; fish fossils, 109, 111; molluscan fossils, 100t, 101; proboscideans, 250; reptile fossils, 117–18, 120, 125; rhinocerotids, 281, 286, 289f; small mammal fossils, 150, 182

Milankovitch cycles, 410, 415, 460, 481

minnows, 110

Miocene Climatic Optimum (MCO), 410, 428, 435, 503, 506–8

Miocene Epoch: mammalian assemblages, 531; paleoecology reconstruction, 409–31, 416t, 419t–20t, 424f–25f, 426f, 427f

Miochoerus, 307

Miomachairodus, 230, 232t; species data, 216t, 234t, 392t, 516t; time range, 218f

Mioparadoxurus, 226–27, 232t, 233; species data, 215t, 234t, 392t, 515t; time range, 218f

Miorhizomys, 179f, 180–81, 180f, 187–90, 458; artistic reconstruction, 426f; featured taxa, 419t; metacommunities, 189t; paleocommunities, 426f, 439t; species data, 167t, 390t, 511t, 514t; time ranges, 173f, 470f–71f, 472f

Miotragocerus, 369, 373–74, 378–79, 458; artistic reconstruction, 413f–14f; astragalus widths, 380, 380f; dental microwear and mesowear, 383, 384f–85f; diet and habitat, 380t, 381–84, 385f, 454–55; featured taxa, 423t; hypsodont scores, 384, 386f; paleocommunities, 413f–14f, 438t; sexual dimorphism, 380; species data, 370t, 394t, 521f; stable isotope data, 381, 382f, 383t, 405f, 450, 450t, 455; time ranges, 371f, 475f–76f

mixed feeders, 464; dental mesowear, 383, 385f; dietary components, 482t; faunal change, 496, 507, 531; functional groups, 506–7; species richness, 499f, 501, 502f

modern ecosystems, 531

Molarochoerus, 308

moles, 153

molluscs, 99–107, 103f, 108, 422, 442; depositional environments, 81–82, 82f; paleocommunities, 437t, 439t; species data, 100t; stratigraphic ranges, 101–3, 102f; taxonomy, 101t

mongooses, 226, 419t, 426, 427f, 428

monkeys, 199, 205, 419t, 427f, 428, 439t, 445

Morgan, Michèle, 8, 533f

Morotochoerus, 313

Moschiola, 332

Munthe, Jens, 149

Muridae, 153, 169, 178, 181–82, 423; change over time, 466f–67f, 492–93; at focus time slices, 471f–72f; proposed relationships, 186–87, 187f; residence time, 459f; species data, 168t, 391t, 511t–12t; species richness, 443f, 448; time ranges, 170f, 174f

Murinae, 164, 181, 183–87, 184f, 190–91, 433; change over time, 463–64, 490, 492f; last appearances, 504; metacommunities, 188, 189t; morphological change, 458; paleocommunities, 425, 428, 438t–39t, 443; proposed relationships, 187f; relative abundance, 461, 463f; replacement of cricetids and gerbils by, 188, 188f, 464; species data, 168t, 391t, 511t–12t; time ranges, 174f, 472f

Muroidea, 178–84, 182f, 442, 461

Murree Formation, 12–15, 14f, 100; chronostratigraphy, 97f; rhinocerotids, 289f, 290; small mammals, 149, 171, 179, 181, 190

Muruyur Beds (Miocene Kipsaramon bonebed), 79

Mus, 185f, 186–87, 187f, 190, 531; metacommunities, 189t; paleocommunities, 439t, 445; species data, 168t, 391t, 512t; time ranges, 174f, 472f

Muséum d'Histoire Naturelle, Toulouse, 281

Muséum National d'Histoire Naturelle in Paris (MNHN), 168t, 216t, 281, 292, 334t

muskrats, 181

Mustela, 231–33

Mustelidae, 213, 222–26, 233; change over time, 490, 492f; paleocommunities, 437t; species data, 215t, 513t; species richness, 493, 498f; time ranges, 217f

Mustelinae, 392t, 515t

Mycteria, 137, 140t

myliobatiformes, 109–10, 109f, 109t

Myocricetodon, 181, 183, 187f, 190, 192; paleocommunities, 436t; species abundance, 448t; species data, 168t, 391t, 512t; time ranges, 170f, 174f, 466f, 471f

Myocricetodontini, 181–83

Myomimus, 172f, 176, 176f, 189, 191; paleocommunities, 438t–39t; species abundance, 448t; species data, 166t, 390t, 513t; time ranges, 469f–70f

Myomininae, 166t, 513t

Myorycteropus, 240, 243t

Myospalax, 178

myosplacines, 178–79

Naduo Formation, 315

Nagpur, Central Deccan Traps, India, 100

Nagri Formation, 12, 14, 14f, 16, 16f, 19f, 88, 413–14; aardvarks, 241; bovids, 377; carbonates, 89–90, 89f, 90t; carnivores, 213; chronostratigraphy, 97f; creodonts, 213, 216–17; depositional environments, 60t, 61f, 440–41; fieldwork, 5, 7–8; fish fossils, 108, 114t; fluvial environments, 22–24, 24f, 25, 28–29, 61; fossil assemblages, 62t–64t, 68–73, 70t–72t, 73f, 74f, 445; fossil localities, 22, 54f, 56, 60t, 61f, 79, 483, 484b; fossil occurrences, 37, 37f; habitat, 428; hippopotamids, 318f; magnetostratigraphy, 17, 18f, 26t–27t; molluscan fossils, 100–101, 100t, 102f, 103–4; paleocommunities, 81, 438t; primates, 202, 207; proboscideans, 250; reptile fossils, 117, 122, 126, 129–30, 421; rhinocerotids, 282f, 286, 289f, 290–92; sediment accumulation, 414; sediment "pods," 79; small mammal fossils, 150, 155–57, 177, 179, 182–83, 186; suid fossils, 299f, 307; taxonomic composition, 67t, 76

Nagri Locality Y311, 79, 207; fossil assemblages, 69–73, 70t–72t, 73f, 445; taxonomic composition, 67t, 76

Nagri Zone (Perim [Piram] Island, India), 262

Naja, 121

Nakalimys, 180

naked catfish, 110–11

Namacerus, 377

Nanger, 375

National Science Foundation, 6, 532

Natricinae, 121–22, 129f

Natural History Museum, London (NHM), 34, 108, 125–26, 213, 260, 281, 314, 332, 353, 368

Natural History Museum of Los Angeles County, 108

near-aquatic mammals, 68t, 81–82, 82f

Necromanis, 245–47

Neoaves, 136

neo-communities, 433

Neognathae, 136–37

Neornithes, 136

Nepal, 531; fossil plant record, 415–17, 416t; primates, 205–6

Nesolagus, 165

Nesomyidae, 181

Nestoritherium, 272–73, 276, 278

new species index (NSI), 435, 446, 447f
Ngorora Formation, 377
Nilssonia, 123–24, 124f, 128, 129f, 131; artistic reconstructions, 424f–25f, 426f; featured taxa, 419t
Nisidorcas, 369, 376, 379–80, 384; species data, 370t, 394t, 522t; time ranges, 370t, 371f, 476f
Notochoerus, 302
Notoungulates, 92
Numulites, 13
Nyanzachoerus, 298, 302, 303f, 308
Nyanzameryx, 357–58
Nycticeboides, 199–200, 200f, 201f; artistic reconstruction, 426f; featured taxa, 419t; paleocommunities, 426f, 436t; species data, 391t, 509t; time ranges, 467f, 472f

ocean circulation, 409–11, 412f
Ochotona, 164
Ochotonidae, 164; faunal change, 465f; paleocommunities, 436t; species data, 166t, 389t, 509t; time ranges, 172f
Odocoileus virginianus, 355t
okapi, 277, 353
Okapia, 353–54, 355t, 357, 361, 364
Okapiinae, 361; paleocommunities, 437t–38t; species data, 355t, 394t, 520t; time ranges, 356f, 475f
Oligopetes, 176
omnivores, 233, 420, 464; dietary components, 482t; faunal change, 490–96, 507; functional groups, 506–7; species richness, 497f, 499f
Opdyke, Neil, 6, 8, 17, 150
Orangemeryx, 354, 358–59; species data, 355t, 394t, 520t; time ranges, 356f, 468f, 475f
orangutans, 199, 203, 204t
Oreinus, 109
Oriensciurus, 176; species data, 166t, 390t, 509t; temporal range, 172f; time ranges, 466f
originations, 457–58; immigrants and, 456, 456f; per-lineage rate, 486, 488b; pulses, 530
Orycteropodidae, 240, 244f; faunal change, 467f; first and last appearances, 493, 504–5; at focus time slices, 473f; paleocommunities, 437t; species data, 392t, 517t; species richness, 493, 498f
Orycteropus, 240–42, 242f, 245; artistic reconstructions, 425, 426f; featured taxa, 419t; paleocommunities, 438t; species data, 392t, 517t; teeth measurements, 242, 243t; time ranges, 244, 244f, 473f
Oryctolagus, 165
Orygotherium, 357; species data, 355t, 394t, 520t; time ranges, 356f, 475f
Oryx beisa, 355t
ossicones, 353–54
Osteichthyes, 109t, 114t
ostracods, 97
ostrich(es), 140t, 141–46, 413f–14f, 420, 423t
Otidae, 420t, 427f
Otididae, 137, 138f, 140t, 141, 142t, 143f, 145–46
Otidiformes, 137, 140t
Otis, 137, 140t
Otomyinae, 181
otters, 108, 233, 419t
Ouzocerus, 376
owls, 152, 165

oxygen isotopes (δ^{18}O): as evidence, 415–17, 416t; fossil enamel values, 395t–407t, 450–55, 450t, 451f–53f, 454f; soil carbonate values, 88–91, 89f, 90f, 90t, 415–17

Paali Nala, Baluchistan, 176
Pabbi Hills, Pakistan, 41, 44t, 165, 187
Pachychilidae, 102f
Pachyportax, 369, 374–75, 378–79, 384, 458; artistic reconstructions, 413f–14f, 426f, 427f; dental microwear and mesowear, 383, 384f–85f; diet and habitat, 381–84, 385f, 386, 454–55; featured taxa, 419t, 423t; hypsodont scores, 384, 386f; species data, 370t, 394t, 522t; stable isotope data, 381, 382f, 383, 383t, 405t–6t, 450t; time ranges, 371f, 476f
Pachytragus, 377
Paguma, 227
PAHs (polycyclic aromatic hydrocarbons), 418
Pakistan Museum of Natural History (PMNH), 9, 34, 150, 213, 281, 332, 368
Palaeanodonta, 245
Palaeochoeridae, 309, 455, 505; at focus time slices, 473f; paleocommunities, 438t–39t, 444; species data, 393t, 517t; stable isotope data, 399t, 450, 450t, 451f; temporal ranges, 297, 298f
Palaeochoerinae, 296–97, 298f, 300f, 517t
Palaeognathae, 136
Palaeohypsodontus, 369, 370t, 384
Palaeomastodon, 249
Palaeomerycidae, 353–59, 364, 504–5; at focus time slices, 474f–75f; paleocommunities, 438t; species data, 355t, 393t–94t, 519t–20t; time ranges, 356f, 474f
Palaeomerycinae, 354–57, 355t, 356f
Palaeomeryx, 355, 358
Palaeoreas, 376
Palaeoryx, 377, 379
Palaeotraginae, 363
Palaeotupaia sivalicus, 157–58
paleobiology: evidence used, 415–17, 416t; Kamlial Formation, 435–40; research topics and references, 48t
paleocommunity(-ies): 7.2 Ma Y581, 424f, 427–28, 427f; 7.4 Ma, 445; 8.7 Ma, 445; 9.2 Ma Y182, 423t, 424f, 425–27, 426f; 10.1 Ma, 444–45; 11.4 Ma, 442–43; 12.4 Ma Y496, 422–25, 424f–25f; 12.7 Ma, 442; 14.1 Ma, 442; composition over time, 528–29; data links, 49; environmental context, 422, 423t; evidence, 416t, 417–22, 419t–20t; inferences from avian fossils, 139–42; inferences from vertebrate fossils, 56–57, 56f; Kamlial Formation, 81, 435, 436t; maps, 410f; post–Kamlial Formation, 442–45; principal characteristics, 436t–39t; reconstructions, 80–82, 409–31, 442–45; resilient, 460; study approach, 433; sub-Himalayan alluvial plain, 411–17, 413f–14f; time slices, 442–45; variance through time, 448–55
paleoecology: reconstruction, 409–31, 416t, 419t–20t
paleoenvironment: alluvial plain, 55, 411–17, 414f; depositional setting, 35–40, 57, 59t, 60t, 413–14, 414f; evidence used, 414–17, 416t; fluvial, 20–30, 21f, 24f, 56–57, 56f, 413–14, 414f; post–Kamlial Formation, 440–42
paleogeography, 409, 411f, 413, 415–17, 416t

paleomagnetic stratigraphy, 8, 12. *See also* magnetostratigraphy
paleosols, 21f, 25–28, 413–15; calcium profile, 88, 88f; carbonate data, 88–91, 89f, 90f, 90t, 412f, 415, 420, 421f; geochemical analysis, 415–17; morphology, 415–17, 416t; post–Kamlial Formation, 440–42; research topics and references, 48t
Paludiavis, 138, 141t, 143f, 144t, 145
pangolins, 79, 242–46, 244f, 246f, 247, 276, 439t
Pangshura, 125
Panopeidae, 97
Panthera leo, 230
Parabrachyodus, 317, 325f
Paraceratherium, 15
Parachanna, 112
Parachleuastochoerus pilgrimi, 305; body sizes, 299f, 300t; species data, 393t, 518t; stable isotope data, 300–301, 301f, 401t; time range, 297, 298f–99f, 474f
Paradakkamys, 183; paleocommunities, 438t; species data, 168t, 391t, 513t; time ranges, 174f, 471f
Paradoxurinae, 227, 233; species data, 215t, 392t, 515t; time range, 218f
Paradoxurine, 515t
Paradoxurus, 227
Paralaucodus, 437t
Paralutra, 226
Paramachaerodus, 230, 230t, 232t; species data, 216t, 234t, 392t, 516t; time range, 218f
Parapelomys, 185f, 186–87, 187f, 190–91; metacommunities, 189t; species data, 168t, 391t, 512t; time ranges, 174f, 472f
Parapodemus, 186, 190–91; species data, 168t, 391t, 512t; time ranges, 174f, 472f
Paratetralophodon, 251f, 254; paleocommunities, 439t; species data, 392t, 517t; stable isotope data, 395t; time ranges, 256f, 477f
Paratragocerus, 379
Paraulacodus indicus, 172f, 173–75, 189; species data, 166t, 390t, 513t; time range, 469f
Parreysia, 100, 100t, 101, 101t, 102f, 103f, 104–5
Passeriformes, 139, 141t, 146
Patnaik, Rajeev, 5
Patriomanis, 245
Pavo, 137, 138f, 139, 140t, 141, 142t, 143f, 144t, 145
Pavoninae, 140t, 143f
peafowl, 137, 139, 140t, 145, 442
Pecarichoerus orientalis, 298f
peccaries, 295
Pecora, 97, 354–63, 394t, 436t, 468f, 523t
Pekania palaeosinensis, 223
Pelecanidae, 139, 141, 141t, 142t, 143f, 146
Pelecaniformes, 138–39, 140t, 141t
Pelecanus, 138f, 139, 143f, 145–46; artistic reconstruction, 426f; featured taxa, 420t; first and last appearances, 144t; temporal range, 141t
pelican(s), 139, 141, 141t, 145–46, 420t
Percrocuta, 227–28, 232t, 233; artistic reconstructions, 426f, 427f; featured taxa, 419t; species data, 215t, 234t, 392t, 515t; time range, 218f
Percrocutidae, 227–28; paleocommunities, 437t; species data, 215t, 392t, 515t; species richness, 493, 498f; time ranges, 218f
Perissodactyla, 74f, 93f, 481, 528; artistic reconstruction, 425, 426f; faunal change, 467f–68f; featured taxa, 419t, 423t; at focus time slices,

473*f*; paleocommunities, 425, 426*f*; species data, 392*t*–93*t*, 523*t*–24*t*; stable isotope data, 92–93, 396*t*, 450

permil notation, 86

persistence index (PI), 435, 446, 447*f*

Peshawar University, 6

Petromus typicus, 172

Phachoerus aethiopicus, 297–98

Phalacrocoracidae, 139, 141, 141*t*, 142*t*, 143*f*, 145–46

Phalacrocorax, 138*f*, 139, 141*t*, 143*f*, 144*t*, 145–46

Phasianidae, 137, 141, 145–46

Phasianinae, 140*t*, 143*f*

Phataginus, 246–47

pheasants, 137, 139, 140*t*, 141–42, 420, 442

Phiomia, 249

Phoberogale, 214, 215*t*, 221, 232*t*

Phoeniconaias, 137, 139, 140*t*, 143*f*, 144*t*, 145–46

Phoenicopteridae, 137, 140*t*, 141, 142*t*, 143*f*, 146

Phoenicopteriformes, 137

Phoenicopterus, 137, 139, 140*t*, 143*f*, 144*t*, 145

Pholidota, 244*f*, 245–47, 392*t*, 473*f*, 514*t*

Phosphatherium escuilliei, 249

photosynthesis: C$_3$, 8, 86–88; C$_4$, 8, 86–88, 91

Phyllotillon naricus, 275

Physidae, 100

Physunio, 100, 100*t*, 101

Pickford, Martin, 7

Pierolapithecus, 207

pikas, 164

Pila, 79, 101, 101*t*, 102*f*, 103, 103*f*, 104–5, 422; artistic reconstruction, 424*f*–25*f*; featured taxa, 420*t*

Pilbeam, David, 313, 533*f*

Pilgrim, Guy, 5, 19, 117, 369

Pinjor Formation, 332, 379

Piram Island, 290, 376

piscivores, 420

Planorbidae, 100–101, 101*t*, 102*f*, 104

plant fossils, 54–55

plant record, 417–18, 417*f*

Platalea, 138, 138*f*, 140*t*, 142*t*, 143*f*

Platybelodon, 253

Plesiaceratherium naricum, 283*t*, 289*f*, 291–92

Plesiogulo, 223, 232*t*; species data, 215*t*, 234*t*, 392*t*, 515*t*; time range, 217*f*

Plesiohipparion, 260, 263

Pleuroceros blanfordi, 283*t*, 289*f*, 291–92

Pliosiwalagus, 165

Polaroid ground photographs, 38*f*, 40

polycyclic aromatic hydrocarbons (PAHs), 418

pondweed, 428

Pongo, 9, 199, 202, 206–7

Pontosmilus indicus, 230

porcupines, 149, 164, 175–76, 189, 192; artistic reconstruction, 427*f*, 428; featured taxa, 419*t*; paleocommunities, 427*f*, 428, 439*t*; species data, 166*t*

Porphyrio, 137

post-*Antemus*, 184–85, 187*f*, 190–91, 461; species data, 168*t*; time range, 174*f*, 471*f*

Potamochoerus, 297–98, 299*f*, 300, 300*t*, 301, 306–8

Potamogeton, 428

Potamotherium, 227

Potwarmus, 181, 183–85, 187*f*, 190–92, 458; metacommunities, 189*t*; paleocommunities, 436*t*; species abundance, 448*t*; species data, 168*t*, 391*t*, 512*t*; time ranges, 170*f*, 174*f*, 467*f*, 471*f*

Potwar Plateau, 2*f*, 13*f*, 35*f*; bioprovince boundaries, 455–56; biostratigraphy, 295; chronostratigraphy, 19–20, 20*f*, 41–44, 97, 97*f*; depositional environments, 35–40, 57, 59*t*–60*t*; fieldwork, 1–2, 4–5; fires, 503; fluvial environments, 20–30, 21*f*, 24*f*; fossiliferous sequence, 4, 7; fossil localities, 37–39, 37*t*, 44–47; fossil occurrences, 35–40, 37*f*. *See also* fossil record; lateral stratigraphy, 22–28; magnetostratigraphy, 17–19, 18*f*, 26*t*–28*t*, 295; mammalian fauna (*see* mammalian fauna); paleogeography, 409, 411*f*, 413; primate faunas, 208; rhinocerotid-yielding localities and areas, 282*f*; sedimentary sequence, 295; Siwalik Group (*see* Siwalik Group); stratigraphy, 35–40; total mammal richness, 497–98; vertical stratigraphy, 28–29. *See also specific formations*

Potwar Silts, 15

precipitation, 416*t*, 503, 506; estimates, 417–418; seasonality, 163, 187, 506–508. *See also* rainfall

predator dens and burrows, 78–79

Pre-Pinjor Beds, 131

pre-*Progonomys*, 185, 187*f*, 190–91, 458; paleocommunities, 438*t*; species data, 168*t*, 391*t*; time range, 174*f*, 471*f*

prey species, 501

Primates, 9, 153, 199–212, 450, 455, 531; artistic reconstruction, 425, 426*f*; depositional environments, 81–82, 82*f*; environmental settings, 68*t*; faunal change, 208, 467*f*; featured taxa, 419*t*, 423*t*; at focus time slices, 472*f*–73*f*; paleocommunities, 425, 426*f*, 436*t*, 439*t*; species data, 391*t*, 509*t*, 514*t*; species richness, 493, 498*f*; stratigraphic ranges, 199, 200*f*

Primus, 167*t*, 169, 170*f*, 181, 190, 192

Pristifelis, 231

Proboscidea, 55, 249–59, 432–33, 455; body sizes, 74*f*, 82, 256; environmental settings, 68*t*; faunal change, 463–64, 468*f*; featured taxa, 419*t*, 423*t*; at focus time slices, 476*f*–77*f*; paleocommunities, 423, 428, 436*t*–39*t*, 440, 442–45; species data, 392*t*, 517*t*; species richness, 443*f*, 447–48, 493, 498, 498*f*, 528; stable isotope data, 92–93, 93*f*, 396*t*, 450, 453*f*; time range, 255–56, 256*f*

Proboscidipparion, 260, 263

proboscis monkeys, 204*t*

Procynocephalus, 206

Prodeinotherium, 250, 251*f*; paleocommunities, 436*t*, 440; species data, 392*t*, 517*t*; stable isotope data, 395*t*; time ranges, 256*f*, 468*f*, 476*f*

Prodendrogale yunnanica, 158

Prodremotherium elongatum, 354

Progiraffa, 358–59, 528; species data, 355*t*, 394*t*, 520*t*; time ranges, 356*f*, 468*f*, 475*f*

progiraffids, 435–40, 436*t*

Progiraffinae, 355*t*, 356*f*, 359–60, 520*t*

Progomphotherium maraisi, 252

Progonomys, 184–87, 185*f*, 187*f*, 190–91, 458–60; metacommunities, 189*t*; paleocommunities, 438*t*; species data, 168*t*, 391*t*, 512*t*; time ranges, 174*f*, 472*f*

Prokanisamys, 179–81, 189, 192, 435; paleocommunities, 436*t*; species abundance, 448*t*; species data, 167*t*, 390*t*, 511*t*; time ranges, 170*f*, 173*f*, 466*f*, 470*f*

Prolibytherium, 354, 355*t*, 357–59

Promellivora punjabiensis, 223

Pronakalimys, 180, 189

Propontosmilus sivalensis, 230, 230*t*

Propotamochoerus hysudricus, 305, 308, 455; artistic reconstruction, 427*f*, 427*f*, 428; body sizes, 299*f*, 300*t*; featured taxa, 419*t*; species data, 393*t*, 517*t*; stable isotope data, 300–301, 301*f*, 400*t*–401, 450*t*, 455; time range, 298*f*–99*f*, 306–7, 473*f*

Prosansanosmilus, 229

Prosantorhinus shahbazi, 283*t*, 289*f*, 291–92

Prosayimys, 171

Prostrepsiceros, 375–76, 379, 384; species data, 370*t*, 394*t*, 522*t*; time ranges, 371*f*, 476*f*

Protaceratherium: body sizes, 283*t*, 289*f*, 291; stratigraphic ranges, 289*f*, 292

Protachyoryctes, 180, 180*f*, 189; paleocommunities, 439*t*; species data, 167*t*, 390*t*, 511*t*; time ranges, 173*f*, 470*f*

Protanancus, 251*f*, 252–53, 256; paleocommunities, 437*t*; species data, 392*t*, 517*t*; time ranges, 256*f*, 476*f*

Protictitherium, 228, 232*t*; species data, 215*t*, 234*t*, 392*t*, 516*t*; time range, 218*f*

Protoryx, 377, 379, 384, 463; species data, 370*t*, 394*t*, 523*t*; time ranges, 371*f*, 476*f*

Protragelaphus, 376, 379, 384; species data, 370*t*, 394*t*, 522*t*; time ranges, 370*t*, 371*f*, 476*f*

Protragelopus, 463

Protragocerus, 372

Proviverrinae, 514*t*

Pseudaelurus sivalensis, 230, 230*t*

Pseudodon, 100–101, 101*t*, 102*f*–3*f*, 104–5, 422

Pteromyinae, 166*t*, 172*f*, 176–77, 509*t*

Ptilocercidae, 156

Ptilocercus, 156, 158

Punjabemys, 182–83, 190; paleocommunities, 436*t*; species abundance, 447–48, 448*t*; species data, 167*t*, 391*t*, 511*t*; time ranges, 174*f*, 466*f*, 471*f*

Punjab University, 332, 368

Python, 120, 121*f*, 129*f*, 131; artistic reconstructions, 424*f*–25*f*, 426, 426*f*, 427*f*, 428; featured taxa, 419*t*

Quade, Jay, 8

Quetta, Pakistan, 6

rabbits, 149, 164–65, 419*t*

rainfall, 417–18, 428, 435, 440. *See also* precipitation

Rajabhat Institute, 281

Rallidae, 137, 139, 140*t*, 141–42, 142*t*, 143*f*, 146

Ramadapis sahnii, 201

Ramapithecus, 6, 9, 202

Ramnagar, India: anthracotheres, 315; primates, 199, 201, 206; small mammals, 153, 157

raptors, 78, 141, 152

Ratitae, 136, 146, 420

Ratufa, 176–77; artistic reconstruction, 426*f*; featured taxa, 419*t*; paleocommunities, 426*f*, 436*t*, 438*t*; species abundance, 448*t*; species data, 166*t*, 390*t*, 510*t*; time ranges, 172*f*, 466*f*, 470*f*

Ratufinae, 166*t*, 172*f*, 176, 177*f*, 510*t*

Rawalpindi Group, 12–15, 14*f*, 16, 97*f*. *See also specific formations*

Raynolds, Bob, 17

rays, 422

Raza, S. Mahmood, 6, 533*f*

Rectidens, 100–101, 101*t*, 102*f*, 103*f*, 104–5

Red Queen model, 4, 460

Redunca, 376

Reduncinae, 522t

Reduncini, 369, 375; paleocommunities, 428, 439t, 445; species accounts, 376, 379–80, 384; species data, 370t; time ranges, 370t, 371f

Reptilia, 69, 117–35, 442; environmental settings, 68t; featured taxa, 419t, 423t; specimens from microvertebrate sites, 129–30, 130t

research: areas, topics and references, 48t, 49; future directions, 532; integration of approaches, 8–9; study approach, 433–35; study materials and methods, 281, 332–33, 368–69; systematic approach, 354

residence time(s) (RT), 434, 456–57, 457t, 458f, 459f, 464, 528, 530; estimation, 482–83

Rhamphosuchus, 126–27, 131, 419t, 424f–25f

Rhinoceros, 281, 285f, 286, 287f, 288f, 290–92; body sizes, 284t, 289f, 290–92; paleocommunities, 438t–39t, 445; species data, 285t, 393t, 524t; stable isotope data, 399t; time ranges, 289f, 468f, 477f

rhinoceroses (rhinos), 30, 55, 290–91, 493, 528, 531

Rhinocerotidae, 276, 281–94, 287f, 288f, 432–33, 455; artistic reconstructions, 423, 424f–25f, 427f, 428; body sizes, 281, 283t–84t, 289f, 291–92; environmental settings, 68t; faunal change, 463–64, 468f; featured taxa, 419t, 423t; at focus time slices, 477f; fossil localities and areas, 282f, 289f; paleocommunities, 423, 424f–25f, 427f, 428, 436t–39t, 440, 442–45; phylogenetic framework, 285f; species data, 285t, 393t, 523t–24t; species diversity, 289f, 291–92; species richness, 443f, 448, 493, 498f; stable isotope data, 398t–99t, 450, 450t, 453f–54f, 455; time ranges, 282f, 286, 289f, 292

Rhinocerotina, 282f, 284–86, 285f, 290–92; species data, 285t, 289f, 524t

Rhinocerotinae, 284–86, 285f, 289f, 292

Rhinolophidae, 156

rhizomycines, 179, 179f, 180f, 426, 447

Rhizomyides, 179, 179f, 180–81, 189–90; paleocommunities, 438t; species data, 167t, 390t, 511t; time ranges, 173f, 470f

Rhizomyinae, 149, 164, 178, 180, 180f, 188, 191, 435, 458; faunal change, 463–64, 493; paleocommunities, 437t, 439t, 445; species data, 167t, 511t; time ranges, 173f

Rhizomyini, 173f, 180–81, 458, 511t, 514t

Rhizomys, 181, 189–90

Rita, 109t, 110

river systems, 20f, 22–23, 24f, 413

Robinson, John T., 9

rocks. See specific formations

Rodentia, 80–81, 149–50, 152–53, 163–87, 192, 278, 433, 464; adaptive radiation, 458; artistic reconstruction, 425–26, 426f, 427f, 428; biogeography, 188–90; Chinji Formation, 65t, 437t–38t; competition among, 461; depositional environments, 65t, 74–75, 75f, 81–82, 82f; environmental settings, 68t, 74; evolutionary trends, 191; faunal change, 465f–67f; featured taxa, 419t; at focus time slices, 469f–72f; isotopic record, 92; micro-vertebrate assemblages, 78; modern, 531; paleocommunities, 187–88, 425–26, 426f, 427f, 428, 436t–39t, 442, 444–45; residence times, 457, 459f; species abundance, 447–48, 448t; species data, 166t–68t, 389t–91t, 435, 509t–13t, 514t; species richness, 498f; temporal and taxonomic resolution, 481; temporal range, 172f–74f

Roe, Lois, 108

Rohtas area, 13, 13f, 17; chronostratigraphic framework, 41, 42t; depositional environments, 442; fluvial environments, 21, 29; fossil localities, 45; magnetostratigraphy, 18f, 26t; stable isotopes, 415–17, 463

rootivores: dietary components, 482t; functional groups, 506–7; species richness, 497f, 499f, 501, 502f

Rudapithecus, 207

ruminants, 353–54, 364, 368, 463

Rupicapra rupicapra, 355t

Ruticeros, 384

Ruticeros pugio, 369, 370t, 374, 384

sabertooth cat, 419t, 423t, 425–26

Saccharum, 427–28

Sahat Sakhan Dinosaur Centre, 281

Salt Range, 13, 13f, 29

Samotheriinae, 355t, 356f, 363, 438t, 521t

Samotherium, 357, 363

Samwal Formation, 12, 14, 14f, 187; chronostratigraphy, 97f; rhinocerotids, 289f; suid fossils, 299f

Sanitheriidae, 97, 393t, 437t, 467f

Sanitherium schlagintweiti, 467f, 517t

Sansanosmilus, 229, 232t, 233; species data, 216t, 234t, 392t, 516t; time range, 218f

Sayimys, 171–73, 171f, 181, 189–92, 458; first and last occurrences, 466f; metacommunities, 189t; paleocommunities, 436t, 438t–39t; species abundance, 448t; species data, 166t, 390t, 513t; time ranges, 170f, 172f, 466f, 469f

scaly-tailed flying squirrels, 178, 187

scaly-tailed gliders, 167t

Scandentia, 153, 156–58, 157f; change over time, 465f; featured taxa, 419t; at focus time slices, 469f; species data, 152t, 389t, 509t

scansorial mammals, 68t, 81–82, 82f

Schizochoerus, 296–97, 306

Schizogalerix, 154, 158–59; paleocommunities, 438t; species data, 152t, 389t, 509t; time ranges, 154f, 469f

Schizopygopsis, 109

Schizotheriinae, 272, 273f, 275, 277–78, 523t

Schizotherium, 272, 275

Schizothorax, 109

Scincidae, 118–19, 419t

Scincinae, 118–19, 119f, 129f

Sciuridae, 169, 176–77, 493, 504; faunal change, 465f–66f, 492f, 494b–96b; at focus time slices, 470f; paleocommunities, 436t–38t, 442, 445; residence times, 457, 459f; species data, 166t, 390t, 509t–10t; species richness, 443f, 448; time ranges, 172f

Sciurinae, 166t, 172f

Sciurotamias, 176–77; paleocommunities, 436t, 438t; species abundance, 448t; species data, 167t, 390t, 510t; time ranges, 172f, 466f, 470f

Sciurotamiini, 166t, 172f

screenwashing, 7; screen(ing), 40, 77–78, 149–151, 153, 155, 163, 176

seasonality, 48t, 49, 412f, 416t; 417–429, 460, 503, 506, 530–32; isotopic indicators, 91; water availability, 23–25, 59t, 80, 103–104, 130, 412–429, 435, 436t, 440–44

sediment "pods," 79

Selenoportax, 369, 373, 378–79, 384, 463; dental mesowear, 383, 385f; diet and habitat, 380t, 381, 454; hypsodont scores, 386f; species data, 370t, 394t, 521t; stable isotope data, 381, 382f, 383t, 405t, 450t; time ranges, 371f, 475f

semi-aquatic mammals, 68t, 81–82, 82f, 108

Semnopithecus subhimalayanus, 206

Semprebon, Gina M., 264, 383

Serengeti National Park, 498

Serpentes, 120–22, 129f

sexual dimorphism, 268, 347, 380

Shah, S. M. Ibrahim, 6, 533f

sheatfishes, 112

sheet deposits. See crevasse splays

Sheikh, Khalid, 17, 533f

shelducks, 137, 139, 140t, 141, 420t, 427f

shellfish, 233

shoebill, 138–39, 141, 141t, 146

Shorea, 423

shovel-tusked proboscideans, 252–53

shrews, 149–50, 153–56, 154f, 155f, 158–59, 192, 439t

siamang, 204t

Siamochoerus, 295

Siamodapis, 201

Siamotherium, 313

Siamotragulus, 343–46, 349; astragalus widths, 336f, 337f; body sizes, 334t; dental microwear, 348, 348f; hypsodonty, 348, 349f; species data, 393t, 519t; time ranges, 334t, 344, 467f, 474f

Siamotragus, 333, 333f, 436t, 443, 474f

Sianpriculus linquensis, 308

Siluridae, 109, 109f, 109t, 110, 112–14

Siluriformes, 109f, 109t, 110, 114t

Simons, Elwyn, 5–6

Sindemys, 182

Sind Province, 14f, 44t, 315, 317

Sinoadapis, 201

Sinolagomys, 164

Sisoridae, 109, 109f, 109t, 110, 112–13, 113f, 114

Sivacanthion, 149, 175

Sivaceros, 369, 372, 377, 379, 384; species data, 370t, 394t, 521t; time ranges, 371f, 468f, 475f

Sivachoerus, 299–300, 299f, 300t, 306–9

Sivaelurus, 230, 230t, 231, 232t; species data, 216t, 234t, 392t, 516t; time range, 218f

Sivahyus, 308–9; body sizes, 299f, 300t; dental morphologies, 301–2, 302f, 303f; species data, 393t, 518t; time ranges, 298f–99f, 306–7, 473f

Sivaladapidae, 199–201, 200f, 207, 493, 504; faunal change, 467f; at focus time slices, 472f; paleocommunities, 436t–37t; species data, 391t, 514t

Sivaladapis, 201; paleocommunities, 436t; species data, 391t, 514t; time ranges, 467f, 472f

Sivalhippus, 260–66, 268–69; artistic reconstructions, 413f–14f, 426f, 427f; dental mesowear and microwear, 264–65, 266f; featured taxa, 419t, 423t; paleocommunities, 413f–14f, 426f, 427f, 439t, 445; species data, 265t, 392t–93t, 523t; stable isotope data, 267f, 396t–98t, 450, 450t, 453–54, 454f; time ranges, 261f, 460, 473f

Sivalictis natans, 231

Sivamellivora, 222

Sivameryx, 316–18, 318f, 319, 323–24, 325f; species data, 393t, 518t; time ranges, 314t, 467f, 474f

Sivaonyx, 224, 225f, 232t, 233; species data, 215t, 234t, 392t, 515t; time range, 217f

Index 551

Sivaophis, 122, 129*f,* 419*t,* 427*f*

Sivapanthera padhriensis, 230–31

Sivapithecus, 6, 9, 126, 202–7, 202*f,* 205*f,* 295, 308, 326; artistic reconstructions, 413*f*–14*f,* 424*f*–25*f,* 425, 426*f;* body sizes, 203, 204*f,* 208; diet and habitat, 203–4, 300; featured taxa, 419*t,* 423*t;* paleocommunities, 413*f*–14*f,* 424*f*–25*f,* 425, 426*f,* 437*t*–38*t,* 442, 445; species data, 391*t,* 514*t;* stable isotope data, 395*t,* 450*t,* 455; time ranges, 199, 200*f,* 208, 442, 473*f*

Sivapterodon lahirii, 214–16

Sivasmilus, 230–31, 232*t;* species data, 216*t,* 392*t,* 516*t;* time range, 218*f*

Sivatheriinae, 354, 361–62, 425; species data, 355*t,* 520*t*–21*t,* 521*t;* time ranges, 356*f*

Sivatherium, 354, 362–65

Sivatupaia ramnagarensis, 157–58

Sivoreas, 373, 379, 381, 384; species data, 370*t,* 394*t,* 521*t;* stable isotope data, 381, 382*f,* 383*t,* 405*t;* time ranges, 468*f,* 475*f*

Siwalik ecosystem, 531

Siwalik fossil collection, 534; documentation of, 39–40, 40*f;* recovery differences, 50

Siwalik fossil enamel stable isotopes: as evidence, 415–17, 416*t;* long-term trends, 91–93, 93*f;* primary data, 395*t*–407*t;* summary statistics, 450–55, 450*t,* 451*f*–53*f,* 454*f*

Siwalik fossil localities, 22, 37–40, 37*f,* 38*f,* 482–83; age estimates, 44–47, 46*f,* 47*f,* 50; Class I, 45; Class II, 45–47; Class III, 47; data links, 49; documentation of, 37, 38*f,* 40, 56; environmental settings, 56–57, 58*f,* 79–80; frequencies, 54, 54*f,* 79–80; with multiple specimens per individual, 78, 78*t*

Siwalik fossil record, 39, 50, 82–83; age estimates, 41–44, 50; birds, 136–48; body sizes, 73–74; Bovidae, 368–88; Chalicotheriidae, 272–80; comparisons with other continental sequences, 530–31; Creodonta and Carnivora, 213–39; data quality, 47–49; depositional environments, 35–40, 57, 59*t*–60*t;* documentation of, 34–53, 40*f;* Equidae, 260–71; fishes, 108–16; Giraffoidea, 353–67; hedgehogs, shrews, bats, and treeshrews, 149–62; highlights, 527–35; Hippopotamoidea, 313–31; lagomorphs and rodents, 163–98; lessons learned, 527; molluscs, 99–107; preservation changes, 79–80; preservation pathways, 55–61; primates, 199–212; Proboscidea, 249–59; reptilia, 117–35; Rhinocerotidae, 281–94; rock formations that preserve, 13–15, 14*f;* sample size, 81; specimen files, 40, 40*f;* Suidae and Palaeochoeridae, 295–312; Tragulidae, 332–52; Tubulidentata and Pholidota, 240–48

Siwalik Group (Siwaliks), 3–8, 12–15, 14*f,* 16–17, 16*f,* 88; bioprovince boundaries, 455–56; chronostratigraphy, 97*f;* climate and vegetation, 409–11, 412*f;* depositional environment, 413–14, 414*f;* fluvial environments, 20–30, 21*f,* 23, 24*f;* fossil localities, 22, 482–83; geochemical analysis, 415–17; lateral stratigraphy, 22–28; magnetostratigraphy, 17–19, 18*f;* mammalian communities, 432–80; mammalian faunas, 481–526; paleoecological reconstruction, 409–31, 410*f,* 413*f*–14*f;* paleosols, 88–91, 88*f,* 89*f,* 90*f,* 412*f;* panel diagrams, 20, 20*f;* physical setting, 409; species data, 389*t*–94*t;*

sub-Himalayan alluvial plain, 411–17, 413*f*–14*f;* vertical stratigraphy, 28–29. *See also specific formations*

Siwalik Hills, 13, 13*f*

Siwalik Hills Indian Expedition (AMNH), 117

Siwalikosorex prasadi, 155

Siwalikomus nagrii, 186

skeletal elements, 62*t,* 69–72, 70*t,* 73*f*

skinks, 419*t,* 426*f*

small mammals, 7, 77–78, 149–50, 192; overview, other than glires, 158–59; analytic approach to, 530; artistic reconstructions, 423, 425, 426*f,* 428; biogeography, 188–90; body sizes, 448–49, 449*f,* 482*t;* competition, 461; conservation insights for, 532; depositional environments, 81–82, 82*f;* diversification rates, 487–90, 489*f;* environmental settings, 68*t;* evolutionary trends, 191; extinction rates, 487–90, 489*f;* faunal change, 463–64, 465*f*–67*f,* 490–96, 492*f,* 494*b*–96*b,* 497*f,* 500, 507–8, 529; feeding habits, 482*t;* first and last appearances, 433, 434*f,* 483, 484*b,* 487, 489*f,* 503–4, 506; at focus time slices, 469*f*–72*f;* fossil ages, 150; fossil localities, 150–52, 151*f;* fossil occurrences, 37*f;* fossil record, 433; functional groups, 506–7; functional richness, 529; joint size-feeding categories, 490, 492*t,* 497*f,* 499*f,* 500; metacommunities, 188, 189*t;* mortality during flood season, 501; orders, exclusive of glires, 153–54; origination rates, 487–90, 489*f;* paleocommunities, 187–88, 423, 425, 426*f,* 428, 436*t*–39*t,* 442–45; residence times, 456–57, 457*t,* 458*f,* 459*f,* 483–86, 486*f,* 507, 528; species change, 446–47, 447*f;* species counts, 440, 441*t,* 443*f,* 498; species data, 152*t,* 509*t*–13*t;* species richness, 446–47, 446*f,* 447–48, 464, 483, 485*f,* 486–90, 489*f,* 492*f,* 496–97, 497*f,* 499–500, 499*f,* 500*t,* 503, 504*f,* 507–8, 529; stable isotope data, 420; taxonomic richness, 503–5, 504*f,* 506, 508, 529; temporal and taxonomic resolution, 481

Smith, G. R., 108

Smithsonian Foreign Currency Program, 6

Smithsonian Institution, 34, 117, 532

Smutsia, 246–47

snails, 420*t,* 422

snakes, 78–79, 108, 117–18, 120–22, 121*f,* 131, 152, 421, 501; artistic reconstruction, 426, 427*f,* 428; depositional environments, 81–82, 82*f;* environmental settings, 68*t;* featured taxa, 419*t;* relative abundance, 130, 130*f*

Soan Formation, 12, 14, 14*f,* 29, 88; rhinocerotids, 289*f;* small mammal fossils, 187; stratigraphy, 17, 97*f;* suid fossils, 299*f*

Soan Synclinorium, 13, 13*f,* 19*f,* 22, 28

soil carbonate stable isotopes, 88–91, 89*f,* 90*f,* 90*t,* 420, 421*f;* Potwar, 412*f,* 415–18

soils, 413–15. *See also* paleosols

songbirds, 139

Soricidae, 153–56, 155*f,* 503–4; at focus time slices, 469*f;* paleocommunities, 439*t;* species data, 152*t,* 389*t,* 509*t;* time ranges, 154*f*

South Asia, 3, 5–6, 8

Spalacidae, 178–81, 445, 457, 504; change in composition over time, 492*f;* faunal change, 466*f;* at focus time slices, 470*f*–71*f,* 472*f;* species data, 167*t,* 390*t,* 511*t,* 514*t;* species richness, 443*f,* 448; time ranges, 173*f*

spalacines, 178–79

Spalax, 178

Spanocricetodon, 167*t,* 181, 190

species change determinants, 4, 460–64

species durations. *See* residence time(s)

species range extensions and contractions, 458–60

species richness, 446–47, 446*f,* 447–48, 483, 528; concept limits, 498–500

specimen files, 40, 40*f*

Sperata (Mystus), 109*t,* 110–11

spoonbill, 138–39, 140*t*

Squamata (squamates), 117–20, 129*f,* 421

squirrels, 149, 169, 176–77, 177*f,* 187, 189, 192, 464; featured taxa, 419*t;* modern, 531; paleocommunities, 438*t,* 442; species data, 166*t*

stable isotopes, 8, 49, 86–95; atmospheric values, 86–87; as evidence, 413–17, 416*t;* fossil enamel values, 91–93, 93*f,* 395*t*–407*t,* 450–55, 450*t,* 451*f*–53*f,* 454*f;* soil carbonate values, 88–91, 89*f,* 90*f,* 90*t*

stasis, coordinated, 528–29

Stegodon, 254–55

Stegodontidae, 249, 254–56, 504; at focus time slices, 477*f;* paleocommunities, 437*t,* 442; species data, 392*t,* 517*t;* stable isotope data, 395*t;* time ranges, 256*f*

Stegolophodon, 251*f,* 254–55; artistic reconstructions, 413*f*–14*f,* 427*f,* 428; featured taxa, 419*t,* 423*t;* paleocommunities, 413*f*–14*f,* 427*f,* 428, 439*t;* species data, 392*t,* 517*t;* time ranges, 256*f,* 477*f*

Stegotetrabelodon, 254

Stenailurus, 231

storks, 137, 139, 141, 146, 442; artistic reconstructions, 424*f*–25*f,* 426*f,* 427; featured taxa, 420*t;* species data, 140*t*

stratigraphy, 35–40; biostratigraphy, 16, 127–30; chronostratigraphy, 19–20, 20*f,* 41–46; high-resolution, 20–21; lateral, 22–28; magnetostratigraphy, 16–20, 18*f,* 20*f,* 26*t*–28*t;* microstratigraphy, 22; paleomagnetic, 8, 12; vertical, 28–29

Strepsiportax, 372, 379–80, 384; diet, 381, 382*f,* 383*t;* species data, 370*t,* 394*t,* 521*t;* stable isotope data, 381, 382*f,* 383*t,* 404*t;* time ranges, 371*f,* 468*f,* 475*f*

Strepsirrhini, 199–201

Struthio, 137, 139, 140*t,* 141–45, 143*f,* 144*t,* 146

Struthionidae, 140*t,* 141, 142*t,* 143*f,* 146

Styriofelis, 231

sub-Himalayan alluvial plain, 411–17, 413*f*–14*f*

Suidae, 55, 295–309; artistic reconstruction, 427*f,* 428; body sizes, 74, 299*f,* 300*t,* 306, 461, 462*f;* bunodont/low-crowned, 299–302; Chinji Formation, 65*t;* dental nomenclature, 295, 296*f;* depositional environments, 76; diet and habitat, 297–98, 305, 455; environmental settings, 68*t;* faunal change, 467*f;* featured taxa, 419*t,* 423*t;* at focus time slices, 473*f*–74*f,* 477*f;* lophodont suoids, 302–5; paleocommunities, 426, 427*f,* 428, 437*t*–39*t,* 443, 445; residence time, 459*f;* species data, 393*t,* 517*t*–18*t,* 518*t;* species richness, 443*f,* 447–48; stable isotope data, 300–301, 301*f,* 399*t*–402*t,* 450, 450*t,* 451*f,* 455; time ranges, 297, 298*f*–99*f,* 473*f*

Suinae, 296–97, 300–301, 455, 458; body sizes, 299*f,* 300*t;* species data, 517*t*–18*t,* 518*t;* temporal ranges, 297, 298*f*–99*f,* 306–7

Sulaiman Lobe, 13, 14*f,* 15–16, 368

Suliformes, 139, 141*t*

Sumatran rhino, 290, 531

Suncus, 155, 155*f*, 159; paleocommunities, 438*t*; species data, 152*t*, 389*t*, 509*t*; time ranges, 154*f*, 469*f*

Suoidea, 295, 302–5, 493, 498*f*, 507

Supramyomorpha, 169, 178–84

Surai Khola area (Nepal): pedogenic carbonate δ¹³C data, 412*f*; plant record, 417–18, 417*f*; vegetation, 440

survey blocks, 37–39, 39*f*

Sus, 297–98, 298*f*, 307–8

Tachyoryctes, 180, 189

Tachyoryctini, 173*f*, 178–80, 180*f*, 439*t*, 511*t*

Tadornini, 137, 140*t*, 142*t*, 420*t*, 427*f*

Tahirkheli, Rashid, 6

Talpidae, 153

Tamias, 177, 189, 191; paleocommunities, 438*t*; species abundance, 448*t*; species data, 167*t*, 390*t*, 510*t*; time ranges, 172*f*, 466*f*, 470*f*

Tamiini, 172*f*

Tamiops, 176–77, 177*f*; artistic reconstruction, 424*f*–25*f*; featured taxa, 419*t*; first occurrences, 466*f*; paleocommunities, 436*t*; species data, 166*t*, 390*t*, 510*t*; time ranges, 172*f*, 466*f*, 470*f*

taphonomy, 54–85; isotaphonomic sampling, 8–9, 55, 80, 83; research topics and references, 48*t*; taphonomic indicators or attributes, 61–76, 62*t*–68*t*, 70*t*–72*t*

Tarsiidae, 199

Tatera indica, 183

Taterillini, 183, 439*t*

Tatrot Formation, 12, 14, 14*f*; bovids, 379; chronostratigraphy, 97*f*; fieldwork, 5; fossil localities, 45; freshwater molluscan fossils, 100, 100*t*, 102*f*; rhinocerotids, 282*f*, 289*f*; small mammal fossils, 165; suid fossils, 299*f*, 307; tragulids, 332, 345

Taucanamo, 304, 307

Tauromeryx, 354–57; species data, 355*t*, 393*t*, 519*t*, 520*t*; time ranges, 356*f*, 474*f*

Tauxe, Lisa, 8, 17–19

taxonomic richness, 503–5, 504*f*

Tayassuidae, 295–96

tectonic setting, 411–13

Teleoceratina, 284, 286–92; phylogenetic framework, 285*f*; species data, 285*t*, 289*f*, 523*t*–24*t*; time ranges, 282*f*, 289*f*, 292

Telmatodon, 317, 323, 325*f*

temperature: cross-over, 87; global, 409–11, 412*f*, 428

Tereulictis, 226

Terminal Miocene, 324–26

terrestrial mammals: environments, 68*t*, 81–82, 82*f*

Testudines, 117, 122–26, 126*f*, 128, 129*f*, 421

Tethys Sea, 409, 411*f*

Tethytragus, 377, 379, 381, 384; hypsodont scores, 386*f*; species data, 370*t*, 394*t*, 522*t*; time ranges, 371*f*, 476*f*

Tetracerus, 345, 531

Tetraconodon, 299–300, 305, 308; body sizes, 299*f*, 300*t*, 306; species data, 393*t*, 518*t*; stable isotope data, 300–301, 301*f*, 401*t*–2*t*; time ranges, 298*f*–99*f*, 306, 477*f*

Tetraconodontinae, 296–97, 299–300, 307; body sizes, 300*t*; species data, 518*t*; time ranges, 297, 298*f*

Tetralophodon, 254

Tetralophodonts, 256

Thalassictis, 228, 228*f*, 232*t*; species data, 215*t*, 234*t*, 392*t*, 516*t*; time range, 218*f*

Thanbinkan, Myanmar, 315

Theobald, William, 117

Thiaridae, 100–101

Thomas, Herbert, 7

Threskiornithidae, 138–39, 140*t*, 143*f*, 146

Thryonomyidae, 149, 164, 173–75, 189; change over time, 465*f*, 492*f*, 493; at focus time slices, 469*f*; species data, 166*t*, 390*t*, 513*t*; time ranges, 172*f*

Thryonomys, 173–75

time slices, 22–23, 530; faunal change between, 465*f*–68*f*; mammal species recorded at, 469*f*–77*f*; post–Kamlial Formation paleocommunities, 442–45

Tomistominae, 126–27, 128*f*, 129*f*, 130–31

tooth enamel. *See* Siwalik fossil enamel

tooth marks, 71*t*

tooth structure, 8

Torolutra, 225–26, 232*f*, 233; species data, 215*t*, 234*t*, 392*t*, 515*t*; time range, 217*f*

tortoises, 122, 413*f*–14*f*; artistic reconstruction, 427*f*, 428; featured taxa, 419*t*, 423*t*

Torynobelodon, 253, 256*f*, 392*t*, 476*f*, 517*t*

total mammal richness, 497–98

Tragelaphini, 375

Tragelaphus, 355*t*, 378

Tragocerus, 373–74

Tragoportax, 373–74, 379–81, 384; artistic reconstruction, 427*f*; featured taxa, 419*t*; paleocommunities, 427*f*, 439*t*; species data, 370*t*, 394*t*, 521*t*; stable isotope data, 382*f*, 383*t*, 405*t*; time ranges, 371*f*, 476*f*

Tragoreas potwaricus, 377

Tragulidae, 9, 30, 55, 80–81, 332–52, 379, 433; artistic reconstruction, 423, 424*f*–25*f*, 427*f*, 428; body sizes, 332, 334*t*, 346–47, 380, 461–62, 462*f*; bone-modification features, 69*f*; Chinji Formation, 65*t*, 437*t*–38*t*; dental microwear, 348, 348*f*; depositional environments, 76, 81–82, 82*f*; diet and habitat, 347–49, 455; environmental settings, 68*t*; faunal change, 467*f*, 493; featured taxa, 419*t*, 423*t*; first and last appearances, 507; at focus time slices, 474*f*; hypsodonty, 348, 349*f*; paleocommunities, 423, 424*f*–25*f*, 426, 427*f*, 428, 436*t*–38*t*, 443, 445; relative abundance, 452–63, 463*f*; residence times, 457, 459*f*; species data, 393*t*, 519*t*; species richness, 443*f*, 448, 462, 493, 498*f*; specimens, 342, 343*t*; stable isotope data, 347–48, 347*f*, 403*t*, 450, 450*t*, 451*f*; time ranges, 332, 333*f*, 334*t*, 349

Tragulus, 332–33, 334*t*, 336*f*, 337*f*, 345, 347

trample damage, 71*t*

trans-Indus region, 13

Trapezoideus, 100, 100*t*, 101

treeshrews, 149–50, 153, 154*f*, 156–59, 192; featured taxa, 419*t*; paleocommunities, 436*t*–39*t*, 445

tree squirrels, 164, 176–77

Triassic Ischigualasto Formation, 79

Triceromeryx, 354–55, 357; species data, 355*t*, 393*t*–94*t*, 519*t*; time ranges, 356*f*, 475*f*

trionychids, 122–23, 130

Trionychinae, 122–23, 124*f*

trionychoids, 122, 123*f*

Tubulidentata, 240–45, 247; change over time, 467*f*; featured taxa, 419*t*; at focus time slices, 473*f*; species data, 392*t*, 517*t*; teeth measurements, 242, 243*t*

Tughlaq, Firuz Shah, 5

Tung Gur, Mongolia, 360

Tupaia, 156–57, 157*f*, 158–59; species data, 152*t*, 389*t*, 509*t*; time ranges, 154*f*, 469*f*

Tupaiidae, 156–59, 504; artistic reconstruction, 424*f*–25*f*; faunal change, 465*f*, 492*f*, 494*b*–95*b*; featured taxa, 419*t*; at focus time slices, 469*f*; paleocommunities, 424*f*–25*f*, 438*t*–39*t*; species data, 152*t*, 389*t*, 509*t*; time ranges, 154*f*, 469*f*

turtles, 78–79, 108, 117, 122–26, 123*f*, 124*f*, 125*f*, 126*f*, 131, 421, 426, 501; Chinji Formation, 65*t*; depositional environments, 81–82, 82*f*; environmental settings, 68*t*, 74; featured taxa, 419*t*; relative abundance, 129–30, 130*f*

Ua pilbeami, 361; species data, 355*t*, 394*t*, 520*t*; time range, 356*f*, 468*f*, 475*f*

U-level mice, 186

ul Haq, Munir, 126*f*

ungulates, 74, 81, 272; body sizes, 444, 529; even-toed (*see* Artiodactyla); faunal turnover, 461–63; isotopic record, 92; odd-toed (*see* Perissodactyla); paleocommunities, 444–45; species abundance, 448; species richness, 498, 531

Unionidae, 100–101, 101*t*, 102*f*, 422

University of Arizona, 6, 34, 117, 213, 332, 368

University of Baluchistan, 150

University of Michigan Museum of Zoology (UMMZ), 108

University of Peshwar, 213, 368

University of the Punjab, Lahore, 5, 9, 34, 213, 273

University of Utrecht, 149

unusual fossil concentrations, 78, 83

Upper Siwalik Group, 14, 14*f*, 88; chalicothere remains, 272, 276; fish fossils, 109; fluvial environments, 29; molluscan fossils, 100*t*, 101; reptile fossils, 117–18, 123–27, 131; rhinocerotids, 289*f*, 291–92; small mammal fossils, 150, 165; suid fossils, 306–7

Urmiornis, 137

Urogale, 157

Uromastycinae, 118, 119*f*, 128–29, 129*f*, 130

Uromastyx, 118

Ursidae, 213, 221, 233; paleocommunities, 439*t*; species data, 215*t*, 391*t*, 515*t*; time range, 217*f*

Urva, 226, 232*t*, 233, 440, 503–4; species data, 215*t*, 234*t*, 392*t*, 513*t*, 515*t*; time ranges, 218*f*

Uttarbaini Formation, 120

Varaniidae, 426, 427*f*, 428

Varanus, 117–20, 119*f*, 129, 131, 421; artistic reconstructions, 413*f*–14*f*, 426*f*, 427*f*; featured taxa, 419*t*, 423*t*

vegetation, 409–11, 412*f*; C₃ grasses, 8, 86–88, 418, 420, 428, 440, 442, 453–55, 463; C₄ grasses, 8, 86–88, 91, 418, 420, 428, 441–42, 445, 450, 453–55, 463–64, 503, 506–8; changes, 503, 530; environmental stress thresholds, 505–8; evidence for, 416t, 417–18, 417f; forested, 445; Late Miocene, 463–64; paleoecological reconstructions, 423, 424f–25f, 425, 426f, 427–28, 427f; post–Kamlial Formation, 440–42; prediction model, 420; sub-Himalayan alluvial plain reconstruction, 413, 413f–14f

vertebrates: body sizes, 73–74; paleoecological reconstructions, 423, 427–28, 427f; taphonomy, 54–85, 56f

vertical stratigraphy, 28–29

vertisols, 25–28

very small mammals. *See* small mammals

Vespertilionidae, 156

Vietnam, 498

Vihowa Formation, 14f, 15–16, 19; bovids, 368–69, 378; carnivores and creodonts, 213, 226; carnivore specimens, 220; chalicotheres, 276; chronostratigraphy, 41; creodonts, 216; Hippopotamoidea, 317; rhinocerotids, 289f; small mammal fossils, 150, 155–57, 181–83; tragulids, 333, 343–44

Vinayakia, 217, 231

Vipera, 120

Viperidae, 120–21, 121f, 129f, 130, 130f

Vishnucyon chinjiensis, 221

Vishnufelis laticeps, 231

Vishnuictis, 226–27, 232t; artistic reconstruction, 426f; featured taxa, 419t; species data, 215t, 234t, 392t, 515t; time range, 218f

Vishnumeryx daviesi, 333, 334t, 345, 348–49

Vishnuonyx, 224–25, 232t, 233; artistic reconstruction, 424f–25f; featured taxa, 419t; species data, 215t, 234t, 392t, 515t; time range, 217f

Vishnutherium, 360, 363

Viverra, 226–27

Viverricula, 226

Viverridae, 213, 226–27, 233; paleocommunities, 436t, 440; species data, 215t, 392t, 515t; species richness, 493, 498f; time ranges, 218f

Viverrinae, 515t

Viviparidae, 100–101, 101t, 102f

Vokes, H. E., 100

volant mammals, 68t, 81–82, 82f, 433

voles, 181

von Koenigswald, Gustaf (Ralph), 5–6

vultures, 139, 141t, 146, 413f–14f, 420, 423t

Wadia Institute of Himalayan Geology, 34, 213, 332, 368

Wadi Moghara, Egypt, 252

Wallago, 109t, 112

waterfowl, 55, 108, 141–42, 146

water snakes, 108

weathering, 61–65

West, Robert M., 149

white rhino, 290

wild boars, 295–97

Willmus maximus, 169–70, 171f, 172f; species data, 166t, 390t, 513t; time range, 469f

Wozny, John, 533f

Xenodermus, 121

Xenokeryx amidalae, 355

Xerini, 166t, 172f

Yale Peabody Museum (YPM), 34, 39–40, 213, 260, 272, 332, 353, 368

Yale University, 7–9, 117, 313

Yale University North India Expedition, 6, 100, 117

Yamuna River, 29

Yunnanochoerus, 296–97, 304–5, 305f, 308; body sizes, 299f, 300t, 306; species data, 393t, 517t; stable isotope data, 300–301, 301f, 305, 399t, 450, 450t; time ranges, 297, 298f–99f, 306, 473f

Zarafa, 354, 359

Zinda Pir Dome, 12–13, 13f, 14f, 15–16, 35f; anthracotheres, 315; birds, 145; bovids, 372, 378; chalicotheres, 275–76, 278; chronostratigraphy, 19, 41, 44t; creodonts, 214; fluvial environments, 29–30; Giraffoidea, 358; Hippopotamoidea, 317; magnetostratigraphy, 17, 19, 26t–28t; primates, 199; rodents, 169, 170f, 181; small mammals, 150, 151f, 155–57, 159, 176, 178–79, 181–83, 190; species documented, 483; suids, 296; tragulids, 344; vertical stratigraphy, 28

Zindapiria, 170f

zokors, 178

Zygolophodon metachinjiensis, 250–51, 251f; artistic reconstruction, 424f–25f; featured taxa, 419t; species data, 392t, 517t; time range, 256f, 476f

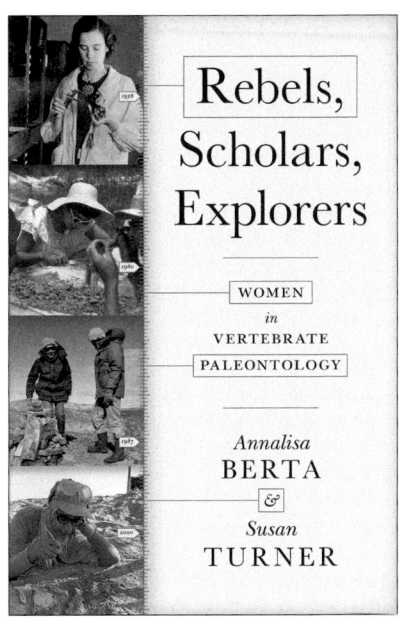
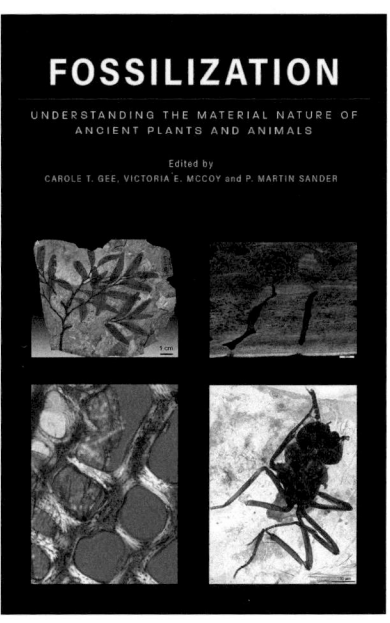

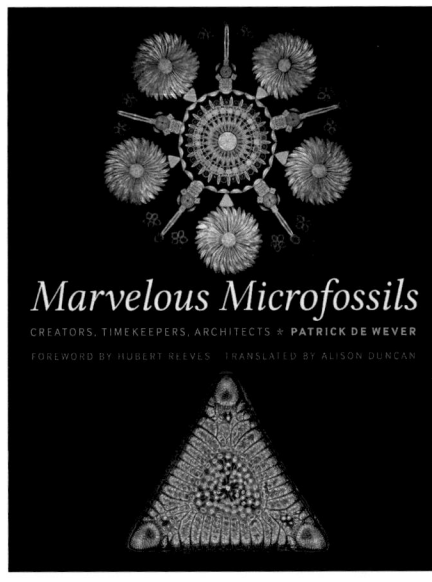
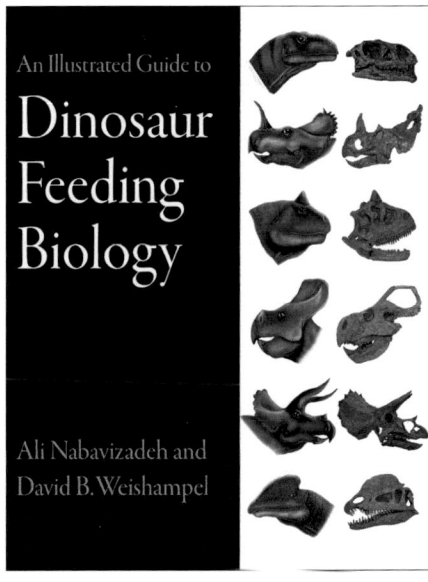